# Harnessing AutoCAD® Civil 3D® 2010

# Harnessing AutoCAD® Civil 3D® 2010

**autodesk® Press**

PHILLIP J. ZIMMERMAN

DELMAR CENGAGE Learning™

Australia • Brazil • Japan • Korea • Mexico • Singapore • Spain • United Kingdom • United States

**Harnessing AutoCAD® Civil 3D® 2010**

**Phillip J. Zimmerman**

Vice President, Career and Professional Editorial: Dave Garza

Director of Learning Solutions: Sandy Clark

Acquisitions Editor: Stacy Masucci

Managing Editor: Larry Main

Senior Product Manager: John Fisher

Senior Editorial Assistant: Dawn Daugherty

Vice President, Career and Professional Marketing: Jennifer McAvey

Marketing Director: Deborah Yarnell

Marketing Manager: Jimmy Stephens

Associate Marketing Manager: Mark Pierro

Production Director: Wendy Troeger

Production Manager: Mark Bernard

Content Project Manager: Angela Sheehan

Senior Art Director: David Arsenault

Technology Project Manager: Joe Pliss

Production Technology Analyst: Thomas Stover



Library of Congress Control Number: 2009934951

ISBN-13: 978-1-4354-9997-3

ISBN-10: 1-4354-9997-2

**Delmar**
5 Maxwell Drive
Clifton Park, NY 12065-2919
USA

Cengage Learning is a leading provider of customized learning solutions with office locations around the globe, including Singapore, the United Kingdom, Australia, Mexico, Brazil and Japan. Locate your local office at: **international.cengage.com/region**

Cengage Learning products are represented in Canada by Nelson Education, Ltd.

For your lifelong learning solutions, visit **delmar.cengage.com**

Visit our corporate website at **cengage.com.**



Printed in Canada
1 2 3 4 5 XX 11 10 09

# CONTENTS

## CHAPTER 3 SITE AND PARCELS 148



## CHAPTER 4 SURFACES 202

## CHAPTER 5 ALIGNMENTS 312



## CHAPTER 6 PROFILE VIEWS AND PROFILES 374

## CHAPTER 7 ASSEMBLIES AND CORRIDORS 430



## CHAPTER 8 CROSS-SECTIONS AND VOLUMES 473



## CHAPTER 9 TRANSITIONS, SUPERELEVATION, AND INTERSECTIONS 523

## CHAPTER 10 GRADING AND VOLUMES 581



## CHAPTER 11 PIPE NETWORKS 636



## CHAPTER 12 CIVIL 3D SHORTCUTS 683

## CHAPTER 13 HYDRAULICS AND PIPE DESIGN 692



## CHAPTER 14 SURVEY BASICS 752



## CHAPTER 15 SURVEY AND TRAVERSE ADJUSTMENTS 803

# INTRODUCTION

Civil 3D is the successor to the Autodesk Land Desktop product line (Land Desktop, Civil Design, and Survey). Like Autodesk Land Desktop (LDT), the program is a powerful drafting tool and unlike Autodesk (LDT), Civil 3D is a dynamic engineering environment. Civil 3D does not create data based on drawing entity snapshots, but drawing objects. The dynamic engineering environment is the single fundamental change that radically alters the CADD design environment. Changes in data produce changes in documentation.

Autodesk (LDT) with its hidden data and definitions disappears in Civil 3D. Prospector and Data Shortcuts provide open access to object data and its properties. Prospector manages an object and provides the interface to the object's data. In Civil 3D, data is only a few clicks away. Data Shortcuts allow data sharing.

The Settings Toolspace presents a unique way of managing a company's "look" that is consistent irrespective of which user develops a design. This is where the journey starts, with Civil 3D styles that accomplish its design drafting and documentation. Where Land Desktop was a take-it or draft-it environment, Civil 3D is a la carte. With Civil 3D you either live with its shipping content or you spend time developing your content. It is a new beginning for the Civil Industry using Civil 3D tools and styles.

Civil 3D's tool sets include points, surfaces, parcels, alignments, profiles, sections, pipes, and survey. To use Civil 3D's tool sets efficiently requires fundamental Civil Design and Survey process knowledge. This book addresses the use of and provides a basic understanding of the tool sets. This book, however, is NOT an engineering or surveying textbook nor is it a tips and tricks book. You can peruse the newsgroups for those needs. This book provides explanations of and exercises helping in the development of a basic understanding of Civil 3D's tools. Many examples are from people using the software while other examples contain difficulties to demonstrate capabilities.

Civil 3D 2010 introduces the ribbon and it is a departure from traditional AutoCAD menus. The ribbon focuses on tasks, whereas the menus focus on object types. I chose to use the ribbon for this book and find its focus on tasks rather than object types a more flowing work process.

When working in Civil 3D you have four potential tasks: creating objects (Home tab), analyzing their values (Analyze tab), editing them if wrong (Modify tab), and labeling (Annotate tab) their values. If using the ribbon and not panicking for menus, the ribbon will become an ingrained habit. The menus remain, as does selecting with the left mouse button, pressing the right mouse button, and selecting a command from a shortcut menu.

Working through the exercises, you will encounter issues with the software. I have not had a single version of Civil 3D complete all of the tasks in this book. With this release, Civil 3D accomplishes most of its goals. I cannot imagine going back to LDT after becoming familiar with Civil 3D.

Drawing files are included on the book's CD for use in exercises.

| Text Element | Example |
|---|---|
| Step-by-Step Tutorials | 1.1 Perform these steps |
| Text Element | Example |
| AutoCAD command | PAN |
| Civil 3D command | ***CREATE ALIGNMENT FROM POLYLINE*** |
| Tab, Icon, or button | ***Save*** |
| User Input | **Bold** |
| Files Names | *Italic* |

## ACKNOWLEDGMENTS

Chapter 2's exercise files are from surveys processed by Jerry Bartels. The manual survey of Chapter 13 is from Gail and Gil Evans of Chicago. Chapter 14's network traverse loop is from Bill Laster. Chapter 3's lot exercise subdivision is from a plat developed by Intech Consultants, Inc. (Tom Fahrenbok, Scott, and Joe).

Thanks to the staff at Delmar Cengage Learning for patiently waiting for the manuscript's completion; especially John Fisher and Angela Sheehan. And Patrick Franzen for Pre-Press PMG's edits and compositing. I want to thank my dear mom (Lorraine), sisters (Sharon and Dawn), and patient friends for their support and help. Also, I want to thank Imaginit, Carl, Len, Angela, Andrea, V, Brent, Jay, Mike Choquette, and many other friends for giving me advice and opportunities. I want to give Mark Martinez special thanks for his joyful and enthusiastic rereading of the manuscript for technical errors. Sadly, no matter how many times he and I reread the manuscript, there will be errors. Let me know where they are so I can fix them.

*Phillip J. Zimmerman*

# CHAPTER 1

# The Beginning

## INTRODUCTION

This chapter introduces AutoCAD Civil 3D 2010, its organization, and its interface. Civil 3D presents a civil design environment addressing traditional design issues. Civil 3D's focus is creating and documenting a design, the same as other civil engineering applications. Civil 3D's tools and design processes are similar to those found in other civil programs, particularly, AutoCAD Land Desktop Companion. Civil 3D's difference is its interactive environment and assumed dependencies between civil design elements (points, surfaces, alignments, profiles, etc.). Civil 3D's design object dependencies are unique.

## OBJECTIVES

This chapter focuses on the following topics:

- Civil 3D Objects
- Prospector Overview
- Prospector's Preview
- Settings
- Data Compatibility and Transfer with Land Desktop

### OVERVIEW

AutoCAD Civil 3D addresses the same civil engineering design issues as Civil 3D AutoCAD Land Desktop Companion (LDT), Inroads, or any other civil engineering application. Civil 3D has tools creating, evaluating, editing, and annotating familiar civil design elements (points, surfaces, alignments, profiles, etc.). Civil 3D's most radical difference is its dynamic environment. Civil 3D's design elements understand relationships and dependencies. When changing an element that has dependencies, Civil 3D updates all of the dependent elements. For example, changing a surface updates contours, profiles, corridors, and sections, reflecting their adjustment to the surface's change.

Civil 3D's development impetus was from perceived and real shortcomings of AutoCAD Land Desktop Companion. The Autodesk civil applications group addressed these issues by developing Civil 3D. Civil 3D's design/drafting environment presents the user with an interactive drafting space, new road design tools, flexible labeling, and an implementation method for standards. Besides learning and becoming comfortable with this design application, other major hurdles include understanding how to document a design, creating Civil 3D implementation content, and learning Civil 3D's behaviors and terminology.

Most company standards are linetypes, layers, pen weights, etc. Civil 3D's implementation focuses on these same standards, but its styles extend to crafting a unified company look. Flexible styles create unique review or submittal profiles, section, parcel, and other design solution elements. When implementing Civil 3D, some questions to ask are: What does a profile or section look like? What types of annotation does an object have? In a sheet, where is the annotation located, what text styles does it use, what information is called out, etc.? This microscopic look at each page is necessary to understand how to implement Civil 3D. A style's power is in creating a consistent and correct document as the designer creates or changes the design solution.

### Unit 1

The first unit reviews the anatomy of Toolspace's Prospector, demonstrates its dynamic qualities, actions occurring in its object and preview areas, and its extensive icon use (signal dependencies, out-of-date status, etc.). This unit also introduces the Civil 3D ribbon.

### Unit 2

The second unit covers Toolspace's Settings, its available styles (object, label, and table), and navigating its tree and branch structure.

### Unit 3

Objects and their behaviors are the topic of this third unit. Civil 3D implements objects with design relationships and dependencies. These objects are a design solution's fundamental building blocks. These objects are linked to styles that produce a look, implement standards, and define object property annotation formats. This object and style environment allows Civil 3D to dynamically manage civil design data and its annotation.

### Unit 4

The last unit reviews transferring data between Autodesk Land Desktop and Civil 3D. Civil 3D reads data directly from a LDT project or imports a LandXML data file (from LDT or other applications). Civil 3D AutoCAD Land Desktop 2009 has routines for extracting Civil 3D data and stores the extracted data in a LDT project.

## UNIT 1: COMMAND RIBBON AND PROSPECTOR

Civil 3D 2010 implements a command ribbon or a user can choose to use the traditional pull down menus (see Figure 1.1). At the Ribbon's top are the various tab names (collections of like tools). For example, the Home tab (top of Figure 1.1) displays icons that display or hide the toolspace, create points, define alignments, etc. The Modify tab (bottom of Figure 1.1) tools edit AutoCAD and Civil 3D objects. Each tab and their commands will be discussed when appropriate in this book.

You hide the Ribbon by typing in the command line, Ribbonclose. To display the traditional menus, click the drop-list arrow at Civil 3D's top left, to the right of redo arrow (see Figure 1.2).

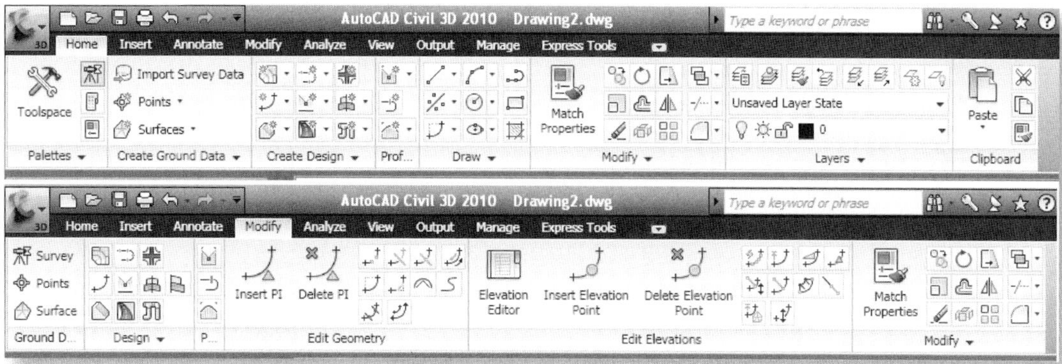


**FIGURE 1.1**

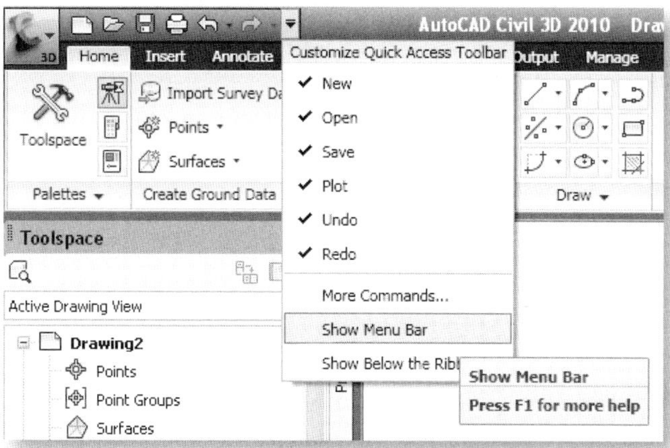


**FIGURE 1.2**

Civil 3D's Toolspace has two tabs: Prospector and Settings (see Figure 1.3). These two tabs manage all objects, styles, references, and data values for a drawing and/or project. The Ribbon's Toolspace icon displays this Toolspace and its tabs. The Toolspace floats or docks and has Auto-hide for when it is not needed. When using survey or generating reports, the Toolspace displays additional tabs. The Ribbon's—Home tab's top left Survey icon displays or hides the Survey tab. In the Ribbon's—View tab, Palettes panel, to the right of the Toolspace icon is the icon that displays or hides the Toolbox tab. The Toolbox tab contains several Civil 3D reports. An alternative, is on the Ribbon's Home tab, Palettes section, click the title to unfold the panel and click Toolbox.

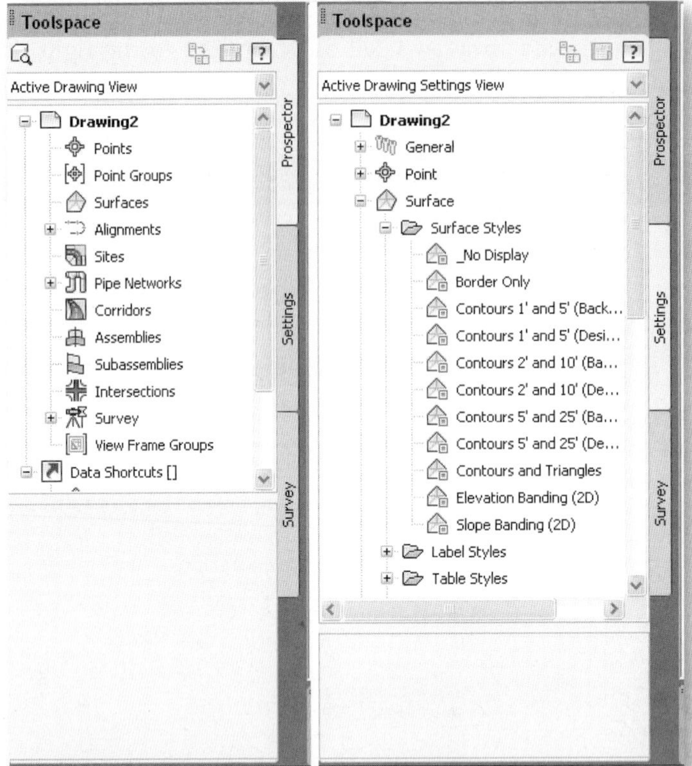


**FIGURE 1.3**

Prospector organizes objects using a hierarchical structure, displaying their data and showing object relationships. Prospector headings identify each object type and branch to reveal specific object information and data. Prospector dynamically manages drawing objects, their listings, and makes their data available to other objects. Prospector updates information and responds to changes and additions to objects and their data. When viewing an object's information, the user interacts with an object's Prospector listing.

Prospector's second area is preview. Whether docked or floating, preview is below or to Prospector's side. This area displays a list or an image of a selected Prospector entry (see Figure 1.4). In Prospector, when selecting a branch heading, a list appears in the preview area (for example, selecting the Prospector's Surfaces heading displays a list of surfaces). When selecting a surface from the named Surfaces' list, preview displays a surface image.

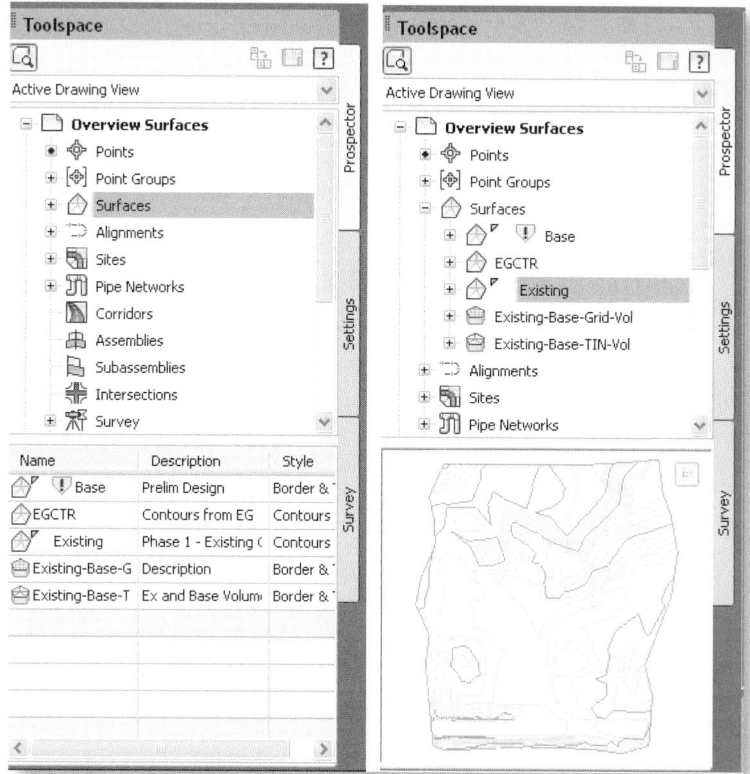


**FIGURE 1.4**

## OBJECTS

In Prospector each object type has a heading and an icon (see Figure 1.4). The object types are Points, Point Groups, Surfaces, Alignments, Sites, Pipe Networks, Corridors, Assemblies, Subassemblies, Intersections, Survey, and View Frame Groups. The Sites branch contains site alignments, grading groups, feature lines, and parcels. A Civil 3D site is the outermost boundary containing alignments and/or parcels. A site can range from a subdivision boundary, or a parcel, to a single alignment. Each Alignment branch includes entries for Profiles, Profile View, and Sample Lines.

When expanding an object type's branch, the first entry is a named instances (occurrences) list for that object type in the drawing. In Figure 1.4, for example, Base and Existing are two surface instances. Adding or deleting surfaces causes Prospector to automatically update its object list. Depending on the object type, each instance has its own branch containing its data or other critical values. For example, the Sites object type contains a sites list (Site 1, Site 2, etc.). Each site instance has its own list of alignments, profiles, sections, and parcels.

Prospector displays object data at some point down an instance's branch. However, there are times when data appears in preview instead of Prospector. For example, the surface Definition branch is a list of assigned surface data types. The data for each entry only shows in the preview area, not the branch. Some objects, such as alignments, profiles, assemblies, and subassemblies, use a Properties dialog box instead of Prospector or preview.

Prospector calls commands using right mouse button shortcut menus. Each heading (object type, object instance, and so on) has a shortcut menu unique to its branch location (see Figure 1.5). A shortcut menu's choices vary with the object type and where the entry is in the branch. For example, Surfaces' shortcut menu commands create, import, export a LandXML file, or refresh the instance list (see the left side of Figure 1.5). A named surface's shortcut menu includes build options, snapshots, zooming, etc. (see the center of Figure 1.5). A specific surface component or data element's shortcut menu has commands for creating, editing, and deleting entries (see the right side of Figure 1.5).

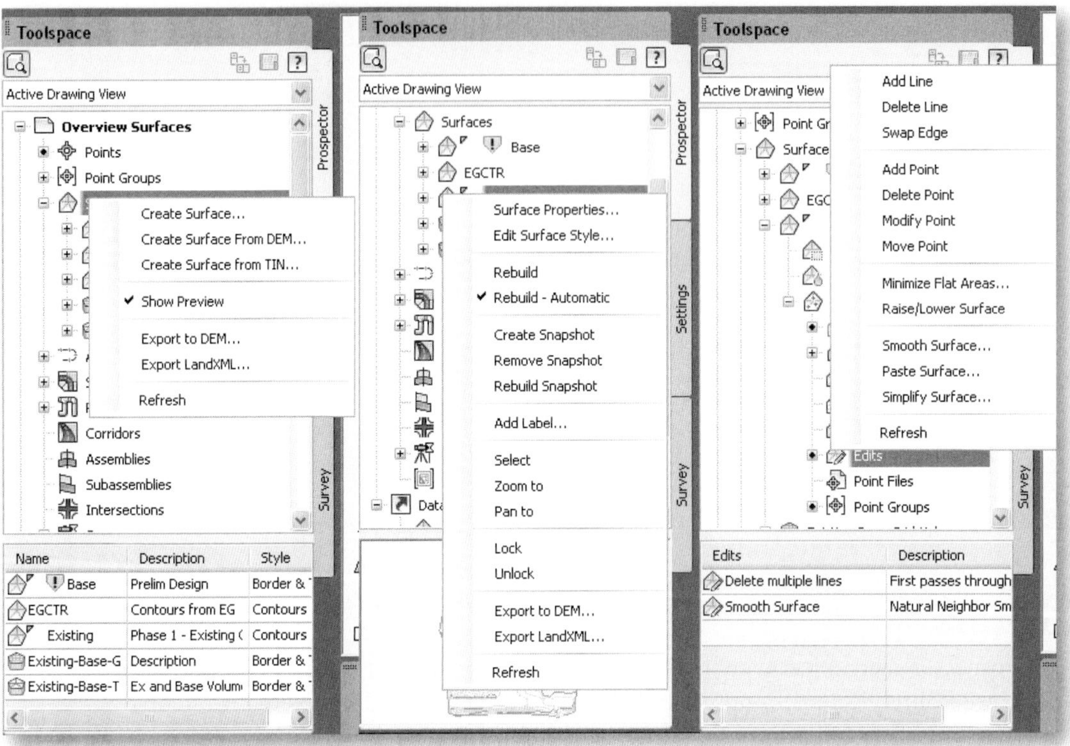


**FIGURE 1.5**

### Prospector Preview

When selecting an object type heading, Prospector responds with an object instances list (see the left side of Figure 1.6). When selecting an object instance, Prospector previews the selected instance (see the center of Figure 1.6). The heading selected determines what preview displays (for example, selecting Sites shows a sites list, selecting Site 1 shows the site's geometry, selecting Parcels shows a parcels list, and selecting a parcel previews its geometry).

- Any value in preview is editable unless it has a gray background.

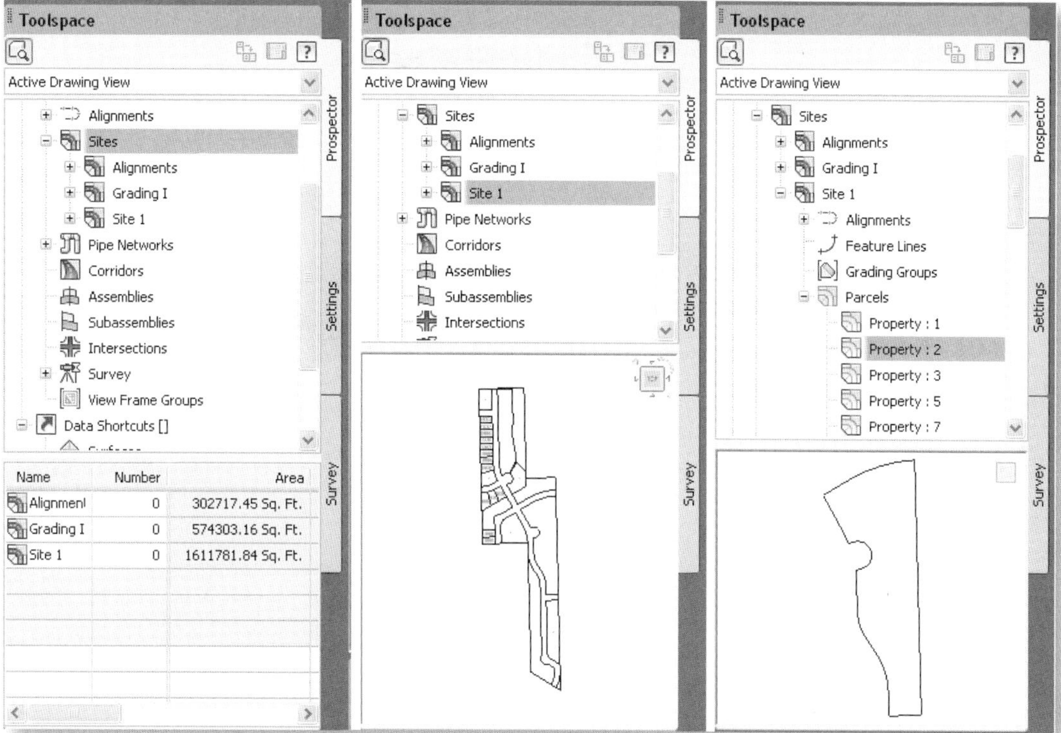


**FIGURE 1.6**

When selecting an item from the preview's list, the user can directly edit the entry or right mouse click it, viewing a shortcut menu specific to the selected item. After completing the preview editing, these changes show in Prospector and the drawing. For example, if selecting Prospector's Surfaces heading, the preview area lists the surface names, descriptions, and their current styles. In the preview area, when clicking on a surface's name, description, or style cell, the user can directly edit the cell's value. After clicking a surface name cell and right mouse clicking, a shortcut menu displays that is the same as if the user had selected a surface name from Prospector's surface instance list (see the center shortcut menu in Figure 1.5). In the preview area, when clicking a style name, a Select Style dialog box appears, listing styles for that object type. The user changes the currently assigned style by selecting another style from the styles list (see Figure 1.7). Again, changes made to a preview entry updates Prospector and the drawing.

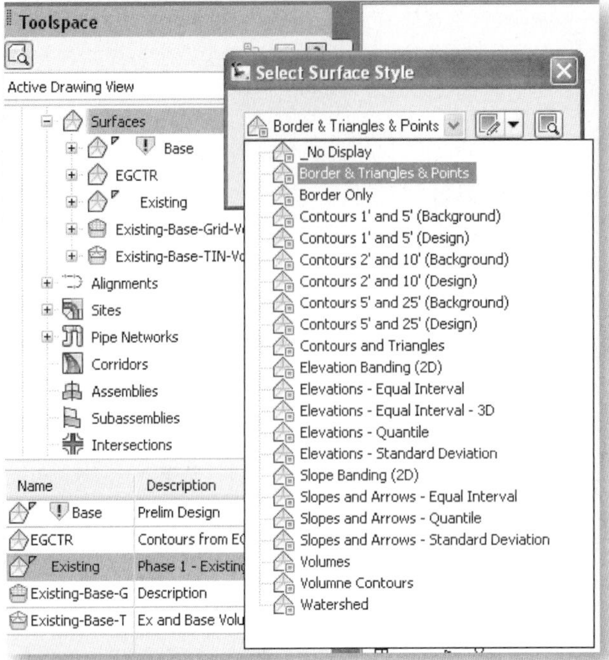


**FIGURE 1.7**

## CIVIL 3D OBJECTS

In AutoCAD Land Desktop (LDT), few, if any, objects interact with each other (the exception being a grading object and a surface). None of the roadway design line work reacts to changes made to any of its elements (a profile reacting to changes in a surface's elevations). It is left to the user to verify what changes need to be made and how to synchronize external data with the entities. In LDT, routines create on-screen line work snapshots and store them as data in external project folders. The drawing's data representation is valid only if nothing changes. If editing or redrawing line work, all related project data files need updating. In many cases the project's final design is the starting point, for fear of out-of-sync data after numerous changes and handoffs between technicians and engineers.

Civil 3D's data schema gives objects knowledge about relationships and dependencies between them. This knowledge allows objects to respond to changes by any object in their group of relationships. So, if a surface's elevations change and an alignment's has an existing surface profile, the profile updates, showing the new surface elevations along the alignment's path. If changing an alignment's location, the profile changes its elevations and either lengthens or shortens, showing the new alignment path. These types of relationships and dependencies are programmed into Civil 3D objects.

## OBJECT DEPENDENCIES AND ICONS

Prospector icons identify object types, their status, and their dependencies (see Figure 1.8). For surfaces, to the left of a surface's name is an icon indicating the surface type. In Figure 1.8, the icons to the left of Base and Existing identify them as Triangular Irregular Network (TIN) surfaces. The icons for the remaining two surfaces identify them as Grid and TIN volume surfaces.

A triangle icon pointing diagonally to the left of a surface's name indicates that another drawing object references that surface's data. When an object is referenced, it cannot be deleted until the reference(s) is removed. Figure 1.8 shows an out-of-date icon (a shield with an exclamation mark), indicating that something changed with the Existing-Base-Grid-Vol surface. To remove out-of-date surface icons, the user must rebuild the surface. Rebuilding surfaces accommodates change(s) and removes the surface out-of-date icon.

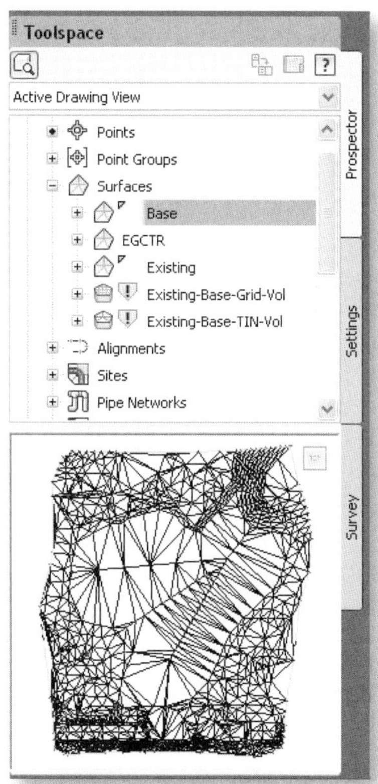


**FIGURE 1.8**

## PROSPECTOR: DATA SHORTCUTS AND PROJECT MANAGEMENT

Civil 3D implements project- and data-sharing capabilities through Prospector, Data Shortcuts, and Autodesk Vault. Data Shortcuts and Project Management are visible only in Prospector's Master View. Master View is set by selecting it from the view list above the drawing's name (see Figure 1.9).

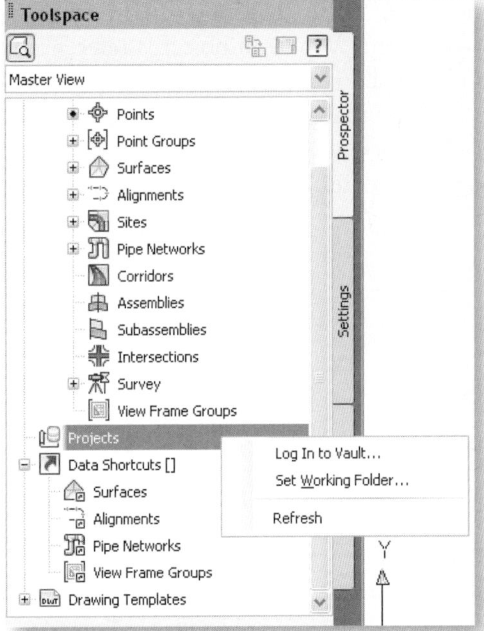


**FIGURE 1.9**

Prospector's Projects branch assists in managing a project's data and drawings. Civil 3D manages project data as a check in or check out process. This process allows users to have data editing control. Users can create data references (consume only) and, if others edit the referenced data, Prospector notifies them that their data is out-of-sync relative to the project's entries. If Vault data changes, the user is notified of the changed data and can synchronize the drawing to the changed data. Again, Prospector uses icons to indicate data's changed status in the current drawing and in the project.

A second data sharing method is a data shortcut. A data shortcut allows a user to share an object with other drawings. If the data shortcuts' object changes, the shortcut notifies the user that the object has changed and after updating the reference, the object present in the drawing matches the current object definition.

Civil 3D Project management and Data Shortcuts are discussed in detail in Chapter 12.

## CIVIL 3D TEMPLATES

Civil 3D ships with two content templates. When in Prospector's Master View, it displays a Drawing Templates branch with a template files list that you can use to start a new drawing. Rather than selecting the New icon, in Prospector's Master View the user can select a template from the Drawing Templates, AutoCAD list, press the right mouse button, and select Create New Drawing (see Figure 1.10).

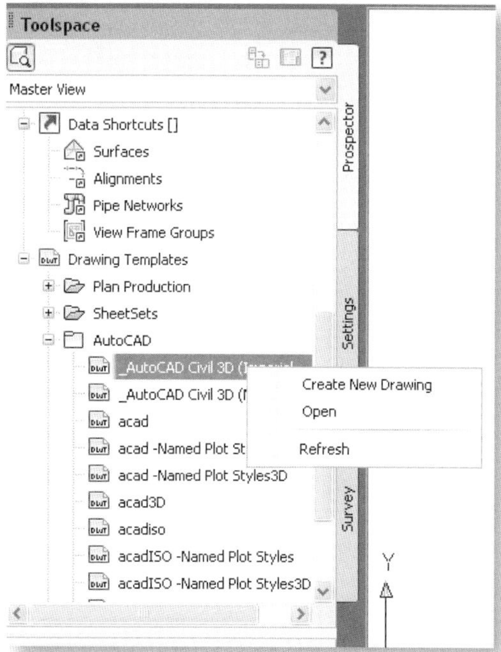


**FIGURE 1.10**

This textbook uses the Autodesk Civil 3D (Imperial) NCS template file. This content template assigns needed layer names and styles. Several exercises will modify, create new styles, and define new layers in addition to those in this template file.

## THE PANORAMA

Civil 3D displays object value editors (grid based) as vistas within a panorama (see Figure 1.11). A panorama can have more than one vista, but only one vista is active at a time.

To close a vista, click the green checkmark at the top of the panorama. To close the panorama, click the X at the top of its mast. If the panorama is closed by clicking its X, it can be redisplayed by selecting the Redisplay Panorama icon to the left of Prospector's Help icon. If the panorama is closed by clicking a vista's green checkmark and it is the only vista in the panorama, you cannot redisplay the panorama because the Redisplay Panorama icon becomes inactive. If this happens, the only way to redisplay the panorama is by executing a command to create a new vista.

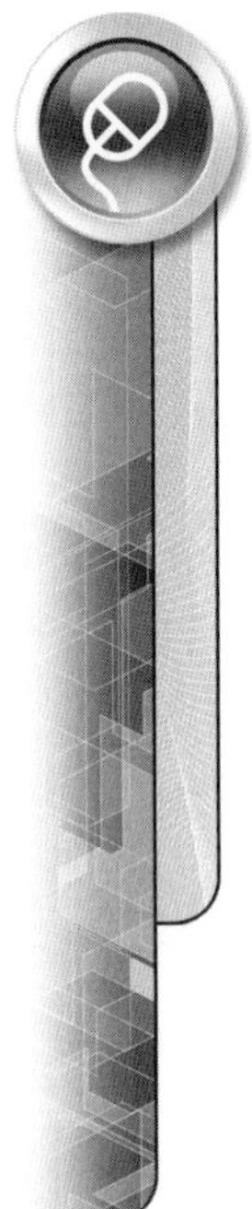

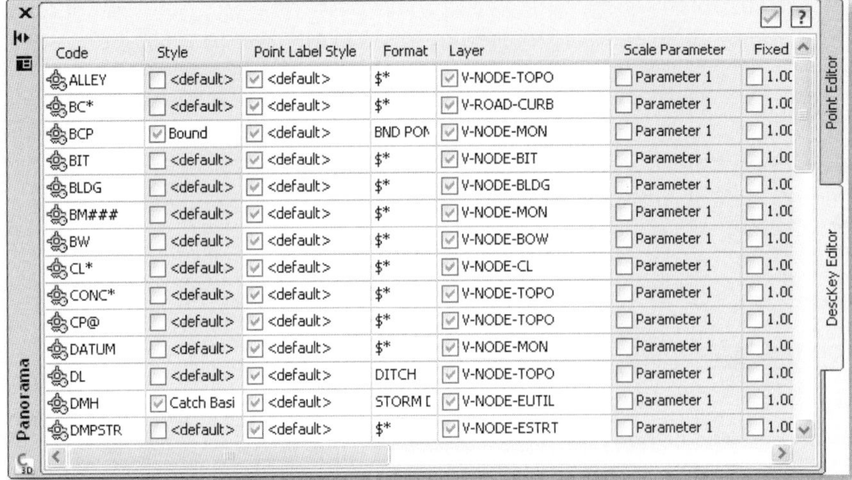


| Code | Style | Point Label Style | Format | Layer | Scale Parameter | Fixed |
|---|---|---|---|---|---|---|
| ALLEY | <default> | <default> | $* | V-NODE-TOPO | Parameter 1 | 1.0C |
| BC* | <default> | <default> | $* | V-ROAD-CURB | Parameter 1 | 1.0C |
| BCP | Bound | <default> | BND POI | V-NODE-MON | Parameter 1 | 1.0C |
| BIT | <default> | <default> | $* | V-NODE-BIT | Parameter 1 | 1.0C |
| BLDG | <default> | <default> | $* | V-NODE-BLDG | Parameter 1 | 1.0C |
| BM### | <default> | <default> | $* | V-NODE-MON | Parameter 1 | 1.0C |
| BW | <default> | <default> | $* | V-NODE-BOW | Parameter 1 | 1.0C |
| CL* | <default> | <default> | $* | V-NODE-CL | Parameter 1 | 1.0C |
| CONC* | <default> | <default> | $* | V-NODE-TOPO | Parameter 1 | 1.0C |
| CP@ | <default> | <default> | $* | V-NODE-TOPO | Parameter 1 | 1.0C |
| DATUM | <default> | <default> | $* | V-NODE-MON | Parameter 1 | 1.0C |
| DL | <default> | <default> | DITCH | V-NODE-TOPO | Parameter 1 | 1.0C |
| DMH | Catch Basi | <default> | STORM I | V-NODE-EUTIL | Parameter 1 | 1.0C |
| DMPSTR | <default> | <default> | $* | V-NODE-ESTRT | Parameter 1 | 1.0C |

**FIGURE 1.11**

## EXERCISE 1-1

After completing this exercise, you will:

- Become familiar with Toolspace and Prospector.
- Observe how Prospector dynamically manages objects.
- Change an alignment's path and view the changes to the road model.
- Dissolve and add parcels to a site design and view the changes to the Parcels data tree.

### Toolspace Basics

1. If not in `CIVIL 3D`, double-click its desktop icon to start the application.
2. If the Toolspace does not show, at the Ribbon's top left (Home tab, Palettes panel) click the ***Toolspace*** icon.
3. Close all open drawings and do not save them.
4. In Toolspace, Prospector tab top right, under the Help icon, select the drop-list arrow and from the list `MASTER VIEW`.
5. In Prospector, expand Drawing Templates and AutoCAD branches, from the list select **_AutoCAD Civil 3D (Imperial) NCS**, press the right mouse button and, from the shortcut menu, select `CREATE NEW DRAWING`.
6. If the Toolspace is not floating, click and hold the left mouse button down near Toolspace's heading, and drag it to the screen's center.
7. On Toolspace's mast, right mouse click and, from the shortcut menu, click Allow Docking to toggle it **OFF**.
8. Again on Toolspace's mast, right mouse click and, in the shortcut menu if it is toggled on, click Auto-hide to toggle it **OFF**.
9. On the Toolspace's mast, press and hold the left mouse button and move the Toolspace to the screen's right and left sides.

Toolspace's mast switches from side to side, but preview remains below Prospector.

10. On the Toolspace's mast, press the right mouse button and, from the shortcut menu, click Auto-hide, toggling it **ON**. Move the cursor away from Toolspace.

The Toolspace hides Prospector under its mast.

11. On the Toolspace's mast, press the right mouse button and, from the shortcut menu, click Allow Docking, toggling it **ON**. At the screen's left side, dock the Toolspace.
12. If necessary, click the ***Prospector*** tab and scroll to the top until viewing the Open Drawings branch.
13. If necessary, to the left of the drawing's name, click the expand tree icon (plus sign), expanding the drawing's object type hierarchy.
14. Adjust preview's size by placing the cursor at the Prospector and preview boundary, pressing and holding down the left mouse button, and sliding it up or down.
15. At the Toolspace's top, under the Help icon, select the drop-list arrow and select ACTIVE DRAWING VIEW.
16. In the display's upper right, select the ***Close*** icon and do not save the drawing.

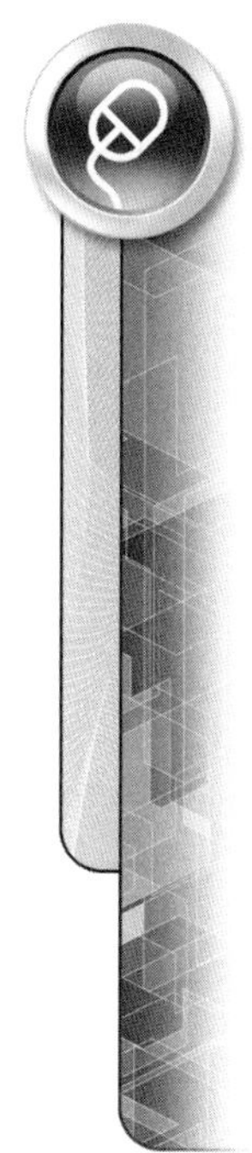

## Prospector Object Management

The remainder of the exercise uses drawings from this textbook's CD, Chapter 1 folder.

Prospector is Civil 3D's control center and it dynamically updates its object list as changes occur in the drawing. Prospector also manages an object's data and makes its data available.

1. At Civil 3D's top left, Quick Access Toolbar, select the ***Open*** icon, browse to this textbook's CD, Chapter 1 folder, select the drawing *Overview Prospector*, and click ***Open***.
2. If necessary, make sure Prospector is showing by selecting the Toolspace's Prospector tab.

## Points

1. In Prospector, select the **Points** heading.

The square with a black dot icon to the left of Points indicates that the drawing has Civil 3D points. When selecting a Prospector object heading, preview displays a corresponding object list (see Figure 1.12). In this case, the preview area lists all of the drawing's points. Any value in preview is editable unless it has a gray background.

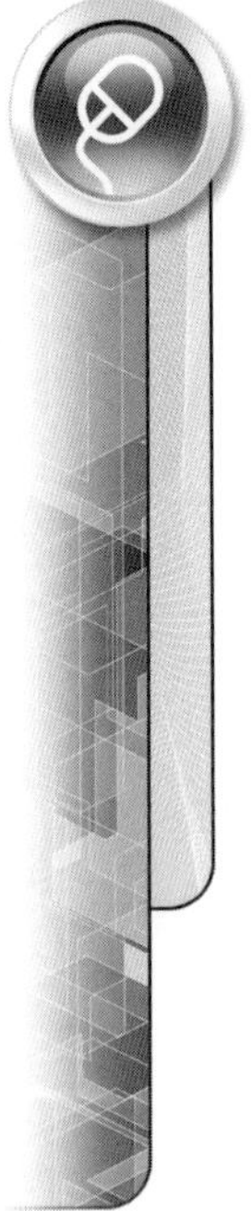

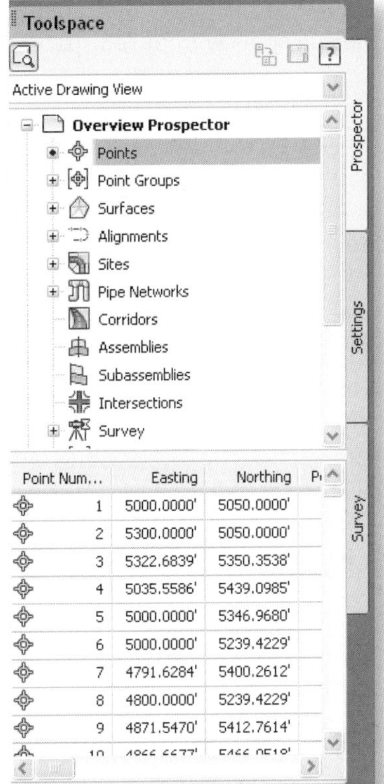


**FIGURE 1.12**

Each Prospector tree heading has a specific shortcut menu (see Figure 1.5). The number of shortcut menu commands depends on an object's complexity or the user's location in the Prospector tree.

2. With **Points** still highlighted, press the right mouse button and, from the shortcut menu, select CREATE..., displaying the Create Points toolbar.
3. To close the toolbar, click the red **X** in its upper-right corner.
4. In preview, double-click a point's number.

By clicking a preview entry, preview allows edits or style changes.

## Point Groups

1. In Prospector, select the **Point Groups** heading and notice that preview lists the drawing's point groups.

Each point group can have a point and/or a label style override. A point and/or a label style override changes the points' display for all point group points.

2. In Prospector, expand the Point Groups branch, viewing the point group list.
3. Select the point group **Existing Ground Points**.

This displays in preview the group's points.

4. With Existing Ground Points still highlighted, press the right mouse button and, from the shortcut menu, select EDIT POINTS....

This displays a panorama with the Point Editor vista containing the point group points. The vista allows point value edits.

5. At the top of panorama's mast, click the **X**, closing it.
6. To the left of Prospector's Help icon, click the ***Redisplay Panorama*** icon.
7. At the top of panorama's mast, click the **X**, closing it.

The Point Groups heading's property is the point groups' display order (see Figure 1.13). The drawing's point groups display begins with the list's lowest point group and continues to the top.

This Display Order property "hides" or changes point markers and labels, enabling you to view or display different point data combinations. In the current drawing, the No Show point group hides all of the points, because it is the last drawn point group (top of the list).

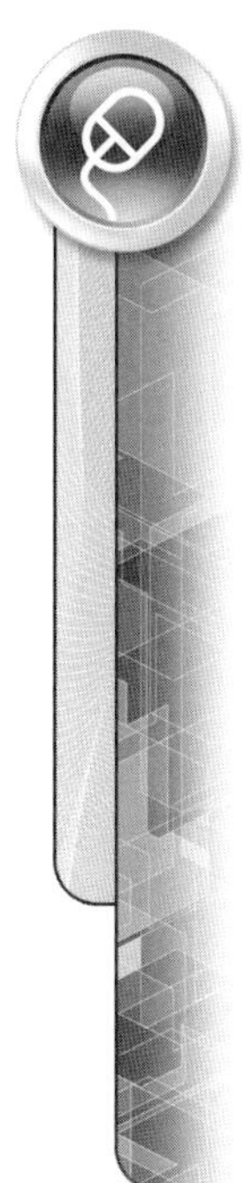

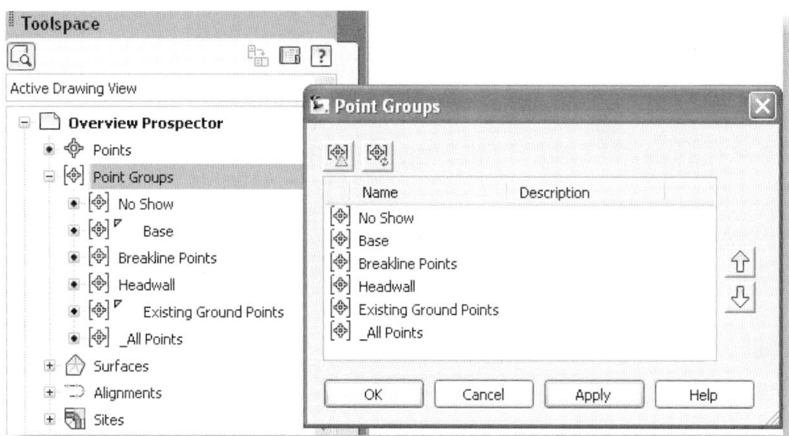


**FIGURE 1.13**

8. In Prospector, select the heading **Point Groups**, press the right mouse button, and, from the shortcut menu, select PROPERTIES....

The top group is No Show and its overrides suppress all point markers and labels (see Figure 1.14). To view different point groups, change their position relative to the No Show group by using the up and down arrow buttons at the right side of the dialog box. The viewable point groups are above No Show.

If the Event Viewer displays, close it by clicking the green check mark in the upper-right of the panorama.

9. In Point Group Properties, select **Existing Ground Points** and move it to the top position by clicking the up arrow on the dialog box's right.
10. If necessary, select **No Show** and move it to the second position.
11. Click ***OK***, viewing the Existing Ground Points point group.

The drawing displays the Existing Ground Points point group points and their assigned point and label styles.

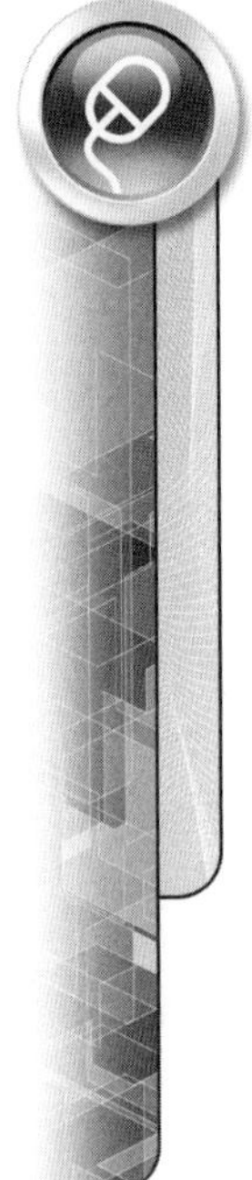

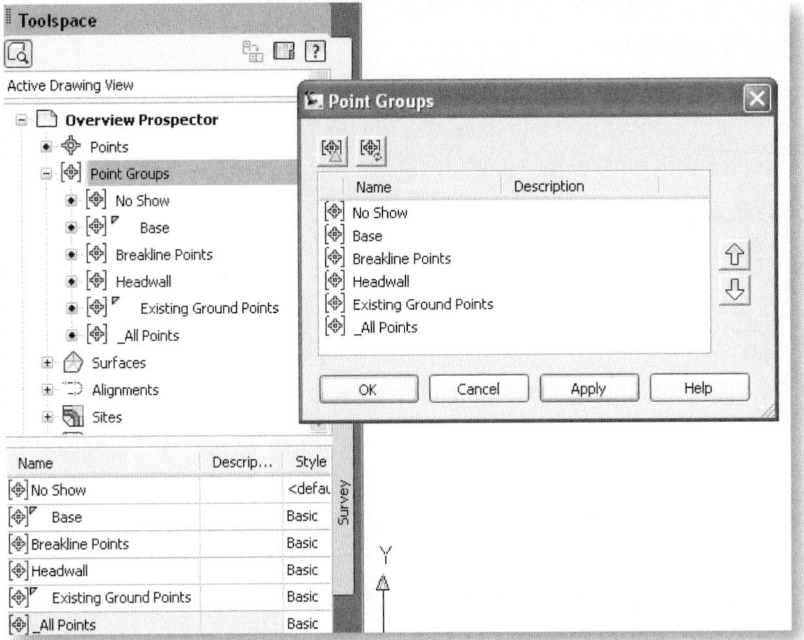


**FIGURE 1.14**

12. Again, in Prospector, select the **Point Groups** heading, press the right mouse button, and, from the shortcut menu, select `PROPERTIES`....
13. In the Point Group Properties dialog box select the **Breakline Points** point group and move it to the top position.
14. Select the **No Show** point group and move it to the second position.
15. Click ***OK***, exiting and displaying the Breakline Points point group. Use the `REGENALL` command to update the display.

This displays the Breakline Points point group (see Figure 1.15). By isolating these points and their surface triangulation you can review the linear objects "successful" triangulation in a surface.

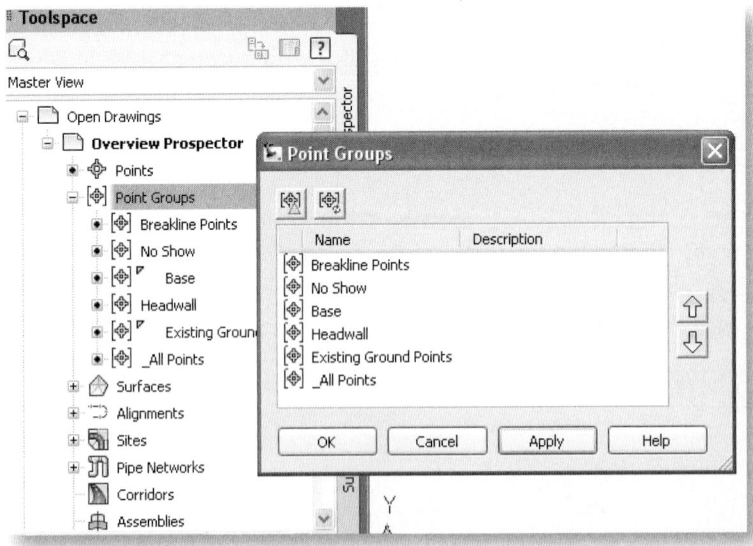


**FIGURE 1.15**

16. At the displays upper right, select ***Close*** icon and do not save the changes.

## Civil 3D Object Reactions and Dependencies — Alignments and Profiles

All Civil 3D objects react to changes affecting their display or information. When changing the data an object depends on, or changing the object itself, all objects dependent on the changed object react and accommodate its change.

1. At Civil 3D's top left, Quick Access Toolbar, select the ***Open*** icon, browse to this textbook's CD, Chapter 1 folder, select the drawing *Overview Road*, and click ***Open***.

When changing an alignment's endpoint, the profile and profile view react to the change by displaying new elevations and profile length.

2. Use the ZOOM and PAN commands and view the profile view's right side elevation annotation.
3. If necessary, click the ***Prospector*** tab.
4. In Prospector, expand the Surfaces branch until you view the surface EG (1).
5. Expand the EG (1) branch until you view the Definition heading and its data entries.
6. From the data list, select the heading **Edits**, press the right mouse button, and select RAISE/LOWER SURFACE (see Figure 1.16).
7. In the Command Line, for the amount to add to the surface's elevation, enter **10** and press ENTER.

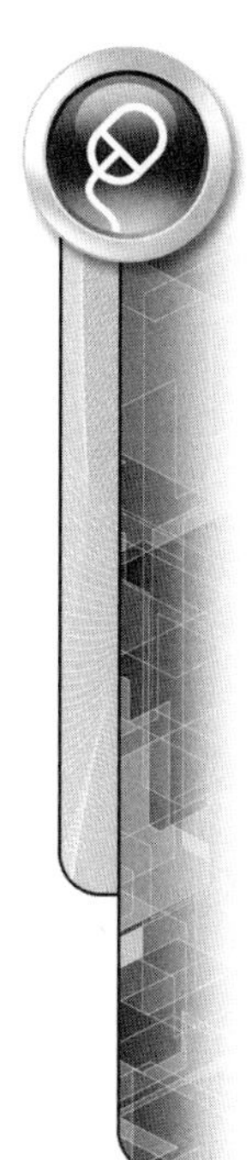

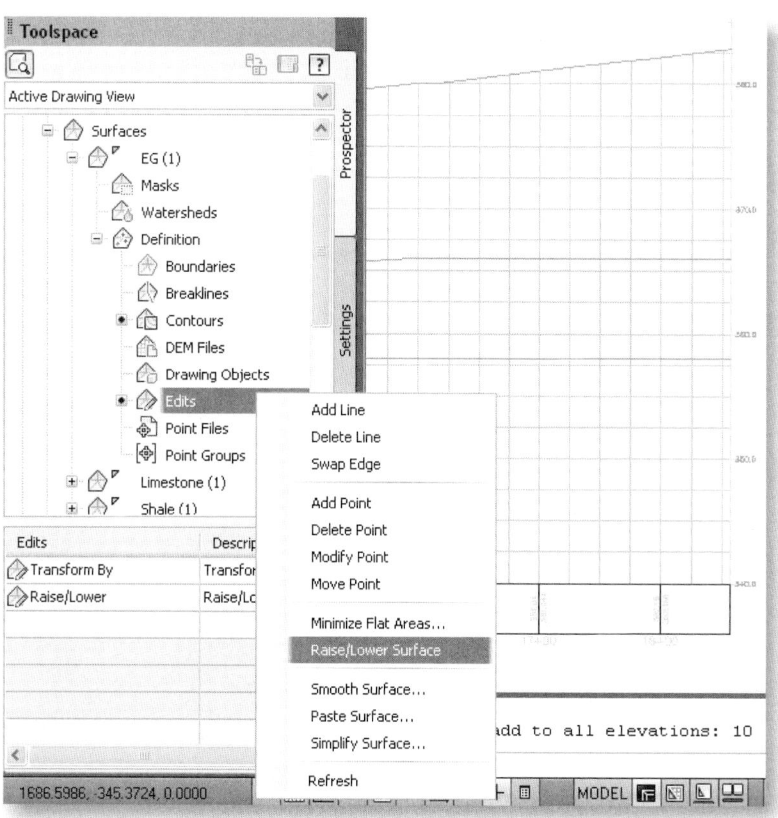


**FIGURE 1.16**

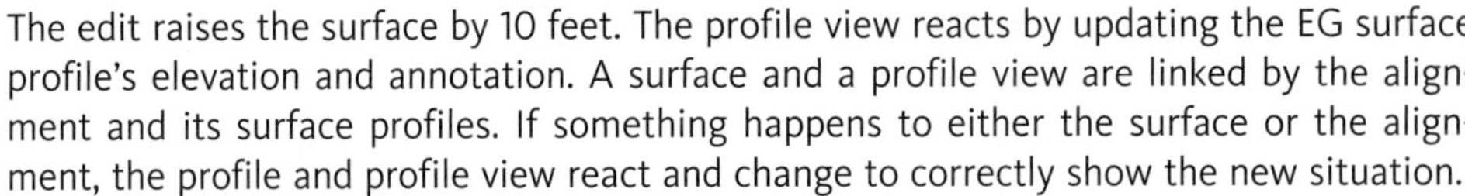

The edit raises the surface by 10 feet. The profile view reacts by updating the EG surface profile's elevation and annotation. A surface and a profile view are linked by the alignment and its surface profiles. If something happens to either the surface or the alignment, the profile and profile view react and change to correctly show the new situation.

8. At the displays upper right, select the ***Close*** icon and do not save the changes.

## Civil 3D Object Reactions and Dependencies — Surface and Contours

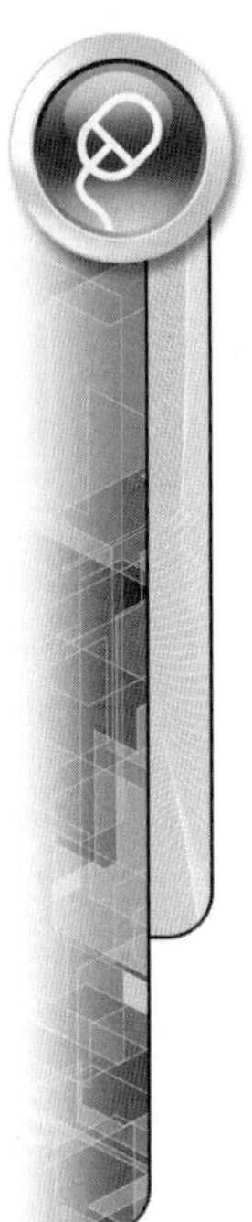

When changing surface data, the surface reacts by updating its definition. If the current style displays contours, the contours change, showing the changed surface elevations.

If the Event Viewer displays, close it by clicking the green check mark in the panorama's upper-right corner.

1. At Civil 3D's top left, Quick Access Toolbar, click the ***Open*** icon, browse to this textbook's CD, Chapter 1 folder, select the drawing *Overview Surfaces*, and click ***Open***.
2. In the drawing, select any surface contour, press the right mouse button, and, from the shortcut menu, select SURFACE PROPERTIES....
3. In the Surface Properties dialog box, click the ***Information*** tab. To the right of Surface style (middle left), view the styles list by clicking the drop-list arrow, and change the current surface style by selecting **Contours and Triangles** (see Figure 1.17).

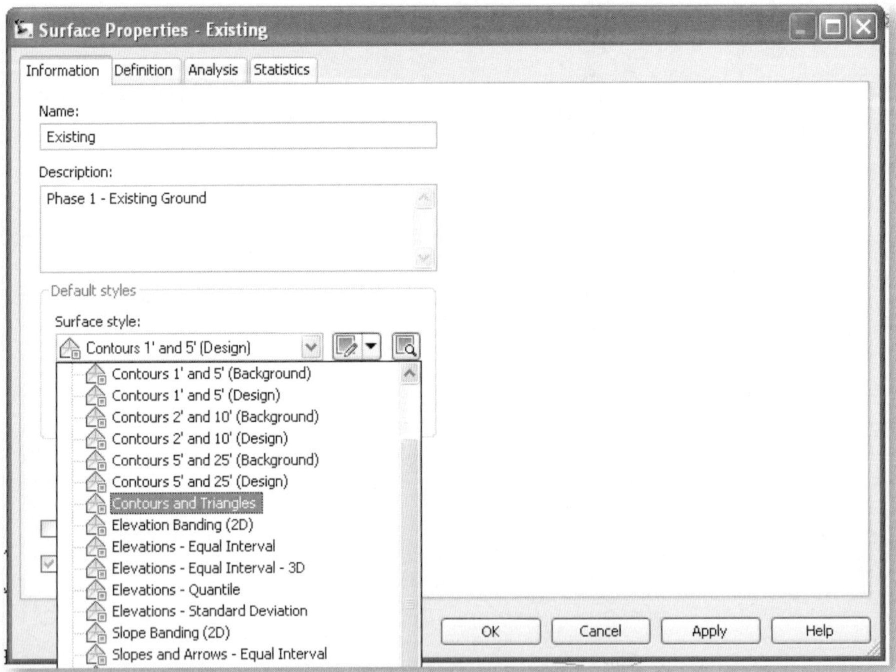


**FIGURE 1.17**

4. Click ***OK***, exiting the dialog box.

The surface now shows contour and triangle surface components. The next step zooms to a point that you are going to modify. After modifying the point, the contours change, reflecting the new surface elevation.

5. In Prospector, at its top, select the **Points** heading, listing all points in preview.
6. In preview, scroll through the point list until you locate point number **71**.
7. In preview, select point number 71's icon, press the right mouse button, and, from the shortcut menu, select ZOOM TO.

The zoom centers the point in the display.

8. In Prospector, expand Surfaces and the Existing surface branches until you view the Definition heading's data list.
9. In the Existing's Definition tree, select **Edits**, press the right mouse button and, from the shortcut menu, select MODIFY POINT.

Your Prospector should look like Figure 1.18.

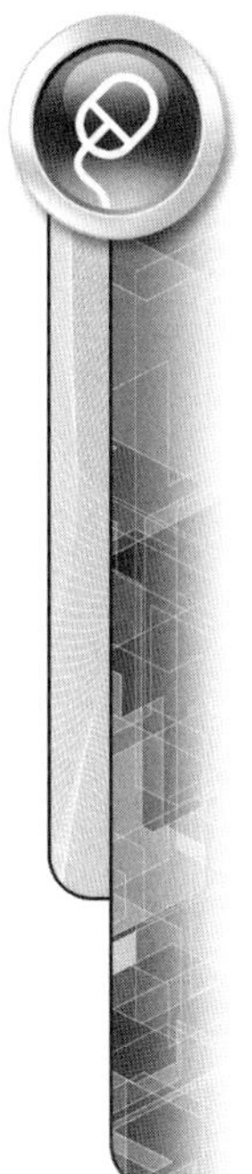

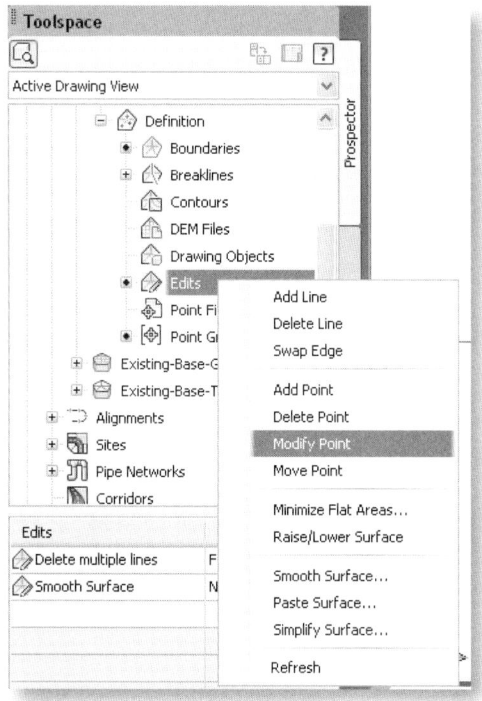


**FIGURE 1.18**

10. In the drawing, at the display's center, select the triangles' intersection, and press ENTER. In the command line, enter a new point elevation of **740** and press ENTER twice, changing the elevation and exiting the command.

This edit changes the surface data, and the contours change, showing the new surface elevation. Prospector's preview adds the edit to the surface's edits list (see Figure 1.19). The surfaces dependent on Existing become out-of-date because of the change made to Existing.

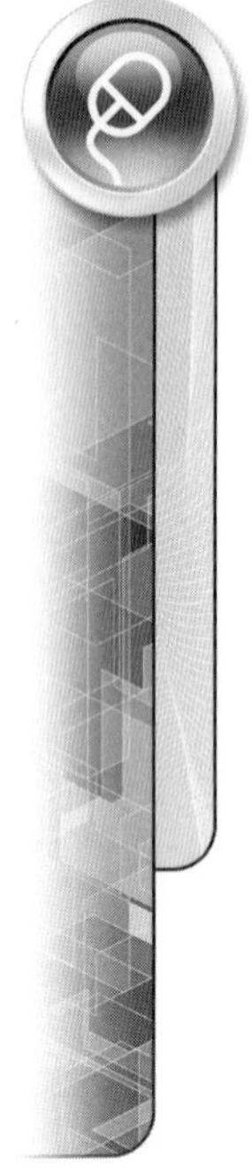

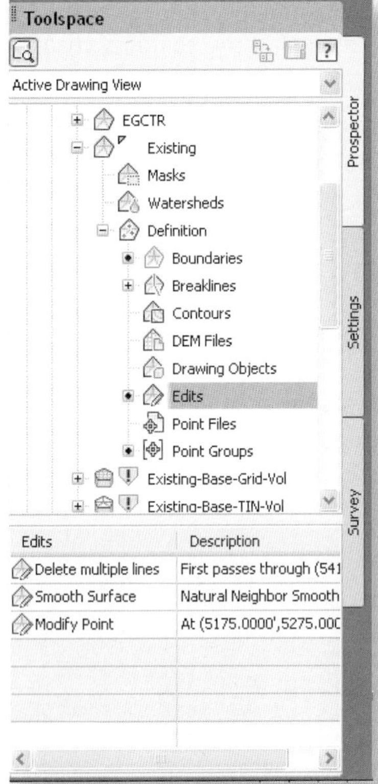


**FIGURE 1.19**

If the previous surface edit is speculative, you can temporarily remove it from the surface. This is a function of the Surface Properties dialog box's Definition panel. If you want to permanently remove the edit, you can delete it either in the Surface Properties dialog box's Definition panel (Remove From Definition) or you can delete it from preview's edits list.

11. In the drawing, select a contour or triangle leg, press the right mouse button and, from the shortcut menu, select SURFACE PROPERTIES....
12. In the Surface Properties dialog box, click the ***Definition*** tab; then in Operation Type (at the bottom of the panel) scroll up or down the list until you locate Modify Point.

The operation has a check mark, indicating it is an active surface edit.

13. Toggle **OFF** the Modify Point edit and click ***OK***, exiting the dialog box.

A warning dialog box displays.

14. In the Warning dialog box, click REBUILD THE SURFACE, rebuilding the surface and exiting the Surface Properties dialog box (see Figure 1.20).

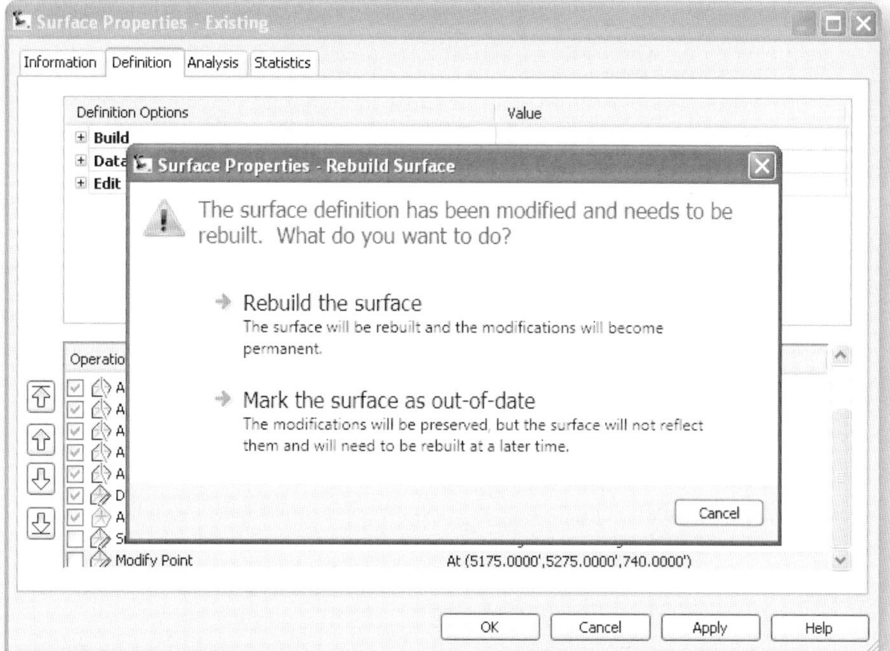


**FIGURE 1.20**

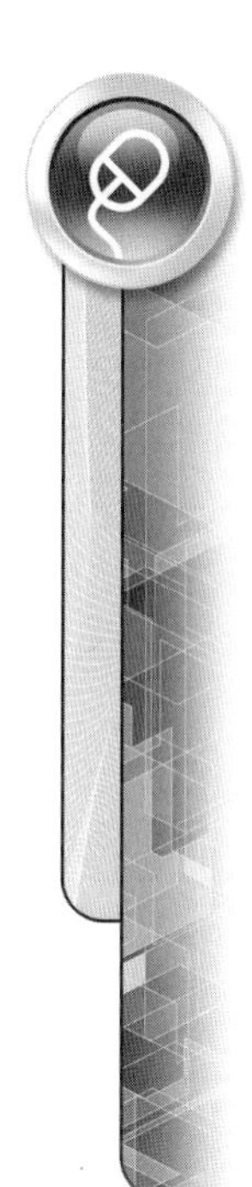

The surface shows the modified point removed from its data by changing the contours.

15. Return to the Surface Properties dialog box (see Step 11). In the Definition panel, Operation Type, toggle **ON** the Modify Point edit, and click ***OK*** to exit the dialog box.
16. In the Warning dialog box, click REBUILD THE SURFACE, rebuilding it and exiting the Surface Properties dialog box.

The edit reappears in the surface.

17. If necessary, in Prospector, in the Existing surface branch, from the Definition heading's list, select **Edits**.

The preview area displays the Existing surface edits.

18. In the preview area, select the entry **Modify Point**, press the right mouse button and, from the shortcut menu, select DELETE... to permanently remove the surface edit.

A Remove From Definition dialog box displays (see Figure 1.21).

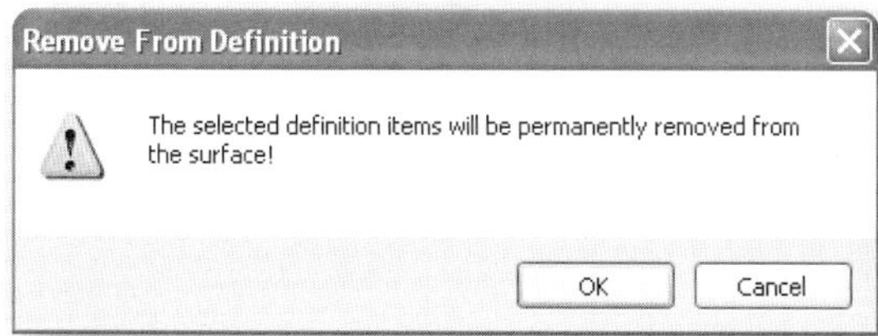


**FIGURE 1.21**

19. In the Remove From Definition dialog box, click ***OK*** to accept the deletion.

This permanently removes the surface edit.

20. At the display's upper right, click the ***Close*** icon and do not save the changes.

## Dynamic Data Management — Parcel Properties

Prospector dynamically manages the drawing's objects list. Each entry is a drawing's object type instance (occurrence). When adding or removing drawing objects, Prospector automatically updates the instance list.

A Site's parcels list identifies each parcel's type and its identifier (usually a parcel number). When defining a parcel, Prospector adds a new parcel to the list and labels it in the drawing.

If the Event Viewer displays, close it by clicking the green check mark in the upper-right of the panorama.

1. At Civil 3D's top left, Quick Access Toolbar, select the ***Open*** icon, browse to this textbook's CD, Chapter 01 folder, select the drawing **Overview Parcels**, and click ***Open***.
2. If necessary, click the **Prospector** tab.
3. In Prospector, expand the Sites branch until you view the Parcels heading and its parcels list. Your screen should be similar to Figure 1.22.

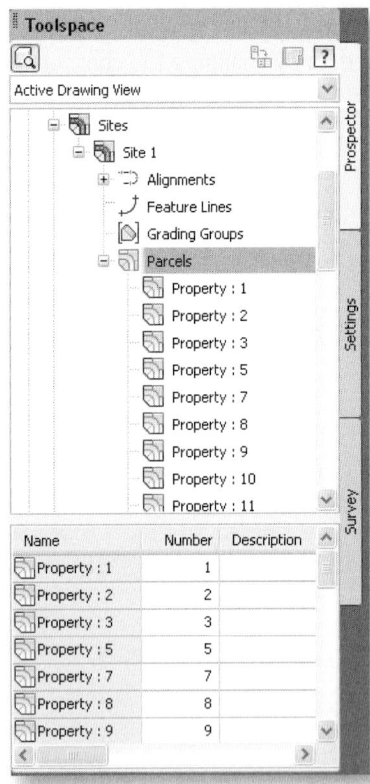


**FIGURE 1.22**

Each parcel's property includes a boundary analysis (see Figure 1.23).

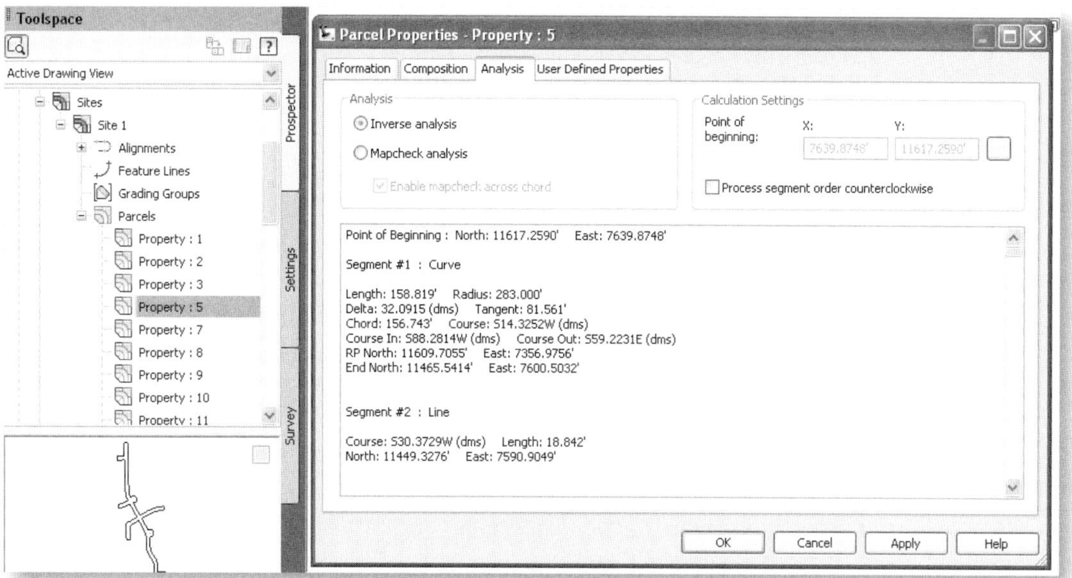


**FIGURE 1.23**

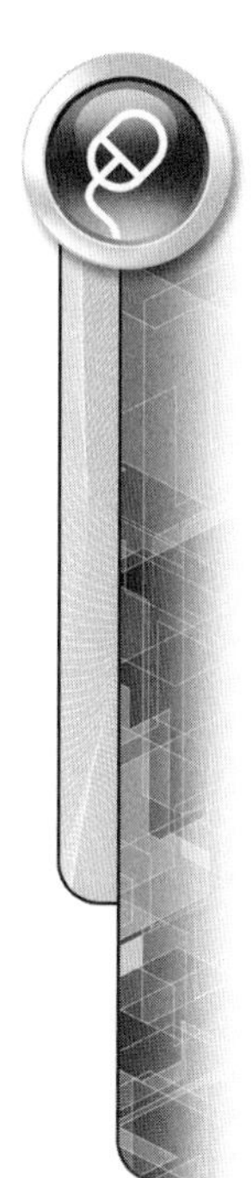

4. In Parcels, select a parcel from the list, press the right mouse button and, from the shortcut menu, select PROPERTIES....
5. In Parcel Properties, click the ***Analysis*** tab and review the selected parcel's **Inverse** and **Mapcheck** reports.
6. Click ***OK***, exiting Parcel Properties.

## Reports Manager — Parcels

1. From the Ribbon headings select the View tab. In the View ribbon, Palettes panel, to the Toolspace icon's right, bottom row, click the middle icon, click ***Toolbox*** to display Toolspace's Toolbox tab.
2. In the Toolbox panel, expand Reports Manager, then expand Parcel until displaying a parcel reports list.
3. From the parcel reports list, select **Inverse_Report**, press the right mouse button and, from the shortcut menu, select EXECUTE....
4. In the Export to XML Report dialog box, deselect everything except for two parcels and click ***OK***.

Internet Explorer displays with a Parcel Inverse Report.

5. Review the report and then close Internet Explorer.

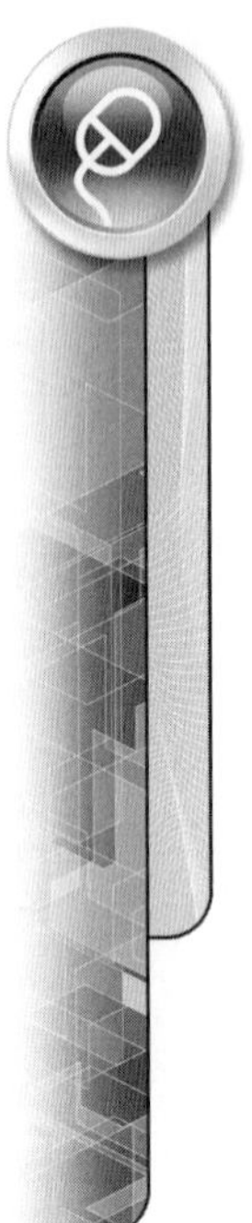

## Adding and Deleting Parcels

Whenever adding or deleting parcels, Prospector modifies its Parcels list.

1. Click the ***Prospector*** tab.
2. From the Parcels list, select **Single-Family : 17**, press the right mouse button and, from the shortcut menu, select ZOOM TO.
3. Use the ERASE command, and in the drawing select the north/south side yard line, dividing parcels **4** and **17**. Press ENTER.

The two parcels merge into one.

4. In Prospector, the Parcels list needs refreshing. Select the **Parcels** heading, press the right mouse button, and, from the shortcut menu, select REFRESH.
5. Scroll through the list, verifying Single-Family : 17's absence from the list.
6. At the displays upper right, select the ***Close*** icon and do not save the drawing.

## SUMMARY

- Prospector is Civil 3D's command center and data and object manager.
- Civil 3D objects are dynamic and update their display when edited.
- Prospector adds and removes objects from its lists as a user creates and deletes Civil 3D objects.
- Prospector displays and manages each object's status (out of date, reference, locked, etc.).
- Prospector displays dependencies and references.
- Civil 3D does not allow objects with dependencies to be deleted.
- Selecting a Prospector heading displays a list or image in the preview area.

## UNIT 2: SETTINGS

The Settings hierarchy manages a drawing's settings and styles. Settings at the drawing name are at the highest level and affect all style values lower in the hierarchy.

Settings' hierarchical tree displays and manages each object type's settings and styles. Each object type is a branch heading and within the branch are the object type's settings and styles. When selecting an object type and pressing the right mouse button, Civil 3D displays a shortcut menu listing editors for feature (object) values, label style defaults, and so on (see Figure 1.24).

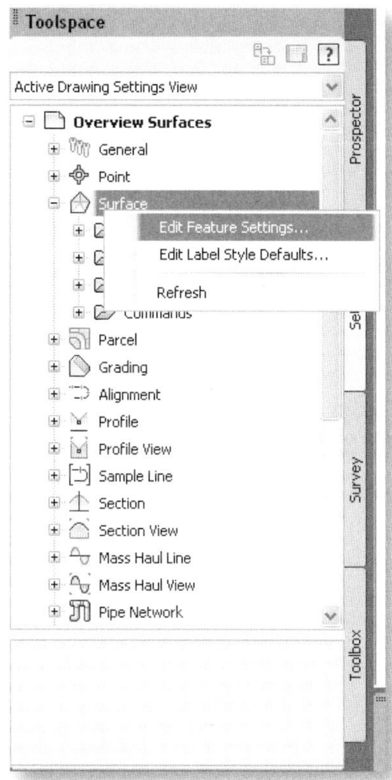


**FIGURE 1.24**

## EDIT DRAWING SETTINGS

Edit Drawing Settings is at the hierarchy's top. This dialog box values affect drawing scale, coordinate systems, label abbreviations, layer assignments, and prompting and listing values.

In Figure 1.25, Edit Drawing Settings' Object Layers panel sets each object type's base layer name, if using a modifier, and the modifier's value. If there is more than one object type instance, the user should define a base layer modifier. A modifier can be a base layer name prefix or suffix with the modifier value being the instance's name (for example, Existing and Base are two TIN surface names that can be appended to the base layer name). An asterisk (*) assigns the object's name to the modifier value. There should be a spacing character between the base layer name and the modifier such as a dash (-) or an underscore (_). If a surface's name is Existing, and the user sets the modifier to Suffix and set its value to -* (a dash and an asterisk), the resulting surface layer name is C-TOPO-Existing. If setting an object label layer, a label style, if its layer is set to 0 (zero) will use the concatenated layer name for its labels. If using the same modifier and value as the TIN Surface Labeling, the resulting label layer would be C-TOPO-TEXT-Existing.

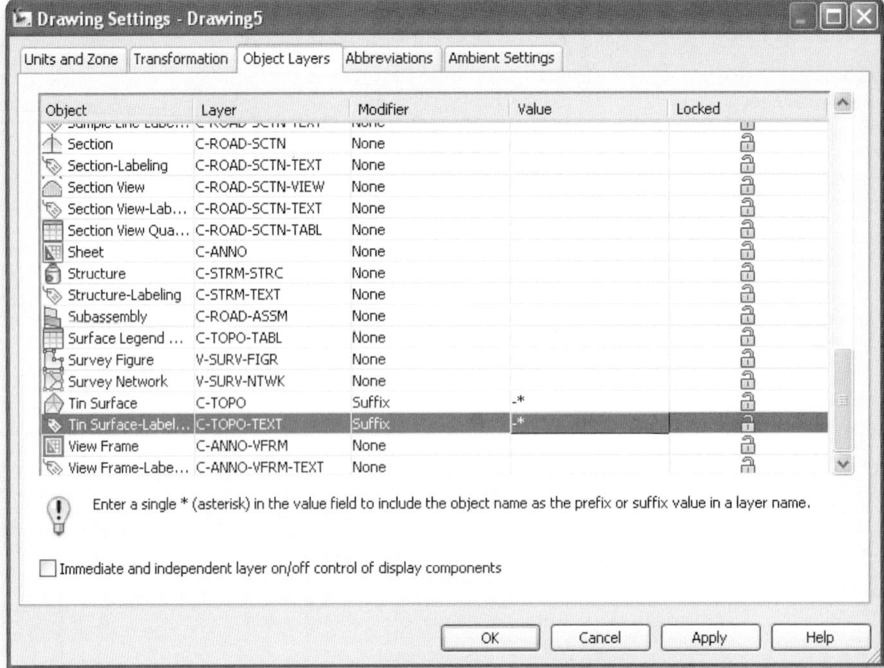


**FIGURE 1.25**

Edit Drawing Settings' Ambient Settings panel sets basic rules for units, rounding, precision, format, etc. For example, to prompt for, list values as, or report cubic feet as the volume unit, in Ambient Settings' Volume section, change the Unit value to cubic feet (see Figure 1.26). The Grade/Slope Section sets slope listing to either percent, rise:run, run:rise, etc., and controls its precision and its rounding.

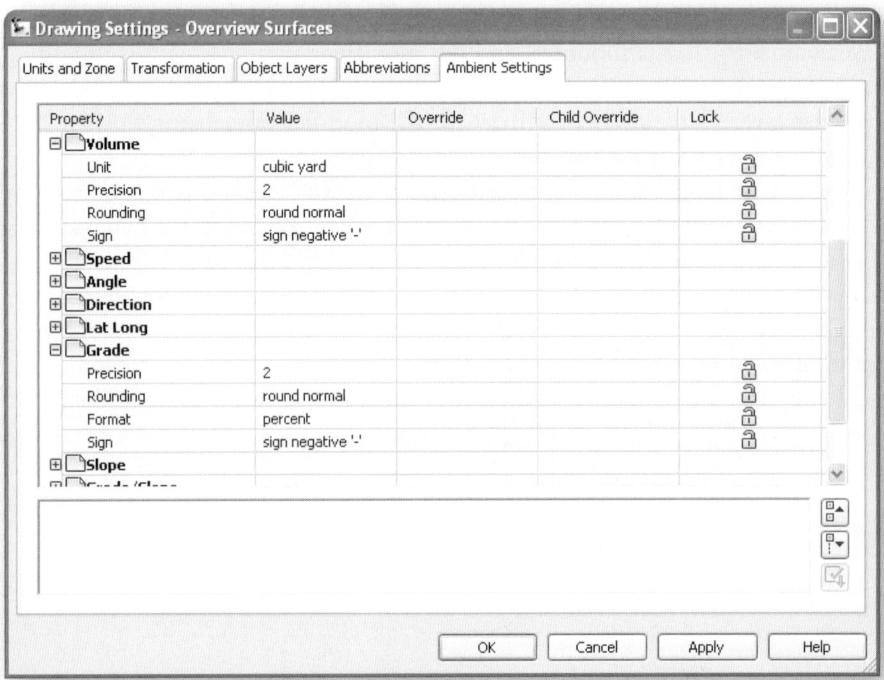


**FIGURE 1.26**

Settings changed in an object type's branch affect only those settings and styles below where the change is made. For example, if you are setting the drawing's ambient settings value to slope (run:rise) and want to change how surface objects report slope, you change the surface branch ambient settings (Edit Feature Settings) to rise:run and override the drawing's preferred value (see Figure 1.27). When overriding a value set at a higher level, an override toggle is set in the dialog box. A down arrow at the higher level indicates the value is changed at a lower level in the hierarchy.

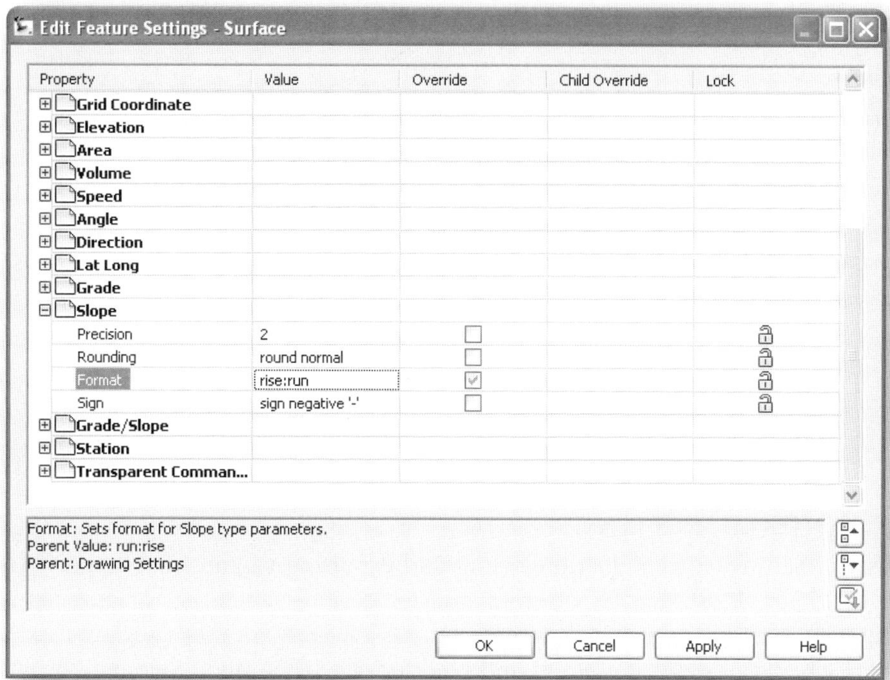


**FIGURE 1.27**

## EDIT LABEL STYLE DEFAULTS

This dialog box's values affect basic label behavior (see Figure 1.28). In the figure, some settings are overridden by "lower" styles (down arrows in the Child Override column). Style overrides occur because a label type behaves, labels, or reports values differently.

To negate overrides, click the Child Override column's down arrow. When clicked, the icon changes to a down arrow with an X. The X indicates the overrides reset to the values in the Edit Label Style Defaults dialog box. A second method for controlling overrides is locking their value. When locked at this level, no lower style can change the value. Again, setting a label style's Label layer to 0 causes the label to use the layer defined in Edit Drawing Settings' Object Layers panel.

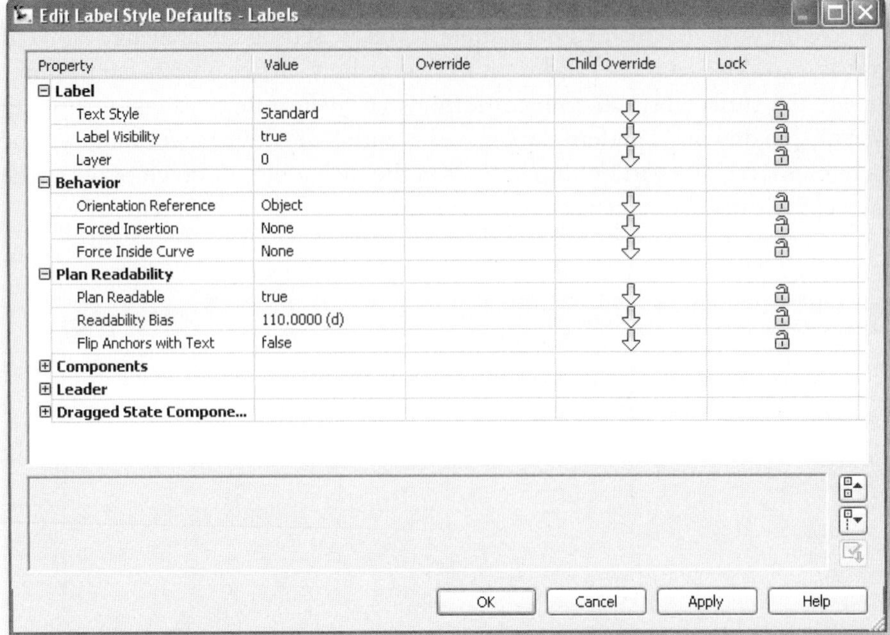


**FIGURE 1.28**

## EDIT LANDXML SETTINGS

LandXML Settings affects data importing and exporting to and from LandXML files (see Figure 1.29). LandXML files transfer design elements (surfaces, alignments, points, etc.) between applications without a loss of fidelity. The LandXML civil data schema allows AutoCAD Land Desktop, Civil 3D, and other applications to transfer data without losing information quality. Civil 3D also uses LandXML files as report data.

When exporting an XML file, the Imperial Units type should be set to match your drawing's value, either International or US Foot. This is set in the Export panel, Data Settings section.

When importing an XML file, you can translate coordinates and elevations. This is done in Import's Translation section. If the drawing's Imperial units are different from the file's units, International (drawing) and US Foot (file), the coordinates for the objects will be transformed.

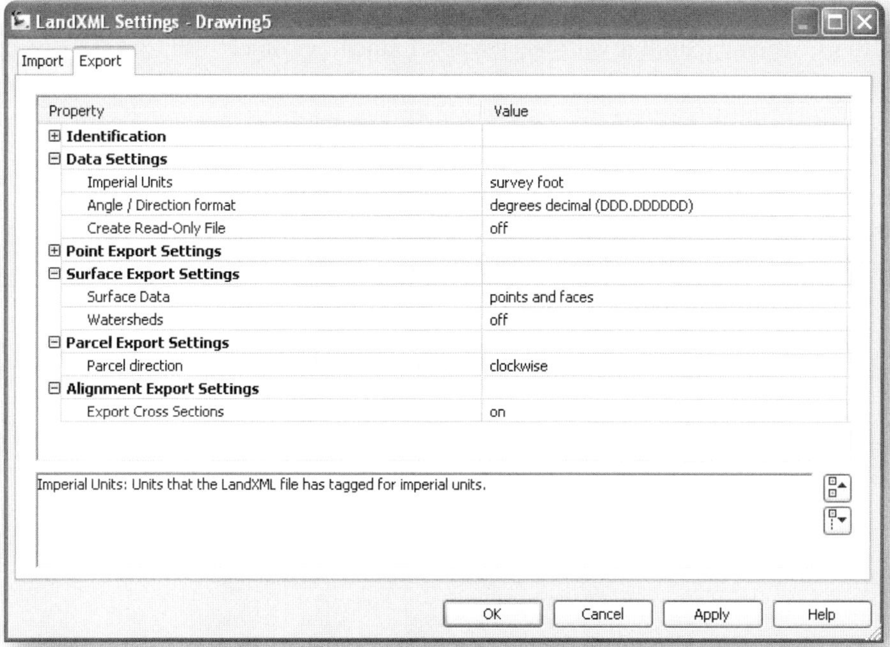


**FIGURE 1.29**

## EDIT FEATURE SETTINGS

In Settings, when selecting an object type heading and pressing the right mouse button, a shortcut menu displays with a call to Edit Feature Settings…. Edit Feature Settings combines Edit Drawing Settings' Ambient Settings values with specific settings for the selected object type (see Figure 1.30). In the figure, there are three sections affecting points: Default Styles, Default Name Format, and Update Points. The number of sections in an Edit Feature Settings dialog box depends on an object's complexity. Changing a ambient value in Edit Features Settings overrides an Edit Drawing Settings value.

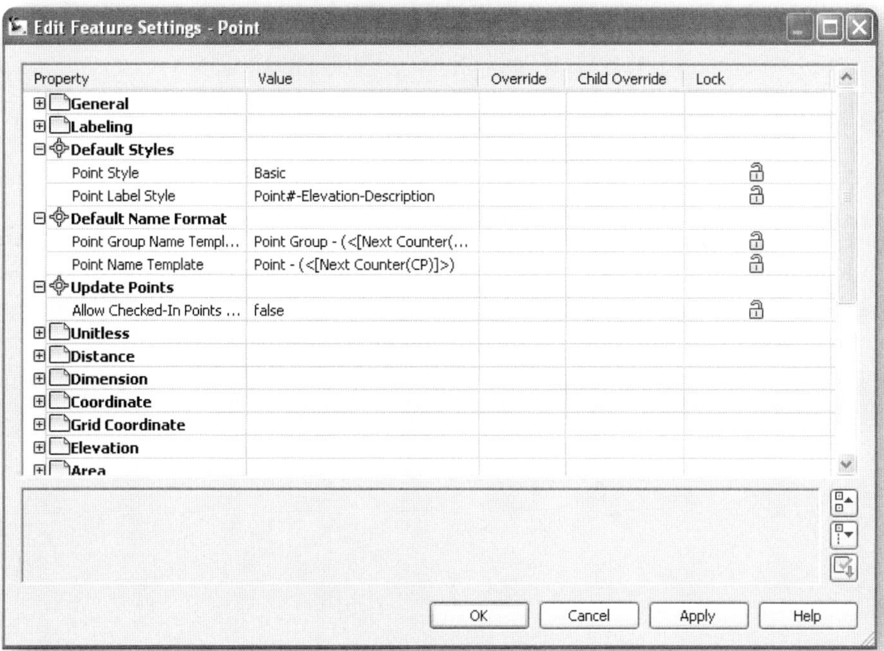


**FIGURE 1.30**

## LABEL AND TABLE STYLE GROUPS

In Settings, when expanding an object type's branch, an object has two style types (Label and Table).

### Label Groups and their Styles

Expanding a Label Styles branch displays label type groups. For example, surface label types are Contour, Slope, Spot Elevation, and Watersheds. The object's complexity determines the label type number. Each heading's shortcut menu contains commands appropriate to the selected heading (see Figure 1.31). For example, Description Key Sets displays a shortcut menu with commands used for creating a new Description Key Set, viewing a set's properties, or refreshing its list. Selecting a named Description Key Set displays a shortcut menu with commands used for viewing the key set's properties, copying it, deleting it, editing its values, or refreshing its list.

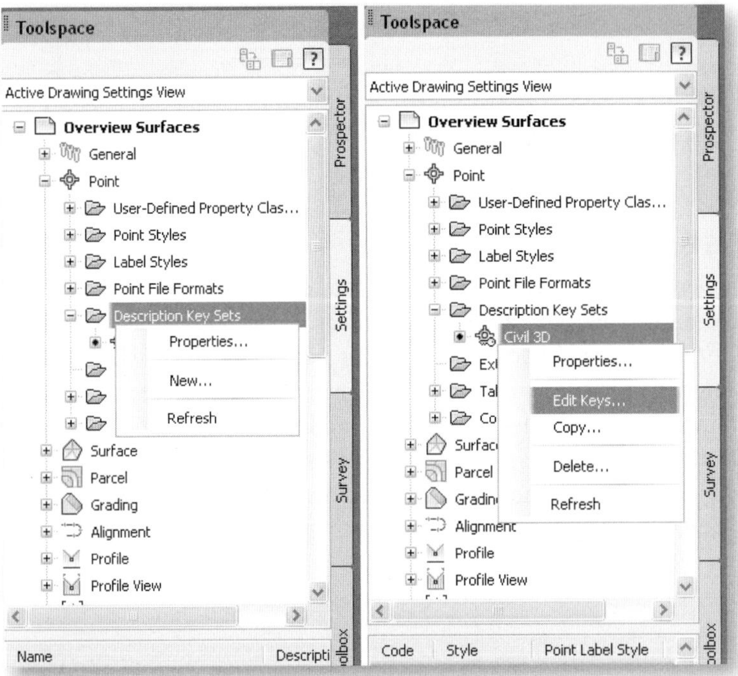


**FIGURE 1.31**

When expanding a Label Styles group, a label styles list displays. The styles create labels, documenting object type properties. For example, a parcel has the label types of area, line, and curve. The area label type styles include Name Area & Perimeter, Name Square Foot & Acres, and Parcel Number. When defining parcels, you can apply one or more area label styles to the parcel. When annotating a parcel's line and curve segments, you use the Line and Curve label type styles.

When selecting a listed style name and pressing the right mouse button, a shortcut menu displays with edit, delete, or copy… commands (see Figure 1.32).

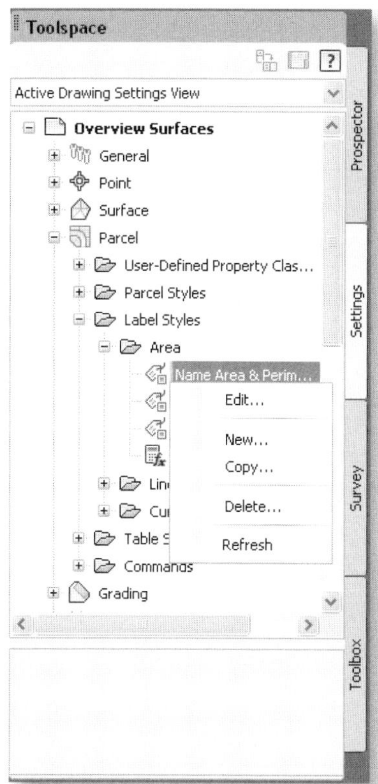


**FIGURE 1.32**

## Transferring Styles between Drawings

To transfer styles to an existing template or drawing, drag and drop them between open files. When you drag and drop a style definition into a file where the style already exists, Civil 3D issues a warning, prompting you to rename, overwrite, or ignore the dropped style.

## EDIT COMMAND SETTINGS

Each object type has commands that use default styles and settings from Edit Drawing Settings or the object's Edit Feature Settings (see Figure 1.33). You can change a command's default styles and settings without affecting other commands or feature settings.

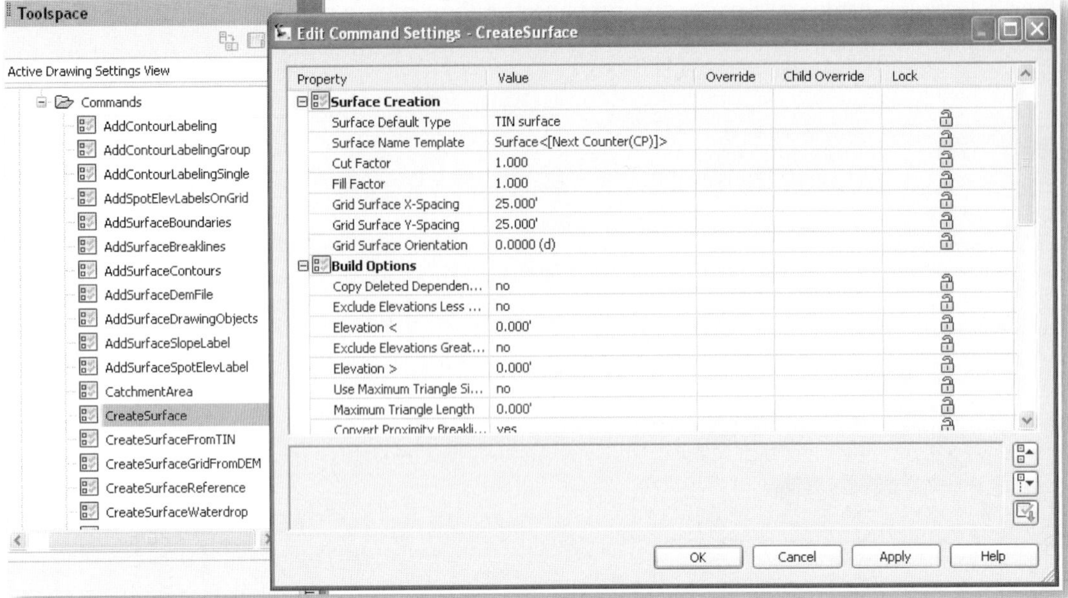


**FIGURE 1.33**

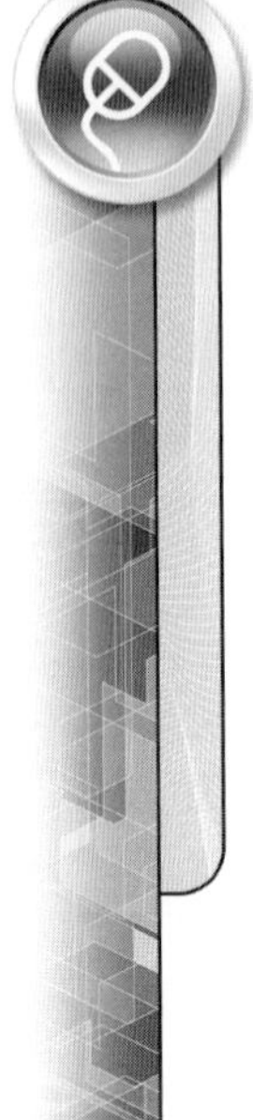
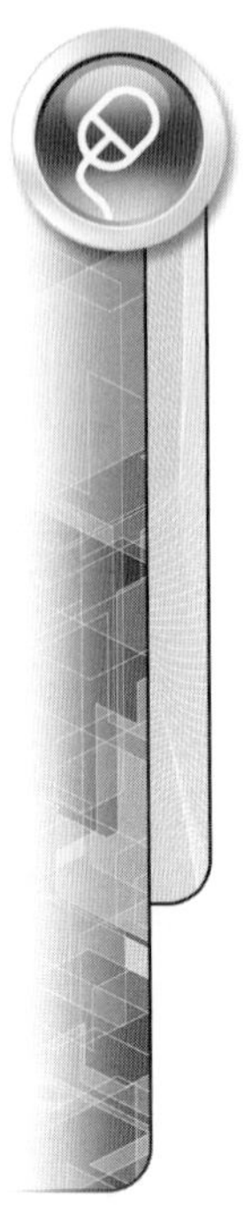

## EXERCISE 1-2

After completing this exercise, you will:

- Review the Prospector data panel.
- Expand and review the Surface data tree.
- Review Prospector icons for status and dependency.
- Review a panorama and its vistas.
- Become familiar with the Settings panel.

### Settings - Edit Drawing Settings

This exercise focuses on Settings and its hierarchical tree management of settings and styles.

1. At Civil 3D's top left, Quick Access Toolbar, click the ***Open*** icon, browse to this textbook's CD, Chapter 1 folder, select the drawing *Overview Surfaces II*, and click ***Open***.
2. Click the ***Settings*** tab.

The Edit Drawing Settings..., Edit Label Style Defaults..., and Edit LandXML Settings... dialog boxes are at the top of Settings' hierarchy.

3. At Settings' top, select the drawing name, press the right mouse button, and, from the shortcut menu, select EDIT DRAWING SETTINGS....
4. If necessary, in Edit Drawing Settings, click the ***Units and Zone*** tab and review its settings and current Zone.
5. Click the ***Transformation*** tab.
6. Click the ***Object Layers*** tab and scroll through the layer list, reviewing layer names and their settings.

7. Click the ***Abbreviations*** tab and review the abbreviations list.

You can change any of the listed abbreviations.

8. Click the ***Ambient Settings*** tab and expand different sections to view their settings.

These values set the general drawing tone. Clicking a padlock icon locks the value so no lower style can override or change the locked value.

9. Click ***OK***, exiting the Edit Drawing Settings dialog box.

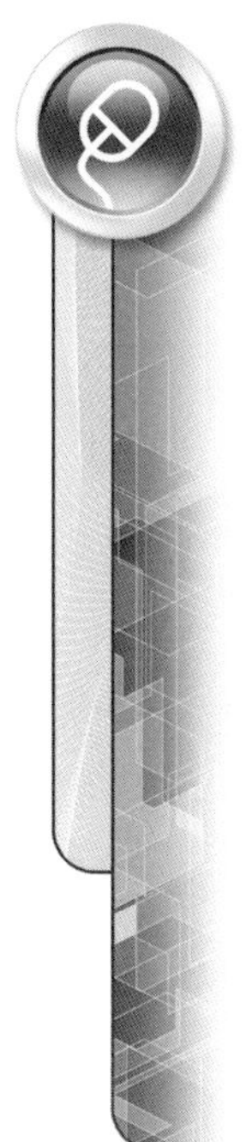

## Edit Feature Settings

1. In Settings, select the **Surface** heading, press the right mouse button and, from the shortcut menu, select EDIT FEATURE SETTINGS....

The Edit Feature Settings dialog box sets the default styles and naming values for Surfaces.

2. Expand the surface sections to review the surface settings, and when done, click ***OK***, exiting the dialog box.

## Edit Label Style Defaults - Drawing

1. In Settings, select the drawing name, press the right mouse button and, from the shortcut menu, select EDIT LABEL STYLE DEFAULTS....
2. Expand and review each section.
3. In the Label section, click the Text Style value cell, and then click the ellipses to view the Select Text Style dialog box.

Selecting a text style here sets it for all non-overridden styles.

4. Close the Select Text Style dialog box and click ***Cancel***.

## Edit Label Style Defaults - Feature

1. In Settings, select the **Surface** heading, press the right mouse button and, from the shortcut menu, select EDIT LABEL STYLE DEFAULTS....

The Edit Label Style Defaults dialog box sets several default values for all surface label styles. The values reflect those set in the drawing name level Edit Label Style Defaults dialog box.

2. Expand each section to review the values and then click ***OK***, exiting the dialog box.

## Object Styles, Label Groups, and Their Styles

1. In Settings, click the plus sign to the left of Surface to expand the Surface branch.

The Surface heading has four branches: Surface Styles, Label Styles, Table Styles, and Commands.

2. Click the expand tree icon to the left of Surface Styles, viewing its object styles list.
3. Select the style **Border & Triangles & Points**, press the right mouse button and, from the shortcut menu, select EDIT.... As an alternative, you can double click the style to edit it.

You modify a style's values in this dialog box.

4. Click the tabs, expand some of the sections in the different panels, and review the values controlling a style.
5. Click ***OK***, exiting the dialog box.
6. In Settings, click the plus sign to the left of Label Styles, displaying the surface label style types list.
7. Expand the Contour branch, viewing its styles.
8. Select a **Contour** style, press the right mouse button and, from the shortcut menu, select EDIT....
9. Click the various tabs and expand some of the sections, reviewing their contents.
10. When you are done, click Cancel to close the label style dialog box.
11. Explore some of the remaining style trees and review their style definitions.
12. CLOSE the drawing and do not save the changes.

## SUMMARY

- The Settings panel manages all styles and command values.
- Settings promote standards and eases implementation with its hierarchical structure.
- In the Settings hierarchy, the higher a setting is, the more styles and values it affects.
- Settings manages, creates, and modifies object and label styles.

## UNIT 3: SETTING CIVIL 3D'S ENVIRONMENT

New terms in AutoCAD Civil 3D describe the drafting environment and its objects and their behaviors (e.g., Prospector, vistas, baselines, assemblies, styles, etc.). Civil 3D sites, baselines, feature lines, and assembly objects are familiar civil concepts and design elements, but these elements are known by more traditional names in other civil design packages (parcels, alignments, grading objects, templates, etc.). The new terms indicate that these familiar design elements have new or expanded capabilities.

Civil 3D uses styles and settings to graphically display a design, set design limits (criteria), produce reports, and create design documentation. Civil 3D ships with a basic style definition content template. However, the template styles may not be right for a user's specific tasks, or they may not meet a user's CAD drafting standards. The biggest initial cost in implementing Civil 3D will be the time spent setting up and modifying styles. As with any Autodesk product, there are several content creation methods, and each strategy has consequences. Harnessing the interplay of the dynamic design environment and the role styles play in creating and finishing a design are the greatest challenges facing Civil 3D implementers and users.

If satisfied with a template's content (styles and layers) as shipped, the user can immediately begin producing a design document. If the content does not reflect specific standards, the user must create new styles, modify existing styles, and define and substitute new layers. When using this new template file, all modified content is present in the drawing.

The interface controlling, displaying, or editing object type information is the same for each object. All object and label styles use similar dialog boxes that have a set structure. By knowing the anatomy and behavior of one object or label style dialog box, a user knows the basics for most of the remaining object type and label styles. What changes is the information available about the object or for the label.

## IMPLEMENTATION

A single Civil 3D template file is an implementation. This file defines all object, label, and miscellaneous styles producing design document. This single file contains the office standards that in LDT took several folders and files. When starting a new Civil 3D drawing with this template, all objects, styles, and settings are there, ready for use. The only remaining settings are the model space plotting scale and a coordinate zone.

A Civil 3D implementation uses a combination of layers and styles. Layers can be used with traditional AutoCAD methods for displaying and hiding drawing elements. Layers are also necessary when implementing AutoCAD Xrefs or sharing data with those without Civil 3D. Styles also control data visibility. A Civil 3D implementation is a combination of styles and layers.

Adding style definitions to a template file means opening the template file and defining the new styles. To add styles from an existing drawing to a template or another drawing, open both files and drag and drop the new styles to the template or other drawing.

The biggest issue with an implementation is layers. If a drawing defines a modifier for an object type, for example the suffix -* for surfaces, Civil 3D creates layers for each surface from the base layer name (C-TOPO-EXISTING and C-TOPO-DESIGN). If assigning both surfaces the same object style, the surfaces use the same layer list. For example, assigning the Border & Contour style to two surfaces causes them to use the same layers to display their borders and contours. You cannot turn off the major contours layer for only one surface because it is the same layer for both surfaces.

In Settings, the four top settings dialog boxes influence all settings and styles below them. These dialog boxes are Edit Drawing Settings, Edit Label Style Defaults, Edit LandXML Settings, and Table Tag Numbering. These dialog boxes' values affect the entire drawing. Any lower style or setting can override these values. If locked at the drawing name level, a lower style cannot change the setting's value.

- Creating a Civil 3D template involves a thorough review and adjustment of values in the four primary dialog boxes; Edit Drawing Settings, Edit Label Style Defaults, Edit LandXML Settings, and Table Tag Numbering.

## OBJECT LAYERS

The Edit Drawing Settings' Object Layers panel defines object type base layer names with an optional modifier (prefix or suffix). Modifiers are necessary when there is more than one instance of an object type (see Figure 1.25).

An asterisk (*) modifier value appends the instance name to the beginning or end of the base layer name. This is the preferred method of modifying a layer's name. One additional option is spacing the object's name from the base layer's name. A dash (-) or underscore (_) is the usual spacing character. For example, setting a surface's name modifier to suffix and setting its value to dash asterisk (-*) creates the layer C-TOPO-EG, the base layer plus the surface name separated by a dash. A second method is entering an explicit value as the modifier (e.g., C-TOPO-SURFACE). In this case, -SURFACE is the modifier value.

For a drawing with two or more object instances, using an object layer modifier is a necessity. Using surfaces as an example, if a drawing has two surfaces, EG and DESIGN, and no layer modifier, both surfaces use the same layer. When objects occupy the same layer, the user cannot independently control their visibility by layers (user must use a style to control the object's display). If the user adds a modifier to the object layer name, Civil 3D creates a new layer for each object from the modifier values. A drawing having two surfaces, EG and DESIGN, would have the layers C-TOPO-EG and C-TOPO-DESIGN (see Figure 1.34).

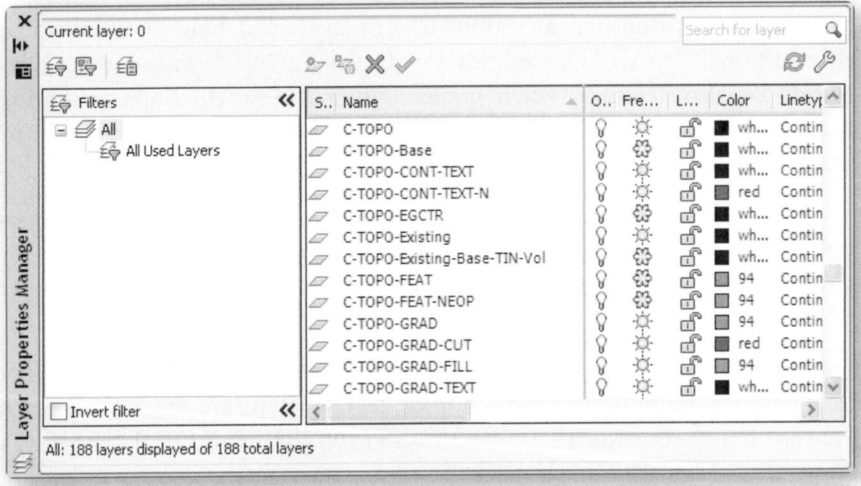


**FIGURE 1.34**

## SETTING THE DRAWING ENVIRONMENT

In Settings, when selecting the drawing's name and pressing the right mouse button, a shortcut menu displays with four settings dialog boxes: Edit Drawing Settings, Edit Label Style Defaults, Edit LandXML Settings, and Table Tag Settings. Because of Settings' hierarchical structure, the values and settings of these editors affect the entire drawing environment. When starting a new drawing, the first step should be reviewing some of the dialog boxes' values.

### Edit Drawing Settings

Selecting Edit Drawing Settings displays a dialog box with five tabs (see Figure 1.35). These tabs affect several areas of the drafting environment.

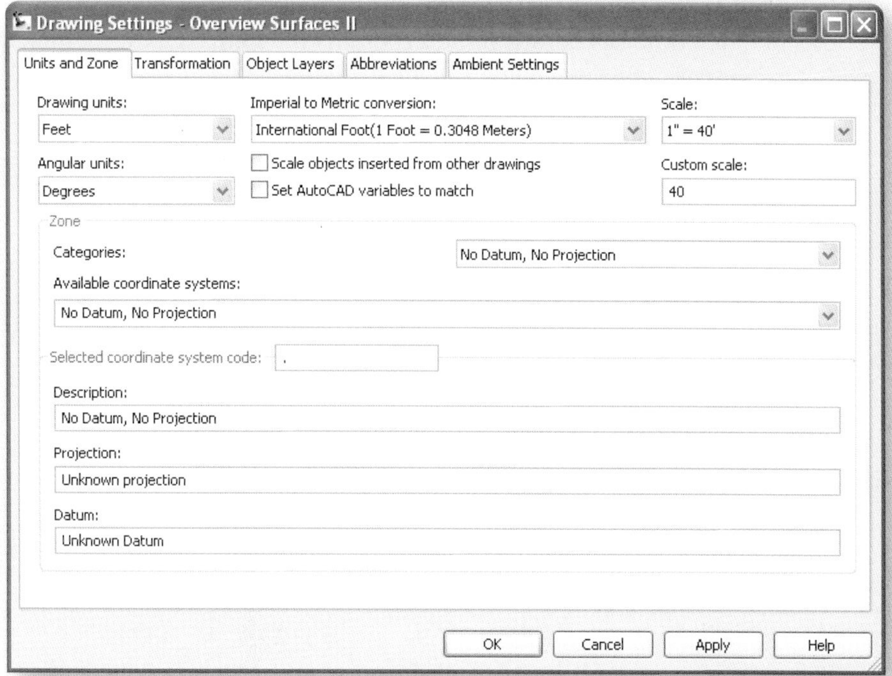


**FIGURE 1.35**

## Units and Zone

Units and Zone top left values set linear and angular base units. There are three angular measurement values; degrees, grads, and radians. The panel's center sets the Imperial units for metric conversion. An Imperial drawing can use International or US Survey feet. You must set this value correctly if you want to move Imperial data to Metric.

US Survey Foot applies to the ratio of feet to meters. In 1866 the ratio was defined as 1200/3937 or 39.37 inches to a meter. In 1959 the ratio was refined when the US changed the definition of a yard to match other country definitions. The yard was redefined to be 0.9144, and the foot was redefined as 0.3048 of a meter. An International Foot is 0.99998 of a US Survey Foot. At the same time, it was decided that data used in Geodetic surveys would use the original foot definition (39.37 = 1 Meter), US Survey Foot. In local coordinates the difference is small. However, when working in coordinates over 1,000,000, the difference is about 2 feet per million coordinates. This value also affects inserted points from Survey.

The Scale objects inserted from other drawings scales the inserted drawing to match the current drawing's units.

The dialog box's top right sets the model space plotting scale. A "standard" scale can be selected here, or a user can enter a custom scale factor below the scale drop-list entry.

The panel's remainder sets and reports coordinate zone settings if set for the drawing. When selecting the Categories' drop-list arrow, a supported coordinate systems list displays. This setting is important for working with state plane or latitude and longitude data.

## Transformation

Transformation values tie local coordinates to a coordinate system. A local coordinate based drawing must have one known point (planar coordinates) and a state plane rotation angle, or two points with known state plane coordinates to relate its local coordinates to state plane coordinates (see Figure 1.36). To correctly determine sea level horizontal distances, you must set elevation (mean elevation of the site), the spheroid radius at sea level (from the coordinate system set in Units and Zone), and Grid scale factor (unity, reference point, Prismodial, and user defined).

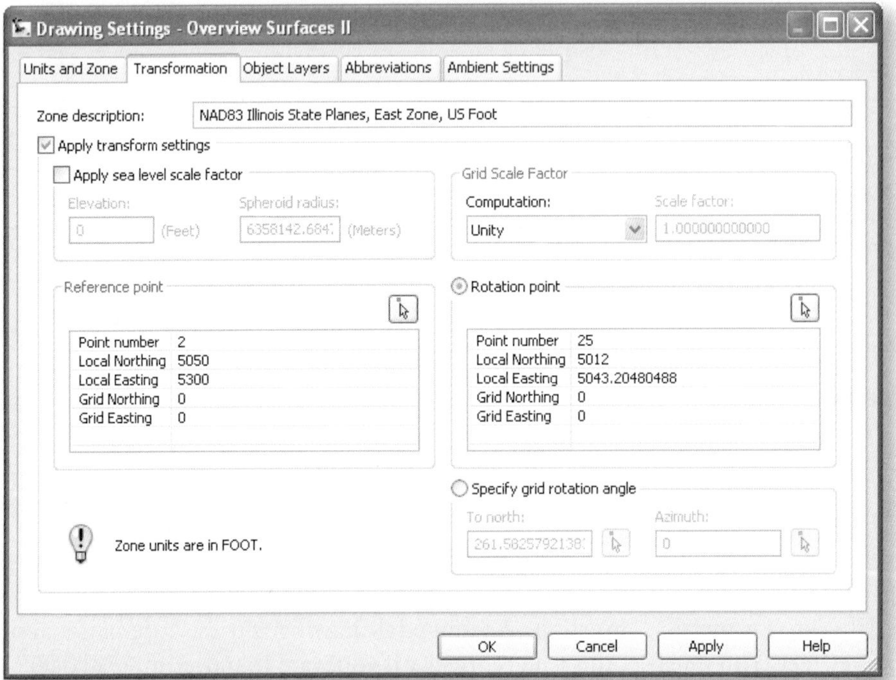


**FIGURE 1.36**

## Object Layers

Object Layers sets base layer names for each object type (see Figure 1.37). The panel's left side lists the object type, and the three columns to the object type's right set layer names if the layer has a modifier, if the modifier is a prefix or suffix, and sets the modifier value. A modifier value can be anything. If the modifier value is an * (asterisk), Civil 3D uses the object's name as the layer modifier. For example, a surface named EG modifies the C-TOPO base layer to C-TOPO-EG when the modifier is a suffix and its value is a dash asterisk (-*). The last column locks the layer name so it cannot be changed. If an object's label styles have their layer set to 0 (zero), the labels will appear on the labeling layer defined in this panel.

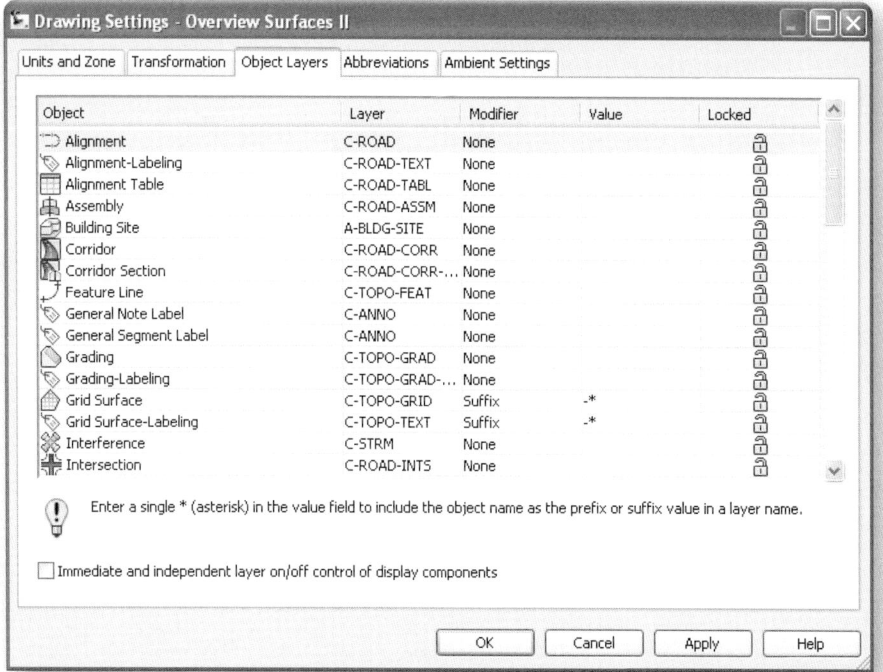


**FIGURE 1.37**

## Abbreviations

The Abbreviations panel affects alignment, superelevation, and profile listing and report values (see Figure 1.38). Many object type labels have regional values and these initial values may need to change.

The Alignment Geometry Point Entity Data values are a set of abbreviations with a complex format string. The format string defines how geometry point's values display. When clicking in a format string cell, at the right of the entry an ellipsis (three dots) appears. Clicking the ellipsis calls the Text Component Editor, and within this editor users edit the abbreviation and its format string.

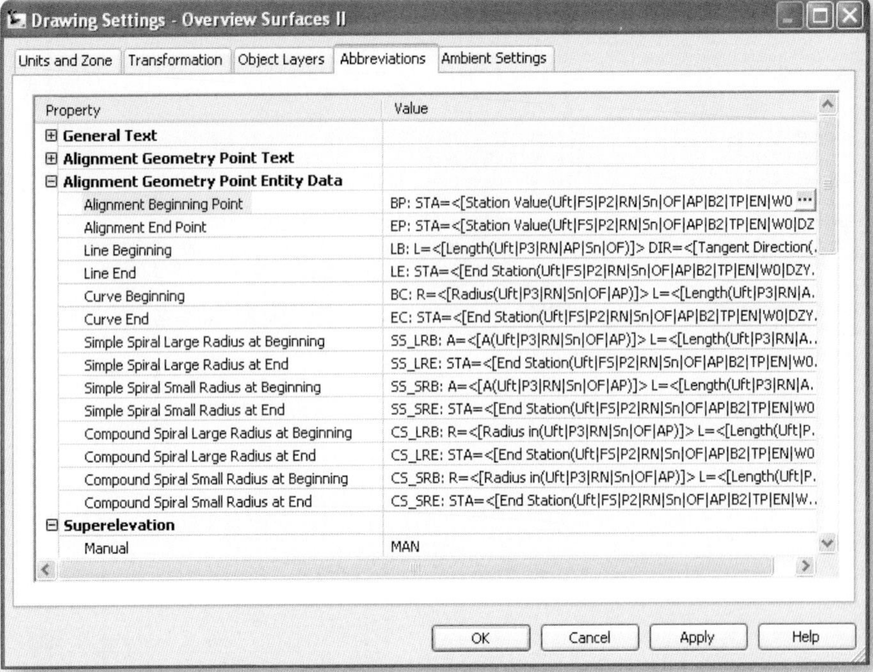


**FIGURE 1.38**

## Ambient Settings

Ambient Settings affects a multitude of drawing values. Each section's values affect units, precision, and rounding; set entry and reporting value formats; and show how to denote a negative value, coordinates and distances, and so on (see Figure 1.39).

The Value column contains the actual setting. A value can be an entered value, a toggle, or a selection from a list of choices.

The Child Override column indicates if any lower settings change the current entry's value. A down arrow indicates a changed value. By clicking on the arrow (adding a red X to the arrow) and clicking Apply, you reset all of Settings' values to the current cell's value.

The Lock column indicates if other values are allowed. If locked, no other values are allowed.

The Angle, Direction, and Lat Long sections can drop the leading 0 (zero) and, if the value is a whole number, becomes an integer.

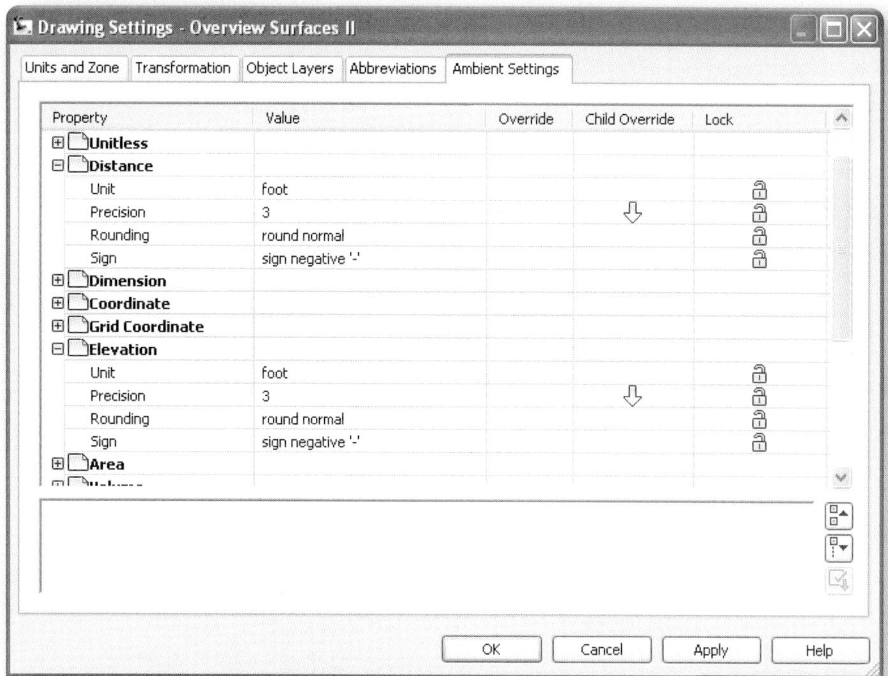


**FIGURE 1.39**

## STYLES

By default every object type has a Basic or Standard style (object, label, etc.). When starting a new drawing without a content template, Civil 3D creates Basic or Standard styles for all Settings tree objects. A Basic or Standard style is a starting point, and Civil 3D expects users to copy and modify them, implementing new styles. If using a Civil 3D content template, additional styles are available to edit or copy to address user needs.

Styles stylize how objects, and what components or characteristics display. A style displays all object components or characteristics, or it can display groupings of components or characteristics. A surface object, for example, has triangles, points, and a border. These components are essential to correctly view and edit a surface under development. Surface slopes and elevations are essential to developing a site's design solution. A style changes an object's information focus by displaying different combinations of components and characteristics.

### Style Layers

A layer's intent is controlling visibility from within Layer Properties Manager. Turning on and off layers determines what is visible in the drawing. This is a typical AutoCAD information display strategy. In Civil 3D, a style controls the component and characteristic visibility. Changing styles changes what components and characteristics display.

An object style's Display panel assigns each object component or characteristic a layer name and properties. The object style's Display panel's left side lists the object's components and characteristics. The number of entries varies by object type; a more complex object has more entries than a simple object (see Figure 1.40 and Figure 1.41).

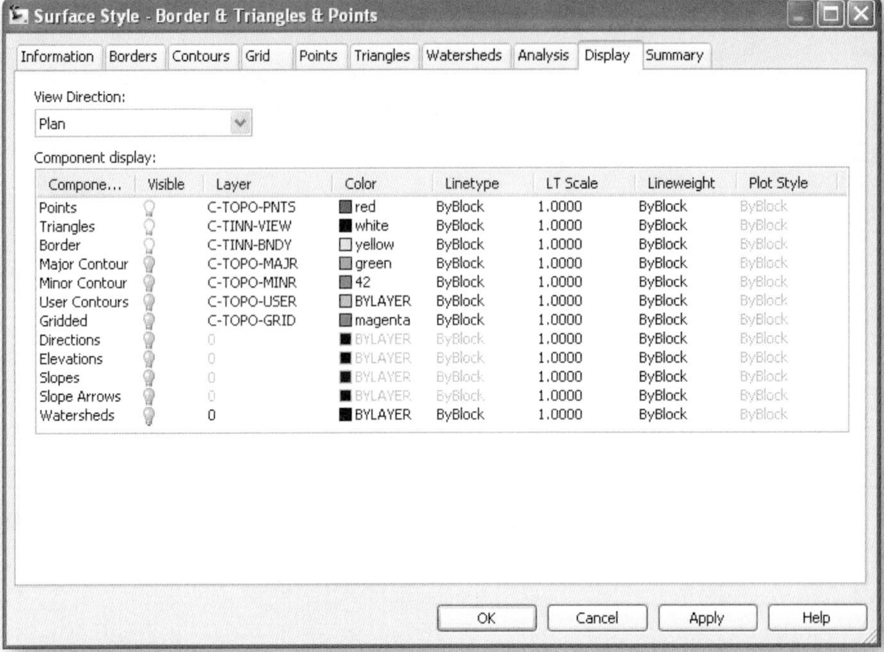


**FIGURE 1.40**

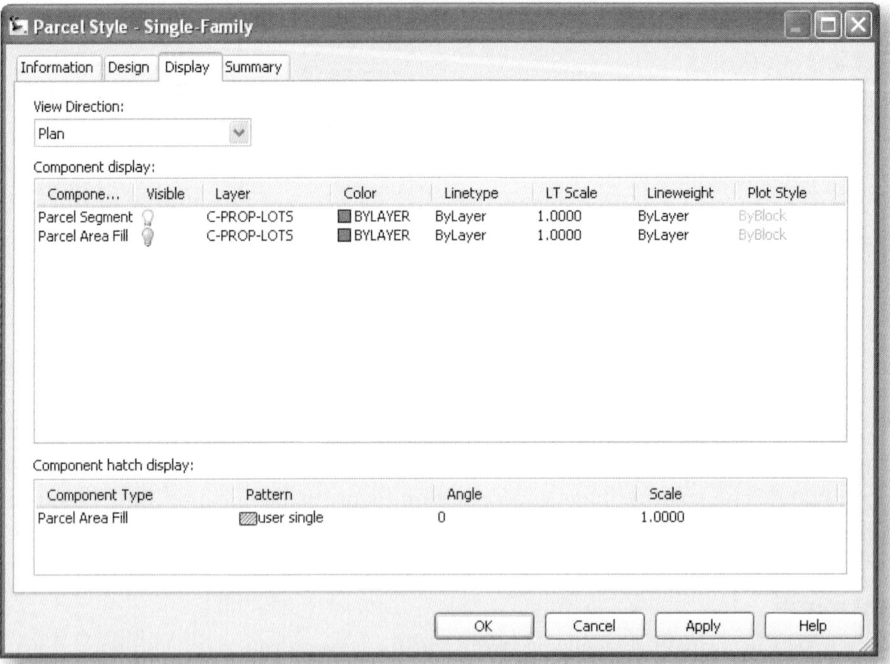


**FIGURE 1.41**

There are three methods for assigning a style layers. The first method uses layers in the current drawing. In the object style's Display panel, clicking a layer name displays the Layer Selection dialog box, and the user selects a layer from the current drawing's list (see Figure 1.42). When returning to the style's Display panel, the component lists the selected layer.

The second method is creating a new layer. When clicking a Display panel's layer name, the Layer Selection dialog box displays. At the dialog box's top right is a New… button that calls the Create Layer dialog box (see Figure 1.43). In this dialog box, users define the new layer. After returning to the Layer Selection box, users select the new layer and return to the Display panel, assigning it to the component or characteristic.

The third method selects layers from another open drawing (see Figure 1.42). Users add to the current drawing's layer list by selecting the drop-list arrow at the Layer Selection dialog box's top left and selecting another open drawing. Layer Selection's list changes, listing the newly selected drawing's layers. Users then select a layer from the list and, when returning to the Display panel, assign the layer to the component or characteristic.

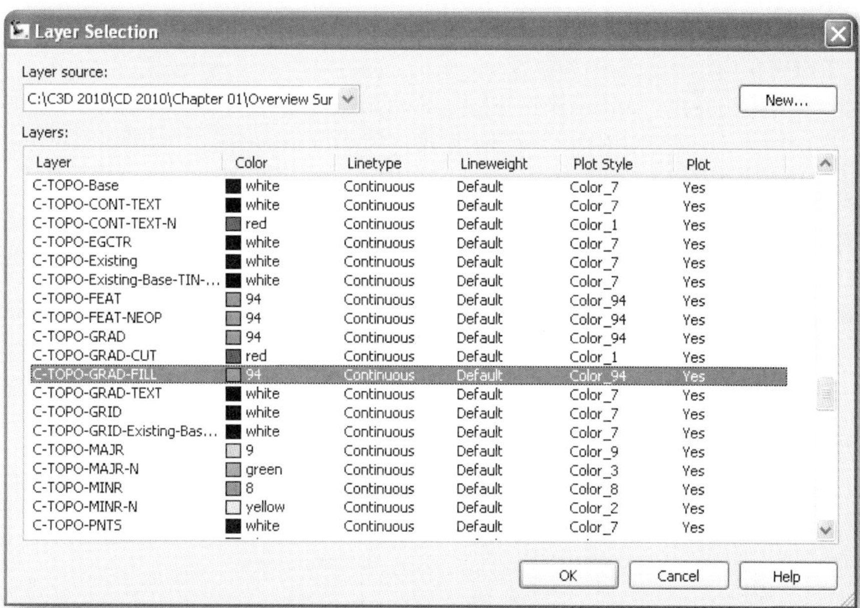


**FIGURE 1.42**

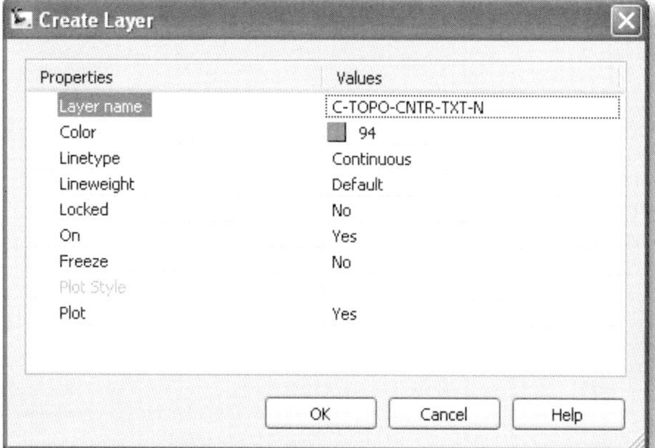


**FIGURE 1.43**

Civil 3D views data from four directions: Plan, Model, Profile, and Section. Plan is a view from directly overhead, and a Model view is from any other angle. Profile and Section apply only to object types displaying in profile and sections views. For example, a surface displays as a line in a profile or section view. Each view group defines component layers and visibility.

Though the Display panel lists all available object type components and characteristics, a style may display only one or only a few components and/or characteristics. A style's purpose determines its components' and/or characteristics' visibility. For example, a surface analysis style focuses on surface elevations. A surface editing/review style shows a surface's border, triangles, and points (see Figure 1.40).

To each component's and characteristic's right is a layer name and properties. The style's layer properties should contain the keyword ByLayer, allowing Layer Properties Manager control of the layer's properties.

## STYLE TYPES

Object styles have special purposes or functions (such as analysis, grouping, or submission documentation). Continuing with the surface objects example, a certain designer has an interest in surface slopes and elevations before starting the design process. To better understand surface slopes, some slope analysis and ranging are necessary. Creating a slope style by ranging slopes and creating down slope arrows lets the style display a more meaningful image of the slopes' spatial distribution. Using an elevation analysis style results in a better understanding of their distribution. Surface styles have color schemes, colorizing their results. Figure 1.44 shows examples of different types of analysis styles output.

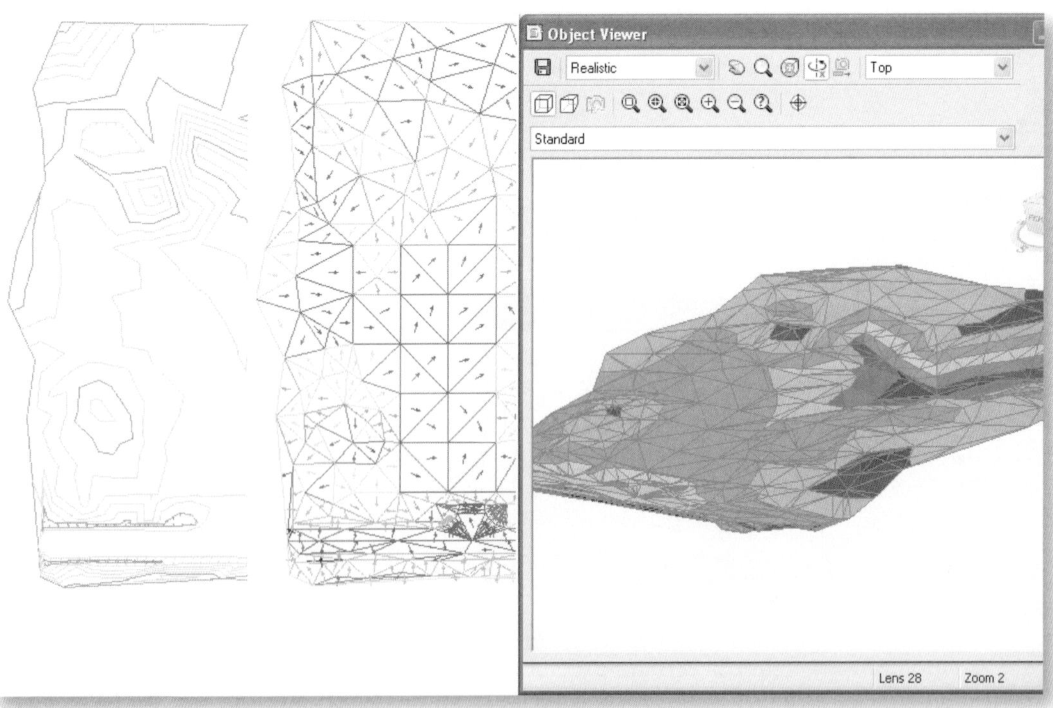


**FIGURE 1.44**

## OBJECT STYLES

Object styles directly control an object's display. Every drawing has a Basic (or Standard) object style. This style has minimal settings and displays a minimal set of object components. Civil 3D ships with several template files containing object styles. These object styles serve as a style type demonstration a user can define for a Civil 3D implementation. Each chapter in this textbook expands, modifies, and creates new styles as needed to document a design solution. The number of styles and their complexity varies depending on the object type. For example, the surface object is complex and has numerous potential styles, whereas parcels have less complexity and fewer styles.

All object types use the same dialog box structure. Fundamental settings, values locations, and components and characteristics use the same basic design. Being familiar with one object's style dialog box means knowing how to navigate the next object's style dialog box.

### Assigning Object Styles

When a new object is created, it is assigned the default object style set in the object type's Edit Feature Settings or the Command Settings dialog boxes. If you want to change an object' style, the change occurs in the object's Properties dialog box, Information panel. Assign a new style by selecting a style from a list of available styles. When exiting the dialog box, the new style changes how an object appears in the drawing (see Figure 1.45).

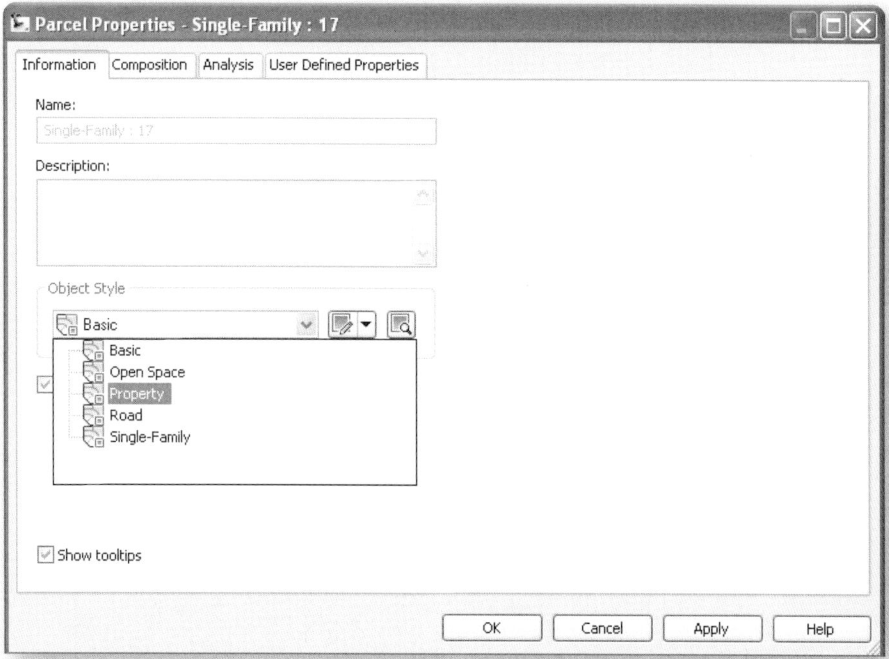


**FIGURE 1.45**

## LABEL STYLES

Label style controls object annotation and its display. Every drawing has a Basic or Standard label style. These styles have minimal settings and annotate a minimal number of object components. Civil 3D ships with several template files containing label styles. These label styles serve as a demonstration of annotation types a user can define for a Civil 3D implementation. Each chapter in this textbook expands, modifies, and creates new styles needed to document a design solution. The number of styles and their complexity varies depending on the object.

When starting a new drawing and using the content template, all styles in the template become part of the new drawing. Displaying a triangle icon to the style's name upper-left means an object in the drawing uses (references) that style.

Label styles are single-purpose, meaning they annotate a specific object facet (see Figure 1.46). The label style types for a parcel are Area, Line, and Curve (see the left side of Figure 1.46). The label style types for an alignment include Station, Station Offset, Line, Curve, Spiral, and Tangent Intersection (see the right side of Figure 1.46).

Each object has label style groups. Each object has specific properties that can be part of a style. All labels have the same general behavior and use the Text Component Editor interface to create new and modify existing label styles. Once familiar with the label editor and its behavior, the user knows how to edit and create labels for all object types.

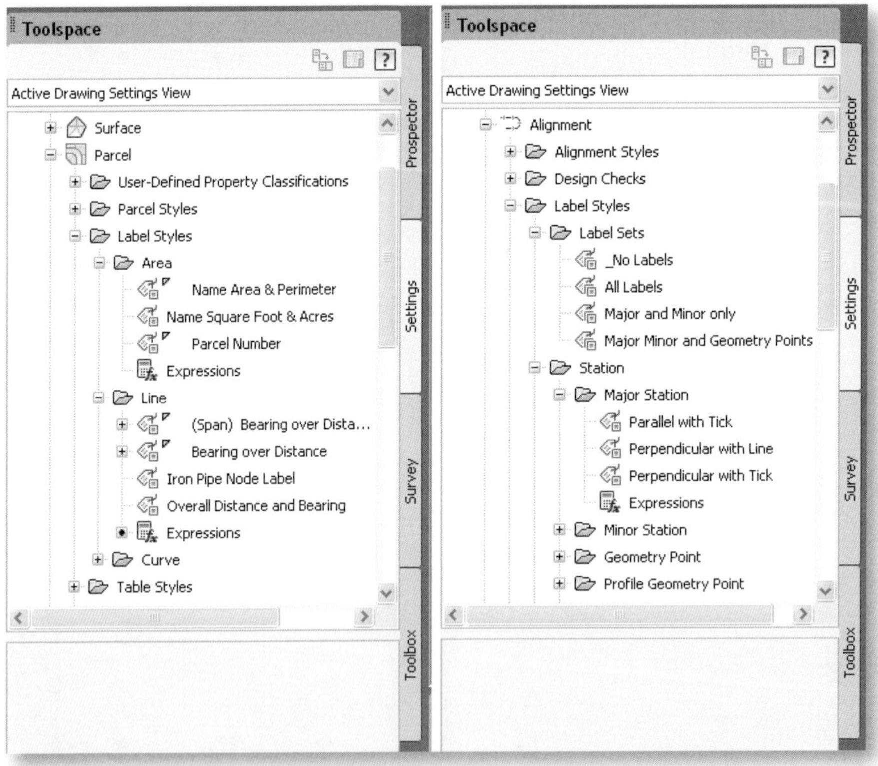


**FIGURE 1.46**

Label styles use the same dialog boxes. The fundamental settings, location of values, and the component and characteristic lists are in the same place for each label type.

Every label has an anchoring point. In Figure 1.47, the parcel's label anchoring point is the parcel's centroid. The label's middle center anchors to the parcel's centroid.

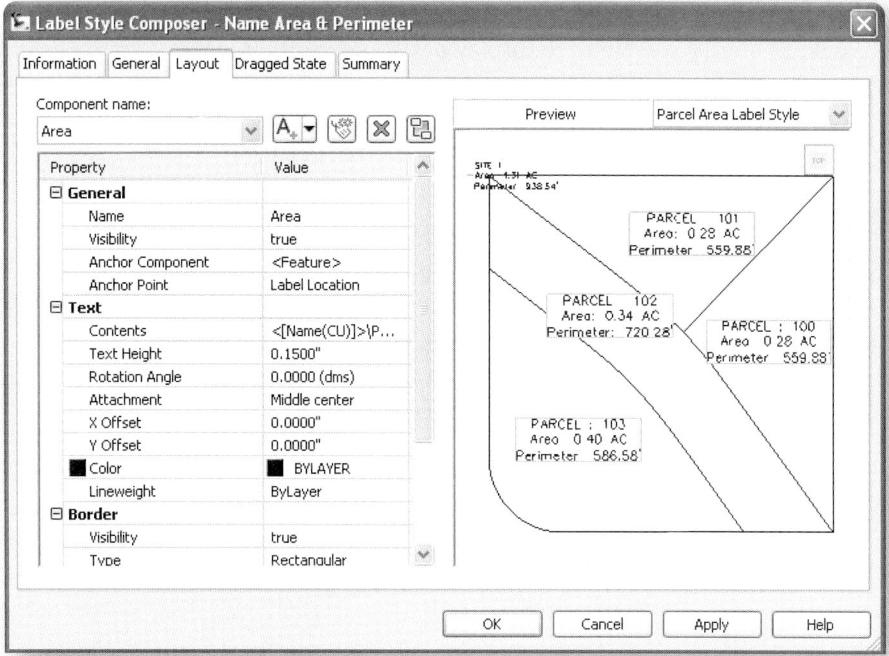


**FIGURE 1.47**

To view a label's content definition, click in the Contents' value cell and click the ellipsis displayed in the cell. This displays the Text Component Editor and the label's content (see Figure 1.48). All styles use both Label Style Composer and Text Component Editor. Label Style Composer's Layout tab defines a label's text component(s), and the Text Component Editor defines and formats each label component's text. The Format tab changes the text's justification, font, color, underline, or adds symbols.

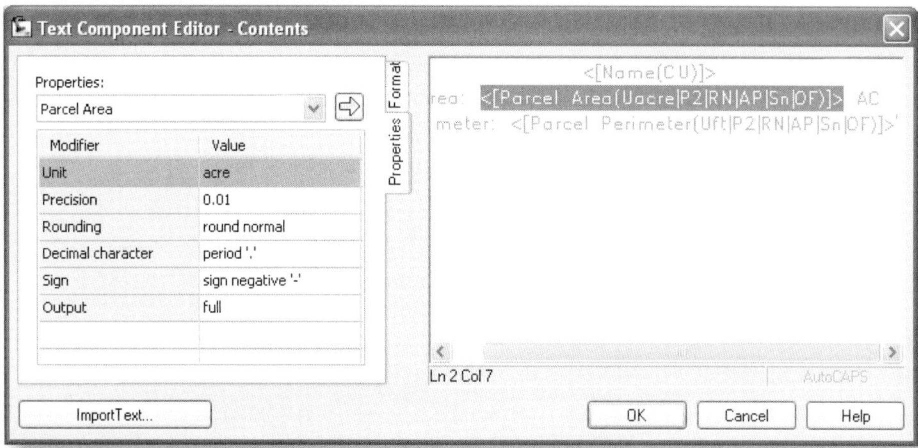


**FIGURE 1.48**

## TABLE STYLES

Some Civil 3D objects use tables to display or document their information. The object types using tables are points, surfaces, parcels, and alignments. Each object type has a Basic or Standard table style. In this and other chapters, exercises access these dialog boxes, evaluate their settings, and, if appropriate, change some values.

## THE PANORAMA

The panorama is important. The palette's structure is similar to a spreadsheet (cell based) and contains vistas (editor, Event Viewer, etc.) (see Figure 1.49). In the figure, the panorama displays a Point Editor vista. Each object type uses the panorama to display its data, and many chapter exercises interact with, edit, and review data within a vista.

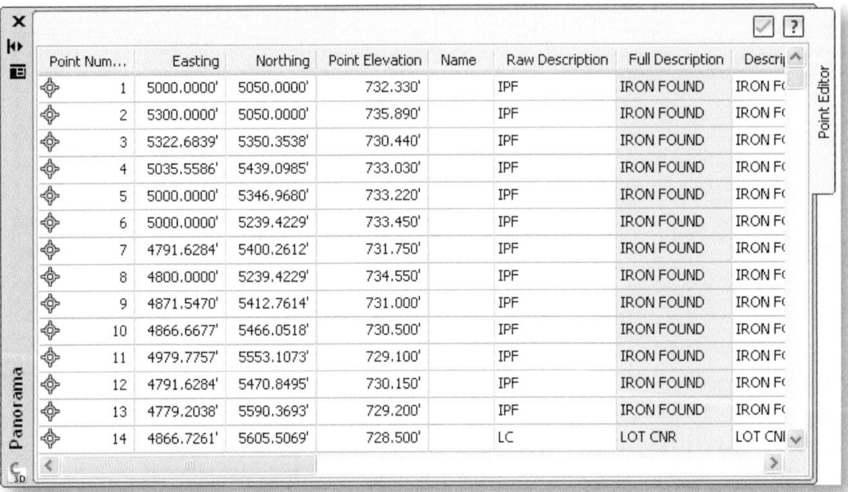


| Point Num... | Easting | Northing | Point Elevation | Name | Raw Description | Full Description | Descri |
|---|---|---|---|---|---|---|---|
| 1 | 5000.0000' | 5050.0000' | 732.330' | | IPF | IRON FOUND | IRON F |
| 2 | 5300.0000' | 5050.0000' | 735.890' | | IPF | IRON FOUND | IRON F |
| 3 | 5322.6839' | 5350.3538' | 730.440' | | IPF | IRON FOUND | IRON F |
| 4 | 5035.5586' | 5439.0985' | 733.030' | | IPF | IRON FOUND | IRON F |
| 5 | 5000.0000' | 5346.9680' | 733.220' | | IPF | IRON FOUND | IRON F |
| 6 | 5000.0000' | 5239.4229' | 733.450' | | IPF | IRON FOUND | IRON F |
| 7 | 4791.6284' | 5400.2612' | 731.750' | | IPF | IRON FOUND | IRON F |
| 8 | 4800.0000' | 5239.4229' | 734.550' | | IPF | IRON FOUND | IRON F |
| 9 | 4871.5470' | 5412.7614' | 731.000' | | IPF | IRON FOUND | IRON F |
| 10 | 4866.6677' | 5466.0518' | 730.500' | | IPF | IRON FOUND | IRON F |
| 11 | 4979.7757' | 5553.1073' | 729.100' | | IPF | IRON FOUND | IRON F |
| 12 | 4791.6284' | 5470.8495' | 730.150' | | IPF | IRON FOUND | IRON F |
| 13 | 4779.2038' | 5590.3693' | 729.200' | | IPF | IRON FOUND | IRON F |
| 14 | 4866.7261' | 5605.5069' | 728.500' | | LC | LOT CNR | LOT CN |

**FIGURE 1.49**

### Event Viewer Vista

The Event Viewer vista lists errors, warnings, and information during command execution. Type icons at the left of the text message indicate the message's severity. A circular red X indicates a failure, a yellow triangle indicates a warning that is not fatal, and an exclamation mark denotes an information message (see Figure 1.50).

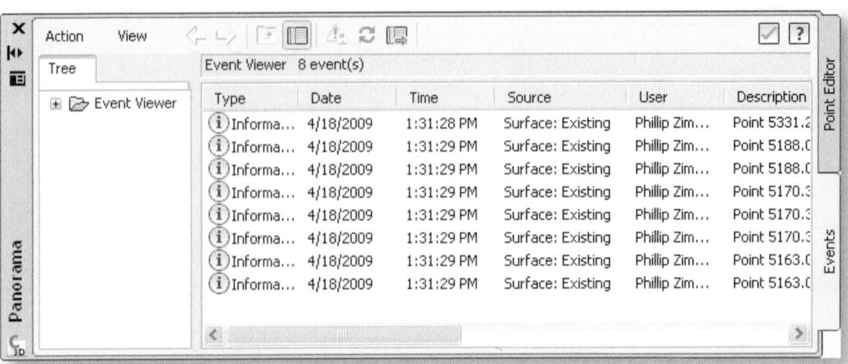


| Type | Date | Time | Source | User | Description |
|---|---|---|---|---|---|
| Informa... | 4/18/2009 | 1:31:28 PM | Surface: Existing | Phillip Zim... | Point 5331. |
| Informa... | 4/18/2009 | 1:31:29 PM | Surface: Existing | Phillip Zim... | Point 5188. |
| Informa... | 4/18/2009 | 1:31:29 PM | Surface: Existing | Phillip Zim... | Point 5188. |
| Informa... | 4/18/2009 | 1:31:29 PM | Surface: Existing | Phillip Zim... | Point 5170. |
| Informa... | 4/18/2009 | 1:31:29 PM | Surface: Existing | Phillip Zim... | Point 5170. |
| Informa... | 4/18/2009 | 1:31:29 PM | Surface: Existing | Phillip Zim... | Point 5170. |
| Informa... | 4/18/2009 | 1:31:29 PM | Surface: Existing | Phillip Zim... | Point 5163. |
| Informa... | 4/18/2009 | 1:31:29 PM | Surface: Existing | Phillip Zim... | Point 5163. |

**FIGURE 1.50**

**EXERCISE 1-3**

After completing this exercise, you will:

- View styles from content templates.
- Review the settings and values in Edit Drawing Settings.
- Create a template file.
- Open an object Properties dialog box.
- Review the anatomy of an object style.
- Review the anatomy of a label style.
- Review the anatomy of a table style.

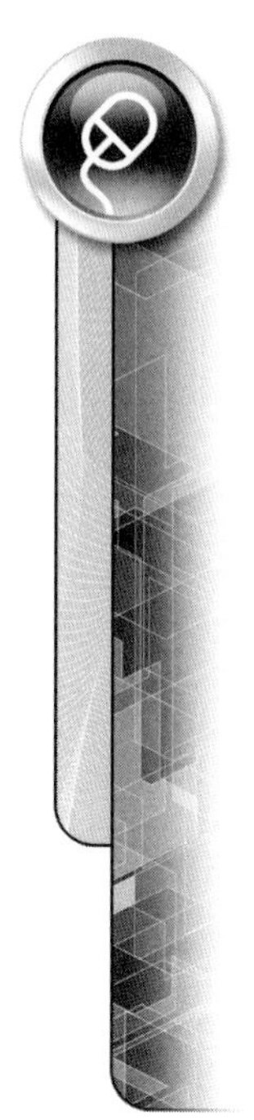

## Drawing Templates

Civil 3D has a content template file that you can use and modify. A template is essential to establishing standard layers, labels, and design documentation.

1. If you are not in **Civil 3D**, start the application and close the open drawing. If you are in Civil 3D, close the current drawing and do not save it.
2. If necessary, click the ***Prospector*** tab.
3. Change Prospector to **Master View**. In Toolspace, top right, under the Help icon, select the drop-list arrow and from the list MASTER VIEW.
4. Expand the Drawing Templates, AutoCAD node, select and right mouse click over *_AutoCAD Civil 3D (Imperial) NCS*, and, from the shortcut menu, select CREATE NEW DRAWING.
5. Click the ***Settings*** tab.
6. In Settings, expand the Surface branch until you view the Surface Styles heading's styles list.
7. Click the **Surface Styles** heading, press the right mouse button, and from the shortcut menu select NEW....
8. In the Surface Style - New Surface Style dialog box, click the ***Display*** tab.

A new style does not reference any layers for its components, but it does assign some specific color and linetype properties (see Figure 1.51).

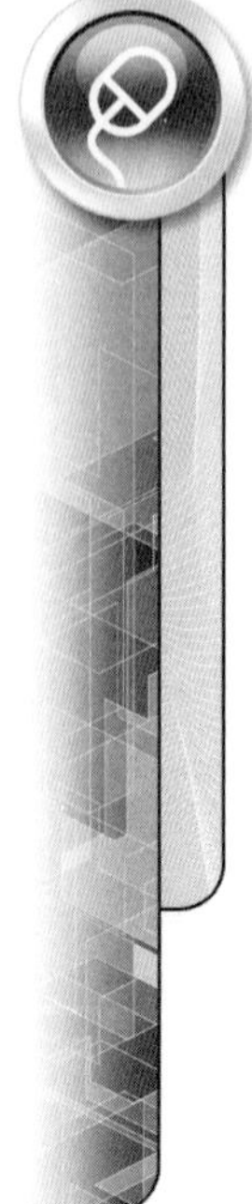

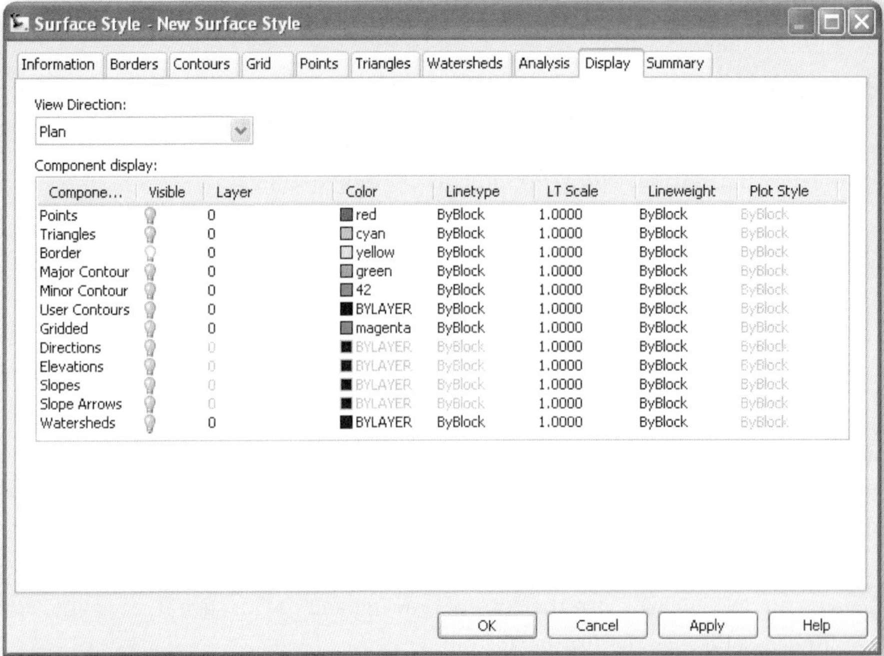


**FIGURE 1.51**

9. For the Border component, click layer **0**.

The Layer Selection dialog box displays, listing the current drawing's layers. If more than one drawing is open, you can select layers from them by clicking Layer Source's drop-list arrow, or you can create new layers by clicking the upper right New... button (see Figure 1.42 and Figure 1.43).

10. At the dialog box's top right click ***New...*** to display the Create Layer dialog box.
11. Click ***Cancel*** until exiting the Surface Style dialog box.
12. In Settings, from the Surface Styles list, select **Contours and Triangles**. Press the right mouse button and, from the shortcut menu, select ***Edit...*** and select the ***Display*** tab.
13. Review the layer assignments and their properties and then click ***OK*** to exit the dialog box.
14. Close the current drawing and do not save the changes.

## Edit Drawing Settings

The Edit Drawing Setting, Edit Label Style Defaults, Edit LandXML Settings, and Table Tag Numbering dialog boxes occupy the Settings' tree top. Their values affect all styles below them.

1. At Civil 3D's top left, Quick Access Toolbar, click the ***Open*** icon, browse to this textbook's CD, Chapter 1 folder, select the drawing *Overview Surfaces II*, and click ***Open***.
2. If necessary, click the **Settings** tab.
3. In Settings, at its top, click the drawing name, press the right mouse button, and, from the shortcut menu, select EDIT DRAWING SETTINGS....
4. In Edit Drawing Settings, click the ***Units and Zone*** tab and review its values.

5. Click the ***Transformation*** tab and review its values.
6. Click the ***Object Layers*** tab, review its settings, and scroll through the layer list, noting its modifier settings and layer names.
7. Click the ***Abbreviations*** tab, viewing Civil 3D's listing and reporting abbreviations. You can change any of them in this panel.
8. Click the ***Ambient Settings*** tab and review its settings for different sections.
9. Click ***OK***, exiting the Edit Drawing Settings dialog box.

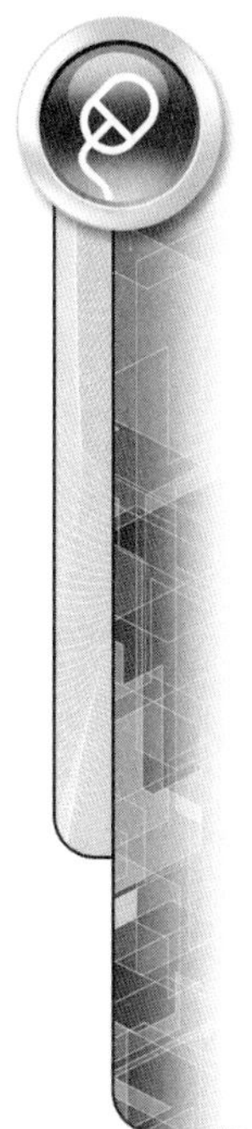

## Object Style

Each object instance has an assigned object style. Civil 3D has several object styles affecting the display of an object's components and characteristics.

This exercise continues in the Overview Surfaces II drawing.

1. In Settings, expand the Surface branch until you view the Surface Styles heading's styles list.

Each surface style has a function or specific surface characteristic to display. The Border & Triangles & Points style shows a TIN surface structure. This surface view allows a user to review if linear features show correctly or if additional editing or data is necessary for a surface to correctly represent its data.

2. From the Surface Styles list, select **Border & Triangles & Points**, press the right mouse button, and, from the shortcut menu, select EDIT....
3. Click the ***Borders***, ***Points***, and ***Triangles*** tabs to review their contents.
4. Click the ***Display*** tab to review what components this style displays.
5. Exit the dialog box by clicking CANCEL.
6. In Settings, in the Surface Styles list, select **Contours 1' and 5' (Design)**, press the right mouse button, and, from the shortcut menu, select EDIT....
7. Click the ***Borders*** and ***Contours*** tabs to review their contents.
8. Click the ***Display*** tab to review what components this style displays.
9. Exit the dialog box by clicking CANCEL.
10. In Settings, expand the Parcel branch until you can view the Parcel Styles heading's styles list.
11. From the styles list, select **Single-Family**, press the right mouse button, and, from the shortcut menu, select EDIT....
12. Click each tab and review its settings and values.

The Display tab settings define what components a parcel displays

13. Exit the dialog box by clicking CANCEL.

## Label Style

1. In Settings, expand the Parcel branch until you view Label Styles' Area heading's styles list.
2. From the list, select **Name Area & Perimeter**, press the right mouse button, and, from the shortcut menu, select EDIT....
3. Click the ***Information*** tab to review its settings and values.

The Information tab displays the style's name, description, and authorship credits.

4. Click the ***General*** tab to view its settings and values.

The General tab displays values from the Edit Label Settings dialog box.

5. Click the ***Layout*** tab, reviewing its settings and values.
6. In the Text section, in Contents, click in the Value cell to display an ellipsis. Click the ellipsis to display the Text Component Editor.
7. Click in the Parcel Area format string on the right side of the dialog box.

This action displays the format string components on the left side of the dialog box. In detail, the left side shows what is in shorthand on the right (i.e., Uacre means the area unit is acre, P2 means two decimal places for areas, etc.). You change values by clicking in their value cell, dropping an options list, and, from the list, selecting a new value.

8. Click in Precision's value cell to display a drop-list, and from the list select a new precision value.
9. Click the **arrow** at the top-center to transfer the new value to the label format.
10. Click CANCEL until exiting all of the dialog boxes.
11. CLOSE the drawing without saving it.

## SUMMARY

- Layer-based drawings have layers for each object type as well as layers for object components and characteristics.
- The Edit Drawing Settings, Edit Label Style Defaults, Edit LandXML Settings, and Tag Table Numbering dialog boxes influence all of the styles and settings in a drawing.
- You can lock values in the Edit Drawing Settings, Edit Label Style Defaults, and Edit LandXML Settings dialog boxes.
- The Edit Drawing Settings, Edit Label Style Defaults, Edit LandXML Settings, and Tag Table Numbering dialog boxes have arrows, indicating that a lower style has changed the value.
- Object styles emphasize a subset of an object's components or characteristics.
- Label styles label a subset of an object's properties.

## UNIT 4: CIVIL 3D AND LAND DESKTOP

Civil 3D works in tandem with Land Desktop. Autodesk gives each application the ability to exchange data, complementing its strengths and making up for its weaknesses. The LandXML file, direct data reading, and data extraction commands transfer points, point groups, description keys, surfaces, alignments, pipe networks, and profile sampling data between the two programs.

### LANDXML SETTINGS

Civil 3D exports and imports LandXML data files from LDT and other civil applications. Setting the proper units for importing and exporting data is critical to successfully using LandXML. You set these values in the Edit LandXML Settings dialog box. The Export panel's Data Settings section is important (see Figure 1.52).

If set to US Foot, the units in the file are tagged to US Foot. When set to International Foot, the units are tagged as International. If you are importing a LandXML file with units tagged as International Foot to a US Foot drawing, the units will be converted upon import.

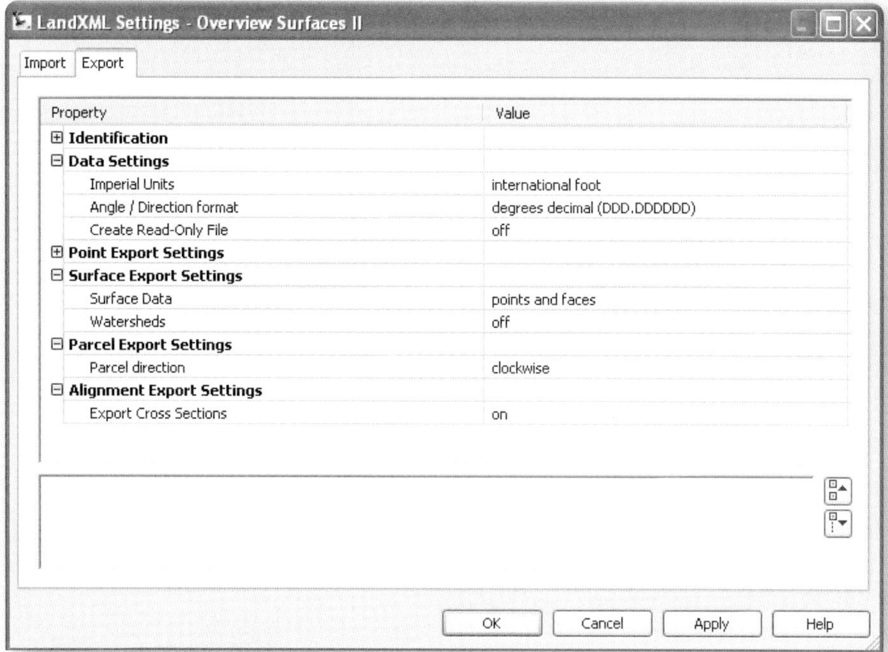


**FIGURE 1.52**

## LandXML Import and Export

The Ribbon's Insert tab, Import panel, LandXML icon starts a LandXML file's import. The command prompts you to select a file and displays a dialog box listing the file's data types (see Figure 1.53). At this time all the data or a subset of data from the file can be selected.

After importing a LandXML file, most information appears in Prospector and in the drawing, except for profile data. When profile data is imported, its entries are placed in the Prospector's Sites branch or in Prospector's Alignments' branch. You have to create a profile view using the profile data to view them in the drawing. You have to create or recreate the remaining roadway design elements (assembly, corridor, section sample group, and section views) to complete a roadway design.

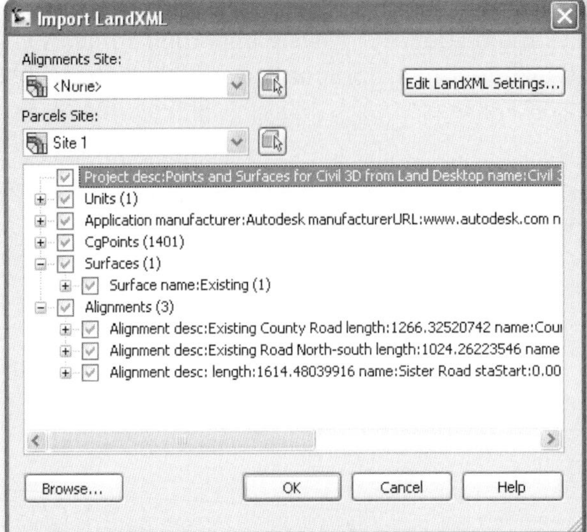


**FIGURE 1.53**

The Ribbon's Output tab, Export panel contains the Export to LandXML icon. When exporting, a user selects data by toggling on the various data types in the Export to LandXML dialog box (see Figure 1.54).

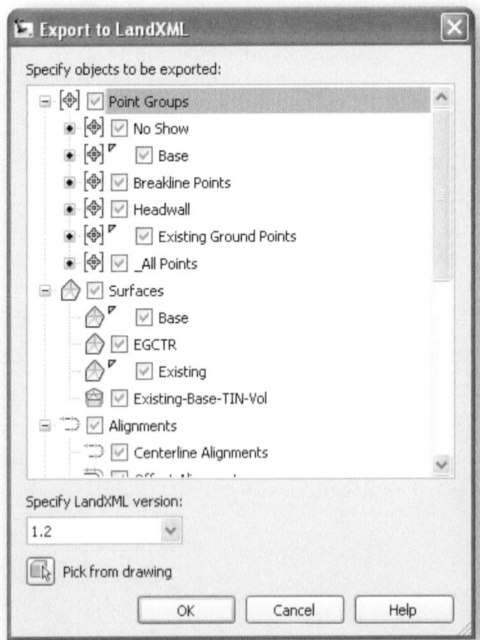


**FIGURE 1.54**

## Importing from LDT Projects

A second method of transferring data from LDT to Civil 3D is directly reading a project's data structure. The Ribbon's Insert tab, Import panel contains the Land Desktop icon which displays a dialog box listing a project's data. To use the dialog box, first identify the LDT project folder and then the project name. After selecting the project, the dialog box populates with the project's data (see Figure 1.55). After selecting the data, click OK and the data is read and converted. You can read project data back to LDT Release 2. If attempting to read earlier versions, users may encounter incompatible file formats.

This method currently does not import points. The best method of importing LDT points is by importing an ASCII file.

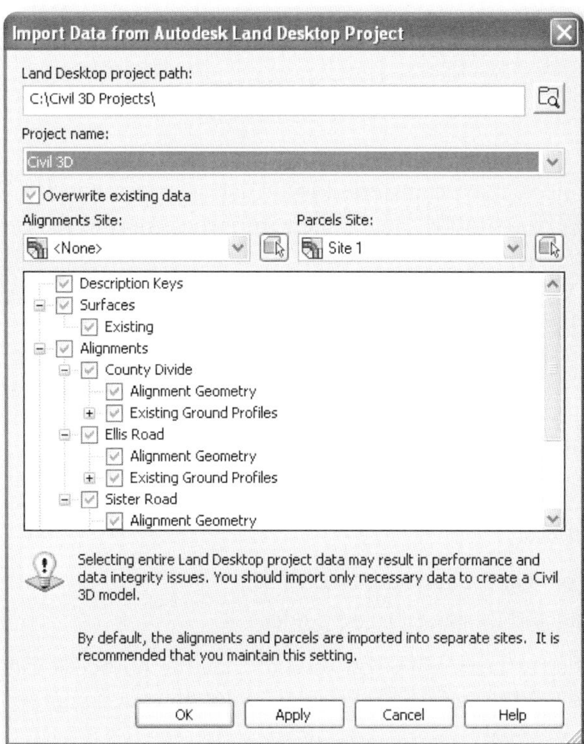


**FIGURE 1.55**

If the LDT project contains pipes, the routine issues a warning about having the proper parts list definition (see Figure 1.56).

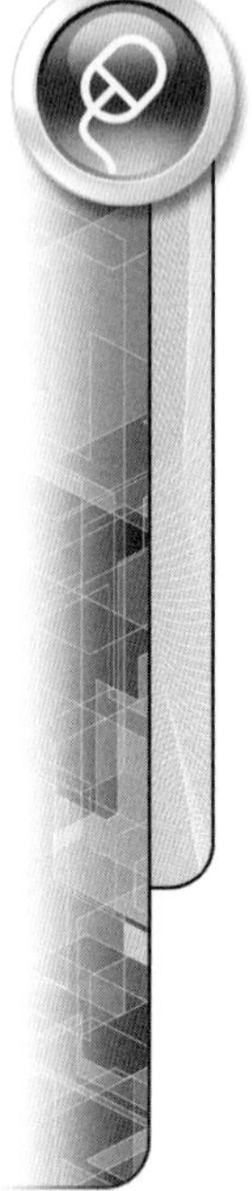

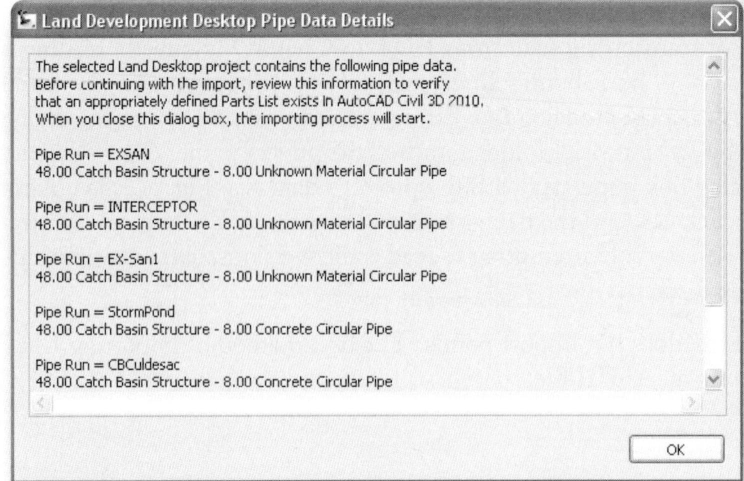
Land Development Desktop Pipe Data Details
The selected Land Desktop project contains the following pipe data.
Before continuing with the import, review this information to verify
that an appropriately defined Parts List exists in AutoCAD Civil 3D 2010.
When you close this dialog box, the importing process will start.
Pipe Run = EXSAN
48.00 Catch Basin Structure - 8.00 Unknown Material Circular Pipe
Pipe Run = INTERCEPTOR
48.00 Catch Basin Structure - 8.00 Unknown Material Circular Pipe
Pipe Run = EX-San1
48.00 Catch Basin Structure - 8.00 Unknown Material Circular Pipe
Pipe Run = StormPond
48.00 Catch Basin Structure - 8.00 Concrete Circular Pipe
Pipe Run = CBCuldesac
48.00 Catch Basin Structure - 8.00 Concrete Circular Pipe
OK

**FIGURE 1.56**

After importing from a LDT project, most transferred information appears in Prospector and in the drawing, except for profile data. When importing profile data, it creates only Prospector entries. You have to create a profile view using the profile data to view them on the screen. You have to create or recreate the remaining roadway design elements (assembly, corridor, section sample group, and section views) to complete a roadway design.

## EXERCISE 1-4

After completing this exercise, you will:

- Import a LandXML file from Land Desktop into Civil 3D.
- Read data directly from a Land Desktop project folder.

This exercise uses files found on the textbook's CD. You need to copy the Autodesk Land Desktop Project, Civil 3D, from the CD to the Civil 3D Projects folder on your computer.

### Exercise Setup

1. Using Windows Explorer, locate the Civil 3D Projects folder on your hard drive.
2. Copy the *Civil 3D project* from this textbook's CD, and place it in the Civil 3D Projects folder (see Figure 1.57). Close Windows Explorer.
3. Click the ***Prospector*** tab.
4. If necessary, change Prospector to **Master View**.
5. Expand the Drawing Templates and AutoCAD node, select and right mouse click over **_AutoCAD Civil 3D (Imperial) NCS**, and, from the shortcut menu, select CREATE NEW DRAWING.

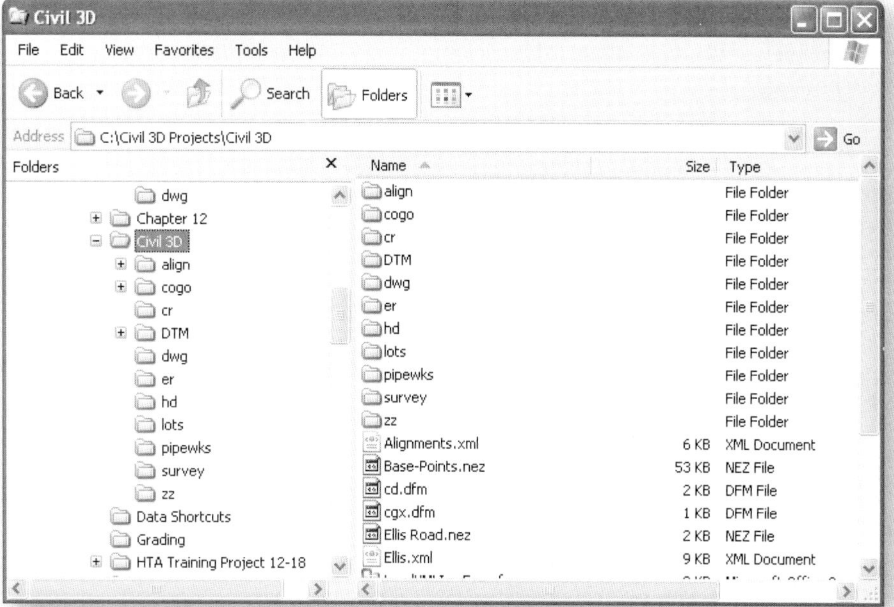


**FIGURE 1.57**

## Importing a LandXML file

A LandXML file contains several data types (surface, alignment, parcel, pipe networks, and profile data). When you view the file in the Import LandXML dialog box, it lists all of the file's data. You select what items to import by toggling them on or off.

1. At Civil 3D's top left click Ribbon's Insert tab.
2. From the Import panel, locate and click the ***LandXML*** icon.
3. In the file Import LandXML dialog box, browse to the folder C:\Civil 3D Projects \Civil 3D, select *Overall Proj.xml*, and click ***Open***, displaying the Import LandXML dialog box.

The Import LandXML dialog box displays, listing the file's data. It is here that you select the data to import (see Figure 1.58).

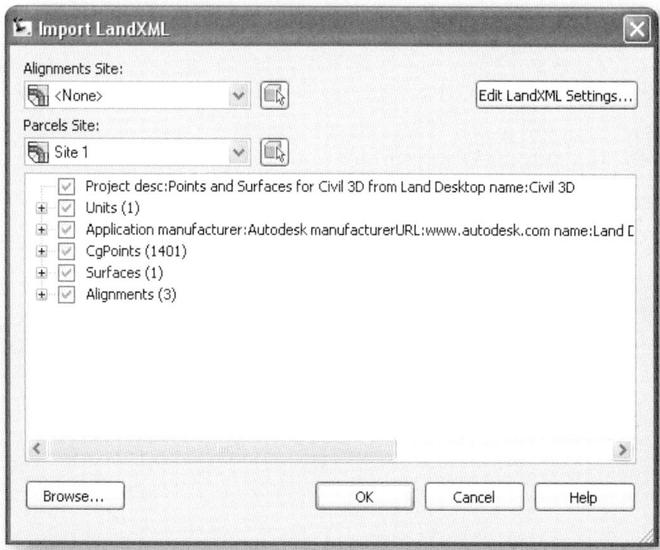


**FIGURE 1.58**

4. Leave all toggles on and click ***OK*** to import the data.

If the Event Viewer displays, close it by clicking the green check mark in the upper-right of the panorama.

5. If necessary, click the ***Prospector*** tab.
6. Expand the Alignments and Centerline Alignments branches to view the LandXML file's alignments.

After reading the points, surface, and alignment data, these objects appear in Prospector under their appropriate object type (see Figure 1.59).

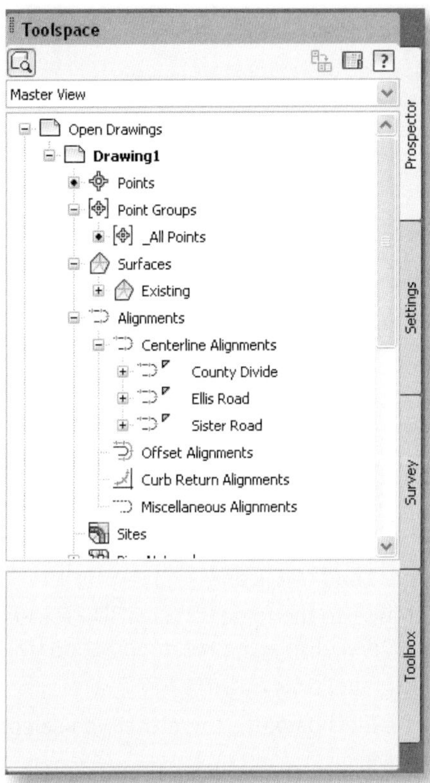


**FIGURE 1.59**

7. In Prospector, click the heading **Points** and, in Prospector's preview, view the imported points.
8. Expand the Surfaces branch to view the surface list.
9. `CLOSE` the drawing, and do not save the changes.

## Importing from a Land Desktop Project

1. In Prospector's Mater View, expand the Drawing Templates and AutoCAD node, select and right mouse click over **_AutoCAD Civil 3D (Imperial) NCS**, and, from the shortcut menu, select CREATE NEW DRAWING.
2. At Civil 3D's top left click the Ribbon's Insert tab.
3. From the Import panel locate and click the ***Land Desktop*** icon.

This displays the Import Data from Autodesk Land Desktop Project dialog box, which sets the projects and named project folder (see Figure 1.55). After setting the project folder and identifying the project, the dialog box displays all of the data that you can transfer from the project.

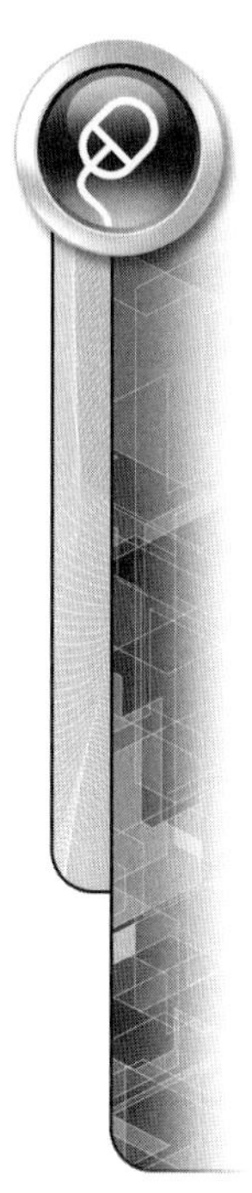

4. In the Import Data from Autodesk Land Desktop Project dialog box, set the project folder to **C:\Civil 3D Projects** and the project name to **Civil 3D**.
5. In the dialog box, click ***OK*** until you return to the command line to import the project data.
6. If necessary, click the ***Prospector*** tab.
7. In Prospector, expand the Surfaces branch to view the surface list.
8. CLOSE the drawing and exit Civil 3D.

Each of the following chapters explores in greater detail an object type and its styles.

## SUMMARY

- LandXML files provide an effective method of transferring data between Civil 3D, Autodesk Land Desktop, and other civil engineering applications.
- Civil 3D reads Land Desktop project data and imports the data directly into a drawing.

# CHAPTER 2

# Points

## INTRODUCTION

Many civil design projects start with coordinate data. Coordinate data comes from surveyors, public records, or a Web site, among other places. It is necessary to import and modify these coordinates into usable point data. Methods for importing and modifying coordinate data include defining standard symbols with descriptions; transforming coordinates to other grid coordinate systems; organizing points into groups based on function, description, and common names; and labeling their coordinates and information as a part of a submitted document.

## OBJECTIVES

This chapter focuses on the following topics:

- Importing Points from External Files
- Defining Point Styles
- Defining Point Label Styles
- Civil 3D Transparent Commands
- Description Key Set
- Point Groups

### OVERVIEW

Coordinate data serves a dual purpose in civil designs. First, it communicates the current site's state to the office. Second, it communicates the design from the office to the field. Coordinates representing the current site's state become points for a survey plat or a surface. A surface influences the strategies for access roads, parcel layout, water control, grading, and so on. This chapter reviews Civil 3D's point implementation.

Points represent several types of data. However, there are three basic point categories that determine how they are processed and assigned marker and label styles. The first point category requires no action other than placing it into a drawing with a marker and label style. An example of this category is a ground shot for a topographic survey. The second category requires a symbol marker style. The symbol is a part of the final product, but the point label may not display. This means the maker and label styles should use different drawing layers. Examples of this category are manholes, signs, power poles, and so on. The third category is points that are a part of a line or arc (a boundary, roadway centerline, edge-of-pavement, etc.). Civil 3D uses this category's points as surface breaklines, for grading feature lines, or line work for a submittal document.

Civil 3D manipulates, represents, and organizes points from the Prospector's Points and Point Groups areas or from the Points ribbon icon (Home, Insert, and Modify—Points). The point tools are flexible, comprehensive, and provide necessary management tools. The Home tab's Points ribbon icon displays routines creating points, Insert tab's Points from File ribbon icon reads external point files, and the Modify Points ribbon displays point editing tools (see Figure 2.1). Prospector's Points shortcut menu creates and transfers points.

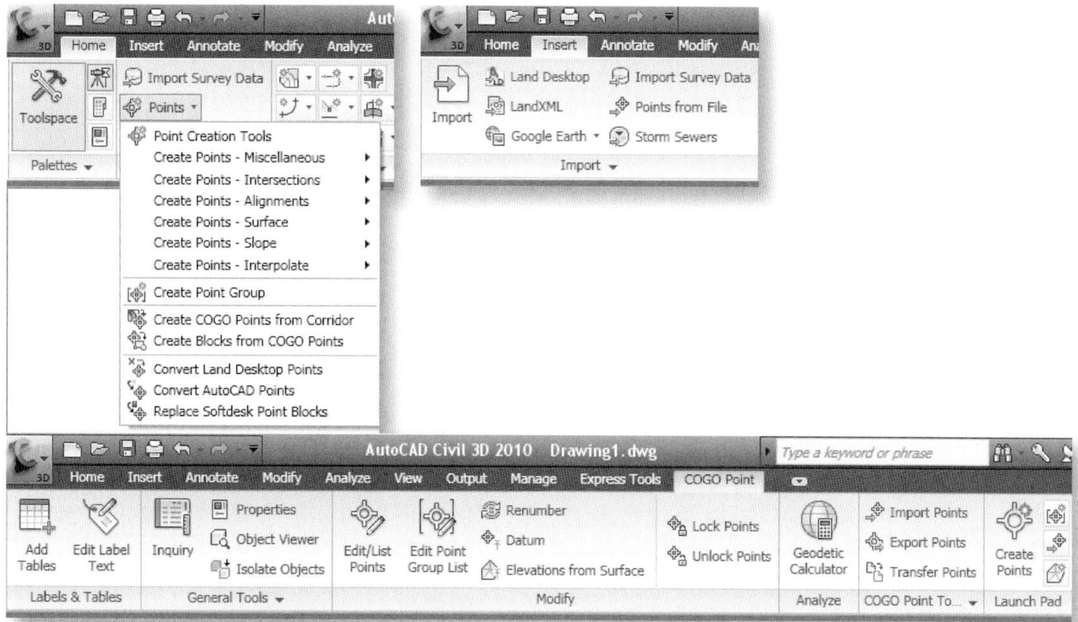


**FIGURE 2.1**

The point creation commands create points from lists, external files, AutoCAD entities, surface elevations, roadway corridors, interpolation between points, from slopes and intersections (distance and direction), and more. The Create Points toolbar organizes its icons by what data source creates the resulting points (see Figure 2.2) and reflects the Home—Points' menu commands (see Figure 2.1).

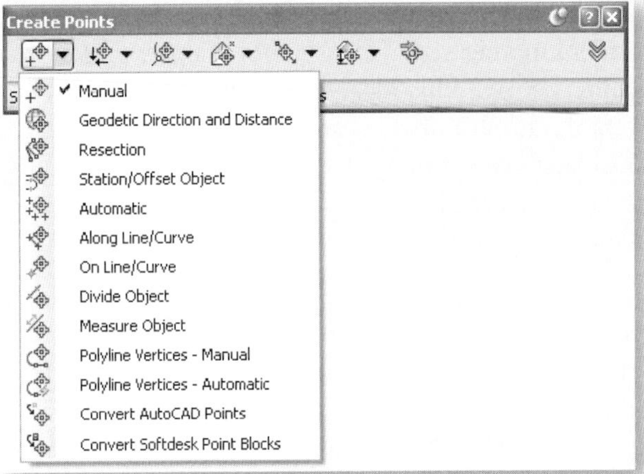


**FIGURE 2.2**

A point represents coordinates. Civil 3D uses a point object that has a marker (its insertion point is the point's coordinate location) and a label. Marker and label styles define the point's look in plan, profile, model, and section. Traditionally, the label displays a point's number, elevation, and description. Users need to develop a basic understanding of style definitions and behaviors when defining and assigning point styles. Default point creation settings (Edit Feature Settings for points) assign default marker and label styles. In addition to these settings, a Description Key Set, if used, affects point style assignments, or a point group can override assigned styles.

If organizing points into point groups, there are two methods of using point group properties to affect their display. First, points in a group retain their original styles. In this situation, display order makes no difference. The second method overrides one or both styles. In this method, display order affects how points display.

If a point belongs to several point groups and each group overrides point and/or label styles, then the last drawn point group determines how the points display. The Point Groups Properties dialog box displays a drawing's point group draw order (lowest being the first drawn). In Figure 2.3, all points belong to the No Show group, and its point and label style overrides hide all point markers and labels. If the Breakline Points group is located at the top of the list and No Show is second, the Breakline point group is drawn last and only its points display.

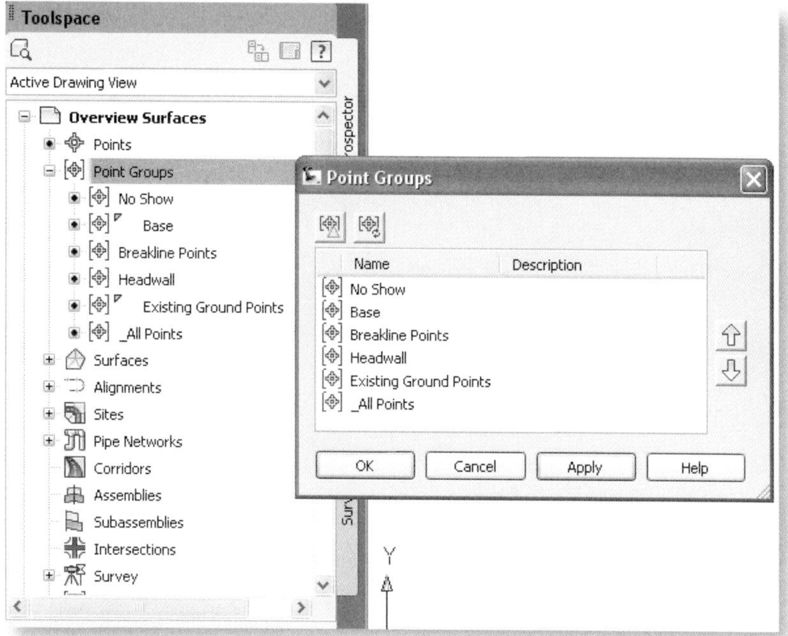


**FIGURE 2.3**

When using the Standard or Basic point style, the marker and label are on the same layer. A point (marker) style can assign a symbol or marker, a behavior (two- and three-dimensional views, at elevation, etc.), and marker and label layers.

If using a Description Key Set, matching a set entry assigns each point a point and label style, marker and label layer, code translation, and scaling.

Prospector locks or unlocks individual points or point groups. Locking points or point groups prevents the `MOVE` command from changing their coordinates. Locking is easily undone, and a vault project does improve on safeguarding point coordinates.

A point represents a point object instance. If using the `ERASE` command to erase a point, Civil 3D deletes the point permanently from the overall point list and any point groups it belongs to.

Points as a part of a vault project are discussed in Chapter 12 of this textbook.

Here are some Civil 3D point rules:

- Points must always be present in a drawing.
- The Properties palette displays and edits all point properties.
- If allowed, the `MOVE` command relocates a point and Civil 3D recognizes the change.

## Unit 1

Settings' Point Edit Feature Settings define the default point and label styles, default point and point group names, and, if checked in points, can be updated. Create Points' command settings define a point's default layer, numbering, information prompting, and handling of duplicate point numbers. These settings and style definitions are the first unit's focus.

### Unit 2

This chapter's second unit reviews routines that create points from printed lists, external ASCII files, or calculated points (such as intersections or interpolations). Other Create Points toolbar routines are discussed in their appropriate context in later chapters of this textbook.

### Unit 3

The subject of this chapter's third unit is point analysis and listing tools. Points are viewed in the preview area, the Properties palette, or in the Panorama's Point Editor vista. The Inquiry tool has some point review commands and the Toolbox has basic point reports.

### Unit 4

Point editing includes changing their number, elevation, or location, and is the subject of the fourth unit. The point editing routines are in Prospector's Points heading shortcut menu, Prospector's preview area, the Properties palette, and the Ribbon's Modify—Points icon group.

### Unit 5

As previously mentioned, some points have symbols. A Description Key Set assigns point styles with a symbol. The Description Key Set, point groups, and Table styles are the subjects of the last unit.

## UNIT 1: POINT SETTINGS AND STYLES

A point object's creation and display is the result of user-defined settings and styles. The point routines use the settings and their values (including default styles) from the Settings' Point branch and the Create Points toolbar. The Point branch sets the default styles, naming convention, if vault project points can be updated, and manages all styles and command properties. The Create Point settings set the next point number, duplicate point number handling, point properties prompting, etc.

### EDIT FEATURE SETTINGS

Displaying Settings' Point heading shortcut menu and selecting Edit Feature Settings… displays the point object default values and specific ambient settings for points. The dialog box values set initial values for all styles and commands in the branch.

#### Default Styles

This section names the default point and label styles. When there are no active Description Key Sets or new point groups default styles, these styles are used (see Figure 2.4).

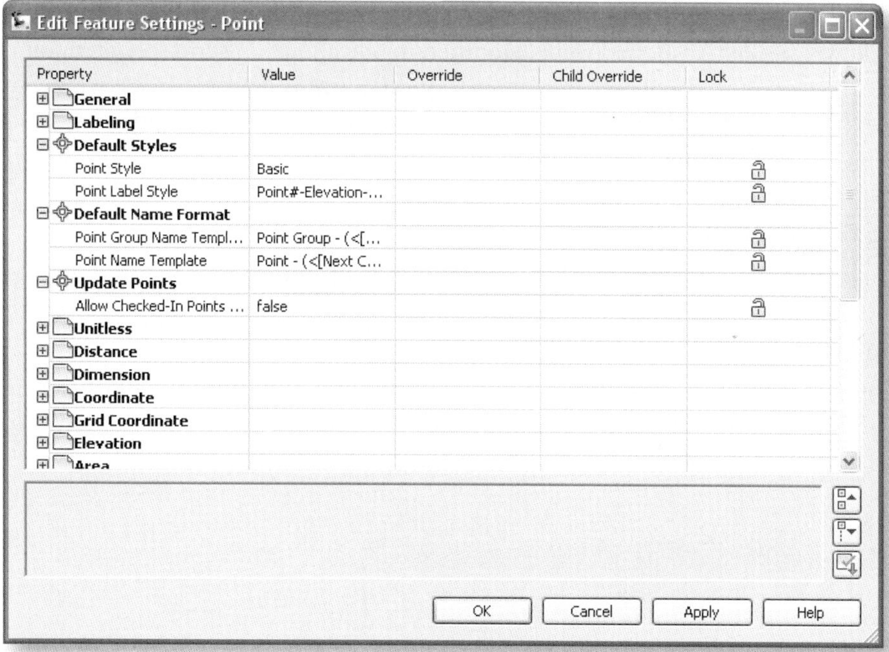


**FIGURE 2.4**

## Default Name Format

This section sets the naming convention for point groups and point names (see Figure 2.4). By default, a point group name is simply the words "Point Group" followed by a sequential number. When defining a point group, a user can assign a more descriptive name. When clicking in the value cell for either format, an ellipsis (...) appears. When you click the ellipsis, the Name Template dialog box displays. In this dialog box, users can change the name prefix (Point Group -) as well as the number style, beginning counter number, and increment values (see Figure 2.5).

Name Template
Name formatting template
Property fields:
Next Counter
Insert
Name:
Point Group - (<[Next Counter(CP)]>)
Incremental number format
Number style:
1, 2, 3...
Starting number:
7
Increment value:
1
OK
Cancel
Help

**FIGURE 2.5**

### Update Points

The Allow Checked-In Points to be Modified toggle allows a user to prevent points checked in from a vault project from being modified (see Figure 2.4). However, by simply changing this value to true, a user can modify point values and then check them back in to the vault project.

## CREATE POINTS

The Create Points command has several values that affect new points.

### Default Layer

If there is no active Description Key Set, this is the default layer for all points created using this toolbar's commands (see Figure 2.6). If using a Description Key Set, this layer is for points whose description does not match any entry in the set.

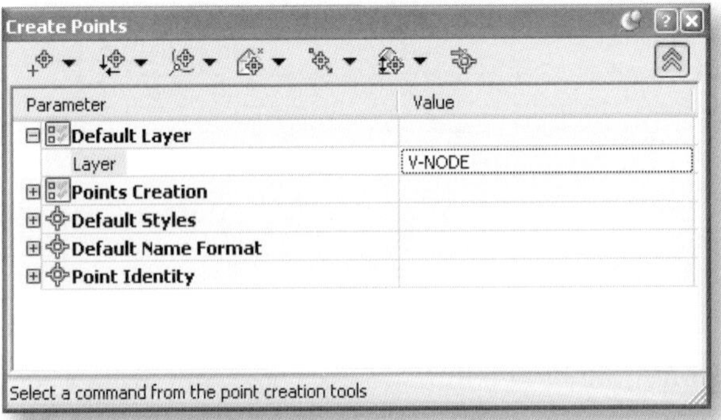


**FIGURE 2.6**

### Points Creation

This section determines how to reference coordinates in a local coordinate system (northing/easting), a state plane system (grid northing/easting), and a geographic system (Long/Lat). It also determines how to assign elevations and descriptions; elevation and description parameters, if using description keys; and determines if the user wants the coordinates echoed to the command line (see Figure 2.7).

There are three possible values for Prompting For Elevations, Descriptions, and Names: None, Manual, and Automatic. None makes a routine not prompt for an elevation, a description, or a name. Manual means that a routine prompts for a value. Automatic uses the default elevation, description, or name, which is farther down in this section's settings. Prompt for Description has one additional method of assigning its value: Automatic – Object. This setting creates a description based on the selected object that is creating the point (e.g., an alignment or parcel).

Set the prompting option by clicking in the value cell for Prompt For Elevations, Prompt for Descriptions, or Prompt for Point Names, then click the drop-list arrow, and, from a list, select the option (None, Manual, or Automatic). Initially, the default elevation is 0 (zero) or blank. After entering an elevation or description, that value becomes the default value for either Default Elevation or Default Description.

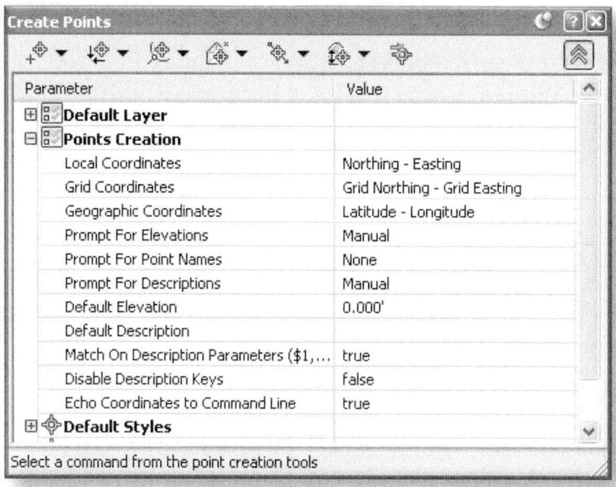


**FIGURE 2.7**

## Point Identity

Whenever the user imports points, there is a possibility of duplicate point numbers or the need to reassign them (see Figure 2.8). The Point Number Offset and Sequence Point Numbers From values are used if referred to by other settings below these two values. The remaining entries display multiple options when you click their drop-list arrows. All point creation routines look to, interact with, and modify this section's values.

The Next Point Number is the next used point number. This number changes when you add or import points. If you are importing points, this number may be the first unused number instead of the first point number after all of the used point numbers. If you decide to erase points, this number does not reset to reflect the erasure. For example, the current point number is 500. After creating 20 additional points, their numbers are 500 through 519. If you erase those 20 points, the current point number remains 520. If you want to reuse the numbers, manually reset the current point number to 500.

Create Points

| Parameter | Value |
|---|---|
| Default Layer | |
| Points Creation | |
| Default Styles | |
| Default Name Format | |
| Point Identity | |
| Next Point Number | 511 |
| Use Sequential Numbering | true |
| Point Number Offset | 1 |
| Sequence Point Numbers From | 100 |
| If Point Numbers Are Supplied | Use |
| Force Names | false |
| If Point Numbers Already Exist | Notify |
| If Point Names Already Exist | Notify |
| If Point Numbers Need To Be Assigned | Use next point number |

Select a command from the point creation tools

**FIGURE 2.8**

### If Point Numbers Are Supplied

Of all the import point settings, If Point Numbers Are Supplied is the critical setting. When you import points, this is the first decision to make. This setting has three options: Use, Ignore, and Add an offset. Each option has its own set of choices and the interactions of these choices are complex because of Settings' dependencies.

**Use.** If Point Numbers Are Supplied is set to Use, the next thing to decide is what to do with file point numbers, some of which may be duplicate numbers. When you are importing and encounter a duplicate point number, the routine checks If Point Numbers Already Exists' value to determine its next step.

**Ignore.** If Point Numbers Are Supplied is set to Ignore, the next thing to decide is how to assign new point numbers to the file's points. If Point Numbers Need To Be Assigned's value sets the assigning method.

**Add an Offset.** If Point Numbers Are Supplied is set to Add an Offset, the next decision is what is the offset value. The Point Number Offset is the value to adjust. Add an Offset's potential problem is it may produce a duplicate point number. When this happens, a Duplicate Point Number dialog box displays and asks for a resolution.

### If Point Numbers Already Exist

If Point Numbers Already Exist has four options: Notify, Renumber, Merge, and Overwrite. Some of these options look to other section settings to resolve a duplicate point number.

**Notify.** This displays the Duplicate Point Number dialog box, which asks for a resolution (see Figure 2.9). By toggling on Apply to all duplicate point numbers, the resolution applies to all duplicate points. If toggled off, each encountered duplicate point number displays this dialog box, asking for a resolution. The Duplicate Point Number dialog box has five possible solutions: Use next point number, Add offset, Sequence from, Merge, and Overwrite.

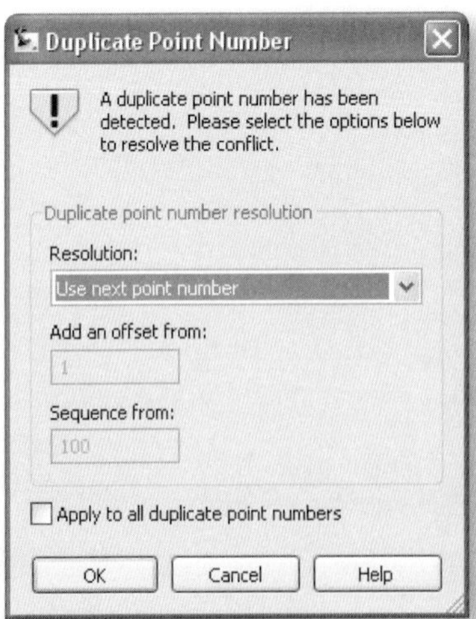


**FIGURE 2.9**

**Use Next Point Number.** Use next point number assigns Next Point Number's value to the duplicate point.

**Add Offset and Sequence From.** Add offset and Sequence from can produce a new point number that is also a duplicate. For example, if you use the offset option and it is set to 1, and the file's point number is 100, this option creates a point number of 101. If that point number is also a duplicate, the next point number is assigned. The same thing can happen using the sequencing option. If the new point number is still a duplicate, the point is assigned the next available point number.

**Merge.** Merge blends a drawing point's data with the source file's values. If the file's point does not contain a value, the current drawing point value remains after import. For example, the source file has point number, northing, and easting data. This source file data will replace the northing and easting values of the same point number in the drawing, but will not change any other values for the point.

Source file values for point 101:

```
101, 5000.0000, 6000.0000
```

Values for point 101 in the drawing:

```
101, 100.0000, 100.0000, 723.84, IPF
```

With Merge on and after importing the file, the new values for point 101 are

```
101, 5000.0000, 6000.0000, 723.84, IPF
```

**Overwrite.** Overwrite replaces a duplicate point number's values with the file's values. When using this option, there may be a loss of data. For example, the source file values for point 401 are

```
401, 5010.0000, 6050.0000
```

The drawing values for 401 are

```
401, 5100.0000, 6500.0000, 723.84, IPF
```

With Overwrite on and after importing the file, the new values for point 401 are

```
401, 5100.0000, 65000.0000, NULL, NULL
```

**Renumber.** This option assigns new point numbers to incoming points. The new point number's value depends on If Point Numbers Need To Be Assigned's value. If Point Numbers Need To Be Assigned is set to Use Next Point Number, the import routine assigns Next Point Number's value. If the value of If Point Numbers Need To Be Assigned is set to Sequence from, the import routine assigns Sequence Point Numbers From's current value. This option's potential problem is also producing duplicate point numbers. When this happens, the Duplicate Point Number dialog box displays and asks for a resolution.

**If Point Numbers Need To Be Assigned.** If Point Numbers Need To Be Assigned is set to Use Next Point Number, the import routine uses Next Point Number's current value. If Point Numbers Need To Be Assigned is set to Sequence from, the import routine assigns Sequence Point Numbers From's current value. This option's potential problem is also producing duplicate point numbers. When this happens, the Duplicate Point Number dialog box displays and asks for a resolution.

## EDIT LABEL STYLE DEFAULTS

This dialog box sets the point labels' general behavior. See this dialog box's discussion in this textbook's Chapter 1, Unit 3.

## POINT STYLES

Standard and Basic point styles are simple marker definitions. A marker is a Civil 3D object that locates a point's coordinates and has four possible views: plan, model, profile, and section.

Settings' Point's Point Styles heading lists the defined styles (see Figure 2.10).

A Point Style dialog box contains five panels, three of which define the marker.

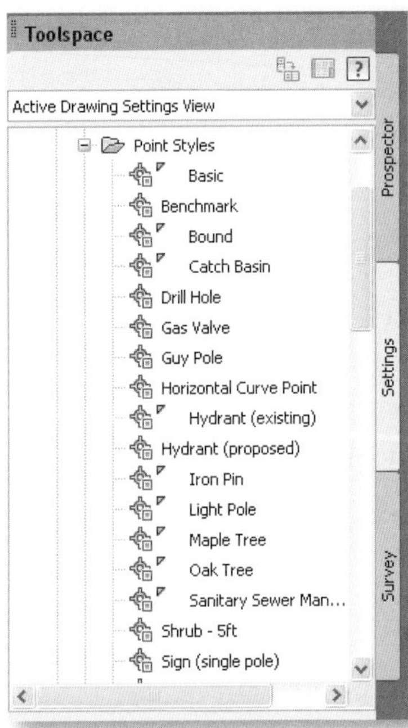


**FIGURE 2.10**

### Information

The Information panel displays the name, description, and who created or modified the style definition (see Figure 2.11).

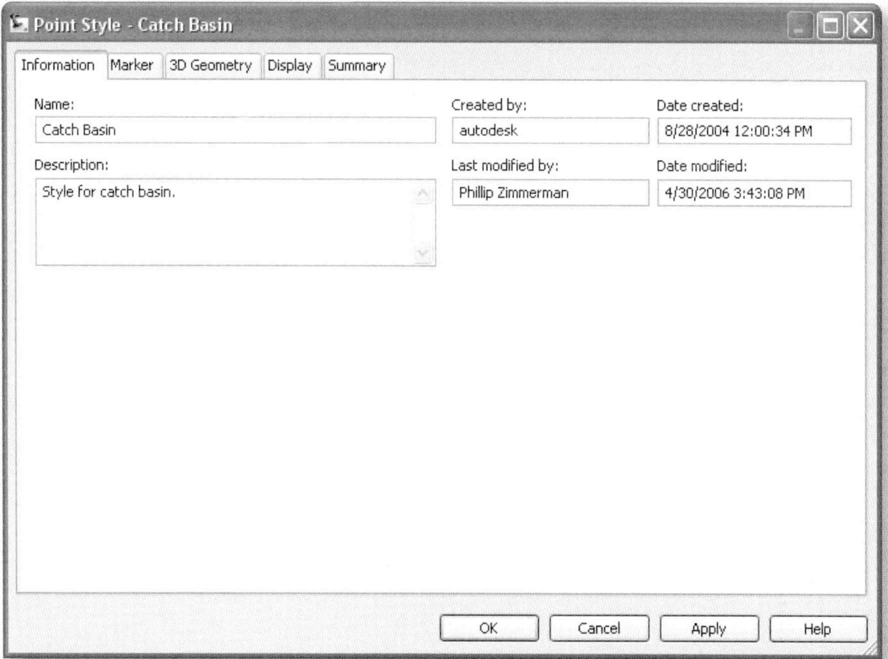


**FIGURE 2.11**

## Marker

The Marker panel defines the marker's appearance (see Figure 2.12). On the panel's left side are three definition options. By default a marker is a Custom marker style. The first option, Use AutoCAD POINT (node), uses the pdmode (appearance) and pdsize (size) system variables. See Civil 3D's help entry for these two AutoCAD system variables.

The second option creates a marker from Use custom marker's list. A custom marker looks similar to an AutoCAD node, but it is an object rather than an AutoCAD node. A custom marker has a base shape (the five leftmost icons) to which users can add two more shapes (a square and/or circle). The base shapes include a node, nothing, a plus symbol (+), an X, or a vertical line.

The third option selects a block definition from Use AutoCAD BLOCK's list. The _Wipeout_Circle marker contains a wipeout that hides line work connected to its coordinates. A marker can be a multi-view block (a block containing different two- and three-dimensional representations). All block definitions must be present in the drawing as a block definition.

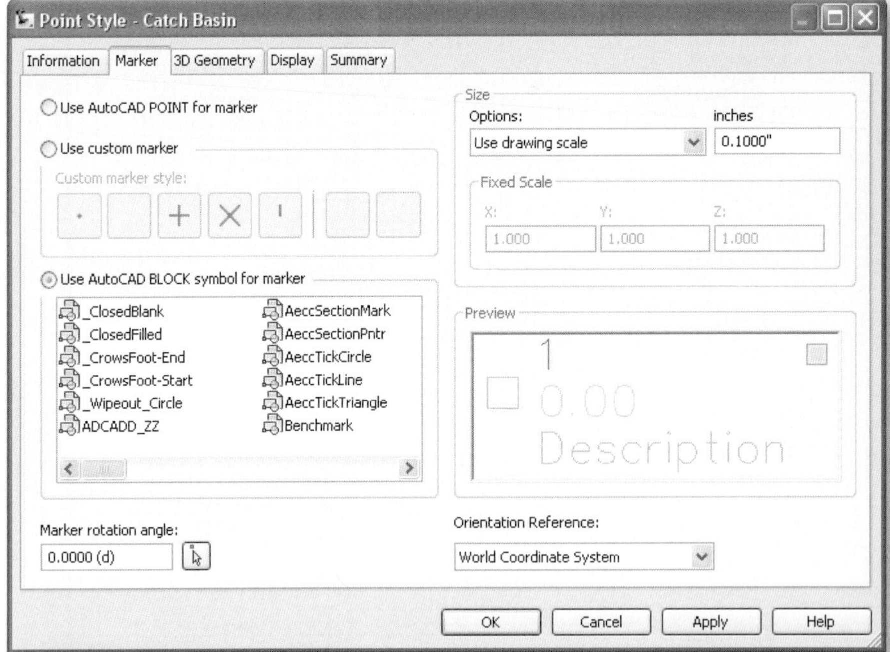


**FIGURE 2.12**

## Marker Rotation Angle

This sets a marker's drawing rotation angle. When in a drawing, the marker displays a rotation grip for rotating the marker.

## Size

The panel's upper-right controls the marker size (see Figure 2.12). By default, a marker's size is the result of multiplying its size by the drawing scale. For example, Use drawing scale multiplies the marker size (0.1) by the drawing's scale (1" = 40') and results in a marker size of 4. Other sizing methods are Use fixed scale, Use size in absolute units, and Use size relative to screen. The Fixed Scale option specifies an X, Y, or Z scaling factor.

## Orientation Reference

By default, a marker orientates itself relative to the World UCS. There are two additional options. The first, the marker's rotation angle, is relative to an object's rotation. The second, the marker's rotation, is relative to the current AutoCAD view direction. This option allows markers to always be plan readable.

### 3D Geometry

When placing points, there are three Point Display mode choices (see Figure 2.13). The default, Use Point Elevation, elevates a point so its AutoCAD Z value matches the entered point's elevation. A second choice is Flatten Points to Elevation, which assigns a user-specified AutoCAD Z elevation. For example, setting Flatten Points to Elevation and entering 0 (zero) creates points with a zero elevation when selecting them with an AutoCAD object snap. The correct elevation is always listed, but selecting it with an object snap produces a zero elevation. Even with Flatten Points to Elevation set, selecting points using a Civil 3D Transparent point filter uses the point's actual elevation.

A Third option is Exaggerate Points by Scale Factor. This option scales a point's AutoCAD Z elevation. For example, setting Exaggerate Points by Scale Factor and setting the scale factor to 2.0 places a point whose elevation is 325.25 at the AutoCAD Z elevation of 650.50. The point displays and lists 325.25, but when you draw a line from the point, the elevation is 650.50.

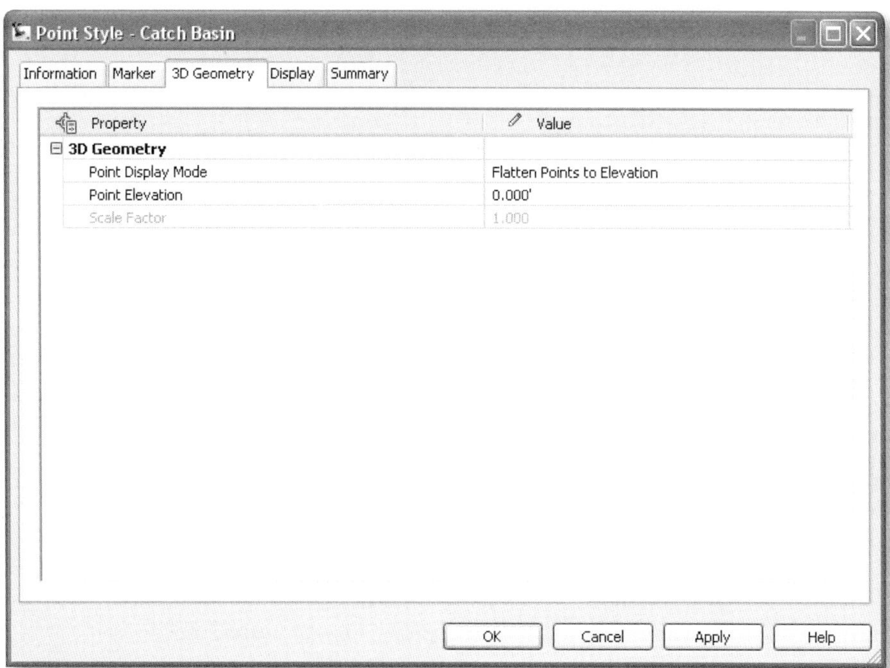


**FIGURE 2.13**

## Display

A point style defines a layer for its marker and associated label components. The Display panel also sets the component's layer state (ON/OFF, and for view directions) and properties (see Figure 2.14).

To change layer assignments, select the layer column's value. A Layer Selection dialog box displays, listing the current drawing's layers. From this list, select an appropriate component layer. If none are appropriate, this dialog box offers the option of creating new layers (by clicking *New...* at the upper-right) or displays other open drawings' layer lists (made visible by clicking Layer source's drop-list arrow).

If using a Description Key Set, the key set assigns the point style's marker layer. In this case the point style marker layer is 0 (zero).

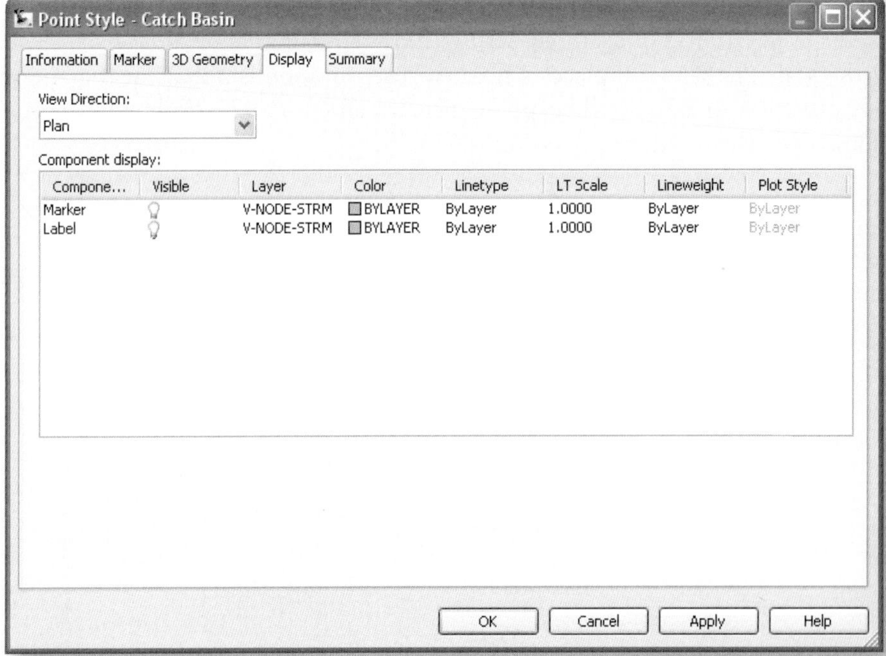


**FIGURE 2.14**

## Summary

The Summary panel reviews all of the style's settings (see Figure 2.15).

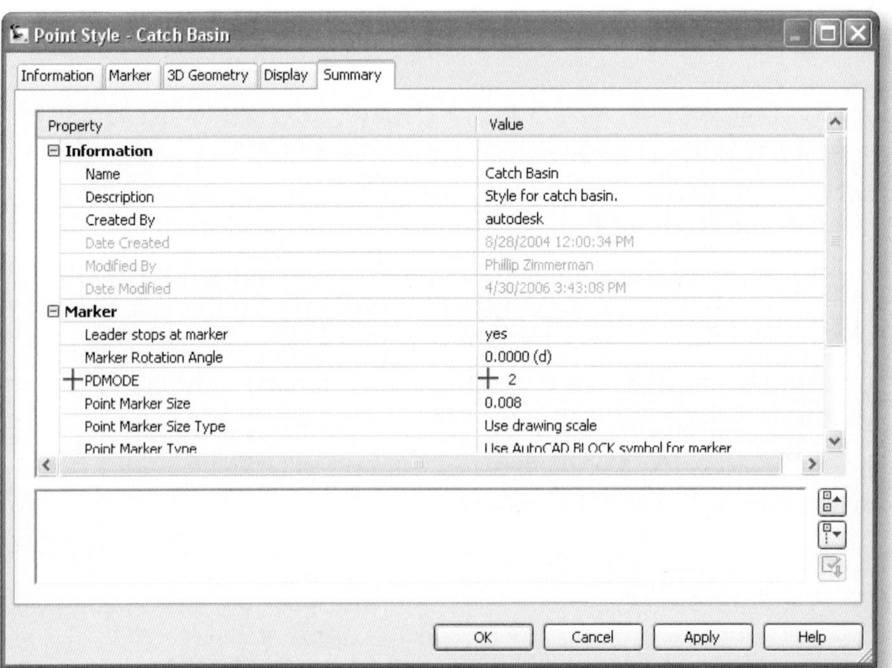


**FIGURE 2.15**

## LABEL STYLES

A label style displays selected point information. It is located relative to the marker and has four possible views: plan, model, profile, and section.

A label style places text and point information at user-specified locations. Label styles are complex, but extremely flexible and easily customized. Labels are sensitive to view rotation and scale and re-orientate or rescale to be "plan readable" and correctly sized (i.e., they read left to right and are the correct size for the plotting scale).

Settings' Point branch, Label Styles heading lists all available label styles (see Figure 2.16). A Point Label Style dialog box contains five panels, three of which affect the label.

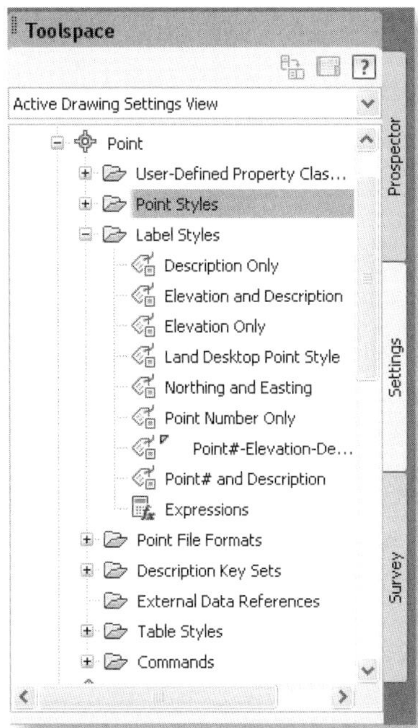


**FIGURE 2.16**

### Information

The Information panel displays the style's name, description, and who created and modified it (see Figure 2.17).

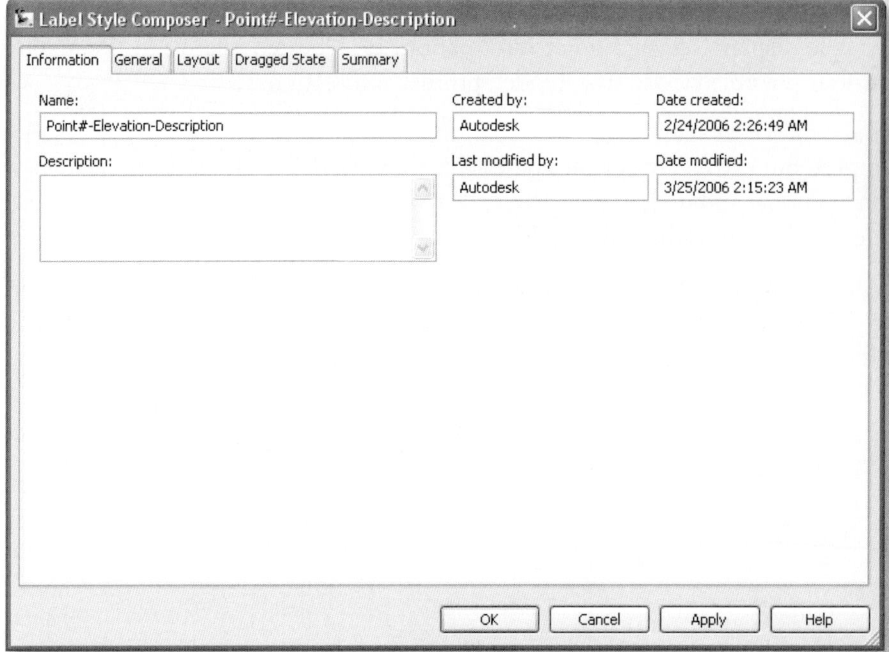


**FIGURE 2.17**

## General

This panel's values reflect the drawing name or point feature level's Edit Label Style values (see Figure 2.18). The Label section identifies the label's text style, its overall visibility, and, if necessary, a layer. If you set this layer to 0 (zero), the associated point style assigns the layer. If Visibility is set to false, all of this style's labels do not show. The Behavior section sets the label's horizontal behavior. Orientation Reference has three possible values: Object, View, or World Coordinate System (WCS). When Orientation Reference is set to Object, the label matches the object's rotation; if set to View, the label is always horizontal regardless of the view's rotation; and when set to WCS, the label is always rotated towards WCS 0 (zero). The last section sets the label's plan readability behavior. The Readability Bias value triggers the label's flip to remain left to right readable, and Flip Anchors with text affects the label's attachment when it flips.

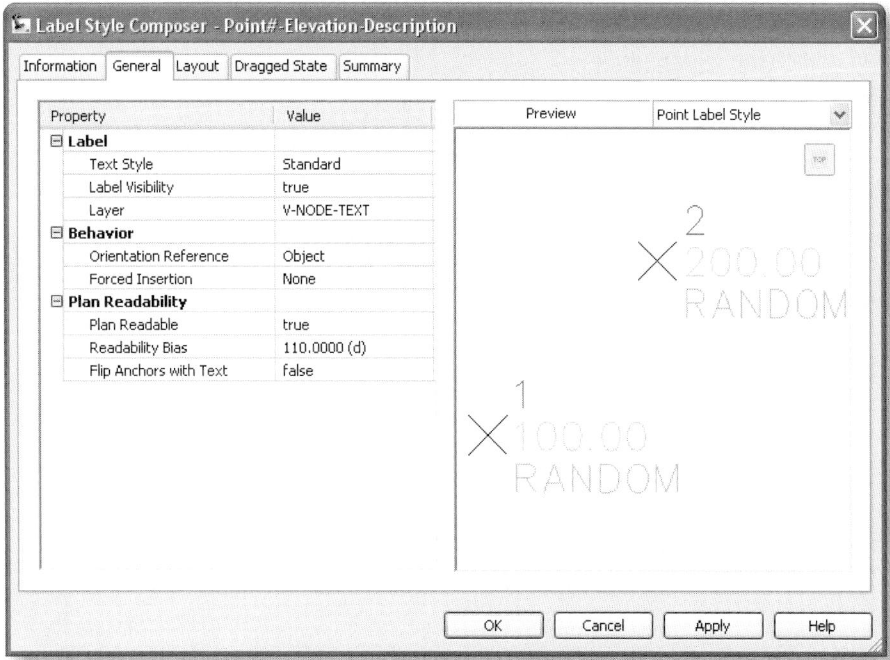


**FIGURE 2.18**

## Layout

Layout settings and values affect each label component (see Figure 2.19). A label component has three sections: General, Text, and Border. General section settings and values name the component and set its visibility and its anchoring point. Text section settings and values set the component's text contents; size; rotation; attachment to the component's anchoring point, where the text is relative to the point's coordinates; offsets; color; and line weight. A label component can be located relative to another label's component rather than the feature, i.e., the elevation component is anchored to the point's coordinates, and the point number component is anchored to the label's elevation component. The Border section settings and values create a border around the text component.

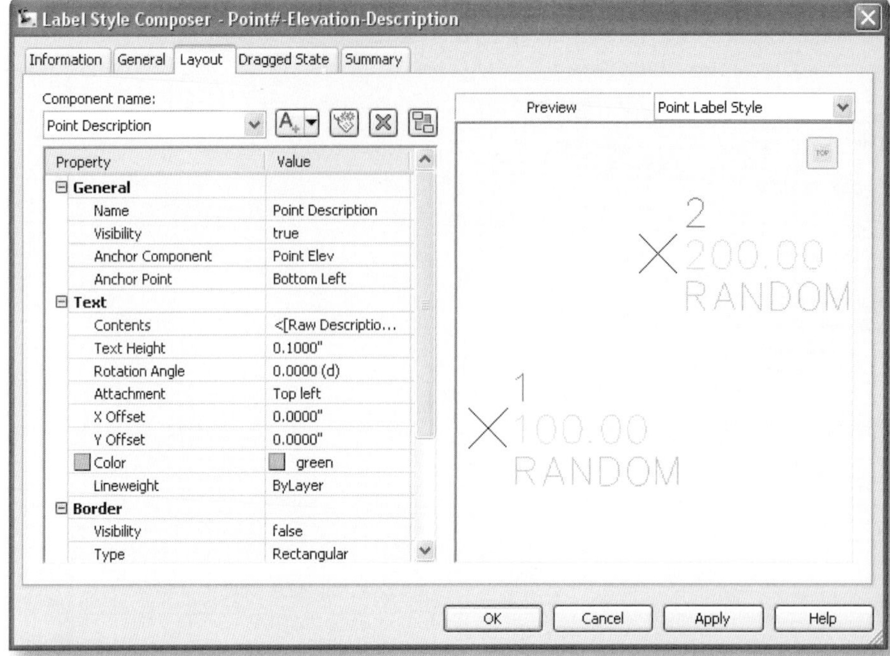


**FIGURE 2.19**

## Anchoring Point

Attaching a label to an object (Feature) is done by anchoring points. A label anchors the main label component to the feature (the point's coordinates). If there are additional label components, they are anchored relative to the main label component. For the Point#-Elevation-Description point label, the main feature anchor point is the elevation label (see Figure 2.20). The label anchors to the coordinates' top-left point. The label's text attaches to the anchor with its text's top-left justification point. To move the label off the point's left side, the style uses an X offset. It would have been simpler to use the feature's middle-right anchor point as the component's middle-left attachment point. Defining it this way removes the need for offsets.

The main component's feature attachment sets the remaining component attachment points (i.e., point number and description). When viewing their anchor and attachment information, they reference the elevation component as the label's Anchor, not the feature. Point Number's bottom-left attaches to elevation's top-left. Point Description's top-left attaches to elevation's bottom-left.

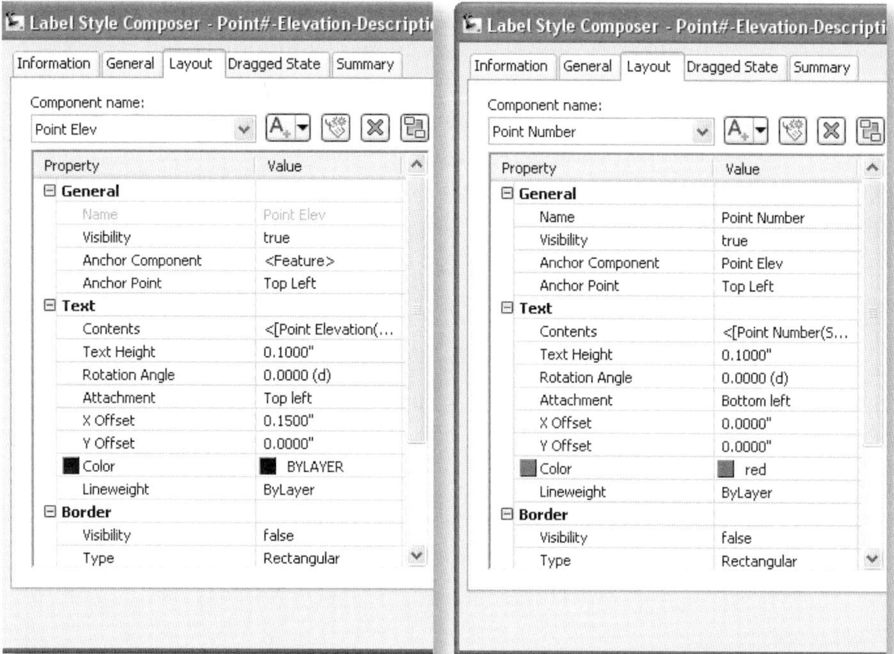


**FIGURE 2.20**

Ideally, anchoring a point and its labels should be set as follows: The right-middle of the feature is the elevation's middle-left attachment point. The elevation's top-left anchor point is the point number's bottom-left attachment point. The elevation's bottom-left anchor point is the point description's top-left attachment point (see Figure 2.21).

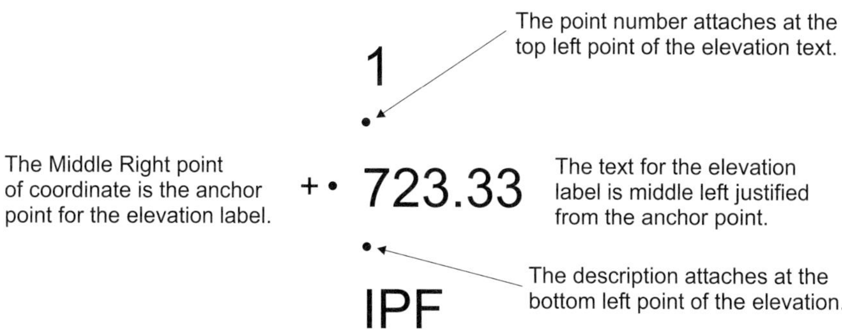


**FIGURE 2.21**

## Contents

Contents' value formats the text component's information. Each named component has its own text format string and its own color, precision, base units, and so on (see Figure 2.22). The Text Component Editor dialog box creates and edits the format string. Access to this dialog box is through Layout's, Text section, Contents' value cell. When clicking in the cell, an ellipsis displays at the cell's right side. Clicking the ellipsis displays the Text Component Editor – Label Text dialog box, which contains the component's format string.

The right side displays the current format string. The left side settings create the right side format string. The left side shows the component's modifiers and their values. On the right, the format string must appear between the < (less than) and > (greater than) signs. A vertical bar separates each modifier. For example, Uft indicates feet, P2 indicates two places of precision, and so on. To change a modifier value, click in the modifier's value column, change the value, and select the arrow, transferring the current left side settings to the right.

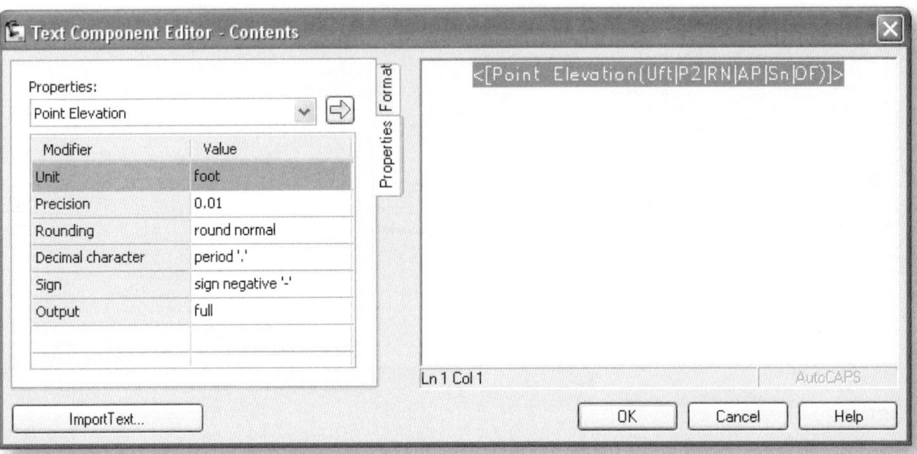


**FIGURE 2.22**

Format string values vary depending on the selected point property and its modifiers. When selecting a property from the list, the Text Component Editor provides all necessary modifiers and settings for a valid format string. An Elevation format string includes units (foot or metric), precision, rounding, the decimal character, sign, and output. Precision can vary from an integer (no decimal places) to seven decimal places. A decimal point can either be a period or a comma. For a signed value, Civil 3D uses a minus sign (−) or brackets to denote negativity, a plus sign, a drop sign (no sign), a left or right parentheses, or the option to display a value to the right of, left of, all of, or only the decimal character.

To modify an existing format string, highlight it on the right side. This action displays on the left side its modifiers and values. Next, change the left side values. Finally, at the top-center of the dialog box, click the arrow, transferring the left side values to the right side, highlighted format string. The right side format string now represents the updated values.

To add a new property to a label component, select its position on the right side, then, at the upper left from the Property drop-list, select a property. On the left appears the property's default values. Modify the values and click the right arrow, transferring them to the right side. When exiting the dialog box the format string becomes the value for the current component's text contents.

## POINT DISPLAY MANAGEMENT

The ERASE command permanently deletes a point. This creates an issue of how to prevent points from displaying on the screen while still having them in a drawing. One method is turning off point marker and label style layers. A second method is defining a point group that hides point markers and labels.

## EXERCISE 2-1

When you complete this exercise, you will:

- Become familiar with Point's Edit Feature Settings.
- Become familiar with the label style defaults.
- Be able to create new point styles.
- Become familiar with the point label style.

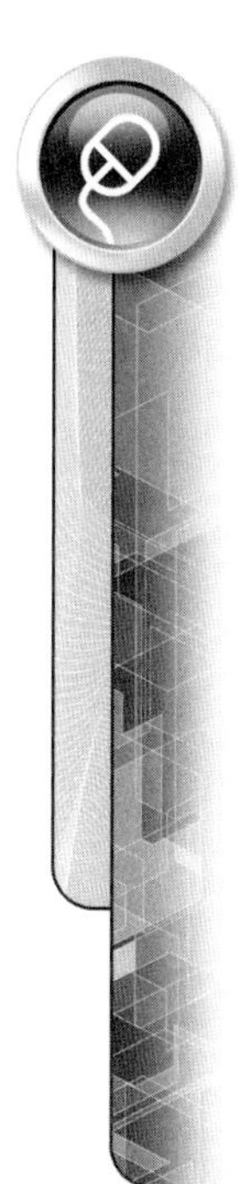

### Exercise Setup

The first task is setting up the point exercises' drawing environment. The drawings use a template file containing several point and label styles. After reviewing these settings, you will create new point and label styles.

1. If you are not in **Civil 3D**, double-click its desktop icon, starting the application.
2. At Civil 3D's top left, to the upper right of the Civil 3D icon, click the black triangle displaying the Application Menu. From the menu, select ***New***, browse to this textbook's CD Chapter 2 folder, select the file *Chapter 2 – Unit 1.dwt*, and click ***Open***.
3. At Civil 3D's top left, to the upper right of the Civil 3D icon, click the black triangle displaying the Application Menu. In the menu highlight Save as, in the flyout menu select AUTOCAD DRAWING, browse to the Civil 3D Projects folder, for the file name, enter **Points-1**, and click ***Save***.

### Review/Edit Drawing Settings

1. Click the ***Settings*** tab.
2. In Settings, at the top, select the drawing name, press the right mouse button, and, from the shortcut menu, select EDIT DRAWING SETTINGS....
3. Click the ***Units and Zone*** tab; if necessary, change the drawing scale to **1" = 40'**, set Imperial to Metric conversion to **US Survey Foot**, and set the zone to **No Datum, No Projection**.
4. Click the ***Object Layers*** tab. Locate the Point Table entry, change its modifier to **Suffix**, and set its value to **-*** (a dash followed by an asterisk).
5. Click the ***Ambient Settings*** tab. Expand the Direction section, click in Format's value cell, and if necessary change the format to **DD.MMSSSS (decimal dms)**.
6. Click ***OK*** to exit the dialog box.

### Edit Feature Settings

Point's Edit Feature Settings dialog box values affect all points. The Default Styles section names point and label styles and the Default Name Format section defines the point groups and point naming format.

1. In Settings, select the **Point** heading, press the right mouse button, and, from the shortcut menu, select EDIT FEATURE SETTINGS....
2. Expand the Default Styles section; if necessary, change the point style to **Basic** and the label style to ***Point#-Elevation-Description***.
3. Expand the Default Name Format section containing the name and sequential counters for both Point Groups and Point Name.
4. Expand the Update Points section and make sure it is set to false.

The Update Points section controls the modification of checked-in points. If this is set to true, you will be able to edit points that should not be changed.

5. Click ***OK*** to exit the dialog box.

## Edit Label Style Defaults

This dialog box's values set the point label styles' general behavior. The Point's Edit Label Style Defaults dialog box can override the drawing's settings. The changes made here only affect the Point branch's labels.

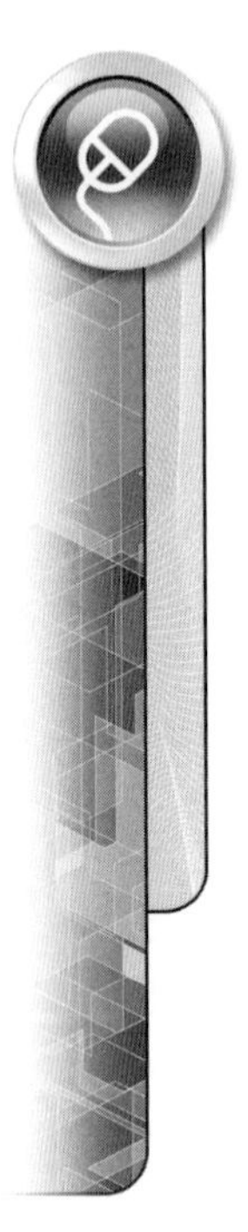

1. In Settings, select the **Point** heading, press the right mouse button, and, from the shortcut menu, select EDIT LABEL STYLE DEFAULTS....
2. Expand each section and review its values.
3. Click ***OK*** to exit the dialog box.

## Point (Marker) Styles

Civil 3D identifies a point's coordinates with a marker. A marker can be a simple AutoCAD node, a custom Civil 3D object, or a drawing block (symbol) (see Figure 2.23).

1. In Settings, go to the Point branch and expand it until you view the Point Styles list.
2. From the Point Styles list, select **Basic**, press the right mouse button, and, from the shortcut menu, select EDIT....
3. If necessary, select the dialog box's ***Information*** tab.

The Information tab contains the name and other data about the style.

4. Select the ***Marker*** tab.

Currently, the marker is a custom object, an X, whose size is a product of the drawing scale and the value 0.1, and its orientation is relative to the WCS (see Figure 2.23).

5. Change the marker to a plus sign (+) and toggle on the *circle*.

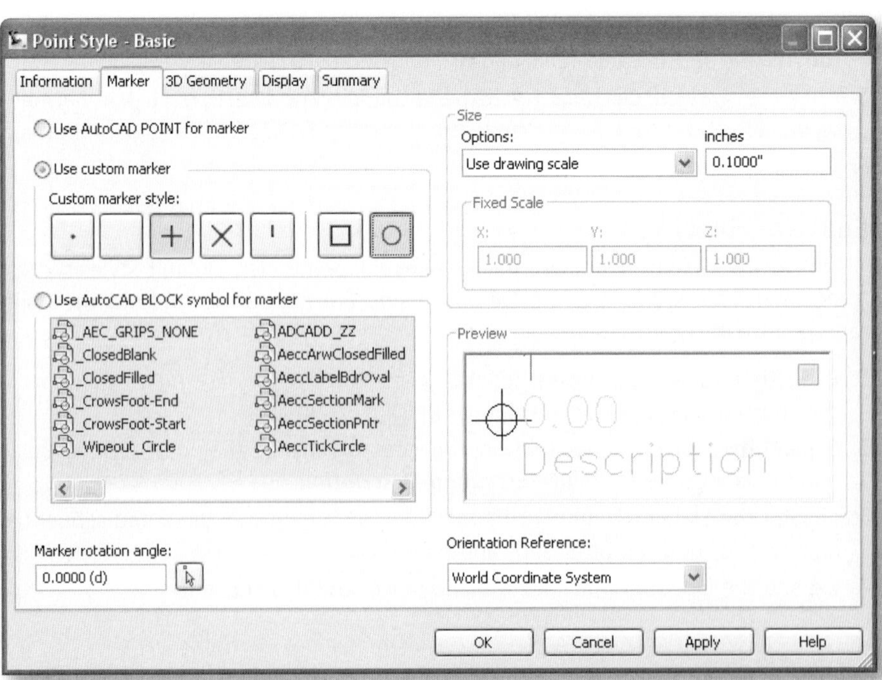


**FIGURE 2.23**

6. Select the 3D Geometry tab; if necessary, change the Point Display Mode to **Flatten Points to Elevation** and the Point Elevation value to **0** (zero).

These settings produce zero elevation objects when selecting a point with an AutoCAD object snap. When selecting a point with a Civil 3D transparent command (point number or point object), the resulting object will have the point's elevation.

7. Select the ***Display*** tab.

The Display tab indicates that the marker and label are on the default point layer, V-NODE and V-NODE-LBL.

8. Click the ***Summary*** tab, expand each section reviewing its contents.

This tab summarizes all of the style's settings and previews the resulting node.

9. Click ***OK*** to exit the dialog box.

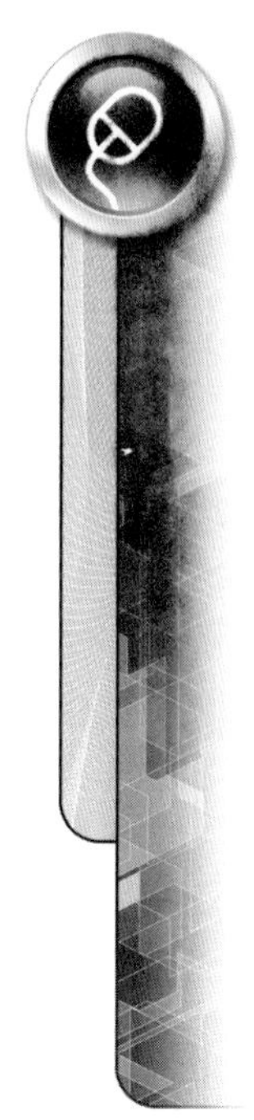

## Create Point Styles

The first new point style is Maple (tree). This style's marker is a maple tree symbol. Use Figure 2.24 and the information in Table 2.1 while defining the point style.

1. In Settings, from the point styles list, select **Basic**, press the right mouse button, and, from the shortcut menu, select COPY....
2. Select the ***Information*** tab; for the name, enter **Maple**, enter a short description, and then select the ***Marker*** tab.
3. In the Marker panel, set the marker to **Use AutoCAD BLOCK symbol for marker**, and from the block list, scroll to and select the **Maple** marker.
4. In the Size area, change the inches value to **0.1**, if necessary.

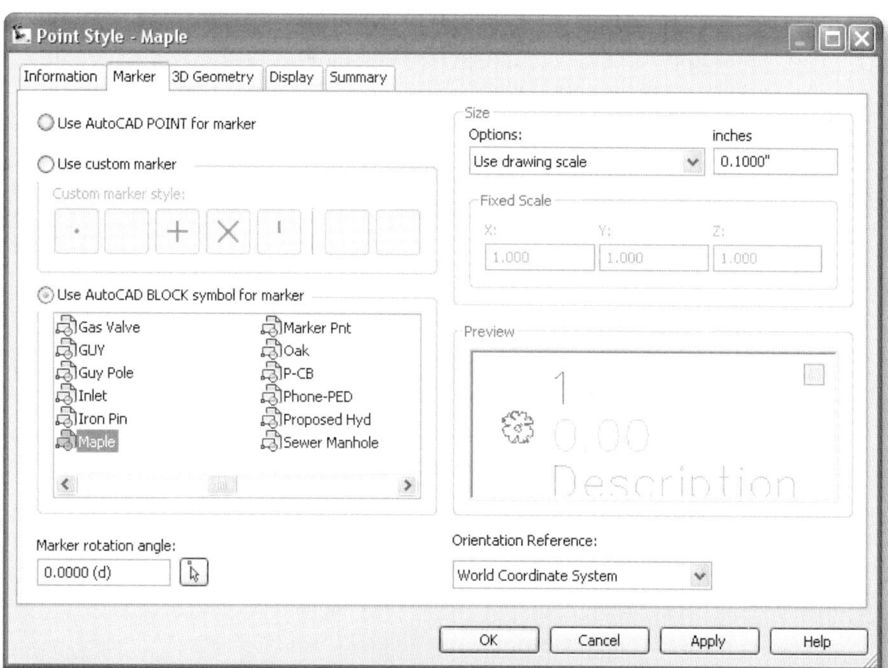


**FIGURE 2.24**

5. Select the ***3D Geometry*** tab; if necessary, change Point Display Mode to **Flatten Points to Elevation** and set the Elevation to **0.0**.

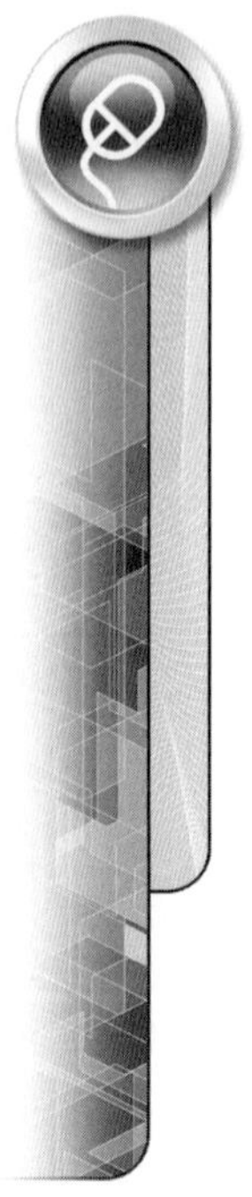

6. Select the ***Display*** tab; click in the cell containing the marker layer name (V-NODE). The Layer Selection dialog box displays.

You select a layer from the current drawing's layer list, from another open drawing (using the Layer Source drop-list at the top-left of the dialog box), or create a new layer (using the New... button at the top right of the dialog box).

7. At the top right of the Layer Selection dialog box, click ***New...*** . This displays the Create Layer dialog box.
8. In the Create Layer dialog box, click in the Layer name value cell and enter **V-NODE-VEG**.
9. Click in the Color value cell, then click the ellipsis (right side of cell). This displays the Select Color dialog box. In the Select Color dialog box, assign a color, and click ***OK*** to return to the Create Layer dialog box.
10. Click ***OK*** to exit the Create Layer dialog box.
11. In the Layer Selection dialog box, if not already selected, locate and, from the layer list, select the layer **V-NODE-VEG**, and click ***OK*** to assign the selected layer to the marker.
12. Repeat Steps 6 through 11, replacing V-NODE-TEXT layer with **V-NODE-VEG-LBL**.
13. Click ***OK***, creating the Maple point style and exiting the dialog box.
14. Repeat Steps 1 through 13 and use the information in Table 2.1 to create an Oak point style. You do not need to create the **V-NODE-VEG** and ***V-NODE-VEG-LBL*** layers; just select them from the layer list. This style's Point Display Mode is also **Flatten Points to Elevation**, and the Elevation is **0.0**.

**TABLE 2.1**

| Style Name | Marker Block | Display Marker | Display Label |
|---|---|---|---|
| Maple | Maple | V-NODE-VEG | V-NODE-VEG-LBL |
| Oak | Oak | V-NODE-VEG | V-NODE-VEG-LBL |

## Review the Point#-Elevation-Description Point Label Style

All points for this exercise use the Point#-Elevation-Description Point Label style.

1. In Settings, from the Point branch, expand Label Styles until you view its styles list.
2. From the list of styles, select **Point#-Elevation-Description**, press the right mouse button, and, from the shortcut menu, select `EDIT...`.

This displays the Label Style Composer dialog box.

3. Select the ***Information*** tab, reviewing its values.

This panel sets the style's name and names the person responsible for creating or modifying it.

4. Select the ***General*** tab, reviewing its settings.

The General panel sets the label's text style, layer, visibility, orientation reference, and plan readability parameters.

5. Select the ***Layout*** tab, reviewing its text component settings.

At the top, the Component name drop-list includes all of the label style's components. The components for this label style are point number, elevation, and description.

6. Click the Component name's drop-list arrow and, from the list, select **Point Elev**.

The Point Elev component's top-left-justified text (Text section's attachment) anchors to the Feature (General section).

7. Click the Component name's drop-list arrow and, from the list, select **Point Number**.

The Point Number component is bottom-left-justified (Text section's attachment value) and is anchored to the top left of the Point Elevation label (General section).

8. Click the Component name's drop-list arrow and, from the list, select **Point Description**.

The Point Description component is top-left-justified (Text section's Attachment value) and is anchored to the Point Elevation's (General section) bottom-left.

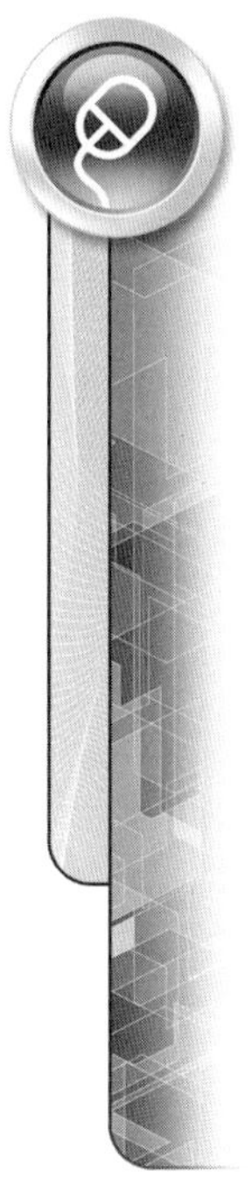

## Formatting Label Text

Point Number, Point Elev, and Point Description all have their own text format string. The string's values format the label's information (see Figure 2.22).

1. With Point Description as the current component, in the Text section, click in the Contents' value cell.
2. At the Value cell's right, click the ellipsis, displaying the Text Component Editor.
3. On the dialog box's left side, click in the Capitalization's value cell and view the capitalization options list.
4. Click CANCEL to exit the Text Component Editor.
5. In Layout, click Component name's drop-list and, from the list, select **Point Elev**.
6. With Point Elev as the current component, in the Text section, click in the Contents' value cell.
7. At the Value cell's right, click the ellipsis, displaying the Text Component Editor.
8. Click the different value cells, viewing their possible formats and options for this label type.
9. Click CANCEL until exiting the Text Component Editor and the Label Style Composer dialog boxes.

## Table Style

Civil 3D has table styles that list point data in table form.

1. In Settings, expand the Point branch until you can view the Table Styles list.
2. From the table styles list, select **Latitude and Longitude**, press the right mouse button, and, from the shortcut menu, select EDIT....
3. Select the ***Data Properties*** tab and review its values.
4. Select the ***Display*** tab and review its values.
5. Click ***OK***, exiting the dialog box.
6. At Civil 3D's top left, Quick Access Toolbar, click the ***Save*** icon to save the drawing.

Civil 3D has several settings that control visibility and how it works with point data. These values should be set in advance; however, Civil 3D is flexible enough to accommodate most unanticipated situations.

## SUMMARY

- Point styles assign both marker and label layers.
- Point styles define a point's marker.
- Point label styles define how and what information displays next to a marker.
- Point label styles have one component feature anchor point.
- The remaining label components anchor relative to the initial component.

## UNIT 2: CREATE POINTS

The Create Points toolbar and the Ribbon's Points menu contain point creation routines. A user displays this toolbar from the Ribbon's Points menu or from Prospector's Points heading shortcut menu. Some point settings are only available in the Create Points toolbar.

The point creation routines create points from printed lists, external ASCII files, or calculated coordinates. This unit does not review all of the point commands. Later chapters will review additional routines in a more appropriate context.

### CREATE POINTS SETTINGS

When using the Create Points toolbar or the Points menu's routines, users need to review and set values. These settings appear in the Create Points toolbar data area or in the Edit Command Settings for the Create Points command. The three settings' groups are as follows: Default Layer, Points Creation, and Point Identity. See the Create Points section of Unit 1.

### RIBBON'S POINTS MENU

The Ribbon's Points menu command Create Point Group displays the Create Point Group dialog box. This is the same dialog box that is displayed when selecting `NEW...` from the Point Groups heading shortcut menu. At the menu's bottom are utilities that convert points from other applications.

#### Convert Land Desktop Points

Convert Autodesk Land Desktop Points creates Civil 3D COGO points from Land Desktop point objects (see Figure 2.25). When running the routine, the Convert AutoCAD Land Desktop Points dialog box displays, containing the current conversion settings. After reviewing the settings, clicking OK starts the conversion process.

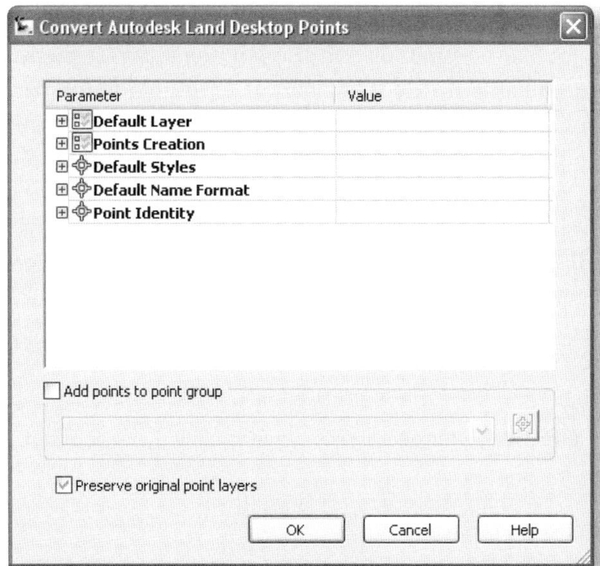


FIGURE 2.25

## Create Blocks from COGO Points

Ribbon's Modify, Points panel, has on its right side in the COGO Point Tools extended ribbon, the command Create Blocks from COGO Points. When you click this command, it displays the Create Blocks from COGO Points dialog box (see Figure 2.26). The dialog box's top selects the points by area (display or window) or by point group. The lower portion identifies the block point block and its layer. After setting the values, clicking OK creates the point blocks.

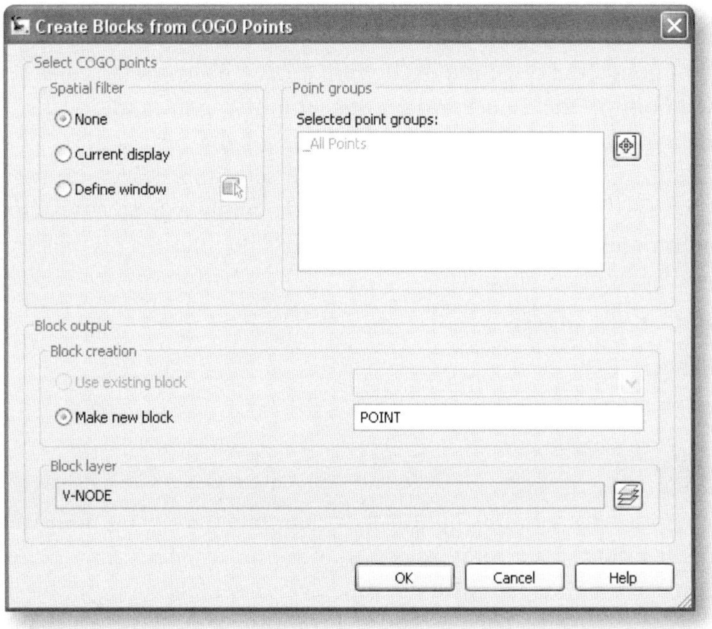

Create Blocks from COGO Points
Select COGO points
Spatial filter
None
Current display
Define window
Point groups
Selected point groups:
_All Points
Block output
Block creation
Use existing block
Make new block
POINT
Block layer
V-NODE
OK
Cancel
Help

FIGURE 2.26

## EXTENDING COMMANDS

In addition to the Create Points toolbar commands and the Ribbon's Points menu, other point creation methods are available in the Transparent Commands toolbar. The Transparent Commands toolbar's overrides change a routine's prompting and allows the routine to use other point placement methods. For example, using the Manual command and entering "'NE" at the command line changes the routine's prompting to northing and easting. Changing Manual to use a direction and distance, enter 'AZ (azimuths) or 'BD (bearing). To mimic field collection techniques, use 'AD (turned angle) and 'SS (side shot). Other transparent commands reference specific Civil 3D objects (points, alignments, and profile views) and AutoCAD objects (lines, arcs, polyline, etc.).

**NOTE**

**To use the transparent commands effectively, Autodesk recommends that dynamic input mode be off when using them.**

### Coordinates

When invoking a transparent command, the routine changes its coordinate prompting. The Grid and Latitude transparent overrides require an assigned coordinate zone. If there is no assigned coordinate system, Civil 3D issues a warning dialog stating that there must be an assigned coordinate system. Set a zone in the Edit Drawing Settings dialog box. The following is a list of Coordinate transparent commands:

- Northing/easting: 'NE
- Grid northing/easting: 'GN
- Latitude/longitude: 'LL

### Angles and Distances

The 'AD, 'DD, and 'SS transparent commands simulate the station (pivot point), backsight, and foresight surveying process. When using this override, users must establish a station and a backsight by selecting a line or two points. If selecting a line, the endpoint nearest to the selection point is the station (pivot point) and the farthest endpoint is the backsight point (the direction of the zero angle). When establishing a station and backsight in point mode, select two points: The first is the station (pivot point) and the second is the backsight point (the direction of the zero angle). The Angle Distance and Deflection/Distance routines use the setup once and the user must define the next station/backsight location. The Side Shot override uses the same pivot and backsight points until the setup is changed. The following is a list of Angle and Distances transparent commands:

- Turned angle and distance: 'AD
- Deflection/distance: 'DD
- Side shot from point: 'SS

### Point Object Filters

Most transparent commands have a Points option. For example, the By Turned Angle transparent command locates a setup by selecting a line or by selecting a point pair. Any command that has a points (coordinate) mode can reference a point object or a northing/easting coordinate set. When a routine prompts for a point, the point can be a command-line entered value or coordinates from an AutoCAD selection (usually with an object snap). If entering .n at the prompt, the prompting changes to northing/easting for the point's coordinates. If entering .p, the prompting changes to point number to determine the point's coordinates. If entering .g, the prompting switches to select a point object to determine the coordinates.

The following are code snippets showing how point object filters affect the prompting for coordinate values when using the Angle Distance override of the transparent commands.

Selecting an AutoCAD line entity for coordinate values:

```
>>Select line or [Points]: (select a line entity)
Selecting AutoCAD coordinates for coordinate values
>>Select line or [Points]: P (toggles to Point mode)
>>Specify starting point or [.P/.N/.G]: (make an AutoCAD
selection)
>>Specify ending point or [.P/.N/.G]: (make an AutoCAD
selection)
```

Specifying a Civil 3D point number for coordinate values:

```
>>Select line or [Points]: P (toggles to Point mode)
>>Specify starting point or [.P/.N/.G]: .P (toggles to Point
number mode)
>>Enter point number: 240
>>Specify ending point or [.P/.N/.G]: .P (toggles to Point
number mode)
>>Enter point number: 250
```

Entering Northings/Eastings for coordinate values:

```
>>Select line or [Points]: P (toggles to Point mode)
>>Specify starting point or [.P/.N/.G]: .N (toggles to NE
coordinates)
>>>>Enter northing <0.0000>: 5500.6395
>>>>Enter easting <0.0000>: 5892.4533
>>Specify ending point or [.P/.N/.G]: .N (toggles to NE
coordinates)
>>>>Enter northing <0.0000>: 5830.3095
>>>>Enter easting <0.0000>: 5942.0523
```

Selecting Civil 3D point object from the display for coordinate values:

```
>>Select line or [Points]: P (toggles to Points mode)
>>Specify starting point or [.P/.N/.G]: .G (toggles to select
a point object mode)
>>
Select point object: (select a point object from the display)
>>Specify ending point or [.P/.N/.G]: .G (toggles to select a
point object mode)
>>
Select point object: (select a point object from the display)
```

## Angle Formats

Civil 3D uses four angle types: decimal degrees and three variations of degrees, minutes, and seconds. A routine prompting for an angle displays the current angle format. Users set the angle format in the Ambient Settings panel of Edit Drawing Settings (see Figure 2.27). The traditional method for surveyors is a decimal format for degrees, minutes, and seconds. The surveyor's convention enters the angle of 34 degrees, 52 minutes, 18 seconds as 34.5218 (dd.mmss). This convention does not allow minute and second values greater than 59.

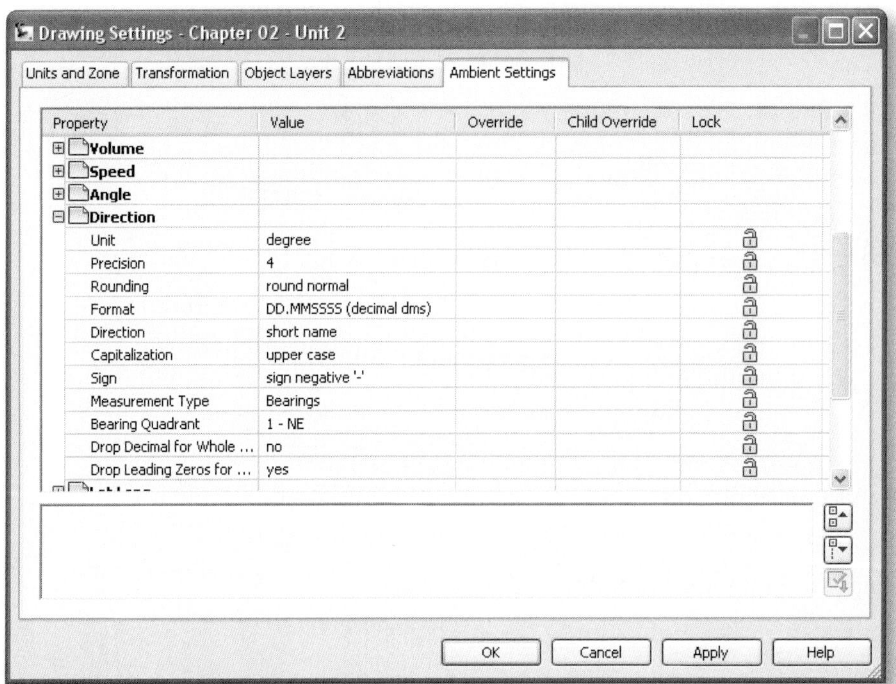


**FIGURE 2.27**

## Directions

When entering directions, Civil 3D supports both the azimuth and bearing systems. The azimuth system assumes north is 0 degrees and measures angles clockwise from 0 to 359.5959 (dd.mmss) (see Figure 2.28). The bearing method divides a circle into four 90-degree quadrants. The top two quadrants, the northeast (quadrant 1) and northwest (quadrant 4), assume that a vector deflects from north to the east or west. So the angle of a vector varies from 0 degrees, a line traveling due north, to 90 degrees, a line traveling due east or west. The bottom two quadrants, the southeast (quadrant 2), and southwest (quadrant 3) assume that a vector deflects from south to east or west. So the angle of a vector varies from 0 degrees, a line traveling due south, to 90 degrees, a line traveling due east or west.

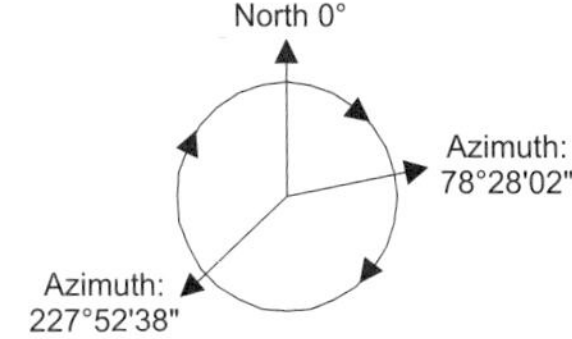


Azimuths:
Assumes North is 0°
Measures Angles Clockwise
Angles are between 0° and 360°

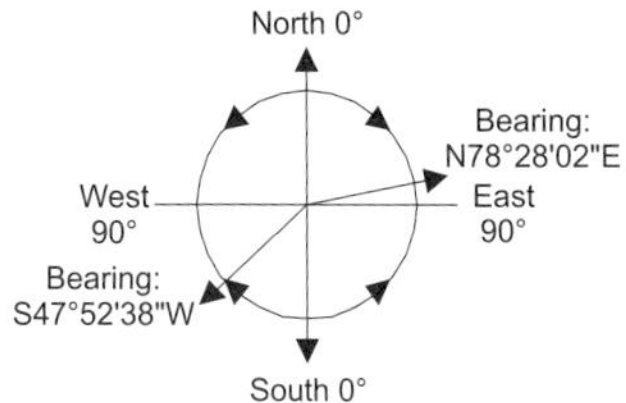


Bearing:
Assumes North or South is 0°
Angles are between 0° and 90°
East or West

**FIGURE 2.28**

## Direction Transparent Commands

After establishing a new point's location or starting from a known point, the two transparent commands ('BD and 'ZD) define a direction to the next point's location. Define the direction by either a bearing ('BD) or an azimuth ('ZD).

- Bearing/distance – 'BD
- Azimuth/distance – 'ZD

The following are code snippets showing how the direction transparent commands affect direction prompting and how a routine indicates the current angle or direction format set in the Edit Drawing Settings dialog box's Ambient Settings panel.

Setting a point by bearing

```
Command:
Please specify a location for the new point: '_BD
>>Select starting point or [.P/.N/.G]: .P
>>Enter point number: 244
Quadrants - NE = 1, SE = 2, SW = 3, NW = 4
>>Specify quadrant (1-4): 1
Current direction unit: degree, Input: DD.MMSSSS (decimal dms)
>>Specify bearing: 57.5652
>>
>>Specify distance: 45.5234
Resuming MODELESSDISPATCH command.
Please specify a location for the new point: (6098.79 5453.6 0.0)
Enter a point description <.>: IPF
Specify a point elevation <.>: 723.44
```

```
N: 5453.5982' E: 6098.7912'
Please specify a location for the new point:
Quadrants - NE = 1, SE = 2, SW = 3, NW = 4
>>Specify quadrant (1-4):
```

Setting a point by azimuth

```
Command:
Please specify a location for the new point: '_ZD
>>Select starting point or [.P/.N/.G]: .P
>>Enter point number: 244
Current direction unit: degree, Input: DD.MMSSSS (decimal dms)
>>Specify azimuth: 73.2644
>>Specify distance: 62.7495
Resuming MODELESSDISPATCH command.
Please specify a location for the new point: (6126.26 5439.78
0.0)
Enter a point description <.>: IPF
Specify a point elevation <.>: 730.76
N: 5439.7831' E: 6126.2615'
Please specify a location for the new point:
Current direction unit: degree, Input: DD.MMSSSS (decimal dms)
>>Specify azimuth:
Resuming MODELESSDISPATCH command.
```

All angle/distance transparent commands can be used with many AutoCAD commands (Line, Polyline, Move, etc.). When using an AutoCAD command with a transparent Civil 3D command, simply click the appropriate transparent command icon after starting the AutoCAD command and follow the prompting until the AutoCAD command is complete.

### Alignments

The Alignment transparent command requires a defined alignment.

- Station/offset: 'SO

### Northing/Easting, Grid, and Longitude/Latitude

When working with local coordinates, users can toggle a command into northing/easting mode by selecting the Northing Easting icon or entering 'NE at the command line. When a drawing has an assigned state plane or Universal Transverse Mercator (UTM) zone, coordinates can be entered as Grid ('GN) or a Latitude/Longitude ('LL). The northing/easting and grid northing/grid easting overrides change the prompting to northing and then easting. Civil 3D never allows these values as a comma-delimited entry. When toggling on Latitude/Longitude, the command first prompts for latitude and then longitude.

### Points

The Point transparent commands are useful when there is a need to reference existing Civil 3D point objects. For example, using the line command 'PN allows users to draw the line by point numbers, and 'PO allows users to select point objects from the display to incorporate their coordinates into the line. 'PA selects points by their names. These commands will use the points' elevation as the elevations at each vertex of the line.

These transparent commands behave exactly like the .p, .n, and .g point filters. However, .p, .n, and .g can be used only when the command line lists them as options. If the dot point filters are not listed as an option, the transparent commands 'PN, 'NE, and 'PO should be used.

- Point number: 'PN
- Point name: 'PA
- Point object: 'PO
- Northing/easting: 'NE

## CREATE POINTS TOOLBAR

The Create Points toolbar's point creation routines divide into six icon stacks (see Figure 2.29). The first six stacks are Miscellaneous, Intersection, Alignment, Surface, Interpolation, and Slope (toolbar stacks from left to right). Each icon stack represents a collection of individual commands. Routines in each stack convert written data, mimic surveying techniques, interpolate between points, and create points from alignment or surface objects. The last icon imports points. This routine creates points while reading ASCII text files.

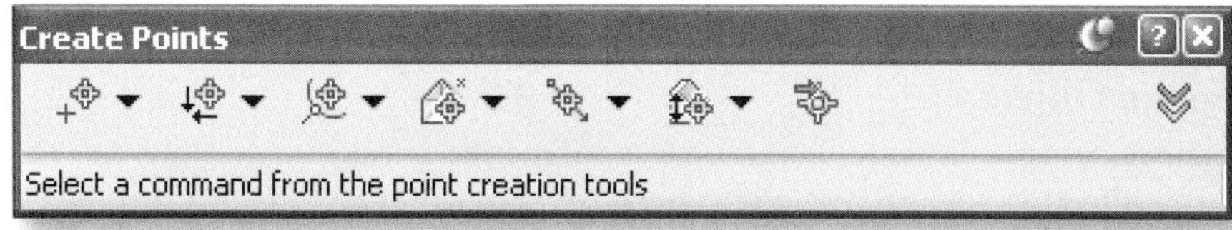


**FIGURE 2.29**

### Miscellaneous

The Miscellaneous icon stack includes creating points manually by referencing existing points (Geodetic Direction and Distance and Resection), by referencing existing AutoCAD objects, and by Converting Softdesk point blocks into Civil 3D point objects (see Figure 2.30).

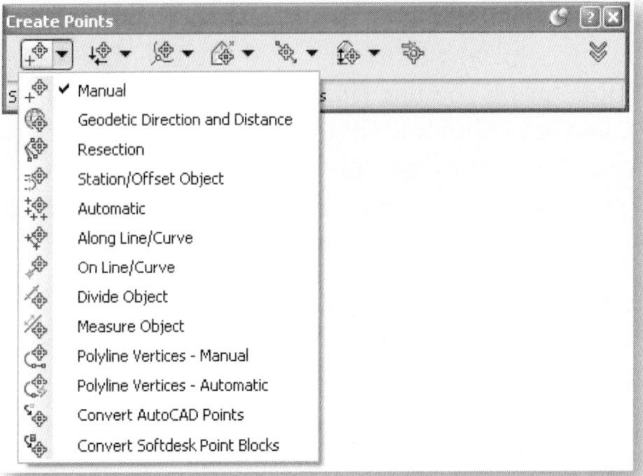


**FIGURE 2.30**

## Manual

Manual places points at command-line entered coordinates (X,Y) or from object snap selection coordinates.

## Geodetic Direction and Distance

This command requires an assigned coordinate system and prompts for grid northings/eastings.

## Resection

Resection creates points whose location is derived from three known points with measured angles between them.

## Station/Offset Object, On Line/Curve, Divide Object, and Measure Object

These routines require existing line and curve segments. If selecting a polyline, the active routine filters it out and asks the user to select another object. These routines measure from the selected object's starting point. The starting point is the endpoint nearest to its selection point. For example, if selecting at a line's southern end, that end is the lowest station or starting measurement point.

## Automatic

Automatic places points at the same locations as the On Line/Curve routine. The difference between the routines is that Automatic works with a selection set of lines and curves rather than individual line and curve segments. Before Automatic places points, it scans the selection set for duplicate (endpoint) coordinates and, if it finds any, places one point at the coordinates. If using On Line/Curve, it places two points at the end coordinates (one point for each selected object).

## Polyline Vertices – Manual

This routine places a point at each vertex with a user-specified elevation.

## Polyline Vertices – Automatic

This routine places a point at each vertex whose elevation is the vertex's elevation.

## Convert AutoCAD Points

This routine converts nodes into Civil 3D point objects. The routine prompts for each point's description.

## Convert Softdesk Point Blocks

This routine converts Softdesk point blocks into Civil 3D point objects.

### Create Points: Intersection

There are three types of intersection routines (see Figure 2.31). The first uses a theoretical intersection, which includes intersections between directions and distances from known points. These are the list's first five commands. The second intersection assumes the existence of line and/or arc entities (Object/Object and Perpendicular). The lines and arcs define directions and/or distances for the computations. The last type works only with Civil 3D alignment objects and comprises the list's last four routines.

A direction can be a bearing, an azimuth, an existing line's direction, a direction between two AutoCAD selections (including object snaps), or a direction between two point objects.

Most intersection routines calculate a single solution. There are, however, some routines that calculate two possible solutions. For example, the Distance/Distance routine calculates two possible solutions. The routine identifies each solution in the drawing with an X. Users have the option of selecting one or both solutions. Select an individual solution by selecting near one X with the cursor, or select All solutions, placing a point at each solution.

Many intersection routines have an offset to the defined direction. The convention is that when at the starting point facing toward the direction, a negative value is to left and a positive value is to the right.

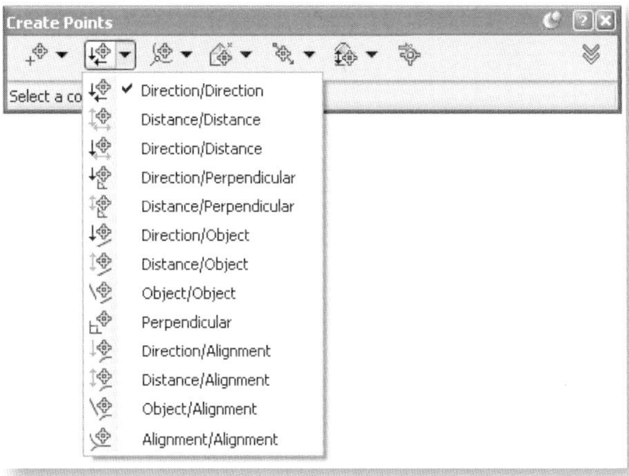


**FIGURE 2.31**

## Alignment/Surface/Interpolate/Slope

This group's routines will be discussed in appropriate chapters later in this textbook.

## Import/Export

The Import Points command is the Create Points toolbar's rightmost icon. If receiving an ASCII file containing coordinate data, the routine reads the file's coordinate records and creates points. The Import Points routine has several predefined formats for reading coordinate file records.

The Import Points dialog box sets the file format and file name, can assign imported points to a point group, and has advanced options for using imported data as modifiers of existing data (see Figure 2.32).

The Import Point dialog box supports reading multiple files in one session.

The Advanced options' first toggle modifies point elevations' Do elevation adjustment if possible. This option is similar to a point's elevation datum adjustment. When importing a point file, a column in the file format named +Z or −Z contains values that adjust the points' elevations as they are read.

The second option transforms a point's file coordinates to new drawing coordinates based on the currently assigned drawing coordinate system. An example of this is importing points having NAD27 US Foot coordinates into a drawing using the NAD83 US Foot system.

The last option, Do coordinate data expansion if possible, forces Civil 3D to calculate longitude and latitude values for the incoming points.

Civil 3D uses the same formats and options when exporting points to ASCII files.

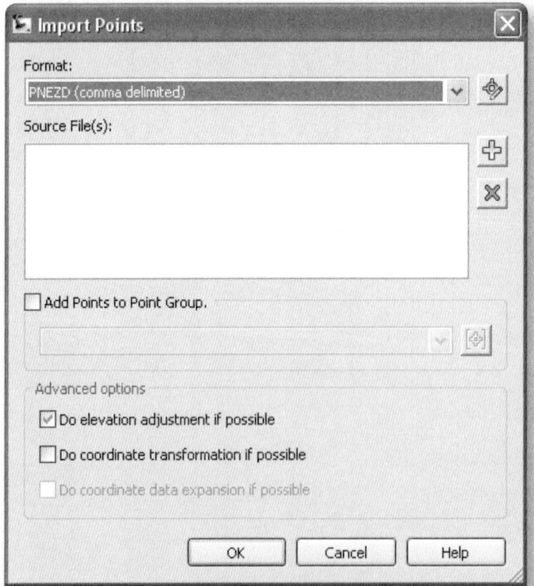


**FIGURE 2.32**

Before importing a file, users must know each line's structure (or a record). Viewing the ASCII file in Notepad helps determine the file's structure. However, it does not reveal one very important piece of information. That piece of information is which fields in a record represent the northing (Y) and easting (X) coordinate values. If using the wrong format, points appear incorrectly in the drawing. The only way to know the file's structure is having notes about the file or talking to persons responsible for the file's creation.

The following is an excerpt from a typical coordinate file:

```
1,5530.83673423,4967.36847364,723.74,PP 60
2,5635.36849278,4952.84689347,722.84,IPF
```

This file is comma-delimited and contains the following fields: point number, northing, easting, elevation, and description. Again, the only way to determine the data order is to have some documentation identifying the fields or a conversation with the person creating the file. Civil 3D contains a format, PNEZD (comma-delimited), which matches this file's structure.

The Import Options section of the previous unit discusses how Civil 3D handles point numbers and how it assigns new point numbers to duplicate points during an import session.

### Export Points

Export Points creates an ASCII file or Access database from selected points or point groups. These routines use the same file formats as Import Points. Ribbon's Points menu, and a point group's shortcut menu contains Export Points.... Alternative locations are Modify tab, on the panel's left click points, Export points or the Output tab's Export section.

### Point File Format

If a file has a format not on the format list or you want to specify a coordinate system with a file format, you can create custom formats. Point File Format defines file formats, accommodating almost any file format and coordinate system. Clicking the icon to the current format's right displays Point File Format (see Figure 2.33).

The type and format of a file's records determines the structure of a format. If setting a coordinate system, the Import routine imports longitude and latitude or state plane coordinates and transforms them to the current drawing's assigned grid system.

Import supports files that are comma-, space-, or column-delimited. The file's structure dictates which format Import Points uses. Users must exercise care when creating a column-delimited format because, in a column-delimited file, a decimal point is a part of the overall field width.

A coordinate file consists of columns and rows. A field is a column of specific information (such as northing, easting, elevation, etc.). A row is known as a record, or a unique occurrence of fields. Field types include point number, northing, easting, elevation, description, latitude, longitude, convergence, and scale factor.

The Point File Format dialog box defines new point file formats or edits existing point file formats. The dialog box's top portion sets the name, extension, delimiter (if applicable), and coordinate zone (if applicable). The dialog box's bottom defines the file record fields (columns).

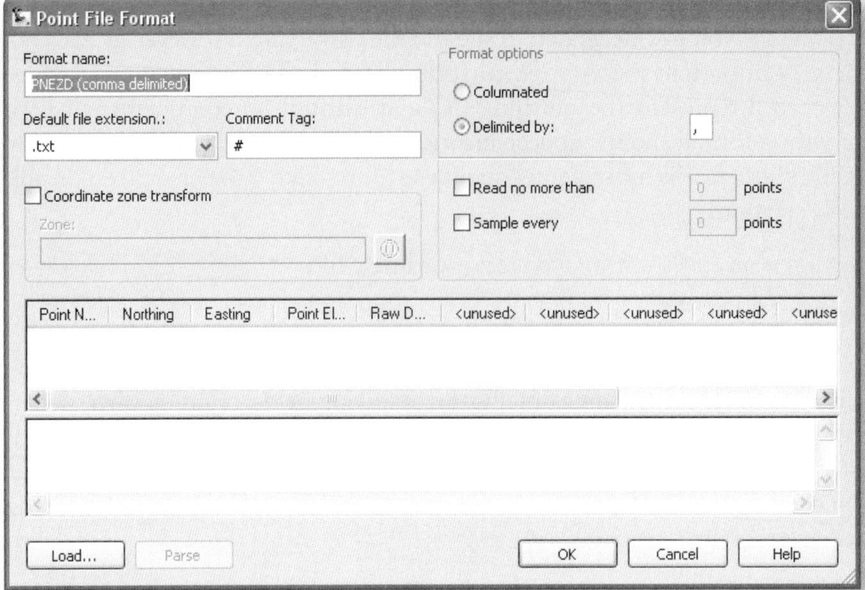


**FIGURE 2.33**

## LANDXML

Another method of creating points is importing a LandXML file. LandXML has a very specific point information schema. The Ribbon's Insert tab's Import panel LandXML routine reads a file and creates points from its data (see Figure 2.1).

The Point Import Settings section has two values that define a point's description: code and desc. A code is a raw description and a desc is a full description. This section has four possible settings: use the code, use the desc, use code first and if no value use the desc, and use desc first and if no value use code. The file snippet below contains both code and desc values. For example, for point 9081, the code is UP and the desc is UTILITY POLE. If using a Description Key Set for this setting, select Code.

```
<CgPoint name=“9080” code=“GR” desc=“GR”>6029.75000000
  4083.87100000 912.132700</CgPoint>
<CgPoint name=“9081” code=“UP” desc=“UTILITY
  POLE”>6026.69100000 4118.11920000 915.166000</CgPoint>
<CgPoint name=“9082” code=“GR” desc=“GR”>5939.18020000
  4113.40820000 911.781700</CgPoint>
```

When exporting points to a LandXML file, the point numbers become a point name (CgPoint name). When importing a LandXML file, Civil 3D assigns a point number that matches the point name as long as it is not a duplicate number (see Figure 2.34). If the number is a duplicate, the settings in Point Identity determine the new point's number.

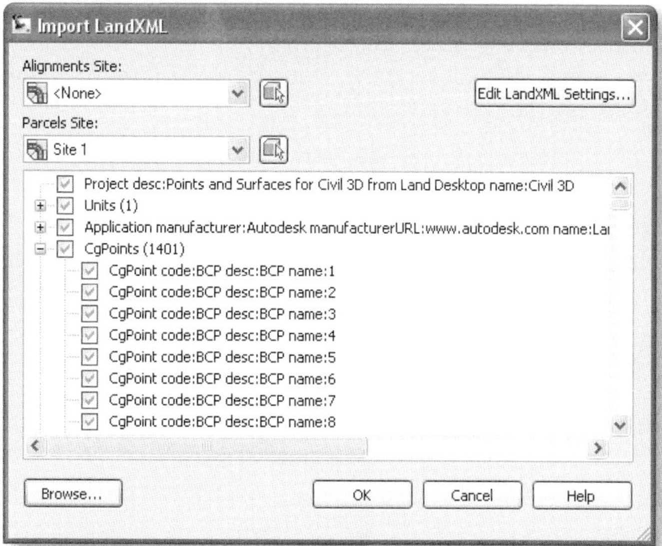


**FIGURE 2.34**

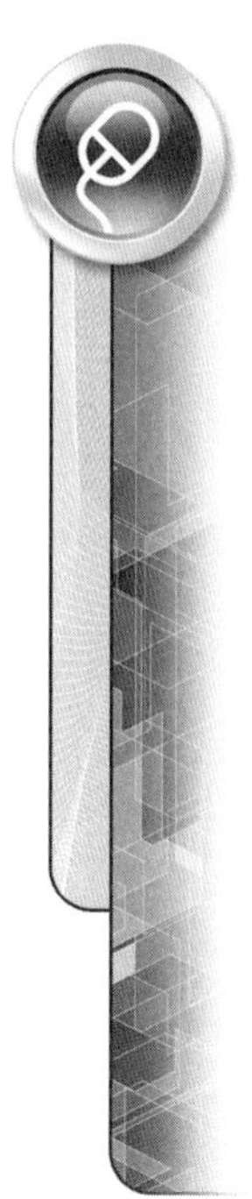

## EXERCISE 2-2

When you complete this exercise, you will:

- Create an Import file format.
- Import an ASCII file.
- Import points using LandXML.
- Be familiar with Create Points toolbar's Miscellaneous icon stack commands.
- Be familiar with Create Points toolbar's Intersection icon stack commands.
- Be able to use transparent commands while creating points.

## Exercise Setup

1. If you are not in the Points-1 drawing file from the previous unit, open it now. If you are starting with this exercise, from Civil 3D's top left, Quick Access Toolbar, select the ***Open*** icon, browse to this textbook's CD, Chapter 2 folder, select the drawing *Chapter 02 – Unit 2*, and click ***Open***.

## Review Point Identity Settings

Before importing points, you need to review the Create Points toolbar's data panel's Point Identity section. This section determines Import points basic values and rules. If the drawing already has points, these settings are critical to not destroy or modify existing point data.

1. Click the ***Prospector*** tab.
2. In Prospector, select the **Points** heading, press the right mouse button, and, from the shortcut menu, select CREATE....
3. Click the chevron icon (Expand the Create Points dialog) at the toolbar's right to display the point settings data panel.

In an empty drawing the current settings are fine. However, if a drawing contains points, the next point number would be something other than 1. When importing points into a drawing containing points, you may want to change the Sequence Point Numbers From value. Make sure your dialog box matches the values in Figure 2.35.

4. Expand the Point Identity section to change Sequence Point Numbers From value to **10000**.
5. Click the chevron icon (Collapse the Create Points dialog) to close the point settings data panel.

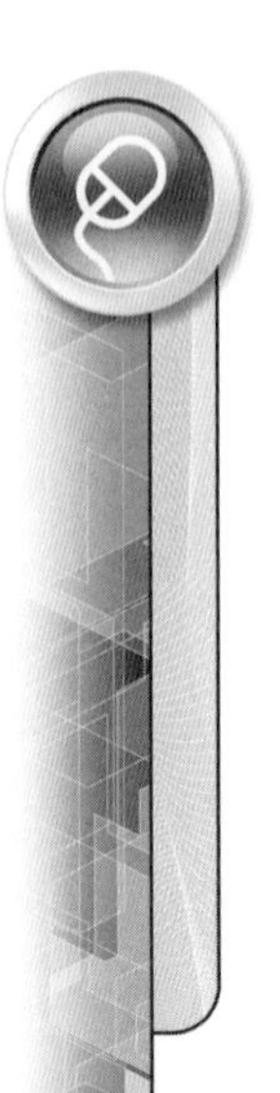

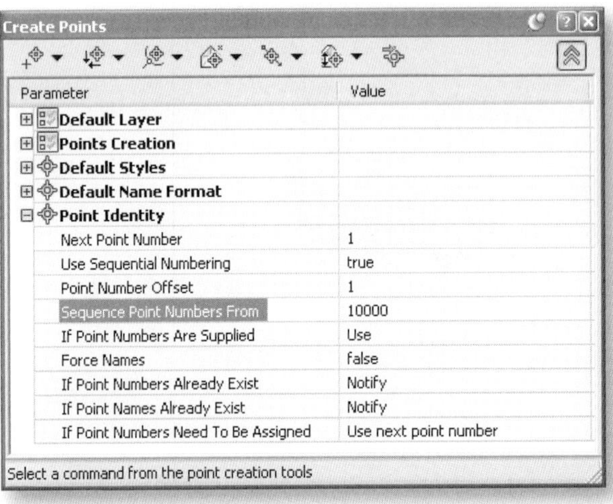


**FIGURE 2.35**

## Defining a File Format

1. Click the various drop-list arrows to review point command icon command stacks.

The Create Points toolbar's Import Points is the rightmost icon. Reading the *Base-Points.nez* file requires a new file format. The file format is point number, northing, easting, elevation, and raw description with a comma delimiter.

2. At the right side of the Create Points toolbar, select the ***Import Points*** icon.
3. In the Import Points dialog box, at the dialog box's top-right, click the ***Point Format*** icon.

This displays the Point File Formats dialog box, listing the current file formats.

4. In the Point File Formats dialog box, click ***New...***, displaying the Point File Formats – Select Format Type dialog box.
5. In the Point File Format – Select Format Type dialog box, select **User Point File**, and click ***OK*** to display the Point File Format dialog box.

To create a new file format, enter a format name, set its extension, set its delimiter, and define a record's data types and order. When you are finished setting the values, your dialog box should look like Figure 2.36.

6. At the top left of the Point File Format dialog box, for the name, enter **HC3D - (Comma Delimited)**, and for the extension, click the drop-list arrow, displaying the extension list and, from the list, select **.nez**.
7. In the Point File Format dialog box, at the top-right in Format Options, select **Delimited by** and, in the box to the right of the toggle, type a comma (**,**).

8. In the Point File Format dialog box, in the middle-left, click the first <unused> heading.

This displays the Point File Formats – Select Column Name dialog box.

9. In the Point File Formats – Select Column Name dialog box, click the drop-list arrow, displaying the data types list and, from the list, select **Point Number**.

When you select Point Number, the Point File Formats – Select Column Name dialog box displays with an entry that defines an invalid point number.

10. Click ***OK***, assigning Point Number to the first field and returning to the Point File Format dialog box.
11. Click the second column in from the left (<unused>).
12. In the Point File Formats – Select Column Name dialog box, click the drop-list arrow that displays the data types list, and select **Northing** from the list.

After selecting Northing, the Point File Formats – Select Column Name dialog box contains northing entries that identify invalid coordinates and the precision coordinates. The precision value is important with column-formatted data, not comma- or space-delimited data. So, for this exercise, you can ignore precision setting. However, if you are working with a columned file or with invalid indicators, these values need to be set.

13. Click ***OK***, returning to the Point File Format dialog box.
14. Repeat Steps 10 through 12 and add **Easting**, **Point Elevation**, and **Raw Description** to the file format. The new headings are to the right of each heading you define.

After adding the headings, your Point File Format dialog box should look like Figure 2.36.

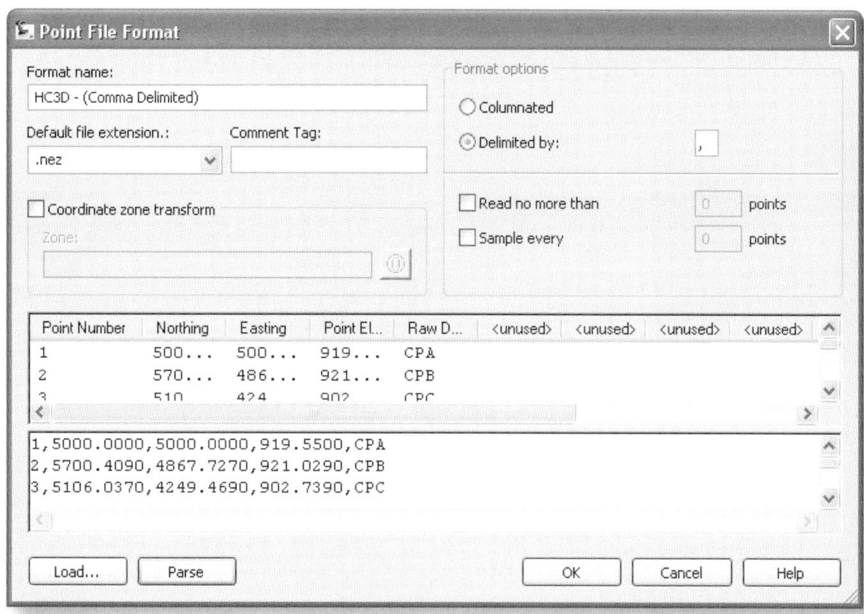


**FIGURE 2.36**

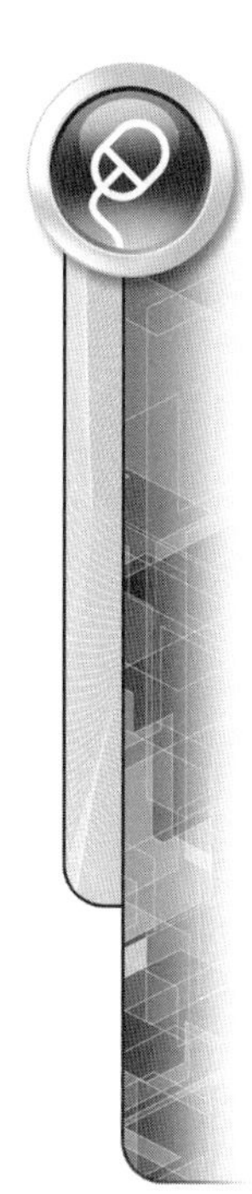

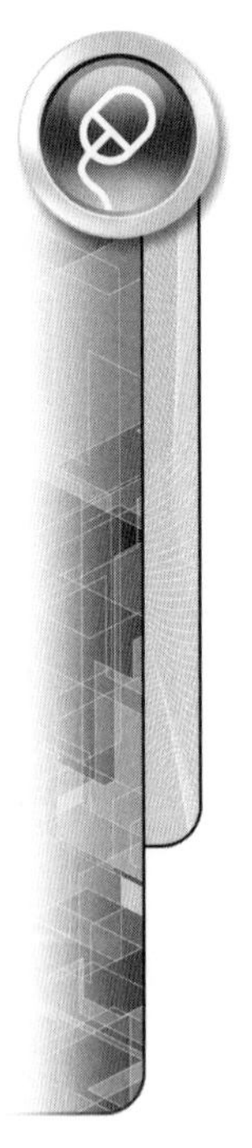

## Testing the Format

Test the new format by loading and parsing *Base-Points.nez*. This file is in this textbook's CD Civil 3D project folder or the Chapter 2 folder.

1. At the bottom-left of the Point File Format dialog box, click ***Load...***, and in the Select Source File dialog box browse to the Civil 3D Projects' Civil 3D folder (C:\Civil 3D Projects\Civil 3D), or browse to this textbook's CD, Chapter 2 folder.
2. In the Select Source File dialog box, at the bottom, change the Files of type to ***.nez**, select the file *Base-Points.nez*, and click ***Open***.
3. In Point File Format dialog box, to read the file, at the bottom-left, click ***Parse***.

The file reads correctly.

4. In the Point File Format dialog box, click ***OK*** and notice that the new format is now listed.
5. Click ***Close*** to return to the Import Points dialog box.

## Importing a File

1. If necessary, in the Import Points dialog box, click the Format: drop-list arrow and select the format **HC3D - (Comma Delimited)**.
2. In the Import Points dialog box, at the right of Source File(s):, click the ***plus sign*** icon, displaying the Select Source File dialog box. If the Civil 3D folder is not the current folder, change the Look in: folder to the local C: drive, double-click Civil 3D Projects, and then the Civil 3D folder to open (C:\Civil 3D Projects\Civil 3D), or navigate to the CD's Chapter 2 folder.
3. Once in the folder, set the Files of type to **.nez*, select the *Base-Points.nez* file, and click Open, returning to the Import Points dialog box.
4. In the Import Points dialog box, do not toggle on any advanced options.

Your Import Points dialog box should look like Figure 2.37.

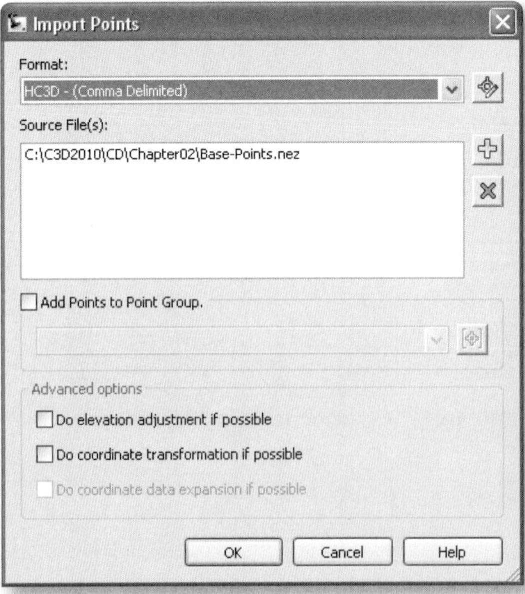


**FIGURE 2.37**

5. Finally, click ***OK***, importing the points.
6. If you cannot see the line work or points, use the ZOOM EXTENTS command to view the entire drawing.
7. In Prospector, expand the Point Groups heading.
8. From the list select **_All Points**, press the right mouse button, and, from the shortcut menu, select ***Properties...*** .
9. Select the Information tab, change the Point label style to **Point#-Elevation-Description**, and click **OK** to exit.
10. In Civil 3D's top left, Quick Access Toolbar, click the ***Save*** icon, saving the drawing.

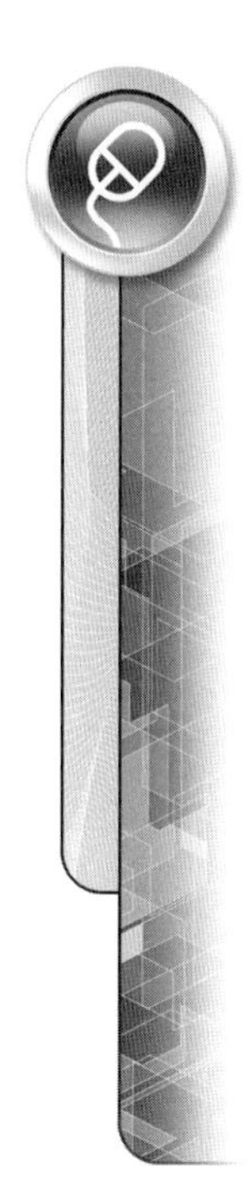

## LandXML — Settings

Before Importing a LandXML file verify and, if necessary, change the LandXML settings. The Settings, Edit LandXML Settings dialog box contains these settings.

1. Click the **Settings** tab.
2. At the Settings panel's top, select the drawing name, press the right mouse button, and, from the shortcut menu, select EDIT LANDXML SETTINGS....
3. In the LandXML Settings dialog box, click the ***Import*** tab, expand the Point Import Settings section, and set Point Description to **use "code" then "desc."**

When exporting to LandXML and using **US Survey Foot**, you need to set Export's base unit in the Data Settings section.

4. Click the ***Export*** tab, expand the Data Settings section, and change the Imperial Units to **US Survey Foot** and the Angle/Direction format to **Degrees decimal dms (DDD.MMSSSS)** (see Figure 2.38).
5. Click ***OK***, exiting the dialog box.

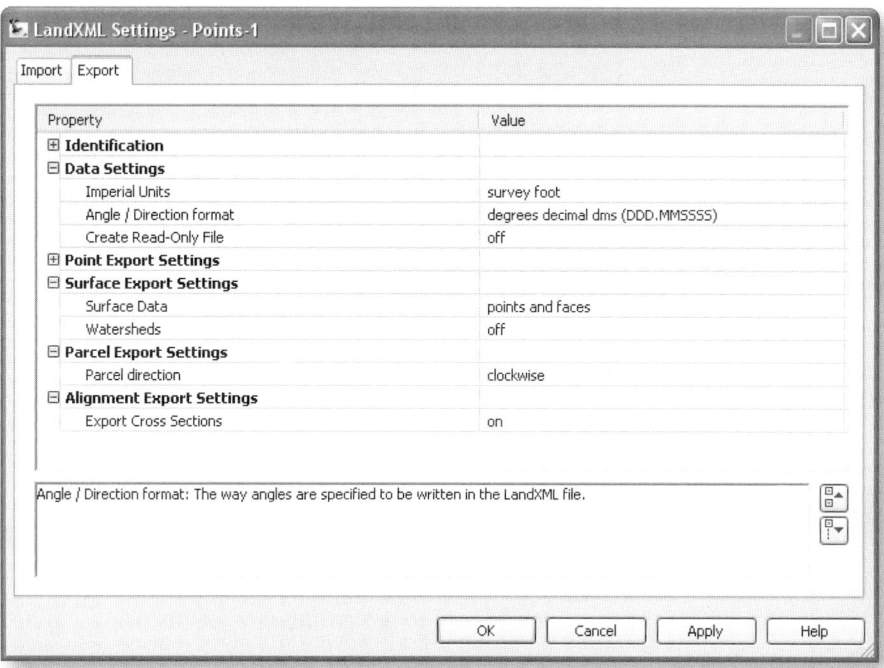
LandXML Settings - Points-1

Import | Export

| Property | Value |
|---|---|
| ⊞ **Identification** | |
| ⊟ **Data Settings** | |
| Imperial Units | survey foot |
| Angle / Direction format | degrees decimal dms (DDD.MMSSSS) |
| Create Read-Only File | off |
| ⊞ **Point Export Settings** | |
| ⊟ **Surface Export Settings** | |
| Surface Data | points and faces |
| Watersheds | off |
| ⊟ **Parcel Export Settings** | |
| Parcel direction | clockwise |
| ⊟ **Alignment Export Settings** | |
| Export Cross Sections | on |

Angle / Direction format: The way angles are specified to be written in the LandXML file.

OK | Cancel | Apply | Help

**FIGURE 2.38**

## LandXML — Import

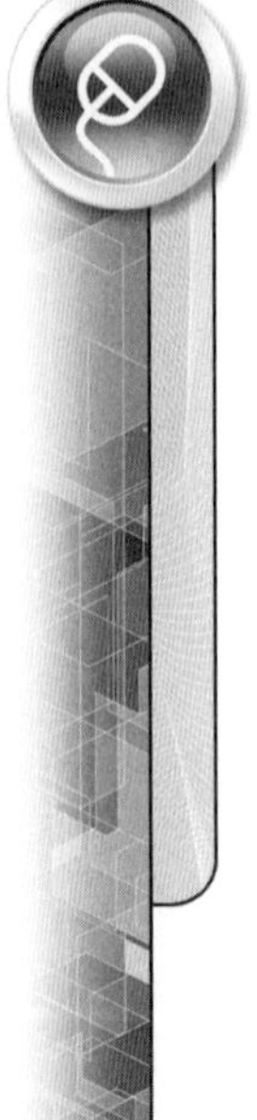

1. In Civil 3D's Ribbon, click the Insert tab. In the Import panel, click the icon ***LandXML*** to display the Import LandXML dialog box.
2. If you are not in the Civil 3D folder, change the Look in: folder to the Local C: drive, double-click Civil 3D Projects, and then the Civil 3D folder to open it (C:\Civil 3D Projects\Civil 3D), or navigate to this textbook's CD Chapter 2 folder.
3. From the list, select *Ellis.xml* and click ***Open***, reading the file.
4. In the Import LandXML dialog box, toggle **OFF** the Alignments and expand CgPoints to review its contents (see Figure 2.39).

Notice that the dialog box lists the point numbers as point names and point codes and descriptions.

5. Click ***OK***, importing the points and closing the dialog box.
6. At Civil 3D's top left, Quick Access Toolbar, click the ***Save*** icon, saving the drawing.

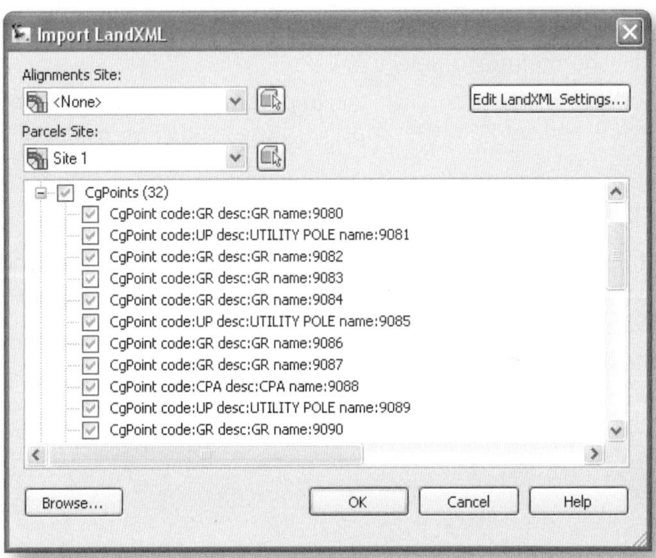


**FIGURE 2.39**

The new points are to the west of the north-south road, at the western edge of the drawing's line work.

## Point Identity — Next Point Number

The Next Point Number is the first unused point number.

1. In the toolbar, click the Expand Create Points chevron, expanding the Create Points toolbar.
2. Expand the Point Identity section, review Next Point Number's value, and change it to **10000**.
3. Close the Point Identity section.

## Create Points Settings

When setting a point, you need to define how to assign the new points' elevation and description.

1. Expand the Points Creation section, review the values, and make sure Prompt For Elevations and Prompt For Descriptions are set to **Manual**.

2. Close the Points Creation section.
3. Click the Collapse the Create Points dialog chevron, collapsing the Create Points toolbar.
4. Click the ***Prospector*** tab.

## Miscellaneous - Manual — Frame Shed and Residence

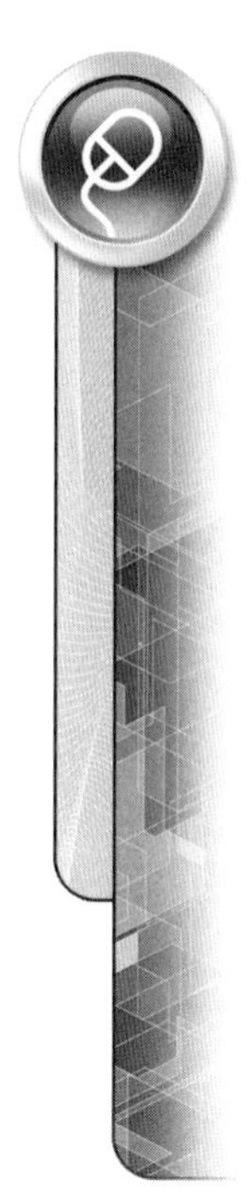

This exercise section uses commands from the Create Points toolbar's Miscellaneous icon stack.

There are a few points missing. In Table 2.2, use the point elevations and descriptions, and for the points' locations, refer to Figure 2.40.

**TABLE 2.2**

| Point Number | Elevation | Description |
|---|---|---|
| 10000 | 920.07 | BLDG |
| 10001 | 917.33 | SWK |

1. Click the Ribbon's View tab, then on the Views panel, select the named view **Frame Shed**.
2. In Create Points toolbar's left side, click the first drop-list arrow and, from the Miscellaneous icon stack, select ***Manual***.
3. Use the **Endpoint** object snap and select the shed's southeast corner, setting point 10000. When prompted for the elevation and description, use the values from Table 2.2.
4. Use the **Endpoint** object snap and select the gravel path's intersection east of the Frame Residence. When prompted for the elevation and description, use the values from Table 2.2.
5. Press ENTER to end the command.
6. At the Civil 3D's top left, Quick Access Toolbar, click the ***Save*** icon, saving the drawing.

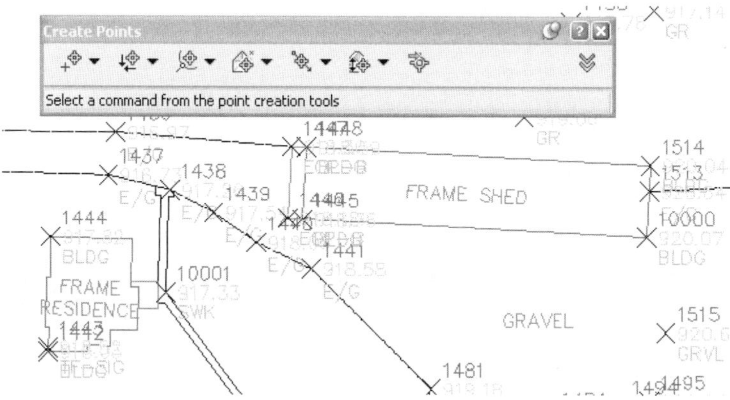


**FIGURE 2.40**

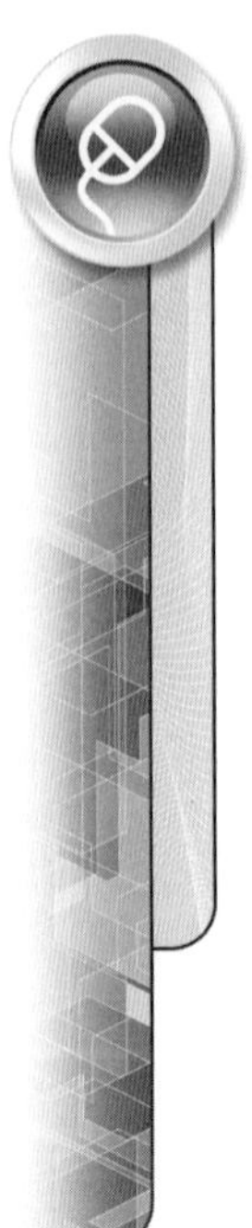

## Miscellaneous – Measure

The next points do not have elevations, but they have the same description. This means changing values in Create Points.

1. From the Ribbon's View tab, Views panel, select the named view **Lot Line**. You will need to click the down arrow to scroll down the list.
2. In the Create Points toolbar, click the Expand the Create Points dialog chevron.
3. Expand the Points Creation section, set Prompt For Elevations to **None**, set Prompt For Descriptions to **Automatic**, and set Default Description to **LLPOINT** (see Figure 2.41).

The Prompt For Elevations and Prompt For Descriptions cells display a drop-list arrow when you click in their value cells.

4. Click the Collapse the Create Points dialog chevron.

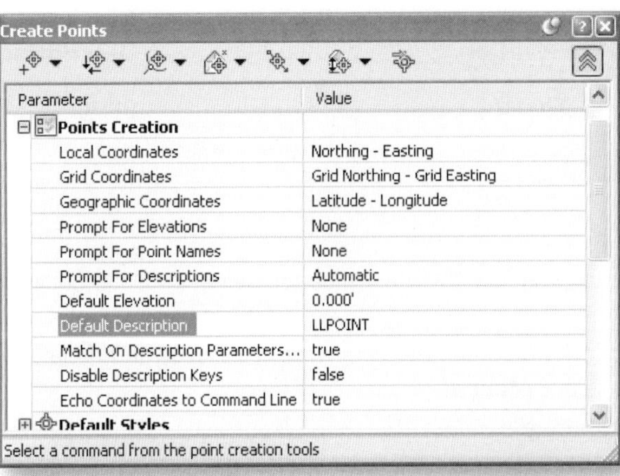


**FIGURE 2.41**

5. In the Create Points toolbar, click Miscellaneous' drop-list arrow (first icon on left) and, from the icon stack, select ***Divide Object***.
6. The routine prompts to select a line, arc, etc. Select lot 16's north lot line, enter **4** as the number of segments, and press ENTER to set the offset to **0.0**.

The routine divides the line into four equal segments with five new points. Each point has no elevation and the description is LLPOINT.

7. Press ENTER, ending the routine.
8. In the Create Points toolbar, click the Miscellaneous drop-list arrow and, from the icon stack, select ***Measure Object***.
9. The routine prompts to select a line, arc, etc. Select lot 16's south lot line, press ENTER twice to accept the beginning and ending stations, set the offset to **0.0**, press ENTER, and for the interval, enter **25** and press ENTER.

The routine creates seven new points with a 25 foot spacing.

10. Press ENTER, exiting the routine.
11. At Civil 3D's top left, Quick Access Toolbar, click the ***Save*** icon, saving the drawing.

Your screen should look like Figure 2.42.

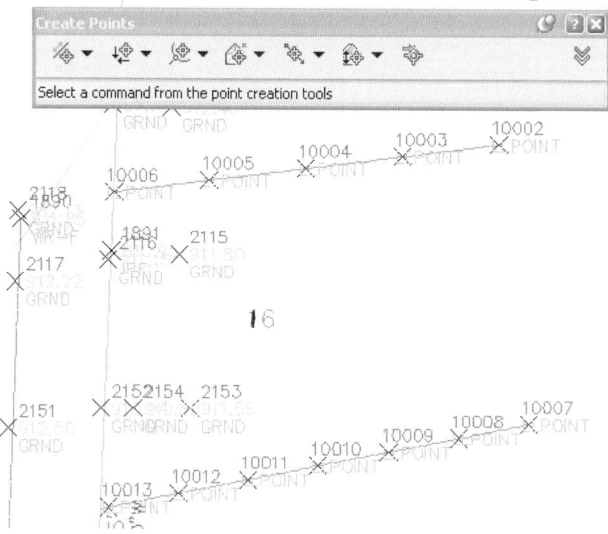


**FIGURE 2.42**

## Polyline Vertices — Automatic

1. In the Create Points toolbar, click the Expand the Create Points dialog chevron. In the dialog box, expand the Points Creation section, set Prompt For Elevations and Prompt For Descriptions to **Automatic**, and set the Default Description to **BERM** (see Figure 2.43).

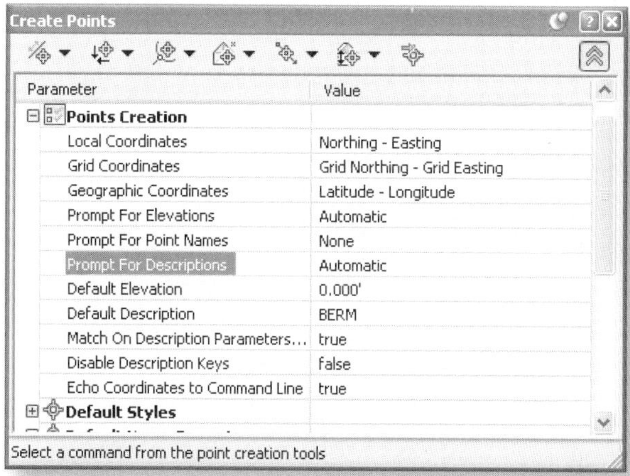


**FIGURE 2.43**

2. Click the Collapse the Create Points dialog chevron.
3. From the Ribbon's View tab, Views panel, select the named view **Berm**. You will need to click the up arrow to scroll up the list.
4. In the Create Points toolbar, click Miscellaneous' drop-list arrow and, from icon stack, select ***Polyline Vertices – Automatic***.
5. The routine prompts for a polyline selection. Select the polyline in the center of the screen.
6. Press ENTER, exiting the routine.

The berm line is a 3D polyline. The routine places a point at each vertex, assigning BERM as the description and the elevation of the vertex to each point (see Figure 2.44).

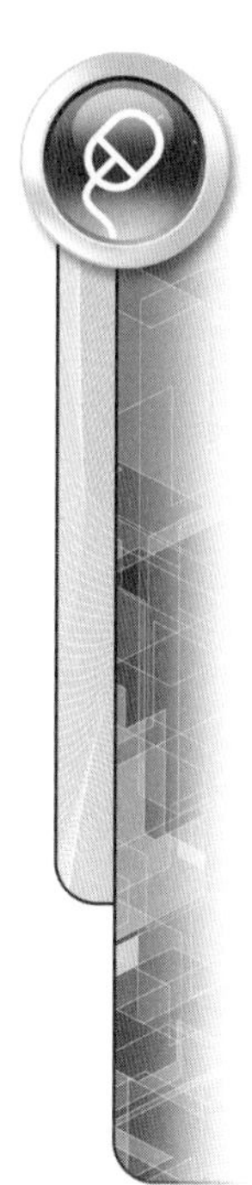

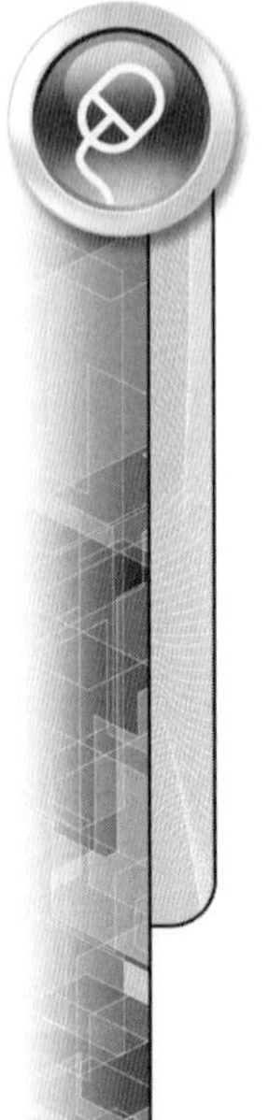

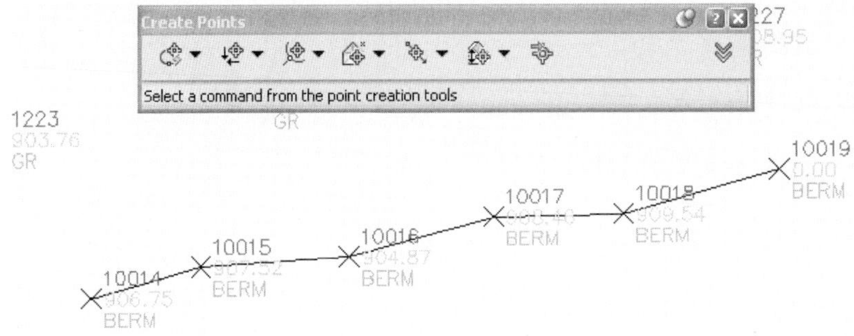


**FIGURE 2.44**

7. At Civil 3D's top left, Quick Access Toolbar, click the ***Save*** icon, saving the drawing.

## Intersection — House Corners

This exercise section uses commands from the Create Points toolbar's Intersections icon stack.

1. If on, toggle **OFF** Dynamic Input mode.
2. In the Create Points toolbar, click the Expand the Create Points dialog chevron.
3. If necessary, expand the Points Creation section and set Prompt For Elevations and Prompt For Descriptions to **Manual**.
4. Expand the Point Identity section, set the Next Point Number to **10050**, and click the Collapse the Create Points dialog chevron.
5. From the Ribbon's View tab, Views panel, select the named view **Lot 11**.

This point series references a point number as the starting point. A field crew measured distances from two points, locating new points that represent the Lot 11 house's south corners. Use the values in Table 2.3 to answer the command prompts.

**TABLE 2.3**

| From Point | Distance | New Point | Elevation | Description |
|---|---|---|---|---|
| 1001 | 88.94 | 10050 | 920.54 | BLDG |
| 1002 | 113.14 | | | |
| 1001 | 116.21 | 10051 | 920.52 | BLDG |
| 1002 | 141.06 | | | |

6. In the Create Points toolbar, click the Intersections drop-list arrow (second icon in from left) and, from the icon stack, select ***Distance/Distance***.
7. In the command line, the routine prompts you to Please specify a location for the radial point location. Click the Transparent Commands' ***Point Number*** filter icon, overriding the current prompt with Enter point number.
8. In the command line, as the first radial point, enter **1001**, press ENTER, enter the distance of **88.94**, and press ENTER, defining the first radial point.
9. The prompt changes, asking for the second point's radial location. Enter as the point number **1002**, press ENTER, enter the distance of **113.14**, and press ENTER, defining the second radial point.

10. The routine then prompts again for a point number. Press ESC, exiting the Point Number prompt and returning to the Point or All prompt.
11. In Lot 11, select a point near the green **X** at the house's southern corner.
12. After selecting the point, the routine prompts for the description and then the elevation. Use the values from Table 2.3 for the description and elevation of the new point.
    After entering the two values, routine places point **10050** at the corner of the house.
13. Repeat Steps 8 through 13 to set point **10051**. Remember to reselect the Transparent Command's **Point Number** filter and to stop it by pressing ESC when selecting the point's location.
14. Exit the *Distance/Distance* command by pressing ESC.
15. In Civil 3D's upper left, Quick Access Toolbar, click the ***Save*** icon, saving the drawing.

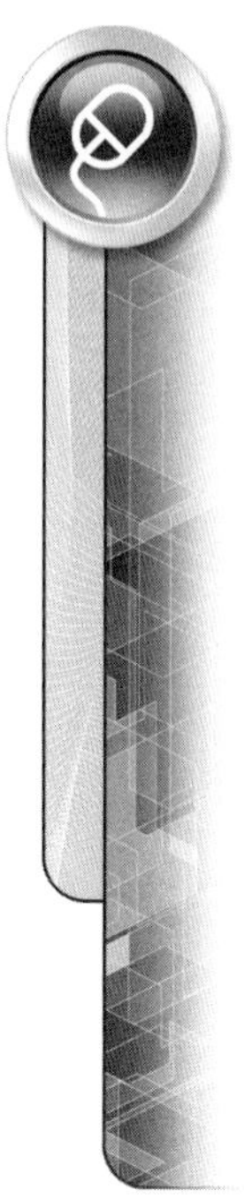

## House Corners by Side Shots

The next point set represents the current building's northern corners and two additional adjacent parcel (Lot 12) building corners. Table 2.4 contains the setup and backsight points, and each new point's side shot angles, distances, elevations, and descriptions.

The Transparent Commands Side Shot override uses point 2341 as the setup point and point 1001 as the backsight point.

Make sure AutoCAD's Dynamic Input mode is off while entering the point values.

**TABLE 2.4**

| Setup | Backsight | Angle | Distance | Description | Elevation |
|---|---|---|---|---|---|
| 2341 | 1001 | 318.2034 | 56.55 | BLDG | 920.54 |
| | | 294.0043 | 67.05 | BLDG | 920.53 |
| | | 252.5419 | 128.29 | BLDG | 920.55 |
| | | 243.5822 | 96.75 | BLDG | 920.53 |

1. In the Create Points toolbar, click Miscellaneous' drop-list arrow and, from the icon stack, select ***Manual***.
2. From the Transparent Commands toolbar, select the override **Side Shot**.

Side Shot changes the prompting to selecting a line or points, establishing a setup and backsight (turned angle and distance from this setup).

3. To establish a setup using points 2341 and 1001, type the letter 'P' (for points mode) and press ENTER.

In points mode, .p changes the prompting to point number and .g changes it to select a point object.

4. In the command line, at the Select line or [Points]: prompt, type **'P'**, and press ENTER.
5. In the command line, at the Specify starting point or [.P/.N/.G]: prompt, type **'.p'** and press ENTER.
6. The prompt changes to Enter Point Number; enter point **2341** and press ENTER.
7. The next prompt is Specify ending point or [.P/.N/.G]: again, type **'.p'** and press ENTER.

8. The prompt changes to Enter Point Number; enter point **1001** and press ENTER.
9. In the command line, the prompt is for an angle; for the angle, enter **318.2034** and press ENTER.
10. In the command line, the prompt is for a distance; for the distance, enter **56.55** and press ENTER.
11. The next two command prompts are for the new point's description and elevation. For their respective prompts, enter **BLDG** and **920.54**.
12. Repeat Steps 9 through 11, creating the three remaining building corner points.
13. Press ESC, exiting the Side Shot override, and press ENTER, ending the command.
14. In the Create Points toolbar, select the red **X** at the right to close it.
15. In Civil 3D's upper left, Quick Access Toolbar, click the ***Save*** icon, saving the drawing.

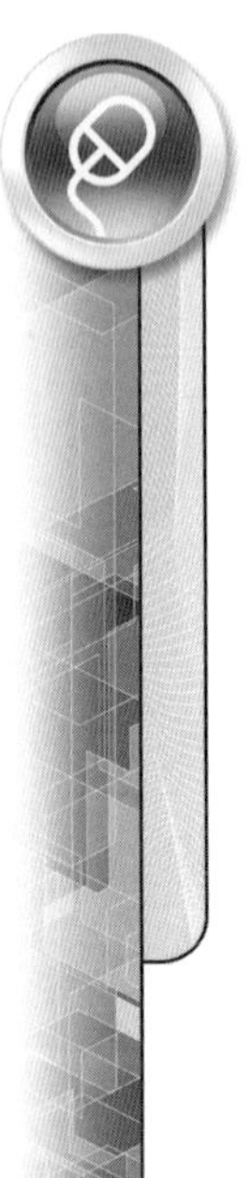

This completes this exercise. After points are created they may need evaluation and possibly editing.

## SUMMARY

- Users can create custom ASCII file formats.
- Adjusting Create Points toolbar's Point Creation prompting options changes how a command prompts for values.
- The Next Point Number is set in Create Points toolbar's or CreatePoints Command's Point Identity section.
- The Polyline Automatic routine assigns the 3D polyline's vertices elevations as the point's elevation.
- Polyline Manual prompts for a new elevation and assigns it to all points.
- Any command prompting for a point can use an AutoCAD selection, point object (.p or .g), or northing/easting coordinate (.n) to establish the point.
- Calculating a new point's coordinates, transparent commands expand the current command's capabilities by using distances, angles, and directions.

## UNIT 3: POINT ANALYSIS

Point analysis and listing tools also function as an editing interface. Prospector's preview and the Point Editor vista are the primary point analysis tools. When selecting Prospector's Points heading, preview displays all points currently in the drawing. When selecting Prospector's Points heading, pressing the right mouse button, and selecting from the shortcut menu Edit Points… or selecting the Ribbon's Modify tab's Points and clicking Edit/List Points command, displays a Point Editor vista (see Figure 2.45). Clicking any column heading in Preview or the Point Editor vista sorts the point list.

The shortcut menu, the Preview, and Point Editor have Zoom to and Pan to commands that change the display to a selected point's location.

To create a LandXML points file, select the Point Groups heading, press the right mouse button, and, from the shortcut menu, select Export LandXML….

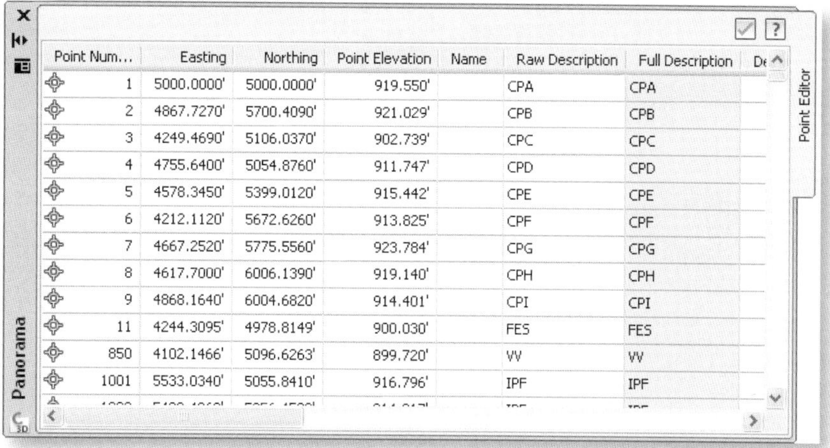

| Point Num... | Easting | Northing | Point Elevation | Name | Raw Description | Full Description |
|---|---|---|---|---|---|---|
| 1 | 5000.0000' | 5000.0000' | 919.550' | | CPA | CPA |
| 2 | 4867.7270' | 5700.4090' | 921.029' | | CPB | CPB |
| 3 | 4249.4690' | 5106.0370' | 902.739' | | CPC | CPC |
| 4 | 4755.6400' | 5054.8760' | 911.747' | | CPD | CPD |
| 5 | 4578.3450' | 5399.0120' | 915.442' | | CPE | CPE |
| 6 | 4212.1120' | 5672.6260' | 913.825' | | CPF | CPF |
| 7 | 4667.2520' | 5775.5560' | 923.784' | | CPG | CPG |
| 8 | 4617.7000' | 6006.1390' | 919.140' | | CPH | CPH |
| 9 | 4868.1640' | 6004.6820' | 914.401' | | CPI | CPI |
| 11 | 4244.3095' | 4978.8149' | 900.030' | | FES | FES |
| 850 | 4102.1466' | 5096.6263' | 899.720' | | VV | VV |
| 1001 | 5533.0340' | 5055.8410' | 916.796' | | IPF | IPF |



**FIGURE 2.45**

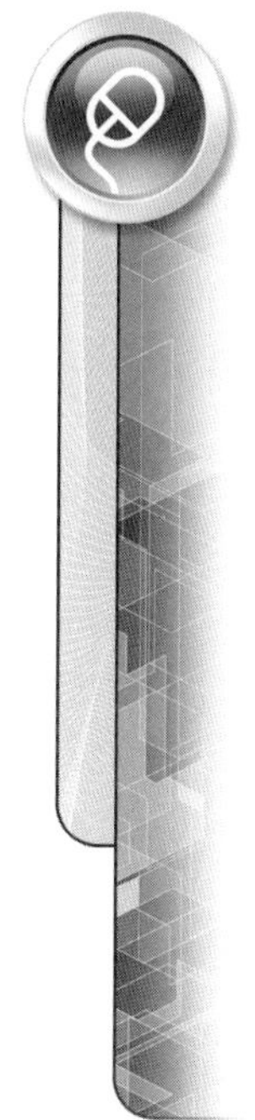

## EXERCISE 2-3

When you complete this exercise, you will:

- Generate a points report.
- Be familiar with the LandXML Report Application.

### Generating a Point Report

1. If you are not in the *Points-1* drawing file from the previous unit, open it now. If you are starting with this exercise, in Civil 3D's upper left, Quick Access Toolbar, click the ***Open*** icon, browse to this textbook's CD Chapter 2 folder, select the *Chapter 02 - Unit 3.dwg*, and click ***Open***.
2. If necessary, click the Ribbon's View tab. To the Toolspace icon's right and in the bottom row, click the ***TOOLBOX*** icon, and click the Toolbox tab.
3. Expand the Reports Manager and Points sections.
4. From the list of Points reports, select **Points_List**, press the right mouse button, and select EXECUTE..., displaying the Export to XML Report dialog box (see Figure 2.46).

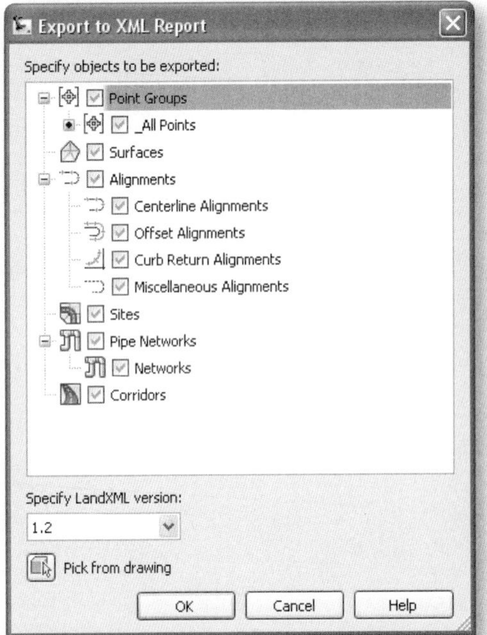


**FIGURE 2.46**

5. If not selected, toggle **ON** the point group _All Points and click ***OK***.

Internet Explorer displays a point list report. You can print or save this report.

6. Review the report and close Internet Explorer.
7. If you want to export an Excel or CSV file, from the Points report list, select **Points_in_CSV**, press the right mouse button, and select ***Execute...***. The Export to XML Report dialog box displays.
8. If not selected, toggle **ON** the point group _All Points and click ***OK***.
9. A File Download dialog box displays, asking to Open or Save the file. Click SAVE, browse to the Civil 3D Projects folder, click SAVE, and CLOSE.
10. If you have Excel, start Excel and view the file.
11. After reviewing the file, exit Excel and return to Civil 3D.

## Zoom to and Pan to

Locating individual points can be difficult. The Points heading and preview area's shortcut menus have the commands ***ZOOM TO*** and ***PAN TO***.

1. Click the ***Prospector*** tab.
2. In Prospector, select the **Points** heading.

The preview area lists all points.

3. In preview, scroll through the points list, select a point, press the right mouse button, and, from the shortcut menu, select ***ZOOM TO***.

The display centers on the point's location.

4. In preview, scroll through the list, select another point, press the right mouse button, and, from the shortcut menu, select ***PAN TO***.

5. Select a few additional points from preview and use the commands ***ZOOM TO*** or ***PAN TO***.
6. In Civil 3D's upper left, Quick Access Toolbar, click the ***Save*** icon, saving the drawing.

## SUMMARY

- Civil 3D exports points to LandXML by selected point groups.
- Toolbox contains point reports.

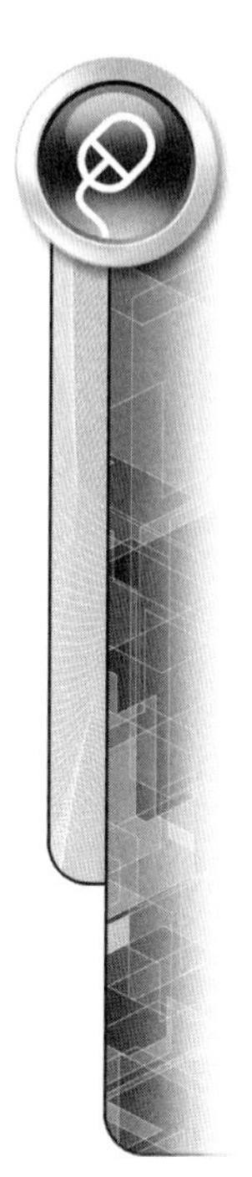

## UNIT 4: EDIT POINTS

Mistakes are inevitable and there needs to be a way to fix them. Prospector's preview or Panorama's Point Editor vista is the point editing interface. Each edits all of the point's information: number, elevation, description, location, styles, and so on. Both Preview and the Points Editor vista have the same shortcut menu commands for editing selected points data.

### PROSPECTOR'S PREVIEW

Clicking a point's cell in Preview activates edit mode. You can edit point numbers, point northings and eastings, point names, elevations, raw and full descriptions, and other point data. When you press the right mouse button over a selected point, a shortcut menu displays with additional point edit commands (see Figure 2.47). See the discussion of these commands in the next heading.

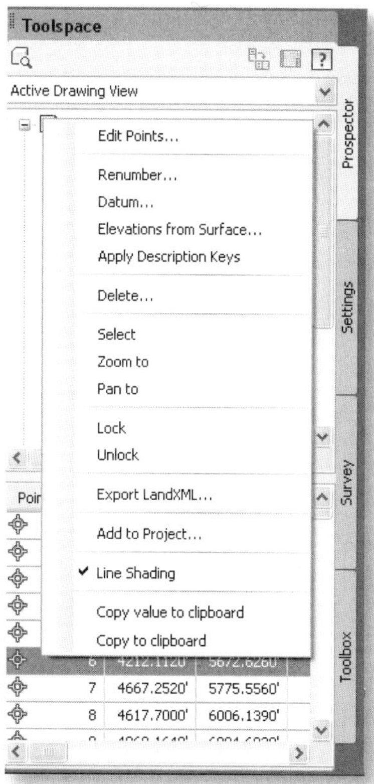

**FIGURE 2.47**

## EDIT POINTS

Edit Points has four starting points: Ribbon's Modify panel, Prospector's Points heading shortcut menu, a right-mouse-click shortcut menu or COGO Point context tab that is available after selecting one or more points on the screen, or a shortcut menu from Prospector's preview.

Edit Points displays the Point Editor vista, listing all or the selected the point values (see Figure 2.45). When selecting a point in Preview and pressing the right mouse button, a shortcut menu displays with additional point editing commands (see Figure 2.47).

### Renumber

Renumber assigns to selected points new point numbers by an additive factor. If the new point number is a duplicate, a dialog box displays, prompting for a solution; assign the next available point number or overwrite the existing point with the newly renumbered point.

### Datum

Datum manipulates selected points' elevations. The amount of elevation change is an absolute (5 or –5) or a reference to an old and new elevation. When using the change in elevation option, a positive or negative value is added to the selected point's elevation. The reference option determines what value to add by taking the difference between two elevations. For example, a field crew assumes a topographic survey benchmark elevation as 100 feet. After researching the benchmark, the surveyor determines that the true benchmark elevation is 435.34. The Datum elevation routine prompts for the reference elevation (the old benchmark elevation) and the new elevation, and then calculates the amount to add to each selected point.

Another situation for using Datum is a blown survey setup elevation. If you are observing points with an incorrect setup elevation, all point elevations are incorrect. If this error is discovered, the datum routine adjusts all affected points.

### Elevations From Surface

Elevations from Surface assigns selected points a surface's elevation at each point's location. The routine selects points by selection, group, numbers, and all.

## DELETE AND AUTOCAD ERASE

The Civil 3D's `DELETE` and `ERASE` commands permanently erase points.

## POINT UTILITIES

The Ribbon's Home tab, Points menu provides tools for converting Land Desktop points, point blocks, and AutoCAD nodes to Civil 3D point objects (see Figure 2.48). The Create Blocks from COGO Points command creates blocks with attributes from Civil 3D cogo point objects.

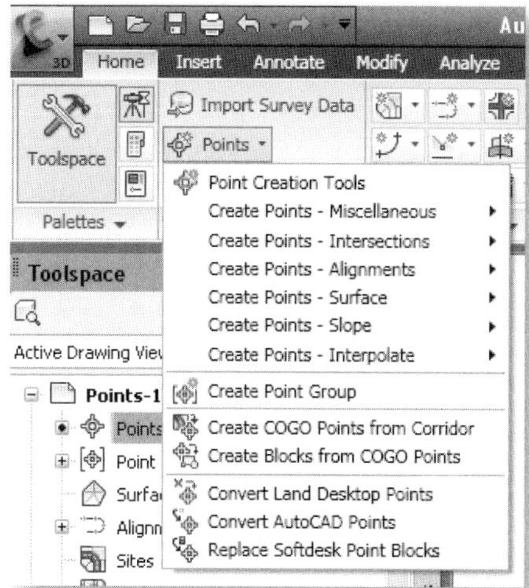


**FIGURE 2.48**

## Transfer and Transfer Points

The Ribbon's Modify tab, Ground Data panel's, Points icon displays point specific editing commands. The Transfer Points routine, in the COGO Point Tools panel, copies point data from an ASCII file or (Access) database to an ASCII file or (Access) database (see Figure 2.49).

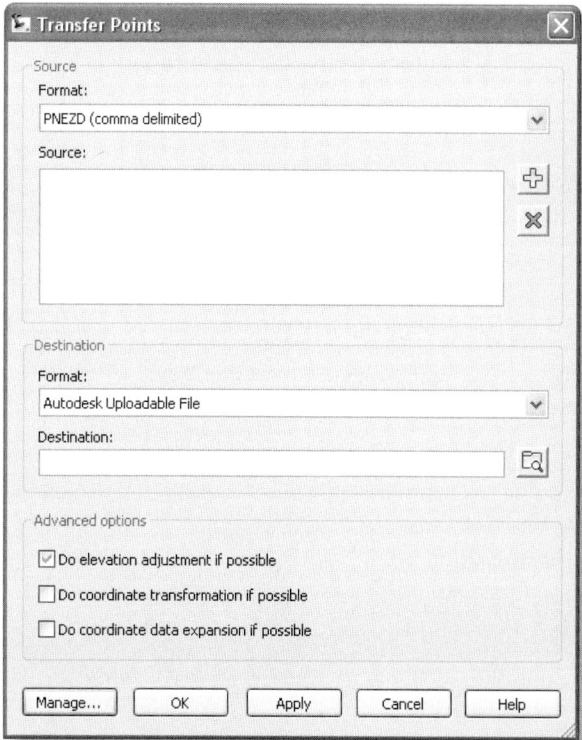


**FIGURE 2.49**

## GEODETIC CALCULATOR

The Ribbon's Modify tab, Points icon displays point specific editing commands. The Geodetic Calculator, in the Analyze panel, creates and calculates a point's grid coordinates and latitude and longitude (see Figure 2.50). To use the calculator, you must have an assigned coordinate system to the drawing.

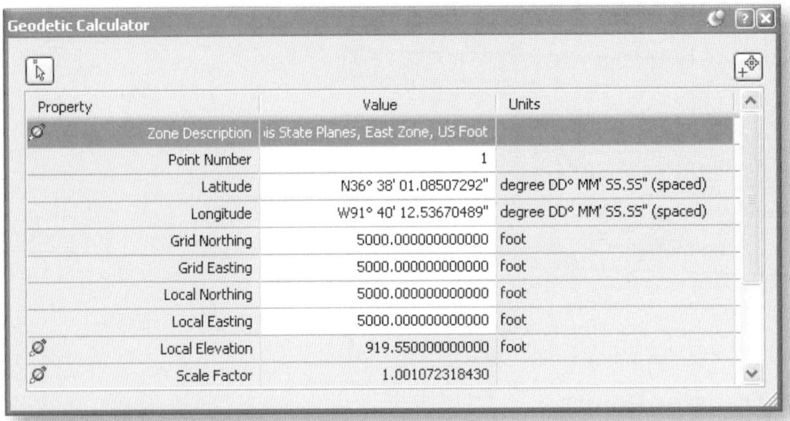

Geodetic Calculator
Property
Value
Units
Zone Description
is State Planes, East Zone, US Foot
Point Number
1
Latitude
N36° 38' 01.08507292"
degree DD° MM' SS.SS" (spaced)
Longitude
W91° 40' 12.53670489"
degree DD° MM' SS.SS" (spaced)
Grid Northing
5000.000000000000
foot
Grid Easting
5000.000000000000
foot
Local Northing
5000.000000000000
foot
Local Easting
5000.000000000000
foot
Local Elevation
919.550000000000
foot
Scale Factor
1.001072318430


**FIGURE 2.50**

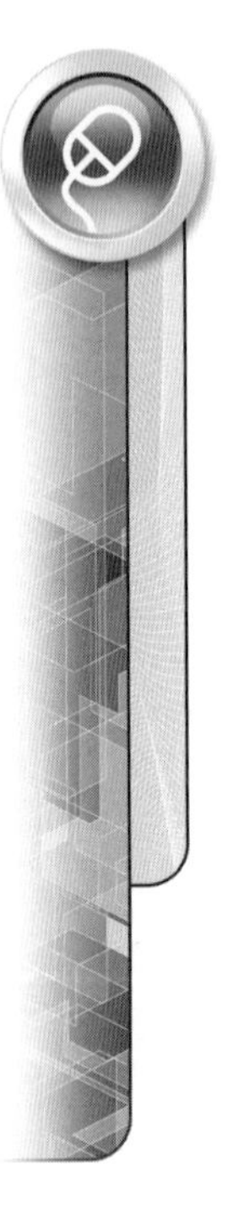

## EXERCISE 2-4

When you complete this exercise, you will:

- Be able to edit points from various locations in the Civil 3D interface.
- Be able to edit points in a vista.
- Renumber selected points.
- Adjust the datum and the elevation of selected points.
- Export points to an ASCII file.

### Exercise Setup

This exercise continues with the drawing from the previous exercise.

1. If you are starting with this exercise and have not done the previous exercises, from the CD that accompanies this textbook, browse to the Chapter 2 folder and open the *Chapter 02 - Unit 4.dwg*.

### Prospector Preview - Edit Points

1. If necessary, click the ***Prospector*** tab.
2. In Prospector, click the **Points** heading, listing the points in preview.
3. In preview, use the scroll bars to view the points.
4. In preview, click the Point Elevation heading, sorting the point elevations.
5. In preview, scroll to and locate point **3001**.
6. In preview, scroll through point 3001's record until locating its elevation (0.0). Click the Point Elevation cell, change the elevation to **898.76**, and accept the new value by pressing ENTER.

7. In Prospector, click the **Points** heading, press the right mouse button, and, from the shortcut menu, select EDIT POINTS... to display the Point Editor vista.
8. In the Point Editor vista, click the Point Number heading, sorting the list.
9. For point 10001, click the Point Elevation cell and change its elevation to **914.15**.
10. For point 10002, click the Point Elevation cell and change its elevation to **914.90**.
11. After editing the points, at the top of Panorama's mast, click the **X**, closing it.
12. In Prospector, select the Points heading, listing the points in preview.
13. In preview, click the Point Number column heading, sorting the points.
14. In preview, scroll thorough the point list, locating point **10003**.
15. In preview, select 10003's point icon, press the right mouse button, and, from the shortcut menu, select ***ZOOM TO***.
16. In the drawing, select point **10003**, press the right mouse button, and, from the shortcut menu, select EDIT POINTS....
17. In the Point Editor vista, click in the Point Elevation cell, enter the value of **916.78**, click the green check in the upper-right to close the vista.
18. In preview, scroll through the point list, locating 3002, select 3002's point icon, press the right mouse button, and, from the shortcut menu, select ***ZOOM TO***.
19. In the drawing, select point **3002**, press the right mouse button, and, from the shortcut menu, select EDIT POINTS....
20. In the Point Editor vista, click in the Point Elevation cell, enter the value of **914.96**, and close the vista.
21. In preview, scroll through the point list, locating number 3003, select 3003's point icon, press the right mouse button, and, from the shortcut menu, select ***ZOOM TO***.
22. In preview, select point **3003**, press the right mouse button, and, from the shortcut menu, select EDIT POINTS....
23. In the Point Editor vista, click in the Point Elevation cell, enter the value of **912.40**, and close the panorama.
24. At Civil 3D's upper left, Quick Access Toolbar, click the ***Save*** icon, saving the drawing.

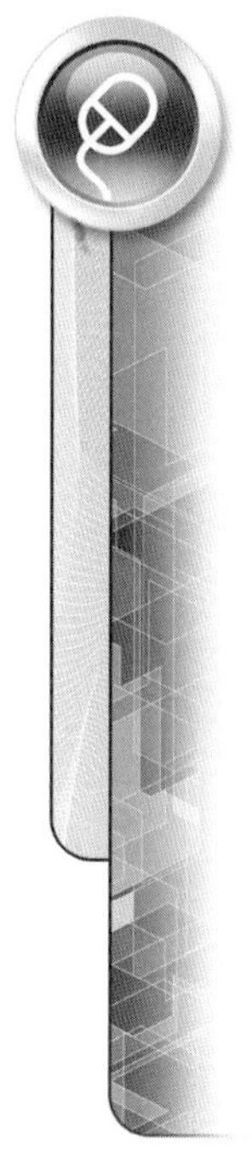

## Renumber Points

There will be times when you will need to renumber points to new numbers.

1. In preview, if necessary, click the Point Number column heading, sorting the points.
2. In preview, scroll through the point list, locating point **3001,** and select the point.
3. While holding down SHIFT, select point number **3003**, selecting points 3001 to 3003.
4. With the cursor over the highlighted points, press the right mouse button, and, from the shortcut menu, select RENUMBER....
5. In the command line, the routine prompts for an additive factor. Enter **2000** and press ENTER to renumber the points to 5001, 5002, and 5003.

The points change to the new point numbers.

6. In preview, scroll until you locate points 5001 through 5003.
7. At Civil 3D's upper left, Quick Access Toolbar, click the ***Save*** icon, saving the drawing.

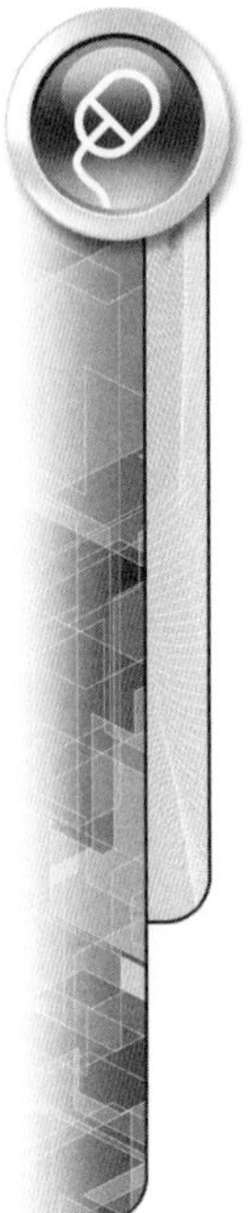

## Edit Datum

There will be times when points have an assumed or incorrect setup elevation. You adjust these points with the Datum routine. The Datum routine prompts for an old and new elevation or an amount (positive or negative) to add to the point's elevation.

1. If necessary, in Prospector, select the Points heading, listing the points in preview.
2. In preview, scroll through the point list, locating point 2025.
3. Select point **2025**, press CTRL, and select point **2024**.
4. With both points highlighted, press the right mouse button, and, from the shortcut menu, select DATUM....
5. In the command line, the routine prompts for an amount of change, enter **5**, and press ENTER.

This adds 5 feet to the two points' elevations.

6. In preview, with the two points highlighted, press the right mouse button, and, from the shortcut menu, select DATUM.... You may have to reselect the two points (2024 and 2025).
7. In the command line, the routine prompts for a change in elevation; for reference, enter the letter '**R**'; as the reference elevation, enter **100**; and for the "new" elevation, enter **95**.

This subtracts 5 feet of elevation from the two points.

8. At Civil 3D's upper left, Quick Access Toolbar, click the ***Save*** icon.

## Export Points

There will be times when you will need to export points to an ASCII file.

1. In Prospector, select the **Points** heading, press the right mouse button, and, from the shortcut menu, select EXPORT....
2. In the Export Points dialog box, select the **HC3D - (Comma Delimited)** format. If you didn't make the format, use the **PNEZD (comma delimited)** format.
3. To the right of Destination File, click the ***folder*** icon, browse to the Civil 3D Projects folder, and name the file Points-1.
4. Click ***Open*** and ***OK***, exporting the points and dismissing the dialog boxes.
5. At Civil 3D's upper left, Quick Access Toolbar, click the ***Save*** icon, saving the drawing and close the file.

This ends the edit point exercise. The next unit covers layer and point and label styles assignments using a Description Key Set.

## SUMMARY

- The point editor is called from a selected point's shortcut menu, the Ribbon, the Points menu, Prospector's preview shortcut menu, or from Prospector's Points heading shortcut menu.
- No matter where the Edit Points... routine is called, it displays the same Point Editor vista.
- The `RENUMBER` command adds a fixed offset to the selected points.
- Select the points first before using the `DATUM` command. If not, the routine assumes you are adjusting the elevations for all points.
- Datum adds a negative or positive amount to the elevations of selected point(s).
- A Datum routine option references an old and a new elevation to determine the value of the additive factor.
- Users can export all points or just the points in a point group when using the `EXPORT` command.

## UNIT 5: ANNOTATING AND ORGANIZING POINTS

A Description Key Set translates raw descriptions into full descriptions. For example, the raw description "IPF" means "Iron Pipe Found." A Description Key Set is a table that translates raw descriptions to full descriptions by matching a code entry (see Figure 2.51). A code represents a list of characters, numbers, or alpha-numeric characters that must be matched from the raw description to use the values in the description key set. Additional Description Key Set functions include assigning point (marker) styles (symbols if appropriate) and label styles, scaling and rotating markers, and assigning point object layers.

A Description Key Set can assign a point style's layer. If the user wants the Description Key Set to assign the point object's layer, the Point Style sets the marker layer to 0 (zero) and defines the label's layer. The Description Key Set references layers in a drawing, and the layers should be a part of the drawing template.

### CODES

As previously mentioned, a code is what a raw description must match to use the values in the Description Key Set. A raw description assignment occurs in the field, in data collectors, in external coordinate files, or in any point creation routine. Many raw descriptions have to be translated to a full description. For example, a MH raw description represents a Manhole or PINE 6 needs to translate to 6" PINE. The only mechanism in Civil 3D that translates a raw description into a full description is the Description Key Set.

Codes do not need to be alphabetic characters. They can be numerical, alphabetical, or a combination of both. When codes are numerical, a Description Key Set is vital in translating the codes into meaningful words.

#### Raw Description Parameters

A raw description may contain spaces (for example, OK 6, MP 3, etc.). Civil 3D considers a space as a separator between parameters. In the example OK 6, the first parameter is "OK" and the second parameter is "6." Civil 3D allows up to 10 parameters in a description and identifies them with a $ (dollar sign) followed by a number.

Civil 3D numbers the parameters starting with 0 (zero). Using the example OK 6, "OK" is parameter 0 ($0) and "6" is parameter 1 ($1).

The Description Key Set's Format column contains the translation format to produce a full description. When translating a raw description, users can rearrange a raw description's parameters to create a full description. With the example OK 6, the Full Description is 6" Oak, so the Format column entry is $1" OAK or $1" $0. This roughly translates to use parameter $1 as the first entry and follow it with an inch sign and the word "OAK." To create a full description from a raw description with no translation, the Format column contains $* (use raw description).

Civil 3D allows parameters as scaling and rotating factors. With the Oak tree example, each measured trunk size can scale the Oak symbol by multiplying it with parameter ($1). Scaling and rotating are Description Key Set parameter columns.

## POINT ANNOTATION METHODS

A point annotation's first method is using default point and label styles from Edit Feature Settings or Command Settings. This method assigns all points the same styles. In a previous exercise, two point styles were defined: OAK and MAPLE. These point styles defined marker and label layers.

When you have several codes, a Description Key Set is very helpful when you want to assign styles other than the default styles (see Figure 2.51).

There should be a Description Key Set entry for all possible point codes. This may be difficult, especially if users do not have any control over the source of the raw descriptions. A text editor with a find-and-replace feature is the only way to substitute the different descriptions with the in-house values. If the work is performed by a contractor, users should encourage the contractor to use the user's point descriptions. Even with this effort, many offices face commonly occurring naming inconsistencies.

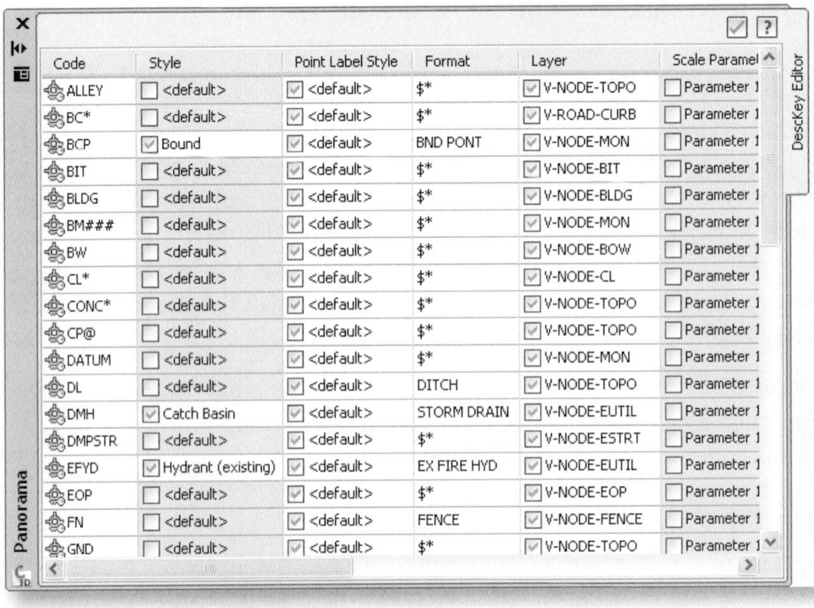

| Code | Style | Point Label Style | Format | Layer | Scale Paramet |
|---|---|---|---|---|---|
| ALLEY | <default> | <default> | $* | V-NODE-TOPO | Parameter 1 |
| BC* | <default> | <default> | $* | V-ROAD-CURB | Parameter 1 |
| BCP | Bound | <default> | BND PONT | V-NODE-MON | Parameter 1 |
| BIT | <default> | <default> | $* | V-NODE-BIT | Parameter 1 |
| BLDG | <default> | <default> | $* | V-NODE-BLDG | Parameter 1 |
| BM### | <default> | <default> | $* | V-NODE-MON | Parameter 1 |
| BW | <default> | <default> | $* | V-NODE-BOW | Parameter 1 |
| CL* | <default> | <default> | $* | V-NODE-CL | Parameter 1 |
| CONC* | <default> | <default> | $* | V-NODE-TOPO | Parameter 1 |
| CP@ | <default> | <default> | $* | V-NODE-TOPO | Parameter 1 |
| DATUM | <default> | <default> | $* | V-NODE-MON | Parameter 1 |
| DL | <default> | <default> | DITCH | V-NODE-TOPO | Parameter 1 |
| DMH | Catch Basin | <default> | STORM DRAIN | V-NODE-EUTIL | Parameter 1 |
| DMPSTR | <default> | <default> | $* | V-NODE-ESTRT | Parameter 1 |
| EFYD | Hydrant (existing) | <default> | EX FIRE HYD | V-NODE-EUTIL | Parameter 1 |
| EOP | <default> | <default> | $* | V-NODE-EOP | Parameter 1 |
| FN | <default> | <default> | FENCE | V-NODE-FENCE | Parameter 1 |
| GND | <default> | <default> | $* | V-NODE-TOPO | Parameter 1 |

Panorama

DescKey Editor

**FIGURE 2.51**

A Key Set assigns a point style layer (marker layer). A Point Style has both marker and label layers. The point style's marker layer should be set to 0 (zero) and the label's layer is the remaining Display panel layer assignment. This strategy places the marker on the Description Key Set layer and the label on the point style's layer. Figure 2.52 shows this strategy for the code CB*. The Description Key Set assigns the marker's layer, and the point style, Catch Basin, defines the label's layer.

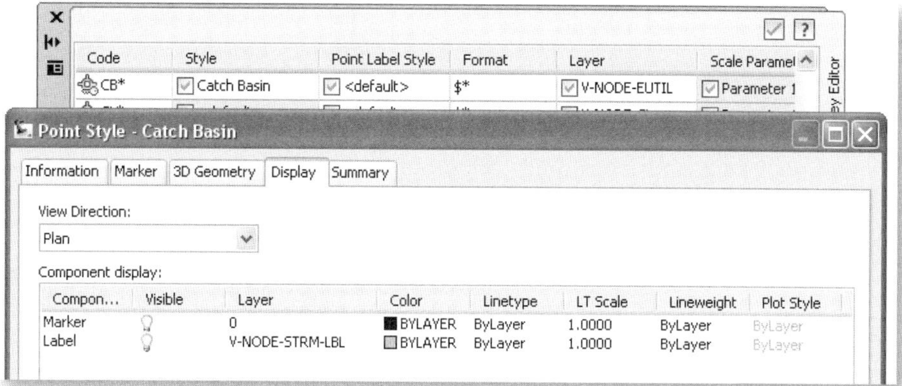


**FIGURE 2.52**

Controlling marker and label visibility occurs in three possible locations. The first place that controls visibility is Layer Properties Manager, by turning on and off the Key Set and point style layers. The second place that controls marker and label visibility is in the point style's Display panel. By turning off the style's Marker and Label component, all of its markers and labels are hidden. The third location used for controlling point visibility is Point Groups.

If you decide to use Description Key Sets, it is best to include them as a part of a template. Adding a Description Key Set after placing points does not make them react to the Description Key Set settings. If editing a point's description, only the description in the label changes not the point style. The behavior of Description Key Sets can be performed manually after imports and edits, by right clicking the _All Points point group and from the shortcut menu, selecting Apply Description Keys.

A Description Key Set is always active. The Create Points toolbar's Points Creation section's Disable Description Keys is the toggle controlling description keys (see Figure 2.53).

In this section there is a second toggle, Match on Description Parameters. This setting indicates whether the numbers in a description are for scaling and rotating the point. If the parameters contain values to scale and rotate the point object, this value needs to be set to true. This also means that spaces in a code must be correct (e.g., Man Hole 48 needs to be Manhole 48). If this is set to false, there can be spaces in the description but there can be no values to scale or rotate the point objects.

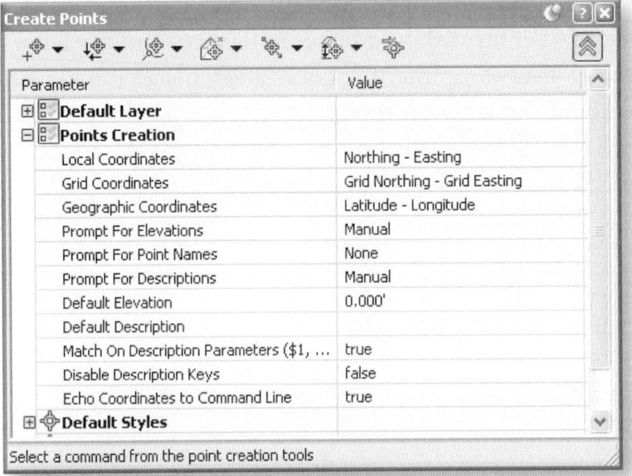


**FIGURE 2.53**

A drawing can have more than one Description Key Set. In this case, the searching order is **down** the key sets list. The Description Key Set heading's property dialog box displays the key sets ordered list and allows manipulation of the key set search order.

## POINT GROUPS

A point group has a name and organizes points by how you think and talk about them (for example, existing utilities, EG surface points, and so on). A point group can assign point, label style, elevation, and description overrides.

If a point belongs to more than one point group, the last displayed point group controls the point's visibility. A point's display can be a complex balance between styles, key sets, point group overrides, and display order.

## DESCRIPTION KEY SET

A Description Key Set is a matrix associating a point style (marker or symbol), label style, layer, translation, and, optionally, a scaling and rotating factor with a point's description. When a point's description matches a code in the key set, the settings for the code in the key set are applied to the point (see Figure 2.54). If the point style is a symbol, the symbol's insertion point is the point's coordinates. For example, the Catch Basin point style uses a block as a symbol for the catch basin and its insertion point represents the point's drawing coordinates.

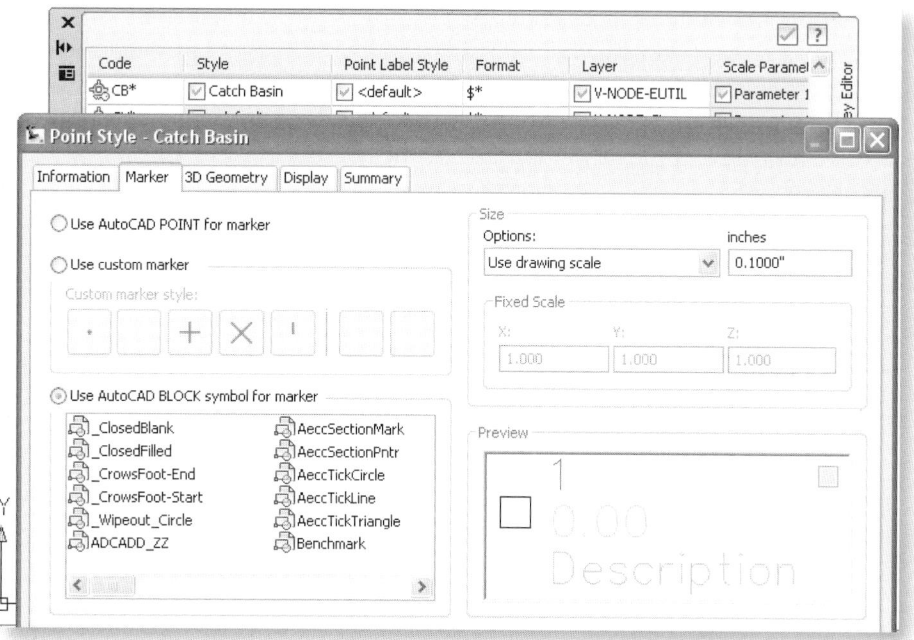


**FIGURE 2.54**

A point label style defines what point information to display and its location relative to the point's coordinates. The separate style layers are used because the symbol is usually part of the final document, while the label may not be.

Civil 3D installs with templates that contain some blocks. If you want to include other symbols, they need to be inserted as block definitions in the template (or current drawing). When using this modified template and starting a new drawing, the symbols appear in the point style's Use AutoCAD BLOCK symbol for marker area (see Figure 2.54).

If making your own symbols, here are some rules:

- Draw and define all symbols by their plotted size.
- All symbol entities should be on layer 0.
- The insertion point of the block should be (0,0) coordinates.

The first rule is draw and define symbols by their plotted size. If a manhole symbol is to be 0.2 inches in diameter when plotted, its block size diameter is 0.2 (foot-based drawing). When placing a point with this point style, it is scaled by the drawing scale. If the drawing scale is 1 inch = 20 feet, the scaling factor for the symbol is 20 and the resulting drawing symbol size is 4 feet in diameter (0.2 * 20).

The second rule refers to the block process. When inserting a block whose entities are on layer 0, they will assume the insertion layer's properties. For example, when placing a shrub symbol on the vegetation layer, the block assumes the vegetation layer's properties (color, linetype, lineweight, etc.).

The last rule is that the assumed block's base coordinates are (0,0) (its insertion point).

A Description Key Set scaling factor exaggerates the symbol's size based on a parameter in the point's code. See the following discussion on setting scale factors.

If the marker and its label need independent visibility; each needs its own layer. When using a Description Key Set, it assigns the point marker layer and the point style assigns the label's layer. With the Description Key Set assigning the marker's layer, the point style has only one layer (0 zero for the marker and a named layer for the label). This type of implementation defines the MAPLE and OAK point styles as those in Figure 2.55.

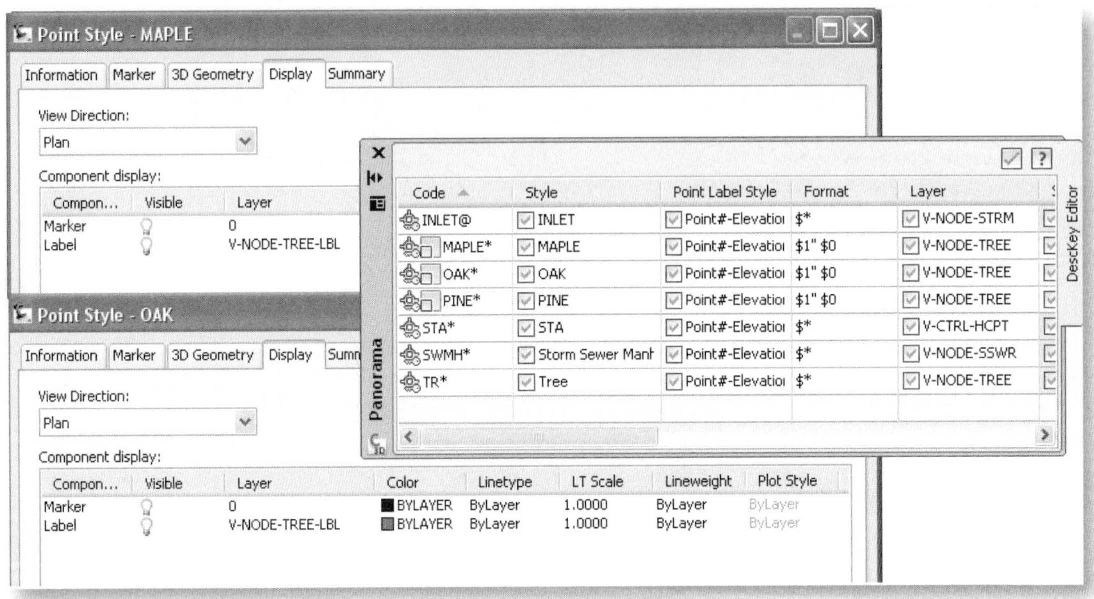


**FIGURE 2.55**

Using a Description Key Set sorts points onto layers, which produces point layer visibility control (for example, all centerline points on the layer V-NODE-RDCL, curb points on the layer V-NODE-CURB, and so forth). Implementing this strategy means that all utilities are on the E-UTIL layer or all survey control are on the Control layer.

If a code does not match any Description Key Set entry, the point is assigned the default point styles and layer.

When using Description Key Sets, remember these three rules:

- The Points Creation's Disable Description Keys option must be set to false.
- The Description Key Set must contain correct code entries.
- A match occurs only when a raw description exactly matches a code in the Description Key Set (including case sensitivity).

## Using Parameters

The Match On Description Parameter option default setting is true. This allows a raw description to contain values separated by spaces. These values can be numeric, for scaling and/or rotating a marker, or text that can be used in the full description or label.

As an example, the OK 6 raw description has two parts: "OK" and "6." The "OK" represents the tree type (oak) and the "6" is its trunk diameter (6 inches). As a result, OK is parameter 0 (zero) and the 6 is parameter 1. To use the code's parameters as a Format entry, reference them in the format's value cell by using \$0, \$1, …\$9. The format of \$1" OAK creates the full description of 6" OAK from the code of OK 6.

This format uses the "6," the second parameter of the raw description ($1), as the first part of the new full description.

## CREATING A DESCRIPTION KEY SET

Civil 3D installs a content template that contains a simple Description Key Set. The expectation is that users will develop their own key sets from scratch or migrate their LDT description keys. The Ribbon's Insert command Land Desktop creates Description Key Sets from existing LDT projects. If migrating an existing Description Key, users still need to edit the resulting key set, adding point and label styles and, if necessary, layer assignments and scaling factors.

Users can drag and drop a key set from one drawing to another to add, update, or overwrite the existing set.

In the Toolspace's Settings tab, in the Point branch, the "Description Key Set" heading is where users create or edit a Description Key Set. To create a new Set, select the heading, press the right mouse button, and, from the shortcut menu, select New.... After selecting New..., a Description Key Set dialog box displays (see Figure 2.56). Enter the Description Key Set's name in this dialog box and exit.

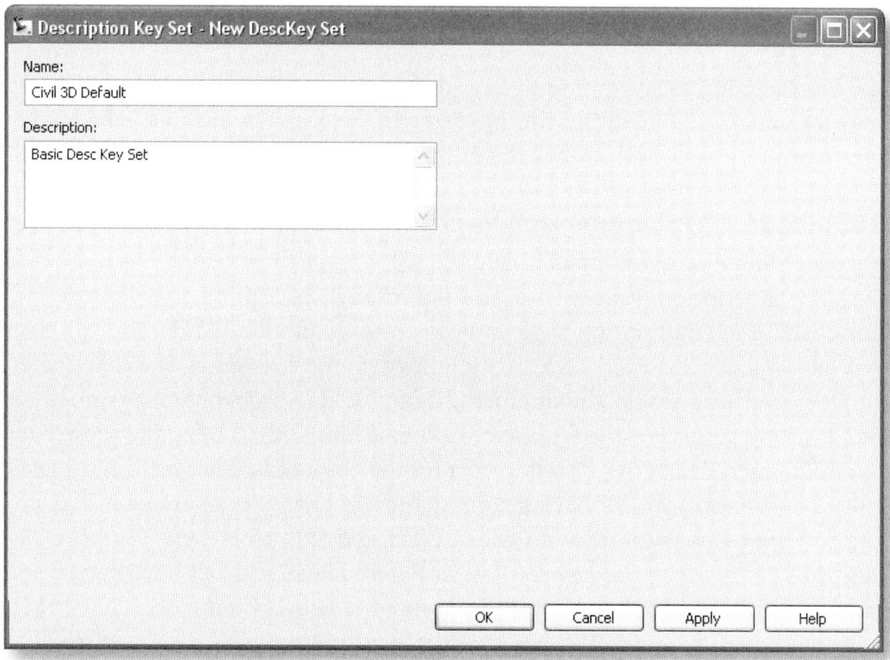


**FIGURE 2.56**

After creating a Description Key Set, expand the Description Key Set branch, displaying the set's list. From the list, select the set name, press the right mouse button, and, from the shortcut menu, select Edit Keys.... This displays a DescKey Editor vista that contains the current Description Key Set (see Figure 2.57). When clicking the first entry's cell, press the right mouse button, and, from the shortcut menu, select New.... This creates a new cell whose code is New DescKey. Change the code name and fill out its remaining information. Adding the next code is repeating the process.

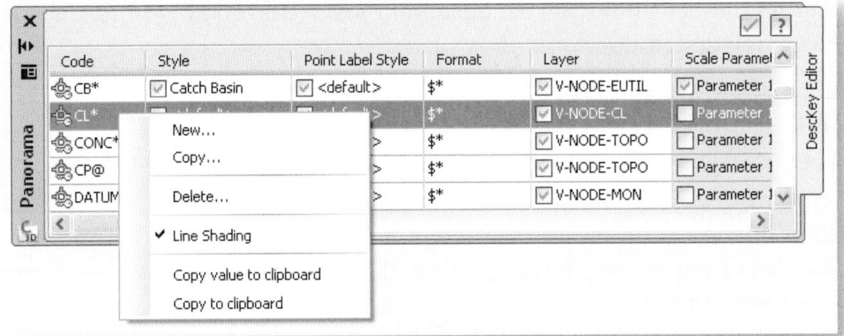


**FIGURE 2.57**

Each key set code has 13 entries. Of the 13 entries, five manipulate the raw description, and the remaining are parameters for scale and rotation. The raw description entries are Code, Point style, Label style, Format (Full Description), and Layer. A description key uses one or all of the entries. If a key only sorts points, the user needs only to identify the code, the point and label styles, and the layer name. If using parameters, the user has to set entries for Format, Scale, and Rotation areas.

## Entry 1: Code

When a routine compares the point's raw description to the Description Key Set Code entries and it matches, the routine assigns the listed point and label styles, does a translation (if set), and assigns the point object a layer (if set).

### Alphabetical and Alphanumeric Description Codes

The process of literal character matching does not work well when adding numerical values to a raw description. Examples of this type of raw description are MH1, MH2, OK 3, or OK 6. Each description has a common root, manholes (MH), and oak trees (OK). What changes for each code is the added numerical value or a parameter. A manhole coded as a single alphanumeric value, MH1, probably means manhole 1, 2, and 3, or in the case of the Oak tree, various trunk sizes. To handle these code types, Civil 3D uses an asterisk (*) wildcard after the common code root. The asterisk wildcard solves the problem of having to match every character or number in a raw description. Using this wildcard with codes, MH and OK creates the Code entries MH* and OK*. This tells the key set that only two characters need to match, and the remainder of the raw description can be ignored. The MH* entry matches MH1, MH2, etc., and the OK* entry matches the raw descriptions of OK 3 and OK 6.

Alphabetical codes matching is case-sensitive; that is, IP is different from Ip, iP, and ip.

In addition to the asterisk wildcard, a Description Key Set can contain the following wildcards: ? (number or character), # (single-digit only), @ (single alphabetical character only), . (any single non-alphanumerical character), [ ] (a list), ~ (logical not), and * (anything).

**NOTE** Wildcards cannot be used at the beginning of a code.

A wildcard in a code is a literal match. If a user defines a code, STA#, it matches STA0 through STA9 (which is STA with a single number following). If the range of station numbers is 1000 to 9999, the Code needs to be STA####. With the code set to STA####, the raw descriptions STA1...STA999 will not match. These values do not match because they do not have enough digits after STA to match the four expected digits in STA####. It is important to remember that when using any wildcard other than * (asterisk), matching is a literal per-character match.

Examples of this type of code are as follows:

- STA#—matches only STA0 through STA9.
- STA##—matches only STA01 through STA99.
- STA###—matches only STA001 through STA999.
- STA*—matches STA0 through STA99999999; also STAKE, STAR12, or anything with STA first.

A code may have multiple raw descriptions that translate to a single full description. For example, in the field, the crew differentiates between recovered iron pins (IPF) and iron rods (IRF). In the office and in the drawing, what is important is that it is iron found. Since IPF and IRF translate to the same full description and have a common beginning and ending (letters I and F), the key set code includes a list of possible character matches between I and F. In this example, the list of possible matching letters, P and R, are bracketed.

- I[PR]F—Matches IPF or IRF

The list of wildcards allows users to include and exclude letters or numbers. Excluding values from matches is done by using a tilde; ~ (not).

- CP[A ... M]—Control points with only the letters A through M matching.
- CP[~N ... Z]—Control points that match that do not include the letters N through Z.

If the code match uses an @ (alpha character only) the following raw descriptions would match CP@:

- CPA
- CPC
- CPM

**Numerical Codes.** Raw descriptions can be numerical. If you are using a numerical description, how would a code describe a 4-inch-diameter tree? The code references a single key set entry for that diameter or a diameter range. For example, the key 1231 indicates a 1-inch- to 2.99-inch-diameter deciduous tree, and 1232 indicates a 3-inch- to 5.99-inch-diameter deciduous tree. There has to be two codes, 1231 and 1232, to handle these raw descriptions.

### Entry 2: Point Style

This entry specifies the matching code's point style (see Figure 2.57). This is how Civil 3D assigns a matching code's symbol.

### Entry 3: Point Label Style

This entry sets the code's point label style.

### Entry 4: Format Full Description

The next code entry is the format or raw description's translation to a full description (see Figure 2.57). The translation's purpose is making the raw description more meaningful to office staff. There are three basic translation types.

The simplest translation from raw description to full description is no translation. When a raw description does not need translation, enter $* (a dollar sign followed by an asterisk) in the Format cell. The $* means use all of the parameters in the raw description as the full description. An example of this is the raw description of TOPO, the full description being the same.

Another translation method is completely replacing the raw description with text in the code's format cell. An example of this is that the raw description is IP, and the Format value is IRON FOUND. When matching the code IP, the point's full description is IRON FOUND.

The last method translates the raw description's parameters order and/or adds information to the full description. An example is translating the raw description OK 6 to a full description. For this translation, the Format value is $1" OAK. This example uses the raw description's second ($1) parameter to create a full description with the tree's diameter.

### Entry 5: Layer

A Code's fifth Description Key Set entry is a point object layer (see Figure 2.57). If you are specifying a layer, the point appears on this layer and the point style has layer 0 (zero) for its marker. If you assign a layer in a point style, you do not need to use this column, because the point style contains the necessary layer.

### Entry 6: Optional Scale Parameter

By default, the point style marker scale setting defines the marker size. The default scale is the drawing plot scale. This toggle adds a second scale factor to marker sizing. The entry identifies which code parameter is a secondary scale factor (see Figure 2.58).

### Entries 7–10: Applied Scales

These four entries scale markers. Depending on the type of symbol, users may want to use various combinations of these toggles and values. Be careful and understand the toggles' consequences when using more than one scaling factor. If all toggles are on, the symbol's scaling is the product of all the checked options.

All Civil 3D point styles have a scaling setting. The default value is Use drawing scale. This makes points the right size for the drawing's plotting scale. The Description Key Set also contains this setting and, if toggled on in the key set, also multiplies the marker or symbol by the scale of the drawing. This creates markers that are too large. Users should toggle off Use drawing scale in the Description Key Set.

The raw description of OK 6 represents a 6-inch-diameter oak tree. If there are other oak trees varying between 2 inches and 12 inches, the user may want differently sized markers to show various tree diameters. To make this happen, set three entries in this area. First, toggle on the Scale Parameter. Next, clicking in the cell to the right of the toggle displays a drop-list arrow. Select the arrow, to display a parameter list, and select the correct parameter (Parameter 1). The last setting is Apply to X-Y.

Toggle this on to scale the symbol in two dimensions only. If the symbols are too large or too small, users can adjust the point style's size in the point style's marker size area.

Another scenario for symbol scaling is scaling a symbol to a specific size. For example, instead of measuring the diameter of each tree, the survey measures the tree's drip line diameters. The desired result is symbols correctly representing the drip-line diameter. If the raw descriptions MP 20, MP 15, and MP 7 are each a maple tree's drip line diameter, the parameter is a scaling factor for sizing a block that represents the diameter.

The first requirement is having a block whose diameter is 1 foot (unit). When the symbol is scaled by parameter, the scaled symbol represents the measured tree drip line. The user would set the scale parameter to parameter 1 ($1) and toggle on Apply to X-Y.

The last scaling option is scaling vertically, Apply to Z. This toggle is useful when you want to display a symbol with a vertical exaggeration. Civil 3D uses multi-view blocks that display a plan view and a second three-dimensional view symbol. For example, this could be used in a survey for recording tree heights. The point's raw descriptions would be OK 15, OK 37, and OK 23. The user would toggle on the scale parameter and set the parameter to 1 ($1), and then would toggle on the Apply to Z. This scales the blocks vertically. The multi-view block would have to have a definition height of 1 to correctly show the different vertical heights.

### Entry 11: Optional Symbol Rotation

The rotation parameter changes a marker's rotation (see Figure 2.58). The default rotation is the USC's zero direction.

### Entries 12 and 13: Fixed Scale and Direction of Rotation

The fixed rotation sets a fixed marker angle.

The default rotation direction is counter clockwise. When using a rotation parameter, this setting defines the direction of the rotation.

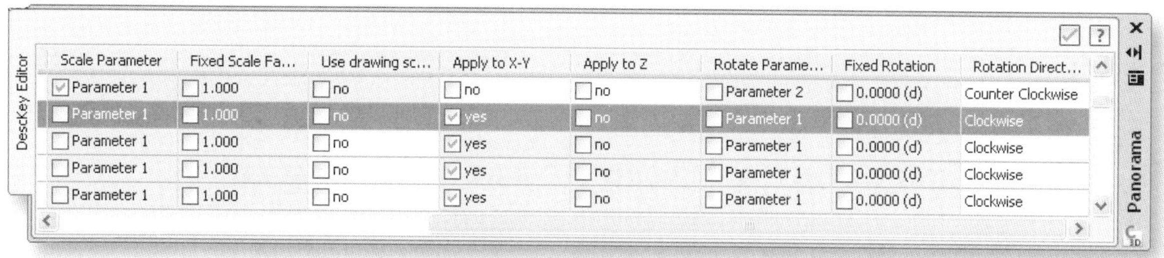

| Scale Parameter | Fixed Scale Fa... | Use drawing sc... | Apply to X-Y | Apply to Z | Rotate Parame... | Fixed Rotation | Rotation Direct... |
|---|---|---|---|---|---|---|---|
| Parameter 1 | 1.000 | no | no | no | Parameter 2 | 0.0000 (d) | Counter Clockwise |
| Parameter 1 | 1.000 | no | yes | no | Parameter 1 | 0.0000 (d) | Clockwise |
| Parameter 1 | 1.000 | no | yes | no | Parameter 1 | 0.0000 (d) | Clockwise |
| Parameter 1 | 1.000 | no | yes | no | Parameter 1 | 0.0000 (d) | Clockwise |
| Parameter 1 | 1.000 | no | yes | no | Parameter 1 | 0.0000 (d) | Clockwise |



**FIGURE 2.58**

## EXAMPLE DESCRIPTION KEY SET

The following is a sample Description Key Set listing:

**TABLE 2.5**

| Code | Point Style | Label Style | Description | Scale |
|---|---|---|---|---|
| BCP | Benchmark | <default> | BND POINT | |
| BLDG | <default> | <default> | $* | |
| BM### | <default> | <default> | $* | |
| CL* | <default> | <default> | $* | |
| DATUM | <default> | <default> | $* | |
| DL | <default> | <default> | DITCH | |
| DMH* | Storm drain | <default> | STORM DRAIN | |
| EOP | <default> | <default> | $* | |
| IN@ | Inlet | <default> | $* | |
| I[PR]F | Iron pin | <default> | IRON FOUND | |
| LC | Iron pin set | <default> | LOT CNR | |
| LP* | Light | <default> | LIGHT POLE | |
| MD | <default> | <default> | MOUND | |
| MH* | Manhole | <default> | MANHOLE | |
| MP* | Maple | <default> | $1" MAPLE | Parameter 1 |
| OK* | Oak | <default> | $1" OAK | Parameter 1 |
| POND | <default> | <default> | $* | |
| PP | Utility pole | <default> | POWER POLE | |
| RP | Survey | <default> | $* | |
| SMH* | Sanitary Manhole | <default> | SAN SEWER | |
| SWALE | <default> | <default> | $* | |
| TOPO | Standard | <default> | $* | |
| WV | WV | <default> | W VALVE | |

## IMPORTING A DESCRIPTION KEY SET FROM LDT

Civil 3D reads the Land Desktop project's description key file and creates a Description Key Set. The LDT description keys do not contain point and label style assignments; these must be set manually. When importing keys from LDT, the point and label styles are set to <default>. After you import the LDT description keys, they should be edited and assigned the correct styles, layers, and toggles.

## POINT GROUPS

Point groups organize points into meaningful groupings. A reason for grouping points is they have similar descriptions, functions, locations, or the user may want to isolate them for further processing. For example, a points group can represent surface data, lots, tree locations, or survey control, and thus, a point group may have different descriptions. Some Civil 3D commands, surfaces for example, require point groups to assign points as surface data.

Point groups are an alternative method of assigning point and label styles to members. This uses the group's point and label overrides, altering what a point displays.

## Point Group Properties

Commands for creating point groups are in the Ribbon's Points flyout menu or the Prospector's Point Groups heading's shortcut menu. There are four methods of selecting points for a point group: Existing Point Groups, Raw Description Matching, Include tab options, and the query builder (SQL statements). At times it is easier to select a large number quickly and then remove a few from the list. To remove points from a group, use the options in the Exclude panel.

### Information

The Information tab contains the point group's name, description, and modification history.

### Point Groups

A point group can be a combination of other point groups. For example, a point group All Breakline can be a combination of the point groups Headwall and Breaklines (see Figure 2.59).

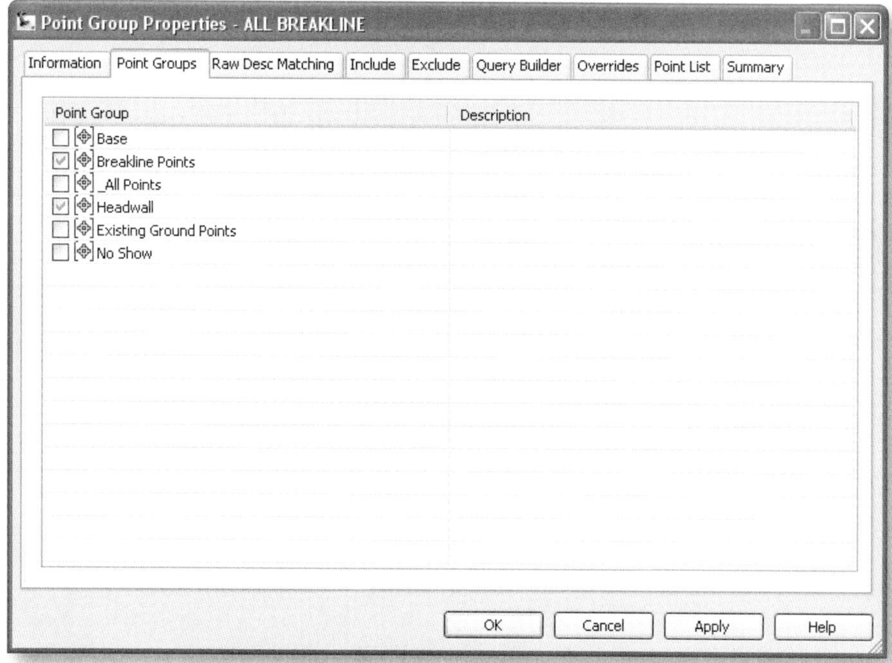


**FIGURE 2.59**

### Raw Description Matching

The Description Key Set(s) supplies this panel a list of raw descriptions. When toggling on a code, a matching point becomes a group's member (see Figure 2.60).

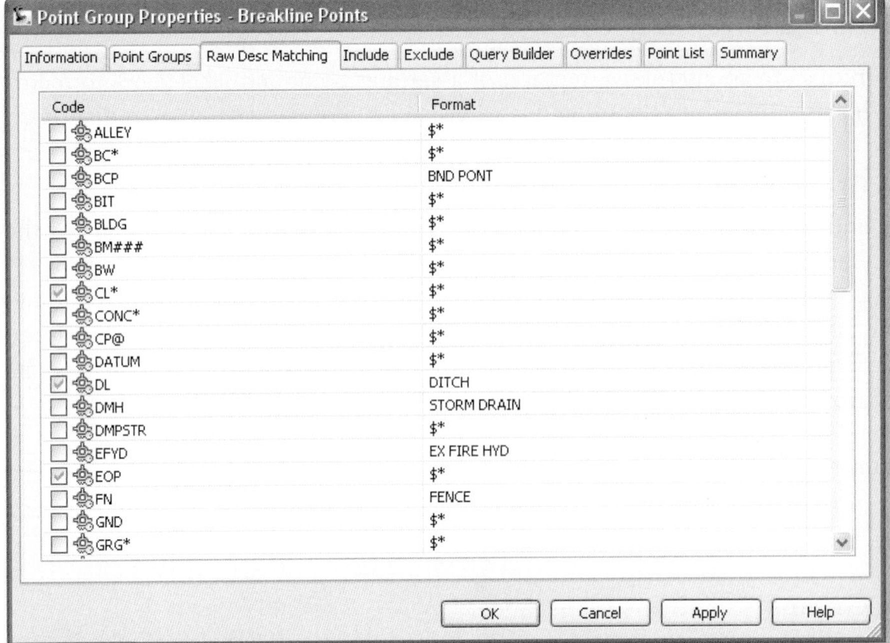


**FIGURE 2.60**

## Include Tab

Include selects points by point number, selection set, raw or full descriptions, point names, and/or elevations (see Figure 2.61). All entries can have multiple values separated by a comma. See the With raw descriptions matching entry in Figure 2.61.

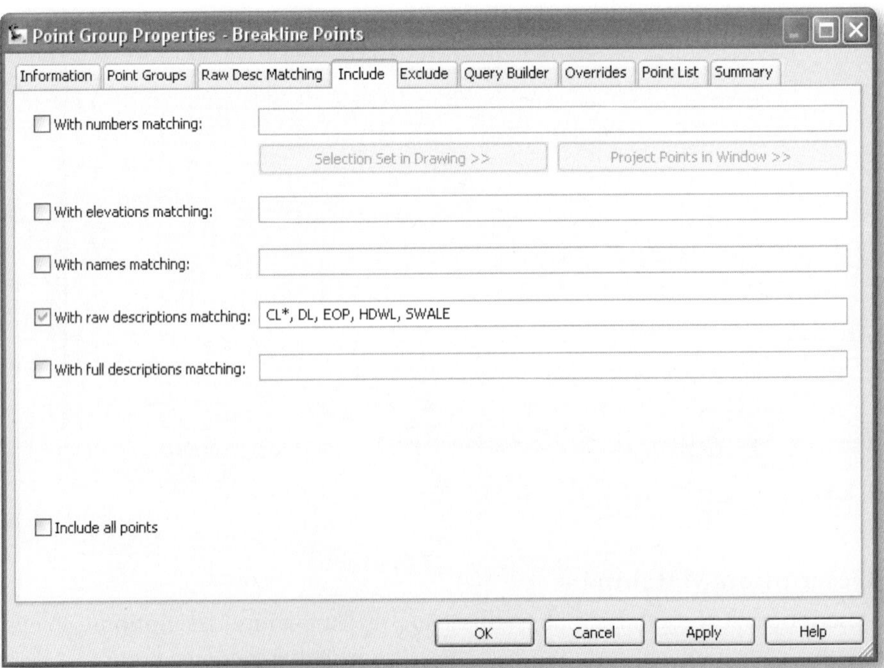


**FIGURE 2.61**

## Exclude

Exclude is exactly like the Include tab; however, it values remove points from the point group. Users can remove points by point number, selection set, raw or full descriptions, point names, and/or elevations (see Figure 2.62).

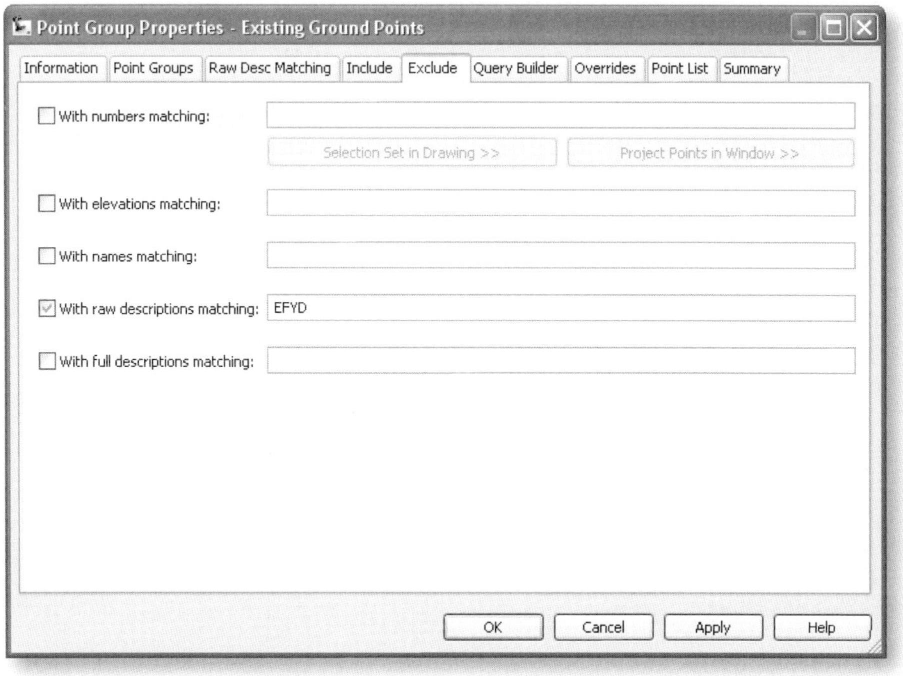

**FIGURE 2.62**

## Query Builder

Query Builder selects points using an SQL (Standard Query Language) query. A query selects points by point number, raw or full description, point names, elevations, and with combinations of logical AND or OR operators (see Figure 2.63).

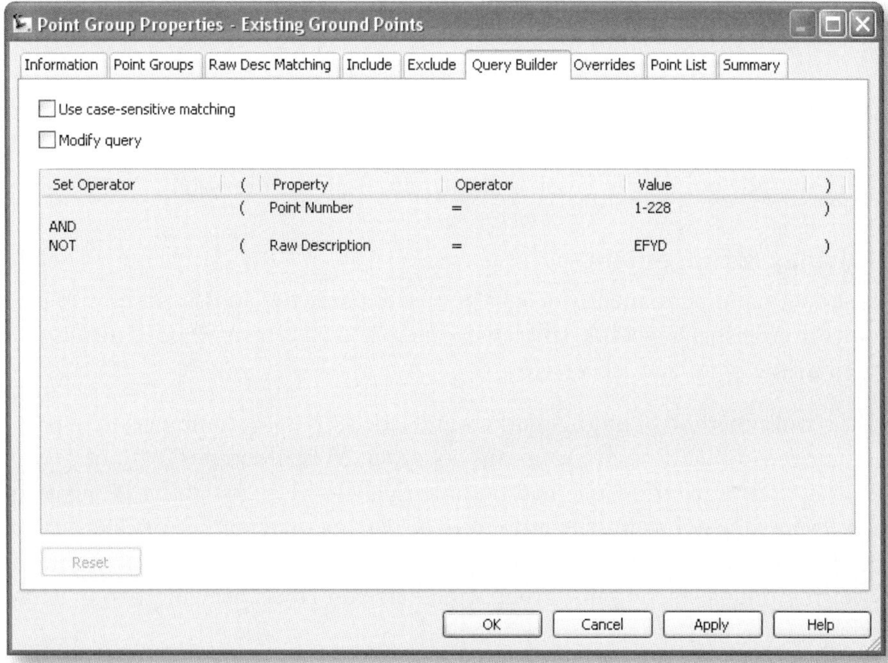


**FIGURE 2.63**

## Overrides

Overrides change the point group's raw description, point elevation, and point style and point label style (see Figure 2.64). When you are defining the overrides, they also appear in the Information tab. The raw description and point elevation overrides change the raw description and elevation for the group. The point style and label overrides assign new markers and labels to the group. You must toggle them on here to activate the override.

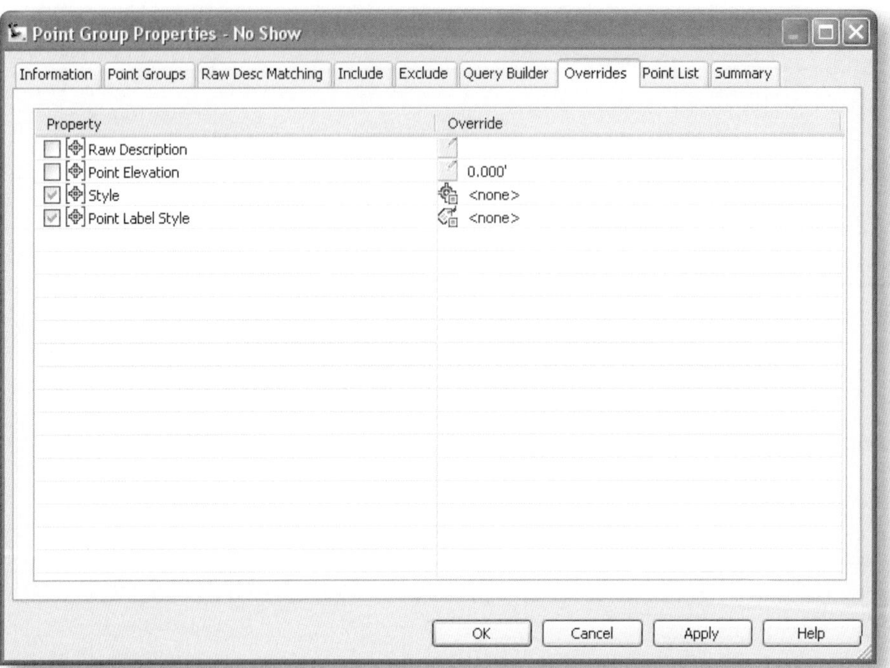


**FIGURE 2.64**

## Point List and Summary Tabs

The Point List tab displays the group's points. The Summary tab lists all of the settings that are defining the group.

### Deleting Point Groups

To delete a group, but not its points, use the group's shortcut menu's Delete command. Deleting the points in a group deletes them from the drawing.

### Point Display Management

The Erase command permanently deletes points. This creates an issue of how to prevent points from displaying while still having them active. One method is turning off layers used by the point and label styles.

There is a second method of hiding markers and labels. It uses a combination of point group style overrides and point group display order. When defining a point group, there is an option to override assigned point and label styles. Assigning None as an override prevents the points in the group from displaying markers and/or labels. Also, Point Groups has display order control. Overrides combined with display control allows viewing only a single point group's markers and labels, viewing two or more point groups, or hiding all markers and labels.

### Point Group Properties

With all of the points in a drawing, and without the ability to remove them, a new strategy for isolating points is a necessity. The point groups display property helps in isolating point groups and controlling their display order. Civil 3D draws the lowest point group first and continues up the list until reaching the top; the last drawn is the top point group. What appears is all of the points in all of the point groups.

If there is a point group using None as an override for point or label styles and it is at the display list's top, all points do not display. A None Point and Label point group override hides the marker and labels. The positioning of this group, relative to other point groups, allows displaying a single or combinations of point groups while hiding other point groups.

If a point belongs to only one point group, it is drawn once. If a point is a member of several point groups, the point's last point group (which is the highest on the list) controls how it displays. For example, the point groups Pine and Oak each display their own symbols. A third point group, Generic Tree, includes the point groups Pine and Oak and overrides the point style with a Generic Tree point style. Placing Generic Tree at the top of the display list displays all pine and oak trees as generic trees (see Figure 2.65). If you change the point group display order and place Pine above Generic Tree, the Pine symbols display and the oak trees display as generic trees (see Figure 2.66).

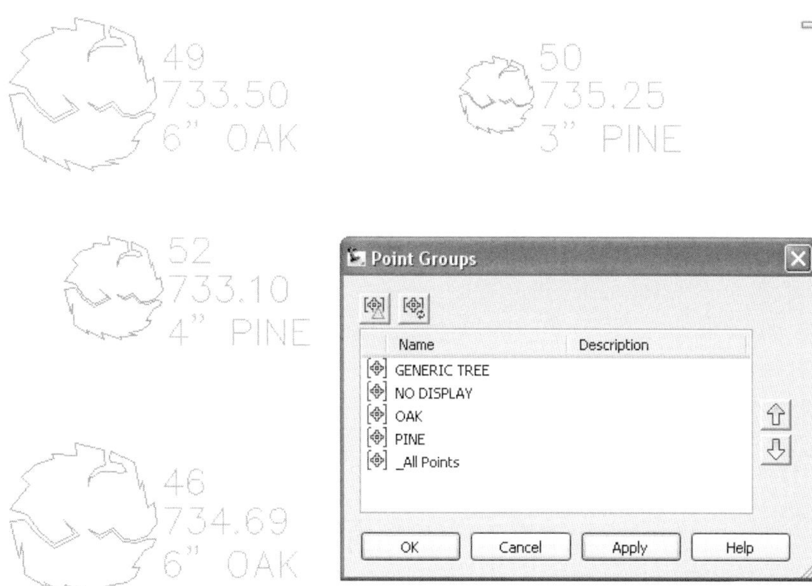


**FIGURE 2.65**

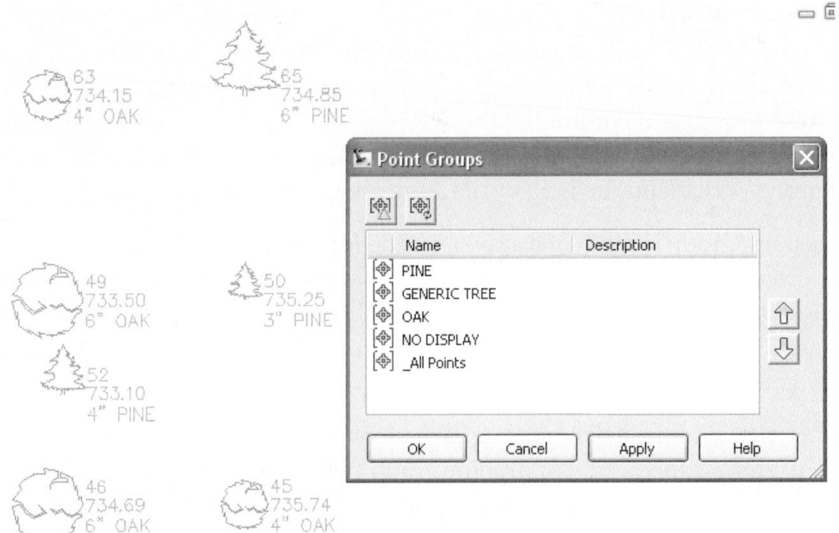


FIGURE 2.66

## POINT TABLE STYLES

Another annotation method is creating a point table. A point table can have all drawing points, a point selection set, or point group points. Point tables are simple to create. The table values are limited to point information: point numbers, point names, elevations, raw or full description, etc. (see Figure 2.67).

The Data Properties panel's plus sign (lower-right) adds new table columns. Double-clicking the column name displays the Text Component Editor, allowing the user to name the column. Double-clicking in the column's value cell displays the Text Component Editor, allowing the user to set the column's text format and property value (see Figure 2.68). The red X deletes a selected column or columns.

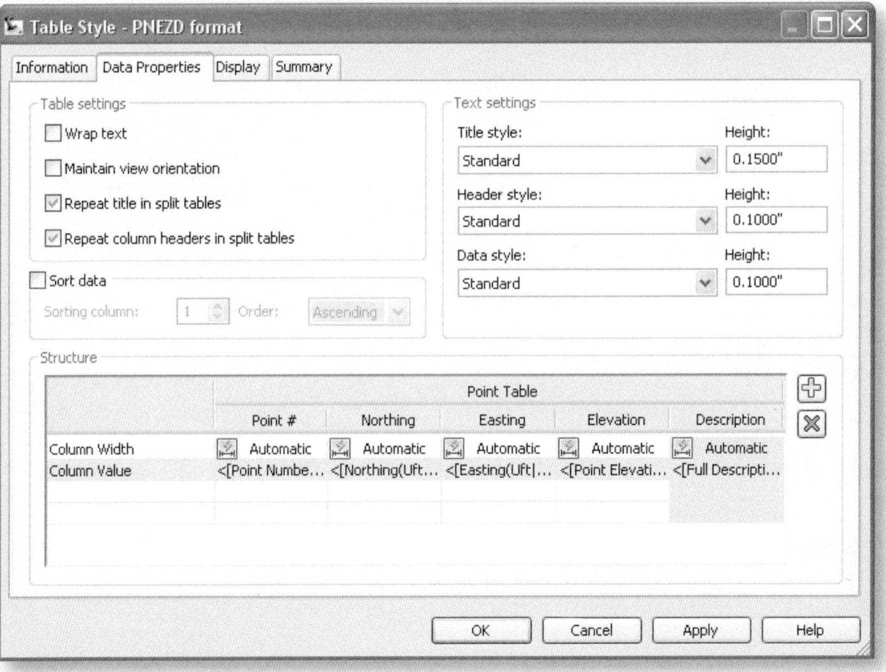


FIGURE 2.67

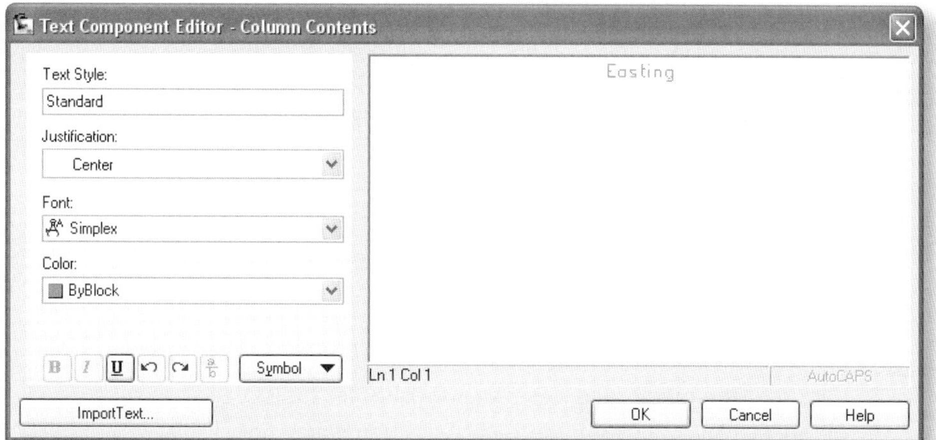


FIGURE 2.68

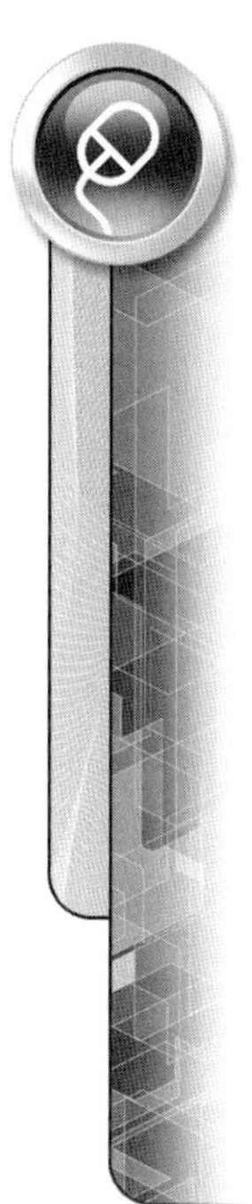

## EXERCISE 2-5

When you complete this exercise, you will:

- Import a Description Key Set.
- Add entries to a Description Key Set.
- Create point groups.
- Assign point group style overrides.
- Adjust the point group display properties.
- Create a point table style.
- Create a point table.

### Exercise Setup

This exercise uses a new drawing template file, Chapter 02 – Unit 5.dwt file. The file is on this textbook's CD, in the Chapter 2 folder.

1. At Civil 3D's top left, click the Civil 3D's icon drop-list arrow, from the Application Menu, select ***New***, browse to the textbook's CD, Chapter 02 folder, select the template file *Chapter 02 – Unit 5.dwt*, and click OPEN.
2. At Civil 3D's top left, click Civil 3D's icon drop-list arrow, from the Application menu, select SAVE AS, in the flyout select AutoCAD Drawing, and save it as *Points-2* in the Civil 3D Projects folder.

### Add Point Styles

There are three missing point styles: Inlet, Maple, and Oak. This exercise creates these point styles, defining a point label layer. The Marker component layer is 0 (zero) for all three point styles because Description Key Set entries assign the point style's layer (see Table 2.6).

**TABLE 2.6**

| Point Style | Marker | 3D Geometry | 2D & 3D Label Layer |
|---|---|---|---|
| Inlet | INLET | Flatten Point to Elevation | V-ESTM-LBL |
| Maple | Maple | Flatten Point to Elevation | V-VEG-LBL |
| Oak | Oak | Flatten Point to Elevation | V-VEG-LBL |

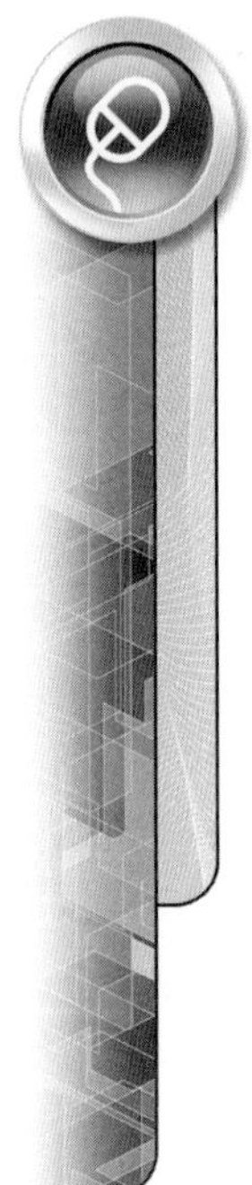

1. Click the ***Settings*** tab.
2. In Settings, expand the Point branch until you can view the Point Styles heading and its styles list.
3. Select the **Point Styles** heading, press the right mouse button, and, from the shortcut menu, select NEW....
4. Click the ***Information*** tab, and for the name, enter **Inlet.**
5. Click the ***Marker*** tab, making it current.
6. Click the **Use AutoCAD BLOCK symbol for marker**, scroll through the blocks, and select **Inlet.**
7. Click the ***3D Geometry*** tab and change the Point Display Mode to **Flatten Points to Elevation.**
8. Click the ***Display*** tab, and leave layer 0 (zero) for the Marker.
9. Click Label layer 0 (zero); the Layer Selection dialog box displays. Scroll down the list, select the layer **V-ESTM-LBL**, and click OK to return to the Point Style dialog box.
10. If necessary, for both the Marker and Label entries, change the Color and Linetype properties to **ByLayer.**
11. Click ***OK***, closing the dialog box.
12. Repeat Steps 3 through 11, making the point styles **Maple** and ***Oak*** by using the information in Table 2.6.
13. At Civil 3D's top left, Quick Access Toolbar, click the ***Save*** icon, saving the drawing.

## Importing a Description Key Set

Civil 3D imports a Land Desktop project's description keys, creating a Description Key Set. If you did not copy it from this textbook's CD Civil 3D project to the Civil 3D Projects folder, please do so now.

1. In Settings, click the **Point** heading, press the right mouse button, and, from the shortcut menu, select EDIT FEATURE SETTINGS....
2. Expand the Default Styles section, reviewing the default Point and Label style names.

Each Description Key Set entry will have these as their initial styles.

3. Click ***OK***, exiting the dialog box.

The Ribbon's LAND DESKTOP command imports the Description Key file from a LDT project.

4. On the Ribbon click the Insert tab. In the Import panel select LAND DESKTOP.
5. Set the Land Desktop Project Path to *Civil 3D Projects*, set the Project Name to *Civil 3D*, and in the data list, except for Description Keys, toggle them **OFF** (see Figure 2.69).

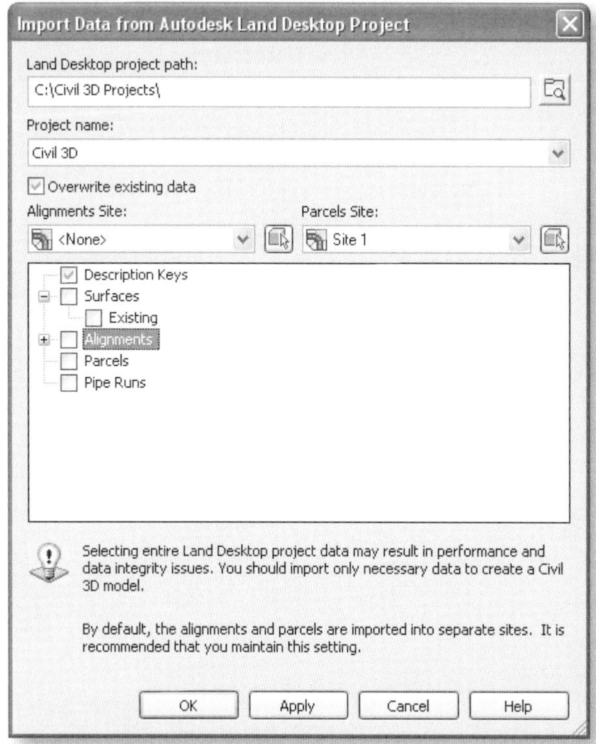


**FIGURE 2.69**

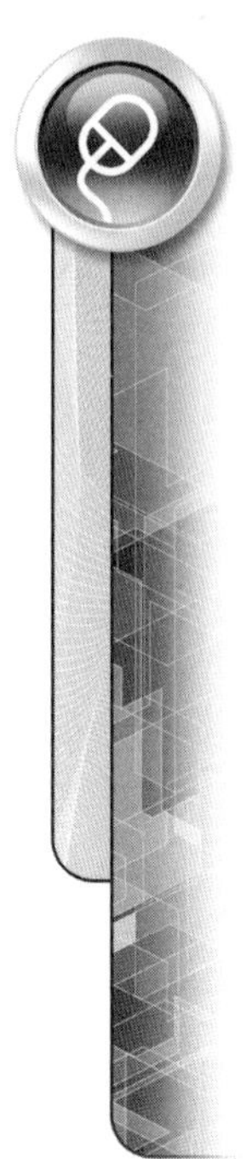

6. Click ***OK***, migrating the description keys.
7. The Description Keys Migration Completed dialog box displays. Click ***OK***, exiting the dialog box and completing the description key migration.
8. At Civil 3D's top left, Quick Access Toolbar, click the ***Save*** icon, saving the drawing.

## Review Imported Description Keys

1. In Settings, expand the Point branch until you can view the Description Key Sets list.

The LAND DESKTOP command imported three description key files: **Civil 3D**, **DEFAULT**, and **HNC3D-DEFAULT**.

2. In Settings, in the Point branch, select the **Civil 3D** description key set, press the right mouse button, and, from the shortcut menu, select DELETE.... The "Are you sure?" dialog box displays, and click ***YES*** to delete the description key set.
3. Repeat step 2 and delete the **Default** description key set.

## Change Codes Point Styles

When importing a Land Desktop Description Key file, each code has the default point style. However, several codes need a point styles with a symbol.

Changing the point style is a two-step process. First, toggle on point style and then select the new point style from a list of styles. When you click in the point style name cell, a Select Point Style dialog box displays with a drop-list of available point styles.

You need to change the current point styles for the codes in Table 2.7.

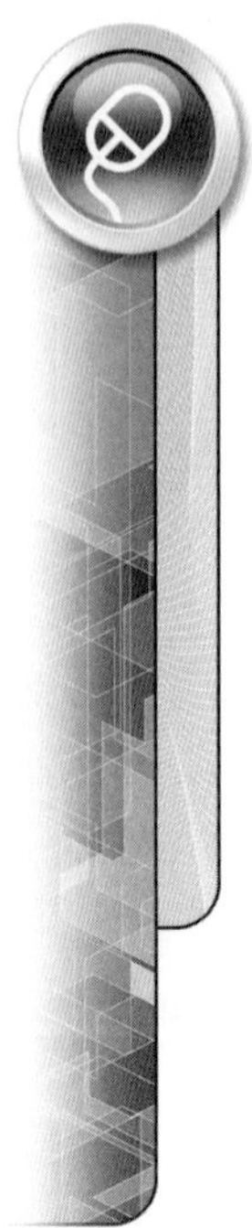

**TABLE 2.7**

| Code | Point Style |
|---|---|
| BM### | Benchmark |
| BP | Iron Pin |
| CB* | Catch Basin |
| CP@ | Benchmark |
| DATUM | STA |
| DMH | Storm Sewer Manhole |
| EFYD | Hydrant (existing) |
| I[PR]F | Iron Pin |
| LC | Iron Pin |
| LP* | Light Pole |
| MH* | Manhole |
| PP* | Utility Pole |
| SIGN* | Sign (single pole) |
| SMH | Sanitary Sewer Manhole |
| STA# | STA |
| TF-SIG | Traffic Signal |
| TR* | Tree |
| VV | Valve Vault |
| WELL | Well |
| WV | Water Valve |

1. In Settings, in the Description Key Set list, select **HNC3D-DEFAULT**, press the right mouse button, and, from the shortcut menu, select EDIT KEYS....

2. In the DescKey Editor vista, for each entry, change the Use Drawing Scale toggle to **NO**. Right-click the column heading, select EDIT..., and a drop-list below the heading will appear, allowing you to select **NO** from the list.
3. In the DescKey Editor vista, in the Point Style column for the BM### entry, indicate the intent to override the default point style by togging it **ON**.
4. In the DescKey Editor vista, in the Point Style column for the **BM###** entry, click <default>, displaying the Point Style dialog box. Click the drop-list arrow, from the list select **Benchmark**, and click ***OK***, selecting the style and closing the dialog box.

The BM### code has the Benchmark point style.

5. Repeat Steps 3 and 4, assigning the remaining keys in Table 2.7 to the appropriate point styles.

## Create New Description Key Set Entries

The Description Key Set has three missing entries: Inlet, Maple, and Oak. The next exercise section creates these keys.

The new keys use related point styles (Inlet, Maple, and Oak). The Maple and Oak codes (MP 2 and OK 6) have a parameter (the tree's trunk diameter). The format entry for these description keys uses the $1 parameter and is the full description's first item. The MP 2's resulting full description is 2" MAPLE and OK 6's is 6" OAK. The format coding is in Table 2.8.

The first parameter is also a symbol multiplier. This displays a 6-inch oak tree as a larger symbol than an oak tree that is only 2 inches in diameter.

The three keys' scaling settings are in Table 2.9. All three keys use the drawing scale to make them the correct size for the drawing. The Maple and Oak entries use Parameter 1 as a secondary scaling factor, which is listed in Table 2.9. This entry makes an MP 4 code's symbol larger than an MP 2's symbol.

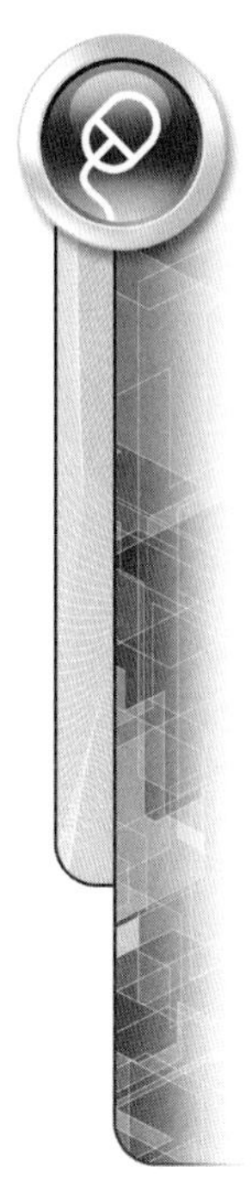

**TABLE 2.8**

| Code | Point Style | Format | Layer |
|---|---|---|---|
| IN@ | Inlet | $* | V-ESTM-PNT |
| MP* | Maple | $1" MAPLE | V-VEG-PNT |
| OK* | Oak | $1" OAK | V-VEG-PNT |

**TABLE 2.9**

| Code | Scale Parameter1 | Drawing Scale | Apply to X-Y |
|---|---|---|---|
| IN@ | ON | OFF | OFF |
| MP* | ON | OFF | ON |
| OK* | ON | OFF | ON |

## Inlet Key

When creating a new Description Key Set entry, you have to select an existing code entry. You should review and adjust all key settings before going on to the next key.

1. In the DescKey Editor vista, select the entry **MH***, press the right mouse button, and, from the shortcut menu, select `NEW...`.

This makes a new key entry, New DescKey, is at the bottom of the list.

2. Locate the New DescKey entry, click in the Code cell, and for the new code, enter **IN@**.

The list resorts and you must located IN@ in the sorted code list.

3. Click the Point Style override toggle to **ON** and click in the point style cell, displaying the Point Style dialog box. Click the drop-list arrow, from the list select **Inlet**, and click ***OK***, returning to the DescKey Editor vista.
4. Make sure the Format cell value is **$***.
5. Click the Layer override toggle, turning it **ON,** and click in the cell, displaying the Layer Selection dialog box. From its list, select the layer of **V-ESTM-PNT**, and click ***OK***, returning to the DescKey Editor Vista.
6. If the Scale Parameter is on, toggle it **OFF**.
7. If the Use Drawing Scale is on, toggle it **OFF**.

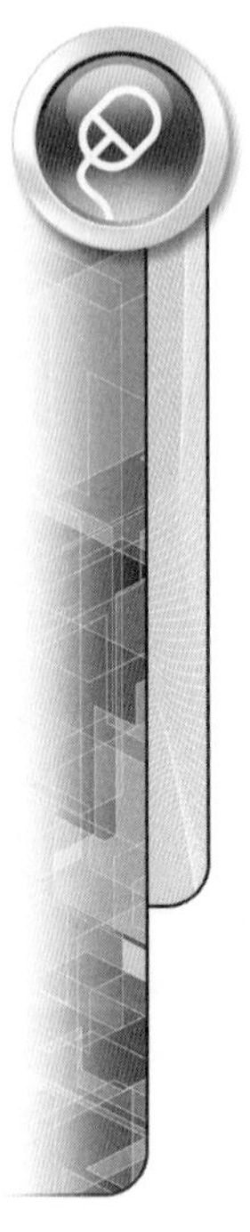

## Maple Key

1. While still in the DescKey Editor, select the entry **ALLEY**, press the right mouse button, and, from the shortcut menu, select NEW....

This makes a new key entry, New DescKey, and sorts it to the bottom of the code list.

2. Locate the New DescKey entry, click in its Code cell and, as the new code, enter **MP***.

The code is again sorted and you must look for the MP* code entry.

3. Click the Point Style override toggle to turn it **ON** and, as in step 8 of inlet (above), select the style **Maple**.
4. Change the Format to **$1" MAPLE**.
5. Click the Layer override toggle to turn it **ON** and, as in step 10 of inlet (above), select the layer **V-VEG-PNT**.
6. Make sure the Scale Parameter is **ON** and is set to **Parameter 1**.
7. If the Use Drawing Scale is on, toggle it **OFF**.
8. Toggle **ON Apply to X-Y**.

## Oak Key

1. Repeat the Maple steps 1 through 8, creating an entry for the **OK*** code, but use the **Oak** point style and for the format use **$1" OAK**.
2. After creating the new codes, close the DescKey Editor vista by clicking the **X** at the top of Panorama's mast.
3. At Civil 3D's top left, Quick Access Toolbar, click the ***Save*** icon to save the drawing.

## Importing Points

Everything is set to import points using the current Description Key Set. The file Chapter 02 - Unit 4.nez is in this textbook's CD, Chapter 2 folder.

1. In the Ribbon click the Insert tab. On the Import panel click POINTS FROM FILE to display the Import points dialog box.
2. In the Import Points dialog box, at its top, click the drop-list arrow and, from the format list, select **PNEZD (comma delimited)**.
3. To select the Source File(s), on the dialog box's right side, click the plus (+) icon.
4. In the Select Source File dialog box, at the bottom, to the right of Files of Type, click the drop-list arrow, and, from the list of file extensions, select ***.nez**.
5. Navigate to the location of the Base-Points.nez file (C:\Civil 3D Projects\Civil 3D), select the file, and click ***Open***, returning to the Import Points dialog box.
6. In the Import Points dialog box, do not set any Advanced Options, and click ***OK***, importing the points.
7. Click the ***prospector*** tab, making it current.
8. In Prospector, expand Point Groups, select **_All Points**, press the right mouse button, and, from the shortcut menu, select PROPERTIES....
9. In _All Points Properties, change the Label Style to **Point #-Elevation-Description** and click ***OK*** to exit.
10. At the drawing's bottom right, the Status Bar, set the Annotation Scale to 1"=30'.
11. In Prospector, select the Points heading, listing the points in preview.

12. In preview, scroll the point list to the right until you can view the Raw Description heading, and click the heading, sorting the list.
13. In preview, scroll through the list until you can view a **WELL** description, and then scroll the listing back until you can view its point number.
14. In preview, select the point number's point icon, press the right mouse button, and, from the shortcut menu, select ZOOM TO.

You should see the Well marker style in the drawing specified by the Description Key Set.

15. At Civil 3D's top left, Quick Access Toolbar, click the ***Save*** icon, saving the drawing.

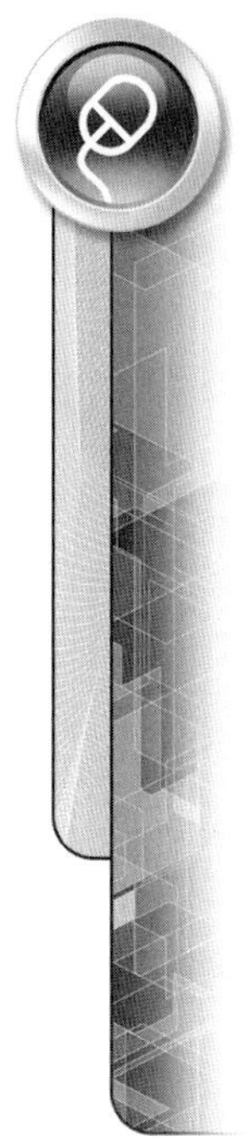

You can zoom to any point from the preview area (Points or Point Group) and the Edit Points vista.

## Creating Point Groups

The easiest way to create a point group is by using raw descriptions. The Description Key Set's raw description entries create the point groups' selection list.

Use Table 2.10's description keys list to define three point groups.

**TABLE 2.10**

| Name | Key(s) |
|---|---|
| Well | WELL |
| Iron | BP, I[PR]F, LC, SIP |
| Site Control | BM###, CP@, STA# |

1. If necessary, click the ***Prospector*** tab, making it current.
2. In Prospector, click the **Point Groups** heading, press the right mouse button, and, from the shortcut menu, select NEW....
3. Select the Information tab and as the name of the point group, enter **Well**.
4. Select the Raw Desc Matching tab, scroll through the list, and toggle **ON** the description **WELL**.
5. To view the selected points, click the ***Point List*** tab.
6. Click ***OK***, exiting the Point Group Properties dialog box.
7. Repeat Steps 2 through 6 and, using key entries from Table 2.10, create the Iron and Site Control groups.

## No Point or Label Point Group — Point and Label Style Overrides

A point group that includes all points, with its point and label styles overridden, set to None, and that is at the point group display's list top, hides all points.

1. If necessary, click ***Prospector's*** tab, making it current.
2. In Prospector, select the **Point Groups** heading, press the right mouse button, and, from the shortcut menu, select NEW....
3. Click the ***Information*** tab and for the name, enter **No Point or Label**.
4. In the Information panel, for the point and label styles, assign <**none**>.
5. Select the ***Include*** tab, and at the panel's bottom-left toggle **ON** include all points.

6. Select the ***Overrides*** tab, toggle **ON** the Style and Point Label Style overrides, and set their Override to <none>
7. Click ***OK***, creating the point group.

All points disappear.

8. In Prospector, select the ***Point Groups*** heading, press the right mouse button, and, from the shortcut menu, select PROPERTIES....
9. In the Point Group Properties dialog box, select the point group **Well**, and move it to the list's top. If necessary, select the point group **No Point or Label**, move it to the second position, and click ***OK***. You will need to REGEN (RE) the drawing.

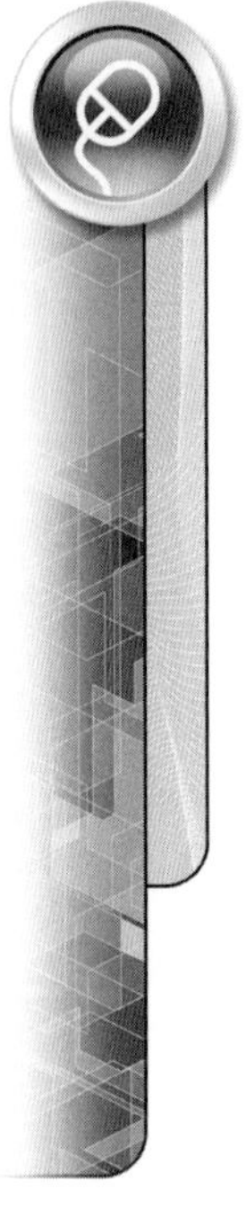

Only the Well points display on the screen.

10. In Prospector, select the **Point Groups** heading, press the right mouse button, and, from the shortcut menu, select PROPERTIES....
11. Finally, in the Point Groups Properties dialog box, select the point group **_All Points**, move it to the list's top, and click ***OK***.
12. At Civil 3D's top left, Quick Access Toolbar, click the ***Save*** icon, saving the drawing.

## Defining a Point Table Style

Many times a site includes a benchmarks and/or survey control points list. A convenient method of documenting them is creating a table. In this exercise you identify the point group Well and create a table containing their coordinates.

The current drawing does not have a table style for the well data.

1. Click the ***Settings*** tab.
2. In Settings, expand the Point branch until you can view the Table Styles' styles list.
3. From the list select the style **PNEZD format**, press ENTER, and from the shortcut menu select EDIT....
4. In the style dialog box, select the tab ***Data Properties***.

This dialog box's panel defines the sorting order, text styles and heights, data options, and the table's structure (see Figure 2.67).

5. Select the Northing heading and drag it to the second column.
6. Select the Easting heading and drag it to the third column.
7. In the Description column, double-click in the Column Value cell displaying the Text Component Editor.
8. On the right side of the dialog box, select Raw Description and delete the entry.
9. In the Text Component Editor's top left, click the Properties drop-list arrow, select **Full Description**, click the arrow (in the top center of the dialog box), and click ***OK***, returning to the Table Style dialog box and setting the cell's value.

Your dialog box should look like Figure 2.67.

10. Click ***OK***, creating the table style.

## Creating a Point Table

1. Click the ***Prospector*** tab, making it current.
2. In Prospector, select the **Point Groups** heading, press the right mouse button, and, from the shortcut menu, select PROPERTIES....

3. In the Point Groups Properties dialog box, select the point group **Well**, move it to the list's top, select and move to the second position the **No Point or Label** group, and click ***OK***, dismissing the Point Groups Properties dialog box.
4. Use the `PAN` and `ZOOM` commands to display a clear area to the east of the site.
5. From the Ribbon, click the Annotate tab. On the Labels & tables panel's left, click the Add Tables drop-list arrow, and from the shortcut menu select `ADD POINT TABLE`.

The Point Table Creation dialog box displays (see Figure 2.70).

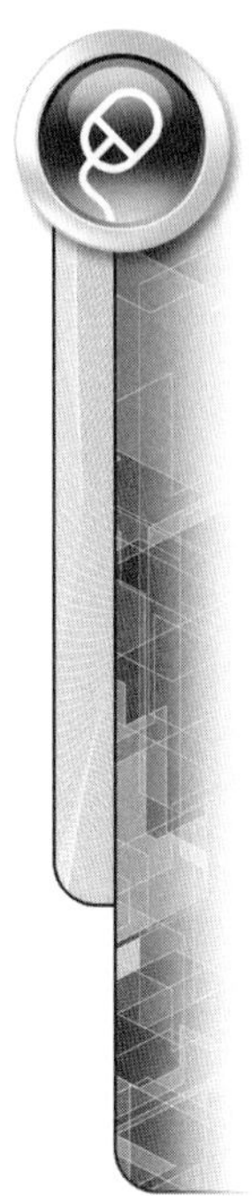

6. If necessary, at the Point Table Creation dialog box's top, change the Table Style to **PNEZD format**.

The dialog box's middle selects the table's points. You select points by Point Label Style Name, Point Groups, or by selecting points from those showing on the screen. The points must be showing on the screen to be a part of a table.

7. At the center-left of the dialog box, click the ***Select Point Groups*** icon.

A list of point groups displays.

8. In the Point Groups dialog box, select the point group **Well** and click OK, returning to the Point Table Creation dialog box.

The dialog box's bottom half defines how many rows are in one table column and how to stack a split table.

9. The current values are correct, including the relationship between the points and their table entry, i.e., Dynamic.
10. Click ***OK***, closing the dialog box.
11. In the drawing, select a point, locating the table.
12. At Civil 3D's top left, Quick Access Toolbar, click the ***Save*** icon, saving the drawing.
13. Use the `ZOOM` command to view the table better.

If you are editing the point data, the table updates to the new values.

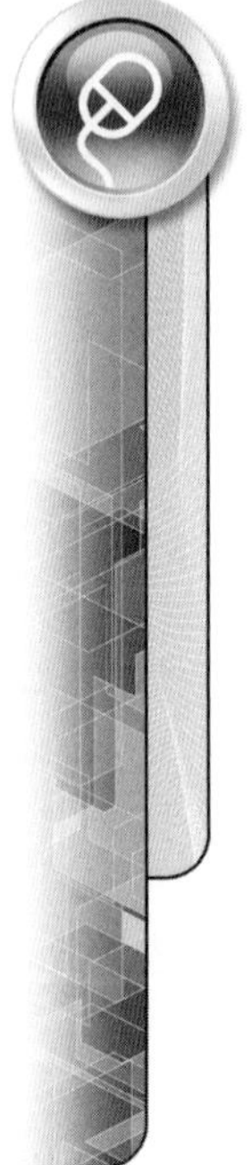

**FIGURE 2.70**

## Dynamic Table Entries

1. In Prospector, click the **Points** heading, previewing the points.
2. In preview, click the Point Number heading to sort the list.
3. In preview, scroll through the points list until you locate point **1473**.
4. Click in the Point Number cell for **1473** and change it to **3050**.

A number of reactions occur. The Well point group goes out of date, displaying an out-of-date icon.

5. In Prospector, expand Point Groups, select the entry Well, press the right mouse button, and, from the shortcut menu, select SHOW CHANGES....

The Point Group Changes dialog box displays that point number 3050 is a new member, while 1473 is to be removed.

6. In the dialog box's upper-left, click ***UPDATE POINT GROUP***, updating the point group and click Close, exiting the dialog box.

The Well group updates, the out-of-date icon disappears, and the new point number (3050) appears in the table.

In Prospector, a reference (orange triangle) icon appears to the left of the point group's name Well. This icon indicates that another object references the point group. Because there is a reference to the group, Civil 3D will not display in the shortcut menu for this group the Delete command.

7. In Prospector, click the point group Well, press the right mouse button, and review the shortcut menu commands; it does not include DELETE....

8. At Civil 3D's top left, Quick Access Toolbar, click the ***Save*** icon, saving the drawing.
9. Exit Civil 3D.

This concludes the point discussion. There are many more point commands in Civil 3D that were not discussed in this chapter. The point commands used for creating points from surface, alignment, profile, and corridor data will be discussed in later chapters.

## SUMMARY

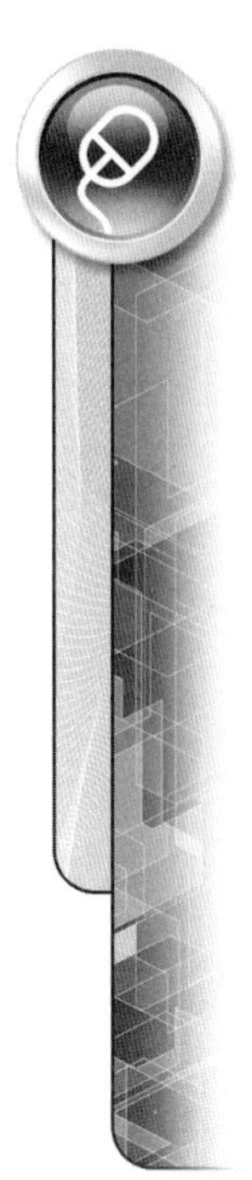

- A Description Key Set assigns a point and label Style and a layer to the point's marker.
- A Description Key Set can translate a point's field code (raw description) to a full description.
- With a Description Key Set assigning a marker layer, the point style should only define a label layer. The marker layer should be 0 (zero).
- An AutoCAD ERASE permanently deletes a point.
- A No Point or Label point group stops points from displaying.
- Point groups properties control what point groups display and in what combination.
- To delete a point group, select it, press the right mouse button, and from the shortcut menu, select Delete... .
- A point table selects points by point label style name, by point groups, or by a selection set.
- A point table is dynamic or static.

The next chapter examines parcel design, annotation, and management tools. Parcels start with an outer boundary, a site. A site is a container for a number of Civil 3D object types, parcels, alignments, profiles, corridors, and sections, and manages their relationships.

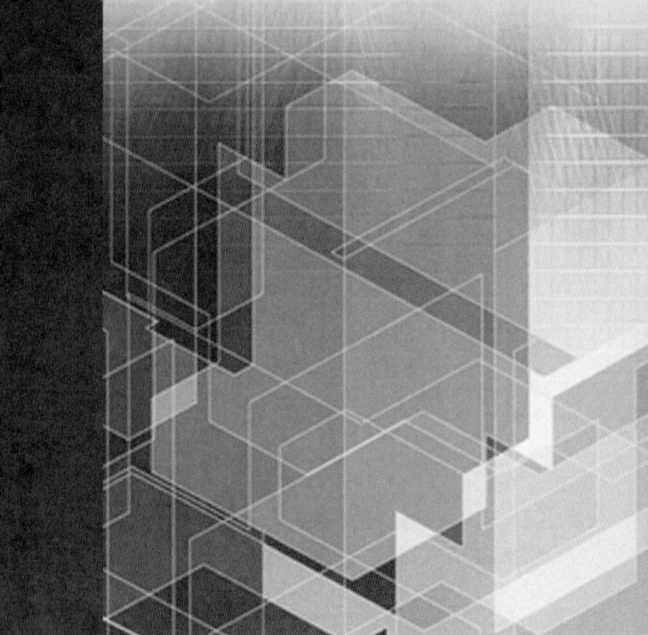

# CHAPTER 3

# Site and Parcels

## INTRODUCTION

Dividing a boundary into specific land-use blocks or resalable parcels is an exercise in practical methods and artistic flair. Parcel's commands provide the tools to subdivide a site. A site boundary is the outermost polygon containing interior homogenous land-use blocks. Subdividing these blocks into small parcels is this chapter's focus.

## OBJECTIVES

This chapter focuses on the following topics:

- Organizing Parcels from the Site to the Individual Parcels
- Defining Parcel Object Styles
- Defining Parcel Area, Line, and Curve Label Styles
- Developing Parcels with the CreateParcelFromObjects and CreateParcel-ByLayout Commands
- Parcel-Sizing Tools in The Parcel Layout Tools Toolbar
- Editing and Renaming Parcel Numbers
- Creating and Modifying Parcel Segments Labels
- Creating Parcel Tables

### OVERVIEW

Civil 3D introduces the concept of an overall site that is carved into smaller and smaller parcels. Both a site and its parcels are custom objects that maintain dynamic relationships to their boundaries and roadway alignments. An example of such a dynamic relationship is creating two parcels from a single site with an alignment. When drawing an alignment through the site, the site will divide into two parcels with the alignment acting as the parcel dividing line. If the dividing alignment is deleted, the parcels merge and return to the original site. Users can also divide a site or parcel into new parcels by using individual parcel segments. Users can also draft parcel lines using the Parcel Layout Tools toolbar; by using line, arc, survey figures, and polyline commands; or by using the line and polyline commands with Civil 3D transparent commands (bearing/distance, direction/distance, etc.). When using lines, arcs, and polylines, the user must define them as parcel segments using the Parcel's Create Parcel From Objects command.

A site is the "container" for parcel objects. Alignments are included in a site because they, too, divide a site and/or parcels into smaller units. There can be more than one site in a drawing, and each site can have its own set of parcels. All members of the same site interact, while those in different sites do not. The interaction rules allow having different subdivision designs within the same drawing, but each subdivision is in its own site.

As mentioned in previous chapters, Civil 3D operates on styles. These styles place objects, their graphical representations (line work), and labeling on predefined layers or within the objects. It is imperative that users define as many styles as possible before using Civil 3D. Each Civil 3D style requires changing a multitude of settings before producing a specific "look" for an office. In the Land Desktop product line, the "office" look was primarily a drafting exercise. However, in Civil 3D, the "look" is produced by a blend of styles, text styles, and basic company blocks (borders, symbols, etc.).

The process of dividing a single parcel into smaller units is an office or personal preference. Some parceling methods use circles, others use frontage measurements, others use rectangular templates, and so on. The parceling methodology implementer believes his or her process produces the best parcel plan. In truth, the rules and regulations governing the site's development are the greatest controls over developing a parcel plan. Each site has rules or covenants that affect the individual parcel's size and shape.

### Unit 1

Civil 3D has settings that affect site and parcel creation and annotation. The unit's focus includes parcel styles and basic drafting tools (lines, arcs, and polylines). The parcel's display and its annotation is a result of currently assigned parcel styles.

### Unit 2

A site and its parcels are drafted and defined by converting existing line work or by using Parcel Layout tools, which are the focus of this unit. The Create Parcel from Objects command converts line work into parcels or parcel segments. The Create Parcel from Objects command requires preexisting lines, arcs, and/or polylines to create the site, parcels, or parcel segments. The Transparent Commands modify how to draft lines—by bearing and distance, azimuth, deflection, or from cogo points. The transparent commands have their own toolbar and work with the line, arc, and polyline commands. When subdividing with the Create Parcel Layout toolbar, the toolbar commands work only with parcel segments.

### Unit 3

This chapter's third unit focuses on evaluating parcels. The parcel-analysis tools are in the Parcel Properties dialog box and Reports Manager. The Parcel Properties dialog box reports the map check and inverse properties for the selected parcel. Reports Manager supplies other necessary reports (metes and bounds, surveyor's certificates, parcel areas, etc.).

### Unit 4

The fourth unit's focus is editing sites and parcels. The Parcel Layout Tools toolbar provides several tools to modify a site and its parcels. Users can modify a site and its parcel segments graphically by grip editing their segments and by erasing individual parcel segments.

### Unit 5

This chapter's fifth unit reviews site and parcel annotation. There are two types of labeling: one for a site's or a parcel's overall values (area, perimeter, etc.) and the other for the site's or parcel's lines and curves (parcel segments). The line and curve labels are either on the parcel's lines and curves or are table entries. All parcel annotation reacts and changes to correctly display newly edited values. All parcel labels are view dependent and will rotate to read correctly in any view orientation. All parcel labels are scale dependent and resize when the viewport plotting scale changes.

## UNIT 1: SITES, PARCELS, AND STYLES

A site and its parcels are defined by one of two methods: from existing objects or by drafting the site and its parcels with parcel-sizing tools. The Create Parcel from Objects command requires preexisting objects (lines, curves, and polylines) to create a site and/or its parcels. Civil 3D has transparent command modifiers that let users draft lines by turned angle, bearing, azimuth, deflection, and distance, or with cogo points. A user can start these commands by selecting a command's icon from the Transparent Commands toolbar. Transparent commands can be used when creating curves; however, their application is limited. The Create Parcel by Layout command presents a toolbar containing parcel segment and curve tools (limited), as well as other parcel-sizing commands. The Parcel Layout Tools toolbar's commands work only with site and parcel segments.

The site and its parcels' display (layer and other object properties) and annotation (label and segments) are a result of assigned parcel styles. Parcel styles define a parcel type or land use and may contain design criteria. The Settings' Parcel Labels section defines parcel label types for parcel information, line and curve segments, and parcel and segment tables (see Figure 3.1). Parcels and their segments can have multiple labels.

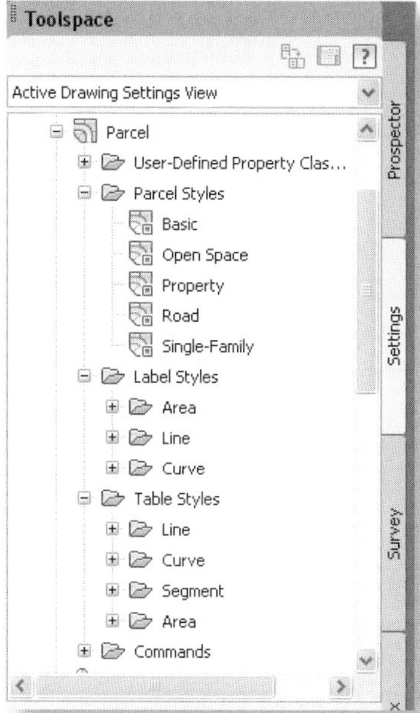


**FIGURE 3.1**

There can be more than one site in a drawing and each site can have its own parcel set. This strategy allows the user to document more than one subdivision strategy in a single drawing.

## PARCEL SETTINGS

The Drawing Settings, Parcel's Edit Feature Settings, and Edit Command Settings affect parcel creation. There are four basic parcel label types: Area, Line, Curve, and Table.

Drawing Settings' Object Layers panel assigns basic parcel objects layers (see Figure 3.2). These layers apply to parcels, their segments, and parcel tables. When assigning a layer modifier, use an asterisk to add the site name as the prefix or suffix.

The Ambient Settings panel contains several critical values that affect parcel commands. These settings affect area, angle, direction, and distances. When defining parcel segments with a direction or angle, the user should change these angle and direction settings to surveyor shorthand: dd.mmssss (Decimal dms). These settings include controls for dropping directions or to angle leading zeros and if the angle is a whole number, the controls allow it to be expressed as an integer.

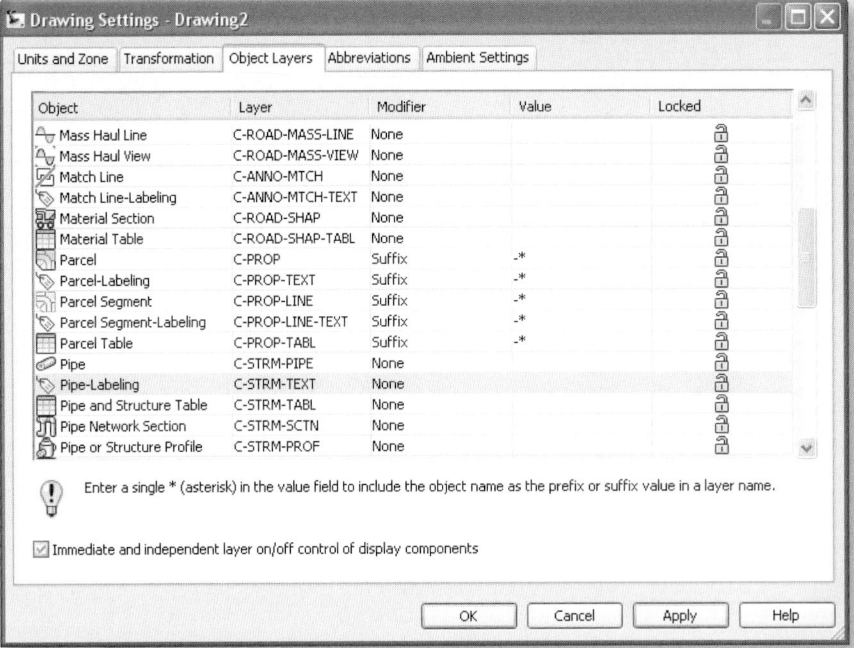


**FIGURE 3.2**

## EDIT FEATURE SETTINGS

The Edit Feature Settings dialog box assigns the default parcel, area, line, and curve label styles (see Figure 3.3). Commands for creating or annotating parcels use this styles list, unless a command overrides this assignment.

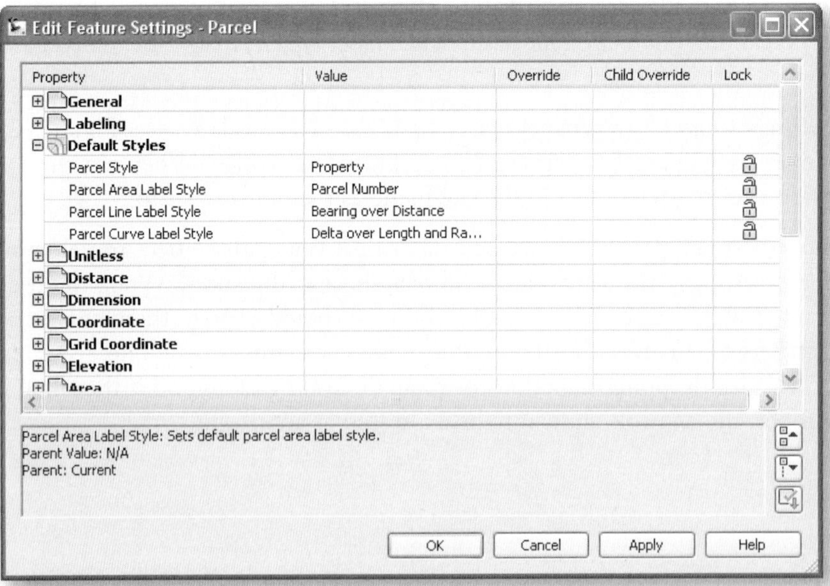


**FIGURE 3.3**

## COMMAND: CREATESITE

CreateSite creates a site. A site is critical to how parcels function. A site is in essence a "container" for parcels, alignments, and grading solutions. Prospector's Sites heading shortcut menu's New ... command creates a site using the command dialog box's settings (see Figure 3.4). A site's default styles (parcel, line, and curve) come from the Parcel Edit Feature settings. The Command's Parcels section sets counter and tag numbers. The Alignment section sets manual and automatic tag values for lines, curves, and spirals. The Feature Line Style Priority setting defines the display and elevation order for intersecting or overlapping feature lines (see Figure 3.5). If using Create Parcel by Layout or Create Parcel from Objects commands, they automatically create a site.

A site's properties define its name, 3D Geometry, construction layers, and parcel area counters.

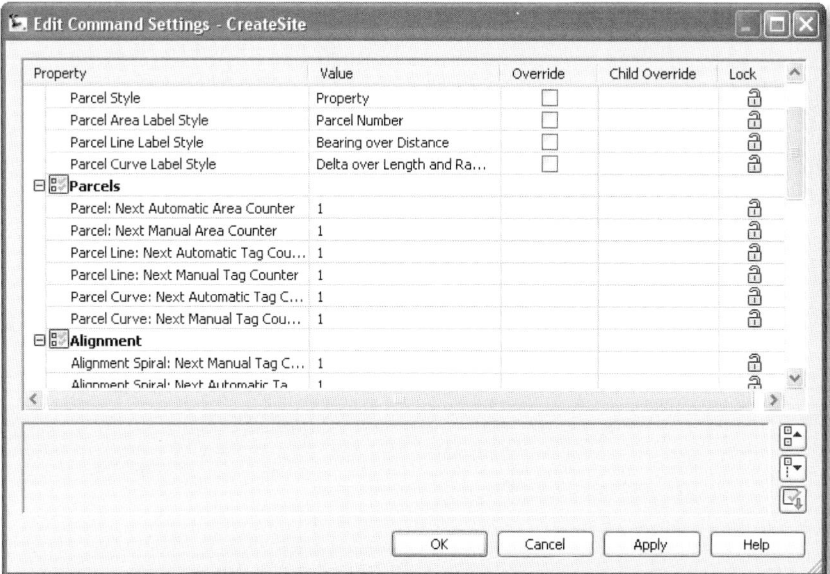


**FIGURE 3.4**

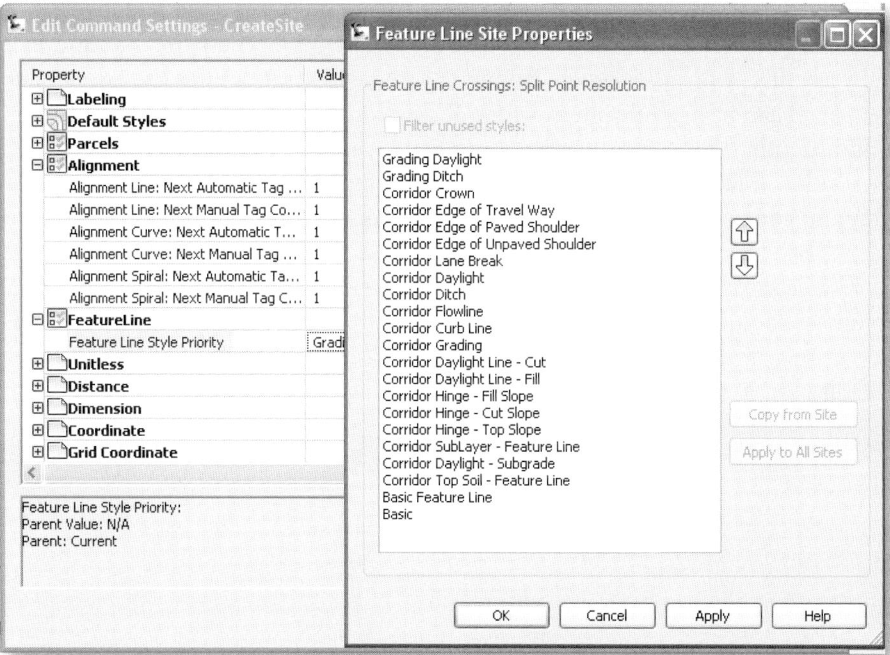


**FIGURE 3.5**

## COMMAND: CREATEPARCELFROMOBJECTS

The Edit Command Settings dialog box's settings affect converted lines, curves, and polylines (see Figure 3.6). The Edit Feature Settings' style settings define the default styles for this dialog box. However, when defining parcels, any style can be assigned to them. The Convert from Entities section controls erasing lines, curves, and polylines after converting them to parcels, and controls whether the parcels are labeled.

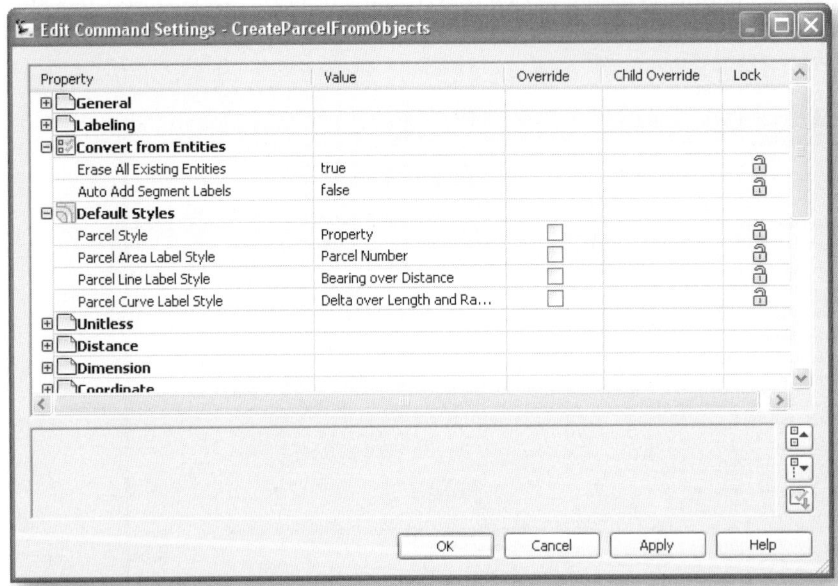


**FIGURE 3.6**

## COMMAND: CREATEPARCELBYLAYOUT

The Edit Command Settings dialog box's four sections affect new parcels created with the CreateParcelByLayout toolbar (see Figure 3.7). The first section, New Parcel Sizing, sets a parcel's minimum area and frontage. Some parcels define their width by an offset from the parcel's frontage. Frontage offset defines the offset value. Also, a parcel can have a minimum width at the frontage offset and a minimum and maximum depth. The second section, Preview Graphics, assigns each parcel's elements a color. The third section, Automatic Layout, toggles on automatic parcel creation and controls how to redistribute the remaining area: Create parcel from remainder, Place remainder in last lot, or Redistribute remainder. When automatically creating parcels, the routine paints the screen with temporary graphics identifying the minimum/maximum and setback values. The fourth section, Convert from Entities, toggles on and off the labeling of the converted parcel segments. The last section sets the default parcel style, and label styles for line and curve segments.

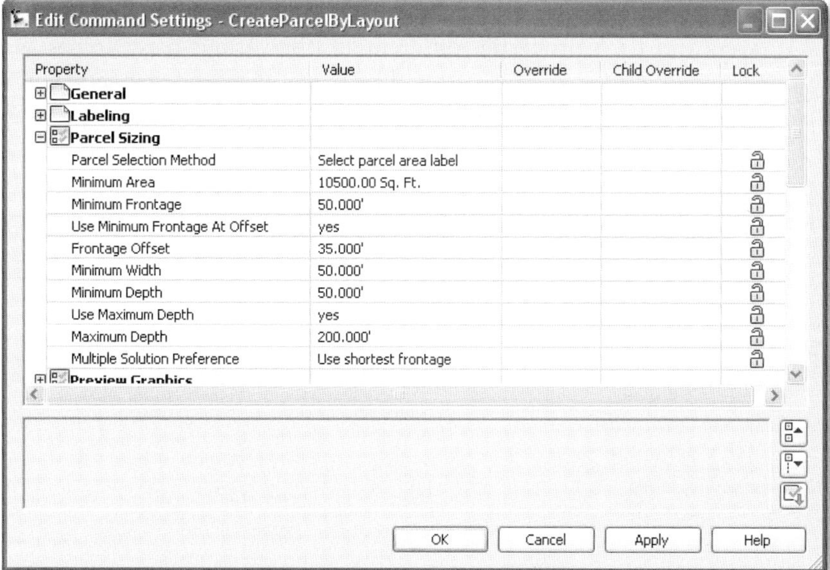


**FIGURE 3.7**

## PARCEL STYLES

Parcel style defines a parcel type, for example, existing, proposed, open space, single family, etc. A parcel's style affects its segment and hatch layers, hatch appearance, and its name template. Each parcel style has four tabs: Information, Design, Display, and Summary.

The Information panel displays the style's name and description. The Design panel defines the parcel-naming template, if it has a fill pattern, and it defines the hatch fill distance, the gap between the parcel's segments and the hatch (see Figure 3.8).

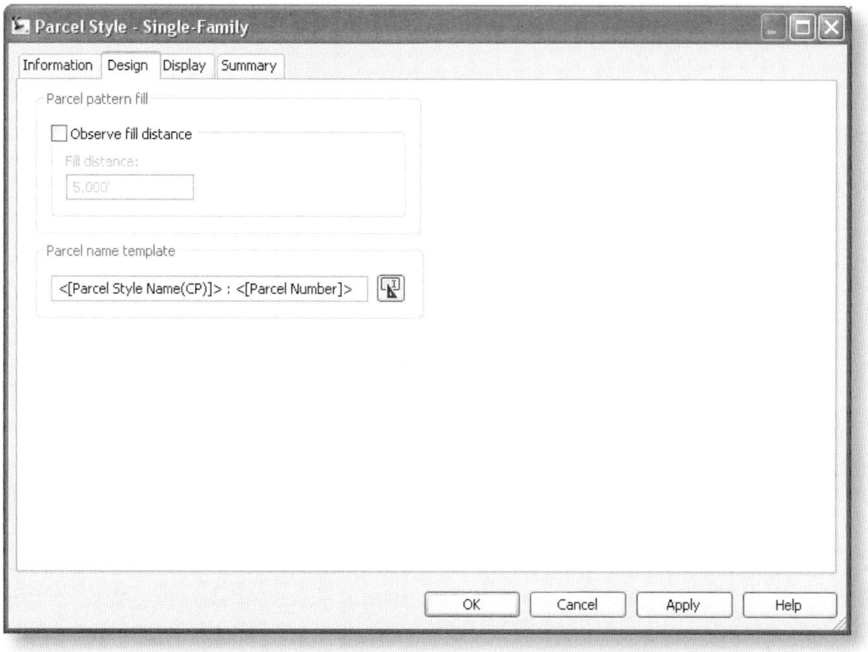

Parcel Style - Single-Family
Information Design Display Summary
Parcel pattern fill
Observe fill distance
Fill distance:
5.000'
Parcel name template
<[Parcel Style Name(CP)]> : <[Parcel Number]>
OK Cancel Apply Help

**FIGURE 3.8**

The display panel sets the parcel style's line and hatch layers (see Figure 3.9). The parcel segment layers and their properties differentiate the various parcel types. As with point groups, parcels have a display order. The parcel display order sets which parcel type covers another parcel type. For example, Site is at the list's top, next is ROW, and last is Single Family. This order implies Site's line work will always be in front of ROW's, and ROW's line work will always be in front of Single Family even when the lines overlap. To plot this display order, toggle off the plotter's line merge option.

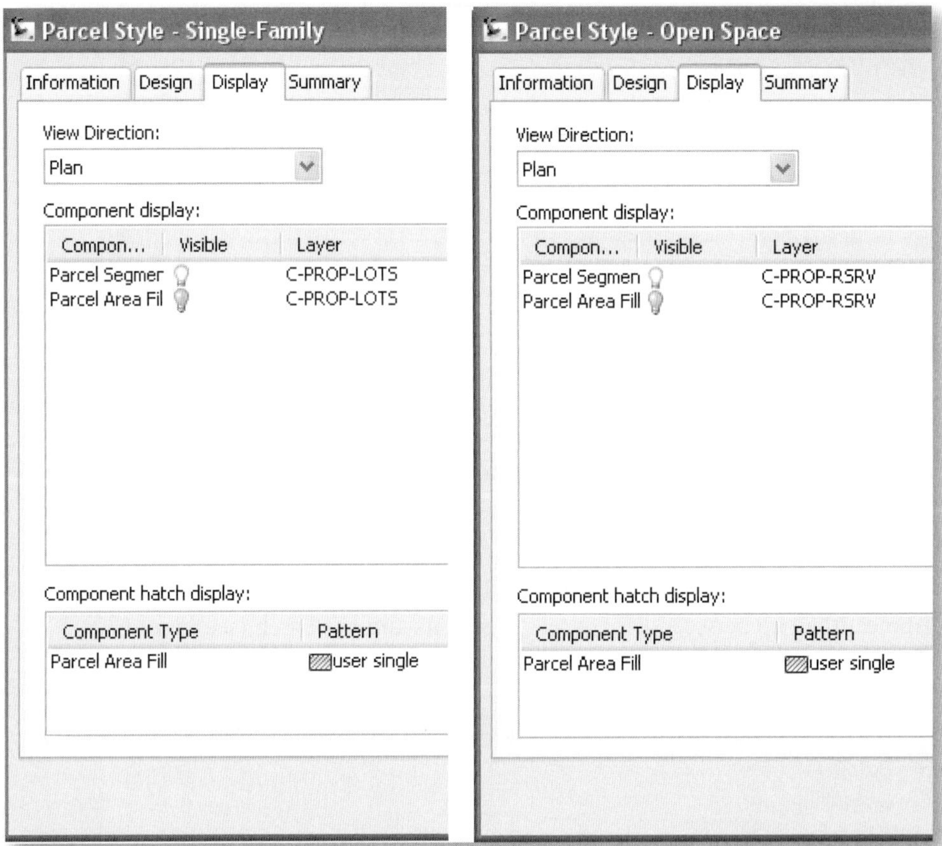


**FIGURE 3.9**

## LABEL STYLES

There are three parcel label styles types: area, line, and curve. Each parcel receives an area label that cannot be deleted.

### Area Label Style

The Label Style Composer interface creates or modifies label styles. The style's General tab defines the text style, its visibility, layer, initial orientation, and plan-readability behavior. The style's Layout tab defines the label's components, their names, contents (the object's property and its format), size, offset, and borders. The label's default location in the parcel is its centroid. Figure 3.10 shows the Name Area & Perimeter area label style. The Dragged State tab values set the label's behavior when moving it from its initial placement point.

A label component has a paper-based text size and each successive component can have a different text size. The model space or viewport plotting scale multiplies this size value, thus creating the correctly sized text for that viewport. Several individual label components cannot have one border around them; there can be a border only around a single component with several properties.

When creating a text component border, its values include color, linetype, lineweight, masks, and gap. When assigning a border's color, the color assignment overrides the label layer's color (ByLayer).

The label's dragged state parameters apply to leaders and text components. All parameters have their own color, lineweight, and linetype properties. A leader has several pointer types to use or you can use your own. When you drag the label, it changes to one of two states: As Composed and Stacked Text. Stacked Text is left-justified text. A dragged label can have a border that is different from the Layout or the assigned color overrides.

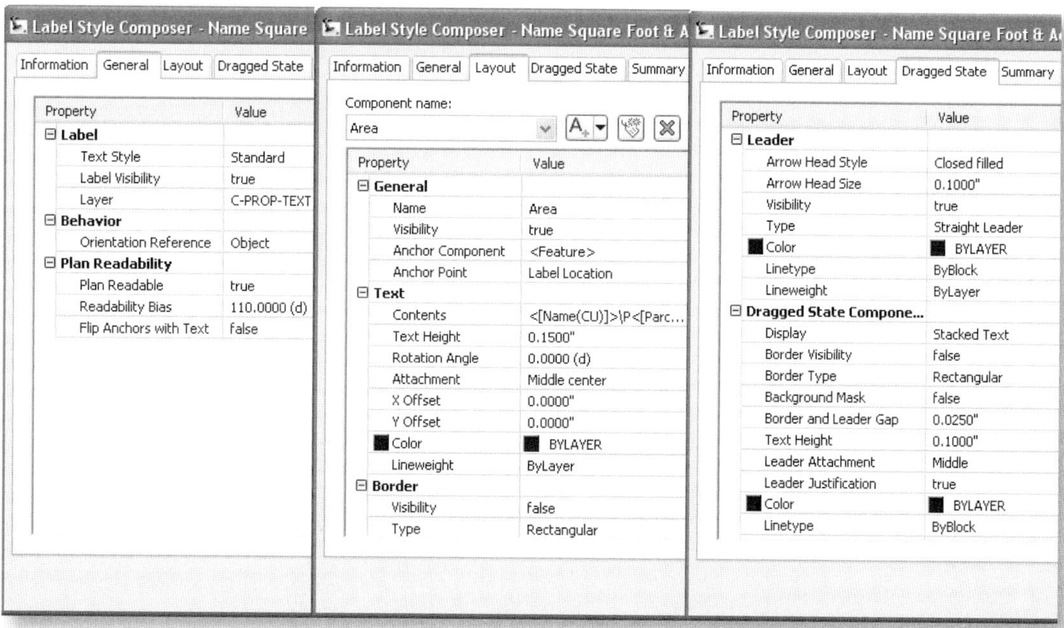


**FIGURE 3.10**

## Line Label Style

A line label style contains a table tag, bearing and/or distance, a direction arrow, and other properties. The three panels, General, Layout, and Dragged State, control the components, visibility, text, form, and shape of the label in its original position (layout) or in a new location if the label has been dragged from its original location (dragged state) (see Figure 3.11). The anchor point can be at the segment's start, middle, and end, which can be used to place any block (crow's feet) or property (Northing/Easting) at their coordinates.

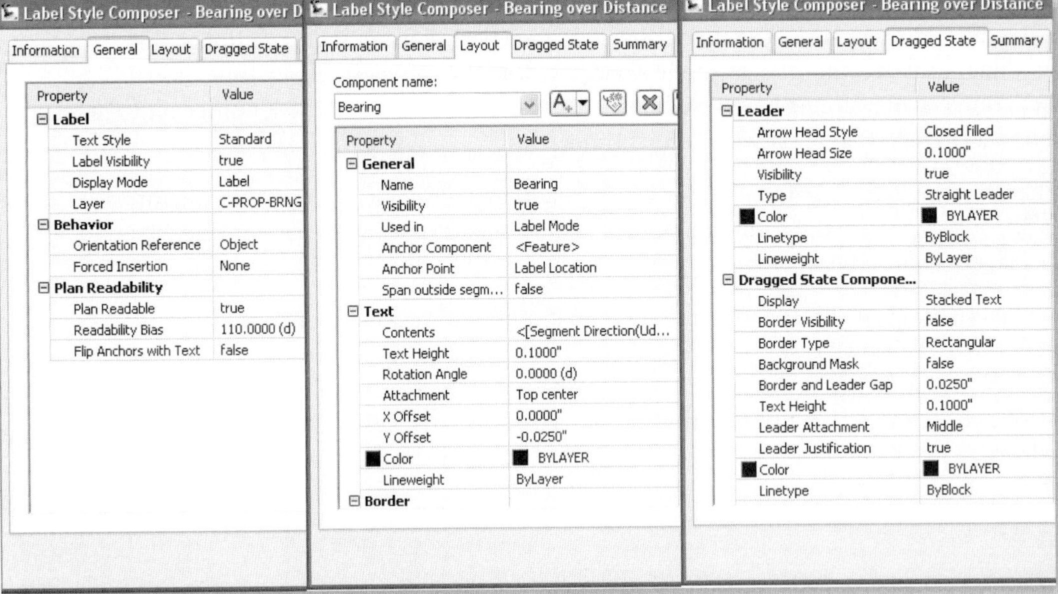


**FIGURE 3.11**

## Curve Label Style

A curve label style contains a table tag, distance, radius, and/or other properties. The three panels, General, Layout, and Dragged State, control the components, visibility, text, form, and shape of the label in its original position (layout) or at a new location if the label has been dragged from its original location (dragged state) (see Figure 3.12). The anchor point can be at the segment's start, middle, and end values and these can be used to place any block (crow's feet) or property (Northing/Easting) at their coordinates.

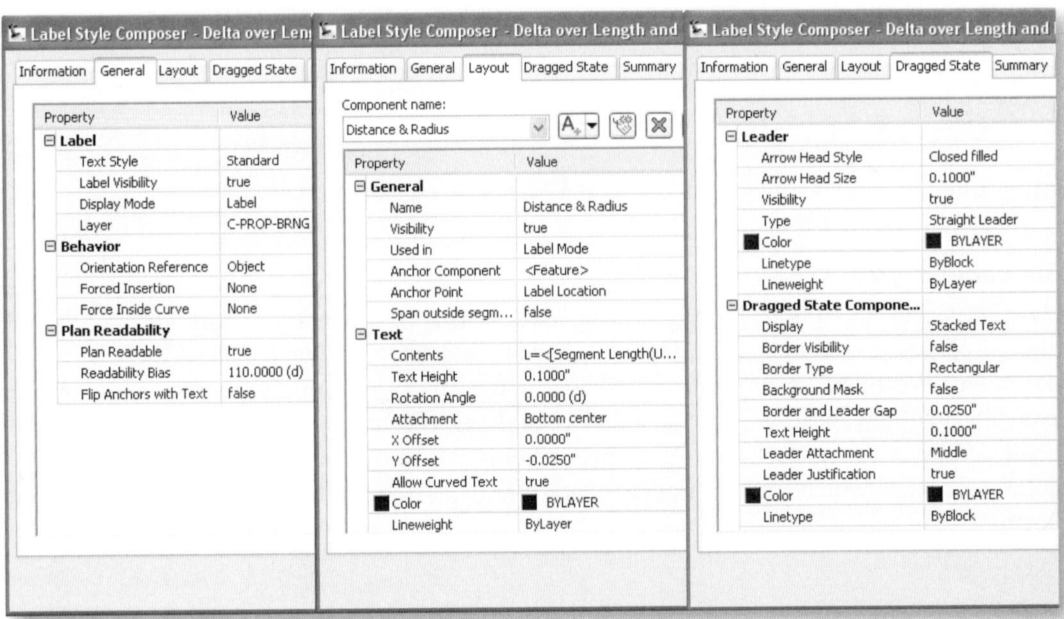


**FIGURE 3.12**

## TABLE STYLES

A table lists parcel information in tabular form. Table types include line, curve, line and curve segments (Line and Curves in the same table), and areas (see Figure 3.13).

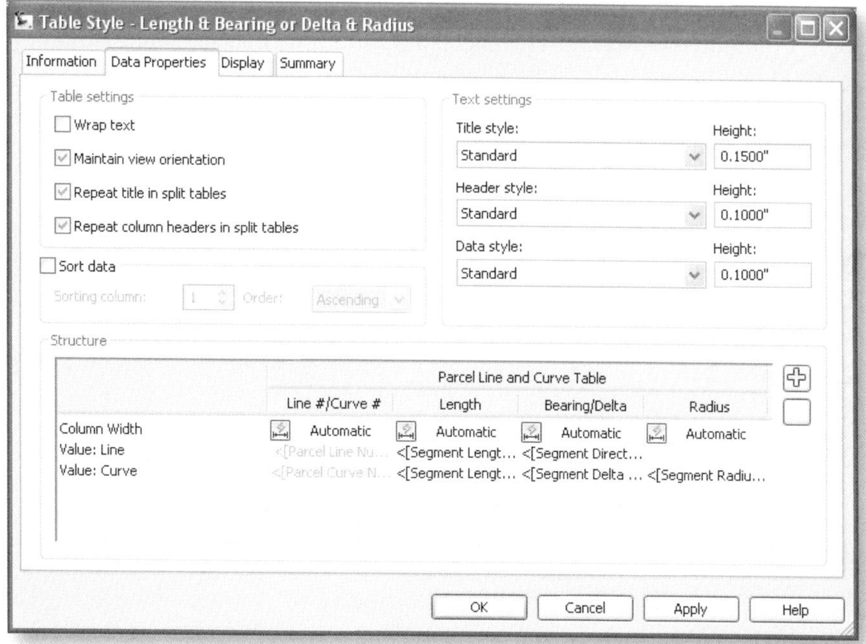


**FIGURE 3.13**

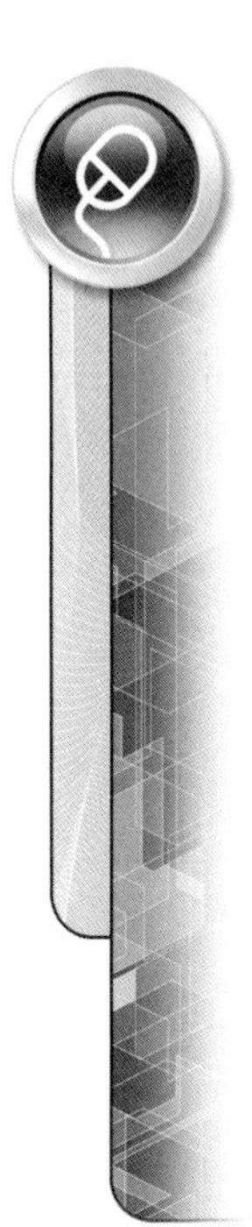

## EXERCISE 3-1

After completing this exercise, you will:

- Be familiar with parcel settings.
- Be familiar with parcel styles.

### Exercise Setup

This exercise's drawing uses a template file, Chapter 03 - Unit 1.dwt. It is found in the Chapter 3 folder of the CD that accompanies this textbook.

1. Start Civil 3D by double-clicking its **Desktop** icon.
2. Close the startup drawing and do not save it.
3. At Civil 3D's top left, click the Application Menu's drop-list arrow, from the shortcut menu select the ***New***, browse to the Chapter 3 folder of the CD that accompanies this textbook, and select and open the file *Chapter 03 – Unit 1.dwt.*
4. At Civil 3D's top left, click the Application Menu's drop-list icon. From the menu place your cursor over SAVE AS..., in the flyout menu select AutoCAD Drawing, save the file to your machine's Civil 3D Projects folder, and name it **Parcels-1**.

### Edit Drawing Settings

1. Click the ***Settings*** tab.
2. In Settings, at the top, select the drawing name, press the right mouse button, and from the shortcut menu, select EDIT DRAWING SETTINGS....

3. Select the ***Object Layers*** tab and scroll down the list until you are viewing the Parcel layer entries.
4. For Parcel, Parcel-Labeling, Parcel Segment, Parcel Segment-Labeling, and Parcel Table, change the modifier to **Suffix** and for each Suffix, enter the value of **-*** (dash asterisk) (see Figure 3.2).
5. Select the ***Ambient Settings*** tab, and expand the Direction section.
6. Click in the Format value cell and, if necessary, set its format to **DD.MMSSSS (decimal dms)**.
7. Click ***OK*** to create the layer suffixes and close the dialog box.

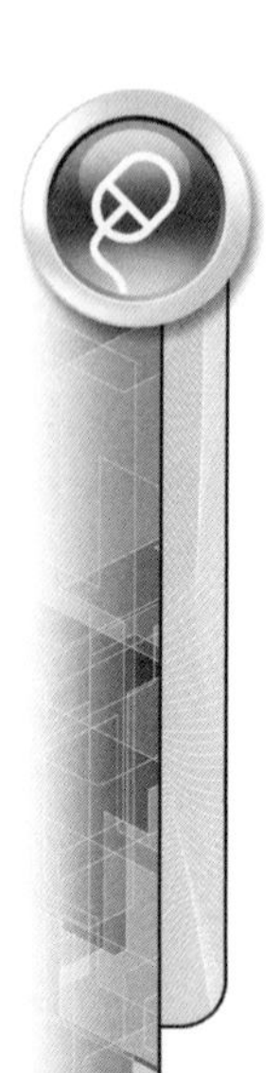

## Edit Feature Settings

The Edit Feature Settings dialog box sets initial parcel and label styles.

1. In Settings, select the Parcel heading, press the right mouse button, and from the shortcut menu, select EDIT FEATURE SETTINGS....
2. Expand the **Default Styles** section, and if necessary, change styles to match those listed in Figure 3.3.
3. Click ***OK*** to close the dialog box.

## Create Site

The CreateSite command creates a site, parcel, tag counters, and increment values.

1. In Settings, expand the Parcel branch until you are viewing the Commands heading's command list.
2. From the Commands list, select CreateSite, press the right mouse button, and from the shortcut menu, select EDIT COMMAND SETTINGS....
3. Review the CreateSite settings and, if necessary, match them to the settings in Figure 3.4.
4. After making any needed changes, click ***OK*** to close the dialog box.

## Create A Parcel from Objects

When converting the parcel's lines, curves, and polylines, Civil 3D wants to know what to do with the selected line work. The choices are to keep or to erase the line work. When defining the parcel, the routine assigns default parcel, area, line, and curve styles.

1. From the Parcel Commands list, select CreateParcelFromObjects, press the right mouse button, and from the shortcut menu, select EDIT COMMAND SETTINGS....
2. Review the Edit Command Settings and, if necessary, match them to those in Figure 3.6.
3. After making any needed changes, click ***OK*** to close the dialog box.

## Create Parcel By Layout

The CreateParcelByLayout command presents the user with a toolbar containing tools to define, size, edit, erase, and review parcel segments or an entire parcel.

1. From the Commands list, select CreateParcelByLayout, press the right mouse button, and from the shortcut menu, select EDIT COMMAND SETTINGS....
2. Review the Edit Command Settings and change them to match those in Figure 3.7.
3. After changing the settings, click ***OK*** to close the dialog box.

## Parcel Styles

Parcel object styles set the layers, hatch pattern, and parcel area label style.

1. In Settings, expand the Parcel branch until you are viewing the Parcel Styles heading and its styles list.
2. From the styles list, select the Property style, press the right mouse button, and from the shortcut menu, select EDIT....
3. Select the ***Information*** tab and review its settings.
4. Select the ***Design*** tab and review its settings.
5. Select the ***Display*** tab and review its settings.
6. Click ***OK*** to close the dialog box.
7. From the list of styles, select the Single-Family style, press the right mouse button, and from the shortcut menu, select EDIT....
8. Select the ***Information*** tab and review its settings.
9. Select the ***Design*** tab and review its settings.
10. Select the ***Display*** tab, review its settings, and, if necessary, match them to those in Figure 3.9.
11. After making any needed changes, click ***OK*** to close the dialog box.

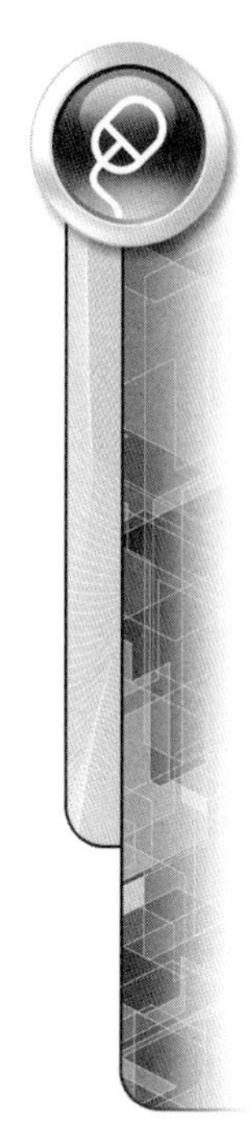

## Area Label Styles

When defining a parcel, all commands creating parcels assign area and segment labels. The segment styles include labels for lines (bearing and distance) and curves (radius, length, chord, etc.). Civil 3D content template files provide basic area label styles.

1. In Settings, expand the Parcel branch until you are viewing the Label Styles' Area label styles list.
2. From the list, select Name Area & Perimeter, press the right mouse button, and from the shortcut menu, select EDIT....
3. Select the ***Information*** tab and review its settings.
4. Select the ***General*** tab and review its settings.
5. Select the ***Layout*** tab and review its settings.

The Name Area & Perimeter label has one named Component: Area. This component has three parcel properties: Parcel Name, Area, and Perimeter (see Figure 3.14).

6. Click in the Component's Text Contents value cell, and then click the ellipsis to display the Text Component Editor – Contents dialog box. This dialog box defines the parcel name, area, and perimeter format.
7. In the dialog box's right side, click a property and review its values on the left side of the dialog box.
8. After reviewing the properties and their format values, click ***OK*** until the dialog boxes are closed and you return to the command prompt.

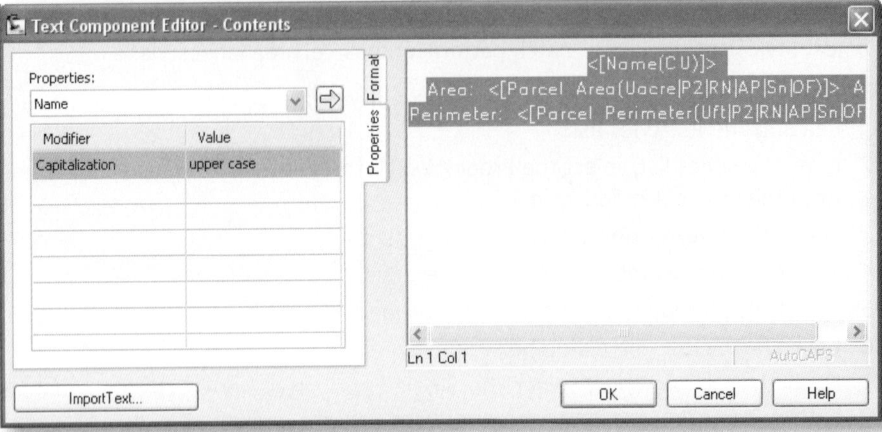


**FIGURE 3.14**

## Line and Curve Label Styles

The Parcel Edit Feature Settings dialog box sets the default line and curve styles.

1. In Settings, expand the Parcel branch until you are viewing the Label Styles' Curve label styles list.
2. From the list, select Delta over Length and Radius, press the right mouse button, and from the shortcut menu, select EDIT....
3. Select and review the settings for the ***Layout*** and ***Dragged State*** tabs.
4. Click ***OK*** to exit the dialog box.
5. Under Label Styles expand Line to view its styles list.
6. From the list, select Bearing over Distance, press the right mouse button, and from the shortcut menu, select EDIT....
7. Select the ***Layout*** tab to review its contents (see Figure 3.11).
8. Click ***OK*** to exit the dialog box.
9. At Civil 3D's upper left, Quick Access Toolbar, click the ***Save*** icon to save the current drawing.

This completes the parcel drawing setup and styles review. The next unit reviews site and parcel drafting and defining. The settings reviewed in this unit directly affect the results of the exercises in the next unit.

## SUMMARY

- The Edit Drawing Settings dialog box's Object Layers tab sets a prefix or suffix for each parcel layer.
- The Create Parcel commands create a site if it does not yet exist.
- Each command to create parcels uses the default parcel style unless the style is changed.
- When creating a parcel, the parcel displays an area label near its centroid.
- Parcel segments labels occur when creating them, or afterward.
- Users can define their own area and segment labels (line and curve).

## UNIT 2: CREATING A SITE AND PARCELS

Creating a property boundary can be a simple drafting exercise. The transparent commands extend the line and polyline commands and allow drafting lines with turned angles, bearings, or azimuths and distances. The data for these lines may be a published plat of a survey, a set of calculated directions and distances, or northings/eastings. Whatever the source, Civil 3D provides all necessary drafting tools to transcribe written values into segments.

Any closed polygon is a potential parcel. If drafting a boundary from survey plat values or a coordinates list, the boundary probably will not close because of calculation rounding errors. When defining a parcel from an open boundary, Civil 3D creates only parcel segments, but not a parcel. To close the boundary, a knowledgeable person needs to be contacted for methods to achieve boundary closure.

Once a boundary parcel is closed, the subdivision design focus first turns to dividing the boundary into smaller homogeneous land-use blocks (residential, common, open, etc.) and finally turns to creating individual parcels (single-family, commercial, etc.). Subdividing strategies vary greatly and at times are more art than science. The subdividing process also involves drafting roadway alignments, which also divide parcels. A final parcel design is a balance between the government's subdivision regulations, developer demands, and engineering constraints. In Civil 3D, the process of subdividing a boundary into parcels includes drawing lines, curves, polylines, and using the Parcel Layout Tools toolbar's sizing commands.

In Civil 3D, a site is an overall parcel boundary that acts as a "container," organizing the internal parcels and alignments. Civil 3D allows multiple sites in a drawing. A second site can be an alternative parcel design or a data set around which to design (wetlands or other protected areas). Objects in different sites do not interact.

### LINE AND POLYLINE COMMANDS

The Line, Fillet, and Polyline commands create potential parcel segments (lines and arcs) from coordinates or object snaps. Civil 3D extends the line and polyline commands by allowing them access to Civil 3D points, thus making them mimic survey (angle and distance) methods. Survey angle and distance methods are similar to the polar toggle. For example, when using the Line command with polar toggled on while dragging toward a direction, entering a distance is all that is required to draw the line. When drafting line work and selecting the bearing and distance transparent command override, the prompt changes and asks for the line segment's bearing and distance values. The transparent command overrides include bearing/distance, azimuth/distance, turned angle, deflection angle, station offset, and referencing Civil 3D points. Users can mix methods to describe the direction and distance by changing the active transparent command. See Unit 2 of Chapter 2 for a discussion on transparent commands.

### CREATE PARCEL FROM OBJECTS

The Create Parcel from Objects command creates a site and parcels from selected lines, curves, and polylines. If there is no defined site, the routine creates a site from the selected line work's perimeter. The command also defines all interior closed polygons as parcels.

A Parcel Feature Settings value specifies whether selected line work is erased. The Create Parcel from Objects routine uses this value as a default and, if on, erases the selected line and arc segments and replaces them with parcel object segments. You can disable the erasure at the bottom of the Create Parcels - From Objects dialog box.

## CREATE PARCEL BY LAYOUT

The Create Parcel by Layout command displays the Parcel Layout Tools toolbar. This toolbar contains tools to create parcel lines and arc segments, sized parcels (manually or automatically), and parcels from user-specified criteria. The main parcel-sizing routines are Slide Angle, Slide Direction, Swing Line, and Free form Create. The routines are an icon stack, fifth in from the toolbar's left. To display the individual commands (icons) within the icon stack, select the drop-list arrow. After displaying the list, select the command for the next parcel segment. Each routine holds the minimum parcel values set in the Parcel Layout Tools toolbar's Parcel Sizing section.

The Slide Line routine creates parcels by drawing the last side yard line at a specified angle to the parcel's frontage line. The Slide Direction routine is the same as Slide Angle, but it uses a direction (bearing, azimuth, or two selected points) drawing for the last side yard line. The Swing Line command draws the last side yard line from a point on the frontage to a point at the back of the parcel. This side yard direction can be perpendicular to the frontage, two user-specified points, a bearing, or an azimuth.

The Parcel Layout Tools toolbar has a roll-down settings panel (see Figure 3.15). The panel's values set the minimum lot size, minimum width and depths, frontage offsets, preferred solution, and settings that affect automatic parcel creation.

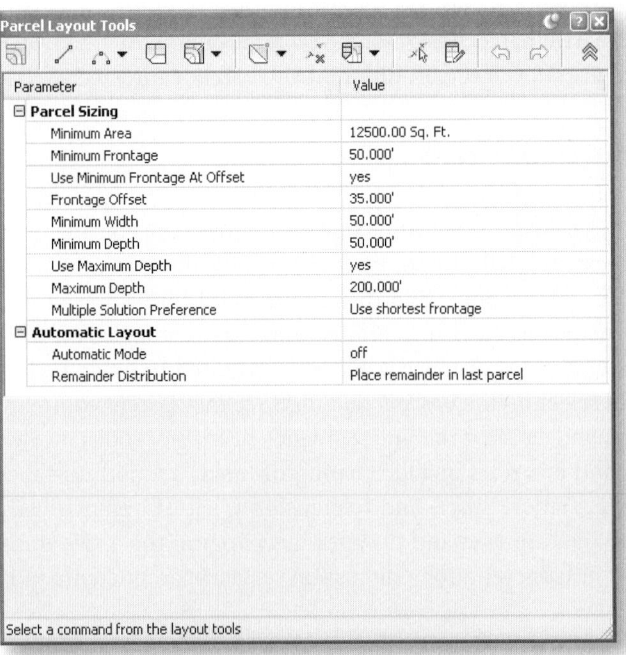


**FIGURE 3.15**

On the toolbar, the second and third icons from the left draw fixed parcel segments (line and curve). You can use the transparent commands to define the segment's direction and length. The sixth icon from the left inserts, deletes, and breaks apart points of intersection (PIs) within parcel segments.

The Delete Sub-entity icon, located to the center right of the toolbar, deletes parcel segments. The Parcel Union icon stack creates a single parcel from two selected parcels. The routine uses the first parcel's properties for the merged parcels.

The Sub-entity Editor, the icon to the left of the undo arrow, displays a selected parcel segment's geometric values. The values shown in black are editable; any grayed-out value cannot be changed. The Sub-entity selection icon to the left of the Sub-entity Editor icon selects the parcel segments to display in the Sub-entity editor.

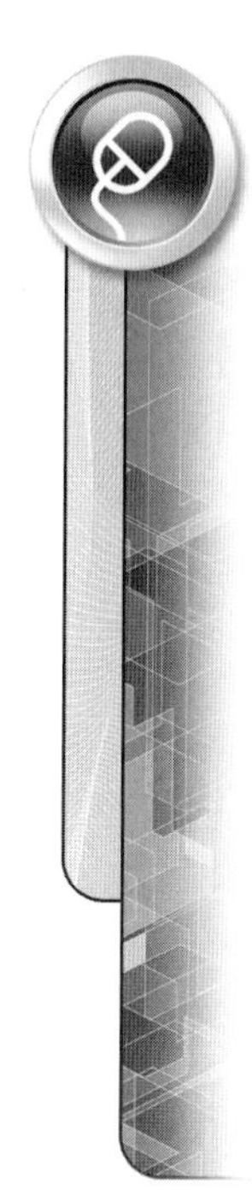

## RIGHT-OF-WAY PARCEL

The Right-of-Way (ROW) parcel uses an alignment to define some of its geometry. The remaining ROW parcel parameters are Offset From Alignment and Cleanup at Alignment Intersections (trim or fillet) (see Figure 3.16).

A ROW parcel, alignments, and adjoining parcels are not dynamically linked. If the alignment moves, the user must delete and redefine the ROW parcel.

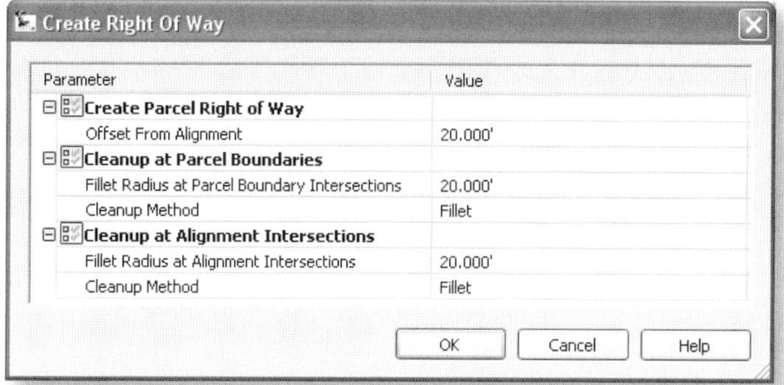


**FIGURE 3.16**

## EXERCISE 3-2

After completing this exercise, you will:

- Be able to draw a boundary using transparent commands.
- Be able to define a site and its parcels using the Create Parcel From Objects command.
- Be able to subdivide parcel blocks using the Create Parcel By Layout command.

### Exercise Drawing Setup

This exercise continues using the Unit 1 drawing, Parcels-1. If you did not do the Unit 1 exercise, you can open the drawing in the Chapter 3 folder of the CD that accompanies this textbook: *Chapter 03 - Unit 2*.

1. If the previous exercise's drawing is not open, open it now, or browse to the Chapter 3 folder of the CD that accompanies this textbook and open the *Chapter 03 - Unit 2.dwg* file.
2. In the Layer Properties Manager, freeze the following layers: **BLDG**, **BNDY**, **BNDYTXT**, **CONT**, **CONT2**, and **EX-ADJLOTS**.
3. Create a New layer, name the layer **BNDY-INPUT**, assign it a color, and make it the current layer.

4. Close Layer Properties Manager.
5. At Civil 3D's top left, Quick Access Toolbar, click the ***Save*** icon to save the drawing.

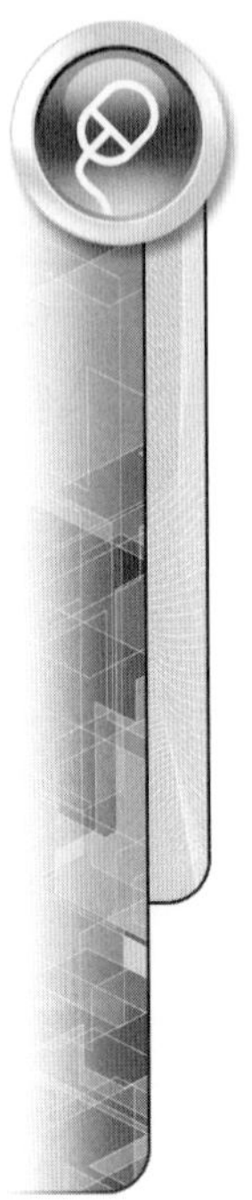

## Drafting a Property Boundary

When drafting a boundary with bearings and distances, a small gap will exist between the last boundary segment's end and the first segment's beginning. This gap represents the closure error. You close this gap using the drafting command's Close option. If you are using a transparent command override, you press ESC once to exit the override and return to the drafting command, and then close the boundary. If you exit before completing the boundary, you will have to pick up where you left off and use an Endpoint object snap to draw the last small segment to the point-of-beginning (POB).

This exercise uses the Line command with the transparent commands Northing/Easting and Bearing and Distance overrides. The *Boundary Calls.txt* file is a Notepad file that has the property boundary's beginning coordinates and its bearings and distances. You can also print the *Property Outline.PDF* file. Both files are in the Chapter 3 folder of the CD that accompanies this textbook.

The property boundary bearings and distances are in Table 3.1. The boundary's northwest corner northing and easting coordinates, 14742.1740 and 6716.9054, are also its POB. The bearings and distances are measured clockwise around the boundary. Using the Table 3.1 values, the boundary almost closes, but not quite.

To complete this exercise, you must have the direction set to Decimal dms (in the Ambient Settings panel). Decimal dms is a surveyor's method of entering bearings and distances. When the direction is set to Decimal dms, the first bearing in the table, N 88d 25'50" E, is entered as a quadrant 1 bearing with an angle value of 88.2550. See the discussion of transparent commands in Chapter 2, unit 2 of this textbook.

```
Point 1:
Northing = 14742.1740
Easting = 6716.9054
```

**TABLE 3.1**

| Segment | From/To Point | Quadrant | Bearing | Distance |
|---|---|---|---|---|
| 1 | 1 to 2 | 1 | N 88d25'50" E | 504.9805 |
| 2 | 2 to 3 | 2 | S 1d34'10" E | 1001.1900 |
| 3 | 3 to 4 | 1 | N 87d59'17" E | 333.0686 |
| 4 | 4 to 5 | 2 | S 1d37'46" E | 1322.0609 |
| 5 | 5 to 6 | 3 | S 88d03'56" W | 1.8090 |
| 6 | 6 to 7 | 2 | S 1d31'46" E | 1028.4872 |
| 7 | 7 to 8 | 4 | N 59d22'31" W | 392.2954 |
| 8 | 8 to 9 | 4 | N 1d33'54" W | 817.3547 |
| 9 | 9 to 10 | 4 | N 1d34'10" W | 614.2499 |
| 10 | 10 to 11 | 3 | S 87d56'21" W | 504.9986 |
| 11 | 11 to 1 | 4 | N 1d34'10" W | 1712.8900 |

1. If necessary, click the ***Settings*** tab.
2. In Settings, at the top, select the drawing name, press the right mouse button, and from the shortcut menu, select EDIT DRAWING SETTINGS....
3. Select the **Ambient Settings** tab and expand the **Direction** section.
4. If necessary, change the Format to **DD.MMSSSS (decimal dms)**.
5. Click ***OK*** to set the format and exit the dialog box.
6. Make sure the Civil 3D Transparent Commands toolbar is visible.
7. Start the LINE command. From the Transparent Commands toolbar, select the ***Northing Easting*** icon, and the command line prompt changes to a starting northing coordinate.

After entering the northing coordinate value, press ENTER for the easting prompt. After entering the easting value, press ENTER to enter both coordinates.

8. At the Northing prompt, enter **14742.1740** and press ENTER.
9. At the Easting prompt, enter **6716.9054** and press ENTER to set the starting point of the line.
10. Press ESC once to end the Northing Easting transparent override and resume the LINE command.
11. Without exiting the LINE command, from the Transparent Commands toolbar select the ***BEARING DISTANCE*** icon and answer the prompts using the information in Table 3.1 for quadrant and bearing and the distance information for each remaining boundary segment.

Civil 3D displays a tripod and prism representing the starting coordinates and each segment's direction. You can graphically (by using selections in the drawing) define a quadrant, direction, and distance while in this override mode.

You will not return to the exact starting coordinates. This is because the distance and angle values contain rounding errors even at four decimal places. If you are looking at the bearings and distances on a plat of survey, the precision is a whole second for bearings and only two decimal places for distances. This precision is less than what you have just entered for this boundary.

12. After entering the last bearing and distance, press ESC only once to exit the Bearing Distance override.
13. At the command line, enter the letter '**C**' and press ENTER to close the boundary by drawing the last line segment to the beginning.
14. At Civil 3D's top left, Quick Access Toolbar, click the ***Save*** icon to save the drawing.

## Create Parcel From Objects

The current line work represents the site definition.

1. In the Ribbon Home tab, Layers panel, click the ***Layer Isolate*** icon, select a line on the BNDY-INPUT layer, and press ENTER to isolate the layer's line work.
2. From the Ribbon's Home tab, Create Design panel, click the ***Parcel*** icon, and from the menu select CREATE PARCEL FROM OBJECTS.
3. Select all of the lines you just drew and press ENTER to display the Create Parcels - From objects dialog box.
4. Match your settings to those in the dialog box in Figure 3.17 and when done, click ***OK*** to exit the dialog box and create the site, parcel, and the parcel area label.

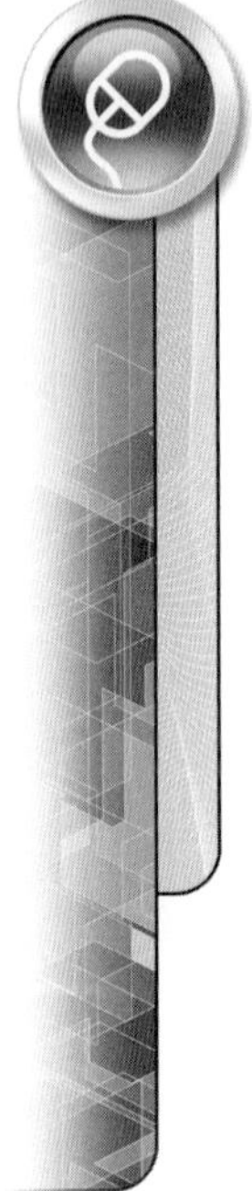

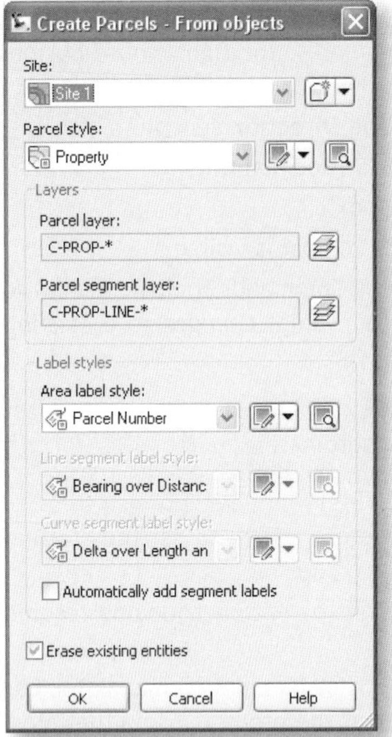


**FIGURE 3.17**

5. In the Ribbon's Home tab, Layers panel, at its middle right, click the UNISOLATE icon to restore the previous layer state.
6. If necessary, open the Layer Properties Manager, toggle **ON** and **Unlock** the **C-PROP-Site1** and **C-PROP-LINE-Site1** layers, and click **X** to exit the Layer Properties Manager palette.
7. If necessary, type REGENALL (REA) at the command prompt and press ENTER.

## Defining the Right-of-Way and the Wetlands Parcels

The ROW and wetlands boundary are polylines on the PR-ROW and Wetlands layers.

1. Isolate the two layers **PR-ROW** (magenta) and **Wetlands** (rust).
2. From the Ribbon's Home tab, Create Design panel, click the ***Parcel*** icon, and from the menu, select CREATE PARCEL FROM OBJECTS, select the line work on the screen, and press ENTER.
3. The Create Parcels - From objects dialog box opens. Adjust the settings to match those in Figure 3.17.
4. After adjusting the settings, click ***OK*** to exit and define the parcels.
5. From the Ribbon's Home tab, Layers panel, click the ***Unisolate*** icon to restore the previous layer state.
6. If necessary, type REGENALL (REA) at the command prompt and press ENTER.
7. At Civil 3D's top left, Quick Access Toolbar, click the ***Save*** icon to save the drawing.

## Creating BackYard Parcel Lines

The previous steps create parcels with uniform land-use types: blocks. These "block" parcels are further subdivided into individual parcels. Some blocks need further subdividing to create front yard and backyard lines. These particular blocks have two frontage lines and no backyard line. Having front yard and backyard lines makes creating individual parcels much easier; all that is left to create is the side yard segment, which, when done, means the final parcel is created. To define the backyard line, you select polylines on the Interior Boundary layer.

1. Isolate the **Interior Boundary** layer.
2. From the Ribbon's Home tab, Create Design panel, click the ***Parcel*** icon, and from the menu, select CREATE PARCEL FROM OBJECTS, select the line work on the screen, and press ENTER.
3. The Create Parcels - From objects dialog box opens. Adjust the settings to match those in Figure 3.17.
4. After adjusting the settings, click ***OK*** to define the parcels.
5. From the Ribbon's Home tab, Layers panel, click the ***Unisolate*** icon to restore the previous layer state.
6. If necessary, type REGENALL (REA) at the command prompt and press ENTER.
7. Click the ***Prospector*** tab.
8. In Prospector, expand the Sites branch until you can see the Site 1 parcel list.
9. At Civil 3D's top left, Quick Access Toolbar, click the ***Save*** icon to save the drawing.

## Creating Single-Family Parcels

The final step is dividing the parcel blocks into individual resalable parcels. You do this using the Parcel Layout Tools toolbar's sizing tools.

1. Click Ribbon's ***View*** tab. In the Views panel's left side select the named view **Residential Parcels 1**.
2. In the Layer Properties Manager (Ribbon's ***Home*** tab, Layers panel), turn on the **Area** layer and click the **X** on the mast to close the palette.
3. From the Ribbon's Home tab, Create Design panel, click the ***Parcel*** icon, and from the shortcut menu, select PARCEL CREATION TOOLS to display the Parcel Layout Tools toolbar.
4. Click the **chevron** at the right side of the toolbar to expand the Parcel Sizing section.
5. If necessary, adjust your settings to match those in Figure 3.15.
6. Go to the toolbar's fifth icon in from the left, and then click the drop-list arrow to the right of the Parcel Sizing Tools icon stack. From the list select SLIDE LINE - CREATE to display the Create Parcels - Layout dialog box.
7. In the dialog box, set the Parcel Style to **Single-Family**, the Area Label Style to **Name Area & Perimeter**, and click ***OK*** to start sizing parcels (see Figure 3.18).

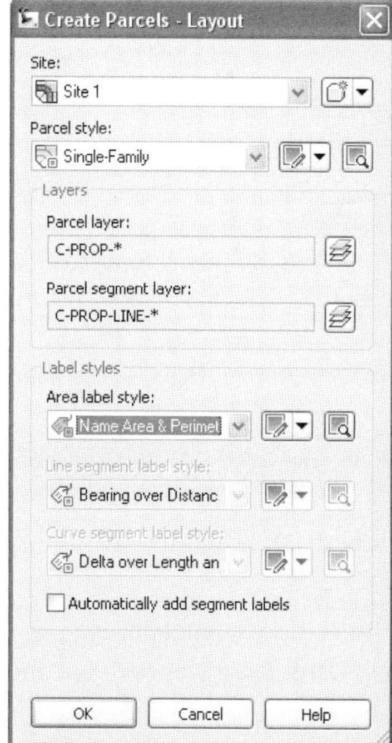
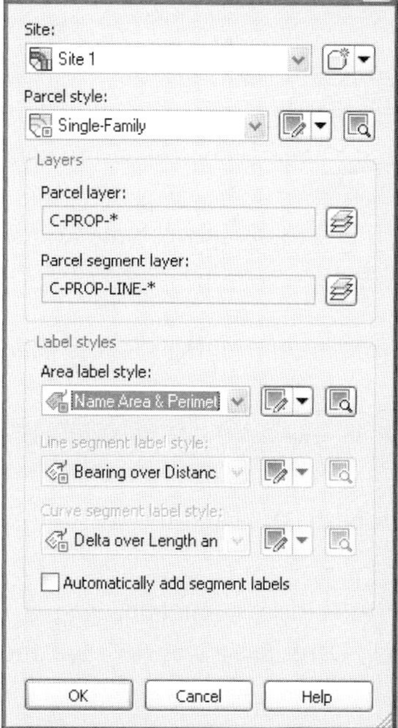

**FIGURE 3.18**

The command prompts you to select a parcel to subdivide.

8. Click the **chevron** at the right side of the toolbar, to collapse the toolbar.
9. If Object Snap is on, toggle it **OFF**.
10. Select the parcel's parcel number.

Next, you are prompted to define the parcel's frontage start and ending points.

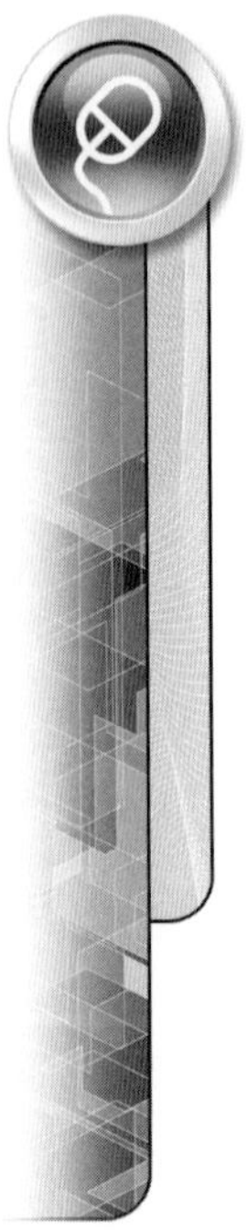

11. In the drawing, using the Endpoint object snap, start the frontage by selecting the parcels' northeast end and again using the Endpoint object snap, end the frontage by selecting the parcels' southwestern end (see Figure 3.19).

While doing this, the routine displays a jig following the frontage geometry. The jig recognizes changes in the frontage's geometry.

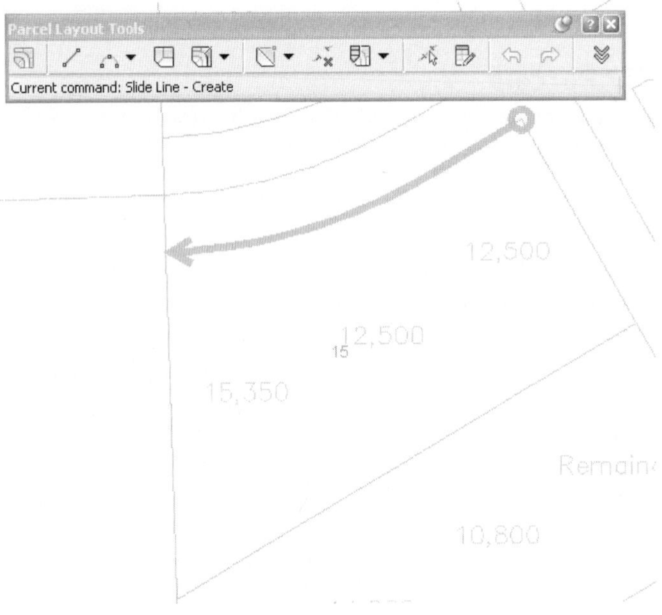


**FIGURE 3.19**

Next, you are prompted to define the frontage line's relative turned angle.

12. At the angle from frontage prompt, enter **90** degrees as the angle.
13. A parcel with 12,500 square feet displays on the screen. The command line prompts to accept the parcel; press ENTER to accept the new parcel and to continue to the next parcel.
14. A parcel with 12,500 square feet displays on the screen. The command line prompts to accept the parcel; press ENTER to accept the new parcel and to continue to the next parcel.
15. A warning dialog box appears stating that the minimum frontage is less than 100', click OK.
16. Press ESC to stop the routine. The command line prompts to select a command or to exit.

The area is now three parcels with one still having the property style and the west most parcel having the block's remaining area.

17. In the command line, enter **'X'** and press ENTER to exit the command.
18. At Civil 3D's top left, Quick Access Toolbar, click the ***Save*** icon to save the drawing.

## Changing a Parcel's Type and Label

When you need to change a parcel's type, change it in the Parcel Properties dialog box, Information panel. When you need to change the area label, change it in the Parcel Properties dialog box, Composition panel.

1. In the drawing, select the central parcel's label, press the right mouse button, and from the shortcut menu, select PARCEL PROPERTIES....
2. Select the ***Information*** tab, click the Object Style drop-list arrow, and from the list, select **Single-Family**.
3. Select the ***Composition*** tab, click the Area Section Label Style drop-list arrow, and select **Name Area & Perimeter**.
4. Click ***OK*** to change the parcel type and label.

The remaining parcel is now single-family with an area label and segments that match the others in the block.

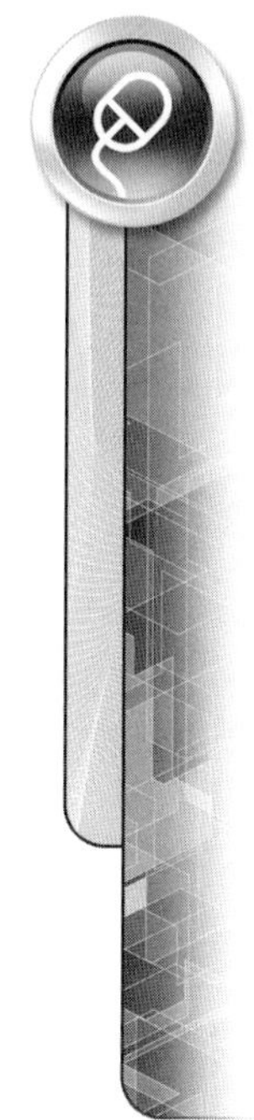

## Slide Line — Create Routine

Many times a side yard segment angle is not 90 degrees to the front yard line, but instead a bearing or perpendicular to the backyard line. The Slide Direction - Create routine is best for this situation.

1. Use the Zoom and Pan commands until you can view the entire adjacent south parcel that shares the backyard line.
2. From the Ribbon's Home tab, Create Design panel, click the ***Parcel*** icon, select PARCEL CREATION TOOLS to display the Parcel Layout Tools toolbar.
3. Click the **chevron** at the right side of the toolbar to expand the Parcel Sizing section.
4. In Parcel Parameters', Minimum Area, set the parcel size to 10500 and at the toolbar's right, click the collapse toolbar chevron.
5. Go to the fifth icon in from the left, and to the right of the icon click the drop-list arrow, and from the Parcel Sizing Tools icon stack, select SLIDE LINE - CREATE.
6. The Create Parcels - Layout dialog box opens. Match the settings to those in Figure 3.18, and when set, click ***OK*** to start the parcel-sizing process.
7. If necessary, close Parcel Parameters by clicking the toolbar's right side chevron.

The command line prompts you to select a label or point within the parcel to subdivide.

8. In the drawing, select the parcel number (in the parcel south center).

Next, you select the parcel frontage's beginning and ending points.

9. In the drawing, at the frontage's southeast end, select its endpoint (Endpoint object snap). Next, select the frontage's northeastern end (Endpoint object snap) to define the parcel's frontage geometry.

While defining the frontage, the routine displays a jig following the frontage geometry. The jig understands the changing frontage geometry (see Figure 3.20).

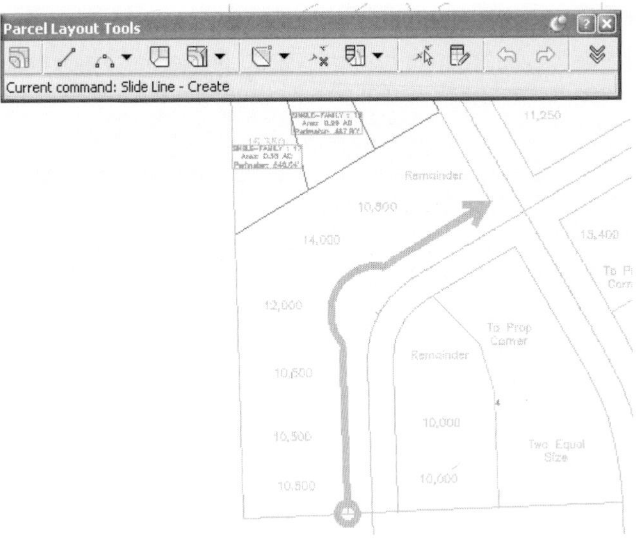


**FIGURE 3.20**

Next, you are prompted for a side yard direction.

10. Using the **Nearest** object snap, select a frontage line point and a second point, using the **Perpendicular** object snap, to the backyard line.

A parcel displays between the front and backyard segments.

11. The command line prompts to accept the parcel; press ENTER twice to accept two parcels with 10,500 square feet.

The next parcel is 12,000 square feet.

12. While the command line prompts to accept the parcel, click the toolbar's Expand Toolbar chevron to display the parcel parameters.
13. In parcel parameters, change the parcel size to 12000, close parcel parameters by clicking on the chevron and press ENTER.

The next parcel is 14,000 square feet.

14. While the command line prompts to accept the parcel, click the toolbar's Expand Toolbar chevron to display the parcel parameters.
15. In parcel parameters, change the parcel size to 14000, close parcel parameters by clicking on the chevron and press ENTER.

The next parcel is 10,800 square feet.

16. While the command line prompts to accept the parcel, click the toolbar's Expand Toolbar chevron to display the parcel parameters.
17. In parcel parameters, change the parcel size to 10800, close parcel parameters by clicking on the chevron and press ENTER.
18. A warning dialog box appears stating that the minimum frontage is less than 100', click OK.

The last parcel is the remainder of the block area.

19. Press ESC to exit the Parcel Sizing command.
20. Change the last parcelin the block to **Single Family** with the label of **Name Area & Perimeter**.
21. At Civil 3D's top left, Quick Access Toolbar, click the ***Save*** icon to save the drawing.

## Manual and Automatic Parcel Sizing

The next area's first parcel uses the manual method, and the remaining parcels use an automatic mode. The last parcel is the block's remainder.

1. Click Ribbon's ***View*** tab and on the Views panel's left, select the named view **Residential Parcels 2**.
2. Go to the fifth icon in from the left, to the right of the Parcel Sizing Tools icon stack, click the drop-list arrow, and select SLIDE LINE - CREATE.
3. The Create Parcels - Layout dialog box opens. In the dialog box, set the Parcel Style to **Single-Family** and the Area Label Style to **Name Area & Perimeter** (see Figure 3.18).
4. Click ***OK*** to exit the dialog box.
5. In the parcel parameters sizing section, set the parcel size to **12500**, press ENTER, and at the toolbar's right side, click the **chevron** to collapse the toolbar.
6. The routine prompts you to select the parcel to subdivide, select the parcel's label.

Next, you are prompted to define the frontage.

7. Using the Endpoint object snap, define the block's eastern frontage by selecting its southern and northern endpoints (see Figure 3.21).

A jig appears, tracing the frontage line and recognizing its geometry.

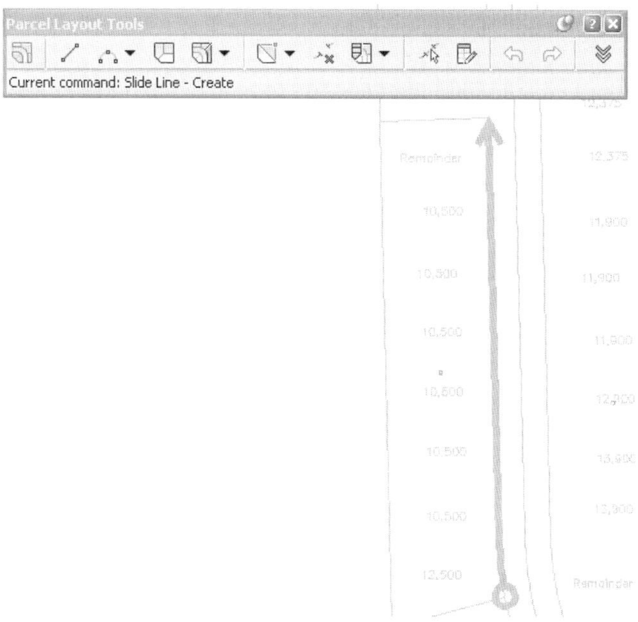


**FIGURE 3.21**

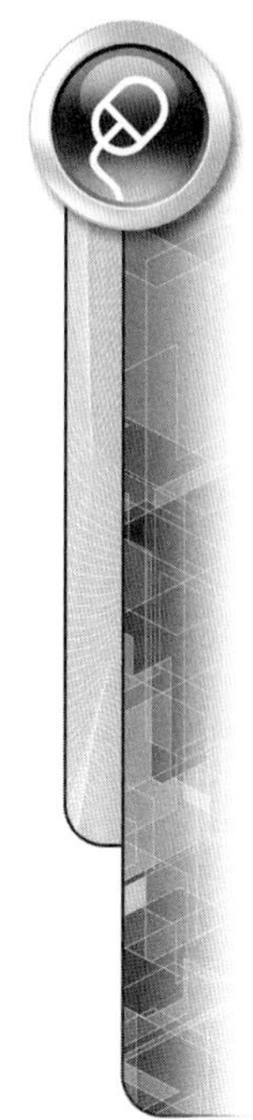

Next, you are prompted for an angle.

8. For the angle enter **90** and press ENTER.
9. A 12,500 square feet displays and the command line prompts to accept the parcel; press ENTER to accept.

The next parcel displays with 12,500 square feet.

10. On the right side of the Parcel Layout Tools toolbar, click the **chevron** to expand the toolbar.

11. In the Parcel Sizing set the parcel size to 10,500 square feet.
12. In the Automatic Layout section, change Automatic Mode to **ON**, and then set Remainder Distribution to **Redistribute remainder**.
13. Click the **chevron** to collapse the toolbar.
14. The 10,500 square feet parcels display and the command line prompts to accept the resulting parcels; press ENTER to accept.

The parcel block is divided into 10,500 square foot parcels with the remaining area distributed evenly to every xparcel.

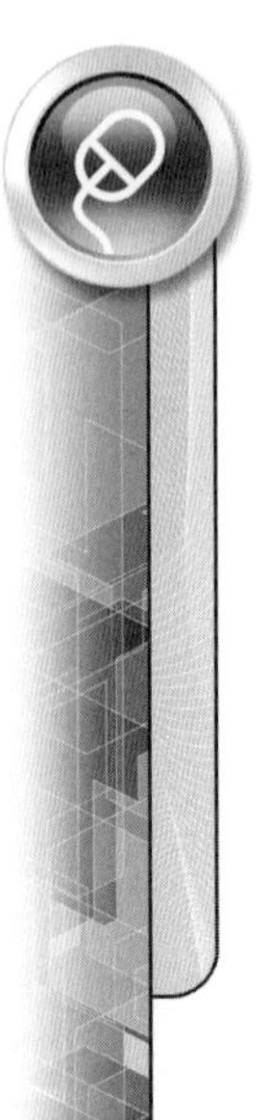

The routine prompts you to select the next parcel to subdivide.

15. Press ESC once, then enter an '**X**' and press ENTER to exit the Parcel Layout Tools toolbar.
16. At Civil 3D's top left, Quick Access Toolbar, click the ***Save*** icon to save the drawing.

The remaining parcel (block's top) needs to have its type and label changed to Single-Family and Name Area & Perimeter.

17. Select the central parcel's label, press the right mouse button, and, from the shortcut menu, select PARCEL PROPERTIES....
18. Select the **Information** tab and change the Object Style to **Single-Family**.
19. Select the **Composition** tab, assign the **Name Area & Perimeter** to the Area selection label style, and click ***OK*** to exit.
20. At Civil 3D's top left, Quick Access Toolbar, click the ***Save*** icon to save the drawing.

The next phase reviews the new residential parcels' areas and boundaries.

## SUMMARY

- Each parcel style defines a parcel segment layer.
- When subdividing a parcel, one strategy is to create parcel blocks (areas of uniform land use) from the overall boundary.
- The Parcel Layout Tools toolbar sizing routines define the last parcel segment (side yard).
- The Parcel Properties dialog box changes a parcel's type.
- When converting line work into parcels, users can preserve or erase the original entities.

## UNIT 3: EVALUATING THE SITE AND PARCELS

Civil 3D has several parcel-evaluation tools. These tools are in the Parcel Properties dialog box, Map Check, and the Toolbox's reports. A parcel's Properties dialog box contains map check and inverse values, Map Check evaluates a parcel's closure, and Toolbox generates several parcel reports, including metes and bounds, surveyor's certificates, parcel areas, and so on. Map Check works best with annotated parcel segments.

### PARCEL PROPERTIES

Parcel Properties contains four panels: Information, Composition, Analysis, and User Defined Properties (see Figure 3.22). The Information panel contains the parcel's name and style. The Composition panel contains the parcel's label style, area, and perimeter. The Analysis panel shows the parcel's Inverse or Map Check values. Inverse uses Civil 3D's precision, 14 decimal places, to calculate each parcel segment's direction, distance, and coordinates. Map Check calculates a parcel's geometry to the distance and angle precisions. This reduced precision introduces distance and angle errors into the perimeter calculations. As a result, the perimeter's starting point coordinates do not match the ending point's coordinates. The distance and angle between the ending and starting points are the Map Check report's parcel "closure error."

All Inverse and Map Check reports traverse a perimeter clockwise. If you want to traverse counterclockwise, reverse the direction by toggling Counterclockwise at the panel's top right.

Each report starts at a point, the POB, which may not be the starting point the user wants. A user can change the starting point interactively by selecting the pick icon to the right of the current POB coordinates (the Analysis panel's top right). When selecting a new POB, Civil 3D displays a glyph at the current POB and gives the option to select a new POB, to traverse the parcel's vertices, or to abort the redefinition process. The Inverse and Map Check reports change to accommodate the new POB and parcel segment order.

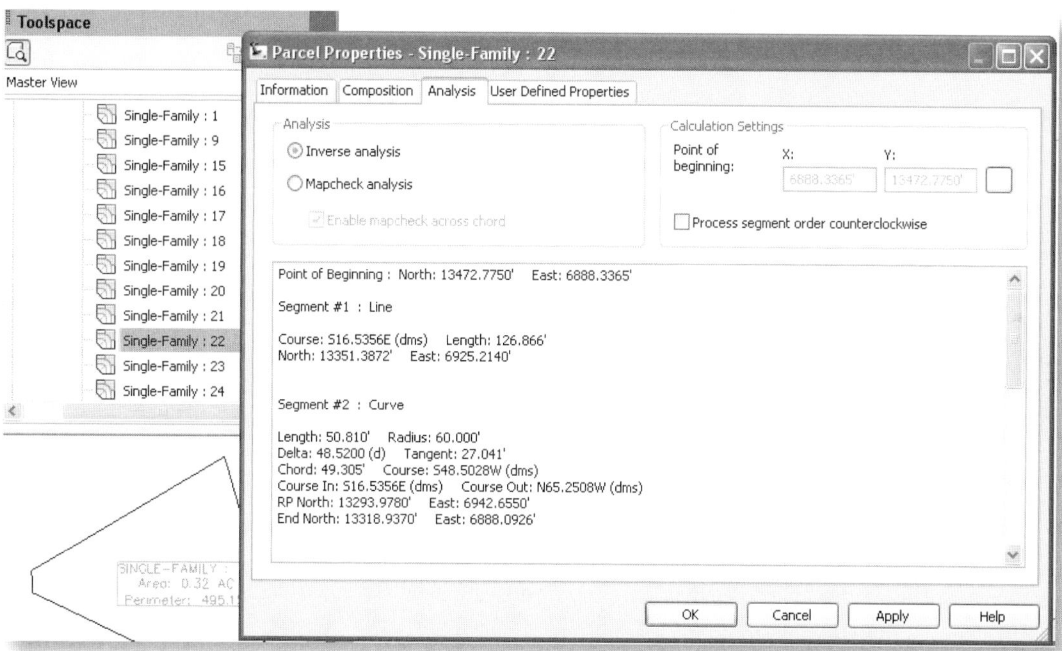


**FIGURE 3.22**

## TOOLBOX/REPORTS MANAGER

Toolbox, or Reports Manager, creates area, inverse, metes and bounds, and surveyor-specific reports (see Figure 3.23). To create a parcel's analysis, select the parcel, press the right mouse button, and select Execute… from the shortcut menu.

Users are able to modify report settings, or if proficient in XSL (eXtensible Style Language) or VBA (Visual Basic), can author their own reports. Visit the LandXML.org Web site to learn more about this data format and its civil industry support. In Ribbon's View tab, at the Toolspace icon's middle right is the Toolbox icon that toggles the Toolspace's Toolbox tab.

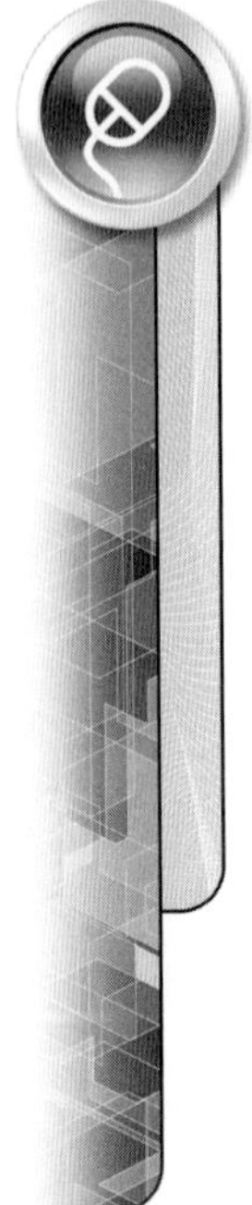

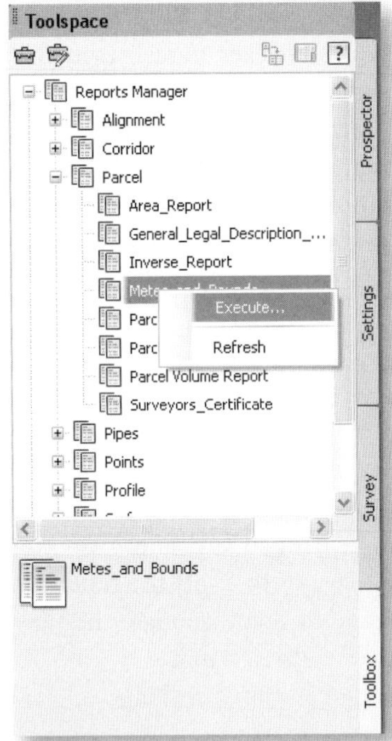


**FIGURE 3.23**

## PARCEL PREVIEW

Preview lists a drawing's object occurrences. Preview also displays a selected object's geometry. Two steps toggle on to display an object's geometry. The first step is to toggle on item previews; the magnifying glass icon is at Prospector's top. The second step is to right-mouse click a prospector heading and select Show Preview from the shortcut menu.

## EXERCISE 3-3

After completing this exercise, you will:

- Be able to review site properties.
- Be able to review parcel properties.
- Be able to create parcel reports with Toolbox.
- Be able to create a parcel data LandXML file.

### Exercise Drawing Setup

This exercise continues with Unit 2's drawing. If you did not complete that exercise, browse to the Chapter 3 folder of the CD that accompanies this textbook and open the *Chapter 03 - Unit 3.dwg* file.

1. If your drawing from the previous exercise is not open, open it now, or browse to the Chapter 3 folder of the CD that accompanies this textbook and open the *Chapter 03 - Unit 3* drawing.

## View Site Properties

Site properties affect a site's layers and the initial parcel's numbers.

1. If necessary, click the ***Prospector*** tab.
2. In Prospector, expand the Sites branch until you are viewing Site 1.
3. Click **Site 1**, press the right mouse button, and from the shortcut menu, select PROPERTIES....
4. Click the ***Information***, ***3D Geometry***, and ***Numbering*** tabs to review their current values.
5. Click **OK** to close the dialog box.

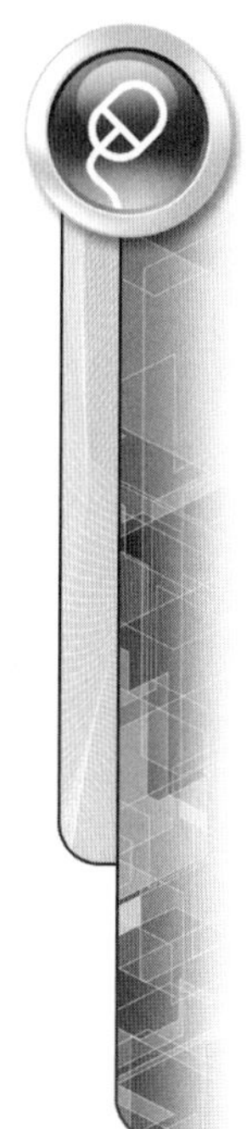

## Preview Parcels

When selecting an item, named site, or individual parcel, if on, Prospector previews the item's geometry.

1. At Prospector's top, select the **magnifying glass** until it displays a border around the icon.
2. In Prospector, expand the **Site 1** branch until you are viewing the Parcels heading and its parcels list.
3. Select the **Parcels** heading, press the right mouse button, and if not on, select SHOW PREVIEW.
4. From the list of parcels, select a parcel to display its vector geometry in the preview area.

## Parcel Properties

Parcel properties include a parcel's name, parcel type, parcel label, and area and perimeter statistics.

1. In Prospector, from the parcels list, select a **Single-Family** parcel, press the right mouse button, and from the shortcut menu, select ***ZOOM TO***.
2. In Prospector, with the parcel entry still highlighted, press the right mouse button, and from the shortcut menu, select PROPERTIES....
3. Click the ***Information*** tab to view its contents.

The parcel name is grayed out and is not editable.

4. Click the ***Composition*** tab to view its contents.

Composition lists the area, perimeter, and currently assigned parcel area selection label style.

5. Click the ***Analysis*** tab to view its contents.

Analysis displays a parcel's inverse or map check values. These reports summarize the boundary's quality based on the line work defining the parcel.

6. Click INVERSE ANALYSIS and review its values.
7. Click MAPCHECK ANALYSIS and review its contents.
8. Click the PICK NEW POB icon to the right of its current coordinates.

The dialog box closes and a glyph indicating the current POB position is displayed on the parcel's perimeter. The default is to move it to the next parcel corner.

9. Press ENTER to move the POB clockwise around the boundary.
10. Type '**S**' and press ENTER to select a new POB.

The dialog box is displayed again with new POB coordinates and a new parcel analysis.

11. Click ***OK*** to exit the Parcel Properties dialog box.
12. At Civil 3D's top left, Quick Access Toolbar, click the ***Save*** icon to save the drawing.

## Toolbox Parcel Report

Toolbox creates printable and customizable parcel reports.

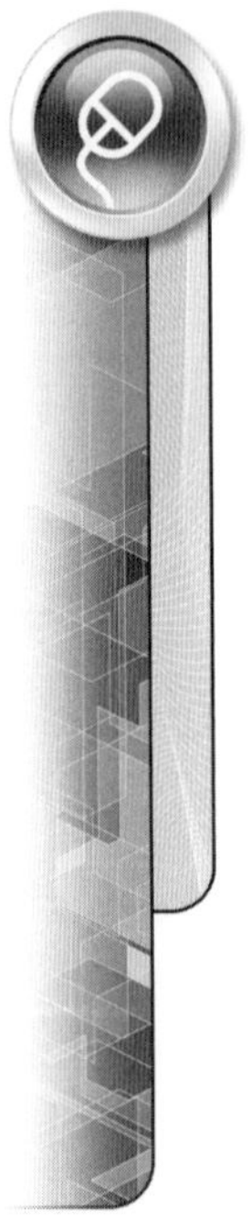

1. If the Toolbox tab is not displaying on the Toolspace, from the Ribbon, click the View tab, and at the Palettes panel's click the ***Toolbox*** icon.
2. In the Toolbox, expand Reports Manager until you are viewing the Parcel report section.
3. Select the METES_AND_BOUNDS report, press the right mouse button, and from the shortcut menu, select EXECUTE....
4. In the Export To LandXML Report dialog box, click ***OK*** to create reports for the selected parcels in Internet Explorer.
5. Scroll through the report and, when finished, from the File menu select SAVE AS ... and save the report.
6. Close **Internet Explorer**.
7. At Civil 3D's top left, Quick Access Toolbar, click the ***Save*** icon to save the drawing.

This completes the analysis of parcels. The next unit reviews the various parcel-editing tools.

## SUMMARY

- Inverse or Map Check report starts at a parcel's POB.
- Users can redefine a parcel's POB and analysis direction (clockwise or counter-clockwise).
- Parcel properties include Inverse and Map Check reports.
- Toolbox has several reports for analyzing parcel values.

## UNIT 4: EDITING A SITE AND PARCELS

The Parcel Layout Tools toolbar has site- and parcel-editing tools. In addition to these tools, users can graphically adjust the parcel segments. All parcel labels change to reflect the modified parcel area and segments.

The Parcel Layout Tools toolbar editing tools include tools to do the following: draw new or delete parcel lines and arcs; merge parcels; add, delete, and break PIs in parcel segments; and change the values (coordinates, radii, etc.) associated with parcel lines and curves. The toolbar also includes the parcel commands undo and redo.

Grip-editing a parcel segment's location is the easiest way to adjust its size. When you want to relocate a parcel's side yard segment, selecting it will display a grip at its frontage intersection. The user can slide this segment along the frontage line to adjust the parcel's area. Users can even transfer the segment to another parcel block simply by selecting a new point on the other block's frontage.

When finalizing a subdivision, the parcel numbers may not be in the desired sequence. The Parcels menu's Renumber and Rename Parcels commands are tools to modify the parcel numbers and names.

The Renumbering parcels process is to select a starting parcel number, the increment value, and the parcels to renumber (see Figure 3.24). For example, renumbering parcels is a block. A block can renumber its parcels starting at 100 with increments of 1. A second block starts its parcel numbers at 200 and also increments by 1.

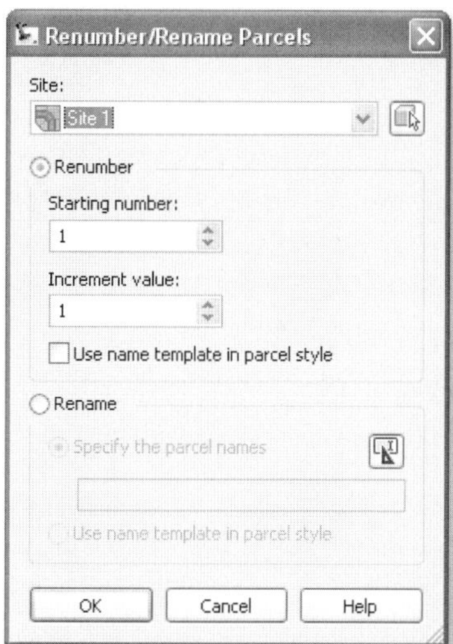


**FIGURE 3.24**

Renaming parcels uses a user-defined prefix and number sequence, or uses a naming template to redefine the parcel name (see Figure 3.25). If users want to change the naming template, the Use name template in parcel style must be toggled off. If this is on, the user cannot change the parcel style's base name (for example, single-family). If toggled off, the user can edit the base name value. However, from that point on, he or she must manually update its value.

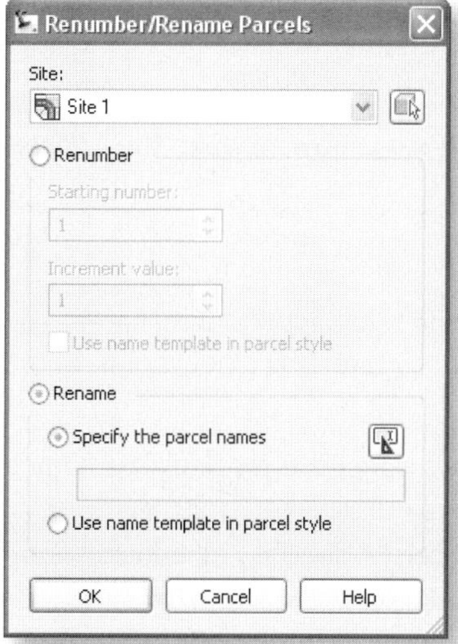


**FIGURE 3.25**

When a parcel name includes a sequential number or a parcel value (address, parcel tax ID, parcel number), the user can define a renaming format in the Name Template dialog box (see Figure 3.26). To display the Name Template dialog box, in the Renumber/Rename Parcels dialog box click the icon to Specify the parcel names' right. It is in the Name Template dialog box that a user redefines the naming template.

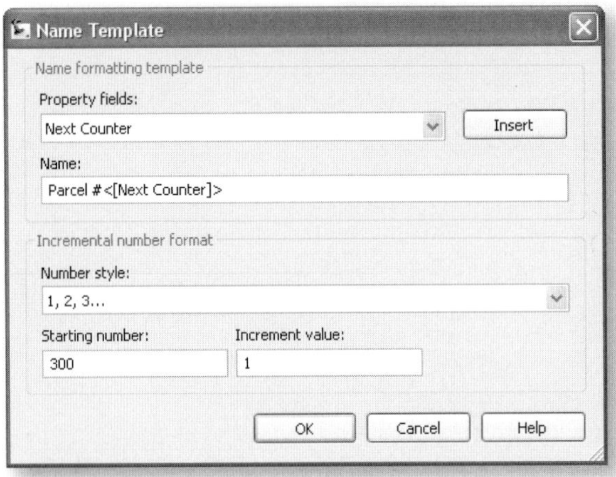


**FIGURE 3.26**

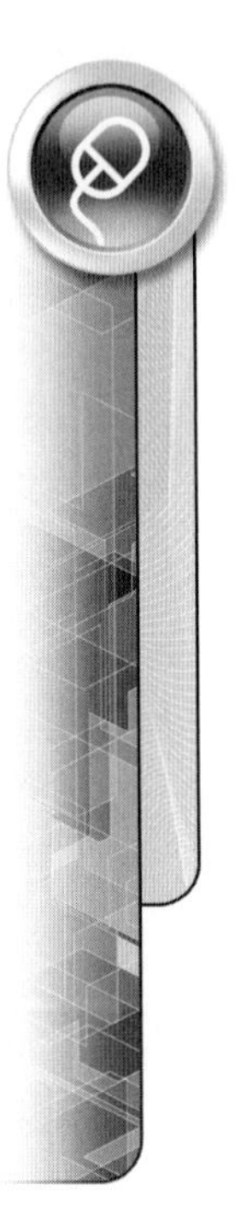

## EXERCISE 3-4

After completing this exercise, you will:

- Be able to edit parcels by manipulating their segments.
- Be able to edit with Parcel Layout Tools toolbar tools.

### Exercise Setup

The exercise continues with the previous exercise's drawing. If you did not complete the previous exercise, start this exercise by browsing to the Chapter 3 folder of the CD that accompanies this textbook and open the *Chapter 03 - Unit 4* drawing file.

1. If the previous exercise's drawing is not open, open it now, or open the *Chapter 03 – Unit 4* drawing file.

### Grip-Editing Parcel Segments

Grips can slide a side yard segment along the frontage line to adjust the parcel area. You can even transfer the segment to an adjacent parcel block by selecting a new point on the block's frontage (see Figure 3.27).

1. On the Ribbon click the ***View*** tab. At the Views panel's left, select the named view **Residential Parcels 1**.
2. Click a side yard line between two adjacent parcels. A special grip appears, indicating that the side yard is "tied" to the two lots' frontage. Select the **grip**.

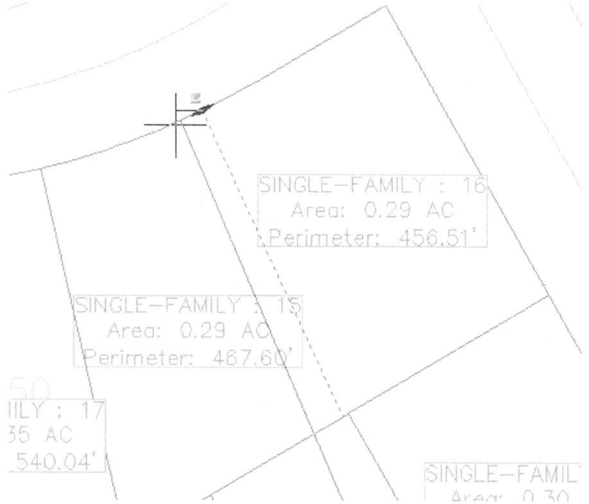


**FIGURE 3.27**

3. If on, toggle **OFF** Osnap and slide the segment along the frontage line.
4. When the side yard line is in the desired position, press the left mouse button to set its location, and press ESC to deselect the object.

If you are dragging a parcel segment into an adjacent parcel block, the segment divides the block. The original parcels sharing the parcel line merge. If moving the segment back to its original location, the newly merged parcels divide and the newly divided block returns to its original form.

## Parcel Layout Tools Toolbar — Delete Parcel Sub-Entity

There are reasons to delete parcel segments and redraw them, to change the segment's angle, or to further divide the parcels. The Parcel Layout Tools toolbar has the tools to perform these actions.

1. In the Ribbon select the Modify tab. At the Design panel's left click the Parcel icon. The Ribbon changes to a Parcels tab. In the Modify panel select PARCEL LAYOUT TOOLS, to display the Parcel Layout Tools toolbar.
2. Near the toolbar's middle, click the ***DELETE SUB-ENTITY*** icon (see Figure 3.28).

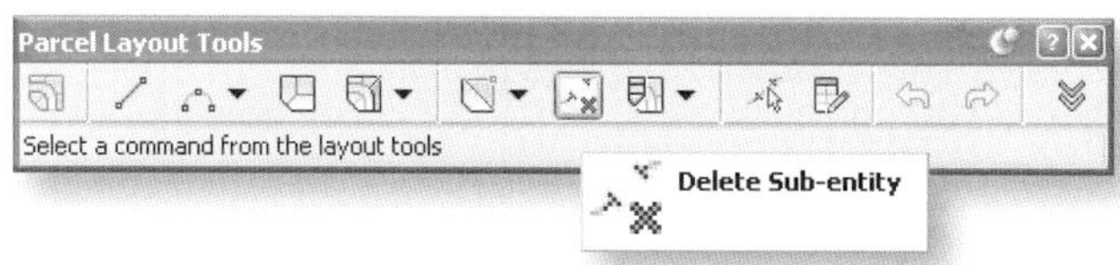


**FIGURE 3.28**

This command deletes parcel segments. If using Delete Sub-entity on frontage, backyard, or boundary segments, this command destroys the parcel definition results in parcel segments. If deleting a side yard segment, the two parcels merge into a single parcel.

3. In the current view, delete one or both side yard segments.

## Parcel Layout Tools Toolbar — Add New Parcel Segment

You can use any of the following parcel-sizing methods to redraft the side yard segment: Slide Line or Swing Line.

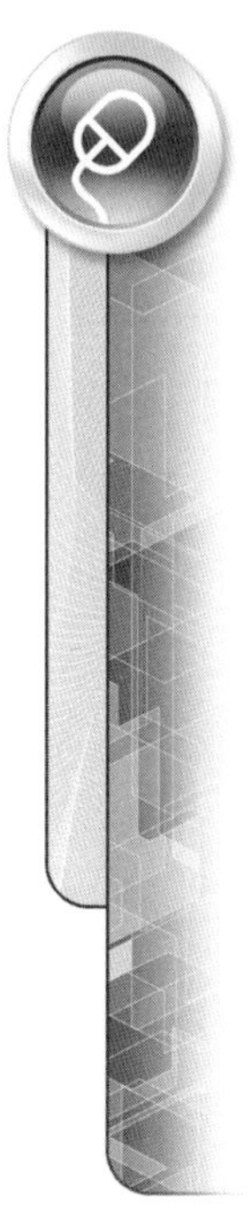

1. Redraft the segment(s) with one of the just-mentioned Parcels Layout Tools toolbar commands. Set Parcel Style to **Single-Family** and use the Area Label Style of **Name Area & Perimeter**.

## Parcel Layout Tools Toolbar — Edit Parcel Segment Angle/Direction

There are times when users want to change the side yard segment's angle or direction. The Slide Line - Edit routine redefines the side yard's angle relative to the frontage geometry or its direction or bearing (Direction).

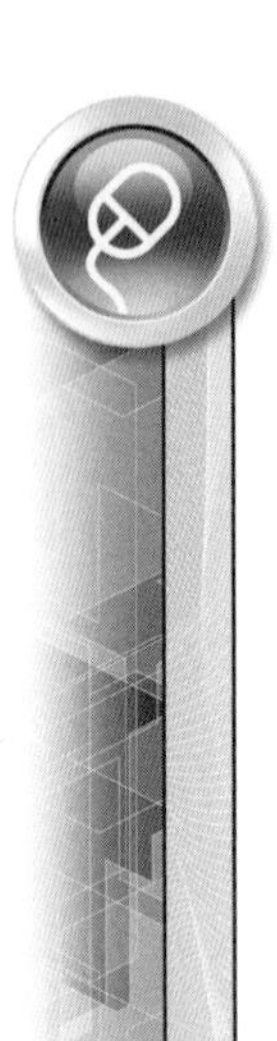

1. From the Parcel Layout Tools toolbar, Parcel Sizing Tools icon stack, select the command ***SLIDE LINE - EDIT***.
2. If the Create Parcel From Layout dialog box opens, set the parcel type to **Single-Family**, the label to **Name Area & Perimeter**, and click ***OK***.
3. Click the collapse toolbar chevron.
4. You are prompted to select a parcel line to adjust. Select a side yard segment between two parcels.
5. Next, select the parcel to edit by placing the highlight over the parcel and clicking in the parcel area.
6. Next, you are prompted to define the parcel's frontage start and ending points.
7. Next, you are prompted for a new angle. For the new side yard line, enter a new angle.
8. After entering a new angle, the command line prompts to accept the adjustment. Press ENTER to accept the new parcel definition.
9. The command line prompts for the next parcel to adjust. Edit a few more parcels in the area.
10. After editing a few parcels, press ESC twice to exit the Parcel Layout Toolbar.
11. At Civil 3D's top left click, Quick Access Toolbar, the ***Save*** icon to save the drawing.

## Renumbering Parcels

After subdividing a site into parcels, the parcel numbers may not represent the desired numbering. The Renumber/Rename Parcels command renumbers selected parcels.

1. If you are not in the Residential Parcels 1 view area, click the Ribbon's View tab. In the Views panel, select and restore the named view **Residential Parcels 1**.
2. In the Ribbon's Parcel tab, Modify panel select RENUMBER/RENAME to display the Renumber/Rename Parcels dialog box (see Figure 3.25).
3. Set the Starting Number to **100** and click ***OK*** to close the dialog box.
4. Next, select a point in the short block's westernmost parcel, and then select a second point in the easternmost parcel.

This selects the parcels for renumbering.

5. Press ENTER twice to end the selection and renumber the parcels west to east, from **100** to **102**.

## Renaming Parcels

There are times when it is desirable to rename the parcels. It may be that the parcel number needs to be a Parcel Identification Number (PIN) or a more general name, such as Parcel #.

1. In Ribbon's Parcel tab, from the Modify panel, select RENUMBER/RENAME to display the Renumber/Rename Parcels dialog box (see Figure 3.25).
2. Toggle **ON** Rename and Specify the Parcel Names and click ***Edit Name Template*** (the icon to the toggle's right).

3. In the Name Template dialog box, Name Formatting Template section, set the Property Fields to **Next Counter**, in the Name area type **Parcel #**, and click **Insert** to put the next counter after "Parcel #."
4. In the Incremental Number Format area, set the Number Style to **1, 2, 3...**, and set the Starting Number to **300**. Your settings should match Figure 3.26.
5. Click ***OK*** to exit the Name Template dialog box and return to the Renumber/Rename Parcels dialog box.
6. Click ***OK*** to close the dialog box.
7. Select the lots for renumbering by first selecting a point in the short block's westernmost parcel and then selecting a point in the easternmost parcel.
8. Press ENTER twice to rename the parcels.
9. Close the Parcel tab by selecting Close on the right.
10. Click the ***Prospector*** tab to make it current.
11. In Prospector, expand the Sites branch until you are able to viewing the Parcels heading and the list of parcels with new names and numbers in Preview.
12. At Civil 3D's top left, Quick Access Toolbar, click the ***Save*** icon to save the drawing.

This completes the parcels editing exercise. The next section reviews parcel segment labels and their annotation.

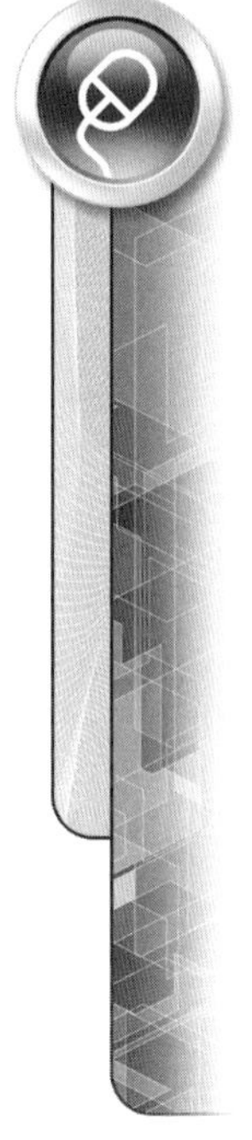

## SUMMARY

- Users can graphically edit a parcel segment's position.
- When graphically moving a segment, the segment follows the attached object's geometry.
- Parcel(s) responds to segment changes (calculates a new area and perimeter).
- If deleting a parcel's boundary segment and it is not a closed boundary, the remaining segments are not labeled.
- When creating a closed polygon from parcel segments, the polygon becomes a parcel with an area label.
- The Parcel Layout Tools toolbar has routines to change a segment's angle or direction.
- The Parcel Layout Tools toolbar has commands to add vertices to parcel segments and add new line and curve segments.

## UNIT 5: PARCEL AND SEGMENT ANNOTATION

Parcel labels an area as part of the parcel-sizing process. When parcels are defined, they receive an area label and optionally label their line and curve segments. If segments are not labeled when the parcels are sized, the Ribbon's Annotate tab, Add Labels drop-list, Parcel flyout, Add Parcel Labels command creates the line and curve labels.

There are two parcel label types: one for the parcel object, the other for the parcel's line and curve segments. The line and curve labels are either on the segments (Label mode) or are entries in a table (Tag mode). All annotation reacts and changes correctly to display the parcel or its segments' values.

Labels can be pinned to a location. If the view rotates, the labels rotate around the pinned point and remain plan-readable.

## PLAN READABILITY AND SCALE SENSITIVITY

By default, all labels are view-dependent and rotate to be plan-readable in any orientation (reading from left to right). Style settings control the angle, thus changing the label's orientation. All area and segment labels are scale dependent. The label style defines the text size as a paper size measurement and it resizes itself by a viewport's scale. For example, a label text height of 0.1 is 5 units tall in a 1"=50' viewport, and 3 units tall in a 1"=30' viewport. If each viewport has a different scale, each viewport's text size will be different, but its paper target size (0.1) will appear in each layout.

## AREA LABELS

When creating a parcel, an area label appears at the parcel's centroid. An area label can be as simple as a parcel number or as complicated as it needs to be. Figure 3.29 shows the list of parcel properties, each of which is available for this label type.

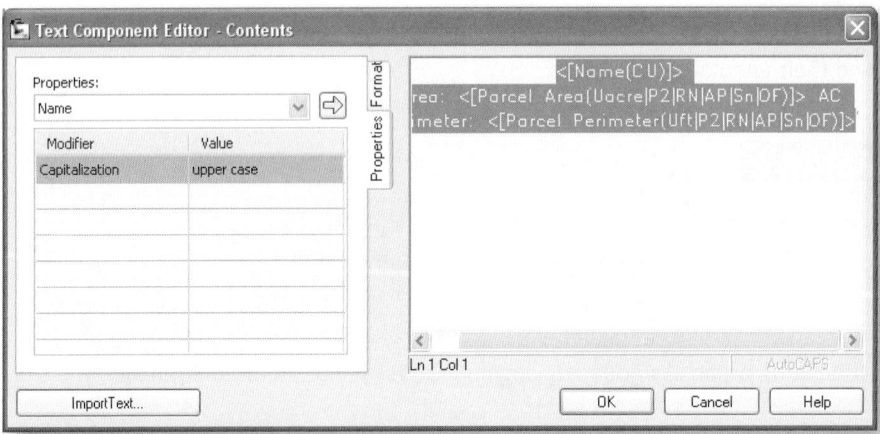


**FIGURE 3.29**

An area label can contain more than one text component. However, when more than one component is included, each has its own border. If users want a single border, the border must be a single multi-line text component. When using a single multi-line component, all text is the same height.

A block can be a label component (see Figure 3.30).

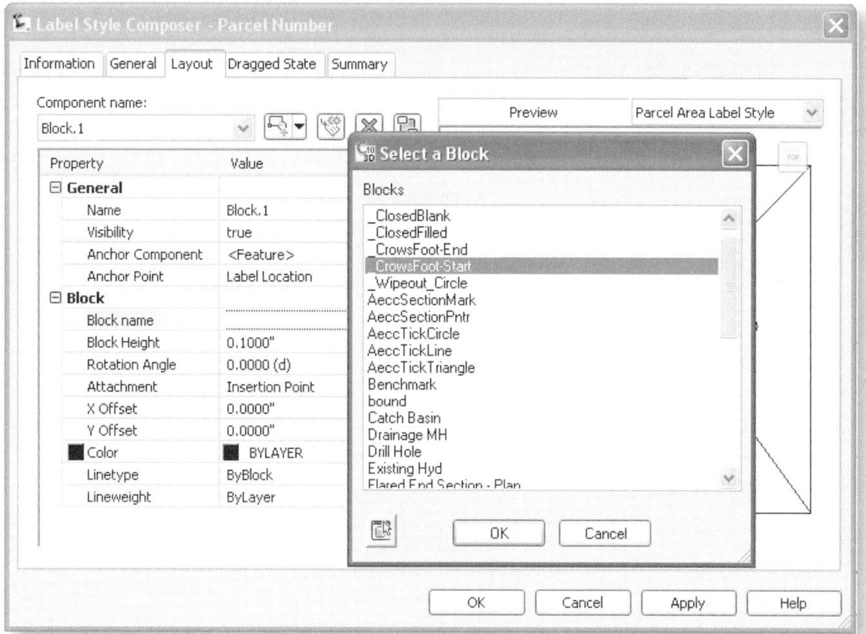


**FIGURE 3.30**

## LABELING SEGMENTS

All parcel segment labels are dynamic and view-dependent. The labels flip to be plan-readable after rotating the view. When a viewport displays labels, the labels resize for the viewport's scale.

When parcels are defined, they can be assigned segment labels. Or, after they are defined, the Add Labels dialog box will label individual or multiple parcel segments (see Figure 3.31). The Add Labels dialog box sets the line and curve segment label styles. When placing multiple segment labels, the routine prompts for clockwise or counterclockwise label placement.

When the labeling is simple and straightforward, the multiple segment option is quickest. However, there are times when more control over the labels' placement is needed, or the labeling needs to span individual lot segments (that is, backyard lines, block parcel lines, etc.). The single segment option places each segment label at the selected label point.

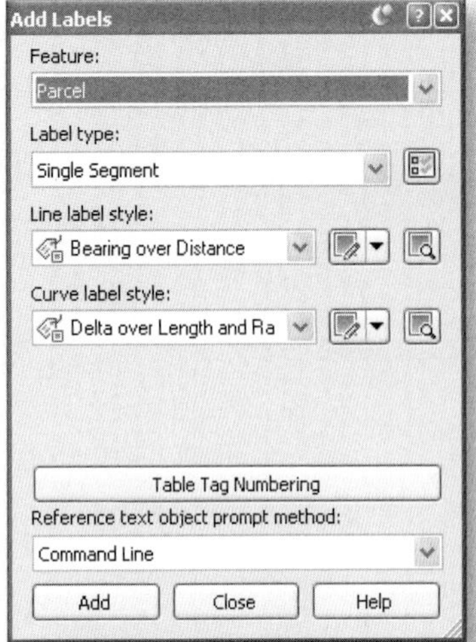


**FIGURE 3.31**

## LINE AND CURVE SEGMENT STYLES

Each segment label style defines a verbose, tag, and dragged state format (see Figure 3.32). The verbose label style annotates the segments with the specified label components. The properties annotated depend on the label's components. The tag variant contains the segment's number corresponding to an entry in a line or curve table. The table's contents depend on the table definition.

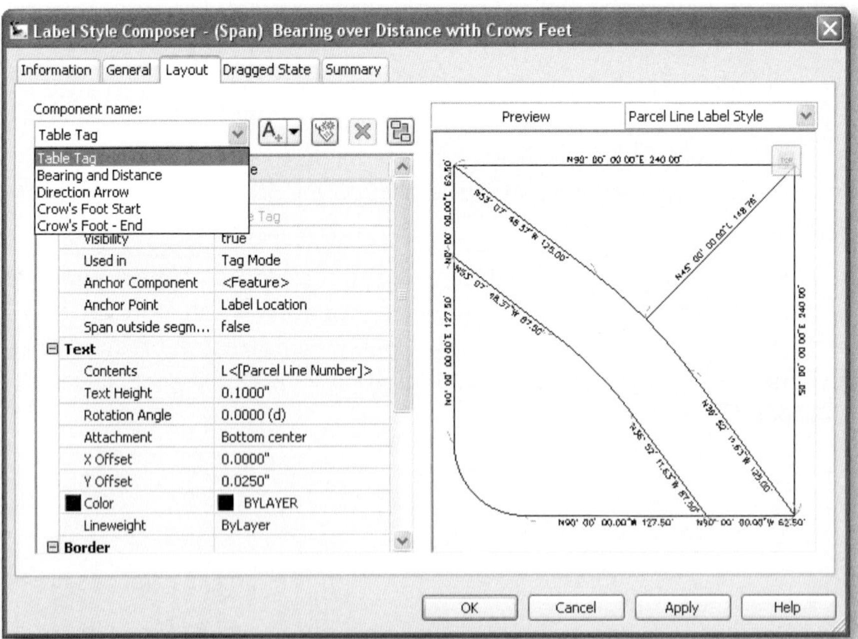


**FIGURE 3.32**

The span line label styles include crow's feet. These styles typically indicate a parcel line segment that has a distance greater than any single segment that abuts the segment. Individual parcels sharing the line have a distance label with no bearing, because the span label provides the direction.

All line label styles include a direction arrow component. Users can erase the Direction Arrow or hide it (see Figure 3.32).

All segment labels attach to a parcel segment. If a user extends a label over several parcels, the segment must be whole and unbroken at the individual side yard line intersections.

The General Plan Readability bias affects how a label orientates itself. This value may need to be adjusted to suit individual user standards.

As with plan readability, labels are viewport-scale sensitive. When setting a viewport's scale, execute a regenall and the labels will resize themselves.

## LABEL NON-PARCEL SEGMENTS

When labeling non-parcel line or curve segments, use the Add Labels' Line and Curve label styles. These styles have the same capabilities as parcel segment labels. However, these styles have different properties. These styles also have crow's feet.

## EDITING SEGMENT LABELS

At times, labels display the wrong direction or positions. Address these situations by selecting the label and pressing the right mouse button to display a shortcut menu with segment label-editing routines (see Figure 3.33). The shortcut menu commands reverse a label's direction, exchange its elements' locations, and adjust the label's properties (different label style, reversed, or flipped). Double-clicking a label displays the Properties dialog box, which lists the label's current settings.

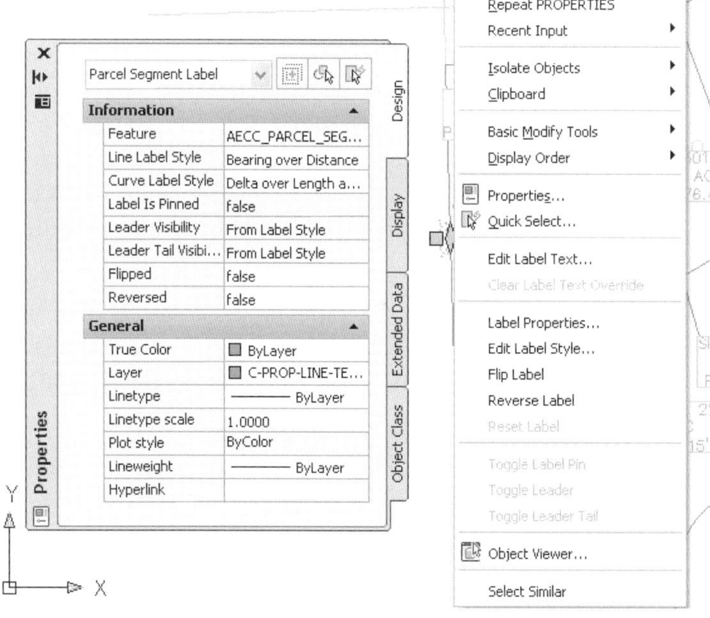


**FIGURE 3.33**

### Flip Label

Flip Label causes a label to exchange its text locations. For example, a label has the direction above and the distance below. When users select the label, press the right mouse button, and select Flip Label, the direction changes to the below position and the distance moves to the above position.

### Reverse Label

Reverse Label causes a selected label to change its bearing by 180 degrees. For example, if a label is N 88° 45' 54" E and reversed, the label changes to S 88° 45' 54" W.

### Reset Label

This command returns the label to its original location. Clicking a dragged label's round grip also will return the label to its original location.

### Dragging a Label

Every label style has two states: layout and dragged. Layout represents the initial label position. When dragging a label from its original position, the layout display changes to represent the label style's Dragged State settings. A label can remain as defined or it can change to stacked text. Reset Label returns a dragged label to its original location. Reset Label is an option on the selected label's shortcut menu or the user can click a label's circular grip (reset label grip) to gain this option.

## LABEL PROPERTIES

When selecting Label Properties from the shortcut menu, the user can assign a new label style, reverse the label's direction (if a line segment), flip the label elements, and pin a label (see Figure 3.33). When Label Is Pinned is set to false, a label is free to move as the associated segment is edited. If Label Is Pinned is set to true, the label remains at its current position even if the associated object is edited. If the view of a pinned label is rotated, the label rotates around the pinned label location.

## TABLE STYLES

Parcels have four basic table styles: lines, curves, segments, and areas. The line and curve tables list line and curve segment geometric values. Each table entry corresponds to a segment tag on the parcel segments. Each segment has a unique number and a corresponding table entry that displays its values. For example, the tag L1 identifies a line segment and has a corresponding L1 table entry. The L1 table record lists the bearing and distance of the segment. A segment table has both line and curve segments. The area table lists the parcels' areas.

Tables apply to general line and curve and parcel segments. To create a table, the segments must be labeled. When creating the table, the label mode changes to tag, the segment labels change to L1... or C1..., and the table lists the corresponding tag and its geometric values.

A Table Style dialog box's Data Properties panel defines the table with three sections: Table settings, Text settings, and Structure (see Figure 3.34).

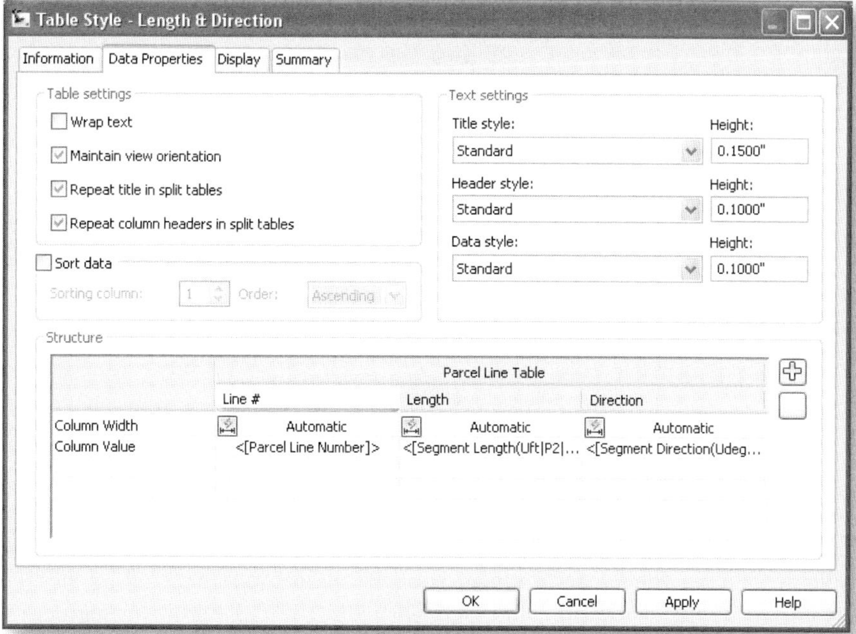


**FIGURE 3.34**

The panel's Table settings section contains toggles for wrapping text, maintaining plan view, repeating the title and column headings in split tables, and sorting table entries. Text settings set the title's text style and size, column headers, and data. The Structure section defines a table record's properties type and its order. Each table column contains object properties information: length, direction, starting northing or easting, ending northing or easting, and a tag number. At the right side of the dialog box are two buttons to add or delete table columns. Edit a column's contents by double-clicking in the column's value cell. This action displays the Text Component Editor. The editor defines which properties (selected from the Properties drop-list) and their formats appear in the table (see Figure 3.35).

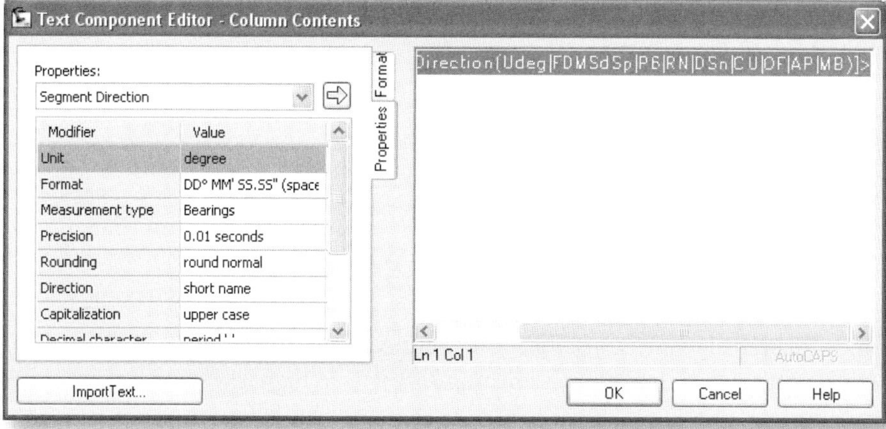


**FIGURE 3.35**

## CREATING A TABLE

To create a table, select the table type and add the necessary information. The table type is selected from the Ribbon's Insert tab, Parcels panel, Tables drop-list. The list has four tables types: line, curve, segment (line and curve), and area.

Selecting a table type from the list opens the Table Creation dialog box. The Table Creation dialog box sets the table style, its layer, what labels to include, the table's layout, and whether it is dynamic or static (see Figure 3.36).

Select the segments by their style or by selecting the segments. The Selection Rule, Add Existing, includes only existing labels in the table. If setting this rule to Add Existing and New, the table is updated and includes any new labeling.

The table's layer is set in the Edit Drawing Settings' Object Layers settings list.

If lengthy, settings at the bottom of dialog box set the next table breakpoint.

When a segment becomes a table entry, it changes the label's visibility mode to Table Tag. When segment labels are viewed, they display L1... or C1.... If you set the labels' display mode (in the General tab of the style definition) back to Label and the table is dynamic, the segments will disappear from the table and the segment's annotation will return to verbose. General labels are for lines and arcs and parcel labels are for parcel segments.

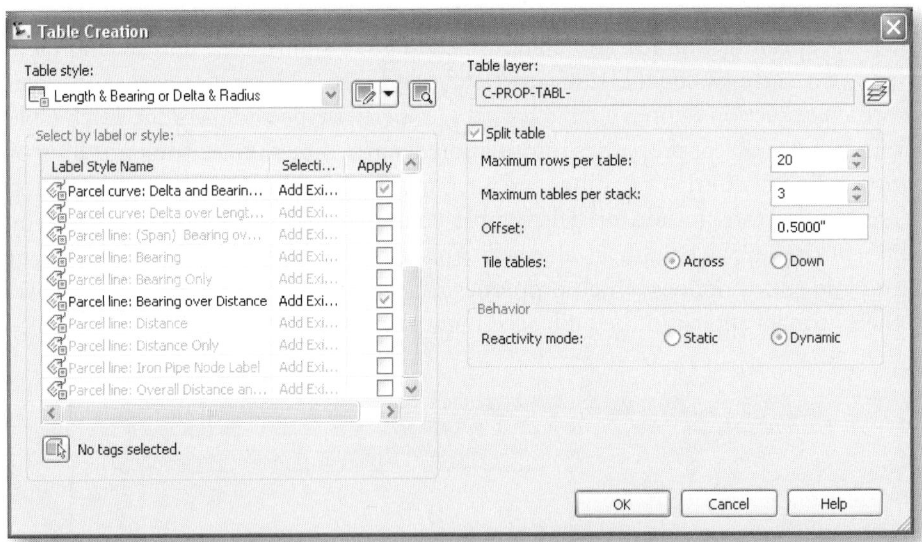


**FIGURE 3.36**

## USER-DEFINED PROPERTY CLASSIFICATIONS

A user can define a parcel property that can be a part of a lot label (for example, defining an Industrial classification). The user-defined property definition is a two-step process. The first is defining the classification name (see Figure 3.37).

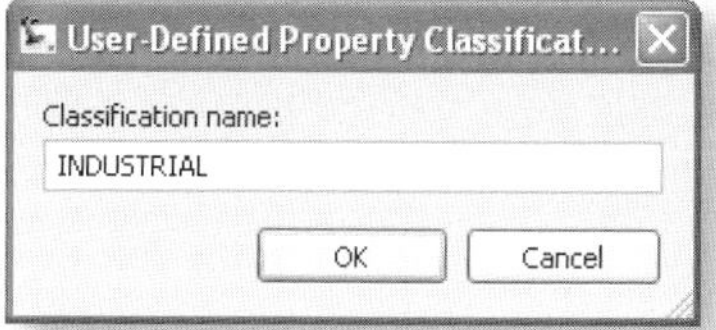


**FIGURE 3.37**

The second is defining the classification's properties (see Figure 3.38). In the case of Industrial, its enumerated types are Light, Medium, and Heavy. These values can be a part of a parcel's label. With the information in the label, the values can be a part of a table or report.

**FIGURE 3.38**

## EXPRESSIONS

Each parcel label type has expressions to build custom calculations (see Figure 3.39). Users first define the expression, and when defining a label, the expression appears as a label property.

**FIGURE 3.39**

## MAP CHECK

Map Check uses selected segment label to create a parcel closure report (see Figure 3.40). First a user defines a map check, then selects the segment labels defining the parcel. Map Check uses the label precision to calculate the closure. The Map Check palette contains tools to create reports, insert values in a drawing, etc. The command is found in Ribbon's Analyze tab, the Survey drop-list, Map Check.

- If a parcel's curve segment is non-tangential, the segment must have a label with the chord's bearing.

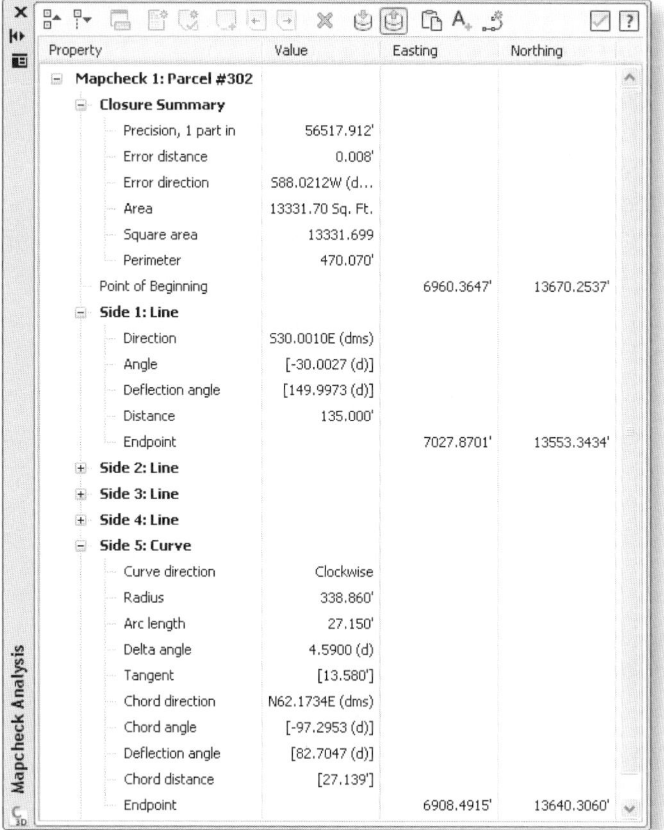

| Property | Value | Easting | Northing |
|---|---|---|---|
| Mapcheck 1: Parcel #302 | | | |
| Closure Summary | | | |
| Precision, 1 part in | 56517.912' | | |
| Error distance | 0.008' | | |
| Error direction | S88.0212W (d... | | |
| Area | 13331.70 Sq. Ft. | | |
| Square area | 13331.699 | | |
| Perimeter | 470.070' | | |
| Point of Beginning | | 6960.3647' | 13670.2537' |
| Side 1: Line | | | |
| Direction | S30.0010E (dms) | | |
| Angle | [-30.0027 (d)] | | |
| Deflection angle | [149.9973 (d)] | | |
| Distance | 135.000' | | |
| Endpoint | | 7027.8701' | 13553.3434' |
| Side 2: Line | | | |
| Side 3: Line | | | |
| Side 4: Line | | | |
| Side 5: Curve | | | |
| Curve direction | Clockwise | | |
| Radius | 338.860' | | |
| Arc length | 27.150' | | |
| Delta angle | 4.5900 (d) | | |
| Tangent | [13.580'] | | |
| Chord direction | N62.1734E (dms) | | |
| Chord angle | [-97.2953 (d)] | | |
| Deflection angle | [82.7047 (d)] | | |
| Chord distance | [27.139'] | | |
| Endpoint | | 6908.4915' | 13640.3060' |



**FIGURE 3.40**

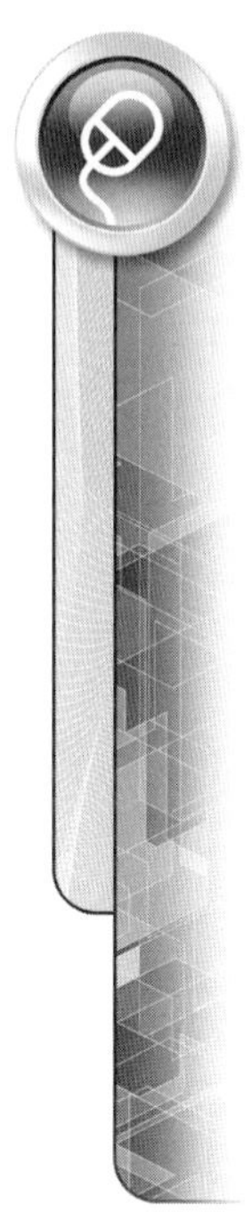

## EXERCISE 3-5

After completing this exercise, you will:

- Be familiar with parcel area label styles.
- Be familiar with parcel segment label styles.
- Be familiar with parcel table styles.
- Be able to create parcel segment labels.
- Be able to create parcel area tables.
- Be able to create parcel segment tables.
- Be able to create a Map Check analysis.

## Exercise Setup

This exercise continues with the previous unit's exercise drawing. If starting with this exercise, browse to the Chapter 3 folder of the CD that accompanies this textbook and open the Chapter 03 - Unit 5 drawing.

1. If the previous exercise's drawing is not open, open it now, or if not completing the previous unit, open the *Chapter 03 - Unit 5* drawing file.

## Parcel Area Labels

Labeling a parcel area or number occurs when creating a parcel.

1. Click the ***Settings*** tab.
2. In Settings, expand the **Parcel** branch until you view the Area label styles list.
3. Click the **Name Area & Perimeter** area label style, press the right mouse button, and from the shortcut menu, select EDIT....
4. If necessary, click the ***Layout*** tab to review its contents. To do this, at the top left click the drop-list arrow and select a component's name from the list.

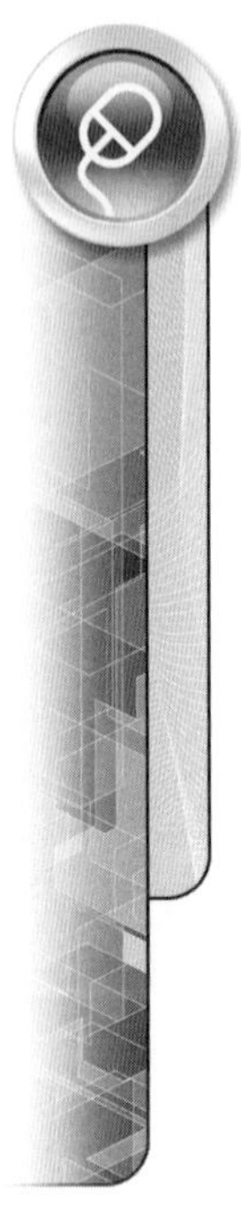

The label component contains three parcel properties. These properties are not individual components, but are one text component entry with three properties.

5. In the Text section, click the **Contents'** value cell displaying an ellipsis.
6. At the cell's right side, click the **ellipsis** to display the Text Component Editor dialog box with the label properties.
7. On the right side, click and highlight each property's format string, and review its settings on the left side.

A border encompasses all three properties because they are a single label component.

8. Click ***Cancel*** to exit the Text Component Editor.
9. Click the ***Dragged State*** tab to review its contents.

If this label is dragged from its original position, it becomes stacked text with a leader. Again, you can change it to remain composed either with or without a leader.

10. Click ***Cancel*** to exit the Label Style Composer.

## Segment Labels — Line Segments

As with area labels, each segment label has a layout and dragged state definition. A line segment label annotates a line's distance, bearing, or both.

1. In Settings, in the **Parcel** branch, expand **Line** label styles until you view its styles list.
2. Select **Bearing over Distance**, press the right mouse button, and from the shortcut menu, select EDIT....
3. Click the **Layout** tab and view the style's components. You do this at the top left by clicking the drop-down list arrow and selecting a component's name from the list.

This style has a table tag and two components: bearing and distance.

4. Click the ***Dragged State*** tab to review its contents.
5. Click CANCEL to exit the style.
6. From the list, select **(Span) Bearing over Distance with Crows Feet**, press the right mouse button, and from the shortcut menu, select EDIT....
7. Click the **Layout** tab and review its components. To do this, at the top left, click the drop-down list arrow and select a component's name from the list. This style has a table tag and four components; bearing and distance, direction arrow, and crow's feet (start and end).

You use this label on segments that extend over individual line segments (backyard line).

8. Click the ***Dragged State*** tab to review its contents.
9. Click ***Cancel*** to exit the style.

## Segment Labels — Curve Segments

A curve segment label annotates a curve's radius and length. A curve label can annotate additional curve properties.

1. In Settings, expand the **Parcel** branch until you are viewing the Curve styles list.
2. From the list, select **Delta over Length and Radius** style, press the right mouse button, and from the shortcut menu, select EDIT....
3. Click the **Layout** tab and change the Component name to **Distance & Radius**.

This label component anchors to the curve's bottom center (the Text Attachment value).

4. Change the Component name to **Delta**.

This label's component anchors to the curve's top center (the Text Attachment value).

5. Click the ***Dragged State*** tab to review its settings.
6. Click ***Cancel*** to exit the dialog box.

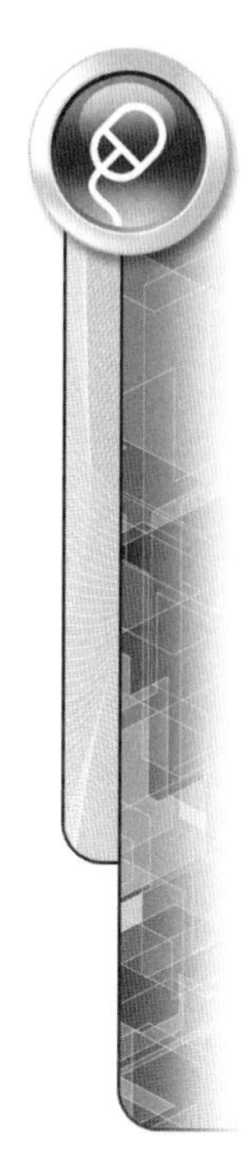

## Adding Segment Labels — Multiple Segment

When defining parcels, you can label their segments. If you are not labeling parcels at this time, the Add Labels routine creates the segment labels. Add Labels has two parcel segment labeling methods: multiple and single segment.

1. If necessary, use PAN and ZOOM to view a few parcels.
2. On the Ribbon, click the Annotate tab, on the Labels & Tables panel's left, click the Add Labels drop-list arrow to display a shortcut menu. Place the cursor over the Parcel entry and in its flyout menu select ADD PARCEL LABELS.
3. In the Add Labels dialog box, set the Feature to **Parcel**, the Label type to **Multiple Segment**, the Line label style to **Bearing over Distance**, and the Curve label style to **Delta over Length and Radius** (see Figure 3.41).
4. Click **Add**.

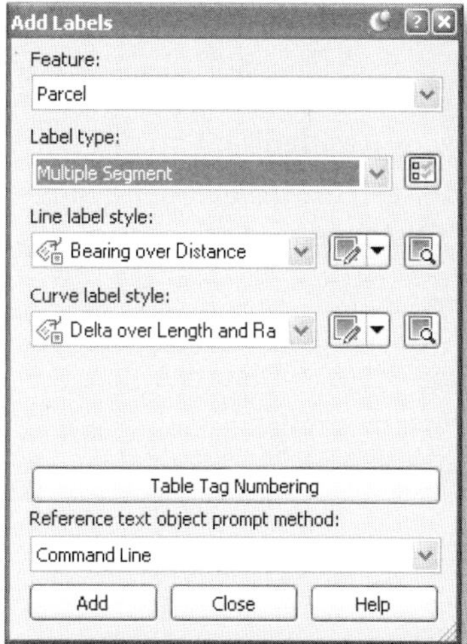


**FIGURE 3.41**

You are prompted to select a parcel label and to identify what segments to label.

5. Select a parcel label to create the segment labels.
6. Press ENTER to specify the clockwise direction for the labels.

There is only one label at each common side yard segment.

7. Undo the labels. You may have to use the ERASE command to remove all segment labels.

## Adding Segment Labels — Single Segment

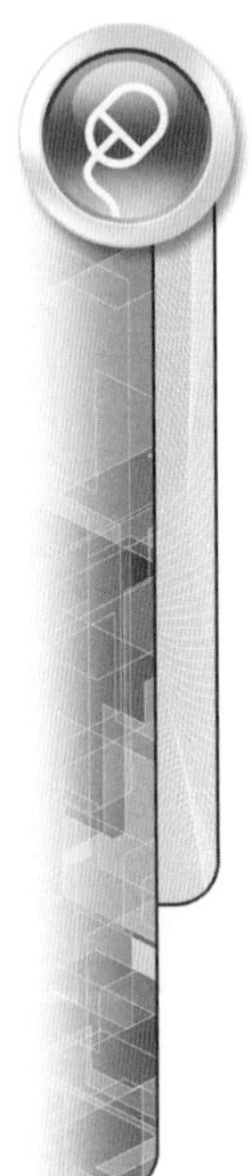

Labeling individual backyard segments is a problem with multiple-segment labeling. It may be necessary to have a label spanning the backyard with a bearing and overall distance, as well as individual parcel backyard distances (see Figure 3.42). In Figure 3.42, the left side uses multiple segment labeling, and the right side shows single segment labeling with the spanning label style.

When using single-segment labeling, the segment's selection point is the label's anchoring point. If it is not in the correct location, you can graphically position the label.

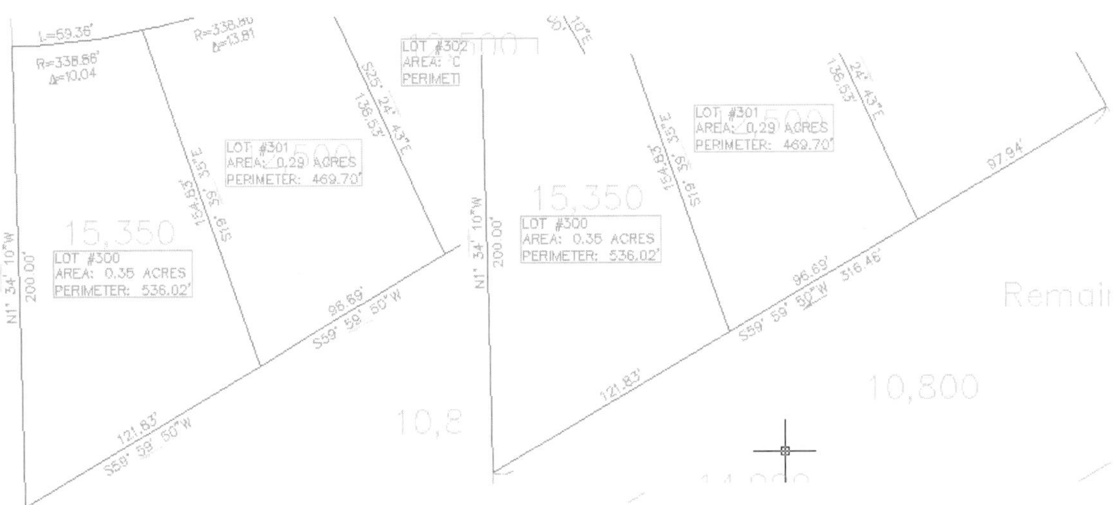


**FIGURE 3.42**

8. If you are not working in the Residential Parcel 1 view area, from Ribbon's ***View*** tab, Views panel, select and restore the named view **Residential Parcel 1**.
9. In the Add Labels dialog box, set the Feature to **Parcel**, the Label type to **Single Segment**, the Line label style to **Bearing over Distance**, and the Curve label style to **Delta over Length and Radius** (see Figure 3.42).
10. Click ***Add*** and select the side and front parcel segments.
11. In the Add Labels dialog box, change the Line label style to **Distance**.
12. Click ***Add*** and label the three parcels' individual backyard distances.
13. Click ***Close*** to exit the Add Labels dialog box.

Your parcel labels should look similar to Figure 3.42.

## Changing Segment Labels

There will be times when a label shows the wrong bearing (north instead of south), or the label is on the segment's wrong side. When you select a label and press the right mouse button, a shortcut menu is displayed with several editing options (see Figure 3.33).

## Reverse Label

If a label's direction is incorrect, use Reverse Label to change its direction.

1. Zoom in to view a few segment labels.

2. In the drawing, select a label, press the right mouse button, and from the shortcut menu, select REVERSE LABEL.

## Flip Label

If a label is on the segment's wrong side, use Flip Label to exchange the label element's location.

1. In the drawing, select a label, press the right mouse button, and from the shortcut menu, select FLIP LABEL.

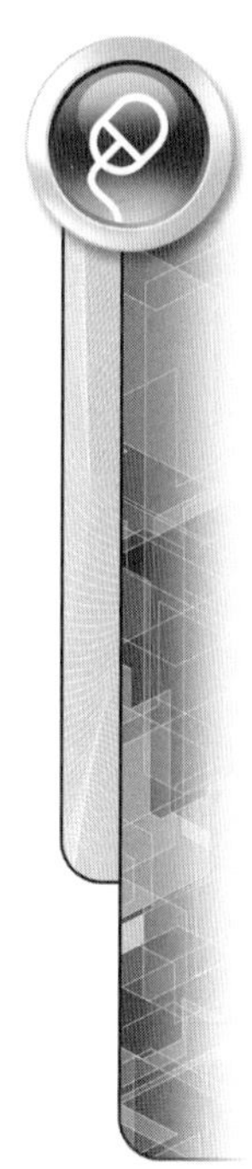

## Drag Label

If selecting and activating a label's grip, you can drag it from its original position. Depending on the label's dragged state settings, the label may dramatically change. Depending on where the label is located, it may use right- or left-justification.

1. In the drawing, select a label, and then select its square grip and move it to a new location.
2. Try different locations and watch the label switch from left- to right-justified.

## Reset Label

If you want to return a label to its original position and composition, use Reset Label.

1. In the drawing, select the dragged label, press the right mouse button, and from the shortcut menu, select RESET LABEL.
2. In the drawing, select a label, and then select its square grip and move it to a new location.
3. In the drawing, select the dragged label and reset the label by clicking its round grip.

## Change Label

Settings in a label's Label Properties dialog box can change its label style.

1. In the drawing, select a label, press the right mouse button, and from the shortcut menu, select LABEL PROPERTIES....
2. In the Properties panel, click in the **Line Label Style** value cell. Click the drop-list arrow to display a list of styles; from the list, select a different label style (see Figure 3.33). Press ESC to deselect the label.
3. Click the **X** on the top of the Properties mast to close the palette.

When you have exited, the new style appears on the segment.

4. At Civil 3D's top left, Quick Access Toolbar, click the ***Save*** icon to save the drawing.

## Label Plan Readibility

All parcel labels are sensitive to view rotation and maintain plan readability (reading from left to right). Whether rotating model space or a layout viewport, the labels react to the rotation and read correctly.

1. Start DVIEW command and select the parcels and their line work. Use the **TWist** option and twist the view until the backyard line is nearly horizontal and select a point.

While rotating the view, the labels remain plan-readable. No matter what the rotation angle is, the area labels and segment annotation are plan-oriented.

2. Press ENTER to exit DVIEW.

3. On the Ribbon click the View tab, in the Views panel, click NAMED VIEWS, click ***New...***, and save the current view as **Parcels Twisted**. Click **OK** until you return to the command line.
4. Use the Plan command and press ENTER twice to return to the World view.

## Scale Sensitivity of Labels

All labels and annotation are scale sensitive. Labels and annotation react to scale's value and size themselves for that scale. Use the Regenall (REA) command to resize the label.

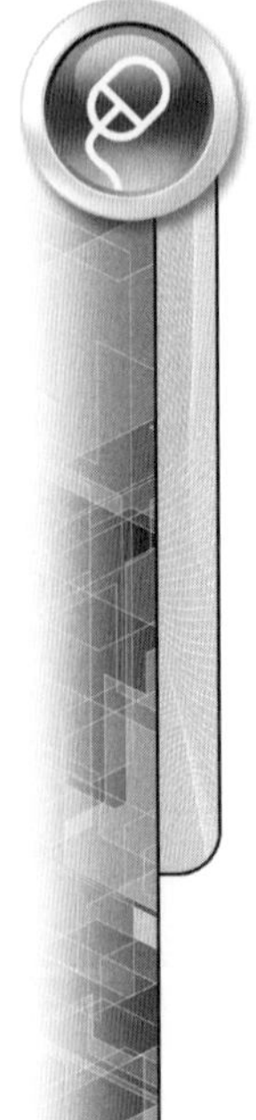

1. Click the **Layout 1** to enter the paper space.
2. Double-click inside the viewport to enter its model space.
3. In Ribbon's Views panel's left, select and restore the named view **Parcels Twisted**.
4. Double-click outside the viewport to return to the paper space.
5. Select the viewport border, press the right mouse button, and from the shortcut menu, select PROPERTIES....
6. In the Properties dialog box, the Misc section, set the Standard Scale to **1''=50'**.
7. Use the Regenall (REA) command to resize the text (if necessary).

The label and annotation text is now correct for the viewport's scale.

8. In the Properties palette, set the viewport's scale to **1''=20'**.
9. Use the Regenall (REA) command to resize the text.
10. Click **X** to close the Properties palette.
11. Click the ***Model*** tab to reenter the model space.

## Parcel Segment Table

Tables list the segments' geometric values. A table is created from existing segment labeling. The Table routine reads the values, creates the table, and toggles the label's visibility to tags (L1, L2..., C1, C2...).

1. On the Ribbon click the ***Annotate*** tab. In the Labels & Tables panel, click the Add Tables drop-list arrow to display a shortcut menu. Place the cursor over Parcel to display its flyout menu and from the flyout, select ADD SEGMENT.

The Table Creation dialog box opens. The Table Style heading at the top lists the current table style. The Select by label or style area lists all label styles.

2. Scroll through the Select by label or style's list, toggling on the styles: Parcel curve; Delta over Length and Radius and Parcel line; Bearing over Distance.

Next, you will set the table type.

3. If necessary, set the Table type to **Dynamic**.

Your dialog box should look like Figure 3.36.

4. Click ***OK*** to close the dialog box.
5. Next, you are prompted to locate the table's upper-left corner. In the drawing, select a point to create the table.

The routine reads the label's values, creates the table, and toggles the styles to tag mode (L1, L2..., C1, C2...).

6. Zoom in to the table and parcels to view their tags.
7. Use undo to remove the tables and return the annotation to verbose mode.
8. At Civil 3D's upper left, Quick Access Toolbar, click the ***Save*** icon to save the drawing.

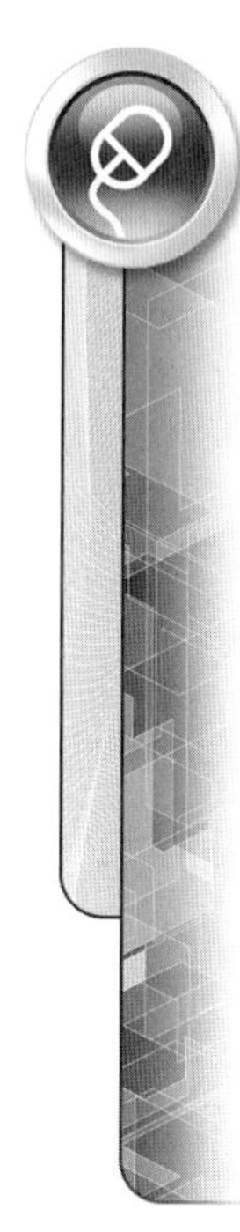

## Parcel Area Table

When submitting documents, you may need to list the parcel areas as a table.

1. From the Labels & Tables panel, click the Add Tables drop-list arrow to display a shortcut menu. Place your cursor over Parcel to display its flyout menu, select ADD AREA.

The Table Creation dialog box opens.

2. In the Select by Label or Style area, toggle **ON** each style.
3. Change Selection Rule to **Add Existing and New**. To do this, double-click in the Selection Rule cell, and select **Add Existing and New** from the list.
4. If necessary, leave Split Table **ON** and set the table to **Dynamic**. Your Table Creation dialog box should look like Figure 3.43.

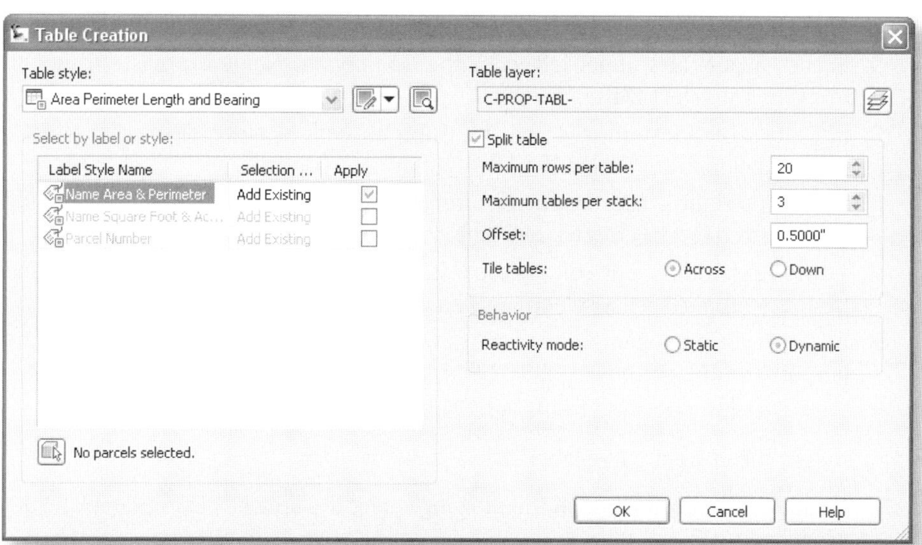

**FIGURE 3.43**

5. Click ***OK*** to close the dialog box.
6. Next, you are prompted for the upper-left location of the table. In the drawing, select a point to create the table.
7. Zoom in to review the table.
8. After reviewing the table, use undo to remove the table.
9. At Civil 3D's top left, Quick Access Toolbar, click the icon to ***Save*** the drawing.

## Map Check

All parcels need their closer documented, the Map Check tool creates this legal documentation.

1. If necessary, click the Prospector tab.
2. In Prospector, click Site 1's parcel heading to list the parcels in the preview area.
3. In preview from the parcel's list select **Parcel #302**, press the right mouse button, and from the shortcut menu, select ZOOM TO.
4. Use the ERASE command and erase all of the parcel segment labels.
5. If necessary, on the Ribbon click the ***Annotate*** tab. On the Labels & Tables panel's left, click the Add Labels drop-list icon and from the shortcut menu's Parcel flyout, select ADD PARCEL LABELS.

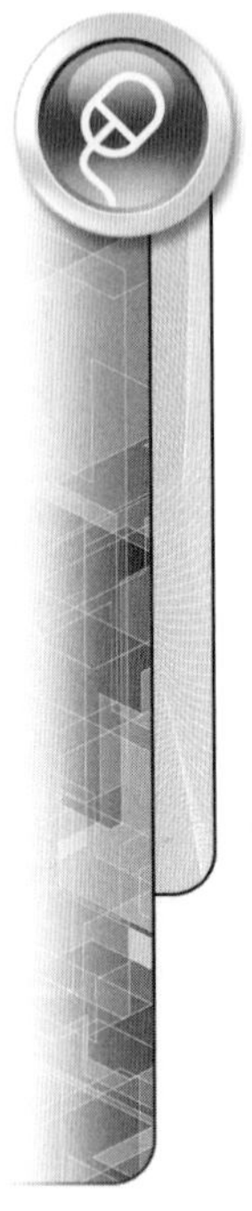

## Create a New Curve Label

Mapcheck must have a curve label with a bearing to correctly follow the brokenback curve, i.e. non-tangential curve on the parcel's west frontage.

1. Set the Label Type to **Multiple Segment** and Line Label Style to **Bearing over Distance**.
2. Click the ***Edit Style*** icon to the left of the Line Label Style.
3. If necessary, click the style's General tab and change the Display mode to **Label**.
4. Click OK to return to the Add Labels dialog box.
5. To the right of the curve label style, click the drop-list arrow and from the shortcut menu select COPY CURRENT SELECTION.
6. In the Information panel, change the style's name to **Delta and Bearing over Radius and Length**.
7. Click the General tab and change the Display mode to Label.
8. Click the Layout tab.
9. In the Layout tab's upper left, click the Component Name drop-list arrow and from the list select **Delta**.
10. In Delta's Text section, click on the Content's value cell to display an ellipsis. Click the ellipsis to display the Text Component Editor.
11. At the top right, place the cursor after the Segment Delta Angle format string and press ENTER to create a new line.
12. At the Text Component Editor's top left, click the Properties' drop-list arrow and from the properties list select **Segment Chord Direction**.
13. At the Text Component Editor's top center, click the transfer **arrow** to place the property on the editor's right side.
14. Click ***OK*** until returning to the Add Labels dialog box.

The new style is the current curve label style.

15. Click the **Add** button and in the drawing select the area labels for Parcels #300, #301, and #302.
16. The command line prompts for clockwise or counterclockwise, press ENTER for clockwise.
17. Close the Add Labels dialog box.
18. Zoom in on Parcel #302 to view its labeling.

## Define a Mapcheck

1. Click Ribbon's **Analyze** tab. On the Analyze tab's left, Ground Data panel, click the Survey drop-list icon and from the shortcut menu, select MAPCHECK.
2. Click the New Mapcheck icon (fifth in from the left) and in the command line for the mapcheck name, enter **Parcel #302** and press ENTER.
3. The command line prompts for a Northing/Easting starting point. In the drawing, using an Endpoint object snap, select the parcel's northeast front yard corner.

The corner's coordinates appear in the Mapcheck Analysis panel.

4. If necessary, toggle off object snaps.
5. The command line prompts for a segment label selection. In the drawing, select parcel #302's east property line label.

An entry for the line appears in the palette and Civil 3D draws a line over the segment.

6. In the drawing select the next two backyard segment labels and then select parcel #302's west side yard line label.

The west side yard was drawn from front to back and the mapcheck bearing arrow points away from the front yard line.

7. In the command line, type the letter '**R**' and press ENTER to reverse the line's bearing to point to the front yard.
8. In the drawing, select the short curve segment label on the front yard's west side.

An arrow should point to parcel's last front yard line segment.

9. In the drawing, select the closing front yard segment label.
10. Press ENTER to end the routine.
11. Review each segment's values for the current mapcheck.
12. At Mapcheck Analysis' top center, click the ***Output View*** icon to display the mapcheck results and review the mapcheck results.
13. At Mapcheck Analysis' top center, click the ***Insert Mtext*** icon and select a point to place the mapcheck values in parcel #302.
14. Close the Mapcheck Analysis palette.
15. At Civil 3D's top left, click the ***Save*** icon to save the drawing.
16. Exit Civil 3D.

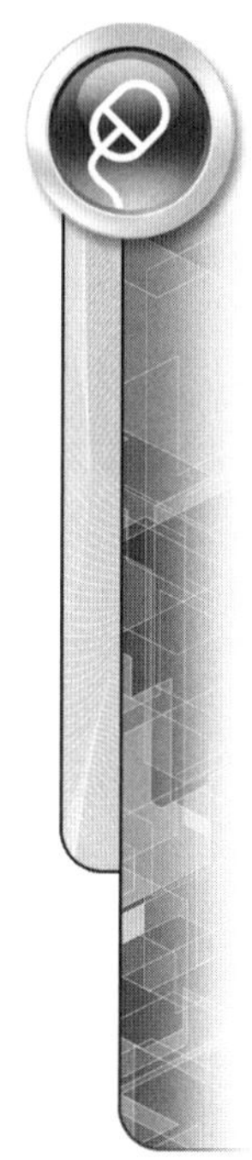

## SUMMARY

- Civil 3D's parcels have three label types: area, line, and curve.
- All labels using the multiple segment option are at each segment's midpoint.
- All labels using the single segment option are at the segment's selection point.
- A label text can flip to opposite segment sides, change direction, be dragged from its original position, be pinned, and be reset to its original position and composition.
- Parcel labels are view-rotation sensitive and if set, rotate to be plan-readable.
- Parcel labels are viewport-plotting-scale sensitive and resize themselves based on the current plotting scale.
- Parcels have segment tables for area, line, curve, or both.

This concludes the parcels chapter. The goal of any subdivision is to create marketable parcels from a site. Parcels may define homogeneous land use (detention ponds, open space, wetlands, etc.) or saleable parcels.

Parcels have tools to design individual parcels, document the subdivided land graphically, and create the necessary reports.

The next chapter is about the Civil 3D surface. The surface is a critical element in the design process.

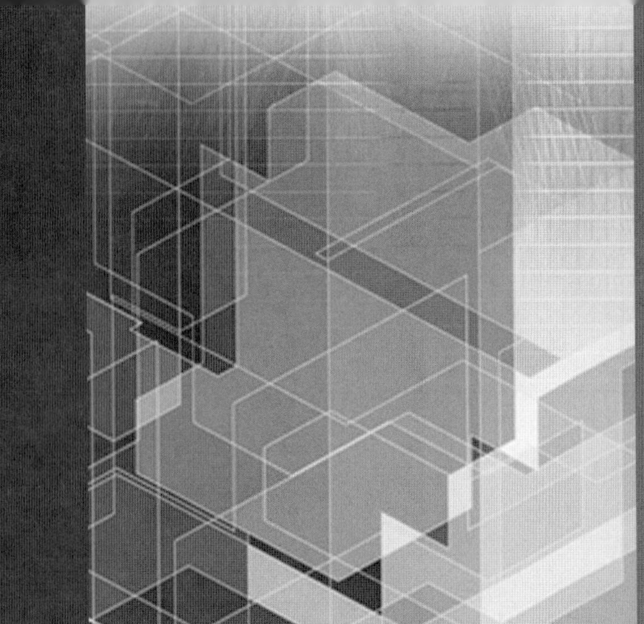

# CHAPTER 4

# Surfaces

## INTRODUCTION

This chapter introduces the surface object. A surface is a triangulated or grid data net and represents a surface's elevations and slopes. Surface data can be points, nodes, and vector objects, Civil 3D cogo point objects, polylines, contours, boundaries, and breaklines. The various surface menus provide tools to create, edit, evaluate, and annotate surface components and characteristics. Surface styles are task orientated, allowing a user to evaluate a surface as it is being built, and to evaluate its slopes and its elevations.

## OBJECTIVES

This chapter focuses on the following topics:

- Basic Surface Settings
- Overview of Surface Objects and Their Label Styles
- Surface Data Types
- How to Evaluate and Edit a Surface
- How to Analyze Slope and Elevation Characteristics
- How to Develop and Assign a Contour Style
- How to Annotate Surface Slopes and Contours
- How to Smooth a Surface
- How to Create Point Objects from Surface Elevations

### OVERVIEW

A surface is fundamental to the civil design process. If surface errors or data misinterpretations are not caught early, they may have grave consequences in the following design phases. It is important that a surface be an accurate data interpretation.

To help understand if a surface correctly interprets its data, displaying specific surface components is invaluable. Displayable surface components are triangle legs, points, borders, boundaries, contours, elevations, slope arrows, etc. In this textbook, surface components are surface border(s), triangles, points, and boundaries. Surface characteristics are relief, water flow (drainage), watersheds, or slopes, and slope directions. The surface object and its label styles display these components and analyze and annotate surface characteristics.

The just-mentioned characteristics influence design strategies and present challenges to each designer. These challenges include how best to define land use, create site access, manage water runoff, or develop manageable earthwork volumes. The solutions applied to the site are in response to these challenges and the solutions may require compromises between cost and best design practices.

To create usable project surfaces, develop a consistent process to produce "good" surfaces (i.e., correct data interpretation). The process includes gathering and evaluating initial data, evaluating each surface build's results, finding and editing errors, analyzing surface characteristics, and, finally, annotating the surface.

Developing consistent surfaces takes seven steps, ranging from using meaningful surface names to documenting everything correctly. The steps occur in related menus or selected object's shortcut menus. However, Prospector's Surfaces shortcut menus call the majority of the needed commands. These commands create, edit, analyze, and annotate surface components and characteristics.

Civil 3D has two terrain surface types: TIN (Triangular Irregular Network or Digital Terrain Model [DTM]) and Grid. A TIN connects all surface data with triangle legs and there is a known elevation at each leg's end. A Grid surface is a mesh structure that has elevations at the mesh's intersections. These elevations are generally interpolated values, not known elevations. A DEM (Digital Elevation Model) is an example of a grid surface.

Civil 3D has two volume surfaces types: TIN and Grid. A volume surface's structure is the same as a TIN or Grid terrain surface; however, the differences in elevation between two terrain surfaces are its elevations. An earthwork volume is a volume surface property.

## SEVEN SURFACE STEPS

- Create the surface object.
- Collect and assign initial data.
- Add breaklines in known problem areas.
- Evaluate resulting triangulation.
- Edit (add, delete, modify) surface triangles, points, breaklines, and/or boundaries.
- Analyze the surface components and characteristics.
- Annotate the surface.

The first step is naming and setting a surface's type. Prospector's Surfaces heading shortcut menu (Create Surface…) or the Ribbon's Surfaces command drop-list Create Surface… entries display the Create Surface dialog box (see Figure 4.1). Create Surface defines a surface type, assigns a layer, name, description, surface type, style, and render material.

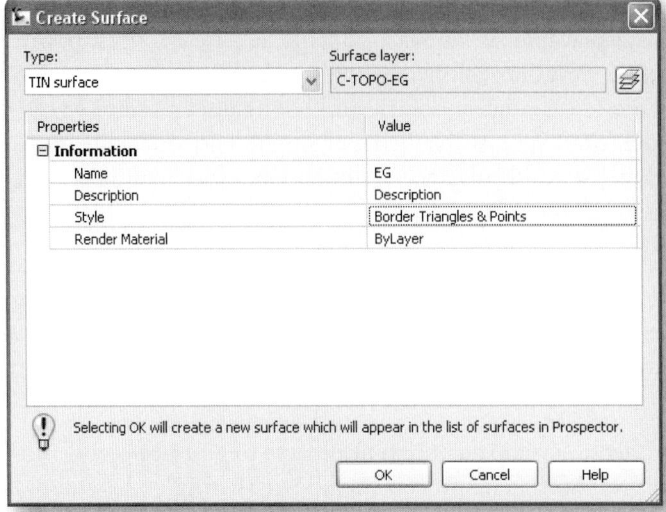


**FIGURE 4.1**

After assigning a name, the next step is collecting the initial data. Surface data comes from diverse sources: drawing objects, Civil 3D cogo points, point files; contours; three-dimensional polylines; boundaries; and/or DEMs (Digital Elevation Models). No matter the data source, before assigning it, review it for obvious errors and omissions. This is because Civil 3D automatically builds a surface from a data assignment.

Another reason for clean data is a surface does not allow crossing breaklines. This prohibition includes contour data. You can toggle this off and allow data crossings, but you risk destabilizing the drawing.

The order of assigning data is important. For example, adding point data outside a boundary makes a surface ignore the boundary, adding breaklines that cross contour data creates data exclusions, adding points after assigning contour data, which limits the scope of the points in the surface. The order of data assignment is as follows:

Data Clip Boundary
Points (Point Groups, Files, drawing objects)
Breaklines and/or Contours (not crossing)
Maximum Triangle Length, Edits, and/or Boundaries

You may have to remove contours from the breakline area to remove potentially overlapping data. If you toggle on, letting a surface have crossing data, you run the risk of destabilizing the drawing.

The third step is adding breaklines to known or obvious problem areas. If you are starting with point data, to correctly triangulate linear features, breaklines are almost always necessary. Examples of breaklines are edges-of-travelway, slope breaks, stream banks, swales, ditches, and other linear features.

Identify breakline areas by reviewing points with similar descriptions. If you are questioning the necessity of a breakline, there are a couple of options. Consult with the original data source, and ask about the area in question. Or, after reviewing the surface and discussing the problem with someone, edit the surface to adjust its triangulation. Editing a surface creates the same effect as adding breaklines. Be sure to understand the consequences of adding new breaklines before editing a surface.

When do you add data, set a maximum triangle length, edit, or add a surface boundary? Some surfaces have lengthy peripheral outside triangle legs. If you can identify the shortest of these long triangles and its length is greater than the remaining surface triangles, setting a maximum triangle length may remove a majority of the unwanted triangles. If you cannot set a maximum triangle length and correcting a problem takes several edits, adding new data or breaklines is advisable. If you are making only a few edits, editing is the preferred method. However, a well designed boundary may eliminate the need for edits.

Breaklines are either two- or three-dimensional polylines, survey figures, or feature lines. When using two-dimensional polylines as breaklines, the only condition is that a point object must be at each vertex. When using three-dimensional polylines, point objects are not necessary.

**NOTE**

**Objects that are to become breaklines must be drawn before assigning them to a surface.**

The fourth and fifth steps review the initial surface triangulation. These steps should use surface-analysis styles that emphasize specific surface components and characteristics which indicate data problems. When completing these steps, it maybe necessary to add data or continue editing.

After adding new data and/or editing a surface, the last two steps are to analyze surface characteristics (elevations and slopes) and create contours.

### Unit 1

Several settings and styles affect surfaces. The first unit reviews and sets the default values.

### Unit 2

The second unit reviews creating a surface, surface data types, and how Prospector manages surface data. Civil 3D builds the surface after each data addition.

### Unit 3

This chapter's third unit covers surface review and editing. There are several options that affect what data a surface uses and how it responds to editing. Surface-editing tools are in the surface's Definition branch's Edits shortcut menu. Surface styles are an aid because they display necessary surface components. For example, displaying triangulation and points is best when looking for incorrect triangulation, seeing a 3-D surface model to view bad elevations, etc.

### Unit 4

The fourth unit reviews developing and assigning surface-analysis styles. Examples of these style types are displaying slopes as ranged values, down slope direction as arrows, and ranged elevations. These styles are displayed as 2D or 3D objects.

### Unit 5

This chapter's fifth unit reviews developing a contour style and setting its values for intervals, layers, and labeling. After reviewing the contours, it may be necessary to smooth the surface and to document their elevations.

### Unit 6

The sixth unit discusses creating points and polylines with elevations from a surface. The Create Points toolbar provides these tools.

## UNIT 1: SURFACE SETTINGS AND STYLES

The Edit Drawing Settings dialog box is the highest level of control for a drawing. The dialog box settings assign basic surface object and label layer names. Surface's Edit Feature Settings dialog box controls three basic surfaces settings: default object and labeling styles; defines the surface name template; and sets the default surface type and its initial settings.

Surface styles determine how a surface displays its components and characteristics. Civil 3D's content templates provide several starter object styles for these components and characteristics. The user creates custom object styles to reflect individual work assignments and information needs.

### EDIT DRAWING SETTINGS

Civil 3D manages surface visibility through layers and/or by styles. The Drawing Settings' Object Layer tab assigns a base layer name with an optional prefix or suffix (see Figure 4.2). C-TOPO is the default surface layer name. When having more than one surface and not changing the default settings, all surfaces are on the same layer, C-TOPO. This situation gives little control over each surface's visibility except by assigning no display to each surface. To remedy this situation, assign a prefix or suffix to the surface's base layer name. Use an asterisk (*) to assign the surface name to the base surface layer name. For example, create a base layer whose name is suffixed by a surface named EXISTING. First, set the Modifier to suffix and enter the value of -*. This creates a layer in the drawing of C-TOPO-EXISTING. The dash is a separator between the base layer name and the name of the surface. This naming convention creates a layer for each named surface.

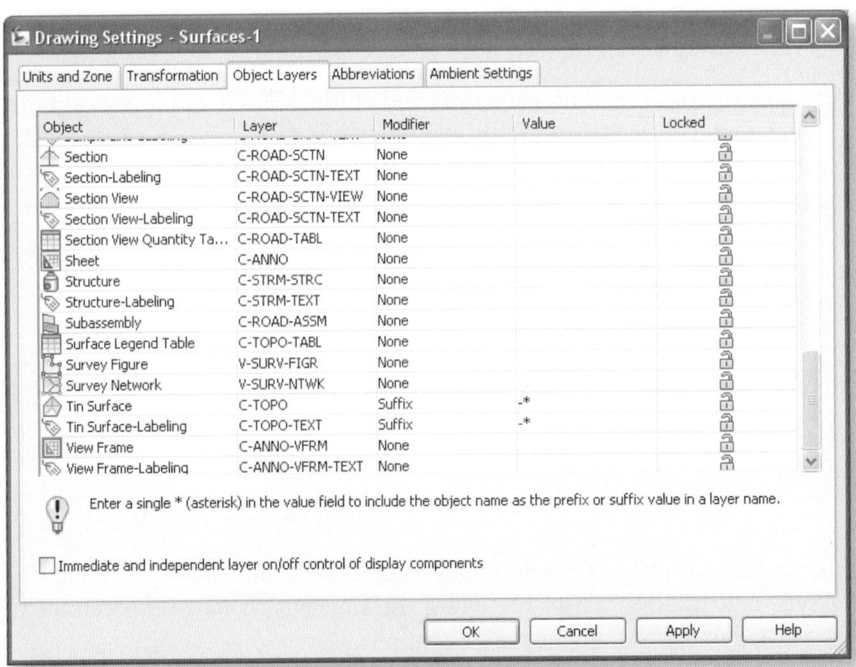


**FIGURE 4.2**

## EDIT FEATURE SETTINGS

Surface Edit Feature Settings sets default styles and defines the surface-naming template. An object style defines how a surface displays its components and characteristics (triangulation, border, breaklines, points, slopes, and elevations) and shows all, one, or any combination of components and characteristics. One style-defining strategy is to create styles showing surface components and characteristics appropriate for an assigned task. While reviewing a newly built surface, the user's interest is the surface's triangulation, breaklines, boundaries, and points. While evaluating a surface for a preliminary design, the styles display surface elevations, relief, and slopes. With specific tasks, styles should emphasize critical surface components and characteristics. Displaying critical information helps make better decisions possible.

### Default Styles

The Default Styles section sets object and the surface annotation styles (see Figure 4.3). The label style types include spot elevation, slope, contours (major and minor), render material, and triangulation markers.

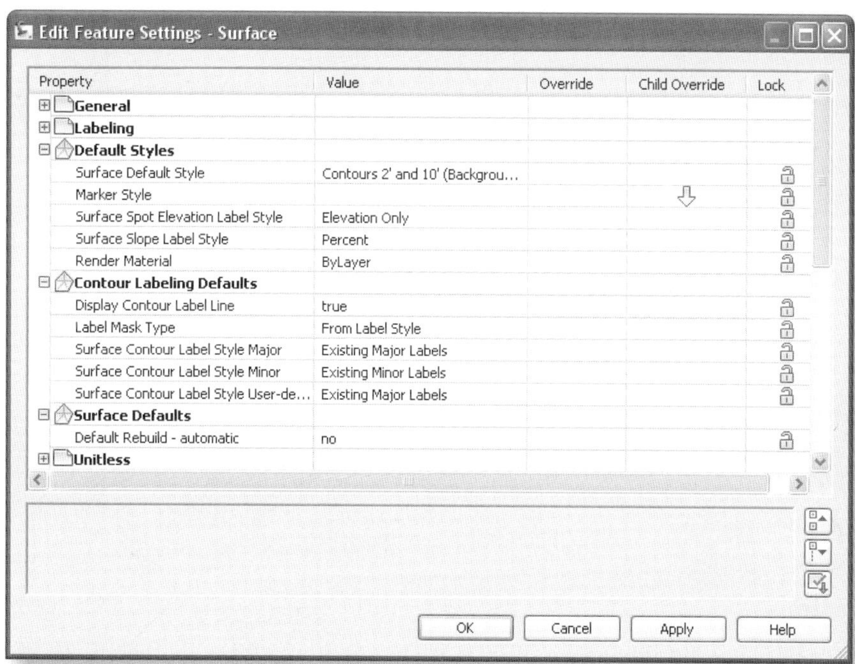


**FIGURE 4.3**

### Contour Labeling Defaults

This section sets the default contour label styles that appear in the Add Labels dialog box. The Display Contour Label Line value should be False. If visible, this line will plot. To display this line to relocate contour labels, select a label and the line will display with grips.

### Surface Defaults

This section sets if automatic rebuild is on when defining a surface.

## COMMAND: CREATESURFACE

The Settings' Surface branch, Commands, CreateSurface sets default values for this command. The CreateSurface command dialog box has two sections: Surface Creation and Build Options.

### Surface Creation

The Surface Creation section sets the default surface type, the surface name format, the grid X and Y spacing, and the grid's rotation (see Figure 4.4).

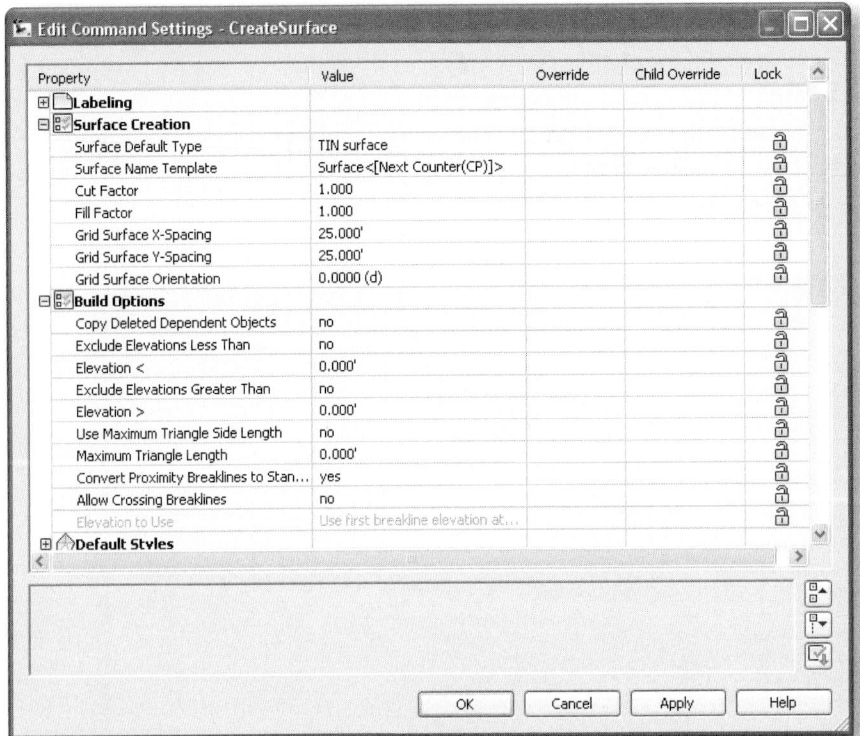


**FIGURE 4.4**

### Build Options

This section sets rules for building a surface: what data to exclude, the maximum triangle length, if proximity breaklines are to become standard breaklines, and how to handle crossing breaklines (see Figure 4.4). These values can be changed at any time.

### Copy Deleted Dependent Objects

Copy Deleted Dependent Objects specifies whether a drawing object's data is preserved in its surface definition after erasing the object(s). The default setting is "no" and assumes the original object(s) are always present in the drawing. When using the default value (no), if the data object is deleted, it is removed from the surface triangulation. If you want to preserve the original data's effect even after erasing it, set this value to "yes." When set to "yes" and you delete the original object, a copy of the object's data is added to the surface's definition and preserves its effect on the surface.

### Exclude Elevations Above/Below

Exclude Elevations Less Than and Exclude Elevations Greater Than exclude data by setting a minimum and maximum elevation. If the elevation data is below the minimum or above the maximum, it is not included in the surface data.

> **NOTE**
> A point with no elevation is not excluded with these settings.

### Maximum Triangle Length

Maximum Triangle Length has two values. First is a toggle to turn this option on or off. Second is the triangle leg's maximum length. This deletes sliver triangles along a surface's periphery. With this setting off, a surface connects points even if there isn't any data supporting the triangle leg. With the setting on and a value set, a surface removes any triangle legs longer than this length. If this value is too low, it may prevent interior triangles, and as a result create interior surface holes with boundaries. Only some surfaces can use this option. Generally, this option is good for surfaces with long peripheral triangles that are longer than any of the surface's body triangles.

### Convert Proximity Breaklines to Standard

When set to "yes," and defining 2D breaklines, the proximity breaklines (2D) convert to standard breaklines (3D). The breakline vertices' elevations are from the point's elevation nearest to each vertex. The 2D object remains in the drawing and is listed as a Standard (3D) breakline. Inserting the breakline into the drawing produces a 3D polyline snapped to the points. Because of this the original polyline and the inserted polyline may not be the same.

### Allow Crossing Breaklines

This toggle and setting allows crossing breaklines in surface data. This includes contours. The default setting is off, or not allowing crossing breaklines. When encountering crossing breaklines their data is added, but the triangulation uses one breakline and does not include the crossing breakline's data where they cross. A Panorama is displayed with an Events vista. Double-clicking the event displays an Event Properties dialog box with the error's details (see Figures 4.5 and 4.6). The Events vista contains a `ZOOM TO` link that takes the user to the offending data. Before including the data, the error must be fixed.

When toggling this setting on, allowing crossing breaklines, the user needs to set a method determining what elevation to use. Including an intersection elevation is one of three methods: use the first breakline's elevation, use the last breakline's elevation, or use the average breakline elevation. If allowing crossing breaklines and their elevations are similar, the effect on the surface is minimal. However, if the elevations vary greatly, it would be better to resolve the crossing breakline issues before using them in a surface.

> **NOTE**
> A surface with this setting on tends to be unstable.

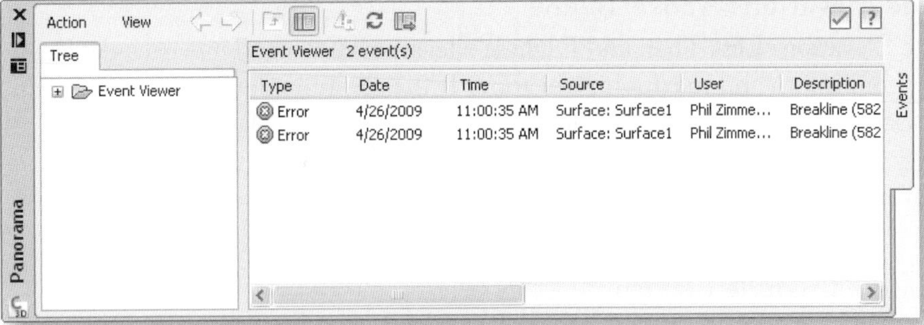


**FIGURE 4.5**

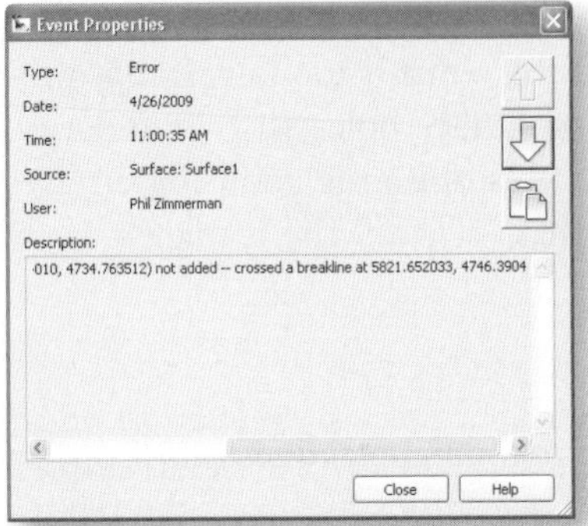


**FIGURE 4.6**

When changing one of the preceding settings after building a surface, the change triggers a surface rebuild. When exiting the surface's Properties dialog box, the dialog box asks whether to rebuild the surface with the new settings.

## SURFACE PROPERTIES — DATA OPERATIONS

Data Operations identifies what surface data types to include (see Figure 4.40). This section's intent is to define surface data. However, this section is also a way to exclude data types already included in a surface. For example, a user has a surface with point groups, breaklines, and boundaries and wants to see the surface triangulation without any breakline data. The user simply goes to this section and toggles off breaklines. When exiting the dialog box, the surface rebuilds without breakline data. You can change these settings before or after building a surface.

## SURFACE PROPERTIES — EDIT OPERATIONS

Edit Operations lists the allowable surface-editing operations (see Figure 4.40). However, this section is also a way to exclude edits already performed on a surface. For example, a user edits a surface with add point, delete line, and surface smoothing and wants to see the surface triangulation without the delete lines. To do this, in this section toggle off Use delete line. When exiting, the surface rebuilds without the deleted line edits.

## SURFACE PROPERTIES — OPERATION TYPE

Operation Type contains a surface's history and data dependencies (see Figure 4.40). Each entry has a toggle allowing it to be included or excluded from the surface. This allows a review of the impact of each entry or a combination of entries on the surface.

Control over the entry extends to reordering the entry's location in the list, i.e., moving an entry up (earlier) or down (later) in the surface build sequence.

## SURFACE STYLE

A surface object style affects how a surface displays its components and characteristics. Surface styles tend to address a task: view triangulation; view contours; view slopes and elevations. A style showing border(s), triangulation, points, and other essential surface components is crucial to evaluating a surface under development (see Figure 4.7). After building a surface, other styles range and display an analysis of surface slopes and elevations (see Figure 4.8). What is displayed by a surface style is set by the style's display panel values. The object has display settings for Plan, Model, and Section.

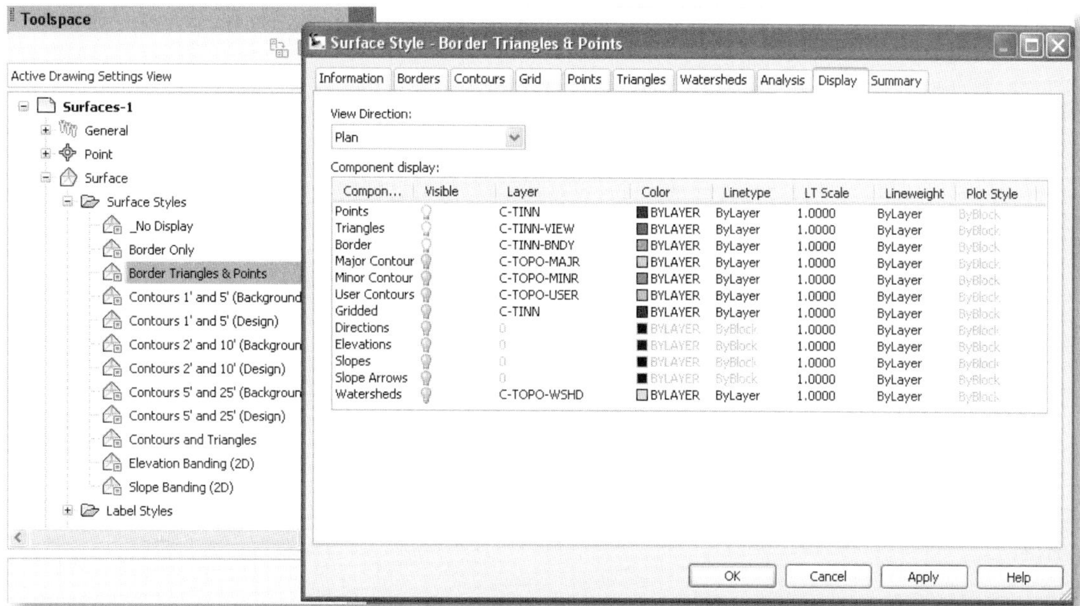


**FIGURE 4.7**

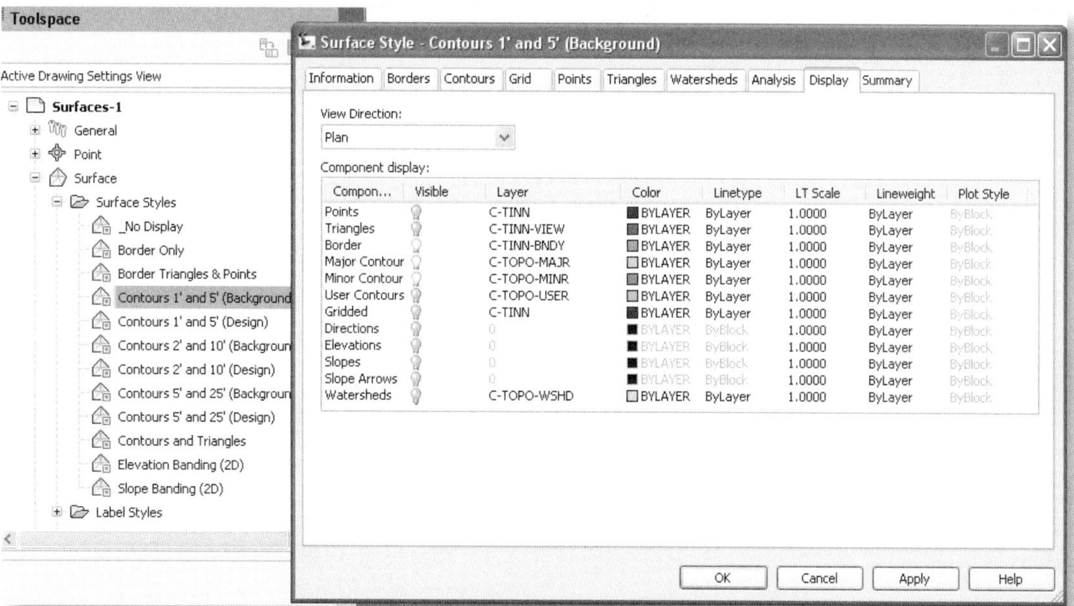


**FIGURE 4.8**

## LABEL STYLES

Surfaces have four label types: Contour, Slope, Spot Elevation, and Watershed (see Figure 4.9).

## TABLE STYLES

Surfaces have seven table types: Direction, Elevation, Slope, Slope Arrow, Contour, Watershed, and User Defined contours (see Figure 4.10).

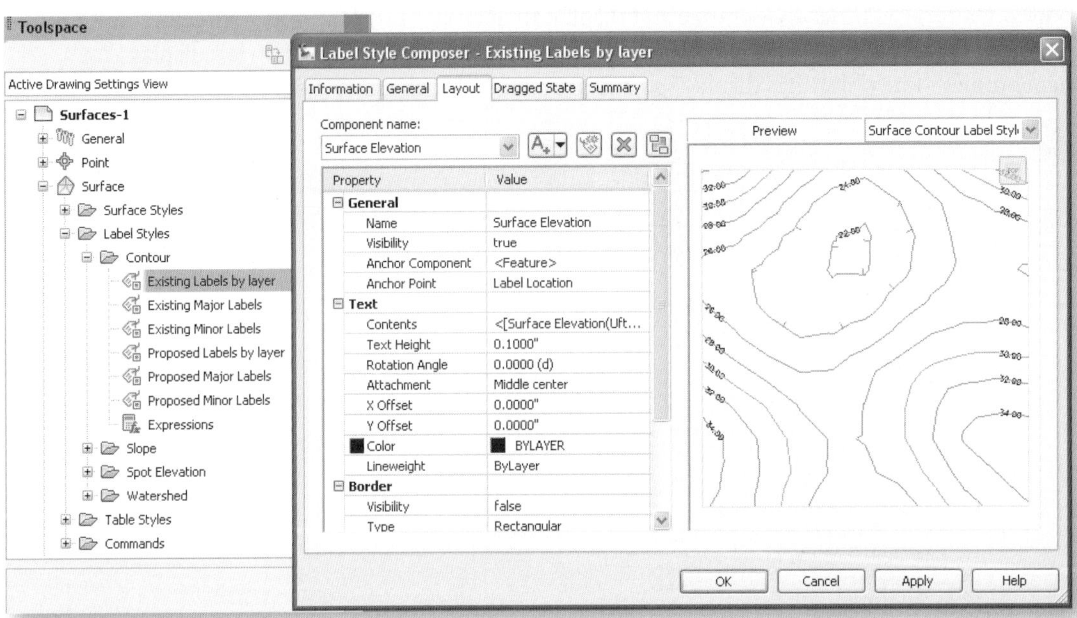


**FIGURE 4.9**

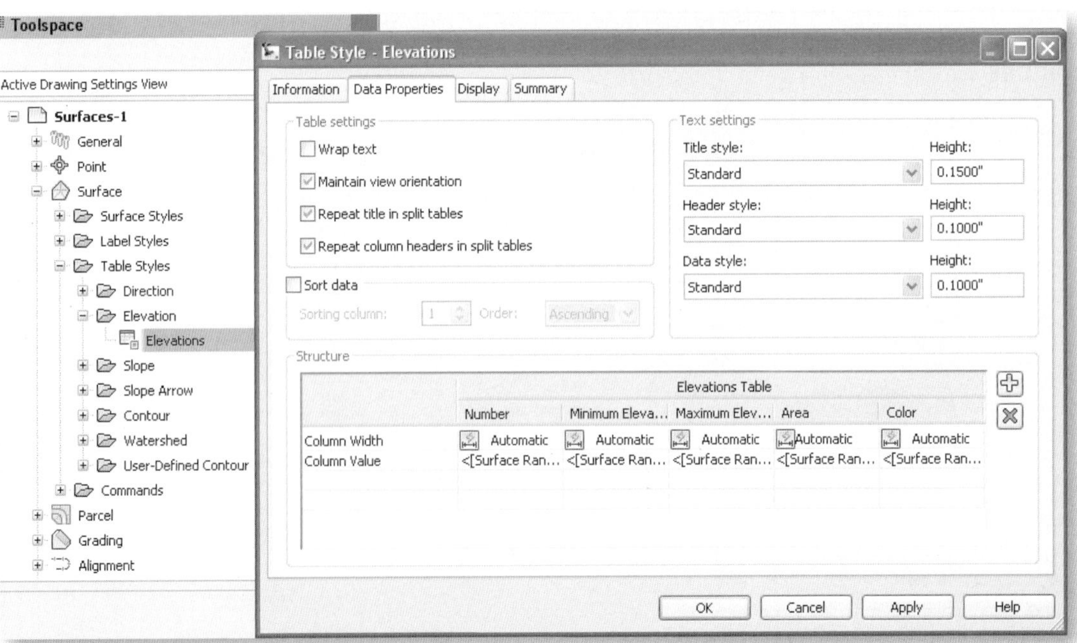


**FIGURE 4.10**

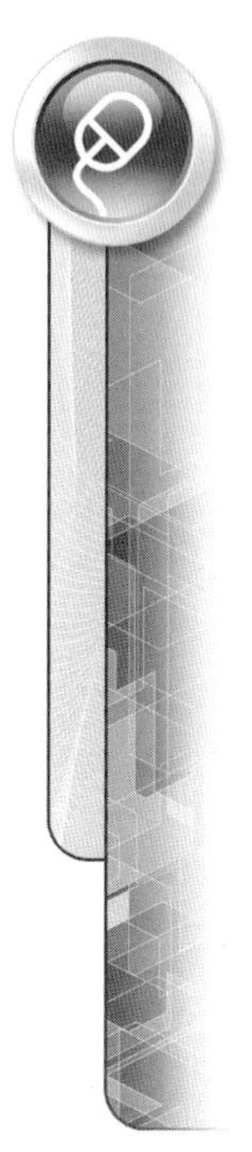

## EXERCISE 4-1

After completing this exercise, you will:

- Be familiar with the settings in Edit Drawing Settings.
- Be familiar with the settings in Edit Feature Settings.
- Be familiar with the settings in Edit Label Default Settings.
- Be familiar with the settings in Surface Object Styles.
- Be familiar with the settings in Surface Label Styles.

### Exercise Preparation

1. If not in Civil 3D, start it by double-clicking the **Civil 3D** desktop icon.
2. Close and do not save **Drawing1**.
3. At Civil 3D's top left, click the Civil 3D drop-list icon and from the Application Menu select NEW.
4. Browse to the Chapter 4 folder of the CD that accompanies this textbook and select the *Chapter 04 – Unit 1.dwt* template file and click ***Open***. Click ***OK*** to create the new file.
5. At Civil 3D's top left, click the Civil 3D drop-list icon and in the Application Menu, place your cursor over SAVE AS, in the flyout menu select AutoCAD Drawing, browse to the Civil 3D Projects folder, name it **Surfaces**, and click ***Save*** to return to Civil 3D.

### Edit Drawing Settings — Review and Set Values

The following steps set a suffix modifier, use the surface's name as part of the layer name, and place a dash separating the surface name from the root layer name, e.g., C-TOPO-EG.

1. Click the ***Settings*** tab.
2. At the top, click the drawing name, **Surfaces**, press the right mouse button, and from the shortcut menu, select EDIT DRAWING SETTINGS....
3. Click the ***Units and Zone*** tab to make it current.
4. If necessary, set the scale to **1"=40'**, set the Imperial to Metric conversion to **US Survey Foot,** and set the Zone Category to **No Datum, No Projection** (see Figure 4.11).

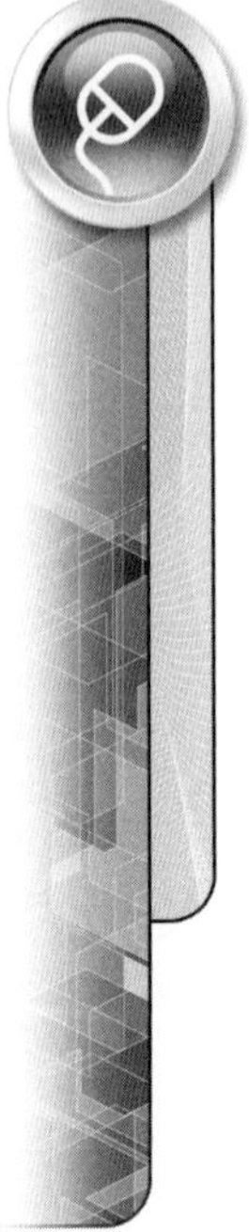

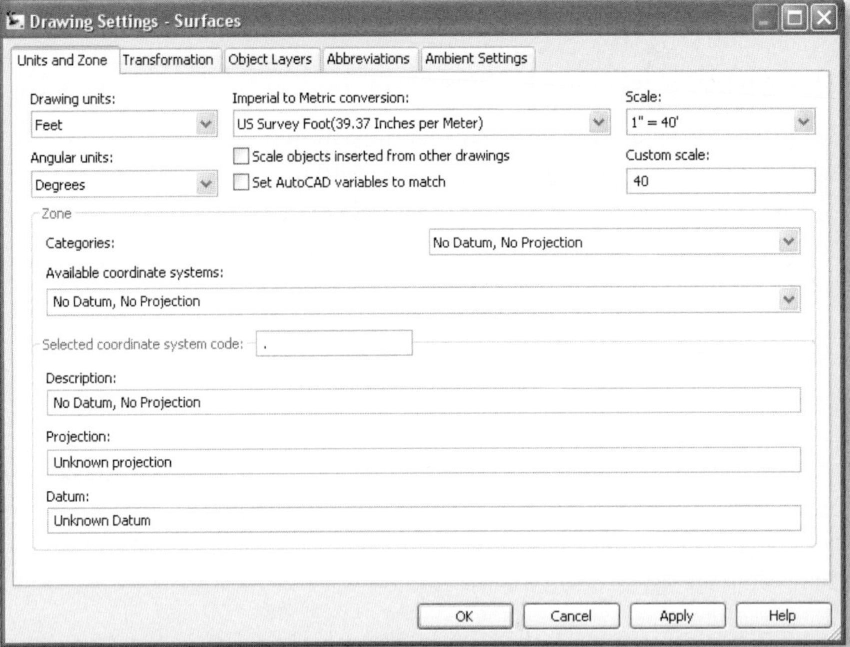


**FIGURE 4.11**

5. Click the ***Object Layers*** tab to view its values.
6. Scroll down the layer list until you are viewing the **Grid Surface** and **Grid Surface-Labeling** entries. Keep the default layer names, but change the Modifier by clicking in the cell and changing its value from None to **Suffix**.
7. Set the Modifier value to **-*** (a dash followed by an asterisk).
8. Continue scrolling down the layer list until you are viewing the **TIN Surface** and **TIN Surface-Labeling** entries. Keep the default layer names, but change the Modifier by clicking in the cell and changing the value from None to **Suffix**.
9. Set the Value of the Modifier to **-*** (dash followed by an asterisk) (see Figure 4.12).

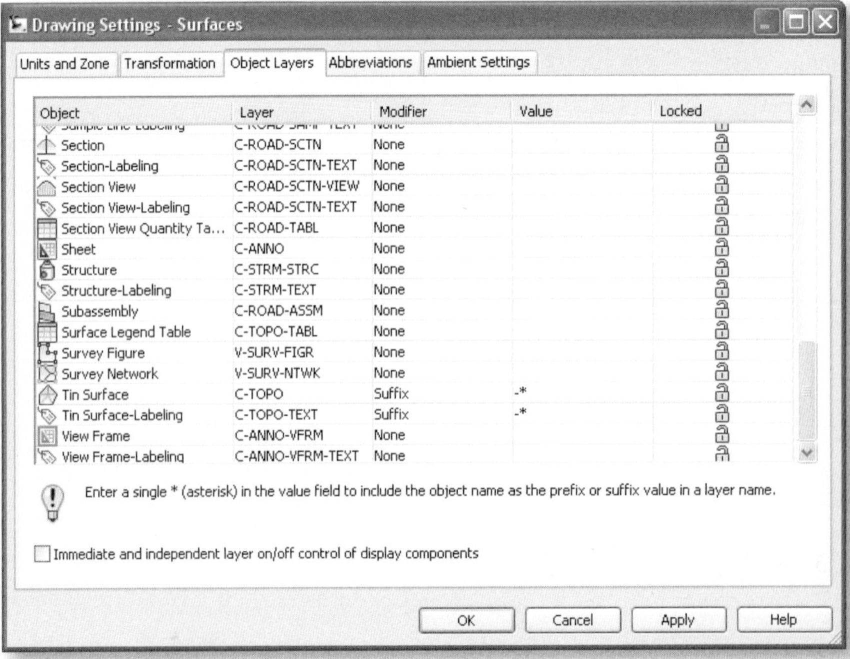


**FIGURE 4.12**

10. Click the ***Ambient Settings*** tab.
11. Expand the **Coordinate** and **Elevation** sections, review their settings, and if necessary adjust them (see Figure 4.13).

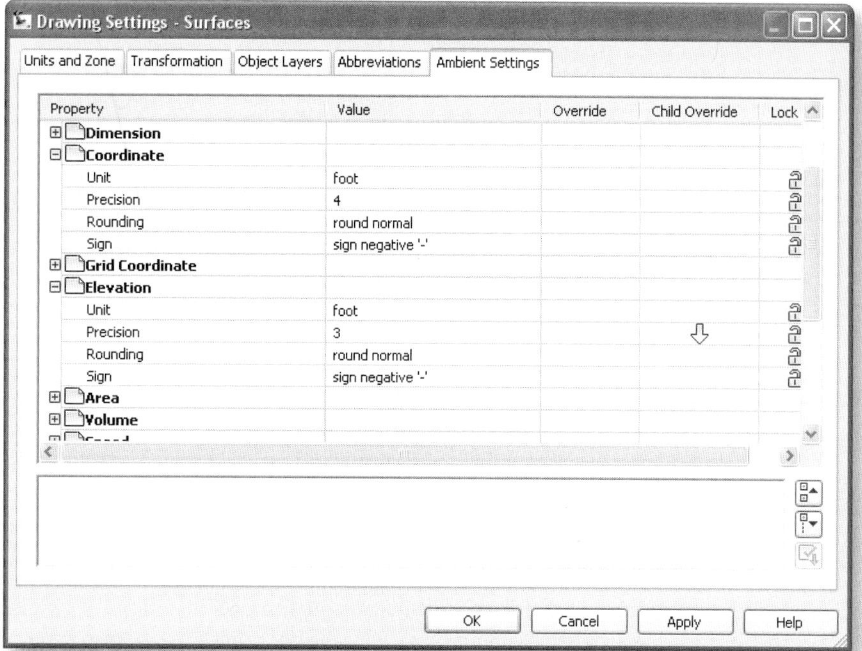


**FIGURE 4.13**

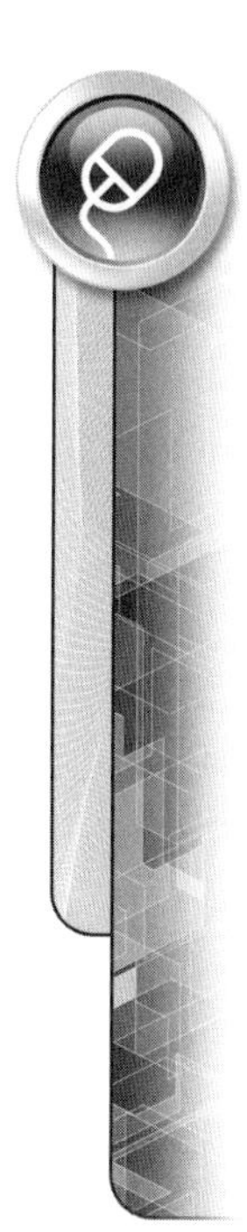

The settings affect the coordinate and elevation display and reporting. The Child Override column's downward-pointing arrow indicates the elevation precision is overridden by a lower style.

12. Click ***OK*** to close the Edit Drawing Settings dialog box.

## Edit Feature Settings Dialog Box

1. In Settings, expand the **Surface** branch by clicking the surface heading's expand tree icon (plus sign).
2. Click the heading **Surface**, press the right mouse button, and from the shortcut menu, select `EDIT FEATURE SETTINGS...`.
3. In Edit Feature Settings, expand **Default Styles** to view the assigned object and Label Styles (see Figure 4.14).

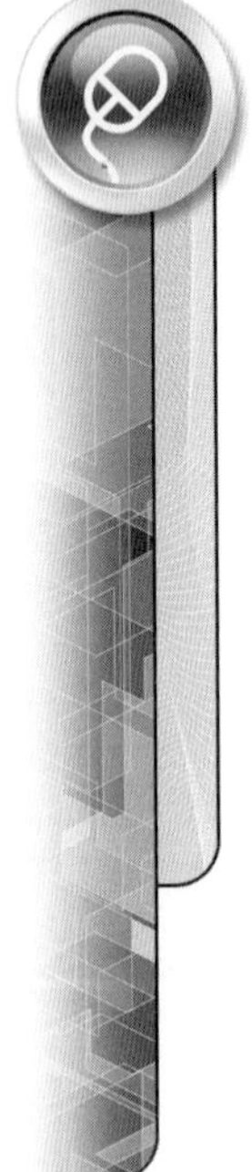

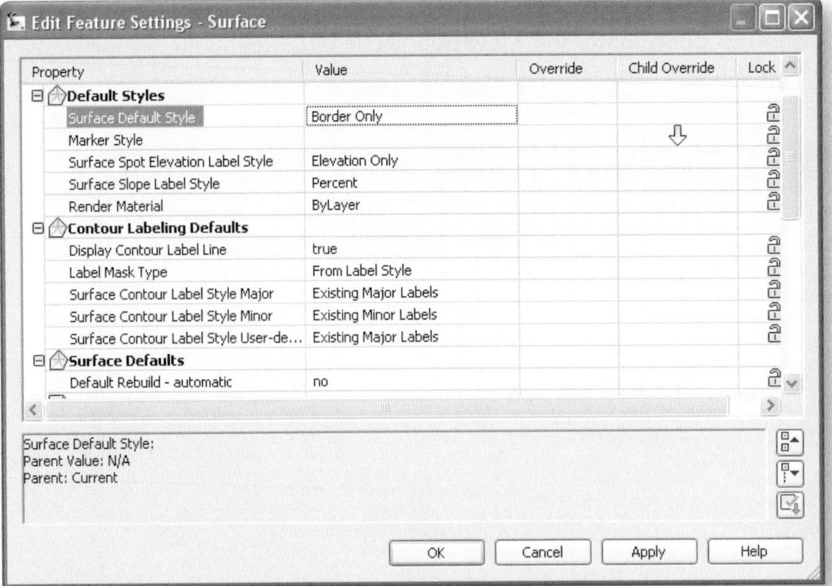


**FIGURE 4.14**

4. Change the Surface Default Style to **Border Only**. This is done by clicking in the Surface Default Style's value cell to display an ellipsis. Clicking the ellipsis displays the Surface Default Style dialog box. Click the drop-list arrow, from the list select **Border Only**, and click ***OK*** to exit the dialog box and assign the style.
5. Click ***OK*** to exit the dialog box.

## Command — CreateSurface

Create Surface's command settings control new surface default values and set surface build options.

1. In Settings, expand the **Surface** branch, until you are viewing its **Commands** branch and commands list.
2. From the list, select **CreateSurface**, press the right mouse button, and from the shortcut menu, select `EDIT COMMAND SETTINGS...`.
3. Expand the **Surface Creation** section to view its values (see Figure 4.4).

The Default Surface Type is TIN surface. The grid spacing and rotation values apply a grid surface type. The default naming convention is Surface1, Surface2, etc.

4. Close the **Surface Creation** section.
5. Expand the **Build Options** section to view its values (see Figure 4.4).

This section contains critical surface-building parameters. The settings and their values are discussed in a later unit's exercise.

6. Click ***OK*** to exit the Edit Command Settings dialog box.
7. Collapse the **Commands** branch.

## Surface Style — Border Triangles Points

The surface style, Border Triangles Points, displays surface components important to building the surface and reviewing the surface quality. Viewing the triangulation allows you to decide whether additional data and/or breaklines are necessary, or if editing the surface suffices.

1. In the Settings, expand the **Surface** branch until you are viewing Surface Styles' list of styles.
2. From the list of Surface Styles, select the style **Contours and Triangles**, press the right mouse button, and from the shortcut menu, select COPY....
3. Click the ***Information*** tab. Change the style name to **Border Triangles Points** and give the style a short description.

The information panel identifies the Name, Description, and the created and modified dates for the style (see Figure 4.15).

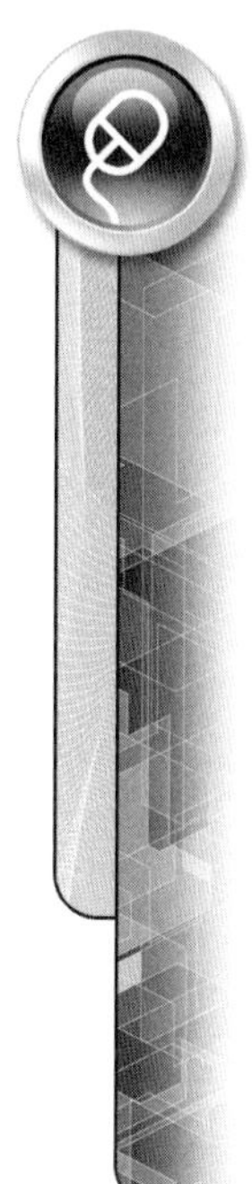

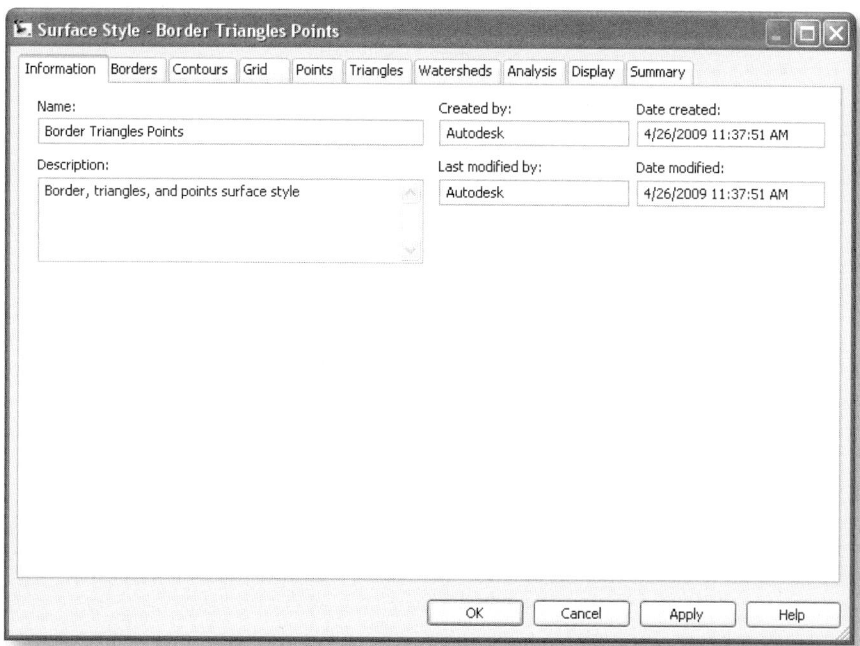


**FIGURE 4.15**

4. Click the ***Borders*** tab.

The 3D Geometry section defines how the surface displays its border (at its true Z-elevation), if it shows an exterior and any interior borders, and if those borders use a datum (see Figure 4.16).

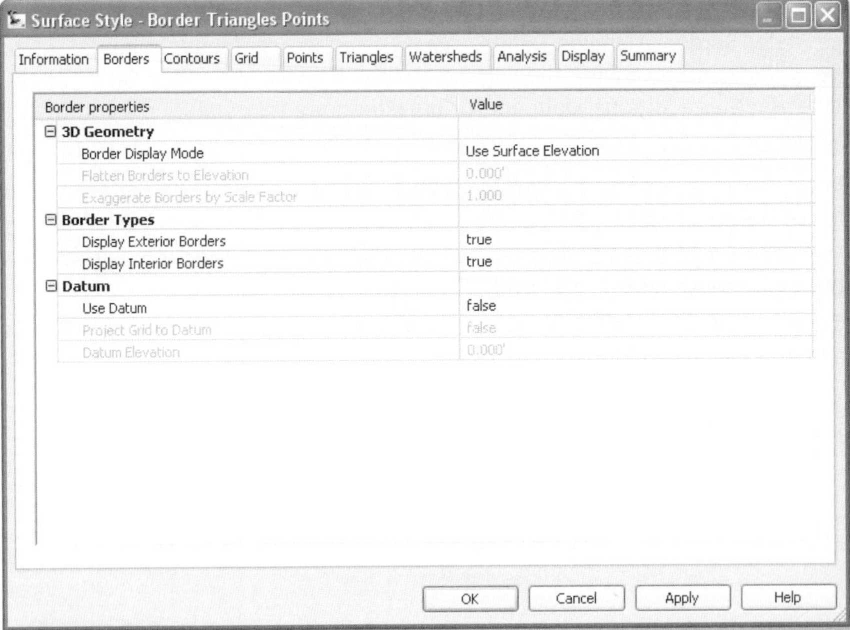


**FIGURE 4.16**

5. Click the ***Points*** tab.

The 3D Geometry section defines how surface points represent surface elevations (they will be at their true Z-elevation). The Point Size section sets the marker's size (this size is set by the scale of the drawing). The Point Display section defines the three surface point types' shape and color: Point Data (actual points from the drawing); Derived Points (smoothing points); and Non-Destructive Points (points on the surface from boundaries) (see Figure 4.17). These settings do not affect cogo points.

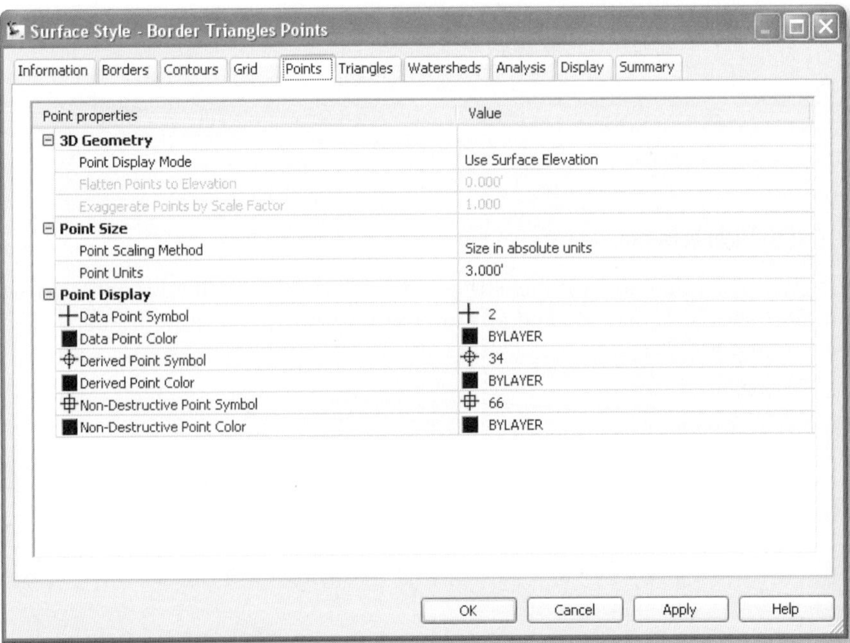


**FIGURE 4.17**

6. Click the ***Triangles*** tab to make it current.

The setting's value affects the surface triangle elevations.

7. If necessary, set the value to **Use Surface Elevation** (see Figure 4.18).

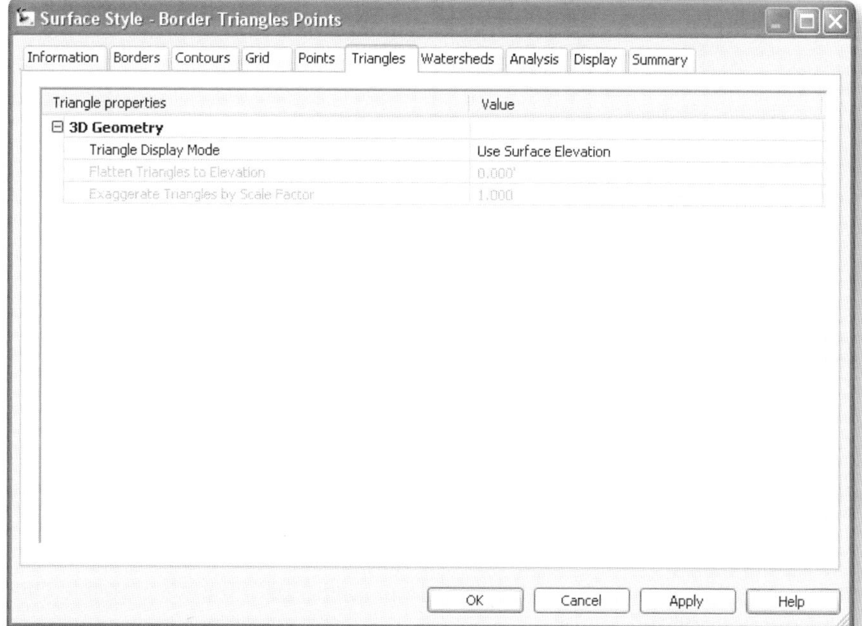


**FIGURE 4.18**

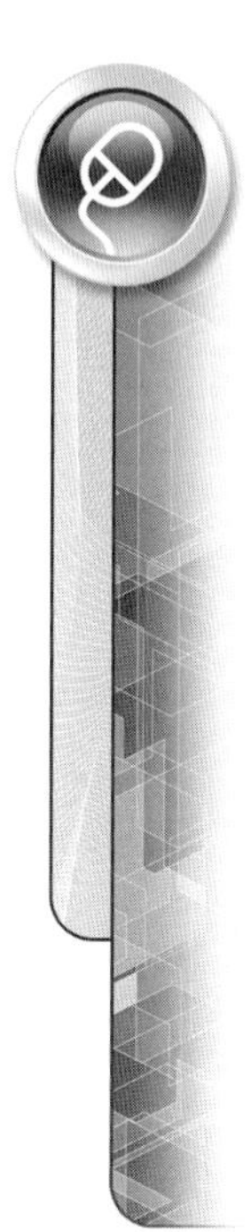

8. Click the ***Display*** tab to make it current.

The Display panel sets surface components and characteristics layers and their properties. The initial settings for plan views (directly over the surface) are surface border, triangles, and points. In the top left, clicking the View Direction drop-list arrow changes the settings to Plan, Model, or Section view (see Figure 4.19).

This surface style's plan view displays a border, triangles, and points, and its Model view shows only triangles.

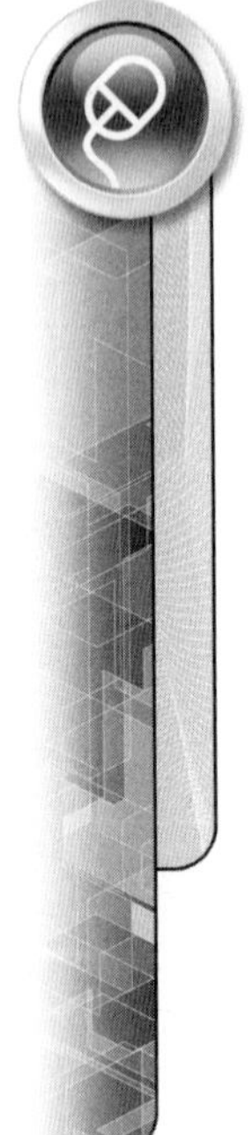

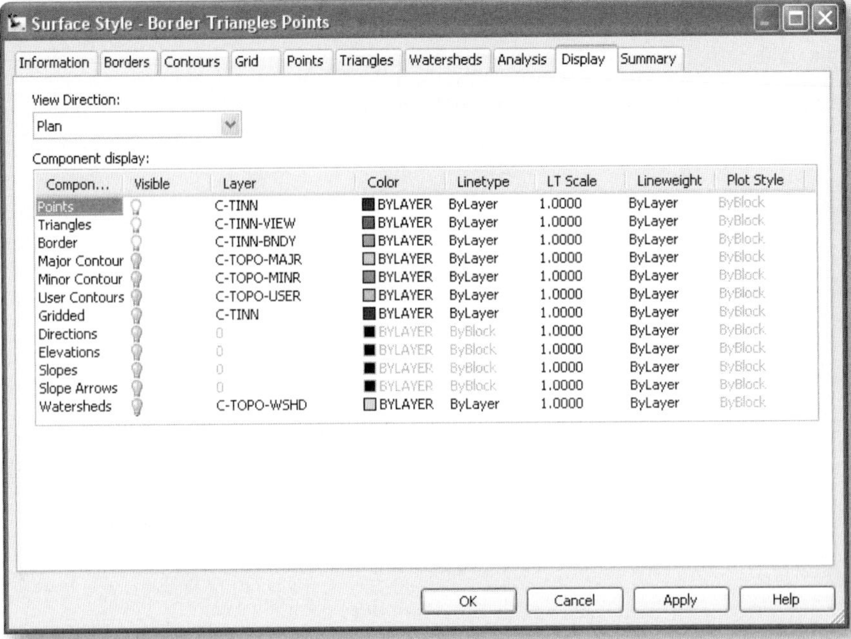


**FIGURE 4.19**

9. In the Display panel for Plan View Direction, toggle **ON Points** and toggle **OFF Contours Minor** and **Major**.
10. In the Display panel, click the **View Direction** drop-list arrow and change the View Direction to **Model**. Only Triangles are visible.
11. In the Display panel, click the **View Direction** drop-list arrow and change the View Direction to **Section**. Only Surface Section is visible.
12. Click ***OK*** to create the Border Triangles Points style.
13. Collapse the **Surface Styles** branch.
14. At Civil 3D's top left, Quick Access Toolbar, click the ***Save*** icon to save the drawing.

## Spot Elevation Label Style – Elevation Only

The Spot Elevation, Elevation Only label style defines surface elevation labels. The drawing contains several additional labeling styles for contours, slopes, watersheds, and spot elevations.

1. While still in the Settings' Surface branch, expand **Label Styles** to view its style's list.
2. Expand the label type **Spot Elevation** to view its label styles list.
3. Select **Elevation Only**, press the right mouse button, and from the shortcut menu, select `EDIT...`.

The Label Style Composer dialog box opens.

4. Click the ***Information*** tab to make it current.

This panel contains the label style's Name and Description. The panel's right side displays the style's creation and modified dates (see Figure 4.20).

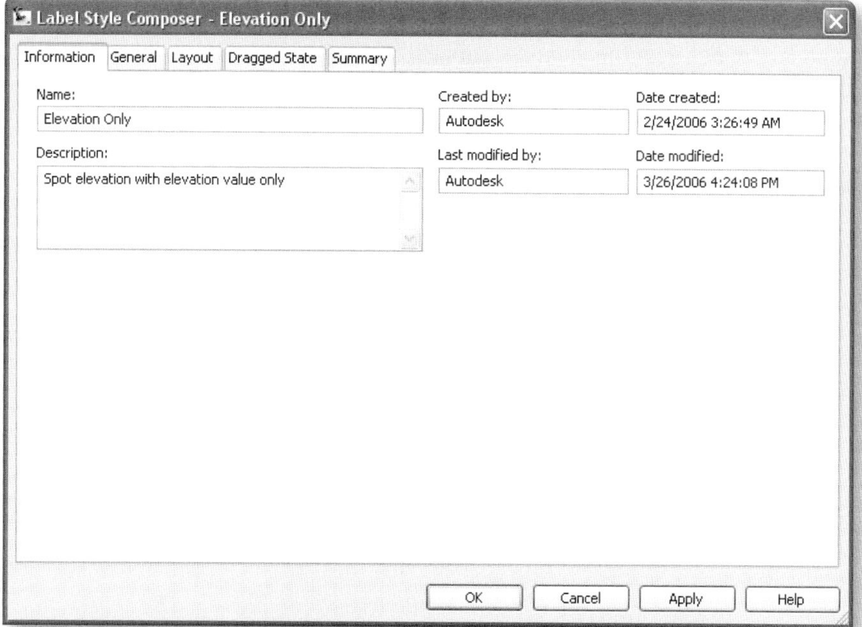


**FIGURE 4.20**

5. Click the ***General*** tab to make it current.
6. If necessary, at the panel's upper right, click the drop-list arrow and set the preview to **Surface Spot Elevation Label Style**.

This panel sets the label style's overall visibility. If you change the Label section's Visibility to false, all of this style's labels are not displayed. If the Label Layer is set to 0, the label will use the label layer defined in Edit Drawing Setting, Object Layer list (Surface Labels — C-TOPO-TEXT). The Orientation Reference sets how the label attaches itself to the surface. The label's initial orientation is the object's orientation. Optional settings include View (displays the label horizontally in all views), or WCS (label reads toward the World Coordinate System's Zero direction). The Plan Readability section defines how the Elevation Only label is displayed in any view (see Figure 4.21).

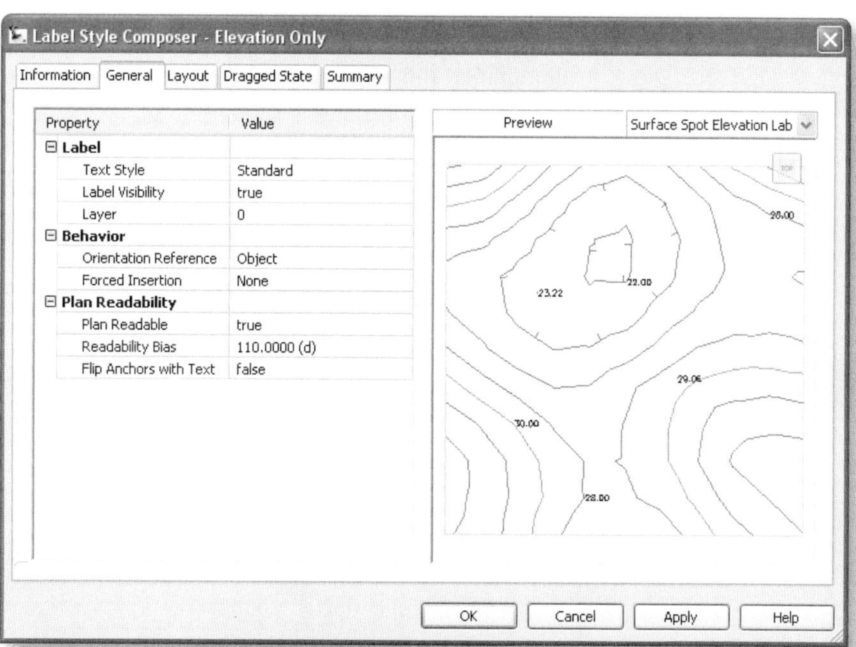


**FIGURE 4.21**

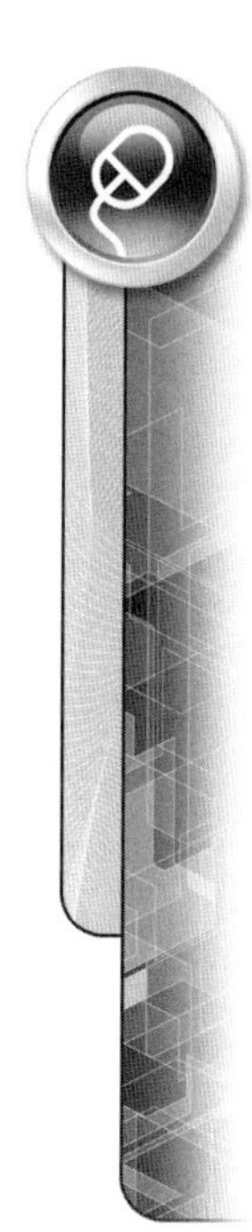

7. Click the ***Layout*** tab to make it current.

The Layout panel defines how the label anchors itself in the object. This label is anchored on the <Feature> (surface) at the selected point's middle right. If toggled on, the Border section defines an enclosing box around the text. The Text section defines the label's text height, rotation, and color. The Attachment setting attaches the text's middle left to the feature's anchor point (see Figure 4.22).

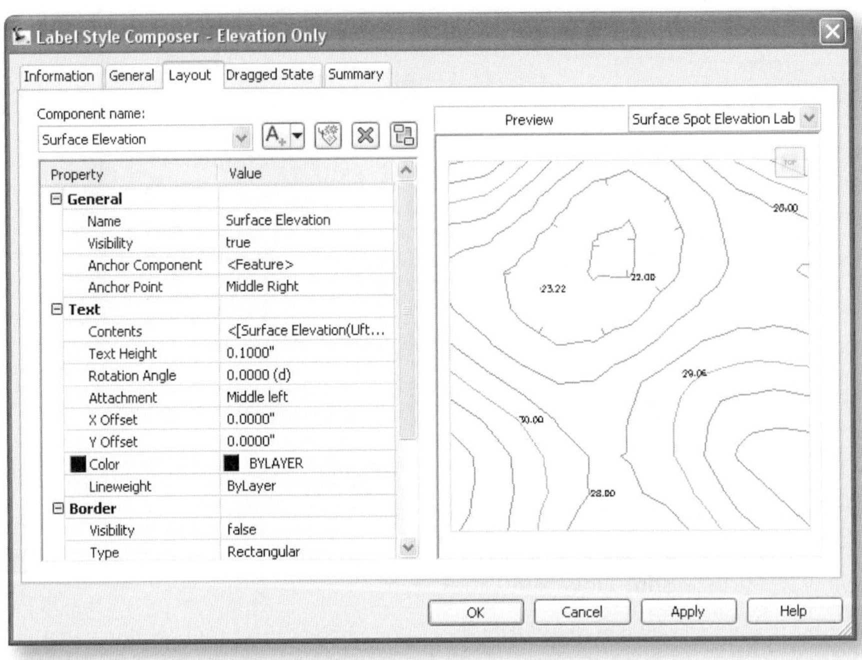


**FIGURE 4.22**

8. In the Text section, in Contents, click in its **Value** cell to display an ellipsis, and click the ellipsis to display the **Text Component Editor**.

The Text Component Editor shows the surface's elevation text labeling format string. The dialog box's right side displays the current format string, which is the sum of the modifiers and their values from the left side. For example, the top entry, Uft, indicates the units are feet, and P2 sets the precision to two decimal places. When highlighting a string on the right, the left side shows each modifier and its value. You change values by clicking in a cell, clicking a drop-list arrow, and selecting a new setting for the Modifier (see Figure 4.23). To transfer the new setting, on the left click the blue arrow; this transfers the setting to the highlighted string on the right.

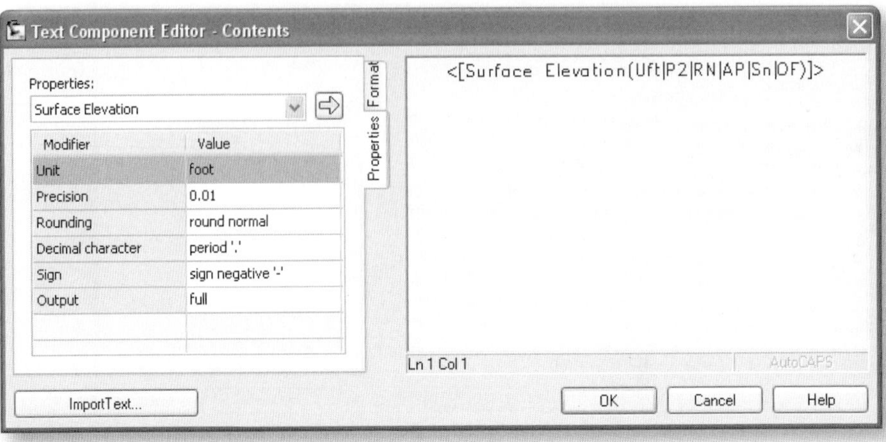


**FIGURE 4.23**

9. Click ***OK*** to exit the Text Component Editor.
10. Click the ***Dragged State*** tab to make it current.

If you are dragging a label from its original location, these settings define what happens to it. In this label style, the dragged label will have a leader and the text will be stacked (left-justified).

11. Click ***OK*** to close the Label Style Composer dialog box.
12. At Civil 3D's top left, Quick Access Toolbar, click the ***Save*** icon to save the drawing.

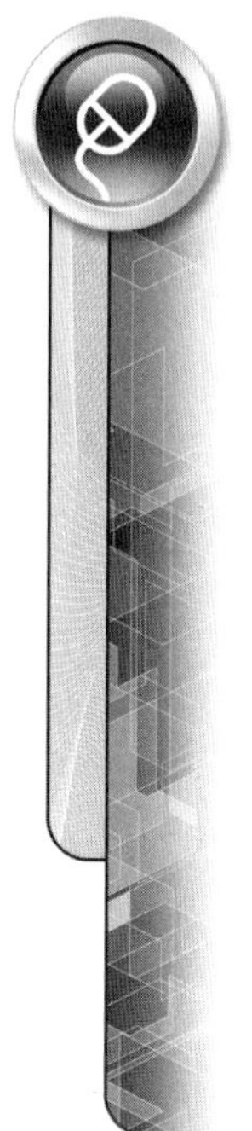

## SUMMARY

- Edit Drawing Settings and Edit Command Settings assign layers, and set default styles, surface types, and build options.
- If a drawing contains multiple surfaces, the Edit Drawing Settings base object layer name should contain a prefix or suffix.
- Edit Feature Settings sets default surface names and styles.
- A command's settings can override default surface styles and parameters.

This completes the surface settings and styles review and the steps to change their values. Next, you will learn to create (name) the surface, add data, and begin the review process.

## UNIT 2: SURFACE DATA

### STEP 1: CREATE A SURFACE

The first step in creating a surface is naming it and assigning it a type and styles. The Ribbon's Home, Surfaces drop-list menu's Create Surface…, or selecting Prospector's Surfaces heading, pressing the right mouse button, and from the shortcut menu, selecting Create Surface… both display the Create Surface dialog box.

#### Create Surface

The Create Surface dialog box defaults are from the values in Edit Drawing Settings (surface layer), Edit Feature Settings (name format and style assignments), and command overrides (see Figure 4.24). The dialog box's top left lists the default surface type and can be changed to any one of the four surface types. The dialog box's top right displays the surface layer (Edit Drawing Settings, Object Layers). If you are using a layer modifier, these affect what is displayed as the layer name.

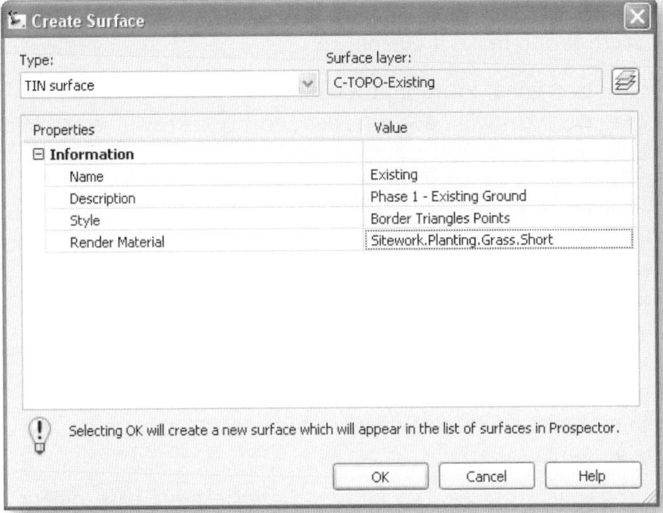


**FIGURE 4.24**

The dialog box's body contains the name, description, and the initial object and rendering material styles. If you want to change the object and render styles, clicking the value cell displays a Select Style dialog box from which you can select an alternative style.

When exiting the Create Surface dialog box, in Prospector under the Surfaces heading, an entry appears with the surface's name. The surface name is a branch heading containing surface information and data. Surface information covers surface masks and watersheds. The data is in a Definition branch listing surface data types (points, point groups, boundaries, etc.) (see Figure 4.25).

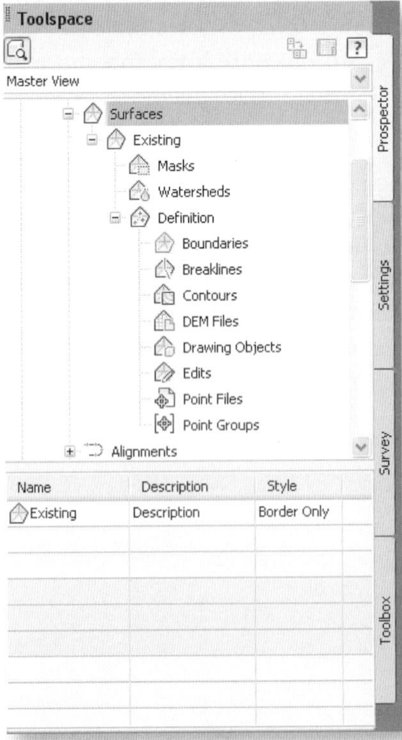


**FIGURE 4.25**

### Step 2: Create Initial Data

A surface's Definition branch lists data added, deleted, and edited, thus creating a surface's triangulation. A right mouse click shortcut menu containing appropriate items to that entry is associated with each heading (see Figure 4.26).

The data assignment order is the following:

Data Clip Boundary
Points (Point Groups, Files, AutoCAD objects)
Breaklines and/or Contours (those that do not cross breaklines)
Setting a Maximum Triangle Length, and/or Edits and/or Boundaries

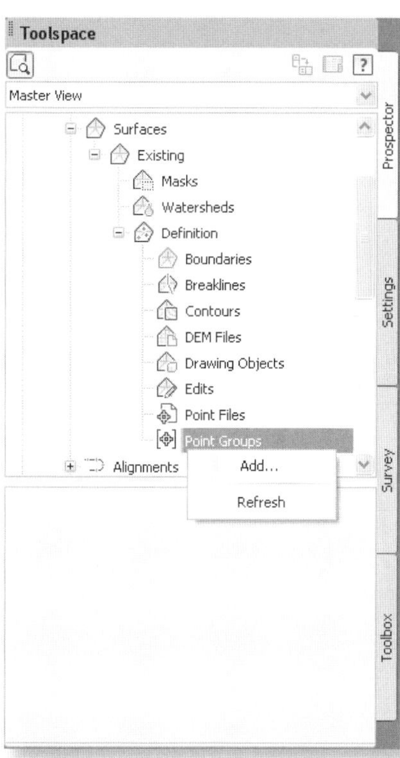


**FIGURE 4.26**

The surface properties dialog box, Definition tab, Operation Type lists, in order, all surface actions (data assignments, edits, etc.). This list toggles their effect on or off, relocates their position in the surfaces history, or can permanently remove them from the surface definition. When changing their status, moving, or removing them from the list, when you exit the dialog box, you are prompted for a surface rebuild. If the operations type list is reordered, the effect each operation has on the surface may change.

## Boundaries

A surface can have four boundary types: Clip, Outer, Hide, and Show.

**Data Clip.** This boundary excludes all data outside its edge. When this is the first surface data entry, all added data outside is not included in the surface. The surface uses all excluded data, if you delete or toggle off this boundary.

**Outer.** An outer boundary forces a surface to use the data only within its edge. The outer boundary focus is limiting surface data to a specific area (i.e., to exclude data outside the closed polyline). The difference between Outer and Data Clip is if you are adding data outside the boundary, the surface expands past the outer boundary's edge. If you are adding data and expanding the surface, in the surface's properties dialog box, reorder the operation types so the boundary is at the bottom of the list. With the boundary at the bottom of the list, it once again excludes data outside its edge, but includes any new data inside its edge.

When an outer boundary crosses triangulation, there are two options to handle the crossed triangulation. The first is to create new triangles from the enclosed data out to the boundary (i.e., Non-destructive breakline) (left side of Figure 4.27). This new triangulation represents as best as possible the elevations of the original surface where the border intersects the triangles. If the boundary contains curves, there is a mid-ordinate setting to adjust the number of new triangles along its curves. The second option is to stop the triangulation at the data. This option does not create triangles between the data and the boundary (right side of Figure 4.27).

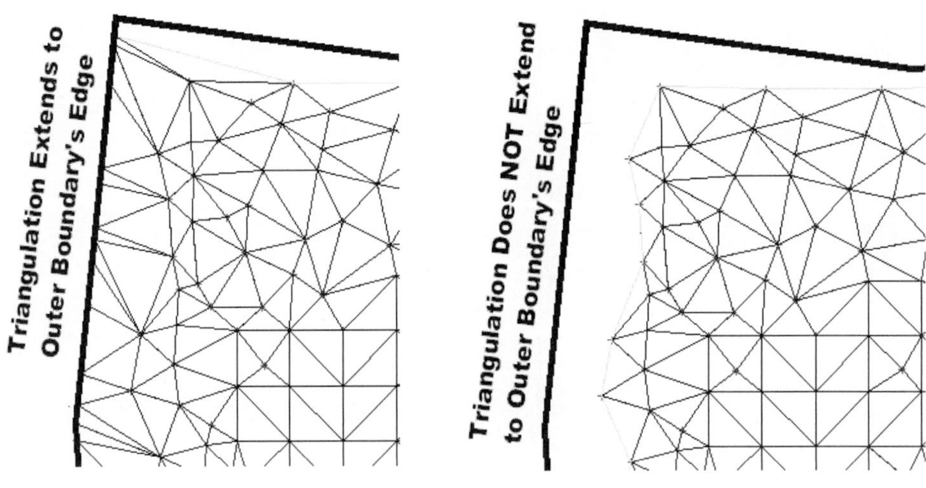


**FIGURE 4.27**

**Hide.** A hide boundary suppresses any surface triangulation within its boundary. The boundary does not need points in its interior to hide the triangulation (see Figure 4.28). It is best to toggle on Non-Destructive for this boundary type.

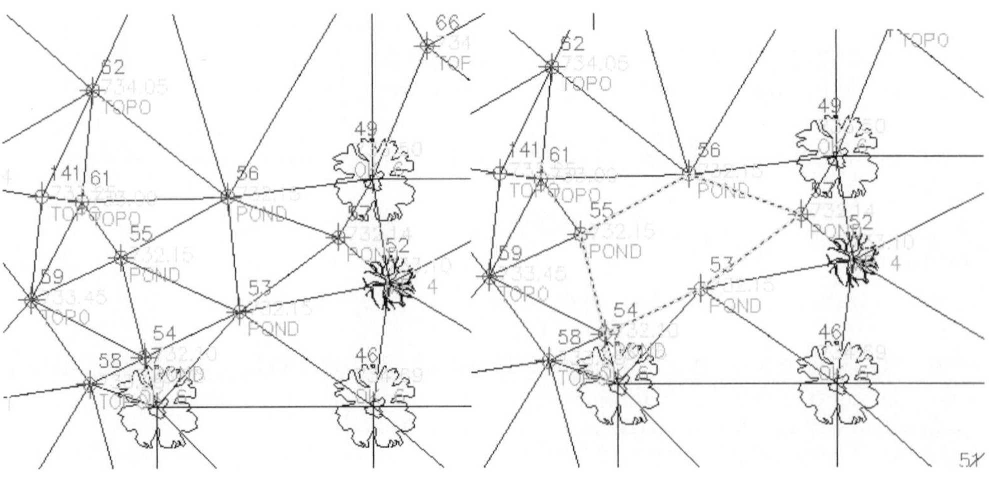


**FIGURE 4.28**

**Show.** A show boundary displays surface triangulation within a hidden boundary. The boundary does not need points in its interior to work when it is set to Non-Destructive.

## Point Data

Points are common surface data. If there are cogo points in a drawing, they must be in a point group. Point data can also be external (ASCII file) and drawing objects.

**Drawing Objects.** The first point data source is drawing objects (see Figure 4.29). When selecting objects, this routine extracts their X, Y, and Z values and uses them as point data. Drawing nodes, 3D lines, blocks, text, 3D faces, and polyface mesh objects are potential surface data. This routine's point data is in addition to point file and/or point group data. The routine does not report or generate any connections between the selected objects' endpoints and vertices. The exception is 3D Faces and toggling on Maintain edges from objects. This preserves their 3D Face edges in the surface.

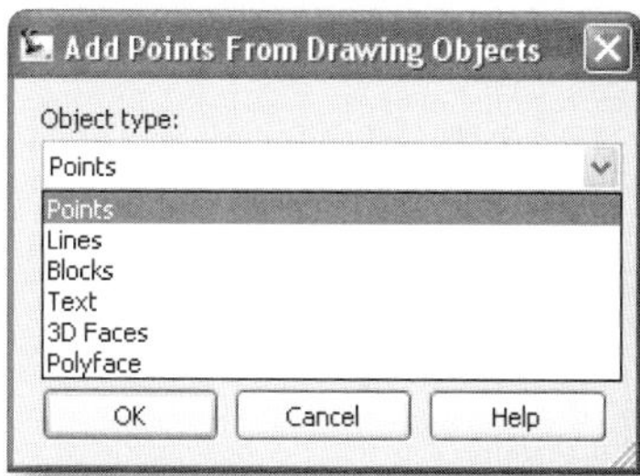


**FIGURE 4.29**

**Point Files.** A second point data source is an external coordinate file. Point data imported by this option is not made into cogo points, but is exclusive to the surface. The routine uses the same Import/Export formats: the file can be ASCII text or an Access database. If you want the data to be active (cogo points), use the Points Import/Export routine.

**Point Groups.** A point group contains all or a selected number of cogo point objects (see Figure 4.30). For information about creating and managing point groups, see the point groups discussion in Chapter 2 of this textbook. Point groups include and exclude points by their number, elevation, description, or selection. Excluding points is necessary because of inaccurate elevation values. For example, points representing fire hydrants are observed by locating their top. These points could be from 1 to 3 feet above the surrounding ground elevations. If you include these points, they create a small hill at each hydrant location. Clearly, they should be excluded from the surface data.

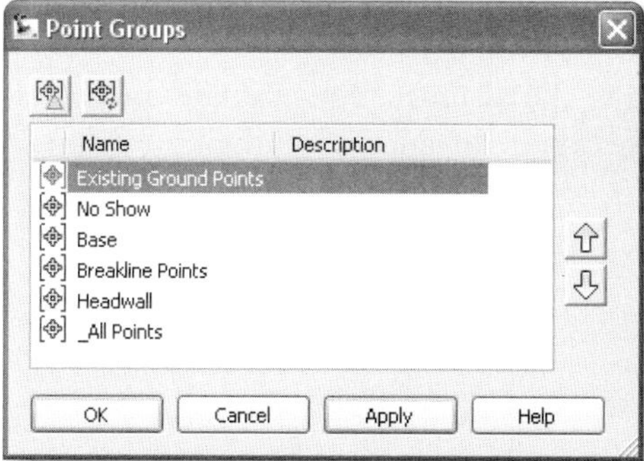


**FIGURE 4.30**

## Contours

Contours represent known surface elevations and are a convenient surface data source. Surfaces from contours do not have the same issues as surfaces from points; which are unable to correctly triangulate linear features. Each contour line is a breakline, which forces the triangulation between contours.

**CAUTION**

If you receive a contour drawing from outside your control, thoroughly check its values before using its data.

**Crossing Contours.** Contour data should never include crossing contours. Even though it occurs in reality, a surface cannot accommodate this situation. When assigning crossing contours to a surface, Civil 3D adds both contours as data, uses the first contour's vertices, issues an error message about the crossing contours, and at the intersection ignores the second contour's data. An error message in the Events vista identifies the intersection point. You should edit the contours so they do not cross (see Figure 4.31). Editing a contour may be as simple as relocating a vertex by using its grip.

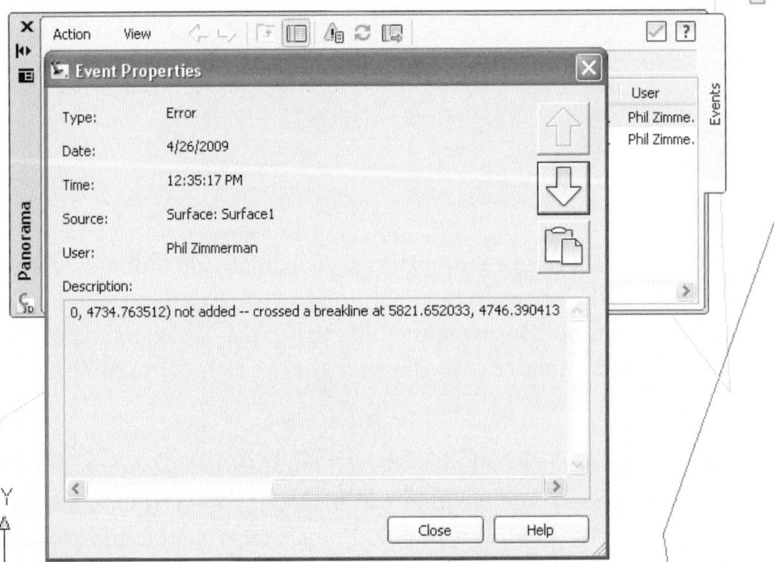


**FIGURE 4.31**

**Weeding Factors.** Weeding removes potentially redundant contour (polyline) vertices. When reading Weeding factors, place a Boolean "AND" between the two values (distance "AND" angle). To remove data (weeding), the vertices must be less than both weeding factors.

The first decision is distance. When evaluating the distance, you evaluate the overall distance between three adjacent vertices. If the overall distance is less than the distance factor, the analysis moves to the second factor, angle (see Figure 4.32).

**FIGURE 4.32**

The angle factor is a right or left turning angle (deflection) made by the contour's three vertices (see Figure 4.32). If the angle factor is 4 degrees, this limits the insignificant turning angle to 4 degrees or less to the right or left of the line from vertex 1 to 2. If the angle is less than 4 degrees, the second vertex is considered redundant and not included. If the angle is greater than 4 degrees, the second vertex is kept as data.

If the distance between three adjacent vertices is greater than the distance weeding factor, the routine moves on to find the next set of three vertices under the Weeding's distance value.

Weeding does not change contours, but instead determines which vertices are data. How much data can be removed before encountering problems? Some experts suggest a 50 percent reduction still produces a viable surface. The number of vertices weeded is not the issue. What is important are the vertices that remain after the weeding process. The only way to evaluate Weeding's effects is to review the resulting triangulation.

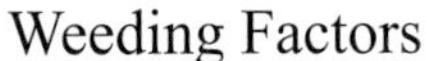

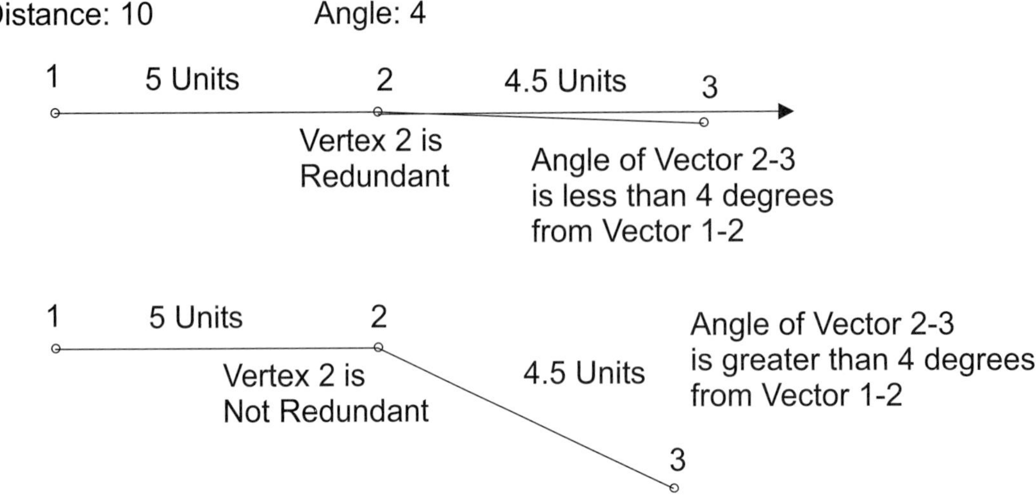


**FIGURE 4.33**

**Supplementing Factors.** Supplementing factors add new data to line and curve contour segments. If Add Contour Data does not encounter a vertex before reaching the distance or mid-ordinate distance value, it calculates new vertices to the data.

The distance factor is the distance along a straight segment, and the mid-ordinate distance is the distance a chord is from an arc. If the distance factor is 20.0, any straight contour segment more than 20 feet between vertices has a data vertex added along its path. If it is an arc, the mid-ordinate distance factor adds vertices representing calculated points along the arc's path. A large mid-ordinate value creates fewer points, and a smaller value creates more points.

Like Weeding, the supplementing factors do not modify the original polylines or contours; they only create more or less surface data. The supplementing distance factor cannot be lower than the Weeding distance factor.

### Contour Data Issues

There are two problems with contour data: bays and peninsulas and the lack of high and low elevations.

**Bays and Peninsulas.** Contours may switch back and as a result create peninsulas and bays (see Figure 4.34). In these loops, a surface switches from triangulating between different contours to triangulating along the same contour. This happens because the nearest triangulation point is along the same contour and not to the next higher or lower contour, or it is impossible to create a triangle between two different contours in the bay or peninsula.

When applying a contour style, the contours cut off the original data's bays and peninsulas. This is because the contour follows the first triangle leg that makes the jump across the bay or peninsula. The result is a surface flat spot.

The errors' net result is usually minimal, and they may even cancel each other out. Civil 3D attempts to optimize the diagonals controlling this problem by using Swapping edges. In many situations, this routine corrects the triangle leg problems; however, this routine does not correct all of the flat spots (see Figure 4.34).

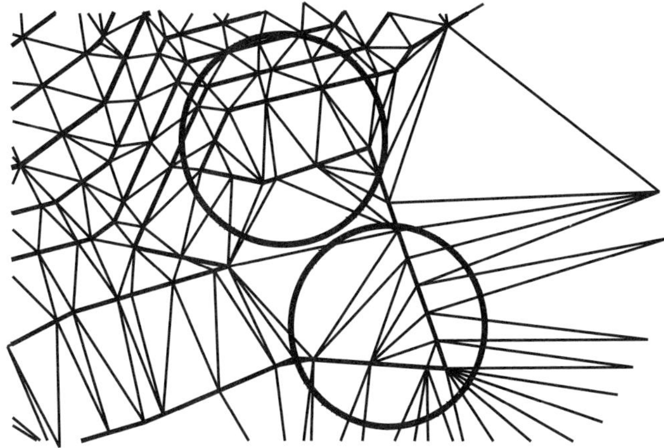


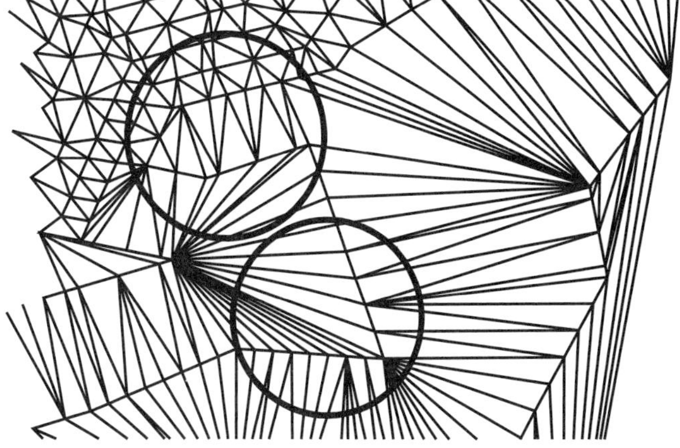


**FIGURE 4.34**

A contour data algorithm resolves the bay and peninsula problem. The algorithm looks at the surface elevation trend and adds elevation data that is slightly higher or lower on the triangle's midpoint where it connects to the same contour. This allows a surface to correctly represent contour data and faithfully re-create the original contours (see Figure 4.35). This does create larger surface files, but is well worth the overhead.

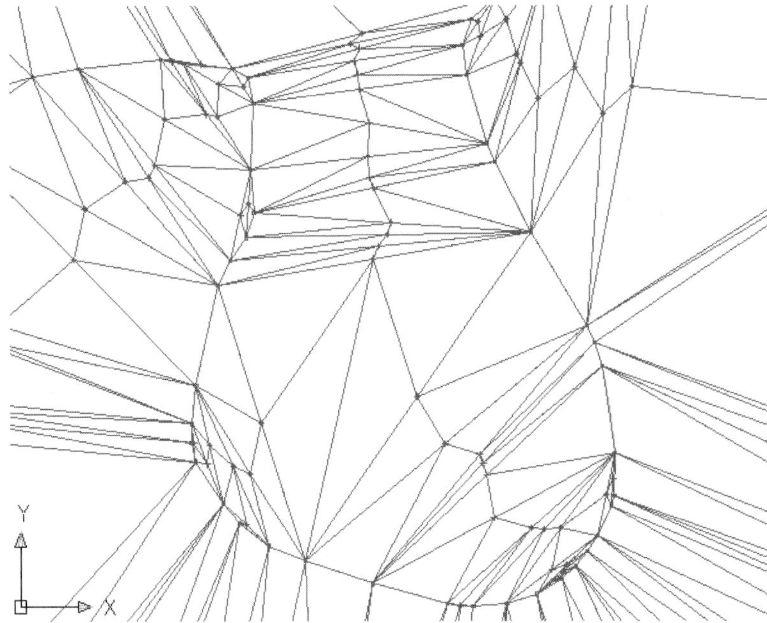


**FIGURE 4.35**

Toggles on Create Contour Data's lower half creates this optimal surface contour data (see Figure 4.32).

## High and Low Points

The second issue is a lack of high and low spot elevations and how to create data representing these missing values. Most aerial firms' contour drawings contain text or blocks inserted at their label's elevation. In addition to text and blocks, other objects can be point data: nodes, 3D lines, and polymesh objects. Including these objects as spot elevations resolves the contour data problem of not having high and low elevations.

## DEM Files

A DEM (Digital Elevation Model) is a USGS standard surface data file (see Figure 4.36). It is a grid with elevations at each intersection. A DEM's area can be quite large and dwarf a project site. Also, the size of DEM cells makes it easy to show a large area surface, but is not very useful for showing a small site's elevations.

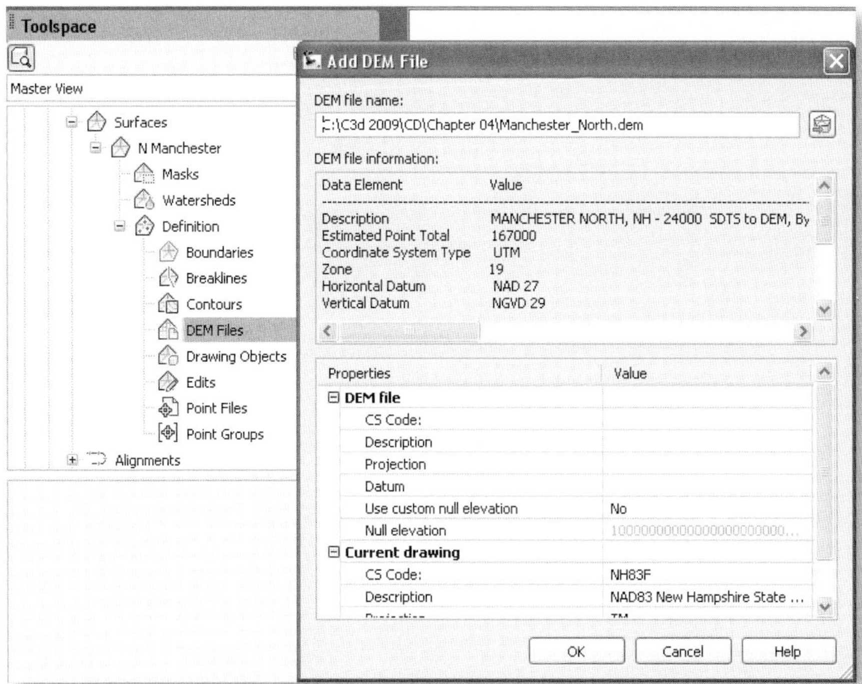


**FIGURE 4.36**

## Step 3: Add Breaklines to Known Problem Areas

A surface contains linear features as points. These features include man-made (edges-of-travelway, curbs, walls, etc.) and natural (swales, ditches, water edges, break-in slopes, etc.) features. The surface triangulation process does not correctly resolve these features. If you want to preserve the points' linear relationships, breaklines must connect the related points. Breaklines represent and preserve the points' linear trend in a surface.

Polylines (2D or 3D) or feature lines representing the points' linear connection first must be drawn before adding them as breaklines. 2D or 3D polylines cannot have curves; only feature lines can contain curves.

### Breaklines

There are five breakline types, four of which control triangulation: Standard, Proximity, From File, and Wall (see Figure 4.37). Standard and wall breaklines are 3D objects, and Proximity breaklines are 2D objects. Even though different (Standard-3D and Proximity-2D breaklines), they produce the same effect. A wall breakline defines a wall's face.

The last breakline type is Non-destructive and it fractures triangulation along its length. A Non-destructive breakline can be a 2D or 3D object. Generally, Non-destructive breaklines are surface-editing tools or are helpful when defining rendering areas.

The lower half of the Add Breaklines dialog box sets weeding and supplementing values for the selected breaklines. These functions affect the breakline's data as they would a contour. See the contour section for an explanation of these values.

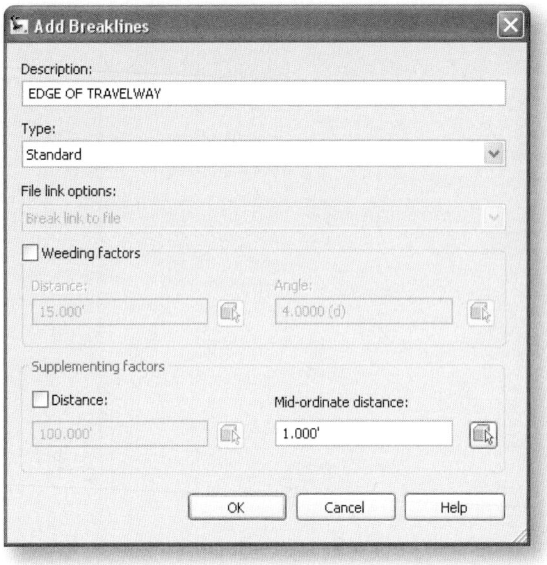

Add Breaklines

Description:
EDGE OF TRAVELWAY
Type:
Standard
File link options:
Break link to file
Weeding factors
Distance: 15.000'
Angle: 4.0000 (d)
Supplementing factors
Distance: 100.000'
Mid-ordinate distance: 1.000'
OK Cancel Help

**FIGURE 4.37**

On any surface, there are features that extend across several points; for example, swales, edges-of-travelway, backs of curbs, berms, etc. Each point's elevation along these features is related to a point before and after it. The only way to guarantee that the TIN triangulation correctly represents this relationship is to define a breakline. A breakline instructs a surface to place triangle legs linking the connected points and not allow any triangle legs to cross the breakline.

Linear feature misrepresentation occurs consistently in point data because of the surface triangulation algorithm. Point data must be dense enough along a linear feature to both describe and delineate it from all other intervening points. In many surveys, this is not the case. A field crew's most difficult task is viewing their survey as a triangulated surface. They see their points within the context of reality. For them, it is hard to understand why a point 100 feet away is not a part of the roadway edge in a triangulated surface. In the field, they see the roadway edge and they do not see their points triangulated. But an edge-of-travelway point and similarly described points separated by 100 feet or more lose their linear connection when creating a surface. When roadway edge points are mixed with surrounding points that do not have the same description (trees, signs, road shoulders, manholes, etc.), the nearest neighbor to a roadway edge point is probably a differently described point (see Figure 4.38). The points surrounding the two edge-of-travelway points (point numbers 17 and 18) form triangles crossing between the two pavement shots, and as a result negate their linear and elevation relationship. This same problem occurs where data is too sparse to support a correct interpretation (see Figure 4.39). The situation shown in Figure 4.39 does not have enough data to correctly resolve the diagonal connection between the 719 or 721 elevations. Without more data or a breakline, the surface creates a diagonal that is always the shortest cross-corner distance.

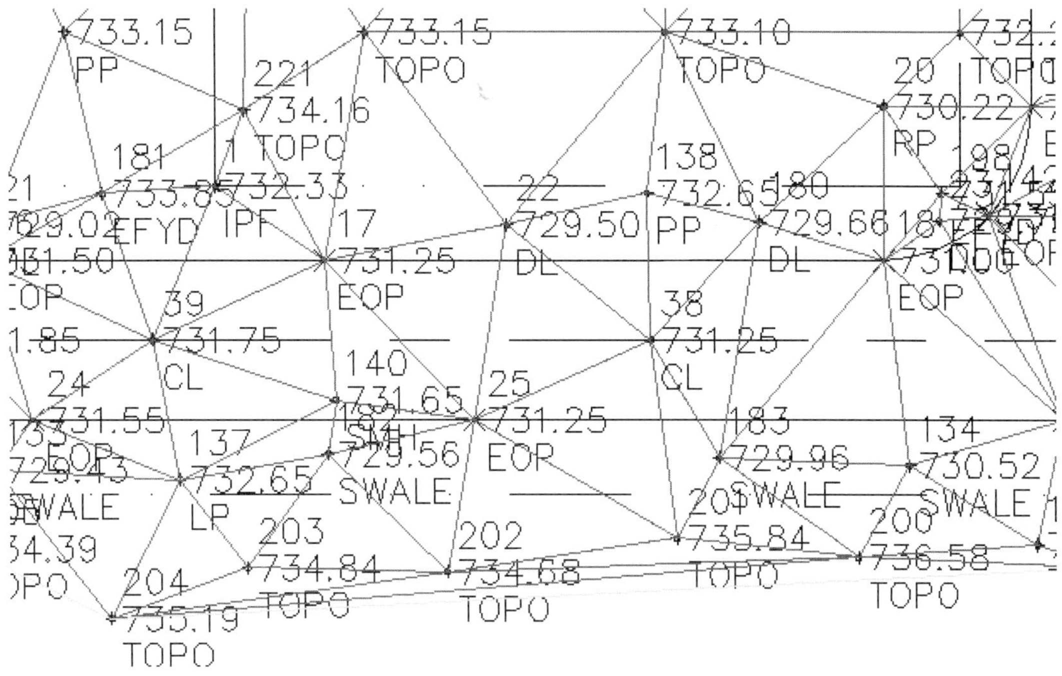

FIGURE 4.38

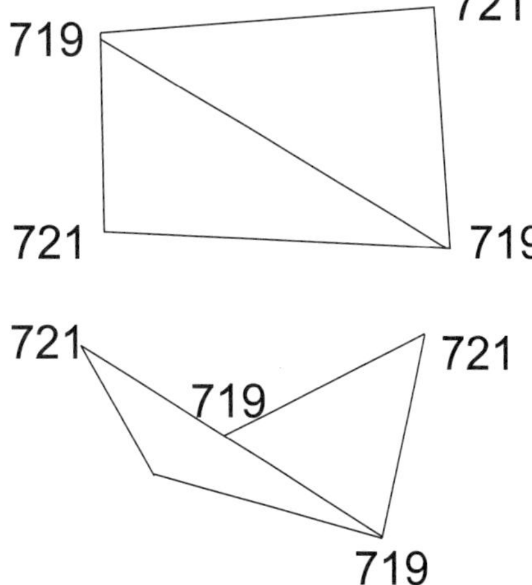
If the diagonal connects the 721 elevations, the area represents a berm.
If the diagonal connects the 719 elevations, the area represents a swale.
719
721
721
719
721
721
719
719
719
721
721
719
721
719
721
719

FIGURE 4.39

The solution to the preceding problem is to create additional triangulation control. The control preserves the points' linear and elevation trends or correctly controls the triangulation in sparse data. Breaklines imply links between points and prevent triangulating crossing the line-linked points. The result is correctly interpreted linear point relationships.

## Drafting Breaklines

Breakline's Add routine expects existing 2D or 3D polylines or feature lines. So, they must exist before using this command. If drawing a 2D polyline, create each vertex near or at a point's location. After drafting the 2D polyline, define it as a Proximity breakline.

If drawing a 3D polyline, use the 3Dpoly command with the Transparent Commands toolbar's point filters (point number ('PN) or point object ('PO)). Without using a transparent command, the 3D polyline command is unaware of a cogo point's elevation and using the filter assigns the point's elevation to the polyline's vertex.

## Feature Line with Arcs

2D and 3D polylines as breaklines cannot contain arcs. If you want breaklines with arcs, create feature lines. When defining feature lines as standard breaklines, they can contain arcs. When defined, each arc is processed by a mid-ordinate value. The mid-ordinate value controls how closely the triangulation follows an arc. The smaller the mid-ordinate value, the closer the triangulation is to the original arc.

## Allow Crossing Breaklines

The default surface build setting does not allow crossing breaklines. If defining breaklines that cross other breaklines, the add breakline routine will add the new breakline and issue an error message about its crossing; the routine does not include some of the breakline's data in the surface.

- Crossing breaklines may destabilize the drawing.

To allow crossing breaklines, change the toggle in the surface's properties dialog box, Definition panel, Build section or in the CreateSurface command (see Figures 4.4 and 4.40). If setting Allow crossing breakline to "Yes," the user needs to set how to resolve the potential elevation difference. The first option uses the first breakline's elevation. The second uses the last breakline's elevation. The third uses the crossing breaklines' average elevation.

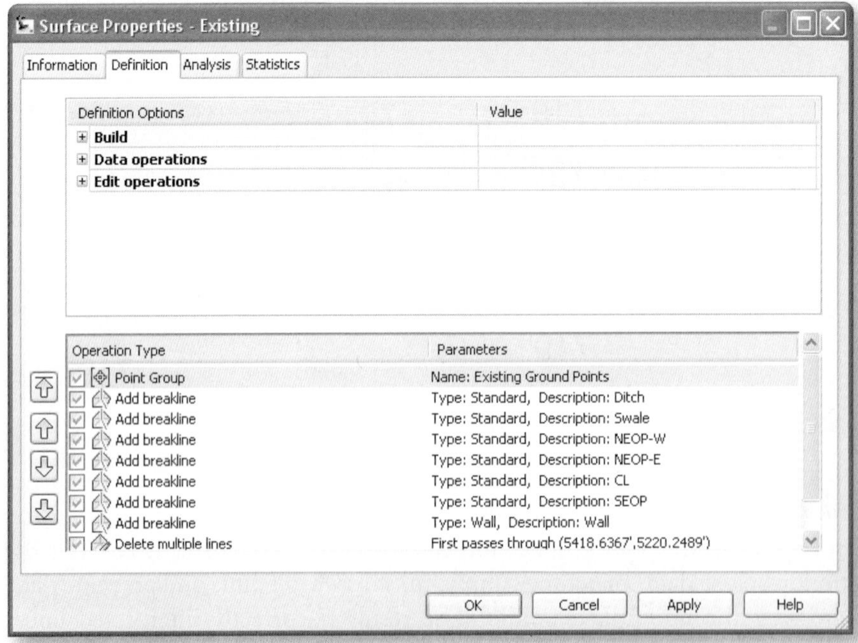


**FIGURE 4.40**

### Standard Breaklines

A standard breakline is a 3D polyline. This polyline contains elevations and it is not necessary to associate cogo point objects with its vertices. The 3Dpoly command creates these objects. If you want a vertex's elevation to be a cogo point's elevation, then use the Transparent Commands toolbar's point number ('PN) or point object ('PO) point filters.

### Proximity Breaklines

A Proximity breakline is a 2D polyline (zero elevation). Because it does not have an elevation, it must have a point object at or near each vertex. The adjacent cogo point objects provide elevations and the polyline provides the line across which no triangle leg can pass. Proximity breaklines cannot have arc segments.

When defining a proximity breakline, the breakline is converted to a 3D polyline snapped to the nearest point. If you were to insert the breakline back into the drawing, it would not match what was originally drawn.

### Wall Breakline

A Wall breakline defines a sheer face (almost) from a 3D polyline. The preexisting 3D polyline is either the wall's top or bottom. After describing the breakline in the Add breakline dialog box, in the drawing select the polyline. After selecting the polyline, specify the offset side by selecting a point to one side or the other of the initial polyline vertex. Take care when selecting the offset side.

After identifying the wall and its offset side, the routine prompts you to set the offset side's elevations. The first option is all. This method prompts for a single value (positive or negative amount) to add to each offset vertex, calculating its elevation. The second option is individual. This method visits each offset vertex and prompts the user for its elevation. To set an elevation, specify an absolute elevation or a delta (a change in elevation).

Introducing a wall into a surface greatly changes its slope's report. Before adding a wall breakline, review its current minimum, maximum, and average slopes.

### Non-Destructive Breakline

A Non-destructive breakline fractures existing triangles along its length while still preserving the elevations found at the intersection of the breakline and the triangulation. The most common situation for this type of breakline is when the user wants to delete triangles to one side of the breakline. The breakline gives the remaining surface triangles a straight or clean boundary.

### From File

Rather than drawing a breakline, create a file to define any of the breakline types. The file has the extension of flt and contains the type of breakline, its name, coordinates, and elevations if it is a standard breakline. See the breakline file entry in Civil 3D's Help file for more information on this type of breakline.

### Feature Lines

The 3DPoly command draws line only segments. If you want a 3D arc, draw a feature line. Feature lines (from Ribbon's Home, Grading icon or the Feature Lines toolbar) are 3D and can contain arc segments. While drafting a feature line, assign elevations to each vertex by typing in the elevation or use a Transparent Commands toolbar point filter ('PO object or 'PN point number) to transfer a point's elevation to a feature line vertex.

Feature lines are powerful when developing design surface data. The Ribbon's Modify panel contains all the necessary tools to create or edit them (see Figure 4.41).

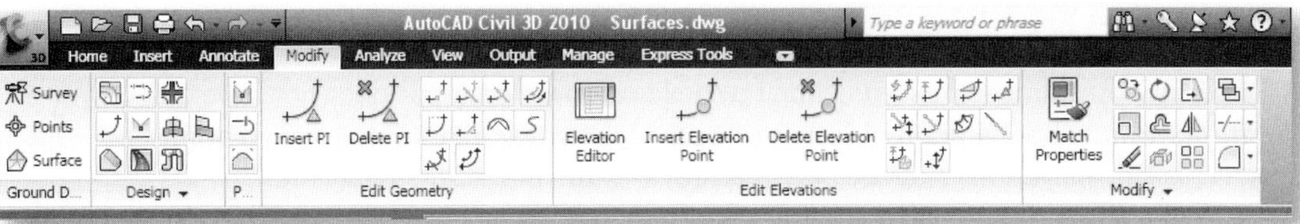


**FIGURE 4.41**

To define standard breaklines from feature lines containing arcs, use a lower mid-ordinate value to better represent the arc in a surface. The lower the mid-ordinate value, the more data points are along its path and this results in triangulation closely following the arc.

This unit's discussion does not cover all of the Feature Lines toolbar icons. Instead, this unit reviews creating feature lines, using Transparent Commands toolbar point filters, editing feature line elevations, and using Quick Profile to evaluate a surface.

**Creating Feature Lines.** The Feature Lines toolbar's leftmost icon draws new feature lines. After selecting the icon, a Create Feature Lines dialog box opens (see Figure 4.42). This dialog box assigns a site, a name, a feature line style, and a layer.

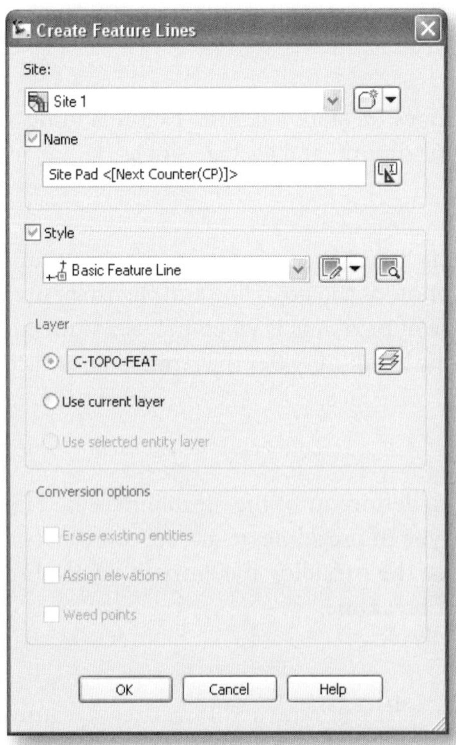


**FIGURE 4.42**

The user accepts or defines a specific layer. At the middle right, if the user clicks the layer icon, an Object Layer dialog box opens to allow the user to define a layer modifier and its value, or a new layer (see Figure 4.43).

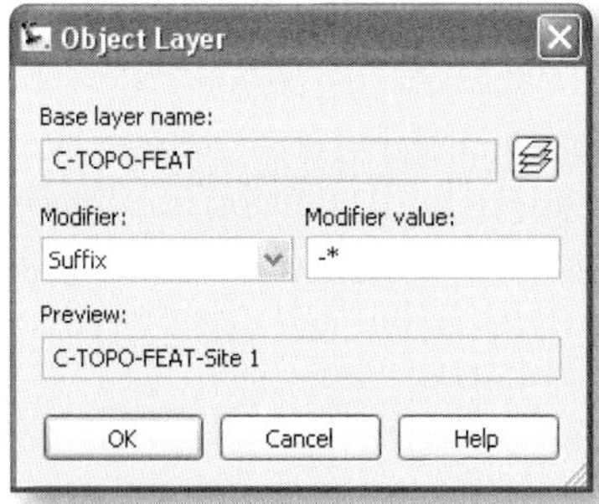


**FIGURE 4.43**

After exiting the Create Feature Lines dialog box, the routine prompts the user for a starting point. If a point is selected, the routine assigns the current elevation (0.00) and prompts the user to accept or to edit its value. If the user selects the first point using a point number ('PN) or point object ('PO) filter, the routine assigns that cogo point's elevation to the feature line's vertex. After selecting the first point, he or she can draft straight or arc segments. After creating the feature line, if the user wants to review or edit its values, he or she can edit it with the elevation editor.

## Elevation Editor

After drafting a feature line, to review or edit its elevations, use the Elevation Editor. To edit a feature line's elevations, select it, press the right mouse button, and from the shortcut menu, select Elevation Editor. This action displays a Panorama with a Grading Elevation Editor vista (see Figure 4.44). Edit the elevations by clicking a vertex's elevation cell and changing its value or by changing the grade between vertices. When you are done, click the Panorama mast's 'X' to close it.

| Station | Elevation | Length | Grade Ahead | Grade Back |
|---|---|---|---|---|
| 0+00.00 | 126.689' | 26.546' | -1.04% | 1.04% |
| 0+26.55 | 126.414' | 20.000' | 1.00% | -1.00% |
| 0+46.55 | 126.614' | 18.692' | 1.05% | -1.05% |
| 0+65.24 | 126.811' | 74.600' | -0.99% | 0.99% |
| 1+39.84 | 126.071' | 76.200' | -1.97% | 1.97% |
| 2+16.04 | 124.571' | 201.200' | 1.99% | -1.99% |
| 4+17.24 | 128.571' | 76.200' | 0.98% | -0.98% |
| 4+93.44 | 129.321' | 74.600' | 0.99% | -0.99% |
| 5+68.04 | 130.061' | 18.692' | -1.05% | 1.05% |
| 5+86.73 | 129.864' | 20.000' | -1.00% | 1.00% |
| 6+06.73 | 129.664' | 26.546' | -16.65% | 16.65% |
| 6+33.28 | 125.245' | | | |

Panorama

Grading Elevation Editor

**FIGURE 4.44**

## Quick Profile

Quick profile is indispensable when evaluating a surface. Creating a mental surface image is difficult by looking only at triangles. A quick profile displays a surface's elevations along a drawing object (lines, arcs, parcel lines, polylines, etc.). If the user moves the drawing object, the profile and its view are updated to show the new elevations. The preferred object is a polyline.

The Feature Lines toolbar's sixth icon in from the left is Quick Profile. A quick profile persists only for the current drawing session. If the user exits or saves the drawing, the quick profile is discarded and has to be redefined.

After selecting the Quick Profile icon, in the drawing select the object representing the quick section. After selecting the object, a Create Quick Profiles dialog box opens (see Figure 4.45). At the dialog box's top, select all or a combination of surfaces. The middle section sets the profile view style. After setting these values and exiting, select an insertion point to create the quick profile (see Figure 4.46).

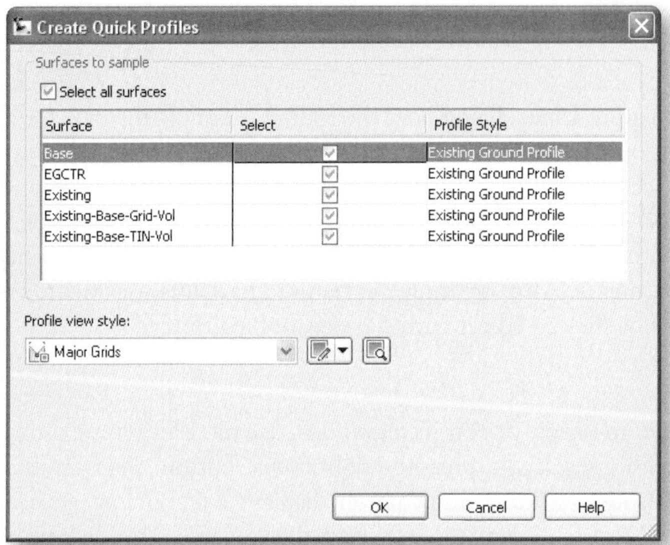


FIGURE 4.45

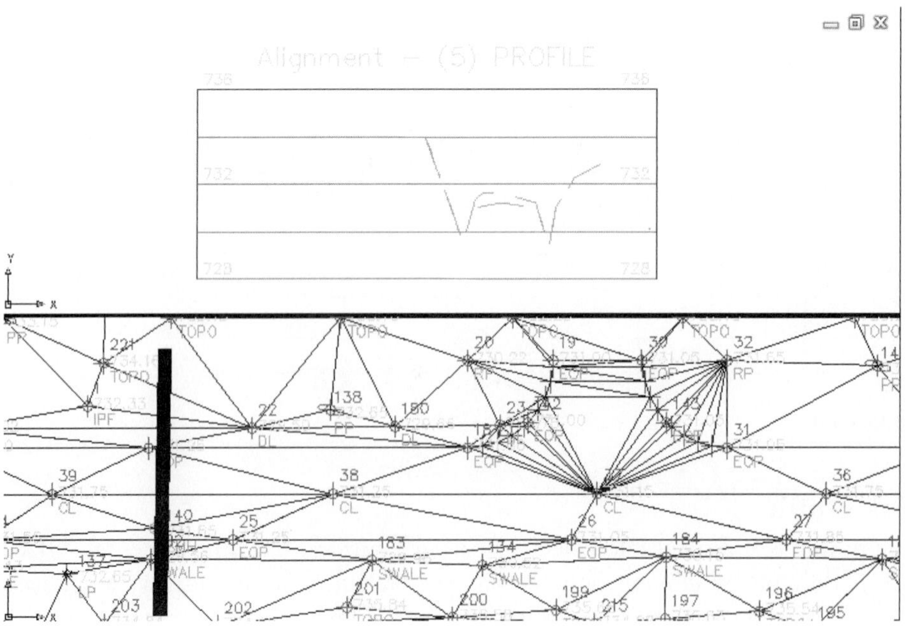


FIGURE 4.46

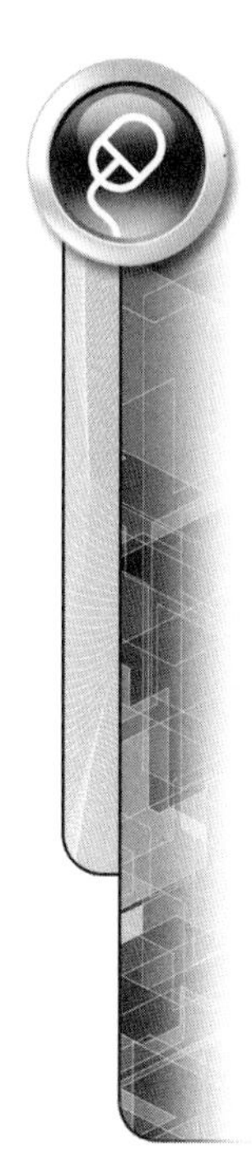

**EXERCISE 4-2**

After completing this exercise, you will:

- Be able to create a surface.
- Assign surface data.
- Change assigned surface styles.
- Review initial surface triangulation.
- Add breakline data.
- Be familiar with the Surface Properties dialog box.
- Be Familiar with the Surface Properties' Surface Statistics panel.

## Exercise Setup

This exercise continues with the previous unit's exercise drawing, Surfaces. If you did not complete the previous exercise or starting with this exercise, browse to the Chapter 4 folder of the CD that accompanies this textbook and open the drawing, *Chapter 04 - Unit 2.*

1. If the previous unit's exercise drawing, Surfaces is not open, open it now, or browse to the Chapter 4 folder of the CD that accompanies this textbook and open the *Chapter 04 - Unit 2.dwg* file.
2. Use the ZOOM EXTENTS command to view the drawing.

## Creating a Surface

The first step is to create a surface from the Ribbon's Home tab, Create Ground Data panel, Surfaces drop-list menu or from Prospector's Surfaces shortcut menu.

1. If necessary, click the ***Prospector*** tab.
2. In the Ribbon's Home tab, Create Ground Data panel, click the Surfaces drop-list arrow, from the menu select CREATE SURFACE or in Prospector, click the **Surfaces** heading, press the right mouse button, and from the shortcut menu, select CREATE SURFACE.... Either action displays the Create Surface dialog box (see Figure 4.24).
3. In the Create Surface dialog box, for the surface name enter **Existing**, for the description enter **Phase 1 – Existing Ground**, and make sure the surface type is **TIN surface**.
4. Set the surface style to **Border Triangles Points**. Click in the Style value cell, and at the cell's right side, click the ellipsis (...) to display the Select Surface Style dialog box, click the droplist arrow, select **Border Triangles Points** from the list, and click ***OK***.
5. Using the same method as the previous step, set the render material to **Sitework.Planting.Grass.Short** and after setting the values, click ***OK*** to create the surface.

The Surfaces branch updates the Existing listing.

6. If necessary, select the ***Prospector*** tab.
7. In Prospector, expand **Surfaces** to view the Existing instance, its surface masks, watersheds, and definition headings.
8. Expand the **Definition** branch to view the surface data tree. You may have to scroll Prospector to view all of the entries.

The surface's Definition branch lists a surface's data types. When assigning a type, Prospector indicates the assignment with an icon (square with a black dot).

## Surface Point Group — Existing Ground Point Group

Next, you will organize point data into point groups. This exercise creates three groups. The first is the surface data. The second consists of points representing possible breaklines. The last hides all point markers and labels. The last group controls point visibility.

The Existing Ground point group contains Existing's surface point data. The group contains all drawing points (point numbers 1–228), except for existing fire hydrants. Fire hydrant elevations are their tops. This creates an incorrect surface elevation at each hydrant. Remove these points from a group by using the group's exclude panel's By Raw descriptions matching option (EFYD).

1. Expand **Point Groups** to view the point group list.
2. Select **_All Points**, and in preview, scroll through its point list (Prospector's lower panel).

The group contains point numbers 1–228, and existing fire hydrants are point numbers 149, 181, 198, and 216.

3. In Prospector, click the **Point Groups** heading, press the right mouse button, and from the shortcut menu, select `NEW...`.
4. In the ***Information*** tab for the point group's name, enter **Existing Ground Points.**
5. Click the ***Include*** tab, toggle **ON With Numbers Matching**, and for the point number range, enter **1–228** (see Figure 4.47).

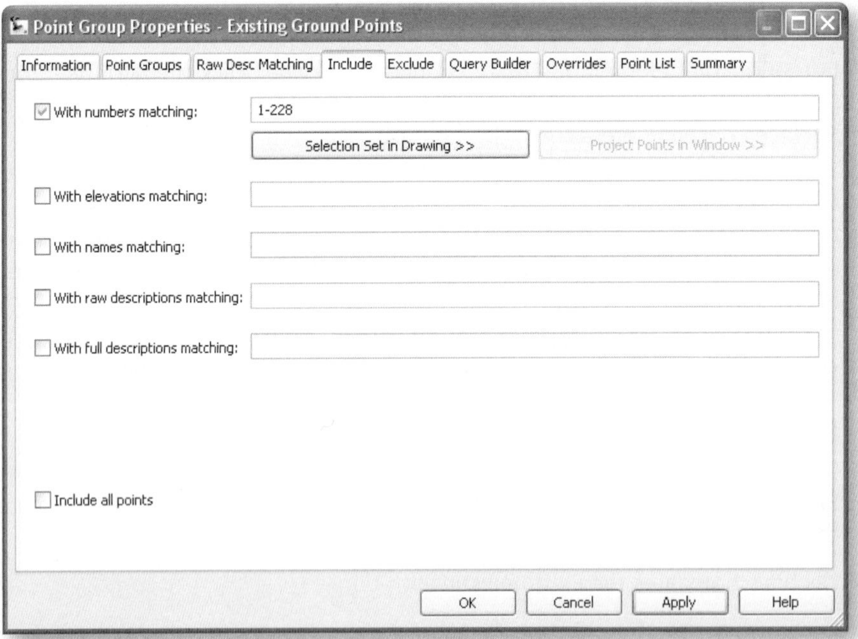


**FIGURE 4.47**

6. Click the ***Exclude*** tab, toggle **ON With Raw Descriptions Matching**, and enter **EFYD** (see Figure 4.48).

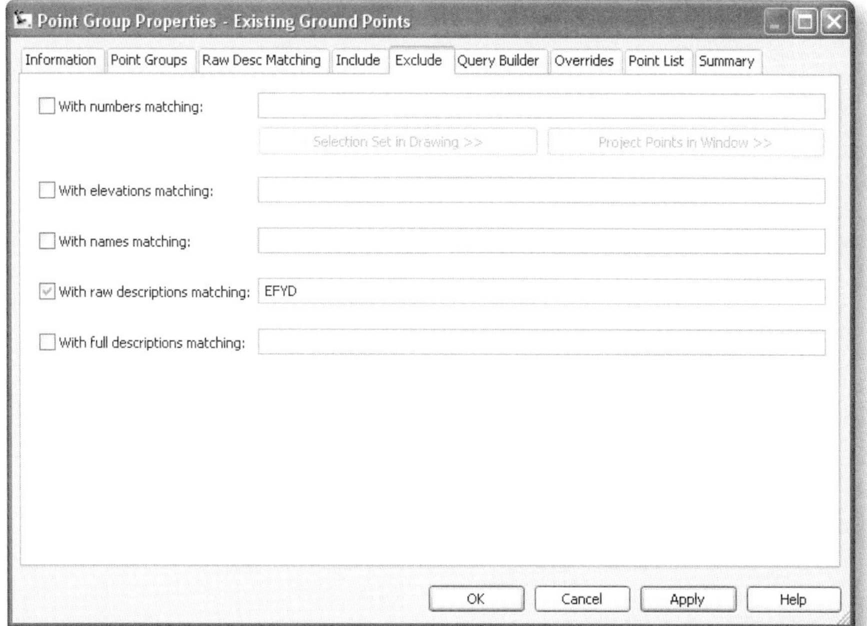


**FIGURE 4.48**

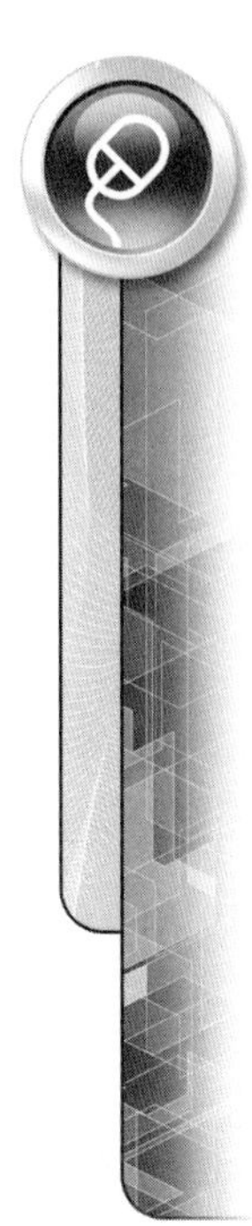

7. Click the ***Point List*** tab and sort the Raw Description list by clicking its heading. Notice the absence of existing fire hydrants (EFYD).
8. Click ***OK*** to create the Existing Ground Points point group.

## Surface Point Group — Breakline Point Group

For some linear features to correctly triangulate, they may need breaklines. These points may include tops and bottoms of slopes or banks (T/S or T/B), edges-of-travelway (EOP), swales (SWALE), ditches (DL), etc. These features rarely have enough data to clearly show their linear relationship.

1. In Prospector, select the **Point Groups** heading, press the right mouse button, and from the shortcut menu, select `NEW...`.
2. In the ***Information*** tab for the point group's name, enter **Breakline Points**.
3. Click the ***Raw Desc Matching*** tab and toggle **ON** the following raw descriptions: **CL***, **DL**, **EOP**, **HDWL**, and **SWALE** (see Figure 4.49).
4. Click ***OK*** to create the Breakline Points point group.

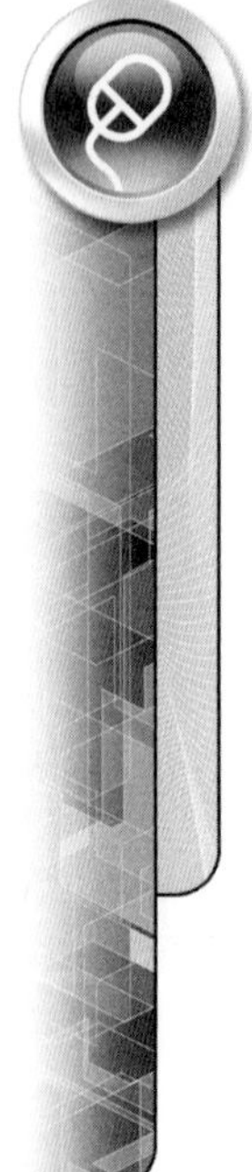

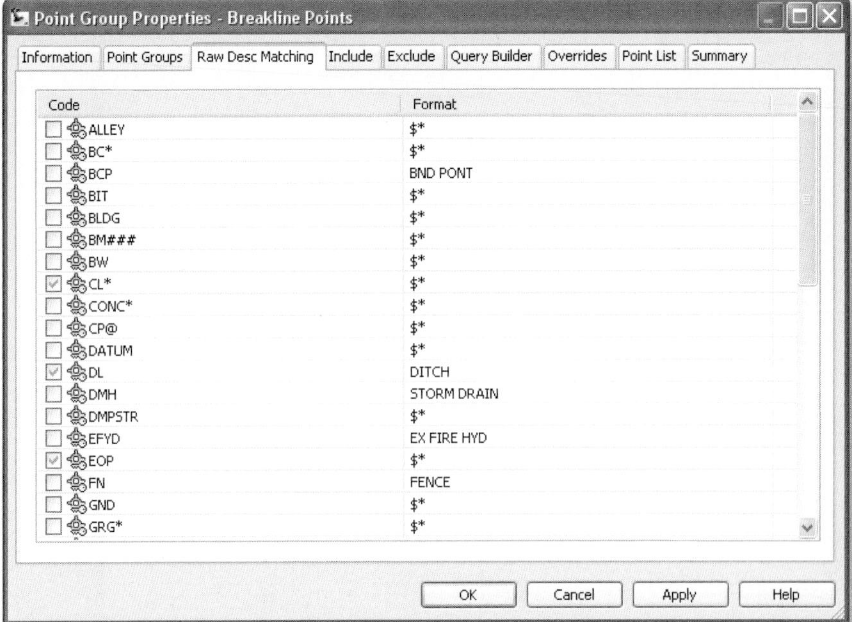


**FIGURE 4.49**

## Point Group — No Show Point Group

There are times when you need to view a subset of all points. You can turn off the appropriate layers or create a point group that hides points. Next, you define a point group that has no point or label style. When placing this point group at the top of the display list, all of the points disappear. The display order list is Prospector's Point Groups heading's property. When changing the display list's order, you control the point's visibility.

No Show is a copy of the _All Points point group with None as the overriding point and label style.

1. In Prospector, from the list of point groups, select the **_All Points**, press the right mouse button, and from the shortcut menu, select COPY....
2. From the list of point groups, select **Copy of _All Points**, press the right mouse button, and from the shortcut menu, select PROPERTIES....
3. In the ***Information*** tab change the point group's name, **No Show**.
4. On the panel's middle left, change the Point Style to **<none>** and the Point Label Style to <**none**>.
5. Click the ***Overrides*** tab and toggle **ON** the overrides for point **Style** and **Point Label Style**.
6. Click ***OK*** to create the No Show point group.
7. In Prospector, Point Groups, select the **Point Groups** heading, press the right mouse button, and from the shortcut menu, select PROPERTIES....
8. In Properties, select **Existing Ground Points** and click the up arrow to move the group to the top of the list.
9. Click ***OK*** to exit and display the points.
10. In Civil 3D's top left, Quick Access Toolbar, click the ***Save*** icon to save the drawing.

## Assigning Surface Data

To assign surface data, you select a data type from the surface definition branch, press the right mouse button, and from the shortcut menu select ADD....

1. If necessary, in Prospector, expand **Existing** until you are viewing its Definition branch.
2. In the Definition branch, select the **Point Groups** heading, press the right mouse button, and from the shortcut menu, select ADD....
3. From the point group list select **Existing Ground Points** and click ***OK*** to close the dialog box and add the group (see Figure 4.50).

The surface builds from the newly assigned data.

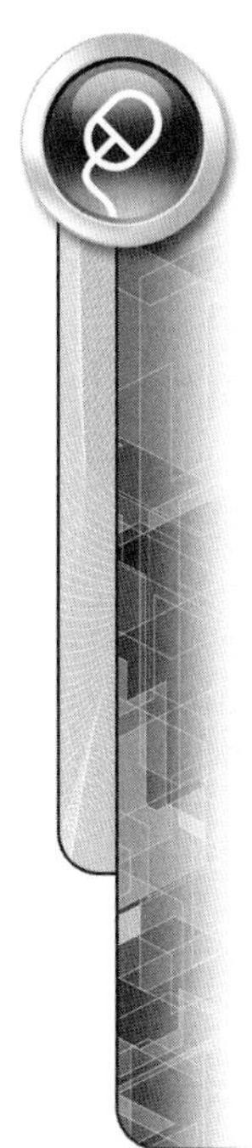

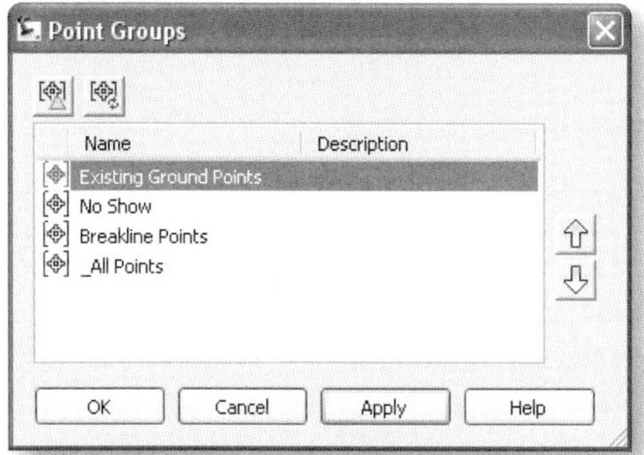


**FIGURE 4.50**

## Review Initial Surface

1. Use the ZOOM and PAN commands to navigate the site to review the surface triangulation.
2. In Prospector, select the **Point Groups** heading, press the right mouse button, and from the shortcut menu, select PROPERTIES....
3. In the Point Groups Properties dialog box, select **Breakline Points**, and move it to the list's top by clicking the up arrow on the dialog box's right side.
4. Select the **No Show** point group and using the up arrow icon on the right, move it to the list's second position (see Figure 4.51).
5. Click ***OK*** to close the dialog box and change the point group display. You may need to use Regenall (REA) to see the update.

The current point group display order hides all points except for Breakline Points.

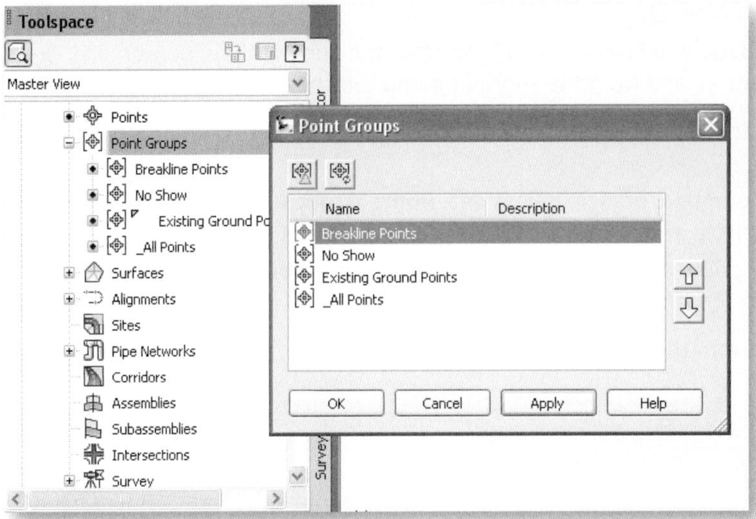


**FIGURE 4.51**

6. In Prospector, click the **Existing** surface name, press the right mouse button, and from the shortcut menu, select REBUILD – AUTOMATIC.
7. Open the **Layer Properties Manager** (Ribbon's Home panel, middle right), create a new layer, for its name enter **Existing-Breakline**, assign it a color, make it the current layer, and click **X** to close the palette.
8. At Civil 3D's top left, Quick Access Toolbar, click the ***Save*** icon to save the drawing.
9. Use the ZOOM command to review the site's southern portion and its triangulation for the points EOP, DITCH, CL, and SWALE.

The initial triangulation does not correctly connect the points representing the north and south edges-of-pavement, the centerline, ditch, or swale (south of the roadway). These points need breaklines to correctly triangulate their elevations.

## Ditch Breakline

The ditch breakline consists of points 21, 22, 180, and 23 and is located just to the road's northwest side (see Figure 4.52).

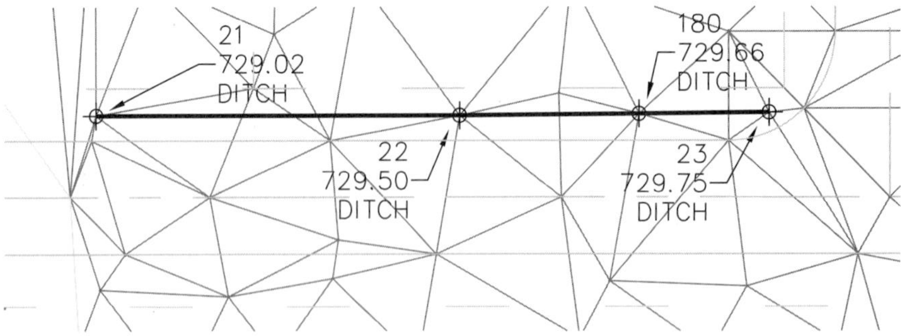


**FIGURE 4.52**

1. Zoom into the area shown in Figure 4.52.
2. Check and, if necessary, display the **Transparent Commands** toolbar.
3. Start the 3DPOLY command, from the Transparent Commands toolbar select either filter, Point Number, or Point Object, and enter or select points **21, 22, 180,** and **23**.

The following are the 3DPOLY command's responses when using the point number transparent command.

```
Command: 3dpoly
Specify start point of polyline: '_PN
>>Enter point number: 21
Resuming 3DPOLY command.
Specify start point of polyline: (4962.18 5043.75 729.02)
Specify endpoint of line or [Undo]:
>>Enter point number: 22
Resuming 3DPOLY command.
Specify endpoint of line or [Undo]: (5048.7 5043.75 729.5)
Specify endpoint of line or [Undo]:
>>Enter point number: 180
Resuming 3DPOLY command.
Specify endpoint of line or [Undo]: (5090.95 5044.15 729.66)
Specify endpoint of line or [Close/Undo]:
>>Enter point number: 23
Resuming 3DPOLY command.
Specify endpoint of line or [Close/Undo]: (5121.48 5044.44
   729.75)
Specify endpoint of line or [Close/Undo]:
>>Enter point number: *Cancel* <Press ESC to exit
   transparent command>
Specify endpoint of line or [Close/Undo]:
Resuming 3DPOLY command.
Specify endpoint of line or [Close/Undo]: <Press ENTER
   to exit>
Command:
```

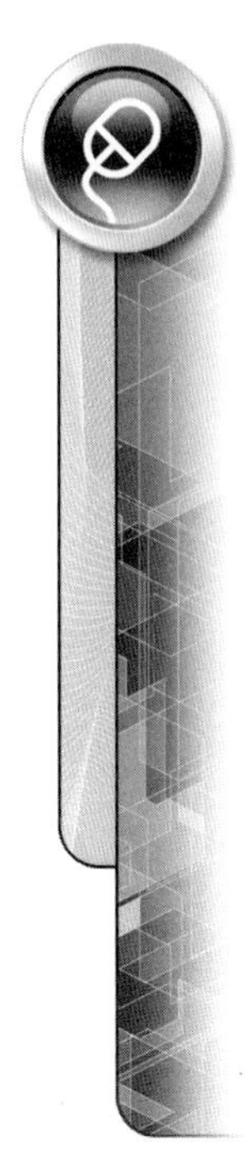

The command echoes the selected point's coordinates and elevations.

4. After drafting the 3D polyline, press ESC to exit the transparent command, and press ENTER to exit the 3DPOLY command.
5. In the Existing surface Definition branch, select the **Breaklines** heading, press the right mouse button, and from the shortcut menu, select ADD....
6. In the Add Breaklines dialog box, for the description, enter **Ditch**, set its type to **Standard**, and click ***OK*** to select the polyline (see Figure 4.53).
7. Select the 3D polyline representing the ditch and press ENTER.
8. If Civil 3D issues a duplicate points warning, close the warning panorama and review the new triangulation.

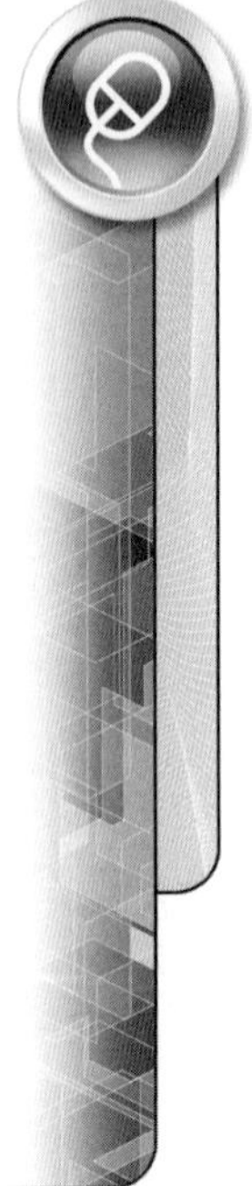

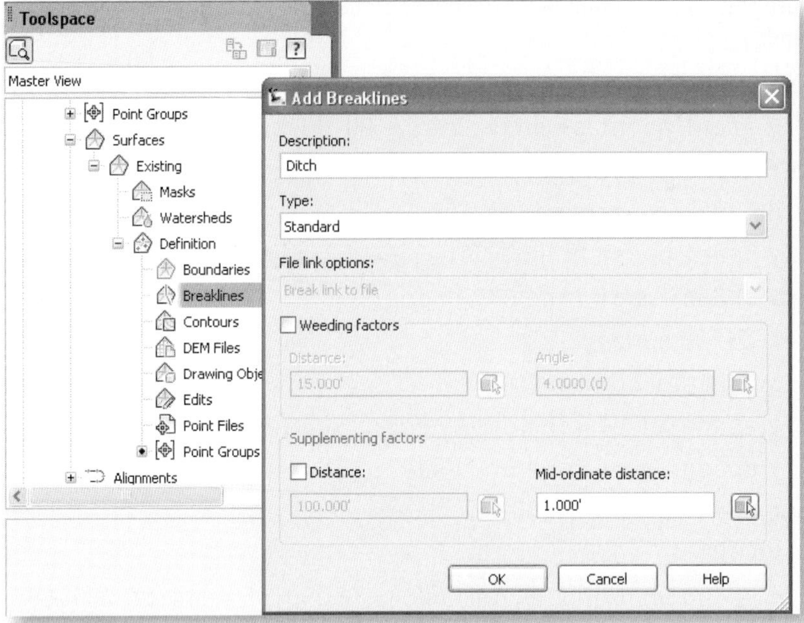


**FIGURE 4.53**

After adding the breakline, the surface updates its definition list and rebuilds its triangulation to resolve the ditch.

9. In the Existing surface Definition branch, expand **Breaklines** to view the Ditch entry.
10. Select the list's **Ditch** entry, press the right mouse button, from the shortcut menu, select PROPERTIES..., and review its coordinate and elevation values.
11. From the list select a vertex, press the right mouse button, and from the shortcut menu select ZOOM TO.
12. Close the Breakline Properties vista by selecting the Panorama's **X**.

Whenever using Zoom to, the object is put at the display's center.

## Swale Breakline

A swale breakline just south of the roadway is next. The swale starts near the road's eastern end and extends all the way to the western edge of the surface. The swale point numbers are (east to west) 190, 135, 191, 185, 184, 134, 183, 182, and 133 (see Figure 4.54).

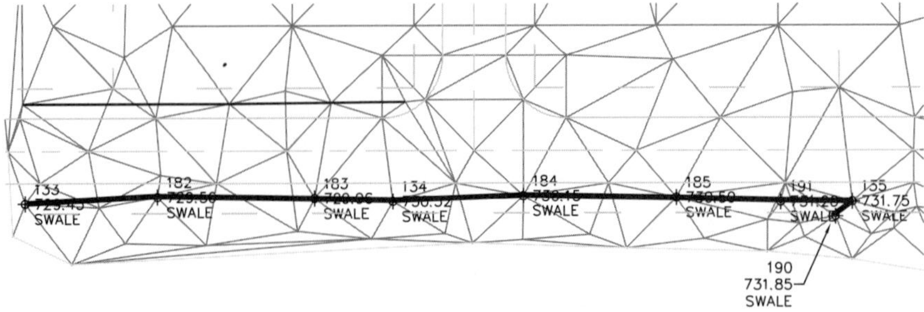


**FIGURE 4.54**

1. Use the ZOOM command and navigate to the area shown in Figure 4.54 to view the swale points.
2. To better view the points, change the Annotation Scale to 1"=20'.

3. Start the 3DPOLY command, from the Transparent Commands toolbar select either filter, Point Number or Point Object, and enter or select points **190**, **135**, **191**, **185**, **184**, **134**, **183**, **182**, and **133**.
4. When you are done, press ESC to exit the transparent command, and press ENTER to exit the 3DPOLY command.
5. In the Existing surface Definition branch, select **Breaklines**, press the right mouse button, and from the shortcut menu, select ADD....
6. In the Add Breaklines dialog box, for the breakline description enter **Swale**, set the type to **Standard**, and click ***OK*** to select the polyline.
7. Select the 3D polyline representing the Swale and press ENTER.
8. If you are warned about duplicate surface points, close the Event Viewer and view the new triangulation.

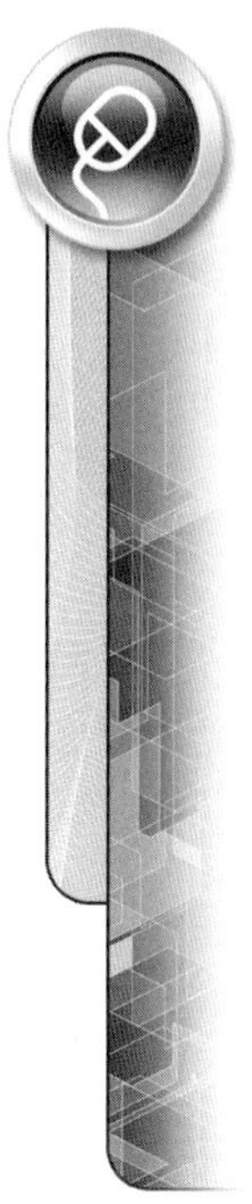

The surface updates its triangulation, correctly resolving the swale.

9. If necessary, in the Existing surface Definition area of Prospector, expand **Breaklines**, to view its list, and notice the Swale breakline is on it.
10. Select the **Swale** entry, press the right mouse button, and from the shortcut menu, select PROPERTIES... and review its coordinate and elevation values.
11. Close the Breakline Properties vista by selecting the Panorama's **X**.
12. At Civil 3D's top left, Quick Access Toolbar, click the ***Save*** icon to save the drawing.

Reviewing a surface's triangulation does not necessarily give you a good idea of what the surface looks like. A Quick Profile is an additional tool to review a surface and its breaklines and edit effects.

## Evaluate Surface with Quick Profile

Reviewing a surface includes looking at its triangulation along the road's southern edge. Few, if any, triangles correctly defined the north ditch and the south swale. As a result of this initial review, two breaklines were added to the surface. Even with these two breaklines, the triangulation representing the roadway is wrong. One triangle even connects a ditch point (180) north of the road to a swale point (183) south of the road (see Figure 4.55).

It is one thing to see the triangulation, but another to understand what the triangles mean as surface elevations or relief. Quick Profile displays elevations along an object as a profile. This exercise portion creates a Quick Profile view that reviews surface elevations and the effects of breaklines.

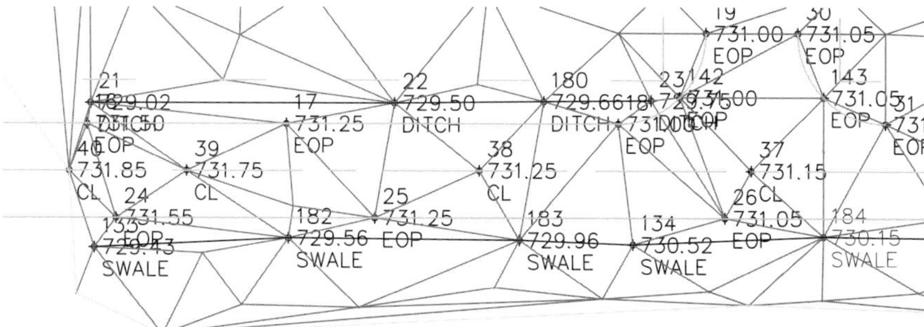


**FIGURE 4.55**

1. From the Ribbon's View tab, Viewports panel, click the Set Viewports icon, from the shortcut menu, select **Two: Horizontal** to split the screen.
2. Click the top viewport to make it current, and use the PAN command to move the site to the left until the screen is clear.

3. Click in the lower viewport to make it current and use the PAN and ZOOM commands to view the roadway's western half (see Figure 4.56).
4. In the lower viewport, start the PLINE command and draw a polyline from south of the swale (near point number 133) to a point north of the ditch (past point 21) at the western end of the roadway (see Figure 4.56).
5. In the Ribbon, click the Analyze tab, Ground Data panel, select the ***Quick Profile*** icon.
6. Select the polyline just drawn, and the Create Quick Profiles dialog box opens. (You can also select the polyline, right-click, and from the shortcut menu, select QUICK PROFILE...).
7. Click ***OK*** to accept the default values and continue drawing the profile.
8. Click in the top viewport to make it current and select a point in its lower-left to locate the profile view.

A Quick Profile appears in the upper viewport and an Event Viewer is displayed to remind you that the profile is only temporary.

9. Click the **X** on Panorama's mast to close the Events vista.

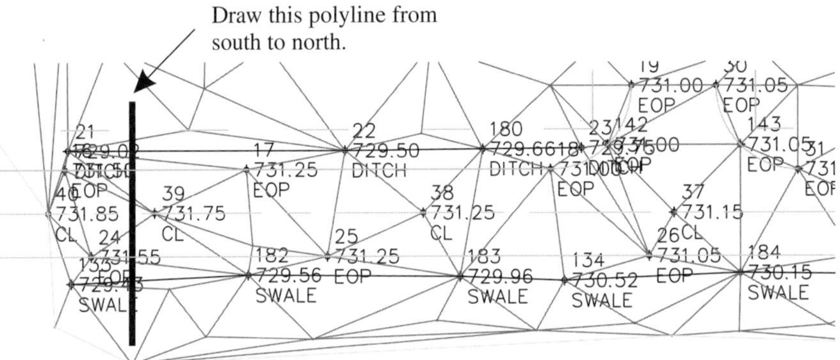


**FIGURE 4.56**

Relocating the polyline segment causes it to update the profile. What should be displayed in the profile is a ditch or swale with a poorly defined road section. Because triangulated points rarely triangulate linear features correctly, you must use breaklines to make them appear correctly in a surface.

10. Use the MOVE command or grips to relocate the polyline segment reviewing various road locations.
11. Finally, relocate the polyline segment back to its original position and press ESC to deselect it.
12. On the Ribbon, select the ***View*** tab. On the Viewports panel, click the Named icon to display the Viewports dialog box.
13. In the Viewports dialog box, click the ***New Viewports*** tab and for the name enter, **A and P**.
14. Click ***OK*** to exit the dialog box.
15. With the lower viewport active from the Viewports panel, click the Set Viewports icon, and from the shortcut menu, select **Single** to reset the drawing to one viewport.
16. At Civil 3D's top left click, Quick Access Toolbar, the ***Save*** icon to save the drawing.
17. If necessary, use the ZOOM and PAN commands to view the roadway from its western end to just east of the intersection.

## Roadway Breaklines

Next, you add four roadway breaklines. The first two will be feature lines and the remaining two repeat the steps you used to create and define the ditch and swale breaklines. The reason for feature lines is the north edges-of-travelway (NEOP-W and NEOP-E) have curve return segments. When defining feature lines containing arcs as breaklines, you process the curves with a mid-ordinate value.

Table 4.1 contains the breakline names, their point numbers, and their breaklines type.

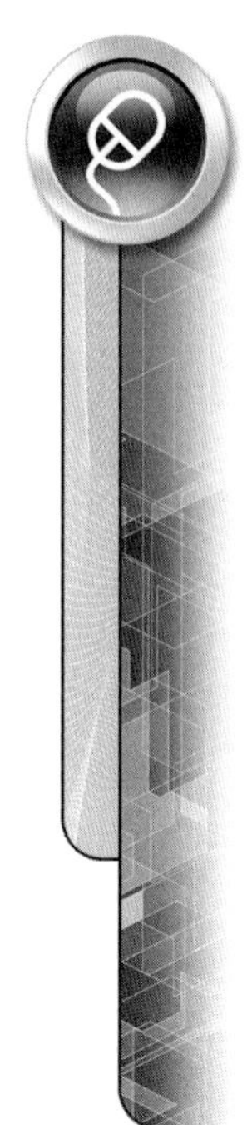

**TABLE 4.1**

| Name | Point Numbers | Type of Breakline |
|---|---|---|
| NEOP-W | 16, 17, 18, 142, and 19 | Standard from Feature Line object |
| NEOP-E | 176, 34, 33, 31, 143, and 34 | Standard from Feature Line object |
| CL | 177, 35, 36, 37, 38, 39, and 40 | Standard |
| SEOP | 24, 25, 26, 27, 28, 29, and 178 | Standard |

1. Click the Ribbon's Home tab. On the Create Design panel's left click the Feature Line icon, and from the menu select the CREATE FEATURE LINE to display the Create Feature Lines dialog box.
2. Match your values to those in Figure 4.57 and click ***OK*** to continue.

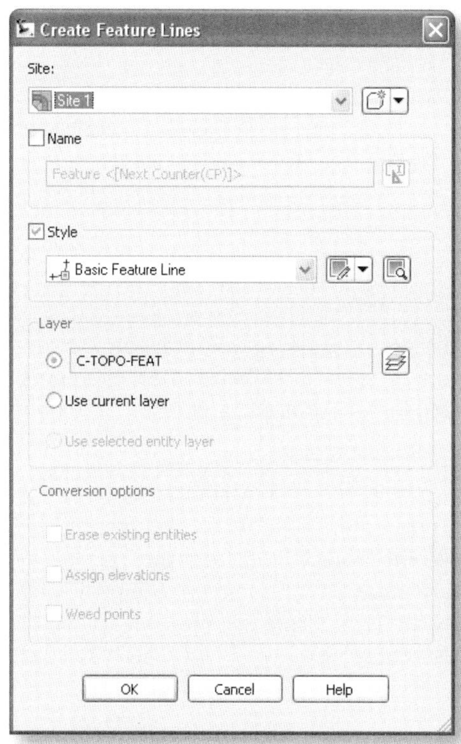


**FIGURE 4.57**

3. From the Transparent Commands toolbar, select the ***Point Object*** override ('PO), and identify the points listed in Table 4.1 for the breakline NEOP-W. When you select the first point object, the routine echoes back its elevation (point 16), and you press ENTER to accept the elevation.

When selecting the second point, the prompt may not be for elevation.

4. Select point 17. Enter the letter '**E**', press ENTER twice using the point's elevation, and the prompt returns to Select point object.

The following segment shows the prompt for the start of the command:

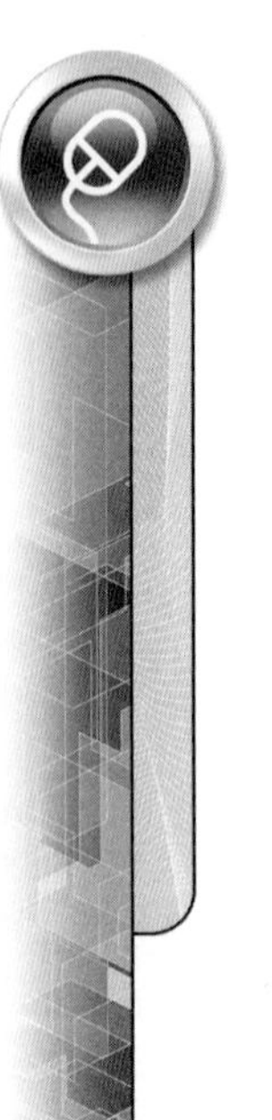

```
Command:
Specify start point: '_PO Start override
>>
Select point object: Select Point 16
Resuming DRAWFEATURELINE command.
Specify start point: (4961.15 5038.0 731.5)
Specify elevation or [Surface] <731.50>: (Press ENTER)
Specify the next point or [Arc]:
>>
Select point object: Select Point 17
Resuming DRAWFEATURELINE command.
Specify the next point or [Arc]: (5018.14 5038.0 731.25)
Distance 56.991', Grade -0.44, Slope -227.96:1, Elevation
731.250'
Specify grade or [SLope/Elevation/Difference/SUrface]
<0.00>: e
  (To set feature line to Elevation)
Specify elevation or [Grade/SLope/Difference/SUrface]
<731.250>:
  (Press ENTER)
Specify the next point or [Arc/Length/Undo]:
>>
Select point object: Select Point 18
```

There needs to be a vertex at each point on the arc. Make sure you draw an arc from point 18 to 142 and from 142 to 19. You will want to zoom in to better view the points. Press ESC to toggle off the transparent command, enter A for arc, and restart the point object override.

5. Select point number **18** and press ENTER to accept its elevation.
6. Press ESC to exit the Point Object override, enter the letter '**A**' and press ENTER to set the Arc option. Toggle **ON** the **Point Object** filter from the Transparent Commands toolbar, select point **142**, and press ENTER to accept the point's elevation. You are now ready to select point **19**, and press ENTER to accept the point's elevation to finish the arc. Press ESC to exit the Point Object override and press ENTER to exit the command.

The following is the command sequence to create the arc segment of the feature line:

```
Select point object:
Resuming DRAWFEATURELINE command.
Specify the next point or [Arc/Length/Undo]: (5112.0 5038.0
  731.0)
Distance 93.857', Grade -0.27, Slope -375.43:1, Elevation
  731.000'
```

```
Specify elevation or [Grade/SLope/Difference/SUrface]
<731.000>:
  (Press ENTER)
Specify the next point or [Arc/Length/Close/Undo]:
>>
Select point object: *Cancel* (Press ESC)
>>
Specify the next point or [Arc/Length/Close/Undo]:
Resuming DRAWFEATURELINE command.
Specify the next point or [Arc/Length/Close/Undo]: a (for
Arc)
Specify arc end point or [Radius/Secondpnt/Line/Close/
Undo]: '_PO
>>
Select point object: Select Point 142
Resuming DRAWFEATURELINE command.
Specify arc end point or [Radius/Secondpnt/Line/Close/
Undo]:
  (5129.68 5045.32
731.0)
Distance 19.635', Grade 0.00, Slope Horizontal, Elevation
  731.000'
Specify elevation or [Grade/SLope/Difference/SUrface]
<731.000>:
  (Press ENTER)
Specify arc end point or [Radius/Secondpnt/Line/Close/
Undo]:
>>
Select point object: Select Point 19
Resuming DRAWFEATURELINE command.
Specify arc end point or [Radius/Secondpnt/Line/Close/
Undo]:
  (5137.0 5063.0
731.0)
Distance 19.635', Grade 0.00, Slope Horizontal, Elevation
  731.000'
Specify elevation or [Grade/SLope/Difference/SUrface]
<731.000>:
  (Press ENTER)
Specify arc end point or [Radius/Secondpnt/Line/Close/
Undo]:
>>
Select point object: *Cancel* (Press ESC)
>>
```

```
Specify arc end point or [Radius/Secondpnt/Line/Close/
Undo]:
Resuming DRAWFEATURELINE command.
Specify arc end point or [Radius/Secondpnt/Line/Close/
Undo]:
  (Press ENTER)
Command:
```

7. Use the PAN command to view the eastern portion of the north edge-of-travelway.
8. Repeat Steps 1–6 and draft the NEOP-E feature line. Start the feature line at point 176. Make sure you draw an arc from point 31 to 143 and from 143 to 30. There must be a vertex at each point on the arc segment.

## Review Feature Lines Elevations

1. Select the **NEOP-E** feature line, press the right mouse button, and from the shortcut menu, select ELEVATION EDITOR....
2. Compare the elevations of the feature to those in Table 4.2, and, if necessary, correct their elevations.

**TABLE 4.2**

| Point | Elevation |
|---|---|
| 176 | 733.75 |
| 34 | 733.45 |
| 33 | 732.35 |
| 31 | 731.05 |
| 143 | 731.05 |
| 30 | 731.05 |

3. Click the Select icon in the upper-left of the vista.
4. Repeat Step 1 and select the **NEOP-W** feature line. If necessary, correct its elevations using those listed in Table 4.3.

**TABLE 4.3**

| Point | Elevation |
|---|---|
| 16 | 731.50 |
| 17 | 731.25 |
| 18 | 731.00 |
| 142 | 731.00 |
| 19 | 731.00 |

5. Click Panorama's **X** to close it.

## Add Feature Lines as Breaklines

1. In the Existing surface's Definition branch, select **Breaklines**, press the right mouse button, and from the shortcut menu, select ADD....
2. In the Add Breaklines dialog box, name the breakline **NEOP-W**, set its type to **Standard**, set the Mid-ordinate value to **0.1**, click ***OK*** to exit the dialog box, select the westernmost feature line, and press Enter.
3. Repeat Steps 1–2 for the eastern feature line using the same values found in Step 2, but describe the breakline as **NEOP-E**.
4. If a Panorama is displayed and informs you about duplicate points, close the Events vista by clicking the **X** and view the new triangulation.

Each time you add a new breakline, the surface rebuilds and the triangulation contains additional data around the curve return arc segments.

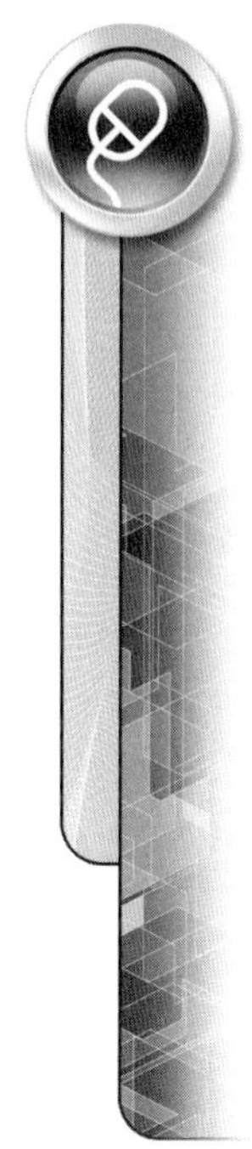

## Draw and Define the CL and SEOP Breaklines

1. Draw 3D polyines for the **CL** and **SEOP** breaklines using the transparent command override of Point Object to reference the points and their elevations.
2. After drafting the 3D polylines, define them as standard breaklines, and assign each one an appropriate name.
3. If a Panorama is displayed and informs you about duplicate points, close the Events vista by clicking the **X** and view the new triangulation.

## View Roadway Quick Profiles

1. Click the Ribbon's ***View*** tab. In the Viewports panel, click NAMED.
2. In the Viewports dialog box's, Named Viewports tab, select the named viewport **A and P**, and click ***OK*** to exit the dialog box.
3. Click the Ribbon's ***Modify*** tab. In the Design panel, click the ***Feature Line*** icon. At the Launch Pad panel's left, click QUICK PROFILE.
4. Select the quick profile polyline drawn earlier, the Create Quick Profiles dialog box displays, click ***OK*** to close it, and in the upper viewport select the profile's insertion point.
5. Close the Panorama to view the quick profile.
6. Review the road by moving the polyline in the lower viewport.
7. After reviewing the road, move the quick profile polyline's southern end to a point just south of point 188 and its northern end just south of the lot line north of the roadway.

A headwall is just south of the road between points 186 and 189.

8. Click the Ribbon's ***View*** tab. In the Viewports panel, click the Set VIEWPORT icon, and from the menu, select SINGLE.
9. At Civil 3D's top left, Quick Access Toolbar, click the ***Save*** icon to save the drawing.

## Wall Breakline

The last breakline is a wall breakline. It starts as a 3D polyline representing the top of the wall. The breakline's offset side (bottom of the wall) is north toward the road.

A wall breakline in a surface greatly changes the surface's slope statistics report. Before adding the wall breakline, review the current slope values for minimum, maximum, and mean.

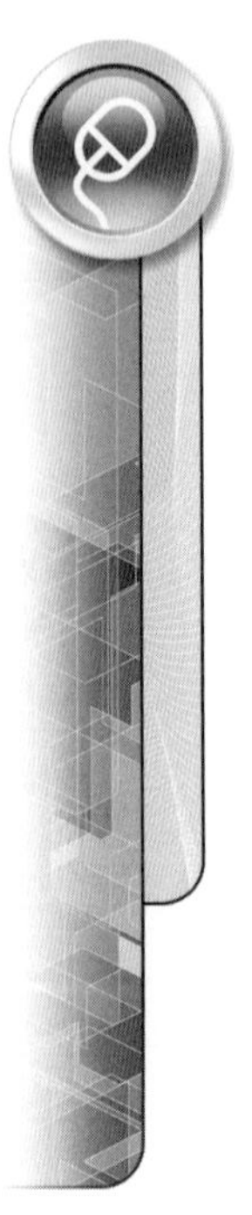

1. In the drawing, select the surface border or a triangle, press the right mouse button, and from the shortcut menu, select SURFACE PROPERTIES....
2. Click the ***Statistics*** tab and expand the **Extended** (statistics) section. Your report should look similar to Figure 4.58. The mean slope is around 6 percent, the lowest slope is zero, and the highest is just over 200 percent.

The mean and maximum slope will change greatly after the wall breakline is defined.

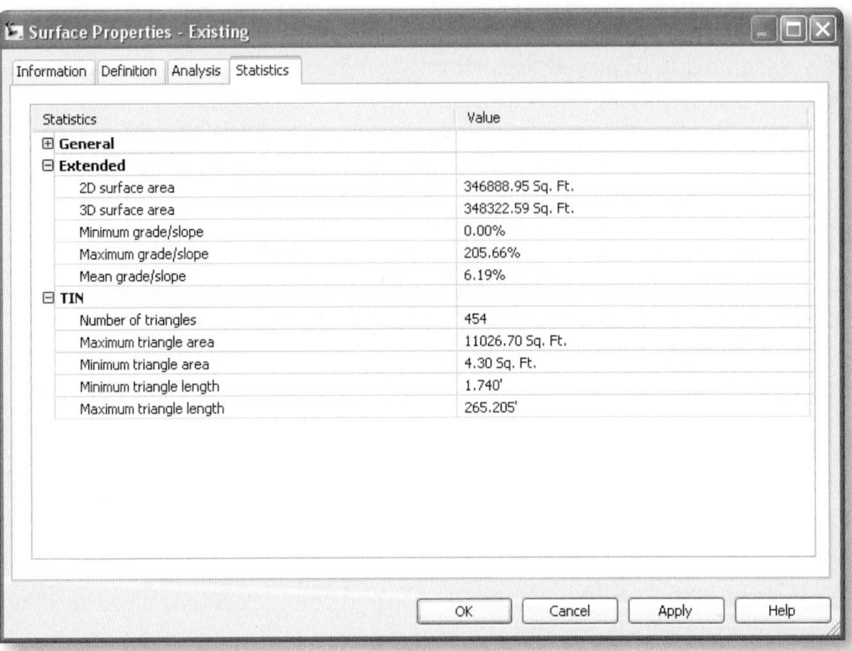


**FIGURE 4.58**

3. Click ***OK*** to exit the Surface Properties dialog box.
4. Make a new point group, **Headwall**, based on the raw description of **HDWL**. Use the ***Raw Desc Matching*** tab to identify the headwall points, and click ***OK*** to exit the dialog box.
5. After defining the Headwall point group, go into the Point Groups properties dialog box and make **Headwall** the top point group, followed by the No Show point group (see Figure 4.59).

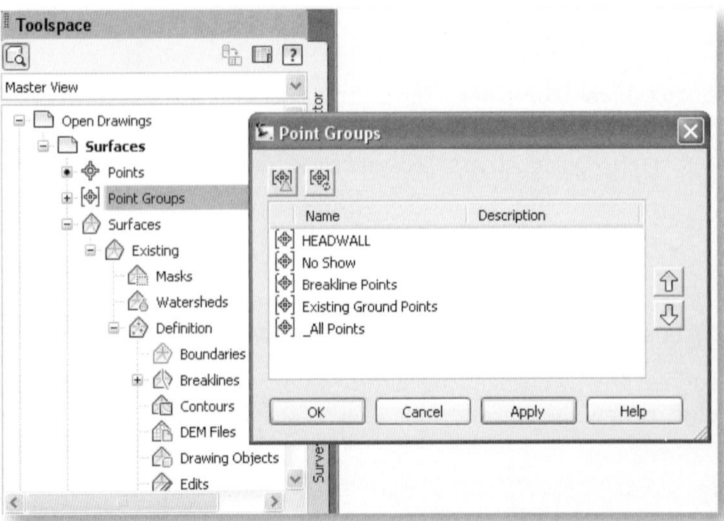


**FIGURE 4.59**

6. Use the 3DPOLY routine with the Transparent Commands toolbar's **Point Object** override and draw a 3D polyline between points 186, 187, 188, and 189.
7. In Prospector, in the Existing surface's Definition branch, select **Breaklines**, press the right mouse button, and from the shortcut menu, select ADD....
8. In the Add Breaklines dialog box, enter the description of **Headwall**, change the breakline type to **Wall**, and click ***OK*** to exit the dialog box.
9. Select the headwall 3D polyline, press ENTER, and select a point to the northeast of point 188 as the offset side.

The routine responds by prompting you for a method to set the offset side elevations.

10. Enter **'I'** for Individual heights and press ENTER to view each headwall vertex.

The survey crew measured a down distance from the wall's top to bottom. Because of this method, the offset elevation for each point is a difference in elevation.

11. At the Specify elevation prompt, for Delta enter **'D'** and press ENTER.
12. The prompt changes to Enter elevation difference for offset point at .... Use the elevation difference entries in Table 4.4 for the difference values.

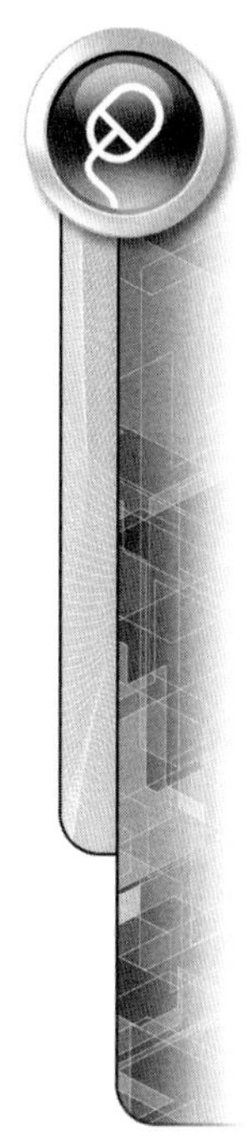

**TABLE 4.4**

| Point Number | Difference in Elevation |
|---|---|
| 186 | -2.5 |
| 187 | -3.5 |
| 188 | -3.5 |
| 189 | -2.5 |

After entering the last value, the routine exits and rebuilds the surface.

13. Select any surface triangle or its border, press the right mouse button, and from the shortcut menu, select SURFACE PROPERTIES....
14. Click the ***Statistics*** tab and expand the **Extended** section. Review the new values. The average for slopes is now nearly 225 and the maximum slope is just over 350,000.

This is a considerable change and the wall skews the surface slope statistic. When reviewing the surface slopes, this has to be taken into consideration.

15. Click ***OK*** to exit the Surface Properties dialog box.

## View the Headwall Profile

1. Click the Ribbon's ***View*** tab. In the Viewports panel, click NAMED.
2. In the ***Named Viewports*** tab, select the named viewport **A and P**, and click ***OK*** to exit the dialog box.
3. Use the ZOOM and PAN commands to place the headwall in the lower viewport.
4. Click the Ribbon's ***Feature Line*** tab, and in the Launch Pad panel, select QUICK PROFILE.
5. Select the quick profile polyline drawn earlier, the Create Quick Profile dialog box displays, click ***OK*** to close it, and in the upper viewport select the profile's insertion point.
6. Close the Panorama to view the quick profile.
7. If necessary, move the quick profile view polyline to view the newly defined headwall.

8. Make the bottom viewport the current viewport.
9. Click the Ribbon's ***View*** tab. In the Viewports panel, click Set VIEWPORT, and from the menu, select SINGLE.
10. Click the Ribbon's **Home** tab.
11. In the Layers panel, use the **Layer Properties Manager** to set layer **0** (zero) as the current layer and freeze the **Existing-Breakline**, **RDS**, **LOT**, **C-TOPO-FEAT,** and **CL** layers to focus on the surface and its triangulation.
12. Use the ZOOM EXTENTS command to view all of the surface and its triangles.
13. In Prospector, select the **Point Groups** heading, press the right mouse button, and from the shortcut menu, select PROPERTIES....
14. Select **No Show**, move it to the top of the list, and click ***OK*** to exit and hide the points.
15. At Civil 3D's top left, Quick Access Toolbar, click the ***Save*** icon to save the drawing.

## SUMMARY

- You create a surface instance by naming, describing, and assigning it a style.
- Each time surface data is added, the surface rebuilds itself.
- Point groups are an effective way of organizing and controlling point display.
- Linear features using point data need breaklines to preserve them on the surface.
- A breakline from a feature line is the only way to represent curvilinear surface features.
- A Quick Profile is an effective way to evaluate a surface under construction.

This exercise and unit covers the initial review and modification of a surface. Civil 3D has other tools to review a surface's properties and edit its data.

## UNIT 3: SURFACE REVIEW AND EDITING

While building a surface, you should evaluate the effects of new data and breaklines. When evaluating each addition, the user may discover missing data, bad data, the need for additional breaklines and/or boundaries, incorrect triangles, and sliver triangles at the periphery—all indicative that a surface needs data tweaking and editing. The process of tweaking a surface is an iterative process involving reviewing the surface, adding or editing data, reviewing the change's effects, and again adding to or editing the surface. It may take several passes before the surface is correct and ready for analysis and annotation.

### STEP 4: EVALUATE RESULTING TRIANGULATION

#### Surface Properties

The initial surface review is conducted by evaluating the values in the Surface Properties dialog box. This dialog box contains surface statistics, data and editing options, editing history, and many other settings and values.

## Information Tab

The Information tab displays the surface name, its object style, assigned rendering material, if the surface is locked, and if the surface should display a tools tip.

## Definition Tab

This tab has two parts: Definition Options and Operation Type. Definition Options has three sections: Build, Data operations, and Edit operations (see Figure 4.60).

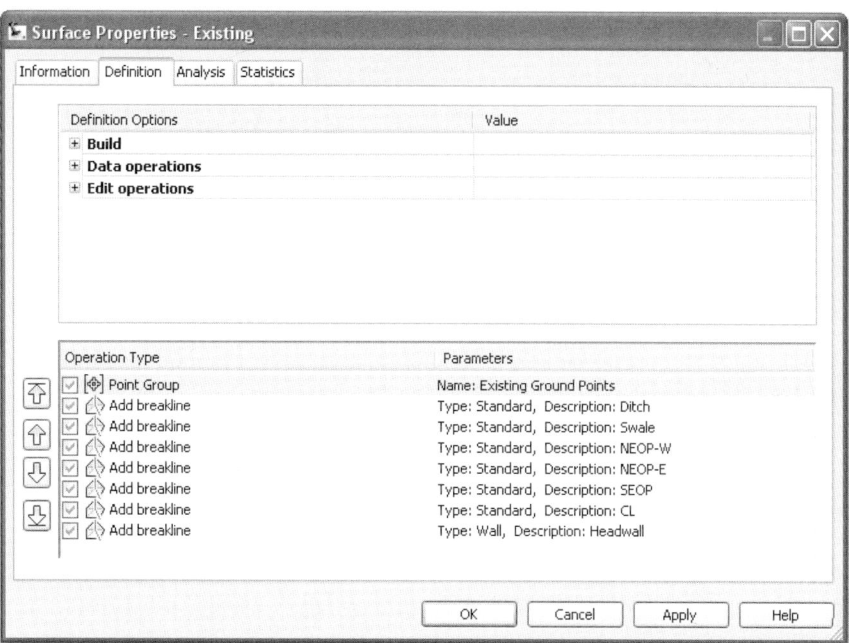


**FIGURE 4.60**

**Definition Options and Operation Type.** This panel maintains settings that affect the building, data use, allowed edits, and the history of the surface. See the discussion in Unit 1 of this chapter.

The Definition area sets the surface data types use, toggles on or off surface data (turning off Use Breaklines removes all of the surface's breakline data), and sets allowed editing operations.

When in Data operations, and you change Use breaklines to No, all breaklines present in the Operation Type (the bottom portion of the panel) are unchecked and are removed from the surface when the change is approved.

Civil 3D then issues a warning dialog box asking if the user wants to rebuild the surface using the new settings (see Figure 4.61). If the user answers Yes, the surface rebuilds with the new settings (rebuilds the surface without breaklines).

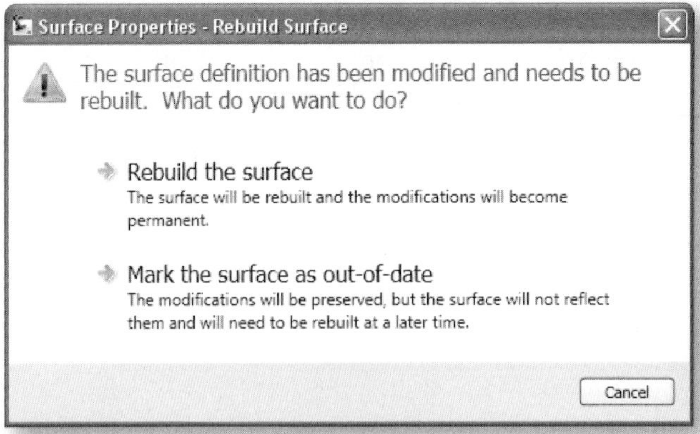


**FIGURE 4.61**

## Statistics Tab

The initial surface review starts in the Surface Properties' Statistics tab. This simple review can identify bad surface data elevations or coordinates. The panel displays three surface properties reports: General, Extended, and TIN. General lists information about the revision number, number of data points, and the minimum, maximum, and mean surface elevation (see Figure 4.62).

General also reports basic surface information: elevation ranges, coordinate ranges, number data points, version, and elevation statistics.

Extended reports the 2D and 3D surface area, and the minimum, maximum, and mean slopes (slopes as a grade). A grade is the rise or fall of elevation expressed as a ratio. So, a 3:1 slope is a 33 percent grade and a 4:–1 slope is a –25 percent grade.

When creating slope range styles, it is important to be aware of any surface feature that distorts the surface statistics (e.g., a headwall distorting the surface slope range). By turning off the headwall breakline in the Operation Type area of the Definition panel of a surface, you could view the surface slopes without the extreme influence of the wall.

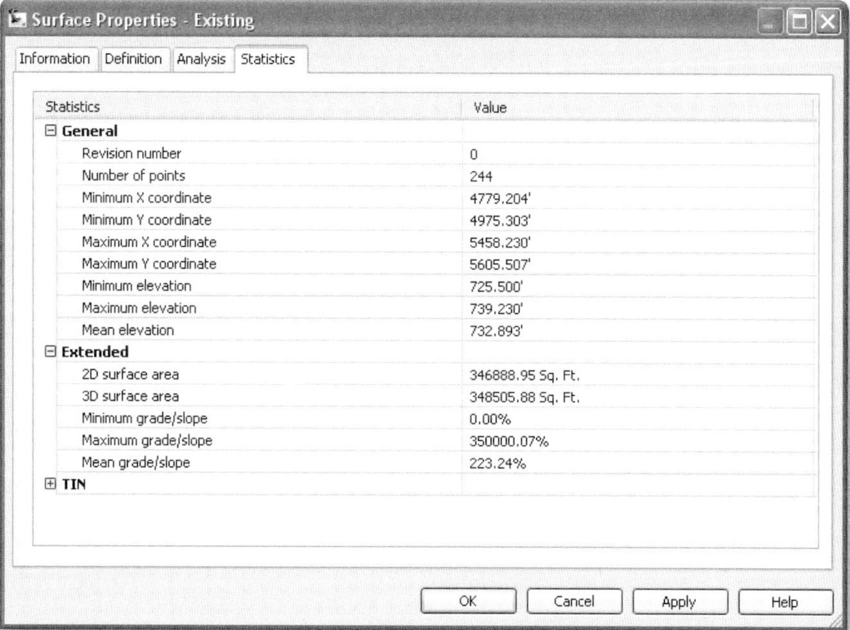


**FIGURE 4.62**

TIN reports a TIN's area, its minimum and maximum triangle sizes, and triangle leg lengths (see Figure 4.63). Excessively long triangle legs occur around a surface's periphery or in areas with sparse data. A surface triangulation review determines whether these triangles need deletion or the points creating the large triangles need exclusion from surface data, or whether there is a need to add or edit data to create a better surface. The user can prevent or control long triangle legs by setting the Maximum triangle length in the Surface Properties Definition tab, Build section (see Figure 4.64).

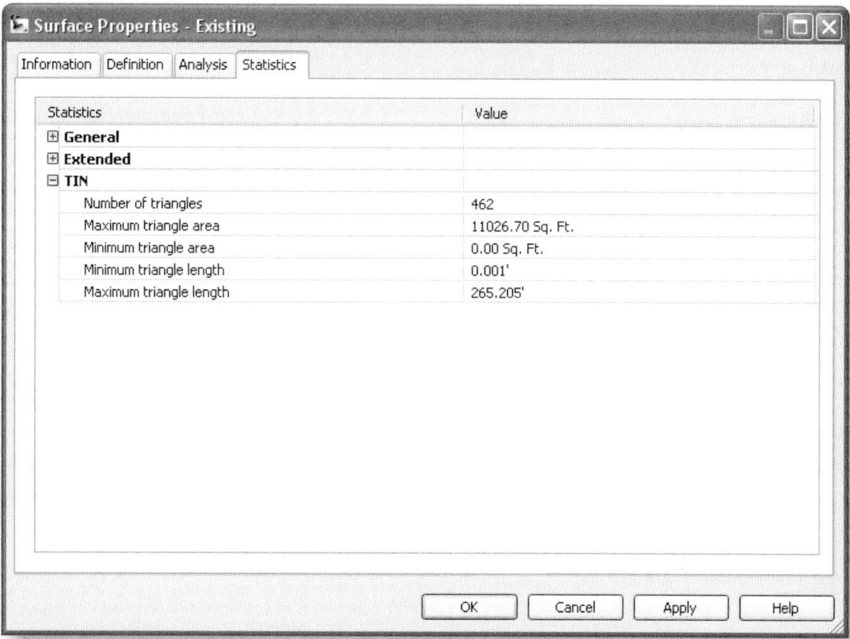


**FIGURE 4.63**

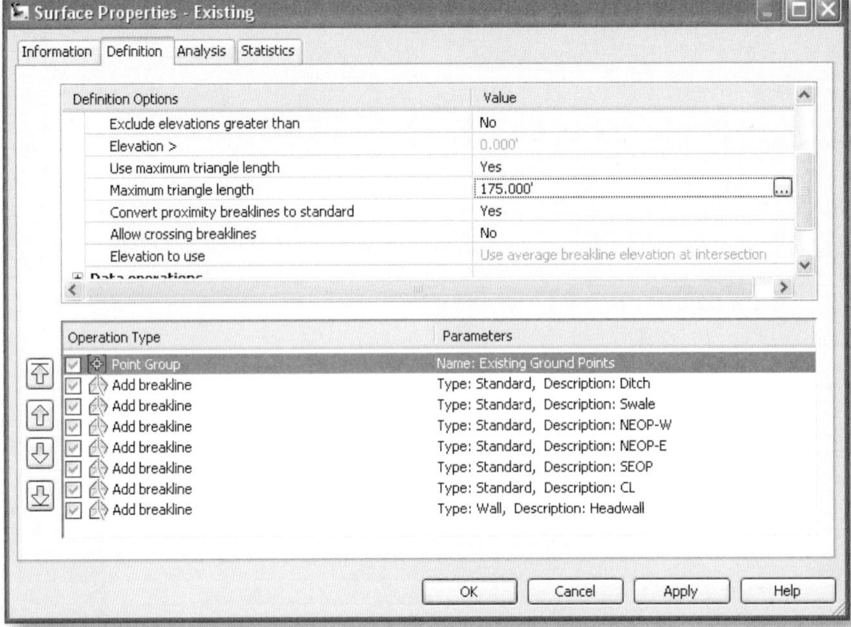


**FIGURE 4.64**

## STEP 5: EDIT SURFACE

The surface-editing tools are in the surface's Definition branch. When you click the Edits heading and press the right mouse button, a shortcut menu displays the editing routines (see Figure 4.65). The shortcut menu is divided into three types of editing tools: lines, points, and overall surface tools. The overall editing tools raise/lower a surface's elevation, minimize flat faces, simplify, and smooth a surface.

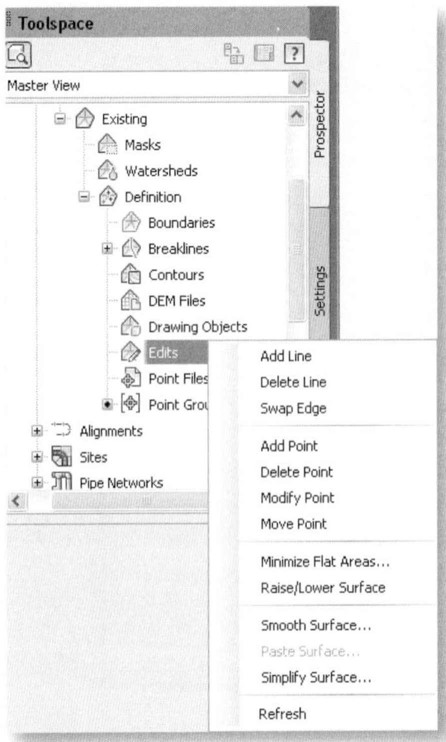


**FIGURE 4.65**

A user can delete any surface edit. When you are in a surface's Definition branch selecting the Edits heading, preview lists the edits. The edits list is from first (at top) to last (at bottom) (see Figure 4.66). Selecting an entry in preview and right mouse clicking displays a shortcut menu listing commands applicable to the selected item. The list includes Delete.... Or, you can delete an edit in the Surface Properties' Definition panel Operation Type list by selecting the edit, right mouse clicking, and from the shortcut menu selecting Remove. When deleting an edit, the surface resolves how the deleted edit changes the surface triangulation. The surface rebuild depends on the Rebuild-Automatic toggle, which is set in a right mouse click shortcut menu when selecting a surface's name. If off, the edit is removed from the list, but the surface does not update its triangulation. What appears is an out-of-date icon to the left of the surface name. When a surface is out-of-date, to update it, select the surface name, press the right mouse button, and from the shortcut menu, select Rebuild. If Rebuild-Automatic is on when deleting an edit from the Preview area, the edit is removed and the surface automatically rebuilds with the new data and edits mix.

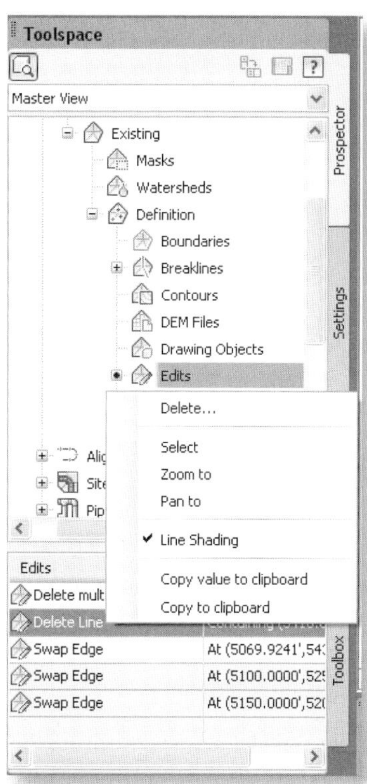


**FIGURE 4.66**

If you prefer not to delete an edit, but rather want to evaluate its impact or view different combinations of edits results, toggle on or off individual entries in Surface Properties, Definition panel's Operation Type area (see Figure 4.67). After evaluating the impact of the edit or edit combinations, the user can return to the Surface Properties dialog box and restore the edits.

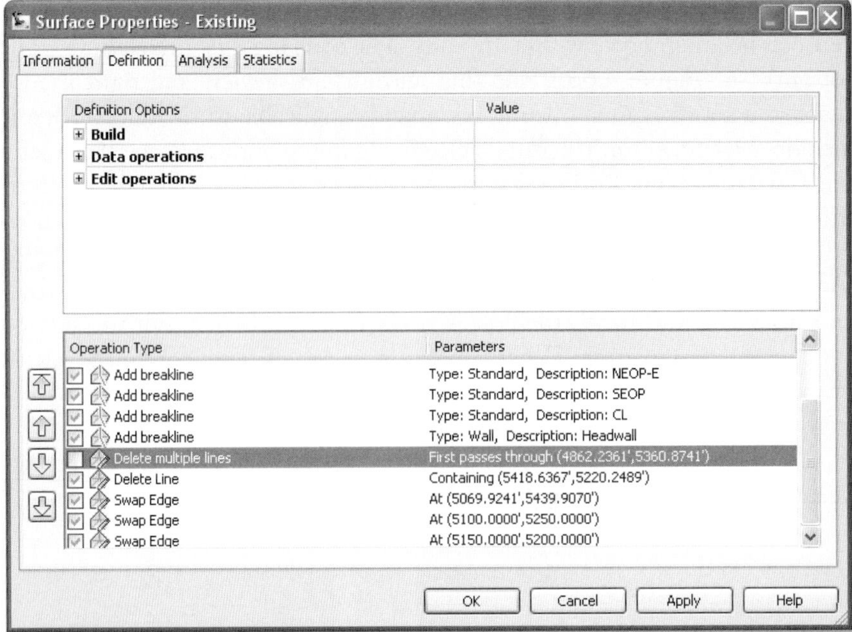


**FIGURE 4.67**

## Line Edits

These edits modify surface triangle legs: delete, add, or swap interior diagonal legs (see Figure 4.65).

Delete Line removes a triangle leg from a surface. Deleting an interior leg creates a hole in the surface. If you delete an interior leg, Civil 3D places a surface border around the hole. If you delete a delete line edit (from preview) or toggle it off in Surface Properties, Civil 3D removes the border and adds the leg back into the surface.

Add Line creates a new triangle leg at the surface's periphery or forces a new diagonal leg in the surface's interior. The first situation is where the surface did not create a triangle and you want to add it to the triangulation. Beware of doing this around the surface edges. The reason the triangle was not made is based on surface settings and it was determined that there is no real data supporting such a triangle. For example, the triangle leg might be greater than the set maximum triangle leg length.

A second use for Add Line is to change an interior diagonal. In a surface's interior, a diagonal leg connects the two nearest points of a group of four neighboring points. The initial diagonal is based on distances and groupings. If it is not the correct solution, add a new leg to the surface to change the diagonal. If the new line crosses an existing diagonal, the surface deletes the original diagonal and replaces it with the new one. Using Add Line for this purpose duplicates the Swap Edge routine.

Swap Edge is a simpler method of changing an interior diagonal. Swap Line changes the diagonal within a four-point group.

**NOTE**

**If edits represent an effort to correctly triangulate a linear feature, use a breakline to correctly resolve the feature instead of editing the triangles.**

## Point Edits

Add Point creates new surface data points (see Figure 4.65). These points are not cogo points or from a directly imported point file. They represent a surface elevation located within a surface's interior. An example of this would be adding high and low elevations to contour data. Add Point only adds a surface elevation.

Delete Point removes a surface elevation. Be aware of how the point is a part of the surface. If the point is from a cogo point that is a part of a surface point group, the deletion removes the elevation from the surface, but does not affect the cogo point. To delete a surface point, select near it, and it is removed from the surface. The cogo point remains on the screen and remains a member of the original surface point group. Preview shows only deleted point coordinates. Delete Point only deletes a surface elevation.

Modify Point changes a surface elevation. If the point is from a cogo point that is a part of a surface point group, the surface elevation changes, but the edit does not affect the cogo point. If you decide to modify the surface elevation, click near the point's location and modify its elevation. The point object remains a point group member and displays its original elevation. Modify Point only modifies a surface elevation.

Move Point changes the location of a surface elevation. If the point is from a cogo point that is a part of a surface point group, the surface elevation moves, but the edit does not affect the cogo point. If you decide to edit a surface elevation, select near the point's location and then select its new location. After moving the surface point, the original point object remains a point group member and does not display a modified location. Move Point only relocates a surface elevation location.

The best practice for adding new points would be as new members of an existing or new point group assigned to a surface or a directly imported point file. This way the user can easily trace their additions and manage their participation in surface data.

When using any of the point edits, you are not editing the cogo point objects. These edits affect only surface points, not the cogo points that create the original data. If you are using these routines, the surface point will be out-of-sync with the point object's values.

**NOTE**

**If you want to modify surface points that are cogo points, use Edit Points ... to edit their values and the surface will rebuild, incorporating their new values.**

A user can delete any surface edit. In a surface's Definition branch, select the Edits heading to display an edits list in preview. From the list, select an edit, right mouse click, and from the shortcut menu, select Delete.... If you don't want to delete the edit, but want to evaluate it or a group of other edits, go to Surface Properties, Definition tab and in the Operation Type area, select and toggle off or on the desired edits.

When selecting a block of entries, select the first entry, hold down SHIFT, and select the last entry. If you want to select several individual entries, hold down CTRL while selecting the individual entries.

### Smooth Surface

Surface smoothing is a cosmetic edit needed for document production. However, when a surface's data is sparse, using one of the smoothing algorithms helps create a more realistic surface. The decision to smooth should wait until after approving the overall surface.

Surface smoothing has two strategies: Natural Neighbor Interpolation (NNI) and Kriging (see Figure 4.68). Both methods create interpolated surface data. The interpolated data creates better transitions between elevations and more data between sparse data points. These interpolation processes result in a "smoother" surface. The NNI method operates best with point data. Kriging extends surface trends to sparser data. Both methods require smoothing regions defined using one of the following methods: select an existing closed polyline, a parcel, draw a polygon; select a surface; or select an existing rectangle.

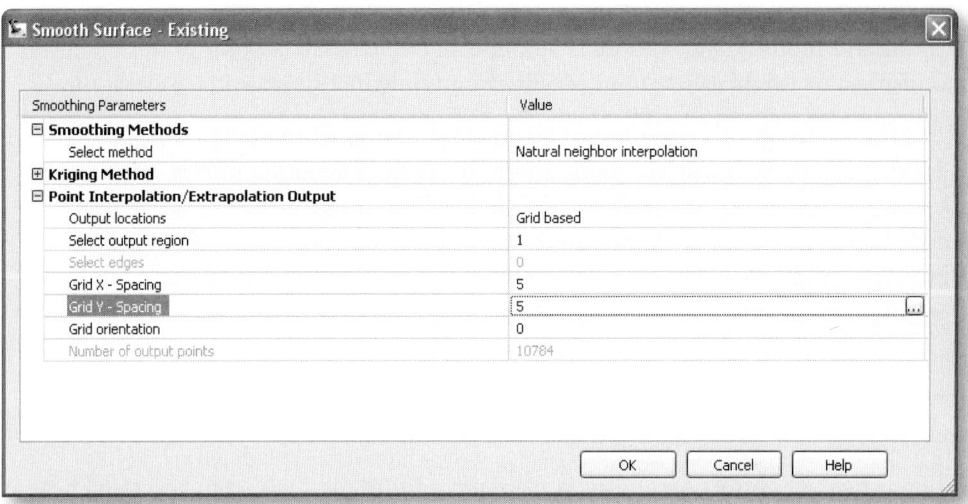


**FIGURE 4.68**

### Natural Neighbor Interpolation (NNI)

Natural Neighbor Interpolation (NNI) uses an NNI algorithm to estimate arbitrary point elevations from a set of points with known elevations. The routine places the resulting interpolated points into the surface by one of four methods: grid, centroids, random points, or as edge midpoints.

**Grid.** Grid smoothing creates new surface elevations from existing elevations within a region based on a grid. Grid smoothing requires a defined interpolation area, grid spacing values, and a grid rotation. The result is new surface data points at the grid spacing and rotation within the smoothing boundary.

**Centroids.** Centroids create new surface elevation data from elevations at the existing triangles' centroids. A region is required to calculate the new data. After calculating new elevation data, the surface creates new triangulation using the existing and newly interpolated points.

**Random Points.** Random Points creates new elevation point data from elevations within the region. The first step is defining the number of new points and the second step is defining the interpolation area. After calculating new data, the surface creates new triangulation between the existing and newly interpolated points.

**Edge Points.** Edge Points creates new surface elevation data from elevation trends found within a region. The new points are slightly higher or lower based on the surface elevation trend. This method allows a contour data surface to correctly interpolate the elevations of bays and peninsulas in the contour data.

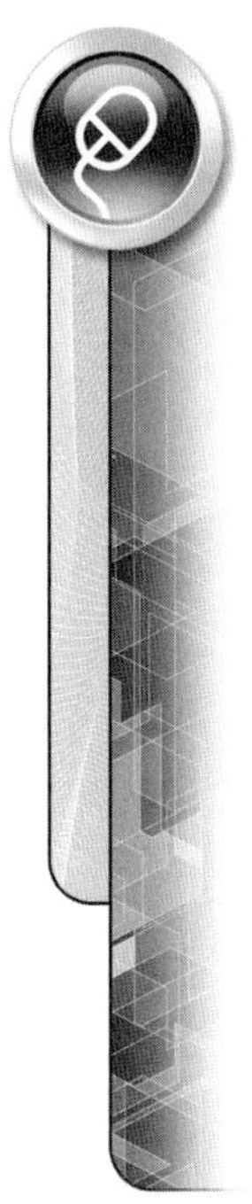

### Kriging

Kriging is a much more speculative method for interpolating surface points. Kriging supplements sparse data and/or expands a surface to sparser data. Examples using Kriging are developing pollution plumes, water flow plumes, or subterranean surfaces. Generally, water flow or pollution plume data is sparse and Kriging's interpolation method is ideal for modifying a surface that has sparse data. An understanding of the mathematics and applications of Kriging is necessary to correctly use this interpolation method.

Any surface smoothing can be deleted. In a surface's Definition branch, select the Edits heading to display an edits list in the preview, select the smoothing, press the right mouse button, and from the shortcut menu, select Delete…. If you don't want to delete the smoothing, but want to evaluate it or a group of other edits, go to Surface Properties' Definition tab and, in the Operation Type area, select and toggle off or on each of the edits.

## EXERCISE 4-3

After completing this exercise, you will:

- Be familiar with the Edit Surface menu.
- Know Surface Properties' Definition tab settings and their effects.
- Be able to delete and control surface edits.
- Review surface edit effects.
- Be familiar with Natural Neighbor Interpolation (NNI) smoothing.

This unit's exercise continues with the previous units drawing. If you did not complete the Unit 2 exercise, or starting with this exercise browse to the Chapter 4 folder of the CD that accompanies this textbook and open the Chapter 04 - Unit 3 drawing.

### Exercise Setup

1. If not open, open the previous exercise's drawing or browse to the Chapter 4 folder of the CD that accompanies this textbook and open the *Chapter 04 – Unit 3* drawing.

### Surface Elevations

A review of simple surface edits can be done using a surface's tool tip. The Surface Properties' Information tab has the tool tip toggle.

1. In the drawing, select the **Existing** surface border, press the right mouse button, and from the shortcut menu, select SURFACE PROPERTIES….

2. Click the ***Information*** tab and at the dialog box's lower-left, notice the Show Tool-tips toggle; make sure it is toggled **ON**.
3. Click ***OK*** to close the dialog box and move the cursor to different surface locations, letting the cursor sit for a moment. The tooltip shows the elevation at the cursor's intersection.

## Reviewing Slopes and Breakline Data

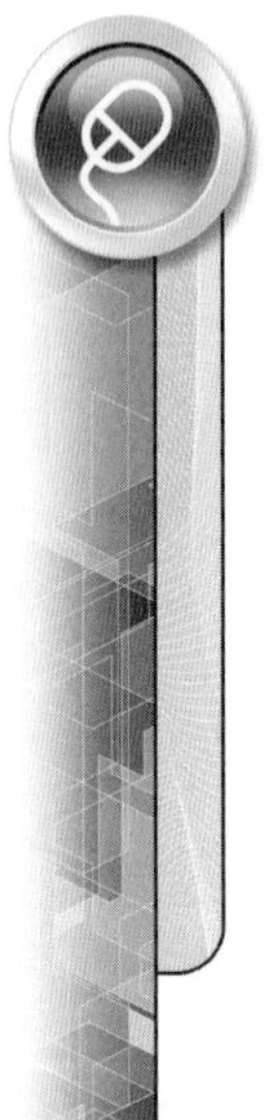

1. In the drawing, select **Existing** surface's border, press the right mouse button, and from the shortcut menu, select SURFACE PROPERTIES....
2. Click the ***Statistics*** tab and expand the **General** section to review its values.

Note the minimum, maximum, and the mean elevation. The mean elevation is about midway between the lowest and highest elevations.

3. Expand the **Extended** section to review its values.

The maximum grade (350,000+) is the result of the wall breakline. This single surface feature changes the average grade from around 6 percent to more than 220 percent.

4. Click the ***Definition*** tab.
5. In Definition Options, expand **Data Operations**, change Use Breaklines to **No**, click ***OK*** to exit the dialog box, and if necessary, click ***Rebuild the Surface*** to rebuild the surface with no breakline data.
6. Use the ZOOM and PAN commands to view the roadway's triangulation.

The surface returns to having only point data.

7. Select a surface triangle or the border, press the right mouse button, and from the shortcut menu, select SURFACE PROPERTIES....
8. In the ***Definition*** tab, expand **Data Operations**, toggle **Use Breaklines** to **Yes**, click ***OK*** to exit the dialog box, and if necessary, click ***Rebuild the Surface***. This restores the breakline data to the surface.

## Temporarily Removing the Wall Breakline

To remove the impact of a single operation, move your attention to the Definition panel's lower half. It is here that one can toggle on or off the effect of one or more operations. By toggling off the wall breakline, it is temporarily removed from the surface. With it removed, the Extended surface statistics return to their values before adding the wall breakline. After reviewing these sans wall statistics, you reenter the surface properties dialog box and toggle it on again.

1. Select a surface triangle or the border, press the right mouse button, and from the shortcut menu, select SURFACE PROPERTIES....
2. In the Operation Type, uncheck the **Wall breakline** toggle.
3. Click ***OK*** to exit the dialog box and click ***Rebuild the Surface*** to rebuild the surface with the new data mix.
4. If the Events vista display contains warnings about duplicate points, click the Panorama's **X** to close it and view the new triangulation.

The headwall triangles change.

5. In the drawing, select the **Existing** surface border, press the right mouse button, and from the shortcut menu, select SURFACE PROPERTIES....
6. Click the ***Statistics*** tab and expand the **Extended** section. Review the maximum and mean slope values.

The mean slopes return to their pre-wall values (they average around 6 percent and have a maximum of just over 200 percent).

7. Click the ***Definition*** tab. In Operation Type, toggle **ON** the **Wall breakline**, click ***OK*** to close the dialog box, and then click ***Rebuild the Surface*** to rebuild the surface.

## Reviewing Build Settings

1. In the drawing, select a surface triangle or the border, press the right mouse button, and from the shortcut menu, select SURFACE PROPERTIES....
2. Click the ***Statistics*** tab and expand the **TIN** section. The minimum and maximum triangle lengths represent some very short and long surface triangles.

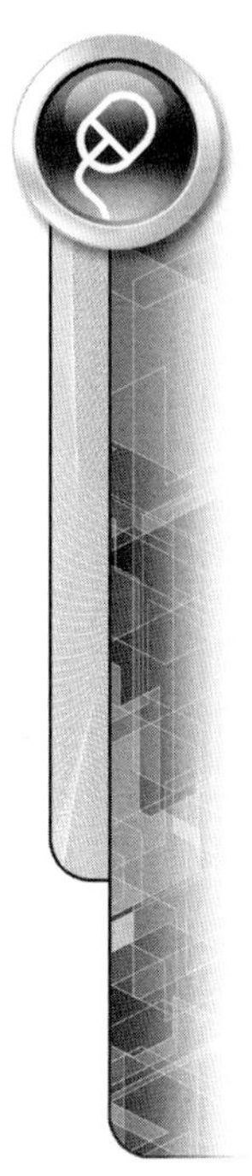

The question is, does a triangle leg of 260 feet represent a valid surface triangle? While building a surface, a surface tries to eliminate some extraneous peripheral triangles. However, if a maximum distance is not specified, a surface may have triangle legs around its periphery without supporting data.

Setting a Maximum triangle to control peripheral triangles may not completely eliminate the problem. After using a few values and viewing the results, you may have to edit the surface with delete lines or use a boundary to control the peripheral triangulation.

3. Click the ***Definition*** tab and in Definition Options, expand the **Build** section. In the Build section, change Use Maximum Triangle Length to **YES**, set the length to **100**, click ***OK*** to close the dialog box, and then click REBUILD THE SURFACE to rebuild the surface.

The surface border and the peripheral triangles change, reflecting the new settings. While triangles are moved in the north, some in the northwest are not removed. It would seem that deleting a triangle and/or defining a boundary is a better solution.

4. In the drawing, select the **Existing** surface border, press the right mouse button, and from the shortcut menu, select SURFACE PROPERTIES....
5. In the ***Definition*** tab Definition Options expand the **Build** section.
6. Toggle the Use Maximum Triangle Length to **NO**, click ***OK*** to close the dialog box, and then click **Rebuild the Surface** to rebuild the surface.

## Deleting Triangles

1. In the Ribbon, click the View tab. In the Views panel's left side, click and restore the named view **Central,** and review the surface's eastern triangulation.
2. From the Existing surface's Definition branch, select **Edits**, press the right mouse button, and from the shortcut menu, select DELETE LINE (see Figure 4.65).
3. In the drawing, select the two triangle legs shown in Figure 4.69 and press ENTER twice to delete the lines and exit the command.

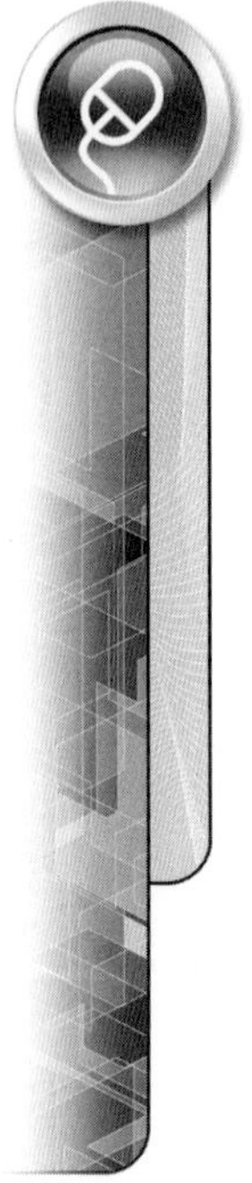

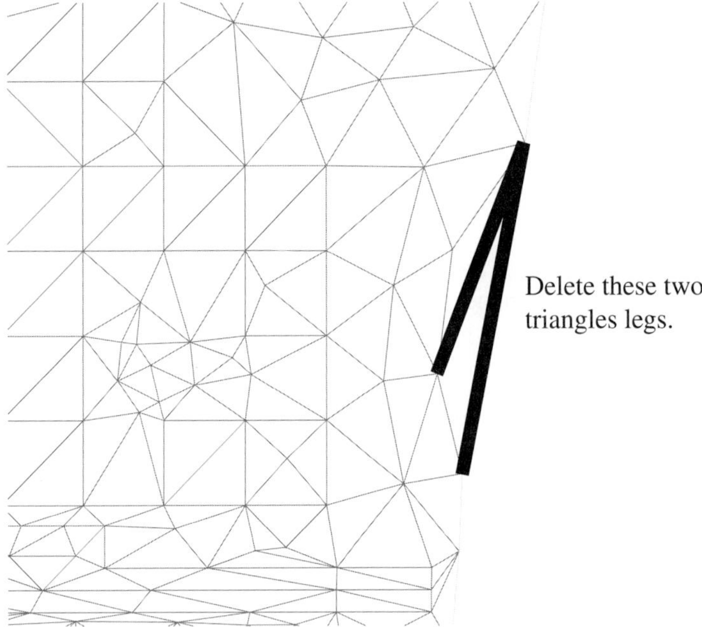


**FIGURE 4.69**

They are removed and surface displays a new surface border.

Prospector updates and indicates the edits by placing an icon to the left of Edits. Select the Edits heading to display a surface listed by the edits preview (see Figure 4.70). The edit also appears in Surface Properties' Definition panel's Operation Type.

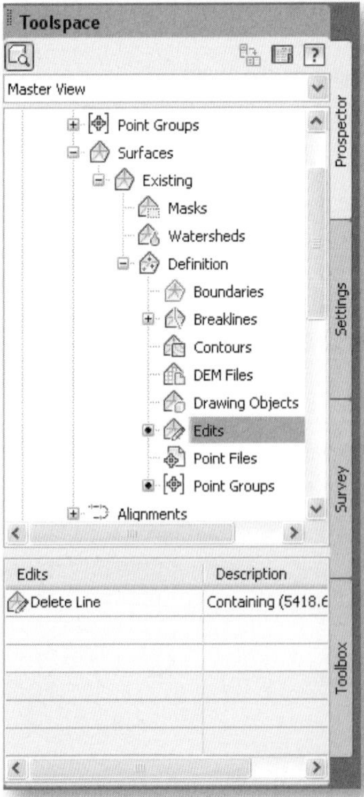


**FIGURE 4.70**

4. From the Definition branch, select **Edits** to display an edit list in preview with the delete line entry.
5. In the drawing, select the surface border, press the right mouse button, and from the shortcut menu, select SURFACE PROPERTIES....
6. If necessary, click the ***Definition*** tab.

Deleting multiple lines is an Operation Type entry at the panel's bottom. There is an option to toggle on or off the deletion's effect or remove it from the surface.

7. From the Operation Type list, select ***DELETE LINE***, press the right mouse button, and review the shortcut menu commands. Remove From Definition... is one of the commands.
8. Press ESC to close the shortcut menu.
9. Click ***OK*** to exit the dialog box.
10. In Prospector's preview, select **Delete Lines** (see Figure 4.70), press the right mouse button, from the shortcut menu, select DELETE..., and in the Remove From Definition dialog box, click ***OK*** to delete the edit.

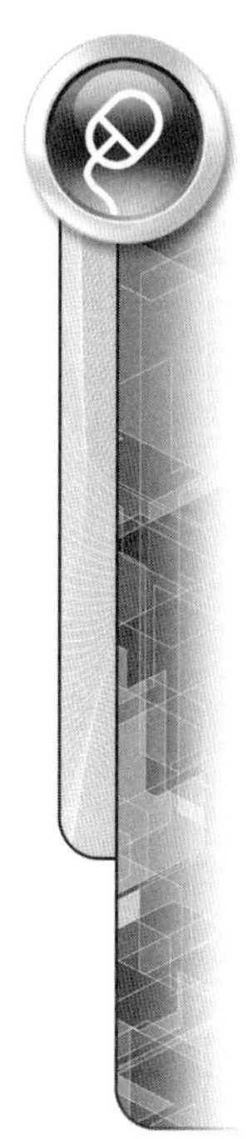

The Remove From Definition dialog box gives you the option of not deleting an edit.

11. If a Panorama is displayed, close the Events vista by clicking the **X** and view the new triangulation.
12. In the Existing surface's Definition branch, select **Edits**, press the right mouse button, from the shortcut menu, select DELETE LINE, and delete the two triangles identified in Figure 4.69.

## Swap Edge

The Swap Edge routine changes a diagonal to the second solution. Use this routine to "tweak" a surface's triangulation in areas that do not need additional data or breaklines. Like Delete Lines, preview displays only the swapped edges' coordinates.

1. In the Existing surface's Definition branch, select **Edits**, press the right mouse button, and from the shortcut menu, select SWAP EDGE.
2. Select two or three diagonals to change their position and press ENTER to exit the routine.
3. In Preview, review the Swap Edge listings.
4. In Preview, select the **Swap Edge** entries, press the right mouse button, from the shortcut menu, select DELETE..., and in the Remove From Definition dialog box click ***OK*** to delete the edits and rebuild the surface.
5. If a Panorama is displayed, close the Events vista by clicking the **X** and view the new triangulation.

## Surface Point Editing — Delete Point

If you want to delete a surface point, the surface point is removed, but the cogo point remains in the drawing and its assigned point group.

1. From the Ribbon's View tab, in the Views panel, select and restore the named view **Northeast**.
2. In Prospector, click the **Point Groups** heading, press the right mouse button, from the shortcut menu, select PROPERTIES..., move **Breakline Points** to the top position and **No Show** to the second position to view the northeastern swale points, and click ***OK*** to exit the dialog box.
3. Use the ZOOM command to better view points 107, 111, and 112.

4. In the Existing surface's Definition branch, select the **Edits** heading, press the right mouse button, and from the shortcut menu, select DELETE POINT.
5. When prompted for a point, click near point number 110 and press ENTER twice to exit the routine.

After deleting the surface point, point 110 remains in the drawing. Preview lists only the deleted point's coordinates.

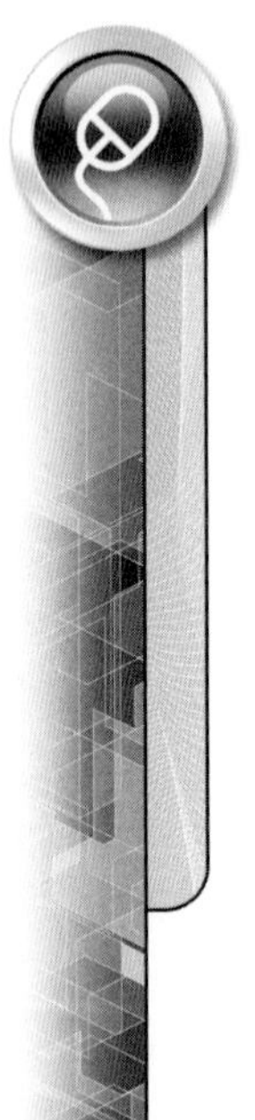

6. In the Existing surface's Definition branch, select the **Edits** heading to preview the edits list.
7. From the Edits list, select the **Delete Point** entry, press the right mouse button, from the shortcut menu, select DELETE..., and in the Remove From Definition dialog box click ***OK*** to restore the deleted point to the surface.
8. If a Panorama is displayed, close the Events vista by clicking the **X** and view the new triangulation.
9. In Prospector, select the **Point Groups** heading, press the right mouse button, from the shortcut menu, select PROPERTIES..., move **No Show** to the top position, and click ***OK*** to exit the dialog box.
10. Use the ZOOM EXTENTS command to view the entire site.

## Boundaries

The surface under review has several points in the west and northwest that seem to be peripheral to the site. These points could be deleted from the surface. However, an outer boundary limits the data to the points inside its boundary.

1. In Ribbon's Home tab, Layers panel, use Layer Properties Manager and create a new layer: **Existing-Boundary**. Make it the current layer, assign it a color, and click **X** to create the layer and exit the panel.
2. Start the POLYLINE routine and draw a closed boundary similar to the one in Figure 4.71.

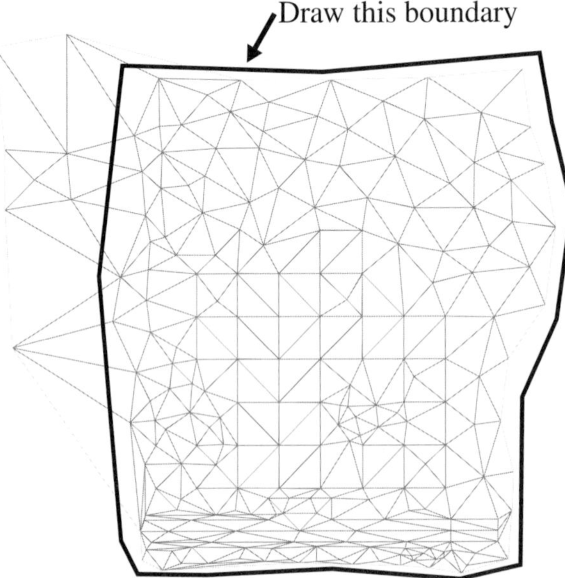


**FIGURE 4.71**

3. In the Existing surface's Definition branch, select **Boundaries**, press the right mouse button, and from the shortcut menu, select ADD....

4. In the Add Boundaries dialog box, name the boundary **Main Site**, set its type to **Outer**, toggle **OFF** Non-destructive breakline, click ***OK***, and select the boundary polyline.
5. Open the **Layer Properties Manager**, make the current layer **0** (zero), freeze the layers **Existing-Boundary** and **V-NODE**, and click **X** to exit the panel.
6. At Civil 3D's top left, Quick Access Toolbar, click the ***Save*** icon to save the drawing.

The surface contains all the needed edits and changes and is ready for analysis. However, it is difficult to decide whether to do any surface smoothing. The decision about smoothing happens after viewing the initial contours.

The next unit reviews the distribution of surface slopes and elevations.

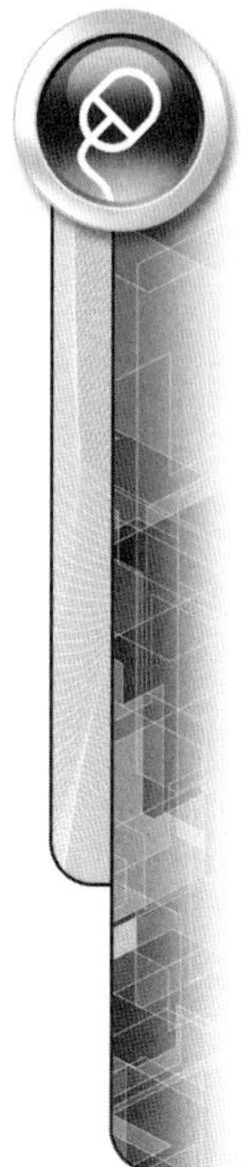

## SUMMARY

- Reviewing and editing a surface is an iterative process.
- To rebuild a surface after each edit, set the Rebuild – Automatic toggle on.
- Delete peripheral triangles with the Delete Line edit.
- If you delete interior triangles, the surface places a border around missing triangles.
- A boundary is an effective method of excluding unwanted data.
- A boundary is an effective method of controlling unwanted peripheral triangles.
- Instead of deleting interior triangles, consider using a hide boundary.

## UNIT 4: SURFACE ANALYSIS

Next in the surface process is how to analyze a surface's slopes and elevations. The analysis is the function of specific-use surface styles. Depending on the analysis type (slope or elevation) and its distribution, each style may use a different range calculation algorithm.

### STEP 6: ANALYZE THE SURFACE CHARACTERISTICS

Because surfaces represent a project's start and end, it is critical that they represent their data correctly. The surface properties dialog box contains a Statistics tab that reports specific information about that surface's slopes and elevations.

However, the best way to understand slopes and elevations is to visualize them in 2D or 3D views. If a surface contains bad data, each view shows it as spikes or wells in the model. By creating analytical surface styles that emphasize these characteristics, one can visualize and communicate a surface in a more understandable way. The surface analysis styles focus mainly on TIN structure, relief, elevation, and slope.

### Slope and Elevation

After focusing on surface triangulation, the next two surface characteristics that need visual review are slope and elevation. Slope is the amount of elevation change over a distance. There are two methods of expressing a slope: as a slope or as a grade. A slope of 3:1 represents a rise of 1 foot over an offset distance of 3 feet. A slope of 4:–1 represents a fall of 1 foot over an offset distance of 4 feet. A grade is the rise and fall of elevation expressed as a percentage. So, a 3:1 slope is a 33 percent grade and a 4:–1 slope is a –25 percent grade.

A slope surface style groups (ranges) triangles by their amount of slope (i.e., 0–2 percent, 2–4 percent, etc.). Each range has a color and all triangles in the range display that color. For example, the range of 0 to 2 percent has red triangles and the range of 2 to 4 percent has yellow triangles.

A slope arrow is another useful slope-analysis tool. A slope arrow displays not only the amount of slope (shows the color of its range), but also shows the direction of down slope. When combined, a slope analysis style displays triangles and slope arrows together.

Slope arrows provide more information than just magnitude of and direction of down slope. When slope arrows point toward each other, they represent a valley or swale (see Figure 4.72). If the slope arrows point away from each other, they represent a high spot (see Figure 4.73). Slope arrows also show the direction of water flow across a surface.

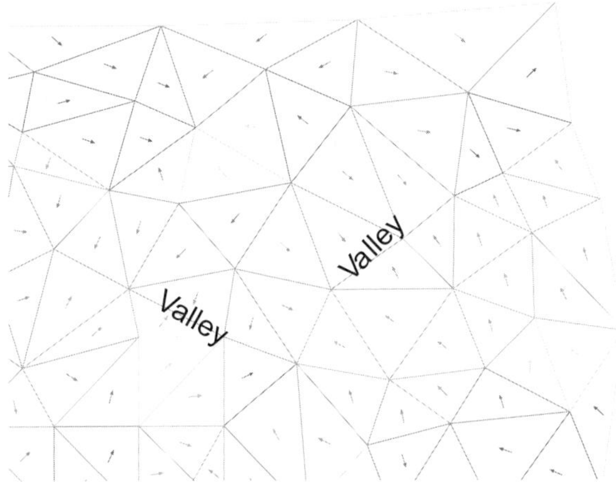


**FIGURE 4.72**

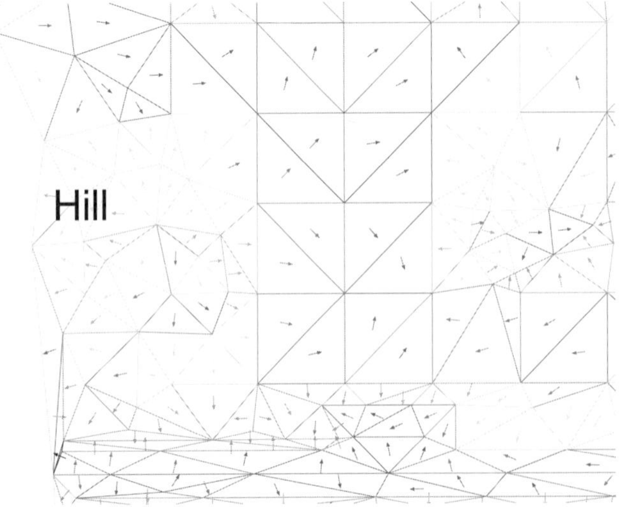


**FIGURE 4.73**

The last surface review styles analyze surface elevations (relief) grouped by elevation. Like slope styles, elevation styles assign colors to each elevation range and each range assigns its color to its triangles. When viewing an elevation style in 3D, the user views the surface's relief.

## Ranging Methods

There are three grouping or ranging methods for slope and elevation data: equal interval, quantile, and standard deviation.

This chapter's surface has a wall occupying a small surface area, but it greatly affects the surface slope statistics. The site has lesser grades and slopes than the wall. The wall slopes are so extreme they bias or skew the statistics to the higher-end values. This was apparent when comparing the slopes before and after adding the wall. The average slope jumped from 6.2 to 225 percent and the maximum slope jumped from 210 to 350,000 percent.

### Equal Interval

Equal interval divides the overall range difference (maximum minus minimum) by a user-specified number of ranges. This method often over generalizes the data. For example, the exercise surface has grades varying from 0.0 to 350,000 percent. Using 10 ranging groups, each range would represent a slope range of 35,000 percent (i.e., 0–35,000, 35,000–70,000, 70,000–105,000, etc.). When evaluating surface slopes using this interval, 0–35,000 for the first range, the range would contain 99 percent of the surface. A skewed data distribution makes the Equal Interval ranging process useless.

The equal interval method is excellent for data that is not skewed. Again, the exercise surface has elevations ranging from 725 to 739 with an average of 732. The average is almost halfway between the minimum and maximum. The equal interval method works well with this elevation data.

### Quantile

The second method is quantile ranging. This method divides the data so that each range interval contains an equal number of members. This methodology focuses on the more frequently occurring values and deemphasizes the few high or low values. With this method, the few high values are put in a range containing lower values, creating a range that has the same number of members as the ranges for the more frequently occurring lower slopes. For example, using the slopes from the exercise surface, this method makes the last slope range contain slopes from 20 to 350,000. This range has the same number of triangles as does the range from 0 to 1.5 percent. This ranging method correctly ranges a skewed slope distribution.

### Standard Deviation

Last is standard deviation. This method calculates and divides data based on how far values differ from the arithmetic mean. This method is most effective when the values approximate a normal distribution (bell-shaped curve). The standard deviation method is used to highlight how far above or below a specific value is in relation to the mean value. So the best use of this method is to show areas of highest and lowest values.

### Watersheds

Surface analysis includes defining surface watersheds. This analysis results in a map representing water collection and discharge areas. Civil 3D creates depressions, single and multiple discharge points watersheds, and hatches each classification with a hatch pattern.

## ASSIGNING AND PROCESSING ANALYSIS STYLES

Viewing an analysis style's effect requires two steps: assign the style to a surface and in the Surface Properties' Analysis tab create the analysis using the style's parameters (see Figure 4.74).

The Analysis panel shows the style's default values. Before running the analysis, the user can change the number of ranges and Legend type. After running the analysis, the user can change the individual range values. When changing the range values, change them from the highest range to the lowest (range 8, then 7, etc.). After changing the values, click Apply to reprocess the data with the new range settings.

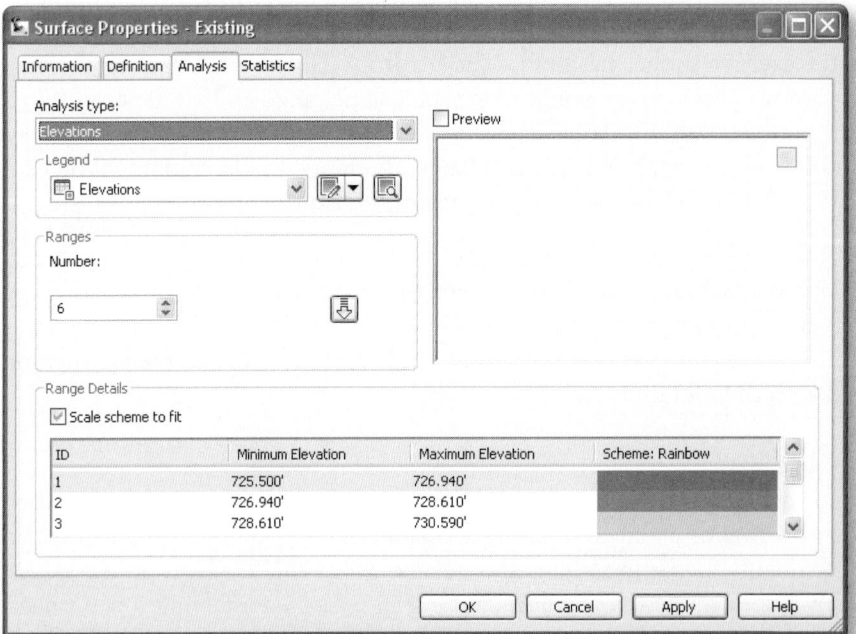


**FIGURE 4.74**

## WATER DROP ANALYSIS

The surface slopes determine and show water flow across a surface. However, the path any one drop takes can be calculated. The Ribbon's Analyze tab's Water Drop command calculates a water drop's surface path (see Figure 4.75).

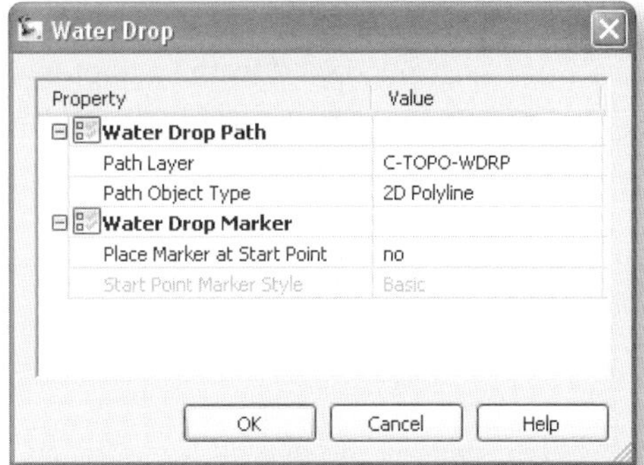


**FIGURE 4.75**

## CATCHMENT AREA

The Ribbon's Analyze tab's Catchment Area command defines a region with a depression low point (catchment point), delineates it with a boundary, and calculates its area. The Catchment Area dialog box sets the drafting properties for a catchment (see Figure 4.76).

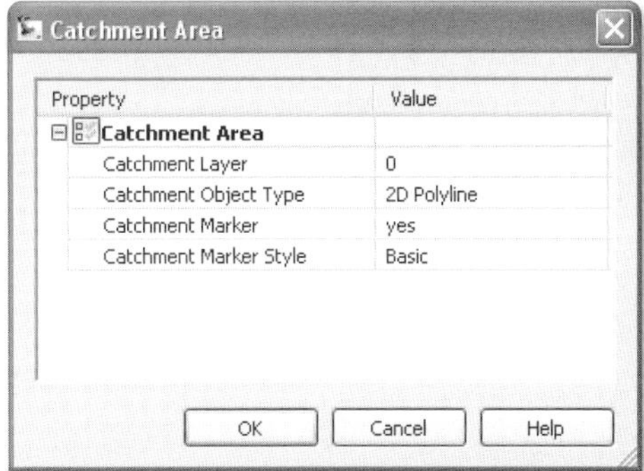


**FIGURE 4.76**

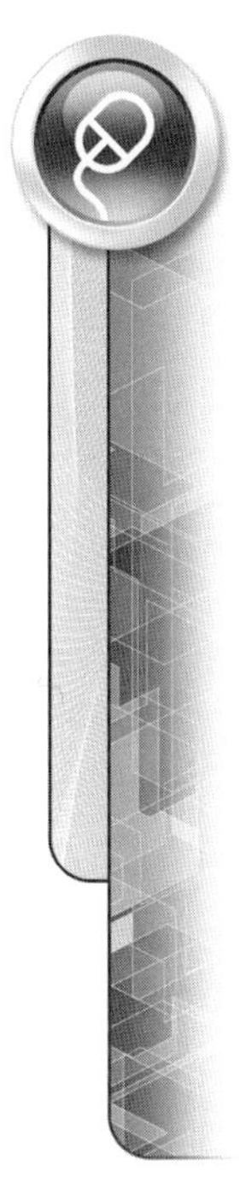

## EXERCISE 4-4

After completing this exercise, you will:

- Define surface slopes and elevations analysis styles.
- Define surface styles using Equal Interval, Quantile, and Standard Deviation parameters.
- Define a watershed surface style.
- Assign and process style settings in the Surface Properties' Analysis tab.
- Create water drop paths across a surface.

This unit's exercise continues with the Unit 3 exercise drawing. If you did not complete the Unit 3 exercise, or starting with this exercise browse to the Chapter 4 folder of the CD that accompanies this textbook and open the Chapter 04 - Unit 4 drawing.

## Exercise Setup

1. If not open, open the previous exercise's drawing or browse to the Chapter 4 folder of the CD that accompanies this textbook and open the *Chapter 04 – Unit 4* drawing.
2. Click the ***Settings*** tab.
3. In Settings, expand the **Surface** branch until you are viewing the Surface Styles list.

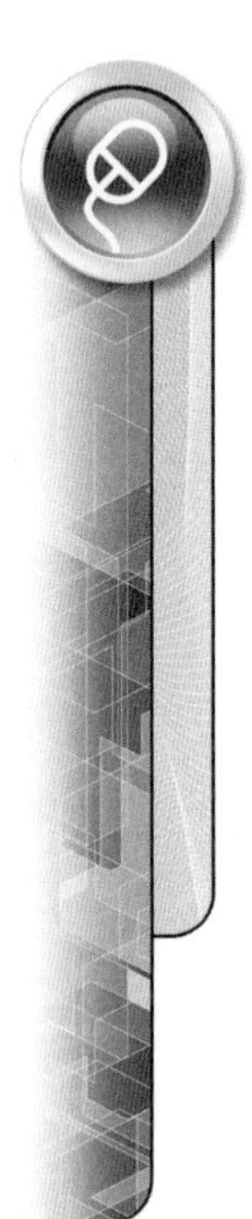

## Slope Surface Styles

The first slope style ranges by Equal Interval. This style divides the overall range by the number of ranges. In the current exercise, the range of slopes is just over 350,000. If a style uses 8 ranges, each range will represent just under 44,000 percent (0–44,000, 44,000–88,000, etc.)

1. From the list of styles select **Slope Banding (2D)**, press the right mouse button, and from the shortcut menu, select COPY....

The Surface style dialog box opens.

2. Click the ***Information*** tab, name the style **Slope and Arrows – Equal Interval**, for the description enter **Slope triangles and arrows by equal interval**, and click ***Apply***.
3. Click the ***Analysis*** tab.
4. Expand the **Slopes** section and adjust your settings to match those in Figure 4.77.
5. Expand the **Slope Arrows** section and adjust your settings to match those in Figure 4.78.
6. Click the ***Display*** tab.
7. At the display list's bottom, turn **ON** the following components: **Border**, **Slopes**, and **Slope Arrows** (see Figure 4.79).
8. Click ***OK*** to exit the dialog box.

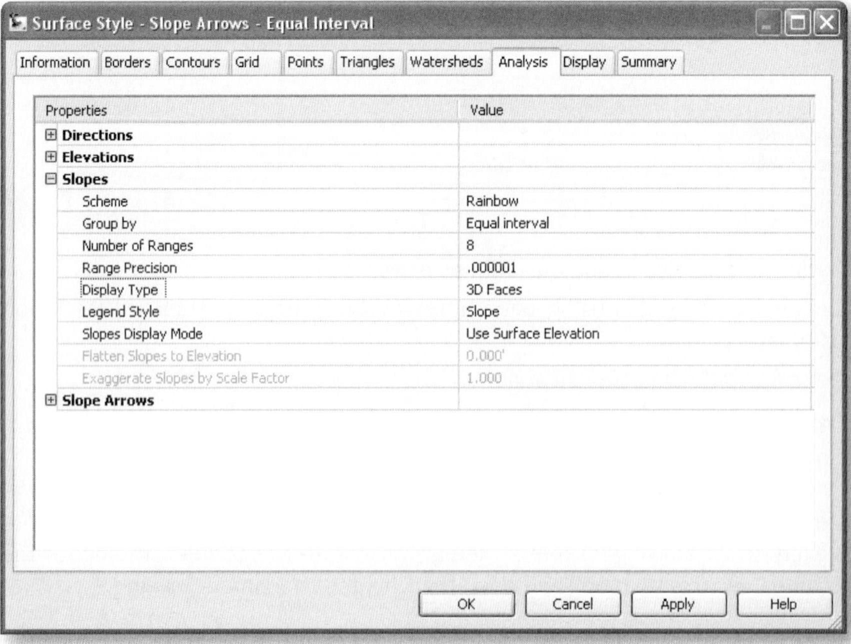

**FIGURE 4.77**

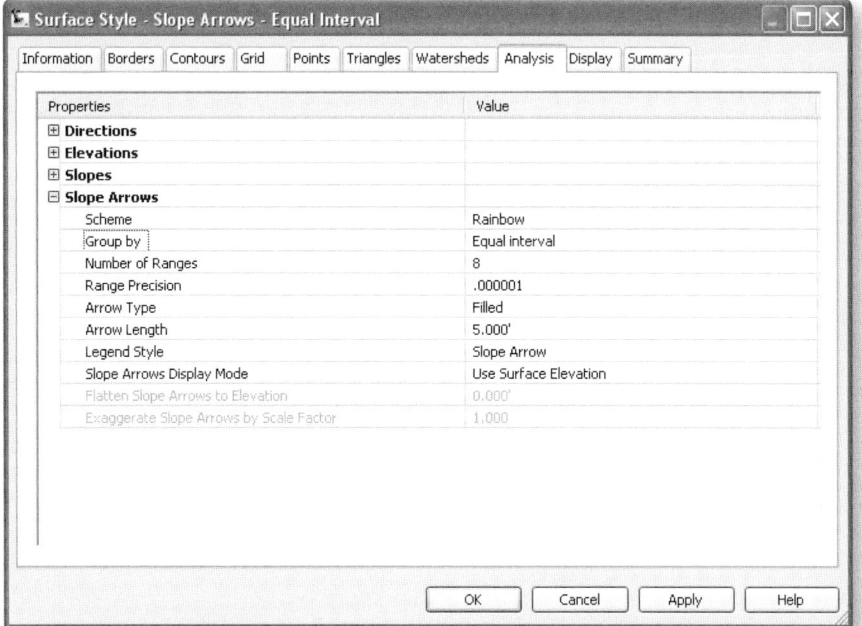


**FIGURE 4.78**

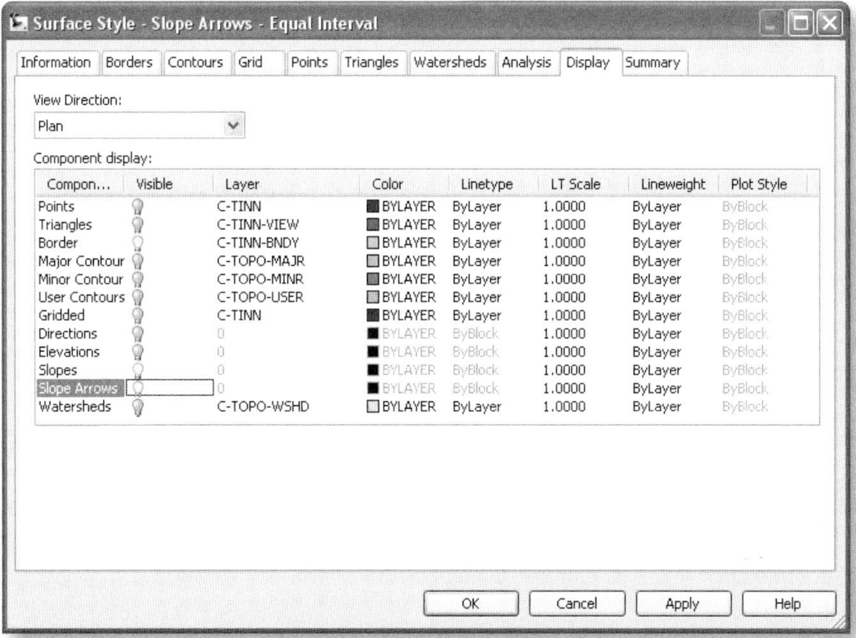


**FIGURE 4.79**

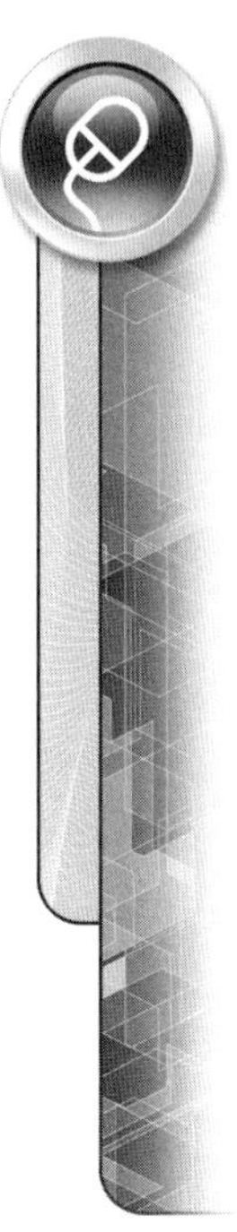

9. Create the remaining two styles, **Slope and Arrows – Quantile** and **Slope and Arrows – Standard Deviation**, by selecting **Slope and Arrows – Equal Interval** from the list of styles, pressing the right mouse button, and selecting `COPY...` from the shortcut menu, and then rename the style and adjust its parameters.

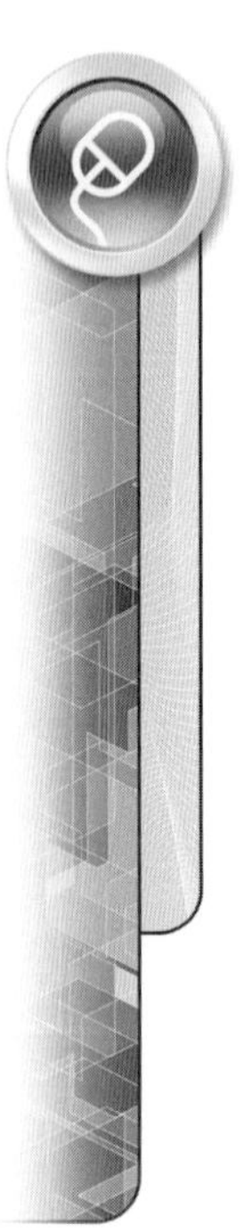

Create the new styles using the following values:

### Slope and Arrows – Quantile

Information tab:

Name: Slope and Arrows – Quantile
Description: Slope triangles and arrows by Quantile.

Analysis tab:

Slope and Slope Arrows: Same settings as Figure 4.77 and 4.78
Group by: Quantile.

Display tab:

Turn on Border, Slopes, and Slope Arrows (see Figure 4.79).

### Slope and Arrows – Standard Deviation:

Information tab:

Name: Slope and Arrows – Standard Deviation
Description: Slope triangles and arrows by Standard Deviation.

Analysis tab:

Slope and Slope Arrows: Same settings as Figure 4.77 and 4.78
Group by: Standard Deviation.

Display tab:

Turn on Border, Slopes, and Slope Arrows (see Figure 4.79).

### Elevation Surface Styles

1. From the list of styles, select **Elevation Banding (2D)**, press the right mouse button, and from the shortcut menu, select COPY....

The Surface Style dialog box opens.

2. Click the ***Information*** tab, for the style's name enter **Elevations – Equal Interval**, and for the description enter **Elevations by Equal Interval**.
3. Click the ***Analysis*** tab.
4. Expand the **Elevations** section and adjust your settings to match those in Figure 4.80.
5. Click the ***Display*** tab, click off all of the components, and click **ON Elevations** (see Figure 4.81).
6. Click the ***OK*** button to exit the dialog box.

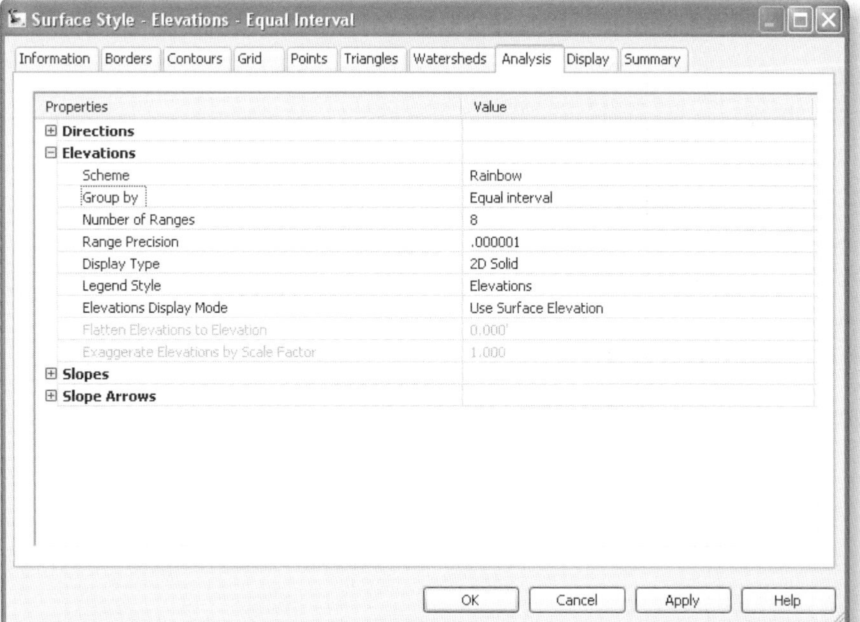


**FIGURE 4.80**

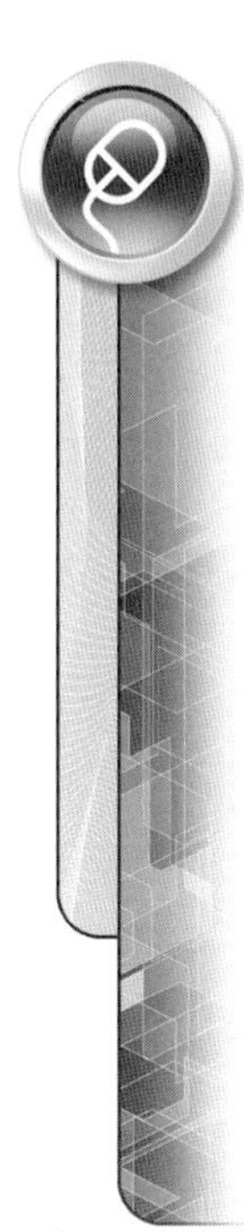

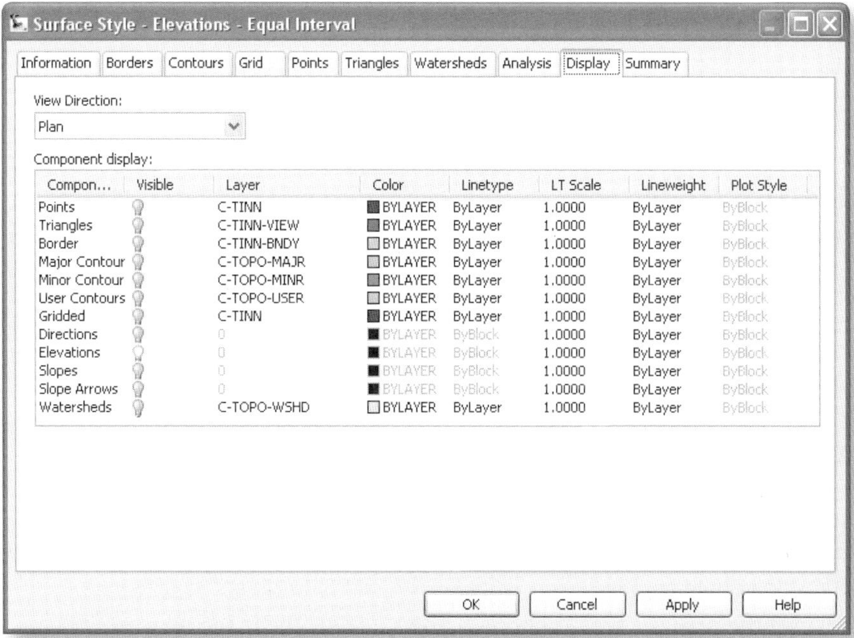


**FIGURE 4.81**

7. Create the remaining two styles, **Elevation – Quantile** and **Elevation – Standard Deviation**, by selecting the **Elevations – Equal Interval** style, pressing the right mouse button, from the shortcut menu, selecting `COPY...`, and renaming the style with its adjusted parameters.

Use the following values to create the styles:

## Elevation – Quantile

Information tab:

Name: Elevations – Quantile
Description: Elevations by Quantile.

Analysis tab:

Set the same settings as Figure 4.80
Group by: Quantile.

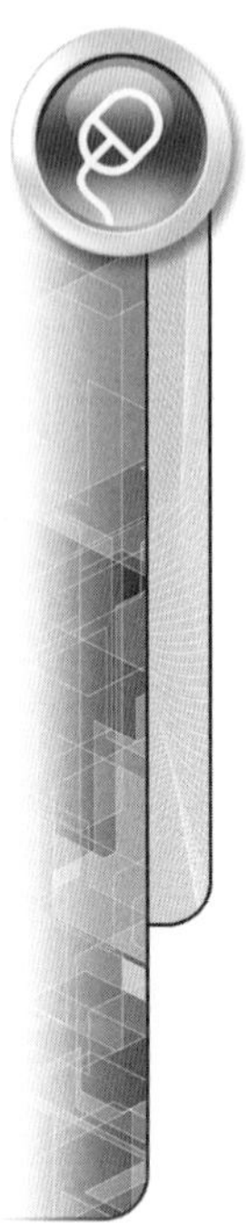

## Elevation – Standard Deviation

Display tab:

Match the settings found in Figure 4.81.
Elevation – Standard Deviation

Information tab:

Name: Elevations – Standard Deviation
Description: Elevations by Standard Deviation.

Analysis tab:

Set the same settings as Figure 4.80
Group: Standard Deviation.

Display tab:

Match the settings found in Figure 4.81.

## Analyzing Surface Slopes

There are two steps to using an analysis style: assign the style and then run the analysis.

1. Click the ***Prospector*** tab.
2. In Prospector, expand **Surfaces** until you are viewing the surface list, from the list select **Existing**, press the right mouse button, and from the shortcut menu, select SURFACE PROPERTIES....
3. Click the ***Information*** tab and change the Surface Style to **Slopes and Arrows – Equal Interval**.
4. Click the ***Analysis*** tab, at the top click the **Analysis type** drop-list arrow, from the list select **Slopes**, review the range number (it should be **8**), and click the **Run Analysis** button (the down arrow to the right of Range number) to produce the slope range details.
5. Change the Analysis type to **Slope Arrows**, review the range number (it should be **8**), and click the **Run Analysis** button (the down arrow to the right of Range number) to produce the arrow range details.
6. Click ***OK*** to view the surface with colorized triangles and arrows.

All of the triangles are red, except for those around the wall. This type of analysis style is not much use with skewed data.

7. Repeat Steps 2–6 and assign and process the **Slope and Arrows – Standard Deviation** for both Slopes and Slope Arrows. Before viewing the new triangles

and arrows, review the range details to see if they better represent the slopes on the surface.

As with the Slopes and Arrows - Equal Interval style, the Slopes and Arrows – Standard Deviation style shows that all but a few of the triangles are red (the first range group). Again, this type of style does not work well with skewed data.

8. Repeat Steps 2–6 and assign and process the **Slopes and Arrows – Quantile** for both Slopes and Slope Arrows. Before viewing the new triangles and arrows, review the range details to see if they better represent the slopes on the surface.

This surface style handles the skewed data better than the Equal Interval or Standard Deviation.

The range details for this style calculate range breaks that fall between integer slope numbers (i.e., 2, 4, etc.). You can edit the range breaks after processing the analysis and adjust each range to a specific minimum and maximum value. The style recalculates the range membership using the new range values.

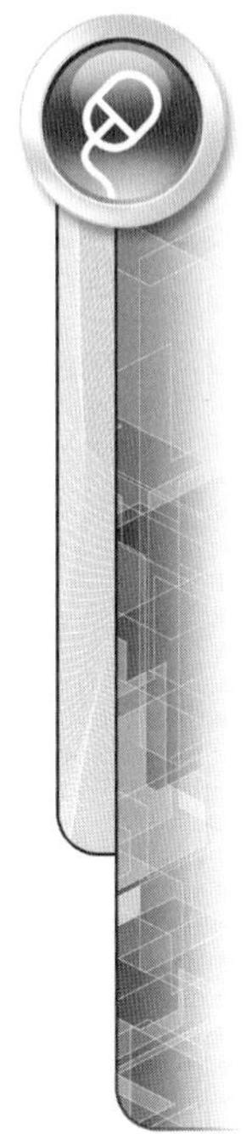

## Modifying Slope Range Values

1. In the drawing, click the surface border or triangle, press the right mouse button, and from the shortcut menu, select SURFACE PROPERTIES....
2. If necessary, click the ***Analysis*** tab.
3. Set the analysis to **Slopes** and in Range Details change the range values to those listed in Table 4.5. Start editing the ranges at the eighth range and work up to the first.

**TABLE 4.5**

| Range ID: | Minimum Slope: | Maximum Slope: |
|---|---|---|
| 1 | 0.0 | 2.0 |
| 2 | 2.0 | 4.0 |
| 3 | 4.0 | 6.0 |
| 4 | 6.0 | 8.0 |
| 5 | 8.0 | 10.0 |
| 6 | 10.0 | 100.0 |
| 7 | 100.00 | 210.00 |
| 8 | 210.00 | 351,000 |

4. After resetting the slope ranges, click the ***Apply*** button to recalculate range membership.
5. Change the analysis to **Slope Arrows** and in the Range Details change the range values to those set in Step 3. Start editing the ranges at the eighth range and work up to the first.
6. After resetting slope arrows for the ranges, click the ***Apply*** button to recalculate the range membership.
7. Click ***OK*** to exit the dialog box and display the range grouping.

## Create a Slope Range Legend

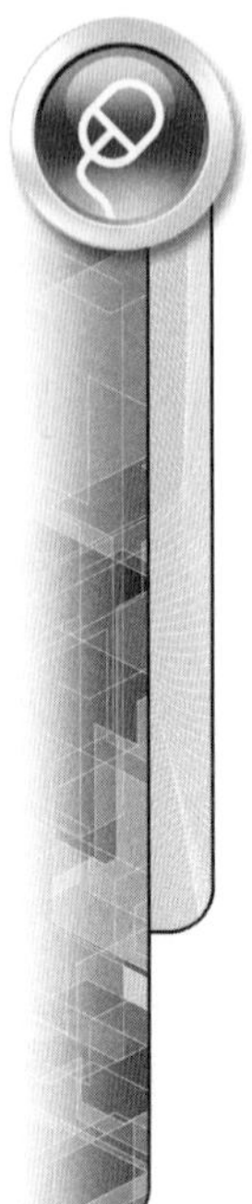

1. On the Ribbon click the Annotate tab. In the Labels & Tables panel, click the Add table drop-list arrow and from the shortcut menu, select, ADD SURFACE LEGEND TABLE. In the command line for the legend type enter '**S**' for slopes, press ENTER, for a dynamic table type '**D**', press ENTER, and select a point to the surface's right to create a legend.
2. Use the Zoom and Pan commands to review the table.
3. Use the ERASE command to erase the slope legend from the drawing.
4. Use the ZOOM EXTENTS command to view the entire site.

## Analyzing Surface Elevations

1. In Prospector, Surfaces branch, select the **Existing** surface, press the right mouse button, and from the shortcut menu, select SURFACE PROPERTIES....
2. Click the ***Information*** tab and change the Surface Style to **Elevations – Equal Interval**.
3. Click the ***Analysis*** tab, change the Analysis type to **Elevations**, check the range number (it should be **8**), and click the **Run Analysis** button (the down arrow to the right of Range number) to produce elevation range details (see Figure 4.82).
4. Click ***OK*** to view the surface with colorized triangles.

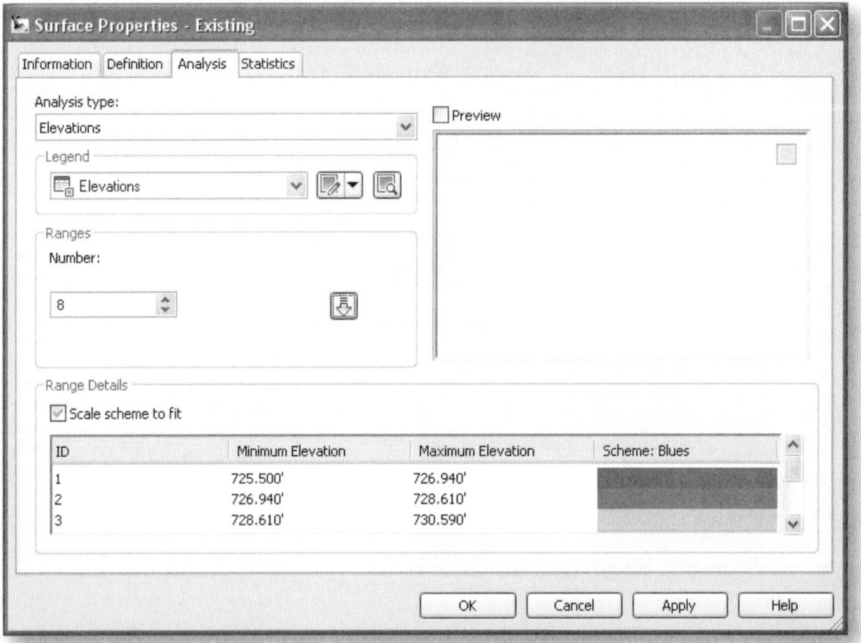


**FIGURE 4.82**

The Elevations – Equal Interval surface style correctly displays the surface elevations.

5. Repeat Steps 1–4 assigning and processing Elevations – Quantile.

Before viewing the display, review the range details to see if they better represent the surface elevations.

The surface elevation map is slightly different from the previous one, but it still displays a good surface elevation map.

6. Repeat Steps 1–4 assigning and processing Elevations – Standard Deviation.
7. At Civil 3D's upper left, Quick Access Toolbar, click the ***Save*** icon to save the drawing.

Before viewing the display, review the range details to see if they better represent the surface elevations.

The surface elevation map is slightly different from the previous one, but it still displays a good surface elevation map.

All three methods correctly represent the elevation information, but in slightly different range values. It would be better if the ranges were at specific whole number elevations (i.e., 725, 728, etc.).

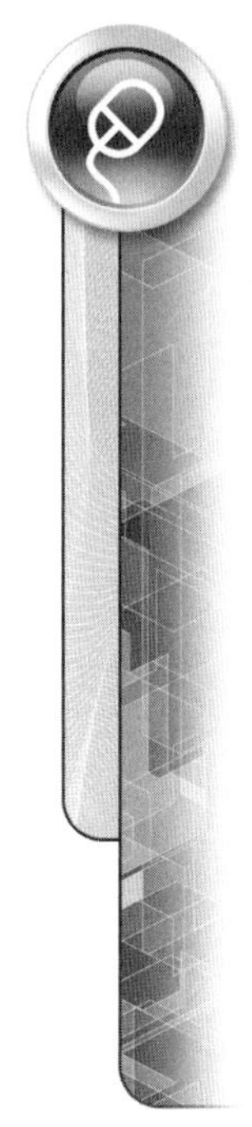

## Modifying Elevation Range Values

1. In Existing's Surface Properties dialog box reassign and reprocess the **Elevation – Equal Interval** surface style.
2. Edit the minimum and/or maximum range values from the last to first elevation range using Table 4.6 as a guide.

**TABLE 4.6**

| Range ID: | Minimum Elevation: | Maximum Elevation: |
|---|---|---|
| 1 | 725.5 | 728.0 |
| 2 | 728.0 | 730.0 |
| 3 | 730.0 | 732.0 |
| 4 | 732.0 | 734.0 |
| 5 | 734.0 | 736.0 |
| 6 | 736.0 | 739.5 |

3. Click ***Apply*** to recalculate the ranges.
4. Click ***OK*** to exit and display a new elevation map.
5. At Civil 3D's upper left, Quick Access Toolbar, click the ***Save*** icon to save the drawing.

All of these styles emphasize and create a 2D elevation map.

## Create a 3D Elevation Surface Style

This surface style creates a 3D elevation view with user-defined elevation ranges.

1. Click the ***Setting*** tab.
2. If necessary, expand the **Surface** branch until you view the Surface Styles list.
3. From the list select **Elevations – Equal Interval**, press the right mouse button, and from the shortcut menu, select `COPY`....
4. Click the ***Information*** tab change the style name to **Elevations – Equal Interval – 3D**.
5. Click the ***Analysis*** tab and expand the **Elevations** section.

6. In the ***Analysis*** tab's Elevations section, change the following values (see Figure 4.83):

   Scheme: Rainbow
   Number of Ranges: 6
   Display Type: 3D Faces
   Elevations Display Mode: Exaggerate Elevation
   Exaggerate Elevation by Scale Factor: 3

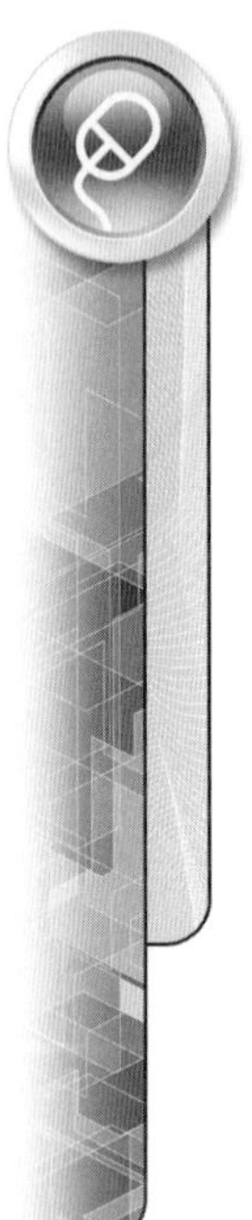

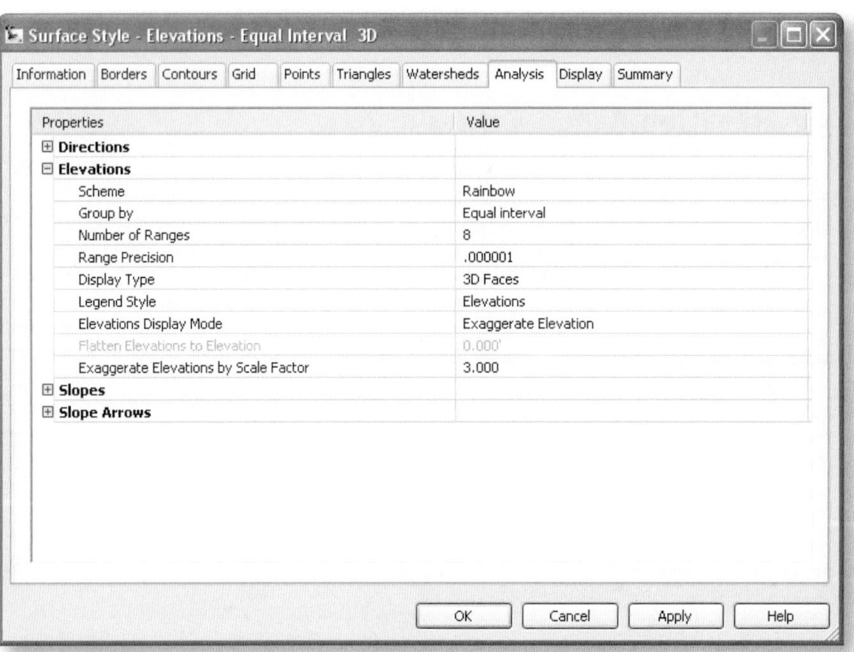


**FIGURE 4.83**

7. Click the ***Display*** tab, at the top left change the View Direction: to **Model**, toggle **OFF** all of the components, and toggle **ON** only **Elevations**.
8. Click ***OK*** to exit the dialog box.
9. In the drawing, select anything representing the surface, press the right mouse button, and from the shortcut menu, select SURFACE PROPERTIES....
10. Click the ***Information*** tab and change the Surface style to **Elevations – Equal Interval – 3D**.
11. Click the ***Analysis*** tab, set the Analysis type to **Elevations**, make sure the range number is **5**, click the **Run Analysis** button (the down arrow to the right of the Range number), and in the Range Details area review the elevation range values.
12. Edit the range minimums and maximums to the values in Table 4.7. Start the editing at the highest range and work down to the lowest range.

**TABLE 4.7**

| Range ID: | Minimum Elevation: | Maximum Elevation: |
|---|---|---|
| 1 | 725.5 | 728.0 |
| 2 | 728.0 | 730.0 |
| 3 | 730.0 | 732.0 |
| 4 | 732.0 | 734.0 |
| 5 | 734.0 | 736.0 |
| 6 | 736.0 | 739.5 |

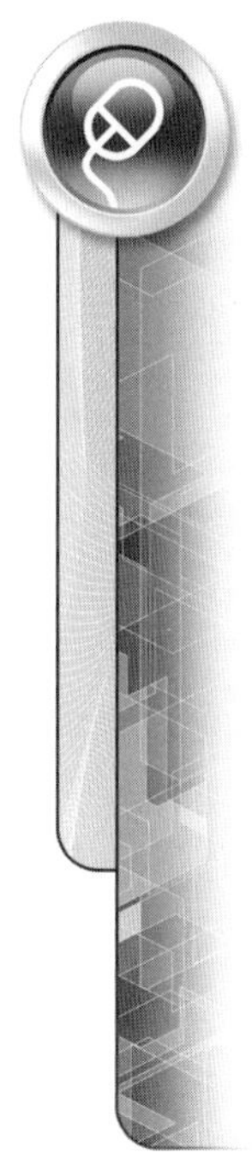

13. Click ***OK*** to exit the dialog box.
14. In the drawing select anything representing the surface, press the right mouse button, and from the shortcut menu, select OBJECT VIEWER....
15. In the Object Viewer, toggle shading to **Realistic** and view the surface from several locations.
16. Exit the Object Viewer.
17. At Civil 3D's upper left, Quick Access Toolbar, click the ***Save*** icon to save the drawing.

## Watershed Surface Style

1. If necessary, click the ***Settings*** tab.
2. If necessary, expand **Surface** until you are viewing the Surface Styles list.
3. From the list select **Elevations – Equal Interval**, press the right mouse button, and from the shortcut menu, select COPY....
4. Click the ***Information*** tab, rename the style **Watersheds**, and for the description enter **Watershed**.
5. Click the ***Watersheds*** tab.
6. Expand and review the watershed sections.
7. Expand the **Surface** section, click in the Value cell, click the ellipsis, and in the Surface Watershed Label Styles dialog box, click the drop-list arrow, and select **Watershed**. Click ***OK*** to exit the dialog box and return to the Surface Style dialog box.
8. Click the ***Display*** tab, scroll to the list's bottom and only toggle **ON Watersheds** and turn off the remaining components.
9. Click ***OK*** to create the style and exit the dialog box.

## Watershed Analysis

1. In the drawing select anything representing the surface, press the right mouse button, and from the shortcut menu, select SURFACE PROPERTIES....
2. Click the ***Information*** tab and set the Surface Style to **Watersheds**.
3. Click the ***Analysis*** tab and adjust your settings to those in Figure 4.84.
4. Click the **Run Analysis** button (the down arrow to the right of Minimum Average Depth number).
5. Click ***OK*** to exit the dialog box and view the watershed analysis.
6. At Civil 3D's upper left, Quick Access Toolbar, click the ***Save*** icon to save the drawing.

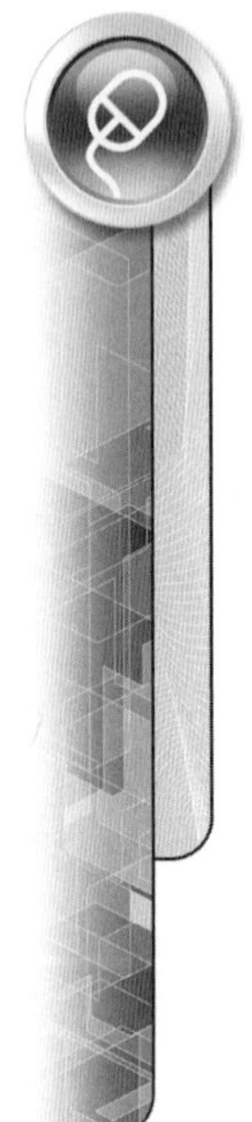

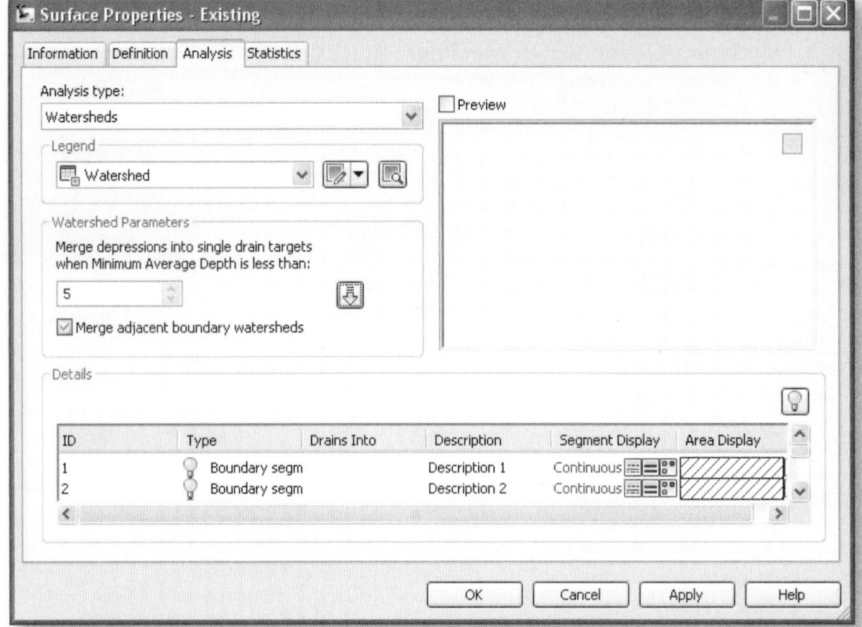


**FIGURE 4.84**

7. Click the ***Prospector*** tab.
8. In Prospector, Surfaces branch, expand **Existing** until you are viewing Masks and Watersheds.
9. Click the **Watersheds** heading and preview the lists of the Existing surface's watersheds.
10. From Preview, select a watershed, press the right mouse button, and from the shortcut menu, select ZOOM TO.
11. Select a couple more watersheds and use the ZOOM TO command to view their drawing location.

## Water Drop Analysis

Water Drop draws water's path across a surface. This path may be a critical review of existing and design conditions.

1. Use the ZOOM EXTENTS command to view the entire site.
2. On the Ribbon, click the Analyze tab. In the Ground Data panel, click ***Water Drop***.
3. In the Water Drop dialog box, click ***OK*** to accept the values and exit the dialog box.
4. Select points in different watersheds and view their water trails over the surface. When done, press ENTER to exit the command.
5. Open **Layer Properties Manager**, freeze the layer **C-TOPO-WDRP**, and click **X** to exit the panel.
6. At Civil 3D's upper left, Quick Access Toolbar, click the ***Save*** icon to save the drawing.

## Surface Elevation – 3D

1. In the drawing, select the surface border, press the right mouse button, and from the shortcut menu, select *SURFACE PROPERTIES....*
2. Click the ***Information*** tab and set the surface style to **Elevation – Equal Interval – 3D**.
3. Click the ***Analysis*** tab and set the analysis type to **Elevations**.
4. Set the range number to **6**, click the **Run Analysis** icon, and click ***OK*** to exit the dialog box.
5. In the drawing, select the surface, press the right mouse button, and from the shortcut menu select OBJECT VIEWER.
6. If necessary, set Object Viewer's mode to **Realistic** and review the surface as a model.
7. Exit the Object viewer.
8. At Civil 3D's upper left, Quick Access Toolbar, click the ***Save*** icon to save the drawing.

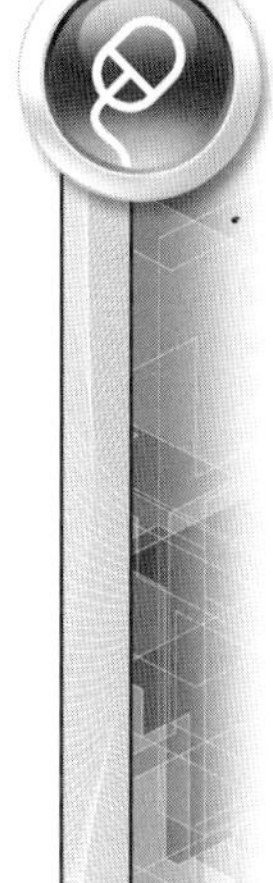

If the current surface components and characteristics are acceptable, then it is time to create the contours that represent the surface elevations.

## SUMMARY

- Slope analysis should include triangles and slope arrows.
- Slope analysis represents the flow of water across a surface.
- Water Drop analysis draws the path of individual water drops as they travel over a surface.
- The Equal Interval and Standard Deviation range methods work best with normally distributed data.
- The Quantile range method works best with skewed data.

## UNIT 5: SURFACE ANNOTATION

Surface annotation consists of contour, spot slope, and/or elevation labels. Contours display a surface's elevation characteristic (hills, valleys, depressions, etc). A contour is a line that has a constant elevation along its entire length. Thus, a contour connects points that have the same surface elevation. There is a dynamic link between a surface and its contours. If a surface changes, it recalculates its triangulation and redraws the contours. How contours are displayed is an effect of a surface object style.

**NOTE**

**If you turn off the surface layer, all surface labels will also turn off. To show surface labels and have the surface not display, assign the _No Display style to the surface.**

### STEP 7: ANNOTATE THE SURFACE

After completing surface editing and analysis, you next create contours and annotate elevations. Additional label styles identify spot slope and elevation at critical surface points.

### Contour Surface Style

You display contours from a surface style assignment containing the appropriate contour settings. A contour style uses three Surface Style tabs: Information, Contours, and Display.

#### Information Tab

The Information panel sets the name and describes the contour style.

#### Contours Tab

The 3D Geometry section contains settings that affect the contours' makeup. This group's settings control the contours' 3D display mode. By default, a contour displays itself as its actual elevation, flattened to an elevation, or as exaggerated elevation showing relief. The default is to display contours at the actual elevation. When flattening contours, specify the flatten elevation. An exaggeration factor exaggerates contours (see Figure 4.85).

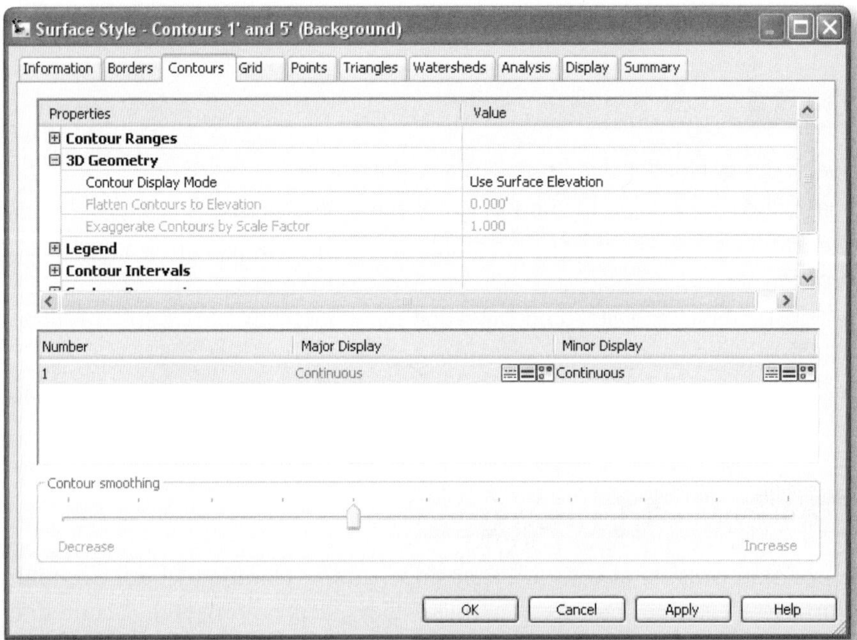


**FIGURE 4.85**

**Contour Interval.** Contour interval sets the major and minor contour interval. The major interval is a multiple of minor (usually a factor of 5). For example, if the minor interval is 0.5, the major is 2.5; if the minor is 1, the major is 5; if the minor is 2, the major is 10, etc. When setting a minor interval, the major value responds with a value five times larger. Always check these settings to make sure they are correct (see Figure 4.86).

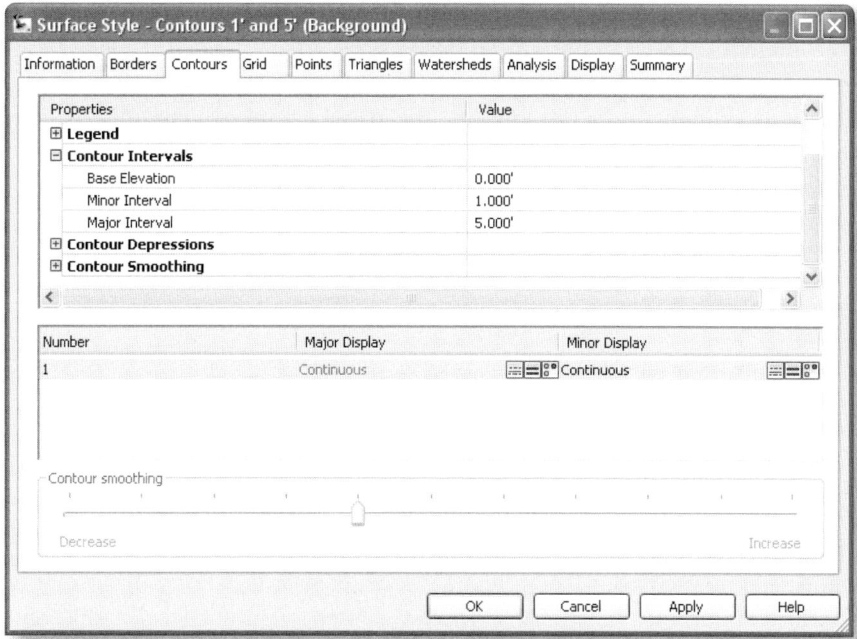


**FIGURE 4.86**

**Contour Ranges.** This section sets contour range groupings. Creating contour ranges assigns different colors to contours. For example, the contour range of 700–730 is blue, 730–760 red, etc. These settings are related to a surface elevations analysis and when using User Contours.

**Depression Contours.** Contour Depression settings set the contour's tick interval (frequency) and size (see Figure 4.87).

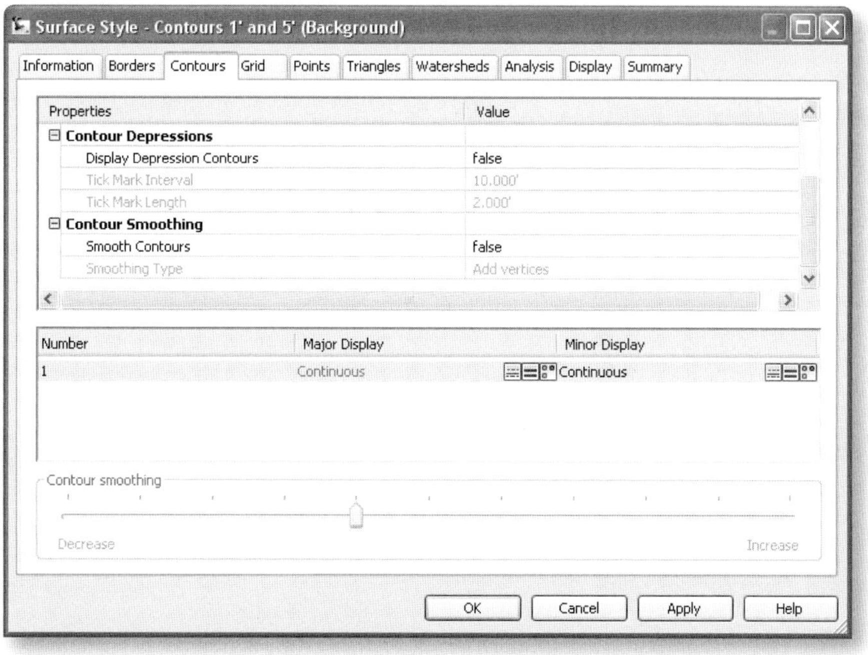

**FIGURE 4.87**

**Contour Smoothing.** Contour Smoothing does not add or modify surface data; it modifies the contours. The two smoothing options are: add vertices and spline. Add vertices smoothes the surface by adding vertices, but preserves as best as possible the surface elevations. When using this option, the panel's bottom contains a slider, which sets the smoothing amount (see Figure 4.87). Spline creates the greatest smoothness by splining the contour. This option also creates the greatest number of crossing contours.

The preferred method of contour smoothing is the edit surface smoothing. See the following discussion.

### Contour Labels

Add Surface Labels annotates contours, spot elevations, and slopes (see Figure 4.88). Contours, spot elevations, and slopes are surface label types and each has an entry under Label Styles in the Settings Surface branch.

Contour labels reflect two basic label strategies; major/minor or existing/proposed. Major/minor assumes a contour label's focus is a major or minor contour no matter what the surface represents. Existing/Proposed assumes a contour label is different when labeling an existing or proposed surface. When implementing Civil 3D, you will have to decide what your office's strategy is. In each label style you can assign a layer or by placing 0 (zero) in the layer name, the label routines will use Edit Drawing Settings, Object Layers' defined layer name.

Surface's Edit Feature Settings set the Add Labels dialog box and Add Surface Labels flyout's default label styles (see Figure 4.89). The Default Styles section sets the Spot Elevation and Slope label style and the Contour Labeling Defaults set default contour label styles.

All contour labels have a label line, and by default it is on. It should be set to false, because when selecting a label the line's grips are displayed (see Figure 4.89). You can select a line grip, stretch and relocate it, and create new labels where contours intersect the line.

**FIGURE 4.88**

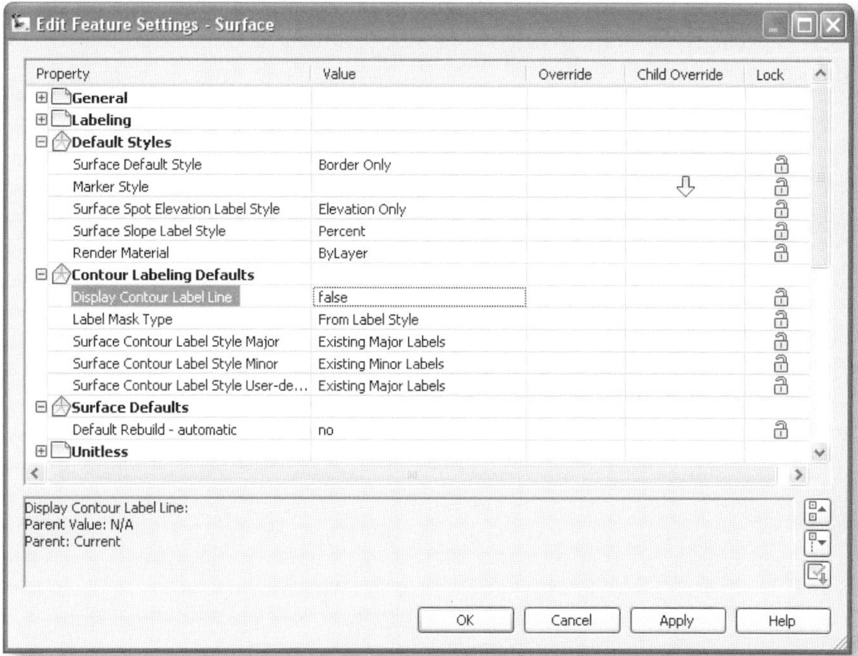


**FIGURE 4.89**

Labeling contours is done by individual, multiple, or multiple at interval. Individual creates a label where a contour is selected. Multiple prompts for a beginning and end of a line, and where the line intersects contours a label appears. If you move the relocating line, the labels shift to the new intersections. Multiple at interval also prompts for a beginning and end of a line and after you select the points, it prompts for a label repeat distance (see Figure 4.90).

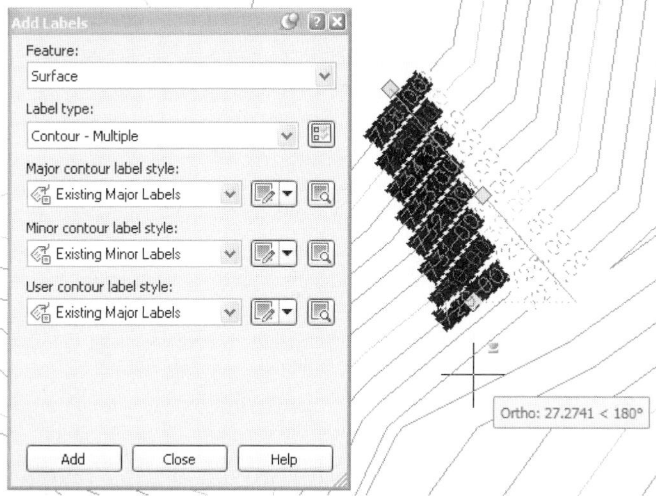


**FIGURE 4.90**

## Spot Slope Labels

Spot Slope labels annotate a surface's slope at the selected point (see Figure 4.91). The best method of placing these labels is with a single pick. When labeling the slope between two points, be careful not to select two points that have several triangles between them. When selecting two points, the routine uses the difference in elevation between the two points to calculate a slope, not the changes that may occur between them.

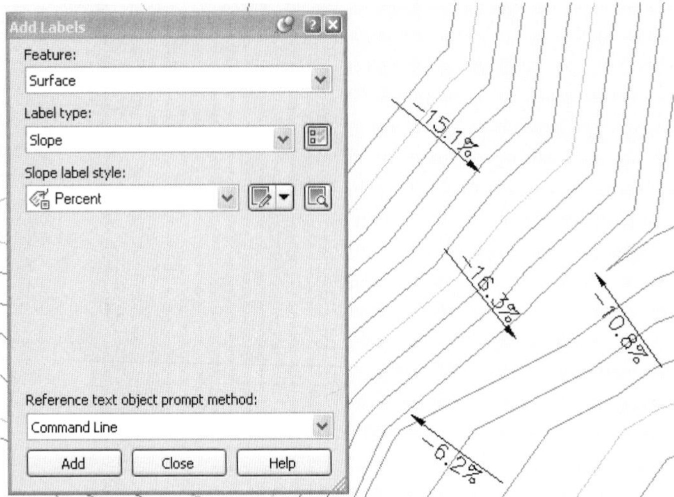


**FIGURE 4.91**

## Spot Elevation Labels

Spot Elevation labels annotate elevations at the selected points (see Figure 4.92).

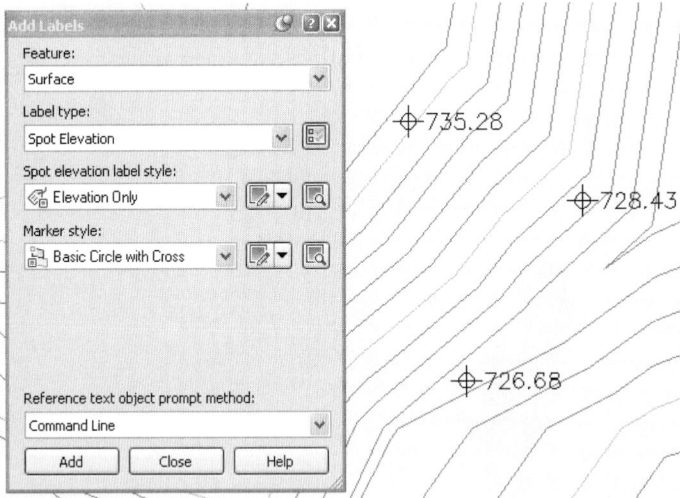


**FIGURE 4.92**

### LABEL GRIPS

When a slope or spot elevation label is selected, it displays two initial grips: relocate the elevation point or drag the label to a new location (see Figure 4.93). The relocate elevation point grip is the diamond grip and the relocate label is the square grip. When the label is dragged to select the label, it shows a dot grip. Clicking the dot grip resets the label to its original position. The plus grip at the midpoint of the leader allows you to add as many vertices to the leader.

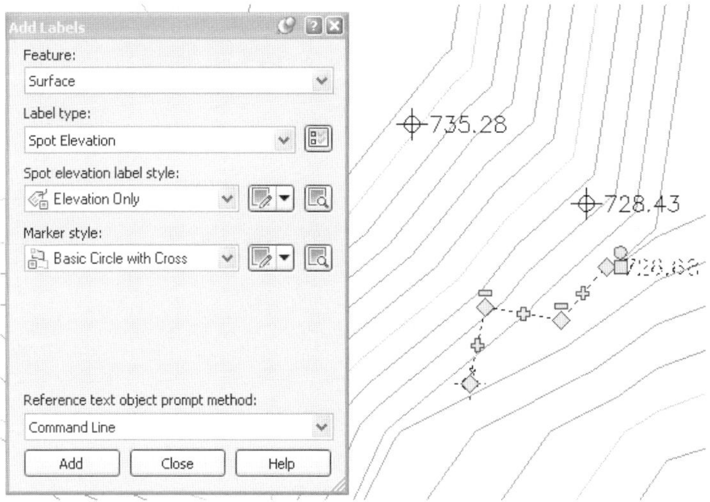


**FIGURE 4.93**

## SURFACE SMOOTHING

Surface smoothing is a surface edit and its options interpolate for a selected region's new surface data. This option is best for areas with great relief over short distances or when there is sparse data. After reviewing the initial surface contours, the user may decide to use smoothing to create more pleasing contours. See the discussion of smoothing in Unit 3 of this chapter.

## EXPRESSIONS

Expressions are user-defined surface properties a label can use. Each label type has its own set of expressions; they label different surface properties. An example expression is top-of-face of curb elevation. The gutter's elevation is from a surface and an expression defines the top-of-face elevation as a surface property.

The expression dialog box presents a calculator-like interface that has access to the label type's object properties and mathematic functions. Once defined, the user assigns a name and description to the expression and when exiting, the named expression becomes a label property (see Figure 4.94).

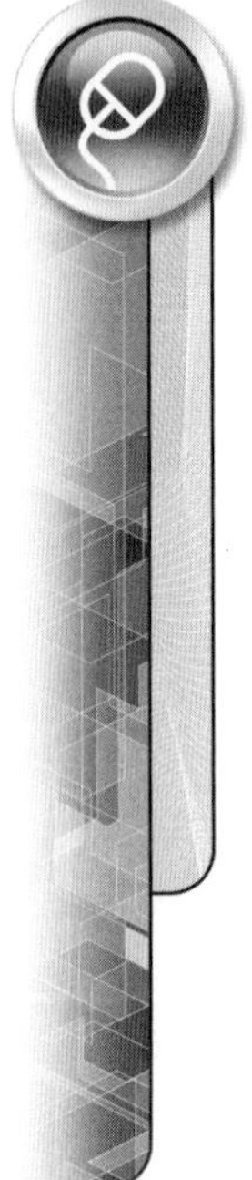

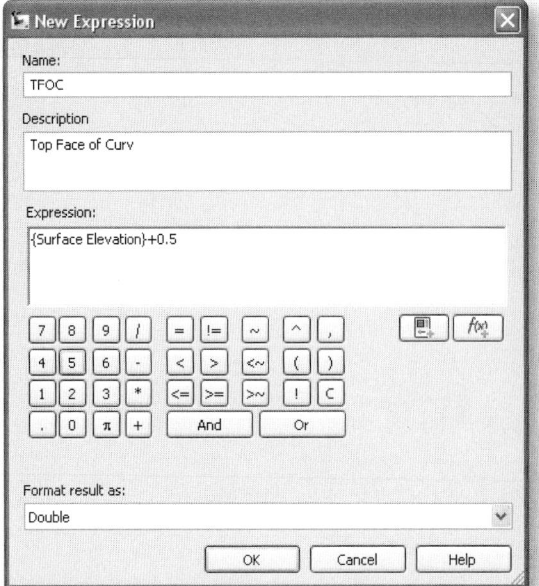


**FIGURE 4.94**

Component Editor lists the named expression and other properties as properties for a label component (see Figure 4.95).

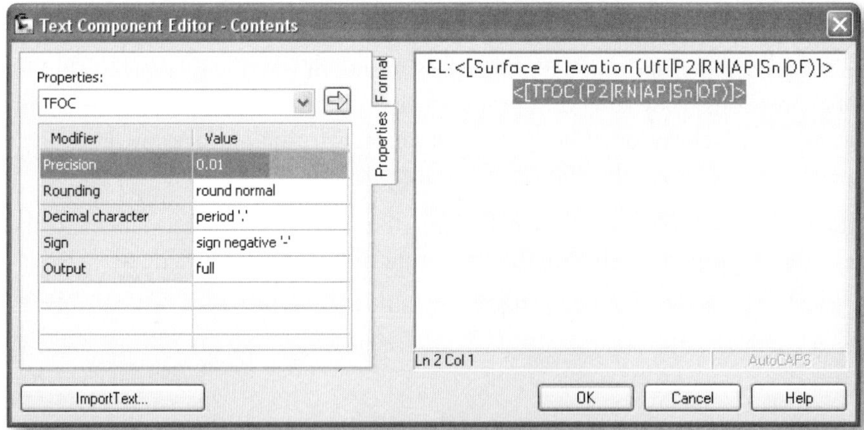


**FIGURE 4.95**

## EXERCISE 4-5

After completing this exercise, you will:

- Be able to modify a contour style.
- Be able to create contour labels.
- Be able to use Natural Neighbor smoothing on a surface.
- Be able to create slope labels.
- Be able to create spot elevation labels.

This exercise continues with the previous unit's exercise drawing. If you did not complete the Unit 4 exercise, or starting with this exercise browse to the Chapter 4 folder of the CD that accompanies this textbook and open the Chapter 04 - Unit 5 drawing.

## Exercise Setup

1. If not open, open the previous exercise's drawing or browse to the Chapter 4 folder of the CD that accompanies this textbook and open the *Chapter 04 – Unit 5* drawing.

## Contour Surface Style

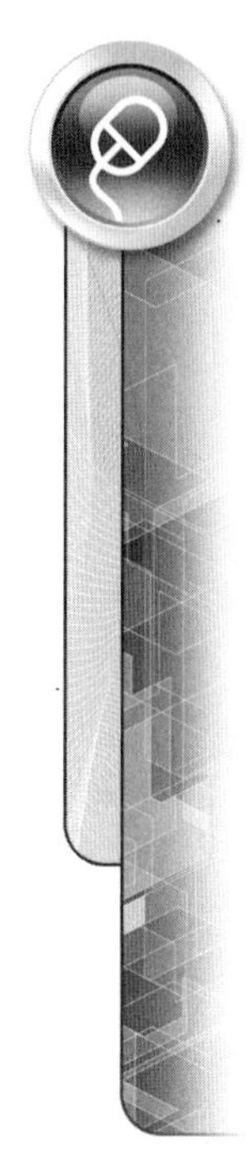

1. Click the ***Settings*** tab.
2. In Settings, expand the **Surface** branch until you view the Surface Styles list.
3. Select the style **Contours 1' and 5' (Background)**, press the right mouse button, and from the shortcut menu, select EDIT....
4. Click the ***Contours*** tab and review the settings for section 3D Geometry, Contour Display Mode.
5. Expand the **Contour Intervals** section and review its values.
6. Expand the **Contour Depressions** section and toggle Display Depression Contours to **True**.
7. Click ***OK*** to exit the dialog box.
8. In Settings, expand the **Surface** branch until you view a list of Contour Label styles.
9. Select the **Existing Major Labels** style, press the right mouse button, and select EDIT... to display the Label Style Composer.
10. Click the ***General*** tab.

The ***General*** tab sets the label's visibility, layer, insertion orientation, and plan readability.

11. Click the ***Dragged State*** tab to review its settings.

The Dragged State settings control how a label reacts if it is dragged from its original location.

12. Click the ***Layout*** tab to review its settings.

The Layout settings define and format the label. The panel's General section defines how the label attaches to a contour, the Text section defines the text's format and its location relative to the contour, and the Border section, if used, defines a box surrounding the label.

13. In the **Text** section, click in the **Contents** value cell, then click the cell's ellipsis to display the Text Component Editor – Contents dialog box (see Figure 4.96).

To change a modifier's value, on the dialog box's right side highlight the format string, on the left side click once in a Value cell, and from the list select the new value. To transfer the new value from the left to the highlighted format on the right, click the arrow at the top center of the dialog box.

14. Click ***Cancel*** until you have exited the dialog boxes and returned to the command line.

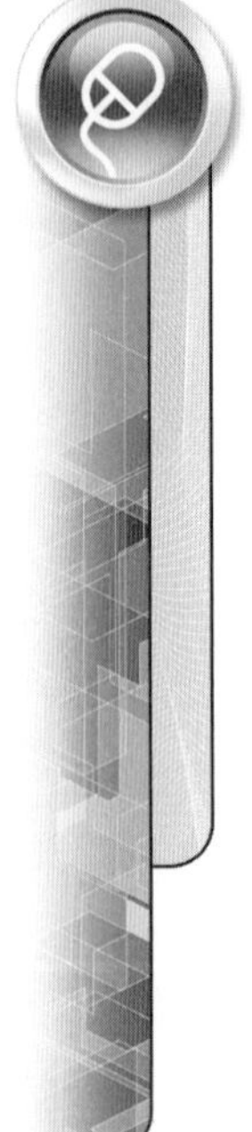

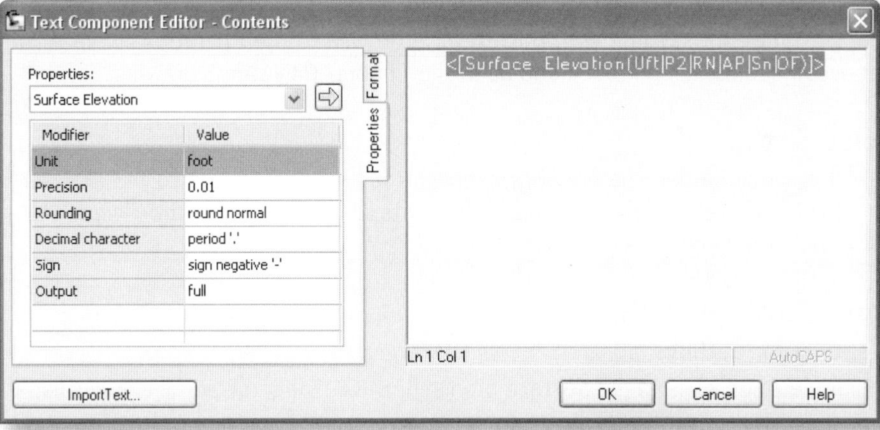


**FIGURE 4.96**

## Creating and Labeling Contours

Create surface contours by assigning a contour surface style.

1. Click the ***Prospector*** tab.
2. In Prospector, expand the **Surfaces** branch until you are viewing the surface list, from the list select **Existing**, press the right mouse button, and from the shortcut menu, select SURFACE PROPERTIES....
3. Click the ***Information*** tab, change the Surface Style to **Contours 1' and 5' (Background)**, and click ***OK*** to exit the dialog box.

Assigning this style makes the surface display elevations as contours.

4. In the Ribbon, select the ***Annotate*** tab. In the Labels & Tables panel's left, click the Add Labels drop-list arrow, select Surface, and in its flyout, select ADD SURFACE LABELS to display the Add Labels dialog box.
5. Click the **Label Type's** drop-list arrow and from the list, select **Contour - Single**.

The dialog box changes to show entries for Major, Minor, and User contour label styles. These styles are from the Edit Feature Settings values.

6. Click ***Add*** and in the drawing select a few contours where there is to be a label. When you are done, press ENTER to exit the routine.

Each contour type (major or minor) receives the correct label type.

7. Use the ZOOM command to better view a label.
8. Click a contour label to view its grips.

A label has a text grip and two line grips.

9. Select one of the grips and drag the line over another contour until you are viewing a new contour label.

A single contour label can become a multiple contour label.

10. Press ESC to remove the grips.
11. Select a label, right mouse click, and from the shortcut menu, select PROPERTIES....
12. In the Properties palette, Labels section, if necessary, set Display Contour Label Line to **false**, and click **X** to close the palette.
13. Press ESC to remove the grips and hide the contour label line.

Selecting a label displays the hidden label line.

14. Use the ERASE command and erase the labels.

## Contour Label – Multiple

1. In the Add Labels dialog box, click the **Label Type** drop-list arrow, and from the list, select **Contour – Multiple**.
2. Click ***Add***, in the drawing select a beginning and ending point of a line intersecting contours, and press ENTER to create the labels.

A label is placed at each intersection of the line and a contour.

3. Select a label and stretch the line so it crosses additional contours.

New labels appear at the line intersections.

4. Erase the labels.

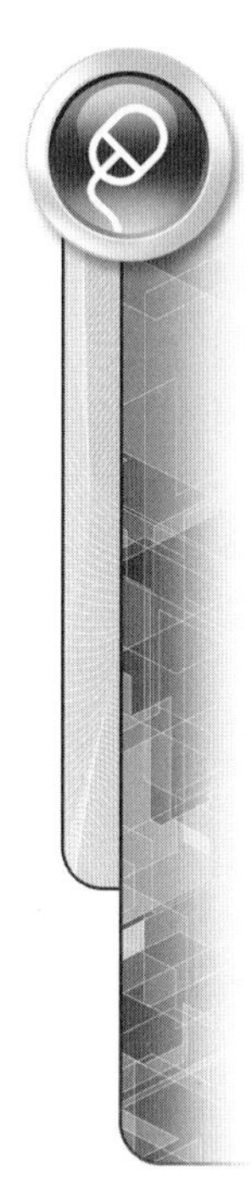

## Contour Label – Multiple at Interval

1. In the Add Labels dialog box, click the **Label Type** drop-list arrow, and from the list, select CONTOUR - MULTIPLE AT INTERVAL.
2. Click ***Add***; in the drawing select a beginning and ending point of a line intersecting contours, in the command line for the interval enter **300**, and press ENTER to create the labels.
3. Use ZOOM command to view the new labels.
4. Activate the grips of some labels and relocate them in the drawing.
5. Close the **Add Labels** dialog box.
6. Use the UNDO command and remove the labels from the drawing.
7. At Civil 3D's top left, Quick Access Toolbar, click the ***SAVE*** icon to save the drawing.

## Surface Smoothing

The surface has several chevron contours. The reason for these contours is the lack of data. The problem with contour smoothing or splining is the potential for creating crossing contours. Edits' Surface Smoothing creates interpolated data, which results in smoother contours.

1. In the drawing, select a contour, press the right mouse button, from the shortcut menu select SURFACE PROPERTIES..., and click the ***Information*** tab.
2. In the ***Information*** tab, set the current surface style to **Border Triangles Points** and click ***OK*** to exit the dialog box.
3. On the Ribbon, click the View tab. In the Views panel's left, select the named view **Central** to set the current view.
4. If necessary, click the ***Prospector*** tab.
5. In Prospector, expand the **Existing** surface until you are viewing its Definition list.
6. Select **Edits**, press the right mouse button, and from the shortcut menu, select SMOOTH SURFACE....
7. Set the smoothing method to **Natural Neighbor Interpolation**, Output Locations to **Grid based**, and set an X and Y spacing of **5**.
8. Click in the **Select Output Region's** value cell; at the cell's right side click its ellipsis, in the command line for the rectangle option enter **'E'**, and select a region covering the surface's central area. After selecting a region, you return to the Smooth Surface dialog box.
9. Click ***OK*** to exit the Smooth Surface dialog box.

The command's effect creates new grid triangulation in the selected region. The interpolation smoothes the surface contours (see Figure 4.97).

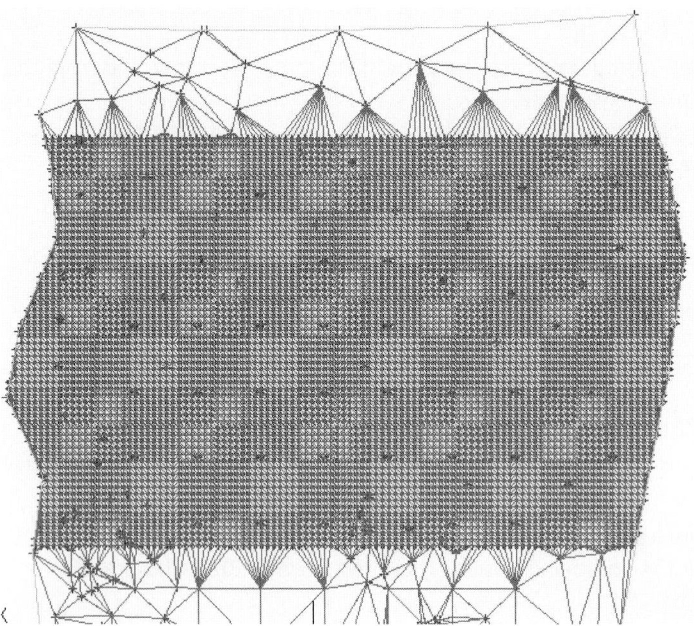

**FIGURE 4.97**

10. In the drawing, select something representing the surface, press the right mouse button, and from the shortcut menu, select SURFACE PROPERTIES....
11. In the ***Information*** tab, set the current surface style to **Contours 1' and 5' (Background)** and click ***OK*** to exit the dialog box.
12. At Civil 3D's top left, Quick Access Toolbar, click the ***Save*** icon to save the drawing.

The contours are considerably smoother than the original ones (see Figure 4.98).

**FIGURE 4.98**

## Slope Labels

Add Labels includes Slope labels. While triangles and slope arrows could be part of a surface's documentation, there is a need to annotate critical surface slopes.

1. In the drawing, select something representing the surface, press the right mouse button, and from the shortcut menu, select SURFACE PROPERTIES....
2. Click the ***Definition*** tab and in **Operations Type** toggle **OFF** surface smoothing.
3. Click ***OK*** to exit the dialog box and in the Rebuild Surface dialog box click ***Rebuild the Surface***.
4. If a Panorama displays a warning about duplicate points, close the **Panorama Event Viewer**.

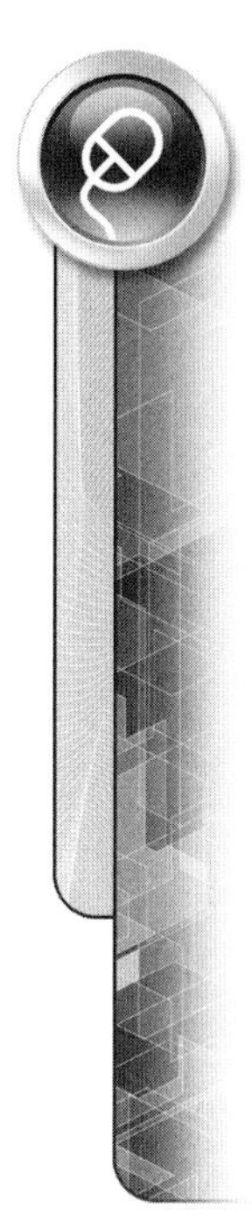

The surface returns to the previous state.

5. Click the ***Settings*** tab and expand the **Surface** branch until you are viewing the list of Slope label styles.
6. From the list of styles select **Percent**, press the right mouse button, and from the shortcut menu, select EDIT....
7. If necessary, click the ***Layout*** tab.
8. At the top left of the ***Layout*** tab, click the **Component name** drop-list arrow, and from the list select **Surface Slope**.
9. In the Text section, click in the **Contents** value cell (an ellipsis appears) and click the ellipsis to display the Text Component Editor. Review the format values for the component.

The dialog box's Properties portion contains the component's format (see Figure 4.99).

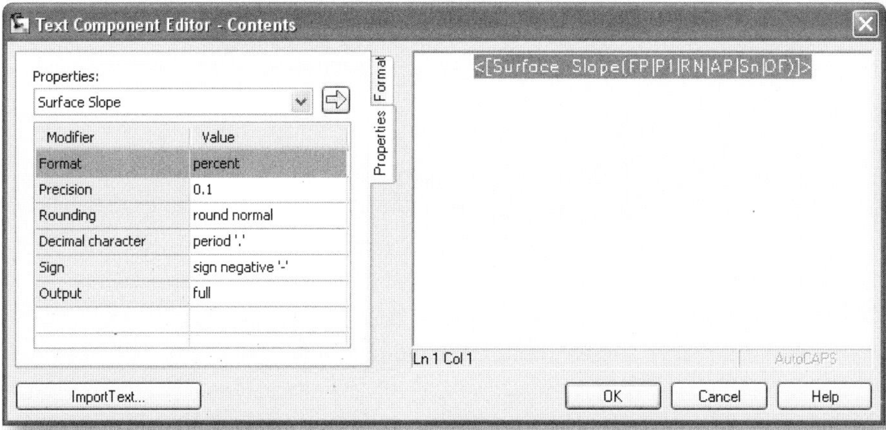


**FIGURE 4.99**

10. Click ***OK*** until you return to the command line.
11. In the Ribbon, select the ***Annotate*** tab. At the Labels & Tables panel's left, click the Add Labels drop-list arrow, select Surface, and in its flyout select ADD SURFACE LABELS to display the Add Labels dialog box.
12. If necessary, with Feature set to **Surface**, change the Label Type to **Slope**, and the Slope Label Style to **Percent**.
13. Click ***Add***.
14. If prompted to select a surface, select the surface border or contour identifying the surface.
15. In the command line make sure the slope method is one-point, press ENTER, and in the drawing select a few points, labeling their slope.

The labels annotate the percentage and direction of the down slope by orientating the text and arrow toward the down slope direction.

16. UNDO until removing the labels.

## Spot Elevation Labels

1. If necessary, click the ***Settings*** tab, and expand the **Surface** branch until you are viewing the Spot Elevation label styles list.
2. Select **EL:100.00**, press the right mouse button, and from the shortcut menu, select ***EDIT***....
3. In the ***Layout*** tab's Text section, click in the **Contents** value cell displaying an ellipsis, and click the ellipsis displaying the Text Component Editor.
4. Highlight the component format on the right to review its settings on the left (see Figure 4.100).

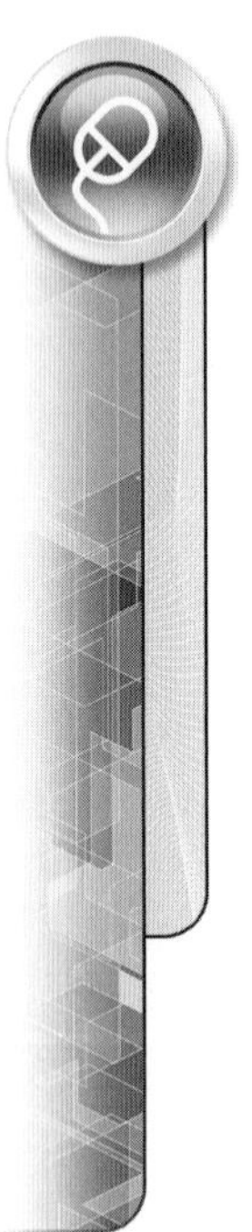

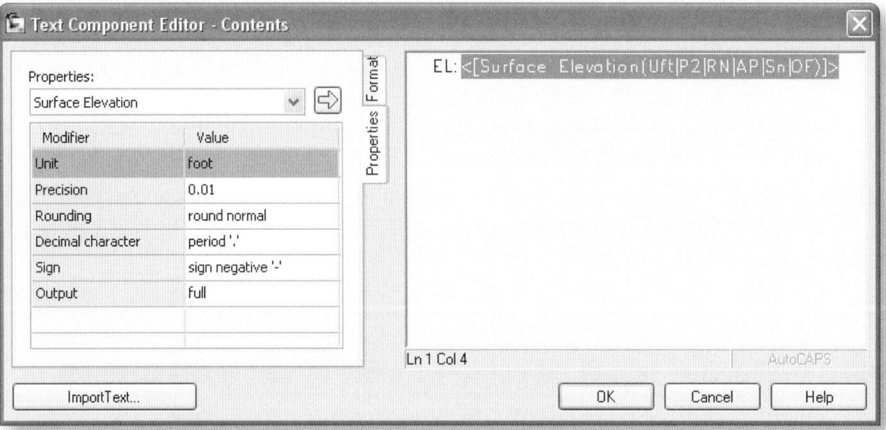


**FIGURE 4.100**

5. Click ***OK*** until you return to the command line.
6. In the Add Labels dialog box, set the Label Type to **Spot Elevation**, set the Spot Elevation Label Style to **EL: 100.00**, and set the Marker Style is **Basic Circle with Cross**.
7. Click ***Add***.
8. If prompted for a surface, select the border or a contour representing the surface.
9. In the drawing, select four or five points to create spot elevation labels.
10. When you are done, press ENTER.
11. In the drawing, select one of the spot elevation labels to display its grips.

The square grip relocates the label.

12. In the drawing, select a square grip and relocate the label.

The diamond grip relocates the elevation point and updates the label's elevation.

13. In the drawing, select a diamond grip and relocate the label's location.

The circular grip of a dragged label resets it to it original location.

14. Click a dragged label's circular grip to reset it to its original label layout.
15. Erase the labeling from the drawing.
16. At Civil 3D's top left, Quick Access Toolbar, click the ***Save*** icon to save the drawing.

## Creating an Expression

An expression uses an object property. The following exercise section creates a spot elevation label using the surface's elevation and adding 0.5 feet.

1. If necessary, click the ***Settings*** tab.
2. If necessary, in Settings, expand the **Surface** branch until you are viewing the Label Styles, Spot Elevations styles list.
3. From the list, select **Expressions**, press the right mouse button, and from the shortcut menu, select NEW....
4. For the Expression Name, enter **TFOC** and for its description enter **Top of Face-of-Curb** (see Figure 4.95).
5. Click the ***Insert Property*** button, (first large button on the right) to display the expression properties list and select **Surface Elevation** from the list.

This adds surface elevation to the expression.

6. On the calculator click the plus sign to add it to the expression.
7. Click the **0** (zero), a **point** (.), and **5** to create a 0.5 entry after the plus sign.
8. Click ***OK*** to exit the dialog box.

The label type's Expression entry has an icon showing that it has content.

9. Click the **Expression** heading for Spot Elevation to list TFOC in preview.

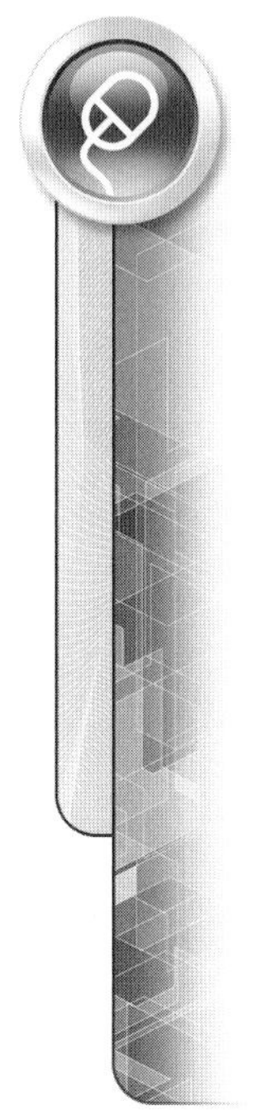

## Adding an Expression to a Label

The expression is now a spot label surface property available as a label component (see Figure 4.96).

1. From the list of Spot Elevation label styles, select **EL:100.00**, press the right mouse button, and from the shortcut menu, select COPY....
2. Click the ***Information*** tab and change the style name to **Gutter and TFOC**. For the description enter **Gutter and Top of Face-of-Curb**.
3. Select the ***Layout*** tab.
4. In the Text section, click in the **Contents** value cell and then click its ellipsis to display the Text Component Editor.
5. On the right side of the dialog box click the text **EL:** and change it to **Gutter:**.
6. Place the cursor after the Gutter's format string (after the >, greater than sign) and press ENTER to make a new line.
7. For text enter **F-TOC:**, leaving a space after the colon.
8. On the left side, click the **Properties** drop-list arrow and from the list select **TFOC**.
9. On the left side, change the modifier Precision by clicking in the **Value** cell, and selecting **0.01** from the list.
10. At the top center click the arrow to create a format string for TFOC.
11. Click ***OK*** until you exit and return to the command line.
12. At Civil 3D's top left, Quick Access Toolbar, click the ***Save*** icon to save the drawing.
13. If necessary, in the Ribbon, select the ***Annotate*** tab. in the Labels & Tables panel's left, click the Add Labels drop-list arrow, select Surface, and in its flyout select ADD SURFACE LABELS to display the Add Labels dialog box.
14. In Add Labels, change Label type to **Spot Elevation**, change Spot Elevation label style to **Gutter and TFOC**, and, if necessary, change the Marker style to **Basic Circle with Cross**.
15. Click ***Add*** and place two labels in the drawing.

16. In the Add Label dialog box, click ***Close***.
17. Erase the labels from the drawing.
18. At Civil 3D's top left, Quick Access Toolbar, click the ***Save*** icon to save the drawing.

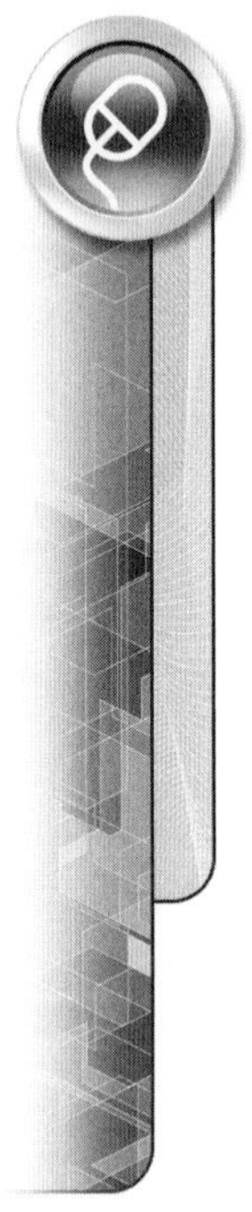

## SUMMARY

- Contour styles fall into two categories: major/minor or existing/proposed.
- You can define depression contours.
- Surface smoothing is effective for areas of sparse data.

Once you have a satisfactory surface, you can create points whose elevations are a surface's elevations. These points can be data for a survey crew or a starting point for a second surface.

## UNIT 6: SURFACE POINTS, LANDXML, AND UTILITIES

The Create Points toolbar creates points from surface elevations. The export LandXML command exports surfaces so other applications can use the surface. The Surfaces', Utilities menu has several routines that are important to images, blocks, volumes, and entity extraction.

### POINTS FROM SURFACES

The Create Points toolbar's Surface icon stack has routines to create new points whose elevations are from a selected surface: Random Points, On Grid, Along a Polyline/Contour, and Polyline/Contour Vertices (see Figure 4.101). The Points menu's Create Points or Prospector's Points heading right-click shortcut menu Create… displays this toolbar.

The Along a Polyline/Contour and (at) Polyline/Contour vertices routines create points whose elevations are at distances along the length of or at the vertices of a polyline or contour. These routines are useful when developing a grading design and create points that represent the intersection of a grading strategy and a surface.

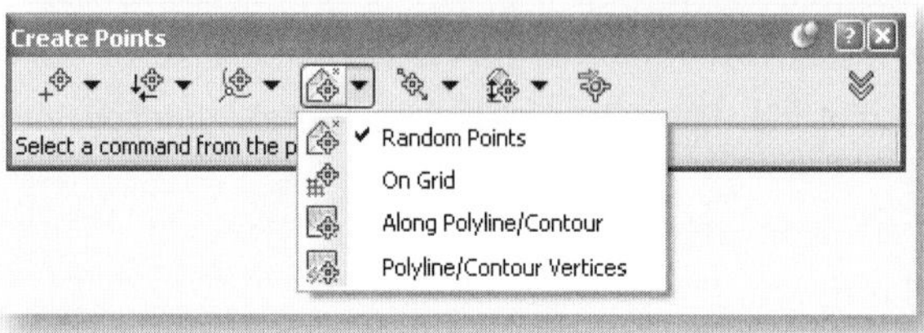


**FIGURE 4.101**

### LANDXML

A method to transfer a surface to other applications, including AutoCAD Land Desktop, is to export a surface as a LandXML file. The File menu, Export flyout or a selected surface's shortcut menu has the Export to LandXML command (see Figure 4.102).

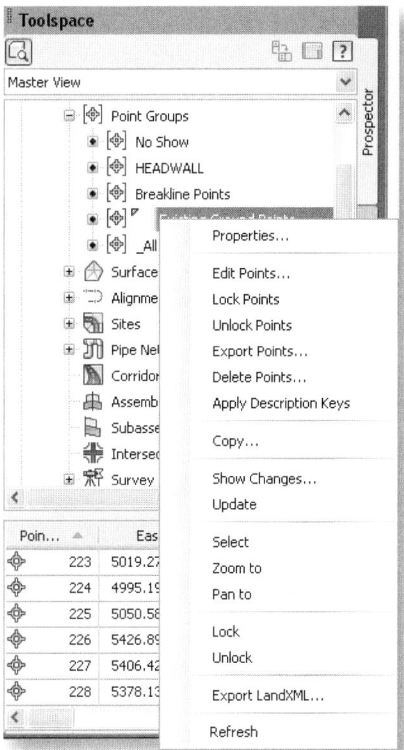


**FIGURE 4.102**

LandXML exporting starts by identifying the export surface. The Export to LandXML dialog box selects the surface(s) or a surface is selected from a drawing (see Figure 4.103).

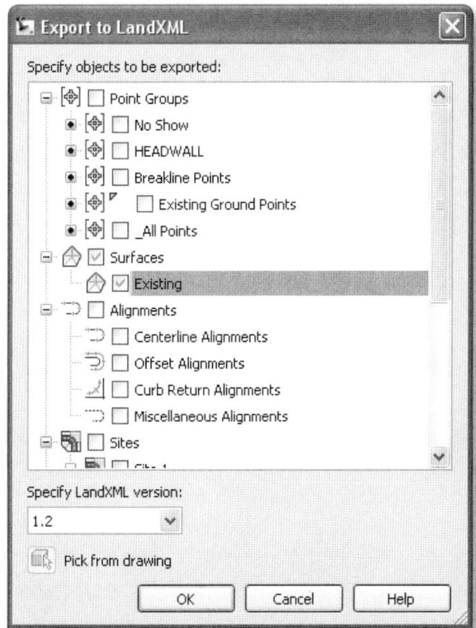


**FIGURE 4.103**

### LandXml Settings

Before exporting to LandXML, check the LandXML settings and make sure they represent the correct units. By default, the LandXML file's Imperial units are International feet and should probably be set to U.S. Foot (see Figure 4.104). The Surface Export Settings section sets the type of surface data (default points and faces) and if to include watersheds, the default number.

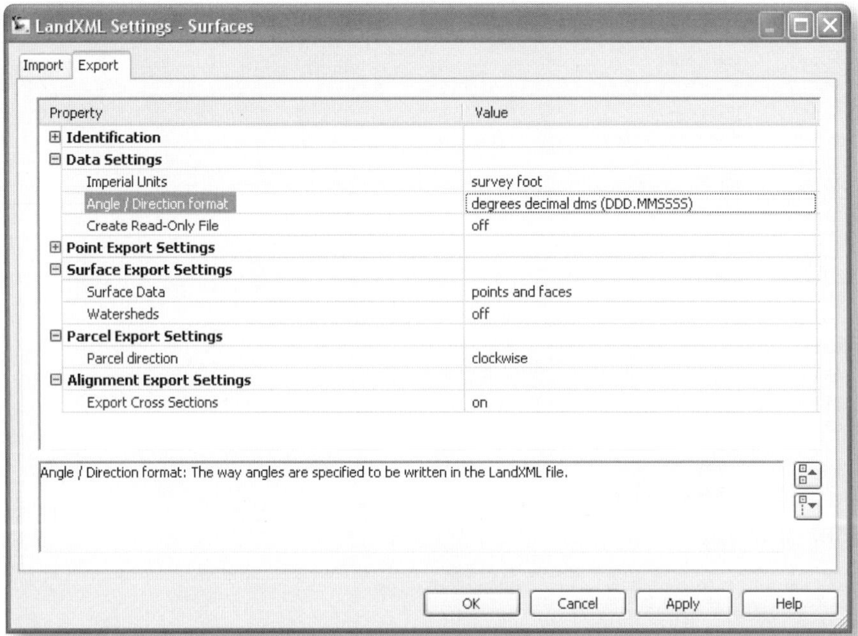


**FIGURE 4.104**

## SURFACE UTILITIES

Surface Utilities tools apply to several specific situations.

### Export as DEM

This routine creates a DEM file (see Figure 4.105). When exporting a DEM, the Export Surface to DEM dialog box opens. The dialog box sets the file name and location, coordinate system, grid spacing, elevation sampling method, and, if necessary, a custom Null elevation.

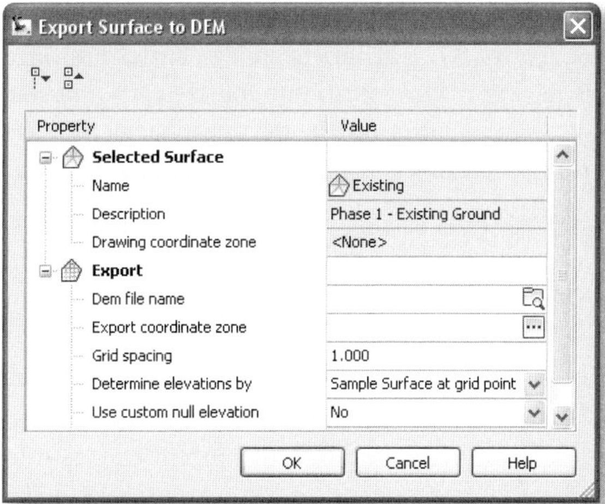


**FIGURE 4.105**

## Volumes and Bounded Volumes

Volumes calculates a TIN volume by comparing two surfaces' triangulation (see Figure 4.106). This is the same as creating a TIN volume surface in Prospector's Surfaces branch. Bounded Volumes calculates an area's volume based on a volume surface (a TIN or Grid volume surface).

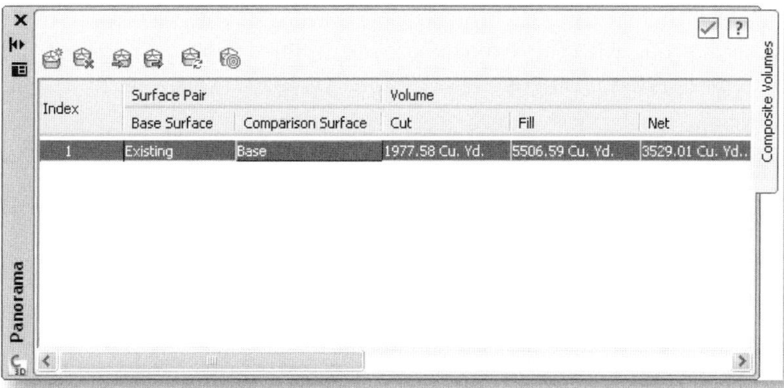


**FIGURE 4.106**

## Water Drop and Catchment Area

Water Drop creates a polyline trail that a water drop takes as it travels over a surface. A water drop is a good way to visualize existing ground surface slopes or to see if the design surface pulls water to drainage catch basins.

A catchment area is a surface low point depression area with a boundary and an area. This is used in hydrological analysis.

## Drape Image

Drape Image places an image on a surface. If the image is larger than the surface, the image is trimmed to the edge of the surface.

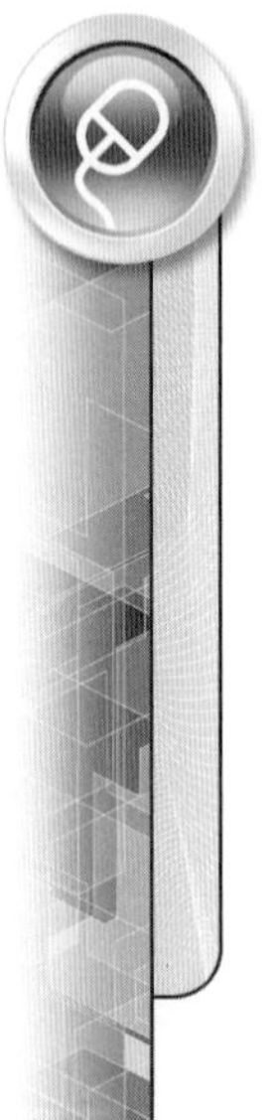

### Extract Objects from Surface

Extract Objects from Surface creates AutoCAD entities from the current surface style. For example, if the surface is displaying major and minor contours, the routine creates polylines at elevations from the surface contours. Or, if the surface is displaying a border and TIN triangles, this routine creates a 3D polyline and 3D lines.

### Move Block to Surface

Move Block to Surface assigns the surface's elevation to a block's Z insertion point at the block's X and Y coordinates.

### Move Block to Attribute Elevation and Move Text to Elevation

These routines use a block's attribute elevation value or the text's content elevation to modify the block's or text's insertion Z value. For example, a text's insertion Z is 0.0 and its content is the elevation 116.13. The routine Move Text to Elevations modifies the text's insertion point to be the Z value of 116.13.

## EXERCISE 4-6

After completing this exercise, you will:

- Be able to place points on a grid.
- Be able to place points on a polyline or contour.
- Be able to export a LandXML file.

This exercise continues with the previous unit's exercise drawing. If you did not complete the Unit 5 exercise, or starting with this exercise browse to the Chapter 4 folder of the CD that accompanies this textbook and open the Chapter 04 - Unit 6 drawing.

### Exercise Setup

1. If not open, open the previous exercise's drawing or browse to the Chapter 4 folder of the CD that accompanies this textbook and open the *Chapter 04 – Unit 6* drawing.

### Point Creation and Point Identity Settings

1. If necessary, click the ***Prospector*** tab.
2. In Prospector, select the Points heading, right mouse click, and from the shortcut menu select CREATE....
3. On the Create Points toolbar's right side, click the **Expand the Create Points dialog** chevron.
4. Expand the **Points Creation** section and set the following values:

   Prompt For Descriptions: **Automatic**
   Default Description: **EXISTING**

5. Expand the **Point Identity** section, set the Next Point Number to **400**, and press ENTER.
6. On the Create Points toolbar's right side, click the **Collapse the Create Points dialog** chevron.
7. In Layer Properties Manager, thaw the **V-NODE** layer, and click **X** to exit the palette.

8. If necessary, click the ***Prospector*** tab.
9. Select the **Point Groups** heading, press the right mouse button, and from the shortcut menu, select PROPERTIES....
10. Move **_All Points** to the list's top, and click ***OK*** to exit the dialog box.

If Civil 3D issues a warning about duplicate points in the surface, close the Panorama's Event Viewer vista.

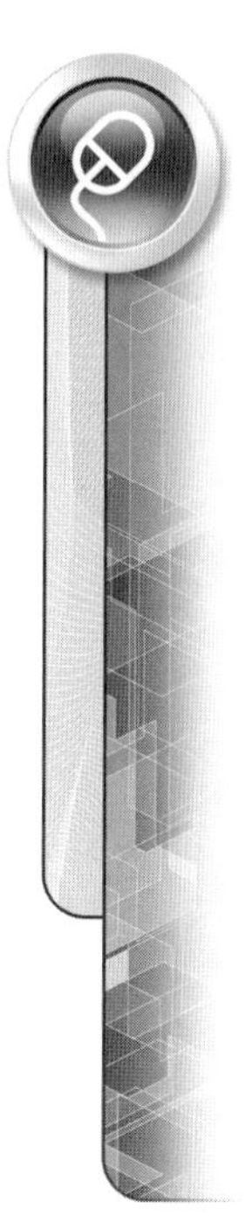

## Points on Grid

Points on Grid creates points on a user-defined grid and assigns surface elevations as their elevations.

1. In the Create Points toolbar, from the Surface's icon stack, click the drop-list arrow and from the list select **ON GRID** (see Figure 4.101).
2. The routine prompts you to select a surface object; select a surface component to identify it.
3. Next, the routine prompts you for a grid base point. Select a point at the surface's lower left.
4. The command line prompts you for a rotation angle; press ENTER to accept zero.
5. The command line prompts you for grid spacings; for grid X and Y spacing enter **30**.
6. Finally, in the drawing select a point and upper-right grid limit within the surface's interior.

A box is displayed to represent the grid area. Within the grid area is another box showing the grid spacing (lower-left corner).

7. The command line asks if you want to change the grid definition. Answer **NO** by pressing ENTER to continue to the next prompt.

The routine creates points using the automatic description (EXISTING).

8. The command line prompts you to select another surface object. Press ENTER to exit the routine and return to the command line.
9. In Prospector, click the **Points** heading to preview the new points.

The new points in preview start with point number **400**.

10. In Prospector, select the **Point Groups** heading, press the right mouse button, and from the shortcut menu, select NEW....
11. In the Point Group Properties dialog box, ***Information*** tab, for the point group name enter **Points From Existing**.
12. Click the ***Include*** tab, toggle **ON With Numbers Matching**, and enter in the point numbers from **400–800** (or whatever ending point number is appropriate for the points in your drawing).
13. Click ***OK*** to exit the Point Group Properties dialog box.

## Points on a Polyline/Contour

1. On the Ribbon, select the View tab. At the Views panel's left, select the named view **Central** to restore the view.
2. Use the ZOOM WINDOW command and zoom in on the surface.
3. On the Create Points toolbar's right side, click the **Expand the Create Points dialog** chevron.
4. In the **Points Creation** section, set the Default Description value to **CONTOUR**.

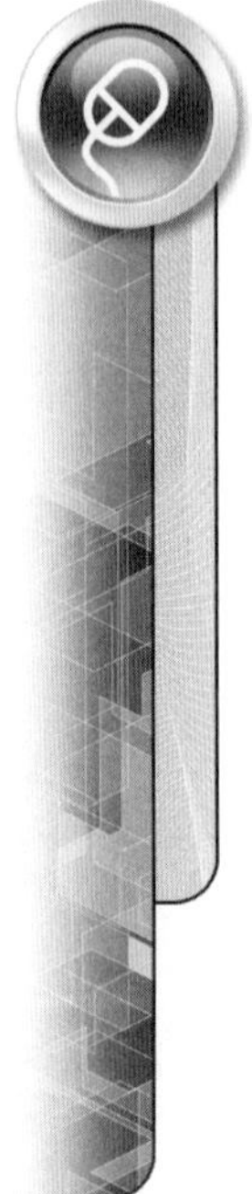

5. On the Create Points toolbar's right side, click the **Collapse the Create Points dialog** chevron.
6. To the right of the Surface icon stack, click the drop-list arrow and select ALONG POLYLINE/CONTOUR from the list.
7. In the drawing, select something representing the surface, press ENTER to accept the default distance (**10**), and select a contour.
8. Press ENTER to exit the routine.
9. In Prospector, select the **Point Groups** heading, press the right mouse button, and from the shortcut menu, select NEW....
10. In the ***Information*** tab, assign the point group the name **From Contour**.
11. Click the ***Include*** tab, toggle **ON With Raw Description Matching**, enter **CONTOUR**, and click ***OK*** to create the point group From Contour.
12. In Prospector, select the heading **Point Groups**, press the right mouse button, and from the shortcut menu, select PROPERTIES....
13. Select **No Show** and move it to the list's second position.
14. Click ***OK*** to exit the dialog box.

Only the contour points are displayed.

15. You may need to expand the Point Groups heading to select any out of date groups, right click and select Update.
16. Select the heading **Point Groups**, press the right mouse button, and from the shortcut menu, select PROPERTIES....
17. Select the point group **No Show** and move it to the list's top.
18. Click ***OK*** to exit the Point Group Properties dialog box.

No points display.

19. Click the Create Points toolbar's red **X** to close it.
20. In Prospector, Point Groups select the **From Contour** point group, press the right mouse button, and from the shortcut menu, select DELETE POINTS.... Click ***OK*** to answer yes to the Are you sure dialog box.
21. In Prospector, select the point group **From Contour**, press the right mouse button, and from the shortcut menu, select DELETE.... Click ***Yes*** to the Are you sure dialog box.
22. In Prospector, select the point group **Points From Existing**, press the right mouse button, and from the shortcut menu, select DELETE POINTS.... Click ***OK*** to answer Yes to the Are you sure dialog box.
23. In Prospector, select the point group **Points From Existing**, press the right mouse button, and from the shortcut menu, select DELETE.... Click ***Yes*** to the Are you sure dialog box.
24. At Civil 3D's top left, Quick Access Toolbar, click the ***Save*** icon and save the drawing.

## LandXML Export Settings

1. Click the ***Settings*** tab.
2. At the top select the drawing name, press the right mouse button, and from the shortcut menu, select EDIT LANDXML SETTINGS....
3. Click the ***Export*** tab, expand the **Data Settings** section, and change the Imperial Units value to **survey foot**.
4. Expand the **Surface Export Settings** section, set Surface Data to **points and faces**, and Watersheds to **off**.
5. Click ***OK*** to exit the dialog box.

## Exporting a Surface and Point Groups

1. Click the ***Prospector*** tab.
2. In Prospector, from the Surfaces branch, select the surface **Existing**, press the right mouse button, and from the shortcut menu, select EXPORT LANDXML....
3. In the Export to LandXML dialog box, toggle **ON** the surface **Existing** and point group **Breakline Points** to include them in the LandXML file.
4. Click ***OK*** to continue the exporting process.
5. In the Export LandXML dialog box, browse to the folder **Civil 3D Projects**, name the file, and click ***Save*** to create the file.

This completes the Civil 3D surfaces chapter. Chapter 10 focuses on designing a second surface and calculating its earthwork volumes (existing ground and a design surface). This chapter focused on surface design tools and the calculating earthwork volumes process.

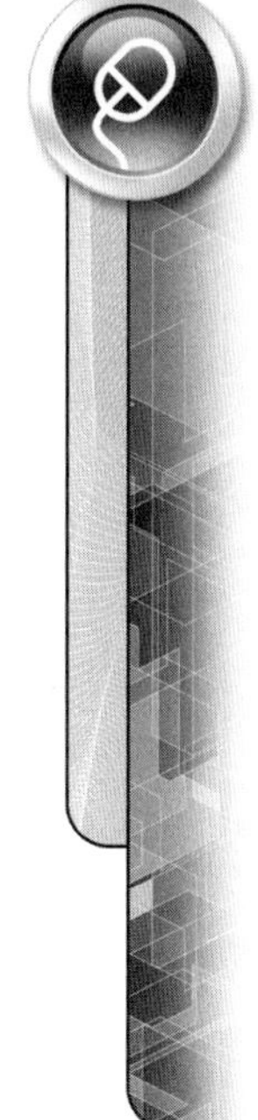

## SUMMARY

- You can create points randomly or on a grid whose elevations are from a surface.
- You can place points whose elevations are from a surface on a polyline at an interval or at its vertices.
- The default Imperial Unit for exporting data to a LandXML file is International feet. Make certain its value represents the measurement value you want.

The next chapter, Chapter 5, starts the discussion of the roadway design process.

# CHAPTER 5

# Alignments

## INTRODUCTION

This and the next four chapters focus on roadway design and its documentation. This chapter concentrates on roadway plan design. The next three chapters focus on the roadway profile and section views and the roadway model (the corridor). As with all Civil 3D objects, styles affect how roadway components display and are annotated.

## OBJECTIVES

This chapter focuses on the following topics:

- Alignment Settings
- Alignment Styles That Affect the Roadway Look and Design Criteria
- Alignment Design Parameters
- Modifying Stationing and Adding Station Equations
- Alignment Reports
- Alignment's Right-Of-Way (ROW)
- Points from Alignments
- Alignment Parcel Subdivision

### OVERVIEW

The Alignments, Profiles, Corridors, and Sections commands create, analyze, edit, and annotate a roadway model. The roadway design process is a sequence of steps using 2D elements and views, alignments (plan view), profiles (side view), and sections (views to the alignment's left and right) to create a 3D road model (a corridor). In Civil 3D a site and its parcels can be related to alignments. If an alignment is a site member, it can subdivide the site and its parcel(s). With this association, alignments, profiles, and sections are a part of the site branch. If you do not want an alignment, its profiles, sections, and corridors to interact with a site and its parcels, do not assign the alignment a site.

An alignment is dynamically linked to all of its dependent elements. When one design element is changed, the change cascades down to the corridor and its sections. The progression follows the Alignments' or Sites' Alignment branch. Changes travel from the list's first dependent element to the last (i.e., from alignment to profile, to the corridor, and finally to the section views). The dependency chain includes surfaces. A surface provides elevations to profile and section. If something should change in a surface, the dependent roadway elements adjust their values to accommodate the surface change.

The roadway design process produces a design model (the corridor) and a documentation set. Civil 3D uses design views, plans, profiles, and sections to document the 2D aspects of this 3D model. The roadway design process reduces the roadway's 3D aspects to three 2D components (see Figure 5.1). This traditional design methodology means projecting the 3D model (the corridor) onto planes showing its 2D road-design portion. This same process occurs when designing a roadway on paper. The paper is the 2D plane on which the designer represents the three respective roadway design model views. It isn't until you create a corridor that the road design appears as model. The road model documentation includes roadway earthwork volume estimations and roadway material volumes.

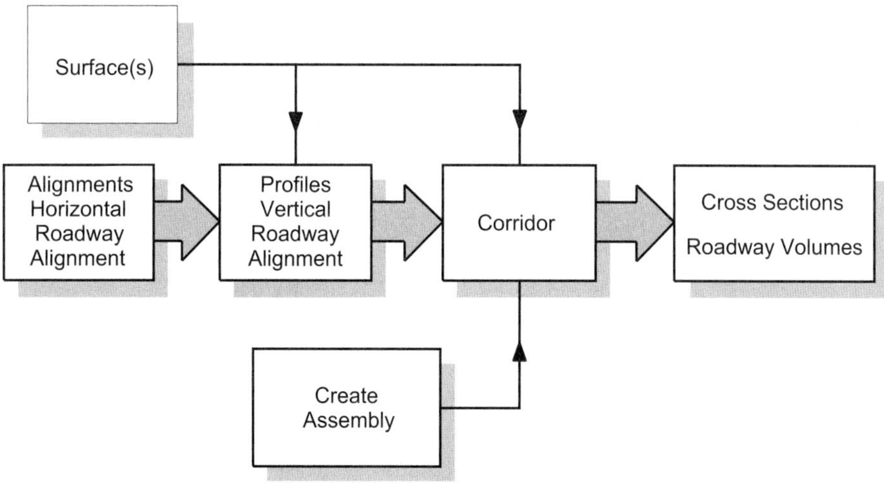


**FIGURE 5.1**

Ribbon's Home tab, Create Design panel's, Alignment drop-list has the horizontal design commands (see Figure 5.2). The Profiles drop-list contains the vertical tools and the cross-sections are in the Sections lines and Section view drop-lists. The Corridors drop-list contains the roadway modeling tools. When developing a vertical frame of reference (profile and cross-section views), the routines sample surface elevations along or to either side of the roadway centerline.

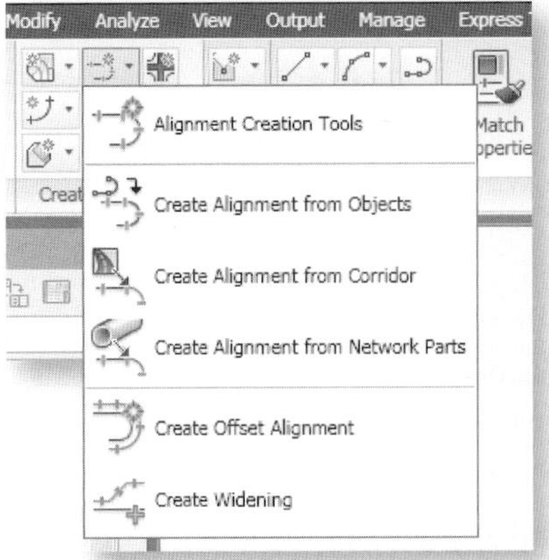


**FIGURE 5.2**

The design process is not overly flexible. Users cannot design a horizontal roadway centerline and then jump to creating a corridor. An assembly following along a centerline also depends on a vertical alignment developed in profile view. What occurs at any point along the road-design process depends on what occurred before the current stage of the process.

The Corridor (roadway model) is beneficial to roadway analysis and editing because of the relationships among the model's design elements. If you want to try an alignment in a new position and view its resulting existing ground profile, all you need to do is grip-edit the alignment, and the profile redrafts itself to the new alignment length, resamples its elevations, and redraws itself within an adjusted profile view. With each alignment change, the vertical design DOES NOT change. You must edit or redraw the vertical design for each alignment relocation.

Roadway design goals vary from project to project. Generally, the goals include to not move more spoil material than necessary and to use the road-design elevations to start overall site grading. Rarely is this process a single pass through the design steps; more likely, it is an iterative process. All initial design element values give the designer a starting point from which to optimize the final model. To modify a design or an element means you must return to a point in the design process and change the values or elements at that point. After making the change, it ripples through the model downward from the change point toward roadway sections.

## Unit 1

The horizontal alignment object styles are the first unit's topic. The alignment object styles control basic design parameters and its component and characteristics' visibility. An alignment object type focuses on the alignment type: existing, proposed, etc. As with other styles, these styles can be saved to a template, and when selected for a new drawing, the styles become a part of the drawing. Design checks notify the user that an alignment has not met minimum or has exceeded maximum design values. Assigning AASHTO values to an alignment guides the designer in creating alignments following AASHTO standards.

It is difficult to separate alignment objects and their labeling. Alignments can be annotated as they are created or after defining them. There may be different styles applied when designing them and when documenting their final values.

### Unit 2

A centerline alignment is a single object composed of tangents, curves, and possibly spirals. An alignment has four creation methods. The first method is to import a LandXML file. The second method is to import from a Land Desktop (LDT) project's alignment definitions. These two methods, LandXML file and LDT project, can import one or more alignments, and when importing them they can select which alignments to import. The third method is to convert objects or polyline into an alignment. The last method is to create an alignment with the Alignment Layout Tools toolbar commands. The toolbar tools draft alignment tangent and curve elements. Drafting and defining the roadway centerlines are the focus of Unit 2.

### Unit 3

Unit 3 focuses on the centerline's design values analysis. Most engineers design by calculations or design criteria. Lines and arcs do not convey their values easily to the designer. The design numbers are found in alignment editors or in alignment reports.

### Unit 4

Many times a design changes or needs additions (station equations or speeds). When an alignment is being edited, the alignment editor vista displays essential data the designer needs to make a design change. An alternative editing method is to graphically relocate alignment elements. An alignment is a collection of fixed, floating, and/or free elements. When an alignment has these segment types, it has a predictable behavior when graphically manipulated. Editing alignments and adding station equations are the fourth unit's subjects.

### Unit 5

Centerline annotation is the topic of unit 5. There are two basic alignment annotation styles: parallel to and offset to, and perpendicular to and on the centerline. Civil 3D provides an extremely flexible alignment annotation system.

### Unit 6

The last unit reviews creating new objects from a centerline, station and offset labels, points, and parcels (ROW).

## UNIT 1: ROADWAY STYLES

An alignment's drawing appearance depends on settings and object and label styles.

### EDIT DRAWING SETTINGS

Editing Drawing Settings' Object Layers panel sets the alignment object's base layer name (see Figure 5.3). The alignment name can be this layer's prefix or suffix. Setting a modifier and entering an asterisk (*) includes the alignment's name in the layer name and creates a unique layer for each alignment.

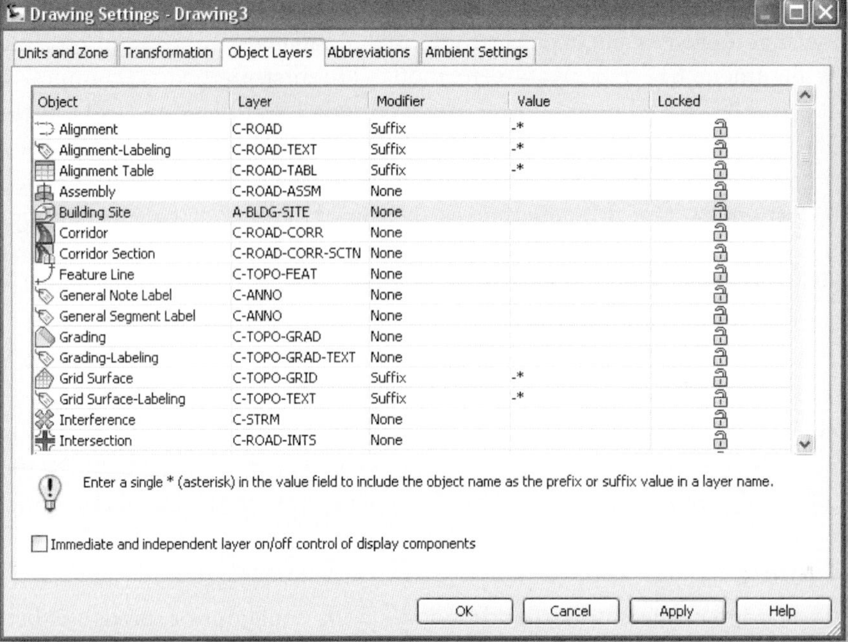


**FIGURE 5.3**

The Abbreviations tab sets critical alignment geometry points codes and value formats (see Figure 5.4). What codes a company uses vary greatly from region to region and it is here that one enters codes specific for his or her region. If you are working with clients using different codes, these codes can be changed to address each client's code values. This section's values affect Civil 3D listing routines.

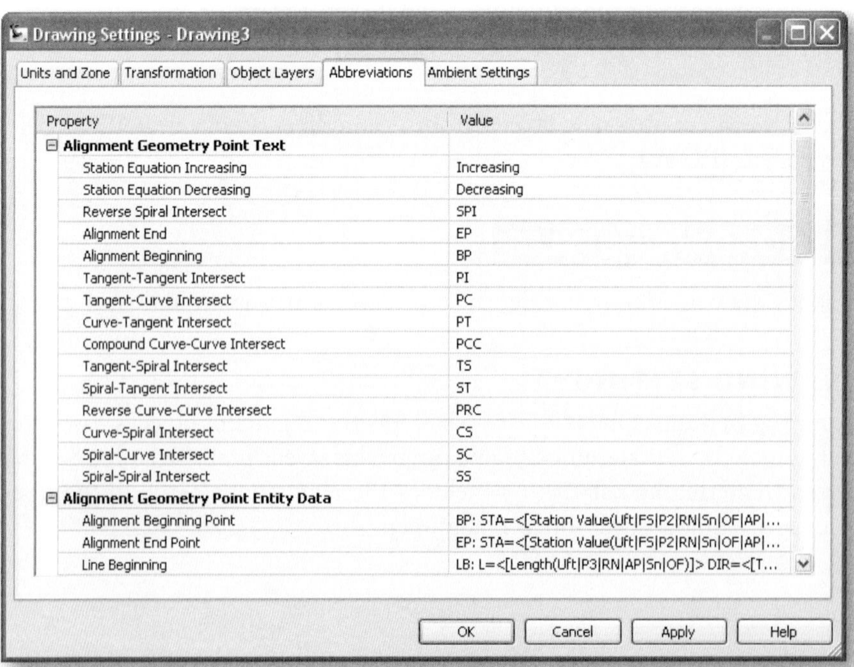


**FIGURE 5.4**

## EDIT FEATURE SETTINGS

Alignment's Edit Feature Settings dialog box identifies default styles and design options (see Figure 5.5). The Default Styles section identifies object and label styles for specific alignment elements. The list includes an alignment label set. A label set is a named group of label styles (e.g., Major Minor and Geometry Point). A label set's effect is labeling specific alignment geometry with a single selection.

The Default Naming Format section defines how an alignment gets its name (alignment followed by a sequentially assigned number). The Station Indexing value sets the interval for stationing annotation.

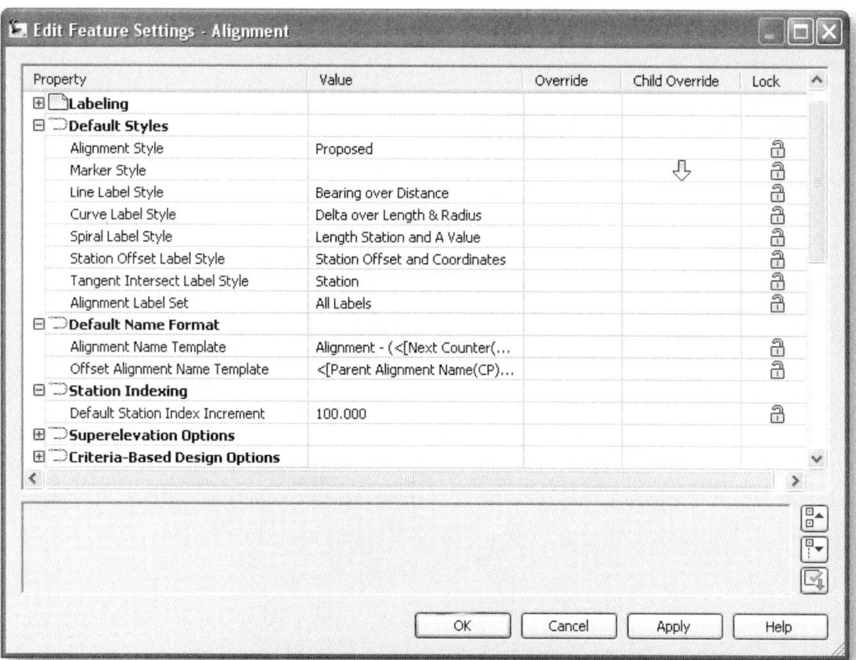


**FIGURE 5.5**

The Superelevation Options section assigns the default parameters, lane and shoulder slopes and methods, to an alignment (see Figure 5.6).

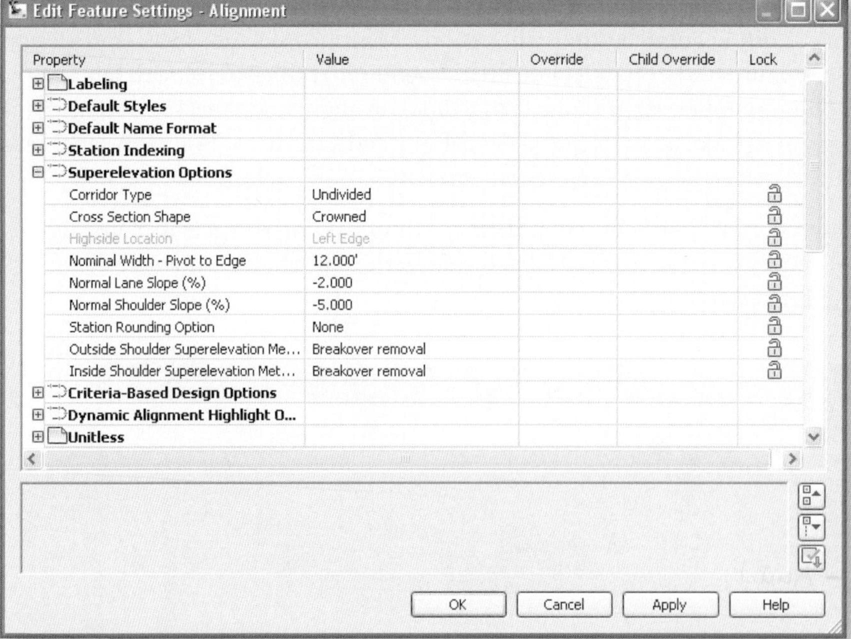


**FIGURE 5.6**

The Criteria-Based Design Option section sets a design speed and toggles on the use of criteria lookup tables (see Figure 5.7). Criteria tables are AASHTO or user-created tables that affect minimum radii, superelevation attainment methods, and superelevation cross slope tables. Design Checks are expressions that affect Tangents, Curves, Spiral, or Tangent Intersections. Check files define minimum radii or minimum tangent lengths for a road.

The Dynamic Alignment Highlight Options, set colors for editor and preview highlighting of alignment types and segments.

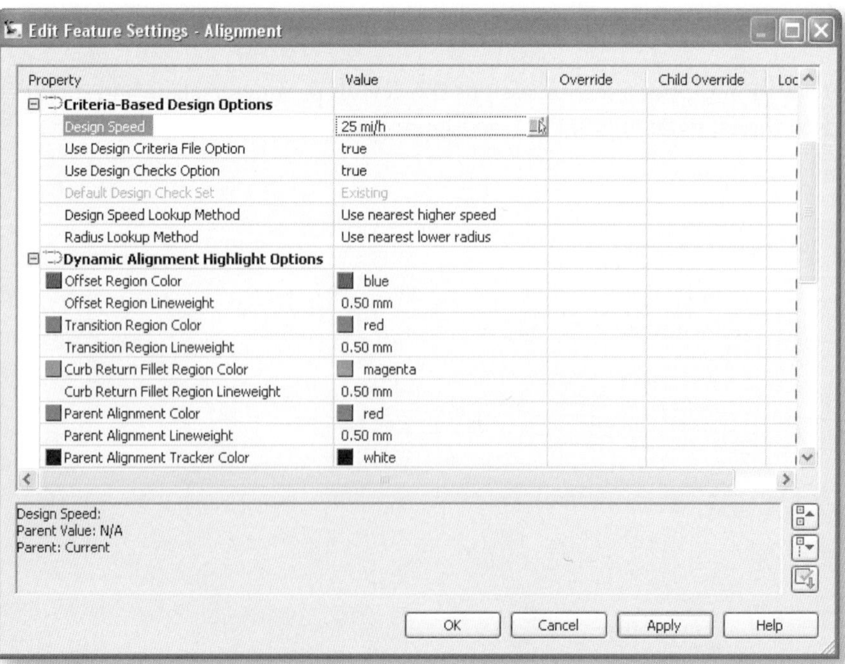


**FIGURE 5.7**

## ALIGNMENT STYLES

The alignment object style sets basic design parameters and visibility for its components. As with all styles, these styles can be saved to an office template, and when selecting the template for a new drawing, its styles become a part of the drawing.

### Object Styles

Alignment object types tend to be types defined by types of alignments. The Proposed style defines a new alignment's display of components.

The Information tab names, describes, and assigns a modification date for the style.

The Design tab's Enable radius snap toggle enables a radius snapping value that affects an alignment's curve graphical editing. When editing an alignment with this toggle on, resulting curves have radii that are divisible by the Radius snap value (see Figure 5.8)

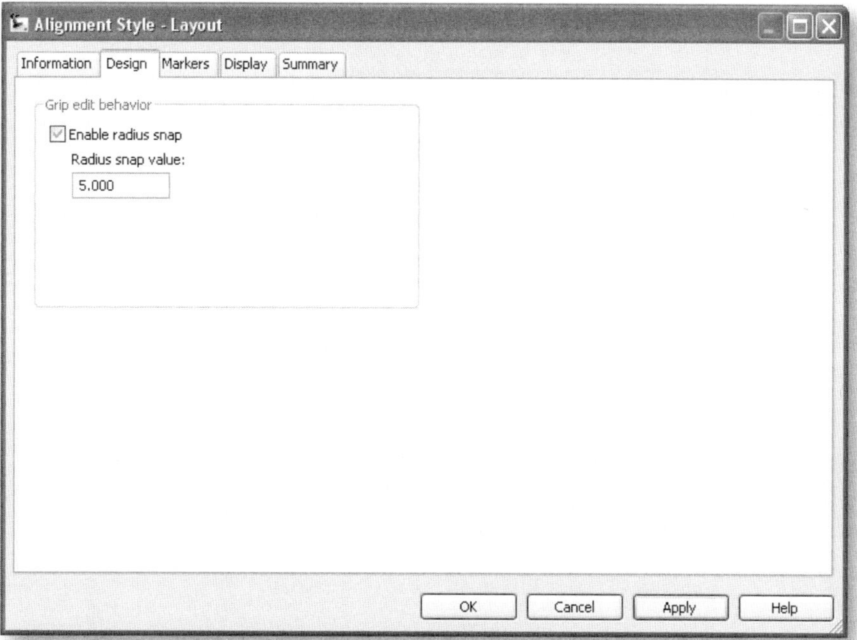


**FIGURE 5.8**

The Markers tab sets marker styles identifying critical alignment points. Marker styles are from Settings, General, Multipurpose Style branch. The panel's bottom defines the alignment direction arrow (see Figure 5.9).

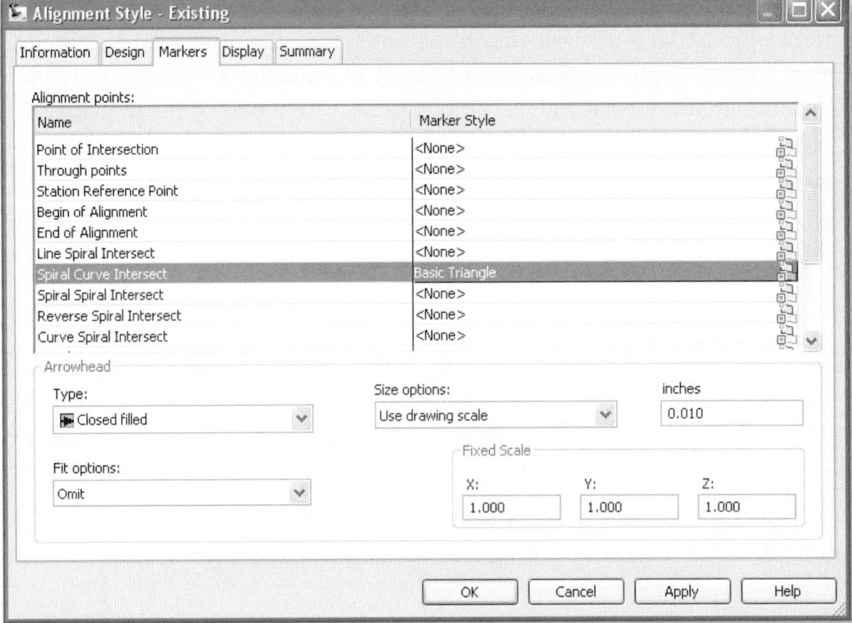


**FIGURE 5.9**

The Display tab assigns layers and their properties to each alignment component (see Figure 5.10).

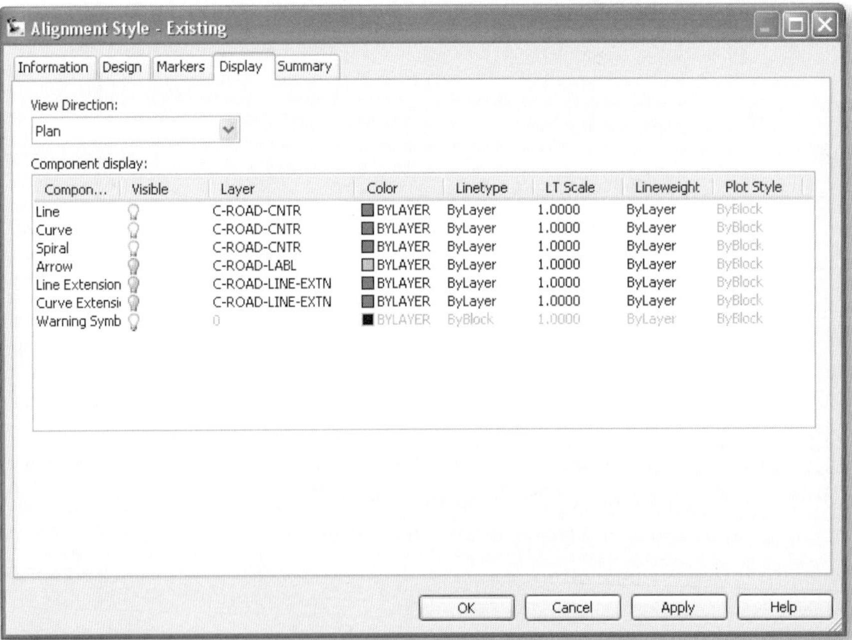


**FIGURE 5.10**

## Label Styles

When creating an alignment, the Create Alignment dialog box also assigns a label set. As with other label styles, they can be saved to a template, and when selecting the template for a new drawing, the styles become a part of the new drawing.

If you want to assign labels, you have the choice of adding individual labels or assigning a label set. A label set is an alias containing several label style types. Assigning a label set to an alignment applies all its styles to the alignment.

## Major Minor and Geometry Points Label Set

The Set's Labels tab assigns the style types and label styles. The Major, Minor, and Geometry Points Label Set labels the major stations with text parallel to centerline and label the minor stations with ticks (see Figure 5.11). The Geometry Points label attaches to a line perpendicular to the alignment point. The increment values set the labels' intervals and ripple through to profile view labeling.

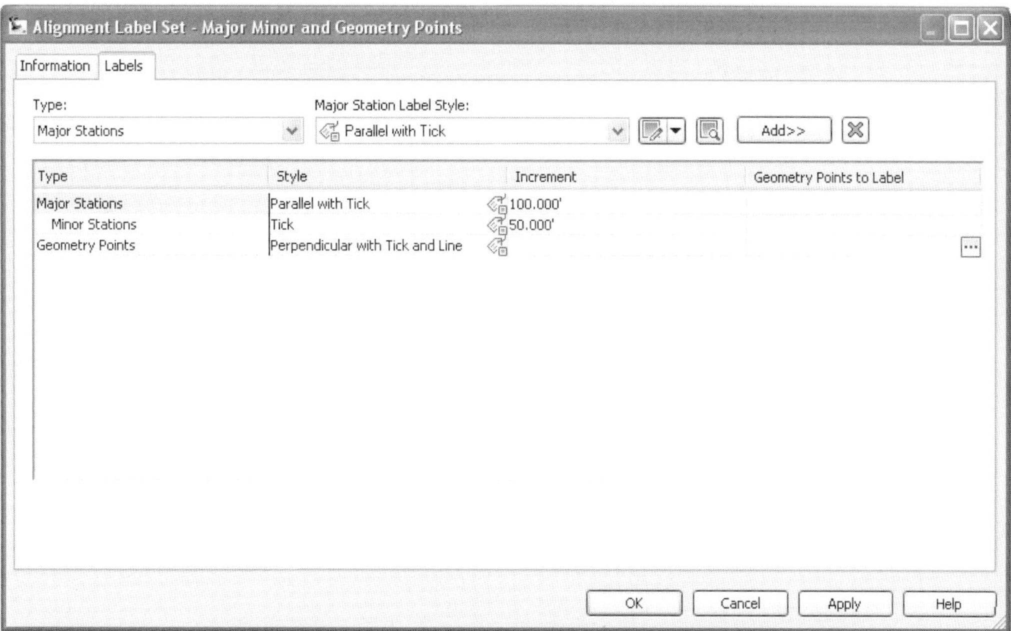


**FIGURE 5.11**

Geometry Points has an ellipsis when clicked, and it displays a geometry points list. By toggling on or off the items, a user sets labeling for the checked alignment points (see Figure 5.12).

Geometry Points

Select geometry points to label:

| Geometry Points | Label |
|---|---|
| Alignment Beginning | ☑ |
| Alignment End | ☑ |
| Tangent-Tangent Intersect | ☑ |
| Tangent-Curve Intersect | ☑ |
| Curve-Tangent Intersect | ☑ |
| Compound Curve-Curve Intersect | ☑ |
| Reserve Curve-Curve Intersect | ☑ |
| Tangent-Spiral Intersect | ☑ |
| Spiral-Tangent Intersect | ☑ |
| Curve-Spiral Intersect | ☑ |
| Spiral-Curve Intersect | ☑ |
| Spiral-Spiral Intersect | ☑ |
| Reverse Spiral Intersect | ☑ |

OK Cancel Help

**FIGURE 5.12**

**Major Stations: Perpendicular with Tick.** Perpendicular with Tick, a major station label, places a tick at an alignment's major station (1+00). The station's value is adjacent to the tick (see Figure 5.13).

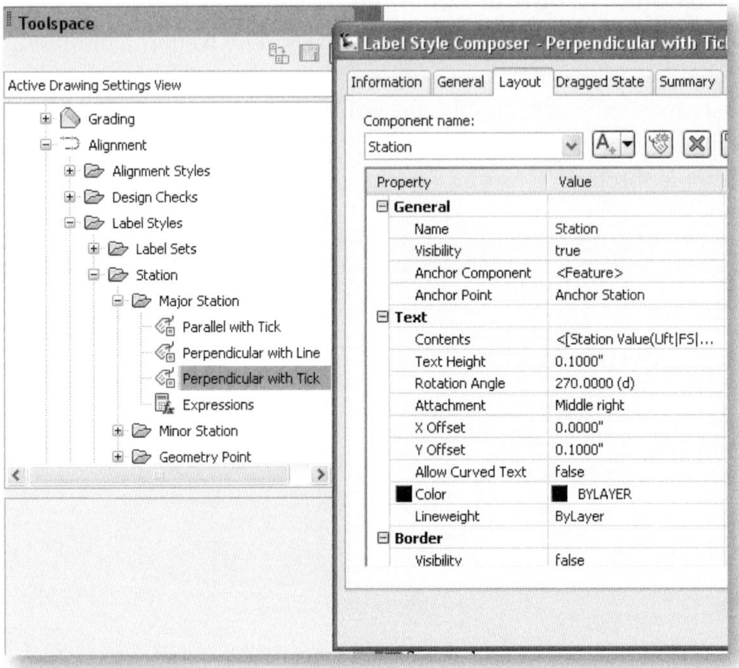


**FIGURE 5.13**

**Minor Station: Tick.** The Tick style places a block (line) at the minor station interval set in the label set (see Figure 5.14).

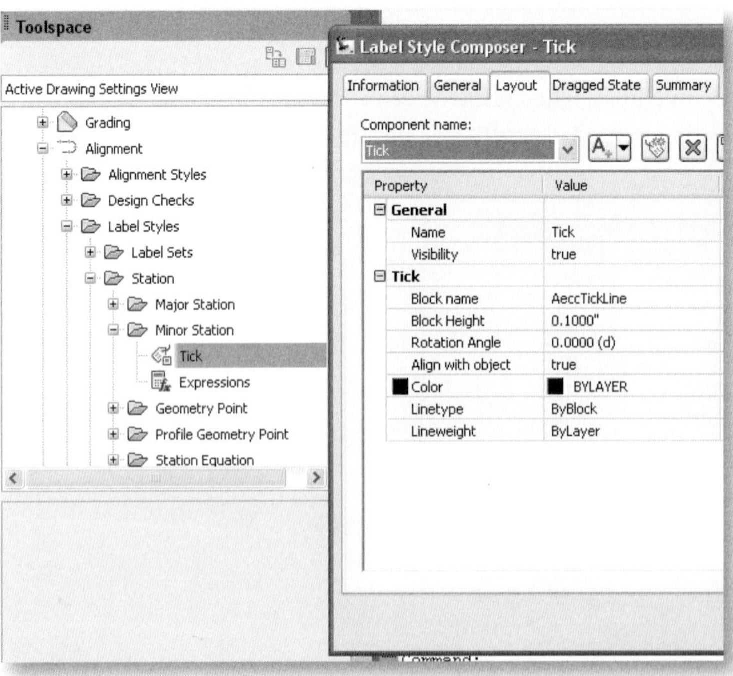


**FIGURE 5.14**

**Geometry Point: Perpendicular with Tick and Line.** This style places a tick, a line, and text anchored to the line labeling centerline geometry points (see Figure 5.15). Geometry points are at a curve's or spiral's beginning or ending point, at a compound curve tangency point, etc.

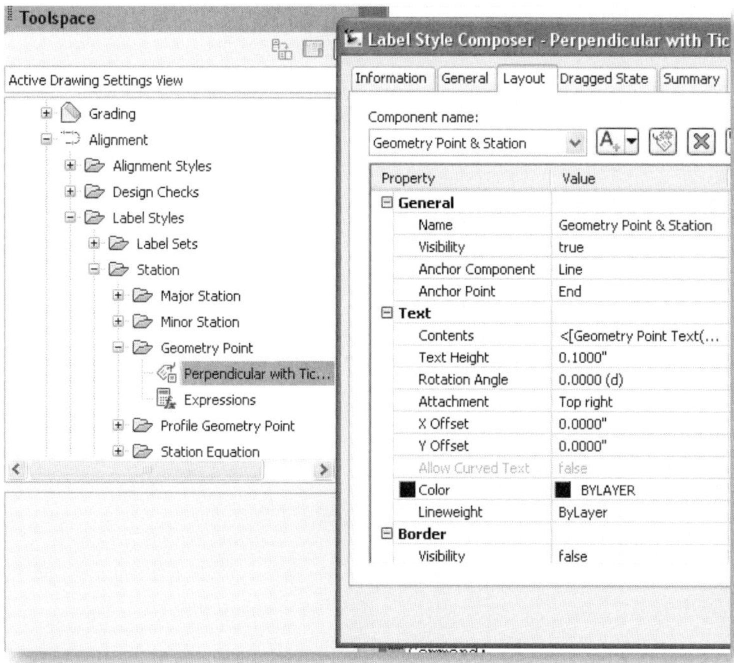


**FIGURE 5.15**

**Profile Geometry Point.** Profile Geometry Point identifies critical vertical design points on an alignment (see Figure 5.16). These points include beginning and end of vertical geometry, beginning and end of a vertical curve, high or low points, etc.

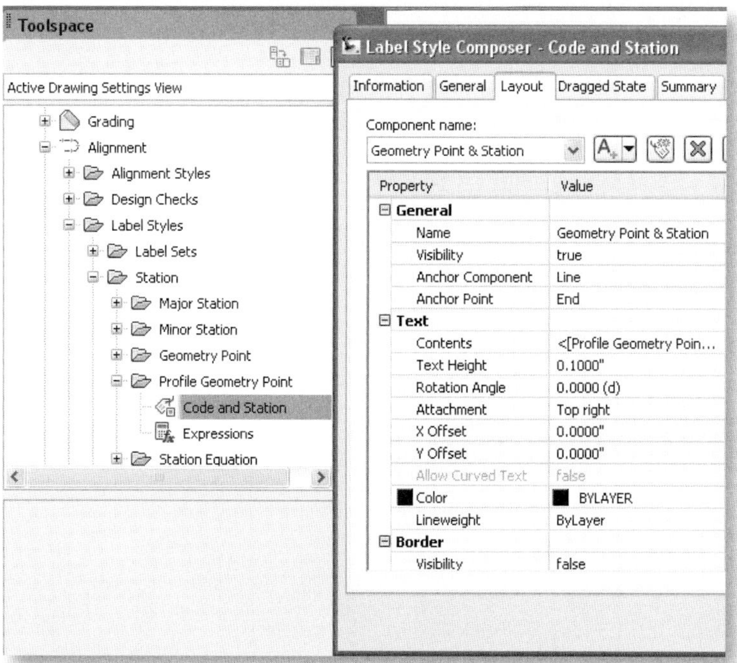


**FIGURE 5.16**

**Station Equation.** Station Equation annotates an alignment point where stationing changes. Usually a station equation is where roadway responsibility changes (a state takes over from a municipality). There is an ahead and back stationing from this point (see Figure 5.17).

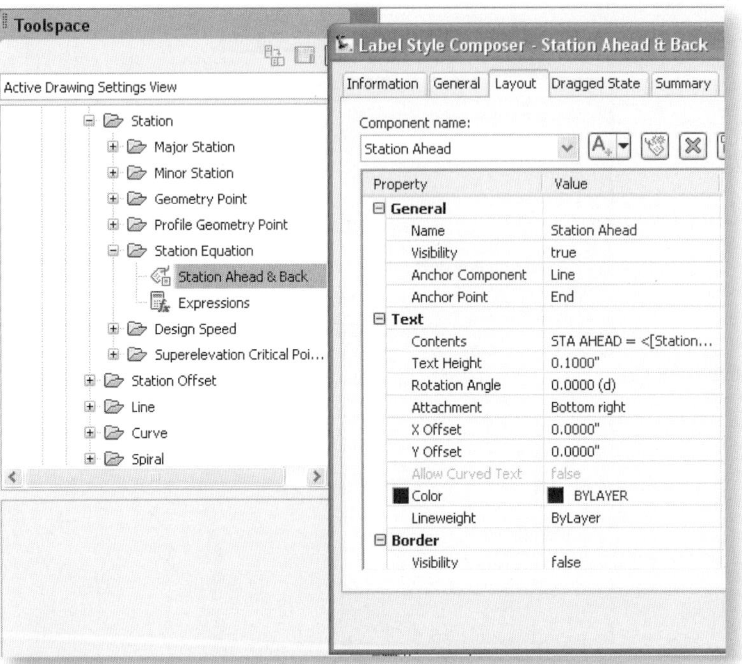


**FIGURE 5.17**

**Design Speed.** Design Speed labels the alignment's segments' design speed (see Figure 5.18).

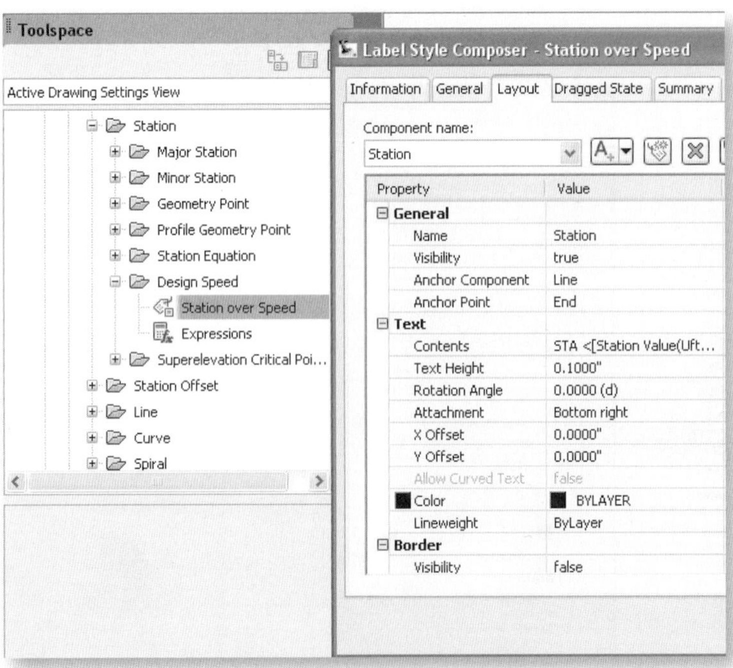


**FIGURE 5.18**

## Add Label Styles

These styles are post-alignment creation labels. The Add Labels dialog box assigns these labels to an alignment. See the Unit 5 discussion on the Add Labels command.

### Station and Offset

The Station Offset labels document station and offsets relative to an alignment. The station and offset - Fixed point labels do not move if alignment stationing changes. The label is updated to show the new station value. Station and Offset labels move, thus retaining the alignment's original station value.

### Line Labels

Line labels annotate tangent segments and are similar to those labeling parcel line segments.

### Curve Labels

Curve labels annotate arc segments and are similar to parcel curve labels.

### Spiral Labels

Spiral labels annotate spiral segments.

### Tangent Intersections

Tangent Intersections labels annotate PI, spiral in and out, and curve parameters (see Figure 5.19).

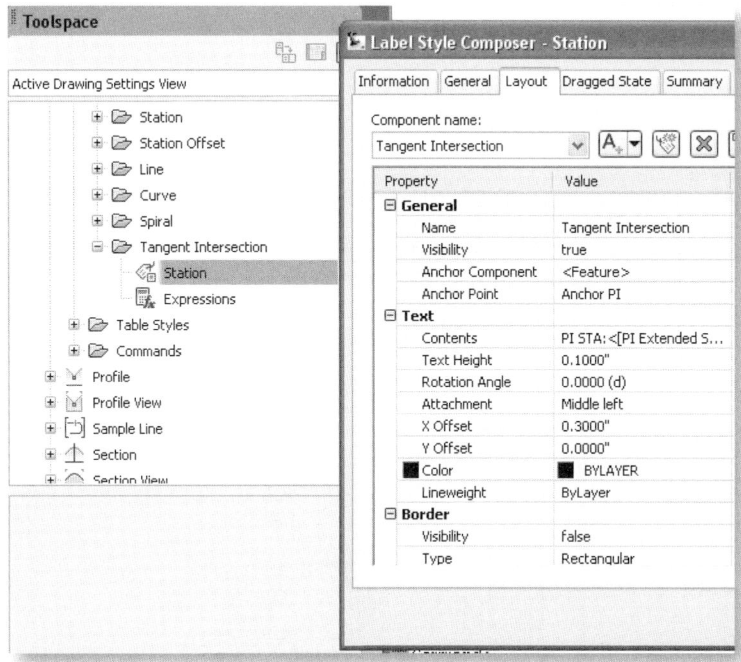


**FIGURE 5.19**

## DESIGN CHECKS

Design Checks set a minimum or maximum tangent, curve, or spiral values defined by an expression. Design Check Sets is an alias for a collection of tangent, curve, or spiral parameters (see Figure 5.20). Each parameter is an expression. For example, a minimum radius is an expression Radius>100, a maximum radius is Radius<300, or a minimum tangent is Tangent>100 (see Figure 5.21). When an alignment segment violates a check, in the drawing and the editing vistas, it is marked with a warning icon.

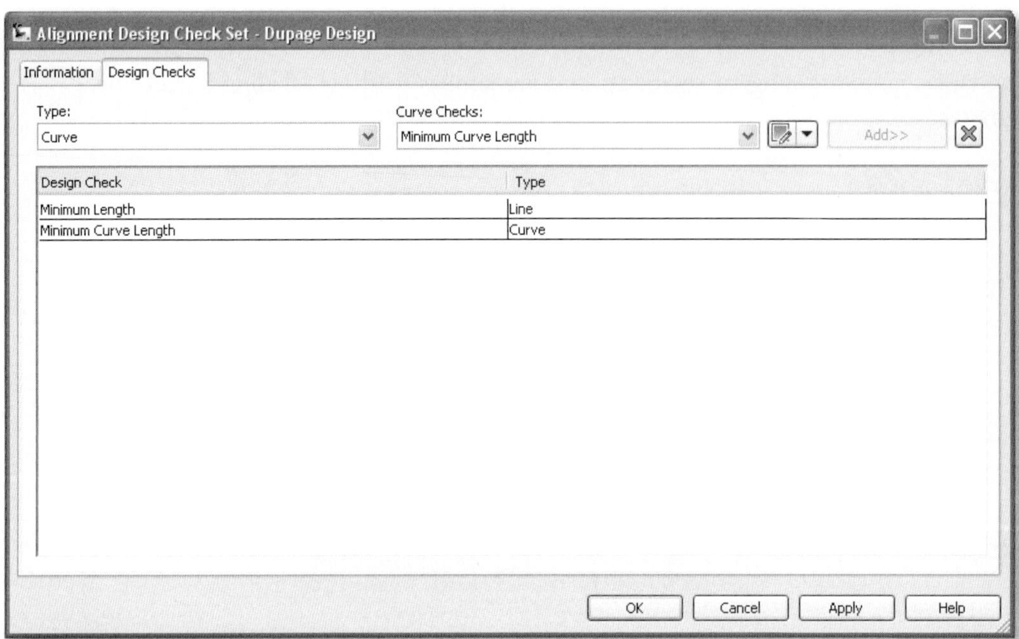


**FIGURE 5.20**

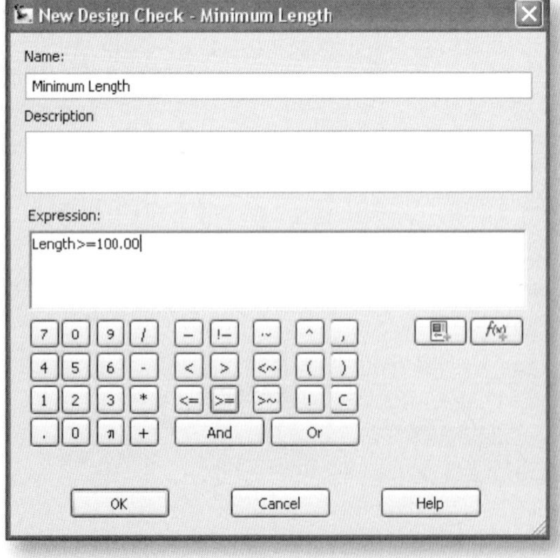


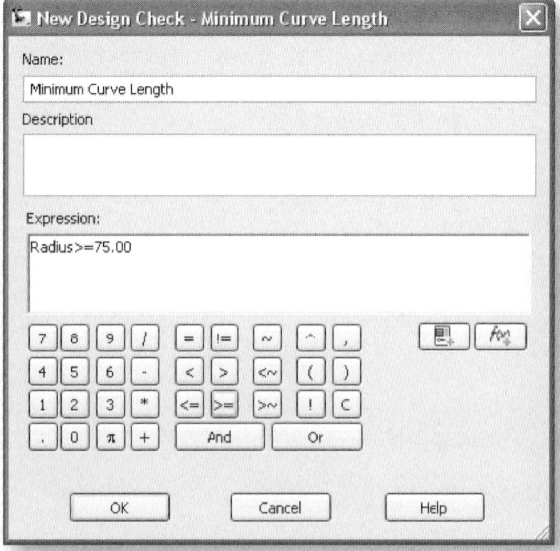


**FIGURE 5.21**

5-1

## EXERCISE 5-1

After completing this exercise, you will:

- Be familiar with alignment settings.
- Be familiar with alignment Edit Features Settings.
- Be familiar with alignment object styles.
- Be familiar with the Major & Minor label set.
- Be familiar with design check expressions and sets.

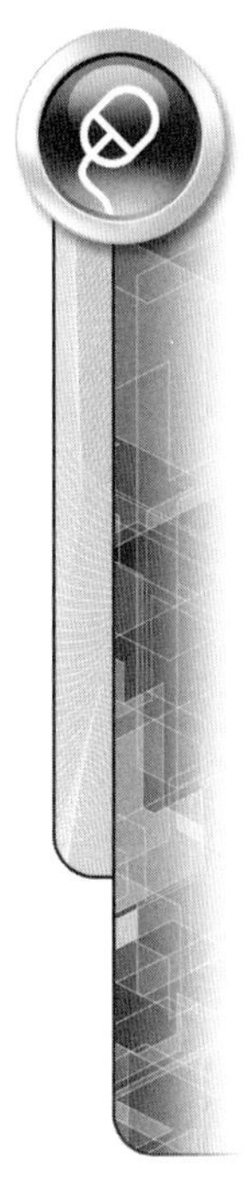

### Drawing Setup

This exercise starts with the new drawing, using the *Chapter 05 – Unit 1.dwt* file, found in the Chapter 5 folder of the CD that accompanies this textbook.

1. If not in Civil 3D, double-click its desktop icon to start the application.
2. Close the drawing and do not save it.
3. At Civil 3D's top let, click Civil 3D's drop-list arrow and from the Application Menu, select NEW. In the Select template dialog box, browse to the Chapter 5 folder of the CD that accompanies this textbook and open the *Chapter 05 - Unit 1.dwt* file. Click ***Open*** to use the template file.
4. At Civil 3D's top left, click Civil 3D's drop-list arrow and from the Application Menu, highlight Save As, from the flyout menu select AutoCAD Drawing, browse to the Civil 3D Projects folder, for the drawing name enter **Alignments**, and click ***Save*** to save the file.

### Edit Drawing Settings

Settings' Edit Drawing Settings values affect the alignment object and its labeling. The settings include layer-naming conventions and abbreviations for alignment geometry points.

1. Click the ***Settings*** tab.
2. In Settings, at its top, select the drawing name, press the right mouse button, and from the shortcut menu, select EDIT DRAWING SETTINGS....
3. Click the ***Object Layers*** tab.
4. In Object Layers, set the Modifier for Alignment, Alignment-Labeling, and Alignment Table to **Suffix**, and for the modifier value enter a dash followed by an asterisk (**-***) (see Figure 5.3).
5. Click the ***Abbreviations*** tab and review its settings for alignment points.
6. Click ***OK*** to save the changes and close Edit Drawing Settings.

### Edit Feature Settings — Alignment

Alignment's Edit Feature Settings assign several initial values and styles for creating and labeling alignments (see Figures 5.5, 5.6, and 5.7).

1. In Settings, select the Alignment heading, press the right mouse button, and from the shortcut menu, select EDIT FEATURE SETTINGS....
2. Expand the Default Styles, Default Name Format, Station Indexing, Criteria-Based Design Options, and Dynamic Alignment Highlight Options sections to review their values.
3. In the Criteria-Based section change the speed to **25** and click ***Apply***.
4. Click ***OK*** to close the dialog box.

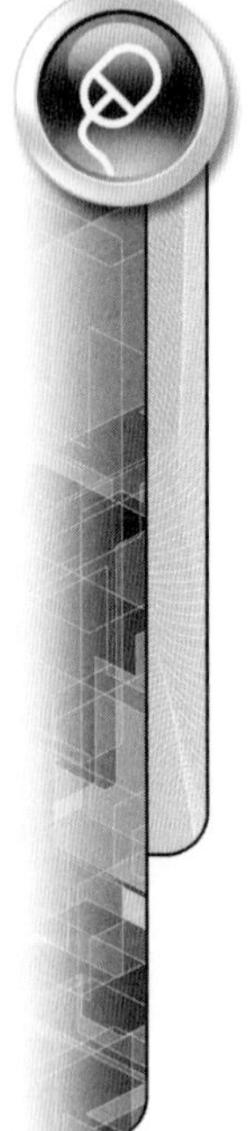

## Alignment Object Styles

This and the following exercises use two alignment object styles: Layout and Existing. The Layout style applies different colors to each alignment segment type, and the Existing style assigns the same color to all alignment segments.

1. In Settings, expand the Alignment branch until you are viewing the Alignment Styles list.
2. From the list, select the style **Layout**, press the right mouse button, and from the shortcut menu, select EDIT....
3. Click the ***Design*** tab. Toggle **ON** and change the Grip edit behavior value to **5.0**.
4. Click the ***Markers*** tab to review its settings and change the arrowhead size to **0.2**.
5. Click the ***Display*** tab to review the components settings for an existing alignment. Each component may have a different color assignment.
6. Click ***OK*** to exit the style.
7. From the list, select the style **Existing**, press the right mouse button, and from the shortcut menu, select EDIT....
8. Click the ***Design*** tab to review its settings.
9. Click the ***Markers*** tab to review its settings.
10. Click the ***Display*** tab to review the components settings for a proposed alignment. Each component is the same color.
11. Click ***OK*** to exit the Existing style.

## Alignment Label Set and Label Styles

The alignment you are going to draft uses the Major Minor and Geometry Points label set. This label set contains five alignment label styles: Major Station – Perpendicular with Line; Minor Station – Tick; Geometry Point – Perpendicular with Tick and Line; Station Equations – Station Ahead & Back; and Design Speeds – Station over Speed (see Figures 5.11 and 5.12). The label styles in this set are all from the Label Styles' Station branch of Alignments.

1. In Settings, expand the Alignment branch until you view Label Styles' Label Sets list.
2. From the list, select the label set Major Minor and Geometry Points, press the right mouse button, and from the shortcut menu, select COPY....
3. In the ***Information*** tab and for the name, enter **Perpendicular – Major Minor and Geometry Points**.
4. Click the ***Labels*** tab.
5. In the Labels panel's upper right, click the red ***X*** to delete all label styles.
6. In the Labels panel, to the right of Type, click the drop-list arrow, and from the list select **Major Stations**.
7. In the Labels panel's top center, to the right of Major Stations Label Style, click the drop-list arrow. From the list select **Perpendicular with Line**, and click ***Add>>*** to add it to the styles list. The increment should be **100.00**.
8. Repeat Steps 6 and 7, and set the label type to **Minor Stations**, the Minor Station Label Style to **Tick**, and add it to the label styles list. The increment should be **50**.
9. Repeat Steps 6 and 7, and set the label type to **Geometry Points**, the Geometry Points Label Style to **Perpendicular with Tick and Line**, and add it to the label styles list. The Geometry Points dialog box is displayed. Review the label toggles and click ***OK*** to add the labels to the list.

10. Repeat Steps 6 and 7, and set the label type to **Station Equations**, the Station Equation Label Style to **Station Ahead & Back**, and add it to the list of label styles.
11. Repeat Steps 6 and 7, and set the label type to **Design Speeds**, the Design Speeds Label Style to **Station over Speed**, and add it to the list of label styles.
12. Click ***OK*** to create the new set.

## Review Label Set Styles

1. In Settings, Alignment Label Styles, expand the Station branch until you are viewing Major Station's styles list.
2. From the styles list, select **Perpendicular with Line**, press the right mouse button, and from the shortcut menu, select EDIT....
3. In the Label Style Composer dialog box, click Layout to review its contents. The text anchors on the Line component.
4. At the panel's top right, click the Component name drop-list arrow and select the Line component to review its anchor point. It is the feature (station).
5. Click ***OK*** to exit the dialog box.

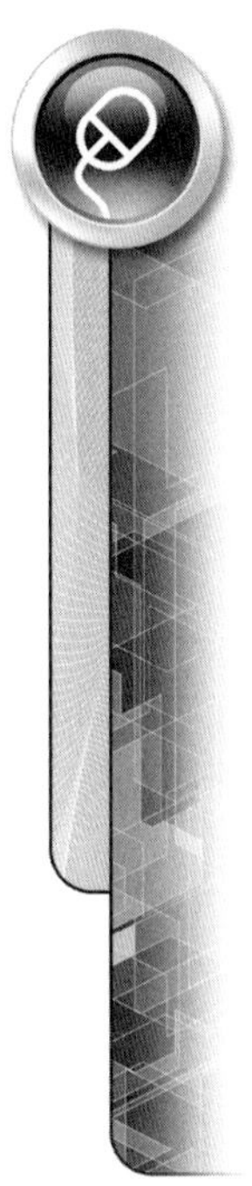

## Design Checks

Design Checks set minimum and maximum values as expressions in a check set.

1. In the Settings, Alignment branch, expand Design Checks until you are viewing the Design Check Sets list.
2. Under Design Checks select Curve, press the right mouse button, and from the shortcut menu, select NEW....
3. In the New Design Check calculator for the check name, enter **Minimum Radius**.
4. In the New Design Check calculator, in the middle right, click the icon Insert property (first button on the right), and from the list, select **Radius**. Next, click the equal to or greater than icon (>=) and then enter **75.00** for the minimum radius. Then review your expression and compare it to Figure 5.21. When you are done entering the values, click ***OK*** to create the check.
5. Under Design Checks select Line, press the right mouse button, and from the shortcut menu, select NEW....
6. In the New Design Check calculator, for the check name enter **Minimum Length**.
7. In the New Design Check calculator, in the middle right, click the icon Insert property, and from the list select **Length**. Next, click the equal to or greater than icon (>=) and then enter **100.00** for the minimum length. Then review your expression and compare it to Figure 5.21. When you are done entering the values, click ***OK*** to create the check.

## Create a Design Check Set

1. In the Settings, Alignment branch, select Design Check Sets, right mouse click, and from the shortcut menu, select NEW....
2. In the Alignment Design Check Set dialog box, click the ***Information*** tab and for the name enter **Dupage Design**.
3. Click the ***Design Checks*** tab.
4. In the upper left, click the Type: drop-list arrow and set the type to **Line**. Line Checks should list **Minimum Length**. Click ***Add>>*** to add it to the check list.
5. In the upper left, click the Type: drop-list arrow and set the type to **Curve**. Curve Checks should list **Minimum Radius**. Click ***Add>>*** to add it to the check list.

6. Click ***OK*** to exit the dialog box.
7. At Civil 3D's top left, Quick Access Toolbar, click the ***SAVE*** icon to save the drawing.

This completes the exercise setup and styles review for this chapter. Next, you learn to define and draft alignments.

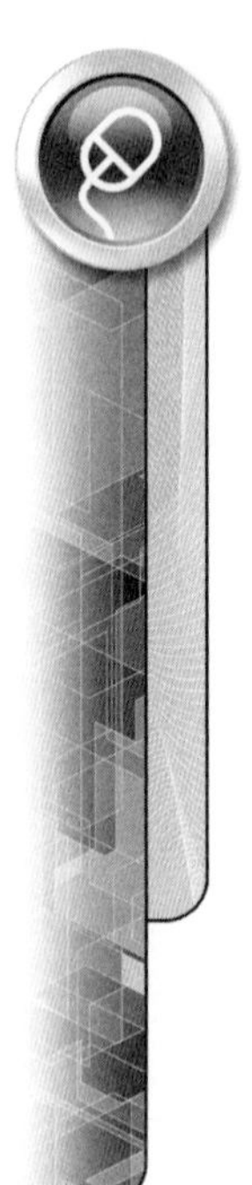

## SUMMARY

- Edit Drawing Settings, Object Layers assigns alignment base layer names.
- If you have more than one alignment, use either a layer prefix or a suffix to place an alignment on a unique layer.
- Alignment annotation is usually assigned when creating an alignment, but the annotation can also be assigned after creating an alignment.
- An alignment label set contains user-specified label styles.
- Alignment labels change immediately when you change or add label set styles.
- Design checks are expressions applied to an alignment.
- Any segment that violates a design check will cause the display of a warning icon in the drawing and in the alignment vistas.

## UNIT 2: CREATING ROADWAY CENTERLINES

A centerline is composed of tangents (lines), curves (arcs), and possibly spirals. There are four methods of creating an alignment. The first method is to import a LandXML file and the second method is to import an LDT project's alignment definition. Both import one or more alignments. Each method displays an alignment list and the user selects which alignment(s) to import.

The third method is to convert objects or polyline into an alignment. A drawback to this method is that the stationing corresponds to the first object's or polyline's direction. However, there is a reverse alignment direction command. A second drawback to these three methods is that the alignments are composed of fixed alignment segments. Fix segments are not dependent on a before or after alignment segment; they are drawn by selecting points or by import. Because they are not dependent on any adjacent existing segments, when editing them, they do not maintain tangency to adjacent segments.

The next method is to draft an alignment using the Alignment Layout Tools toolbar. The Alignment Layout Tools toolbar routines create tangent, curve, and spiral segments. Each segment can be fixed, floating, or free. Another alignment design method is to sketch controlling geometry and use Alignment Layout Tools commands to trace or convert them to alignment segments (see Figure 5.22).

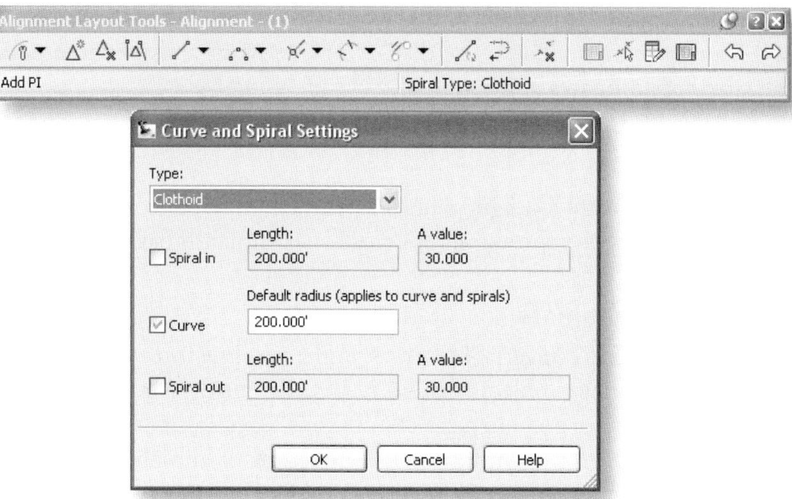


**FIGURE 5.22**

Whether creating an alignment from objects, polyline, or from the tools in the Alignment Layout Tools toolbar, each method displays the Create Alignment dialog box. The General tab sets the alignment name, its description, starting station, site, style, layer, and alignment label set (see the left side of Figure 5.23). The Design Criteria tab sets the design speed, toggles criteria checking, sets AASHTO (superelevation parameters), and sets the design check set name (see the right side of Figure 5.23).

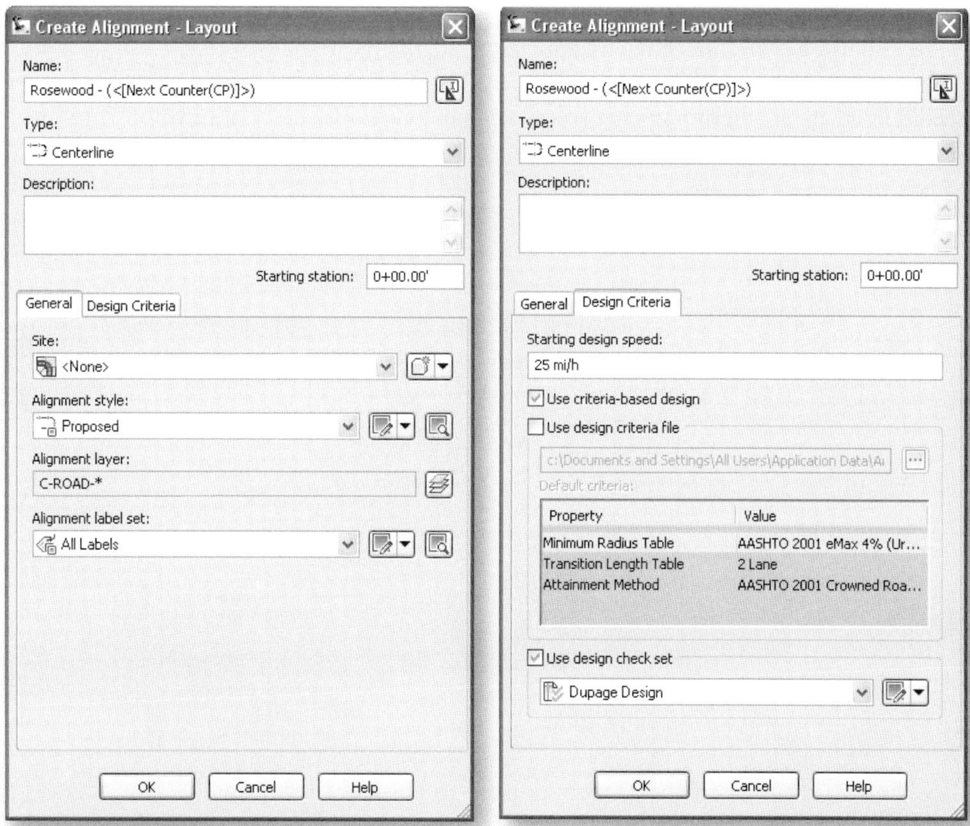


**FIGURE 5.23**

How the alignment is displayed is a result of two associated styles: the alignment and label set styles. The alignment style sets the display of line work and its curve radii constraints. The label set annotates alignment line work.

## ALIGNMENT CREATION

The Ribbon's Home tab, Create Design panel has the Alignment icon and when clicking its drop-list arrow displays the alignment creation commands (see Figure 5.2).

### Create Alignment from Objects

Create Alignment from Objects converts lines, arcs, and a polyline into an alignment that has all fixed segments. These objects can be in the current drawing or selected from an xrefed drawing. With a polyline, the alignment's direction is the polyline's direction. The command prompting indicates the direction of the alignment and gives you the opportunity to reverse the alignment's direction. If you do not reverse the alignment's direction at this time, the Ribbon's Modify, Alignment panel's Modify section drop-list menu Reverse Direction command will reverse its direction. At the bottom of the Create Alignment from Objects dialog box are settings affecting the creation of curves at each PI. When drafting an alignment, all segments must have the same direction. If you draft individual segments, even if separated, they all must have the same direction to connect with other segment types. If a segment has the wrong direction, the Alignment Layout Tools toolbar has a reverse segment command.

### Alignment Creation Tools

The Alignment Layout Tools toolbar commands create tangent and curve segments (see Figure 5.24). The tangent and curve segments can be fixed, floating, or free. These segment types define an alignment with the user's desired behaviors and when edited, create an alignment that solves several constraint and relationship issues.

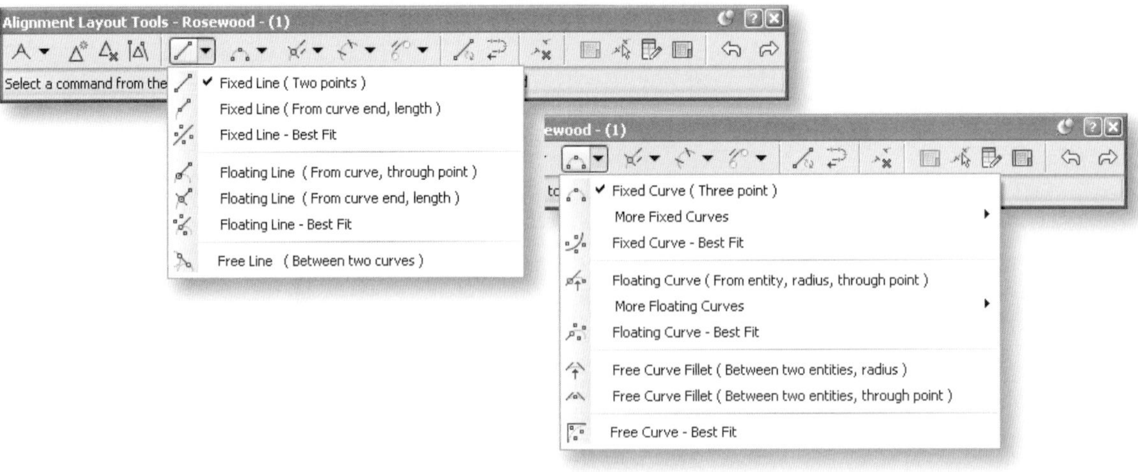


**FIGURE 5.24**

At times, certain segment properties need to be held (for example, always tangent, always attached to two entities, etc.). Fixed, floating, or free segment types can maintain a complex alignment that has specific segment connection relationships.

### Tangent Lines Only and Tangent with Curves

The first icon stack at the toolbar's left side draws centerlines with or without curve segments (see Figure 5.24). When you click the icon stack drop-list arrow, it displays the command choices. All tangents are fixed segments (two selected points with no before or after dependencies) and when you are using Tangents with Curve, the curves are free segments (dependent on the before and after tangent). Curve and Spiral Settings set the curve radius or spiral lengths for each free curve or spiral. If you are using spirals, Civil 3D supports Bloss, Clothoid, Cubic (JP), and Sine Half-wave Diminishing Tangent spiral types.

### PI Management

The toolbar's next three icons in from the left insert, delete, and break apart PIs. These commands add a PI to a tangent segment where there is none, relocate an existing PI, and separate tangents at a PI to create space for a new tangent.

### Individual Tangent and Curve Entities

The Tangent and Curve icon stacks provide tools to create specific tangent and curve segment types (see Figure 5.24). Each segment can reference existing alignment segments, use the transparent command direction overrides (bearing, azimuth, etc.), or use object snaps. When connecting to a segment, use an object snap.

Alignments have three segments types: fixed, floating, and free.

A fixed segment:

- Is defined by selecting points and, if a curve, possibly by specifying a radius.
- Is editable only by adjusting its segment coordinates (line endpoints or the three curve points).
- Does not depend on another alignment segment for its geometry or tangency.
- Does not maintain tangency when edited.

The first rule states that two points (line), three points (curve), two points and a radius, etc., specify a fixed tangent. The second rule implies you can change only the coordinates of the line or curve (three points define the curve). The Alignment Editor vista or sub-entity editor presents the segment's coordinates (see Figure 5.25). The blackened values indicate editable values.

The third rule states that the entity is not attached to any alignment segments (tangent to, from the endpoint of, etc.). The fourth rule states that if drawn initially tangent to another object, editing the entity breaks the original tangency between the objects.

**NOTE**

**When thinking about fixed segments, think zero dependencies.**

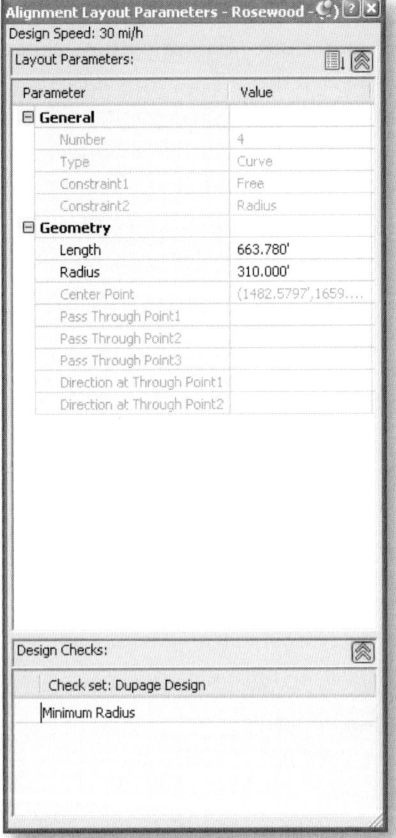


**FIGURE 5.25**

A floating segment is:

- Dependent on and always attached to the before alignment segment.
- Always tangent to the attached before entity.
- Defined by user parameters.
- Attachable only to another fixed or floating entity.

The first and second rules state that a floating segment attaches to a before alignment segment and is tangent. The third rule states that after attaching to the before segment, the user must specify other parameters to complete the segment's definition. The last rule states that the segment attaches only to fixed or floating alignment segments.

**NOTE**

When thinking about floating segments, think one dependency.

A free segment is:

- Dependent on two alignment segments for its geometry.
- Always tangent to the before and after segments.
- Connected between two fixed, two floating, or one fixed and one floating segment.
- Defined by before and after segment geometry and additional user-specified parameters.

The first and second rules state that a free entity requires two existing entities and is tangent to both. The third rule states that a free entity connects only to combinations of fixed and floating segments. The last rule states that after attaching a free segment, the user enters additional parameters.

NOTE

**When thinking about free segments, think two dependencies.**

### Spiral Segments

The Alignment Layout Tools toolbar has three icon stacks for creating spirals: Floating Line and Spiral, Floating and Free Spiral-Curve-Spiral, and Fixed and Free Compound or Reverse Spirals. A fixed spiral does not depend on any alignment segments and when edited does not retain tangency. A floating spiral attaches to an existing segment's end and remains tangent when edited. The default method is spiral, curve, and spiral (S-C-S). Any spiral's length can be 0 (zero). This makes the curve tangent to its connecting segment. The default in and out spiral settings are set in the Tangent/Tangent Curve icon stack at the toolbar's left, Curve and Spiral Settings....

## STATION EQUATIONS

Alignments may change stationing at one or more points along their path; this is a station equation. The alignment stationing changes at this point, and from this point, the stationing may increase or decrease. An alignment may have several station equations. Station equations are defined and managed as an alignment property.

## DESIGN SPEEDS

When defining an alignment, it potentially has design speeds. The Alignment Properties dialog box manages design speeds. A design speed in a necessary value when designing roads with superelevation.

## IMPORTING LANDXML

LandXML files can contain alignment definitions (see Figure 5.26). The file contains the alignment's coordinates and segment geometry. This allows transferring design data between applications without losing design fidelity. If the file contains more than one alignment, the user can select which alignments to import.

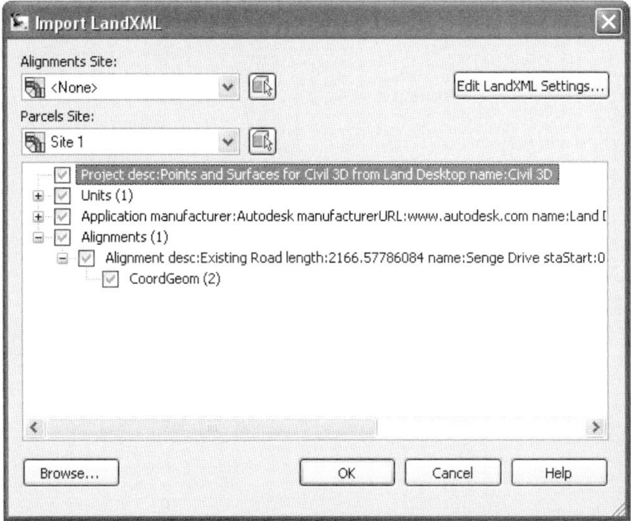


**FIGURE 5.26**

## IMPORTING FROM LAND DESKTOP

Importing an LDT project's alignment is similar to importing a LandXML file. The Land Desktop command, found in Ribbon's Insert tab, Import section, displays a dialog box where the user selects a project path and a project name (see Figure 5.27). After selecting the project path and project name, the dialog box displays the LDT project data as a list, including alignments. From the list, select the alignment(s) and, when you exit, the alignments are imported to the drawing.

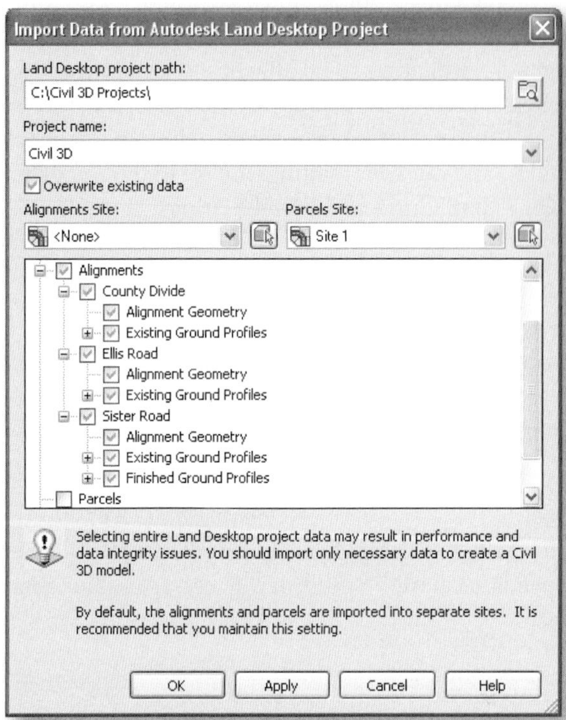


**FIGURE 5.27**

## ALIGNMENT TYPES

Civil 3D defines several alignment types; Centerline, Offset, Curb Return, and Miscellaneous. The Offset alignment's purpose is to control the location of a subassembly point. For example, an offset can widen the travelway or widen a shoulder. Curb return alignments are for intersection design.

An offset alignment can be static or dynamically linked to the centerline. These alignment types are discussed and used in the Chapter 9 of this textbook.

## EXERCISE 5-2

After completing this exercise, you will:

- Be able to create an alignment by importing a LandXML file.
- Be able to create an alignment from a polyline.
- Be able to create an alignment using the Alignment Layout Tools toolbar.
- Be able to use transparent commands while laying out a centerline.

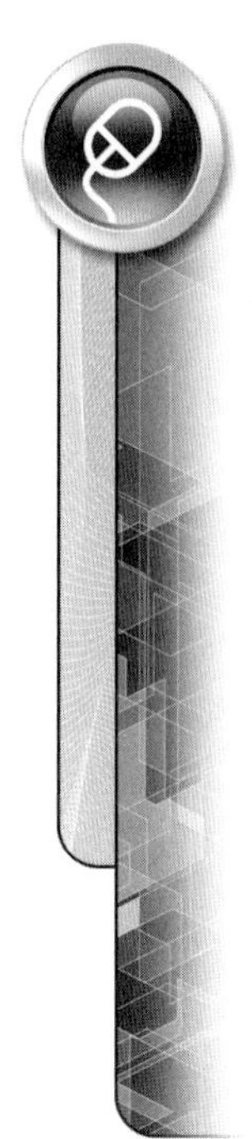

### Exercise Setup

This exercise continues with the previous exercise's drawing, Alignments. If you did not complete the previous exercise, browse to the Chapter 5 folder of the CD that accompanies this textbook and open the *Chapter 05 – Unit 2.dwg.*

1. If not open, open the previous exercise's Alignments drawing, or browse to the Chapter 5 folder of the CD that accompanies this textbook, and open the *Chapter 05 – Unit 2* drawing.
2. In Layer Properties Manager, or by using the Layer Freeze icon, freeze the contour and boundary layers **3EXCONT**, **3EXCONT5**, and **Boundary**, and click **X** to exit the palette.

### Import an Alignment

The LandXML file contains alignment definitions and can be imported to create alignments. The LandXML files for this section are in the Chapter 5 folder of the CD that accompanies this textbook.

1. On the Ribbon, click the ***Insert*** tab. On the Import panel, select LANDXML.
2. In Import LandXML, browse to the Chapter 5 folder of the CD that accompanies this textbook, and from the files select **Existing Road.xml**, and click OPEN.
3. In the Import LandXML dialog box, click ***OK*** to import the Senge Drive alignment.

### Create Alignment From Objects

Create Alignment From Objects converts a polyline into an alignment.

1. On the Ribbon, click the ***View*** tab. At the left side of the Views panel, select and restore the named view **Existing Road**.
2. Click Ribbon's ***Home*** tab, Create Design Panel, click the drop-list arrow for Alignment, from the shortcut menu select CREATE ALIGNMENT FROM OBJECTS, in the drawing near its southern end, select the red polyline (C-ROAD-CTLN), and press ENTER to continue.
3. The command line prompts to accept or reverse the alignments direction. Press ENTER to accept the direction and to continue.

4. The Create Alignment from Objects dialog box opens. Using Figure 5.28 as a guide, enter the following values: for Name, leaving the counter, enter **OMalley Phase 2**; for the type set it to **Centerline**, for Site, click the Create New icon on the right, for the name enter **Site 1**, click ***OK***; for the Alignment Style assign **Existing**; for the Alignment Label Set assign **Perpendicular – Major Minor and Geometry Points**; toggle **OFF** Add curves between tangents, and toggle ON Erase existing entities.

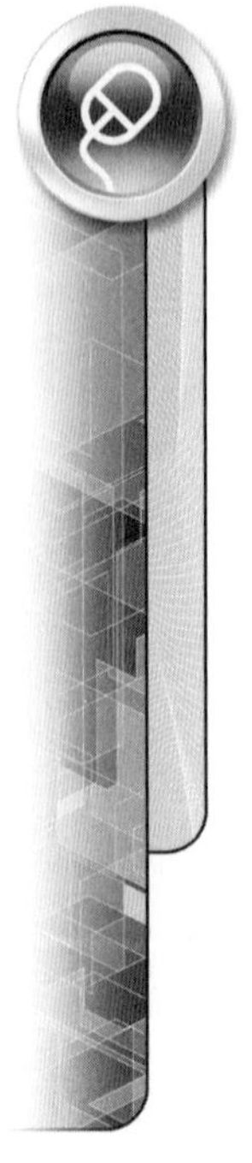

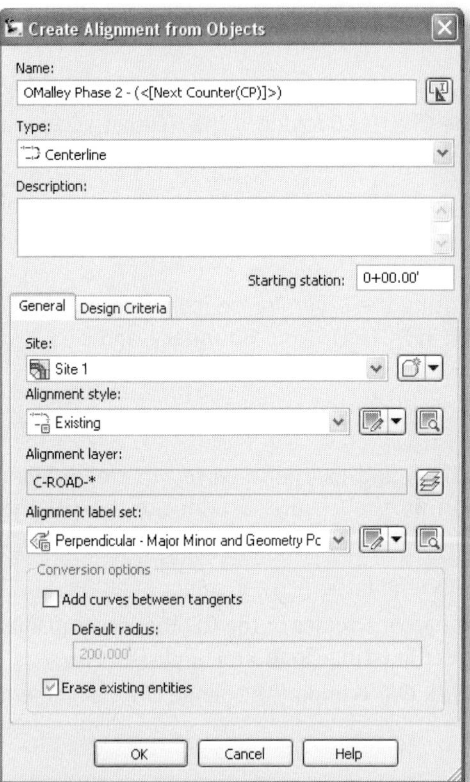


**FIGURE 5.28**

5. Click the ***Design Criteria*** tab.
6. Using Figure 5.29, set the following values. Set the Starting design speed to **25**, toggle **ON** Use criteria-based design, toggle **OFF** Use design criteria file, toggle **ON** Use design check set, click the check set drop-list arrow, and from the list select **Dupage Design**.
7. Click ***OK*** to exit and define the alignment.

**FIGURE 5.29**

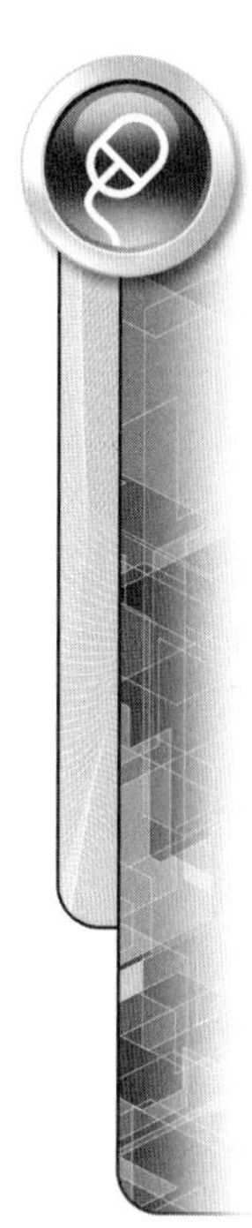

The alignment violates the 100 foot minimum tangent rule. 

8. Place the cursor over the violation icon and a tooltip lists the violation.

The stationing should start at the alignment's northern end and increase toward the south.

9. If the stationing increases from the south to north, select the Alignment. From the Ribbon's Alignment: Omalley Phase 2 – (1) tab, select the Modify panel's title to unfold the panel and select Reverse Direction, and in the warning dialog box click ***OK*** to accept the change and reverse its direction.
10. Use the `ZOOM EXTENTS` command to view the entire site.
11. At Civil 3D's top left, Quick Access Toolbar, click the ***Save*** icon to save the drawing.

## Offsetting Preliminary Centerline Tangents

The engineer wants the alignment to start at two points at the site's southwestern corner (point numbers 1 and 2). The centerline should be about 185 feet in from the Phase I and II boundary north of the points. Lines on the P-CL layer represent lines to be offset to create centerline line segments (see Figure 5.30).

1. On the Ribbon, click the ***View*** tab. In the Views panel's left, select and restore the named view **Proposed Starting Point**.
2. Click Ribbon's ***Home*** tab. In the Layers panel, click `ISOLATE`, in the upper right of the view, select the entity on the **P-CL** layer (objects in beige) and press `ENTER`.

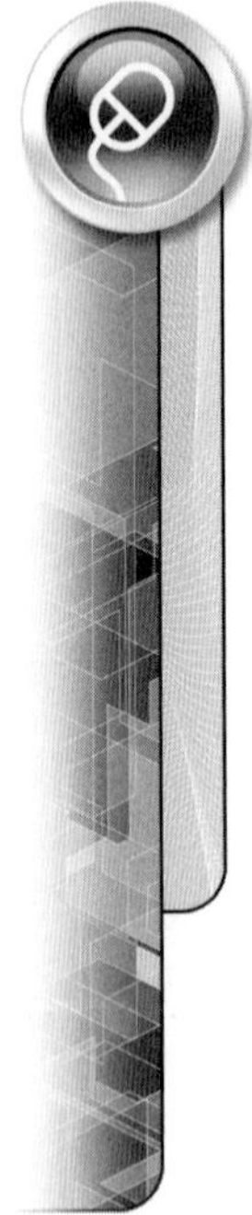

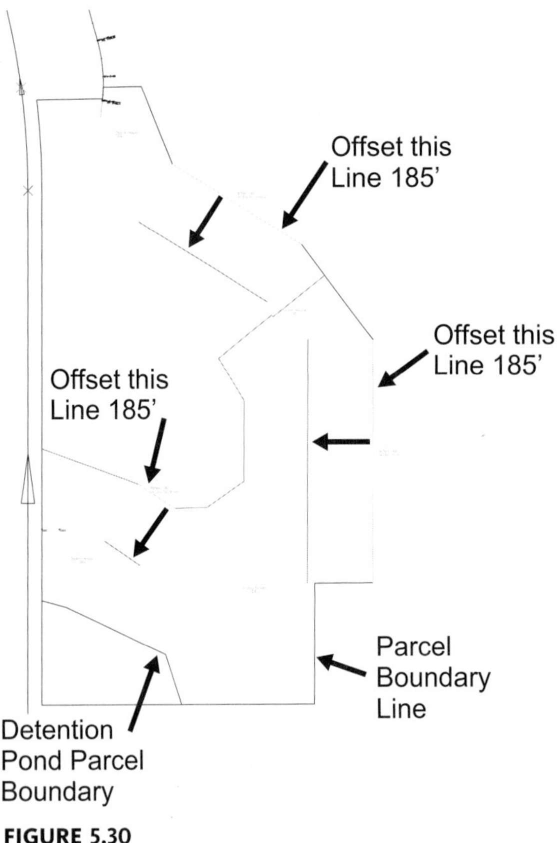


**FIGURE 5.30**

3. Use the OFFSET command, set the distance to **185**, offset the line in the upper right, southwest, and exit the offset command.
4. Use the ZOOM and PAN commands to view the remaining two lines.
5. Use OFFSET command and offset the eastern line, west; the northeastern line, southwest; and then exit the command.
6. Click the Ribbon's ***Home*** tab. In the Layers panel's right side, click unISOLATE.

## Define the Site Boundary

Next, you define the boundary as a parcel (site).

1. In Ribbon's Home tab, Create Design panel's left side, click the drop-list arrow of the ***Parcel*** icon. From the shortcut menu select CREATE PARCEL FROM OBJECTS, select the parcel's outer boundary (see Figure 5.30), press ENTER, and click *OK* to accept the defaults in the Create Parcels – From Objects dialog box.

## Define Detention Pond Parcel

1. In Ribbon's Create Design panel's left side, click the drop-list arrow of the ***Parcel*** icon. From the shortcut menu select CREATE PARCEL FROM OBJECTS, select the Detention Pond boundary located in the site's southwestern corner (see Figure 5.30), press ENTER, and click ***OK*** to accept the defaults in the Create Parcels – From Objects dialog box.
2. At Civil 3D's upper left, Quick Access Toolbar, click the ***Save*** icon to save the drawing.

## Drafting the Rosewood Alignment

The Rosewood alignment starts at the site's lower-left (southwest) side (point number 1) and winds its way north to its connection with the OMalley centerline. Begin drafting the alignment with points 1 and 2 using transparent command point filters. In the Alignment Layout Tools toolbar other commands convert the previously offset lines to fixed tangent segments. Finally, free curves connect the converted tangent segments creating the finished alignment.

1. Make sure the Transparent Commands toolbar is visible.
2. Click the Ribbon's ***View*** tab. At the Views panels left, select and restore the named view **Proposed Starting Point**.
3. Click the Ribbon's Home tab. In the Create Design panel, click the ***Alignment*** icon, and from the shortcut menu select `ALIGNMENT CREATION TOOLS`.
4. In the Create Alignment – Layout dialog box, for the alignment Name and leaving the counter, enter **Rosewood**, set the alignment type to **Centerline**, create the alignment in **Site 1**, set the Alignment Style to **Layout**, and set the Alignment Label Set to **Perpendicular – Major Minor and Geometry Points.**
5. Click the ***Design Criteria*** tab and set the Starting design speed to **25**, toggle **ON** Use criteria-based design, toggle **OFF** Use design criteria file, toggle **ON** Use design check set, click the check set drop-list arrow, and from the list select **Dupage Design**.
6. Click ***OK*** to continue.

## Tangent 1 – Fixed Tangent

The first segment is a fixed segment drawn with Fixed Line – Two Points and references points 1 and 2 with a transparent command point filter.

1. From the Alignment Layout Tools toolbar, click the ***Fixed Line – (Two points)*** icon (fifth in from the left).
2. Next, in the Transparent Commands toolbar click the point filter ***Point Object*** ('PO), and in the drawing select anywhere on point 1 and then anywhere on point 2.
3. Press `ENTER` twice to return to the command line.

A fixed alignment segment appears between the points with an arrow pointing from left to right, and it violates the minimum tangent length criteria (see Figure 5.31).

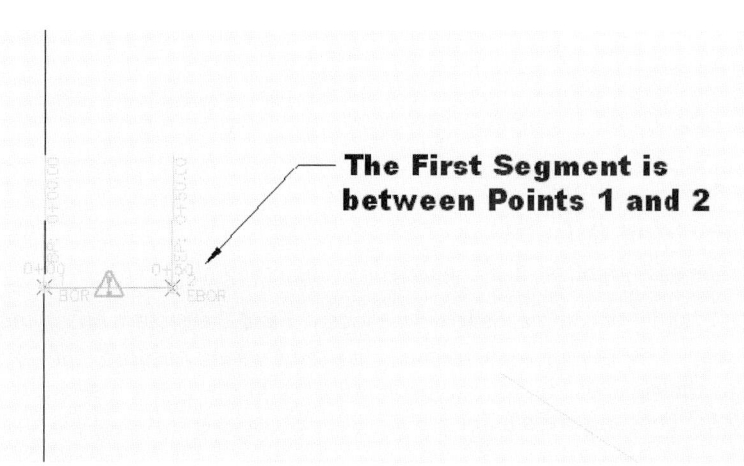


**FIGURE 5.31**

## Tangent 2 - Converted Offset Line

The second tangent is the southern offset line converted to a fixed segment.

1. From the Alignment Layout Tools toolbar, click the ***Convert AutoCAD line and arc*** icon (to the right of the Spiral type icon stack), in the drawing select the offset line's western end, and press ENTER.

The converted segment's direction is the same and it is the first tangent.

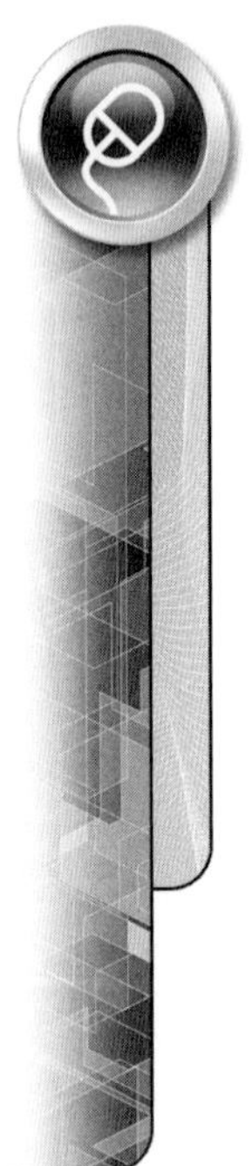

## Curve 1 - Free Curve

The first curve has two dependent segments, before and after, and is a free curve segment.

1. Use the ZOOM command to better view both entities' ends.
2. From the Alignment Layout Tools toolbar, click the Curve drop-list icon (sixth icon in from the left), and from the Curve icon stack, select FREE CURVE FILLET (BETWEEN TWO ENTITIES, RADIUS).
3. The command line prompts you to select the first entity. Select Tangent 1's eastern end, and when you are prompted for the next entity, select Tangent 2's western end.
4. The command line prompts you if the curve is Less than 180 degrees. Press ENTER to accept the value and continue.
5. The command line prompts you for the last parameter, radius. For the radius type **285** and press ENTER twice to create the curve and exit to the command line.

The Curve 1 is tangent at Tangent 1's end and slightly down from Tangent 2's western end (see Figure 5.32).

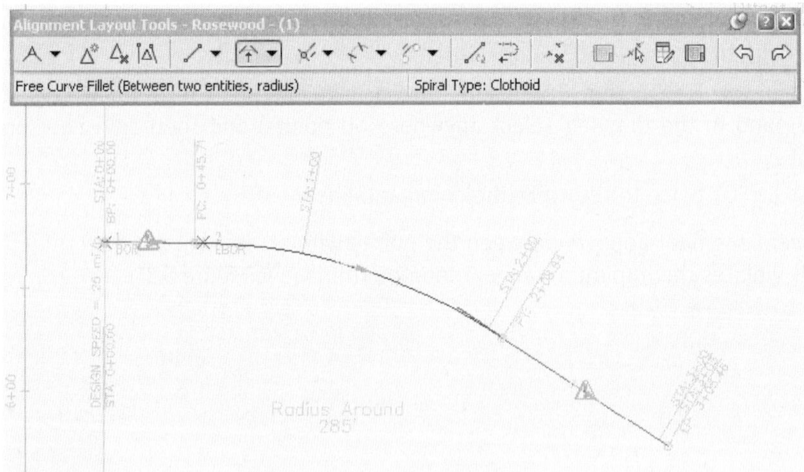


**FIGURE 5.32**

## Tangent 3 - Converted Offset Line

The third tangent is the eastern offset line converted to a fixed segment.

1. Use the ZOOM and PAN commands to view Tangent 2 and the eastern offset line.
2. From the Alignment Layout Tools toolbar, click the ***Convert AutoCAD line and arc*** icon (to the right of the Spiral type icon stack), in the drawing select the line's southern end, and press Enter.

The converted segment's direction should be the same as the alignment.

3. If necessary, on the Ribbon, click the Home tab. Click Layer Control's drop-list arrow, and then scroll to and turn off the layer P-CL.

## Curve 2 – Free Curve

The second curve has two dependent segments, before and after, and is a free curve segment.

1. From the Alignment Layout Tools toolbar, click the Curve drop-list icon (sixth icon in from the left), and from the Curve icon stack, select FREE CURVE FILLET (BETWEEN TWO ENTITIES, RADIUS).
2. The command line prompts you to select the first entity. Select Tangent 2's eastern end, and when prompted for the next entity, select Tangent 3's southern end.
3. The command line prompts you if the curve is Less than 180 degrees. Press ENTER to accept the value and continue.
4. The command lines prompts you for the last parameter, radius. For the radius type **310** and press ENTER twice to create the curve and exit to the command line.

The Curve 2 is tangent at Tangent 2's end, up from Tangent 3's southern end (see Figure 5.33).

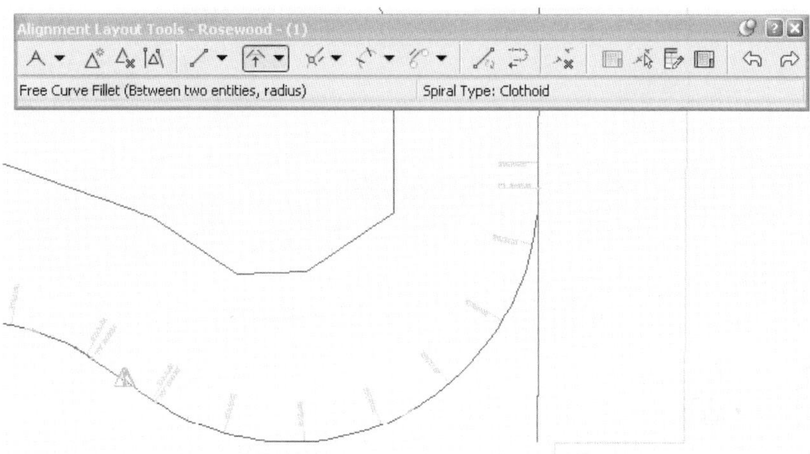


**FIGURE 5.33**

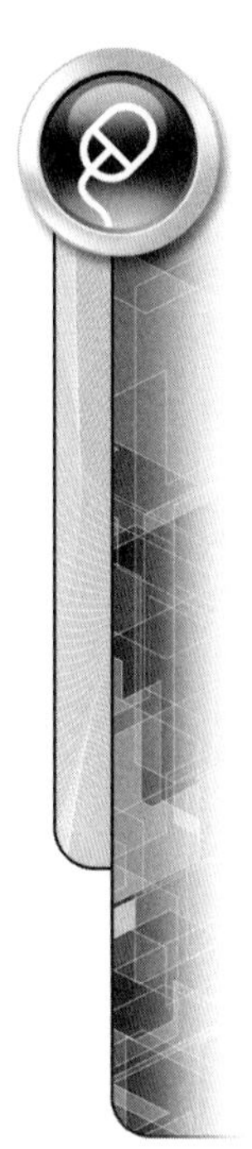

## Tangent 4 – Converted Offset Line

The fourth tangent is the northern offset line converted to a fixed segment.

1. From the Layers panel, click the Previous icon to turn on the **P-CL** layer.
2. Use the ZOOM and PAN commands to view Tangent 3 and the northern offset line.
3. From the Alignment Layout Tools toolbar, click the ***Convert AutoCAD line and arc*** icon (to the right of the Spiral type icon stack), in the drawing select the line's western end, and press ENTER.

The converted segment's direction is in the opposite direction of the third tangent.

4. Click the Layer Control's drop-list arrow, and then scroll to and turn off the **P-CL** layer.
5. From the Alignment Layout Tools toolbar, click the ***Reverse Sub-entity Direction*** icon (the next icon to the right), in the drawing select the line's southern end, and press ENTER.

The converted segment's direction changes to be the same as the alignment.

## Curve 3 – Free Curve

The third curve has two dependent segments, before and after, and is a free curve segment.

1. From the Alignment Layout Tools toolbar, click the Curve drop-list icon (sixth icon in from the left), and from the Curve icon stack, select FREE CURVE FILLET (BETWEEN TWO ENTITIES, RADIUS).
2. The command line prompts you to select the first entity. Select Tangent 3's northern end, and when prompted for the next entity, select Tangent 4's southern end.
3. The command line prompts you if the curve is Less than 180 degrees. Press ENTER to accept the value and continue.
4. The command lines prompts you for the last parameter, radius. For the radius type **310** and press ENTER twice to create the curve and exit to the command line.

The Curve 3 is down from Tangent 3's northern end and slightly up from Tangent 4's southern end (see Figure 5.34).

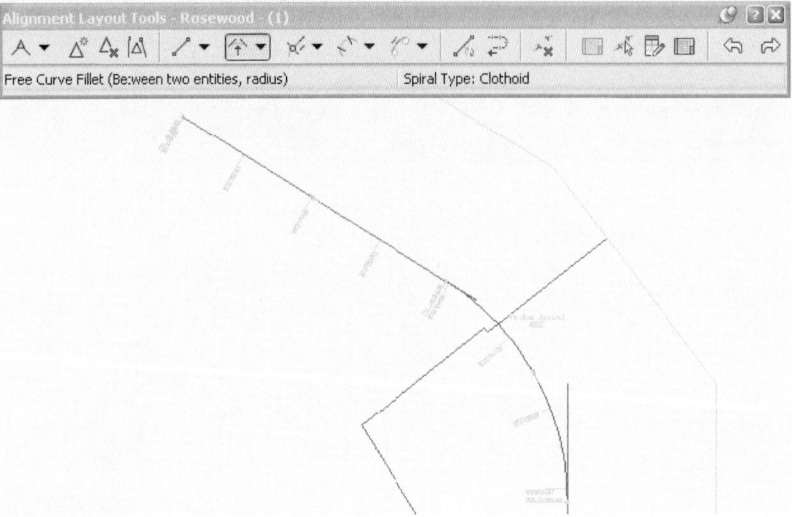


**FIGURE 5.34**

## Tangent 5 – Converted Line

The fifth tangent is a line converted to a fixed segment.

1. Use the ZOOM and PAN commands to view Tangent 4 and the line just south of the OMalley alignment.
2. From the Alignment Layout Tools toolbar, click the ***Convert AutoCAD line and arc*** icon (to the right of the Spiral type icon stack), in the drawing select the line's southern end, and press ENTER.

The converted segment's direction is in the opposite direction of the third tangent and violates Minimum Length.

3. From the Alignment Layout Tools toolbar, click the ***Reverse Sub-entity Direction*** icon (the next icon to the right), in the drawing select the line's southern end, and press ENTER.

The converted segment's direction changes to be the same as the alignment.

## Curve 4 – Free Curve

The third curve has two dependent segments, before and after, and is a free curve segment.

1. From the Alignment Layout Tools toolbar, click the Curve drop-list icon (sixth icon in from the left), and from the Curve icon stack, select FREE CURVE FILLET (BETWEEN TWO ENTITIES, RADIUS).

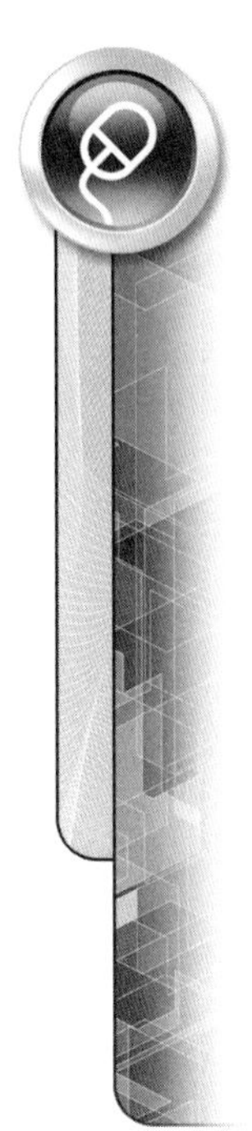

2. The command line prompts you to select the first entity. Select Tangent 4's northern end, and when prompted for the next entity, select Tangent 5's southern end.
3. The command line prompts you if the curve is Less than 180 degrees. Press ENTER to accept the value and continue.
4. The command lines prompts you for the last parameter, radius. For the radius enter **335** and press ENTER twice to create the curve and exit to the command line.

The Curve 4 is tangent down from Tangent 5's southern end and tangent to Tangent 4 (see Figure 5.35).

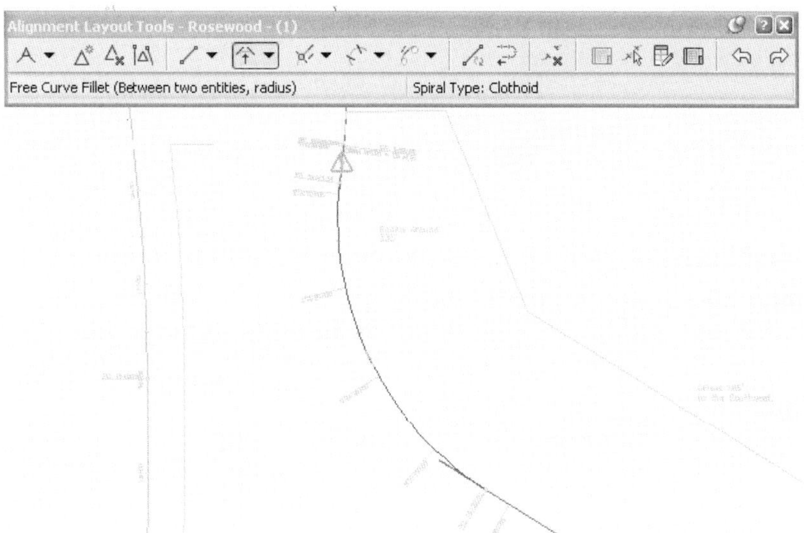


**FIGURE 5.35**

5. Click the red **X** on the Alignment Layout Tools toolbar to close it.
6. At Civil 3D's top left, Quick Access Toolbar, click the ***Save*** icon to save the drawing.

This completes importing and drafting alignment segments. When using the Alignment Layout Tools toolbar, the resulting alignment is a combination of fixed, floating, and free segments. You can also convert objects or polylines into alignments, or import them from LandXML files or LDT projects. The next step is to review their segment values to see if they meet the minimum requirements.

## SUMMARY

- One or more alignments can be imported when reading a LandXML file or an LDT project.
- The Create Alignment From Objects routine converts objects or a polyline into an alignment.
- Imported and Converted entities are fixed alignment segments.
- If a segment has the wrong direction, the Reverse Direction command changes its direction.
- The Alignment Layout Tools toolbar icons draw tangents and curves.
- Alignment segments do not need to be connected to be recognized as segments of the same alignment.

## UNIT 3: CENTERLINE ANALYSIS

Engineers design by calculations or design criteria. Alignment graphics do not reveal all of these numbers. An alignment's values are displayed in its properties dialog box or in its Alignment Layout Tools toolbar editors. To display a drafted alignment's editors, select an alignment in a drawing, press the right mouse button, and from the shortcut menu, select Edit Alignment Geometry.... Another method of editing an alignment is by selecting it in the drawing and the ribbon changes to an alignment editing panel. Or, from the Ribbon's Modify tab, Design panel, select the Alignment icon and a new panel displays the Geometry Editor command. To identify the alignment to edit, select an alignment segment in the drawing. Two toolbar editors display the alignment's numbers: a Sub-entity editor (a selected segment) or Alignment grid view (an overall view).

As seen in the previous unit's exercise, design criteria icons that mark violations are a visual review tool. Also, the Toolbox publishes alignment reports to the screen or to files.

### EDIT ALIGNMENT GEOMETRY – ALIGNMENT GRID VIEW

The Alignment Layout Tools toolbar's Alignment Grid View icon, displays all of an alignment's critical geometric values in the Alignment Entities vista (see Figure 5.36). Only the values in black are editable.

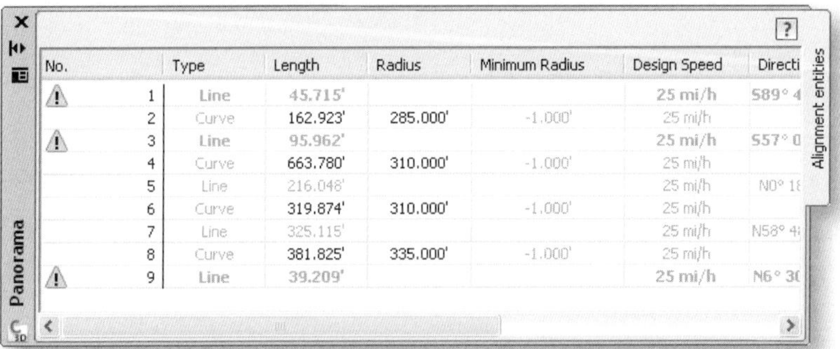


| No. | Type | Length | Radius | Minimum Radius | Design Speed | Directi |
|---|---|---|---|---|---|---|
| 1 | Line | 45.715' | | | 25 mi/h | S89° 4 |
| 2 | Curve | 162.923' | 285.000' | -1.000' | 25 mi/h | |
| 3 | Line | 95.962' | | | 25 mi/h | S57° 0 |
| 4 | Curve | 663.780' | 310.000' | -1.000' | 25 mi/h | |
| 5 | Line | 216.048' | | | 25 mi/h | N0° 1 |
| 6 | Curve | 319.874' | 310.000' | -1.000' | 25 mi/h | |
| 7 | Line | 325.115' | | | 25 mi/h | N58° 4 |
| 8 | Curve | 381.825' | 335.000' | -1.000' | 25 mi/h | |
| 9 | Line | 39.209' | | | 25 mi/h | N6° 3 |

**FIGURE 5.36**

### EDIT ALIGNMENT GEOMETRY – SUB-ENTITY EDITOR

The Alignment Layout Tools toolbar's Sub-entity Editor icon displays a selected alignment's segment's values. To use the Sub-entity Editor, first display it by clicking its toolbar icon (second icon in from the right), click the Pick Sub-entity icon (third in from the right), and select an alignment segment (sub-entity). When selecting a sub-entity, its values appear in the Alignment Layout Parameters dialog box (see Figure 5.37). To view another sub-entity's values, select another alignment segment and its values replace those currently in the dialog box. Only the values in black are editable.

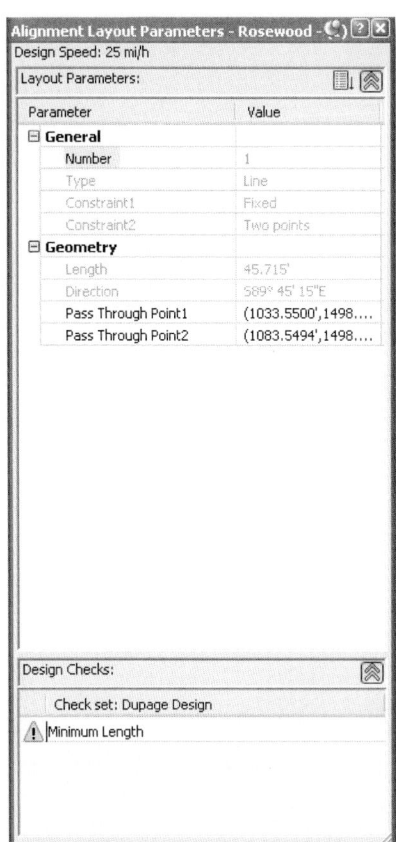


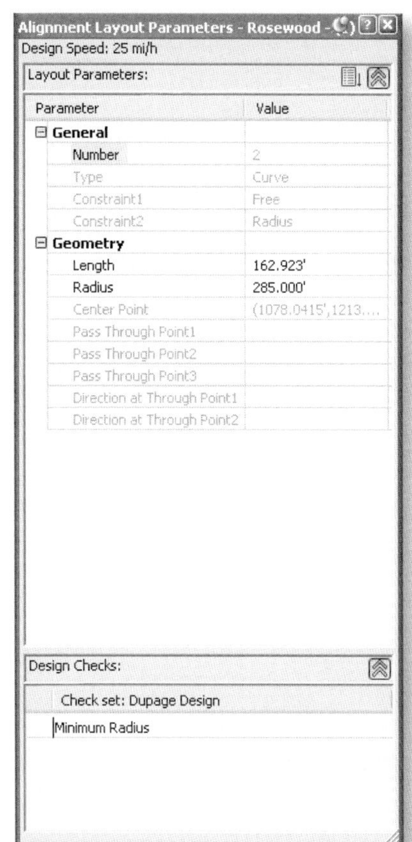


**FIGURE 5.37**

## DESIGN CHECK

Design Check icons display tooltips that report the criteria violated (see Figure 5.38).

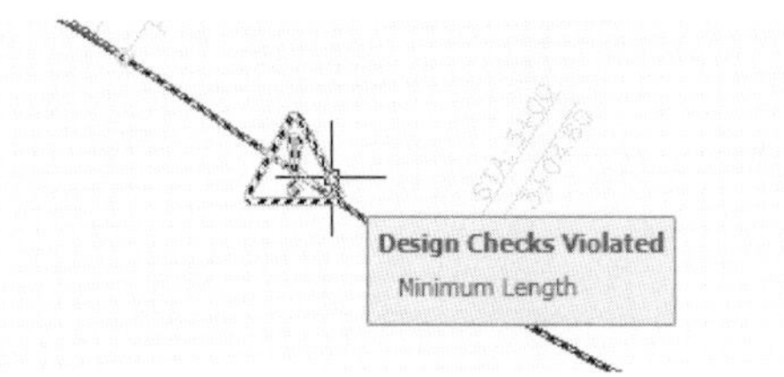


**FIGURE 5.38**

## TOOLBOX REPORTS

The Toolspace's Toolbox creates several alignment reports (see Figure 5.39). The reports include station, station and curve, stakeout, design criteria, etc. Some reports prompt for additional information before creating a report (see Figure 5.40).

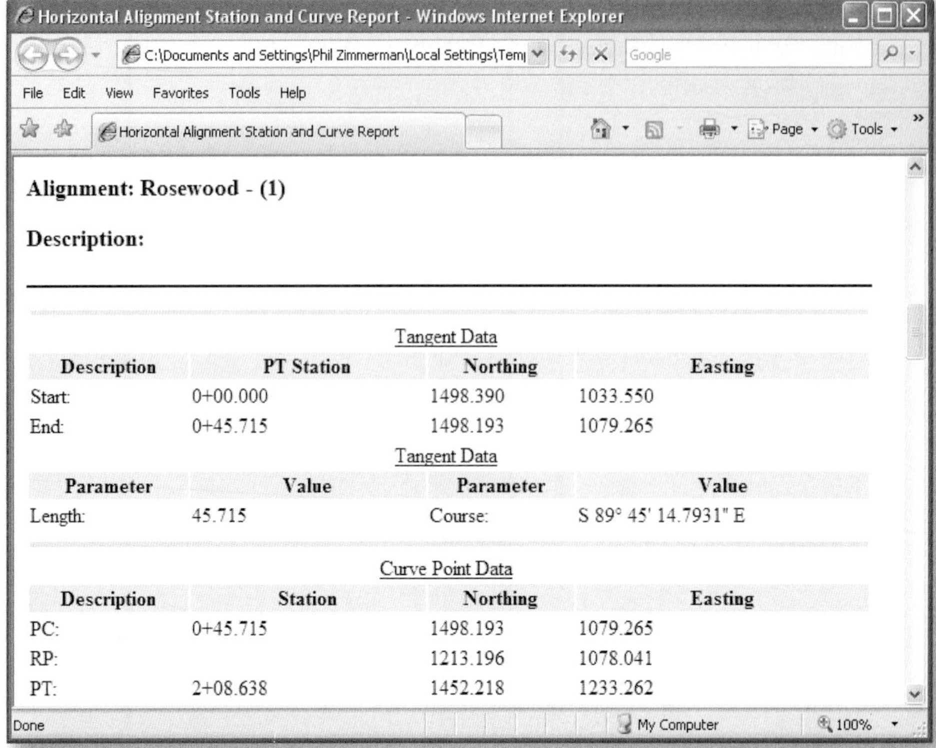


**FIGURE 5.39**

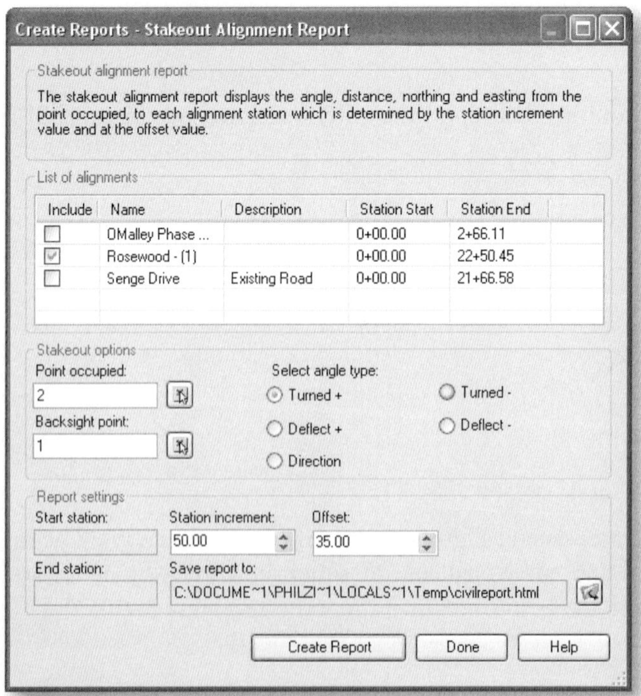


**FIGURE 5.40**

## INQUIRY TOOLS

The Ribbon's Analyze tab, Inquiry panel's Inquiry Tool command displays an Inquiry Tool palette and its alignment inquiries report station, offset, and elevation values (see Figure 5.41).

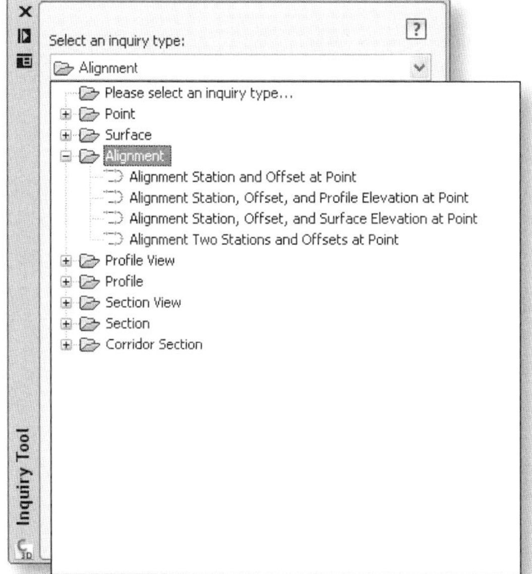


**FIGURE 5.41**

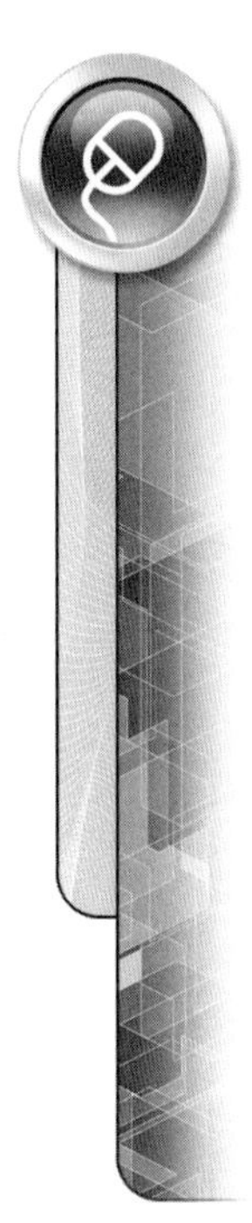

## EXERCISE 5-3

After completing this exercise, you will:

- Review an entire alignment's values.
- Review an alignment's sub-entity values.
- Select and view various alignments reports.

### Exercise Setup

This exercise continues with the previous exercise's drawing. If you did not complete the previous exercise, browse to the Chapter 5 folder of the CD that accompanies this textbook and open the *Chapter 05 – Unit 3.dwg* file.

1. If not open, open the previous exercise's drawing, or browse to the Chapter 5 folder of the CD that accompanies this textbook and open the *Chapter 05 – Unit 3* drawing.
2. If you have closed the Alignment Layout Tools toolbar, in the drawing select the Rosewood alignment and from the Ribbon's Alignment: Rosewood – (1) tab, Modify panel, select GEOMETRY EDITOR.

### Alignment Entities Vista

The Alignment Grid View icon displays all critical alignment values in a single vista.

1. In the Alignment Layout Tools toolbar, click the ***Alignment Grid View*** icon (right-most icon). Your Alignment Entities vista should look similar to Figure 5.36.

The Panorama is displayed with the Alignment Entities vista displaying all alignment values, including any criteria violations. Editable values are black.

2. Scroll horizontally through the Panorama to review the vista's values.
3. Click the Panorama's **X** to close it.

## Sub-entity Editor

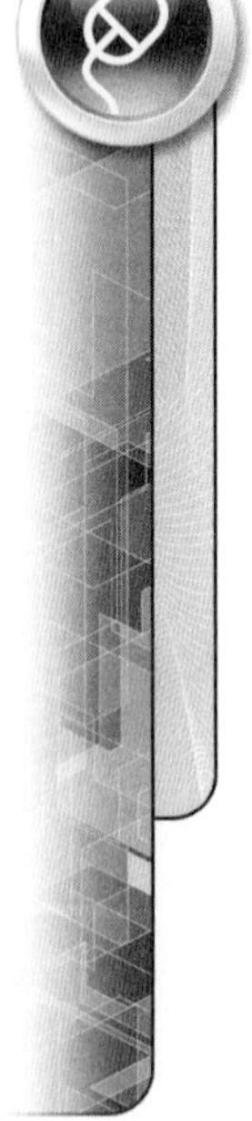

The Sub-entity Editor's Alignment Layout Parameters dialog box displays a selected segment's values.

1. On the Ribbon, click the ***View*** tab. At the Views panel's left, select and restore the named view **Proposed Starting Point**.
2. In the Alignment Layout Tools toolbar, select the ***Sub-entity Editor*** icon (the second icon in from the right-hand side) to display the Alignment Layout Parameters dialog box.
3. In the Alignment Layout Tools toolbar, click the ***Pick Sub-entity*** icon to the left of the Sub-entity Editor icon and select Curve 1. Your Editor should look similar to Figure 5.37.

The Alignment Layout Parameters dialog box displays the selected segment's values, including any criteria violations. Editable values are black. You may need to select the Show More and Expand All Categories button at the upper-right corner of the Alignment Layout Parameters dialog box to view more details.

4. Select Tangent 2 and review its values; notice its criteria violation.
5. Close the Sub-entity Editor by clicking its red **X**.
6. Close the Alignment Layout Tools toolbar by clicking its red **X**.

## Toolbox Reports

Toolbox creates assorted reports for selected alignments.

1. If the ***Toolbox*** tab is not displayed in the Toolspace, in the Palettes panel, to the right of the Toolspace icon, click the ***TOOLBOX*** icon.
2. If necessary, click the ***Toolbox*** tab.
3. In Toolbox, expand Reports Manager and the Alignment section until you are viewing its reports.

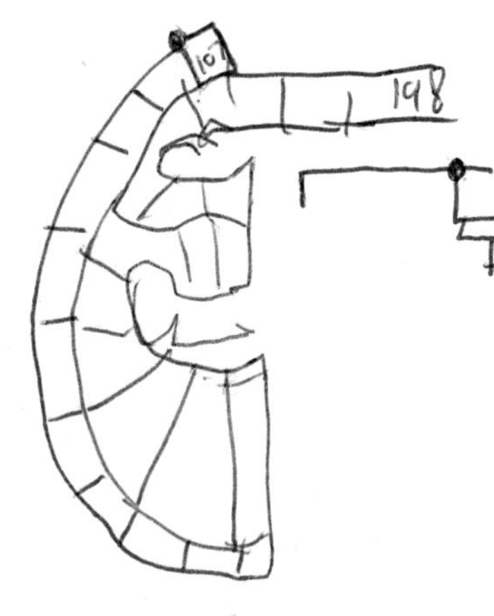


4. From the list of reports select ***Station_and_Curve***, press the right mouse button, and from the shortcut menu, select EXECUTE....
5. The Export to XML Report dialog box is displayed and lists all alignments. Click ***OK*** to create reports for each alignment.

You can toggle off the alignments that do not need a report.

Internet Explorer is displayed with a report for each alignment.

6. Scroll through the report until you can view Rosewood's values, and then review them.
7. Close Internet Explorer.
8. From the list of reports, select ALIGNMENT DESIGN CRITERIA VERIFICATION, press the right mouse button, and from the shortcut menu, select Execute....
9. The Create Reports – Alignment Design Criteria Verification Report dialog box opens. Toggle OFF all alignments except for Rosewood – (1), and click CREATE REPORT.

Internet Explorer displays a report marking each violation.

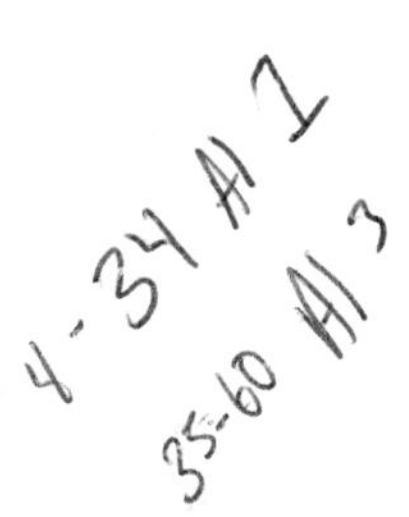

10. Scroll through the report to review Rosewood's values.

11. Close Internet Explorer.
12. In the Create Reports – Alignment Design Criteria Verification Report dialog box, click ***Done*** to close the dialog box.
13. At Civil 3D's top left, Quick Access Toolbar, click the ***Save*** icon to save the drawing.

## Inquiry Tools

1. In the Ribbon, click the Analyze tab. In the Inquiry panel, select INQUIRY TOOL.
2. In the Inquiry Tool palette, click the Select an inquiry type's drop-list arrow, expand the Alignment section, and from the list select **Alignment Station and Offset at Point.**
3. In the Select Alignment dialog box, select the alignment Rosewood – (1), and click ***OK*** to continue.
4. In the drawing, select a point near the Rosewood alignment.

The Inquiry Tool palette displays the selected point's coordinates, station, and offset.

5. Select a few more locations and note the values listed in the Inquiry Tool palette.
6. In the Inquiry Tool palette, click the Select an inquiry type's drop-list arrow and from the list select ***ALIGNMENT TWO STATIONS AND OFFSETS AT POINT.***
7. In the Select First Alignment dialog box, select Rosewood – (1), and click ***OK*** to continue.
8. In the Select Second Alignment dialog box, select Senge Drive and click ***OK*** to continue.
9. In the drawing, select a point near the Rosewood alignment.

The Inquiry Tool palette shows the selected point's coordinates, station, and offset relative to the two alignments.

10. Select a few more locations and note their Inquiry Tool palette values.
11. Click the Inquiry Tool palette's **X** to close it.
12. At Civil 3D's top left, Quick Access Toolbar, click the ***Save*** icon to save the drawing.

This completes an overview of alignment review methods. The Grid View and Sub-Entity vista editors display all alignment segment values. Toolbox's reports create various alignment reports.

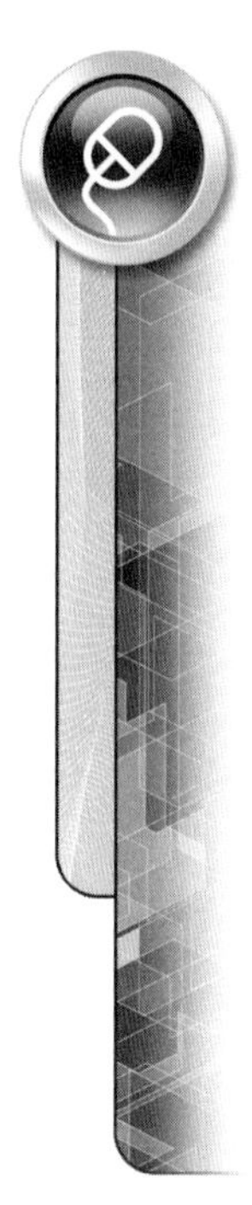

## SUMMARY

- Edit Geometry displays the alignment's Alignment Layout Tools toolbar, having an Alignment Grid View and Sub-entity Editor icons.
- The Alignment Entities vista and Alignment Layout Parameters dialog boxes display all alignment segment values.
- In an alignment editor, editable values are in black; when changing their value, update the alignment.
- Toolbox creates the alignment reports.

## UNIT 4: EDITING A HORIZONTAL ALIGNMENT

Designs change, need tweaking, or need new properties (station equations). You edit an alignment by using three methods: graphical, sub-entity, or grid (overall). Graphical uses alignment grips and relocates or changes a segment's parameters (see Figure 5.42).

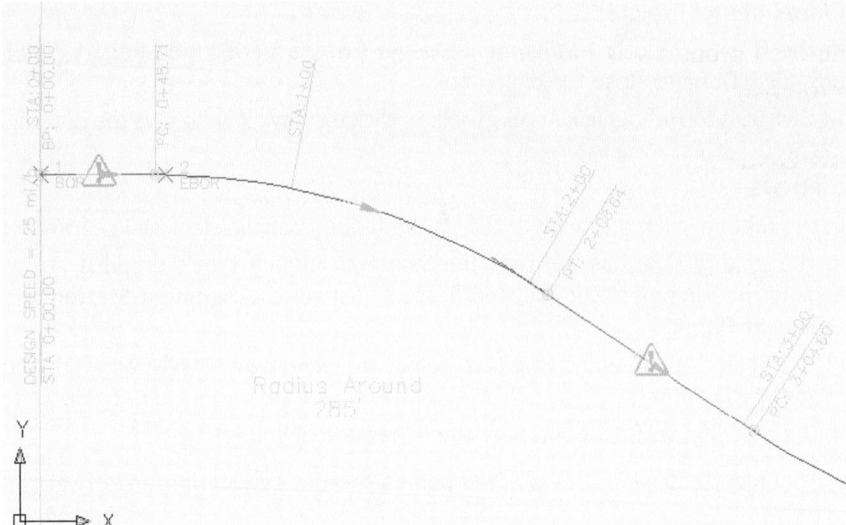

**FIGURE 5.42**

The Edit Alignment Geometry routine displays the Alignment Layout Tools toolbar. The Alignment Layout Tools toolbar displays the Alignment Layout Parameters dialog box or Alignment Entities vista. The Alignment Layout Parameters dialog box requires a selected alignment segment, and, after selecting a sub-entity, will display its values. The Alignment Editor vista contains all alignment values. Both editors display editable values in black and update the alignment after edits have been made.

## GRAPHICAL EDITING

When graphically editing an alignment, first select the alignment to display its grips, select a grip, and then manipulate its location by moving it. Depending on the type of tangent (fixed, floating, or free), it has grips at its endpoints and possibly at its midpoint. Curve segments have grips at the beginning (PC) and end (PT), midpoint, and PI. If locating a grip that the alignment cannot correctly resolve, the alignment will not allow that point to be selected.

The sub-entity types (fixed, floating, and free) influence how an alignment "drags" and resolves at the selected location. The fixed, floating, and free entity rules and the entity type before and after the active grip all affect the resulting edit. For example, fixed entities do not guarantee tangency, but floating and free do. Users need to be aware of each segment's behavior before or during edits. If these behaviors are not considered, the edit may create non-tangent segments and curves, unanticipated tangent extensions, and other unforeseen effects.

If you select a floating curve's PC or PT grip, the only edit is to lengthen the curve. If you select its midpoint grip, the adjustment affects its radius and changes the attached tangent's bearing.

When you are graphically editing an alignment converted from a polyline, the editing options are limited because all tangents and curves are fixed segments. If you are manipulating fixed segments, they do not retain tangency.

## EDIT ALIGNMENT GEOMETRY

The Alignment Layout Tools toolbar has two editors: Sub-entity and Alignment Grid View (overall).

### Alignment Grid View

The Alignment Entities vista displays all values critical to alignment segment in a single vista. Only values in black can change; editing a value and pressing ENTER cause the alignment to update.

### Sub-Entity Editor

There are two steps to using the Sub-entity Editor. The first step is to display the Alignment Layout Parameters dialog box and the second step is to select a segment with the sub-entity selection tool. Only parameters in black can change; after changing its value and pressing ENTER, the alignment is updated to match the new values (see Figure 5.37).

When selecting a curve segment, the Sub-entity Editor displays all of its values. Depending on the curve type (fixed, free, floating), users can change a curve's length, radius, chord, mid-ordinate, and the external ordinate or secant values. Changing a value and pressing ENTER updates the curve (see Figure 5.37).

When selecting a spiral segment, the vista displays values pertinent to a spiral. Edits available include length, the in and out A value, and the circle's radius (S-C-S) (see Figure 5.43).

**FIGURE 5.43**

## DELETING AND INSERTING SEGMENTS

Tangents, curves, and spirals cannot be deleted in the editors. They can be deleted only by using the Alignment Layout Tools toolbar's sub-entity delete command (fifth icon in from the right). The segment types affect what sub-entities can be deleted. For example, if you want to delete a tangent attached to a free curve, the free curve must be deleted first, and then the tangent can be deleted.

Alignment Layout Tools toolbar PI routines include inserting, deleting, and breaking apart a PI, so additional or new sub-entities can be added.

## ALIGNMENT PROPERTIES

Alignment properties adds, modifies, and reports stationing, station equations, design speeds, labeling sets, and superelevation parameters. It also reports which profiles and profile views depend on the alignment.

### Station Control

Station Control sets an alignment's location (reference point) and stationing information. By default, the starting station is at reference coordinates. In Reference point, users can define a new beginning station or alignment beginning point (see Figure 5.44). If a new beginning station value is assigned, the station information area is updated.

### Station Equations

At times, an alignment's stationing changes at a station equation (see Figure 5.44). This happens for many reasons, such as a new road connecting to an existing one, a change of jurisdiction controlling the centerline, and so on. A station equation point signals the end of one stationing system and the beginning of another. The stations on the other side of the equation can be totally different and can even decrease along the remainder of the roadway.

The two Station Equation icons create and delete equations. When you select an Add station equation, the dialog box hides, the command line prompts for a station, and a station jig appears, sliding along the alignment. To select a station, select a point in the drawing or enter a station in the command line. After identifying the station, the Properties dialog box reopens and lists the station. To finish the equation, modify the station back to its earlier state (if necessary), enter the station ahead, and set the increase/decrease value. If you are defining a station equation, the original stationing is known as Raw stationing.

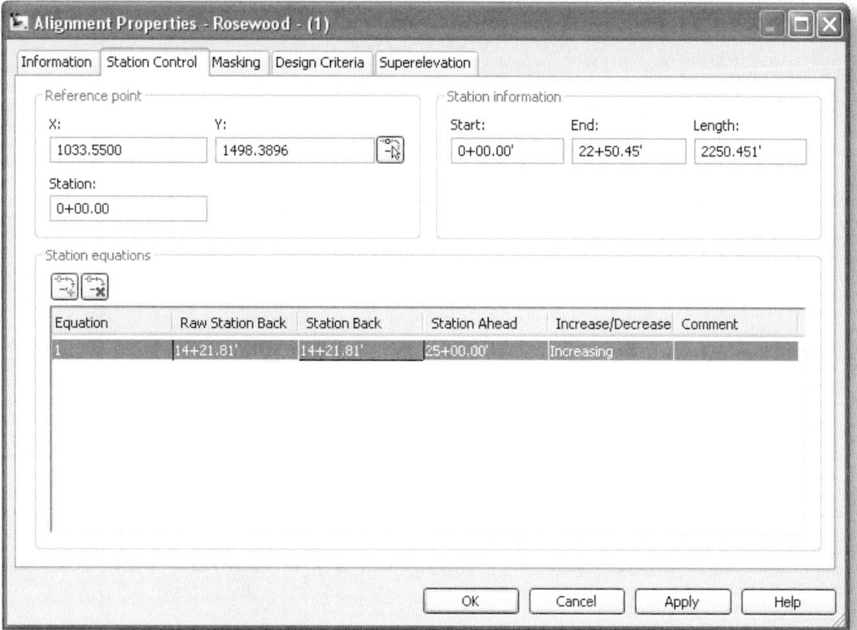


**FIGURE 5.44**

## Mask

The Mask panel lists the masks used in conjunction with alignment intersections.

## Design Criteria

If you are using design criteria, the initial alignment has a uniform design speed. If you want to add speeds for different alignment sections, at the panel's top left click the Add design speed icon to create a new design speed entry and modify its values. You can also delete any design speed entry.

The panel's right half toggles on and off design criteria and design check set; it also resets the criteria files and the check set name (see Figure 5.45).

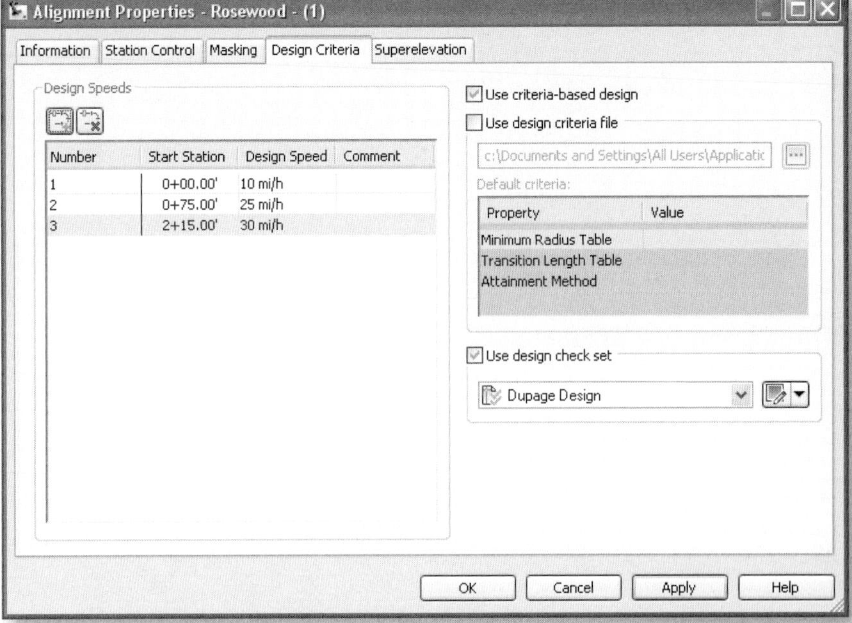


**FIGURE 5.45**

## Superelevation

This panel defines the superelevation regions within an alignment (each curve of the alignment is a region). The panel's values reflect the Design Criteria panel's selected tables and the superelevation regions (each curve is a superelevation region) (see Figure 5.46 and Figure 5.47). Arrows mark areas where minimums or needed lengths are not available.

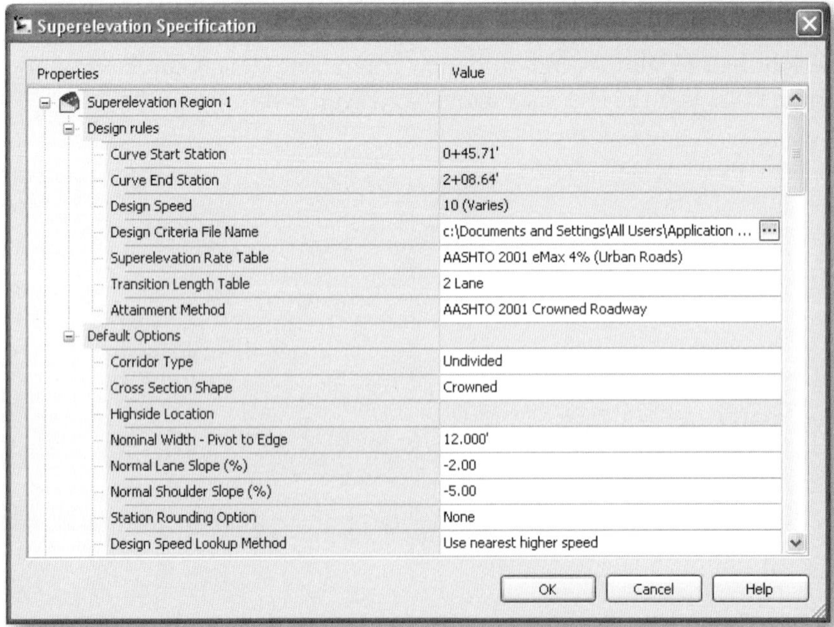


**FIGURE 5.46**

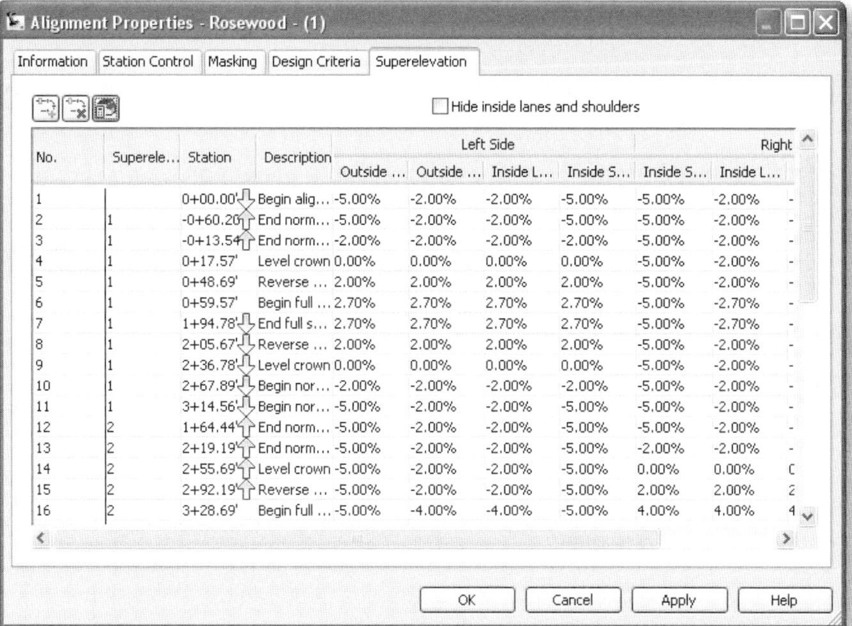


| No. | Superele... | Station | Description | Left Side | | | | Right | | |
|---|---|---|---|---|---|---|---|---|---|---|
| | | | | Outside ... | Outside ... | Inside L... | Inside S... | Inside S... | Inside L... | |
| 1 | | 0+00.00' | Begin alig... | -5.00% | -2.00% | -2.00% | -5.00% | -5.00% | -2.00% | - |
| 2 | 1 | -0+60.20' | End norm... | -5.00% | -2.00% | -2.00% | -5.00% | -5.00% | -2.00% | - |
| 3 | 1 | -0+13.54' | End norm... | -2.00% | -2.00% | -2.00% | -2.00% | -5.00% | -2.00% | - |
| 4 | 1 | 0+17.57' | Level crown | 0.00% | 0.00% | 0.00% | 0.00% | -5.00% | -2.00% | - |
| 5 | 1 | 0+48.69' | Reverse ... | 2.00% | 2.00% | 2.00% | 2.00% | -5.00% | -2.00% | - |
| 6 | 1 | 0+59.57' | Begin full ... | 2.70% | 2.70% | 2.70% | 2.70% | -5.00% | -2.70% | - |
| 7 | 1 | 1+94.78' | End full s... | 2.70% | 2.70% | 2.70% | 2.70% | -5.00% | -2.70% | - |
| 8 | 1 | 2+05.67' | Reverse ... | 2.00% | 2.00% | 2.00% | 2.00% | -5.00% | -2.00% | - |
| 9 | 1 | 2+36.78' | Level crown | 0.00% | 0.00% | 0.00% | 0.00% | -5.00% | -2.00% | - |
| 10 | 1 | 2+67.89' | Begin nor... | -2.00% | -2.00% | -2.00% | -2.00% | -5.00% | -2.00% | - |
| 11 | 1 | 3+14.56' | Begin nor... | -5.00% | -2.00% | -2.00% | -5.00% | -5.00% | -2.00% | - |
| 12 | 2 | 1+64.44' | End norm... | -5.00% | -2.00% | -2.00% | -5.00% | -5.00% | -2.00% | - |
| 13 | 2 | 2+19.19' | End norm... | -5.00% | -2.00% | -2.00% | -5.00% | -2.00% | -2.00% | - |
| 14 | 2 | 2+55.69' | Level crown | -5.00% | -2.00% | -2.00% | -5.00% | 0.00% | 0.00% | C |
| 15 | 2 | 2+92.19' | Reverse ... | -5.00% | -2.00% | -2.00% | -5.00% | 2.00% | 2.00% | 2 |
| 16 | 2 | 3+28.69' | Begin full ... | -5.00% | -4.00% | -4.00% | -5.00% | 4.00% | 4.00% | 4 |

**FIGURE 5.47**

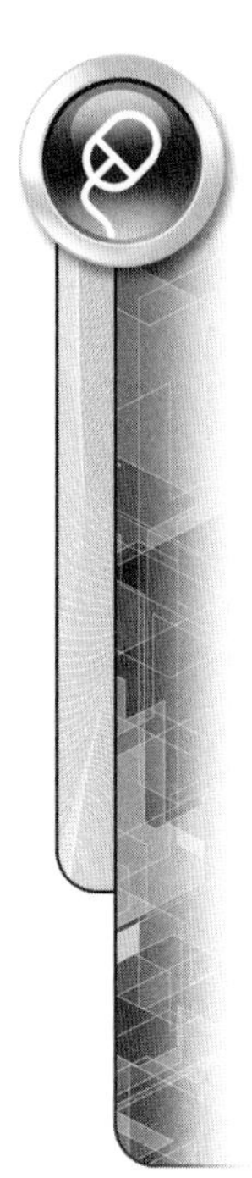

## EXERCISE 5-4

After completing this exercise, you will:

- Graphically edit an alignment.
- Edit alignment values in the Grid View vista.
- Edit an alignment's values in the Sub-entities Editor dialog box.
- Review and edit an alignment from its Properties dialog box.
- Modify design speeds for an alignment.

### Exercise Setup

This exercise continues with the previous exercise's drawing. If you did not complete the previous exercise, browse to the Chapter 5 folder of the CD that accompanies this textbook and open the *Chapter 05 – Unit 4.dwg* file.

1. If not open, open the previous exercise's drawing or browse to the Chapter 5 folder of the CD that accompanies this textbook and open the *Chapter 05 – Unit 4* drawing.

### Graphical Editing — Tangent Segments

When you have selected any alignment segment, the entire alignment displays its editing grips. To edit the alignment, click a grip and move it to a new location. If you locate a grip that does not correctly resolve, a solution will not be displayed and you cannot pick that point.

There are four fixed tangents in the alignment. You can move all of the tangents by selecting their middle grip or by adjusting their bearing by selecting their end grip.

1. Use the `ZOOM` command to better view the eastern north/south tangent.
2. In the drawing, click the alignment to activate its grips.

3. Select the eastern tangent's middle grip (square) and move the alignment east and west.

As the tangent moves, the two free curves change to accommodate the tangent's motion.

4. Press ESC to release the tangent.
5. In the drawing, click the eastern tangent's southern extension grip and move the cursor east and west.

This grip changes the tangent's direction from its southern end while holding its northern point. As the grip relocates, the free curves adjust to maintain valid attachments.

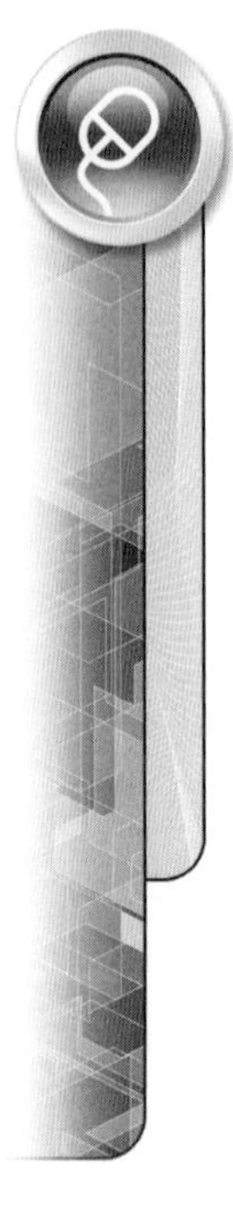

6. Press ESC to release the tangent.

### Curve Segments

When you select any free curve's grip and move its location, you adjust the curve's radius.

1. Use the ZOOM and PAN command and view Curve 2.
2. If necessary, click the alignment to activate its grips.
3. Activate the curve's middle grip and slowly move it toward the southeast.

Both tangents lengthen to accommodate the curve's changing radii.

4. Press ESC to release the curve.

### Alignment Grid View

1. While the alignment is highlighted, from the Modify palette, select GEOMETRY EDITOR to display the Alignment Layout Tools toolbar.
2. On the toolbar's right, click the ***Alignment Grid View*** icon to display the Panorama with the Alignment Entities vista.
3. In the Alignment Entities vista, review each curve's radii. All radii are divisible by 5.
4. Click the **X** to close the Alignment Entities vista.
5. At Civil 3D's top left, Quick Access Toolbar, click the ***Save*** icon to save the drawing.

### Sub-entity Editor

The Sub-entity Editor's – Alignment Layout Parameters dialog box displays a selected segment's information.

1. In the Alignment Layout Tools toolbar, click the ***Sub-entity Editor*** icon to display the Alignment Layout Parameters dialog box.
2. In the Alignment Layout Tools toolbar, click the ***Pick Sub-entity*** icon. In the drawing, select the alignment Curve 2.

The curve's data appear in the editor. The current radius is **310** feet and is editable.

3. Click the Alignment Layout Parameters dialog box's red **X** to close it.
4. Click the Alignment Layout Tools toolbar's red **X** to close it and return to the command line.

### Alignment Properties — Station Control

1. On the Ribbon, click the ***View*** tab. At the Views panel's left, select and restore the named view Station Equation.
2. In the drawing, select the alignment, from the Modify panel, select ALIGNMENT PROPERTIES icon.
3. Click the ***Station Control*** tab to review its contents.
4. In the dialog box, at the middle left click the ***Add station equation*** icon.

The dialog box disappears, and in the drawing a jig attaches to the alignment reporting stations.

5. Using an Intersection object snap, select the alignment's intersection with the Phase 1 line and return to the Alignment Properties dialog box.
6. In the dialog box, for station ahead enter **2500**, and set Increase/Decrease to Increasing.
7. Click ***Apply*** to modify the alignment's stationing.

The alignment's beginning station and length remain the same; however, the ending station changes to a higher value.

8. Review the new station values.

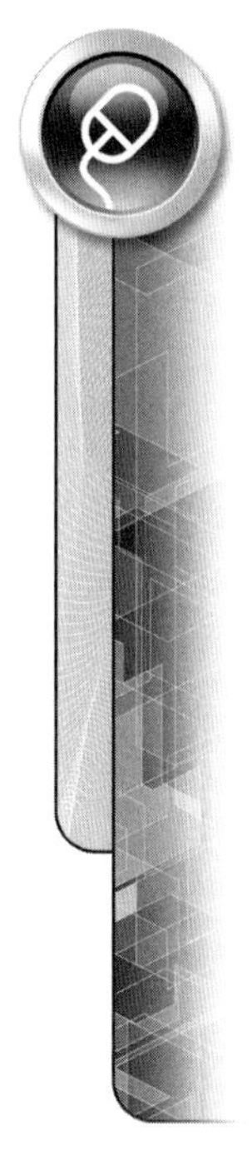

## Alignment Properties — Design Speed

1. In the Alignment Properties dialog box, click the ***Design Criteria*** tab.
2. At the panel's top left, click the ***Add design speed*** icon twice.
3. For station 0+00 change the design speed to **10**.
4. In the new speed entry, for the start station enter **75** and for its speed enter **25**.
5. In the second new speed entry, for the station enter **215** and for its speed enter **30**.

Your Design Criteria panel should now look like Figure 5.45.

6. Click ***OK***.
7. At Civil 3D's top left, Quick Access Toolbar, click the ***Save*** icon to save the drawing.

This ends the exercise on alignment editing. Alignment edits occur graphically, in a sub-entity or an overall editor, and in the Alignment Properties dialog box. The Alignment Properties dialog box is the only place where you can define station equations, design speeds, superelevation, and other important alignment values.

## SUMMARY

- Users can graphically relocate alignment segments.
- When graphically editing an alignment, segments react to any changes and will not allow an incorrect solution.
- When graphically editing an alignment, the segment types (fixed, floating, and free) determine how they react to the changes.
- The Alignment Grid View vista displays all alignment geometric numbers.
- The Sub-entity Editor displays only the selected segment's geometric numbers.
- When editing in a vista, values in black are editable.
- The Alignment Properties dialog box tabs affect stationing, station equations, design criteria, design speeds, and superelevation settings.
- The Alignment Properties dialog box lists all profiles and profile views referencing the alignment.

## UNIT 5: ALIGNMENT ANNOTATION

When designing or creating an alignment, it is automatically annotated based on the assigned label set. A label set is an alias for a collection of label style types. Users can change assigned label styles by changing a set's labels list or by adding or removing label types and styles with Edit Alignment Labels. A label set's label styles are specifically for annotating an alignment's segment geometry. All label sets and their styles are listed in the Label Sets and Stations sections under the Alignment Label Styles heading (see Figure 5.48).

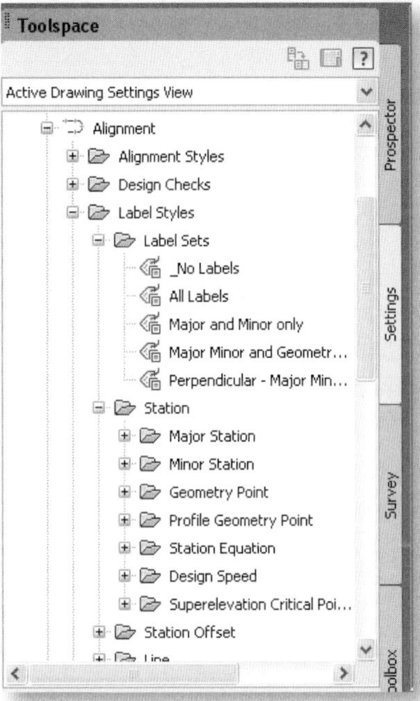


**FIGURE 5.48**

A second labels group annotates alignments or locations near them after creating alignments. These labeling types appear in the Add Labels dialog box. The styles appearing in Add Labels are in Settings' Alignment Label Styles branch after the Label Sets and Station headings. These labels include Station Offset, Line, Curve, Spiral, and Tangent Intersection.

All labels adjust their size when they change or set a viewport plot scale. All labels react to alignment changes (geometry and stationing). Civil 3D templates provide several "starter" styles; however, users can modify them to accommodate company standards.

## ALIGNMENT LABEL SETS AND STYLES

Label sets are a collection of label style types grouped under an alias. When assigning a label set, the alignment uses the styles appropriate for its current definition. Users can change the assigned set or assign new styles to an alignment. See the discussion of these styles in Unit 1 of this chapter.

## ADD LABELS

The Add Labels dialog box adds labels to an existing alignment (see Figure 5.49). Alignment's Edit Feature Settings dialog box sets the default label styles. See the discussion of Edit Feature Settings in Unit 1 of this chapter.

To change a label type, click its drop-list arrow and select the desired label type. When you change the label type, the dialog box changes to show the available label types and any other appropriate selections.

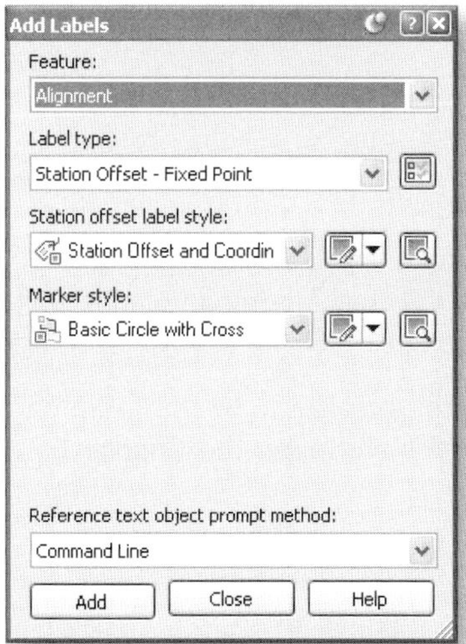


**FIGURE 5.49**

## ALIGNMENT TABLE

Alignment tables list line, curve, or spiral segments (see Figure 5.50). You must label the alignment first, because each table entry corresponds to its alignment segment label (line, curve, and spiral).

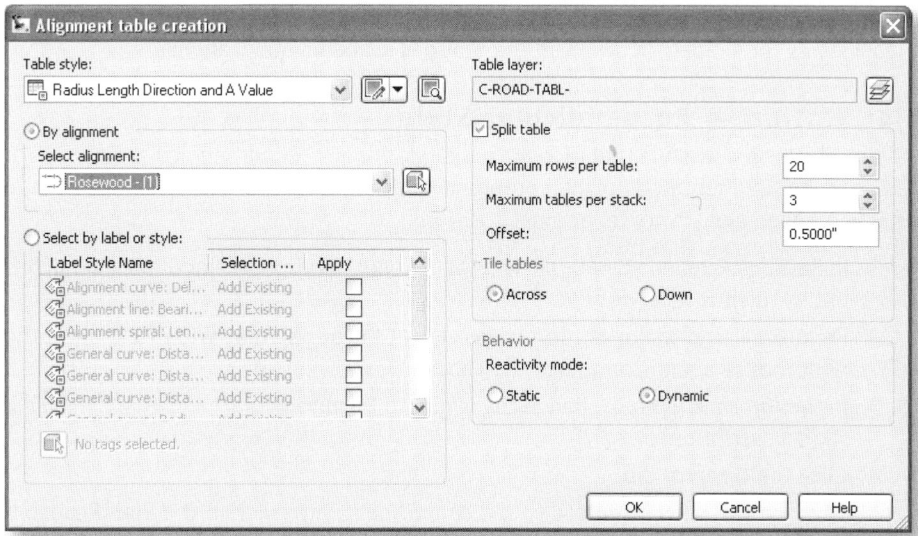


**FIGURE 5.50**

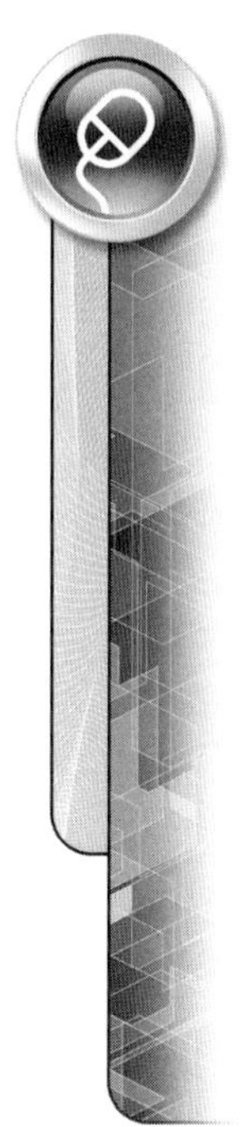

## EXERCISE 5-5

After completing this exercise, you will:

- Be able to review alignment label sets.
- Be able to review alignment label styles.
- Be able to create new label styles.
- Be able to define a new label set.

- Be able to apply a label set to an alignment.
- Be able to add station and offset labels.
- Be able to add segment labels.

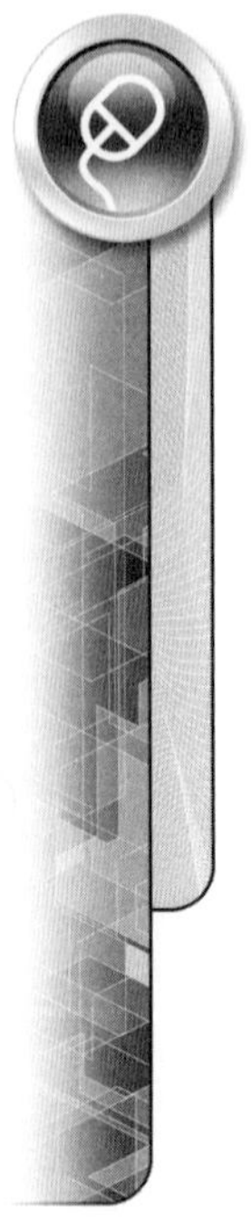

## Exercise Setup

This exercise continues with the previous exercise's drawing. If you did not complete the previous exercise, browse to the Chapter 5 folder of the CD that accompanies this textbook and open the *Chapter 05 – Unit 5.dwg* file.

1. If not open, open the previous exercise's drawing or browse to the Chapter 5 folder of the CD that accompanies this textbook's and open the *Chapter 05 – Unit 5* drawing.

## Label Sets — All Labels

When creating an alignment, the Create Alignment dialog box has an entry for assigning a Label Set (see Figure 5.28).

1. Click the ***Settings*** tab.
2. In Settings, expand the Alignment branch until you view the Label Styles, Label Sets list.
3. From the list of label sets, select ***All Labels***, press the right mouse button, and from the shortcut menu, select EDIT....
4. In the Alignment Label Sets dialog box, click the ***Labels*** tab and review the list of assigned label styles.
5. Click ***Cancel*** to exit the dialog box.

The label styles for a label set are from the Stations branch below Label Sets.

## Major Station — Perpendicular Style

The label set uses a new style that places a major station's label perpendicular to an alignment. The label is orientated to the centerline direction and the text does not change when it is dragged away from its original position.

1. In Settings, expand the Alignment, Label Styles', Station, Major Station branches until you view its styles list.
2. From the list of Major Station styles, select **Perpendicular with Tick**, press the right mouse button, and from the shortcut menu, select COPY....
3. In the ***Information*** tab, change the Name to **Perpendicular**, and give the style a short description.
4. Click the ***General*** tab.
5. To make the label orientate to the centerline's path, click in the Plan Readability's Plan Readable value cell to display its drop-list arrow. Click the drop-list arrow, and change its value to **False**.
6. Click the ***Layout*** tab.
7. At the panel's top left, click the Component name drop-list arrow and select **Tick** from the list. Delete it by selecting the red **X** in the panel's upper-middle portion.

The current component name is now Station.

8. In the Text section, click in the Attachment's value cell to display its drop-list arrow. Click the drop-list arrow, and from the list select **Middle Center**.
9. In the Text section, click in the Y Offset value cell and change the offset value to **0** (zero).

Your Layout panel should look similar to Figure 5.51.

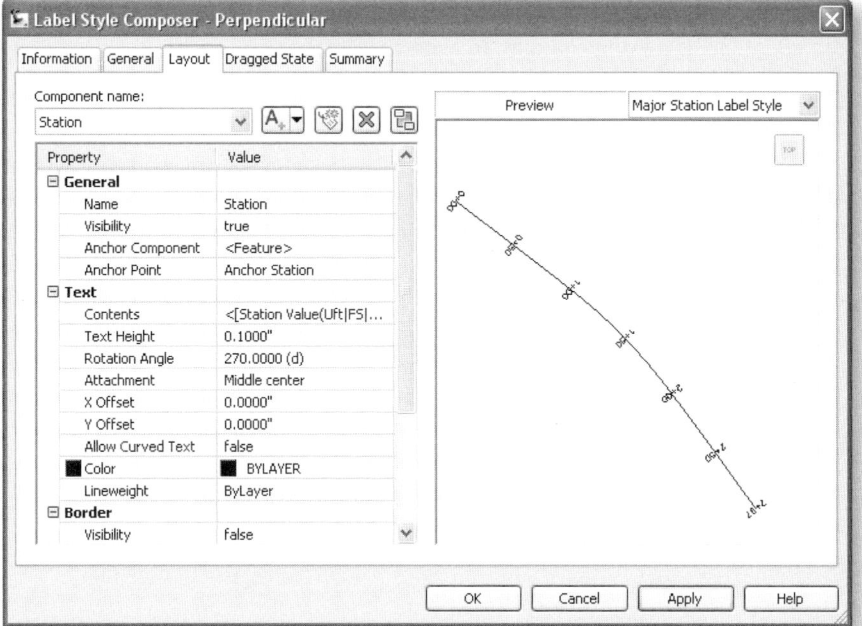


**FIGURE 5.51**

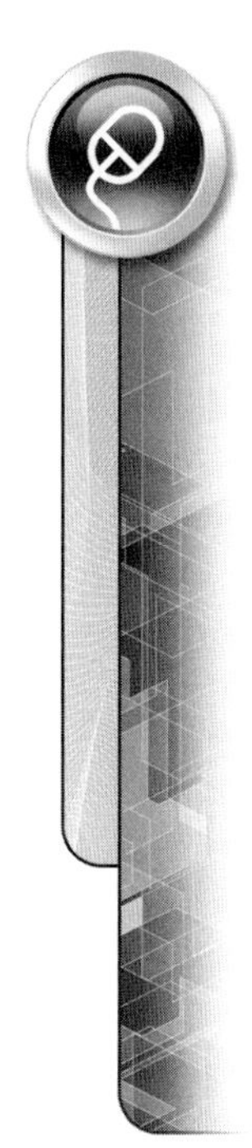

10. Select the ***Dragged State*** tab.
11. In the Leader section, click in the Visibility value cell to display its drop-list arrow. Click the drop-list arrow, and change it to **False**.
12. In the Dragged State Components section, click in the Display value cell to display its drop-list arrow. Click the drop-list arrow, and change the value to **As Composed**.
13. Click ***OK*** to create the style.
14. At Civil 3D's top left, Quick Access Toolbar, click the ***Save*** icon to save the drawing.

## Perpendicular Label Set

The All Labels label set contains parallel label station styles. The new label set is a copy of All Labels except for the Major Station label.

1. If necessary, expand the Alignment, Label Styles, and Label Sets branches until you view the Label Sets list.
2. From the label sets list, select **All Labels**, press the right mouse button, and from the shortcut menu, select COPY....
3. Click the ***Information*** tab and for the name, enter **All Labels – Perpendicular**.
4. Click the ***Labels*** tab.
5. In the labels list under the Type heading, Major Stations type, at the right of the Style's cell, click the blue label icon to display the Pick Label Style dialog box.
6. In the dialog box, click the drop-list arrow, from the list select **Perpendicular**, and click ***OK*** until you exit the dialog boxes.

This changes the Major Stations label to Perpendicular and the Label Set is completed.

## Changing Label Assignments

The Edit Alignment labels setting changes an alignment's labels.

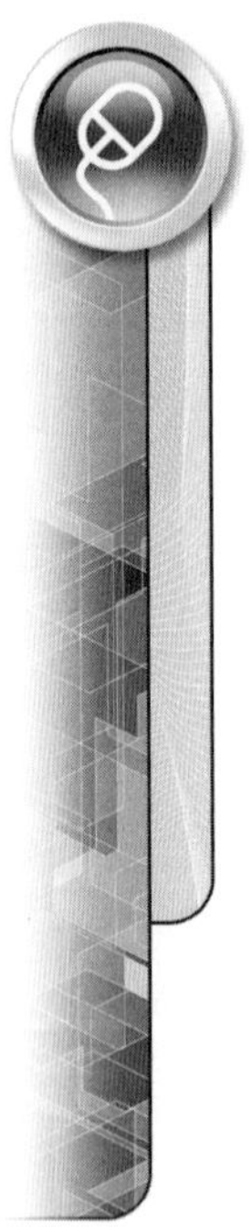

1. In the drawing select the Rosewood – (1) alignment, press the right mouse button, and from the shortcut menu, select EDIT ALIGNMENT LABELS....
2. In the labels list under the Type heading, select Major Stations, and to the right of Add>>, click the red **X** to remove the label type and its style.
3. Repeat the previous step until all Types are removed.
4. At the bottom of the dialog box, click IMPORT LABEL SET....
5. In the Select Style Set dialog box, click the drop-list arrow. From the list select All Labels – Perpendicular, and click ***OK*** to return to the Alignment Labels dialog box (see Figure 5.52).
6. Click ***OK*** to exit the dialog box.
7. Use the ZOOM and PAN commands to better view the alignment labeling.

The major stationing labels should now be perpendicular to the alignment.

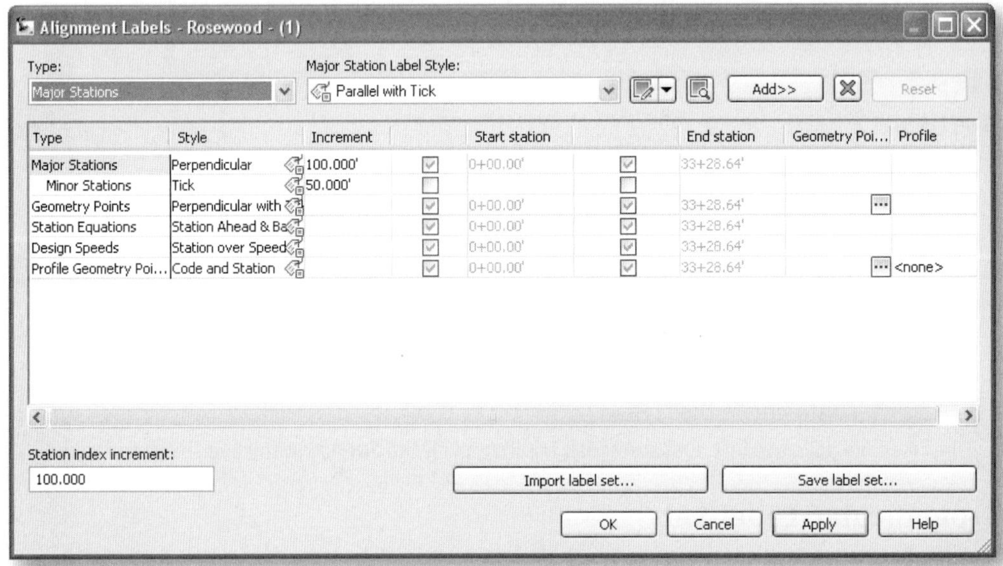


**FIGURE 5.52**

## Add Labels – Station and Offset – Fixed Point

There are critical points along the centerline's path that need annotation: lot corners, existing trees, existing utilities, driveway intersections, and so on.

The file for this exercise is in the Chapter 5 folder of this textbook's CD.

1. Make sure the Transparent Commands toolbar is displayed.
2. On the Ribbon, click the ***Insert*** tab. On the Import panel, select LANDXML.
3. In the Import LandXML dialog box, browse to Station *Offset.xml*, select it, and click Open. The Import LandXML dialog box opens. In the Import LandXML dialog box, click ***OK***.

The file is read and the points appear in the drawing.

4. On the Ribbon click the ***View*** tab. On the Views panel's left, select and restore the named view **Station Equation**.

5. On the Ribbon, click the ***Annotate*** tab. On the Labels & Tables panel's left, click Add Labels' drop-list arrow, on the shortcut menu, click Alignment, and on the flyout select ADD ALIGNMENT LABELS.
6. If necessary, in Add Labels, change the Label type to **Station Offset – Fixed Point** and set the Station offset label style to **Station Offset and Coordinates**.

Your Add Labels dialog box should match Figure 5.53.

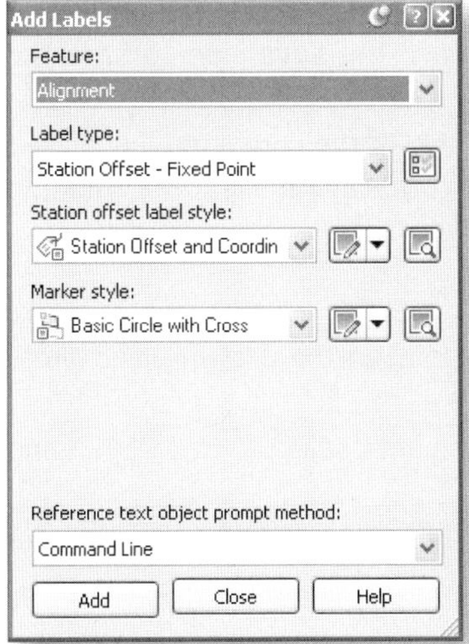


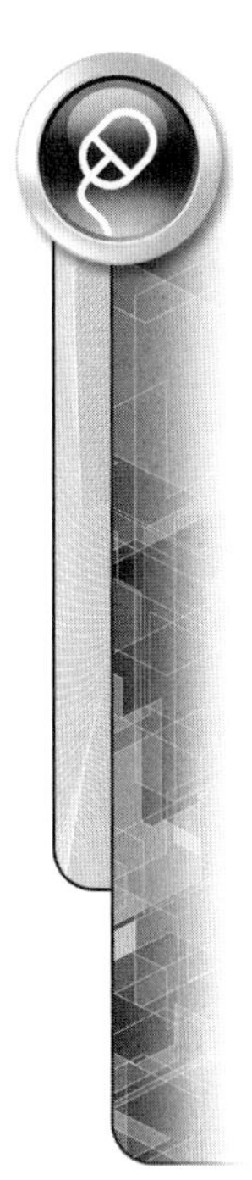

**FIGURE 5.53**

7. Click ***Add*** to start the labeling process.
8. The Add Labels command prompts you to select an alignment. In the drawing, select the Rosewood alignment.

The routine displays a station jig, and asks you to select a point.

You want a label that references the cogo points. Use the transparent command Point Object override. The transparent override affects only one selection. You need to reselect the Point Object icon for each point selection.

9. From the Transparent Commands toolbar, click the ***Point Object*** icon and in the drawing select one of the points.
10. Repeat the previous step, and annotate the remaining points.
11. When you finish annotating, press ENTER to end the command.

The labels may be on top of one another.

12. In the drawing, select a station/offset label to display its grip; select the grip (cyan diamond), and drag the label to a new position.

The label's Dragged State changes the label to stacked text with a leader.

13. Use the PAN command and pan the drawing to view more of the alignment.

## Add Labels – Station and Offset

1. In the Add Labels dialog box, change the Label type to **Station Offset** and the Label style to **Station Offset and Coordinates**.
2. Click ***Add*** to start the labeling process.
3. The Add Label command prompts you to select an alignment. In the drawing, select the Rosewood alignment.

A station jig appears ready to select a station and offset.

4. In the drawing that is using the station jig, identify the station by selecting a point near the alignment. To identify the offset, select a second point to one side or the other of the alignment.
5. Repeat the previous step to create a few more station/offset labels. You can also type in values for the Station and Offset.
6. Press ENTER to exit the labeling, and in the Add Labels dialog box, click ***Close*** to exit it.
7. At Civil 3D's top left, Quick Access Toolbar, click the ***Save*** icon to save the drawing.
8. In the command line type DVIEW and press ENTER. In the command line, select object, enter **All**, and press ENTER twice. In the command line, enter **TWist** and press ENTER. Slowly move the cursor to rotate your view of the drawing.

All labels rotate to be plan-readable (horizontal).

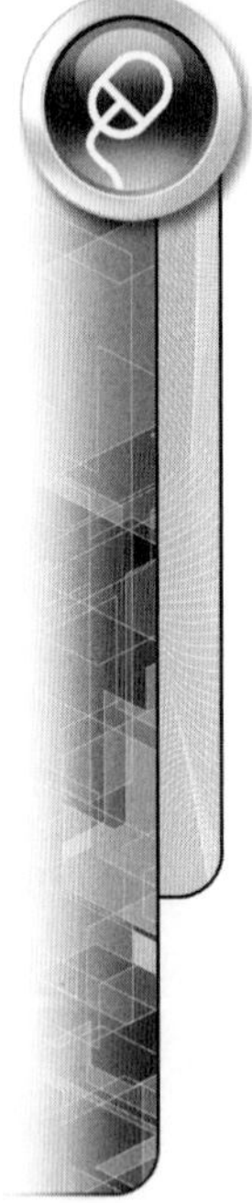

9. Move the mouse to specify a rotation, pick a point in the drawing, and press ENTER to exit Dview.
10. In the Ribbon, select the ***View*** tab. On the Views panel, select NAMED VIEWS and create a New view named **Twisted**.
11. In the command line, type **Plan** and then press ENTER twice to restore the previous display.

## Labels and Layouts

1. On the Status Bar, click the ***Layout1*** icon.
2. In the layout, to enter the viewport's model space by double-clicking in it.
3. From the Views panel, select and restore the named view **Twisted**. You may need to scroll down the list by clicking the down arrow.
4. On the Status Bar, set the Viewport Scale to **1″=60′**.
5. If necessary, in the command line, type REGENALL (REA) and press ENTER to resize the station offset labels.

The labels are resized appropriately for the current scale.

6. On the Status Bar, set the Viewport Scale to **1″=30′**.
7. If necessary, in the command line, type REGENALL (REA) and press ENTER to resize the station offset labels.

The labels are resized appropriately for the current scale.

8. From the Status Bar click the Model icon.
9. At Civil 3D's top left, Quick Access Toolbar, click the ***Save*** icon to save the drawing.

## Alignment Segment Labels

There are times when you need to label an alignment's tangents and curves. You use the Add Labels dialog box to create these labels.

1. Click the ***Annotate*** tab. On the Labels & Tables panel's left, click Add Labels' drop-list arrow, on the shortcut menu, click Alignment, and on the flyout select ADD ALIGNMENT LABELS.
2. In Add Labels, Label type, click the drop-list arrow and from the list select **Multiple Segment**.
3. Click Table Tag Numbering to review its current numbers. After reviewing the numbers, click ***OK*** to return to the Add Labels dialog box.

4. Click ***Add***. The command line prompts you to select an alignment; in the drawing, select the Rosewood alignment.

Segment labels appear along the alignment's tangents and curves.

5. In the Add Labels dialog box, click ***Close***.
6. Use the ZOOM and PAN commands to view labels.

## Alignment Table

Segment labels are the basis for an alignment's lines and curves table.

1. Use the ZOOM and PAN commands to view an empty area to the right of the site.
2. If necessary, click the ***Annotate*** tab. On the Labels & Tables panel, click Add Tables' drop-list arrow, on the shortcut menu, click Alignment, and on the flyout select ADD SEGMENT.
3. In the Alignment Table Creation dialog box, set the table to By alignment, click its drop-list arrow, and from the list select Rosewood – (1). All remaining settings stay the same. Click ***OK*** to close the dialog box.
4. As you exit, a table hangs on the cursor. In the drawing, select a point to locate the table.
5. At Civil 3D's top left, Quick Access Toolbar, click the ***Save*** icon to save the drawing.
6. Use the ZOOM command to better view the table.
7. Use the ZOOM and PAN commands to better view the alignment labeling.

The labels are now tags. When you create a table, the labels change their display mode to tag mode.

This completes alignment annotation. The next unit reviews how to create new entities from an existing alignment.

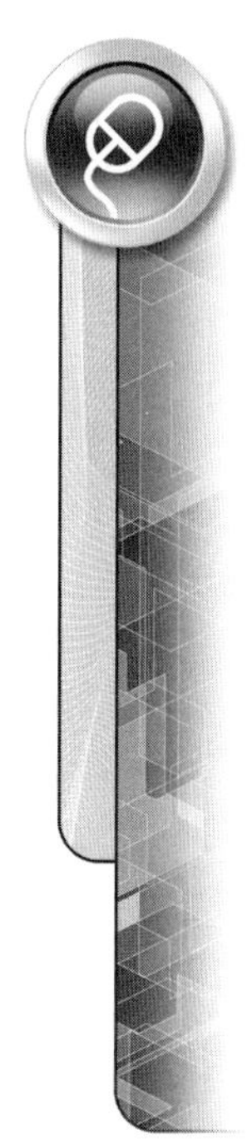

## SUMMARY

- When creating an alignment, the label set automatically annotates critical alignment values.
- All labels are scale- and rotation-sensitive.
- All station and offset labels update their station and offset values in response to alignment changes.
- Alignment segment labels are similar to parcel segment labels, but are defined in the Alignment branch.
- Users can create an alignment table listing line, curve, and spiral segment values.

# UNIT 6: OBJECTS FROM ALIGNMENTS

After defining alignments, Civil 3D has routines to create new objects to reference those alignments. Most routines create points and are in the Create Points toolbar.

### POINT SETTINGS

The Create Points' toolbar Expand Create Points Dialog can affect many new point values. The dialog's Elevations and Descriptions affect how points are assigned descriptions and elevations. If description and elevation are set to Manual, a routine prompts for these values before the point is created. If descriptions and elevations are set to Automatic, a routine assigns the current value to each point without prompting the user.

When you create points and you want to use the alignment's name as the description, set Prompt for Description to Automatic - Object. By selecting this option, the alignment's name becomes the point's description.

## CREATE POINTS—ALIGNMENT

The Create Points toolbar—Alignment icons stack references an existing alignment (see Figure 5.54).

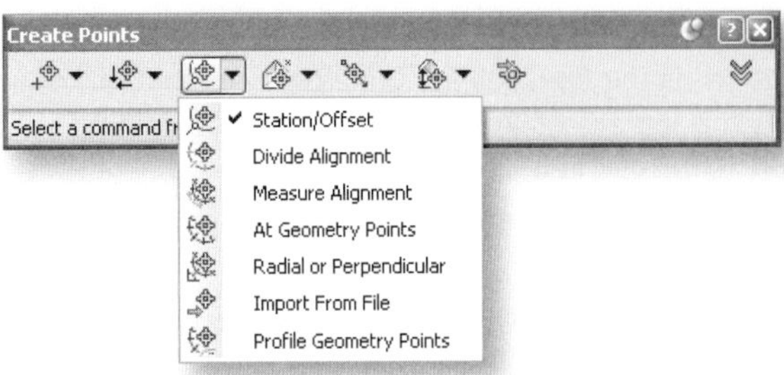


**FIGURE 5.54**

### Station/Offset

Station/Offset places points at specific stations and offsets along a selected alignment. When using this routine, you are prompted for an alignment. After selecting the alignment, a station jig appears and reports the stations and offsets. You use this jig to locate the station and then locate the offset. Instead of using the jig to locate a station and offset, users can enter the station and offset values.

### Divide Alignment and Measure Alignment

Divide Alignment and Measure Alignment place points on the alignment at intervals. Divide places points to reflect the number of user-specified segments (for example, 10 equal segments). Measure places points at a user-specified distance (for example, every 20 units).

### At Geometry Points

At Geometry Points places points at critical alignment locations. The critical points are PC (point of curve beginning), PT (point of curve ending), SC (point of spiral curve intersection), RP (radial point), CC (compound curve points), and spiral curve intersections.

### Radial or Perpendicular

Radial or Perpendicular routine places points on an alignment that are radial or perpendicular to user-selected points away from the alignment. Radial and Perpendicular use a station jig to locate the points.

### Import from File

Import from File reads an external file and creates points using an alignment's stationing. The file formats contain station and offset values and other values. The following file formats are used by the routine:

```
Station, Offset
Station, Offset, Elevation
Station, Offset, Rod, Hi
Station, Offset, Description
Station, Offset, Elevation, Description
Station, Offset, Rod, Hi, Description
```

The Rod and Hi are for level surveys. The Rod is the prism elevation, and the Hi is the instrument height. The file can be space- or comma-delimited.

## CREATE POINTS — INTERSECTION

Create Points toolbar's Intersection icons stack contains routines to place points at the intersection of alignments and directions, distances, other alignments, and other objects in the drawing (see Figure 5.55).

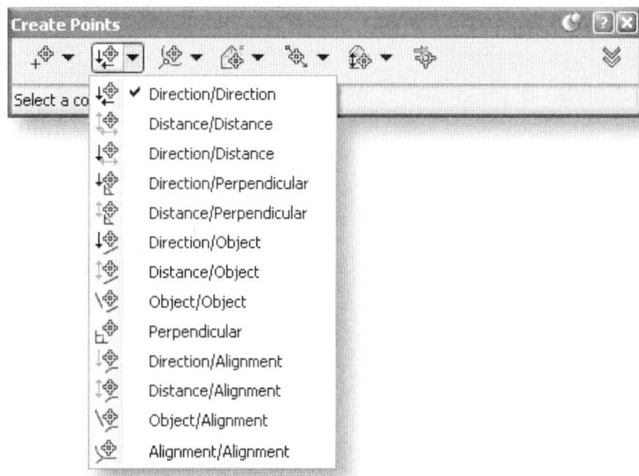


**FIGURE 5.55**

### Direction/Alignment

This routine places points that are at the intersection between an alignment and a direction from a known point. The intersection can be an offset from both the direction and alignment or either the direction or alignment.

### Distance/Alignment

The Distance/Alignment routine creates points at the intersection of a distance from a known point and an alignment. There can be an offset from the alignment.

### Object/Alignment

The Object/Alignment routine creates an intersection point at the intersection of an object (line, circle, or spiral) and an alignment. Both the object and the alignment can have an associated offset.

### Alignment/Alignment

The Alignment/Alignment routine creates a point at the intersection of two alignments. There can be an offset from both or either alignments.

## LANDXML OUTPUT

LandXML Export creates a file that describes an alignment's geometry (see Figure 5.56). This file transfers an alignment's geometry to other applications without loss of coordinate and design fidelity.

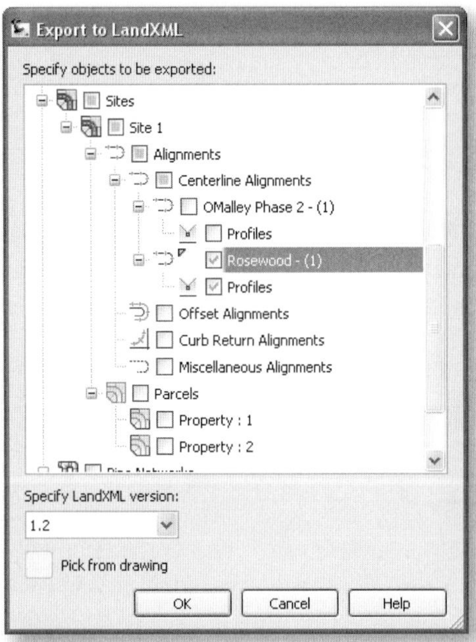


FIGURE 5.56

## CREATE ROW (RIGHT-OF-WAY)

Create ROW is in the Parcels menu and creates right-of-way parcels (right and left side of the alignment) that encompass the alignment and any intersecting alignment. When running the routine, Create ROW displays a dialog box that sets fillet or chamfer values for intersecting ROW lines from intersecting alignments (see Figure 5.57).

Most Civil 3D objects react to or are linked to other objects, allowing them to interact with one another. Unfortunately, the ROW parcel is not linked to its alignment and as a result will not change its geometry to match any changes to the alignment definition.

A ROW parcel can be a parcel frontage boundary.

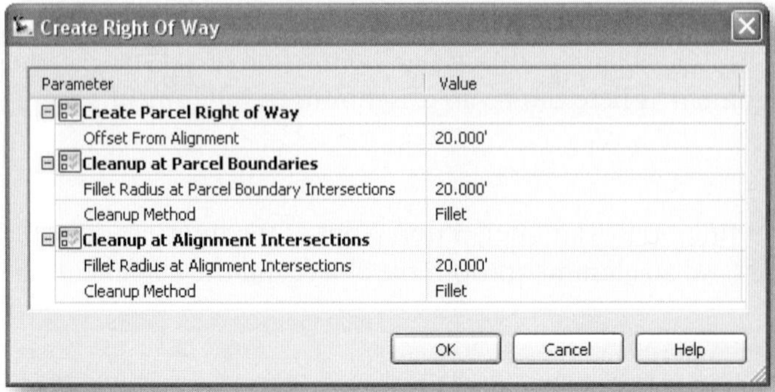


FIGURE 5.57

## EXERCISE 5-6

After completing this exercise, you will:

- Be able to create points by intersection with an alignment.
- Be able to create points from the geometry with an alignment.
- Be able to import points with station and offset values.
- Be able to create an alignment ROW.

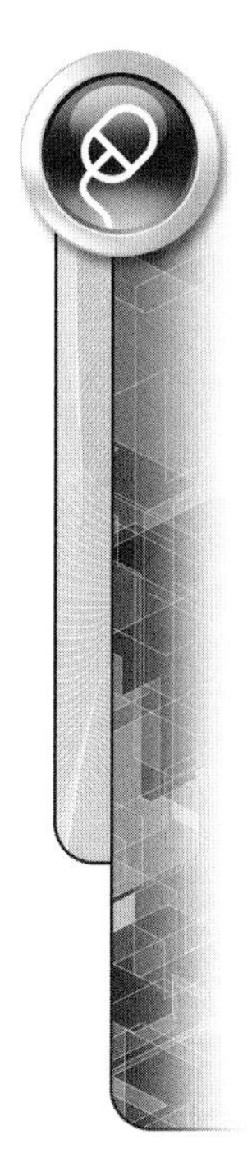

### Exercise Setup

This exercise continues with the previous exercise's drawing. If you did not complete the previous exercise, browse to the Chapter 5 folder of the CD that accompanies this textbook and open the *Chapter 05 – Unit 6.dwg* file.

1. If not open, open the previous exercise's drawing or browse to the Chapter 5 folder of the CD that accompanies this textbook and open the *Chapter 05 – Unit 6* drawing.

### Alignment Points — Station/Offset

Civil 3D creates points whose locations are critical centerline points, reference station and offset, measure lengths along it, or divide it into equal-length segments.

Setting point by Station and Offset is similar to labeling alignment stations and offsets.

1. On the Ribbon, click the ***View*** tab. On the Views panel's left, select and restore the named view **Proposed Starting Point**. You may need to scroll up the list by selecting the up arrow.
2. On the Ribbon, click the ***Home*** tab. On the Create Ground Data panel, click the Points drop-list arrow and on the shortcut menu, select `POINT CREATION TOOLS`.
3. On the toolbar, click (chevron on the toolbar's right side) the Expand the Create Points dialog.
4. Expand the Points Creation section, set the values in Table 5.1.

**TABLE 5.1**

| Section | Setting | Value |
|---|---|---|
| Points Creation | Prompt For Elevations | None |
| Points Creation | Prompt For Descriptions | Automatic - Object |
| Points Creation | Default Description | CL |

Your toolbar should look like Figure 5.58.

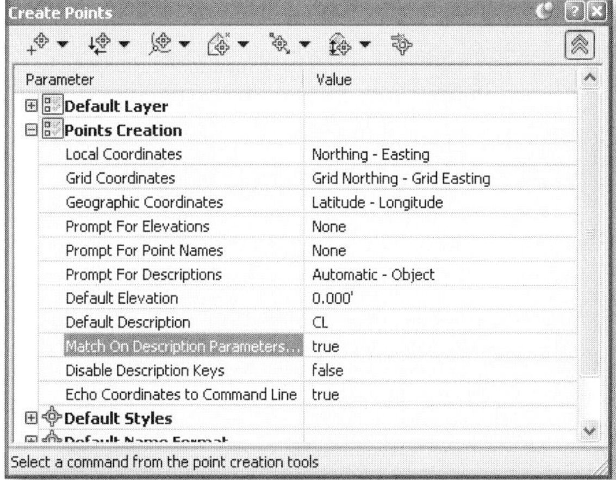

**FIGURE 5.58**

5. Click the Collapse the Create Points dialog chevron to close it.
6. In the Create Points toolbar, click the ***Alignment*** icon stack drop-list arrow, and from the list select AT GEOMETRY POINTS.
7. In the command line, the routine prompts you to select an alignment; select the Rosewood alignment.
8. In the command line, the routine prompts you for the starting and ending stations; set the starting station to 0+00, press ENTER, set the ending station to 1+00, and press enter twice to exit the routine.

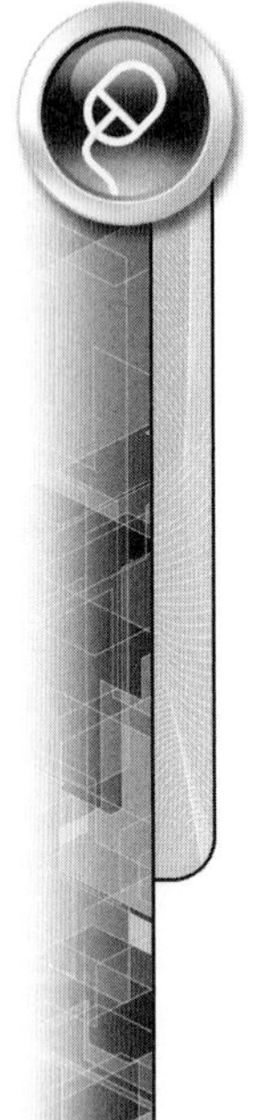
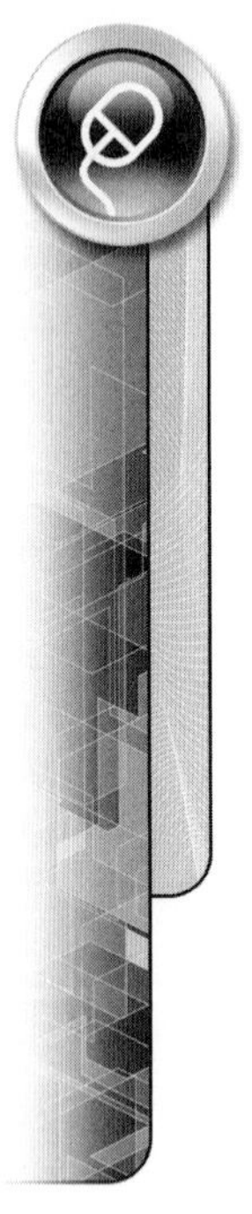

The routine creates points representing the alignment's curve RPs, PCs, and PTs.

## Import from File

You can import points from a file that references an alignment's stations and offsets.

The file for this exercise is in this textbook's CD, Chapter 5 folder.

1. From the Create Points toolbar, to the right of the Alignment icon stack, click the drop-list arrow, and from the list select IMPORT FROM FILE.
2. In the Select File dialog box, browse to the CD that accompanies this textbook, and from the Chapter 5 folder, open the *Station Offset.txt* file.
3. In the command line, the routine prompts you for a file format number (you may have to press F2 to see the format list). For the file format enter **1** (Station, Offset), and press ENTER to continue.
4. In the command line, the routine prompts you for a delimiter; enter **2** (comma), and press ENTER twice to use a comma and accept the default invalid station/offset indicator.
5. In the command line, the routine prompts you to select an alignment; in the drawing, select the Rosewood alignment.

The points reference stations 1+00 to 2+00.

6. Use the ZOOM command to better view the points in the area of stations 1+00 and 2+00.
7. At Civil 3D's top left, Quick Access Toolbar, click the ***Save*** icon to save the drawing.

## Points from intersections with Alignments

1. On the Ribbon, click the ***View*** tab. In the Views panel select and restore the named view **Station Equation**.
2. If the area is crowded with labels, select a station/offset label to display its grip; select the grip (cyan diamond), and drag the label to a new position, or use the ERASE command to remove some labels from the drawing.
3. In the Create Points toolbar, click the Expand the Create Points dialog chevron. In the Points Creation section, change the Default Description to **INTERSECTION**.
4. Click the Collapse the Create Points dialog chevron to hide the panel.
5. In the Create Points toolbar, to the right of the Intersection icon stack, click the drop-list arrow and from the list, select DIRECTION/ALIGNMENT.
6. The routine prompts you for an alignment; in the drawing select the Rosewood alignment. Next, the routine prompts you for an offset; for the offset, enter **-25** and press ENTER.
7. In the command line, the routine prompts you for a direction starting point; in the drawing, select a point near the Phase 1 line.

8. After selecting the starting point, the routine prompts you for a second point to set a direction (a drag helps you define the direction). In the drawing, select a second point to define a direction that intersects the Rosewood alignment.
9. In the command line, the routine prompts you for an offset; for the offset enter **0** (zero) and press ENTER. The routine marks a point that represents the intersection of an offset from the alignment and the direction from the Phase 1 line. Press ENTER to exit the routine.
10. In the Create Points toolbar, click the toolbar's red **X** to close it.
11. At Civil 3D's top left, Quick Access Toolbar, click the ***Save*** icon to save the drawing.

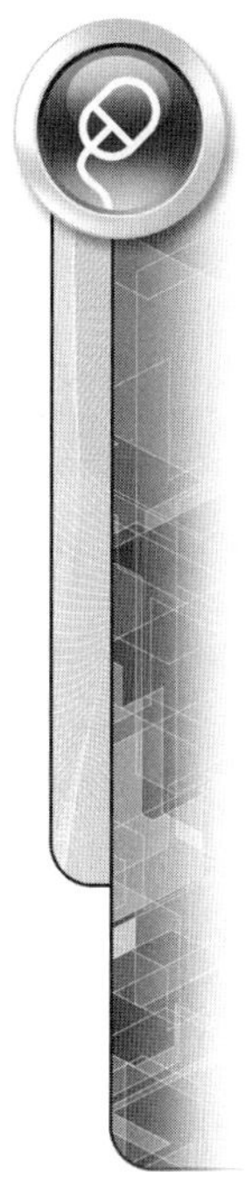

## Creating a ROW

The Rosewood alignment divides the property into two new parcels. In each subdivision, there is a buffer that extends to either side of the alignment. This buffer is the ROW parcel. Civil 3D creates a ROW from all of the defined centerlines of a drawing.

1. Use the ZOOM EXTENTS command to view the entire drawing.
2. On the Ribbon, click the Home tab. On the Create Design Panel, click Parcel's drop-list arrow and from the shortcut menu select CREATE RIGHT OF WAY.
3. In the command line, the routine prompts you to select the parcels that are adjacent to the alignment. In the drawing, select the parcel labels to the east and west of the Rosewood alignment, and press ENTER to display the Create Right Of Way dialog box.
4. In the Create Parcel Right of Way section, set the Offset From Alignment value to **33** and set the cleanup methods for Cleanup at Parcel Boundaries and Alignment Intersections to **None** (no fillet or chamfer) (see Figure 5.57).
5. Click ***OK*** to create the ROW.
6. At Civil 3D's top left, Quick Access Toolbar, click the ***Save*** icon to save the drawing.

This completes the alignment exercises.

## SUMMARY

- Users can create points to reference an alignment's geometry or stations and offsets.
- Users can create points that intersect an alignment by using other alignments, directions, or distances.
- Users can import points from a file containing station and offset values (plus additional data).
- Users can create a Right-of-Way parcel based on an alignment's definition.
- A Right-of-Way parcel is not linked to an alignment and will not react to any changes in the alignment's geometry.

An alignment is a roadway's horizontal centerline. An alignment is a collection of segment types that understand basic rules about their connection with segments before and after them. If you graphically adjust or manually edit segment values, the entire alignment adjusts to accommodate the changes. When you create alignments, they can include station and geometry annotation.

The next chapter focuses on the elevations along a centerline's path, the profile.

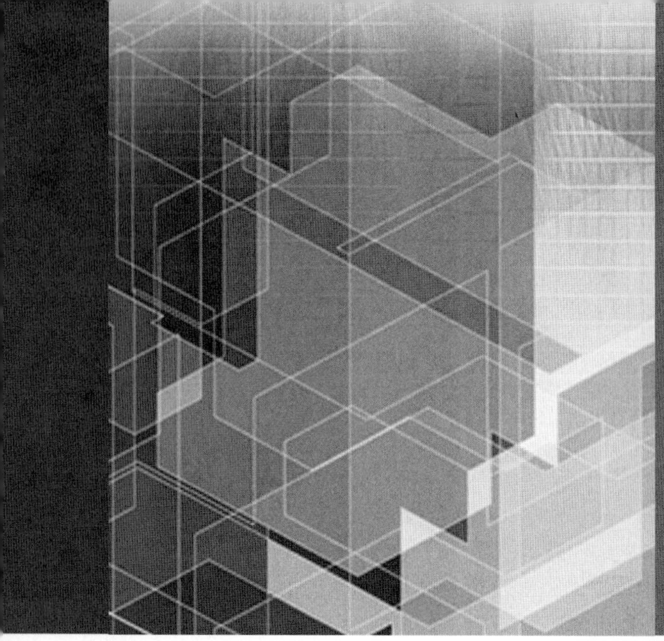

# CHAPTER 6

# Profile Views and Profiles

## INTRODUCTION

Road design's second step views an alignment from its side, showing the elevations along its path. In Civil 3D, a profile view is a grid that displays and annotates stations and elevations. A profile is the elevations from a surface or line work representing a roadway's vertical design within a profile view. Civil 3D dynamically links the profile view (grid) and the surface profiles to the alignment. If you are editing the alignment (move, shorten, or lengthen), the profile view and the surface profiles change to show the modified alignment's length and elevations. Changing a surface causes its profile to change within a profile view. The vertical design is unaffected by a changing alignment.

Styles affect the grid format, the grid annotation and types, profiles, and their display properties. A complicated web of dependencies and styles makes up the final profile view and its profiles.

## OBJECTIVES

This chapter focuses on the following topics:

- Introducing and Creating a Profile Grid
- Creating a Simple Profile Grid Style and Modifying Existing Grid Styles
- Introducing and Creating a Profile
- Creating a Simple Profile Style and Modifying Existing Grid Styles

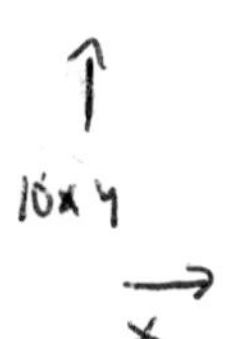

### OVERVIEW

This chapter covers roadway design's second phase, profile view and profiles. A profile view is a graph that represents an alignment's stationing and elevations along that alignment's path. A profile is line work within a profile view that represents surface elevations along an alignment's path or a proposed roadway's vertical design. A profile view and its surface profiles are the backdrop for the proposed roadway vertical design.

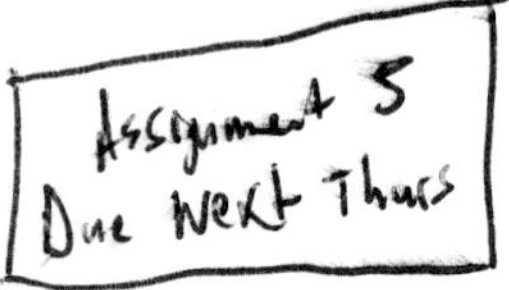

It is in the profile view that you start developing a feel for a road design's impact on earthworks volumes. The proposed roadway's height above or below existing ground's elevations begins by giving visual feedback as to amounts of earthwork or what problems need to be resolved to build the road (see Figure 6.1).

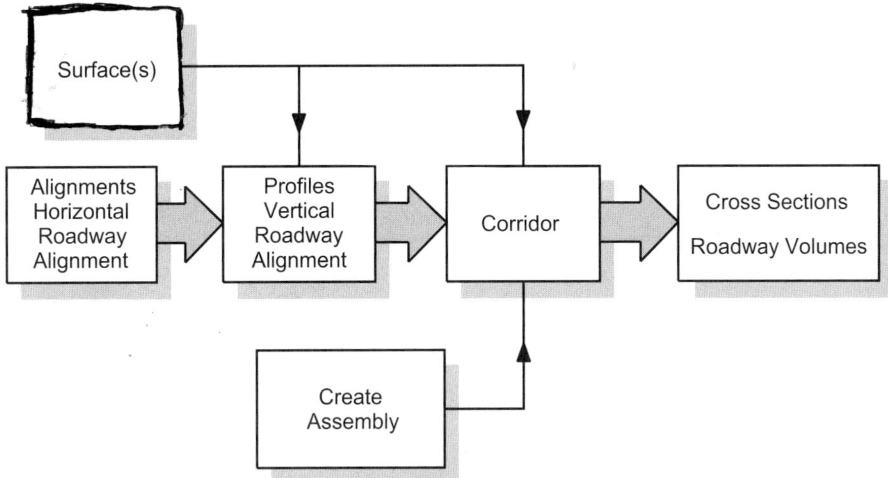


**FIGURE 6.1**

There are three steps to creating a profile and profile view. First, there must be data; an alignment and its profile data which can be a surface or a file containing alignment stations and their elevations. The second step is creating the profile (sampling surface elevations, assigning styles, and adjusting other values), and the last step, if you chose to do so, is immediately creating a profile view containing the profile(s).

Profile and Profile View steps:

1. Have an alignment and a surface or read profile data from a file
2. Create a Profile
3. Create a Profile View

A profile view can contain multiple profiles (surfaces and proposed vertical designs). Each profile can have a different style, thus allowing it to display its information uniquely in the profile view.

### Unit 1

A Profile View is a graph in which profiles are drawn. A profile represents the existing ground, subsurfaces, or one or more proposed vertical designs. The styles that affect a profile view and its profiles are the focus of this first unit.

### Unit 2

The second unit of this chapter reviews the steps needed to create a profile and its Profile View.

### Unit 3

Within a profile view and an existing ground profile, a designer creates a vertical (alignment) roadway design. A vertical alignment contains tangents and vertical curves. Civil 3D has three vertical curves types: circular, symmetrical, and asymmetrical parabolic curves. Creating a vertical alignment is the topic of the third unit of this chapter.

### Unit 4

The analysis and editing of a vertical alignment (profile) is the focus of this chapter's fourth unit. This unit reviews analyzing a vertical alignment using Toolbox reports, the Inquiry Tools, profile properties, and vistas within the vertical editor.

### Unit 5

The fifth unit of this chapter reviews profile and vertical alignment annotation and projecting 3D objects to a profile view. The annotation is the result of label sets applied when a profile is created or is the result of labels that are added after the profile is created.

## UNIT 1: PROFILE VIEW AND PROFILE STYLES

A Profile View is a graph in which profiles exist. Profiles within a view represent existing ground and other surfaces along an alignment and one or more proposed vertical designs.

### PROFILE VIEW

Profile View is a grid that represents alignment stations and elevations along its path. Stations are along the graph's top or bottom and create vertical lines to mark the stations in the profile view's elevation area (see Figure 6.2). The station interval has a major and minor increment with station annotation at the major stations (minimally). All values are user specified and set in the styles applied to the profile view.

Elevations are measured from a graph's datum (lowest elevation) upward to the highest elevation. The line interval is an even elevation (every 2 or 5 feet, for example) and has secondary tick marks at minor increments with annotation at major elevations (minimally). All values are user-specified values set in the styles applied to the profile view.

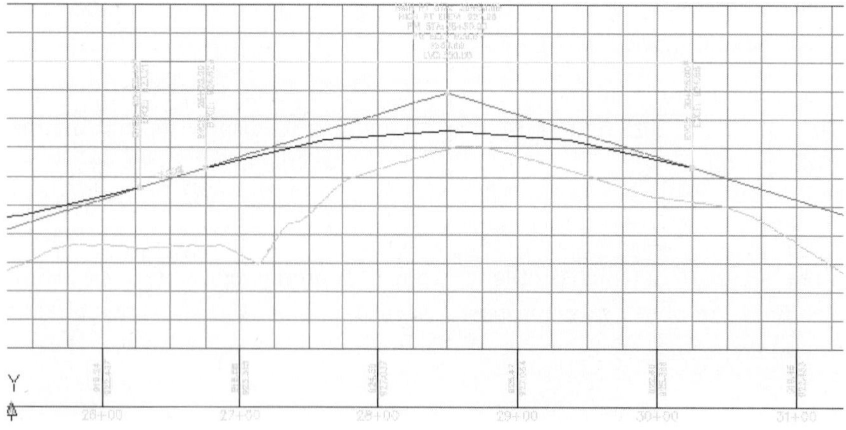


**FIGURE 6.2**

Traditionally a profile view is 1/10th of the horizontal scale. If you are working with a drawing that has a 1"=40' scale, the vertical scale is 1"=4' (1 inch of paper represents 4 feet of relief).

## PROFILE VIEW STYLE — FULL GRID

A Profile View Style defines values that affect titles and defines station and elevation annotation. A Profile View Style is a multi-tabbed dialog box with each tab affecting different aspects of the view.

### Information

As with all styles, there is an Information tab. This tab contains the name, description, and details on who created or modified the style as well as when that style was created or modified.

### Graph

The Graph tab affects the vertical scale and profile view direction (see Figure 6.3). The panel's top half lists the current vertical and horizontal scales, and the amount of vertical exaggeration. By default, a vertical scale is 1/10th of the horizontal scale. If the horizontal scale is 1"=40', then the vertical scale is 1"=4'. The vertical scale at the top left sets traditional scales and just below it, a user can set a custom scale value.

The panel's lower half sets the profile view's direction (left to right or right to left).

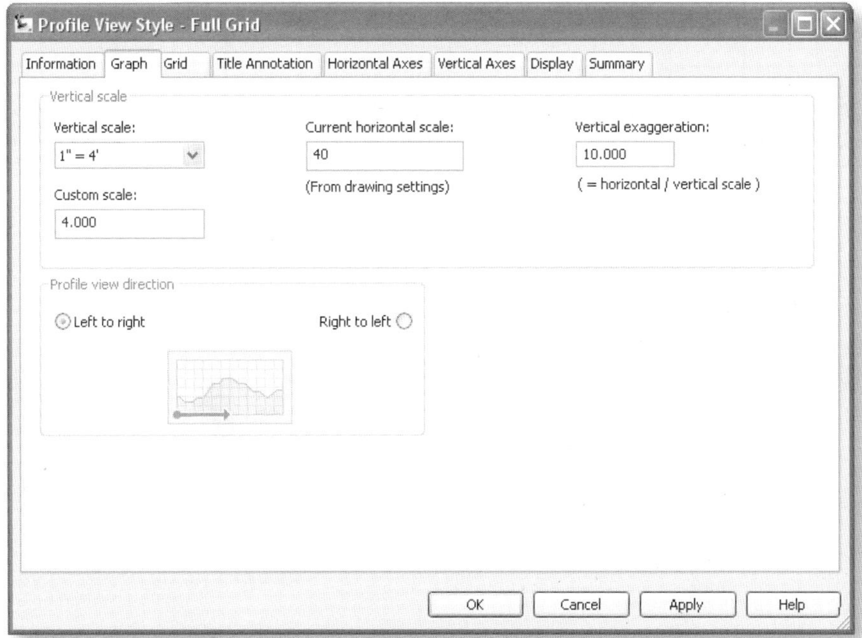


**FIGURE 6.3**

### Grid

Grid's top sets a view's clipping parameters. There are four possible toggle combinations: all on; all off; clip horizontally; and clip vertically. The effect of toggling on all clip toggles is seen in the right side of Figure 6.4. The left side of Figure 6.4 shows what occurs when all the toggles are off.

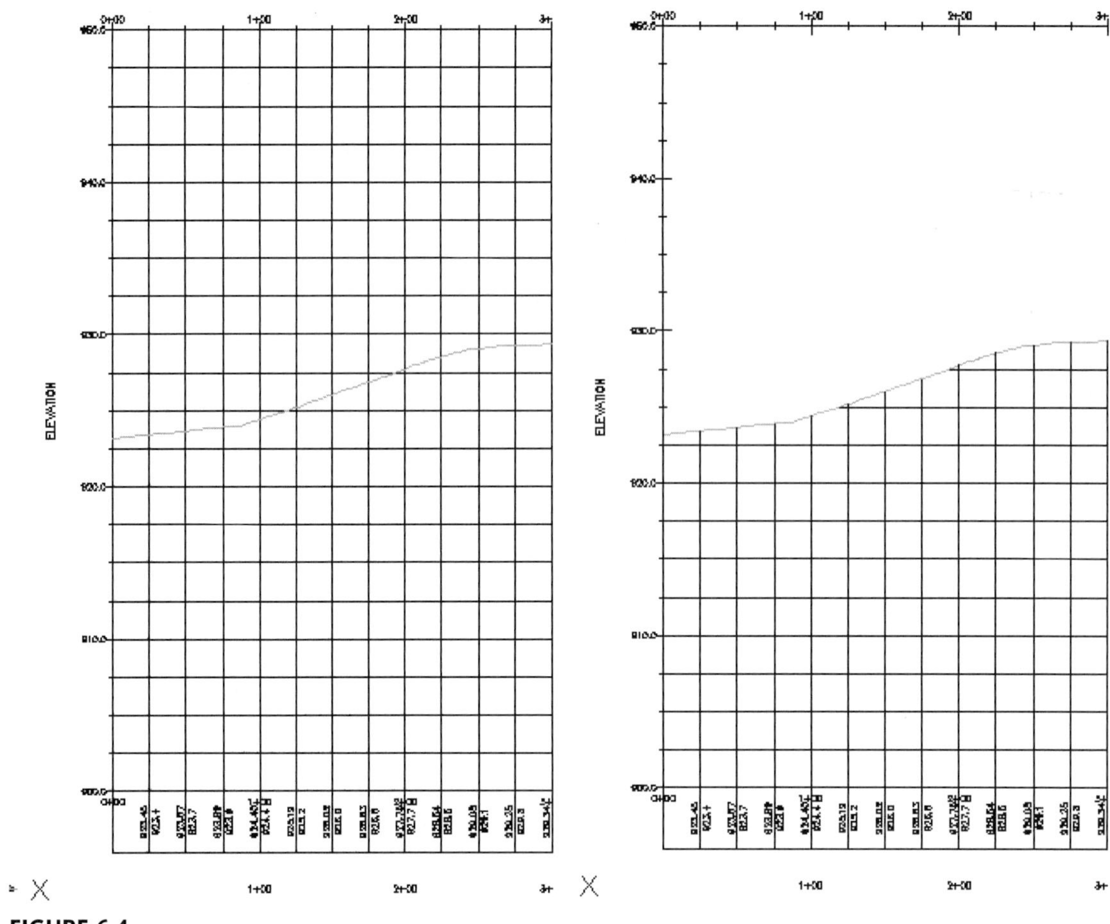

FIGURE 6.4

If you are toggling on only horizontal clipping, the right side of Figure 6.5 shows the result. If you are toggling on only vertical clipping, the left side of Figure 6.5 shows the result.

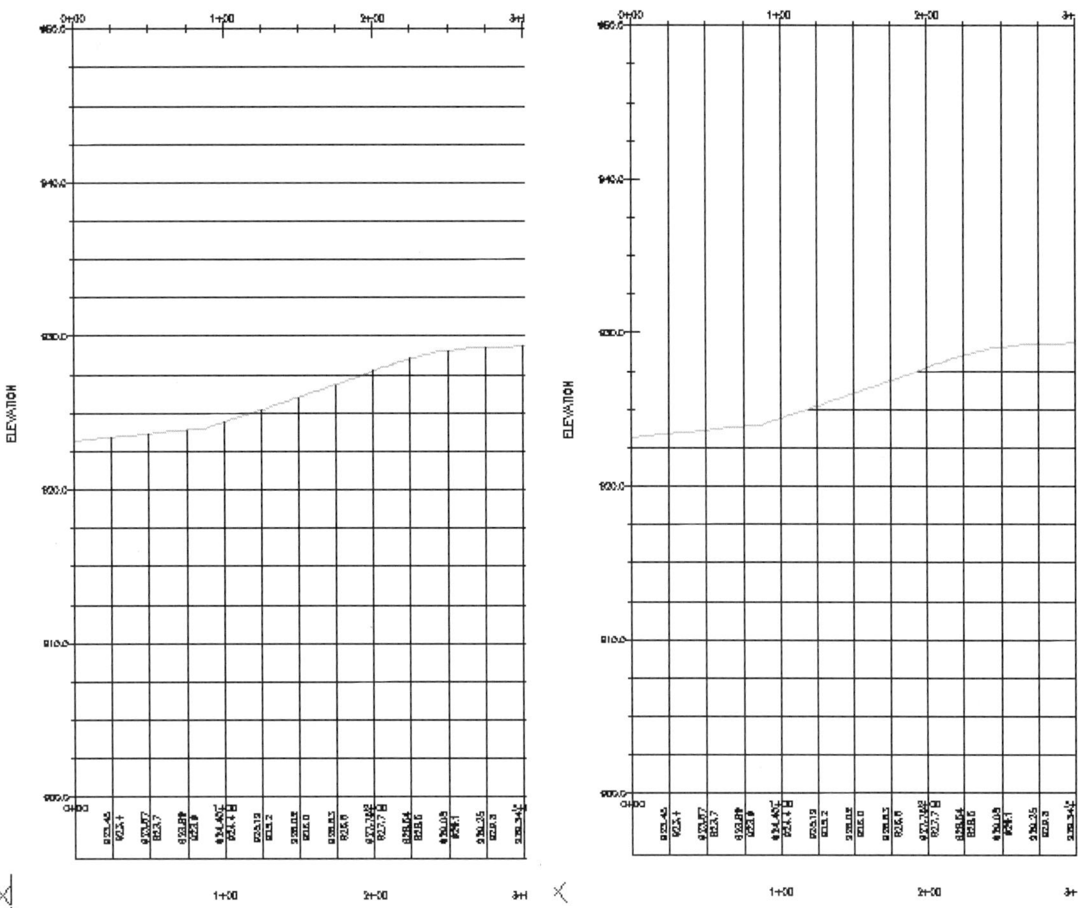


**FIGURE 6.5**

The panel's lower left sets the major grid padding around a grid. The panel's lower right sets an axis offset. When the axis offset has a positive value, a gap appears between the axis and the grid lines (see Figure 6.6).

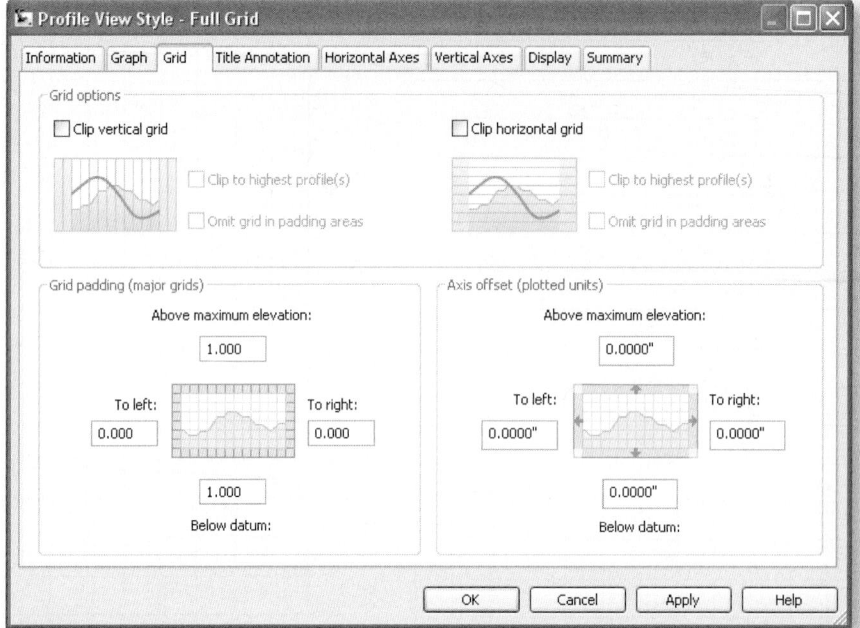


**FIGURE 6.6**

## Title Annotation

The panel's left side sets the title's text, text style, text size, and location (see Figure 6.7). The title has a justification setting and a border toggle. The panel's right side sets the title text, justification, text style, size, and content, for each of the four axes.

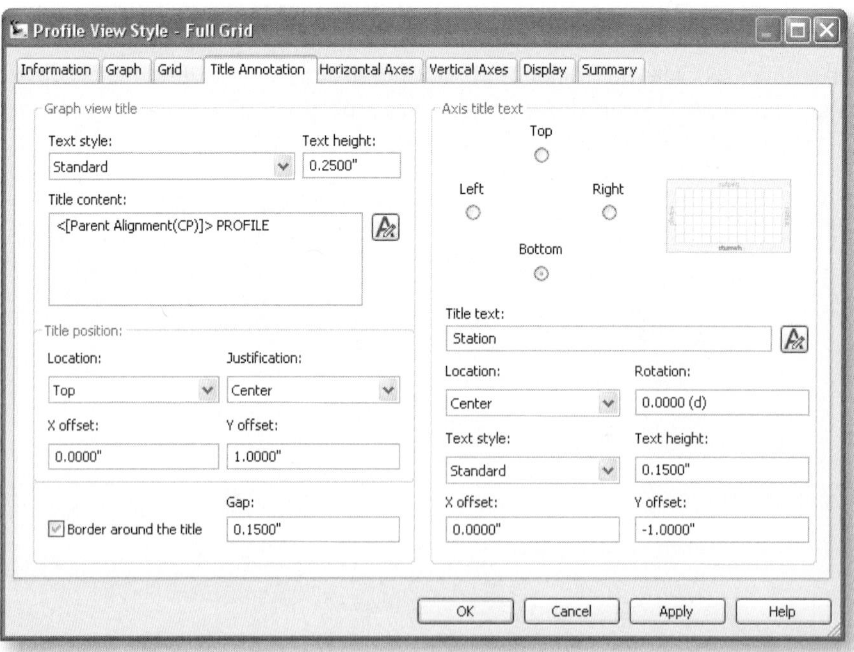


**FIGURE 6.7**

## Horizontal Axes

Horizontal Axes defines major and minor station text and tick intervals, text styles and size, and titles (see Figure 6.8). Station labeling can be at the top, bottom, or both. At the panel's top is the control for displaying top or bottom settings and if the first grid line is annotated. The rightmost column defines ticks that represent the horizontal geometry that is displayed in the view.

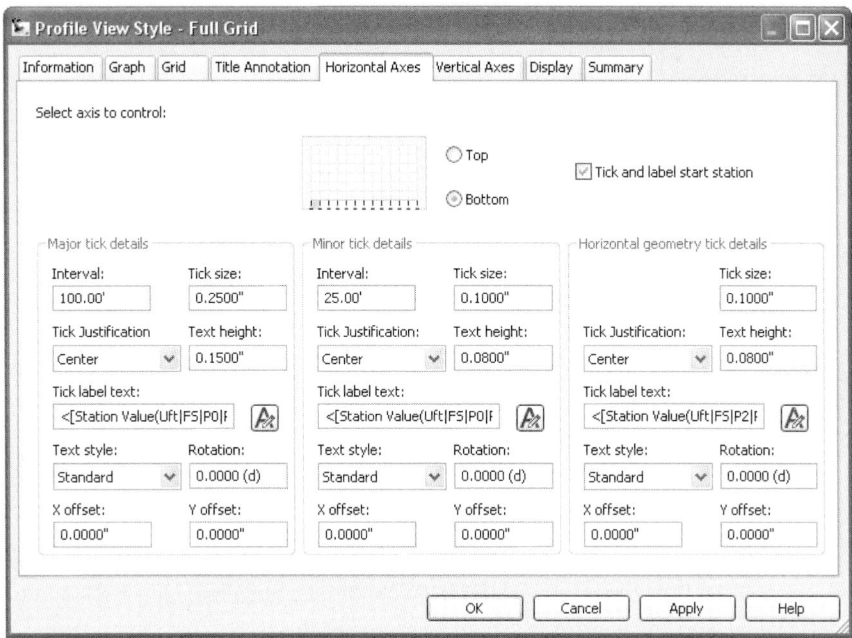


**FIGURE 6.8**

## Vertical Axes

Vertical Axes defines the elevation annotation's major and minor text and tick intervals and text styles, justification, rotation, and size (see Figure 6.9). Elevation labeling can be at the right, left, or both. At the panel's top center is the control for the side to which the settings apply and for whether the first elevation grid line has annotation.

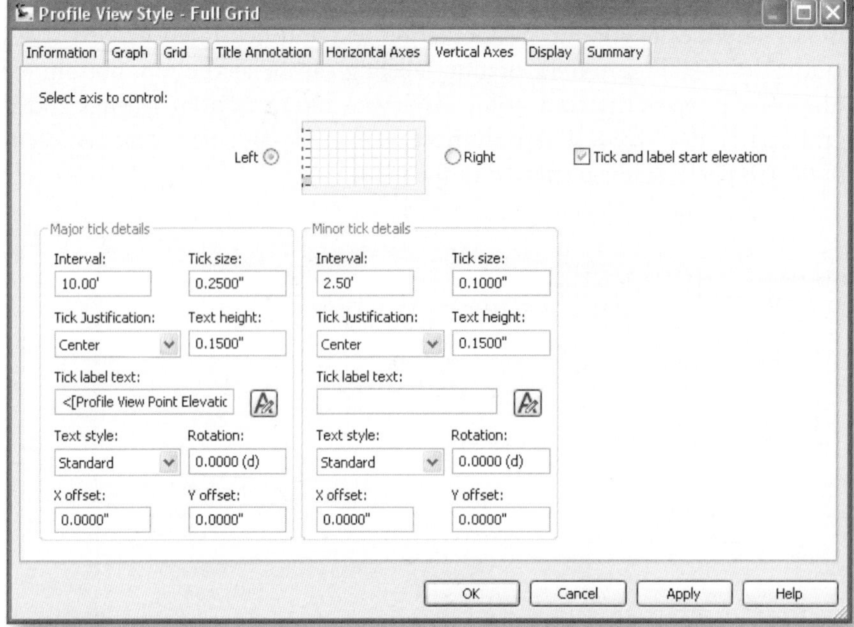


**FIGURE 6.9**

## Display

Even though they define profile view axes' annotation, the Display tab settings also determine what is displayed by the view style. The Display tab lists all of the view's components, their layer, layer properties, and their visibility (see Figure 6.10). Even though every possible tick, grid, or label is defined, they will show *only* if they are turned on in this panel.

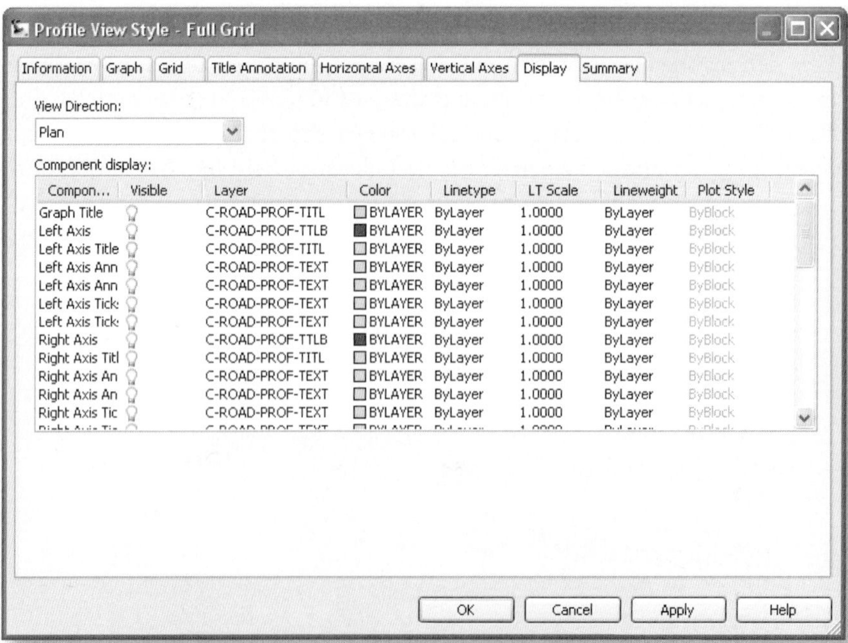


**FIGURE 6.10**

## BAND SETS

A Band Set name is an alias for a collection of Band Styles and depends on specific (band) styles to create their definition. Band Sets place vertical and horizontal information as a band at the profile view's bottom (or top) (see Figure 6.11). Traditionally, a band is at the profile view's bottom. A band's information may include the following: elevation or station values; graphics representing vertical, horizontal, section, or superelevation geometry; or a pipe network. When there is more than one band, the bands stack below the view with a gap between them. The Create Profile View dialog box assigns band sets when you create a view. Profile view's Properties dialog box, Bands panel manages bands already assigned to a view (add, delete, or modify).

If the band set contains station and elevation data, it displays the same data as a profile view style. You need to turn off the profile view annotation to remove it from the view, otherwise it will show in the band.

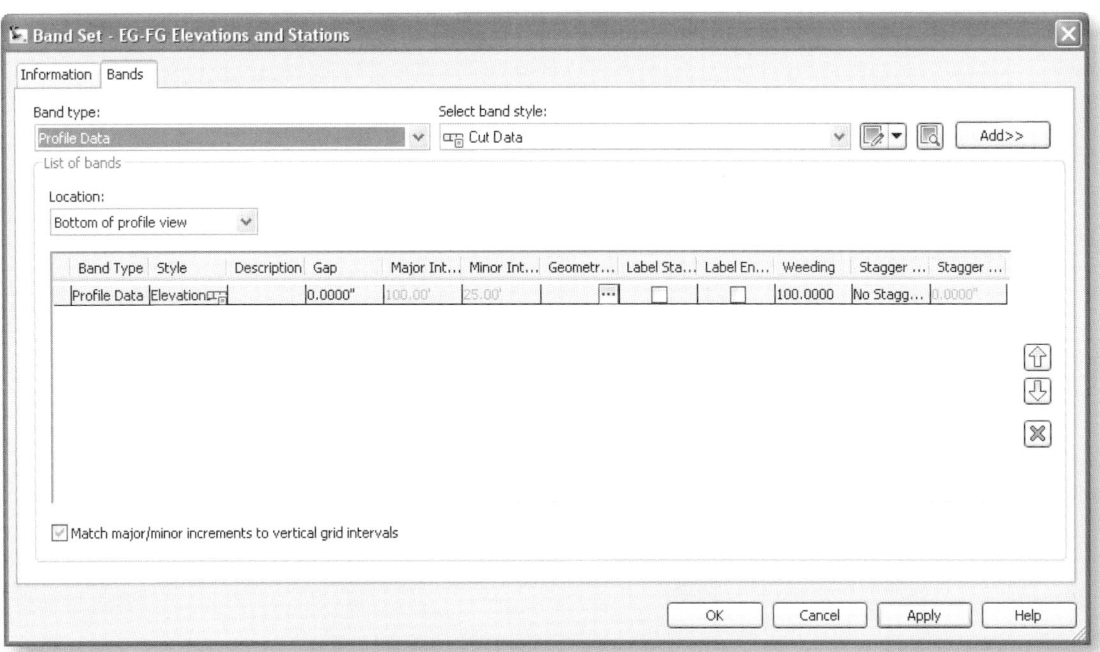


**FIGURE 6.11**

A Band Set can have one or any combination of six band styles types: Profile Data; Horizontal and Vertical Geometry; Sectional Data; Pipe Network Bands; and Superelevation data. These six style types are themselves a collection of values from their respective object type values. For example, the Profile Data styles emphasize labels for major and minor stations, horizontal and vertical geometry points, and station equations. The Vertical Geometry styles emphasize Up and Down hill tangent and sag and crest curve labeling, etc.

Profile View Properties' Bands panel manages its assigned band sets: add, delete, or modify (see Figure 6.12). When you are creating a Profile View from a surface, Civil 3D automatically assigns the surface to Profile 1 and Profile 2. The assignment remains until a specific profile is reassigned to Profile 2. Generally, this assignment is after the proposed vertical design has been drafted.

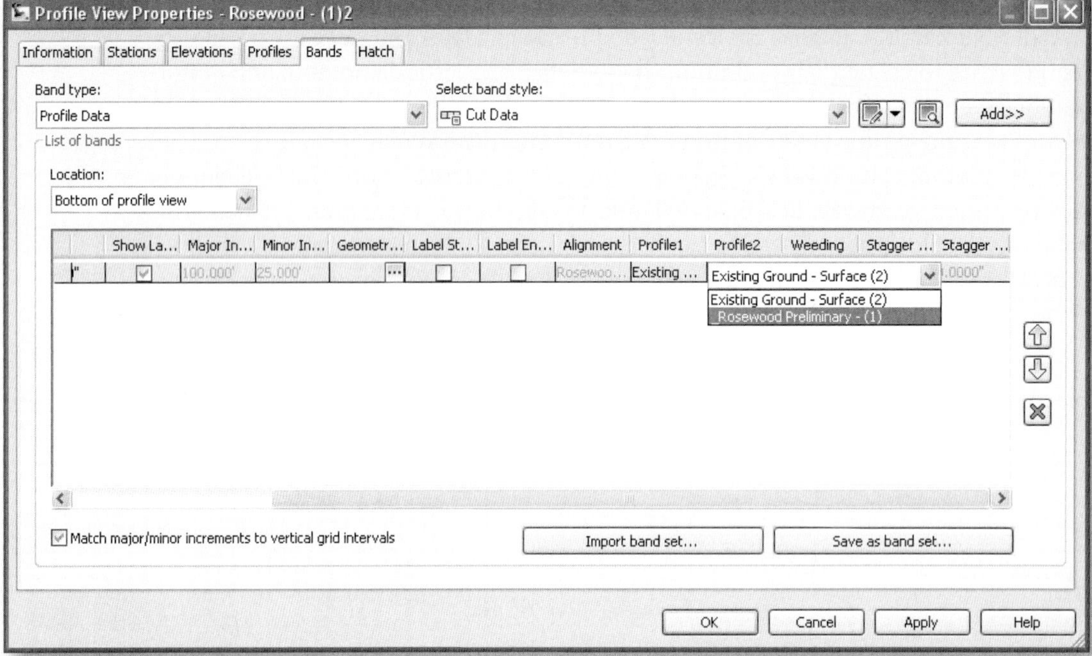


**FIGURE 6.12**

## Band Styles

All band style dialog boxes have four tabs: Information, Band Details, Display, and Summary (see Figure 6.13). The Band Details panel defines all potential style labels and the Display panel defines what label components are displayed.

The Information and the Summary panels are the same for all styles. Information names and records the style's creation and modification dates. The Summary panel reviews only the basic settings for each label type and tick and does not review specific label values. Viewing specific label values is the function of the Band Details tab.

### Profile Data Band Style

Profile Data Band style annotates major and minor stations, stations and elevations, cut and fill, profile elevations, etc. Each band type style will have a different list of values.

### Band Details

Band Details defines the band's title text style (top left), title content, its size, and location (middle left), and the band's general layout (see Figure 6.13). The layout area (bottom left) includes the gap between the band and the profile bottom (or top) and the title's text box width and height.

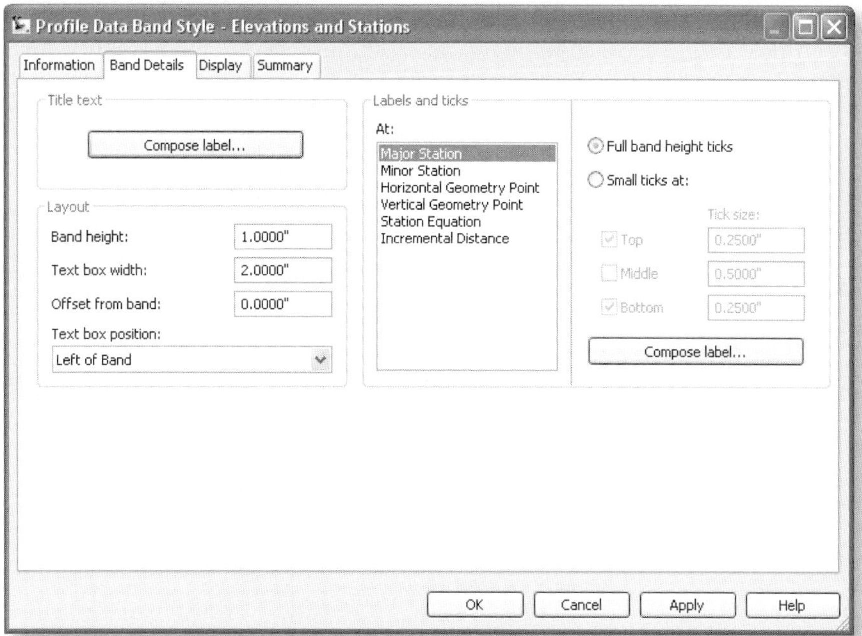


**FIGURE 6.13**

Band Details panel's critical part is what is not visible. What is hidden are listed label Types label definitions: At Major Station; At Minor Station; etc. When you highlight a label type and click Compose Label..., the Label Style Composer dialog box opens and names the label type you are reviewing or creating (see Figure 6.14).

When the Component Name drop-list arrow is clicked, it lists all of the label's components. Figure 6.14 lists all of Profile Data style's At Major Station label components. By selecting each label type (At Major Station, At Minor Station, etc.), clicking Compose Label..., and listing its Component Name, the style reveals what it labels.

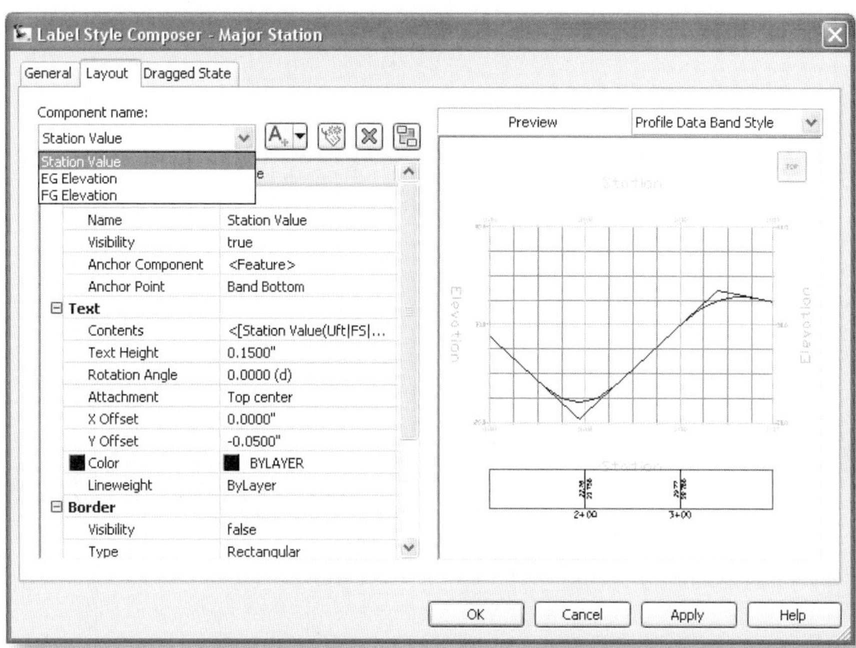


**FIGURE 6.14**

## Display

Display's settings assign each component its layer, properties, and visibility for a band's title, ticks, and lines (see Figure 6.15).

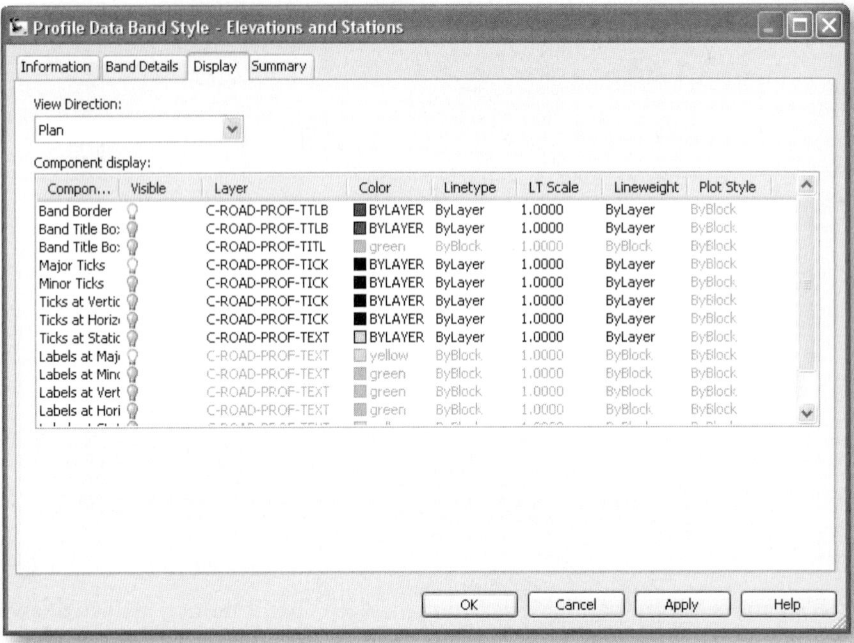


**FIGURE 6.15**

## Vertical Geometry Band Style

Vertical Geometry draws and annotates critical proposed vertical design values (see Figure 6.16). This style uses an assigned profile as data for its labels and this style also sketches its geometry in the band.

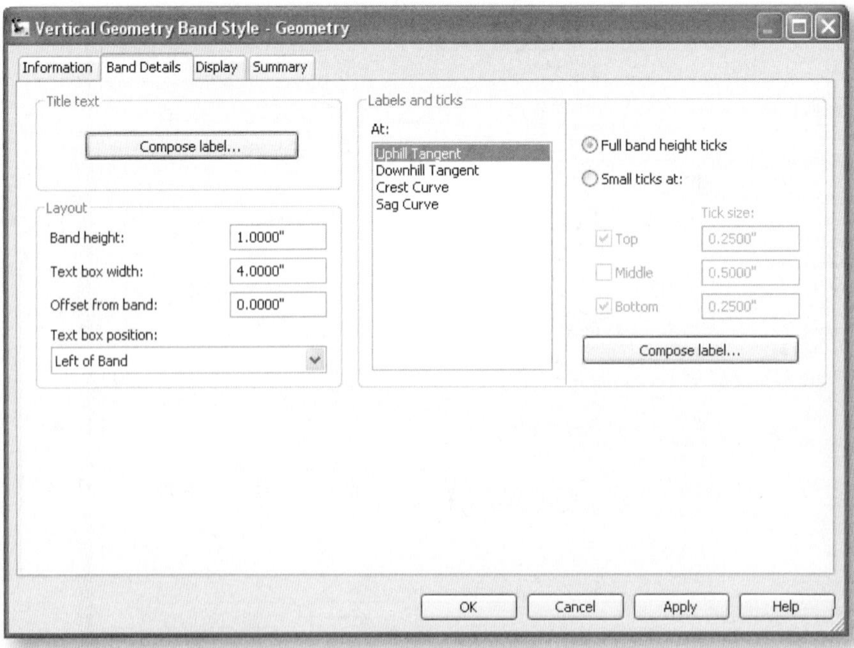


**FIGURE 6.16**

## Horizontal Geometry Band Style

Horizontal Geometry draws and annotates critical horizontal alignment values (see Figure 6.17). This style uses an assigned alignment as data for its labels and then it sketches its geometry in the band.

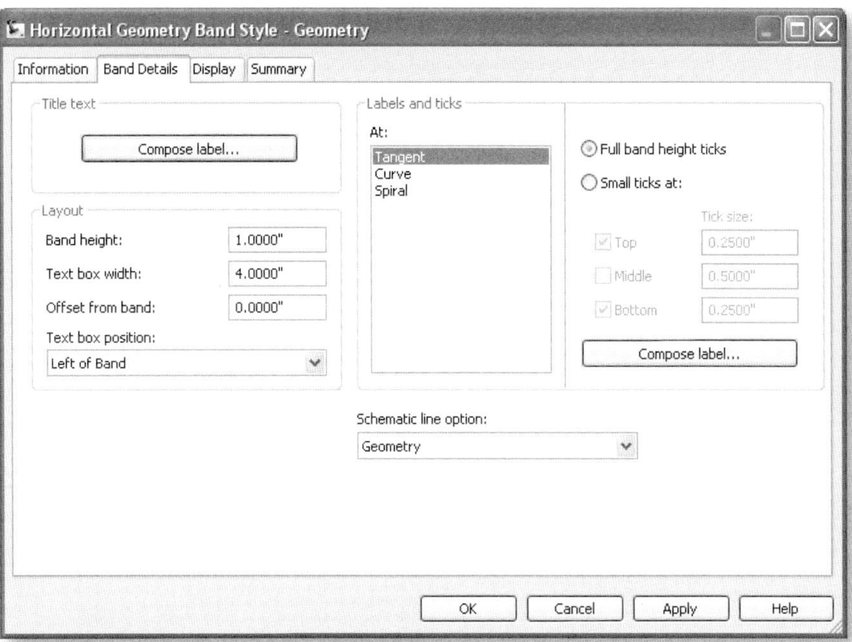


**FIGURE 6.17**

## Superelevation Band Style

Superelevation Band style annotates proposed superelevation critical points (see Figure 6.18). This style uses an assigned alignment as data for its labels and sketches the superelevation in the band.

Superelevation Band Style - Undivided - No Shoulders
Information
Band Details
Display
Summary
Title text
Compose label...
Layout
Band height:
1.0000"
Text box width:
4.0000"
Offset from band:
0.0000"
Text box position:
Left of Band
Labels and ticks
At:
Normal Crown
Level Crown
Reverse Crown
Full Super
Shoulder Critical Points
Slope Transition Region
Full band height ticks
Small ticks at:
Tick size:
Top
0.2500"
Middle
0.5000"
Bottom
0.2500"
Compose label...
OK
Cancel
Apply
Help

**FIGURE 6.18**

## Sectional Data Band Style

Sectional Data Band style creates annotation that represents data from a sample line group: sample line stations (At Sample Line Station) and the distance from the previous sample line (see Figure 6.19). These styles use an assigned sample line group as data for its labels.

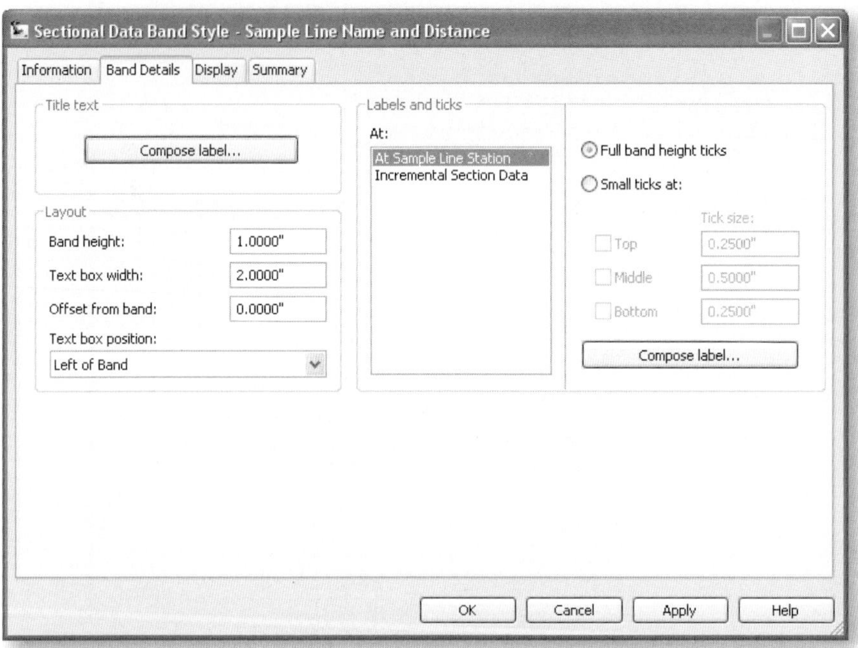


**FIGURE 6.19**

## Pipe Network Band Style

Pipe Network Band style creates annotation from an assigned pipe network (see Figure 6.20). The band labels structure and pipe values: station, inverts, slope, length, etc.

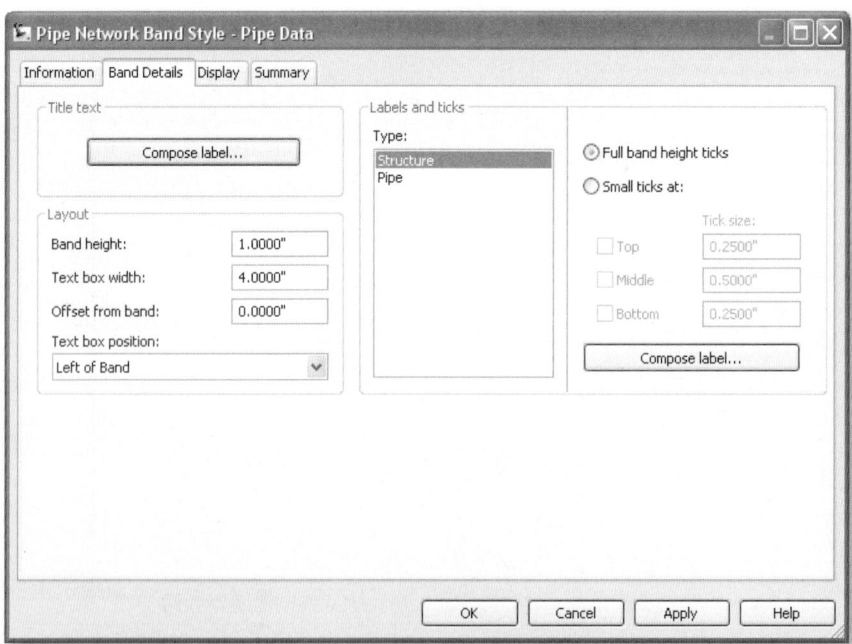


**FIGURE 6.20**

## PROFILE STYLES

Profile styles affect the way a profile is displayed in a profile view. In any implementation, there will be at least four profile styles: existing ground: right and left; and design. These styles visually differentiate the profile types in a profile view (see Figure 6.21).

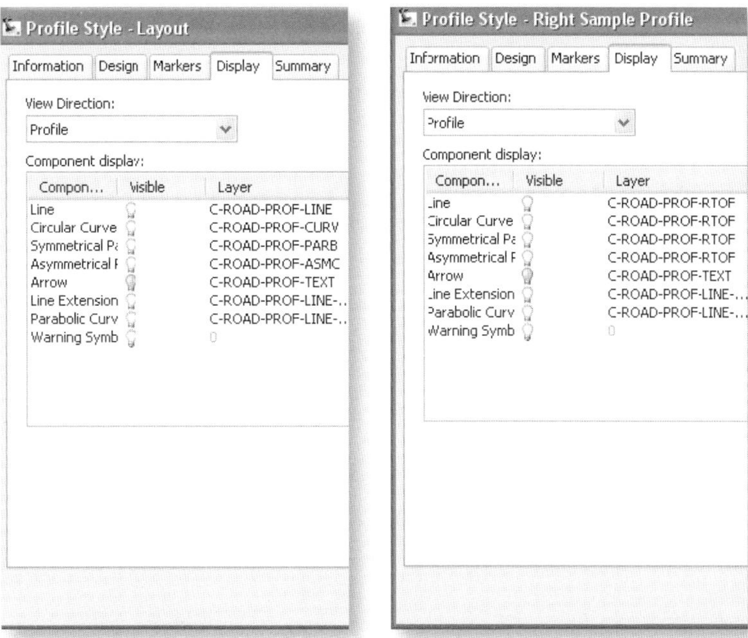


FIGURE 6.21

## PROFILE LABEL SETS AND STYLES

Profiles have Label Sets. A Label Set name is an alias for a collection of profile label styles. Profile label styles focus on Major and Minor Stations, Horizontal Geometry Points, Lines, Grade Breaks, and Sag and Crest curves. The label set most often used is the one assigned to a vertical design. You can later change a profile's labels in its Properties dialog box (see Figure 6.22).

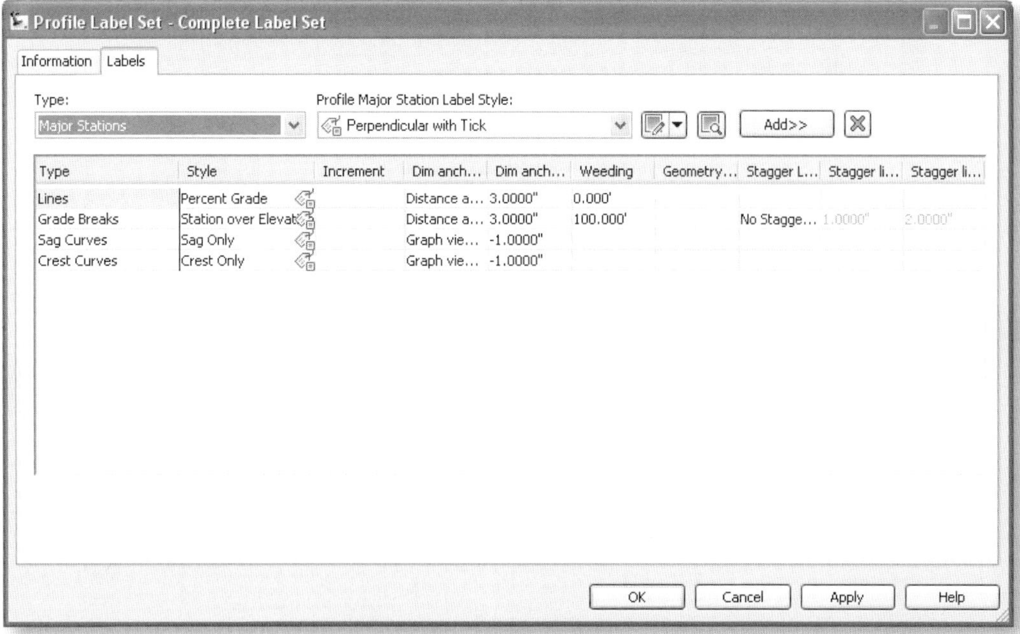


FIGURE 6.22

## Station

Station styles label vertical designs with major, minor, and horizontal geometry points.

### Major, Minor, and Horizontal Geometry Styles

The Major and Minor Station styles are interval label styles. The Major and Minor label styles frequency is set by the profile view stationing parameters. The Horizontal Geometry Points labels attach to the vertical design. The data types for the label styles include alignment, profile, and superimposed profile data (see Figures 6.23 and 6.24).

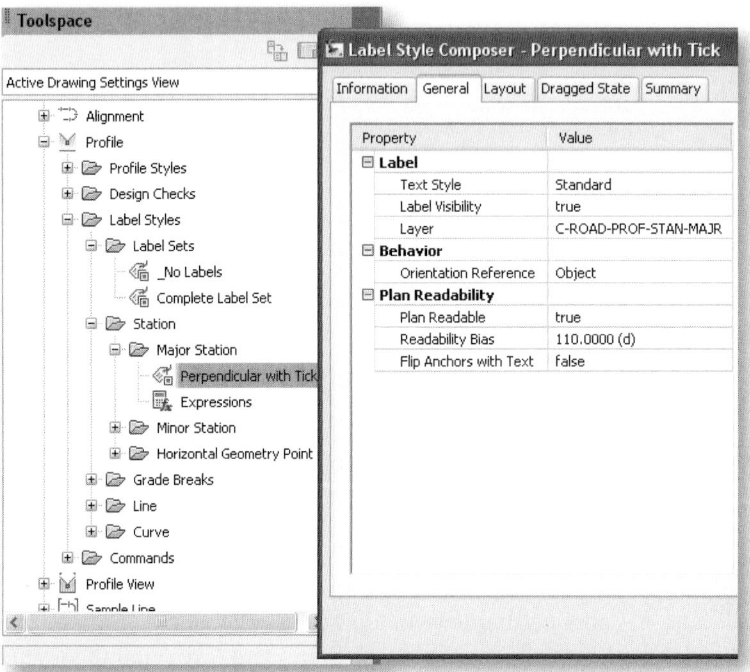


FIGURE 6.23

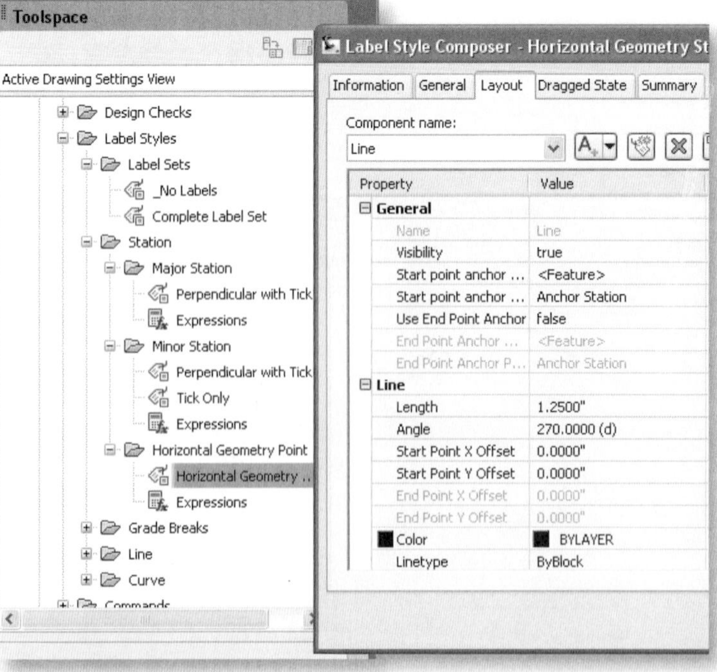


FIGURE 6.24

## Grade Breaks

Grade Breaks labels a profile tangent's beginning and ending station. Using this label on surface profiles creates a multitude of labels. Like all Profile labels, the focus of Grade Breaks labels should be the vertical design, not surface profiles (see Figure 6.25).

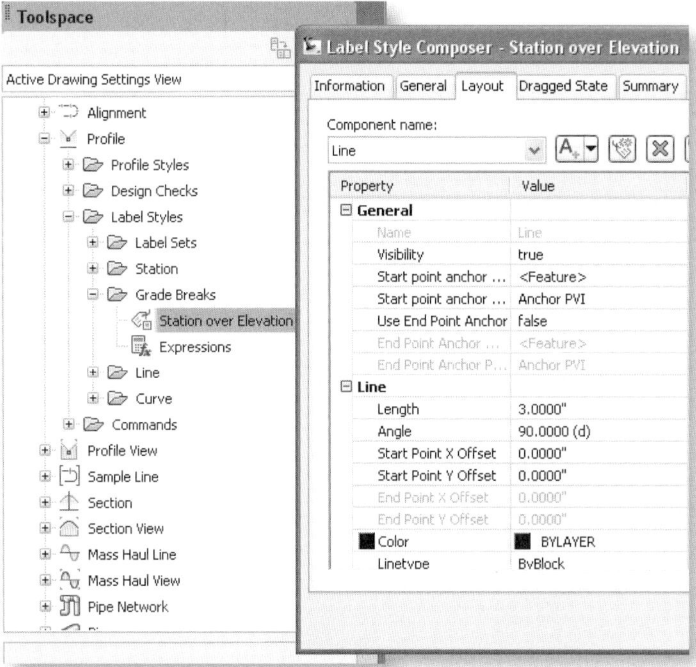


**FIGURE 6.25**

## Line

Line label styles annotate a profile tangent's grades and/or slopes. In addition to tangent information, additional alignment and profile data is available to this style (see Figure 6.26).

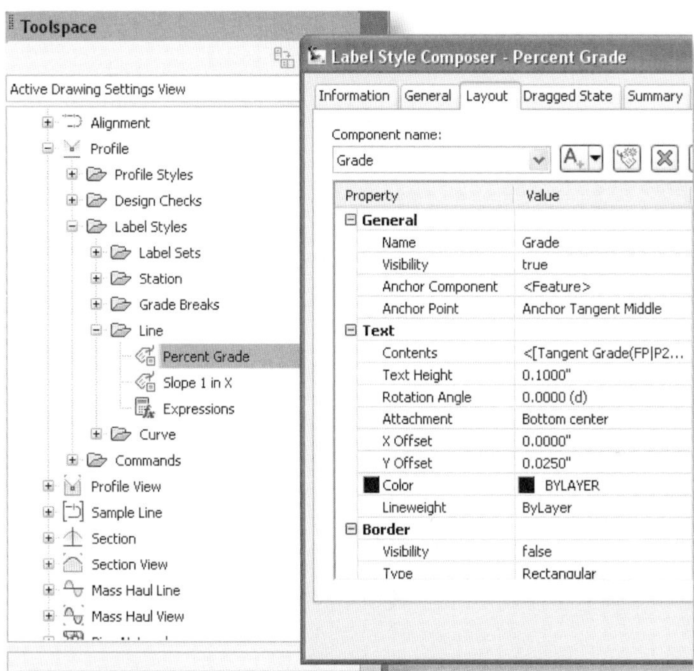


**FIGURE 6.26**

## Curve

Curve label styles annotate a profile's vertical curve's critical values. In addition to curve information, additional alignment and profile is available to this style (see Figure 6.27).

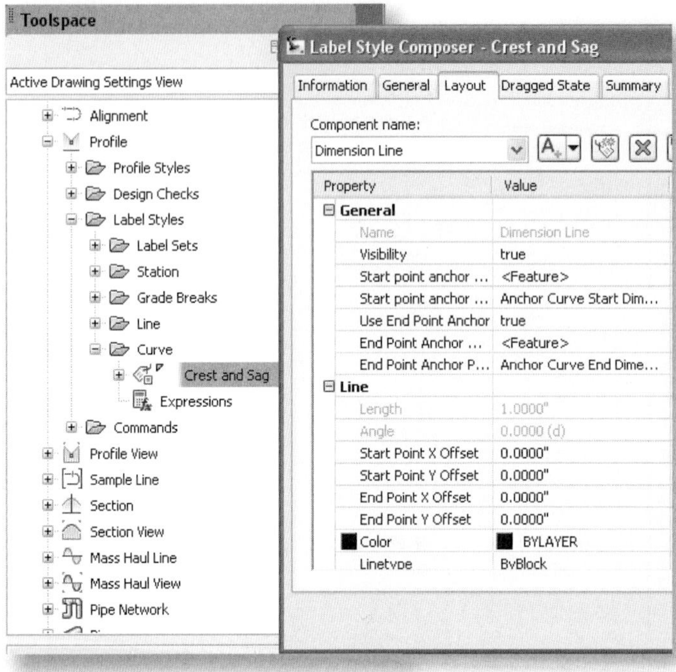


**FIGURE 6.27**

## ADD LABELS LABEL STYLES

The Add Labels command adds labels to a profile after creating it. These labels include two point profile view label styles: Station and Elevations; and Depths. Figure 6.28 shows a Station and Elevation label style example. These labels appear in the profile view's grid.

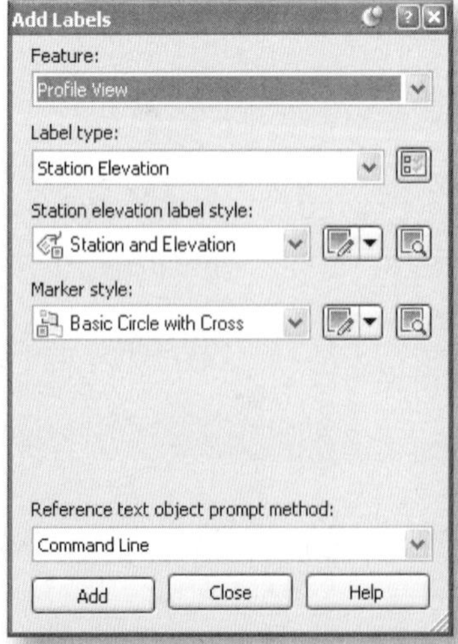


**FIGURE 6.28**

## PROFILE VIEW AND PROFILE SETTINGS

The Edit Drawing Settings dialog box contains several values that affect profiles, profile views, and their labels. Any style in the Profile settings branch can override these initial settings. However, if the Edit Drawing Setting values are locked, the lower styles that reference the locked values cannot change the values.

### Edit Drawing Settings — Object Layers

Edit Drawing Settings sets profile object base layer names. Each layer can have a modifier (prefix or suffix) and its value is the profile's name (see Figure 6.29).

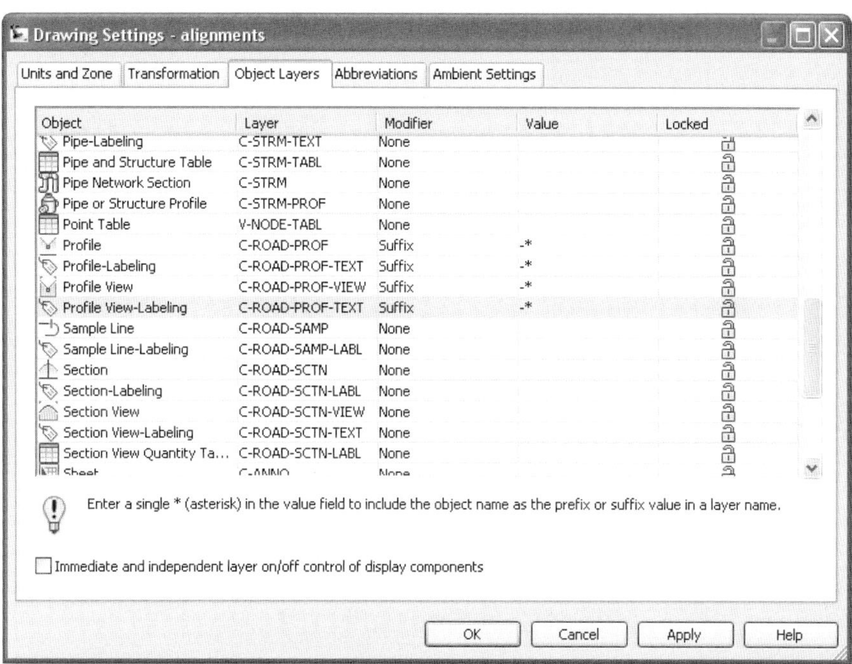


**FIGURE 6.29**

### Edit Drawing Settings — Abbreviations

This section defines critical points profile abbreviations (see Figure 6.30). You can change these values to reflect area conventions. These abbreviations are for reporting or listing critical profile points.

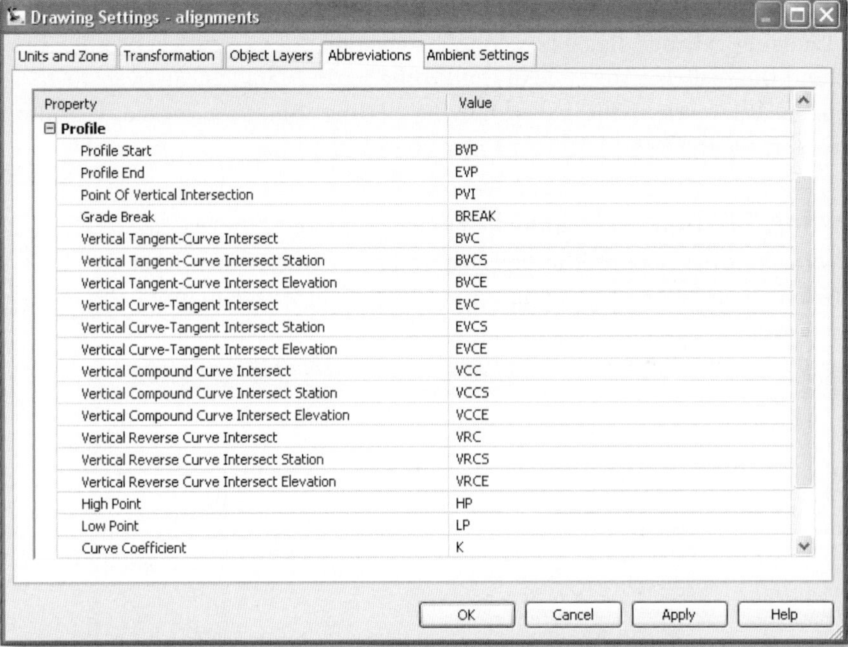


**FIGURE 6.30**

## Edit Feature Settings — Profile View

Edit Feature Settings has four sections that assign initial profile view values. Default Styles assigns view, marker, label set, and Add Label styles (see Figure 6.31). Default Name Format defines how to create a profile file view name.

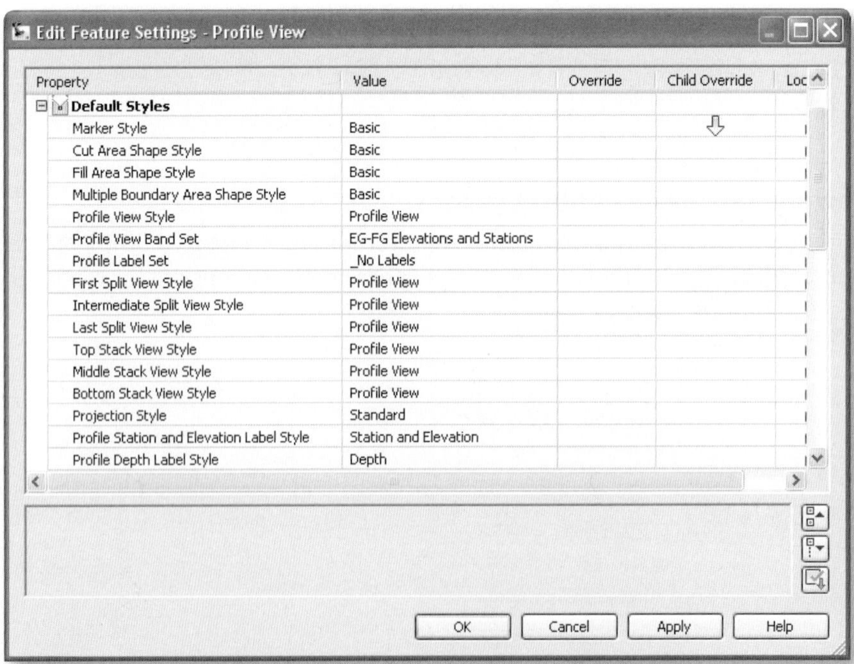


**FIGURE 6.31**

Profile View Creation sets whether a file is to be split when created, whether to set elevation values manually, and what parts of a pipe network to include (only selected segments or the whole network). Split Profile View Option assigns whether a profile should be split when it is made and how to determine the split view section's datum elevation (see Figure 6.32). Stacked Profile View Options determine if profile view are stacked or set side by side. Default Projection Label Placement settings specify the default placement of labels for objects projected to profile views.

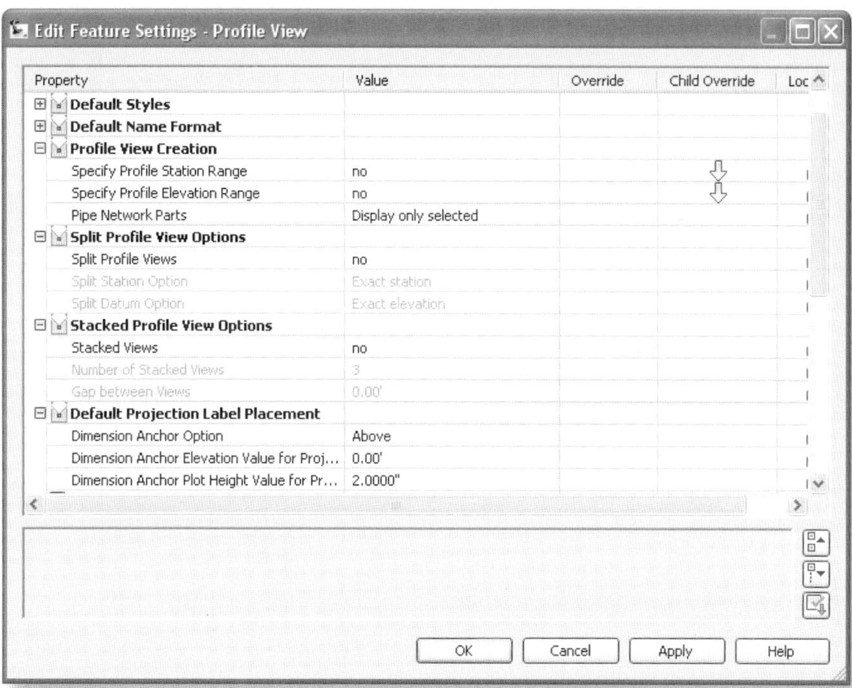


**FIGURE 6.32**

## Edit Feature Settings — Profile

Profile's Edit Feature Settings dialog box assigns initial styles and a label set for all Profiles (see Figure 6.33). Default Styles assigns what style appears when a new profile is created. Default Name Format defines the Profiles, Offset Profiles, Superimposed, and the 3dEntity Profile Name Template naming convention.

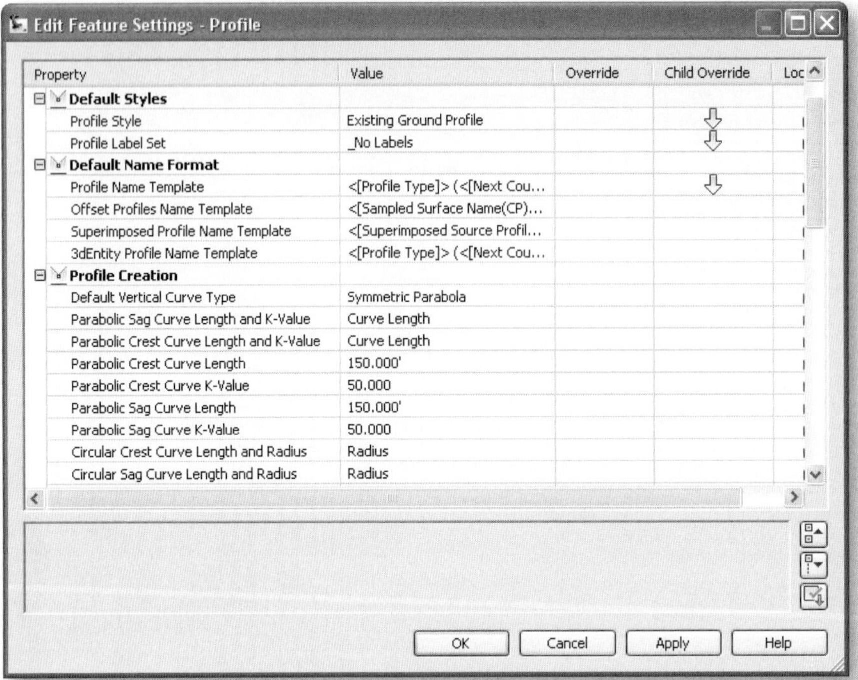


**FIGURE 6.33**

The Profile Creation section sets the default vertical curve type (parabolic) and the initial values for each vertical curve type (circular, parabolic, and asymmetrical parabolic), and vertical curve design parameters (eye height, stopping height). Criteria-Based Design Options toggle on their use and also set the name of the Default Design Check Set (see Figure 6.34).

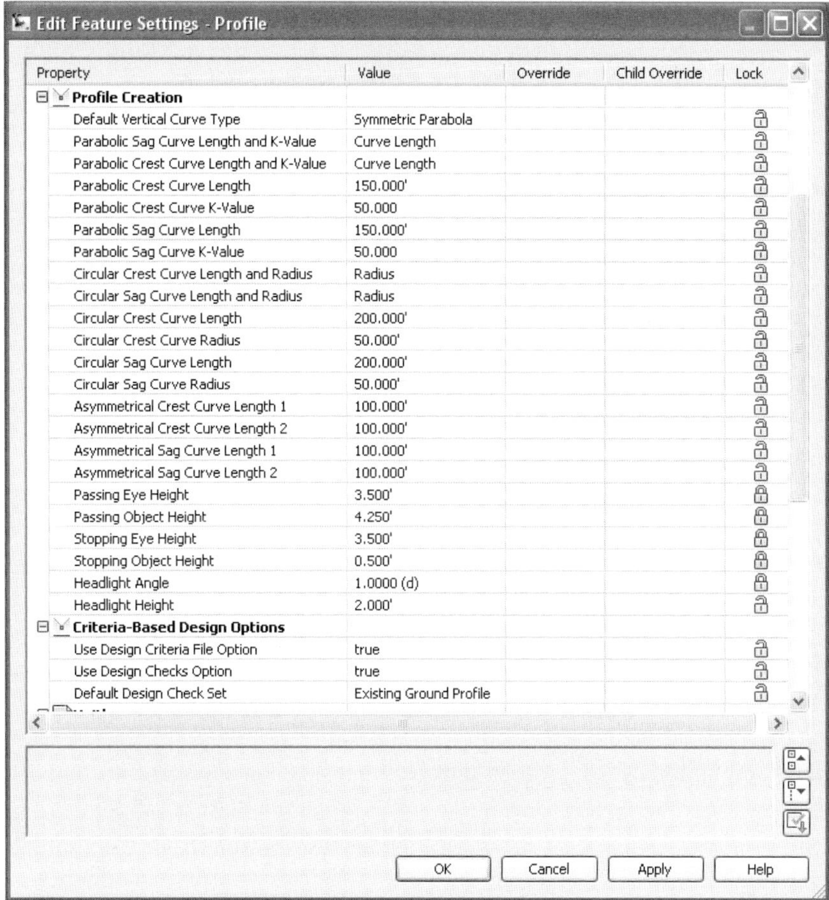


**FIGURE 6.34**

## VERTICAL SEGMENT CHECKS

A user evaluates a design by comparing AASHTO charts to a design. If a user does not want to use criteria design values, then the user can define design checks for vertical tangents and curves. Defining these checks is similar to defining an expression (see Figure 6.35). Figure 6.35 defines a tangent and curve check.

New Design Check - DuPage 5% Grade
Name: DuPage 5% Grade
Description
Expression: {Tangent Grade}<5.0

New Design Check - Minimum Curve Length
Name: Minimum Curve Length
Description
Expression: {Profile Curve Length}>350.00

**FIGURE 6.35**

## EXERCISE 6-1

After completing this exercise, you will:

- Be familiar with Profile and Profile View Settings.
- Be familiar with Profile View Styles.
- Be familiar with Profile View bands.
- Be familiar with Profile Styles.

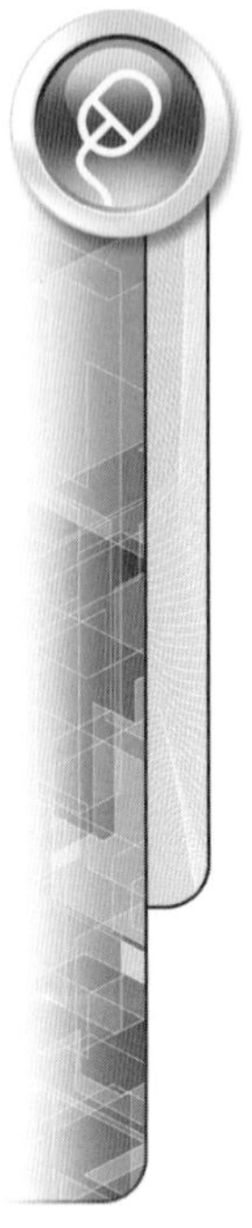

### Exercise Setup

This exercise starts with the previous Chapter's drawing. If you did not complete the previous exercise, browse to the Chapter 6 folder of the CD that accompanies this textbook and open the *Chapter 06 – Unit 1.dwg* file.

1. If ***Civil 3D*** is not open, double-click its desktop icon to start the application, and close the current drawing.
2. Open your drawing from the previous chapter or browse to the Chapter 6 folder of the CD that accompanies this textbook and open the *Chapter 06 – Unit 1* drawing.
3. If necessary, click the ***Prospector*** tab.
4. In Prospector, select the **Points** heading, press the right mouse button, and from the shortcut menu, select EDIT POINTS....
5. In the Point Editor vista, click the **Point Number** heading and sort the point list.
6. Select point number 3, hold down SHIFT, and select the last point number.
7. With the points highlighted, press the right mouse button, from the shortcut menu, select DELETE..., and then click ***Yes*** in the Are you sure dialog box.
8. Close the Panorama.
9. In Layer Properties Manager, freeze the layer **V-NODE**, and click **X** to exit the Layer Properties Manager palette.
10. At Civil 3D's top left, Quick Access Toolbar, click the ***Save*** icon to save the drawing.

### Edit Drawing Settings

Edit Drawing Settings affects all styles and settings below it (see Figure 6.29).

1. Click the ***Settings*** tab.
2. At the Settings' top, select the drawing name, press the right mouse button, and from the shortcut menu, select EDIT DRAWING SETTINGS....
3. Click the ***Object Layers*** tab.
4. In Object Layers, scroll to Profile, Profile-Labeling, Profile View, and Profile View-Labeling, set their modifier to **Suffix**, and for their value enter, **-***.

The –* (dash followed by an asterisk) appends the layer names with the object's name.

5. Click the ***Abbreviations*** tab.
6. Collapse the Alignment and Superelevation sections to view the Profile section and review their values.
7. Click the **Ambient *Settings*** tab.
8. Expand the Grade/Slope and Station sections to review their values.
9. Click ***OK*** exiting the dialog box.

## Profile View – Edit Feature Settings

Profile View's Edit Feature Settings set initial styles, band set assignments, naming formats, and other profile view creation settings (see Figure 6.31).

1. In Settings, click the **Profile View** heading, press the right mouse button, and from the shortcut menu, select EDIT FEATURE SETTINGS....
2. In Edit Feature Settings, expand Default Styles, Default Name Format, Profile View Creation, Split Profile View Options, Stacked Profile View Options, Default Projection Label Placement, and review their values.
3. Click ***OK*** to exit the dialog box.

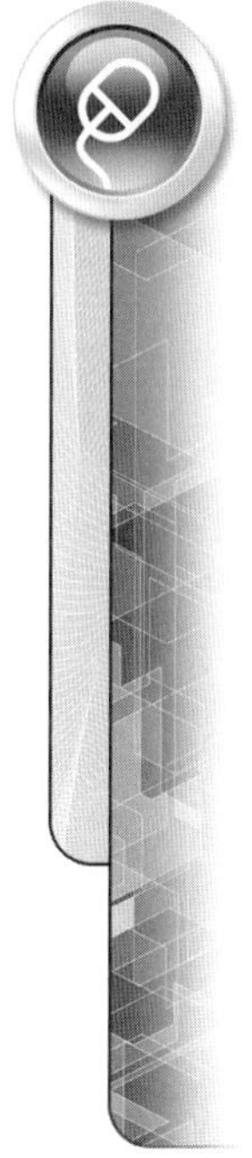

## Profile – Edit Feature Settings

Profile's Edit Feature Settings dialog box sets initial styles, label assignments, and other critical profile values (see Figure 6.33).

1. In Settings, click the **Profile** heading, press the right mouse button, and from the menu, select EDIT FEATURE SETTINGS....
2. In Edit Feature Settings, expand each Profile section and review its settings.
3. Click ***OK*** to exit the dialog box.

## Profile View Style – Full Grid

Profiles appear within a profile view. A profile view has vertical lines to represent horizontal alignment stations and horizontal lines to represent elevations along the alignment. A Profile View Style defines the grid and its annotation (see Figures 6.3, 6.6, and 6.7).

1. In Settings, expand the Profile View branch until you are viewing Profile View Styles' list of styles.
2. From the list, select **Full Grid**, press the right mouse button, and from the shortcut menu, select EDIT....
3. Click the ***Graph*** tab.

This panel controls a profile view's vertical exaggeration and direction.

4. Click the ***Grid*** tab to review its settings.

This panel controls if the grid is clipped, and if there are any extra vertical or horizontal grid segments.

5. Click the ***Title Annotation*** tab to review its settings.

This panel controls the title, the title's location (left half), Axis Title, and location (right half).

6. Click the ***Horizontal Axes*** tab.

This tab controls the stationing annotation and interval.

7. In Major tick details, to the right of the Tick label text, and click the ***Text Component Editor*** icon.

The Text Component Editor opens and contains the label's format.

8. Click ***OK*** to exit the Text Component Editor.
9. Click the ***Vertical Axes*** tab.

This tab controls the elevation annotation and interval.

10. Click the ***Display*** tab to review its component list and display values.
11. Click ***OK*** to exit the dialog box.

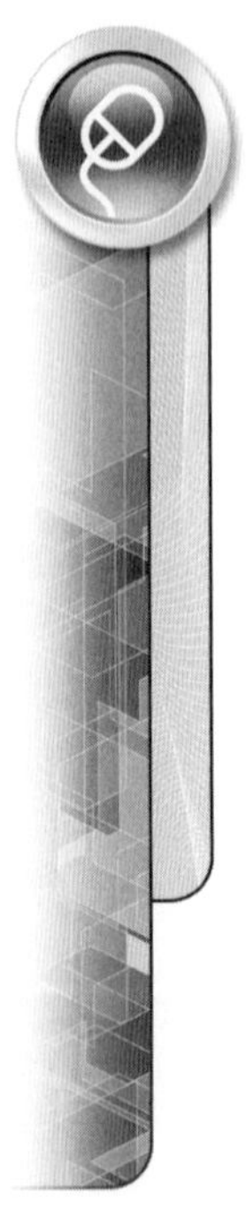

## Profile View - Band Set

Civil 3D bands place information above or below a profile view. A band can contain profile data, horizontal and vertical geometry, superelevation, section sample line data, and pipe network values.

1. In Settings, expand the Profile View branch until you are viewing the Band Styles heading and its styles list.
2. In the Band Styles branch, expand Band Sets, from the styles list select **Profile Data with Geometry and Superelevation**, press the right mouse button, and from the shortcut menu, select EDIT....
3. Click the ***Bands*** tab to view its list of band types for the top and bottom of the profile view and label styles. Note the styles referenced in this style.
4. Click ***OK*** to exit the Band Set - Profile Data with Geometry and Superelevation dialog box.

## Profile - Existing Ground

Existing Ground is a style that affects the display of elevations along an alignment. This style assigns each component a layer and properties.

1. In Settings, expand the Profile branch until you are viewing the Profile Styles list.
2. From the styles list, select **Existing Ground Profile**, press the right mouse button, and from the shortcut menu, select EDIT....
3. Click the ***Display*** tab to review the component layer and property assignments.
4. Click ***OK*** to exit the dialog box.

## Profile - Label Sets

A Profile Label Set is an alias for a collection of label styles.

1. In Settings, expand the Profile branch until you are viewing the Label Styles' branch Label Sets list.
2. From the list, select **Complete Label Set**, press the right mouse button, and from the shortcut menu, select Edit....
3. If necessary, click the ***Labels*** tab and note the label types and label styles list.
4. Click ***OK*** to exit the dialog box.

## Profile Label Set Styles — Line

Label sets annotate a profile view's profile and are used for design alignment labeling, and not for surface profile labeling.

The line styles label a profile tangent with a grade percent format (see Figure 6.26).

1. In the Label Styles branch, expand the Line styles branch until you are viewing its styles list.
2. From the list, select **Percent Grade**, press the right mouse button, and from the shortcut menu, select EDIT....
3. Click the ***Layout*** tab.
4. In the Text section, click the Contents' value cell to display an editing ellipsis.
5. Click the ellipsis to display the Text Component Editor, and in the Editor, review the tangent label's format.
6. At the dialog box's top left, click Properties' drop-list arrow to view all of the alignment and profile properties available to this label type.
7. Click ***OK*** to exit the dialog boxes and return to the command line.

## Profile Label Set Styles — Curve

Curve places a multi-component label on a vertical curve. This label is the most complex profile label. Its components range from arrow blocks to formatted text (see Figure 6.27).

1. In Profile, Label Styles, expand the Curve branch until you are viewing its styles list.
2. From the list of styles, select **Crest and Sag**, press the right mouse button, and from the shortcut menu, select `EDIT...`.
3. If necessary, click the ***Layout*** tab.
4. In the top-left of the Layout panel, click the Component name's drop-list arrow to view its label components.
5. From the list, select **PVI Sta and Elev**.
6. In the Text section, click the Contents' value cell to display an editing ellipsis.
7. Click the ellipsis to display the Text Component Editor.
8. In Text Component Editor, review the component's format.

This label component has four properties.

9. At the top left, click the Properties drop-list arrow to view all of this label type's label properties.
10. Click ***OK*** to exit the dialog boxes and return to the command line.

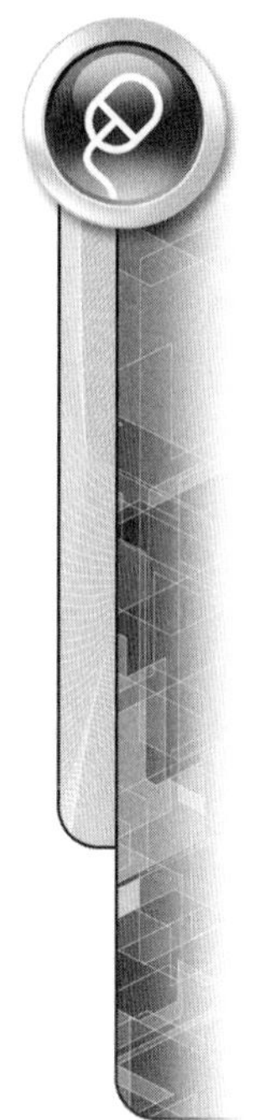

## Profile Label Set Styles — Station and Elevation

Station and Label labels show profile spot elevation within a profile view. If you are moving the label from its original location, its dragged state settings define how it is displayed.

1. In Settings, expand the Profile View branch until you are viewing Label Styles' Station Elevation labels list.
2. From the list, select **Station and Elevation**, press the right mouse button, and from the shortcut menu, select `EDIT...`.
3. If necessary, click the ***Layout*** tab.
4. If necessary, at its top right, change the Preview to Station Elevation Label Style.
5. In the Text section, click the Contents' value cell to display an editing ellipsis.
6. Click the ellipsis to view the Text Component Editor.
7. At the top left, click the Properties drop-list arrow to view all of this label type's label properties.
8. Click ***OK*** to exit the dialog boxes and return to the command line.

## Profile Label Set Styles — Depth

Depth labels place labels between two user-selected points, which results in a label with a slope or grade between the selected points.

1. Expand the Profile View branch until you are viewing Label Styles' Depth styles list.
2. From the styles list, select **Depth**, press the right mouse button, and from the shortcut menu, select `EDIT...`.
3. Click the ***Layout*** tab.
4. If necessary, at the panel's top right, change Preview to Depth Label Style.
5. In Layout, at the top, click the Component name drop-list arrow to view the label's components list.
6. Set the Component name to **Depth**.

7. In the Text section, click the Contents' value cell to display an editing ellipsis.
8. Click the ellipsis to display the Text Component Editor, and review the format string for Depth.
9. At the top left, click the Properties drop-list arrow to view all of this label type's label properties.
10. Click ***OK*** to exit the dialog boxes and return to the command line.
11. At Civil 3D' top left, Quick Access Toolbar, click ***Save*** icon to save the drawing.

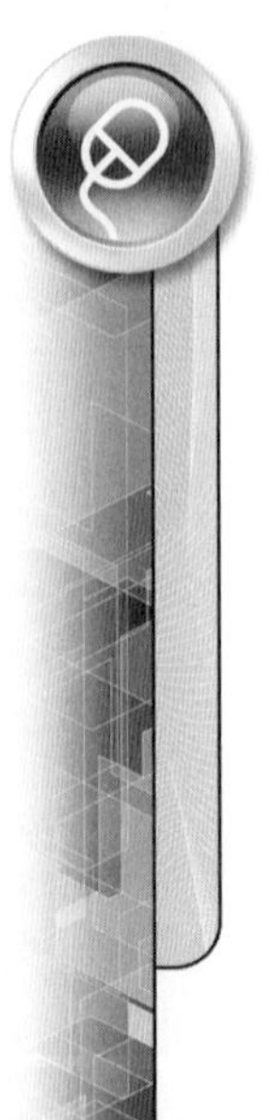

## Tangent and Curve Design Checks

Design checks are like expressions and identify any alignment segment that does not pass their test. All tangents must be less than 5 percent, and vertical curves cannot be less than 350 feet.

1. Expand the Profile branch until you are viewing Design Checks and its check type list.
2. From the check type list, select **Line**, press the right mouse button, and from the shortcut menu, select `NEW...`.
3. For the check name, enter **DuPage 5% Grade**.
4. In the New Design Check dialog box, in the lower-middle right, click the ***Insert Property*** icon (first large button on the right), and from the list of properties, select **Tangent Grade**.
5. In the dialog box, click the less than sign (<) and then click **5**.

Your check should look like Figure 6.35.

6. Click ***OK*** to create the check.
7. From the check type list select *Curve*, press the right mouse button, and from the shortcut menu, select `NEW...`.
8. For the check name, enter **Minimum Curve Length**.
9. In the New Design Check dialog box, in the lower-middle right click the ***Insert Property*** icon, and from the list of properties, select **Profile Curve Length**.
10. In the dialog box, click the less than sign (<) and then click **145**.

Your check should look like Figure 6.35.

11. Click ***OK*** to create the check.

## Creating a Vertical Design Check Set

1. Select the **Design Check Sets** heading, press the right mouse button, and from the shortcut menu, select `NEW...`.
2. Click the Information tab for the name, enter **DuPage Vertical Checks**.
3. Click the ***Design Checks*** tab.
4. At the panel's top left, click the Type drop-list arrow, set it to **Line**, set the Line check to **DuPage 5% Grade**, and click ***Add>>*** to add it to the list.
5. At the panel's top left, click the types drop-list arrow, set it to **Curve**, set the Curve check to **Minimum Curve Length**, and click ***Add>>*** to add it to the list.
6. Click ***OK*** to create the Design Check Set.
7. At Civil 3D's upper left, Quick Access Toolbar, click the ***Save*** icon to save the drawing.

This ends the Profile View and Profile styles and settings review. The next unit covers creating an existing ground profile and a profile view.

## SUMMARY

- Edit Drawing Settings sets initial values used by all Profile View or Profile styles and commands.
- Even though they define all possible profile view annotation, the display settings control what is visible for the profile view style.
- Profile Styles assigns component layers and their properties.
- Profile label styles are primarily for vertical alignment labeling.
- A Profile Label Set places labels in a profile view, and a Profile View Band Set places data at a Profile View's top or bottom.
- Add Labels creates Station and Elevation and Depth labels after creating profiles and their view.
- Instead of using criteria-based design, a user can define a series of design checks.

## UNIT 2: CREATING A PROFILE AND ITS VIEW

Creating a profile and its profile view is a two-step process: sample elevations and create a profile view. These two steps can be executed as a single command sequence or as two separate steps. You can create a profile view without a surface; however, you will have to add elevations to the profile view to view its grid.

### SURFACE DATA

First, you determine the elevations along the alignment's path. To do this, Civil 3D samples a surface or reads a station and elevation data file. This step associates surface or file elevations to the alignment stationing. The easiest method is to sample surface elevations. If you have multiple surfaces, you can sample one or all of them.

### CREATE PROFILE FROM SURFACE

The Create Profile from Surface dialog box displays values necessary to sample a surface (see Figure 6.36). The dialog box's top left sets the alignment, and its top right sets the surface(s) to sample. The dialog box's middle left sets the beginning and ending sampling stations. By default, they are the alignment's beginning and ending stations. At the middle right, the Sample offset is toggled on for use. The box to the toggle's right lists offset sampling distances (to the alignment's right and/or left side). Each offset appears as a separate entry in the Profile list with its assigned offset value. On the dialog box's right side, the Add button places the selected and set items in the Profile list (dialog box's bottom).

The Profile List displays each profile's sample information as a ribbon. If you have multiple entries, multiple profiles are created. Each profile entry displays the surface name, type, update mode, profile style, stations, and minimum and maximums elevation. Each profile can have a different Profile style and values.

At the dialog box's bottom are two important buttons: Remove and Draw in profile. Remove deletes unwanted Profile list entries. Draw in profile view calls the Create View Wizard, and creates a profile view (grid) for the listed profile(s). If you click OK, then creating the profile view is a separate step.

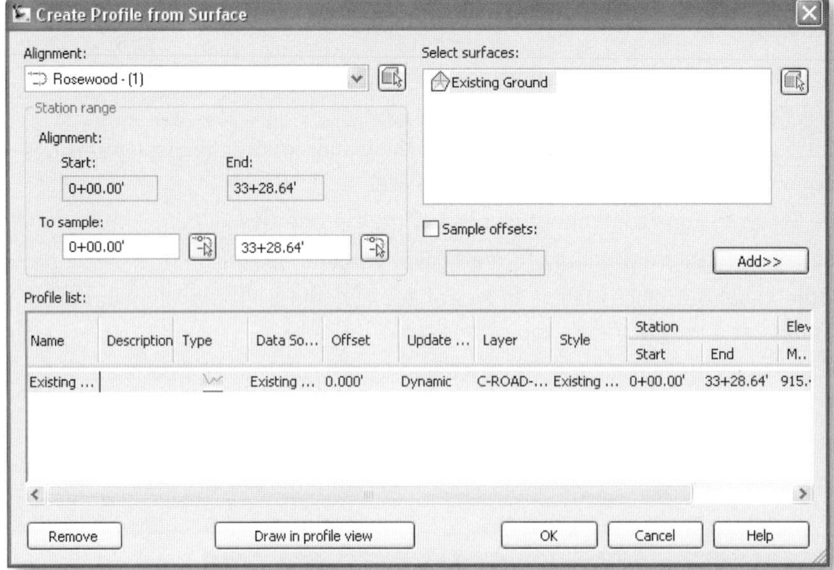


**FIGURE 6.36**

## CREATE PROFILE VIEW

Create Profile View displays a wizard. The wizard defines the name, style, band set, stations, elevations, station splits, and view contents.

### General

General sets the alignment's name and sets the profile view's name, profile view style, and its base layer, and if there are multiple profiles, they are stacked (see Figure 6.37). If multiple profiles are not stacked, then all profiles appear in a single profile view.

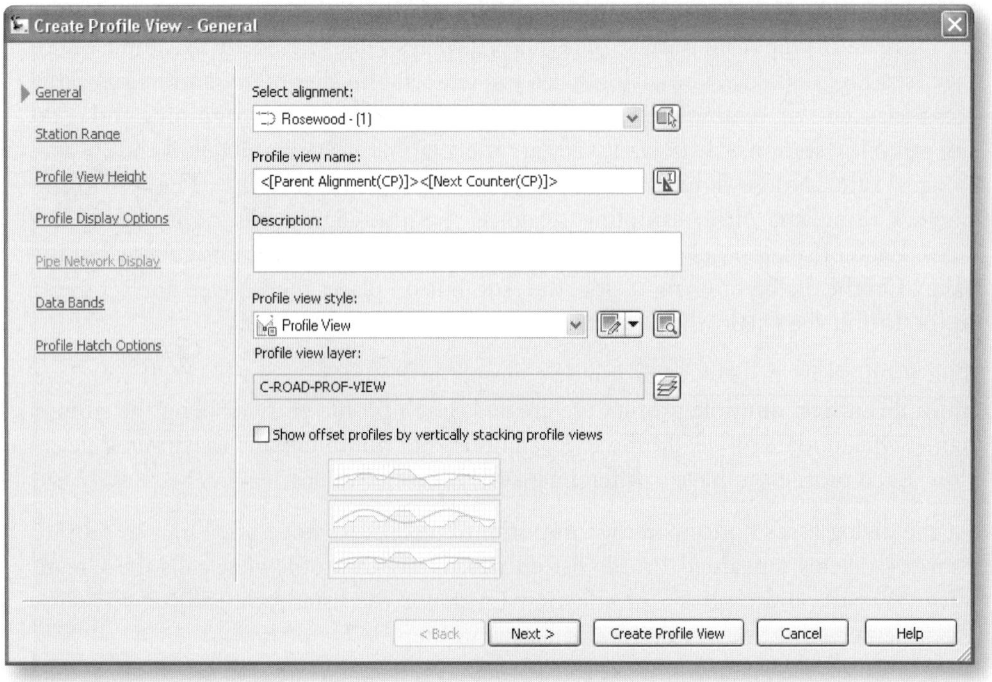

**FIGURE 6.37**

## Station Range

Station Range sets the profile view's station range; it may be different from the sampled range (see Figure 6.38). If set to Automatic, the values are the sampled station range. When set to manual, the values are a user-specified range.

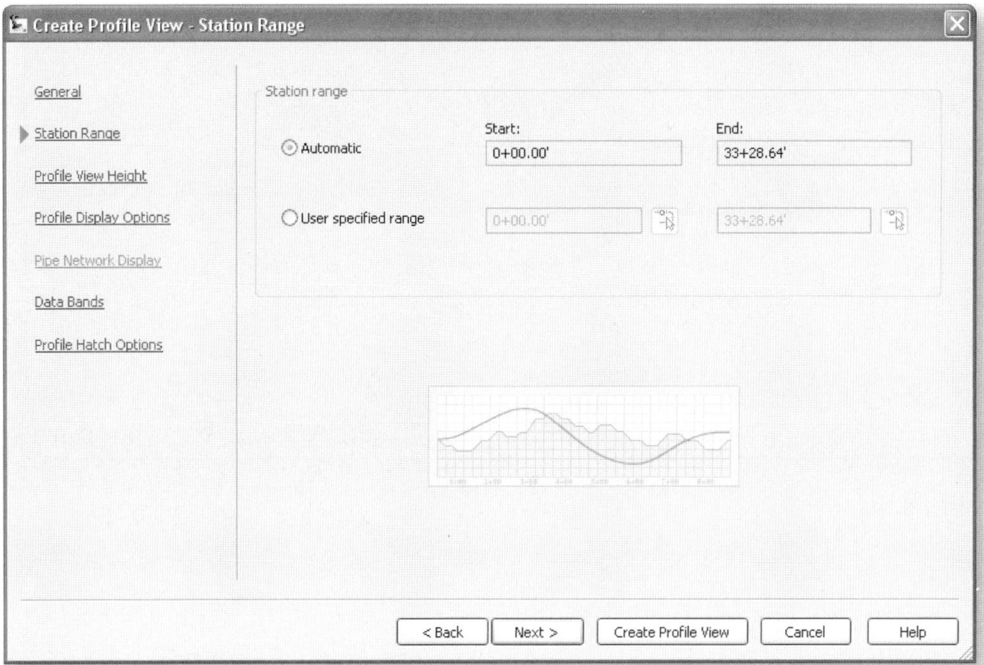


**FIGURE 6.38**

## Profile View Height

Profile View Height sets the profile view's elevation range (see Figure 6.39). If set to Automatic, the values are the sample elevation range. If set to manual, they are a user-specified range. When User specified is toggled on, the lower half of the dialog box becomes active. It is here that a user can split a profile into three segments using station and elevation settings with each having their own profile view style. Splitting a profile view creates a profile view with limited vertical height. This is not usually necessary when a user creates an initial design profile; however, when a user creates plan and profile sheets, splitting the profile may be necessary.

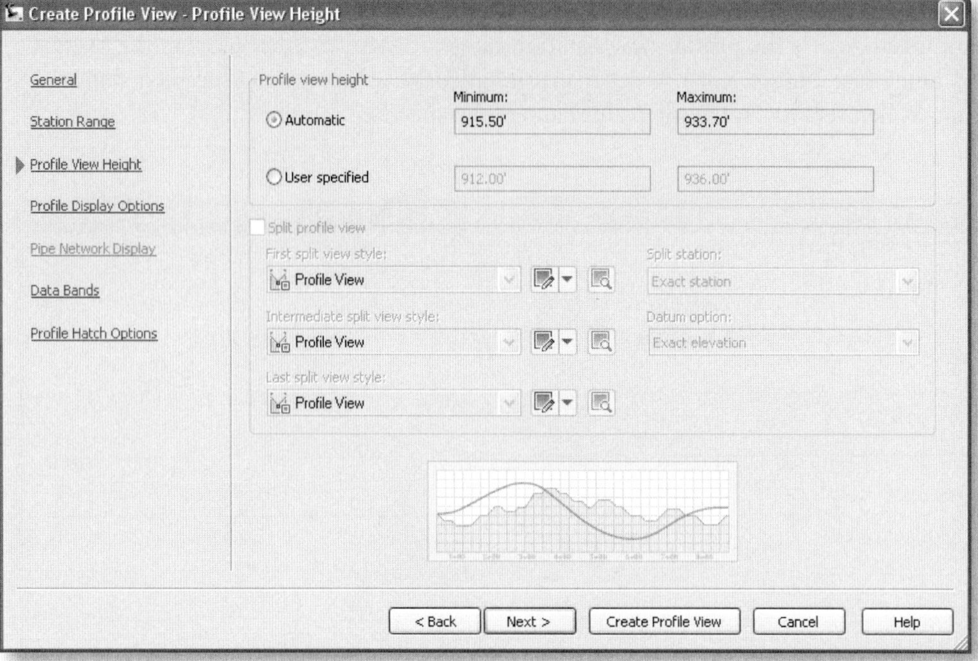


**FIGURE 6.39**

## Profile Display Options

Profile Display Options determine which profiles to display and their mode, it sets their object and label styles, and it reviews their station and elevation values (see Figure 6.40).

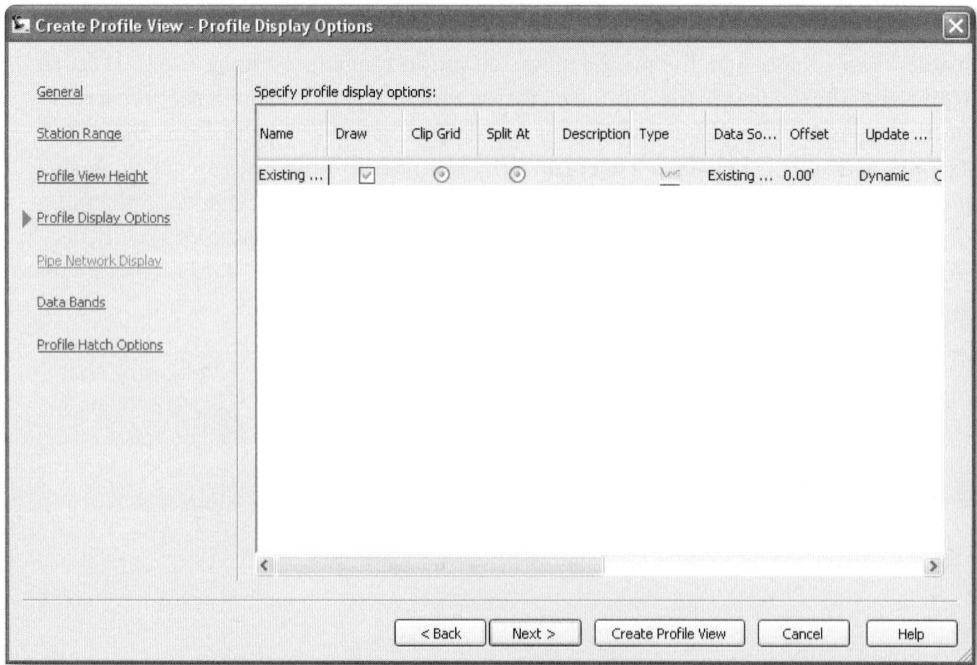


**FIGURE 6.40**

## Pipe Network Display

Pipe Network Display selects entire pipe networks or their pipes and structures to draw in a profile view (see Figure 6.41). By selecting a network, all of its pipes and structures are selected. The Show only parts selected to draw in profile view draws only selected parts in the profile view.

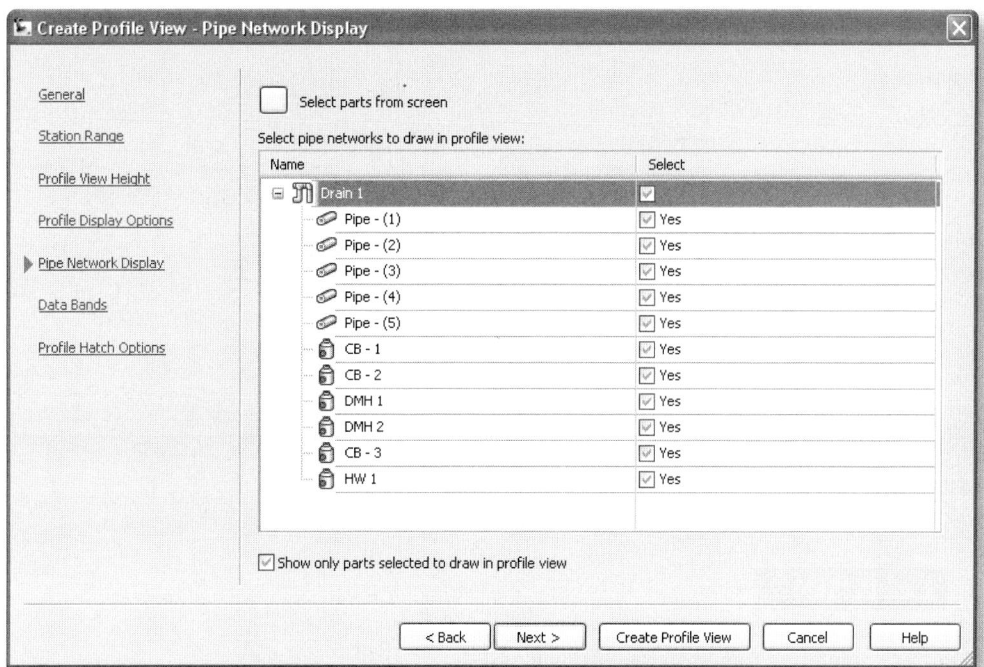


**FIGURE 6.41**

## Data Bands

Data Bands sets the Band Set (top), its location (middle), and its properties (bottom) (see Figure 6.42). When you are creating the initial profile view, there is usually only one profile. In this case, the wizard assigns the surface profile to Profile1 and Profile2. When you are defining the second profile, you must change Profile2's assignment to the new profile.

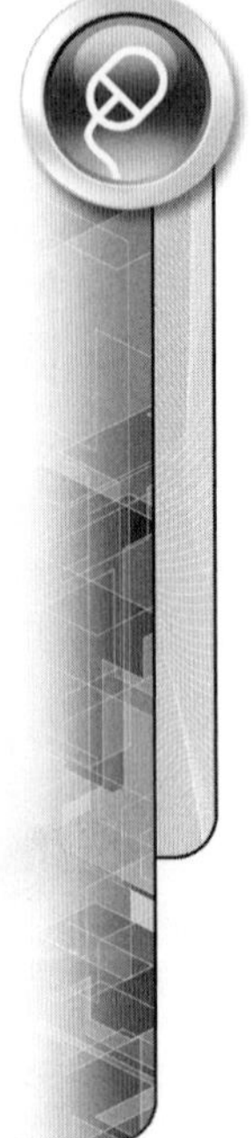

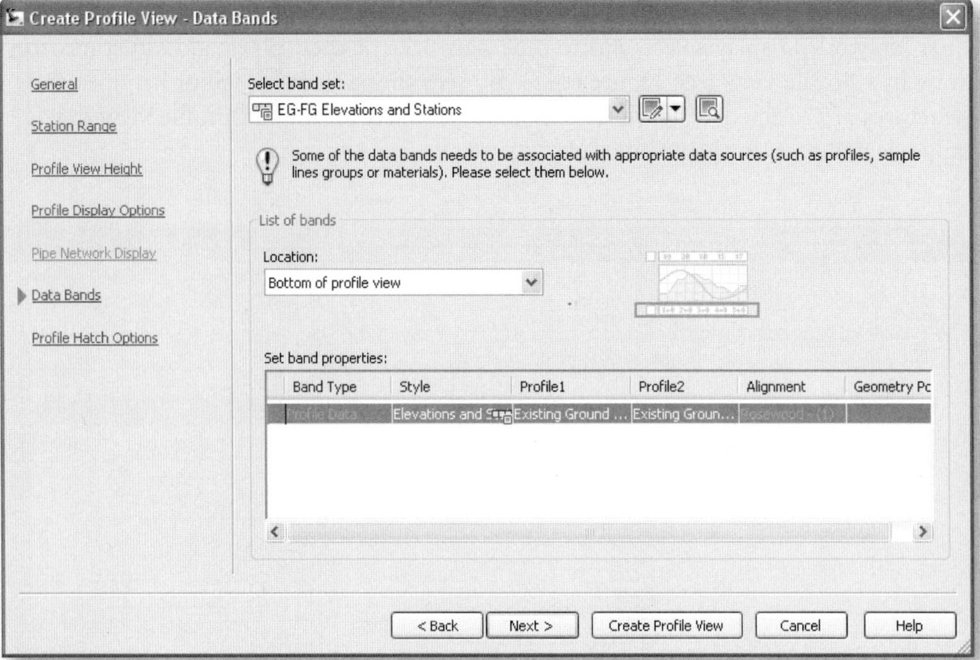


**FIGURE 6.42**

## EXERCISE 6-2

After completing this exercise, you will:

- Be able to create a Profile.
- Be able to create a Profile View.
- Be able to change an assigned Profile View.
- Be able to modify Profile View annotation.

## Exercise Setup

This exercise continues with the previous Unit's exercise drawing. If you did not complete the previous exercise, browse to the Chapter 6 folder of the CD that accompanies this textbook and open the Chapter *06 – Unit 2.dwg* file and start the exercise.

1. If not open, open the previous exercise's drawing or browse to the Chapter 6 folder of the CD that accompanies this textbook's and open the *Chapter 06 – Unit 2* drawing.

## Create an EG Surface

First, there needs to be a surface. If you haven't created a surface before, you should review Chapter 4 – Surfaces.

1. In the Home tab's Layers panel, select Layer Properties, thaw and turn on the layers **3EXCONT** and **3EXCONT5**, and click **X** to exit the Layer Properties Manager.
2. From the Home tab's Layers panel, select ***Isolate*** and select one contour layer from each of the two contour layers.
3. Click the ***Prospector*** tab.

4. In Prospector, click the **Surfaces** heading, press the right mouse button, and from the shortcut menu, select CREATE SURFACE....
5. In the Create Surface dialog box, set the type to **TIN surface**, for the Name, enter **Existing Ground**, set the style to **Border only**, and click ***OK*** to make the surface.
6. In the Layers panel, click the Layer Control drop-list arrow, and then scroll to and unlock the **C-TOPO-Existing Ground** layer.

The C-TOPO-Existing Ground layer contains the surface object.

7. In Prospector, expand Surfaces, then expand Existing Ground until you are viewing its Definition branch and its data list.
8. In Existing Ground's Definition list, select **Contours**, press the right mouse button, and from the shortcut menu, select ADD....
9. In the Add Contour Data dialog box, for the description enter **Aerial Contours** and set the Weeding and Supplementing values to match those in Figure 6.43.

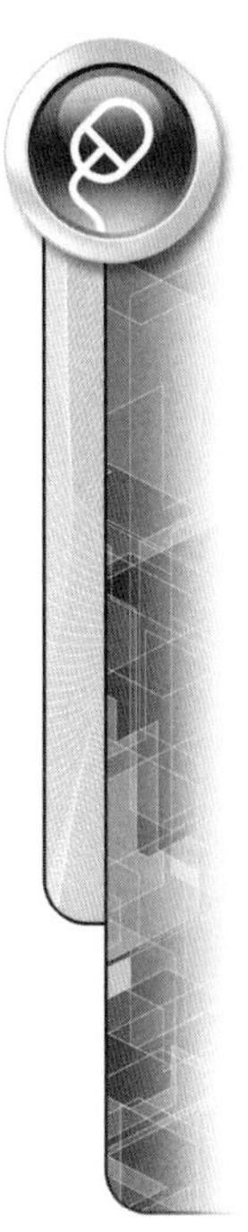

**FIGURE 6.43**

10. Click ***OK***, in the drawing select the contours, and then press ENTER.
11. From the Layers panel, click the ***Unisolate*** icon to restore layer visibility.
12. In Layers panel, Layer Properties Manager, freeze the layers **3EXCONT** and **3EXCONT5**, thaw the layer **Boundary**, and click the **X** to exit.
13. In Existing Ground's Definition branch, select **Boundaries**, press the right mouse button, and from the shortcut menu, select ADD....
14. In the Add Boundaries dialog box, for the description enter **Outer**, set the type to **Outer**, toggle **OFF** Non-destructive breakline, and click ***OK*** to continue.
15. In the drawing, select the polyline boundary to add the boundary to the surface.
16. Reopen Layer Properties Manager, freeze the layer **Boundary**, and click the **X** to exit the palette.
17. If you still do not see the surface boundary, in the command line, enter REA and press ENTER.

18. Use the ZOOM and PAN commands to place the site at the drawing's left side.
19. At Civil 3D's top left, Quick Access Toolbar, click the ***Save*** icon to save the drawing.

## Create a Surface Profile

After creating a surface, next you sample the surface along the alignment to create the existing profile.

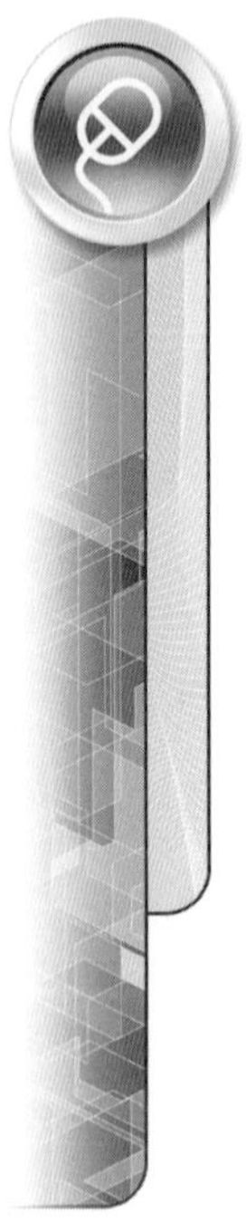

1. From the Home tab, on the Create Design panel, click the Profile icon, select CREATE SURFACE PROFILE.

The drawing has three alignments.

2. In the upper left of the Create Profile from Surface dialog box, set the alignment to **Rosewood - (1)**, and in the upper right select the surface **Existing Ground**.

Next, you decide the station sample range. By default, sampling is from the alignment's beginning to end. If you want a right and left profile, toggle ON Offsets and enter their values.

If you decide to sample the alignments' entire lengths, do not adjust their values.

3. If it is not already off, toggle **OFF** offset sampling.
4. In the dialog box's middle right, click ***Add>>*** to place the current values in the Profile List.

If you click OK to exit, you next need to select from the Profile & Sections Views panel, Profile View's Create Profile View command to create a profile view with the current profile. A second option is to continue to Create Profile View by clicking Draw in Profile view.

## Create a Profile View

Create Profile View reads the sampled data and displays the Create Profile View Wizard.

1. At the bottom of the Create Profile from Surface dialog box, click ***Draw in Profile View*** to display the Create Profile View wizard.
2. Using the values in Figures 6.37, 6.38, 6.39, 6.40, and 6.42, click ***Next*** until you are in the last panel, click ***Create Profile View***, and in the drawing to the right of the site, select a point to locate the profile view.
3. At Civil 3D's upper left, Quick Access Toolbar, click the ***Save*** icon to save the drawing.
4. Use the ZOOM and PAN commands to better view the profile.

## Change Profile View Styles

A Profile View style defines the view's grid and annotation. Changing the profile view style changes the view's display and shape.

1. In the drawing, select the profile view, press the right mouse button, and from the shortcut menu, select PROFILE VIEW PROPERTIES....
2. In the Profile View Properties dialog box, click the ***Information*** tab, click Object Style's drop-list arrow, and from the styles list, select **Major Grids**.
3. Click ***OK*** to exit the Profile View Properties dialog box.
4. Repeat the previous three steps, but set the Profile View Style to **Full Grid**.

## Modify a Profile View Style

Modifying a profile view style changes how it is displayed.

1. Click the ***Settings*** tab.
2. In Settings, expand the Profile View branch until you are viewing the Profile View Styles list.
3. From the list, select **Full Grid**, press the right mouse button, and from the shortcut menu, select EDIT....
4. In the Profile View Style dialog box, click the ***Grid*** tab.
5. In Grid Options, toggle **ON** both Clip Vertical Grid and Clip Horizontal Grid.
6. Click ***OK*** to view the profile view changes.
7. In the drawing, select the profile view, press the right mouse button, and from the shortcut menu, select EDIT PROFILE VIEW STYLE....
8. In the Profile View Style dialog box, click the Grid tab and toggle **OFF** both Clip Horizontal Grid and Clip Vertical Grid.
9. Click ***OK*** to make the profile view changes.
10. At Civil 3D's top left, Quick Access Toolbar, click the ***Save*** icon to save the drawing.

This completes the exercise on creating a profile and its view.

The next unit describes how to create a proposed vertical alignment for Rosewood.

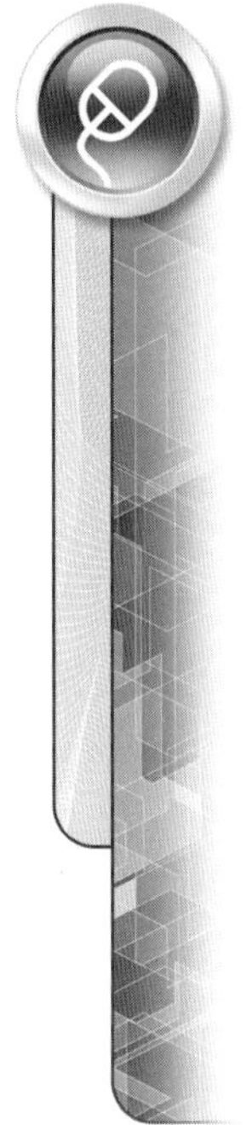

## SUMMARY

- The Create Profile from Surface command samples one, all, or any combination of surfaces.
- The Create Profile or Create Profile View dialog box select which profiles appear in a profile view.
- A Profile View can pad the grid area around a profile.
- A Profile View can clip the horizontal or vertical view grid.
- A Profile View's annotation is around its grid perimeter.
- A Profile style assigns a component's layer name and properties.
- A Profile's annotation is within a profile view's grid.
- Profile View label styles are placed after the profile(s) and the view are created.

## UNIT 3: DESIGNING A PROPOSED PROFILE

Within the profile view and its existing ground profile, a designer creates a vertical road design. This design goes by many names: vertical alignment, finished ground, etc. A vertical alignment contains proposed tangents (lines) and vertical curve segments.

Vertical curves transition a vehicle from one grade to the next. There are two types of vertical curves: crest (transitioning from an up to a down grade) or sag (transitioning from a down to an up grade). There are three vertical curve types: circular, asymmetrical (parabolic), and parabolic. The parameters to create these curves are curve length or a K Value. A K Value represents the horizontal distance along which a 1 percent change in grade occurs on the vertical curve and is a measure of abruptness.

## PROFILE CREATION TOOLS

The Home tab's Create Design panel's Profile – Profile Creation Tools toolbar is the only toolset for creating a vertical alignment's tangent and curve segments (see Figure 6.44). After starting the command and identifying the profile view, the Create Profile dialog box opens with two tabs assigning site, styles, and design criteria.

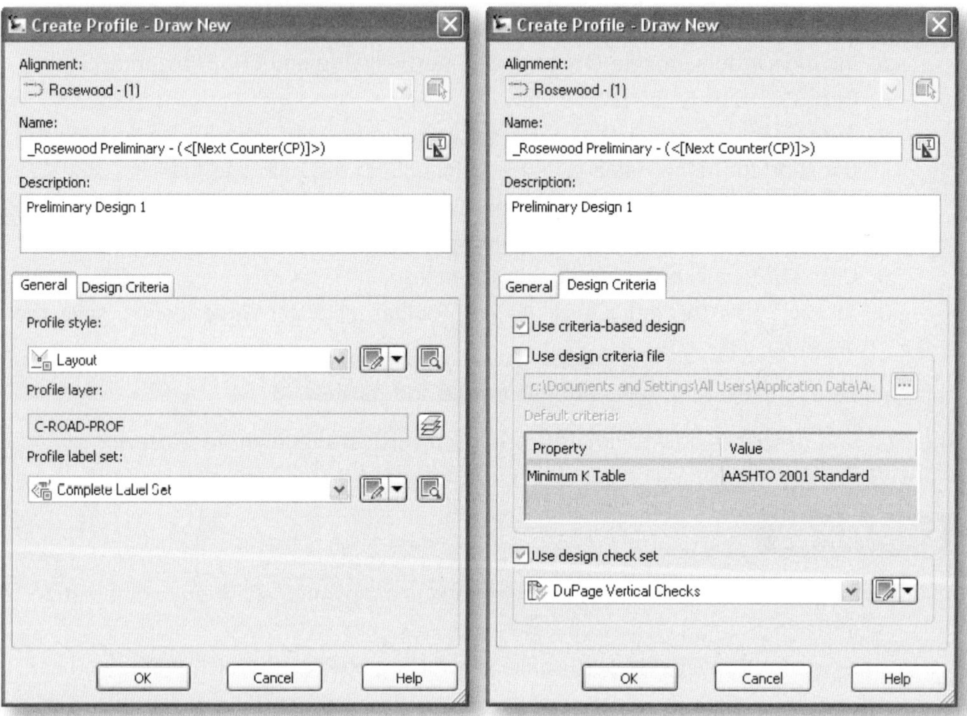


**FIGURE 6.44**

After setting the values in the Create Profile dialog box, click OK to display the Profile Layout Tools toolbar. Minimally, you can draft tangent (line) segments and later add the vertical curves to the alignment.

Clicking the toolbar's leftmost icon's drop-list arrow lists its available drafting modes: draw tangents only and tangents with vertical curves. The last item listed sets the vertical curve default values: curve length or K Value. The default type, length, and design values are from Profile's Edit Feature Settings (see Figure 6.45).

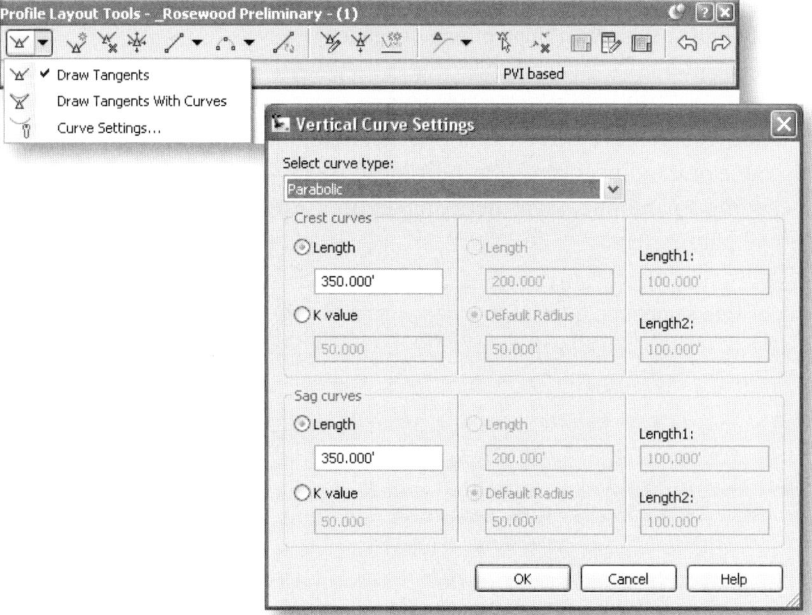


**FIGURE 6.45**

The toolbar's next three icons insert, delete, or move tangent Points of Vertical Intersection (PVIs).

## Tangents

Tangents' icon stack creates three tangent types; fixed, floating, and free (see Figure 6.46). You draft a fixed tangent by selecting two points, a floating tangent attaches to the end of an existing vertical curve, and a free tangent is drawn between two existing vertical curves. Fixed and Floating also use a best fit drawing method.

## Curves

The Curves' icon stack drafts fixed, floating, and free vertical curves.

### Fixed

A fixed curve segment does not connect to or depend on preexisting segments. Fixed vertical curves have several variations: three points; best fit; selecting an endpoint and identifying a pass-through point; selecting two points and entering in the tangent grade at the start of the vertical curve; selecting two points and entering in the tangent grade at the end of the vertical curve; and selecting two points and entering a parameter (K or minimum radius).

**NOTE**

When thinking of fixed segments, think of zero dependencies.

**Floating.** Floating vertical curves have three variations: selecting an existing segment to attach to and identifying a pass-through point and a K value; selecting an existing segment to attach to and identifying a pass-through point and a grade; or using a best-fit method.

**NOTE** When thinking of floating segments, think of one dependency.

**Free.** Free vertical curves have three variations: selecting two existing segments and specifying a K value or a radius; using a best-fit method; or selecting two existing segments and specifying a parabolic, asymmetrical parabolic, or circular vertical curve and setting additional parameters.

**NOTE** When thinking of free segments, think of two dependencies.

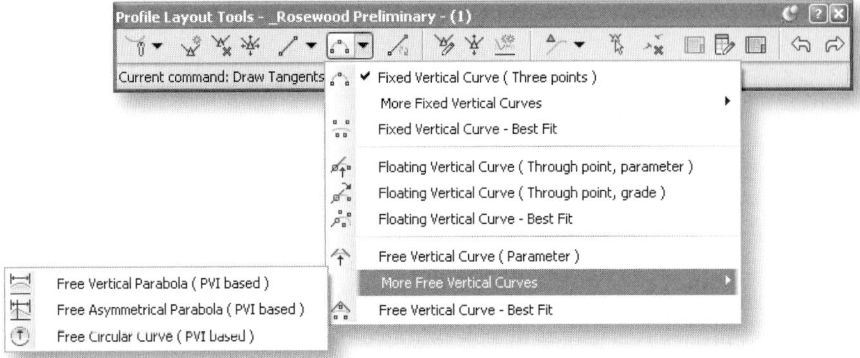


**FIGURE 6.46**

The next icon converts AutoCAD line work into profile segments.

The central three icons (left to right) create PVIs by entering their values in a table, raise or lower a PVI, and copy a profile.

The sixth icon in from the right toggles vertical design between PVI and entity-based commands.

The four icons on the right side are used to select sub-entities (tangents or vertical curves) for the Sub-Entity Editor, Delete Sub-Entities, and display a Grid View editor (overall alignment editor).

Vertical Design Check Sets marks tangents and curves that do not pass the Sets' scrutiny.

## TRANSPARENT COMMANDS

Six transparent commands affect drafting a vertical alignment. The first three use a selection in plan view to set a value in profile view: Profile Station from plan; Profile Station and Elevation from Plan; and Profile Station and Elevation from COGO Point. Each of these methods prompts for a plan view selection and transfers the selection's values to a profile view. The remaining three methods work inside a profile view: Station and Elevation; Profile Grade Station; and Profile Grade Length. Station and Elevation prompts for a station and then prompts for its elevation. Profile Grade Station prompts for a grade, freezes the cursor at the specified grade, and uses the cursor to draft the tangent line with the grade. The last method prompts for a grade, a starting point, and the length of the grade.

### PROFILE2 ASSIGNMENT

Creating the initial profile view assigns the surface to both Profile1 and Profile2. Profile2 is traditionally the proposed vertical design and its elevations appear in the profile view's data band (see Figure 6.12). Profile2's assignment is a manual step and is done in the Profile View's Properties dialog box.

### ASSIGNING DESIGN CHECKS

When creating a profile, it can be assigned a design checks set. When a segment violates a design rule, it receives a tag indicating it did not clear the check.

### EDIT LABELS...

If you are not adding labels while designing a profile, Edit Labels... adds or changes labels assigned to a profile. In the drawing, select a profile, press the right mouse button, and from the shortcut menu, select Edit Labels.... The Profile Labels dialog box lists currently assigned label types and styles (see Figure 6.47). The Add control allows for label additions to the current list and the Delete button deletes any assigned label type and style. The dialog box's top right sets the label type and style to add, and at the bottom are controls to import a profile label set or to create one from the current label list.

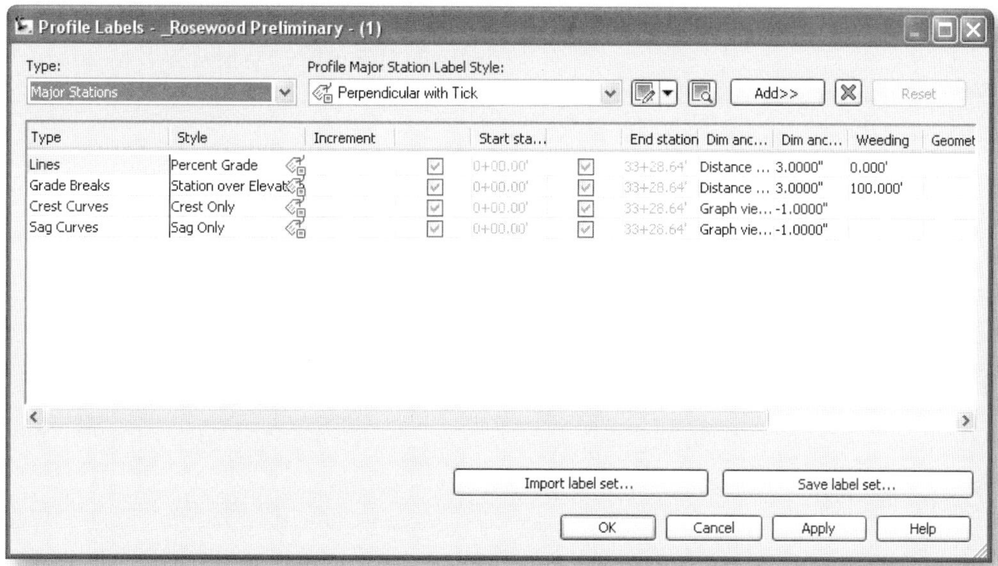


**FIGURE 6.47**

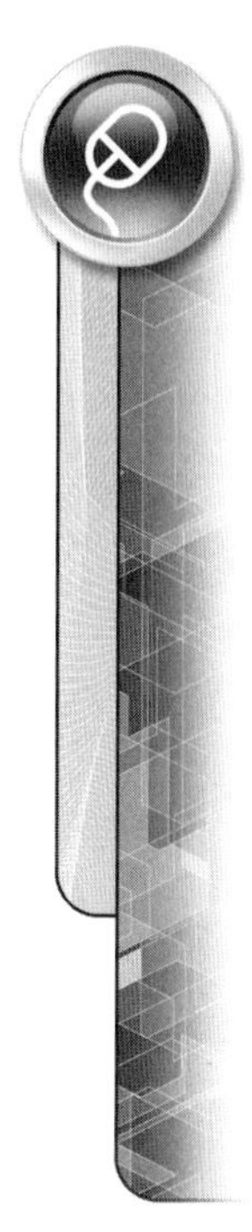

## EXERCISE 6-3

After completing this exercise, you will:

- Be familiar with the Create By Layout Command Settings.
- Be able to create a Profile using the Profile Layout Tools toolbar.

### Exercise Setup

This exercise continues with the previous Unit's exercise drawing. If you did not complete the previous exercise, browse to the Chapter 6 folder of the CD that accompanies this textbook and open the *Chapter 06 – Unit 3.dwg* file.

1. If not open, open the previous exercise's drawing or browse to the Chapter 6 folder of the CD that accompanies this textbook and open the *Chapter 06 – Unit 3* drawing.

## Edit Feature Settings

The Profile Edit Feature Settings dialog box settings affect vertical curves.

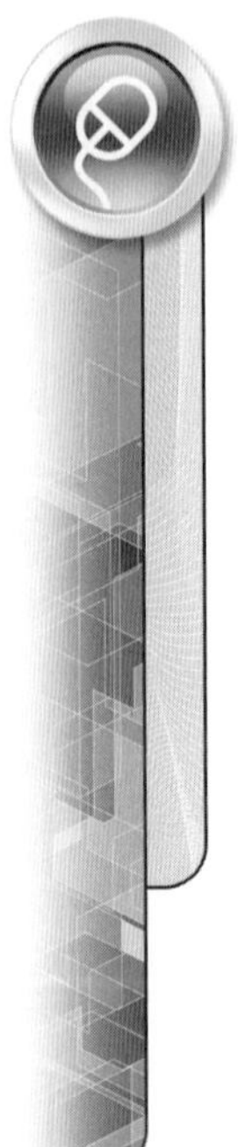

1. If necessary, click the ***Settings*** tab.
2. In Settings, click the *Profile* heading, press the right mouse button, and from the shortcut menu, select EDIT FEATURE SETTINGS....
3. Expand the Profile Creation section and review its values.

The first value sets the default vertical curve type. The remaining values are vertical curve type settings.

4. If necessary, set the Default Vertical Curve Type to **Symmetric Parabola**.
5. If necessary, set the parabolic criteria to **Curve Length**.

Your values should match those in Figure 6.48.

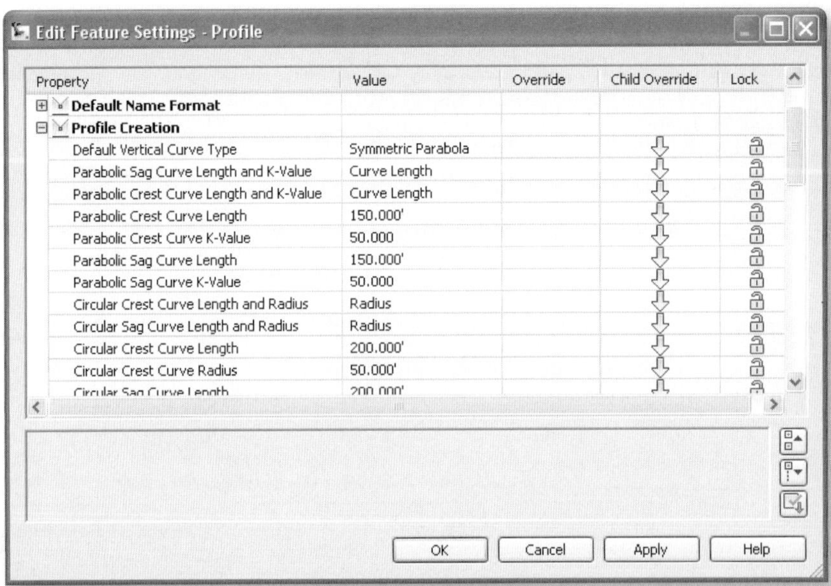


**FIGURE 6.48**

6. Click ***OK*** to set the values and to exit the dialog box.

## Drafting a Proposed Centerline

Creating a vertical design is not always a simple single pass process. The vertical design may have grade, curve length, K Value, or distance between restrictions that affect the design. In addition to these design issues, you may have to design a road that balances or meets some targeted earthworks amount.

1. Use the ZOOM and PAN commands to view the profile view.
2. On the Ribbon, click the ***View*** tab. From the Views panel, select NAMED VIEWS, and create a New view, naming it **Profile**.

3. At Civil 3D's top left, Quick Access Toolbar, click the ***Save*** icon to save the drawing.

Make sure that the Transparent Commands toolbar is visible.

4. On the Ribbon, click the ***Home*** tab. From the Create Design panel, click the ***Profile*** icon, select `PROFILE CREATION TOOLS` and in the drawing, select the Rosewood profile view.

The Create Profile – Draw New dialog box opens.

5. In the dialog box, for the name replace <[Profile Type]> with **_Rosewood Preliminary -**, give the profile a short description, set the Profile Style to **Layout**, and set the Label Set to **Complete Label Set**.

Your dialog box should look similar to Figure 6.44.

6. Click the Design Criteria tab, toggle **ON** Use criteria-based design, toggle *OFF* Use design criteria file, toggle **ON** Use design check set, and change the set to **DuPage Vertical Checks**.
7. Click ***OK*** to exit the Create Profile – Draw New dialog box and to display the Profile Layout Tools toolbar.

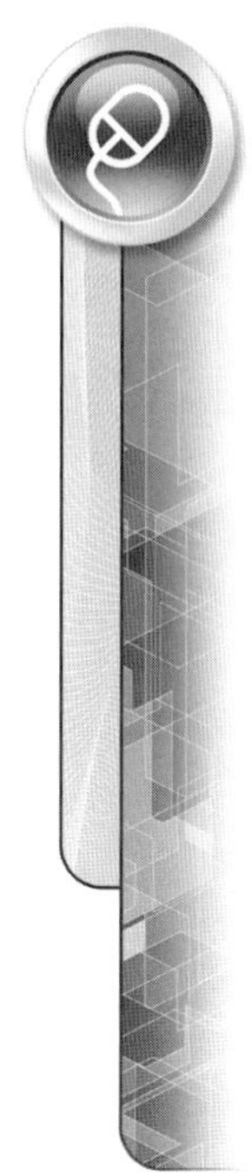

The Rosewood vertical design criteria are the following:

No tangent grades over 5%.

No vertical curves less than 350.

## Drafting Tangent Segments with Vertical Curves

The vertical design uses the Draw Tangents With curves icon at the toolbar's left end. Table 6.1 has the tangents' From and To Stations and their grades. The first and last vertical alignment ends are Endpoint object snap selections of the existing ground profile.

1. On the Profile Layout Tools toolbar, in the first icon stack on the left, click its drop-list arrow and from the list, select Curve Settings....
2. In the Vertical Curve Settings dialog box, if necessary, at the top click the Select curve type drop-list arrow and select the **Parabolic curve** type from the list.
3. Set the Curve length to **350** for crest and sag curves.

Your screen should look like Figure 6.45.

4. Click ***OK*** to close the dialog box.
5. On the Profile Layout Tools toolbar, the first icon stack on the left, click its drop-list arrow and from the list, select ***Draw Tangents With Curves***.

The command line prompts for a starting point.

6. Use the `ZOOM` command to better view the profile's beginning.
7. Use an **Endpoint** object snap and select the intersection of existing ground and the 0+00 profile station.
8. The command line prompts you for an ending point, but from the Transparent Commands toolbar, select the icon ***Profile Grade Station***.
9. The command line prompt changes to Select a Profile View; select the profile view.
10. The command line prompts you for a Grade (positive up and negative down). For the grade, enter **1** and press `ENTER`.

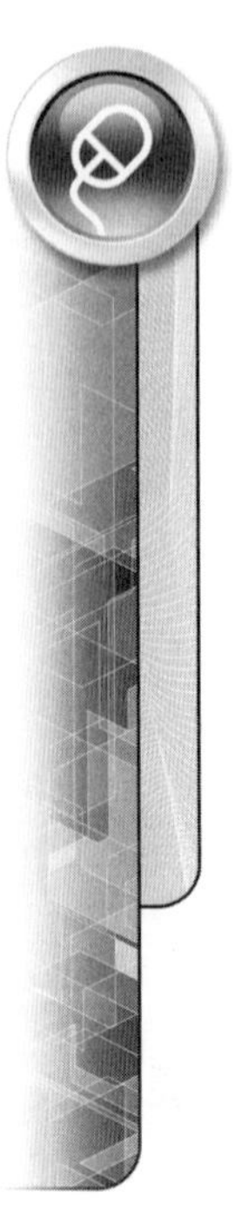

The crosshairs display a 1 percent grade, a station jig, and a tooltip that reports a station and elevation.

11. In the profile view, select a point near station **900**, or in the command line enter **900**. If you are entering a command line value, you must press ENTER to assign the value.
12. Use the mouse wheel and pan and zoom to view the profile past station 900.
13. The command line prompts you for the next Grade. Set the grade to -3%, enter **-3** (and press ENTER), drag the cursor to 13+75, and select that point or enter **1375** (and press ENTER).
14. Use the mouse wheel and pan and zoom to view the profile past station 2000.
15. The command line prompts you for the next grade. Set the grade to 3%, enter **3**, and select a point near 28+50 or enter **2850**.
16. Press ESC once to end the Transparent Command. The prompt changes to Specify End Point. Using an Endpoint object snap, select the intersection of the existing ground and the profile's last vertical grid line, and press ENTER to complete the profile.
17. Close the Profile Layout Tools toolbar.

**TABLE 6.1**

| From Station | To Station | Grade |
|---|---|---|
| Endpoint | 9+00 | 1% |
| 9+00 | 13+75 | -3% |
| 13+75 | 28+50 | 3% |
| 28+50 | Endpoint | No specific grade |

Your profile should look similar to Figure 6.49.

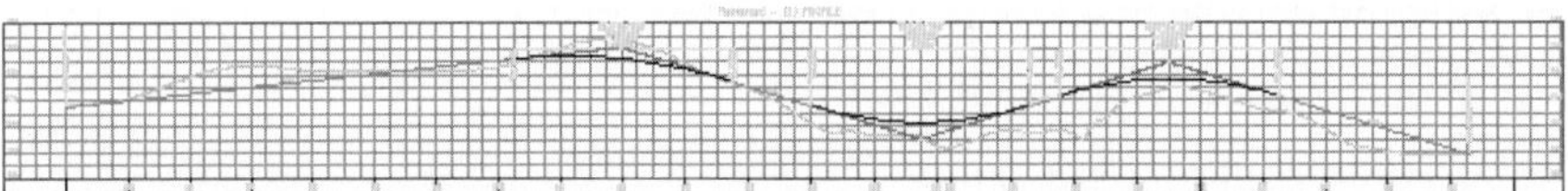

**FIGURE 6.49**

## Assigning the Profile to Profile2

The Profile2 assignment is done in Profile View Properties Band Set values.

1. Use the ZOOM and PAN commands to better view the band's Major and Minor Station annotation.

Currently, both values represent existing ground elevations.

2. In the drawing, select the Profile View, press the right mouse button, and from the shortcut menu, select PROFILE VIEW PROPERTIES....
3. Click the ***Bands*** tab.
4. In the Bands panel, scroll it completely to the right to view the Profile1 and Profile2 assignments.

5. Under the Profile2 entry, click in the cell to display the drop-list, and from the list select **_Rosewood Preliminary - (1)**.
6. Click ***OK*** to exit the dialog box.
7. At Civil 3D's top left, Quick Access Toolbar, click the ***Save*** icon to save the drawing.

This ends the drafting vertical alignment tangents and vertical curves exercise. Next, you focus on evaluating and editing vertical alignments.

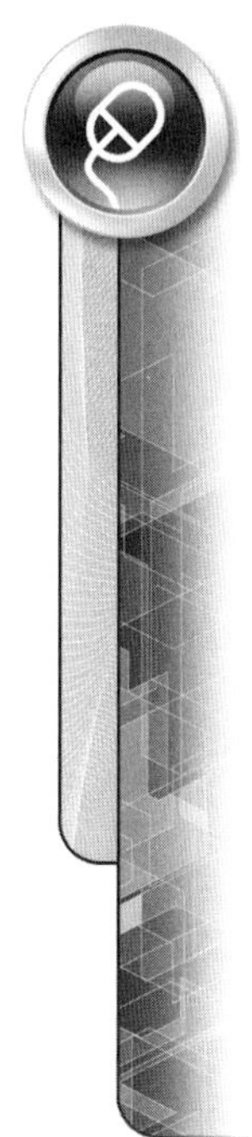

## SUMMARY

- Profile's Edit Feature Settings dialog box values set the vertical curve's type, criteria, and calculation values.
- There are three vertical curve types: circular, symmetrical, and asymmetrical parabolic curves.
- There are three plan view vertical design transparent commands: Profile Station from Plan; Profile Station and Elevation from Plan; and Profile Station and Elevation from COGO Point.
- There are three profile view vertical design transparent commands: Profile Station Elevation; Profile Grade Station; and Profile Grade Length.
- You can create an initial vertical design with or without vertical curves.
- If you are creating an initial design with only tangent segments, you can add vertical curves at anytime.
- If you are creating overlapping vertical curves, the routine issues cannot resolve error and start prompting you for new values.
- If you are drafting an initial vertical design with tangents and vertical curves and there is no solution for a vertical curve, the routine will not draw a vertical curve.

## UNIT 4: ANALYZING AND EDITING A VERTICAL DESIGN

Vertical alignment analysis and editing is this unit's focus. The vertical design analysis uses Toolbox reports or information displayed in the Profile Layout Tools toolbar's Grid View or Sub-entity editors. Also, grips graphically edit a design profile. The Inquiry Tool palette has profile and profile inquiry commands.

### PROFILE REPORTS

Toolbox has several profile reports (see Figure 6.50). In Toolbox, select the report type, right mouse click, and from the shortcut menu, select Execute.... Depending on the report type, a user selects the profile or enters a Create Reports dialog box. Here, a user selects the profile and continues to generate the report.

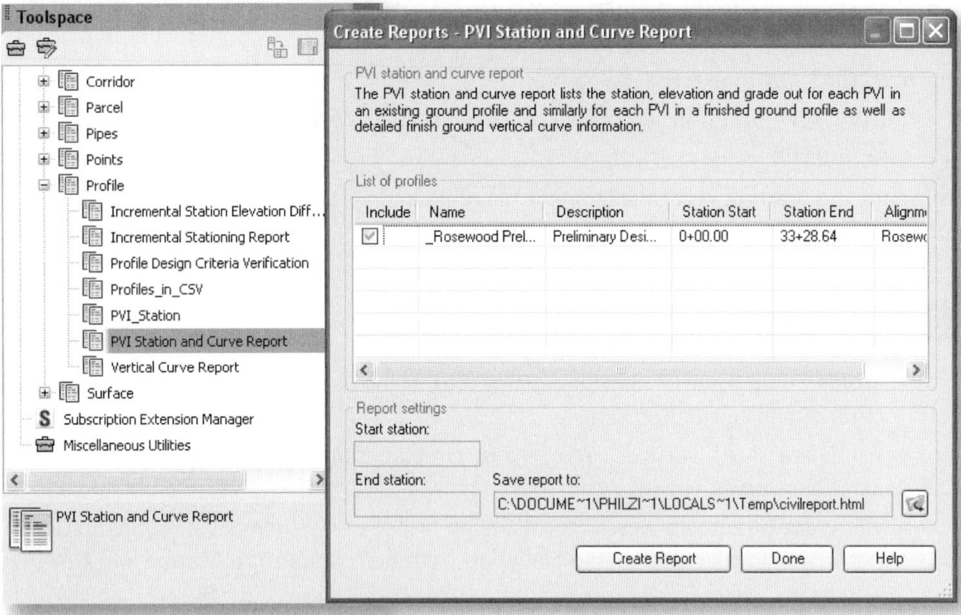


**FIGURE 6.50**

## INQUIRY TOOL

Inquiry Tool has two profile view and profile inquiries (see Figure 6.51). The profile view inquiries are Profile View Station and Elevation at Point and Profile View Elevation and Grade between Points. The Profile inquiries are Profile Station and Elevation at Point and Profile Elevation Difference at Station. Both query sets prompt for the surface and profiles. After selecting a point or point(s), the Inquiry Tool palette populates and displays the query results.

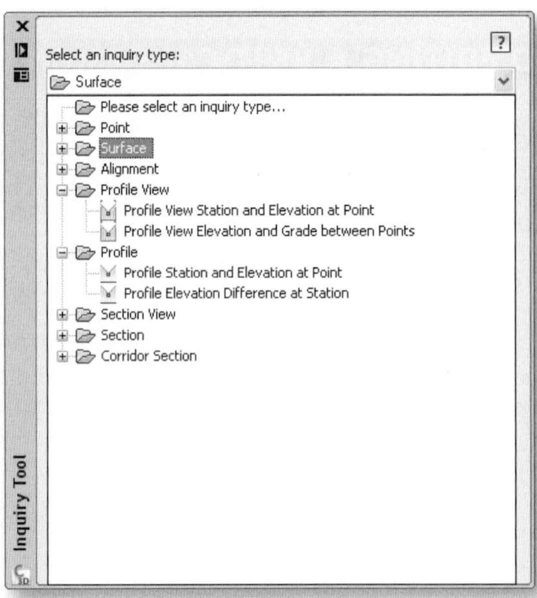


**FIGURE 6.51**

## GRAPHICAL EDITING

Profile grips graphically manipulate its PVIs and segments. When selecting a profile, Profile displays grips to edit a vertical curve's length, change its PVI location and elevation, or move a PVI by holding one tangent's grade and changing the second tangent's grade. When manipulating a PVI's grip, all connected tangents change to accommodate its new location (see Figure 6.52).

Each grip type has a specific editing function. For example, round grips at a vertical curve's end and midpoints lengthen the curve. PVI right and left arrow grips hold one tangent's grade while moving the PVI and changing the second tangent's grade. A PVI red triangle grip adjusts both tangent grades, which changes the PVI's location.

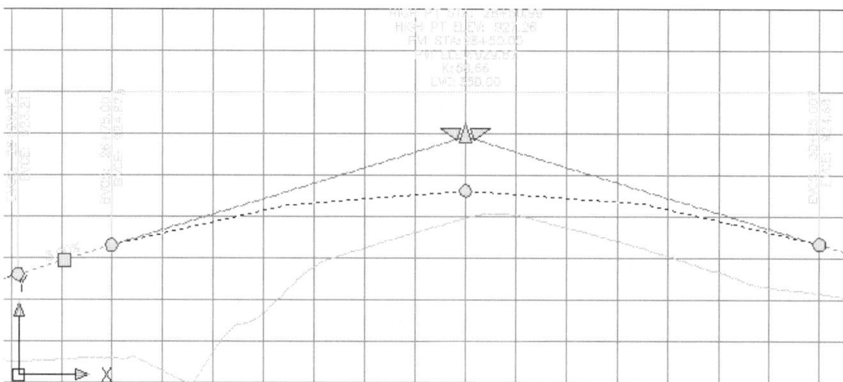

**FIGURE 6.52**

## PROFILE LAYOUT TOOLS EDITORS

The Profile Layout Tools toolbar provides two editors for reviewing and editing profile tangent and vertical curve values: Grid View displays all vertical alignment values; and Profile Layout Parameters displays selected vertical segment values. Figure 6.53 shows the Profile Layout Parameter dialog box.

Making any adjustments in either editor causes the profile to recalculate and it is then redisplayed with its new values. Each editor indicates editable values with black print.

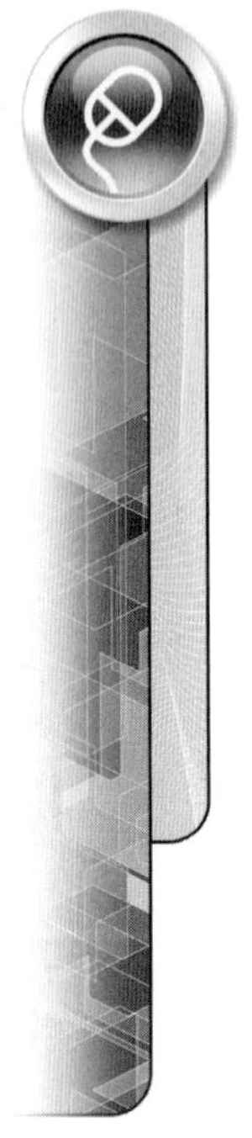

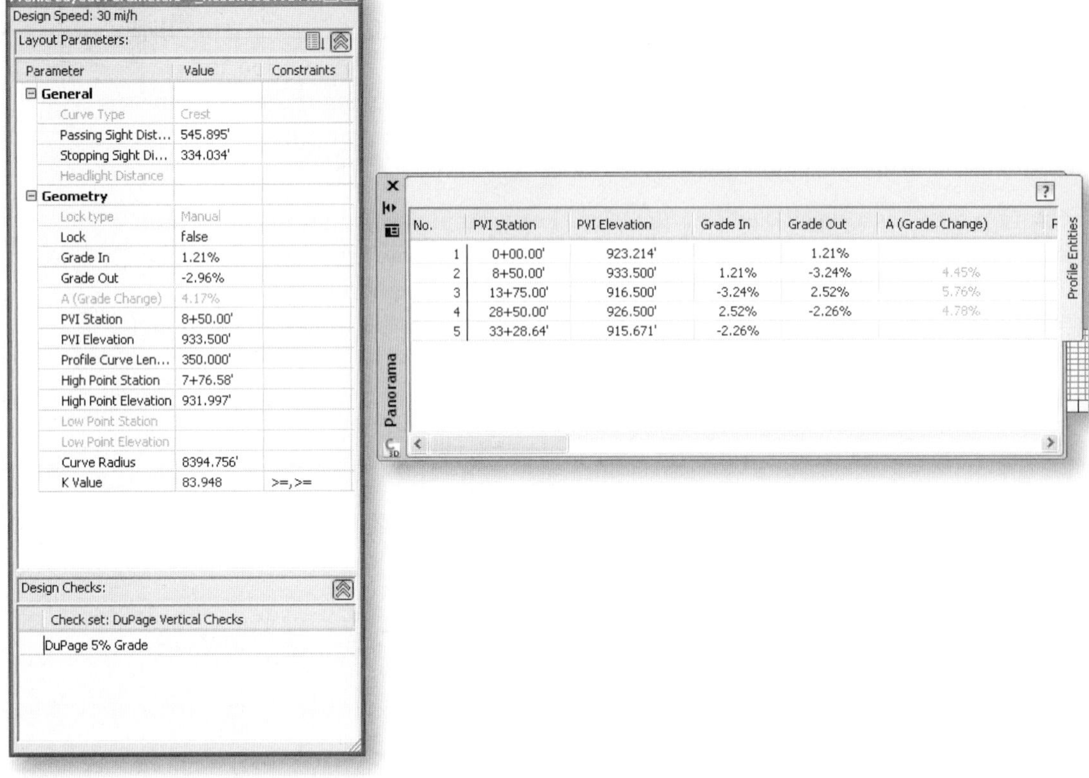


| No. | PVI Station | PVI Elevation | Grade In | Grade Out | A (Grade Change) |
|---|---|---|---|---|---|
| 1 | 0+00.00' | 923.214' | | 1.21% | |
| 2 | 8+50.00' | 933.500' | 1.21% | -3.24% | 4.45% |
| 3 | 13+75.00' | 916.500' | -3.24% | 2.52% | 5.76% |
| 4 | 28+50.00' | 926.500' | 2.52% | -2.26% | 4.78% |
| 5 | 33+28.64' | 915.671' | -2.26% | | |

**FIGURE 6.53**

## EXERCISE 6-4

After completing this exercise, you will:

- Be familiar with profile reports.
- Be able to create and view a vertical alignment report.
- Be able to graphically edit a vertical alignment.
- Be able to edit the vertical alignment in a grid view or sub-entity editor.

### Exercise Setup

This exercise continues with the previous Unit's exercise drawing. If you did not complete the previous exercise, browse to the Chapter 6 folder of the CD that accompanies this textbook and open the *Chapter 06 – Unit 4.dwg file.*

1. Open the drawing from the previous exercise or browse to the Chapter 6 folder of the CD that accompanies this textbook and open the *Chapter 06 – Unit 4* drawing.

### Profile Review

The first profile report is a tangents and curves review.

1. Click the Toolbox tab. If necessary, click the ***View*** tab. At the Palettes panel's right, select the ***Toolbox*** icon displaying the Toolbox tab on the Toolspace.
2. In Toolbox expand the Reports Manager's Profile section, and from the list, select `PVI STATION AND CURVE REPORT`.

3. After selecting the report, press the right mouse button, and from the shortcut menu, select EXECUTE....
4. In the Create Reports dialog box, click ***Create Report*** to create the report.

The report is displayed.

5. Review the report, close Internet Explorer, and click ***Done***.

## Inquiry Tool

Inquiry Tool reports profile and profile view values in a panel.

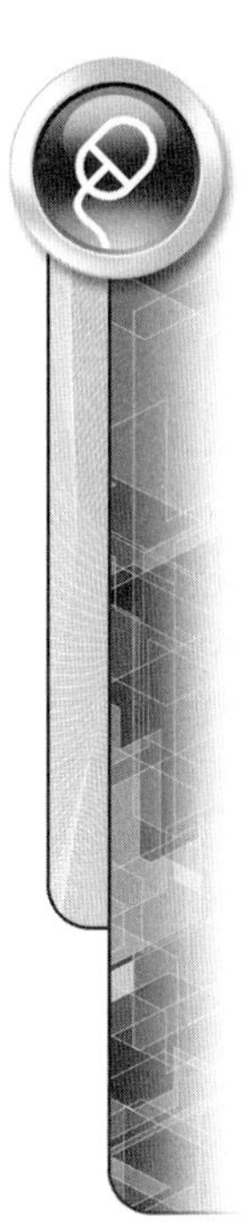

1. On the Ribbon, click the ***Analyze*** tab. On the Inquiry panel, select INQUIRY TOOL
2. At the palette's top, click the Inquiry type drop-list arrow, expand the Profile View section, and from the inquiries list, select PROFILE VIEW STATION AND ELEVATION AT POINT.
3. In the drawing, select a few points within the profile view and review the Inquiry results in the tool palette.
4. At the palette's top, click the Inquiry type drop-list arrow, expand the Profile section to change the Inquiry type to **Profile Station and Elevation at Point**. In the Select Profile dialog box, select the profile _Rosewood Preliminary - (1), and click ***OK***.
5. In the profile view, select some points and review their values in the Inquiry Tool palette.
6. Close the Inquiry Tool palette by clicking its **X**.

## Graphically Editing a Proposed Profile

You can graphically edit a vertical alignment.

1. At Civil 3D's top left, Quick Access Toolbar, click the ***Save*** icon to save the drawing.
2. Use the ZOOM and PAN commands to better view the first PVI.
3. Select the profile _Rosewood Preliminary - (1) to display its grips.
4. Adjust the vertical curve by selecting its round grip and relocating it, and then press ESC.
5. Adjust the PVI by selecting one tangent arrow grip and moving its location, and then press ESC
6. Adjust the PVI by selecting the opposite tangent arrow and moving its location, and then press ESC.
7. Adjust the PVI by selecting it and moving it, and then press ESC.

## Sub-entity Editor Dialog Box

When making any adjustments in an editor, all associated changes accommodate the edit.

1. In the drawing, select the _Rosewood Preliminary - (1) profile, press the right mouse button, and from the shortcut menu, select EDIT PROFILE GEOMETRY....

The Profile Layout Tools toolbar is displayed.

2. On the toolbar's right side, click the ***Profile Layout Parameters*** icon to display the Profile Layout Parameters dialog box.

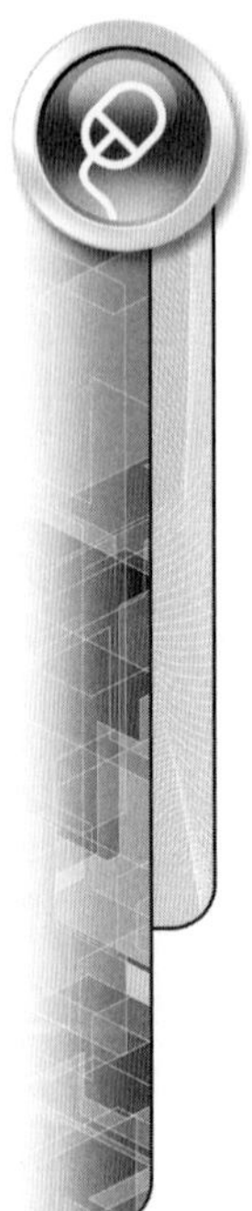

3. If necessary, use the ZOOM and/or PAN commands until you view the first PVI.
4. From the toolbar, click the ***Select PVI*** icon, and in the profile view, select a point near the first PVI.
5. In the Editor, change the PVI Station to **8+50** and change its elevation to **933.50**.
6. Use the ZOOM and/or PAN commands to view the second PVI.
7. Select near the second PVI.
8. Change its PVI Station to **13+75** and raise its elevation to **916.50**.
9. Use the ZOOM and/or PAN commands until you are viewing the last PVI.
10. Select near the last PVI.
11. In the Profile Layout Parameters dialog box, change the PVI's location to **28+50** and change its Elevation to **926.5**.
12. Click the **X** in the Profile Layout Parameters dialog box to close it.

## Profile Grid View Vista

Profile Grid View displays all of the profile values in a single Panorama vista.

1. In the Profile Layout Tools toolbar, on its right side, click the ***Profile Grid View*** icon.
2. Scroll through the values and note which values can be edited.
3. Close the Panorama and close the Profile Layout Tools toolbar.
4. At Civil 3D's top left, Quick Access Toolbar, click the ***Save*** icon to save the drawing.

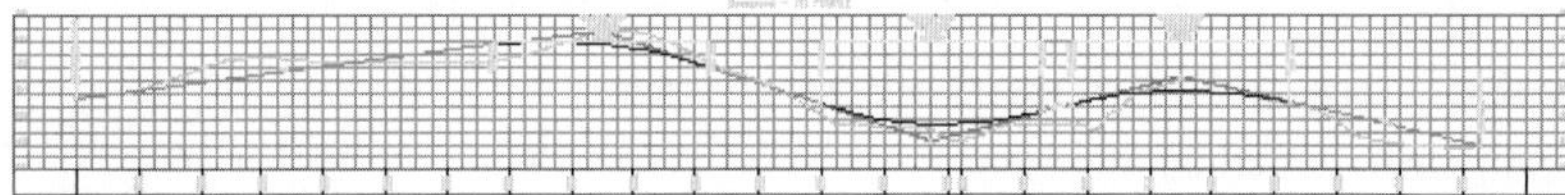

**FIGURE 6.54**

Your profile and profile view should look similar to Figure 6.54.

This completes the vertical alignment evaluation and editing exercise. Editing is done by two methods: graphically manipulating the alignment's grips; or by editing the vertical design in the profile editors.

## SUMMARY

- Toolbox reports a profile's station, elevation, grades, and length of vertical curves.
- When you are graphically editing a vertical curve, manipulate its grips linked to specific geometric PVI and vertical curve points.
- When you are graphically editing a profile, round grips affect vertical curves.
- When you are graphically editing a profile, arrow grips affect the PVI's location by holding the first or second grade.
- When you are graphically editing a profile, by selecting the triangular PVI grip, you change its location, the in and out grades, the PVI elevation, and/or the PVI station.

## UNIT 5: PROFILE ANNOTATION

This unit reviews profile view annotation. A design profile can be annotated as it is created or after it has been created. This concept was reviewed earlier in this chapter.

### SPOT PROFILE LABELS

A second profile annotation type is Profile View labels to annotate stations and elevations. The Add Labels, Profile Views menu, Add Profile View Labels command creates these label types: Station and Elevation and Depth. The Station and Elevation label annotates its namesake at user-selected locations (see Figure 6.55). After identifying the profile view, a jig appears connecting the stationing along the profile's bottom to the cursor. After selecting the station, the jig freezes at the station and it switches to identify the elevation.

Depth annotates the distance between two selected points. If you are selecting a lower then a higher point, the label is a positive grade. If you are selecting a higher and then a lower point, the label is a negative grade.

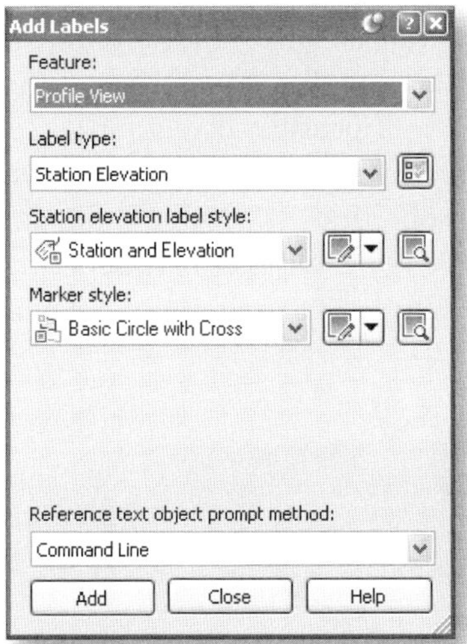


**FIGURE 6.55**

### PROJECTING OBJECTS TO PROFILE AND SECTION VIEWS

Civil 3D projects AutoCAD points, blocks, 3D solids and 3D polylines, Civil 3D COGO points, feature lines, and survey figures to a profile and/or section view. The objects must exist in the drawing before projecting them to a view and **CAN NOT** start and end after the profile stationing. If the object's type is not correct, the routine removes it from the projection selection set. A projected object may represent a waterline, cable line, electric lines, ROW lines for a profile and/or section, etc.

After drafting the object, in the Profile or Profile View ribbon tab, the Launch Pad panel's right side, you select Project Objects to Profile View. After selecting the objects to project and the profile/section view, the Project Objects to Profile/Section View dialog box displays (see Figure 6.56). In the Project Objects to Profile View dialog box, for each selected object, you select a style, an elevation option, and if desired, a label style.

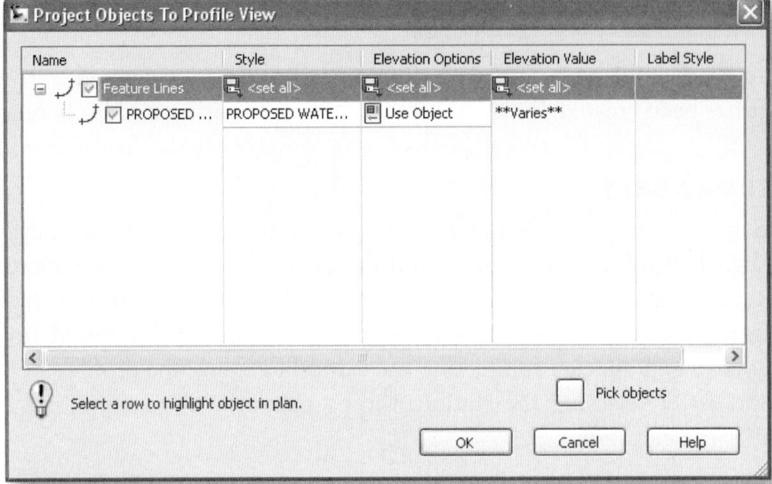


**FIGURE 6.56**

## Projection Styles

A projected object style defines both profile and section behavior and is found in Setting's Multipurpose section. The Information tab supplies the style's name. The Profile and Section panels define the behavior for all possible object types (see Figure 6.57). You select the object type from a list by clicking the drop-list arrow at the panel's top left. Each object type has a unique properties list. Both Profile and Section have the same object type and properties list. The Display panel sets the profile and section layers and their properties. The Display tab sets the projection object's layers (see Figure 6.58).

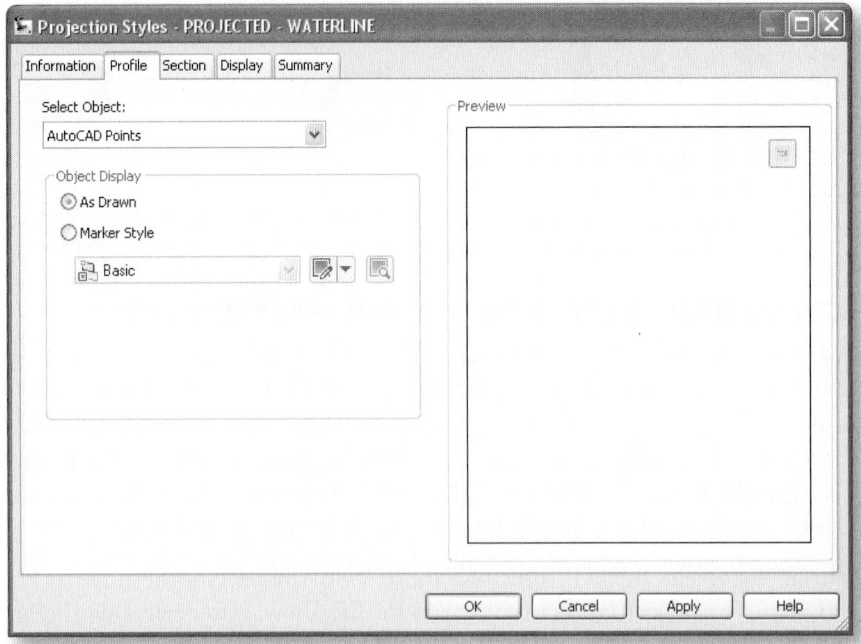


**FIGURE 6.57**

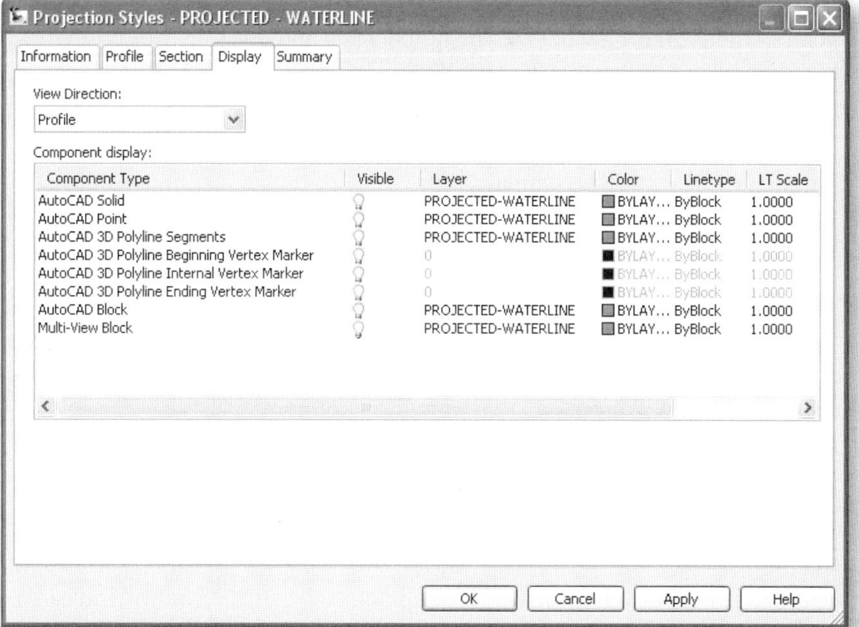


**FIGURE 6.58**

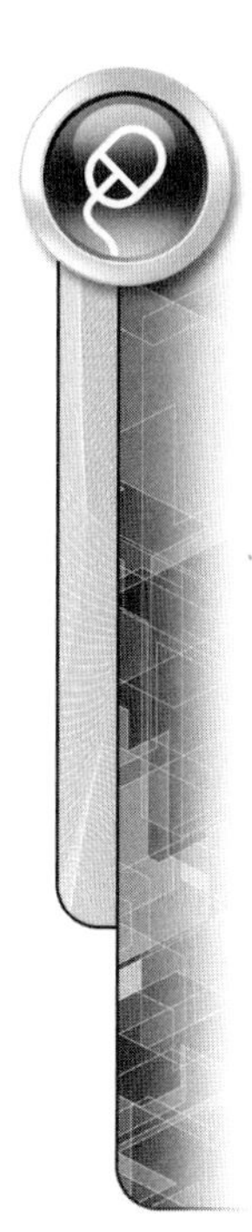

## EXERCISE 6-5

After completing this exercise, you will:

- Be able to label a profile view.
- Be able to apply different profile label styles.

### Exercise Setup

This exercise continues with the previous Unit's exercise drawing. If you did not complete the previous exercise, browse to the Chapter 6 folder of the CD that accompanies this textbook and open the *Chapter 06 – Unit 5.dwg* file.

1. If not open, open the previous exercise's drawing or browse to the Chapter 6 folder of the CD that accompanies this textbook and open the *Chapter 06 Unit 5* drawing.

### Spot Profile Elevations — Station and Elevations

The Add Labels dialog box creates two types of spot profile labels: a station and elevation and depth label.

1. Click Ribbon's ***Annotate*** tab. On the Labels & Tables panel's left, click the Add Labels drop-list arrow to display a shortcut menu. Place your cursor over the menu's Profile View entry and in the flyout select `ADD PROFILE VIEW LABELS`.

The Add Labels dialog box opens.

2. In the Add Labels dialog box, if necessary, change the Label Type to **Station Elevation** and click ***Add***.
3. The command line prompts you for a Select a Profile View. In the drawing, select the **Rosewood Profile View**.

4. The command line prompts you for a Station. Using the station jig, select a point in the profile view or enter a Station value and press ENTER.
5. The station jig freezes and a second one appears prompting you for an elevation. Select a point in the profile view or enter an elevation and press ENTER.
6. Place two more labels in the profile view.
7. Press ENTER to exit the command.
8. Use the ZOOM command to better view the labels.

## Depth

1. In the Add Labels dialog box, change the Label Type to **Depth** and click ***Add***.
2. The command line prompts you to Select a Profile View. In the drawing, select the **Rosewood Profile View**.
3. The command line prompts you to Select First Point. In the profile view, select a point. A jig appears connecting the cursor to the selected point.
4. The command line prompts you to Select a Second Point. In the profile view, select a second point.

The routine draws a line between the two selected points, labels the line with a distance, and exits to the command line.

5. Create two more Depth labels, and ZOOM in to better view the labels.
6. Click the Add Labels' ***Close*** button.
7. In Civil 3D's upper left, Quick Access Toolbar, click ***Save*** icon to save the drawing.

## Changing Profile Labels

Label sets apply label types and their styles to a profile. Edit Labels... changes or assigns new labels to an existing profile.

1. Use ZOOM command and zoom in to better view the first vertical curve.
2. In the profile view, select the **_Rosewood Preliminary - (1)** profile, press the right mouse button, and from the shortcut menu, select EDIT LABELS....
3. In the Profile Labels dialog box, review the currently assigned labels.
4. Delete all of the labels by selecting each type from the list and clicking the red **X**.
5. In the panel's upper left, change the label type to **Grade Breaks**, set the style to **Station over Elevation**, and click ***Add>>***.
6. In the panel's upper left, change the label type to **Crest Curves**, set the style to **Crest Only**, and click ***Add>>***.
7. For Crest Curves, click in its Dim anchor opt cell and change it to **Graph view top**.
8. Click ***OK*** to label the profile.
9. In the profile view, select the **_Rosewood Preliminary - (1)** profile, press the right mouse button, and from the shortcut menu, select EDIT LABELS....
10. In the Profile Labels dialog box, review the currently assigned labels.
11. At the panel's bottom, click IMPORT LABEL SET....
12. In the Select Style Set dialog box, click the drop-list arrow, from the list select **Complete Label Set**, and click ***OK*** to return to the Profile Labels dialog box.
13. Click ***OK*** to assign the label set to the profile.
14. At Civil 3D's top left, Quick Access Toolbar, click ***Save*** icon to save the drawing.

## Projecting a Watermain to a Profile View

1. Select Ribbon's ***Home*** tab. On the Layers panel, select Layer Properties, locate and turn on the layer **C-TOPO-PUTIL-WATER**, and exit the Layer Properties Manager.

The watermain is adjacent to the Rosewood alignment.

2. Use ZOOM and PAN to better view the watermain.
3. Select the Ribbon's Modify tab and then on the Profile & Section Views panel, click the ***Profile View*** icon.
4. On the Profile View tab's, Launch Pad panel's right, select Project Objects To Profile View.
5. The command line prompts you to select an object to project. In the drawing select the watermain near the Rosewood alignment and press ENTER.
6. The command line prompts you to select a profile view. Select the Rosewood profile view and the Project Objects To Profile View dialog box displays.
7. In the Project Objects To Profile View dialog box, set the style to Proposed Watermain, Elevation Options to Use Object, and click OK to project the watermain to the profile view.
8. Select Ribbon's ***Home*** tab. On the Layers panel, select Layer Properties, locate and turn OFF the layer **C-TOPO-PUTIL-WATER**, and exit the Layer Properties Manager.
9. In the drawing, select the profile view, press the right mouse button, and from the shortcut menu, select Profile View Properties....
10. In Profile View Properties, click the Projections tab, toggle off Feature Lines, and click OK to exit the Profile View Properties.
11. At Civil 3D's top left, Quick Access Toolbar, click ***Save*** icon to save the drawing.

This completes the Profile labels exercise.

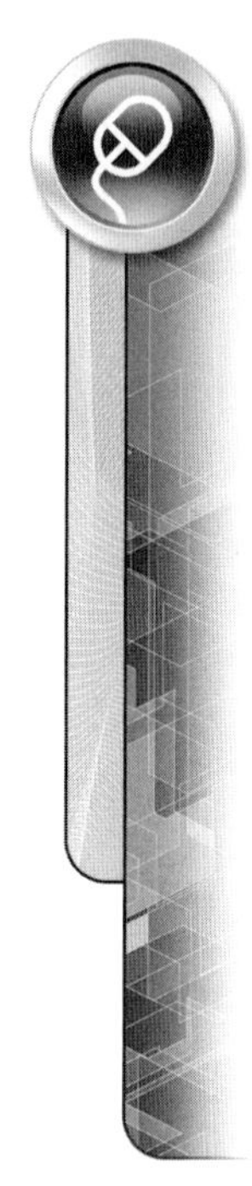

## SUMMARY

- Station and Elevation are profile view labels.
- Depth labels the distance between select points within a profile view.

This ends the Profiles and Profile Views Chapter. The next chapter focuses on the Assembly, its subassemblies, and the roadway model, the corridor.

# CHAPTER 7

# Assemblies and Corridors

## INTRODUCTION

In Civil 3D, the roadway design process is the same as any other application. What is beneficial about the process is that the design model elements interact while you are designing the roadway. Chapters 5 and 6 focus on a road design's first two elements: horizontal and vertical alignments. Chapter 7 covers the third element, the assembly. An assembly is similar to an LDT template, but with greatly enhanced capabilities. Assemblies contain subassemblies with parametric controls and "intelligence" behind them, making them more than simple templates.

A corridor is the roadway 3D model that results from the combination of a horizontal and vertical alignment, a surface, and an assembly.

## OBJECTIVES

This chapter focuses on the following topics:

- The Corridor Modeling Catalogs (Imperial)
- Subassemblies, Their Behaviors, and Parameters
- Assembly Creation and its Modification
- Creating a Simple Corridor from a Surface, an Alignment, and an Assembly
- Reviewing Subassembly and Assembly properties
- Use of the Simple Corridor Command versus the Corridor Command

### OVERVIEW

After designing the horizontal and vertical roadway, next you will define a roadway cross section, or an assembly. An assembly is also known as a template. An assembly has enhanced capabilities, or native intelligence. Subassembly parameters define rules that affect how an assembly solves a roadway design.

An assembly is a vertical line that represents the horizontal (baseline) and vertical alignment attachment point (see Figure 7.1). Subassemblies attach to the assembly to make a roadway cross section. Each subassembly has a set function (curb, slope to daylight, travelway, etc.), attaches to an assembly or another subassembly at a connection point, and has design constraints or parameters that govern its behavior. Connection points locate the attachment point for the next outward subassembly. Most assemblies are built from the inside out.

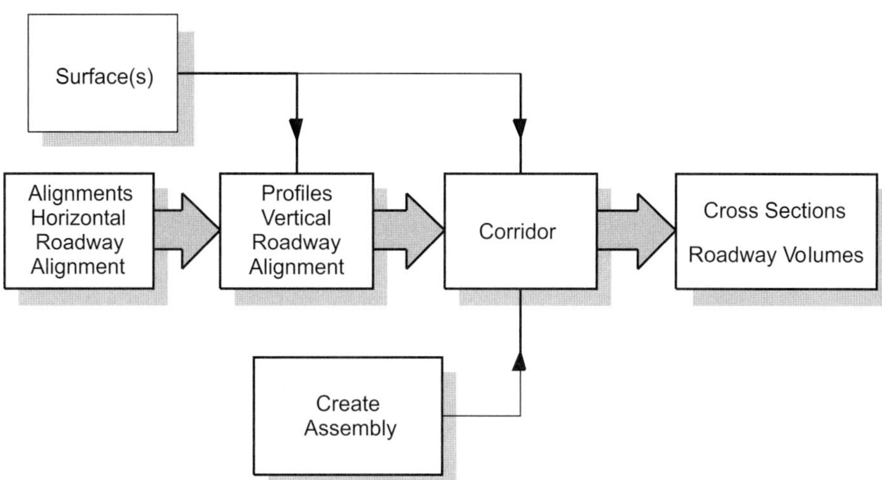


**FIGURE 7.1**

Each subassembly has right and left behaviors set by its parameters list. Parameters allow a subassembly to react to varying design conditions. In Figure 7.2, the right side shows the grading slope parameters for a basic cut slope ditch. Using a more complex slope subassembly, the parameters can automate a more complex design solution (left side of Figure 7.2).

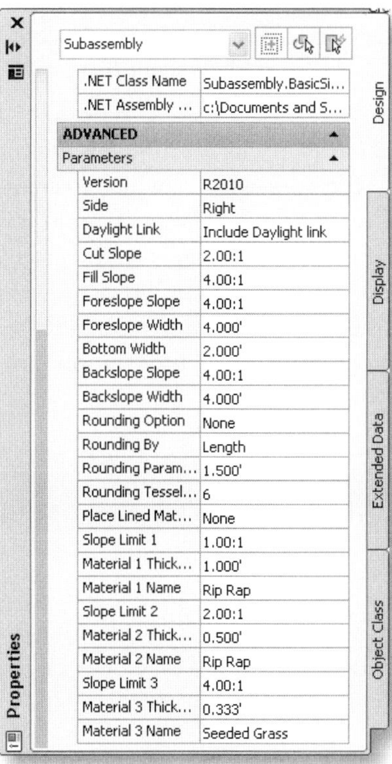


**FIGURE 7.2**

After creating one or more assemblies, you next create a corridor. A corridor combines the horizontal and vertical alignments, the assembly, and even a surface into the model. A roadway model provides several review opportunities and creates new data. Users can slice the model diagonally and see real roadway sections. Users can also create feature lines (base grading objects) from this model. Feature lines become starting points for grading solutions that surround the roadway. The model's elevations are data for developing piping networks and roadway sections.

Corridors create surfaces, and they use surface object and label styles. Corridor surfaces use subassembly links and feature lines for data. A contour surface style displays its contours. A corridor surface appears in Prospector's Surfaces list. A subassembly shape has an area that allows it to be material volume data, and a corridor surface is a comparison surface for general roadway earthworks.

Road design goals vary from project to project, but generally the goal is to not move more soil material than necessary. Occasionally, achieving this goal means editing the design. This editing may start at the very beginning: editing the horizontal alignment. Or, the modifications occur in the proposed profile or with changes to the assembly. Wherever the editing takes place, Civil 3D moves the changes forward from that point to the corridor. All roadway design elements are dynamically linked.

### Unit 1

The first unit focuses on corridor styles and settings. Each subassembly has associated parameters and styles. These styles affect the subassembly connection points, fill patterns, outlines, and labeling of critical subassembly points.

### Unit 2

Next, you create the assembly. An assembly is an anchor point to which subassemblies attach. Subassemblies represent roadway cross-section elements (pavement, curbs, sidewalks, links, etc.). Each subassembly has a right, a left, and a list of parameters.

### Unit 3

The third unit focuses on creating a corridor (the roadway model). After defining the roadway elements, you next create the corridor. There are two ways in which to create a corridor: Create Simple and Create Corridor. Create Simple makes assumptions about section spacing and critical point sampling. If these assumptions are incorrect, users edit the corridor's properties to assign correct values. Create Corridor makes fewer assumptions and lets the user control assignment values to more complex station and assembly relationships, targets, and alignment, profile, and user station sampling intervals.

### Unit 4

After creating a corridor, you evaluate and possibly edit its values. This is done in the Section Editor.

### Unit 5

This unit explores corridor data that generate new data. This data includes grading feature lines, points, transitioning offset baselines, volumes, or other useful roadway design elements.

## UNIT 1: CATALOGS, PALETTES, AND STYLES

The catalog concept comes from the building systems world and Civil 3D uses it to organize subassemblies.

### CORRIDOR MODELING CATALOG (IMPERIAL)

There are six Imperial Corridor Modeling catalogs: Channel and Retaining Wall Subassembly, Generic Subassembly, Getting Started Subassembly, Rehab Subassembly, Subdivision Roads Subassembly, and Transportation Road Design Subassembly (see Figure 7.3). Some of the catalogs are more than one page. Each catalog has similar design elements, but each also has specific functionality or constraints that set it apart. Users can add to or customize the existing roadway elements.

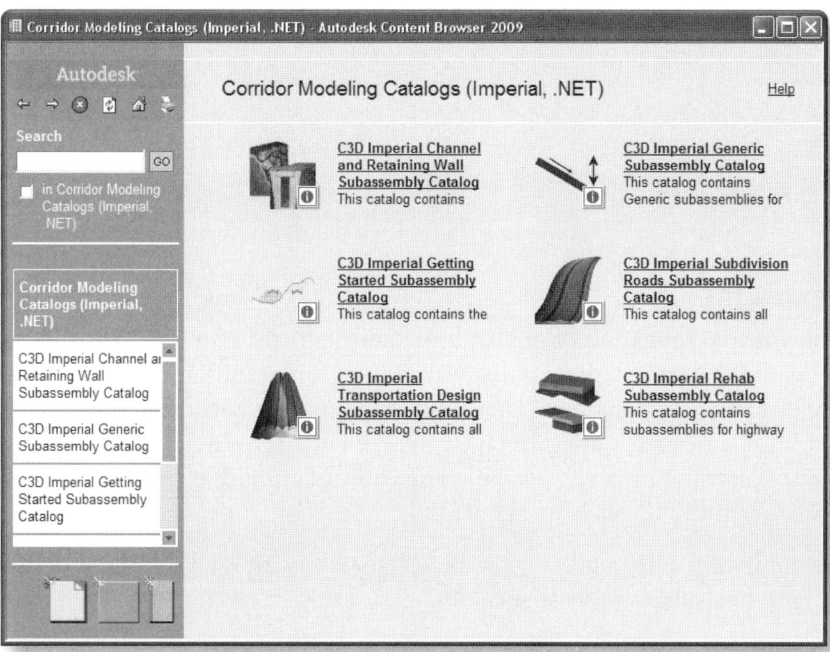


**FIGURE 7.3**

### CIVIL 3D IMPERIAL PALETTE

The Civil 3D - Imperial tool palette displays the cataloged subassemblies in groups, based on their function (see Figure 7.4). Each tab represents function-related subassemblies (basic or complex pavement, sidewalks, sod, shoulders, rehabilitation strategies, and links between subassemblies). The palette includes a tab with predefined assemblies.

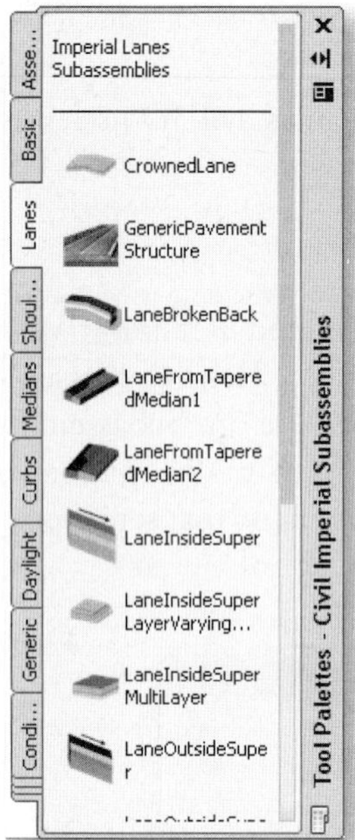


**FIGURE 7.4**

## SUBASSEMBLIES

A subassembly is a combination of points, links, and shapes. Point codes, links, and shapes label corridor data, integrate a roadway with grading solutions, and use their data for slope staking, as construction staking data, and for corridor visualization. Point codes, links, and shapes are skeletal elements that allow subassemblies to react to parameters. Customizing a subassembly requires understanding roadway design needs and VBA scripting. This is no simple task.

Each subassembly vertex has a point code. Between each point is a link. Links define a closed polygon or a subassembly shape.

Each point, link, and shape has a style. A point code style defines a marker and/or color. Link styles (lines between point codes) define link layer and color properties. Shape styles define a shape's fill color and its outline. Marker, Link, and Shape styles are in Settings' General, Multipurpose Styles branch.

### Edit Drawing Settings

The assembly object's base layer is C-ROAD-ASSM. If you have more than one assembly, it is best to add a suffix or prefix to the base layer name so each assembly has its own layer.

### Edit Feature Settings

Settings' Subassembly branch has an Edit Feature Settings dialog box with values that affect the subassemblies of an assembly.

### Subassembly Points (Codes)

Subassemblies contain a series of points that have specific functions. A subassembly's first point (P1) is its connecting point. It connects the subassembly to the assembly or a more centrally located subassembly. A subassembly can connect to any adjacent subassembly point. Attaching point code P1 to an adjacent subassembly point is a matter of selecting the "correct" subassembly point. Users may have to zoom in to better view the specific adjacent subassembly connection points. The P1 point switches to the subassembly's right side when changing its side parameter to left. For example, the curb or a sidewalk subassembly will mirror itself to position the P1 point to the correct side. A symmetrical assembly only needs the right side defined, and then it is mirrored to the left to complete its definition.

Each subassembly point has unique point codes. Figure 7.5 lists the basic lane subassembly codes.

**NOTE** Civil 3D Help is an excellent resource for subassembly information.

**NOTE** Civil 3D's Help folder has a C3DStockSubassemblyHelp PDF with the same information as the help file.

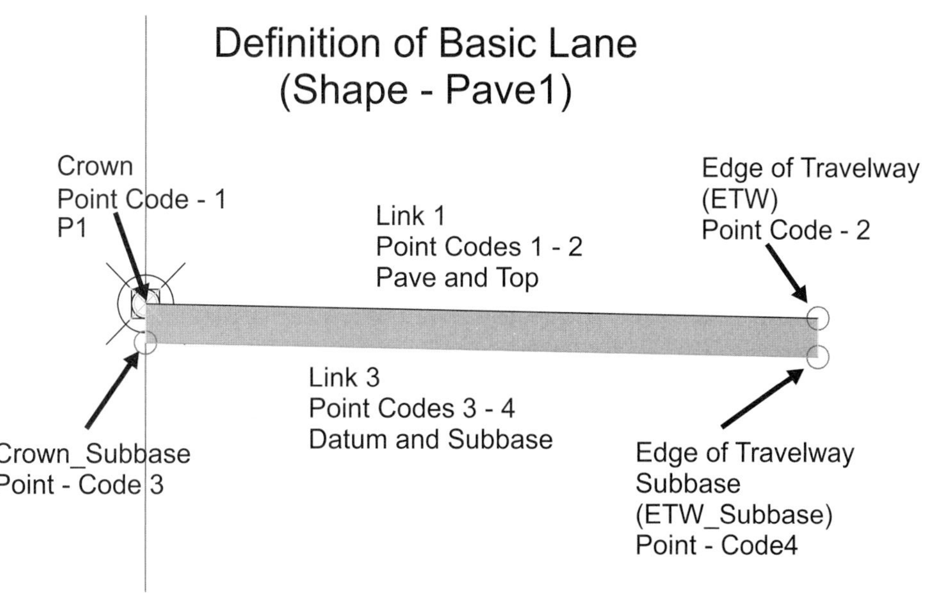


**FIGURE 7.5**

Point codes provide section label data (offset and elevation) and create corridor feature lines.

## Subassembly Links

Between each subassembly vertex (point code) is a link. A link creates a segment of a subassembly's "shell" (see Figure 7.5). Each link has a type, and all subassemblies have a top-of-subassembly and datum link.

Links provide slope and distance data. A label uses these values as data: pavement cross slope; daylight slope; and so on. Links are critical data for a corridor surface.

### Subassembly Shapes

Points (codes) and links create a subassembly shape. In Figure 7.5, Pave1 is the subassembly shape and its name is BasicLane.

A shape provides a name for a label and an area for a material volume calculation.

## CORRIDOR

A corridor is a 3D model from the processing of horizontal and vertical alignments and the assembly. The point codes create 3D strings representing the assembly's points along the horizontal and vertical alignments. These strings and the links combine to create corridor surfaces.

### Edit Drawing Settings

Corridor's base and sections layers are C-ROAD-CORR and C-ROAD-CORR-SCTN. If there is more than one corridor, it is best to add a suffix or prefix to the base layer name so each corridor has its own layer.

### Edit Feature Settings

Settings' Corridor branch has an Edit Feature Settings dialog box with values that affect the corridor's creation, style assignments, and region highlight graphics. Edit Feature Settings has three sections: Default Styles, Default Name Format, and Region Highlight Graphics (see Figure 7.6). The Create Corridor commands use these values when calculating a corridor.

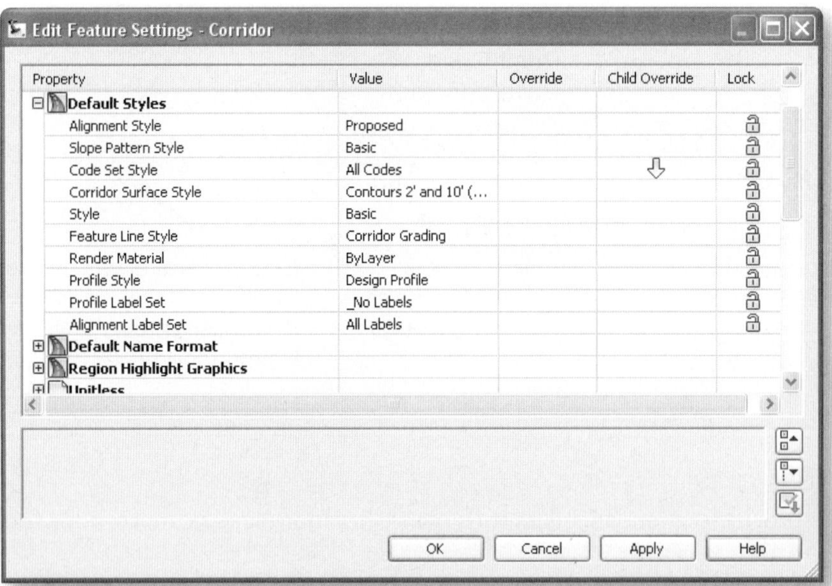


**FIGURE 7.6**

### Corridors – Create Simple Corridor

Create Simple Corridor creates a corridor that has no regions and the vertical design stationing matching the alignment stationing (see Figure 7.7). A region is an alignment segment that uses a different assembly. Examples of regions are modeling intersecting roadways, cul-de-sacs, assigning an different assembly for a portion of a corridor, and knuckles.

The command's Assembly Insertion Defaults section defines the assembly (section) frequency along tangents, curves, and spirals. Also, this section defines whether the corridor includes critical geometry points from horizontal, superelevation, sampling frequencies, and profile elements or parameters.

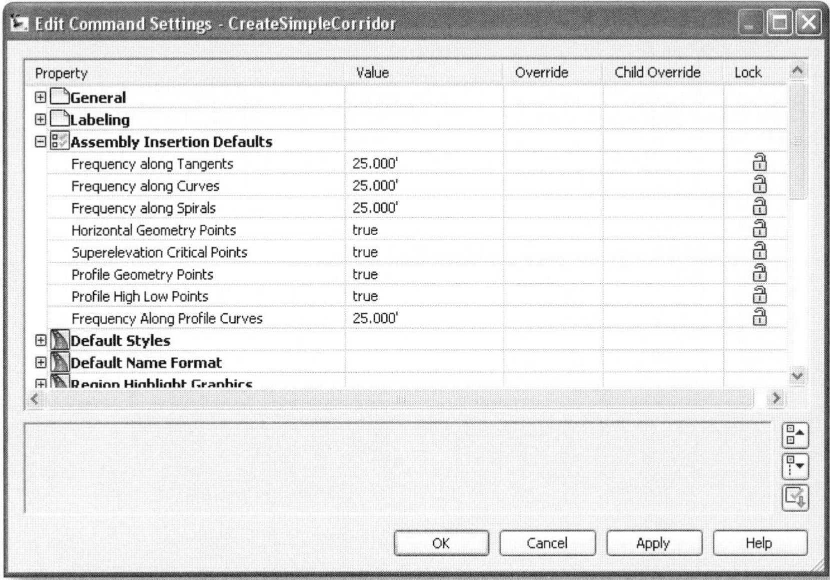


**FIGURE 7.7**

The Default Styles and Default Name Format sections set styles for each corridor element and its naming format. These values reflect the Corridor's Edit Feature Settings and are changeable at the command level.

### Corridors – Create Corridor

Create Corridor displays the same setting as Create Simple Corridor. The difference between the commands is their order of prompts for information. Create Corridor displays more dialog boxes and is for more complicated corridors. These corridors have vertical design stationing that does not match the horizontal alignment stationing or have regions with different assemblies.

## CORRIDORS – MULTIPURPOSE STYLES – CODE SET

A Code Set assigns several important values to a corridor (see Figure 7.8). A code set assigns an assembly and its point, links, and shapes point, link, or shape label styles, render materials, area fill hatches, feature line styles, and Pay Item codes. Each value displays at different stages of the design process. The label styles display in corridor sections and cross sections, the pay item codes affect the calculation of material list costs, etc.

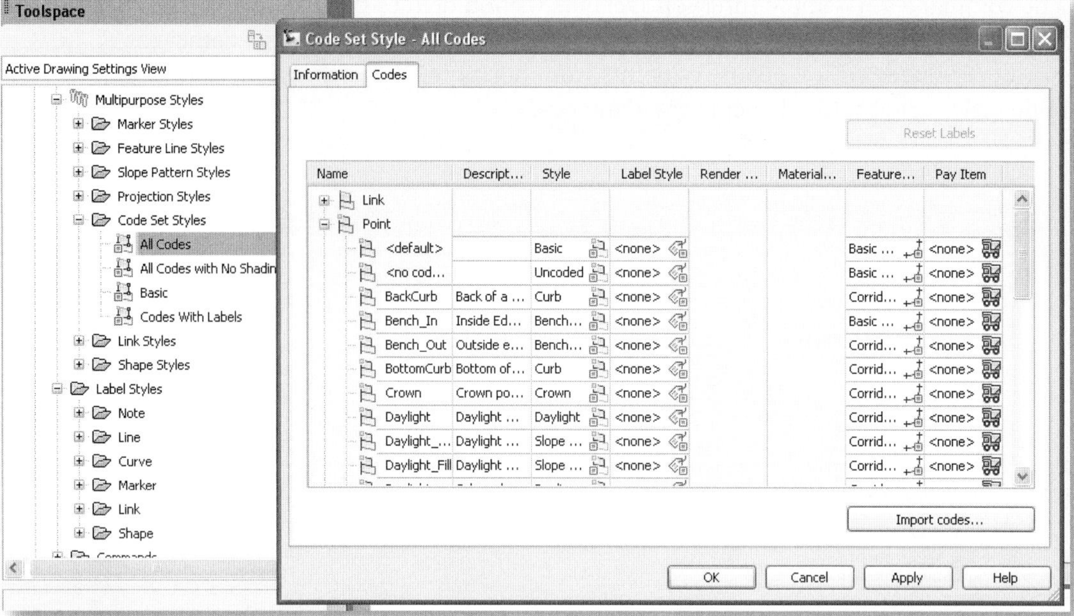


**FIGURE 7.8**

## Point, Link, and Shape Label Styles

A code set's point label styles annotate offset and elevation values (see Figure 7.8). These labels include back-of-curb, daylight, and other elevations that are a point in a cross section. Link label styles annotate slopes that appear in an assembly's subassemblies. These labels include travelway cross slopes, daylight slopes, etc. A shape label style annotates the name of a subassembly shape. These labels include pave1, base, curb, etc.

## Render Material and Material Area Fill

The Render Material assignment places a material on a subassembly link. When you view the corridor as a model with realistic or conceptual visual styles, these links display the assigned material. When viewing corridor or creating cross sections, the material area fill defines if the shapes are hatched and with which hatch pattern.

## PAY ITEM

Pay items allow a user to assign a lump sum or per unit value to a project item. For example, the cost of pipes, structures, curbing, asphalt, granular base, etc. are all items that can be assigned a pay item. A code set contains a column where this value is assigned to corridor assemblies (their subassemblies) points and links. The material lists assign costs for assembly shapes and volumetric calculations. In Pipes, a pay item is assigned in the parts lists. After assigning and creating a corridor or pipe network, a user can create a estimation report based on the costs and assigned objects.

## Code Set Style

In the Code Set, a user assigns the pay item value to a named subassembly of the assembly. See Figure 7.9. The pay item assignment allows you to calculate a cost with a material quantity, e.g. asphalt, concrete, base material, etc. The quantity take off routines calculate the material quantities and costs as a detailed or summary report.

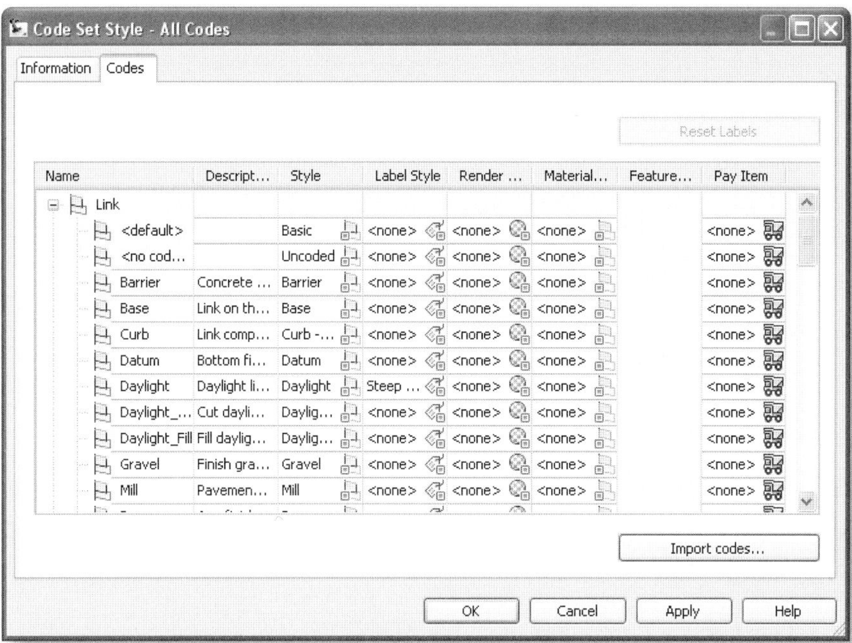


**FIGURE 7.9**

## Pay Item External Files

Pay items are lump sum or per unit costs you attach to drawing and/or Civil 3D objects. After attaching the values to the objects, you can extract a detailed or general report. You must have at least a pay item file to use this feature.

## Pay Item file

The pay item file contains the item's code, description, and payment type, per unit or lump sum, and is a comma delimited file with a CSV extension. The following is an excerpt from the file.

```
15101-0000,MOBILIZATION,LPSM,0,0,N,2003,,,,

15201-0000,CONSTRUCTION SURVEY AND STAKING,LPSM,0,0,
N,2003,,,,

15205-0000,"SLOPE, REFERENCE, AND CLEARING AND GRUBBING
STAKE",LPSM,0,0,N,2003,,,,

15206-0000,"SLOPE, REFERENCE, AND CLEARING AND GRUBBING
STAKE",STA,3,3,N,2003,,,,
```

The file contains ten fields, Pay Item, Item Description-USC, Unit_E, Bid_Dec, Pay_Dec, Pay Item Type, FP-YR, Date, added/modified, Division, and Comments. Quantity Take Off (QTO) manager loads and manages the pay item list.

## Pay Item Category File

A second file, Pay Item Category, organizes the pay items with headings and relationships. This and the Pay Item file must be loaded before using the routines of

QTO. The categorization file is an XML file with the extension of FOR, defining the categories as groups of pay item number codes and has. The following is an example of a category file.

```
<payItemCategorizationRules>
  <Properties>
    <Company>Autodesk</Company>
    <Product>AutoCAD Civil 3D</Product>
    <Description>Quantity takeoff pay item list
      categorization file</Description>
    <Version>1.0</Version>
  </Properties>
  <payItemIDLocation>
        <start>1</start>
        <end>5</end>
      </payItemIDLocation>
      <categories>
    <category type="value" start="" end="" title="Division
      100" description="General Requirements">
    </category>
    <category type="value" start="" end="" title="Division
      150" description="Project Requirements">
      <category type="value" start="" end="" title="Group
        151" description="Mobilization">
        <category type="value" start="15101" end="" title=
          "Section 15101" description="Mobilization"/>
    </category>
    <category type="value" start="" end="" title="Group
      152" description="Construction Survey and Staking">
      <category type="value" start="15201" end="" title=
        "Section 15201" description="Construction survey
        and staking"/>
```

The category file titles the groups, assigns the groups description, and identifies what pay item numbers are in each category. Figure 7.10 shows the QTO manager with a pay item list with an applied category file.

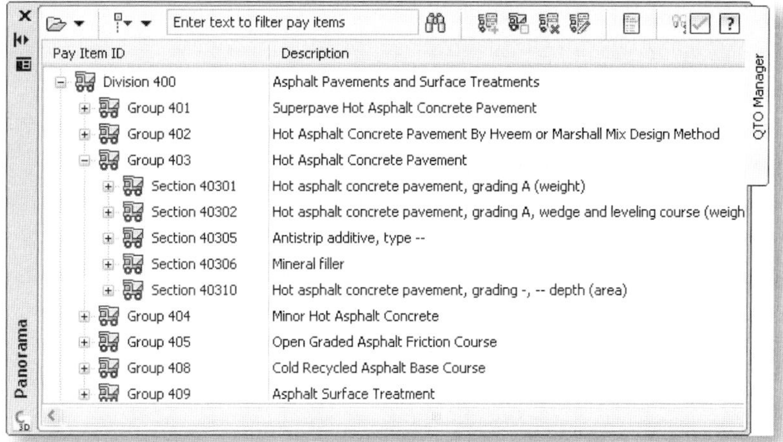


**FIGURE 7.10**

## Pay Item Assignment

After loading in a pay item file and categorizing it, the next step is assigning pay item entries to entities, code set styles, or material lists. The entities can be Civil 3D objects or AutoCAD entities and each object or entity can have more than one assigned pay item entry. In Figure 7.11, the icons at QTO's top center from left to right assign pay items to objects in the drawing, closed area, removes assigned pay items from selected objects, and edits assigned pay items.

At QTO manager's top right, the second icon in from the right is an icon stack with commands to highlight entities with and without pay item assignments and objects with a specific pay item. When hovering the cursor over an object with a pay item, Quick Properties displays the object's assigned pay item(s).

Enter text to filter pay items
Pay Item ID | Description | Unit Type
Highlight objects with pay items
Highlight objects without pay items
Highlight objects with selected pay items
✔ Clear highlight
Division 400 | Asphalt Pa…
Group 401 | Superpave…
Group 402 | Hot Asphalt … Met…
Group 403 | Hot Asphalt Concrete Pavement
Section 40301 | Hot asphalt concrete pavement, grading A (weight)
Section 40302 | Hot asphalt concrete pavement, grading A, wedge and leveling course…
Section 40305 | Antistrip additive, type --
Section 40306 | Mineral filler
Section 40310 | Hot asphalt concrete pavement, grading -, -- depth (area)
Group 404 | Minor Hot Asphalt Concrete
Group 405 | Open Graded Asphalt Friction Course
Group 408 | Cold Recycled Asphalt Base Course
Group 409 | Asphalt Surface Treatment
Group 410 | Slurry Seal
QTO Manager
Panorama

**FIGURE 7.11**

## Pay Item Reports

Objects with pay item assignments are available for reports using the item's values. There are two types of pay item reports; summary and detailed. A summary report lists the pay item ID, its description, total quantity, and unit of measure for each pay item. A detailed report contains a line of information for each selected object.

- A pay item report's scope is a drawing, sheet, or selected objects.
- A summary report does not calculate values for corridor codes.
- A detailed (itemized) report calculates Corridor codes assignments.
- If the pay item relates to an alignment, the report can be limited by alignment station values.

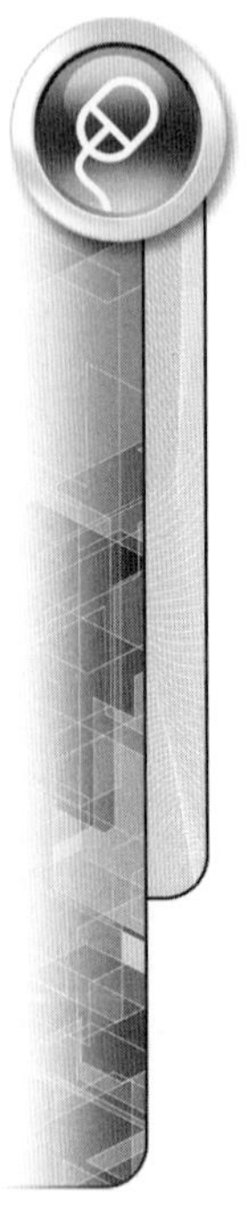

**EXERCISE 7-1**

After completing this exercise you will:

- Be familiar with Marker, Link, and Shape styles.
- Be familiar with the settings for the CreateSimpleCorridor command.

## Exercise Setup

This exercise starts with the previous chapter's exercise drawing. If you did not complete the previous chapter's exercise, browse to the Chapter 7 folder of the CD that accompanies this textbook and open the *Chapter 07 – Unit 1.dwg* file.

1. If you are not in ***Civil 3D***, double-click its desktop icon to start the application.
2. When you are at the command prompt, close the open drawing and do not save it.
3. Open the drawing from the previous exercise's chapter or browse to the CD that accompanies this textbook and open the *Chapter 07 – Unit 1* drawing.

## Edit Drawing Settings

The assembly, corridor, and corridor section layers need a suffix and a dash asterisk (-*).

1. Click the ***Settings*** tab.
2. At Settings' top, select the drawing name, press the right mouse button, and from the shortcut menu, select EDIT DRAWING SETTINGS....
3. Click the ***Object Layers*** tab.
4. In Object Layers, change the Modifier for Assembly, Corridor, and Corridor Section to **Suffix**, and change their Value to **-*** (a dash followed by an asterisk).
5. Click ***OK*** to set the values and exit the dialog box.

## Subassembly — Edit Feature Settings

It is important that each subassembly name include its side. In Civil 3D 2010, this is a default value. In this exercise this value has not been set. You will change the Subassembly feature settings' value to use the side as a part of the subassembly name. The side property makes it easier to correctly assign controlling alignments and profiles to complicated corridors.

1. In Settings, from the headings list, select **Subassembly**, right mouse click, and from the shortcut menu, select EDIT FEATURE SETTINGS....
2. Expand the Subassembly Name Templates section.
3. Click the Create From Macro's value cell that is displaying an ellipsis.
4. Click the ellipsis to display the Name Template dialog box.
5. Click in the Name cell, after Macro Short Name's dash, add two spaces, and an additional dash.

6. Place the cursor between the two dashes, click the Properties field's drop-list arrow, from the list select **Subassembly Side**, and click ***Insert*** to add the side property to a Subassembly's name.
7. Click ***OK*** until the dialog boxes are closed.

## Point Code Styles

Points, links, and shapes styles define their symbols, layers, and visibility.

1. In Settings, expand General and Multipurpose Styles until you are viewing the Marker Styles list. From the list, select **Crown**, press the right mouse button, and from the shortcut menu, select EDIT....
2. If necessary, in the Marker Style dialog box, click the ***Marker*** tab.

Marker defines what shape a roadway crown marker displays.

3. Click the ***Display*** tab.

The Display tab defines marker's visibility, layer, and properties.

4. Click ***OK*** to exit the dialog box.
5. In Settings, collapse the Marker Styles branch.

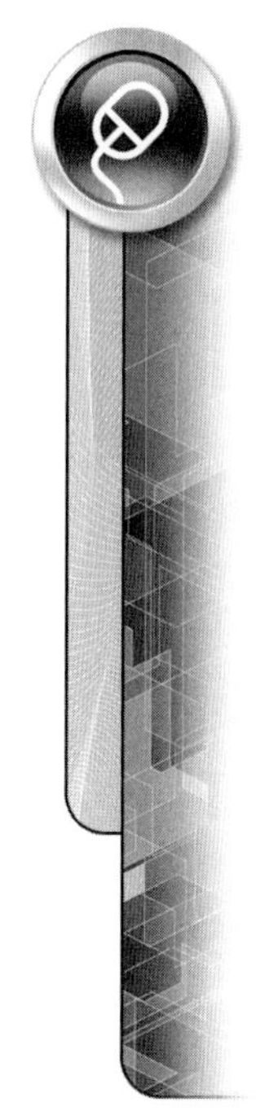

## Link Styles

A link style defines a link's visibility, layer, and layer properties.

1. In Settings, General, Multipurpose Styles branch, expand Link Styles until you view its styles list. From the list, select **Pave1**, press the right mouse button, and from the shortcut menu, select EDIT....
2. Click the ***Display*** tab to review its values.
3. Click ***OK*** to exit the dialog box.
4. In Settings, collapse the Link Styles branch.

## Shape Styles

A shape style defines visibility, outline and fill layers, their properties, and the fill pattern.

1. In Settings' Multipurpose Styles branch, expand Shape Styles until you view its styles list.
2. From the list, select **Pave1**, press the right mouse button, and from the shortcut menu, select EDIT....
3. In the Shape Style dialog box, click the ***Display*** tab to review its contents.
4. Click ***OK*** to exit the dialog box.
5. In Settings, collapse the General branch.

## Create Simple Corridor

CreateSimpleCorridor assigns several default object styles and corridor parameters.

1. In Settings, expand the Corridor branch until you view the Commands list.
2. From the list, select **CreateSimpleCorridor**, press the right mouse button, and from the shortcut menu, select EDIT COMMAND SETTINGS....
3. Expand the Assembly Insertion Defaults section.

These values set the corridor section interval (every 25 feet), what critical geometry points to include (horizontal, vertical, and superelevation), and how often to sample a vertical curve.

4. Collapse the Assembly Insertion Defaults section and expand the Default Styles section.

All styles in this section are from Corridor's Edit Feature Settings.

5. Collapse the Default Styles section and expand the Default Name Format section.

The Default Name Format section sets a corridor naming for corridors, surfaces, and feature lines.

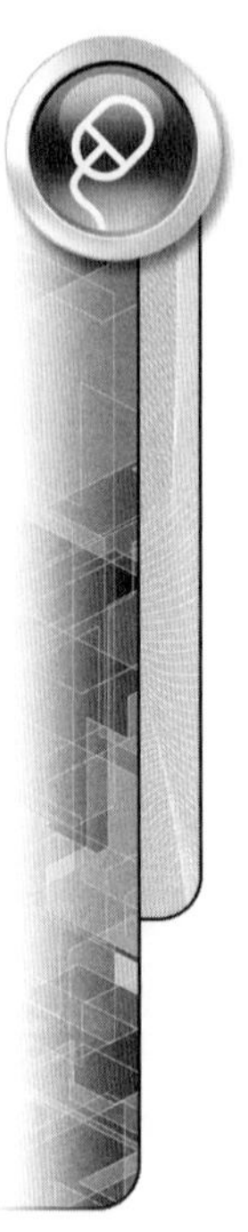

6. Collapse the Default Name Format section and expand the Region Highlight Graphics section.

The Region Highlight Graphics section sets a corridor region's appearance and display.

7. Click ***OK*** to close the dialog box.

## Imperial Subassembly Catalogs

The Roadway Catalogs provide content for the Civil 3D Imperial and Metric subassembly palettes.

1. On the Ribbon, click the View tab. In the Palettes panel's middle right, click the ***Content Browser*** icon.
2. The Imperial and Metric Catalogs are displayed.
3. In the catalog library, select the Corridor Modeling Catalogs (Imperial) icon.
4. Select and review each catalog's contents.
5. Close Corridor Modeling Catalogs dialog box.
6. At Civil 3D's top left, Quick Access Toolbar, click the ***Save*** icon to save the drawing.

## The Imperial Roadway Palette

Imperial Roadway is a multi-tabbed palette containing subassemblies that address several road design issues. Each tab represents the subassemblies categories.

1. In the Palette panel, to the left of the Content Browser icon, select the ***Tool Palettes*** icon.
2. Click each palette tab to review each subassembly collection.

This completes the exercise that reviews corridor settings, styles, catalogs, and palettes. The next unit creates an assembly by using various subassemblies.

## SUMMARY


- Point codes are critical subassembly vertices.
- Links and shapes use point codes as their endpoints or vertices to define their shape.
- Point code labels create station, offset, and elevation annotation.
- Links define a shape's edge, provide slope/grade labeling data, and provide surface data.
- Shapes provide the name and an area for material volumes.
- The CreateSimpleCorridor settings affect how a corridor model is made and the styles it uses.

## UNIT 2: ASSEMBLIES AND SUBASSEMBLIES

An assembly is a roadway cross-section anchor for defining a road section. The assembly is the section's midpoint and all subassemblies attach to and outward from it. Subassemblies represent discreet cross-section elements (for example, pavement, curbs, and shoulders).

The horizontal and vertical alignments pass through the assembly's central eyelet. As these alignments move, they pull the assembly's eyelet to the right, left, up, or down. The subassembly shapes and their locations along this path create the roadway corridor model.

Each subassembly has a marker set surrounding its shape. Each marker has a point code and a link connecting it to the next shape point. Each link has a name (for example, top or datum).

Creating or customizing subassemblies requires a familiarity with VBA scripting and .NET. The Help file documents the necessary steps to create a custom subassembly.

### SUBASSEMBLY PROPERTIES

A subassembly attaches to the assembly's right and left. The side relative to the assembly on which a subassembly is located should be a part of the subassembly's name. This is done because subassemblies by default receive only a subassembly name and number. When subassembly names become complicated, the names pay dividends when assigning alignments and profiles in the Create Corridor or Corridor Properties Target Mapping dialog box (see Figure 7.12).

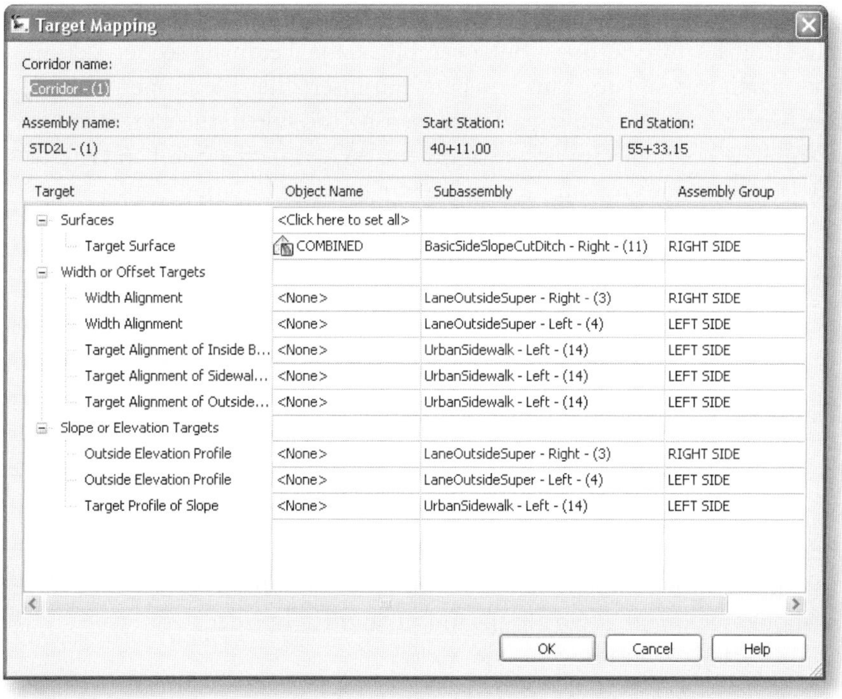

Target Mapping

Corridor name: Corridor - (1)

Assembly name: STD2L - (1) Start Station: 40+11.00 End Station: 55+33.15

| Target | Object Name | Subassembly | Assembly Group |
|---|---|---|---|
| Surfaces | <Click here to set all> | | |
| Target Surface | COMBINED | BasicSideSlopeCutDitch - Right - (11) | RIGHT SIDE |
| Width or Offset Targets | | | |
| Width Alignment | <None> | LaneOutsideSuper - Right - (3) | RIGHT SIDE |
| Width Alignment | <None> | LaneOutsideSuper - Left - (4) | LEFT SIDE |
| Target Alignment of Inside B... | <None> | UrbanSidewalk - Left - (14) | LEFT SIDE |
| Target Alignment of Sidewal... | <None> | UrbanSidewalk - Left - (14) | LEFT SIDE |
| Target Alignment of Outside... | <None> | UrbanSidewalk - Left - (14) | LEFT SIDE |
| Slope or Elevation Targets | | | |
| Outside Elevation Profile | <None> | LaneOutsideSuper - Right - (3) | RIGHT SIDE |
| Outside Elevation Profile | <None> | LaneOutsideSuper - Left - (4) | LEFT SIDE |
| Target Profile of Slope | <None> | UrbanSidewalk - Left - (14) | LEFT SIDE |

OK Cancel Help

**FIGURE 7.12**

A subassembly's Parameters panel contains its values and allows changes to them after attaching the subassembly to the assembly (see Figure 7.13).

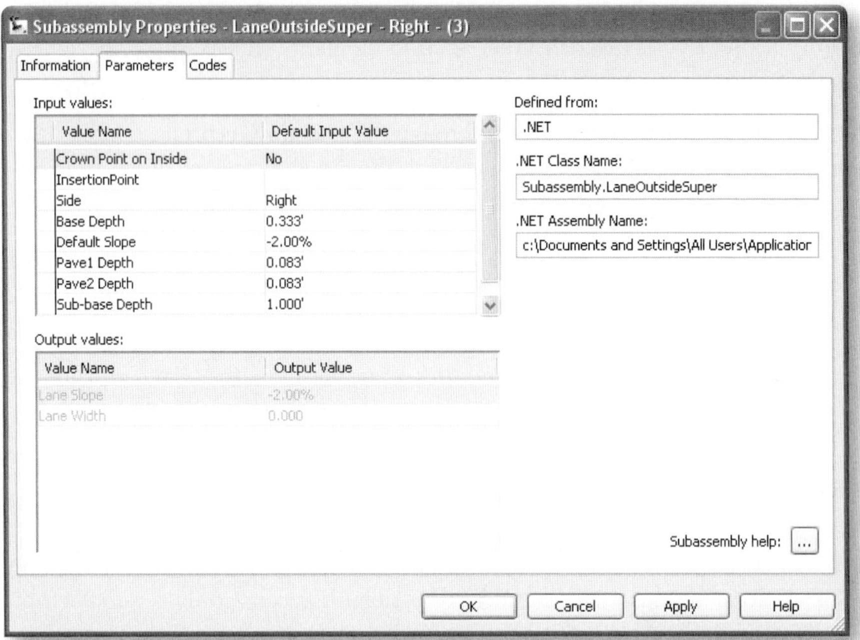


**FIGURE 7.13**

### Edit Feature Settings — Subassembly

Subassembly's Edit Feature Settings, Subassembly Name Template section defines a subassembly's naming template. This is where you insert the side property into the Create From Macro naming template. After inserting the side in the name template, each subassembly will contain its side assignment. When in the Set Targets dialog box, the subassemblies will contain their side as a part of their name.

## ASSEMBLY PROPERTIES

Assembly properties show its construction, subassembly dependencies, and their parameters. These dependencies are especially important in a complex assembly. It is here that subassembly naming pays off. Also, users should rename the assembly groups, giving them more meaningful names (see Figure 7.14).

In the Construction tab, each subassembly parameter is editable by selecting the subassembly on the panel's left side and then editing its parameters on the right.

An assembly and its subassemblies are displayed in tree form. The main branches are right and left side, and below each branch are the attached subassemblies. For convenience, you should rename the two groups to Left and Right side.

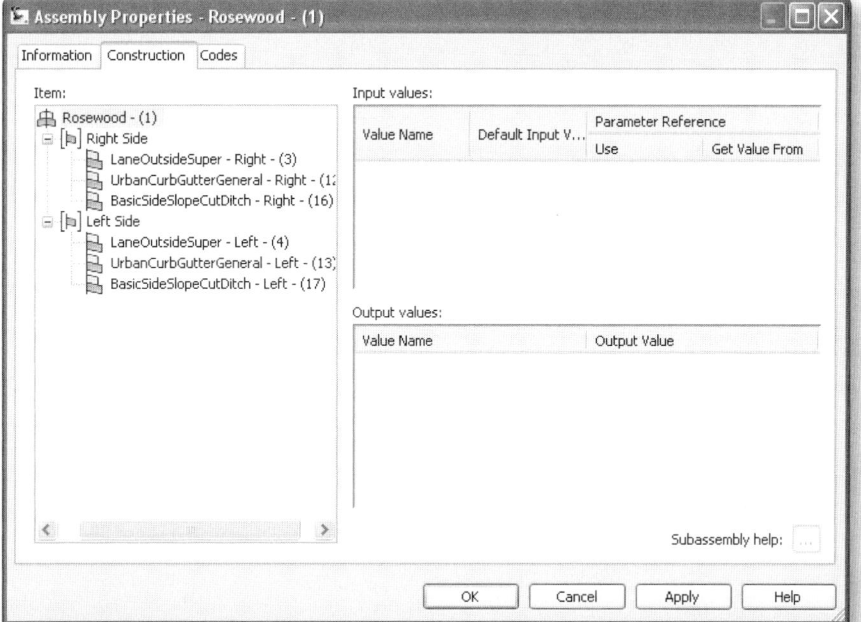


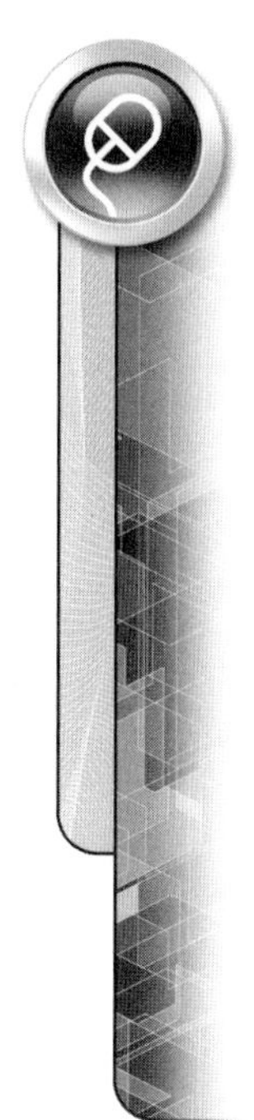

**FIGURE 7.14**

## EXERCISE 7-2

After completing this exercise, you will:

- Be able to create an assembly.
- Be familiar with various subassemblies.
- Be able to select and attach a subassembly to an assembly.
- Be able to edit a subassembly's properties.
- Be able to review an assembly's properties.

### Exercise Setup

This exercise continues with the previous exercise's drawing. If you did not complete the previous exercise, browse to the Chapter 7 folder of the CD that accompanies this textbook and open the *Chapter 07 – Unit 2.dwg* file.

1. If not open, open the previous exercise's drawing, or browse to the Chapter 7 folder of the CD that accompanies this textbook and open the *Chapter 07 – Unit 2* drawing.

### Create the Assembly

Creating an assembly requires naming it, assigning styles, and placing it in the drawing. After placing the assembly in the drawing, you next add the appropriate subassemblies.

1. In the Home tab's, Create Design panel, click the ***Assembly*** icon and from the drop-list, select CREATE ASSEMBLY.
2. In the Create Assembly dialog box, for the assembly name replace "Assembly" with **Rosewood**, leave the Assembly Style as **Basic** and the Code Set Style to **All Codes**, and click ***OK*** to close the dialog box (see Figure 7.15).
3. The command line prompts you for the assembly's location. Select a point just to the left of Rosewood Profile View's lower-left corner.

The assembly is a vertical line with connection symbols at its midpoint.

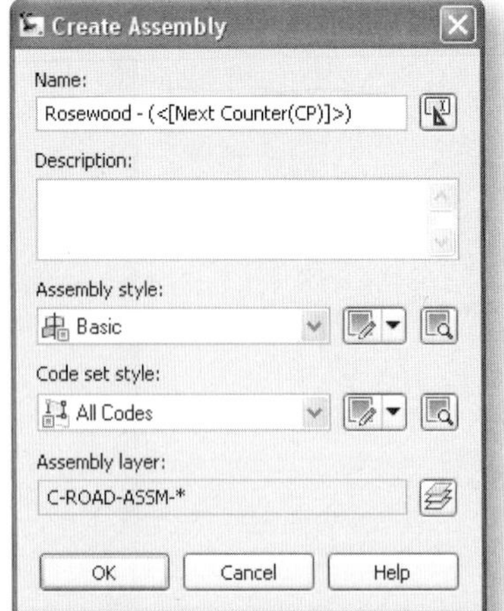


**FIGURE 7.15**

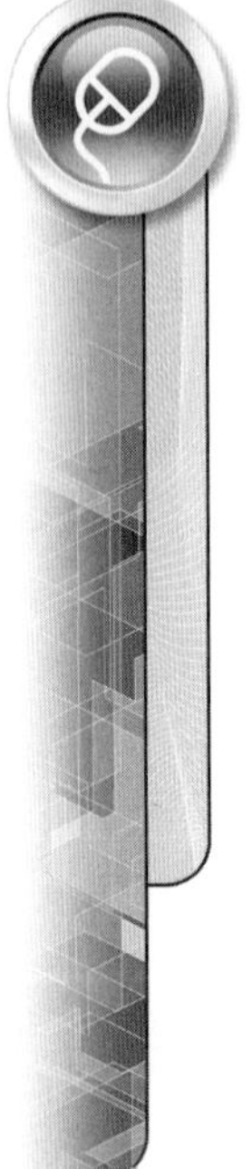

## Add Subassemblies — Travel Lanes

Subassemblies create the roadway section and are from the Civil 3D – Imperial palette.

Travel lanes are 12 feet wide, have four materials with varying thicknesses, and have a −2 percent cross slope. You adjust these parameters before attaching the subassembly to the assembly (see Figure 7.16).

Most subassemblies have right- and left-side property. When you attach a subassembly to the assembly or to an inner subassembly, take care to set the correct side parameter and to be able to view the correct attachment code.

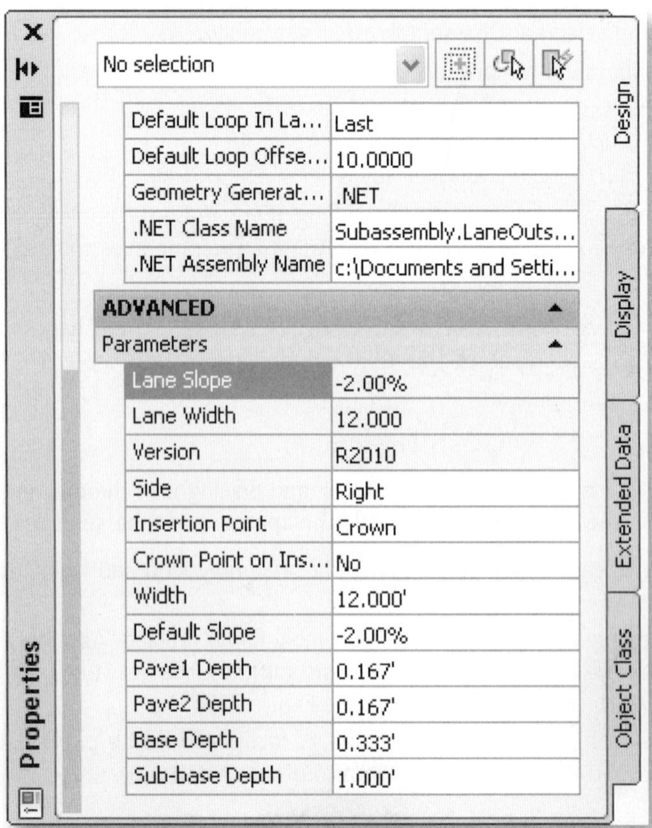


**FIGURE 7.16**

1. If necessary, click the View tab. On the Palettes panel, right of the Toolspace icon select the **Tool Palettes** icon.
2. Click the ***Imperial - Lanes*** tab to display its subassemblies.
3. From the Imperial - Lanes palette, select ***LaneOutsideSuper*** and the Properties palette displays (see Figure 7.16).
4. In Properties, set the Side to **Right**, change the Pave1 and Pave2 depths to **0.167**, and in the drawing, select the assembly to attach the subassembly.

The subassembly attaches the right travelway to the assembly.

5. In the Properties, change the Side to **Left**, and in the drawing, select the assembly to attach the subassembly to its left side.
6. Press ENTER twice to end the routine.
7. Select the ***Prospector*** tab.
8. In Prospector, expand the Subassemblies branch, select the first subassembly from the list, press the right mouse button, and from the shortcut menu, select PROPERTIES....

The Information tab displays a name with the side parameter (see Figure 7.9).

9. Click the ***Parameters*** tab to review its values.
10. Click ***OK*** to exit the dialog box.

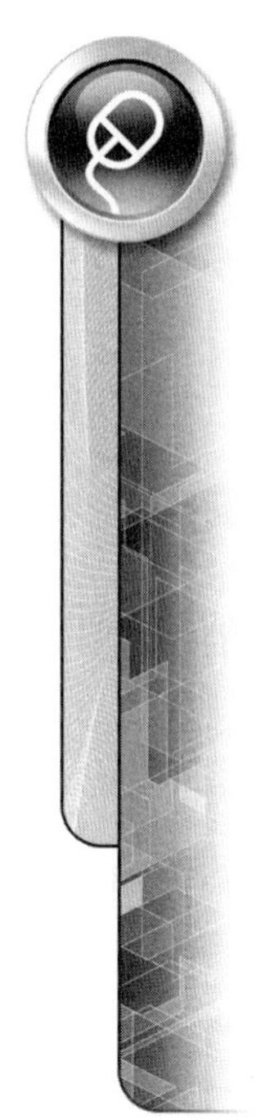

## Add Subassemblies — Curbs

The Curb subassembly attaches to the top of the pavement's red ringlet on the left side.

1. Click the ***Imperial - Curbs*** tab, and from the panel click ***UrbanCurbGutter-General***.
2. In Properties, change the Side property to **Right**. In the drawing, pan to the assembly's right side, and attach the curb to the travelway's top code (red circulate) (see Figure 7.17).

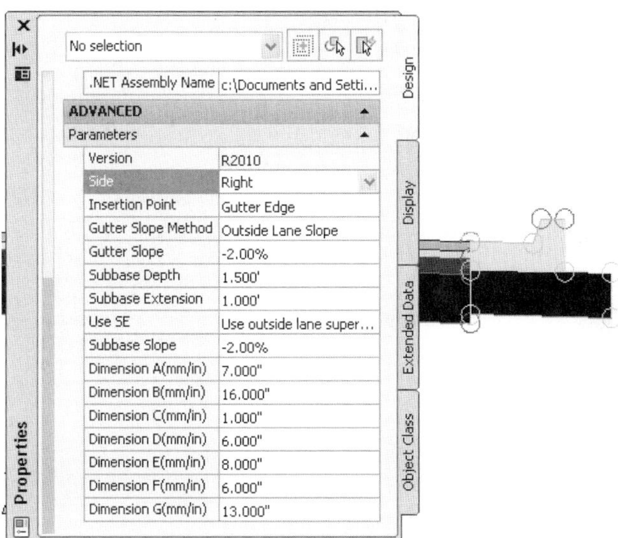


**FIGURE 7.17**

3. In Properties, change the Side to **Left**. In the drawing, pan to the assembly's left side and attach the curb to the top outside travelway red ringlet.
4. Press ENTER twice to end the routine.
   Your subassembly should look like Figure 7.18.

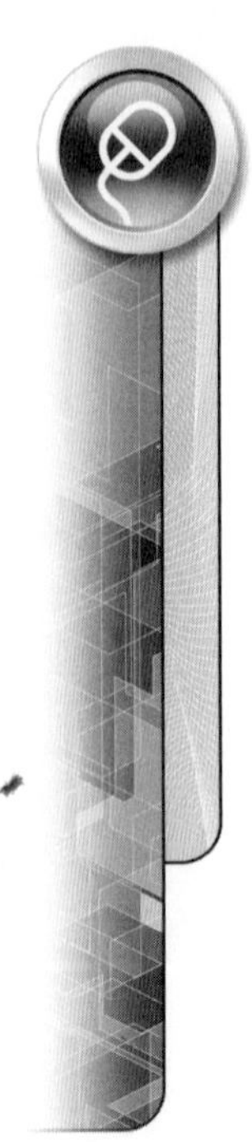

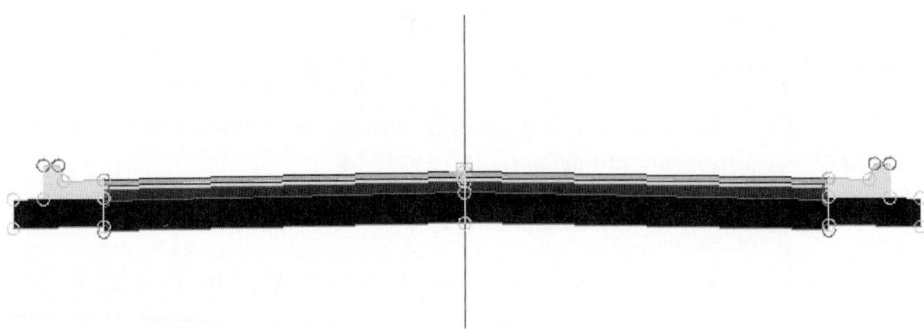

**FIGURE 7.18**

## Add Subassemblies — Ditch and Daylight Slopes

BasicSideSlopeCutDitch daylights to a surface. When attaching the daylight subassembly to the curb's outside back edge, it shows as a sideways "V." When processed, it creates the expected daylight.

1. In the Civil 3D – Imperial tool palette, click the ***Imperial - Basic*** tab, and from the palette, select ***BasicSideSlopeCutDitch***.
2. In the Properties palette, review its settings, change the side to **Right**, and in the drawing, attach the subassembly to the curb's upper-right side.
3. In Properties, change the Side to **Left**, and in the drawing, attach the subassembly to the back of the curb's upper-left side.
4. Press ENTER twice to exit the command.

Your subassembly should look like Figure 7.19.

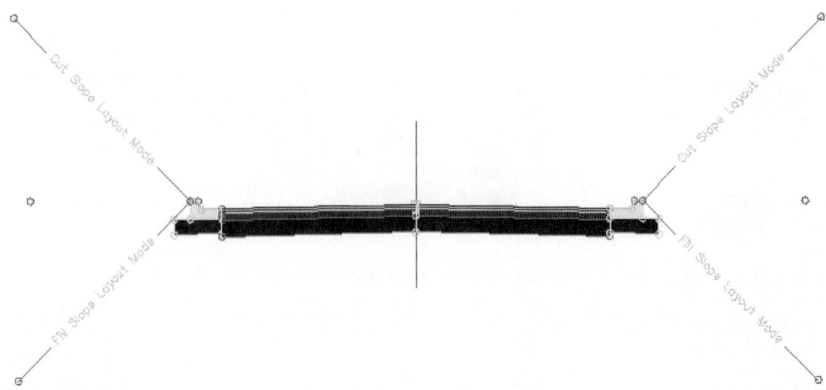


**FIGURE 7.19**

5. At Civil 3D's upper left, Quick Access Toolbar, click the ***Save*** icon to save the drawing.

## Subassembly Properties

The curb's Subbase depth does not match the travelway's depth.

1. If necessary, select the ***Prospector*** tab.
2. If necessary, expand the Subassemblies branch until you view the subassembly list.
3. From the list, select **UrbanCurbGutterGeneral – Right**, press the right mouse button, and from the shortcut menu, select PROPERTIES....
4. In the Subassembly Properties dialog box, select the ***Parameters*** tab, and in the lower right, select the ellipsis for Subassembly Help.

In Help, the Subbase depth represents edge-of-travelway depth. This value needs to be set to 1.6666.

5. Close Help to return to the Subassembly Properties dialog box.
6. Scroll down the Default Input Values list, locate the value for **Subbase Depth**, and change its value to **1.6666**.
7. Click ***OK*** to exit the dialog box.
8. Repeat Steps 3 through 7 and update the values for **UrbanCurbGutterGeneral – Left**.
9. At Civil 3D's upper left, Quick Access Toolbar, click the ***Save*** icon to save the drawing.

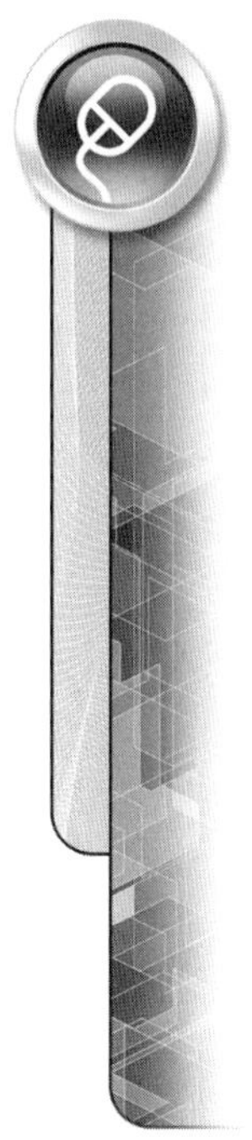

## Assembly Properties

Assembly Properties' Construction tab lists the assembly's subassemblies and their side.

1. In Prospector, expand the Assemblies branch.
2. From the list, click **Rosewood - (1)**, press the right mouse button, and from the shortcut menu, select PROPERTIES....
3. In the Properties dialog box, click the ***Construction*** tab.
4. Click the heading Group – (1), press the right mouse button, and from the shortcut menu, select RENAME. Rename the group **Right Side**.
5. Click the heading Group – (2), press the right mouse button, and from the shortcut menu, select RENAME. Rename the group **Left Side** (see Figure 7.14), and click ***OK***.
6. At Civil 3D's upper left, Quick Access Toolbar, click the ***Save*** icon to save the drawing.

This ends the assembly exercise. Next, you will create a corridor from the horizontal and vertical alignment data and the assembly.

## SUMMARY

- The first step in a road section is to create an assembly.
- After creating an Assembly, you next attach subassemblies from the assembly outward.
- Each subassembly has a Right or Left property. Set this property before you attach the subassembly to an assembly or to a more central subassembly.
- Each subassembly has parameters. Set them before you attach the subassembly to an assembly or to a more central subassembly.
- Each subassembly has properties and all of its parameters are available for editing.
- Each assembly has properties that contain the right and left subassembly attachments. Assembly properties displays all subassemblies and makes their parameters available for editing.

## UNIT 3: CREATING A SIMPLE CORRIDOR

The final step is creating the corridor with one of the Create Corridor routines. A corridor combines each roadway element's settings and parameters and produces a model. When using the Create Simple Corridor command, a Create Simple Corridor dialog box is displayed. It sets the corridor's name, style, and layer (see Figure 7.20). When exiting the Create Simple Corridor dialog box, the user selects the three corridor elements: horizontal and vertical alignment, and the assembly.

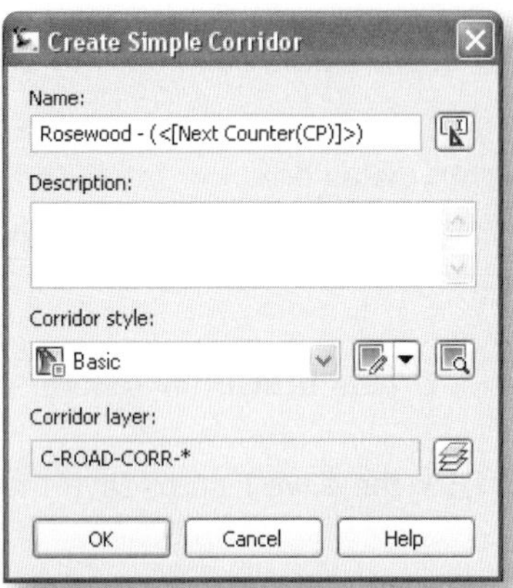


**FIGURE 7.20**

Next, the Target Mapping dialog box opens and prompts you for a review of corridor elements (alignment(s), profile(s), and assemblies) and sets target values. Most simple corridors require that the daylight surface be named (see Figure 7.21).

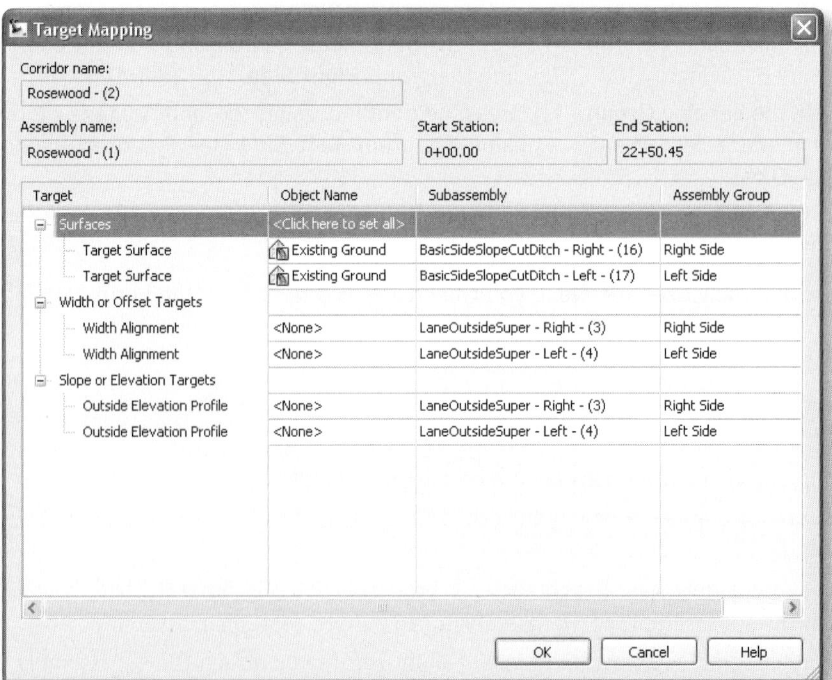


**FIGURE 7.21**

When you are viewing a corridor with Object Viewer or 3D Orbit, Civil 3D displays the corridor sections as assemblies with strings connecting them (see Figure 7.22).

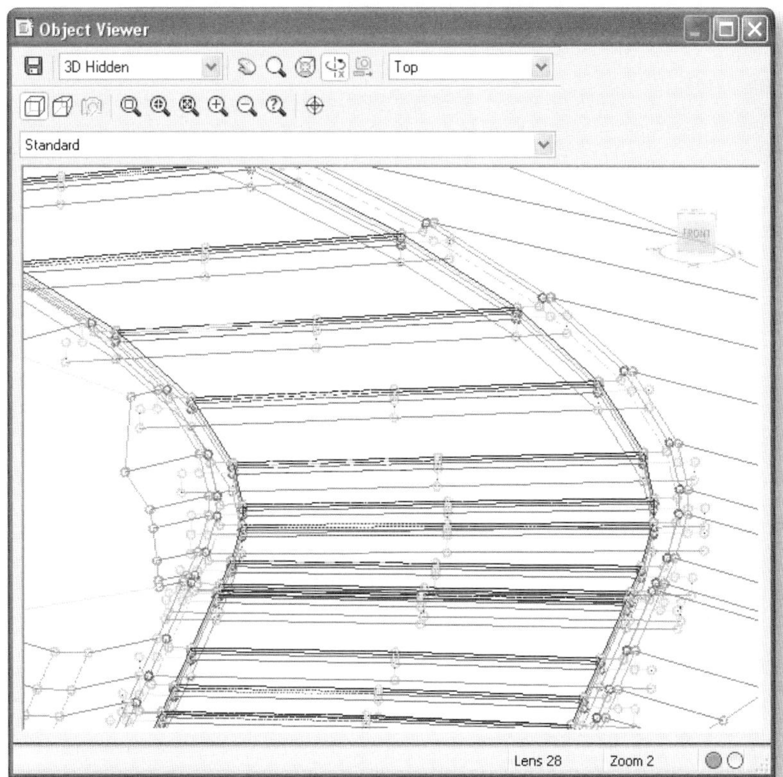


**FIGURE 7.22**

It is evident from Figure 7.22 that points (crown, gutter, etc.) act as eyelets (the subassembly point codes) through which threads are strung (feature lines). These threads are data for many new Civil 3D objects, including feature lines, design annotation, points, and surfaces.

## CORRIDOR PROPERTIES

Corridors have properties that represent their construction.

### Parameters

Parameters contains the corridor regions (see Figure 7.23). There may be times when you want to use different assemblies at various locations along a corridor. This is where roadway regions are made. Regions represent alignment transitions, mergers, or portions of a cul-de-sac. Each region can have its own assembly, frequency, targets, and overrides.

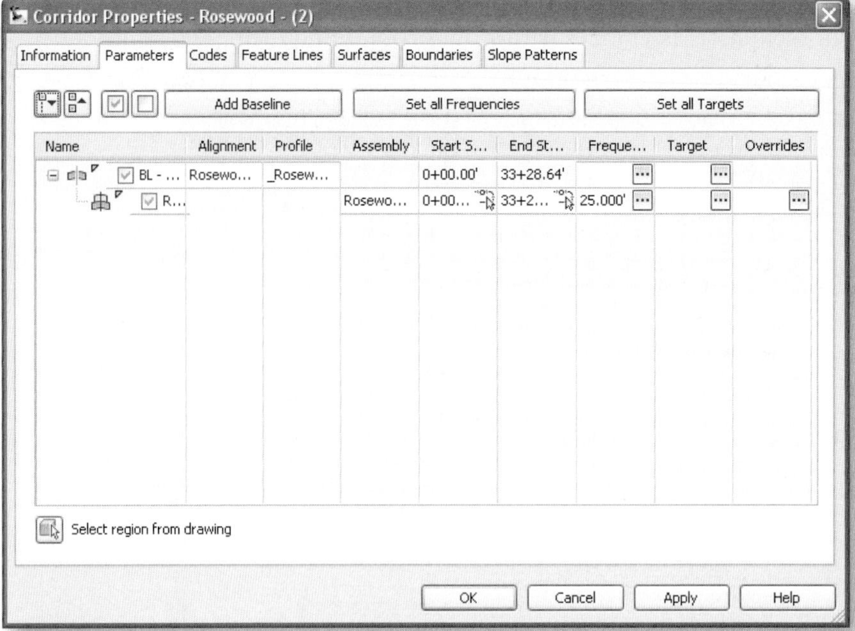


**FIGURE 7.23**

## Codes

Codes lists all points, links, and shapes in a corridor. The entries reflect the assembly's subassemblies codes (see Figure 7.24).

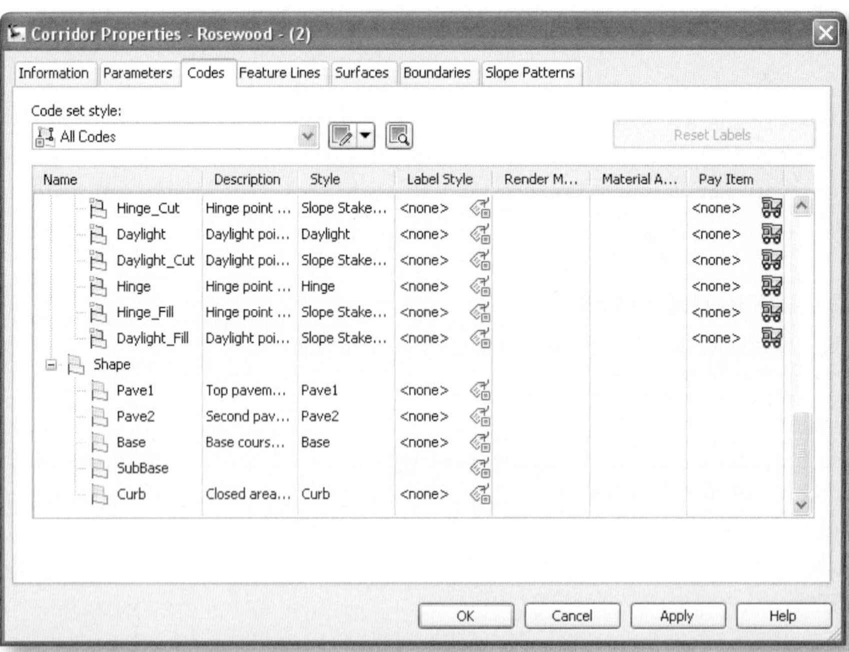


**FIGURE 7.24**

## Feature Lines

Feature Lines are strings or threads that pass through subassembly point codes. Their names are the same as the points through which they pass (see Figure 7.25). At the panel's bottom are two settings that affect complex sections that use a point code multiple times. Branching Inward and Outward affect how these lines merge. Merging Inward means the feature lines, if they are not present in the next section, should merge to the next interior point of the same description. Outward is the opposite of inward: If more points are present in the next section, they connect to the outside points of the same name. Connect extra points is a toggle that tells feature lines to join between sections when there are varying numbers of the same point code between two sequential sections.

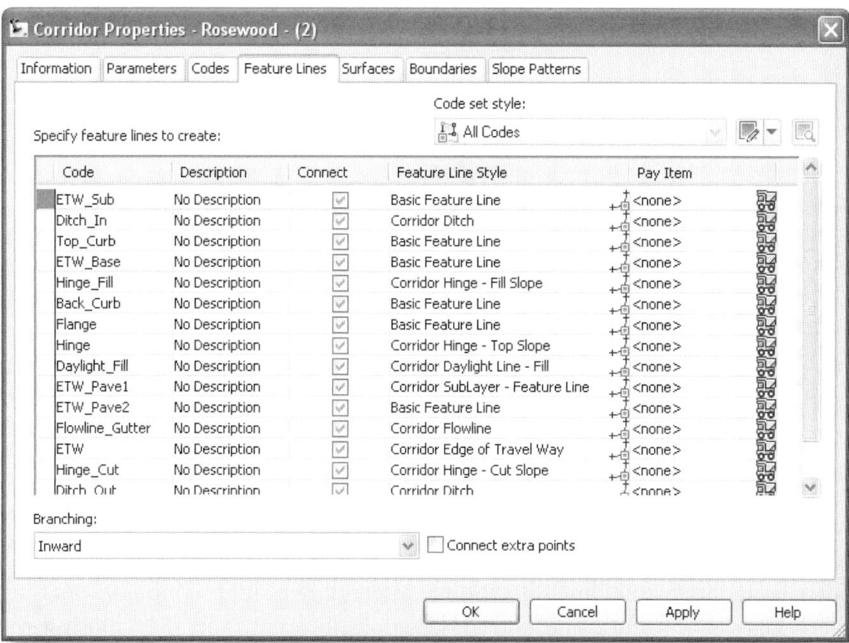


**FIGURE 7.25**

## Surfaces

A corridor model contains surface data (see Figure 7.26). Surface data comes from any section link or feature line. For example, the corridor top link represents the road's top surface. A surface that represents the road's top uses the subassembly's top link as its data source. This surface would represent the final road product. Another important corridor surface is datum, which represents the limit of cut and fill. When comparing datum and the existing ground surfaces, an earthworks quantity is the result.

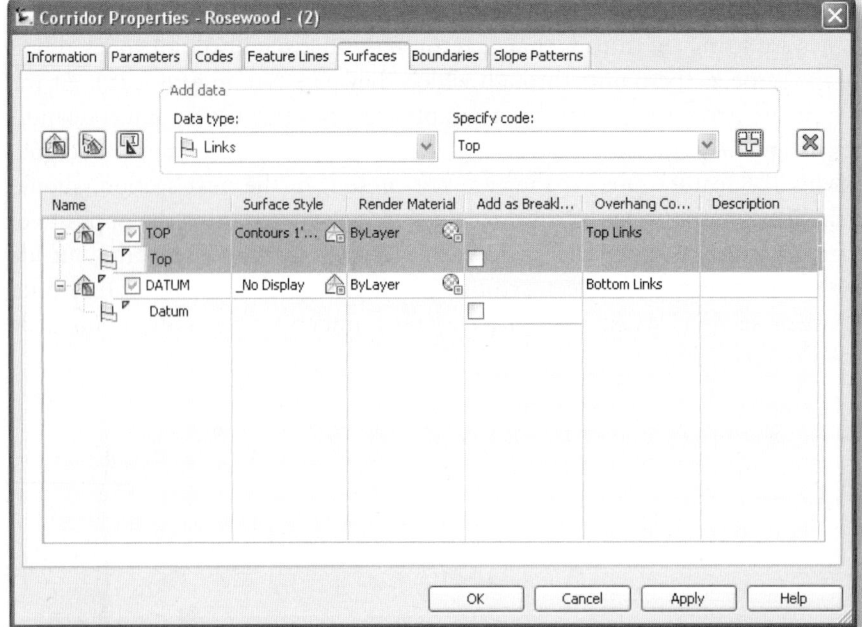


FIGURE 7.26

## Boundaries

As with all surface data, there is a need to control spurious triangulation around a surface's periphery. Corridor feature lines are viable boundaries for roadway surfaces. The most common boundary is daylight (see Figure 7.27). This boundary focuses the triangulation within the daylight boundary and produces a clean corridor surface.

If the assembly is asymmetrical, there is no one feature line enclosing the corridor. To create a boundary for an asymmetrical assembly use the Add Interactively ... command. By selecting a boundary feature line's beginning and ending point on one side, crossing over to the other side's feature line end and beginning point, and then closing the boundary, you create an interactive boundary from two different feature line codes.

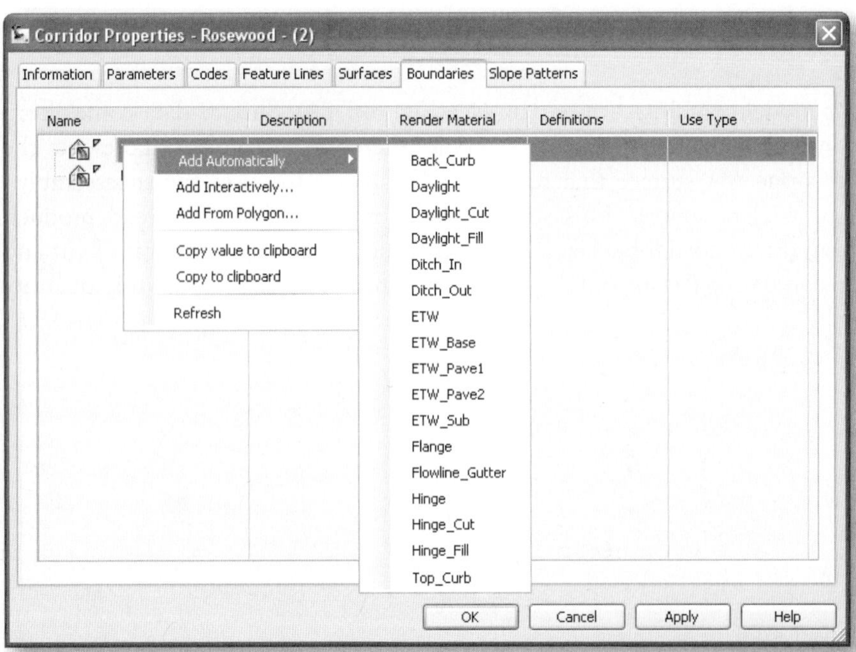


FIGURE 7.27

### Slope Patterns

Slope Patterns are symbols that indicate the slope type along a corridor's path (see Figure 7.28). Settings' Multipurpose Styles define their patterns.

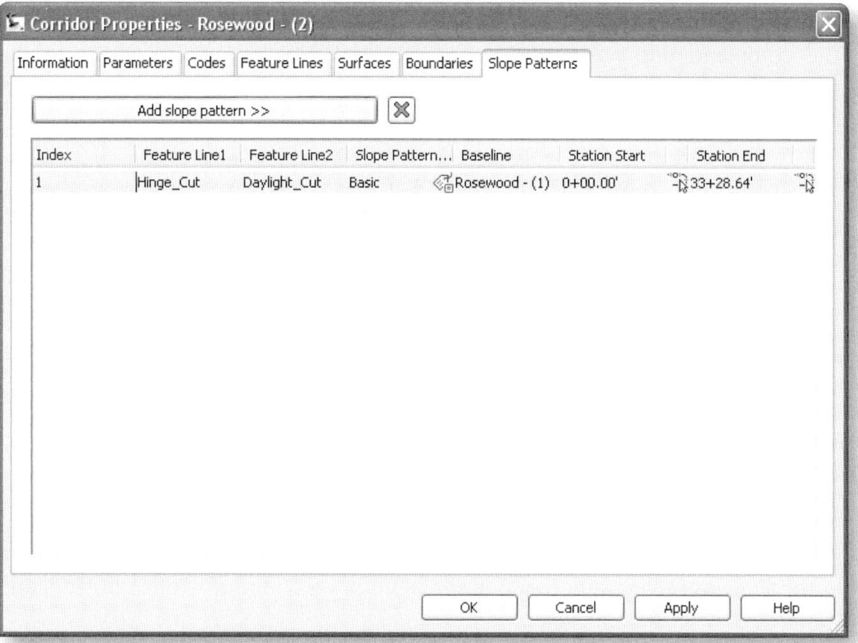


**FIGURE 7.28**

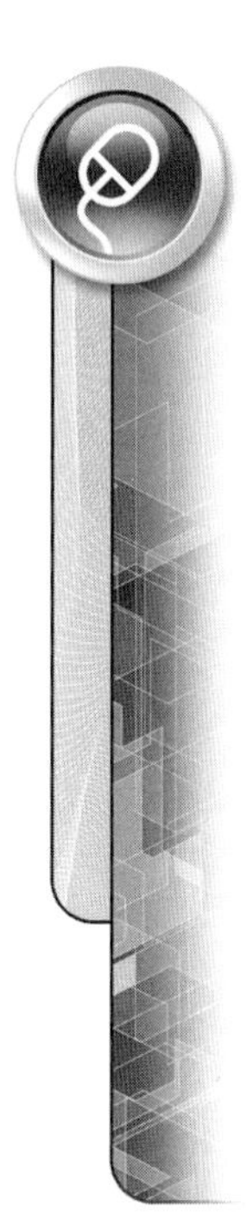

## EXERCISE 7-3

After completing this exercise, you will:

- Be able to create a simple corridor.
- Be familiar with corridor properties.

### Exercise Setup

This exercise continues with the previous exercise's drawing. If you did not complete the previous exercise, browse to the Chapter 7 folder of the CD that accompanies this textbook and open the *Chapter 07 – Unit 3.dwg* file.

1. If not open, open the previous exercise's drawing or browse to the Chapter 7 folder of the CD that accompanies this textbook and open the *Chapter 07 – Unit 3* drawing.

### Create Simple Corridor

1. If necessary, click the ***Home*** tab. At the Create Design panel's middle, click the ***Corridor*** icon and from the drop-list, select CREATE SIMPLE CORRIDOR.
2. In the Create Simple Corridor dialog box, leave the counter and replace "Corridor" with **Rosewood**. Enter a short description, and click ***OK*** to begin identifying the corridor components.

3. The command line prompts you for an alignment. If you are able to select the alignment from the drawing, select it or right mouse click, from the list select **Rosewood - (1)**, and click ***OK*** to continue.
4. The command line prompts you for a profile. If you are selecting the profile from the drawing, select it or right mouse click, from the list select **_Rosewood Preliminary - (1)**, and click ***OK*** to continue.
5. The command line prompts you for an assembly. If you are selecting the assembly from the drawing, select it or right mouse click, from the list select **Rosewood - (1)**, and click ***OK*** to continue.
6. The Target Mapping dialog box opens. For Surfaces, Object Name click in the cell <Click here to set all>. The Pick a Surface dialog box opens. From the list, select the **Existing Ground** surface, and click ***OK*** to specify the surface.
7. Click ***OK*** to build the corridor.

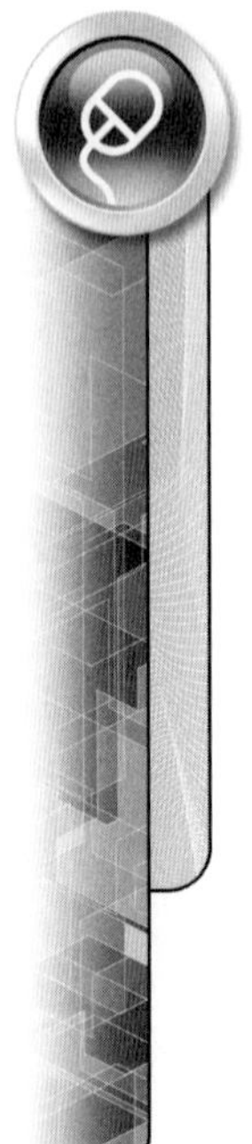

The corridor builds and is displayed as a mesh.

8. At Civil 3D's top left, Quick Access Toolbar, click the ***Save*** icon to save the drawing.

## Viewing a Corridor

The corridor is a 3D model and can be viewed with Object Viewer or with 3D Orbit.

1. Use the ZOOM and PAN commands to view the site's corridor.
2. In the drawing, select any corridor segment, press the right mouse button, and from the shortcut menu, select OBJECT VIEWER....
3. In Object Viewer, begin viewing by clicking and holding the left mouse button down in the area of curve 2, and slowly moving the cursor toward the center of the object viewer. This tilts the roadway toward your point of view.
4. Just before you are viewing the corridor edge-on, release the left mouse button.
5. After viewing the corridor, select the Object Viewer's **X** to close it.

## Corridor Properties

Corridor properties range from alignments, parameters, codes, surfaces, boundaries, slope patterns, and feature lines.

1. In the drawing, select any corridor segment, press the right mouse button, and from the shortcut menu, select CORRIDOR PROPERTIES....
2. In the Corridor Properties dialog box, click the ***Parameters*** tab.

Parameters list the alignment, profile, and assembly assignments for the corridor. You change the name of any of these elements by clicking the named element and selecting a new alignment, profile, etc., from a list in a dialog box.

3. Click **Set All Targets** at the panel's top right.

This panel reports the Right and Left target surface (existing ground) and any alignment or profile assignment.

4. Click ***OK*** to exit the Target Mapping dialog box and return to the Corridor Properties dialog box.
5. Select the ***Codes*** tab.

Codes lists all corridor links, points, and shapes.

6. Click the ***Feature Lines*** tab.

Feature lines are threads that pass through the assembly point codes. Each code has a name and a specific function. You can import feature lines and use them to design a surface or use them as grading objects.

7. Click the ***Surfaces*** tab.

This panel creates surfaces from feature lines and links.

8. Click the ***Boundaries*** tab.

This panel defines surface boundary control.

9. Click the ***Slope Patterns*** tab.

If you want to include a corridor slope pattern, define it here. You identify where it occurs in the corridor and what pattern to use.

10. Click ***Cancel*** to exit the Corridor Properties dialog box.

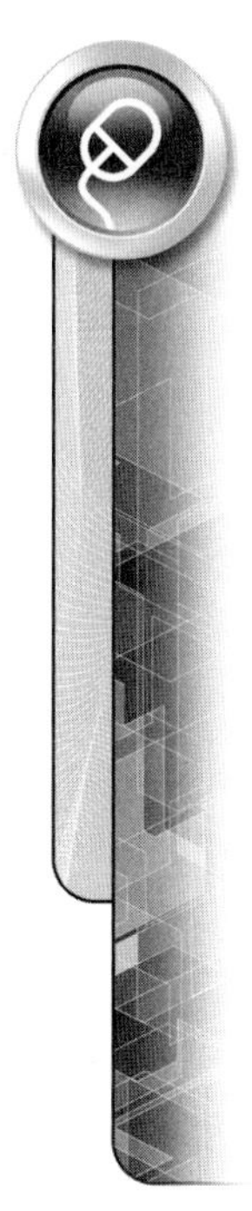

## Feature Lines

The All Codes style assigns each feature line a style.

1. On the Ribbon, click the ***View*** tab and in the Views panel, select and restore the named view **Proposed Starting Point.**
2. Use the ZOOM and PAN commands to better view the corridor (see Figure 7.29).

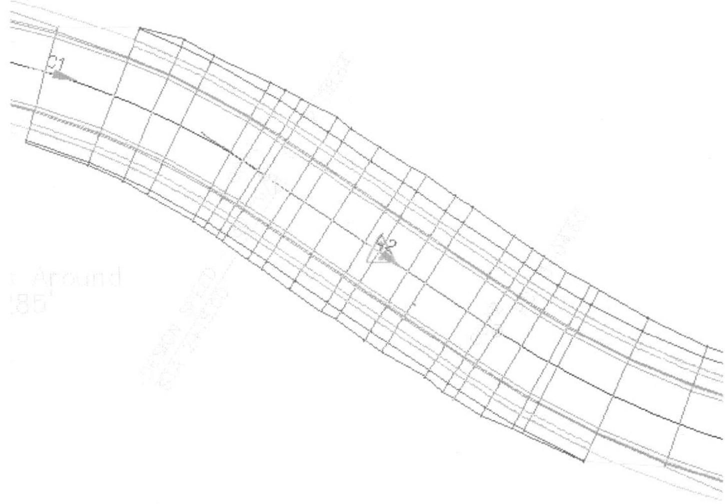

**FIGURE 7.29**

3. In the drawing, select any corridor segment, press the right mouse button, and from the shortcut menu, select CORRIDOR PROPERTIES....
4. In the Corridor Properties dialog box, click the ***Feature Lines*** tab to review its contents.

This panel lists all corridor feature lines. Each has a name, is visible, and has styles.

5. Click ***OK*** to exit the Corridor Properties dialog box.
6. At Civil 3D's top left, Quick Access Toolbar, click the ***Save*** icon to save the drawing.

This ends the corridor exercise on creating and viewing its properties. Once you have a corridor, you can create new objects from it. The next unit reviews and edits a corridor's values.

## SUMMARY

- Creating a simple corridor is a one-step process.
- The alignments, profiles, assemblies, and surfaces all contain data and constraints, and the Create Corridor command blends these elements together to create a corridor.
- A corridor has an extensive properties list.
- The Corridor Properties dialog box creates corridor surfaces, surface boundaries, and adds corridor slope markings.

## UNIT 4: CORRIDOR REVIEW AND EDIT

The Corridor Properties dialog box displays values that describe a corridor's overall character. However, it may be necessary to review and possibly edit some corridor values. The Corridors' Section Editor provides these tools.

The Corridor - Section Editor tab controls the currently viewed station, sets section overrides, extends section edits to a station range, and has a Corridor Parameters palette that edits a section's values (see Figure 7.30). When you select the corridor from the screen, the Corridor Section Editor tab displays when you select Corridor Section Editor from the Modify panel. If you select the corridor from a list, the editor displays the first corridor section. The Station Selection panel's middle identifies the current section and has controls that affect which section is displayed. The panel's left side identifies the baseline name, and its center identifies the current section. The icons on the station's right control the location to which the changes are extended: this station only; a range of stations; or the entire corridor. The Parameter Editor, adds or deletes points, links, or shapes or modifies a subassembly's parameters.

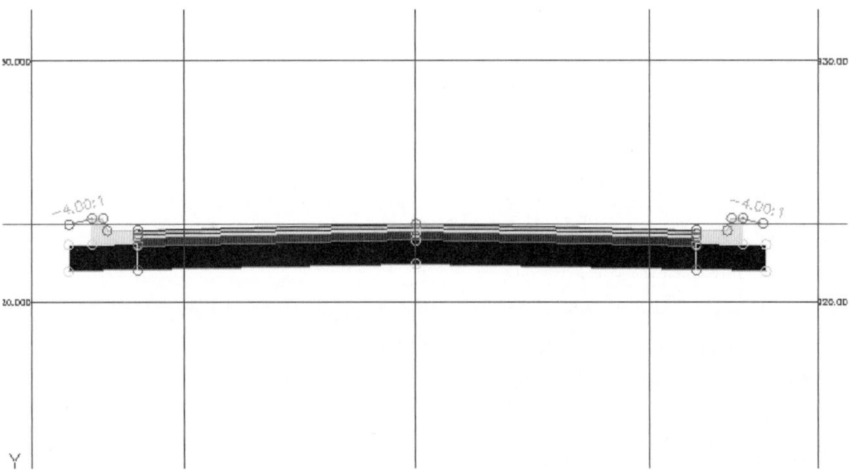


**FIGURE 7.30**

The Section Editor and Parameter Editor are interactive. When you make a parameter change, the section reacts to the change. Changes can apply to the current section or extend over a station range. You can measure the assembly distances in the editor.

## VIEW/EDIT CORRIDOR SECTION OPTIONS

The Edit/View Options affect the view's grid, labeling, vertical scale, colors, and text (see Figure 7.31). The Code Set Style affects the labeling and assembly display properties. It would be best to create a style just for section viewing. Most Code Set Styles use text that is too big to be useful.

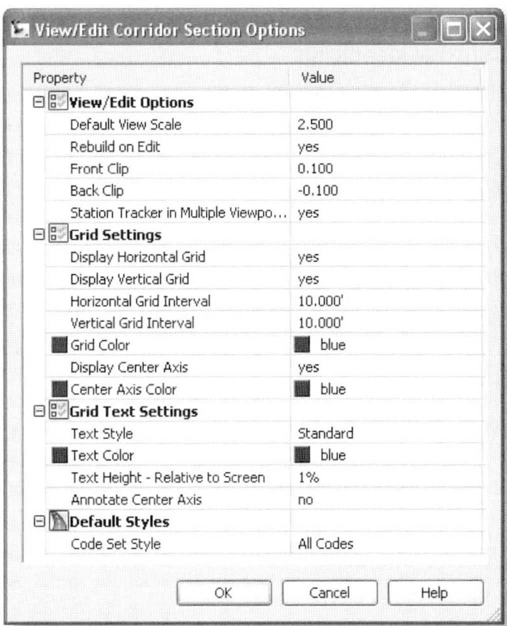


**FIGURE 7.31**

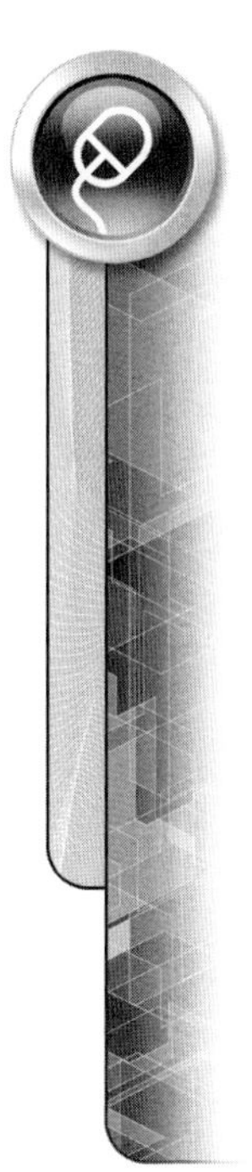

## EXERCISE 7-4

After completing this exercise, you will:

- Be able to use the View/Edit Section Editor.
- Be familiar with section editing.

### Exercise Setup

This exercise continues with the previous exercise's drawing. If you did not complete the previous exercise, browse to the Chapter 7 folder of the CD that accompanies this textbook and open the *Chapter 07 – Unit 4.dwg* file.

1. If not open, open the previous exercise's drawing, or browse to the Chapter 7 folder of the CD that accompanies this textbook and open the *Chapter 07 – Unit 4* drawing.

### Viewing Corridor Sections

Viewing sections is one method of road design review.

1. In the drawing select the corridor. The Ribbon displays the Corridor panel and from the Modify panel, select CORRIDOR SECTION EDITOR.

The Section Editor ribbon panel displays with the section. In the Ribbon, to the current station's right and left, single arrows display the next ahead or back stations.

2. On the Station Selection panel, click the **right single arrow** a few times to move to higher stations.
3. On the Station Selection panel, click the **left single arrow** a few times to move to lower stations.
4. On the Station Selection panel, to the right of the current section, click the **barred-arrow** icon to view the corridor's last station.
5. On the Station Selection panel, to the left of the current section, click the **barred-arrow** icon to view the first corridor station.
6. On the View Tools panel, select the ***Edit/View Options*** icon.
7. In the View/Edit Corridor Section Options dialog box, View/Edit Options section, set the Default View Scale to **2.0**, adjust any colors, Default Styles section, set the Code Set Style to **All Codes with No Shading**, and click ***OK*** to return to the section view.

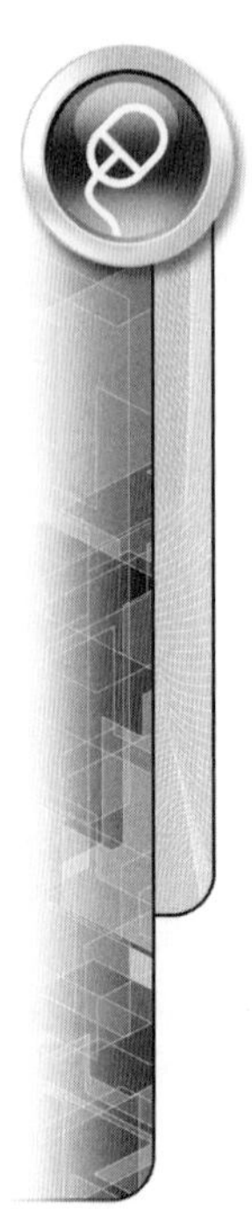

## Editing Sections

Edit each subassembly's parameters in an Parameters Panorama. Any changed value is applied to the current section or to a range of stations.

1. In the Station Selection panel, in the panel's middle, click the stations drop-list arrow and select station **9+25**.
2. With 9+25 as the current station and in the Corridor Edit Tools panel, click the ***Parameter Editor*** icon to display the Corridor Parameters palette. You may need to expand the palette to view its contents.
3. In Corridor Parameters palette, locate the Right Lane's Default Slope entry. Click in the Value cell for slope and change it to **6** (percent).

This changes the right lane slope to 6 percent. After you make the change, a check appears in the override box, an information icon appears, and the section responds by raising its right pavement to the 6 percent slope.

4. In Corridor Parameters, click the **X** to close the palette.
5. In the Corridor Edit Tools panel, at its middle right, select ***Apply to a Station Range***.
6. In the Apply to a Range of Stations dialog box, set the station range from **925** to **1350** and click ***OK*** to apply the change.
7. In the Station Selection panel, to the right of the current station, click the **single arrow** and view the changes to station **14+00**.
8. To remove the change and restore the original value, in the Station Selection panel, click the stations drop-list arrow, and select station **9+25** from the list.
9. In the Corridor Edit Tools panel, click the ***Parameter Editor*** icon. In the Right Lane Default Slope entry, **uncheck** the override (the value returns to −2%), and close the Parameter Editor.
10. In the Corridor Edit Tools panel, click the ***Apply to a Station Range*** icon, apply the change to the full station range, and click ***OK*** to exit the Apply to a Range of Stations dialog box.
11. Verify the change by reviewing the stations between 9+50 and 13+50.
12. At the Section Editor tab's right, click the **X** to close the panel.

This exercise ends corridor review and editing. The Corridor Section Editor tweaks values for any corridor parameter. After each edit, a section reacts to the change. Changes can apply to the current station or to a range of stations.

## SUMMARY

- The Corridor Section Editor displays each section with all of its subassembly parameters.
- When changing a subassembly parameter, the section responds by showing the change in the view, and when you exit the viewer, the change is shown in the corridor.
- Changes apply to the current section or to a range of stations.

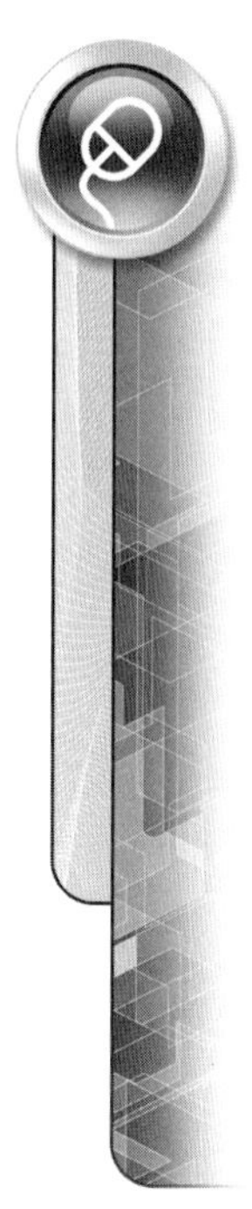

## UNIT 5: OBJECTS FROM A CORRIDOR

Feature lines (the strings or threads) that appear between corridor sections connecting the subassembly point codes are an integral part of a corridor's definition. Feature lines can become additional alignments, profiles, polylines, grading feature lines, and surface data.

To create objects from a corridor, use the commands from the Ribbon's Modify - Corridor tab. These commands create feature lines, points, alignments, profiles, and polylines from the corridor (see Figure 7.32).

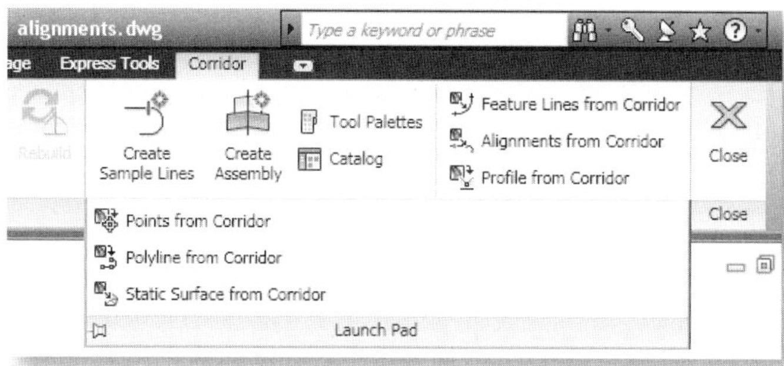


**FIGURE 7.32**

Each corridor feature line displays a tooltip that identifies its name. The tooltip is an effective way to determine which feature line to export (see Figure 7.33).

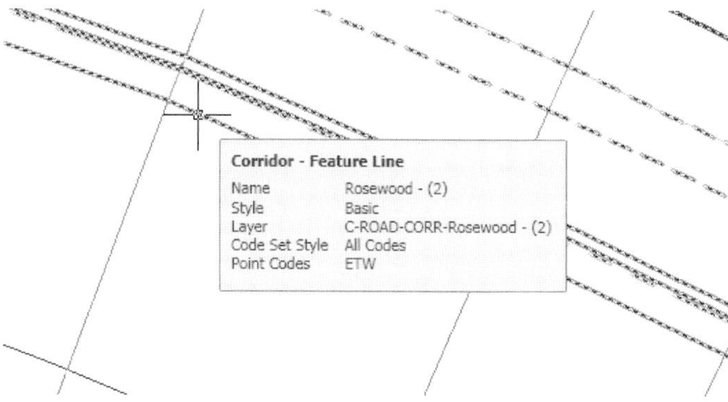


**FIGURE 7.33**

If you are selecting where there is more than one feature line, a Select a Feature Line dialog box opens. From the list, select the desired feature line (see Figure 7.34).

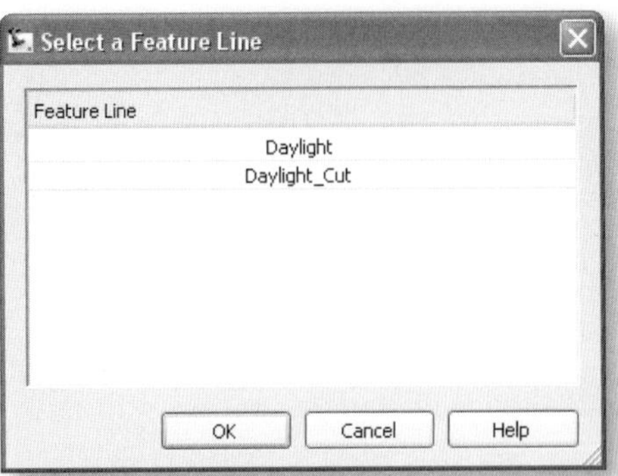


**FIGURE 7.34**

## CREATE POLYLINE FROM CORRIDOR

This routine creates a 3D polyline whose elevations are corridor feature line elevations. The polyline resides on the current layer.

## CREATE GRADING FEATURE LINE FROM CORRIDOR

The resulting feature line has corridor feature line elevations. The routine places the resulting feature line on the feature line style's layer.

## CREATE ALIGNMENT FROM CORRIDOR

This routine creates an alignment object whose path is the same as the corridor feature line. The routine displays the Create Alignment – from Objects dialog box and names the alignment using the feature line's name (see Figure 7.35). After defining an alignment, the routine prompts you to create a profile.

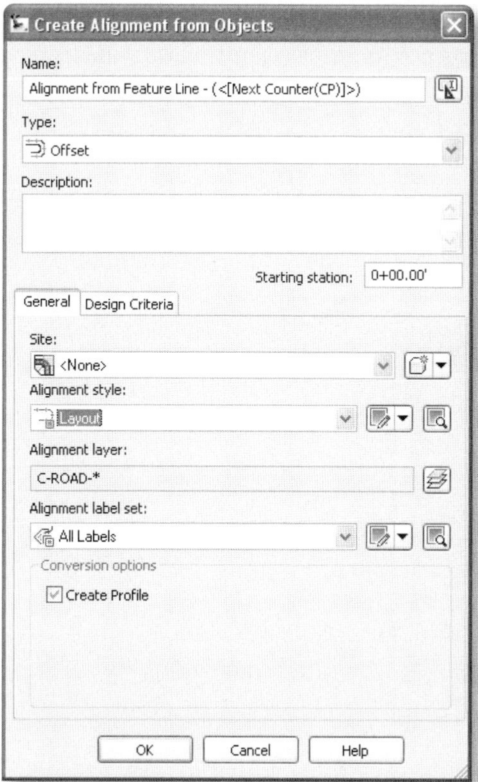


**FIGURE 7.35**

## CREATE PROFILE FROM CORRIDOR

This routine creates a profile whose path and elevations are the same as the corridor feature line. The routine displays the Create Profile – Draw New dialog box (see Figure 7.36).

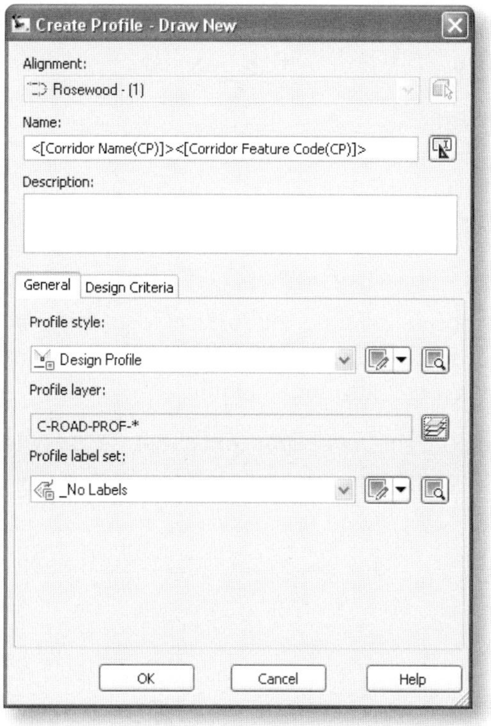

**FIGURE 7.36**

## CREATE COGO POINTS FROM CORRIDOR

Create COGO Points from Corridor exports points whose elevations are the feature line elevations. The points are at each corridor section. After selecting a corridor, the Create COGO Points dialog box opens (see Figure 7.37). This dialog box lists all of the corridor feature lines and creates points only for those toggled on. The routine can create a point group from the exported points. Users can export points for the entire corridor's length, or for a range of stations.

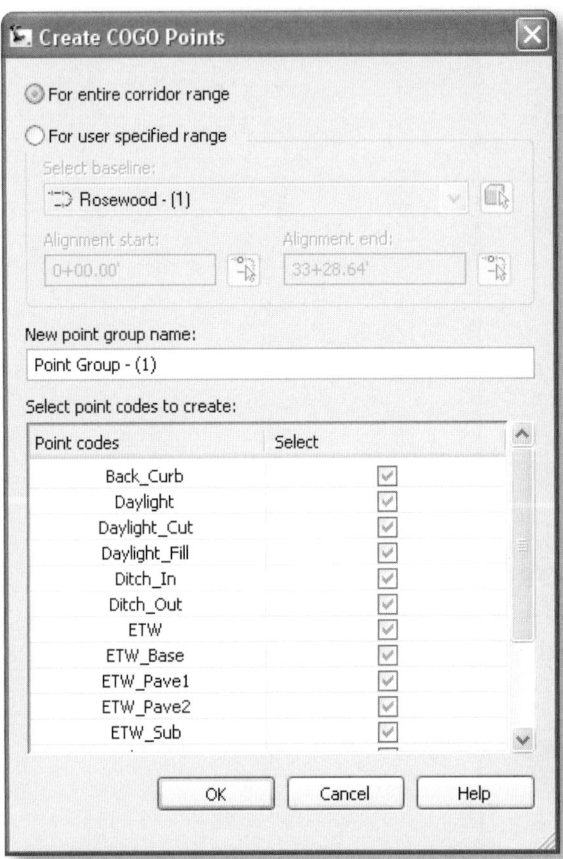


FIGURE 7.37

## CORRIDOR SURFACES

Creating corridor surfaces is a three-step process: naming the surface, identifying its data, and defining boundary control. A corridor's surface is displayed in Prospector's Surfaces list and is dynamically updated when the corridor is rebuilt.

Corridor Properties' Surfaces panel assigns the surface's name and the surface's data (see Figure 7.38). You should turn on both top and bottom overhang corrections.

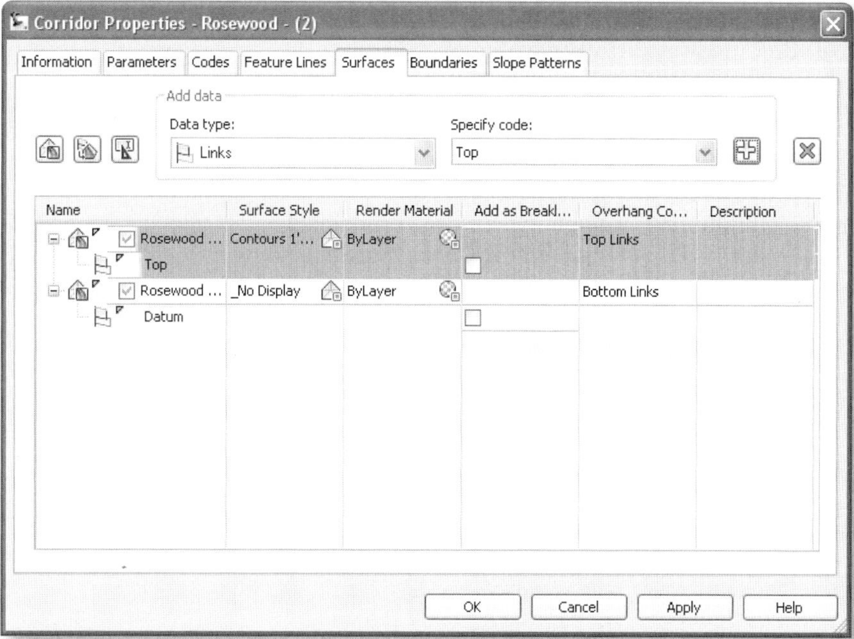


**FIGURE 7.38**

The Boundaries panel has a corresponding entry for each surface. A boundary limits triangulation to the data between the two outermost feature lines (see Figure 7.39). If a boundary is not automatically defined, the Add Interactively routine uses a jig to draw the boundary around the corridor.

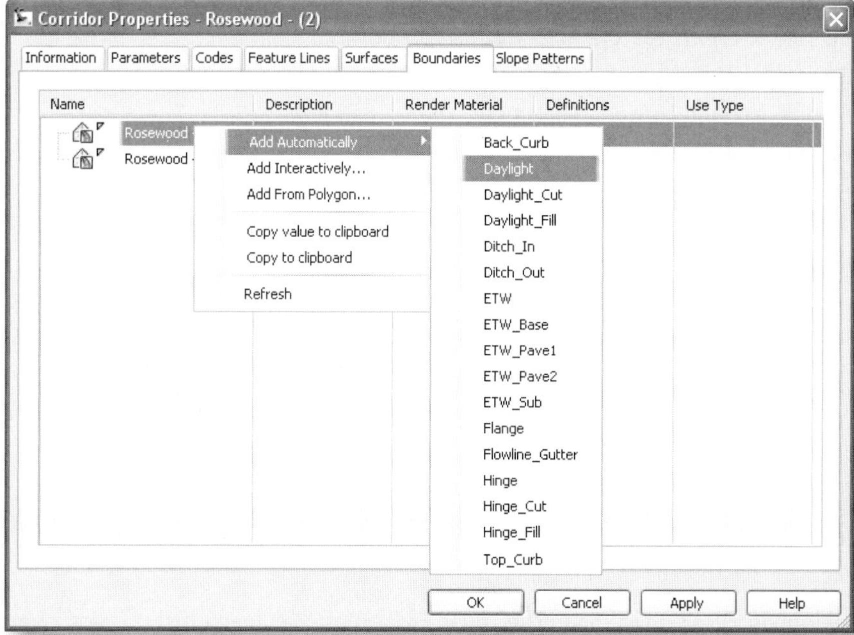


**FIGURE 7.39**

## CALCULATING OVERALL EARTHWORK VOLUMES

Civil 3D has three roadway earthwork volumes calculators. The first is a simple comparison between two surfaces. The Ribbon's Modify - Surfaces tab has the Volumes command. This command displays a Composite Volumes calculator vista (see Figure 7.40). To start the calculation process, create a volume entry, assign the two surfaces, and click in a Cut or Fill output cell. This calculator calculates an up-to-the-minute volume without having to define a volume surface.

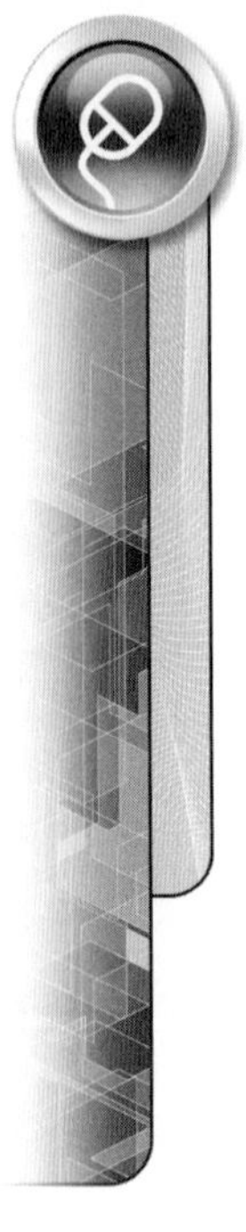

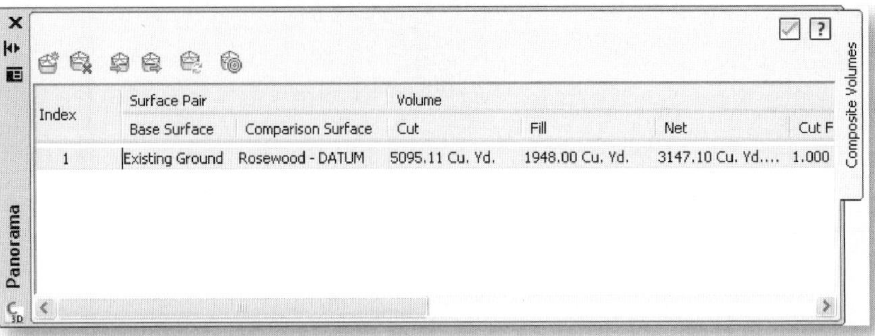


**FIGURE 7.40**

The second volume method defines a volume surface to calculate earthworks. The third method creates a quantity takeoff report from the corridor sections and is a part of the next chapter.

## EXERCISE 7-5

After completing this exercise, you will:

- Be able to create 3D polylines from a corridor.
- Be able to create feature lines from a corridor.
- Be able to create points from a corridor.
- Be able to build corridor surfaces.
- Be able to add a corridor surface boundary.
- Be able to add a corridor slope pattern.

### Exercise Setup

This exercise continues with the previous exercise's drawing. If you did not complete the previous exercise, browse to the Chapter 7 folder of the CD that accompanies this textbook and open the drawing *Chapter 07 – Unit 5.dwg* file.

1. If not open, open the previous exercise's drawing or browse to the Chapter 7 folder of the CD that accompanies this textbook and open the *Chapter 07 – Unit 5* drawing.

### Polyline

You can export a 3D polyline by selecting a corridor feature line.

1. In Ribbon's Home tab, open the Layer Properties Manager, create a new layer, and name it **3D poly**. Make it the current layer, assign it a color, and click the **X** to exit the palette.

2. If necessary, from Ribbon's View tab, Views panel, select and restore the named view **Proposed Starting Point**.

The Daylight feature line represents the intersection of the ditch slope out to an intersection with Existing Ground.

3. Click the ***Modify*** tab, on the ***Design*** panel, click the ***Corridor*** icon. From the ***Corridor*** tab, click the Launch Pad panel's drop-list arrow, and from the shortcut menu, select POLYLINE FROM CORRIDOR.
4. In the drawing, select the corridor's north **Daylight Line** (yellow), the Select a Feature Line dialog box displays, select **Daylight**, click ***OK*** to create the polyline, and press ENTER to exit the routine.
5. Select the new object, press the right mouse button, and from the shortcut menu, select PROPERTIES....

The new object is a 3D polyline with daylight feature elevations.

6. Click the **X** to exit the Properties palette.
7. Erase the just-created 3D polyline.

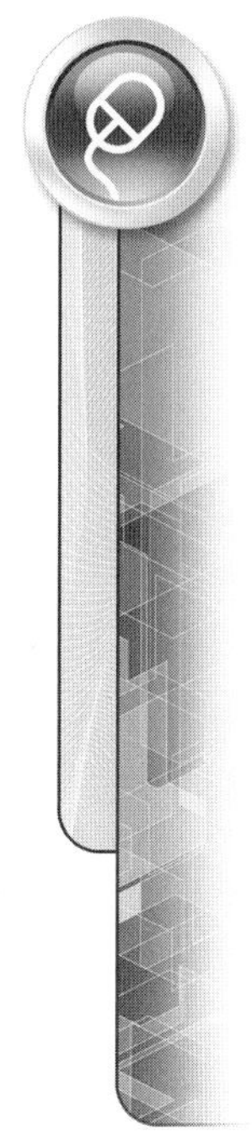

## Feature Lines

Feature Lines are similar to 3D polylines. They have varying elevations at each vertex, but feature lines are custom Civil 3D grading objects. The Feature Line export routine works exactly like the Polyline from Corridor command.

1. From the Launch Pad panel, at its top right, select FEATURE LINES FROM CORRIDOR.
2. Select the corridor's north daylight line, if the Select a Feature Line dialog box displays, select **Daylight**, click ***OK*** to display the Create Feature Line from Corridor dialog box.
3. Change the style to Corridor Daylight, toggle **OFF** the Smoothing option, click ***OK***, to continue, and press ENTER to exit.
4. Select the new object, press the right mouse button, and from the shortcut menu, select PROPERTIES....

The object is listed as an Auto Corridor Feature Line. The feature line is on C-TOPO-FEAT, the default feature line layer.

5. Click **X** to close the Properties palette.
6. Erase the just-created feature line.

## Points

Points that represent corridor elevations are critical to placing a design in the field. The point numbers should be offset from other existing points and they should have their own point group.

1. Click the ***Prospector*** tab.
2. In Prospector, select the **Points** heading, press the right mouse button, and from the shortcut menu, select CREATE ... to display the Create Points toolbar.
3. At the toolbar's right, click the Expand the Create Points dialog chevron.
4. Expand the Point Identity section, change the **Next Point Number** to **10000**, and press ENTER.
5. At the toolbar's right, click the Collapse the Create Points dialog chevron.
6. Close the Create Points toolbar.

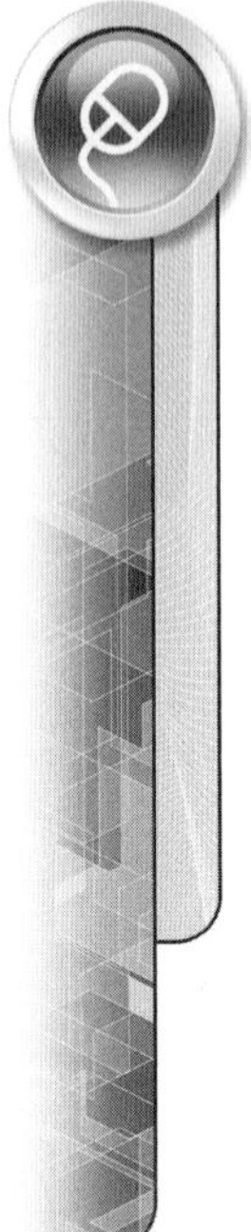

7. Click the Ribbon's ***Corridor*** tab, click the Launch Pad panel's drop-list arrow, and select POINTS FROM CORRIDOR.
8. The command line prompts you to identify a corridor. In the drawing, select a Rosewood – (1) corridor segment.
9. In the Create COGO Points dialog box, set the station range to the entire corridor length, for the point group name enter **Design Points**, and export the following point codes:

   Back_Curb (Back of Curb)

   Daylight

   ETW

   Flowline_Gutter
10. Click ***OK*** to create the points and their point group.

The points are displayed and are in the Design Points point group.

11. In Prospector, expand Point Groups, select the **Design Points** point group, press the right mouse button, and from the shortcut menu, select EDIT POINTS....
12. After reviewing the points, click the Panorama's **X** to hide it.
13. In the point groups list, select the **Design Points** point group, press the right mouse button, and from the shortcut menu, select DELETE POINTS....
14. Click ***OK*** in the Are you sure? dialog box to delete the points.
15. With the **Design Points** point group still highlighted, press the right mouse button, and from the shortcut menu, select DELETE... to delete the point group.
16. Click ***Yes*** in the Are you sure? dialog box.

## Slope Patterns

When preparing for a submission or a presentation, you may want to show slope patterns along the corridor's path.

1. If necessary, on the Ribbon, click the ***View*** tab. In the Views panel, select and restore the named view **Proposed Starting Point**.
2. If necessary, use the PAN and ZOOM commands until you are viewing the corridor cut area (stations 1+25 to 2+50).
3. Use the ZOOM command to better view the northern corridor section. You will be identifying the **Ditch_Out** and **Daylight_Cut** feature lines around station 2+50. Use the tooltip to identify the location of feature lines.
4. In the drawing, select any corridor segment, press the right mouse button, and from the shortcut menu, select CORRIDOR PROPERTIES....
5. Click the ***Slope Patterns*** tab.
6. At the dialog box's top, click ***Add Slope Pattern*** >>.
7. The command line prompts you for the first feature line. On the corridor's north side, select the **Ditch_Out** line just north of the Ditch_In feature line.
8. The command line prompts you for the second feature line. On the corridor's north side, select the **Daylight_Cut**. The Select a Feature Line dialog box opens. Select **Daylight_Cut** and click ***OK***.

The Corridor Properties dialog box again opens and lists the first and second feature lines.

9. In the dialog box, in the Slope Pattern Style column, click the ***Slope Pattern Style*** icon, and in the Pick Style dialog box, select the style, **Slope Schemes**. Click ***OK*** to return to the Corridor Properties dialog box.
10. Again, at the panel's top, click ***Add Slope Pattern >>.***

You may have to pan to the corridor's southern side to select the next two lines.

11. In the drawing on the corridor's southern side, select the same two lines, **Ditch_Out** and **Daylight_Cut**.

The Corridor Properties dialog box again opens, listing the first and second feature lines.

12. For the new entry, click the ***Slope Pattern Style*** icon, in the Pick Style dialog box select the style, **Slope Schemes**, and click ***OK*** to close it and return to Corridor Properties dialog box.
13. Click ***OK*** to exit the dialog box and assign the slope patterns to the corridor.
14. Use the ZOOM and PAN commands to better view the pattern.
15. At Civil 3D's top left, Quick Access Toolbar, click the ***Save*** icon to save the drawing.

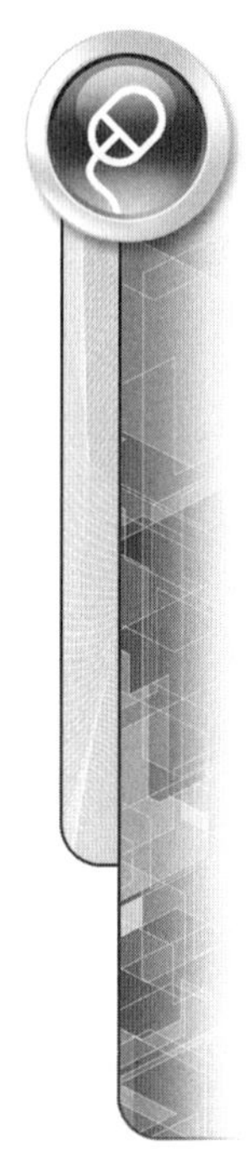

## Corridor Surfaces

Calculating corridor earthworks compares existing ground and the corridor datum elevations.

1. In the drawing, select any corridor segment, press the right mouse button, and from the shortcut menu, select CORRIDOR PROPERTIES....
2. Click the ***Surfaces*** tab.

This tab defines the surface names and assigns their data.

3. At the dialog box's top left, click the ***Create a Corridor Surface*** icon.

A surface entry appears.

4. Click in the Name column and change the surface name to **Rosewood - Top**.
5. In the dialog box top center, Add Data area, set the Data Type to **Links**. Set Specify Code to **Top**, and click the **+** (plus sign) to assign the link data.
6. Assign Rosewood - Top the surface style **Contours 1' and 5' (Design)**.
7. Click in the Overhang Correction column and select **Top Links** (see Figure 7.38).
8. To make a second surface, at the dialog box's top left, click the **Create a Corridor Surface** icon.
9. Click in the second surface's Name column and change the surface name to **Rosewood - Datum**. At the top center of the dialog box, Add Data area and set the Data Type to Links. Set Specify Code to **Datum** and click the **+** (plus sign) to add the link data.
10. Assign Rosewood - Datum the surface style **_No Display**.
11. Change Overhang Correction to **Bottom Links**.
12. Click the ***Boundaries*** tab.
13. Select **Rosewood - Top**, press the right mouse button, and in the Add Automatically flyout menu from the boundaries list select **Daylight**. Make sure the Use Type is **Outside Boundary** (see Figure 7.39).
14. Repeat the previous step and add the same boundary to **Rosewood - Datum**.
15. Click ***OK*** to create the surfaces and exit the dialog box.
16. Place your cursor over the corridor and review the corridor's station and surface elevations.

17. At Civil 3D's top left, Quick Access Toolbar, click the ***Save*** icon to save the drawing.

Any corridor changes automatically update the corridor surfaces.

## Calculate an Earthworks Volume

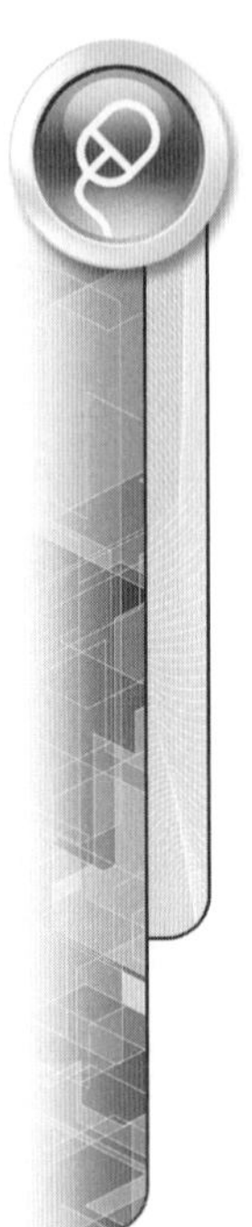

1. If necessary, click the Ribbon's ***Modify*** tab. On the Ground Data panel, click the ***Surface*** icon. In the Surface tab's Analyze panel, select VOLUMES.
2. In the Composite Volumes vista's upper left, click the ***Create new volume entry*** icon.

Clicking in the Base and Comparison Surface cells displays a surface drop-list.

3. In Base Surface, click twice in <select surface>, and from the surfaces list, select **Existing Ground**.
4. In Comparison Surface, click in <select surface>, and from the surfaces list, select **Rosewood - Datum**.
5. If necessary, click in the Cut cell to calculate a volume.
6. Click the Panorama's **X** to hide it.
7. Click the Tool Palettes's **X** to close the panel.
8. At the Surface tab's right, click the **X** to close the panel.
9. At Civil 3D's top left, Quick Access Toolbar, click the ***Save*** icon to save the drawing.

## SUMMARY

- A simple corridor combines alignment, profile, and assembly data and parameters to create a corridor (roadway model).
- A corridor is dynamic and changes if any one of its dependent objects (alignment, profile, or assembly) change.
- Corridor Properties assigns feature line styles to each corridor feature line.
- Corridor Properties assigns slope patterns to corridor cut and fill areas.
- Corridor Properties creates surfaces, surface data, and surface boundaries.
- The Surfaces menu, Utilities flyout Volumes ... command compares any two surfaces to calculate an earthwork volume.

This ends the simple corridors chapter. A corridor is a dynamic roadway model and reacts to any change to its data. Next, you learn how to document critical roadway cross sections.

# CHAPTER 8

# Cross-Sections and Volumes

## INTRODUCTION

This chapter is the last one that discusses Civil 3D roadway design basics. Chapters 5, 6, and 7 cover creating the horizontal alignment, profiles, and corridor. This chapter covers creating cross-sections that document the design. Cross-sections document the roadway assembly and the existing ground to the alignment's right and left.

## OBJECTIVES

This chapter focuses on the following topics:

- Roadway Cross-Section Settings
- Sample Line Groups
- Roadway Volumes Review
- Importing Cross-Sections
- Annotating Design Results
- Creating Section Sheets
- Creating Plan and Profile Sheets
- Assigning Pay Items
- Creating Pay Item Reports

### OVERVIEW

The design itself is not necessarily the end of the design process. While a design solves engineering issues, it must be reviewed and documented and it must have calculated volumes (see Figure 8.1). The alignment and profile view document the design's horizontal and side views. The last view, cross-sections, completes the design documentation. This view looks from the design's starting station toward the highest station with an offset from left to right of the centerline. A left offset is always a negative value, and a right offset is always a positive value.

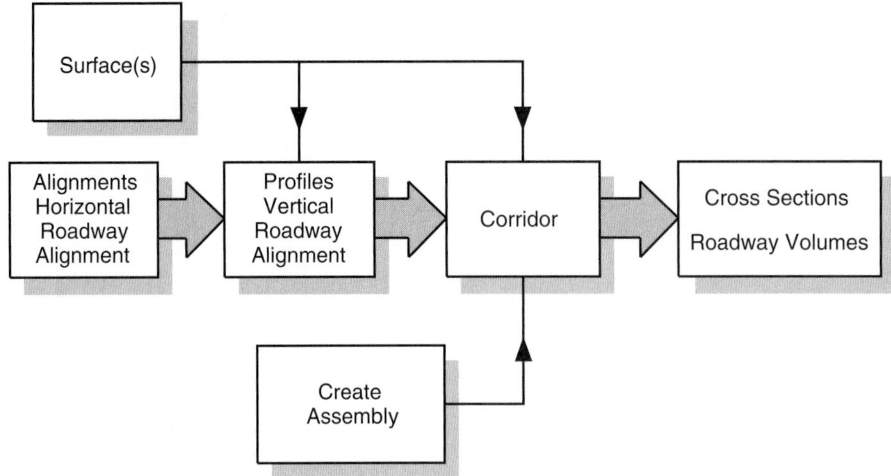


**FIGURE 8.1**

Offset's overall left-to-right distance is known as swath width, and it is wider than the Right-of-Way (ROW). ROW distances may vary along a road's path, but the swath width stays the same. Swath width distance is an office standard or mandated value.

Civil 3D samples a corridor with groups of sample lines. A corridor can have any number of sample line groups documenting a corridor. By applying different styles and sampling different data, each group produces completely different-looking sections. If you add new data to a corridor—for example, a pipe network—the sample lines need to be resampled.

After you create sample lines, next you will create section views. To create section views, you use three methods: individual, as a page(s), or all sections. A section view is a grid with basic annotation. Much like profile views, a section view can have bands. Even though it defines annotation for all grid axes and extra band annotation, it is the Section View style's Display settings that control what is visible.

The process of creating a section view also creates a section, for example, existing ground and the corridor. The section contains a surface and a corridor. Other sampled sections may not be displayed, but are used for material and earthwork volumes (corridor shapes and Datum).

Section annotation comes from two different style groups: Section and General's Multipurpose. The section styles annotate a sampled surface's offsets and elevations, grade breaks, etc. These label types are not frequently used in section views. General's Multipurpose styles use the corridor as their label data: grades/slopes; offsets; and elevations from the roadway assembly points, links, and shapes. These styles are the typical sections view labels.

## Unit 1

Section's settings and styles are the first unit's focus. There are three style types that affect sections: sample lines, section views, and sections. Each type has its own Edit Feature Settings that contain values for each object type. These styles control layer names, sampling, annotation (spot and grade), and page layout.

### Unit 2

The second unit's focus is a sample line group and its sections.

### Unit 3

After sampling the corridor, the section commands create views that contain sampled sections. Unit 3 covers importing sections and their annotation. This unit also discusses section view properties.

### Unit 4

Unit 4 reviews General's Multipurpose styles and what they annotate within a section view.

### Unit 5

This unit focuses on earthwork calculations, material estimates for quantity takeoffs, and pay items.

## UNIT 1: CROSS-SECTION SETTINGS AND STYLES

There are myriad settings and styles for sections and section views. Of the styles, section and section view styles are the most complex.

### EDIT DRAWING SETTINGS

The Edit Drawing Settings dialog box affects basic object layers and their initial values (see Figure 8.2). If a project contains more than one section sample line group, it is best to assign either a prefix or suffix to the base object layer name. This places sample lines, sections, and section views on their own layer.

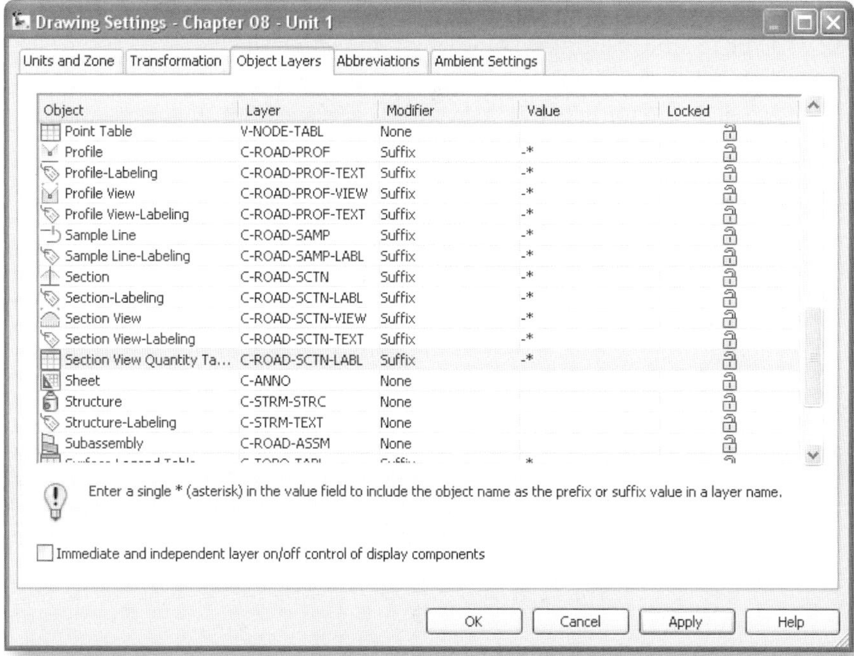


**FIGURE 8.2**

## EDIT FEATURE SETTINGS

Each object type's Edit Feature Settings dialog box assigns initial values to sample line, section view, and section elements. These values affect styles and commands that are lower in their respective branches.

### Sample Line

Sample Line's Edit Feature Settings dialog box includes assigning object and label styles and defining their naming convention (see Figure 8.3).

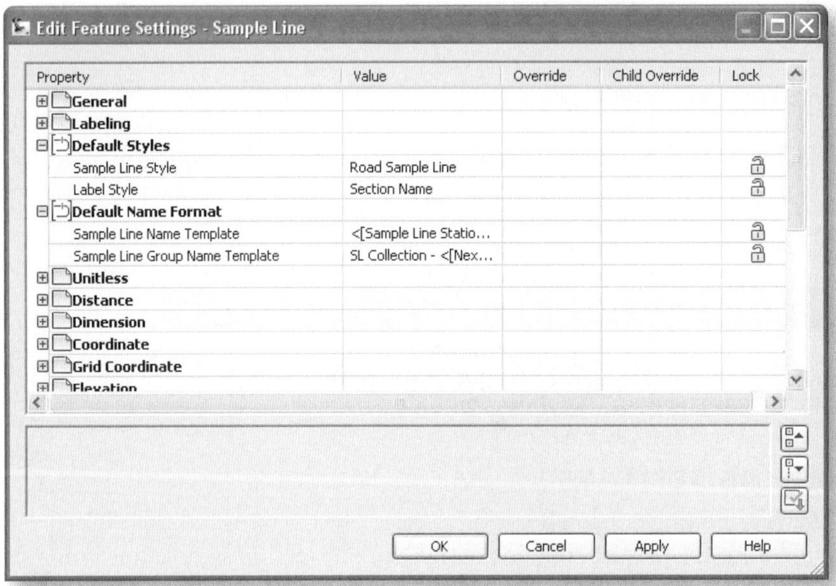


**FIGURE 8.3**

### Section View

Section View's Edit Feature Settings dialog box sets section view object styles, label styles, plotting styles, and the default Add Labels command styles (see Figure 8.4). The settings also assign the initial band style and label placement. Whether working with band or label sets, their names are an alias for a collection of styles.

A Section Label Set defines what section labels appear in the section view. Group Plot Style defines how to organize the sections in a drawing. The last step is defining the section view naming convention.

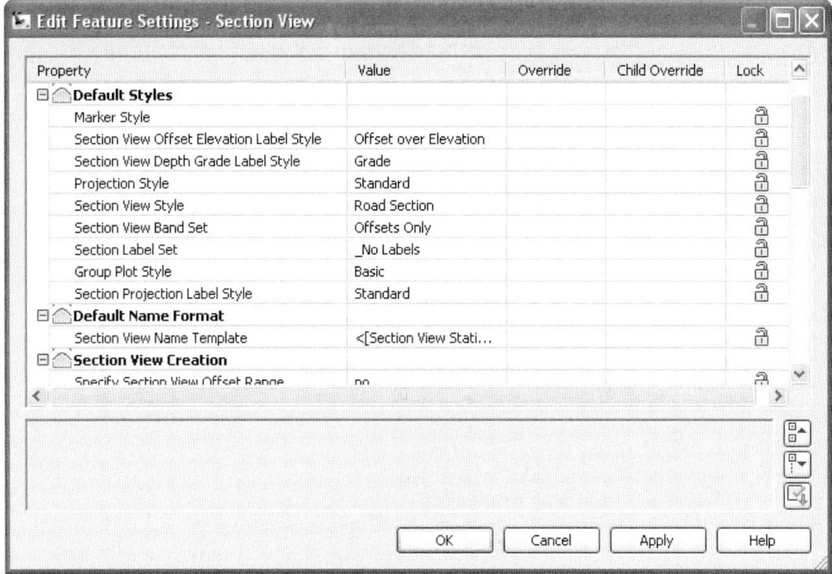


**FIGURE 8.4**

## SECTION

Section's Edit Feature Settings dialog box defines initial styles for a section and their naming convention (see Figure 8.5).

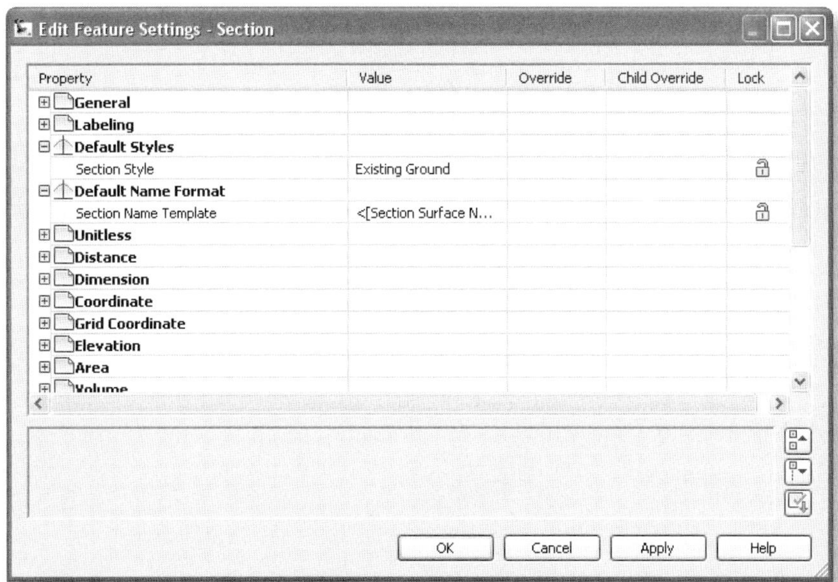


**FIGURE 8.5**

## OBJECT STYLES

Each sample line, section, or section view has an object style. The style defines its shape and component layer names and display properties.

### Sample Line

Sample line object styles assign layers and their properties to the sample line object (see Figure 8.6).

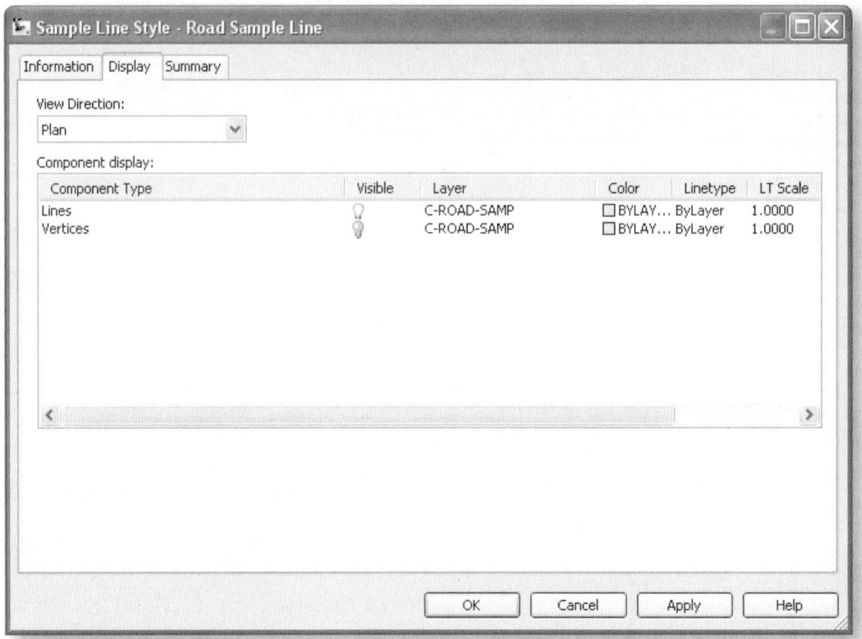


**FIGURE 8.6**

## SECTION VIEW

Section view object styles assign grid properties, title, basic annotation, and layers and their properties.

### Graph Panel

This panel controls section's vertical exaggeration. Traditionally, a vertical scale is a 10:1 ratio (horizontal scale/vertical).

### Grid

Grid defines its clipping parameters, station padding, and axes offset. Clipping removes one or both grids above their line work. Padding places one or more grid lines around the sections, and axes offset pushes the axes outward from the grid (see Figure 8.7).

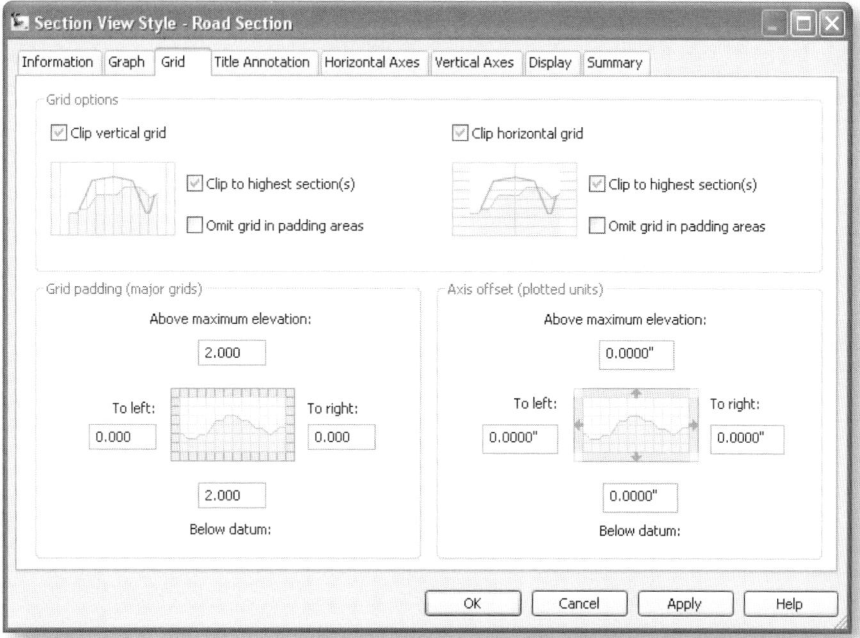


**FIGURE 8.7**

## Title Annotation

Title Annotation's left side defines a section view's title, size, location, text style, justification, and if it has a border. Title Annotation's right side defines axes titles, their content, and properties. See Figure 8.8.

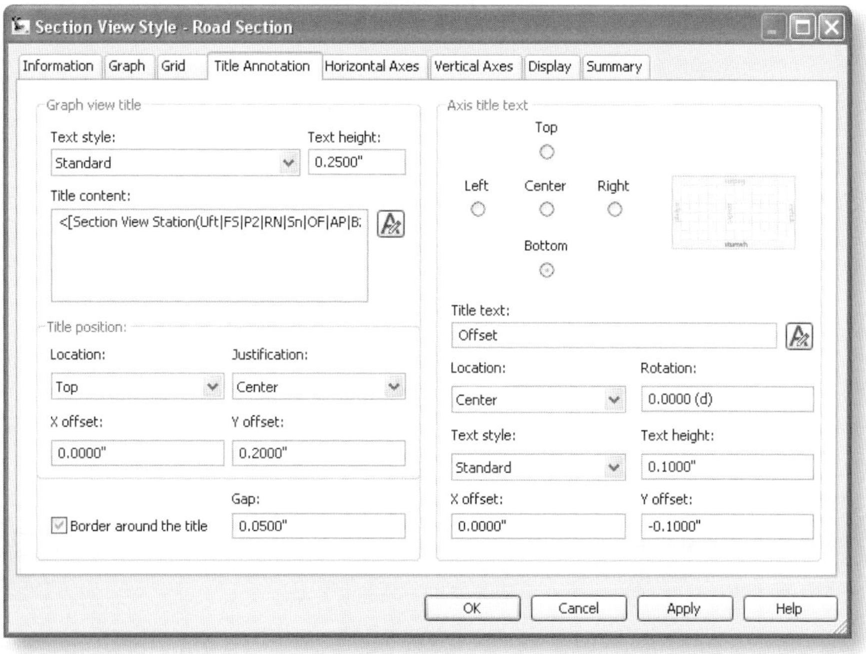

**FIGURE 8.8**

## Horizontal Axes

Horizontal Axes defines a section view's major and minor stationing annotation (see Figure 8.9). A section view has vertical lines that demark and annotate horizontal section offsets. The panel's left side defines the major offset distance and its annotation for a grid's top and bottom. The panel's right side defines minor station interval and its annotation. Major and minor settings are for size, text style, height, and other properties. The text icons display the Text Component Editor, thus giving access to the offset label's format (precision, sign, and so on).

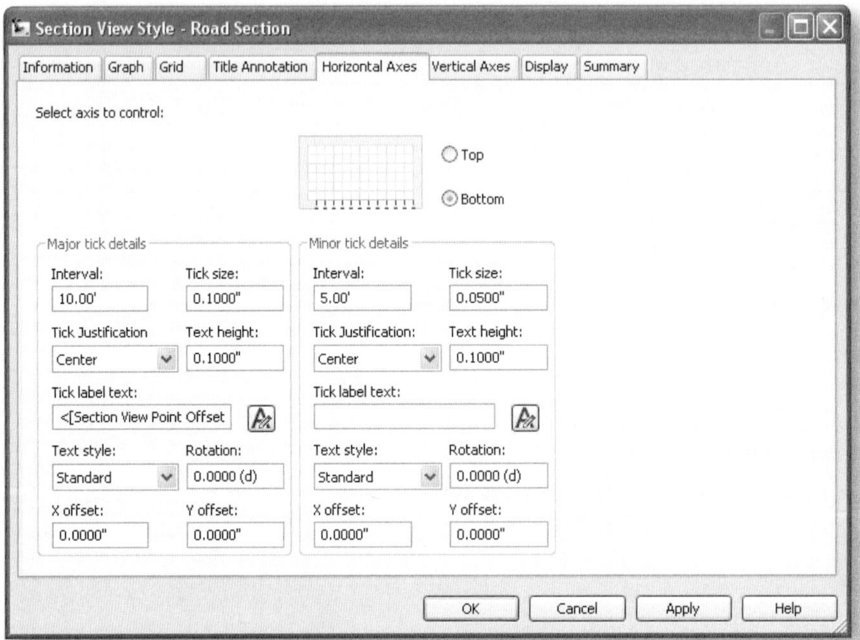


**FIGURE 8.9**

## Vertical Axes

Vertical Axes defines a section view's elevation annotation. A Section view has horizontal lines that demark and annotate elevations. This panel's controls are similar to those on the Horizontal Axes panel, except they affect the annotation of a view's elevations.

## Display

Even though a view style has tick, label, and title definitions for each axis, the Display panel settings control what the style actually displays.

## SECTION

A section style assigns the component's layer names and properties (see Figure 8.10).

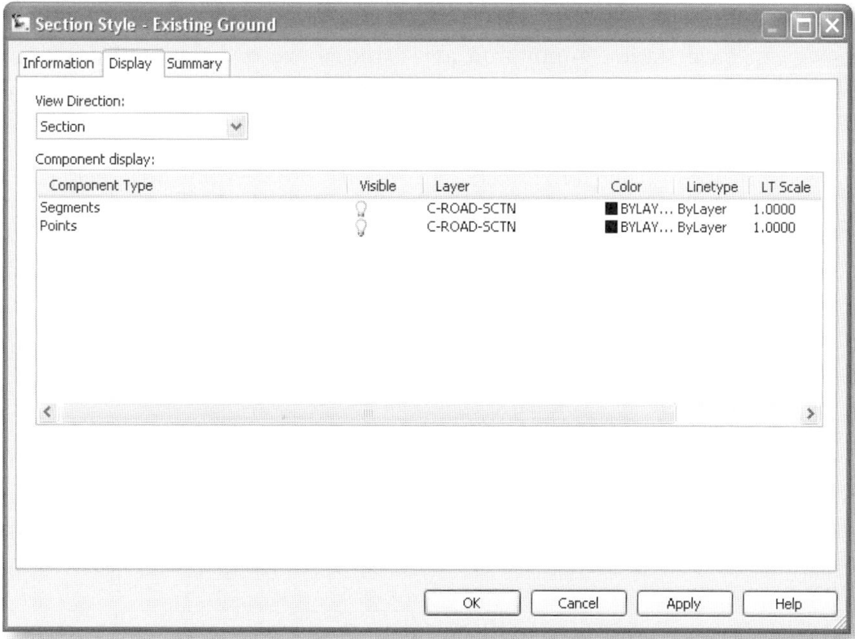


**FIGURE 8.10**

## LABEL STYLES

Civil 3D has label styles for sample lines, section views, and sections.

### Sample Lines

A sample line label style defines the sample line, its look, and any notations (generally a station value) (see Figure 8.11).

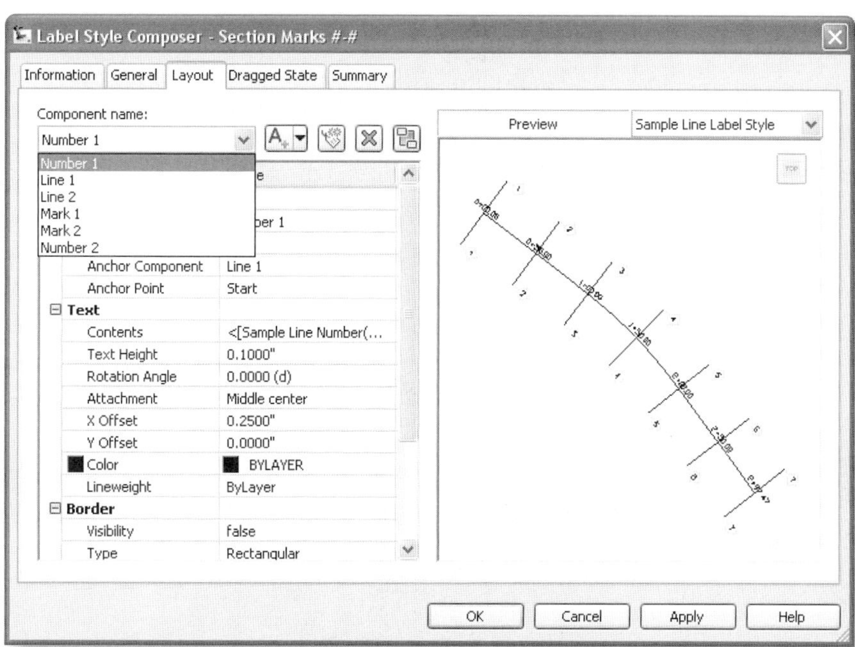


**FIGURE 8.11**

## Section View Band Set

Section views can have station annotation as a band above or below the grid (see Figure 8.12). A band set is an alias that includes several individual styles. By changing a section's band set, a section view's annotation may radically change.

In a section's Properties dialog box, users can modify the section's assigned label styles. This is done by clicking the Band Type drop-list arrow, selecting the desired style, and adding it to the styles list.

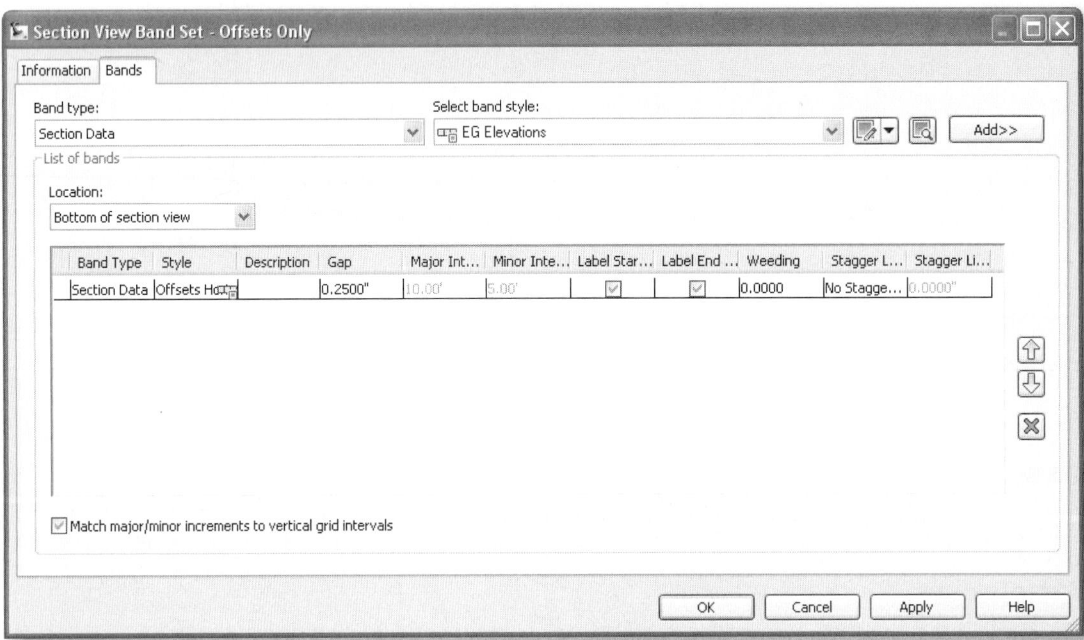


**FIGURE 8.12**

## Band Set Styles

A band set style's Band Details panel defines the title and it properties: text style, height, location, etc. (see Figure 8.13).

The panel's right side defines each label type listed at the dialog box's center. To view, create, or modify a label's values, from the list label type, select a type and click the Compose Label... button. The Label Style Composer displays a selected label's definition (see Figure 8.14).

The panel's upper right defines if the label has any ticks and, if it does, their size.

Even though the Display panel contains all possible label components and ticks, the Display panel settings determine what is actually displayed.

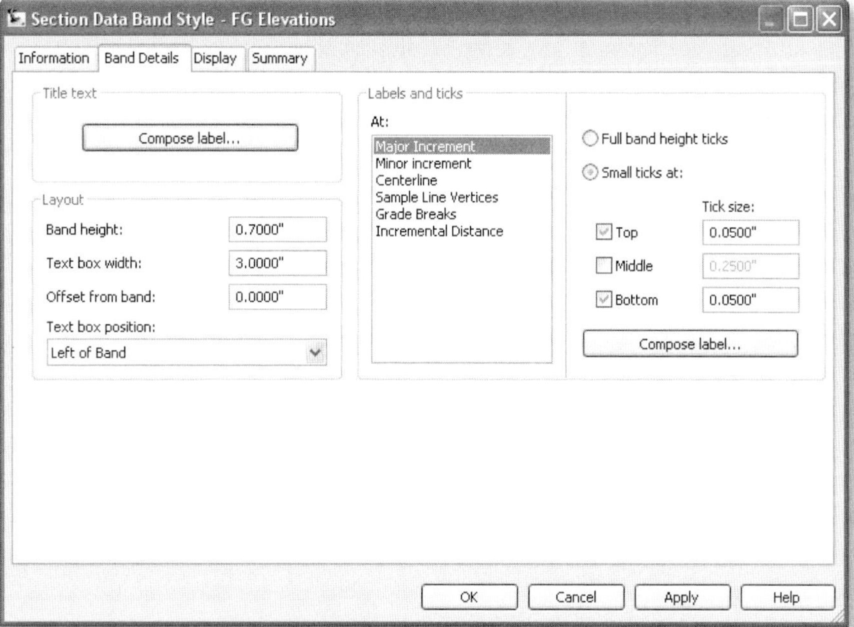


FIGURE 8.13

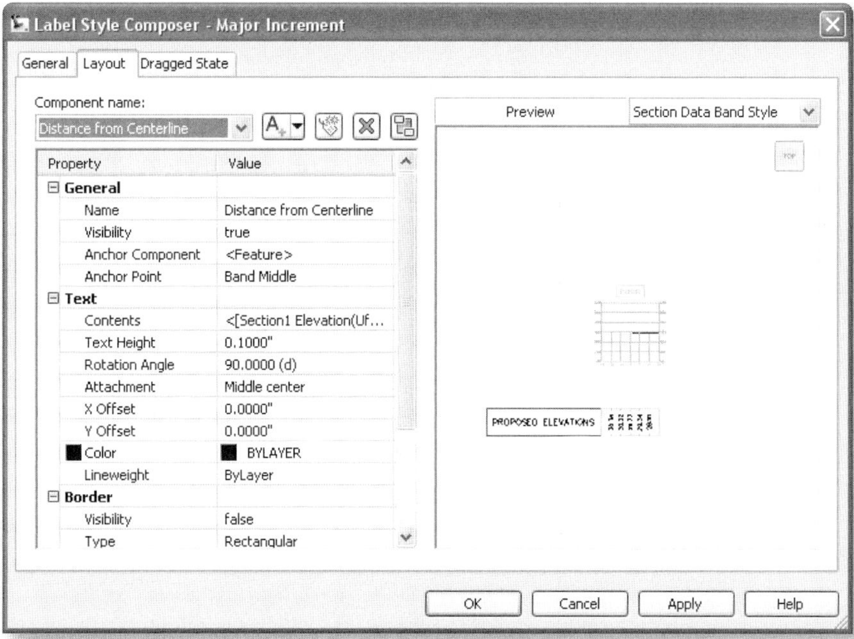


FIGURE 8.14

## Section Data

These styles have six possible functions: major and minor increment, centerline, sample line vertices, grade breaks, and incremental distance. The properties used by these labels include distance from centerline, offset from centerline, and elevations from Section1 (generally Existing) or Section2 (proposed). The style types include major station offset, EG elevation, and FG elevations.

### Section Segment

These styles annotate section segment lengths, slopes, elevations, offset sides, etc.

### Section View — Spot Label Styles

These styles label selected points within a section view. The Add Labels dialog box lists these styles and places them in a drawing, and are discussed in unit 4 of this chapter.

### Section Label Styles

Section label styles annotate section offset and elevation, grade break elevations, and segment values (length, grade, etc.). These styles are not roadway section annotation. Corridor section annotation comes from the Multipurpose code set styles. These are a part of the unit 4 discussion of section annotation.

### Section Label Sets

A section label set is an alias for a collection of several individual styles (see Figure 8.15). Changing the section label style's list may completely change a section's annotation.

Label sets use a combination of the following label type styles: Major and Minor Offsets, Grade Breaks, and Segments.

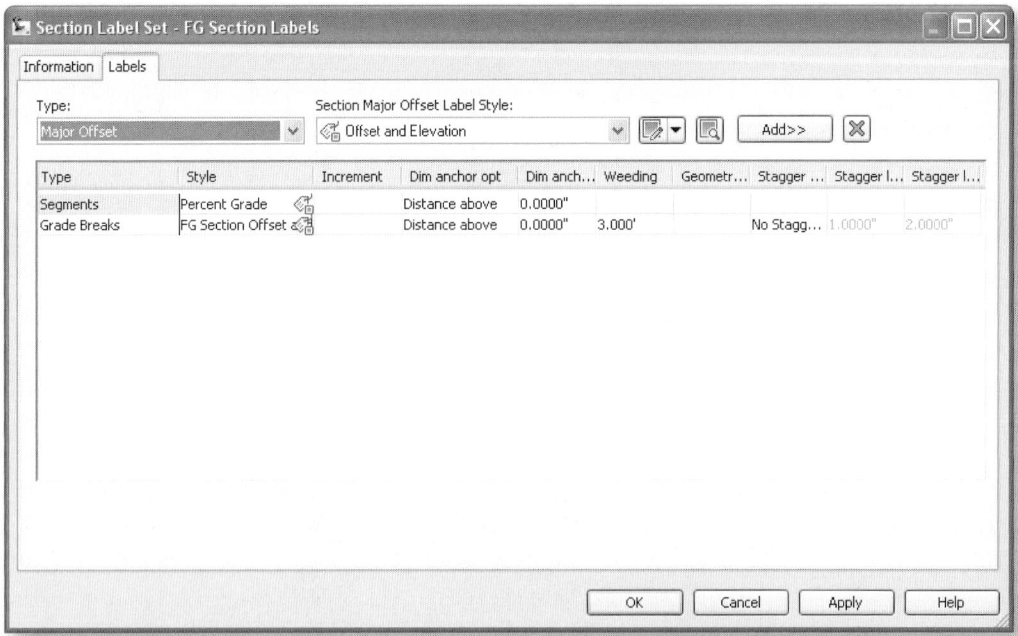


**FIGURE 8.15**

### Major and Minor Offsets

Major and Minor Offset styles label a section's offset and elevation at a major and minor increment.

### Grade Break

Grade Break styles label a grade break's offset, grade, and length.

### Segment

Segment styles label a section's segment length and grade.

## COMMAND SETTINGS

Sample lines, section view, and sections commands contain critical values that determine swath widths, section spacing, and default ranges, default styles, etc.

### Create Sample Lines

The Ribbon's Home tab, Profile & Section View panel's, Sample Lines command settings affect how the corridor is sampled (see Figure 8.16). The Default Swath Widths section defines a sample line's maximum left and right offset. The Sampling Increments section defines if the sampling is incremental, and, if incremental, the sampling interval for tangents, curves, and spirals segments. Additional Sample Controls section settings determine which additional sample lines to create: beginning and ending alignment stations, horizontal geometry points, and critical superelevation stations.

The Miscellaneous section Lock to Station setting, when true, tells the sample lines to update if the alignment changes.

After creating a sample line group, Prospector displays all sample line data under the alignment. This list includes the surface names, corridors, and pipe networks that are a part of the section's data.

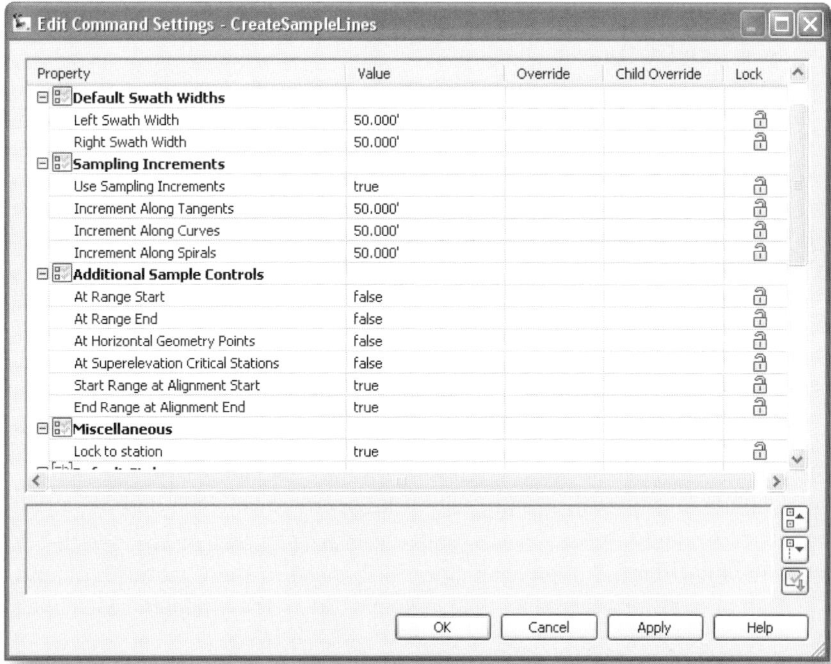


**FIGURE 8.16**

### Section View — Create Section View

Create Section View or Create Multiple Section View command settings are the same (see Figure 8.17). The Default Styles section assigns section view, band sets, section labels styles, section view group plot, and default Add Labels styles. The Default Name Format section defines the section views naming convention.

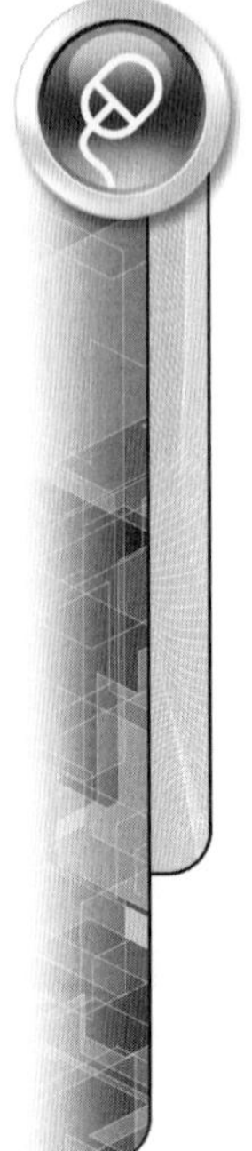

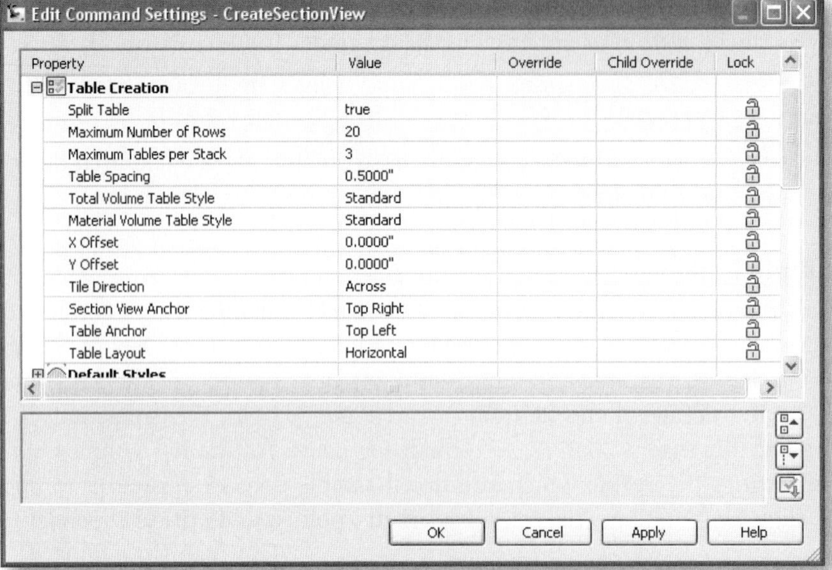


**FIGURE 8.17**

## EXERCISE 8-1

After completing this exercise, you will:

- Be familiar with Edit Drawing Settings.
- Be familiar with Sample line, Section, and Section View Edit Feature Settings.
- Be familiar with sample lines and sections Label Styles.
- Be familiar with sample lines and sections Edit Command Settings.

### Exercise Setup

This exercise starts with the previous chapter's exercise drawing. The drawing contains an alignment, profile, assembly, and corridor. If you did not complete the Chapter 7 exercises, browse to the Chapter 8 folder of the CD that accompanies this textbook and open the *Chapter 08 - Unit 1.dwg* file.

1. If you are not in **Civil 3D**, double-click the ***Civil 3D*** desktop icon to start the application.
2. When you are at the command line, close the open drawing and **do not save it**.
3. At Civil 3D's top left, Quick Access Toolbar, click the Open icon to open the exercise drawing from the previous chapter or browse to the Chapter 8 folder of the CD that accompanies this textbook and open the *Chapter 08 - Unit 1* drawing.

### Edit Drawing Settings

Edit Drawing Settings affect sample lines, sections, and section views.

1. Click the ***Settings*** tab.
2. At Settings' top, click the drawing name, press the right mouse button, and from the shortcut menu, select EDIT DRAWING SETTINGS....
3. Select the ***Object Layers*** tab.

If you have multiple sample line groups and section objects, their layers entries should have a modifier and a value (see Figure 8.2).

4. In the Object Layers panel, change the modifier for **Sample Line**, **Sample Line-Labeling**, **Section**, **Section-Labeling**, **Section View**, **Section View-Labeling**, **Section View Quantity Takeoff Table**, and **Sheet** to **Suffix**. For the value, type (**-***) (a dash followed by an asterisk).
5. Click ***OK*** to exit the dialog box.

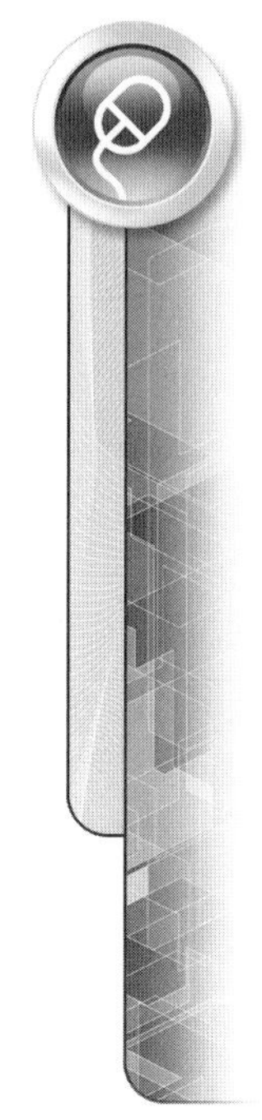

## Edit Feature Settings — Sample Line

Sample Line's Edit Feature Settings dialog box sets default styles and name formats (see Figure 8.3).

1. In Settings, select the Sample Line heading, press the right mouse button, and from the shortcut menu, select EDIT FEATURE SETTINGS....
2. In Edit Feature Settings, expand the Default Styles and Default Name Format sections to review their values.
3. Click ***OK*** to close the dialog box.

## Edit Feature Settings — Section View

Section View's Edit Feature Settings assign default view styles, its section label sets, plotting styles, label styles, and default name formats (see Figure 8.4).

1. In Settings, select the Section View heading, press the right mouse button, and from the shortcut menu, select EDIT FEATURE SETTINGS....
2. In Edit Feature Settings, expand the Default Styles, Default Name Format, Section View Creation, and Default Projection Label Placement sections to review their values.
3. Click ***OK*** to close the dialog box.

## Edit Feature Settings — Section

Section's Edit Feature Settings set the default style and naming format for a section (see Figure 8.5).

1. In Settings, select the Section heading, press the right mouse button, and from the shortcut menu, select EDIT FEATURE SETTINGS....
2. In Edit Feature Settings, expand the Default Styles and Default Name Format sections to review their values.
3. Click ***OK*** to close the dialog box.

## Object and Label Styles — Sample Line

Sample lines, section views, and section styles specifically label their critical values. Section view band sets and section label sets are aliases for a collection of styles with specific purposes.

The sample line object style sets its drawing layers. The sample line label styles affect its appearance and labeling.

1. In Settings, expand the Sample Line's Label Styles branch until you view its styles list.
2. From the styles list, select Section Name and Marks, press the right mouse button, and from the shortcut menu, select EDIT....
3. If necessary, click the ***Layout*** tab.

4. At Layout's top, click the Component name drop-list arrow to display the label component list. Select each one and review each component's values.
5. From the Component name drop-list, select **Sample Name**. In its Text section, click in the Contents value cell to display an ellipsis, and click the ellipsis to display the Text Component Editor.
6. Review this component's format string.
7. Click ***Cancel*** until you have returned to the command line.

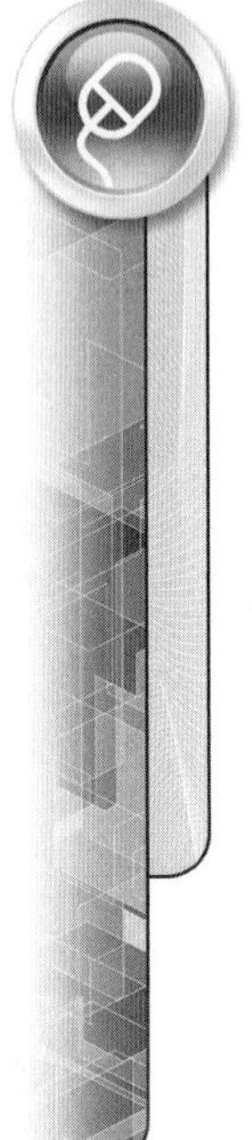

## Object and Label Styles — Section View

Section view is a grid that encloses the section station and elevations. The view styles provide basic station and elevation annotation, grid, exaggeration, and title content.

1. In Settings, expand the Section View branch until you view Section View Styles' styles list.
2. From the list, select **Road Section**, press the right mouse button, and from the shortcut menu, select EDIT....
3. In turn, select the ***Graph***, ***Grid***, ***Title Annotation***, ***Horizontal Axes***, ***Vertical Axes***, and ***Display*** tabs to review their contents.
4. Click ***OK*** to exit the Road Section view style.

## Band Sets and Band Set Styles

Band sets are aliases for style groups that appear below or above a section view (see Figure 8.12).

1. In Settings, expand the Section View and Band Styles branches until you are viewing the Band Sets list.
2. From the list, select **Major Stations Offsets and Elevations**, press the right mouse button, and from the shortcut menu, select EDIT....
3. Click the ***Bands*** tab.
4. In the dialog box's top left, click the Band Type drop-list arrow, and from the list, select **Section Data**.
5. At the dialog box's middle top, click the Select Band Style drop-list arrow to display the styles list.
6. In the dialog box's top left, click the Band Type drop-list arrow and from the list, select **Section Segment**.
7. At the dialog box's top middle, click the Select Band Style drop-list arrow to display a styles list.
8. Click ***Cancel*** to close the dialog box.

The band set styles are below the Band Set heading, Section Data, and Section Segment headings list.

## Section Data

1. In Settings, expand the Section View, Band Styles branch until you view the Section Data's styles list.
2. From the Section Data styles list, select **Offsets**, press the right mouse button, and from the shortcut menu, select EDIT....
3. Click the ***Band Details*** tab.

The panel's left side defines the band's title. At the top right, Labels and Ticks sets their size. At the center are the label's annotation locations. The Compose Label... button displays the Label Style Composer dialog box.

4. In the Band Details top middle, from the At: area, select **Centerline** and click ***Compose Label...*** to display the Label Style Composer.
5. In the Label Style Composer's Text section, click in the Contents value cell to display an ellipsis, and then click the ellipsis to display the Text Component Editor.
6. At the Text Component Editor's left side, click the Properties drop-list arrow to view the properties list.
7. Click ***Cancel*** until you have returned to the command line.

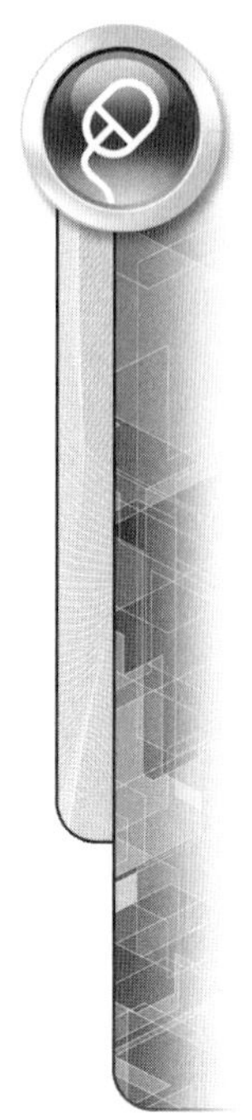

## Section Segment

This band label type annotates section segment lengths and grades.

1. In Settings, expand the Section View, Band Styles branch until you view the Section Segment styles list.
2. From the Section Segment styles list, select **Segment Length**, press the right mouse button, and from the shortcut menu, select EDIT....
3. If necessary, click the ***Band Details*** tab.

The panel's left side defines the band's title. At the center, the Labels and Ticks area lists all label types and tick sizes and locations. Compose Label... displays the Label Style Composer dialog box.

4. In Band Details' top middle, in the At: area, select **Segment Labels** and click ***Compose Label...*** to display the Label Style Composer.
5. In Label Style Composer's Text section, click in the Contents' value cell displaying an ellipsis, and then click the ellipsis to display the Text Component Editor.

This band label annotates the surface segment lengths.

6. In the Text Component Editor, at its top left, click the Properties drop-list arrow to view the label's list properties.
7. Click ***Cancel*** until you have returned to the command line.

## Section Labels — Section Label Set

Section labels annotate surface elevations and stations within a section view.

Section label sets are aliases containing one or more style types: Major and Minor Offsets; Grade Breaks; and Segments style. A set also specifies the label's location and a weeding factor. Weeding removes overlapping labels.

1. In Settings, expand Section's Label Styles branches until you view the Label Sets' styles list.
2. From the styles list, select **FG Sections Labels**, press the right mouse button, and from the shortcut menu, select EDIT....
3. If necessary, click the ***Labels*** tab.
4. At the dialog box's top left, click Type's drop-list arrow to view the style types list.
5. From the types list, select **Segments**.
6. In the dialog box's middle top, click the Section Segment Label Style's drop-list arrow to view its styles.
7. Click ***Cancel*** to return to the command line.

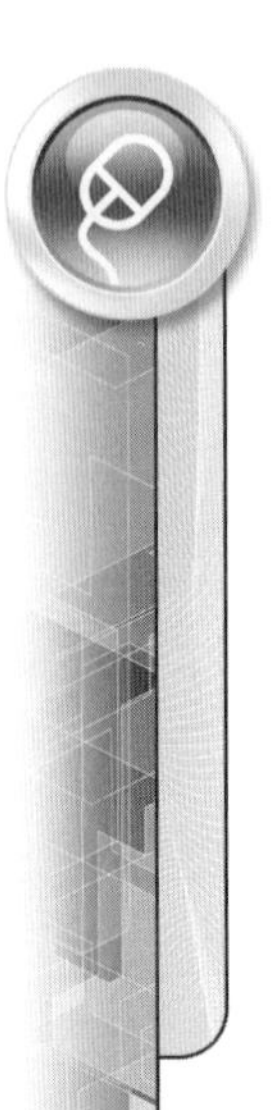

## Section Labels — Major and Minor Offset

Major and Minor Offset label styles annotate offsets and elevations. A section view defines the major and minor styles intervals.

1. In Settings, expand the Section branch until you view the Major and Minor Offset styles list.
2. From the Major Offset styles list, select **Offset and Elevation**, press the right mouse button, and from the shortcut menu, select EDIT....
3. Click the ***Layout*** tab.
4. In the Layout tab, in the Text section, click in the Contents' value cell to display an ellipsis, and then click the ellipsis to display the Text Component Editor.

The label at the major station intervals annotates the section view.

5. In the Text Component Editor, at its top left, click the Properties drop-list arrow to view the label's list properties.
6. Click ***Cancel*** until you return to the command line.

## Section Labels — Grade Break

Grade Break styles label a surface or design section grade break using EG or FG. The labeling appears in the section view.

1. In Settings, expand the Section branch until you view the Grade Break styles list.
2. From the styles list, select **FG Section Offset and Elevation**, press the right mouse button, and from the shortcut menu, select EDIT....
3. If necessary, click the ***Layout*** tab.
4. Set the Component name to txtOffset. In its Text section, click in the Contents' value cell to display an ellipsis, and then click the ellipsis to view its format string.
5. Click ***Cancel*** until you return to the command line.

## Section Labels — Segment

This label type annotates a surface section segment length and cross slope.

1. In Settings, expand the Section branch until you view the Segment styles list.
2. From the styles list, select the **Percent Grade**, press the right mouse button, and from the shortcut menu, select EDIT....
3. Click the ***Layout*** tab.
4. In the Text section, click in the Contents' value cell to display an ellipsis, and then click the ellipsis to display its label format string.
5. In the Text Component Editor, at its top left, click the Properties drop-list arrow to view the label's list properties.
6. Click ***Cancel*** until you return to the command line....

## Command — Create Sample Lines

Commands create sample lines, section views, and sections default settings in addition to those set by Edit Feature Settings.

The Create Sample Line command samples a corridor at an interval, at specific stations, or at a station range.

1. In Settings, expand the Sample Line branch until you view the Commands list.
2. From the list, select **CreateSampleLines**, press the right mouse button, and from the shortcut menu, select EDIT COMMAND SETTINGS....
3. Expand the Default Swath Widths section.

A swath width is the distance sampled to the roadway's left and right.

4. Expand the Sampling Increments section.

This section sets the initial corridor sampling frequency.

5. Expand the Additional Sample Controls section.

This section sets the sampling of additional critical corridor points.

6. Expand the Miscellaneous section.

If Lock To Station is true, if any change occurs to the alignment geometry or properties, then the sections resample.

7. Expand the Default Styles and Default Name Format sections.

Sample Line Edit Feature Settings sets these styles, and you can change them here.

8. Click ***Cancel*** to return to the command line.

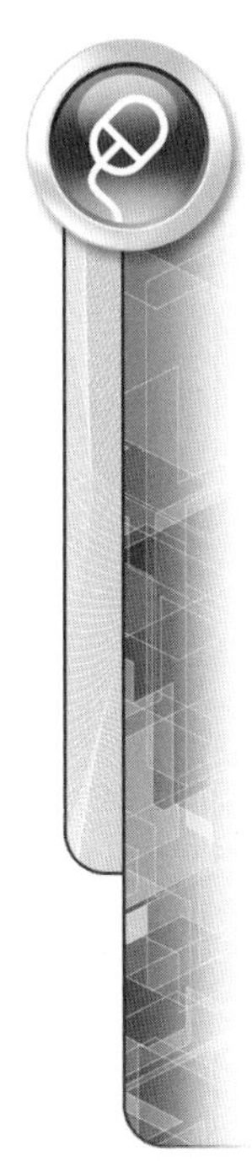

## Command — Create Section View

Create Section View (single section) and Create Multiple Section Views (multiple sections) commands use the same settings.

1. In Settings, expand the Section View branch until you view the Commands list.
2. From the list, select **CreateSectionView**, press the right mouse button, and from the shortcut menu, select EDIT COMMAND SETTINGS....
3. Expand the Table Creation, Default Styles, Default Name Format, Section View Creation, and Default Projection Label Placement sections and review their values.

Table Creation controls the format and behavior of quantity takeoff tables. The Default Styles lists the style for a new section view. The Default Name Format specifies a new view's naming format. Section View Creation governs prompting for Offset Range and Height, and how to display the section group's elevation range. Default Projection Label Placement controls the placement of labels for objects projected to section views.

4. Click ***Cancel*** to return to the command line.
5. At Civil 3D's top left, Quick Access Toolbar, click the ***Save*** icon to save the drawing.

This ends the exercise that reviews objects, styles, and commands that affect sample lines, section views, and sections. The next unit reviews creating section sample lines.

## SUMMARY

- The section view is the most complex Civil 3D object.
- A section samples surface, corridor stations, and/or pipe networks along an alignment's path.
- If you are designing pipe networks, it is best to design them before you create sample lines.
- A section view is a grid with sections within it.
- A section view has annotation on all four axes, plus it has bands that appear at its top and bottom.
- When you create a section view, the command assigns the section label styles and/or sets.

## UNIT 2: CREATE SAMPLE LINES

Sample lines are the link between a corridor and a section view and its sections. It is easiest to first define all corridor content before you create sample lines. For example, if you are creating sample lines and then you add a piping network, the sample line group must be resampled with the pipe network data.

### CREATE SAMPLE LINES

The process starts by identifying which roadway elements to sample (see Figure 8.18). The Create Sample Line Group dialog box reports the sample line group name format (upper left), the current sample line style, label style, and layer (upper right), and the alignment in the middle left.

The dialog box's lower half lists all elements available for sampling along the named alignment. There are four types of section data: surface(s), corridors, corridor surface(s), and pipe network(s). Each data type has a unique icon. The dialog box identifies each potential data source, if sampled, and uses the style, layer, and updating mode listed to the source's right. At this point, it is best to set the styles for each data type.

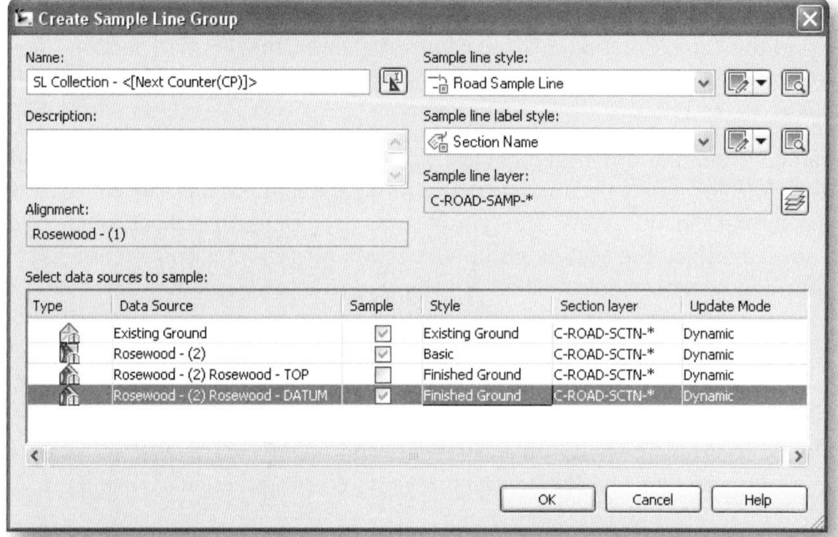


**FIGURE 8.18**

After you identify sample elements and click OK, the Sample Line Tools toolbar is displayed. The Sample Line Tools toolbar creates sample lines, views their content, and deletes individual sample lines (see Figure 8.19). Users can define sample lines by stations, by selecting drawing points, by selecting existing polylines, by a station range, or from the corridor sections. When you define sample lines, you can use a combination of sampling methods (for example, by station range and by selecting points along the centerline of the corridor). After you create sample lines, the toolbar reverts to defining sample lines by stations.

At the toolbar's center is the current sample line group (SLG), at its lower left is the current definition method, and at the lower right is the current alignment. Creating sample line methods are in an icon stack on the toolbar's middle right.

The toolbar has editing tools and tools to delete existing groups, create new groups, and review a sample line's values.

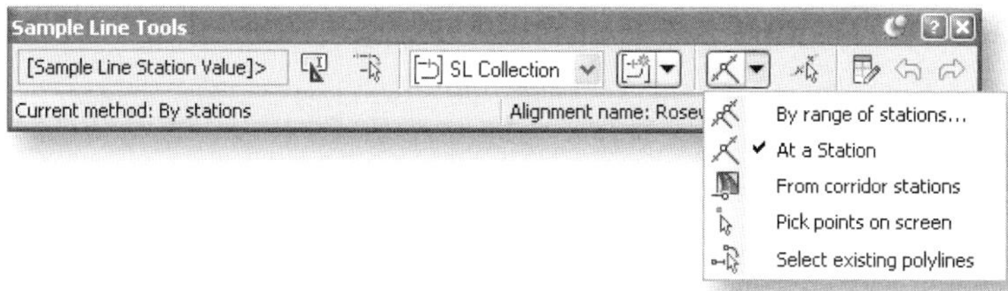


**FIGURE 8.19**

The By Station method uses a station jig to identify a station and then prompts for the left and right swath width. When you use By range of stations… or From corridor stations, the methods display the Create Sample Lines – By Station Range dialog box (see Figure 8.20). This dialog box sets right and left swath width and the sampling increments.

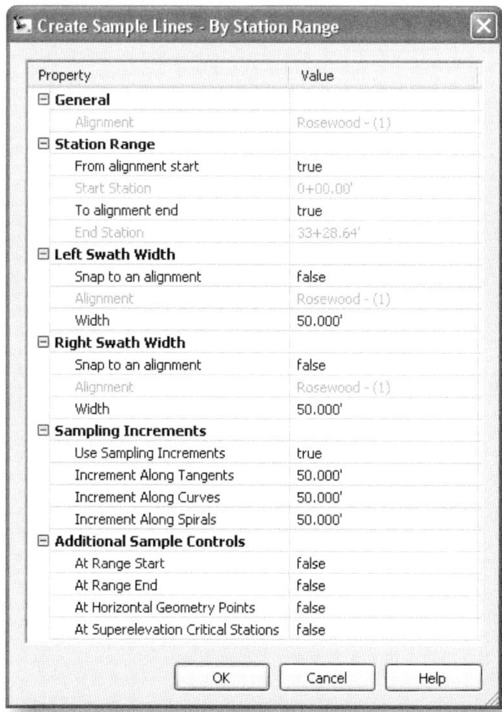


**FIGURE 8.20**

If you are creating sample lines where existing sample lines are duplicated, a warning dialog box opens and has options to resolve the sample line duplication (see Figure 8.21). You can resolve duplicates by adding them as new, overwriting old sample line data, or ignoring new sample line data. You can also use the duplicates to append to the current sample line group or you can delete the existing list and replace it with the duplicate data.

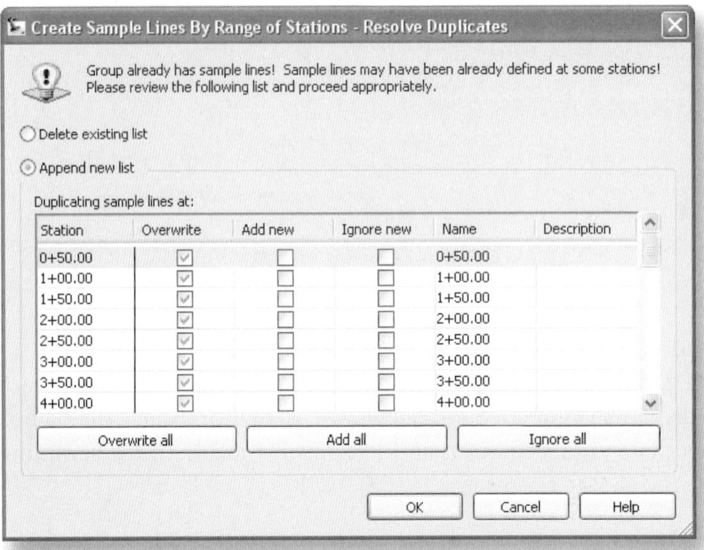


**FIGURE 8.21**

The Sample Line Tools toolbar has a cell-based viewer that displays a section's properties (see Figure 8.22). To view a section's Sample Line properties editor, click the Select/Edit Sample Line icon, and select a sample line. At the Editor's bottom are buttons that display the section's previous or next vertex's information (Center, Left 1, and Right 1).

**FIGURE 8.22**

## EDITING AND REVIEWING SAMPLE LINES

Reviewing and editing sample lines occurs in the Sample Line Tools toolbar. The toolbar's SampleLine Entity view icon displays the Edit Sample Line dialog box. The Ribbon's Modify, Section panel's Edit Sample lines command displays the Sample Line Tools toolbar and presents the Edit Sample Line dialog box.

## SAMPLE LINE GROUP PROPERTIES

Each Sample Line Group has a multi-tabbed Properties dialog box. The Information tab sets the group's name. The Sections tab lists all sample lines, and Edit Group Labels modifies the sample line group's labeling. The Sections tab lists the sampled corridor components. The panel lists the objects sampled when creating the sample lines (see Figure 8.23). If you want to calculate quantities, this list must include surfaces and/or a corridor for the material lists defined in the Material List tab. Material lists are the focus of unit 5 of this chapter. Clicking the Sample more sources... button displays the Section Sources dialog box (see Figure 8.24). Section Sources adds and removes components from section sampling. After you change the components list and click OK, Section Sources resamples the sections.

The Section Views tab lists all sections that are using the sample line group's data. As mentioned before, the Material List tab contains the definitions for quantities.

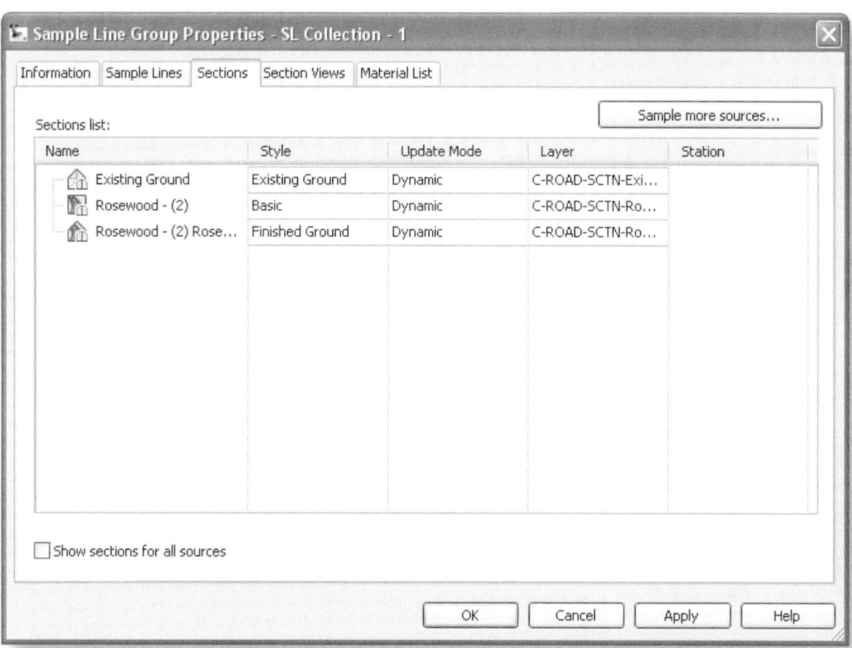


**FIGURE 8.23**

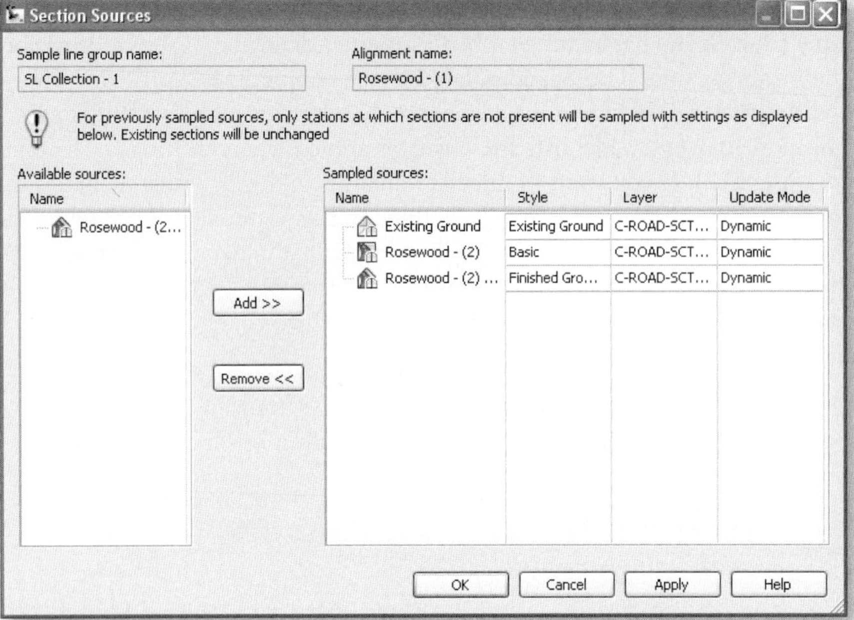


**FIGURE 8.24**

## SAMPLE LINE PROPERTIES

The Sample Line Properties dialog box's tabs lists data (name, alignment, number, and station), sampled components, and views using the section's data (see Figures 8.25 and 8.26). In Prospector's Preview, select a sample line and press the right mouse button to display a shortcut menu with a Properties… pick. The menu also includes zooming to and deleting the selected sample line (see Figure 8.27).

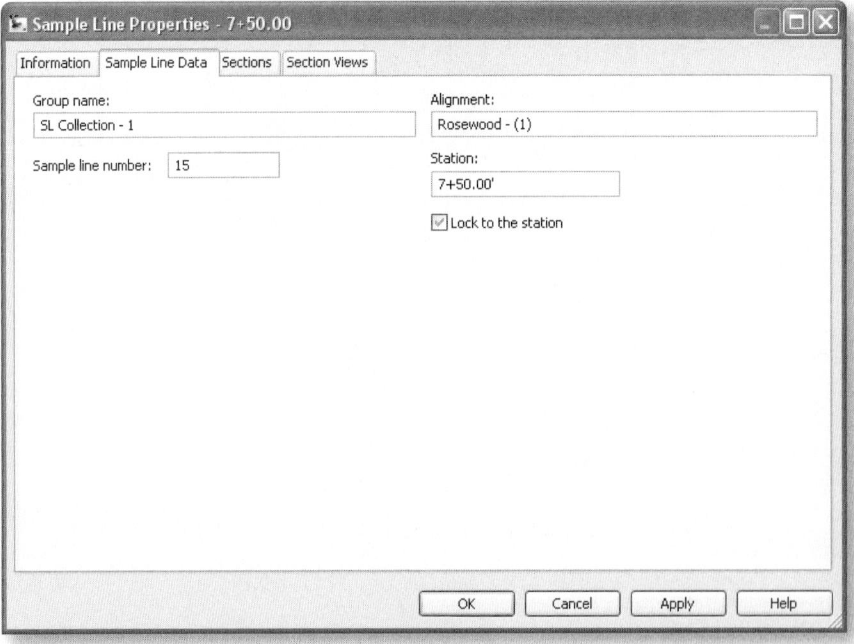


**FIGURE 8.25**

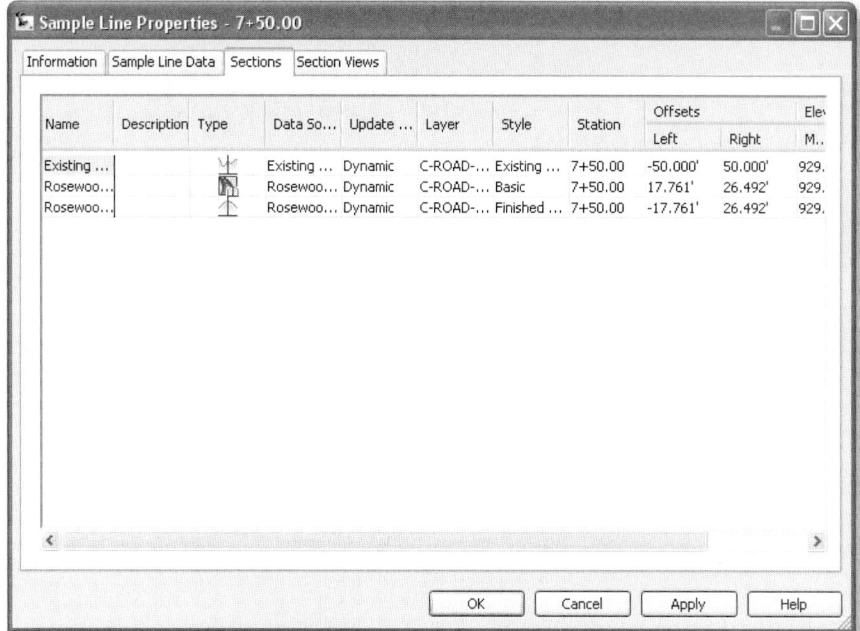


FIGURE 8.26

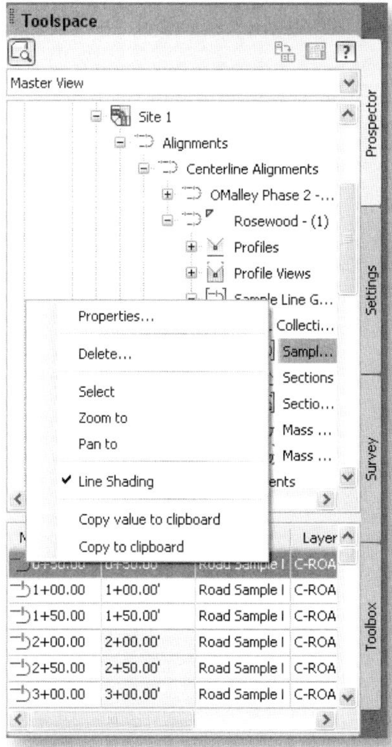


FIGURE 8.27

EXERCISE 8-2

After completing this exercise, you will:

- Be able to create sample lines.
- Be able to create sample lines by different methods.
- Be familiar with the grid view of a section.
- Be able to view sample line properties.

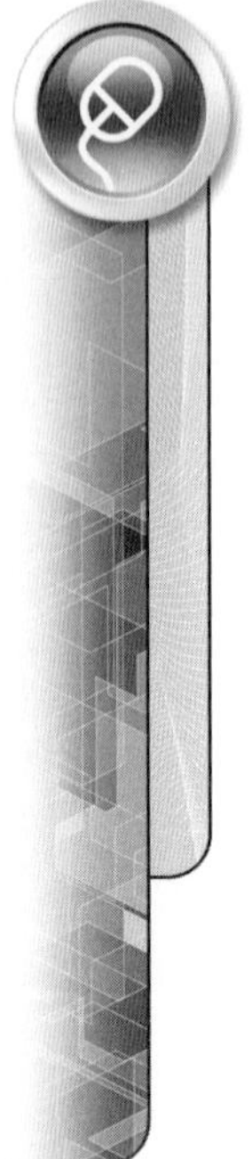

## Drawing Setup

This exercise continues with the previous exercise's drawing. If you did not complete the previous exercise, browse to the Chapter 8 folder of the CD that accompanies this textbook and open the *Chapter 08 – Unit 2* drawing.

1. If not open, open the previous exercise's drawing or browse to the Chapter 8 folder of the CD that accompanies this textbook and open the *Chapter 08 – Unit 2* drawing.

## Create Sample Lines

Create Sample Lines displays a Create Sample Line Group dialog box. This dialog box identifies what components to include in the sampling. After the sample component has been identified, the Sample Line Tools toolbar is displayed. The Sample Line methods icon stacks sets and changes the sampling methods.

1. On the Ribbon, click the ***Home*** tab. On the Profile & Section Views panel, click the ***Sample Lines*** icon.
2. The command line prompts you for an alignment. If you are able to select the alignment from the drawing, select it or right mouse click, from the list select **Rosewood - (1)**, and click ***OK*** to continue displaying the Create Sample Line Group dialog box.
3. In the dialog box, toggle off Rosewood – Top, change the styles for Rosewood – TOP and DATUM to **Finished Ground**, and click ***OK*** to display the Sample Line Tools toolbar with a jig attached to the alignment that reports its current station.
4. Move your cursor around the corridor to view its station reporting. If necessary, zoom or pan to better view the corridor.
5. In the Sample Line Tools toolbar, to the Sample Line Group name's right, click the Sample line creation method's icon stack drop-list arrow, and from the list, select `FROM CORRIDOR STATIONS`.

The Create Sample Lines dialog box opens and lists the beginning and ending stations and the left and right swath width.

6. Click ***OK*** to create the sample lines.

The sample lines appear, and creation mode returns to At a Station.

7. Without changing the creation mode, in the drawing, select a couple of stations and, when you are prompted, for both swath widths enter **50**.

These stations are added to the sample line list's end.

## Reviewing Sample Line Data

1. On the Sample Line Tools toolbar's right, click the ***SampleLine Entity View*** icon to display the Edit Sample Line dialog box editor.

2. In the Sample Line Tools toolbar to the left of the SampleLine Entity View icon, click the ***Select/Edit Sample Line*** icon, and in the drawing, select a sample line.

The sample line's properties are displayed in the editor. You can change the station name, swath width, or review its station and elevations (center, left, and right).

3. Click the ***Next Vertex*** or ***Previous Vertex*** to review the section values.
4. Click the editor's red **X** to close it.
5. Press ENTER to exit the Create Sample Lines command.
6. At Civil 3D's upper left, Quick Access Toolbar, click the ***Save*** icon to save the drawing.

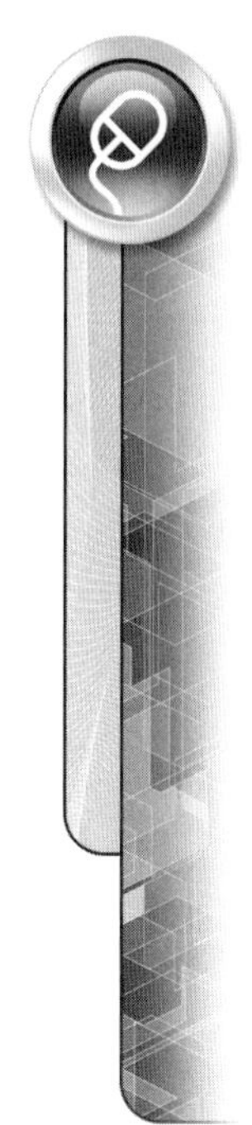

## Sample Line Group Properties

Each SLG has properties: Sample Line Properties include sample line data, what has been sampled, what section views use the section's data, and material list data.

1. Click the ***Prospector*** tab.
2. In Prospector, expand the Sites branch until you are viewing the Rosewood - (1), Sample Line Groups' SL Collection – 1 branch.
3. Select the SL Collection – 1 heading, press the right mouse button, and from the shortcut menu, select PROPERTIES....
4. Click the ***Sample Lines*** tab to review its information.

This lists the group's sample lines, their layer, style, and swath width offset.

5. Click the ***Sections*** tab.

This displays the sample groups sample components list.

6. Click ***OK*** to exit the dialog box.

## Sample Line Properties

1. If necessary, in Prospector, expand the Site1 branch until you are viewing the Sample Lines heading under SL Collection – 1.
2. Select the Sample Line heading to display, in Preview, the sample line list.
3. Scroll the sample line list until you locate line **2+75**, select it, press the right mouse button, and from the shortcut menu, select ZOOM TO.
4. In Preview, select **2+75**, press the right mouse button, and from the shortcut menu, select PROPERTIES....
5. Click the ***Sample Line Data*** tab.

This panel's content lists the section's location, number, alignment, and if the sample line is station locked.

6. Click the ***Sections*** tab.

This panel lists the section's sampled components.

7. Click ***Cancel*** to exit the dialog box.
8. At Civil 3D's upper left, Quick Access Toolbar, click the ***Save*** icon to save the drawing.

This ends the unit on creating sample lines and their properties. Next, you will create section views with sections.

## SUMMARY

- Sample lines sample what exists along an alignment.
- It is best to have all data present before you create the sample lines.
- If you add data to a corridor, you must resample the sample lines.
- Define sample lines first by station range, by corridor sections, by selecting multiple points, by existing polylines, and by points, and then specify their left and right offsets.
- Sample line's properties list individual section vertices, the sample line's station, and what section views use its data.
- A sample line lists its sections and their styles.
- Corridor sections use styles from the General, Multipurpose style list.

## UNIT 3: CREATING ROAD SECTIONS

The last step in documenting a roadway design is to create the section views. Commands import sections one at a time, all at once (as an array), or as organized section pages.

Section views contain section offset and elevation data. Sections contain within them all of their data and can be relocated anywhere in the drawing without affecting their data. To create section sheets, you must take care when planning and experimenting with settings that affect their creation.

### CREATE SECTION VIEW

Create Section View displays the Create Section View wizard. The General panel sets the alignment, section and its station, sample line group, the section view name, the section view style, and its layer (see Figure 8.28).

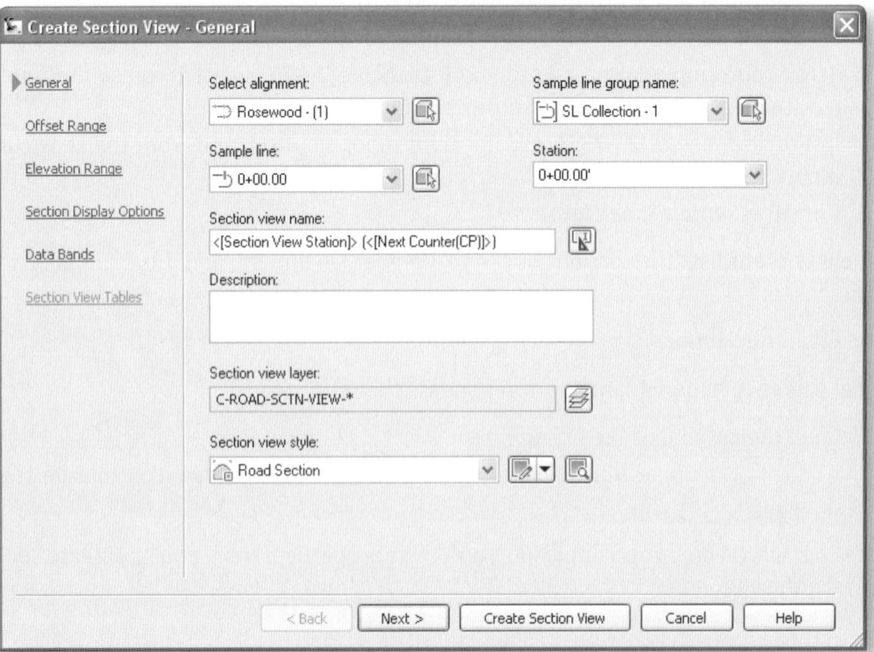


**FIGURE 8.28**

The Offset Range panel sets the sampled offset or a user-specified value for the section view.

The Elevation Range panel sets the sampled elevation range or a user-specified range for the section view.

Section Display Options list the section's components with drafting, styles, and labeling options (see Figure 8.29). You may have to change settings that do not carryover from previous commands.

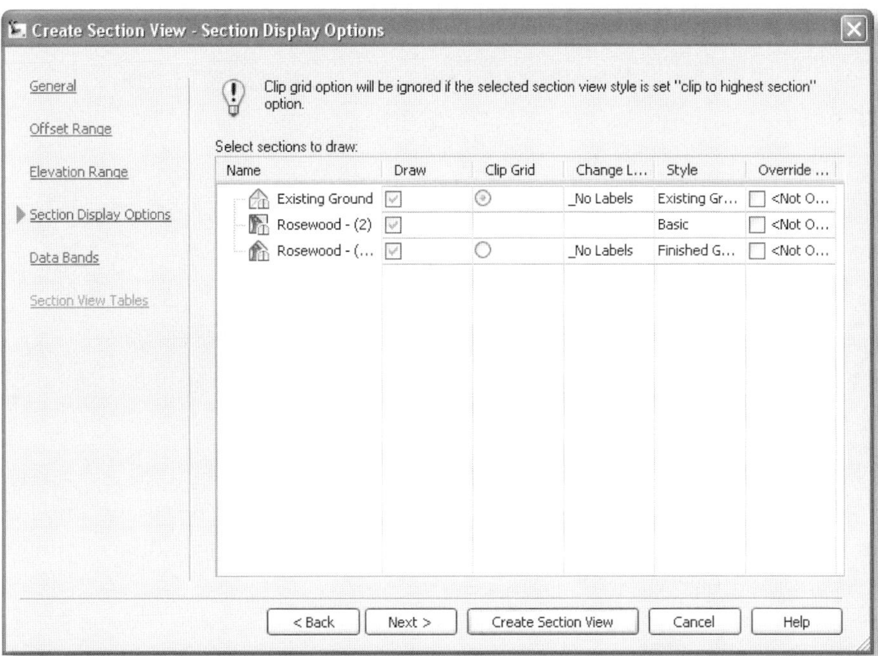


**FIGURE 8.29**

Data Bands assigns the data band type and its location (see Figure 8.30).

Section View Tables assign tables associated with material lists. This will be the focus for unit 5 of this chapter.

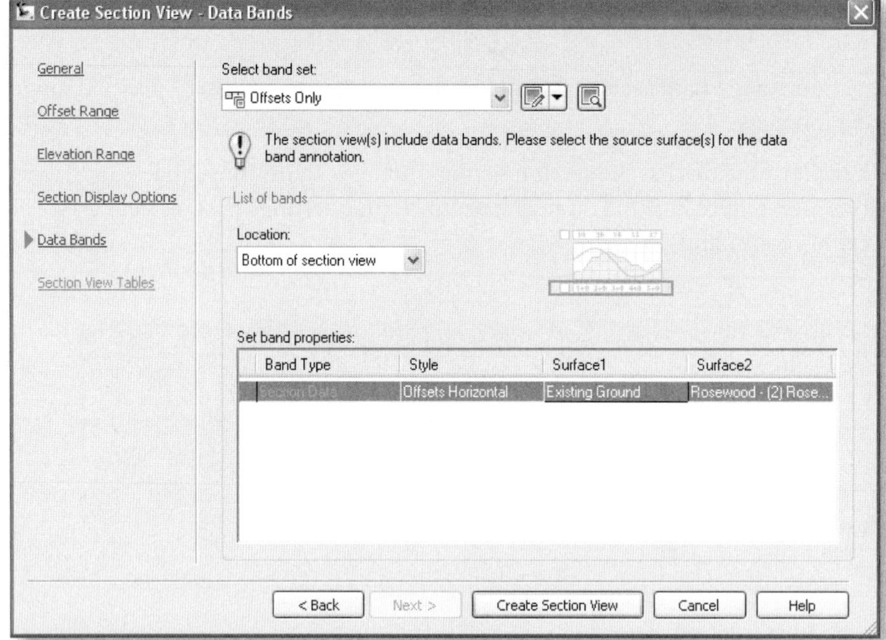


**FIGURE 8.30**

## CREATE MULTIPLE VIEWS

Create Multiple Views uses the same interface as Create Section View, but has changes to the first panel. These changes include the station range and assigning a group plotting style (see Figure 8.31).

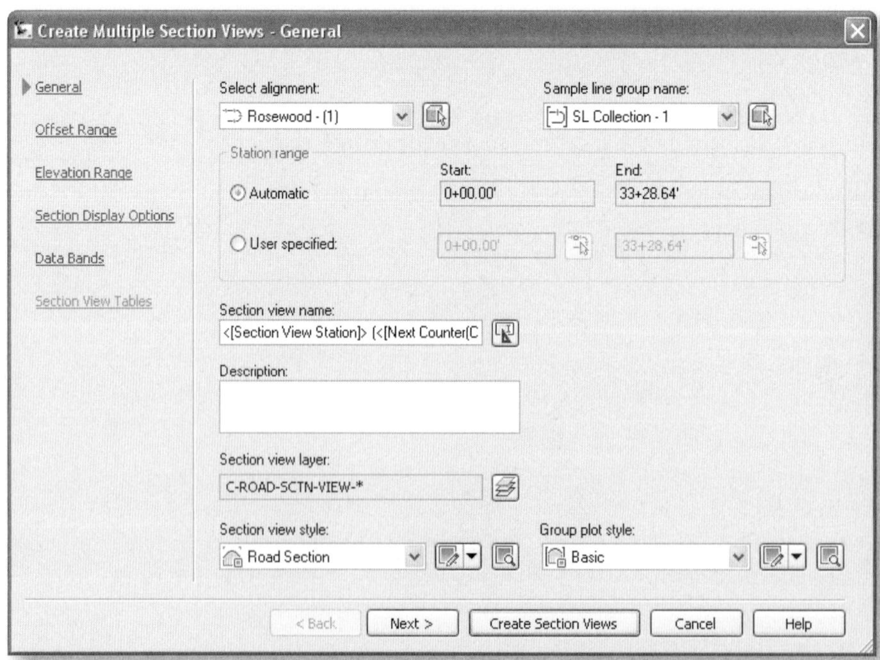


**FIGURE 8.31**

## GROUP PLOT STYLES

When plotting more than one section, a group plotting is necessary. A group plot's styles define section view sheets or arrays.

### Plot by Page

A page style defines a sheet for sections.

#### Array Panel

Array defines the section's distribution (rows or columns), number of sections (per row/column), and the rows/columns spacing between the sections (see Figure 8.32). The spacing distance (on the left) is the number of grid lines between views. A sheet style defines the grid size.

The panel's lower left sets the array's beginning point (such as the Lower Left corner), section justification (Centerline), and if section cells are Uniform Per Row, Column (variable width), or All (same width for all). Uniform per row or column means the sections in the column or row have the same number of horizontal or vertical cells. Uniform for all means all sections have the same number of horizontal and vertical cells.

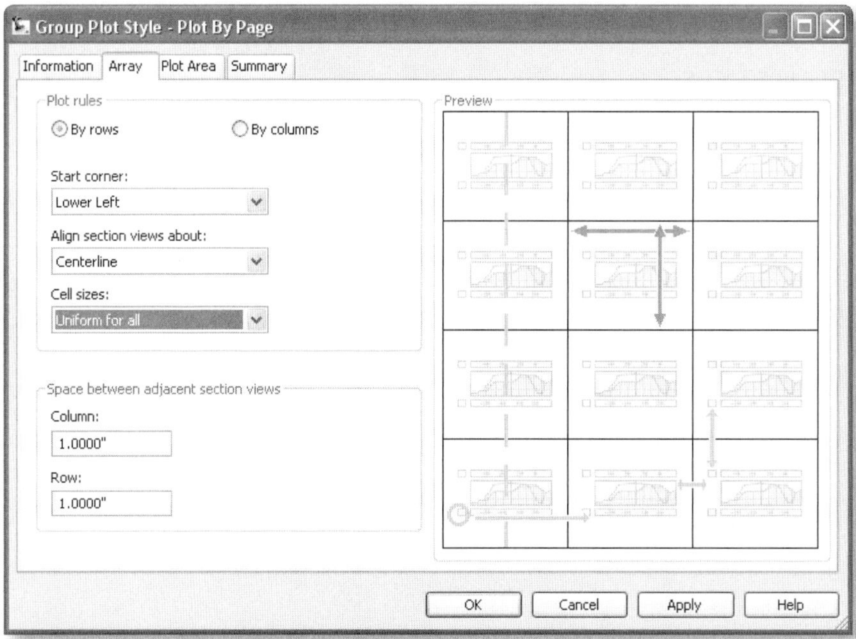


**FIGURE 8.32**

#### Plot Area Panel

Plot Area defines the Plot by Page's sheet size. The sheet style defines the sheet size and its margins. The dialog box's top left sets the successive sheets gap (see Figure 8.33).

**NOTE**

**AutoCAD's model space sheet plot size must be the same sheet size set in this dialog box (see Figure 8.33).**

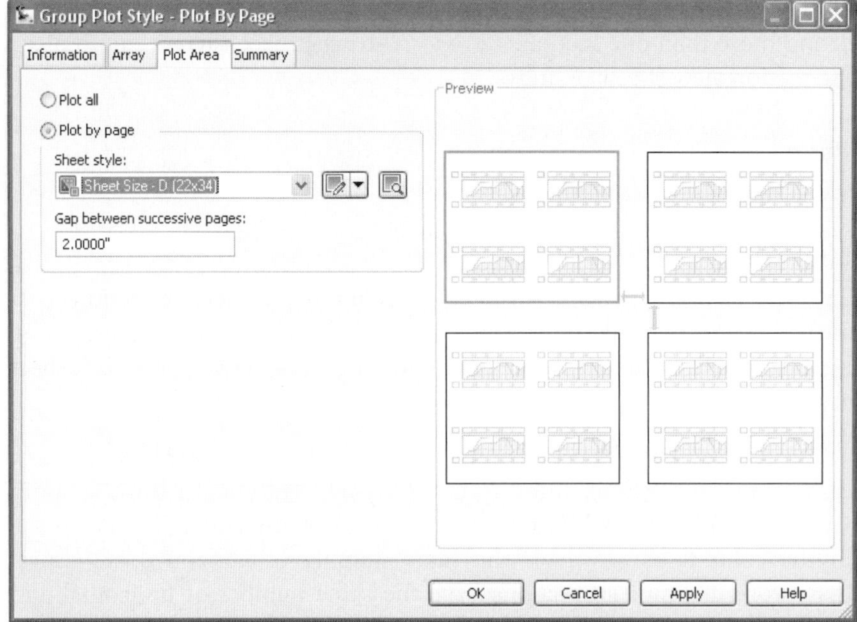


**FIGURE 8.33**

## SHEET SIZE STYLE

A sheet size style sets a section sheet's plotting area (see Figure 8.34). The sheet sizes and margins should reflect the plotter's specifications.

The margins define space around the sheet's edge, thus allowing a border to be anywhere on the sheet. The panel's left side defines a horizontal and vertical spacing grid for the sheet's printable area. The grid acts as a section location "snap." Group plot styles use these values to determine a section's horizontal and vertical spacing location.

When you are setting a sheet size and the page layout is set to Default (Model), the model space page setup *must* be set to the same sheet size. For example, if you set the sheet size to D (24×36), make sure the model space tab plot setup is for a D size sheet. The same is true if using Default (Layout). The paper space layout must match the size that existed prior to creating the section views.

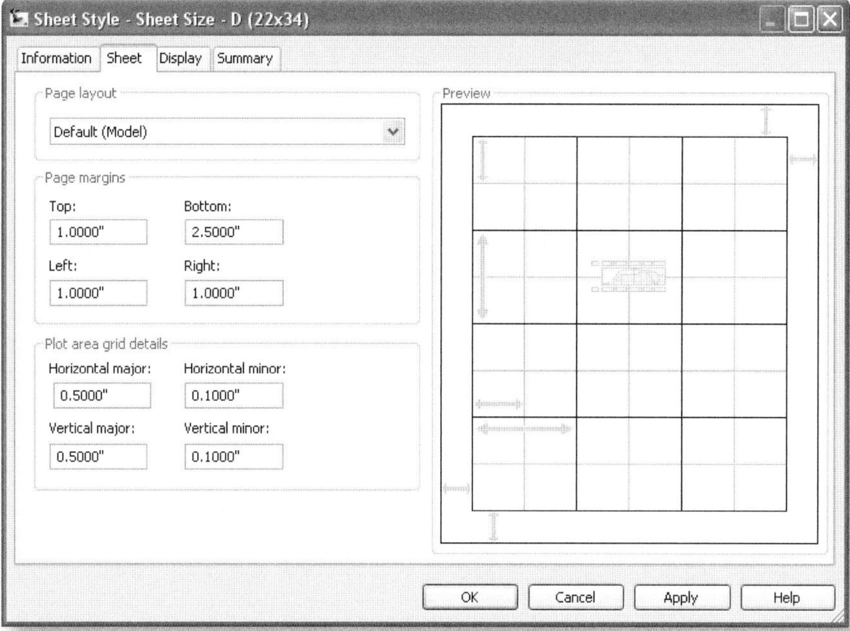


**FIGURE 8.34**

## Plot All

Plot All plots all sections in an array defined by the Array tab's settings (see Figure 8.35).

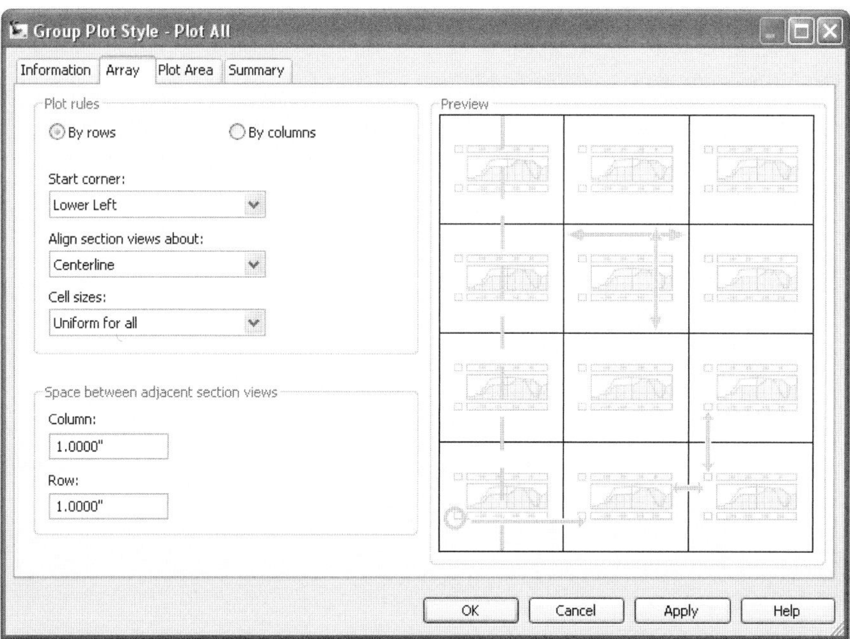


**FIGURE 8.35**

## SECTION VIEW BAND SET

When you create section views, a traditional section band labels only offsets. A Section View Band set may also include EG (existing) and FG (top of corridor) surface elevations. If you are using surface elevations, they must be assigned in the Create Section View or Create Multiple Section View commands. The Data Bands panel makes the assignments (see Figure 8.30).

## ERASING EXISTING SECTION VIEWS

If you are erasing and you want to re-create the existing section views, the sample line group must be resampled. Resampling is a Section Line Groups' property. A better strategy is to save the drawing, import the sections and if not correct, undo the section import.

## EXERCISE 8-3

After completing this exercise, you will:

- Be familiar with the settings for paper styles.
- Be familiar with the plot page and plot all styles.
- Be able to adjust and plot pages of sections.
- Be able to plot all of the sections.
- Be familiar with section view properties.
- Be familiar with section properties.

### Exercise Setup

This exercise continues with the previous exercise's drawing. If you did not complete the previous exercise, browse to the Chapter 8 folder of the CD that accompanies this textbook and open the *Chapter 08 - Unit 3.dwg* file.

1. If not open, open the previous exercise's drawing or browse to the Chapter 8 folder of the CD that accompanies this textbook and open the *Chapter 08 - Unit 3* drawing.

### Review Page Sheet Style

When you plot with a page group style, the style defines the sheet size, its printable area, and its grid spacing. The Model or Layout space default layout must make the sheet style size definition.

1. Click the ***Settings*** tab.
2. In Settings, expand the Section View branch until you view the Sheet Styles' styles list.
3. From the list, select **Sheet Size - D (22×34)**, press the right mouse button, and from the shortcut menu, select EDIT....
4. Click the ***Sheet*** tab.

The settings allow for a border with its information around the sections.

5. Click the ***Display*** tab.

This panel controls the page boundary, its printable area, and its grid visibility.

6. Click ***Cancel*** to exit the dialog box.

## Plot By Page

Plot By Page plots sections as an array (see Figure 8.32). Currently, the array is four sections wide and has as many rows as necessary to plot all of the sections.

1. In Settings, expand the Section View branch until you view the Group Plot Styles' styles list.
2. From the Styles list, select **Plot By Page**, press the right mouse button, and from the shortcut menu, select EDIT....
3. Click the ***Array*** tab.
4. If necessary, match your panel's settings to those in Figure 8.32.
5. Click the Plot Area Tab. When you are done, click ***OK*** to exit the dialog box.
6. Use the PAN and ZOOM commands to move the site west.
7. At Civil 3D's upper left, Quick Access Toolbar, click the ***Save*** icon to save the drawing.

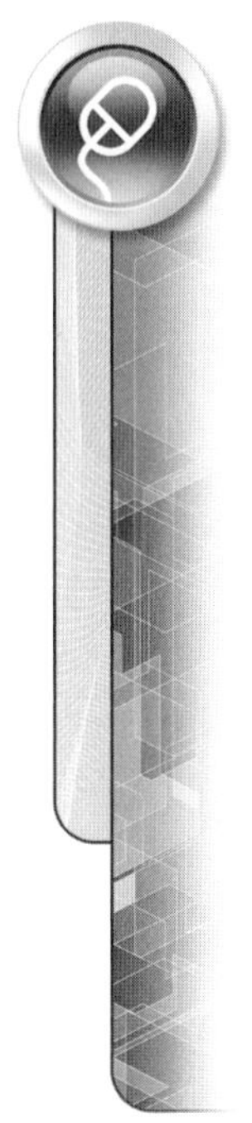

## Set Model Space Plotting

The model space paper size must match the section view sheet size.

1. Click the Ribbon's Output tab, from the Plot panel, select PAGE SETUP MANAGER.
2. With Model highlighted, click MODIFY... to display the Plot Setup – Model dialog box.
3. Click the Paper size's drop-list arrow, and from the list of paper sizes, select **ANSI expand D (34.00×22.00 inches)**.
4. Click ***OK*** and then click Close to return to the command line.
5. At Civil 3D's top left, Quick Access Toolbar, click the ***Save*** icon to save the drawing.

## Creating Section Views

1. Click the ***Prospector*** tab.
2. In Prospector, expand the Corridors branch.
3. Pan the drawing so you have empty space for the sections.
4. If the Rosewood corridor is out-of-date, select Rosewood - (1), press the right mouse button, and from the shortcut menu, select REBUILD.
5. Click the ***Modify*** tab, on the Profile & Section Views panel, and click the ***Sample Line*** icon. From the Ribbon's Sample Line tab's Launch Pad panel, select the ***Create Section View*** icon displaying a shortcut menu and from the shortcut menu, select CREATE MULTIPLE SECTION VIEWS to display the Create Multiple Section Views wizard.
6. In the dialog box's lower right, set the Group Plot Style to PLOT BY PAGE, and click ***Next*** to continue.
7. In Offset Range, review its values, and click ***Next*** to continue.
8. In Elevations Range, review its values, and click ***Next*** to continue.
9. In Section Display Options, toggle **OFF** Rosewood - (1) Rosewood - Datum.
10. For the Rosewood - (1) corridor Overrides column, toggle it **ON** and assign **All Codes**.
11. Click ***Next***.
12. In Data Bands, set Surface1 to **Existing Ground**, set Surface2 to **Rosewood - (1) Rosewood - Datum**, and click ***Create Section Views*** to continue.
13. Select a point in the lower left of the screen.

14. Use the PAN and ZOOM commands to review the sections.
15. Close the drawing and **do not save it**.

The only annotation remaining is the section labels from Multipurpose style, All Codes. This concludes creating section views. Next, you will review Multipurpose styles that annotate assembly values.

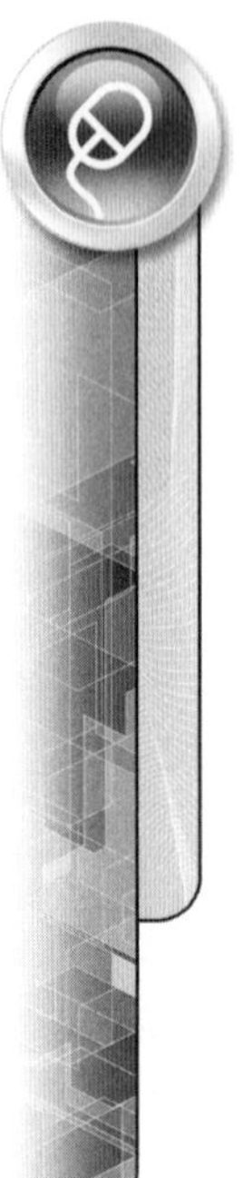

## SUMMARY

- Create Multiple Section Views imports section views and sections in an array or page format.
- When you create sections using a sheet size, the page setup for model or paper space must be set to the specified sheet style's size.
- The Create Section View command creates one section view at a time.
- Section View Properties' Sections panel assigns, removes, or changes a section's label styles.

## UNIT 4: MULTIPURPOSE CODE STYLES

Section view and section label styles focus annotation on surfaces. Corridor assembly annotation comes from General's Multipurpose Style, Code sets and their styles. The code styles reference subassembly points, links, and shape properties for their label data. Even though an assembly has more than one curb, there is only one code for back-of-curb, flange, and so on. Code set label style assignments create labels for all code occurrences.

### CODE SET STYLES

Code Set Style – All Codes has all possible point, link, and shape assembly codes (see Figure 8.36). Each entry has a description, style, label style, and render material. The Style columns define how the subassembly's points, links, and shapes are displayed. The Label Style column assigns each entry a label style. If you assign a label style to an entry, the label appears in the drawing.

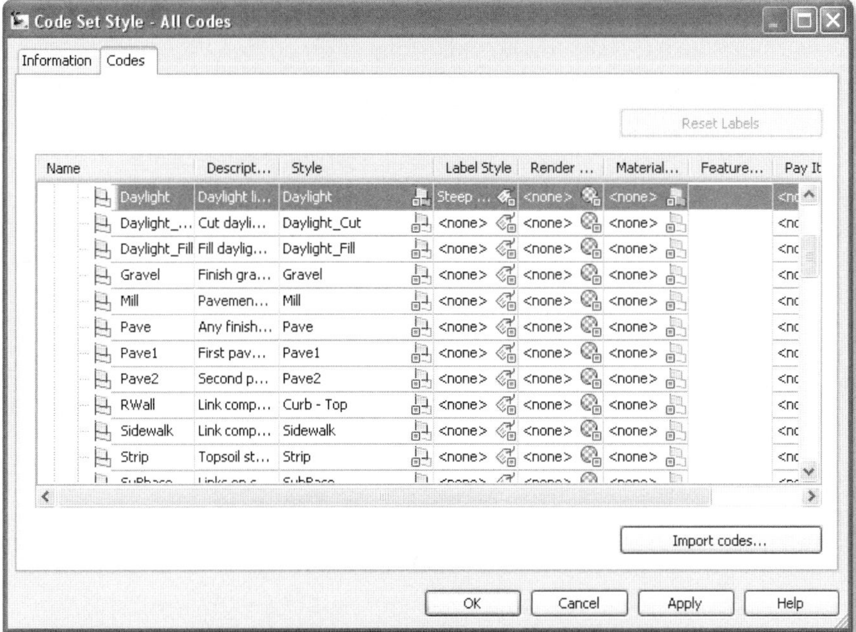


**FIGURE 8.36**

## CODE LABEL STYLE

Code label styles affect point, link, or shape objects. The style's Layout panel focuses on the object type and its properties.

### Marker Label Styles

Marker label styles annotate a subassembly's point code offset and elevation of (see Figure 8.37).

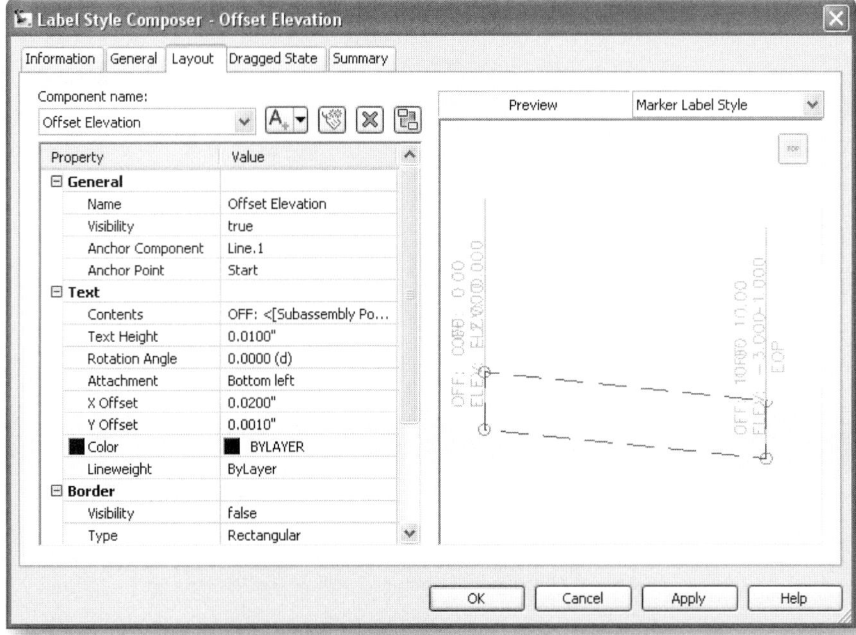


**FIGURE 8.37**

## Link Label Styles

Link label style annotates a subassembly's link slope or grade (see Figure 8.38).

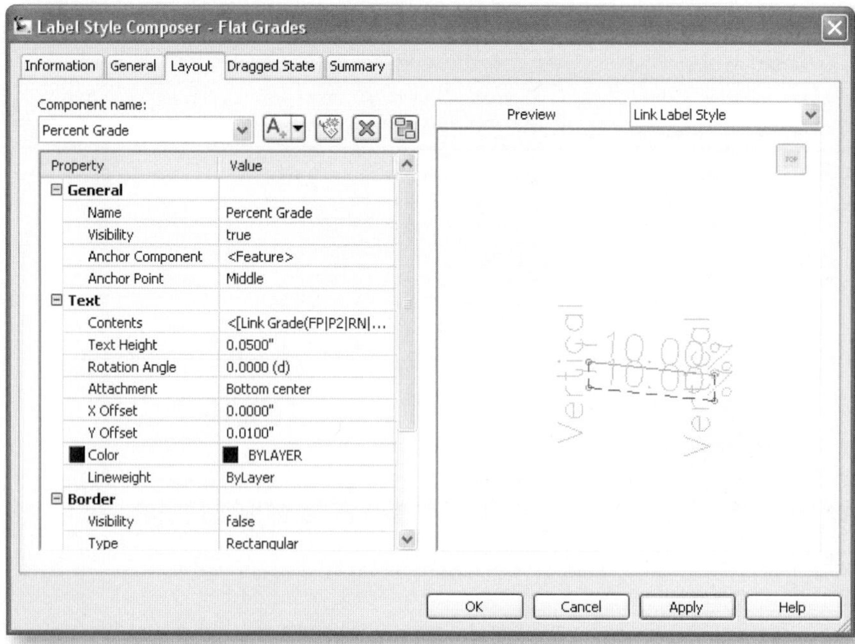


**FIGURE 8.38**

## Shape Label Styles

Shape label style annotates a subassembly's shape (see Figure 8.39).

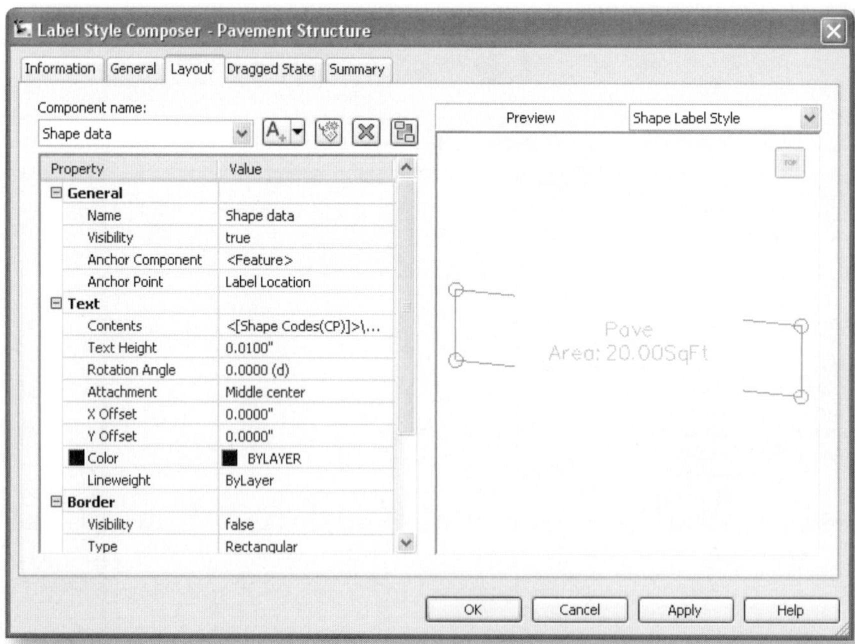


**FIGURE 8.39**

## PROJECTING OBJECTS TO SECTION VIEWS

See the discussion on projecting objects to profile and section views in Chapter 6, Unit 5 of this textbook.

## EXERCISE 8-4

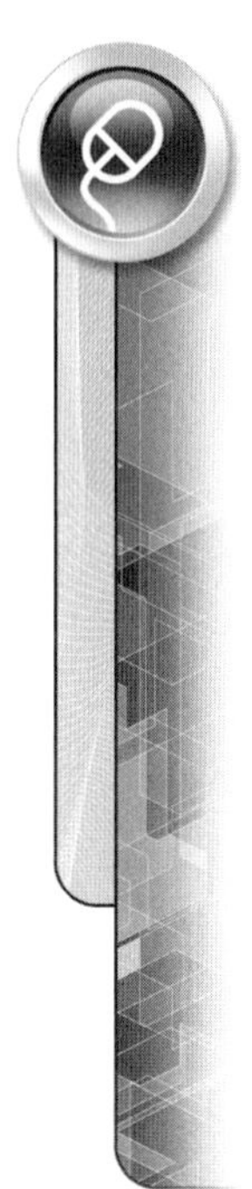

After completing this exercise, you will:

- Be familiar with the code set label styles.
- Be familiar with the code label styles.
- Be able to assign a code label style for a code label set.

### Exercise Setup

This exercise continues with the previous exercise's drawing. If you did not complete the previous exercise, browse to the Chapter 8 folder of the CD that accompanies this textbook and open the *Chapter 08 - Unit 4.dwg* file.

1. Open the previous exercise's drawing or browse to the Chapter 8 folder of the CD that accompanies this textbook and open the *Chapter 08 - Unit 4* drawing.

### Review All Codes Code Set Style

All Codes Set style contains all possible subassembly codes, links, and shapes.

1. Click the ***Settings*** tab.
2. In Settings, expand the General, Multipurpose Styles branch until you are viewing the Code Set Styles' style list.
3. From the styles list, select **All Codes**, press the right mouse button, and from the shortcut menu, select EDIT....
4. If necessary, click the ***Codes*** tab.
5. Review the Link label style assignments.

The only assigned label style is Daylight (Steep Grades).

6. Review the Point section.
7. In the Point section, for BackCurb and Ditch_Out in their Label Style column assign the **Offset Elevation** style, and click ***OK*** until you return to the command line.

### Modifying Labels

The Daylight slope and Offset and Elevation labels are too large and need to be adjusted.

1. Expand the General, Labels Styles branch until you view the Link styles list.
2. From the styles list, select **Steep Grades**, press the right mouse button, and from the shortcut menu, select EDIT....
3. If necessary, click the ***Layout*** tab.
4. In the Label Style Composer - Steep Grades, Text section, click in the Text Height value cell, change the size to **0.05**, and click ***OK*** to exit the dialog box.
5. Expand the Label Style's Marker branch to display its styles list.
6. From the styles list, select **Offset Elevation**, press the right mouse button, and from the shortcut menu, select EDIT....

7. In the Label Style Composer - Offset Elevation, Text section, click in the Text Height value cell, and change the size to **0.05**.
8. Click the Component name drop-list arrow, and select Point Code from the list. Scroll to the Border section, set Background Mask to **False**, and click ***OK*** to exit the dialog box.
9. At Civil 3D's top left, Quick Access Toolbar, click the ***Save*** icon to save the drawing.

## Create Multiple Section Views

1. Click the ***Prospector*** tab.
2. In Prospector, expand the Corridors branch.
3. If the Rosewood corridor is out-of-date, select **Rosewood - (1)**, press the right mouse button, and from the shortcut menu, select REBUILD.
4. Use the ZOOM EXTENTS command to view an open area of the drawing.
5. Click the Ribbon's Sample Line tab, on the Launch Pad panel, select ***Create Section View*** displaying a shortcut menu and from the shortcut menu, select CREATE MULTIPLE SECTION VIEWS to display the Create Multiple Section Views wizard. If the Sample Line tab, is not displayed on the Ribbon, click the ***Modify*** tab and on the Profile & Section Views panel, click the ***Sample Line*** icon.
6. In the dialog box's lower right, set the Group Plot Style to **Plot By Page**, and click ***Next*** to continue.
7. In Offset Range, review its values, and click ***Next*** to continue.
8. In Elevations Range, review its values, and click ***Next*** to continue.
9. In Section Display Options, toggle **OFF** Rosewood - (1) Rosewood - Datum.
10. For the Rosewood - (1) corridor Style column, assign the style **All Codes**.
11. Click ***Next***.
12. In Data Bands, set Surface1 to **Existing Ground**, set Surface2 to **Rosewood - (1) Datum**, and click ***Create Section Views*** to continue.
13. Select a point in the lower left of the screen.
14. Use the PAN and ZOOM commands to review the sections.
15. Close the drawing and **do not save it**.

This concludes this unit's discussion on labeling corridor components. The next unit covers calculating earthwork and material quantities.

## SUMMARY

- Section view and section styles do not label the corridor components.
- Code set styles label assembly markers, links, and shapes.
- Adding a label to the code set updates sections using the set.

## UNIT 5: QUANTITY TAKEOFFS

This last unit covers calculating corridor and subassemblies quantities: earthworks and materials. A volume surface or quick comparison of the Existing Ground and Rosewood - Datum, calculate roadway earthwork volumes. However, a third calculation method creates section material lists for formatted earthwork and materials volume reports.

Before you calculate a volume, you must define the takeoff criteria. This is done in the Settings, Takeoff Criteria branch. There can be multiple material lists for earthwork or material volumes.

## MATERIALS

When calculating subassembly material volumes, the focus is on subassembly shapes (Pave1, Pave2, Subbase, Base, Curb, etc.) (see Figure 8.40).

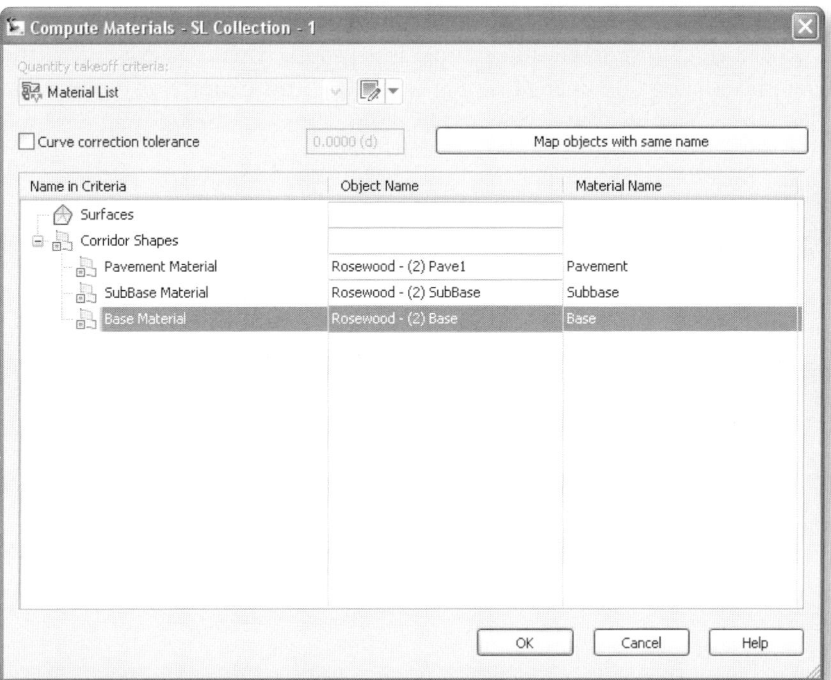


**FIGURE 8.40**

## EARTHWORKS

Earthworks is a quantity takeoff criteria and is a comparison of two surfaces' elevation differences: a base and comparison surface. Civil 3D uses this information, first, to hatch cut and fill areas (see Figure 8.41) and second, to calculate volumes. Reading the cut and fill criteria definition may be confusing. The above and below is not a position reference, but rather is an area reference, that is, the area below EG and the area above Datum define the cut area.

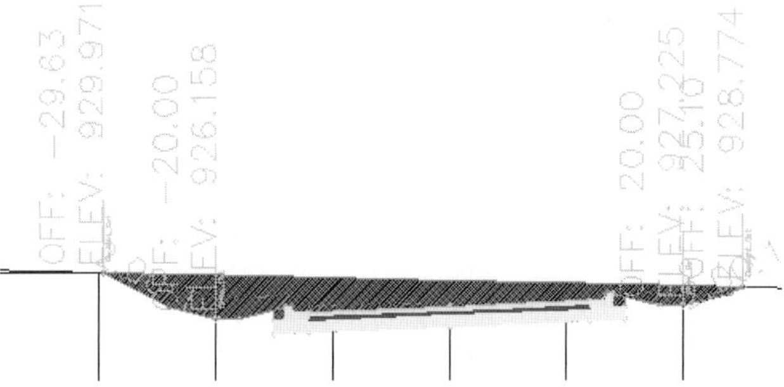


**FIGURE 8.41**

## CUT AND FILL

Cut and fill defines respective areas for each section, and this data is used for hatching these areas in roadway sections.

## QUANTITY TAKEOFF

After defining criteria sets, you next assign corridor components to the criteria values. There are two places to define the material list: Sample Group Properties and from the Ribbon's Modify, Sample Line panel, Compute Materials command. When you use Sample Line's command, a Select a Sample Line Group dialog box opens. After setting the correct alignment and sample line group, when you click OK, the Compute Materials dialog box opens (see Figure 8.42). This dialog box defines one material list at a time. Sample Group Properties defines any number of material lists.

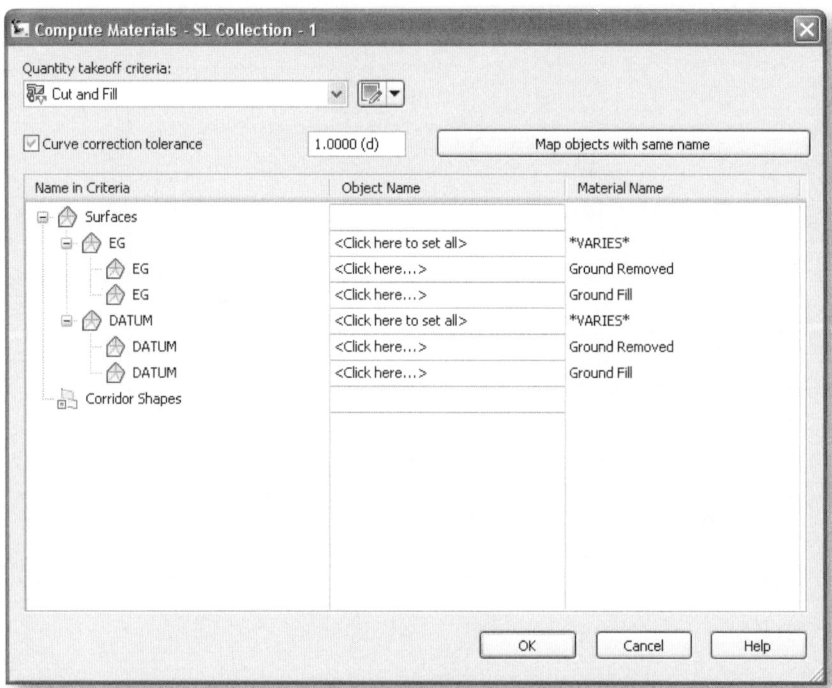


**FIGURE 8.42**

When defining the material list in Sample Line Group Properties dialog box, you add new criteria and its materials here. At the dialog box's lower right, clicking Import another criteria opens a Select Quantity Takeoff Criteria dialog box to set the new criteria (see Figure 8.43). After you select the Criteria, the Compute Materials dialog box opens (see Figure 8.44). Compute Materials sets the initial material list and its assigned components. The list is set and cannot be modified until you click OK and add the new material list to the Compute Materials dialog box. Once added, at Compute Materials' top left, click Add new material to add a material. The new material must be renamed and assigned a corridor component. You add a component at the dialog box's top center. If this is a surface assignment, change the Data type to Surface, and from the Select surface drop-list, select the new surface and click the plus sign, (+). If it is a Corridor Shape, change the Data type to Corridor Shape, and from the Select corridor shape drop-list, select the appropriate corridor shape (see Figure 8.45).

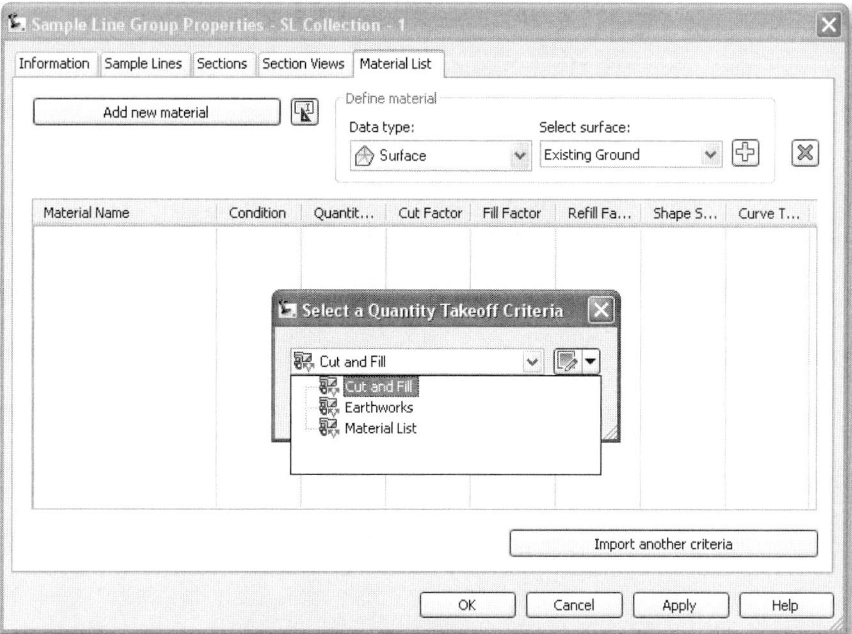


**FIGURE 8.43**

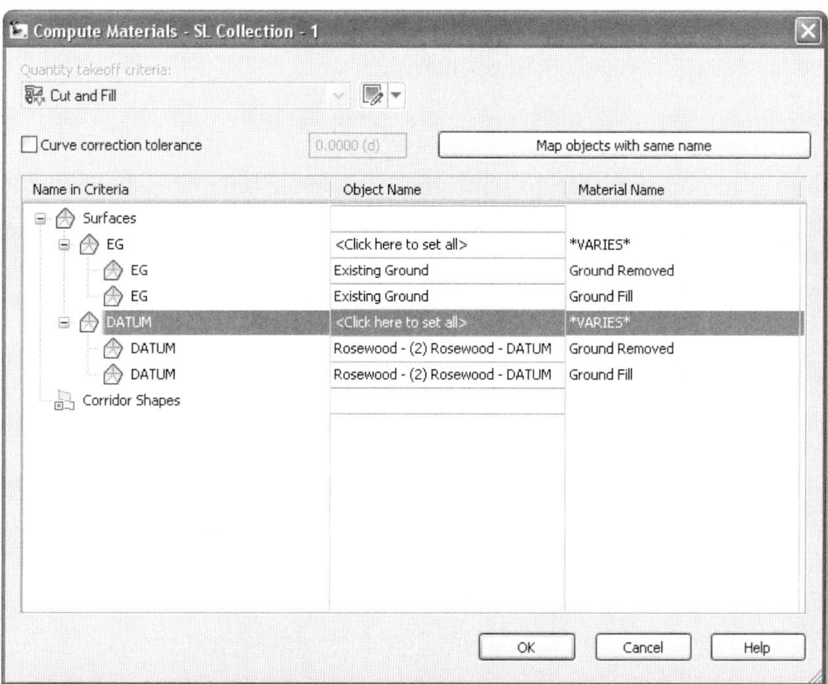


**FIGURE 8.44**

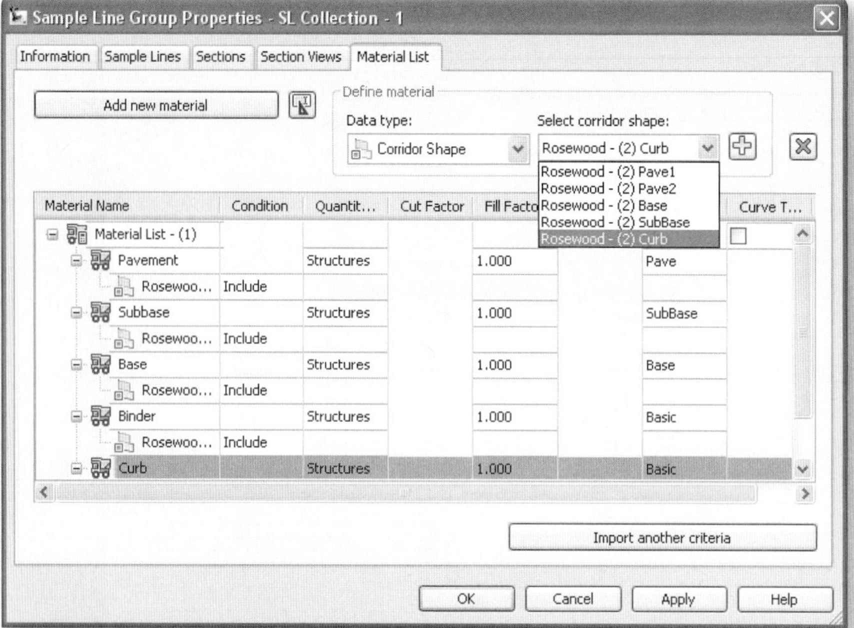


**FIGURE 8.45**

## QUANTITY REPORTS

After defining material lists, next you create a report. The most important step in the report process is having a correctly defined material list.

## MASS HAUL DIAGRAM

A mass haul diagram is a chart that diagrams the status of cut/fill materials and a design's overall balance. A mass haul diagram has free haulage parameters (from points of balance) and borrow and dump parameters. Borrow adds available fill material and Dump subtracts from the fill material.

## PAY ITEM REPORTS

Objects with pay item assignments are available for reports using the item's values. There are two types of pay item reports; summary and detailed. A summary report lists the pay item ID, its description, total quantity, and unit of measure for each pay item. A detailed report contains a line of information for each selected object.

- A pay item report's scope is a drawing, sheet, or selected objects.
- A summary report does not calculate values for corridor codes.
- A detailed (itemized) report calculates Corridor codes assignments.
- If the pay item relates to an alignment, the report can be limited by alignment station values.

## EXERCISE 8-5

After completing this exercise, you will:

- Be familiar with the quantity criteria takeoff.
- Be able to create a quantity takeoff earthwork and material report.
- Be able to create a quantity takeoff table.

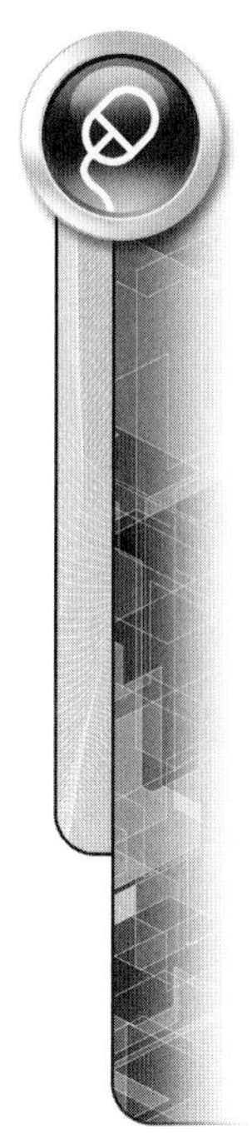

### Exercise Setup

This exercise continues with the previous exercise's drawing. If you did not complete the previous exercise, browse to the Chapter 8 folder of the CD that accompanies this textbook and open the *Chapter 08 - Unit 5.dwg* file.

1. Open the previous exercise's drawing or browse to the Chapter 8 folder of the CD that accompanies this textbook and open the *Chapter 08 - Unit 5* drawing.

### Material List — Criteria Styles

First, you define a cut and fill and then an earthworks material list. Sections use cut and fill data for hatching an assembly's cut and fill areas and earthworks is used as a section label or a detailed earthworks report.

1. Click the ***Settings*** tab.
2. Expand the Quantity Takeoff branch until you are viewing the Quantity Takeoff Criteria list.
3. From the list, select **Cut and Fill**, press the right mouse button, and from the shortcut menu, select EDIT....
4. Click the ***Material List*** tab.
5. Expand the Material Name sections and review the values.

The Condition settings are confusing. The settings do not define the surfaces' relative positions, but do define the areas that are cut and fill. For example, the area below EG and above Datum creates a cut area. The area below Datum and above EG creates a fill area.

6. Click ***OK*** to exit the dialog box.
7. From the Quantity Takeoff Criteria list, select **Earthworks**, press the right mouse button, and from the shortcut menu, select EDIT....
8. Click the ***Material List*** tab.
9. Expand the Material Name and review the values.

The criteria compare a base to a comparison surface, just like the Ribbon's Modify, Surface panel's, Volumes routine.

10. Click ***OK*** to exit the dialog box.

The assembly has five potential volume materials: Pave1, Pave2, Base, Subbase, and Curb.

11. From the Quantity Takeoff Criteria list, select **Material List**, press the right mouse button, and from the shortcut menu, select EDIT....
12. Click the ***Material List*** tab.
13. Expand the Material Name sections and review the values.

The materials list contains three materials: Pavement, Base, and SubBase. The list needs two more materials: Binder and Curb.

14. In the dialog box's top left, make a new material entry by clicking ***Add New Material***.

15. Click in the Material Name cell, and for its name enter **Binder**.
16. Click in Binder's Quantity Type cell, and from the drop-list select **Structures**.
17. At the Panel's top center, change the Data Type to **Corridor Shape**. Adjacent to the Data Type, click Select Corridor Shape's drop-list arrow, from the list select **Pave2**, and click the plus sign (+), to add Pave2 to Binder.
18. Repeat Steps 14–17 and add **Curb** to the material list. The Curb material is a **structure** using the corridor shape of **Curb**.
19. For the Pavement entry, click its Pavement shape (Pavement Material) and click the red **X** to delete it.
20. With Pavement still highlighted, at the top center, change the Data Type to **Corridor Shape**, from the Select Corridor Shape list select **Pave1**, and click the plus sign (+), to add Pave1 as Pavement's data type.
21. Change the Shape Style for Binder to **Pave2**, and the Shape Style for Curb to **Curb**.

    Your values should match those in Table 8.1 and Figure 8.46.

**TABLE 8.1**

| Material Name | Quantity Type | Data Type | Corridor Shape and Style |
|---|---|---|---|
| Pavement | Structures | Corridor Shape | Pave |
| SubBase | Structures | Corridor Shape | Subbase |
| Base | Structures | Corridor Shape | Base |
| Binder | Structures | Corridor Shape | Pave |
| Curb | Structures | Corridor Shape | Curb |

22. Click ***OK*** to exit the Quantity Takeoff Criteria - Material List dialog box.
23. At Civil 3D's top left, Quick Access Toolbar, click the ***Save*** icon to save the drawing.

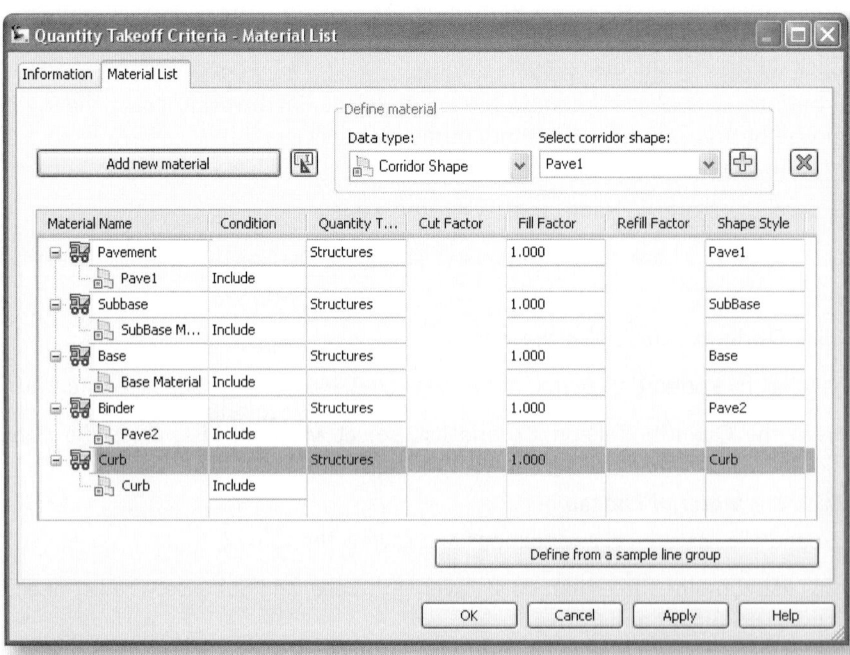


**FIGURE 8.46**

## Sample Line Properties Material Lists — Cut and Fill

1. Use the ZOOM and PAN commands to view the corridor and its section lines.
2. Click the ***Prospector*** tab.
3. In Prospector, expand the Sites branch until you are viewing the Rosewood - (1) Sample Line Group groups list.
4. From the sample line list, select **SL Collection – 1**, press the right mouse button, and from the shortcut menu, select PROPERTIES....
5. Click the ***Material List*** tab.
6. At the panel's lower right, click Import Another Criteria to display the Select a Quantity Takeoff Criteria dialog box. From its criteria list , select **Cut and Fill**, and click ***OK*** to display the Compute Materials dialog box.
7. In the dialog box's top center, in EG's Object Name cell, click (<Click here to set all>), and from the surface list, select **Existing Ground**.
8. Repeat the previous step for the Datum entry and select **Rosewood - (1) Rosewood - Datum**.
9. Click ***OK*** to compute the materials and return to the Sample Line Group Properties dialog box.

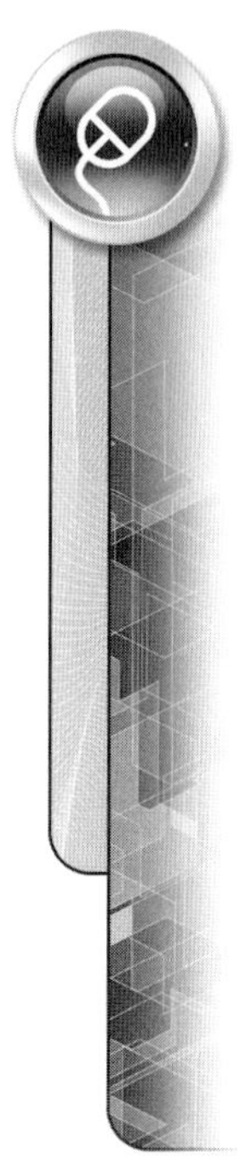

## Sample Line Properties Material Lists — Earthworks

1. At the panel's lower right, click ***Import Another Criteria*** to display the Select a Quantity Takeoff Criteria dialog box. Click the drop-list arrow, from its list of criteria, select **Earthworks**, and click ***OK***.
2. The Compute Materials dialog box opens. In the top center, in Existing Ground's Object Name cell, click (<Click here>), and from the surface list, select **Existing Ground**.
3. In Datum's Object Name cell, click (<Click here>), and from the surface list, select **Rosewood - (1) Rosewood - Datum**.
4. Click ***OK*** until you return to the command line.
5. At Civil 3D's top left, Quick Access Toolbar, click the ***Save*** icon to save the drawing.

## Material List — Sample Line Panel

1. From the Sample Line tab's Launch Pad panel, select COMPUTE MATERIALS. If the Sample Line tab, is not displayed on the Ribbon, click the Modify tab and on the Profile & Section Views panel, click the Sample Line icon.
2. In the Select a Sample Line Group dialog box, set the alignment to **Rosewood - (1)**, the Sample line group to **SL Collection – 1**, and click ***OK*** to continue.
3. In the Edit Material List's bottom right, click ***Import Another Criteria*** button.
4. In the Select a Quantity Takeoff Criteria dialog box, click the drop-list arrow, select **Material List**, and click ***OK***.
5. Using Figure 8.47 as a guide, set the object types in Compute Materials. Click Map objects with same name to fill in some of the values.

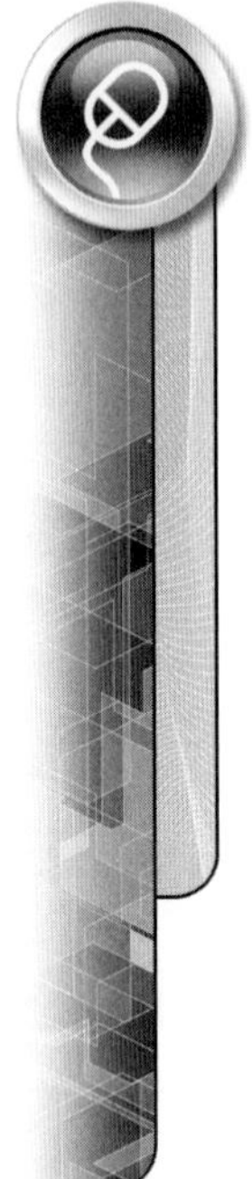

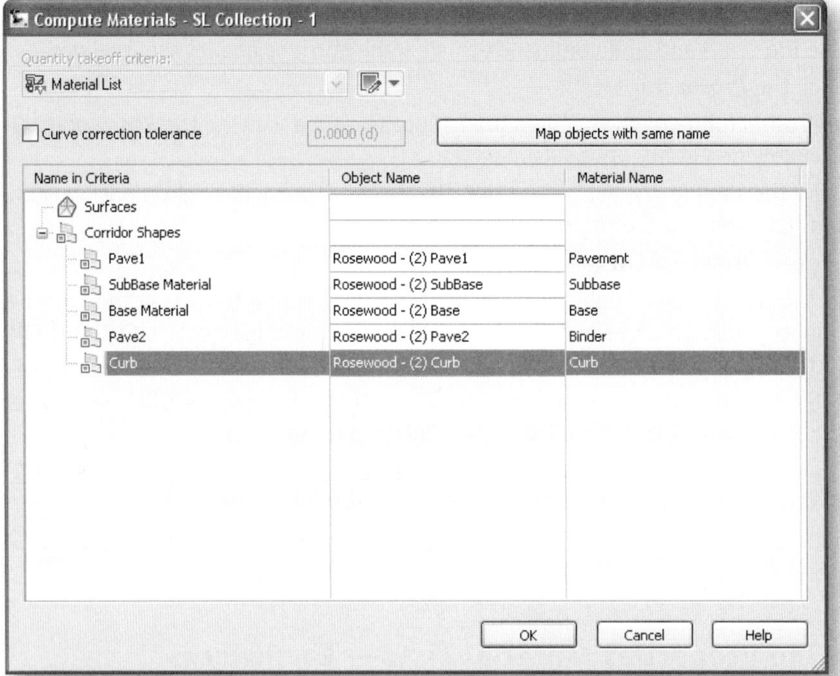


**FIGURE 8.47**

6. Click ***OK*** twice to add the materials to the list and to exit the dialog boxes.
7. At Civil 3D's top left, Quick Access Toolbar, click the ***Save*** icon to save the drawing.

## Calculating Earthwork Volumes

1. From the Sample Line tab's Launch Pad panel, select GENERATE VOLUME REPORT.

The Report Quantities dialog box opens.

2. In the Report Quantities dialog box, set the Alignment to **Rosewood - (1)**.
3. In the Report Quantities dialog box, set the Material List to **Material List - (2)**.
4. In the Report Quantities dialog box, to the right of Select a style sheet, click its icon, in the Select Style Sheet dialog box, select **Earthwork.xsl**, and click the ***Open*** button.
5. Click ***OK*** to create an earthwork volume report.
6. If you get a scripts warning, click ***Yes*** to continue.
7. Review the values, and after reviewing them, click the Internet Explorer **X** to close the report window.

## Calculating Material Volumes

1. From the Sample Line tab's Launch Pad panel, select GENERATE VOLUME REPORT.

The Report Quantities dialog box opens.

2. In the Report Quantities dialog box, set the Alignment to **Rosewood - (1)**.
3. In the Report Quantities dialog box, set the Material List to **Material List - (3)**.

4. In the Report Quantities dialog box, to Select a style sheet's right, click its icon, in the Select Style Sheet dialog box, select **Select Material.xsl**, and click ***Open***.
5. Click ***OK*** to create a materials volume report.
6. If you get a scripts warning, click ***Yes*** to continue.
7. Review the values, and after reviewing them, click the Internet Explorer **X** to close the report.

## Creating a Volume Table

1. Use the ZOOM EXTENTS command to view an open area of the drawing.
2. From the Sample Line tab's Labels & Tables panel, click the Add Tables icon, and from the drop-list, select TOTAL VOLUME.
3. In the Create Total Volume Table dialog box, set the Alignment to **Rosewood - (1)**, set the Material List to **Material List - (1)**, and click ***OK***.
4. The command line prompts you for a table location. In the drawing, select a point below the profile view to locate the table.
5. Use the ZOOM and PAN commands to better view the table.
6. At Civil 3D's top left, Quick Access Toolbar, click the ***Save*** icon to save the drawing.

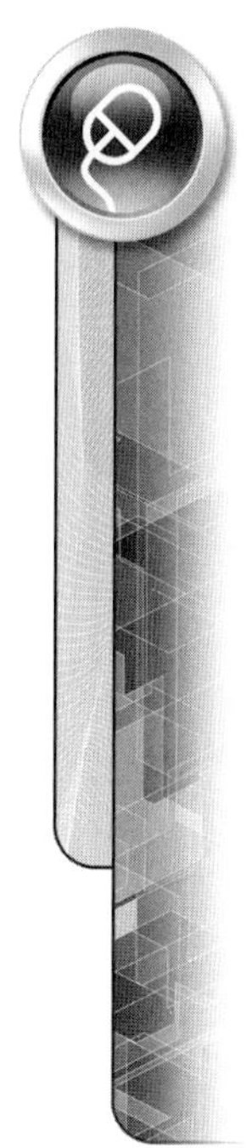

## Create a Mass Haul Diagram

1. Use the PAN and ZOOM commands to view an open area in the drawing.
2. From the Sample Line tab's Launch Pad panel, select CREATE MASS HAUL DIAGRAM.
3. The Create Mass Haul Diagram wizard opens. Review its contents and click ***Next***.
4. Set the Material list to **Material List - (1)**, set the Material to display as mass haul to **Total Volume**, and click ***Next***.
5. Review the values in the Balancing Options panel, and, finally, click ***Create Diagram***.
6. The command line prompts you for a diagram location. In the drawing, select a point to locate the Mass Haul View.
7. Use the ZOOM and PAN commands to better view the graph.
8. At Civil 3D's top left, Quick Access Toolbar, click the ***Save*** icon to save the drawing.

## Creating Sections with Volume tables

1. Use the PAN and ZOOM commands to view an open area in the drawing.
2. From the Sample Line tab's Launch Pad panel, click the ***Create Section View*** icon and from the drop-list, select CREATE MULTIPLE SECTION VIEWS.
3. In Create Multiple Section Views – General, change the Group plot style to **Plot by Page**, and click ***Next***.
4. In Offset Range, click ***Next***.
5. In Elevation Range, click ***Next***.
6. In Section Display Options, toggle OFF **Rosewood - (1) Rosewood - Datum**.
7. For the Rosewood - (1) corridor Style column assign **All Codes**.
8. Click ***Next***.
9. In Data Bands, set Surface1 to **Existing Ground**, and click ***Next***.

10. In Section View Tables, set Type to **Total Volume**, set Table Style to **Standard**, and click ***Add>>***.
11. In Section View Tables, Position of Tables, set X-Offset to **0.5**, and click ***Create Section Views***.
12. Select a point in the lower left of the screen.
13. Use the PAN and ZOOM commands to review the sections and tables.
14. At the Sample Line tab's right, click the **X** to close the panel.
15. At Civil 3D's top left, Quick Access Toolbar, click the ***Save*** icon to save the drawing.

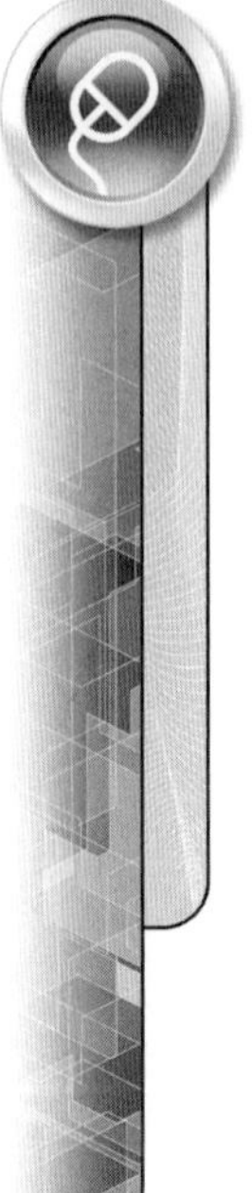

This ends the review of corridor section and volumes. Sample Line Groups are the basis for cross-sections, their annotation, and volume reports. Material lists compute section data for reports or tables that display volume values.

## SUMMARY

- Quantity takeoff criteria define the base and comparison surfaces.
- Quantity takeoff criteria use assembly shapes to calculate material volumes.
- A materials list links a quantity takeoff criteria entry subassembly shape or surface.
- Materials list are the basis of volume-calculation tables and reports.

This ends the section view and section chapter. A section is the roadway design cross-section view. It is an important part of design evaluation and documentation.

# CHAPTER 9

# Transitions, Superelevation, and Intersections

## INTRODUCTION

Roadway designs must accommodate a variety of issues. These issues include road widening, cut or fill designs, and sustaining constant speeds. Civil 3D subassemblies react to changes in depth above or below the existing ground surface, transitions to widen or narrow lane widths, and transitions to tilt the lanes to one side or another to safely maintain speeds through a design.

## OBJECTIVES

This chapter focuses on the following topics:

- Creating a Cul-de-Sac Knuckle
- Transiting a Road to Different Lane Widths
- Superelevation Using Alignment Parameters and Design Criteria
- Creating an Intersection Design

### OVERVIEW

This chapter's topics cover how Civil 3D uses a subassembly's parameters to create solutions. Additional transition alignments widen and narrow a lane's width. Superelevation is a combination of alignment properties and design criteria. The Intersection editor creates an engineering design based on user parameters (see Figure 9.1).

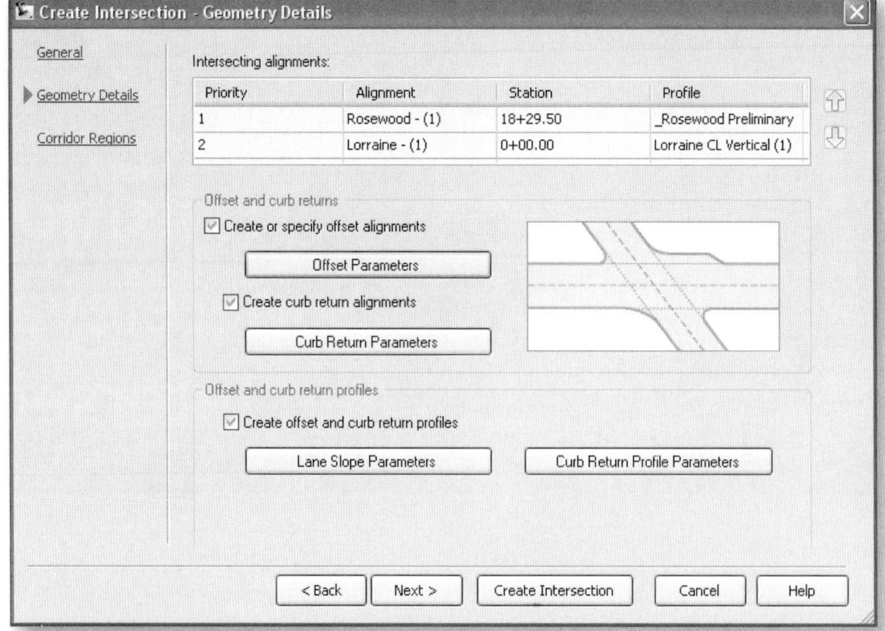


**FIGURE 9.1**

When creating transitioning pavement, there must be additional alignment type (offset alignments) to control the edge-of-travelway (width). In addition to horizontal alignments, some subassemblies require a vertical alignment for each horizontal transition alignment. The Create Corridor dialog box's Set Target panel creates the alignment – assembly attachments (see Figure 9.2). Subassemblies that transition must be attached to an alignment. A target can be an alignment, polyline, feature line, or survey figure. Civil 3D has types of alignments, this allows the user to specify the function of an alignment by specifying its type. For example, a widening alignment controls a subassembly point. A widening alignment is a separate alignment or is associated with the centerline or other widening alignment. Widening alignment has other properties that need definition to correctly work.

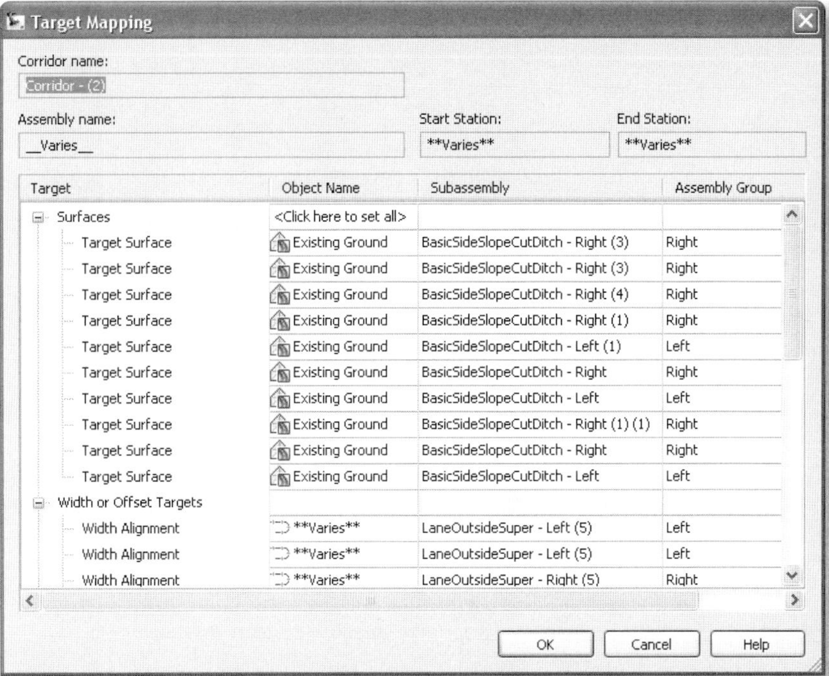


**FIGURE 9.2**

When designing superelevations, design speeds and superelevation table values must be a part of the alignment's properties (see Figures 9.3 and 9.4).

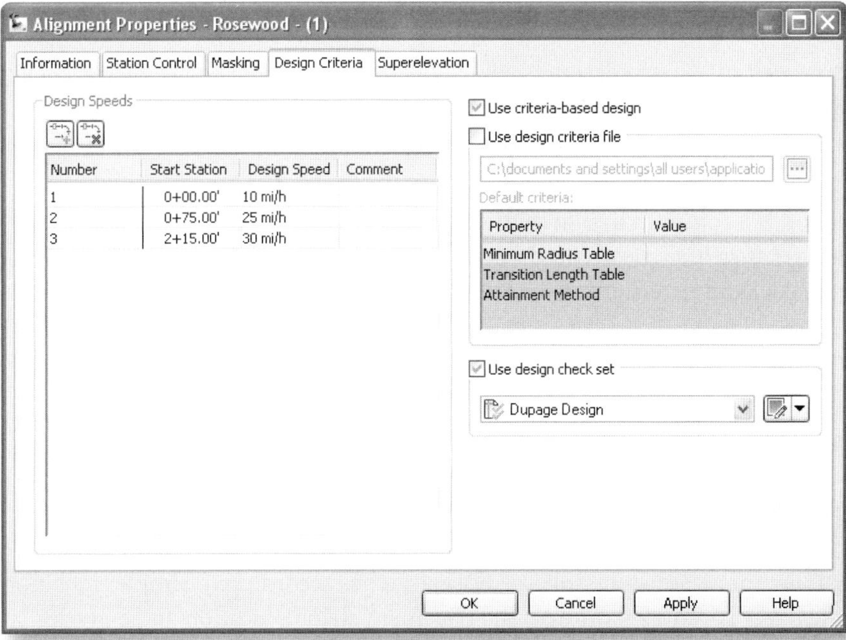


**FIGURE 9.3**

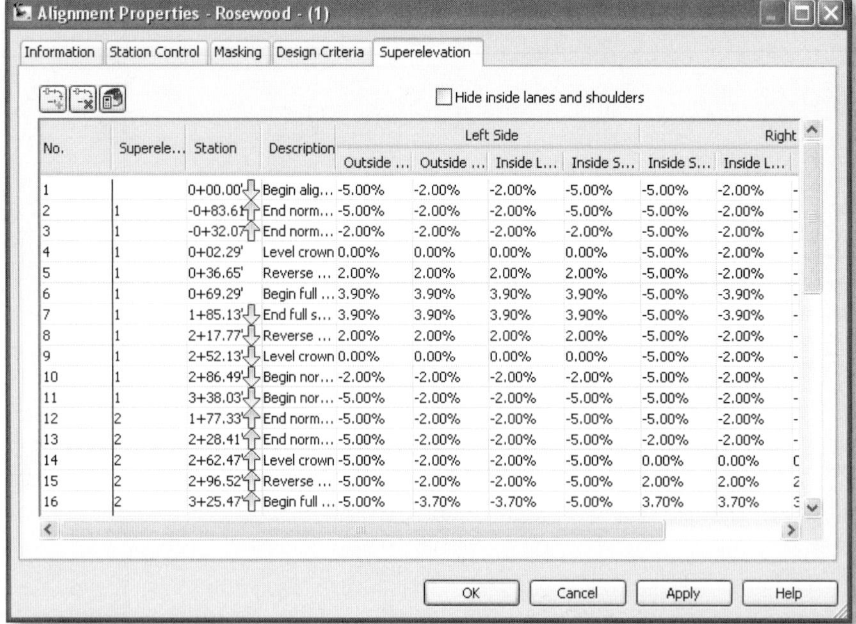


**FIGURE 9.4**

### Unit 1

The focus of the first unit is roadway transitioning. Civil 3D requires a second alignment to widen a lane. The unit's example is a subdivision street knuckle. This unit introduces alignment types.

### Unit 2

This unit reviews cul-de-sac design. Civil 3D widens pavement (edge-of-travelway) with additional horizontal alignments attached to the assembly.

### Unit 3

Superelevation is a common design method with roadways that carry significant traffic, have higher traveling speeds, and have safety concerns. Selecting design criteria, assigning speeds, and determining regional superelevation rules are the focus of this unit.

### Unit 4

The focus of Unit 4 is designing a simple intersection.

## UNIT 1: SIMPLE TRANSITIONS

The Basic Lane transition stock subassembly stretches according to simple transition parameters. These parameters address the pavement shape's location and elevation as it transitions to wider or narrower pavement widths. The basic lane transition subassembly uses this simple approach to transitioning (see Figure 9.5). The subassembly's transition value controls its behavior when an alignment affects the subassembly's location. The values for transitioning a basic lane transition are the following: Hold elevation, change offset; Hold grade, change offset; Change offset and elevation; Hold offset and elevation; and Hold offset, change elevation.

**Hold elevation, change offset**—The offset alignment changes the pavement's width, and the edge-of-travelway elevation is held. The pavement's cross slope changes to maintain the pavement's elevation.

**Hold grade, change offset**—The lane's grade is held, and as the offset alignment increases the pavement width, the pavement's edge lowers in elevation. If the alignment moves the pavement edge closer to the centerline, the pavement edge will rise in elevation to maintain the grade.

**Change offset and elevation**—The offset alignment (horizontal location) and its profile (vertical location) control the edge-of-travelway's location.

**Hold offset and elevation**—This parameter prevents a subassembly from responding to any attached horizontal or vertical offset alignment.

**Hold offset, change elevation**—This parameter holds the pavement width as specified, but an offset profile controls the edge-of-travelway's elevation.

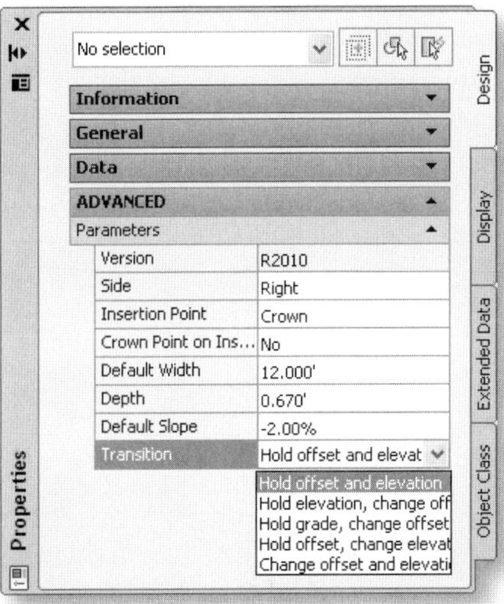


**FIGURE 9.5**

Sections from an assembly that uses simple transition parameters do have issues. The main issue is the sections on the edge-of-travelway. They are always perpendicular to the centerline, but are not perpendicular to any arc in the transitional alignment (see Figure 9.6).

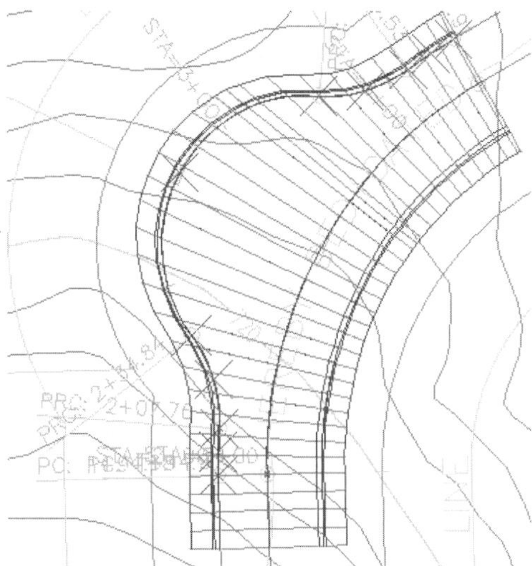

**FIGURE 9.6**

What should appear in the sections are curbs and sidewalks perpendicular to the offset alignment's curve (see Figure 9.7).

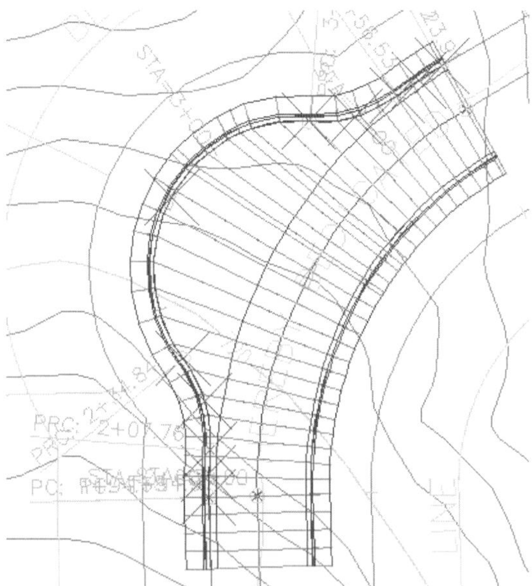

**FIGURE 9.7**

To create perpendicular sections (the curb and sidewalk of Figure 9.7) for subassemblies outside the transition alignment, there must be an assembly offset. The subassemblies attached to an offset are perpendicular to the offset, not to the roadway centerline.

An assembly's Construction tab shows the subassemblies' attachment construction (see Figures 9.8 and 9.9).

An assembly using an offset alignment must have the outside subassemblies attached to that offset alignment, not to the assembly. When attaching a subassembly, the user must select the offset marker, not the assembly marker. An example of an assembly using an offset is in Figure 9.19.

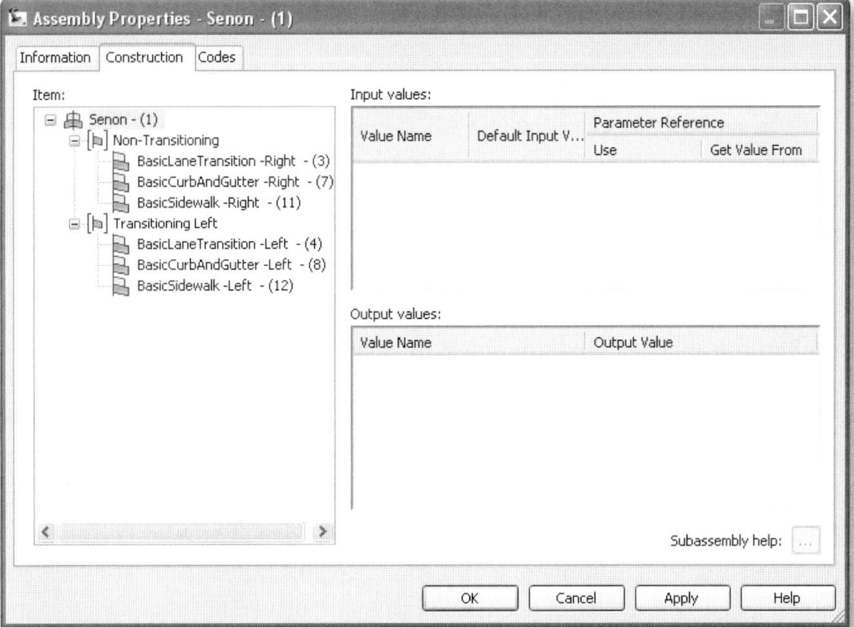


**FIGURE 9.8**

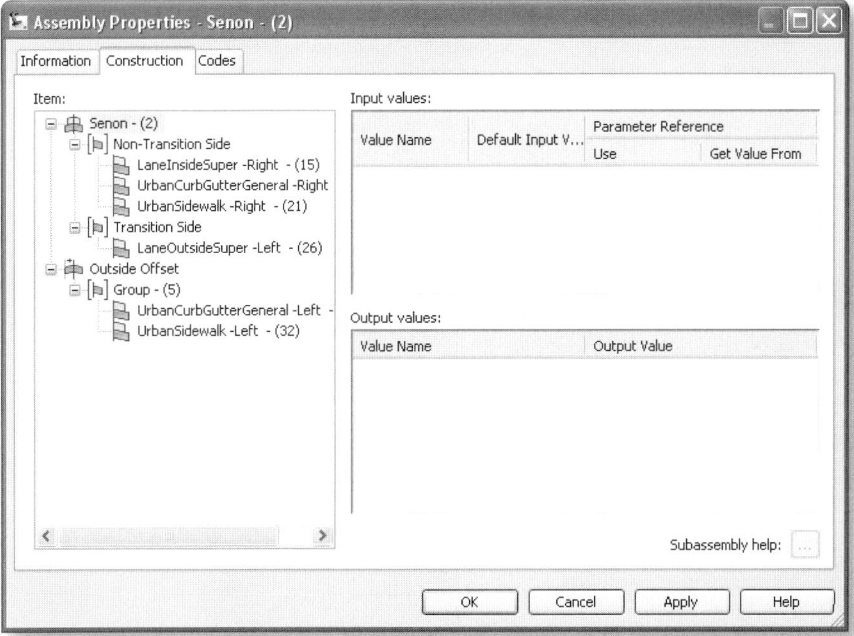


**FIGURE 9.9**

As mentioned, a transition can also have a vertical alignment that controls the transition point's vertical location. The basic lane transition does not need a vertical alignment, except when the transition property is set to Change Elevation.

**NOTE**

**When using an offset in an assembly, Civil 3D may require a vertical alignment that provides elevations for the subassemblies outside the offset.**

**NOTE**

All subassemblies, assemblies, and assembly groups should have descriptive names.

**NOTE**

Add the subassembly side to a subassembly's name by adding it in the Subassembly Edit Feature Settings.

When creating an assembly, the user should assign meaningful names to each subassembly. When the subassembly groupings are viewed in the Assembly Properties dialog box, the subassembly names easily identify the important grouping in a potentially complex assembly. This naming rule should apply to any grouping the assembly defines (right side, left side, or offset attachments).

## IDENTIFYING OFFSET ALIGNMENTS OR OBJECTS

An assembly offset defines a control point and, often, an attachment point for a second alignment (transitional alignment) or specified object. All subassemblies attached to the offset's outside are perpendicular to the alignment's path that controls the offset.

When you attach the first outside subassembly, it must be attached to the offset, not the assembly. The Assembly Properties dialog box indicates if the subassemblies are attached to the offset or the assembly.

Offset alignment or object attachments determine what corridor creation command to use. If you are creating a simple corridor with the basic lane transition, you can use Create Simple Corridor. More complex corridors require that you use Create Corridor. The Create Simple and Create Corridor commands both use a dialog box to set the names of the offset horizontal and vertical alignments or objects (see Figure 9.10).

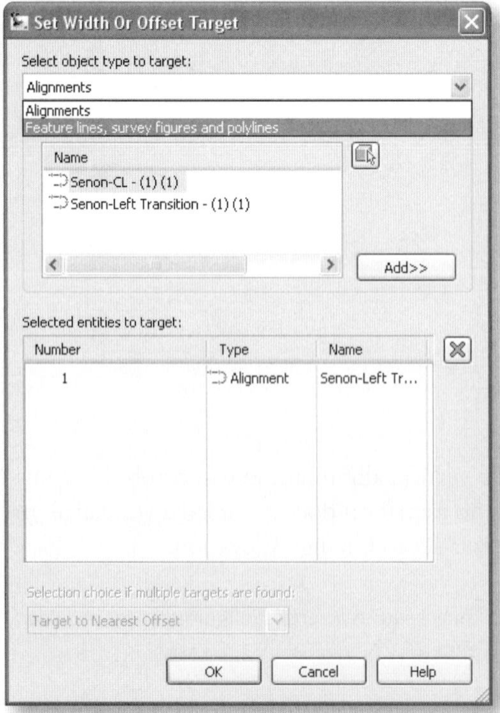


**FIGURE 9.10**

## CORRIDOR PROPERTIES

After creating the corridor, users can edit the corridor properties by adding or changing attached offset alignments (target name mapping) and/or the frequency to apply assemblies (see Figure 9.11).

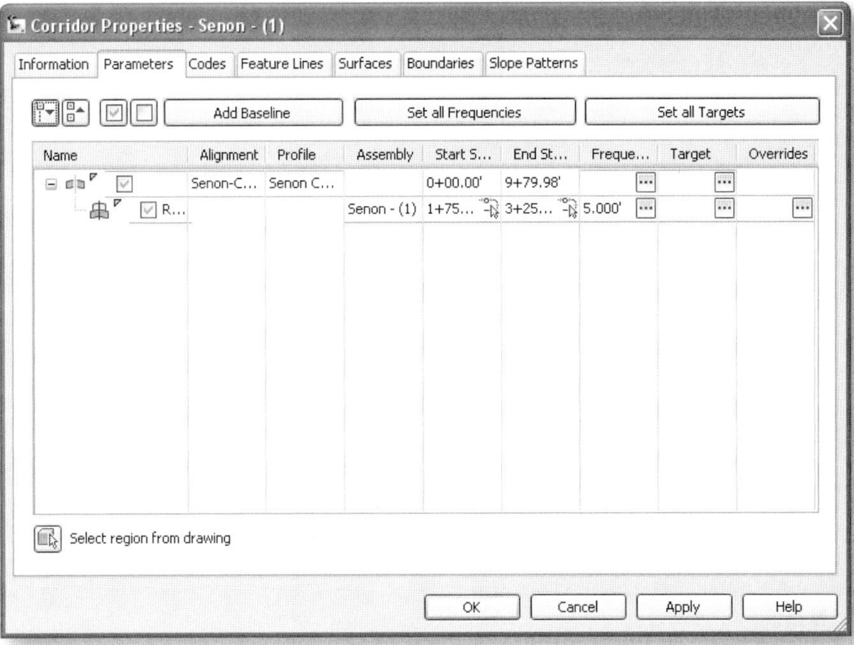


**FIGURE 9.11**

## CORRIDOR SECTION FREQUENCY

With a more complex corridor (containing regions, multiple offset alignments, and so on), users should increase the number of sections that create the corridor. This is done in the Frequency to Apply Assemblies dialog box of the Corridor Properties dialog box (see Figure 9.12).

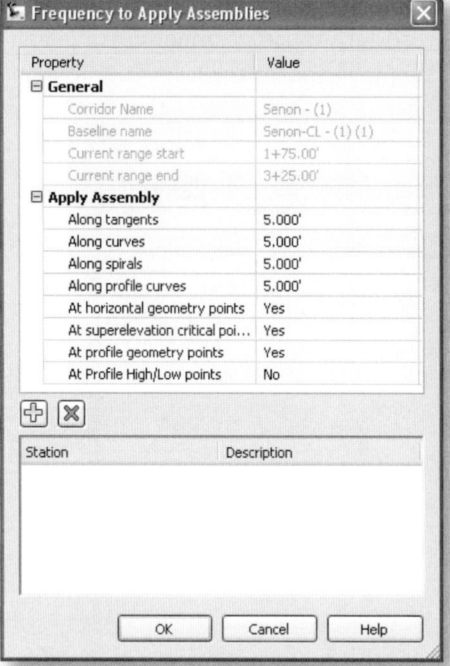


**FIGURE 9.12**

## WIDENING

A roadway may have a widening to accommodate a bus stop or an intersection's turning lane. The first step to create a widening is to identify the alignment and the widening's starting and ending stations. Next, the routine prompts for the widening's width and side. After entering the widening parameters, an Offset Alignment Parameters palette displays with the widening's current values (see Figure 9.13). The widening routine uses the default length of 75 feet to enter and exit the widening. All widening parameters are editable in the palette and display interactive graphics as you change the widening parameters. You can adjust the default values for the Alignment Widening command in Settings' Alignment's command branch.

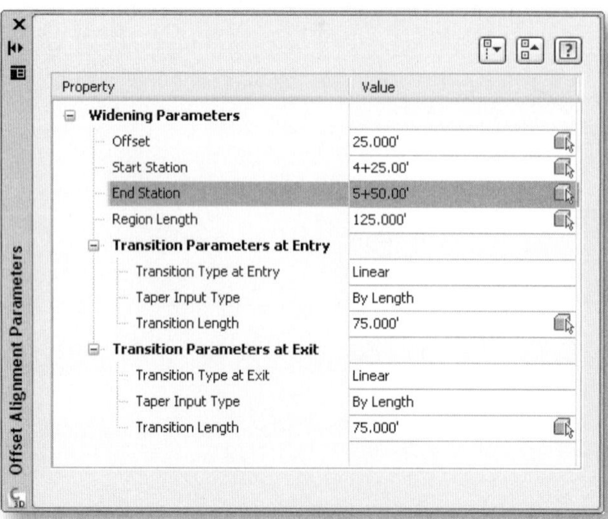


**FIGURE 9.13**

**EXERCISE 9-1**

After completing this exercise, you will:

- Be familiar with creating a transition assembly.
- Be familiar with the BasicLane transition subassembly.
- Be able to transition a lane using an offset horizontal and vertical alignment.
- Create a widening.

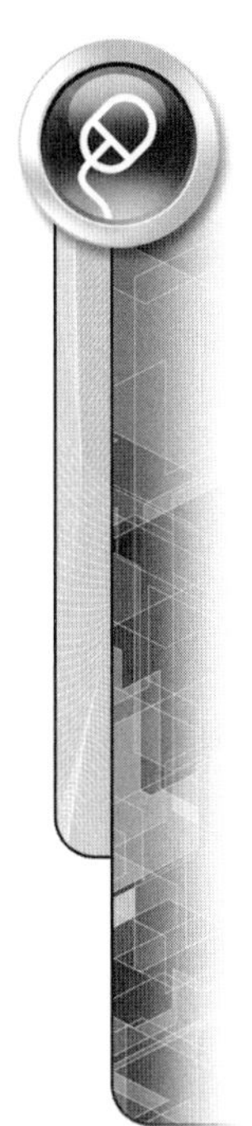

## Exercise Setup

This exercise uses the *Chapter 09 - Unit 1.dwg* file. To find this file, browse to the CD that accompanies this textbook and open the *Chapter 09 - Unit 1.dwg* file.

1. If you are not in **Civil 3D**, double-click the ***Civil 3D*** desktop icon to start the application.
2. When you are at the command line, CLOSE the open drawing and **do not save it**.
3. At Civil 3D's top left, Quick Access Toolbar, click the Open icon, browse to the Chapter 9 folder of the CD that accompanies this textbook and open the *Chapter 09 - Unit 1* drawing.
4. At Civil 3D's top left, click Civil 3D's drop-list arrow, from the Application Menu, highlight SAVE AS, from the flyout select AUTOCAD DRAWING, browse to the Civil 3D Projects folder, for the drawing name enter **Basic Transition**, and click ***Save*** to save the file.
5. If necessary, click Ribbon's Home tab, in the Palettes panel, click the ***Tool Palettes*** icon to display the Civil 3D Imperial tool palette.

## Add Side to Subassembly Name

1. Click the ***Settings*** tab.
2. In Settings, select *Subassembly*, press the right mouse button, and from the list, select EDIT FEATURE SETTINGS....
3. Expand the Subassembly Name Templates section, click the **Create From Macro** value cell to display an ellipsis, and click the ellipsis to open the Name Template dialog box.
4. Place the cursor after the dash between the two name values, press the Spacebar **twice**, add a dash (-), and press the back arrow to place the cursor between the two spaces.
5. Click the Property Fields drop-list arrow, and from the list, select **Subassembly Side**.
6. Click ***Insert*** to place the Subassembly side in the name template (see Figure 9.14).
7. Click ***OK*** until you have returned to the command line.

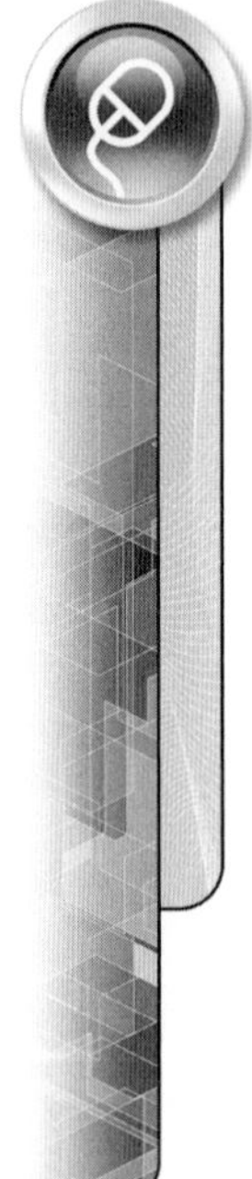

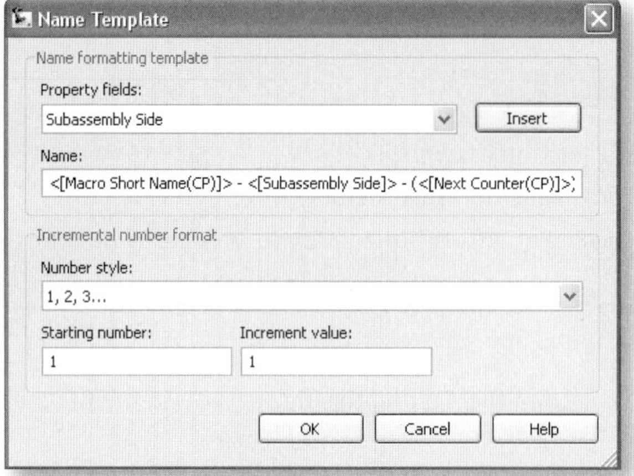


**FIGURE 9.14**

## Create a Basic Lane Transition Assembly — Senon

This assembly uses the basic lane transition, basic curb and gutter, and urban sidewalk subassemblies (see Figure 9.15).

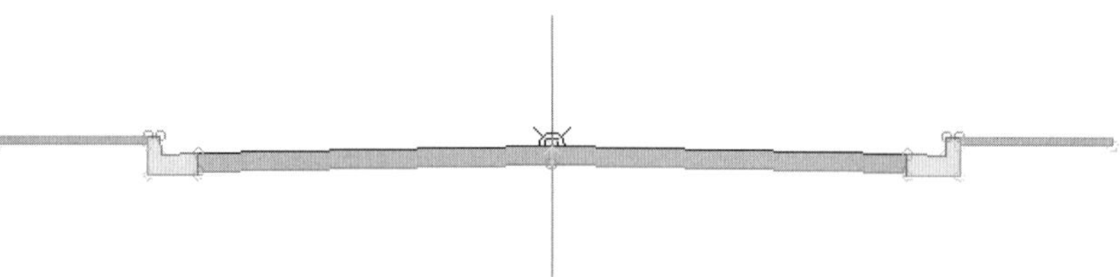

**FIGURE 9.15**

1. From Ribbon's Home tab, Create Design panel, click the Assembly icon, and from the shortcut menu, select CREATE ASSEMBLY.
2. In the Create Assembly dialog box, for the assembly name, enter **Senon**, leaving the counter. Click ***OK***, and in the drawing, select a point to the profile's left.

## Basic Lane Transition — Right Side

1. On the tool palette, select the ***Imperial - Basic*** tab, and from the palette, select ***BasicLane*** to display its properties dialog box.
2. In Properties, the Parameters area, if necessary, change Side to **Right**, set the lane Width to **14** feet, in the drawing select the assembly to attach the subassembly, and press ENTER twice to exit.
3. On the tool palette, select the ***Imperial - Basic*** tab, and from the palette, select ***BasicLaneTransition*** to display its properties dialog box.
4. In Properties, the Parameters area, set the Side to **Left** and the Transition property to **Hold grade, change offset**. In the drawing, attach the subassembly by selecting the assembly and press ENTER twice to exit the routine.

## Curb, Gutter, and Sidewalk

1. Repeat the process of attaching subassemblies and add to the right and left sides a ***BasicCurbAndGutter***. The curbs attach to the BasicLaneTransition's outside top.
2. Repeat attaching the subassemblies and add a **BasicSidewalk** to the right and left sides. The sidewalks attach to BasicCurbAndGutter's outside top.

3. At Civil 3D's top left, Quick Access Toolbar, click the ***Save*** icon to save the drawing.

Your assembly should now look like Figure 9.15.

## Assembly Properties

1. Click the ***Prospector*** tab.
2. In Prospector, expand the Assemblies branch. From the list, select **Senon - (1)**, press the right mouse button, and from the shortcut menu, select PROPERTIES....
3. Click the ***Construction*** tab.

There should be two groups: one (right side) and two (left side) of the assembly.

4. In the dialog box, select the Group - (1) heading that represents the assembly's right side, and press the right mouse button. From the shortcut menu, select *RENAME*, and rename the group **Non-Transitioning Right**.
5. In the dialog box, select the Group - (2) heading that represents the assembly's left side, and press the right mouse button. From the shortcut menu, select *RENAME*, and rename the group **Transitioning Left**.

Your Assembly Properties dialog box should now look like Figure 9.16.

6. Click ***OK*** to exit the Assembly Properties dialog box.

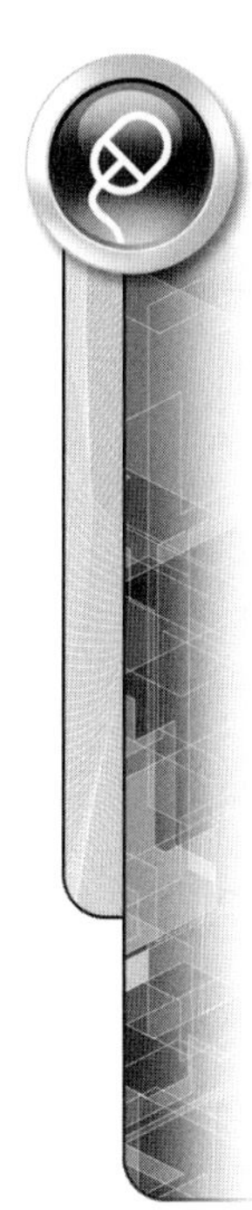

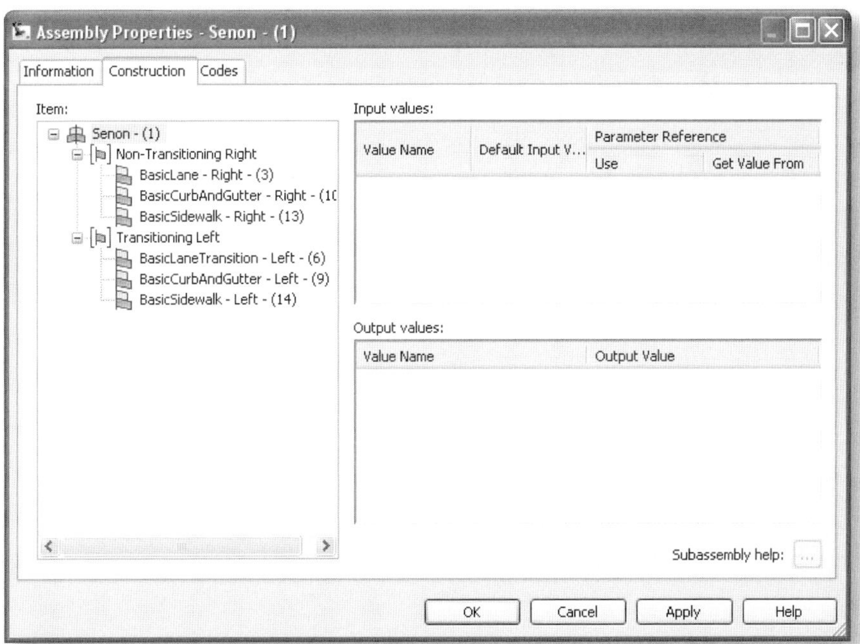


**FIGURE 9.16**

## Create a Simple Corridor

1. From Ribbon's Home tab, in the Create Design panel, click the ***Corridor*** icon and from the shortcut menu, select CREATE SIMPLE CORRIDOR to open the Create Simple Corridor dialog box.
2. In Create Simple Corridor, keeping the counter for the corridor, for the name enter **Senon**, and click ***OK*** to continue.

3. The command line prompts you for an alignment. Press ENTER key, from the list select **Senon-CL – (1) (1)**, and click ***OK*** to continue.
4. The command line prompts you for a profile. Press the right mouse button, click drop-list arrow, and from the list, select **Senon Centerline (1)**. Click ***OK*** to continue.
5. The command line prompts you for an assembly. Press ENTER, click the drop-list arrow, and from the list, select **Senon - (1)**. Click ***OK*** to continue.

The Target Mapping dialog box opens.

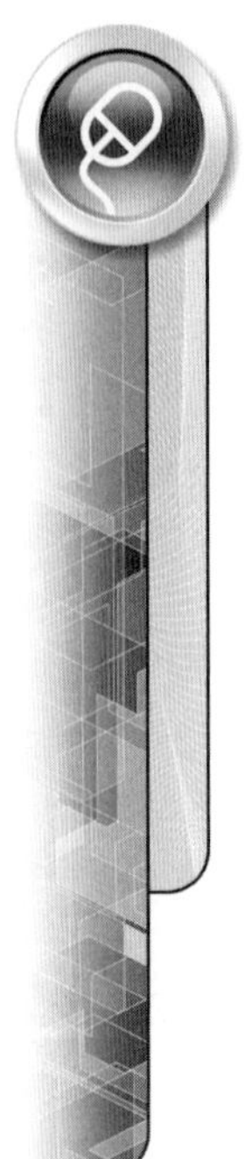

6. In the dialog box, for the Transitioning Left assembly group, set the Width or Offset Targets section by clicking in the Transition Alignment's Object Name cell (<None>). A Set Width or Offset Target dialog box opens. In the Select Alignments list, select **Senon-Left Transition - (1) (1)**, click ***Add>>***, and, making sure the alignment appears in Selected Entities to target, click ***OK*** to return to the Target Mapping dialog box (see Figure 9.17).

You do not need to set a profile because the transitioning subassembly parameter (Hold grade, change offset) sets the edge of the travelway's elevation.

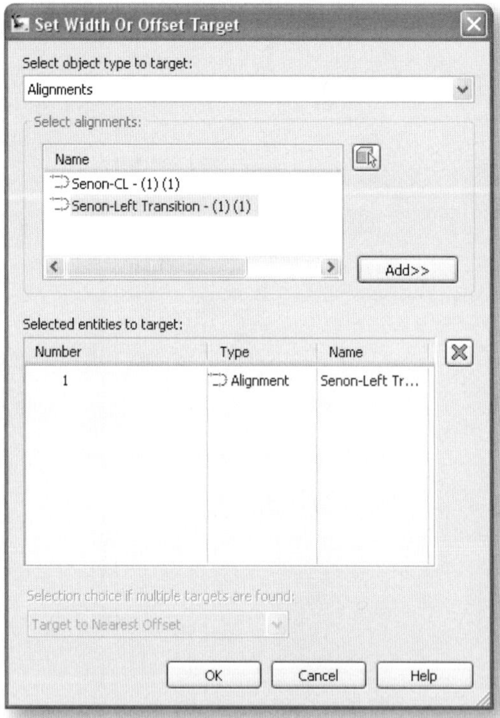


**FIGURE 9.17**

7. Click ***OK*** to create the corridor.
8. If an event panorama is displayed, close it.
9. At Civil 3D's top left, click the ***Save*** icon to save the drawing.

## Review Corridor

1. Use the ZOOM and PAN commands to view the new corridor.

The section frequency does not create a good transition representation in the knuckle. The section frequency needs to be increased, and the station range should be reduced.

2. In the drawing, select the corridor, press the right mouse button, and from the shortcut menu, select CORRIDOR PROPERTIES....
3. In Corridor Properties, select the ***Parameters*** tab, and for Region (1) of RG-Senon - (1) - (1), change its range Start Station to 1+75 (**175**), and its End Station to 3+25 (**325**) (see Figure 9.18).

You could graphically select start and end stations by clicking the pick station icon in each cell, and, in the drawing, by selecting a station along the corridor.

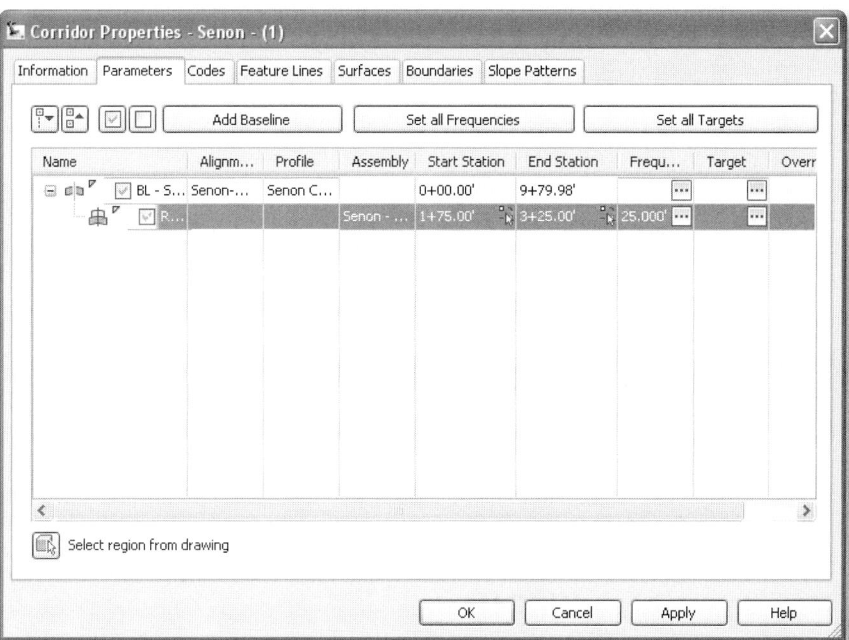


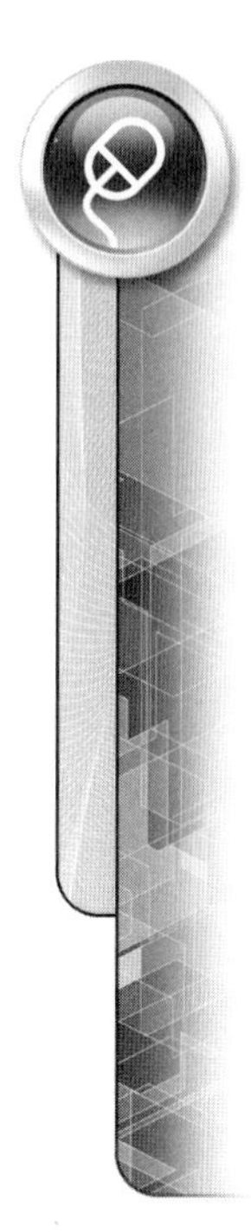

**FIGURE 9.18**

4. For Region (1), in the Frequency cell, click the ellipsis to open the Frequency to Apply Assemblies dialog box.
5. In the Apply Assembly section, change each Along entry value to **5** (see Figure 9.12).

You can graphically select a frequency by clicking the ellipsis in a value cell and selecting a distance from the drawing.

6. Click ***OK*** until you have exited the dialog boxes and rebuilt the corridor.
7. At Civil 3D's top left, Quick Access Toolbar, click the ***Save*** icon to save the drawing.
8. Use the ZOOM and PAN commands to better view the knuckle.
9. Use the Object Viewer to view the corridor in 3D.

The transition looks all right. However, the curb and sidewalk subassemblies are not perpendicular to the roadway edge. When you are using a basic lane transition, the curb and sidewalk sections will not be perpendicular to the transition alignment. Also, the constant −2 percent grade is from the centerline to the knuckle's edge-of-travelway. This subassembly has little control over its slope.

10. Exit Object Viewer after you have reviewed the corridor.
11. Open the Layer Properties Manager, freeze the layer containing the corridor **(C-ROAD-CORR-Senon - (1)),** and click **X** to exit the palette.

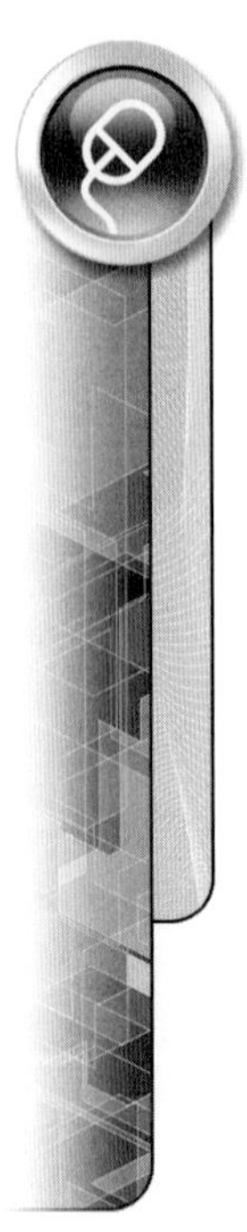

To make the curb and sidewalk perpendicular to the transition curve, the assembly needs an offset point and two additional alignments: a width and a vertical.

## Assembly with Offset

Previously, an offset alignment defined the transitional control point for the BasicLane-Transition (the edge-of-travelway) and all of the subassemblies attached to that point. Making the curb and sidewalk perpendicular to the edge-of-travelway requires an assembly offset and an alignment to control its horizontal location and a profile to control its elevations.

Figure 9.19 shows an assembly using an offset (the vertical line between the edge-of-travelway and the curb flange). You locate the offset by selecting PAVE1's upper-left endpoint. The left curb subassembly attaches to the OFFSET, not the assembly. The sidewalk attaches to the top back-of-curb.

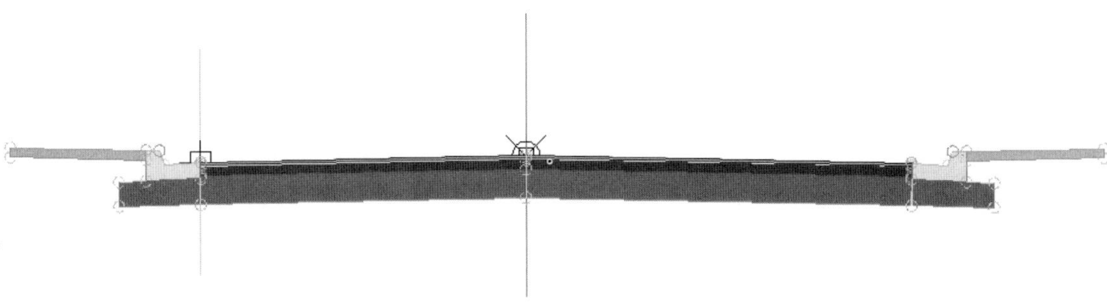

**FIGURE 9.19**

## Create New Assembly with Offset

1. If necessary, click the ***Prospector*** tab.
2. Expand Assemblies, from the list select Senon - (1), click the right mouse button, and from the shortcut menu, select ZOOM TO.
3. PAN the assembly so there is a clear working space beneath the previous assembly.
4. From Ribbon's Home tab, Create Design panel, click the ***Assembly*** icon, and from the shortcut menu, select CREATE ASSEMBLY.
5. In the Create Assembly dialog box, for the assembly name enter **Senon**, leaving the counter. Click ***OK***, and just below the Senon - (1) assembly, select a point.

## Create the Assembly's Right Side

The right side uses a transitional lane subassembly, but there will be no attached alignment.

1. If necessary, click Ribbon's Home tab, in the Palettes panel, click the Tool Palettes icon to display the Civil 3D Imperial tool palette.
2. Click the ***Lanes*** tab. From the tool palette, select ***LaneOutsideSuper***.
3. In Properties, in the Parameters area, set the Side to **Right**, change the Width to **14** feet, and in the drawing, select the assembly.
4. Press ENTER twice to exit the routine.
5. On the palette, click the ***Curbs*** tab and select ***UrbanCurbGutterGeneral***.
6. In Properties, the Parameters area, set the Side to **Right**. In the drawing, attach the curb and gutter to the right lane's upper-right outside marker (edge-of-travelway).

7. Press ENTER twice to exit the routine.
8. From the same palette tab, select ***UrbanSidewalk***.
9. In Properties, the Parameters area, set the Side to **Right**. In the drawing, attach the sidewalk to the curb and gutter's upper-right marker (back-of-curb), and press ENTER twice to exit the routine.
10. At Civil 3D's top left, Quick Access Toolbar, click the ***Save*** icon to save the drawing.

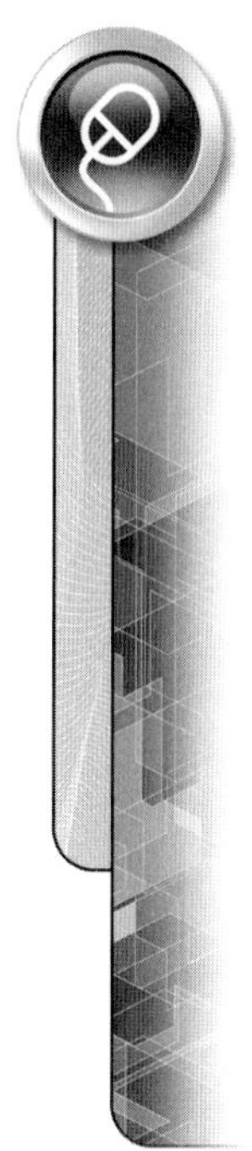

## Create Assembly Left Side

1. Click the ***Lanes*** tab. From the palette, select **LaneOutsideSuper**.
2. In the Properties palette, the Parameters area, change the Side to **Left**, the Width to **14** feet, and attach it to the assembly. Press ENTER twice to exit the routine.

## Add Assembly Offset

1. Use the ZOOM and PAN commands to view the assembly's left half.
2. In the drawing, select the assembly (red vertical line), press the right mouse button, and from the shortcut menu, select ADD OFFSET....
3. Use the ZOOM command to better view PAVE1's upper-left vertex and using the Endpoint object snap, select its upper endpoint to set the offset (see Figure 9.19).

An offset vertical line appears at the selected point.

## Attach the Curb and Sidewalk

The left side curb subassembly must attach to the offset, not the assembly. The curb and its attached subassemblies are perpendicular to the offset's (transition) alignment.

1. Click the ***Curbs*** tab. From the tool palette, select ***UrbanCurbGutterGeneral***.
2. If necessary, in Properties, Advanced Parameters, change the Side to **Left**. In the drawing, attach the curb and gutter to the offset by selecting the **offset** (left vertical line). Press ENTER twice to exit the routine.
3. From the same palette, select ***UrbanSidewalk***. If necessary, in Properties, Advanced Parameters, change the Side to **Left**. Select the curb's upper-left marker (back-of-curb) and press ENTER twice to exit the routine.
4. At Civil 3D's top left, Quick Access Toolbar, click the ***Save*** icon to save the drawing.
5. Click the Subassembly tool palette's **X** to close it.

## Edit Assembly Properties

1. In Prospector, expand Assemblies, select **Senon - (2)**, press the right mouse button, and from the shortcut menu, select PROPERTIES....
2. In the dialog box, click the ***Construction*** tab.

Construction indicates the UrbanCurbandGutterGeneral - Left and UrbanSidewalk - Left subassemblies attach to the offset, not to the main assembly (see Figure 9.20).

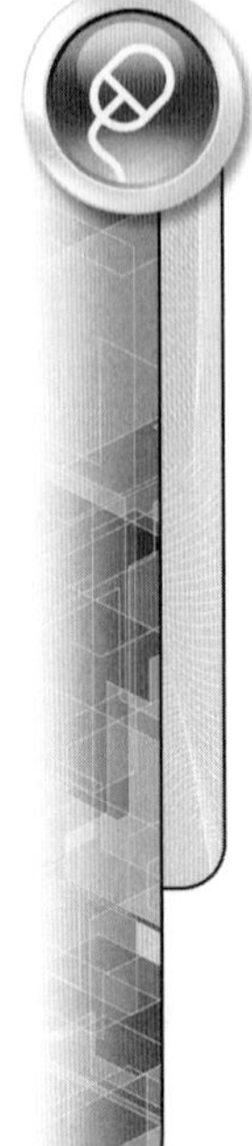

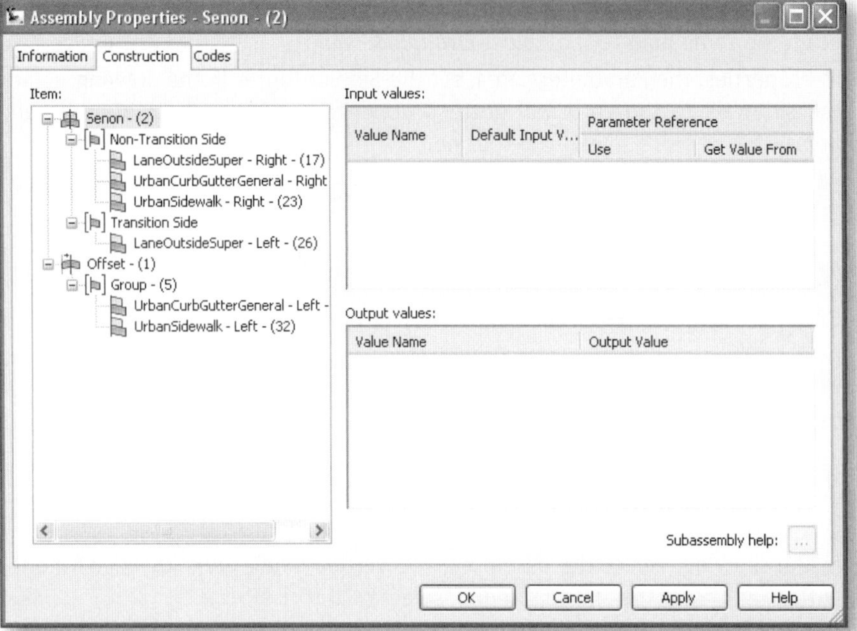


**FIGURE 9.20**

3. Rename the subassembly groups using Figure 9.20 as a guide.
4. In Construction, click the right side subassemblies group name, press the right mouse button, from the shortcut menu, select RENAME, and for the group name enter **Non-Transition Side**.
5. In Construction, click the left side of the LaneOutsideSuper group name, press the right mouse button, from the shortcut menu, select RENAME, and for the group name enter **Transition Side**.
6. In Construction, under Offset – (1) containing the Left UrbanCurbGeneral - Left and UrbanSidewalk - Left subassemblies, click the group name, press the right mouse button, from the shortcut menu, select RENAME, and for the group name enter **Outside Offset**.
7. Click ***OK*** to exit Assembly Properties.
8. At Civil 3D's top left, Quick Access Toolbar, click the ***Save*** icon to save the drawing.

## Create a Corridor

1. In Ribbon's Home tab, Create Design panel, click the Corridor icon and from the shortcut menu, select CREATE CORRIDOR.
2. The command line prompts you for an alignment name. Press the right mouse button, and from the alignment list, select **Senon-CL – (1) (1)**, and click ***OK*** to continue.
3. The command line prompts you for a profile name. Press the right mouse button, click the drop-list arrow, and from the profile list, select **Senon Centerline (1)**. Click ***OK*** to continue.
4. The command line prompts you for an assembly name. Press the right mouse button, click the drop-list arrow, and from the assembly list, select **Senon - (2)**. Click ***OK*** to continue, and the Create Corridor dialog box opens.
5. Replace "Corridor" with **Senon**, leaving the counter, in the Corridor Name area at the top left of the dialog box.

The first items to set are the corridor stations for Region 1 (see Figure 9.29).

6. In the dialog box, click in the Region (1) RG – Senon - (2) – (2) Start Station cell and set it to **125**.
7. In the dialog box, click in the Region (1) End Station cell and set it to **375**.

## Set Offset Alignment and Profile Names

1. In the Alignment cell for outside Offset (1), click (<Click here...>). In the Pick Horizontal Alignment dialog box, click the drop-list arrow, select **Senon-Left Transition - (1) (1)** from the list, and click ***OK*** to return to the Create Corridor dialog box (see Figure 9.21).

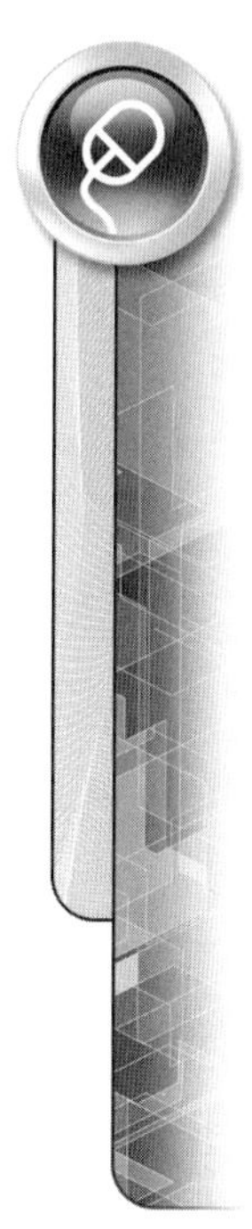

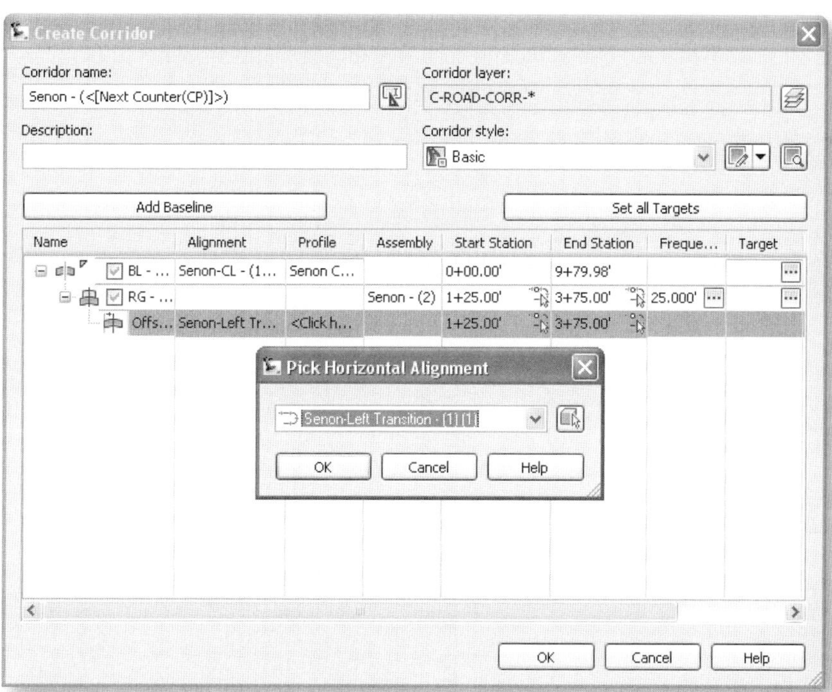

**FIGURE 9.21**

The Alignment name appears in the cell and the Profile cell now contains the text "<Click here...>".

2. In outside Offset (1)'s Profile cell, click (<Click here...>). In Select a Profile, for the Alignment, select **Senon-Left Transition - (1) (1)**, from the Profile drop-list arrow, select **Senon-Left Transition Vertical (1)**, and click ***OK*** to return to the Create Corridor dialog box.

This sets the offset's horizontal and vertical alignments from station 1+25 to 3+75.

## Set All Targets

The Set All Targets dialog box links the transitional alignment to a subassembly point.

1. In the dialog box, at the top right, click ***Set All Targets***.
2. Widen the Object Name and Subassembly columns so that you can see their complete names.

The right side width is the subassembly's width parameter. The left side's LaneOutside-Super uses the transition alignment.

3. In the Width or Offset Targets (Alignments) section, LaneOutsideSuper – Left, Transition Side, click in the Width Alignment's Object Name cell (<None>). In the Set Width or Offset Target dialog box, select **Senon-Left Transition - (1) (1)**, click ***Add≫***, and click ***OK*** to return to the Target Mapping dialog box.

When you are using an assembly offset, a profile is required. The Profiles section sets edge-of-travelway elevation (see Figure 9.22).

4. In the Slope or Elevation Targets (Profiles) section, for LaneOutsideSuper – Left, click in the Outside Elevation Profile's Object Name cell (<None>), in the Set Slope Or Elevation Target dialog box, set the Alignment to **Senon-Left Transition - (1) (1)**, in Select Profiles, select **Senon-Left Transition Vertical (1)**, click ***Add≫***, and click ***OK*** to return the Target Mapping dialog box.

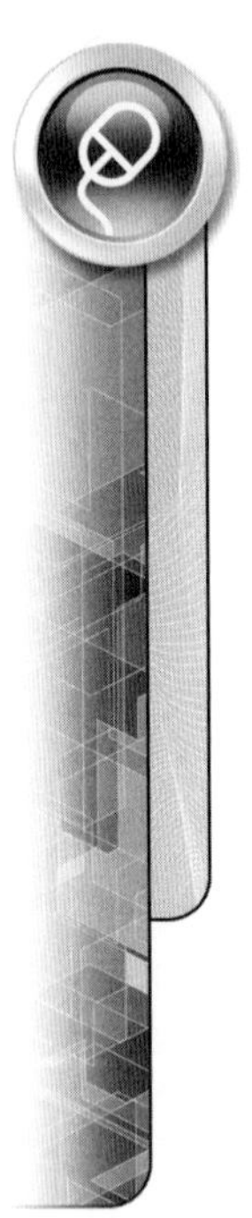

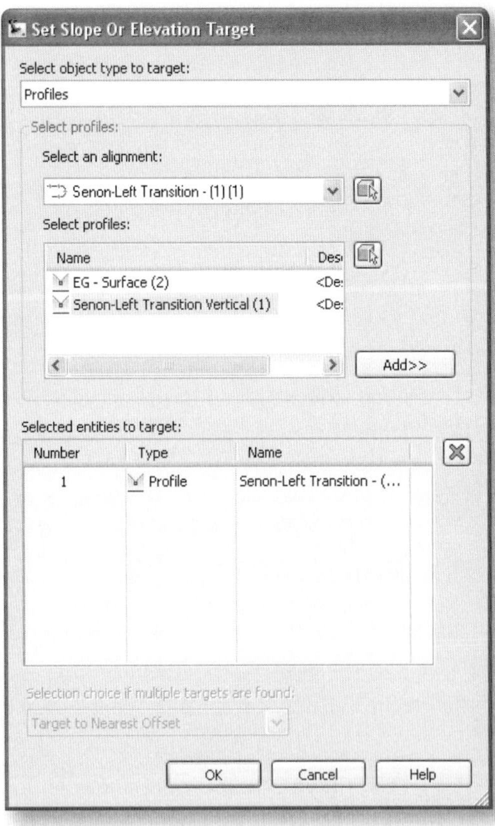


**FIGURE 9.22**

5. Click ***OK*** to exit the Target Mapping dialog box and return to Create Corridor dialog box.

## Change Region Frequency

1. In the Region (1) Frequency cell, click the ellipsis.
2. In Frequency to Apply Assemblies, change the Along Tangents and Along Curves frequency to **5**, and click ***OK*** to return to the Create Corridor dialog box.

3. Click ***OK*** to create the corridor.
4. At Civil 3D's top left, Quick Access Toolbar, click the ***Save*** icon to save the drawing.
5. Use the ZOOM and PAN commands to view the new Corridor.

The Curb and Sidewalk subassemblies are perpendicular to the offset alignment's path.

6. Use the *OBJECT VIEWER* to view the corridor in 3D.
7. Exit the Object Viewer after you review the corridor.
8. At Civil 3D's upper left, Quick Access Toolbar, click the Save icon to save the drawing.

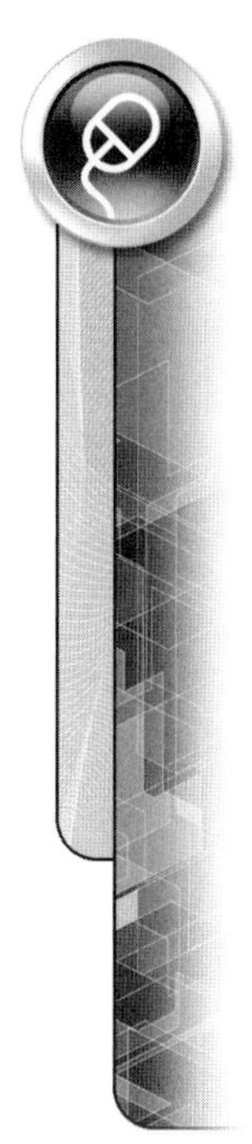

## Widening

Senon needs a widening alignment to create a corridor with a bus stop. The widening starts at 5+50 and ends at 6+25 and the transition lengths are both 55 feet.

1. In Prospector, expand the Site's Site1 branch until viewing the Senon alignments.
2. In the Centerline Alignment branch, select Senon-Left Transition – (1) (1), right mouse click, and from the shortcut menu, select Move to site.
3. In the Move to Site dialog box, at the top set the Destination site to None and click OK to transfer the alignment.
4. In Prospector, expand the Alignments branch, select the alignment Senon-Left Transition – (1) (1), right mouse click, and from the shortcut menu, select PROPERTIES....
5. Click Senon-Left Transition – (1) (1)'s ***Information*** tab, set the alignment type to **Offset**, and click OK.
6. From the Ribbon's Home tab, Layers panel's right side, click the ***Freeze*** icon and in the drawing select the Senon - (2) corridor.

## Define the Widening

1. On the Home tab, Create Design panel, click the Alignment icon, and from the shortcut menu, select Create Widening.
2. The command line prompts for an alignment and in the drawing select the Senon-Left Transition – (1) (1) alignment and a station jig displays prompting for the starting station. Move the jig near station 5+50 and select the point.
3. The routine now prompts for the ending station. Move the jig near station 6+25 and select the point.
4. The command line prompts for a width; press enter to accept 25.
5. The command line prompts for a side; type in '**L**' and press enter to display the Offset Alignment Parameters palette.
6. In the palette, change the start station to 5+50 (**550**) and the end station to 6+25 (**625**).

As you edit the widening, the drawing contains interactive graphics showing the changes.

7. Change both transition lengths to **55**.
8. Close the Offset Alignment Parameters palette.

## Create a Offset Alignment Profile for the Widening

The new offset alignment needs to have a vertical design profile. You can create a vertical design that is a superimposed copy Senon-Left Transition – (1) (1)'s vertical alignment.

1. If necessary, click the Home tab. In the Profile & Section Views panel, click the Profile View icon and from the shortcut menu, select Create Profile View.
2. In the Create Profile View wizard, the General panel, set the alignment to **Senon Left Transition – (1)(1)-Left 0.000**, the profile style to **Major Grids**, click Create Profile View, and in the drawing select the location of the profile view.

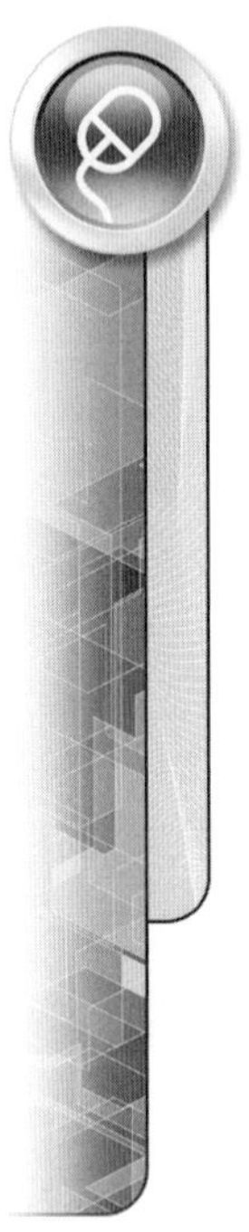

## Superimpose Profile

1. In the Create Design panel, click the Profile icon and from the shortcut menu, select CREATE SUPERIMPOSED PROFILE.
2. The command line prompts for a profile. In the Senon Left Transition profile view select the Senon-Left Transition Vertical (1) vertical design.
3. The command line prompts for a profile view. In the drawing select the new offset alignment's profile view.
4. The Superimpose Profile Options dialog box opens. Toggle ON select Start and End stations and click OK.
5. The profile transfers to the new profile view.

## Create New Corridor

1. In Ribbon's Home tab, Create Design panel, click the Corridor icon and from the shortcut menu, select CREATE CORRIDOR.
2. The command line prompts you for an alignment name. Press the right mouse button, and from the alignment list, select **Senon-CL – (1) (1)**, and click ***OK*** to continue.
3. The command line prompts you for a profile name. Press the right mouse button, click the drop-list arrow, and from the profile list, select **Senon Centerline (1)**. Click ***OK*** to continue.
4. The command line prompts you for an assembly name. Press the right mouse button, click the drop-list arrow, and from the assembly list, select **Senon - (2)**. Click ***OK*** to continue, and the Create Corridor dialog box opens.
5. In the dialog box's top left, replace "Corridor" with **Senon**, leaving the counter.

The first items to set are the corridor stations for Region 1.

6. In the dialog box, click in the Region (1) RG – Senon - (2) – (2) Start Station cell and set it to **125**.
7. In the dialog box, click in the Region (1) End Station cell and set it to **800**.

## Set Offset Alignment and Profile Names

1. In the Alignment cell for outside Offset - (1), click (<Click here...>). In the Pick Horizontal Alignment dialog box, click the drop-list arrow, select **Senon-Left Transition - (1) (1)-Left 0.000** from the list, and click ***OK*** to return to the Create Corridor dialog box.

The Alignment name appears in the cell and the Profile cell now contains the text "<Click here...>".

2. In outside Offset - (1)'s Profile cell, click (<Click here...>). In Select a Profile, for the Alignment, select **Senon-Left Transition - (1) (1)-Left 0.000**, from the Profile drop-list arrow, select **Senon-Left Transition Vertical (1) - [Senon-Left Transition - (1) (1)] - (1),** and click ***OK*** to return to the Create Corridor dialog box.

This sets the offset's horizontal and vertical alignments from station 1+25 to 8+00.

## Set All Targets

The Set All Targets dialog box links the transitional alignment to a subassembly point.

1. In the dialog box, at the top right, click ***Set All Targets***.
2. Widen the Object Name and Subassembly columns so you can see their complete names.

The right side width is the subassembly's width parameter. The left side's LaneOutsideSuper uses the transition alignment.

3. In the Width or Offset Targets (Alignments) section, LaneOutsideSuper – Left, Transition Side, click in the Width Alignment's Object Name cell (<None>). In the Set Width or Offset Target dialog box, select **Senon-Left Transition - (1) (1)-Left 0.000**, click ***Add>>***, and click ***OK*** to return to the Target Mapping dialog box.

When you are using an assembly offset, a profile is required. The Profiles section sets edge-of-travelway elevation.

4. In the Slope or Elevation Targets (Profiles) section, for LaneOutsideSuper – Left, click in the Outside Elevation Profile's Object Name cell (<None>). In the Set Slope Or Elevation Target dialog box, set the Alignment to **Senon-Left Transition - (1) (1)**, in Select Profiles, select **Senon-Left Transition Vertical (1)**, click ***Add>>***, and click ***OK*** to return the Target Mapping dialog box.
5. Click ***OK*** to exit the Target Mapping dialog box and return to Create Corridor.

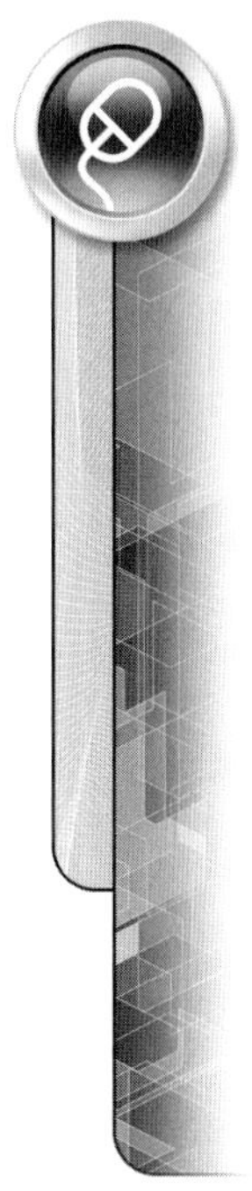

## Change Region Frequency

1. In the Region (1) Frequency cell, click the ellipsis.
2. In Frequency to Apply Assemblies, change the Along Tangents and Along Curves frequency to **5**, and click ***OK*** to return to the Create Corridor dialog box.
3. Click ***OK*** to create the corridor.
4. At Civil 3D's top left, Quick Access Toolbar, click the ***Save*** icon to save the drawing.
5. Use the `ZOOM` and `PAN` commands to view the new Corridor.
6. Exit Civil 3D.

This completes the simple transitions and widening exercise.

## SUMMARY

- BasicLane with transition creates simple lane-widening designs.
- BasicLane with transition subassembly has settings to control the transition point's offset and elevation.
- When BasicLane with transition's Transition parameter is set to Change offset, change elevation, you must assign both an offset horizontal and vertical alignment.
- When you are using an assembly offset, the offset point must have assigned horizontal and vertical alignments.
- The first subassembly outside the assembly's offset must attach to the offset.
- Widenings are parameter driven transitions.

## UNIT 2: CREATING A CUL-DE-SAC

Creating a cul-de-sac is similar to transitioning a roadway. In a cul-de-sac, an alignment controls the edge-of-travelway's outside edge, and the main roadway's centerline provides the pavement width (see Figure 9.23). When going around the cul-de-sac, the controlling alignment switches to the edge-of-travelway and the centerline assumes the role of pavement widening.

Having perpendicular curb and sidewalk sections around the cul-de-sac means the curb and sidewalk are to the left of the edge-of-travelway and an alignment attaches to the edge-of-travelway (see Figure 9.23).

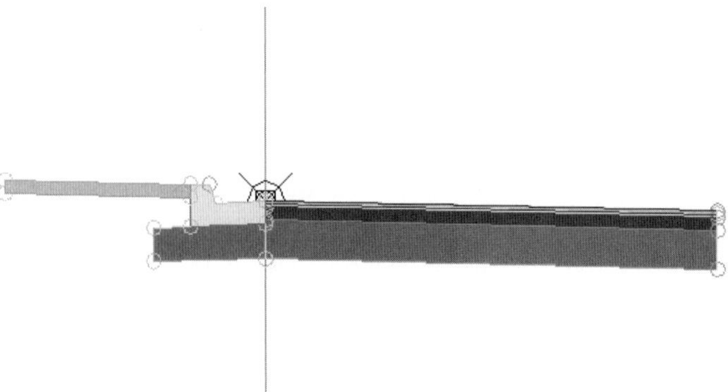

**FIGURE 9.23**

### CORRIDOR SURFACES

Using more than one region for a corridor introduces corridor surface boundaries issues. When you are creating a corridor that has a single region, the boundary selection is a single feature line (daylight). When there is more than one region, selecting the boundary segments may be a manual selection process. When manually identifying boundary segments, a boundary jig appears to help define the boundary (see Figure 9.24).

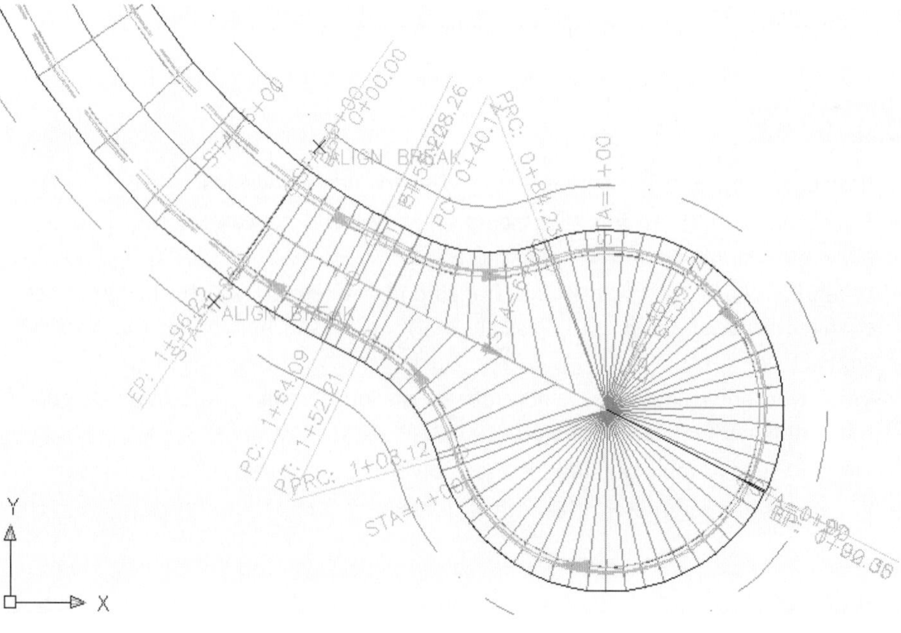


**FIGURE 9.24**

**EXERCISE 9-2**

After completing this exercise, you will:

- Be able to create a cul-de-sac.
- Be familiar with the Create Corridor command.
- Be able to define corridor regions.
- Be able to create a corridor surface.
- Manually define a corridor surface boundary.

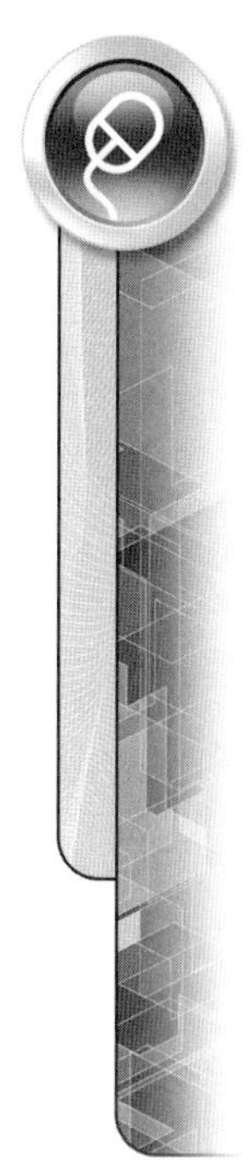

## Exercise Setup

This exercise uses the *Chapter 09 - Unit 2.dwg* file. To find this file, browse to the CD that accompanies this textbook and open the *Chapter 09 - Unit 2.dwg* file.

1. If you are not in **Civil 3D**, double-click the ***Civil 3D*** desktop icon to start the application.
2. When you are at the command line, CLOSE the open drawing and **do not save it**.
3. At Civil 3D's top left, Quick Access Toolbar, click the Open icon, browse to the Chapter 9 folder of the CD that accompanies this textbook and open the *Chapter 09 - Unit 2* drawing.
4. At Civil 3D's top left, click Civil 3D's drop-list arrow, from the Application Menu, highlight SAVE AS, from the flyout menu, select AUTOCAD DRAWING, browse to the Civil 3D Project folder, for the name enter **Cul-de-sac**, and click ***Save*** to save the file.

There are three alignments: one entire roadway baseline (Lorraine - (1)); and two offset (transition) alignments for the cul-de-sac (Lorraine Left and Right Cul-de-sac – (1)). There are two roadway assemblies: one non-transitioning assembly applied to stations 0+00 to 5+25; and one transition assembly.

The two offset alignments and their profiles already exist. You would have had to calculate or determine their profile elevations so they initially match the main alignment's elevations and then reflect cul-de-sac drainage design. One way of developing a cul-de-sac design is to use an ETW feature line that determines the starting elevations. Creating Feature lines from a corridor are in the Ribbon's Modify, Corridors' panel: Feature Lines from Corridor. Alignments from Corridor also prompt to create a profile.

## Create the Transition Assembly

Use Figure 9.24 as a guide when you are creating the transitional assembly.

1. If necessary, click the ***Prospector*** tab.
2. Expand the Assemblies branch. From the assembly list, select **Lorraine – No Transition – (1)**, press the right mouse button, and from the shortcut menu, select ZOOM TO.
3. From Ribbon's Home tab, Create Design panel, click the ***Assembly*** icon and from the shortcut menu, select CREATE ASSEMBLY. For the name, replace "Assembly" with **Lorraine – Transition**, keeping the counter. Click ***OK***, and in the drawing, select a point to locate the assembly.

## Right Side Transition Subassembly

1. If necessary, click Ribbon's ***Home*** tab. In the Palettes panel, click the Tool Palettes icon to display the Civil 3D Imperial tool palette.
2. On the tool palette, click the ***Lanes*** tab.

3. From the Imperial – Lanes palette, select ***LaneOutsideSuper***.
4. In Properties, the Parameters area, set the Side to **Right**, and set the Width to **12** feet. In the drawing, place the subassembly by selecting the assembly, and press ENTER twice to exit the routine.

## Left Side Curb and Sidewalk

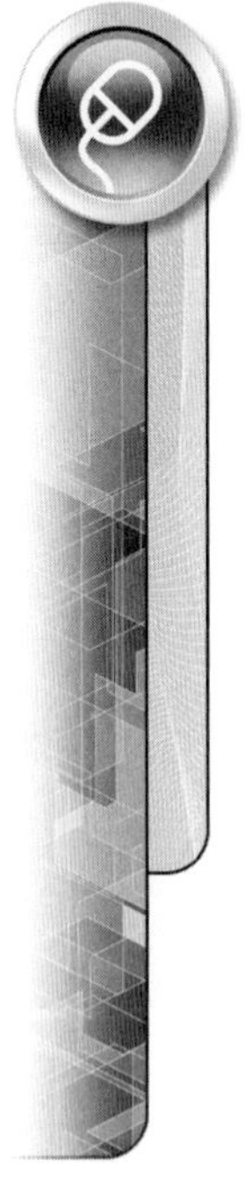

1. On the tool palette, click the ***Curbs*** tab.
2. From the Imperial – Curbs palette, select ***UrbanCurbGutterGeneral***.
3. In Properties, the Parameters area, set the Side to **Left**, and in the drawing, select the assembly, and press ENTER twice to exit the routine.
4. From the Imperial – Curbs palette, select ***UrbanSidewalk***.
5. In Properties, the Parameters area, set the side to **Left**, and in the drawing, select the upper back-of-curb marker to place the sidewalk, and then press ENTER twice to exit the routine.
6. Close the tool palette.
7. At Civil 3D's top left, Quick Access Toolbar, click the ***Save*** icon to save the drawing.

## Edit Assembly Properties

1. If necessary, expand the Prospector Assemblies branch until you view the assembly list.
2. From the list, select **Lorraine – Transition – (1)**, press the right mouse button, and from the shortcut menu, select PROPERTIES....
3. Click the ***Construction*** tab, and select and rename the group with right Lane-OutsideSuper to **Transition Subassembly**.
4. Select and rename the group with Curb and Sidewalk to **Non-Transition Subassemblies**.
5. Click ***OK*** to exit the dialog box.
6. At Civil 3D's top left, Quick Access Toolbar, click the ***Save*** icon to save the drawing.

## Create the Non-Transition Corridor Region

The process starts with a simple corridor, and later adds to its complexity by editing its properties.

1. From Ribbon's Home tab, Create Design panel, click the Corridor icon and from the shortcut menu, select CREATE SIMPLE CORRIDOR. For the name, replace "Corridor" with **Lorraine**, and click ***OK*** to continue creating the corridor.
2. The command line prompts you for an Alignment. Press the right mouse button, and from the list, select **Lorraine - (1)**. Click ***OK*** to continue.
3. The command line prompts you for a profile. Press the right mouse button, click the drop-list arrow, and from the list, select **Lorraine CL Vertical (1)**. Click ***OK*** to continue.
4. The command line prompts you for an assembly. Press the right mouse button, click the drop-list arrow, and from the list, select **Lorraine – No Transition – (1)**. Click ***OK*** to continue.

The Target Mapping dialog box opens. There are no target names to set at this time.

5. Click ***OK*** to create the corridor.
6. If the Event Viewer displays, close it by clicking the green check mark in the panorama's upper-right corner.

7. Use the ZOOM and PAN commands to view the corridor.
8. Use the ZOOM and PAN commands to view the cul-de-sac from station 5+25 to its end.
9. At Civil 3D's top left, Quick Access Toolbar, click the ***Save*** icon to save the drawing.

The corridor goes from the beginning to the end (center of the cul-de-sac). The non-transitional corridor region should stop at station 5+25.

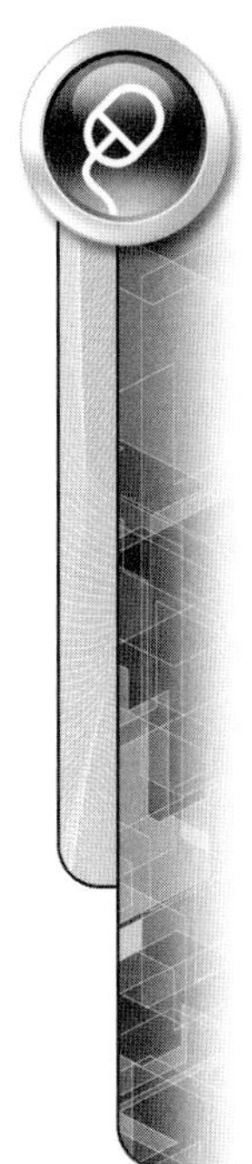

## Edit the Corridor Properties

1. In the drawing, select the corridor, press the right mouse button, and from the shortcut menu, select CORRIDOR PROPERTIES....
2. If necessary, scroll right until you are viewing the beginning and ending stations. Click in the Region (1) RG – Lorraine – No Transition – (1) – (1) ending station cell and change its station to 5+25 (**525**).
3. Click ***OK*** to modify the corridor's ending station.

## Define and assign Baseline (2)

The corridor control around the cul-de-sac's northern edge passes to a second baseline (centerline) at station 5+25.01. This alignment's vertical setting controls the edge-of-travelway's vertical location. The Lorraine - (1) alignment stretches the pavement to create the cul-de-sac's paved area.

1. In the drawing, select the corridor, press the right mouse button, and from the shortcut menu, select CORRIDOR PROPERTIES....
2. In the Parameters panel, in the panel's upper-center, click ***Add Baseline*** to create a new baseline entry in the Create Corridor Baseline dialog box.
3. In Create Corridor Baseline dialog box, click the Horizontal alignment drop-list arrow. From the list, select **Lorraine Left Cul-de-Sac – (1)**, and click ***OK*** to return to the Corridor Properties dialog box.
4. For Baseline (2) Lorraine Left Cul-de-Sac – (1) – (1), in its Profile cell, click (<Click here...>). In the Select a Profile dialog box, click the drop-list arrow. From the list, select **Lorraine Left Transition - Vertical (1)** and click ***OK*** to return to the Corridor Properties dialog box.

## Create Baseline (2)'s Region 1

1. In the Corridor Properties dialog box, with the Baseline (2) highlighted, press the right mouse button, and, from the shortcut menu, select ADD REGION... to open Create Corridor Region dialog box.
2. In Create Corridor Region, click the Assembly drop-list arrow. From the list, select **Lorraine - Transition – (1)**, and click ***OK*** to return to the Corridor Properties dialog box (see Figure 9.25).

The preceding steps created Region (1) for Baseline (2) with the selected assembly.

3. Click ***OK*** to create the modified corridor.
4. Use the ZOOM and PAN commands to inspect the intersection of the two corridor regions.

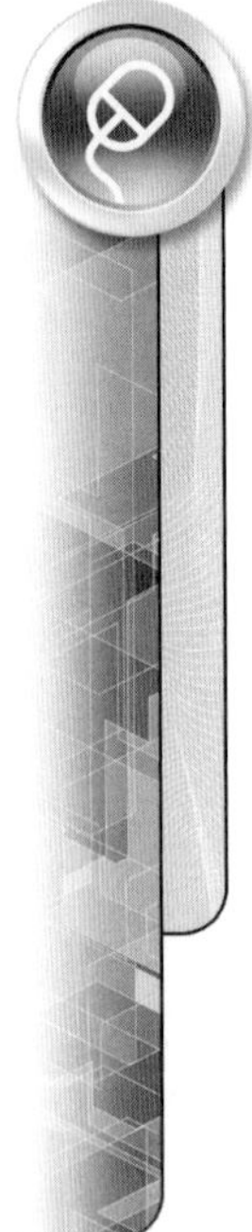

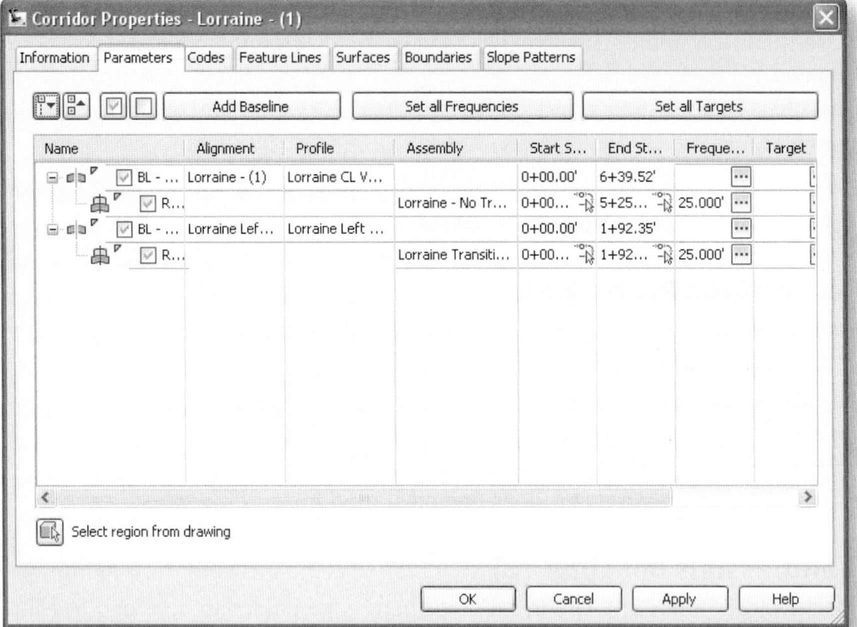


**FIGURE 9.25**

The new baseline creates sections that follow the new alignment and its vertical design. However, the right side of the assembly does not reach the centerline of the road or the cul-de-sac.

To create the correct width, in Target Mapping, assign the Lorraine - (1) centerline and vertical as the lane width target alignment and profile.

## Assign Transition Targets

1. In the drawing, select the corridor, press the right mouse button, and from the shortcut menu, select CORRIDOR PROPERTIES....
2. If necessary, click the ***Parameters*** tab.
3. If necessary, click Baseline (2)'s Region (1) entry to highlight it, scroll to the right, and in the Target column, click the ellipsis.
4. Under the Width or Offset Targets (Alignments) section, locate the Transition Subassembly's Width Alignment entry, and click in the Object Name cell (<None>). The Set Width Or Offset Target dialog box opens. In Select alignments, select **Lorraine - (1)**, click ***Add>>***, and click ***OK*** to return to the Target Mapping dialog box (see Figure 9.26).
5. In the Slope or Elevation Targets (Profiles) section, for Transition Subassembly locate its Outside Elevation Profile entry. Click in its Object Name cell (<None>) to open the Set Slope Or Elevation Target dialog box. With Alignment set to Lorraine - (1), from the Select Profile list, select **Lorraine CL Vertical (1)**, click ***Add>>***, and click ***OK*** to return to the Target Mapping dialog box (see Figure 9.26).

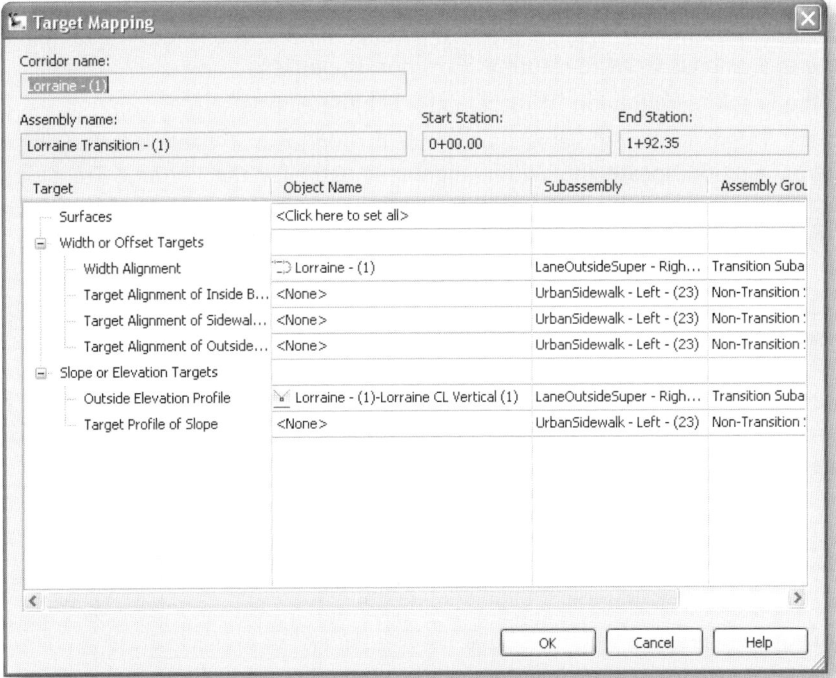


**FIGURE 9.26**

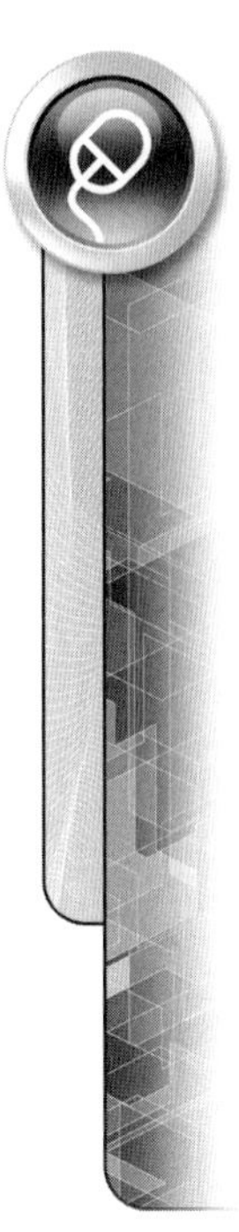

6. Click ***OK*** to exit the Target Mapping dialog box.
7. In Baseline (2), the Region (1)'s Frequency column, click the ellipsis, change the Along Tangents and Along Curves sampling rate to **5**, and click ***OK*** to return to the Corridor Properties dialog box.
8. Click ***OK*** to exit the dialog box and update the corridor.
9. Use the `ZOOM` and `PAN` commands to inspect both corridor baselines and their regions.

The new baseline and its region stretch the pavement to Lorraine - (1)'s centerline.

## Define and Assign Baseline (3)

1. In the drawing, select the corridor, press the right mouse button, and from the shortcut menu, select `CORRIDOR PROPERTIES`....
2. If necessary, click the ***Parameters*** tab.
3. In the panel's upper-center, click ***Add Baseline*** and create Baseline (3).
4. In Create Corridor Baseline, click the Horizontal Alignment drop-list arrow. From the list, select **Lorraine Right Cul-de-sac – (1)** and click ***OK*** to return to Corridor Properties.
5. For Baseline (3), click in its Profile cell (<Click here...>). In the Select a Profile dialog box, click the drop-list arrow, and from the list, select **Lorraine Right Transition Vertical (1)**. Click ***OK*** to return to Corridor Properties.
6. With Baseline (3) still highlighted, press the right mouse button, and from the shortcut menu, select `ADD REGION`....
7. In Create Corridor Region, click the drop-list arrow. From the list, select **Lorraine Transition – (1)**, and click OK to return to Corridor Properties.
8. Expand Baseline (3) to view Region (1). Scroll the corridor panel to the right until you view the Frequency and Target columns.

9. In the Baseline (3) Region (1) Frequency cell, click the ellipsis. In the Frequency to Apply Assemblies, change the Along Tangents and Along Curves sampling rate to **5**, and click ***OK*** to return to Corridor Properties.
10. In the Baseline (3) Region (1) Target cell, click the ellipsis.
11. In the Target Mapping dialog box, under the Width or Offset Targets (Alignments) section, locate the Width Alignment entry for the Lorraine Transition Subassembly, and click in the Object Name cell (<None>). The Set Width Or Offset Target dialog box opens. From the list of alignments, select **Lorraine - (1)**, click ***Add>>***, and click ***OK*** (see Figure 9.26).
12. In the Slope or Elevation Targets (Profiles) section, locate the Transition Subassembly Outside Elevation Profile entry. Click in its Object Name cell (<None>) to open the Set Slope Or Elevation Target dialog box. Making sure the Alignment is Lorraine - (1), from the Select profiles list, select **Lorraine CL Vertical (1)**, click ***Add>>***, and click ***OK*** until you have returned to Corridor Properties (see Figure 9.27).

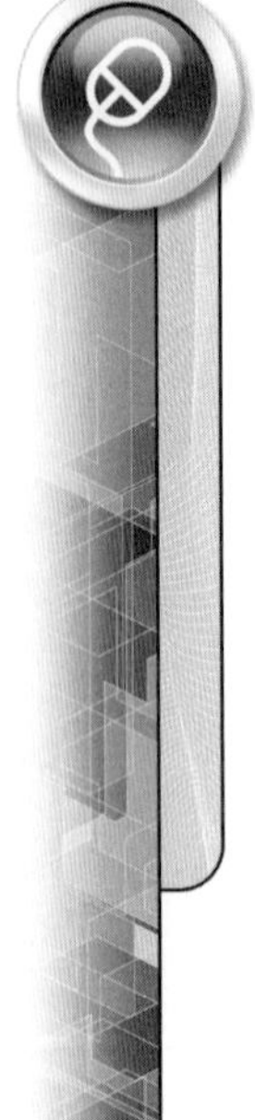

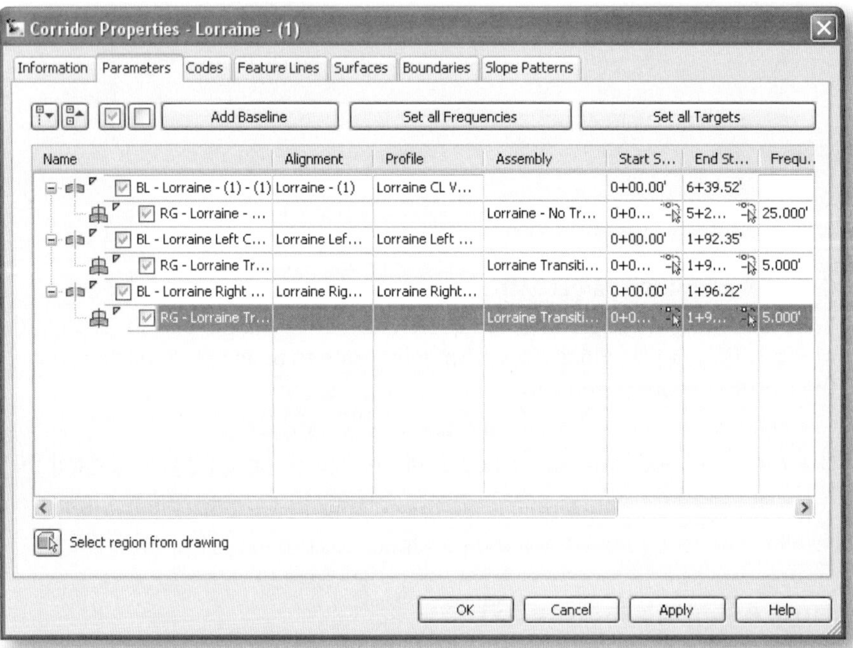


**FIGURE 9.27**

13. Click ***OK*** to exit and update the corridor.
14. Use the ZOOM and PAN commands to inspect the cul-de-sac.
15. At Civil 3D's top left, Quick Access Toolbar, click the ***Save*** icon to save the drawing.

## Corridor Surface

1. In the drawing, select the corridor, press the right mouse button, and from the shortcut menu, select CORRIDOR PROPERTIES....
2. Click the ***Surfaces*** tab.
3. In the panel's upper-left, click the ***Create a corridor surface*** icon.
4. Rename the surface **Lorraine – Top**, assign it the style **Border & Triangles & Points**, and set Overhang correction to Top Links.

5. From the panel's top center, set the Data type to **Links**, from the list of links select **Top**, and click the plus sign (+) to add it as surface data.

Each Baseline contributes to an overall surface and their 'extents' define the surface boundary.

6. Click the ***Boundaries*** tab.
7. In the panel, select Lorraine - (1) Surface - Top, right mouse click, and from the shortcut menu, select CORRIDOR EXTENTS AS OUTER BOUNDARY....
8. Click ***OK*** to build the corridor surface.
9. Select a surface triangle, and use the Object Viewer to review the surface.
10. At Civil 3D's top left, Quick Access Toolbar, click the ***Save*** icon to save the drawing.

This ends the exercise on cul-de-sac transitions.

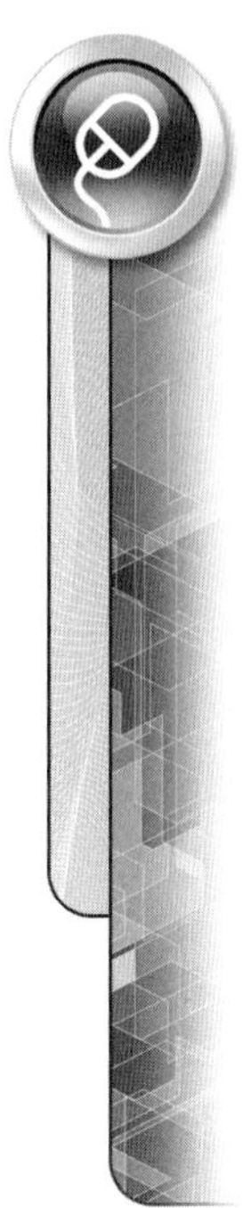

## SUMMARY

- The key to cul-de-sacs is defining an assembly and attaching the curb and sidewalk subassemblies to the left side so that they are perpendicular to the cul-de-sac arc.
- The centerline horizontal and vertical alignments are the width and height for the cul-de-sac's center.

## UNIT 3: SUPERELEVATION

Superelevation allows a roadway to carry higher speeds around its horizontal curves. To maintain these higher speeds, the roadway assembly includes subassemblies that change the lanes' cross slope (superelevate), and possibly the shoulders, toward the horizontal alignment curve's center. The lanes' rotation takes advantage of centrifugal force and stabilizes a vehicle passing through the curve. This design methodology is found in both highway and railway designs (see Figure 9.28).

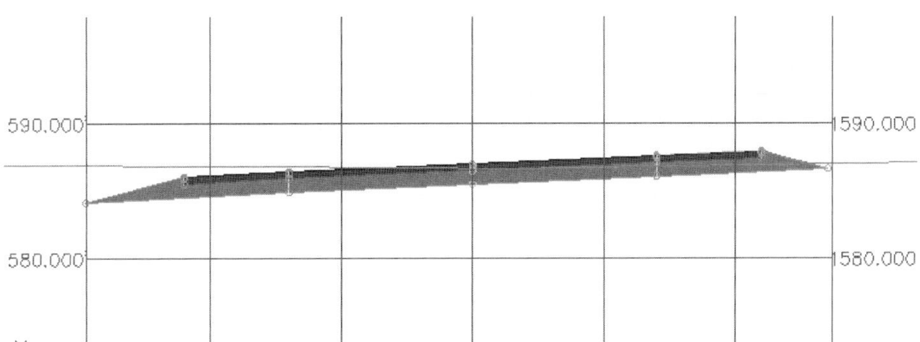


**FIGURE 9.28**

The pavement rotation methods vary greatly, and there is no uniform standard governing their use or design. Pavement rotation can occur about the centerline, around its inside or outside lane, or, if it's a divided highway, around the lane edges or centerline. There is usually a regulatory body document that defines the necessary distances, radii, and lengths to achieve maximum superelevation.

The American Association of State Highway & Transportation Officials (AASHTO) publishes documents that define one highway design standards set. The AASHTO "Green Book" standards are incorporated in Civil 3D as Design Criteria. These standards are modifiable and a user can create an entirely new set. The standards affect both horizontal and vertical design.

## DESIGN STANDARDS FILE

The Corridor Design Standards file contains tables with critical superelevation design values. This file's values calculate roadway cross-section rotation and check for minimum design values. The Design Standards file is an XML-based file and can be customized to accommodate differing design standards. Customization occurs in Alignment's Design Criteria Editor.

The Design Criteria Editor sets how it evaluates a design. First, it evaluates the minimum curves along an alignment. Second, it understands the attainment method for the type of road (with or without a crown). This includes lengths between critical superelevation points expressed as formulas. Third, is for a design speed and curve radius, what lane cross slope is necessary to maintain the design speed through the superelevated curve. Related to this value (the necessary cross slope) is a length to transition from normal crown to superelevation.

### Horizontal — Base Units

The design's base units are critical to all computation. The Design Criteria Editor lists and modifies its default values (see Figure 9.29).

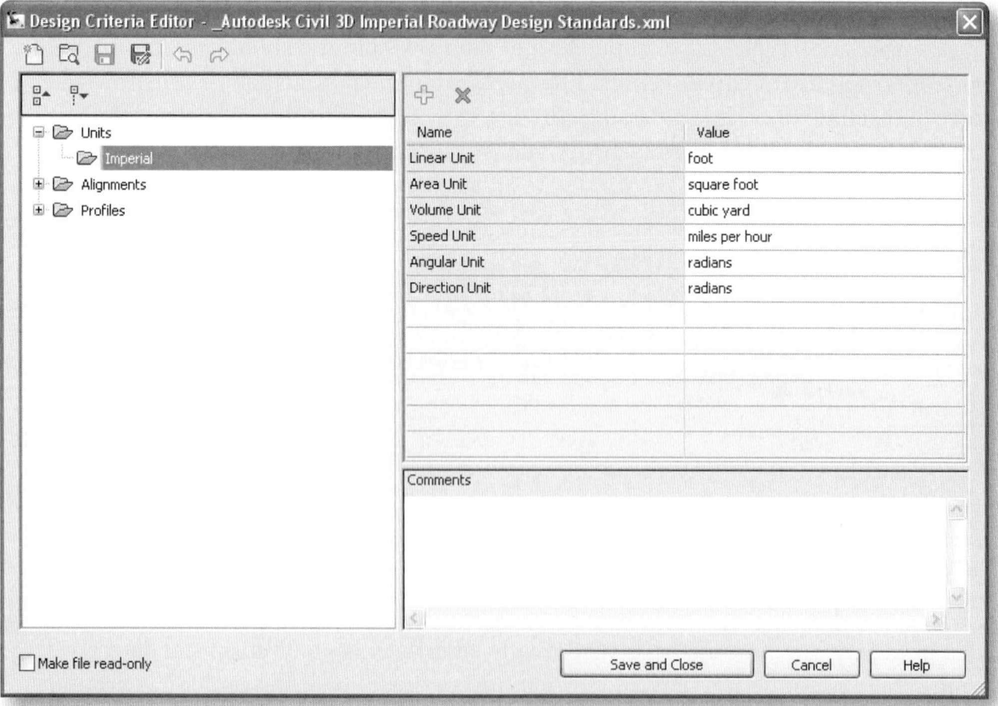


**FIGURE 9.29**

### Horizontal — Minimum Radius

Design Standards contains minimum curves for design speeds and superelevation. The following is an excerpt from the file for a 4 percent superelevation, its design speeds, and the recommended minimum radii. To maintain a higher speed with a 4 percent pavement cross slope, the minimum road radius needs to lengthen to safely handle the greater speeds.

```
<MinimumRadiusTables>
<!--======================================== -->
<!-- Defines minimum radii for road type and design speed -->
<MinimumRadiusTable name="AASHTO 2001 eMax 4%">
<MinimumRadius speed="15" radius="70"/>
<MinimumRadius speed="20" radius="125"/>
<MinimumRadius speed="25" radius="205"/>
<MinimumRadius speed="30" radius="300"/>
<MinimumRadius speed="35" radius="420"/>
<MinimumRadius speed="40" radius="565"/>
<MinimumRadius speed="45" radius="730"/>
<MinimumRadius speed="50" radius="930"/>
<MinimumRadius speed="55" radius="1190"/>
<MinimumRadius speed="60" radius="1505"/>
```

Figure 9.30 shows how this area of the Design Standards file is displayed in the Design Criteria Editor.

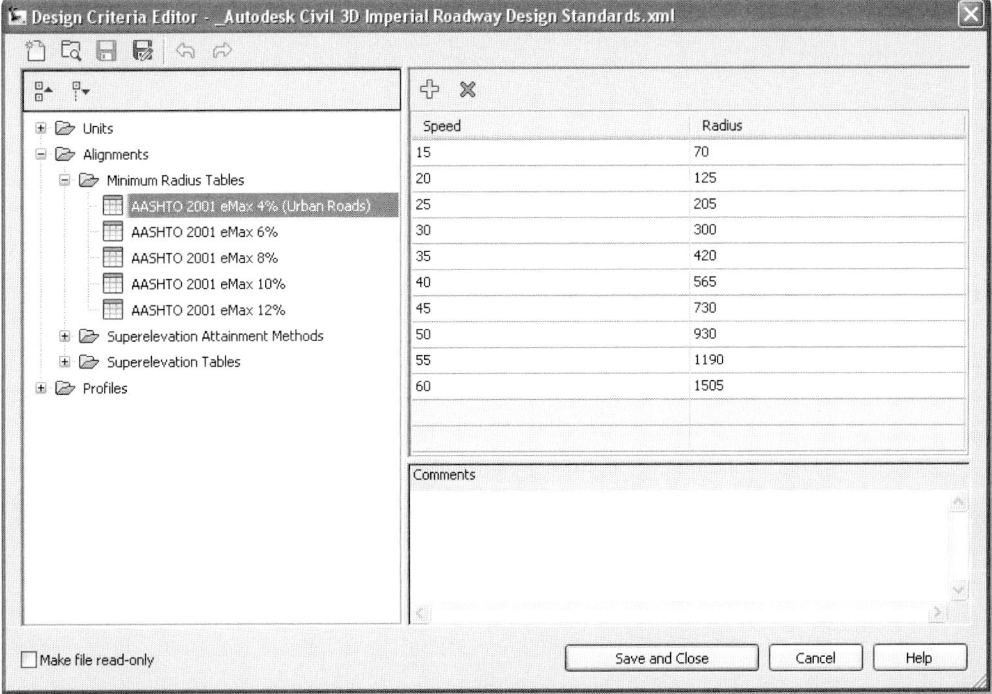


**FIGURE 9.30**

### Horizontal — Superelevation Attainment Methods

A transition length section defines the length of transition for any design speed. The length varies for each speed and radius of curve. For example, a road with a design speed of 20 and a radius of 150 needs 72 feet of transition. If the same road has a radius of 500, it needs only 37 feet of transition.

### Transition Formulas

Civil 3D uses formulas to calculate superelevation values. The superelevation parameters' variables are the following:

{e} - The superelevation rate (from tables)

{t} - The superelevation length (from tables)

{c} - The normal crown lane slope (from alignment settings)

{s} - The normal shoulder lane slope (from alignment settings)

These variables are a part of formulas that define different methods of superelevating a road. Civil 3D supports the following five key transition formulas:

LC to FS — Level crown station (LC) to full super (FS) station (runoff)

LC to BC — Level crown station (LC) to beginning of curve (BC)

NC to LC — Normal crown station (NC) to level crown station (LC) (Runout)

LC to RC - Level crown station (LC) to reverse crown station (RC)

NS to NC - Normal shoulder station (NS) to normal crown station (NC)

The Design Criteria Editor lists and modifies these variables' values. By default, Civil 3D uses the two-thirds rule—that is, two-thirds of the transition length is along the tangent, and the remaining one-third is along the path of the curve (see Figure 9.31).

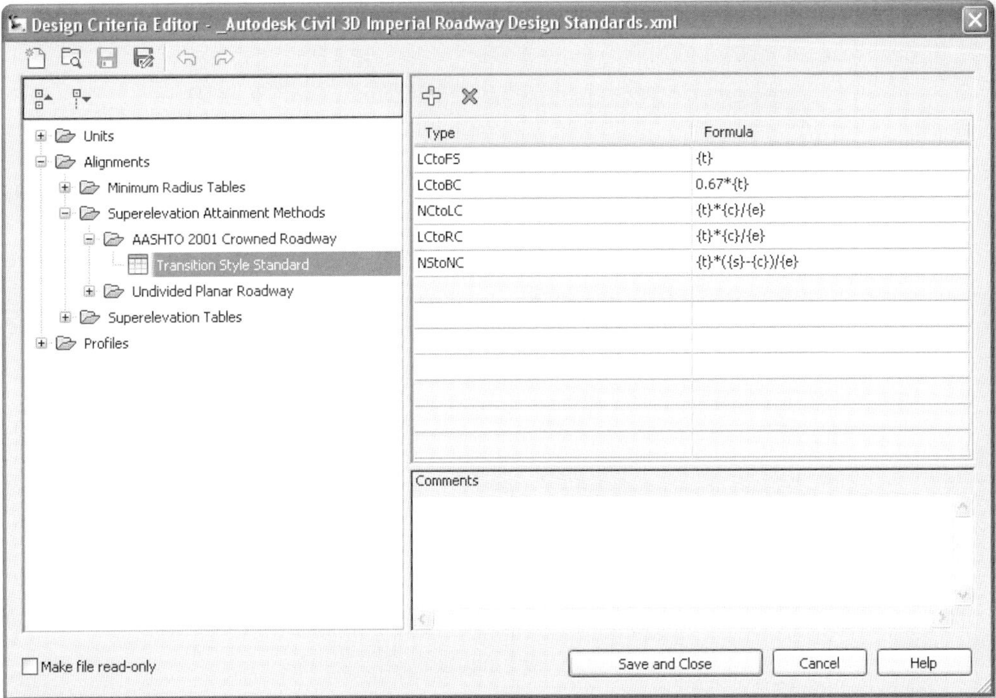


**FIGURE 9.31**

## Horizontal – Superelevation Tables — Speed and Cross Slope

A minimum radius table defines the smallest horizontal radius that maintains the design speed. If the roadway maintains the same speed, but contains different radii, what cross slopes does each have? What roadway lengths are needed for each curve superelevation? To answer these questions, Civil 3D references two additional tables. The first specifies the cross slope amount based on various curve radii for a fixed speed. The second specifies the required transition length. The following Design Standards file excerpt is for an urban road with a design speed of 20 and a maximum cross slope of 4 percent (eMax). NC stands for "normal crown" and RC stands for "reverse crown."

```
<DesignSpeed speed="20">
<SuperelevationRate radius="1400" eRate="NC"/>
<SuperelevationRate radius="1200" eRate="RC"/>
<SuperelevationRate radius="1000" eRate="RC"/>
<SuperelevationRate radius="900" eRate="2.1"/>
<SuperelevationRate radius="800" eRate="2.2"/>
<SuperelevationRate radius="700" eRate="2.3"/>
<SuperelevationRate radius="600" eRate="2.5"/>
<SuperelevationRate radius="500" eRate="2.6"/>
<SuperelevationRate radius="450" eRate="2.7"/>
<SuperelevationRate radius="400" eRate="2.9"/>
<SuperelevationRate radius="350" eRate="3.0"/>
<SuperelevationRate radius="300" eRate="3.2"/>
```

```
<SuperelevationRate radius="250" eRate="3.4"/>
<SuperelevationRate radius="200" eRate="3.7"/>
<SuperelevationRate radius="150" eRate="3.9"/>
<SuperelevationRate radius="125" eRate="4.0"/>
```

Figure 9.32 shows how this area of the Design file is displayed in the Design Criteria Editor.

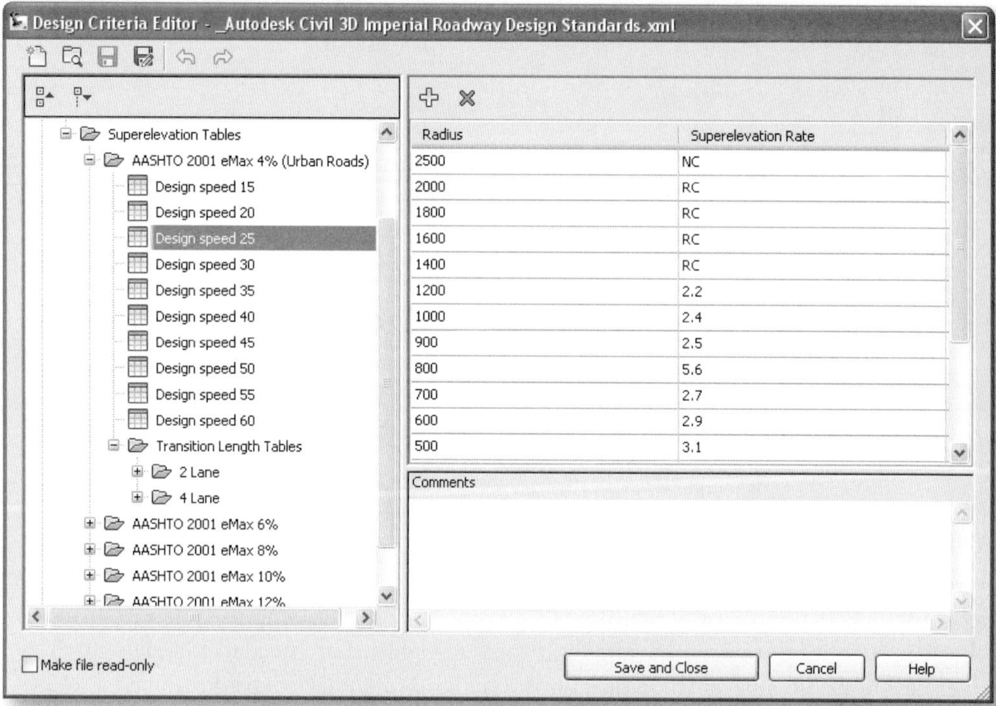


**FIGURE 9.32**

The maximum rotation of the roadway pavement occurs only at the minimum curve radius for the design speed. The cross slope amount changes with the radius of the curve—the shorter the curve radius, the greater the amount of cross slope needed to maintain the speed. In the previous excerpt, a curve with a radius of 125 has a cross slope of 4 percent, and a curve with a radius of 400 will need a cross slope of only 2.9 percent to maintain the speed.

## Horizontal – Transition Length Tables

Transition Length has two tables: 2 and 4 lanes. Each table has a speed and links a radius with an overall transition length (see Figure 9.33).

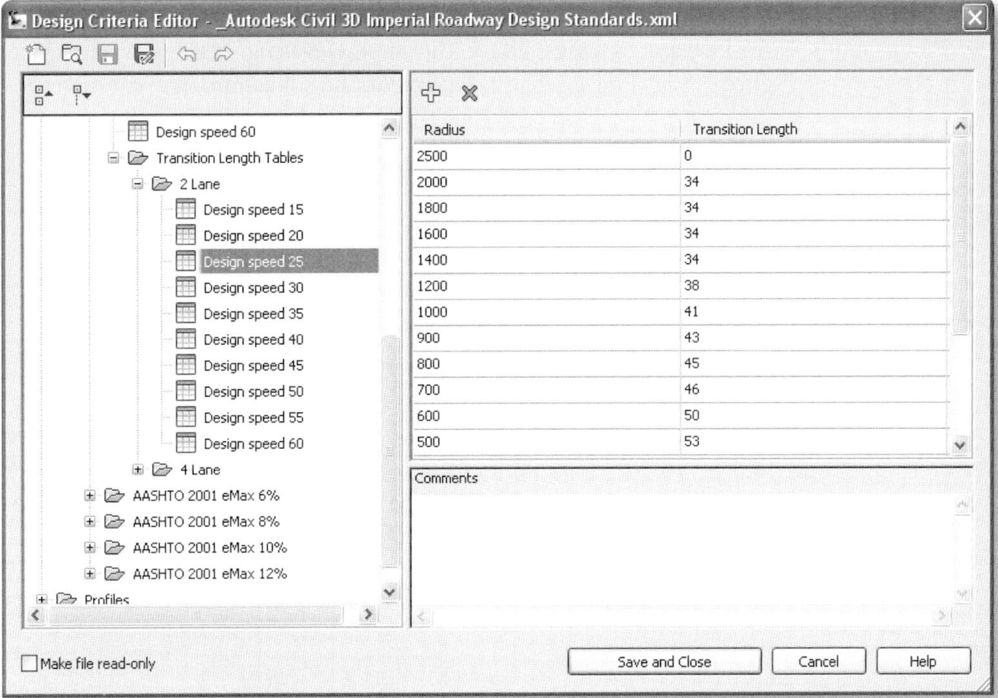


**FIGURE 9.33**

## Vertical Design Criteria

Design Criteria's last section defines vertical curve standards by one of three methods: Stopping Sight Distance, Passing Sight Distance, or Headlight Sight Distance. Each table uses a speed to define a minimum K value (see Figure 9.34).

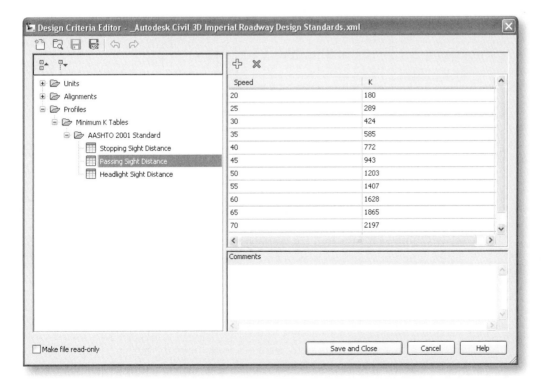

Design Criteria Editor - _Autodesk Civil 3D Imperial Roadway Design Standards.xml

Units
Alignments
Profiles
Minimum K Tables
AASHTO 2001 Standard
Stopping Sight Distance
Passing Sight Distance
Headlight Sight Distance

| Speed | K |
|---|---|
| 20 | 180 |
| 25 | 289 |
| 30 | 424 |
| 35 | 585 |
| 40 | 772 |
| 45 | 943 |
| 50 | 1203 |
| 55 | 1407 |
| 60 | 1628 |
| 65 | 1865 |
| 70 | 2197 |

Comments

Make file read-only
Save and Close
Cancel
Help

**FIGURE 9.34**

## EDIT FEATURE SETTINGS

Alignment's Edit Feature Settings affect the superelevating lanes and shoulders behavior (see Figure 9.35). These settings supersede the subassembly's parameters.

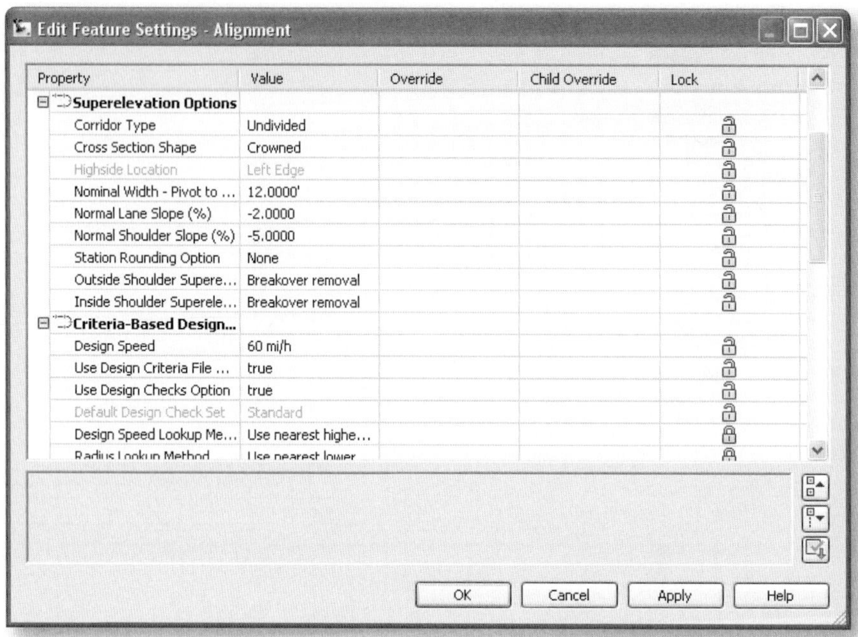


**FIGURE 9.35**

### Superelevation Options

### Corridor Type and Cross Section Shape

Corridor Type and Cross Section Shape show and set the Design Criteria settings.

### Subassembly Lanes and Shoulder Parameters

Nominal Width – Pivot to Edge, Nominal Lane Slope %, and Nominal Shoulder Slope % supersede the subassembly parameters.

### Outside/Inside Shoulder Superelevation Method

There are three shoulder superelevation methods. The first method is breakover removal. This forces the shoulder slope to match the roadway cross slope before beginning superelevation. This method also introduces additional transitional length to rotate the shoulder until it matches the lane's slope. After matching the lane's slope, the lane begins its transition.

The second method is to match lane slopes; the shoulder's slope always matches the lane slopes. The third method is to maintain the shoulder's default slope through the entire superelevation.

### Criteria-Based Design Options

This section sets the default design speed, if you are using Design criteria and Design checks, and explains how to resolve speeds or radii that are not in the table (see Figure 9.36).

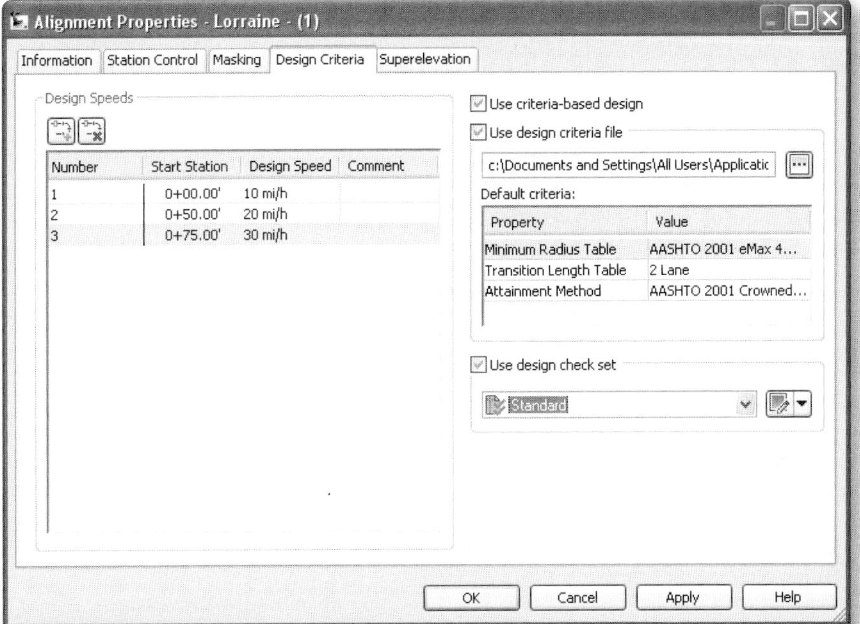


**FIGURE 9.36**

## GENERAL TERMS

When working with superelevations, some basic terms to understand are the following:

**Runout**—The distance over which a lane on one side of the pavement rotates from a normal crown (NC) to no crown (level crown, or LC). Tangent runout distance is another name for runout, because the rotation occurs along the tangent before entering the curve.

**Runoff**—The distance over which one side of the pavement rotates from a level crown (LC) to full superelevation (FS).

**Percentage of Runoff**—The runoff percentage represents the amount of superelevation that occurs along the tangent before entering the curve. In Civil 3D, the default is that two-thirds of the rotation occurs along the tangent, and the remainder occurs along the entering curve.

**E value**—The maximum superelevation rate. The E value is either ft/ft or m/m ratio. A 0.10 E value is a 10 percent grade.

**Superelevation Regions**—Civil 3D considers each curve a superelevation region. When setting the superelevation parameters, users need to set them for each region (curve) of the alignment.

## DESIGN RULES

The first option of the Superelevation Region is the Design rules section (see Figure 9.37). This section identifies the starting and ending station of the curve, the design speed, what Design Standards file to use, the superelevation rate table, the transition length table, and the method of attaining superelevation. The last three settings are tables in the Design Standards file.

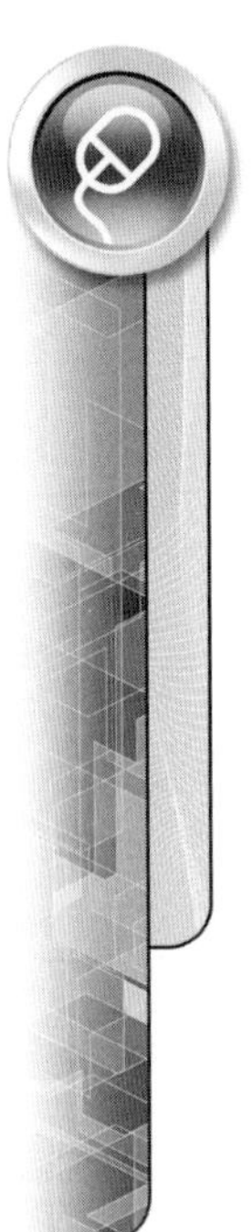

## DEFAULT OPTIONS

The Default Options section sets corridor, shoulder, and calculation rules. The first part of the section sets the corridor type and cross-section shape (see Figure 9.37). The next three entries set the lane width and grades for the lane and shoulders. The next group of values sets which assumptions to use when you have to calculate a value when the speed and curve radius is not in the Design Standards file. The last entries specify how to remove the shoulder cross grade.

## WARNINGS AND ERROR MESSAGES

Civil 3D issues a warning in the Event Viewer if the curve radii and speeds exceed minimums in the Superelevation Design file. These design issues should be dealt with immediately—before you continue with the design process.

The Superelevation panel of an alignment's Properties dialog box will indicate problems with a design by using arrows to point to the full superelevation of the region.

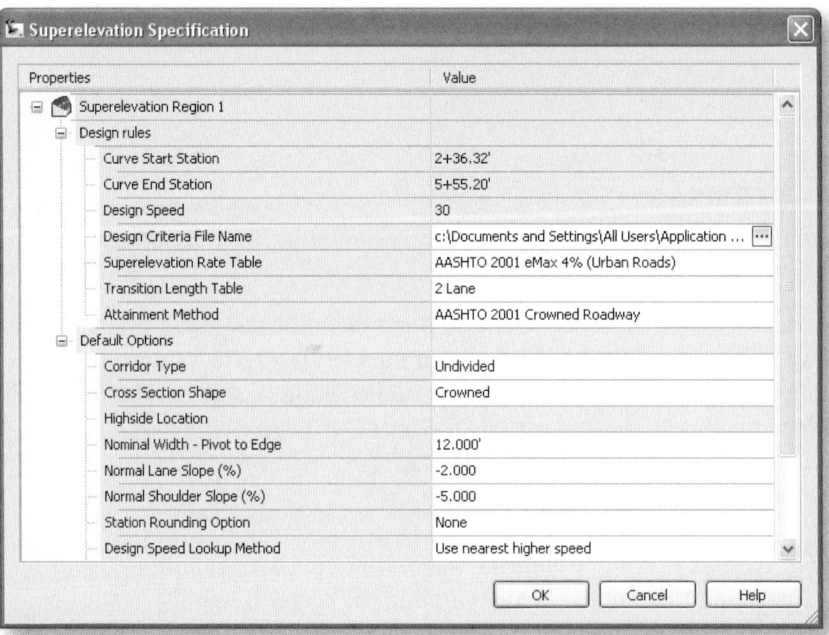


**FIGURE 9.37**

## EXERCISE 9-3

After completing this exercise, you will:

- Be able to review Design Criteria values.
- Be able to use Design Criteria.
- Be able to superelevate a two-lane crowned road.
- Be able to superelevate a four-lane divided highway.

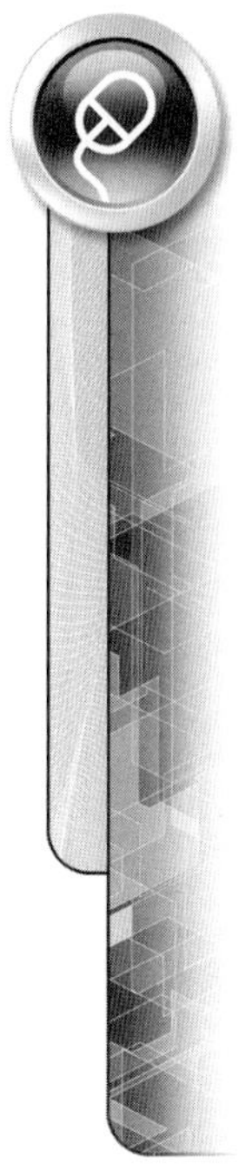

## Exercise Setup

This exercise uses the *Chapter 09 - Unit 3.dwg* file. To find this file, browse to the CD that accompanies this textbook and open the *Chapter 09 - Unit 3.dwg* file.

1. If you are not in **Civil 3D**, double-click the ***Civil 3D*** desktop icon to start the application.
2. When you are at the command line, CLOSE the open drawing and **do not save it**.
3. At Civil 3D's top left, Quick Access Toolbar, click the Open icon, browse to the Chapter 9 folder of the CD that accompanies this textbook and open the *Chapter 09 - Unit 3* drawing.
4. At Civil 3D's top left, click Civil 3D's drop-list arrow, from Application Menu, highlight SAVE AS, from the flyout menu, select AUTOCAD DRAWING, browse to the Civil 3D Projects folder, for the drawing name enter **Superelevation**, and click ***Save*** to save the file.

## Alignment — Edit Feature Settings

1. Click the ***Settings*** tab.
2. Select the Alignment heading, press the right mouse button, and from the shortcut menu, select EDIT FEATURE SETTINGS....
3. Expand the Superelevation Options section and review its values.
4. Expand the Criteria-Based Design Options section. If necessary, using Figure 9.35 as a guide, set the values to match the figure.
5. Click ***OK*** to exit the dialog box.

## Import the Briarwood Alignment and Profile

The Briarwood Alignment and its vertical design profile are in a second LandXML file. When you import an alignment and profile, the alignment is drawn, but the profile data is read. You must create a profile view, sampling the EG surface, and when you draw the view, you must include the Briarwood vertical alignment.

1. In the Import panel, select LANDXML.
2. In the Import LandXML dialog box, browse to the Chapter 9 folder of the CD that accompanies this textbook, select the file *Briarwood.xml*, and click ***Open***.
3. In the Import LandXML, click ***OK*** to import the file.
4. Use the ZOOM and PAN commands to better view the alignment (the surface's western center).

The Briarwood centerline is the small subdivision road at the surface's middle left.

## Assign Design Speeds

1. In the drawing, select the Briarwood alignment, press the right mouse button, and from the shortcut menu, select ALIGNMENT PROPERTIES....
2. Click the ***Design Criteria*** tab.
3. Click the ***Add Design Speed*** icon to set the first station and its design speed.
4. Click the ***Add Design Speed*** icon two more times.
5. Using Table 9.1 as a guide to set the following stations and their design speeds.

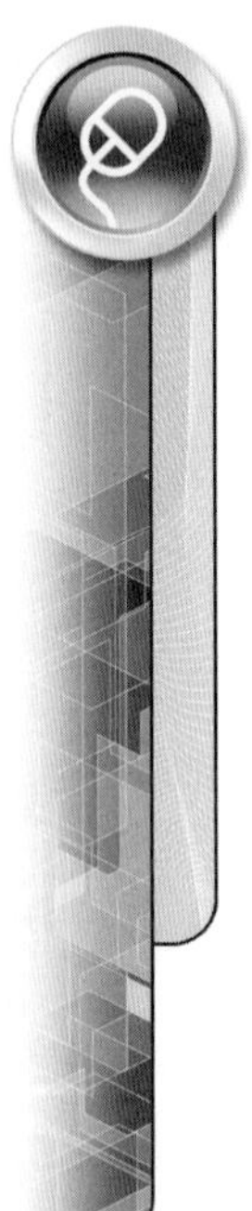

**TABLE 9.1**

| Station | Speed |
|---|---|
| 1+05 | 10 |
| 5+50 | 20 |
| 27+50 | 10 |

## Assign Criteria

1. In the Design Criteria tab, in its upper right, toggle **ON** Use criteria-based design and Use design criteria file.
2. Click ***Apply***.

## Set Superelevation Properties

The alignment's properties maintain superelevation information and assignments.

1. Click the ***Superelevation*** tab.
2. At the dialog box's top, if necessary, toggle **ON** Hide Inside Lanes and Shoulders.

Briarwood is a two-lane road with shoulders.

3. In the dialog box's top left, click the ***Set Superelevation Properties*** icon (the right icon) to display the Superelevation Specification dialog box.
4. Set the Region 1 and 2 design rules as follows:

   Superelevation Rate Table: AASHTO 2001 eMAX 6%

   Transition Length Table: 2 Lane

   Attainment Method: AASHTO 2001 Crowned Roadway
5. Use Figure 9.38 as a guide and if necessary, set the Default Options section's values for all regions.
6. Set the Region 3 design rules as follows:

   Superelevation Rate Table: AASHTO 2001 eMAX 8%

   Transition Length Table: 2 Lane

   Attainment Method: AASHTO 2001 Crowned Roadway
7. Set the Region 4 design rules as follows:

   Superelevation Rate Table: AASHTO 2001 eMax 12%

   Transition Length Table: 2 Lane

   Attainment Method: AASHTO 2001 Crowned Roadway

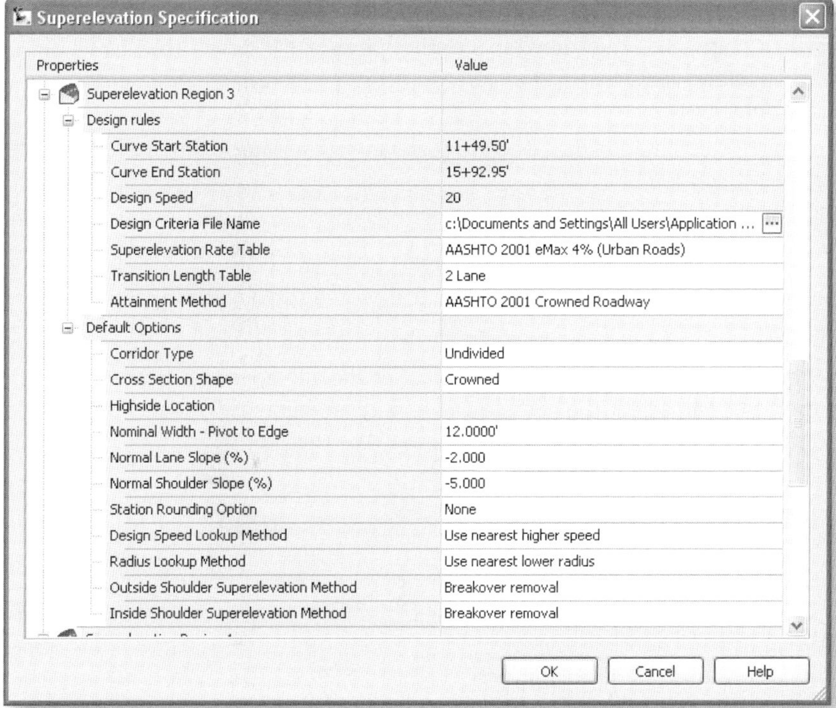


**FIGURE 9.38**

8. Click ***OK*** to set the specifications and return to the Alignment Properties dialog box.
9. Scroll the panel until you are viewing the Region 2 stationing.

The arrows indicate that some alignment lengths do not meet minimum AASHTO standards, and the parameters need to be reviewed and possibly changed (see Figure 9.39).

10. Click ***OK*** to exit the dialog box.
11. At Civil 3D's top left, Quick Access Toolbar, click the ***Save*** icon to save the drawing.

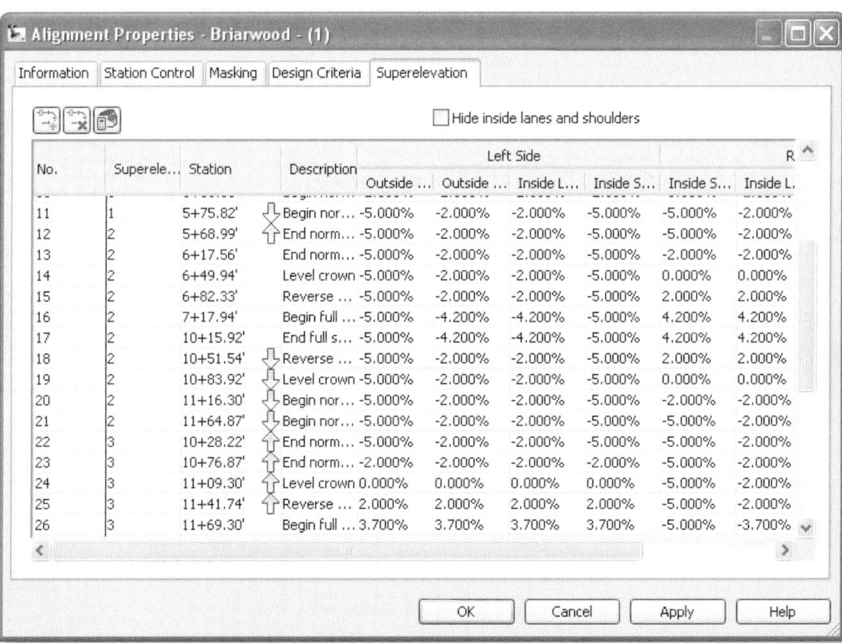

**FIGURE 9.39**

## Create the Briarwood Profiles and View

1. On the Ribbon, click the ***View*** tab. On the Views panel, select NAMED VIEWS, create a New view using the current screen display, and name the view **Briarwood**.
2. Use the PAN command and place the roadway to the right to create open space to the surface's left.
3. Click the Ribbon's ***Home*** tab. In the Create Design panel, click the ***Profile*** icon, and from the shortcut menu, select CREATE SURFACE PROFILE.

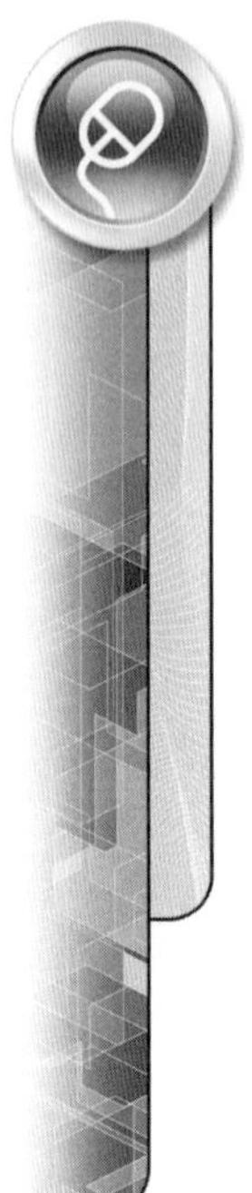

In Create Profile from Surface, the Briarwood vertical profile should already be in the Profile list area.

4. In Create Profile from Surface dialog box, set the alignment to **Briarwood – (1)**. At the top right, select the surface **EG**, and in the middle right, click ***Add≫*** to add the list EG – Surface (1) (see Figure 9.40) to the profile.

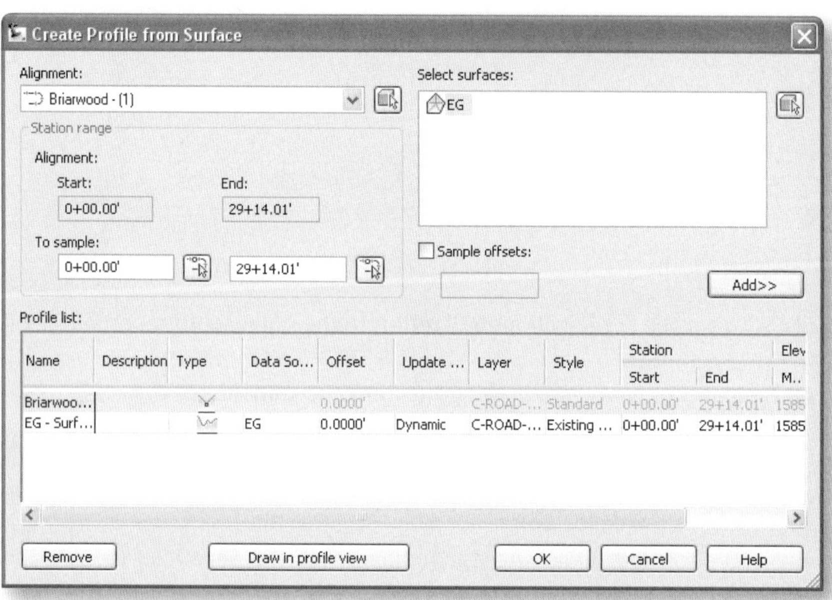


**FIGURE 9.40**

5. At the dialog box's bottom left, click ***Draw in Profile View*** to open the Create Profile View wizard.
6. In the Create Profile View wizard – General panel's middle, set the Profile View Style to **Clipped Grid**.
7. Click ***Next*** three times, and in the Create Profile View – Profile Display Options panel, change the Briarwood Vertical (1) style to **Design Style** and click Next.
8. In Data Bands set Profile 1 to EG – Surface (1).
9. Click ***Create Profile View***, and in the drawing, select a point at the screen's left side.
10. If you need to move the profile, use the MOVE command to relocate the profile.

Your drawing should now look like Figure 9.41.

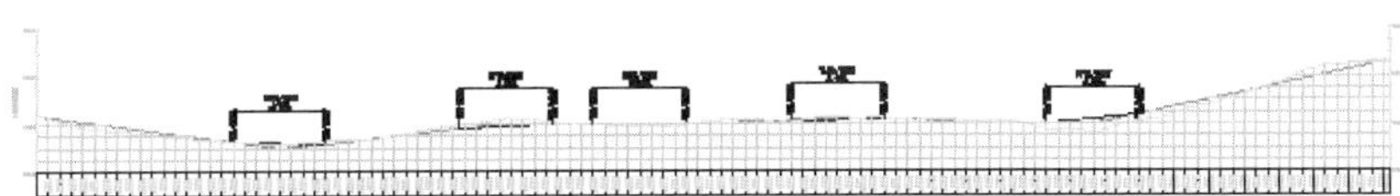

FIGURE 9.41

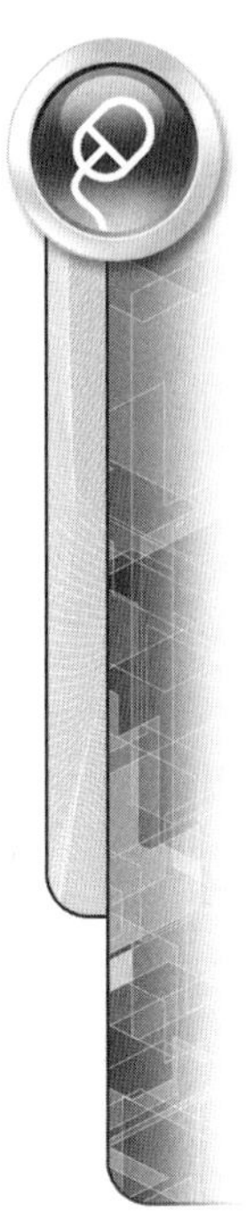

11. At Civil 3D's top left, Quick Access Toolbar, click the ***Save*** icon to save the drawing.

## Create the Assembly

Figure 9.42 is the Briarwood assembly.

1. On the Home tab, in the Create Design panel, click the Assembly icon, and from the shortcut menu, select CREATE ASSEMBLY.
2. In the Create Assembly dialog box, name the assembly **Briarwood**, leaving the counter. Click ***OK***, and select a point to the left of the Briarwood profile.

## Add Subassemblies

1. From the Ribbon's ***Home*** tab. In the Palettes panel, click the Tool Palettes icon to display the Civil 3D Imperial tool palette.
2. On the tool palette, click the ***Lanes*** tab.
3. From the Imperial – Lanes palette, select ***LaneOutsideSuper***.
4. In Properties, the Parameters area, change the Side to **Right**, change the Width to **14** feet, and in the drawing, select the assembly.
5. In Properties, the Parameters area, change the Side to **Left**, create the lane by selecting the assembly, and press ENTER twice to exit the routine.
6. On the tool palette, click the **Shoulders** tab.
7. From the Imperial – Shoulders palette, select ***ShoulderExtendSubbase***. In Properties, the Parameters area, set the Side to **Right**, and in the drawing, select the lane's upper-right marker.
8. In Properties, the Parameters area, change the Side to **Left**. In the drawing, select the lane's upper-left outer marker and press the ENTER key twice to exit the routine.
9. At Civil 3D's top left, Quick Access Toolbar, click the ***Save*** icon to save the drawing.

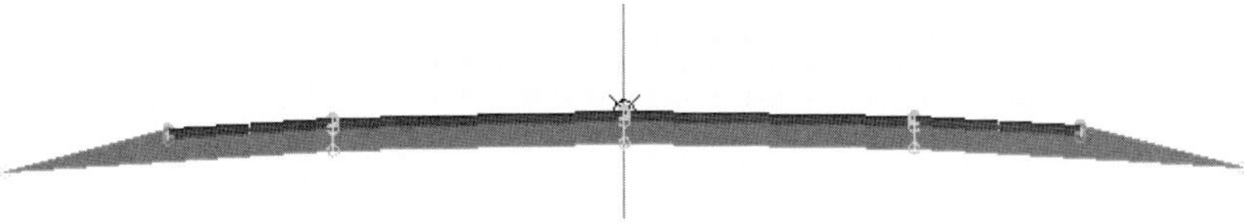

FIGURE 9.42

## Create the Corridor

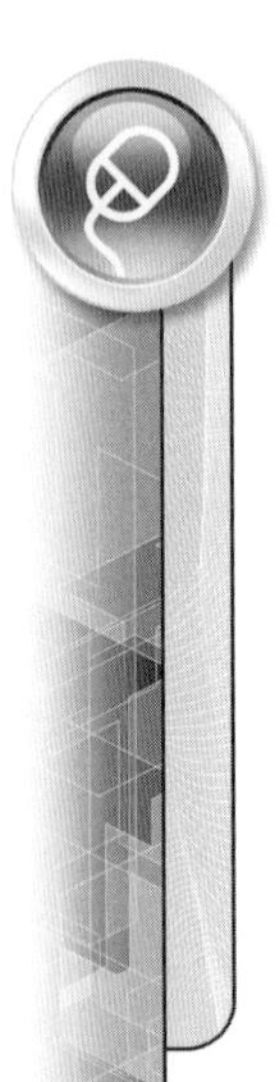

1. On the Ribbon, click the ***View*** tab. On the Views panel's left, select and restore the named view **Briarwood**.
2. On the Ribbon, click the Home tab. On the Create Design panel, click the Corridor icon, from the shortcut menu, select CREATE SIMPLE CORRIDOR, and in Create Simple Corridor, keeping the counter, enter **Briarwood** for the name, and click ***OK*** to continue.
3. The command prompts you for an alignment. Press the right mouse button, select **Briarwood – (1)**, and click ***OK*** to continue.
4. The command prompts you for a profile. Press the right mouse button, and in the Select a Profile dialog box, click the drop-list arrow. From the list, select **Briarwood Vertical (1)**, and click ***OK*** to continue.
5. The command prompts you for an assembly. Press the right mouse button, and in the Select an Assembly dialog box, click the drop-list arrow. From the list, select **Briarwood – (1),** and click ***OK***.
6. The Target Mapping dialog box opens. There are no mappings. Click ***OK*** to create the corridor.

## View Corridor Sections

1. Click the Ribbon's Modify tab. On the Design panel, click the Corridor icon, and in the Modify panel, select CORRIDOR SECTION EDITOR. In the drawing, select the Briarwood corridor to view the sections.
2. After viewing the sections and their superelevation, exit the routine by clicking the Section Editor panel's ***Close*** icon.
3. At Civil 3D's top left, Quick Access Toolbar, click the ***Save*** icon to save the drawing.

## Divided Highway Corridor

A divided highway corridor starts at the site's top left and continues to its middle right. The LandXML file for this highway is in the Chapter 9 folder of the CD that accompanies this textbook.

1. Use the ZOOM EXTENTS command to view the entire drawing.

## Import Route 7380 Alignment and Profile

1. On the Ribbon, click the Insert tab. On the Import panel, select LANDXML. Browse to the Chapter 9 folder of the CD that accompanies this textbook, select the file *Route 7380.xml*, and click ***Open***.
2. In Import LandXML, if necessary, set the Alignment site to **None**, and then click ***OK***.

The alignment appears in the site's northern half.

3. Use the ZOOM and pan the site to the left.
4. If necessary, click the ***Prospector*** tab.
5. On the Ribbon, click the Home tab. On the Create Design panel, click the Profile icon, and from the shortcut menu, select CREATE SURFACE PROFILE.
6. In Create Profile from Surface, change the alignment to **Route 7380 – (1)**, select the surface **EG**, and click ***Add>>*** to add the surface to the profile list. At the bottom of the dialog box, click ***Draw in Profile View*** to display the Create Profile View wizard.
7. In the Create Profile View – General panel, change the style to **Clipped Grid**, and click ***Next*** three times.

8. In the Create Profile View – Profile Display Options panel, change the Route 7380 Vertical (1) design vertical alignment style to **Design Style**, and click Next.
9. In the Data Bands panel, change Profile 1 to EG – Surface (2), click ***Create Profile View***, and in the drawing, to the east of the site, select a point.
10. At Civil 3D's top left, Quick Access Toolbar, click the ***Save*** icon to save the drawing.

## Create the Assembly

Figure 9.43 represents the exercise's assembly.

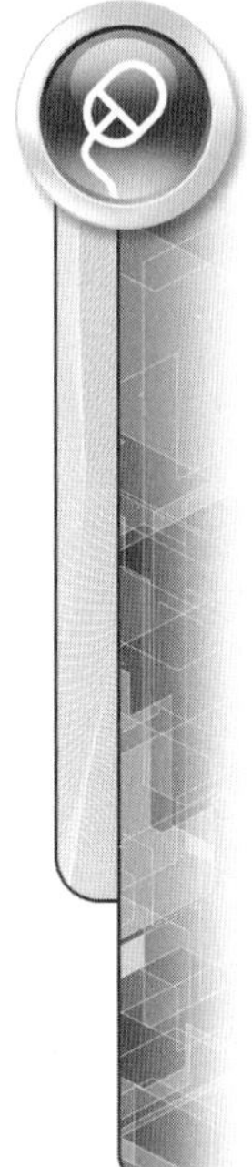

1. From the Ribbon's Home tab, Create Design panel, click the Assembly icon, and from the shortcut menu, select *CREATE ASSEMBLY*. In the Create Assembly dialog box, for the assembly name, enter **Route 7380**, keeping the counter, click ***OK***, and place the assembly near the Route 7380 profile.
2. If necessary, from the Ribbon's Home tab, on the Palettes panel, click **Tool Palettes**.
3. Click the ***Medians*** tab.
4. From the Imperial – Medians palette, select ***MedianDepressedShoulderVert***.
5. In Properties, the Parameters area, change the Hold Ditch Slope to **Hold Ditch at Center, adjust sideslope on high side**. Change the Paved Shoulder Width to **8** feet, and change the Unpaved Shoulder Width to **3** feet. In the drawing, attach the subassembly by selecting the assembly, and press ENTER twice to exit the routine.

The assembly pivots around the median's centerline.

6. Click the ***Lanes*** tab.
7. From the ***Imperial - Lanes*** palette, select ***LaneOutsideSuper***.
8. In Properties, the Parameters area, change the Side to **Right**, and change the Width to **26** feet. In the drawing, select the median's upper-right marker to place the lane.
9. In Properties, in the Parameters area, change the Side to **Left**. In the drawing, select the median's upper-left marker to place the lane.
10. Press ENTER twice to exit the routine.
11. Click the ***Shoulders*** tab.
12. From the ***Imperial - Shoulders*** palette, select ***ShoulderVerticalSubbase***.
13. In Properties, the Parameters area, set the Side to **Right**, set the Paved Width to **8**, and set the Unpaved Width to **3**. In the drawing, select the lane's upper-right marker to place the subassembly.
14. In Properties, the Parameters area, change the Side to **Left**. In the drawing, select the lane's upper-left outer marker to place the subassembly, and press ENTER twice to exit the routine.
15. Close the Civil 3D - Imperial tool palette.
16. At Civil 3D's top left, Quick Access Toolbar, click the ***Save*** icon to save the drawing.

**FIGURE 9.43**

## Assign Design Speeds

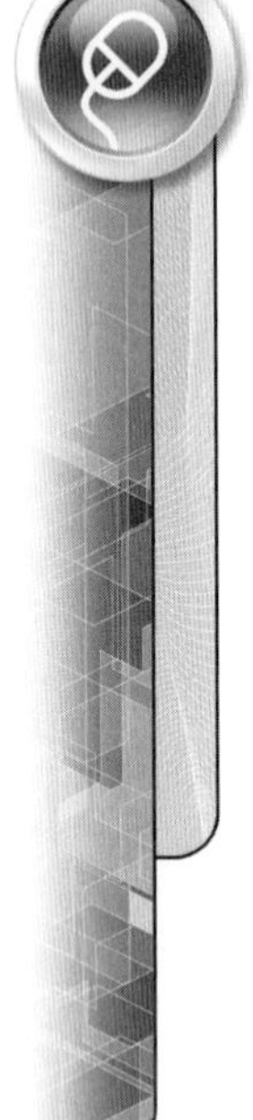

1. Click the ***Settings*** tab.
2. In Settings, select Alignment, press the right mouse button, and from the shortcut menu, select EDIT FEATURE SETTINGS....
3. Expand the Criteria-Based Design Options, set the Design Speed to **70**, and click ***OK*** to exit the dialog box.
4. Use the ZOOM and PAN commands to view the Route 7380 alignment.
5. Click the ***Prospector*** tab.
6. Expand the Alignments branch until you are viewing Route 7380 – (1).
7. Select **Route 7380 – (1)**, press the right mouse button, and from the shortcut menu, select PROPERTIES....
8. Click the ***Design Criteria*** tab.
9. In Design Criteria, click the ***Add Design Speed*** icon, set the Start station to **10**, and set the Design speed to **70**.
10. In Design Criteria, click the ***Add Design Speed*** icon, set the Start station to **10250**, and set the Design speed to **70**.

If warnings appear indicating that speeds are too great, select Manually Change the Speeds to close them. The warning stems from the current superelevation table, max 4%.

11. In the panel's upper right, toggle **ON** Use criteria-based design and Use design criteria file.
12. In Default criteria, change the Minimum Radius Table to **AASHTO 2001 eMax 10%**, change the Transition Length Table to **4 Lane**, and toggle Off Use Design Check set.
13. Click ***Apply*** to set the speeds and criteria.

## Assign Superelevation Properties

1. Click the ***Superelevation*** tab.
2. Click the ***Set Superelevations Properties*** icon to set the Route 7380 region values (see Figure 9.44).
3. Set the following values for Superelevations Region 1 and 2:

Design rules:

Superelevation Rate Table: AASHTO 2001 eMax 10%
Transition Length Table: 4 Lane
Attainment Method: AASHTO 2001 Crown Roadway

Default Options:

Corridor Type: Divided
Cross Section Shape: Planar
Nominal Width - Pivot to Edge: 26
Normal Shoulder Slope (%): −6

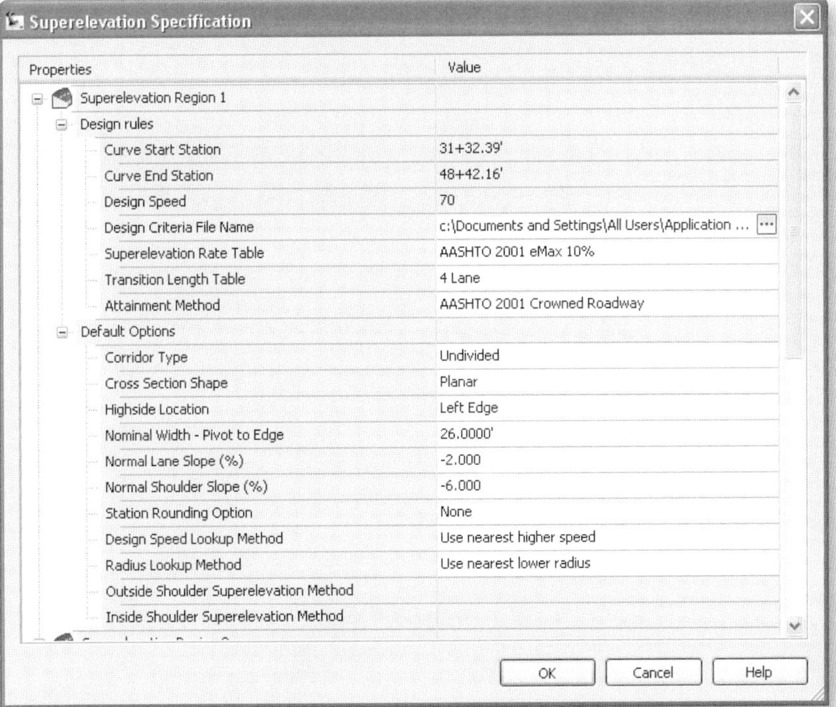


**FIGURE 9.44**

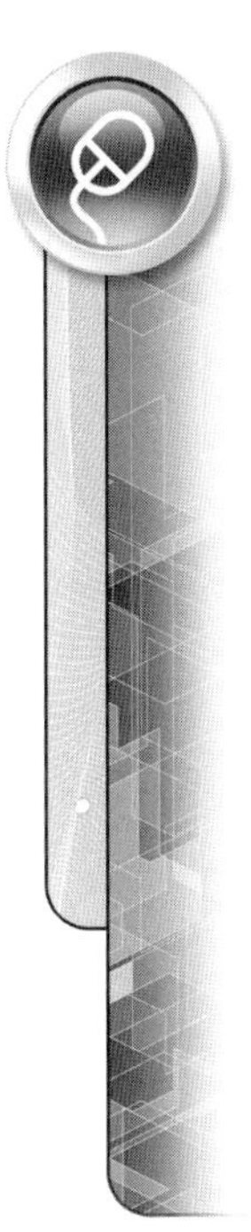

4. Click ***OK*** to set the Route 7380 superelevation parameters.
5. Scroll through the superelevation data and note the stations of maximum or full superelevation.
6. Click ***OK*** to change the alignment properties.
7. At Civil 3D's top left, Quick Access Toolbar, click the ***Save*** icon to save the drawing.

## Create the Corridor

1. If necessary, click the Ribbon's Home tab. On the Create Design panel, click the Corridor icon and from the shortcut menu, select CREATE CORRIDOR.
2. The command line prompts you for an alignment. Press the right mouse button, select **Route 7380 – (1)**, and click ***OK*** to continue.
3. The command line prompts you for a profile. Press the right mouse button, and in the Select a Profile dialog box, click the drop-list arrow. From the list, select **Route 7380 Vertical (1)**, and click ***OK*** to continue.
4. The command line prompts you for an assembly. Press the right mouse button, and in the Select an Assembly dialog box, click the drop-list arrow. From the list, select **Route 7380 – (1)**, and click ***OK*** to continue.
5. In the Create Corridor dialog box, name the corridor **Route 7380**, keeping the counter.
6. At the dialog box's top right, click ***Set All Targets***.
7. In Target Mapping, for the Surfaces, Object Name cell, click (<Click here to set all>). In Pick a Surface, select **EG**, and click ***OK*** (see Figure 9.45).

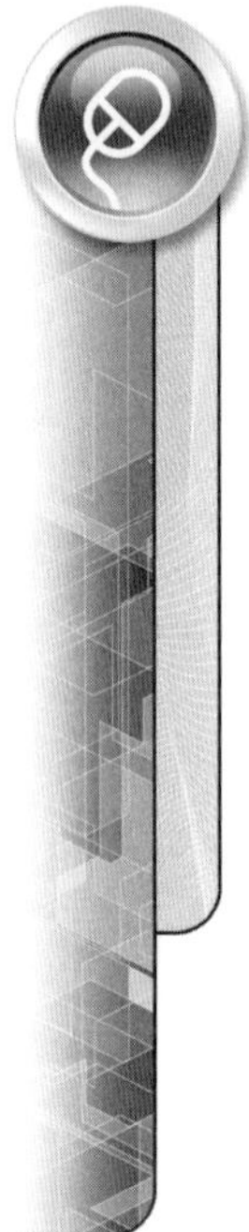

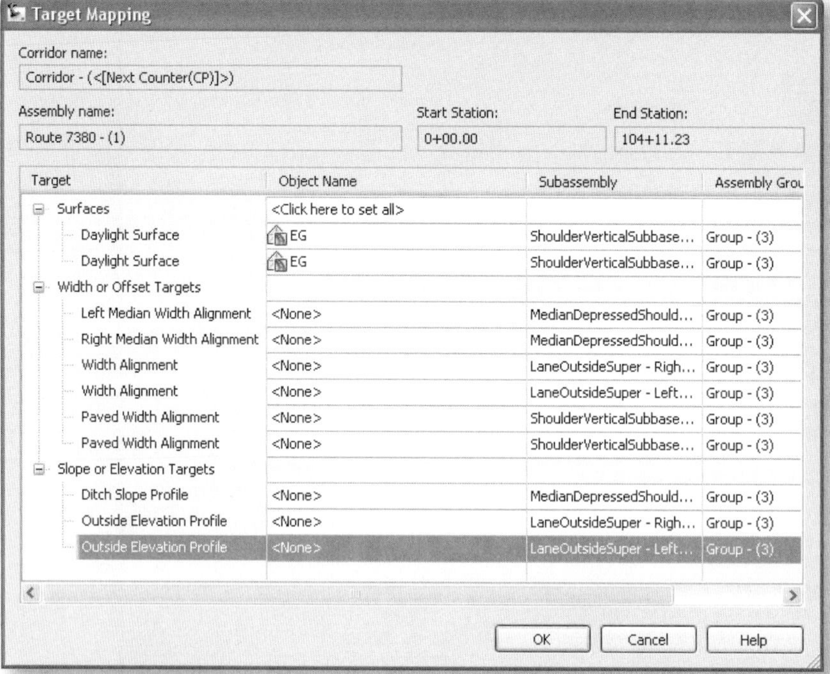


**FIGURE 9.45**

8. Click ***OK*** until you have exited the dialog boxes and built the highway.
9. At Civil 3D's top left, Quick Access Toolbar, click the ***Save*** icon to save the drawing.
10. Use the `ZOOM` and `PAN` commands to view the new corridor.

The maximum superelevation occurs at stations 32+00 to 47+00 and 71+00 to 80+00. Start viewing the first region's sections at station 25+00 and the second region at station 54+00.

11. Click the Ribbon's Corridor tab. In the Corridor tab's Modify panel, select `CORRIDOR SECTION EDITOR`, pick any line representing the corridor, and view the assembly behavior through the superelevated curves. When you are finished reviewing the curves, click Close to exit the Section Editor tab.
12. Click Close to exit the Corridor tab.

This completes the transitions and superelevations exercise. The following Unit focuses on Civil 3D's Intersection Wizard.

## SUMMARY

- The Design Criteria shipped with Civil 3D reflects the values published in the AASHTO "Green Book."
- The Design Criteria is an XML-based file and can be copied and edited to suit customer needs.
- The Design Criteria contains variables and formulas to accomplish roadway superelevation and transition.
- The two basic superelevation design parameters are the design speeds and roadway curve radii.
- Design Criteria evaluates a roadway design horizontally and vertically.

- If a superelevation design does not meet the criteria, the alignment properties dialog box will show blue arrows.
- Design Criteria marks horizontal and vertical tangents and curves that do not meet the current criteria.
- Design speeds and superelevation properties are in Alignment Properties.
- The superelevation properties should be appropriate for the assembly that is creating the roadway corridor.

## UNIT 4: INTERSECTION WIZARD

Roadway intersections are essential to a roadway design. Civil 3D intersection wizard creates intersection designs. The wizard is simple to use and produces an intersection that can be edited to the user's needs.

2D intersection rules are:

- Two alignments intersecting only ONCE.
- No vertical alignment or corridor objects.

3D intersection rules are:

- Two alignments intersection only ONCE.
- Both with vertical design profiles.
- Both with a corridor, but not intersecting.

The resulting intersection has an entry in Prospector's Corridor branch, but the Intersection collection lists edits the intersection.

### INTERSECTION WIZARD

Civil 3D's Intersection Wizard appears after selecting the intersection of two alignments. The Wizard has three panels; General, Geometry Details, and Corridor Regions.

#### General Panel

The General Panel's top sets the Intersection's name and description (see Figure 9.46). The panel's middle sets the intersection style and intersection type. The two intersection types are Primary Road Crown Maintained and All Crowns Maintained. The Primary Road Crown Maintained method holds the primary roads crowned cross section while flattening the secondary road's crown to match the elevations along the primary roads edge-of-travelway. The All Crowns Maintained method intersects the two road crowns and blends the two road's edge-of-travelway elevations so they match around a circular arc, multiple arcs, or chamfer.

#### Geometry Details

The Geometry Details panel sets the primary and secondary alignment. You change the alignments status by selecting the alignment's name and clicking the Up/Down arrows on the panel's top right.

The Offsets and Curb Returns section define pavement widening values (offsets) for the intersection. The offsets values widen one or both sides of one or both intersection alignments (see Figure 9.47).

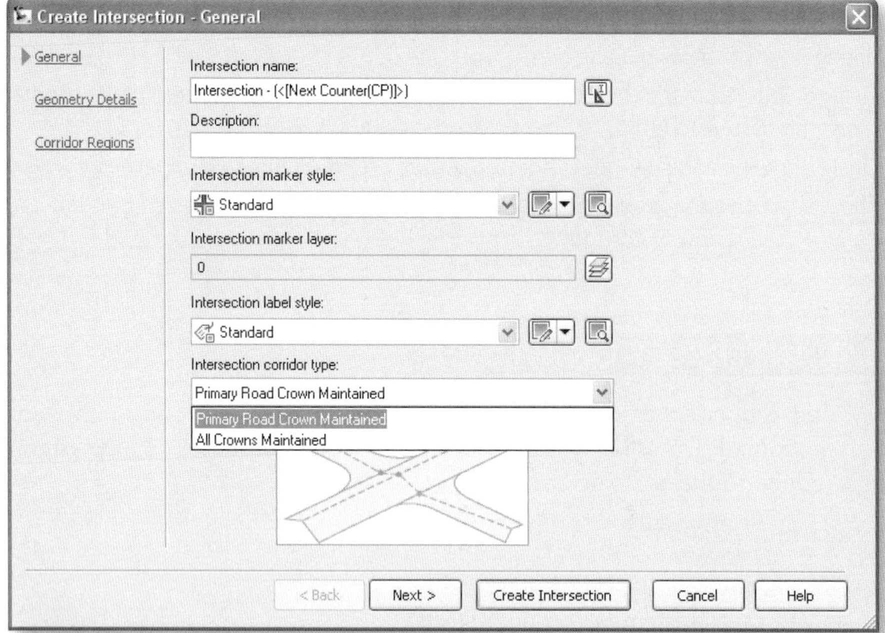


**FIGURE 9.46**

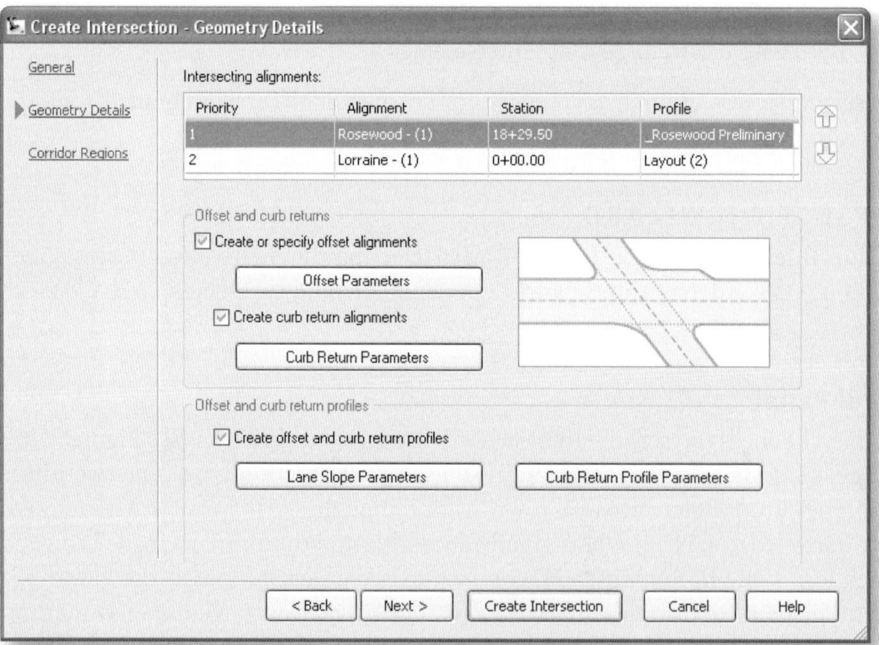


**FIGURE 9.47**

## Offset Parameters

When clicking the Offset Parameter button, the Intersection Offset Parameters dialog box displays (see Figure 9.48). This dialog box sets the offset alignment length (it can be for the entire length of both intersecting alignments or only the length of the intersection), use of an existing alignment, a name format of the offset alignment, and the offset distance from the centerline. The toggle just below these settings defines the offset's length, for intersection length this toggle is off, for the entire alignment length toggle it on. When selecting each item, the item highlights in the dialog box preview area or in the drawing.

- Even if not using offset alignment, you must toggle them on to create curb return alignments.

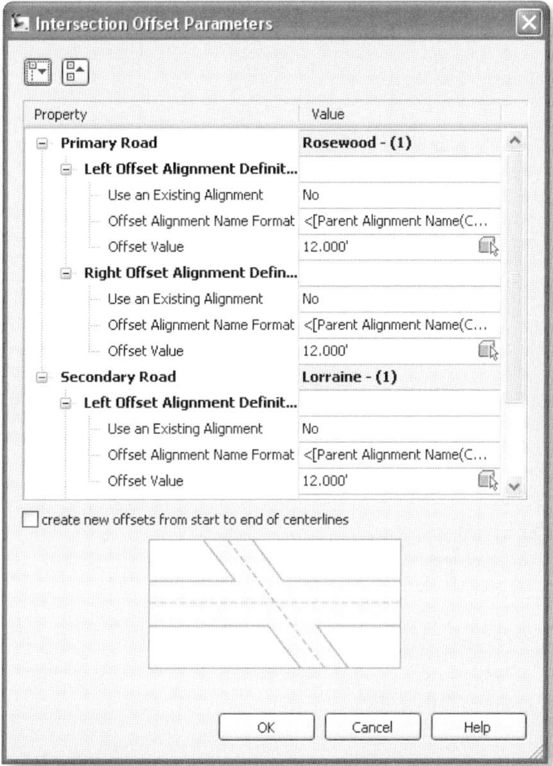


**FIGURE 9.48**

## Curb Return Parameters

The Intersection Curb Return Parameters dialog box sets the curb return type and its radii. At the dialog box's top is a quadrant designation identifying each curb return's location with its parameters and displays the quadrant's location in the drawing. The Next and Previous button at the top focuses the dialog box parameters to all intersection quadrants (see Figure 9.49). In the main portion of the dialog box are the curve type and its radii. A curb return can be a single circular radius, a chamfer, or a 3-Centered arcs. 3-Centered arcs are three compound curves with specified lengths and radii. While specifying these values, the drawing displays graphics relative to the values being set.

At the dialog box's top are the Widen turn lane toggles. When toggled on, the editor uses the offset values from the Intersection Offset Parameters dialog box.

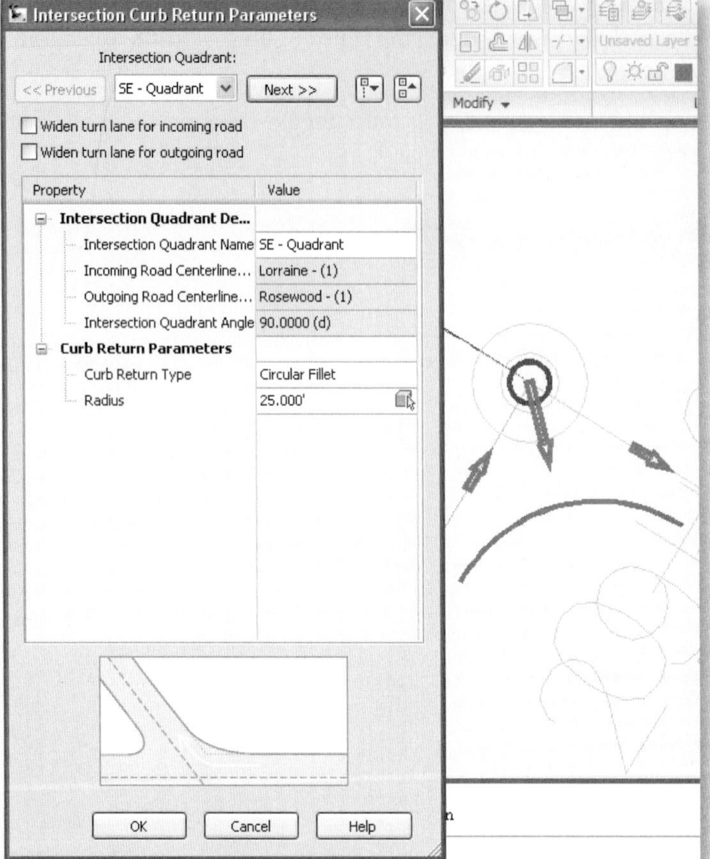


**FIGURE 9.49**

## Offset and Curb Profiles

The Offset and curb profile section defines initial profiles for the offset and return design between the intersecting roadways.

## Lane Slope Parameters

The Lane Slope Parameters dialog box defines the profile that transitions from one roadway's cross slope to the intersecting roadway's cross slope (see Figure 9.50). By Default it is the lane subassembly's default cross slope.

## Curb Return Parameters

The Curb Return Parameters dialog box sets values for the vertical tangents defining the curb returns profile. The incoming and outgoing tangent length set the length of the tangent transitioning from one roadway to the next. The Next and Previous buttons at the top display each intersection design quadrant.

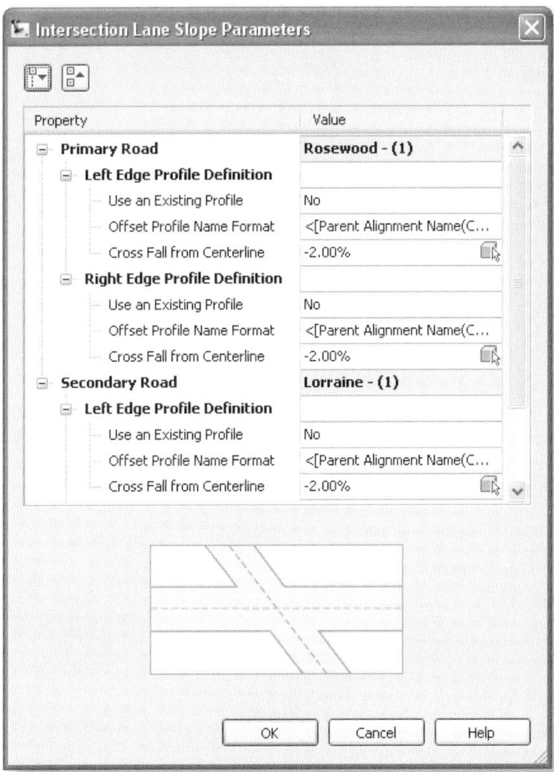


**FIGURE 9.50**

## Corridor Regions

This panel defines how to create the intersection and what assemblies to use (see Figure 9.51). The panel's top toggles the intersection's creation as a corridor or as alignments with profiles. An intersection can be added to an existing corridor or be its own new corridor. If you are adding to an existing corridor, you select the corridor's name from a drop-list. In this area, you must also identify which surface is the daylight surface.

The middle portion set the assembly set, its path, or creates a new assembly set. An Assembly set is a external file containing the name and path to drawings containing specific task assemblies; curve fillet assembly, road part section, etc.

You can create your own sets from assemblies in the current drawing. Autodesk suggests using their set until your are comfortable with the intersection design process. Your assemblies should reside in a drawing dedicated to intersection design and not production. When saving the set, Civil 3D writes an XML file to Civil 3D's data area.

The Assembly set assigns specific assemblies to different intersection regions. As you select each region, the dialog box previews the region and the assembly.

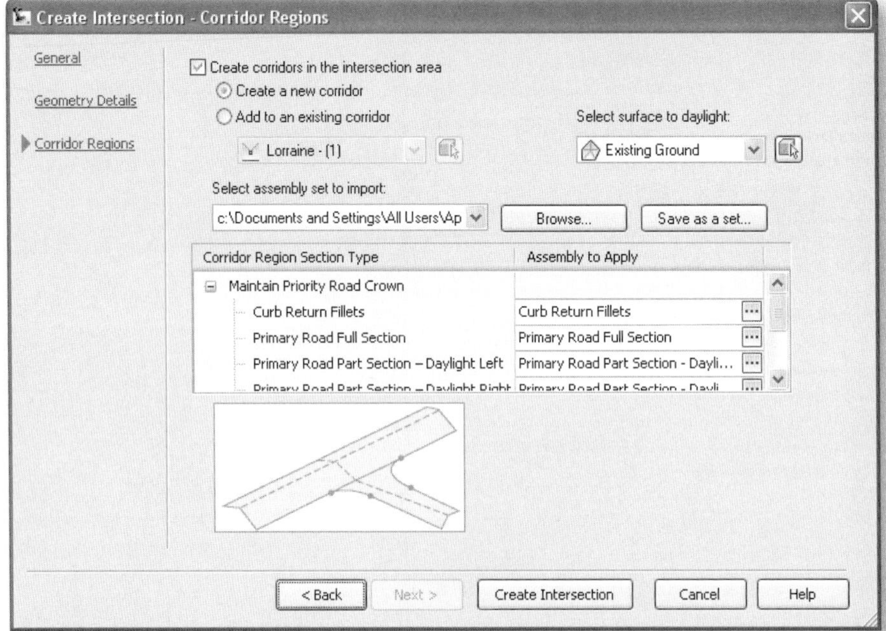


**FIGURE 9.51**

## INTERSECTION SETTINGS

An intersection has many Civil 3D setting values.

### Edit Drawing Settings — Object Layers

The Edit Drawing Settings - Object Layers dialog box lists two layers for intersections. The first layer, Intersections is the object's layer in a drawing. The second layer is Intersection Labels and is used if the intersection label layer is set to 0 (zero). If the label has a specified layer, that layer is used instead of this one.

### Edit Feature Settings

The intersection feature settings set the default object and label styles and the naming formats.

### Intersection Styles

The intersection style defines the marker type and layer for an intersection.

### Intersection Label Styles

Currently, an intersection label is the intersection's name and the alignments creating the intersection.

### Command Settings — Create Intersection

The CreateIntersection command settings define all of the values for the Intersection Wizard (see Figures 9.52 and 9.53).

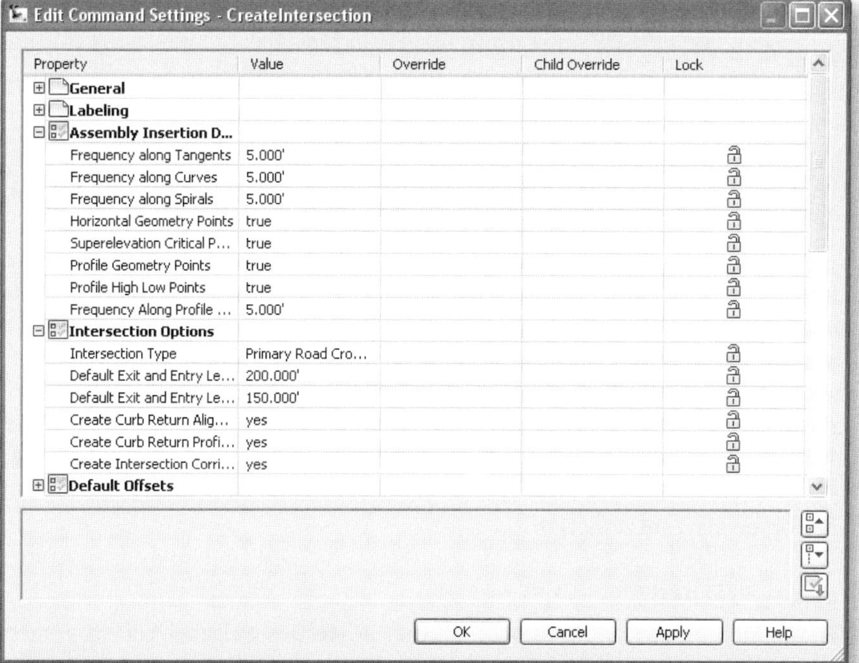


**FIGURE 9.52**

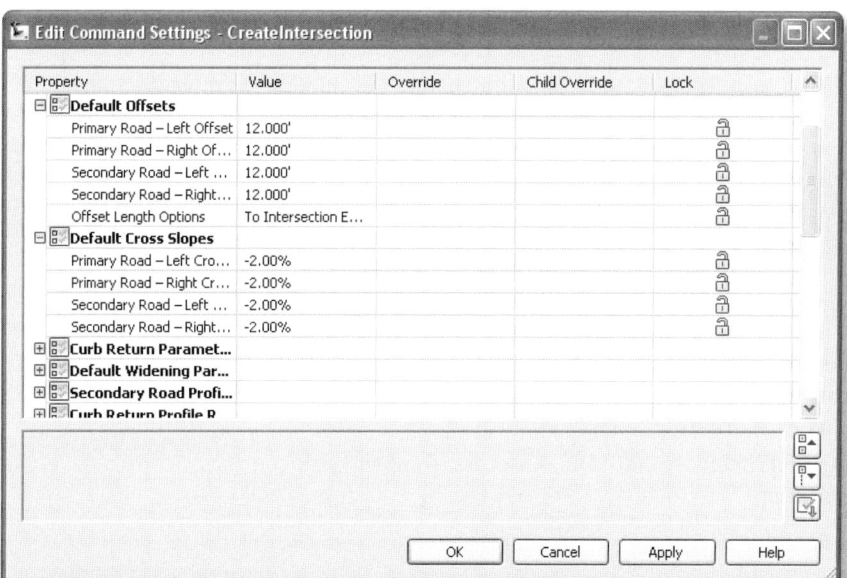


**FIGURE 9.53**

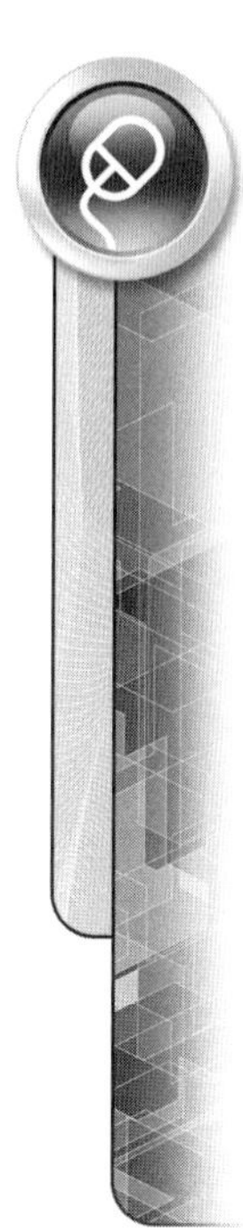

## EXERCISE 9-4

After completing this exercise, you will:

- Be able to review Design Criteria values.
- Be able to create an intersection.

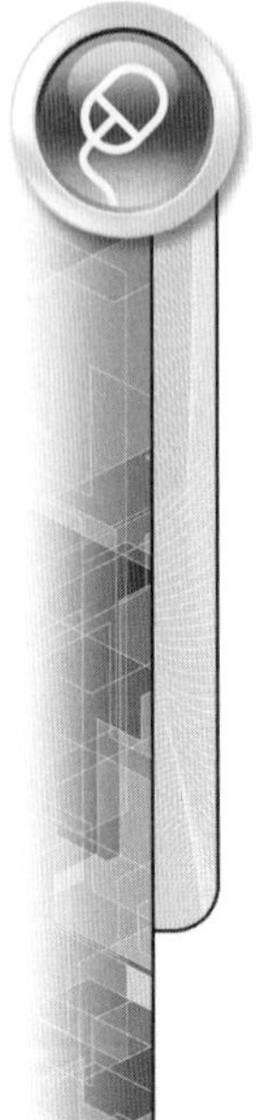

## Exercise Setup

This exercise uses the *Chapter 09 - Unit 4.dwg* file. To find this file, browse to the CD that accompanies this textbook and open the *Chapter 09 - Unit 4.dwg* file.

1. If you are not in **Civil 3D**, double-click the ***Civil 3D*** desktop icon to start the application.
2. When you are at the command line, CLOSE the open drawing and **do not save it**.
3. At Civil 3D's top left, Quick Access Toolbar, click the Open icon, browse to the Chapter 9 folder of the CD that accompanies this textbook, and open the *Chapter 09 - Unit 4* drawing.
4. At Civil 3D's top left, click Civil 3D's drop-list arrow, from Application Menu, highlight SAVE AS, from the flyout menu, select AUTOCAD DRAWING, browse to the Civil 3D Projects folder, for the drawing name enter **Intersection**, and click ***Save*** to save the file.
5. On the Ribbon, click the ***View*** tab. On the Views panel's left, select and restore the named view **Intersection**.

## Intersection Wizard

1. On the Ribbon, click the Home tab. In the Create Design panel, click the Intersection icon.
2. The command line prompts for an intersection point. In the drawing select the intersection between the two roadways.
3. The Create Intersection Wizard displays, review its values, and click Next.
4. Even though you will not use any offsets, click Offset Parameters and review its settings. When done, click OK to return to the Create Intersection wizard.
5. Click Curb Return Parameters and change the quadrants to view their preview graphics. When done, click OK to return to the Create Intersection wizard.
6. Click Lane Slope Parameters and review the intersection slope settings. When done, click OK to return to the Create Intersection wizard.
7. Click Curb Return Profile Parameters and change the quadrants to view their preview graphics. When done, click OK to return to the Create Intersection wizard.
8. Click Next.
9. At the panel's top left, toggle Create corridors in the intersection area to Add to an existing corridor, click the drop-list arrow, from the list select Lorraine - (1), and click Create Intersection.
10. At Civil 3D's top left, Quick Access Toolbar, click the Save icon to save the drawing.

This ends the exercise on intersection design.

## SUMMARY

- The Intersection Wizard uses parameters to define an intersection.
- Curb returns can be a circular arc, chamfer, or a combination of three radii.
- An intersection can be its own corridor or integrated in one of the intersection roadway's corridor.

The next Chapter focuses on the grading tools in Civil 3D.

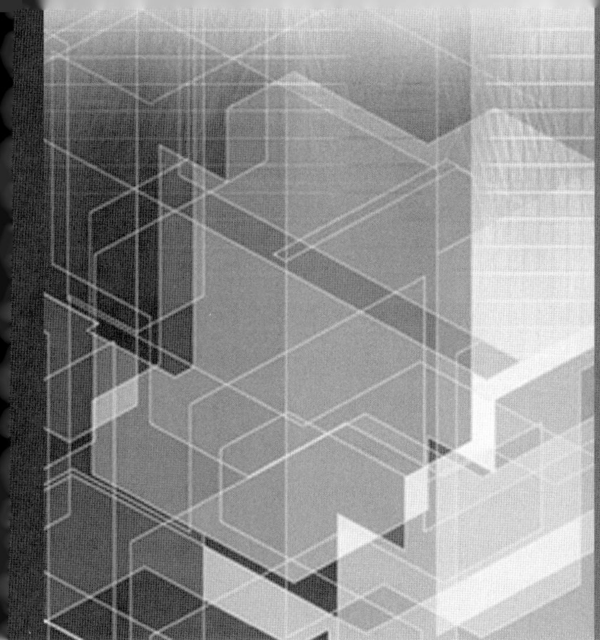

# CHAPTER 10

# Grading and Volumes

## INTRODUCTION

AutoCAD Civil 3D uses surface breaklines, cogo points, contours, feature lines, and grading objects to create a surface design. There are numerous ways to grade a site, so there are numerous strategies to grade a site, and there are numerous object types to use when grading a site.

When using breaklines, cogo points, and contours, feature lines are the most powerful design tool. Feature lines can be converted lines, arcs, or polylines. The most powerful feature line ability is that of containing a curve and preserving elevations along that curve's length. Feature lines are the basis for grading solutions. Another use for feature lines is as a surface breakline. When processing a feature line as surface data, a mid-ordinate variable samples the changing elevations along the feature line's curve.

## OBJECTIVES

This chapter focuses on the following topics:

- Using Various Grading Tool Types
- Creating Designs Using Points, Feature Lines, and Grading Solutions
- Designing with a Grading Object
- Calculating an Earthworks Volume

### OVERVIEW

This textbook's Chapter 4, Surfaces, introduced basic site design tools (i.e., points, feature lines, breaklines, and 3D polylines). In this chapter, the point routines described assign elevations to points from surfaces, 3D polylines, or from interpolating elevations between controlling points. Figure 10.1 shows the Interpolation routines from the Create Points toolbar.

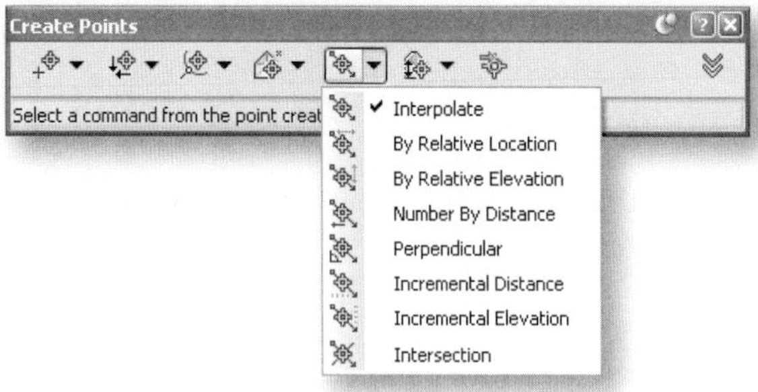


**FIGURE 10.1**

Grading routines focus on feature lines and the grading object. A grading object sets blending strategies, matching a design to an existing surface, projecting a surface to a distance using a grade or slope, or projecting a surface to a specific depth using a grade or slope. A grading object does not need to have a surface to which to grade. It can grade for a distance or to an elevation, relative to or absolute from itself. A grading object can automatically create a surface and calculate a volume.

A feature line creates linear shapes that a grading object uses to create a grading solution. Developing key feature lines that later contribute to creating a surface solution is an essential step in developing an overall grading solution.

After grading a site or developing a design surface, the next step is to calculate an earthworks volume. Unit 4 of this chapter describes a design process to create a surface and calculate an earthworks volume.

### Unit 1

The first unit uses the Create Points toolbar point tools to create a solution from a series of points. The points' elevations are from a surface or from point grade/slope interpolations.

### Unit 2

The second unit focuses on the Feature Line, its creation, settings, and modification.

### Unit 3

The third unit focuses on grading its settings and styles. Whether estimating a stockpile or pond, or designing a pad or parking lot, the grading object presents the user with several opportunities to evaluate different grading scenarios.

### Unit 4

The fourth unit focuses on designing a second surface with contours, points, and feature lines. After designing the surface, the next step is to calculate the earthwork volume.

## UNIT 1: GRADING WITH POINTS

Create Points toolbar commands create new points for a surface that represents a grading plan or a proposed site design. Some routines assign point elevations from a surface. Others assign point elevations by interpolating a grade or slope either from one point or between two controlling points. When you are creating a surface from point data, you may need to add breaklines to correctly triangulate the point data. Instead of using points and breaklines, it may be simpler to use contours, feature lines, and 3D polylines. When you are creating a surface with these three objects, each object type can be a breakline. No matter what object types are used to create the surface, it is necessary to review their effects and resolve any conflicts or crossings.

The Create Points toolbar has three icon stacks that affect grading: Interpolation, Slope/Grade, and Surface.

## INTERPOLATION

The Interpolation icon stack represents routines that calculate elevations from or between points (see Figure 10.1). Most routines require two points (cogo or selected). Most routines also place points in a direction for a distance at a slope or grade from one point or between two controlling points. To create new points, routines may prompt for one or more of the following values: distance, elevation, grade, offset, and slope.

Why create interpolated points? A wide gap between points along a linear feature, for example, a swale, will allow triangles to cross the swale feature. The crossing triangle legs indicate that the points that represent the swale are not related and those points break the assumed constant slope between the swale points. This triangulation incorrectly interprets the related swale points' elevations. To control this problem, there are two options: placing additional points between the swale controlling points, or placing a breakline between the swale control points. In either case, the purpose is to make the surface create triangles that correctly interpret the swale point data.

### Interpolate

Interpolate places points between two existing control points. The new points' elevations are a straight slope calculation of a change in elevation over a distance. The routine places a specified number of points between two controlling points. Each intervening point has a different elevation that reflects a straight slope calculation from the difference in elevation of the controlling points.

### Relative Location/Relative Elevation

These two routines use a starting and ending point. The new points are interpolated elevations between the starting and ending points' elevation. The Location routine places points at a distance measured from the first point. The Elevation routine places points along a line from the first to second point, whose elevations are calculated at a constant slope or grade. Both routines have an optional offset. If the elevation or distance is greater than the original distance between two points, the routines continue placing points until they reach their specified distance or elevation.

### Intersection

This routine places a point at the intersection of two directions and grades or slopes.

### Perpendicular

This routine places a point perpendicular to an object or direction. The point's elevation is calculated by the distance and grade/slope perpendicular to the intersection with the object.

### Slope

The Slope icon stack's two routines are similar to and less complicated than the point interpolation routines: Slope/Grade - Distance and Slope/Grade - Elevation (see Figure 10.2). The Distance and Elevation routines require only one starting point, and the second point does not need to be an existing point. The new points can include a point that represents the distance end.

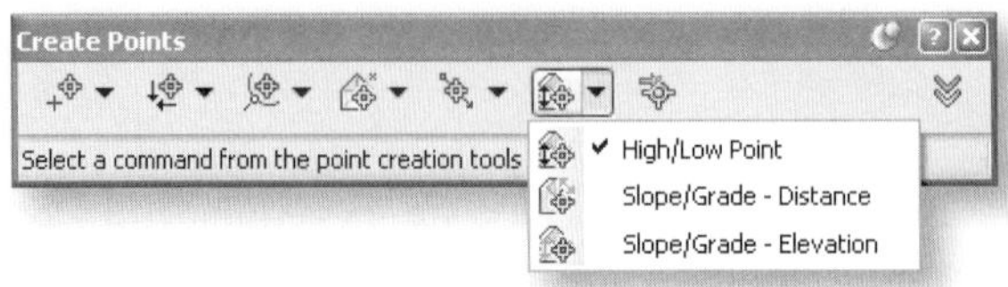


**FIGURE 10.2**

### Slope/Grade - Distance

Slope/Grade - Distance prompts you for a starting point, a direction, a distance, and a slope or grade. It also prompts you for a selected distance and, before proceeding, gives you an opportunity to adjust the distance. The routine then prompts you for the number of intermediate points and whether the farthest distance also has a point.

```
Command:
Specify start point: '_PO (Transparent Command: Point Object)
>> Select point object:
Specify start point: (9655.92 8982.44 681.0)
Elevation <681.000'> (Press ENTER)
Specify a point to define the direction of the intermediate
points:
  (Select a point from the selected point)
Slope (run:rise) or [Grade] <Horizontal>: 5
Slope: (run:rise): 5.00:1
Grade: (percent): 20.00
Distance: (Select a distance from the selected point)
```

The routine responds with the slope and grade, prompts you for the number of intermediate points, and prompts you to answer if the end point also should receive a point.

```
Enter the number of intermediate points <0>: 3
Specify an offset <0.000>: (Press ENTER)
Add ending point [Yes/No] <Yes>: (Press ENTER)
Command:
```

### Slope/Grade - Elevation

Slope/Grade - Elevation works much like the Distance routine, except that the primary values are elevations. If the ending elevation over the default distance is not attained, it will continue placing points in the direction until the elevation is achieved. The routine prompts are as follows:

```
Command:
Specify start point: '_PO (Transparent Command: Point
Object)
```

```
>> Select point object:
Resuming CREATEPOINTS command.
Specify start point: (9655.92 8982.44 681.0)
Elevation <681.000'> (Press ENTER)
Specify a point to define the direction of the intermediate
points:
  (Select a point from the selected point)
Slope (run:rise) or [Grade] <Horizontal>: 7
Slope: (run:rise): 7.00:1
Grade: (percent): 14.29
Ending Elevation: 687.5
Enter the number of intermediate points <0>: 4
Specify an offset <0.000>: (Press ENTER)
Add ending point [Yes/No] <Yes>: (Press ENTER)
Command:
```

Again, if the elevation in the specified distance (the distance between points 1 and 2) is not attained, the routine continues placing points until the elevation is reached.

## SURFACE

Surface point routines require a surface and use the surface's elevations to assign point elevations (see Figure 10.3). If the surface type is terrain, the points represent surface elevations. If the surface type is volume, the point elevations represent the difference in elevation between the two surfaces at the point's location.

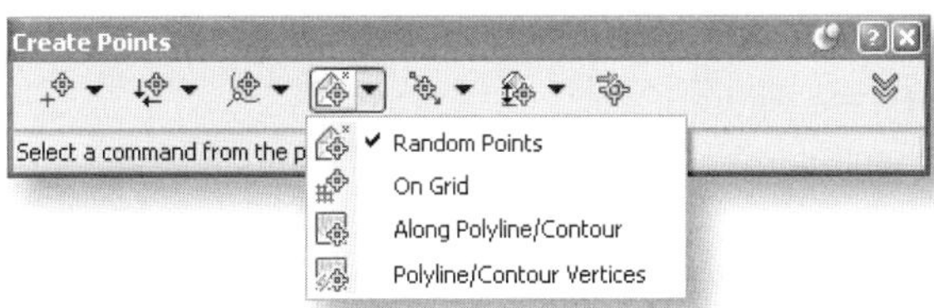


**FIGURE 10.3**

### Random Points

Random Points creates a point whose elevation is a surface's elevation at the selected coordinates.

### On Grid

On Grid places points in a user-defined X and Y spacing. The point elevations are the surface's elevations at the grid intersections.

### Along Polyline/Contour

Along Polyline/Contour places cogo points at a specified distance along a polyline or contour. This is a measure type of command.

### Polyline/Contour Vertices

Polyline/Contour Vertices places cogo points at each polyline or contour vertex. The point elevations are the surface elevation at the polyline vertex or the contour's elevation.

## EXERCISE 10-1

After completing this exercise, you will:

- Be able to set points with elevations from a surface.
- Be familiar with placing points between two existing points.
- Be able to place points on a polyline or contour.
- Be able to place points at a distance or elevation.

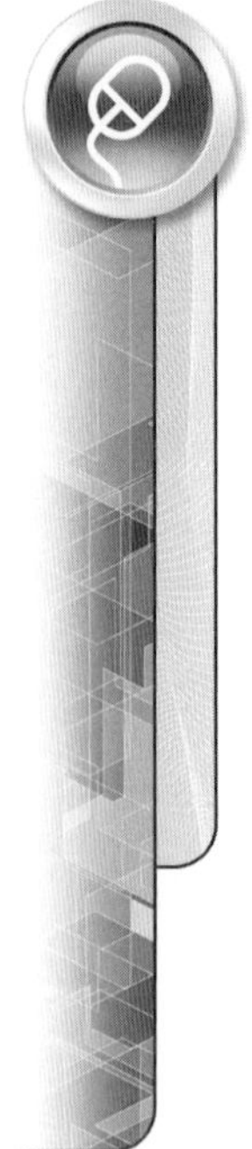

### Exercise Setup

This exercise uses the *Chapter 10 – Unit 1.dwg* file. To find this file, browse to the CD that accompanies this textbook and open the *Chapter 10 – Unit 1.dwg* file.

1. If you are not in Civil 3D, double-click the **Civil 3D** desktop icon to start the application.
2. When you are at the command prompt, CLOSE the opening drawing and do not save the file.
3. At Civil 3D's top left, Quick Access Toolbar, click the ***Open*** icon, browse to the Chapter 10 folder of the CD that accompanies this textbook, and open the *Chapter 10 – Unit 1* drawing.
4. At Civil 3D's top left icon, click Civil 3D's drop-list arrow, from the Application Menu, highlight SAVE AS, from the flyout select AutoCAD Drawing, browse to the Civil 3D Projects folder, for the drawing name **Surface Points**, and click ***Save*** to save the file.
5. In Prospector, select the Points heading, press the right mouse button, and from the shortcut menu, select ***CREATE...*** to display the Create Points toolbar.
6. On the Create Points toolbar's right, click Expand the Create Points dialog (chevron).
7. Expand Points Creation, change Prompt For Elevations to **Manual**, change Prompt For Point Names to **None**, change Prompt For Descriptions to **Automatic**, and set the Default Description to **GP** (see Figure 10.4).
8. At the toolbar's right, click the chevron Collapse the Create Points dialog.

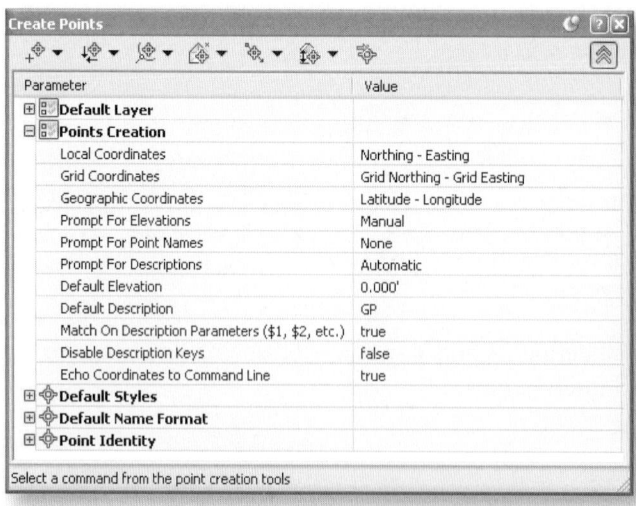


**FIGURE 10.4**

### Interpolate — Relative Location

The Interpolate routines use two points between which they calculate elevations for new points. The second point can be an existing or a selected point that you assign an elevation. The new points can have an offset; negative is left and positive is right. You can select two points from the drawing, assign them elevations, and continue using the command. You do not need to select a cogo point for either point.

1. Make sure you can view the Transparent Commands toolbar.
2. From the Miscellaneous icon stack, select ***Manual***, and select a point to the west of the building points. For the point's elevation, enter **675.00**, and press ENTER until you have exited the routine.
3. From the Create Points toolbar, Interpolation icon stack, select its drop-list arrow and select ***By Relative Location***.
4. To select a cogo point as a starting point, select the ***Point Object*** filter from the Transparent Commands toolbar.
5. The command line prompts you for the first point. In the drawing along the building's edge, select the northernmost point and press ENTER to accept the elevation of 681.00.
6. The command line prompts you for the second point object. Select the just-created new point and press ENTER to accept its elevation (675.00).

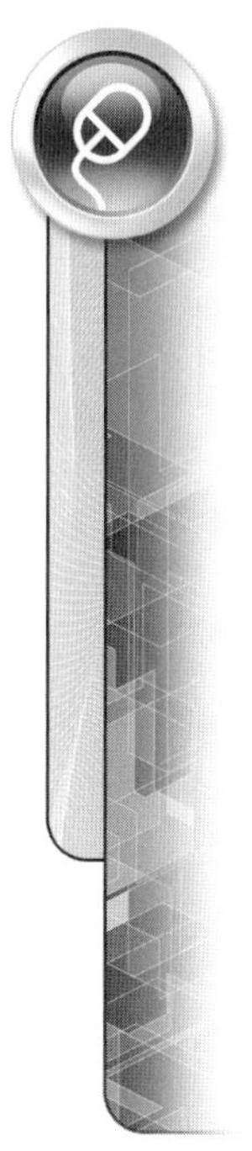

The routine reports the elevation, distance, and grade between the selected points, a direction arrow, and a distance jig from the first point. If necessary, press F2 to view the reported information.

7. The command line prompts you for a distance. Enter **25** and press ENTER.
8. The command line prompts you for an offset. Press ENTER for no offset distance.

A new point appears 25 feet from the first point (building 681.00), whose elevation is interpolated from the distance and elevation between the two points.

9. The command line prompts you for another distance. Enter **45** and press ENTER.
10. The command line prompts you for an offset. Press ENTER for no offset.
11. Press ESC to end the routine.
12. At Civil 3D's top left, Quick Access Toolbar, click the ***Save*** icon to save the drawing.

## Interpolate — Incremental Distance

This routine works only with the selected points. The routine's focus is placing points either at a distance or elevation increment between the control points.

1. From the Create Points toolbar, Interpolation icon stack, click its drop-list arrow and select ***Incremental Distance***.
2. From the Transparent Commands toolbar, select the point filer ***Point Object***.
3. The command line prompts you for the first point. Select the point object, select the second point down from the building's north end, and press ENTER to accept the 681.00 elevation.
4. The command line prompts you for the second point. Select the point object, select the point you placed in the drawing (elevation of 675.00), and press ENTER to use its elevation.
5. The command line prompts you for a distance between points. Enter **25** and press ENTER.
6. The command line prompts you for an offset. Press ENTER for no offset.

The routine places points between the control points at 25-foot increments. Their elevations are calculated using a constant slope between the two control points.

7. Press ESC to exit the routine.
8. Use the ERASE command and erase the new points, leaving the points along the building's westerly side.
9. At Civil 3D's top left, Quick Access Toolbar, click the ***Save*** icon to save the drawing.

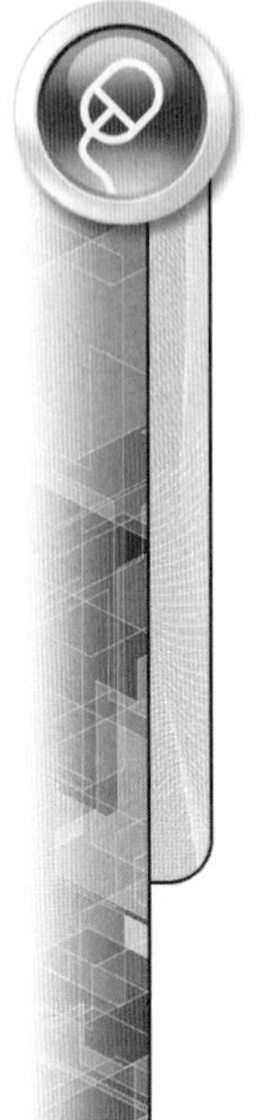

## Slope — Slope/Grade – Elevation

This routine creates points in a direction from an existing or selected point. The elevation is either the point's elevation or one you specify.

1. From the Create Points toolbar, Slope icon stack, click its drop-list arrow and select ***Slope/Grade – Elevation***.
2. From the Transparent Commands toolbar, select the ***Point Object*** override.
3. The command line prompts you for the first point as a point object. Select the northernmost point of the building edge.

The routine prompts you for the next point, to set a direction and default distance.

4. In the drawing, select a point to the west.
5. The command line prompts you for a slope. Enter **8** (for 8:1) and press ENTER.

The command line reports the slope and the grade.

6. The command line prompts you for an ending elevation. Enter **675.00** and press ENTER.
7. The command line prompts you for the number of intermediate points. Enter **4** and press ENTER.
8. The command line prompts you for an offset. Press ENTER for no offset.
9. The command line prompts you about including a point at the end of the distance. Press ENTER for Yes.

The routine creates points that radiate from the control point. The last point's elevation is 675.

10. Press ESC to end the routine.
11. Use the ERASE command to erase the new points from the drawing, leaving those points along the building's side.
12. At Civil 3D's top left, Quick Access Toolbar, click the ***Save*** icon to save the drawing.

## Surface — Random Points

Random Points requires a surface. Any surface elevation can be assigned to a point by selecting it, pressing the right mouse button, and from the shortcut menu, selecting Elevations from Surface....

1. In the Layer Properties Manager, thaw the layer **C-TOPO-EXISTING GROUND** and click **X** to exit the palette.
2. From the Create Points toolbar, Surface icon stack, click its drop-list arrow, and from the list, select ***Random Points***.
3. The command line prompts you for a surface. In the drawing, select a surface contour to identify the surface.
4. In the drawing, select a few random points, and, after creating them, press ENTER twice to exit the routine.

The routine creates points whose elevations are from the selected surface.

5. ERASE the points you just made and leave the points along the building.
6. At Civil 3D's top left, Quick Access Toolbar, click the ***Save*** icon to save the drawing.

## Surface — Along Polyline/Contour

This routine creates points whose elevations are from a selected surface at a measured distance along a polyline's path.

1. If necessary, in the status bar, toggle **OFF** Object Snaps.
2. Start the POLYLINE command, and in the drawing, draw two polylines, each with multiple vertices.
3. From the Create Points toolbar, Surface icon stack, click its drop-list arrow and select ***Along Polyline/Contour***.
4. The command line prompts you to select a surface. In the drawing, select a contour to identify the surface.
5. The command line prompts you for a distance. For distance, enter **15** and press ENTER.
6. The command line prompts you to select a polyline or contour. In the drawing, select a polyline and press ENTER.

The routine creates points at 15-foot intervals along the polyline, ignoring the polyline vertices.

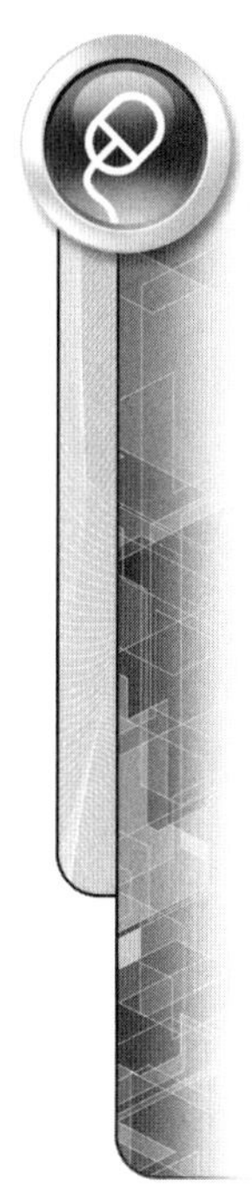

## Surface — Polyline/Contour Vertices

This routine creates points whose elevations are from a selected surface at each polyline or contour vertex.

1. From the Create Points toolbar, Surface icon stack, click its drop-list arrow, and from the list, select ***Polyline/Contour Vertices***.
2. The command line prompts you for a surface. In the drawing, select a contour to identify the surface.
3. The command line prompts you to select a polyline or contour. In the drawing, select the remaining polyline.

The routine places a point at each vertex. When you are using a polyline, the point's elevation is the surface's elevation.

4. Press ENTER to exit the routine.
5. Close the Create Points toolbar.
6. At the top right of the drawing display, click the ***Close*** icon, exit the drawing, and do not save the changes.

This ends the exercise on points as grading data. The next unit reviews feature lines and grading objects.

## SUMMARY

- The Create Points toolbar's Surface icon stack creates new points that use surface elevation as the point's elevations.
- Surface icon stack routines create points on a grid, along a polyline or contour, or randomly.
- The Create Points toolbar's Interpolation icon stack uses two points to create new points.
- Interpolates' two points are point objects or points that have been assigned an elevation.
- When using an interpolate point command, the resulting points generally occur between the two control points.
- The Create Points toolbar's Slope icon stack creates new points from a single control point at a slope or grade for a distance, or until a specified distance or elevation has been reached.
- The Interpolation and Slope routines prompt you for an optional offset.

## UNIT 2: FEATURE LINES AND GRADING OBJECTS

Civil 3D takes a programmatic approach to solving grading design issues. Traditional grading solves the issue of blending a design to the existing conditions as offsetting and blending contours or elevations. The Civil 3D Grading Object can be a linear or closed object and can daylight to a surface or just grade at a distance or a slope. The end result is a solution that uses the assigned grades and distances. After creating a solution, you may need to calculate a volume or stitch together other grading objects and their solutions into a single solution.

The feature line is fundamental to grading objects. A feature line has many applications and uses. It can be a surface breakline, a corridor string, and a grading object. Feature lines have two ribbon areas (see Figure 10.5). The first location is the Feature Line icon in the Home panel (top of Figure 10.5). Assigning or editing elevations and modifying the feature line are found in the Feature Line panel called from the Modify panel.

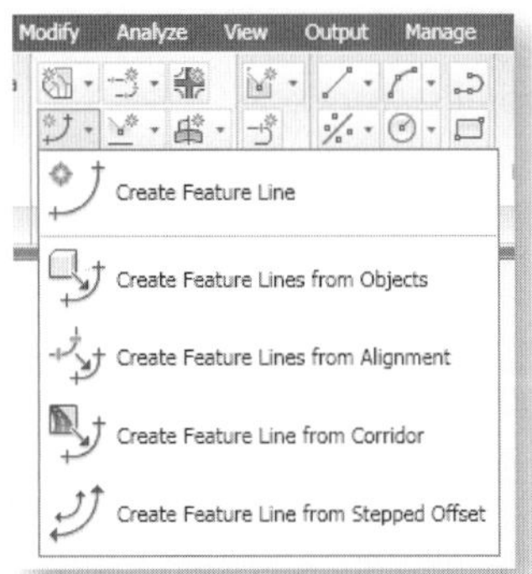


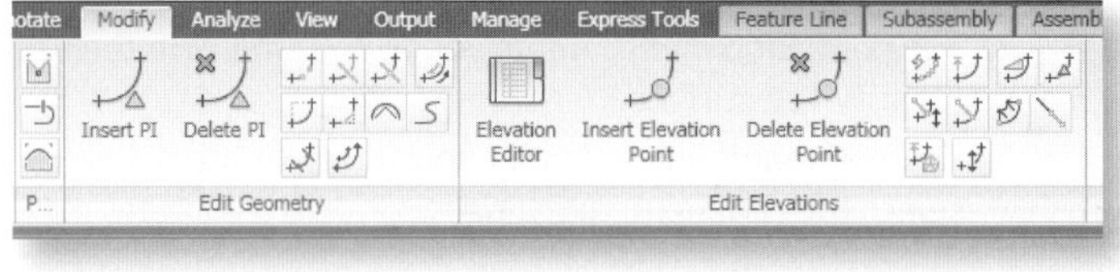


**FIGURE 10.5**

A feature line is the only surface breakline that supports arc segments. Because of this, they are ideal for designing curvilinear features such as entrance returns, parking islands, ponds, etc.

### EDIT DRAWING SETTINGS

Feature lines have an entry in the Object Layers list and can be assigned a modifier and value.

### FEATURE LINE STYLES

Feature line styles are found in the Settings, General, Multipurpose Styles branch. Feature Line Styles assign a layer and its properties (see Figure 10.6). Feature lines also use marker styles, which are in the General, Multipurpose Styles branch.

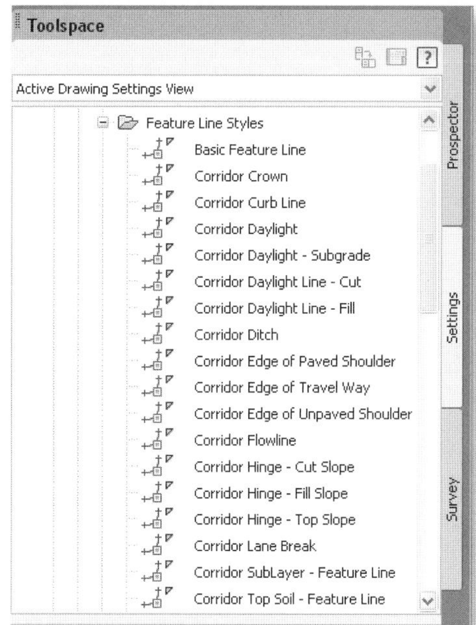


**FIGURE 10.6**

## CREATING FEATURE LINES

Feature lines originate from four sources: drawn, converted entities, an alignment, and from a corridor. The four icons on the Feature Lines toolbar's left create feature lines. Feature line creation commands are also in the Ribbon's Modify - Corridor panel.

When you create feature lines, a Create Feature Lines dialog box opens. A feature line must belong to a site and can have a style and a layer, can be weeded for redundant data, and can be assigned elevations (see Figure 10.7). When toggling on Assign elevations, a second dialog box opens with elevation assignment options (see Figure 10.8). Elevations can be a user-assigned value, from a grading object, or a surface. Inserting intermediate grade break points produces elevations between control points where the feature line crosses surface triangulation.

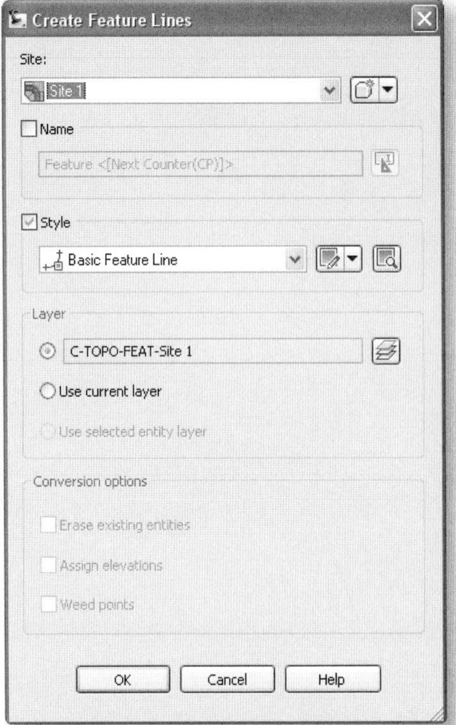


**FIGURE 10.7**

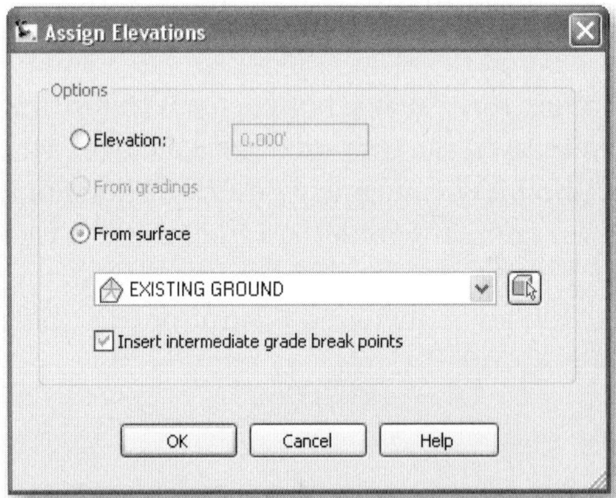


**FIGURE 10.8**

**NOTE**

**Feature line tools affect 3D polylines, survey figures, and alignment feature lines.**

## SITES

A site is a collection of related objects, parcels, alignments, feature lines, and gradings. All items in a site interact with each other. For example, you want to show a site's soil types (as parcel polygons) and parcels. If both are in the same site, they interact. If they are in different sites, they do not interact.

## DRAFTING

To draft a feature line, you start with the assignment of a site, style, name, and layer. When you select the first feature line point, the command line prompts you for an assignment method: a user-specified elevation or a surface elevation. After you select the second point, you need to select the second point's elevation assignment method: grade, slope, elevation, difference, surface, or transition.

**NOTE**

The method used to create the feature line determines the elevation-assignment options.

### Grade or Slope

Grade or Slope assigns the just-selected point's elevation by calculating a grade or slope by the distance between the first and second point. For example, the first point's elevation is 100, the second point is 100 feet distant, and a 2 percent grade assigns the second point the elevation of 102. A slope of 10:1, ten feet raises the height 1 foot, which means the second point's elevation is 110.

### Elevation

Elevation assigns a user-entered value as the second point's elevation.

### Difference

Difference adds a positive or negative value to the first point's elevation to determine the second point's elevation. If the first point is 100 and the difference is 8, the second point's elevation is 108. If the difference is −3.5, the second point's elevation is 96.5.

### Surface

Surface assigns the surface's elevation to the second vertex.

### Transition

Transition starts with the first point and its initial elevation. When you Enter T, and then press ENTER, and you make a series of picks to define transition vertices, no elevation is assigned until you specify an assignment method. This option creates a multi-segment (tangents and/or arcs) that from the first to the last segment has a constant grade or slope, a specific starting and ending elevation, a single elevation difference, or a starting point and ending point with specific surface elevations. When you use the surface option with transitions, the first and last vertices set the starting and ending elevation. The intermediate elevations are a constant grade or slope based on the difference between the first and last elevations.

## CONVERTING

When converting an object to a feature line, Create Feature Lines enables an option to assign elevations. There are three elevation-assignment methods. The first method is to enter a single elevation for the entire feature line: the Elevation option. The second method is to assign an elevation from a grading: From Grading. This creates a temporary surface from a selected grading and then assigns the feature its elevations. The third method is from a surface. From surface assigns surface elevations to the feature line. When you assign elevations from a surface, there is an option: Insert intermediate grade break points. This option creates elevation points where the feature line crosses a TIN triangle. See the following discussion about control and elevation points.

## CREATE FROM ALIGNMENT

Create from Alignment draws a feature line whose elevations are from the alignment's vertical profile. This quickly creates a feature line that can be graded to represent other critical road points: edge-of-travelway, gutter, top-face-of-curb, etc.

## CREATE GRADING FEATURE LINE FROM CORRIDOR

From Corridor extracts a feature line from a corridor point code threads and assigns the corridor elevations to the resulting line.

## CONTROL POINT

Control points are major points, for example, a rectangle's beginning and ending points (each vertex has the same elevation) or all vertices on a linear feature line (each vertex has a different elevation) (see Figure 10.9). They are displayed as green solid fill triangles. When you edit a feature line in the Elevation Editor, and when you do not select a vertex, all vertices show a triangle. When you select a vertex, just that vertex's triangle is displayed (see Figure 10.10).

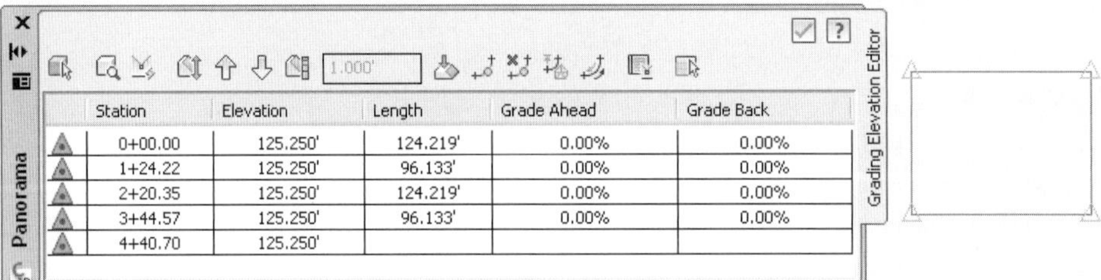


| Station | Elevation | Length | Grade Ahead | Grade Back |
|---|---|---|---|---|
| 0+00.00 | 125.250' | 124.219' | 0.00% | 0.00% |
| 1+24.22 | 125.250' | 96.133' | 0.00% | 0.00% |
| 2+20.35 | 125.250' | 124.219' | 0.00% | 0.00% |
| 3+44.57 | 125.250' | 96.133' | 0.00% | 0.00% |
| 4+40.70 | 125.250' | | | |

**FIGURE 10.9**

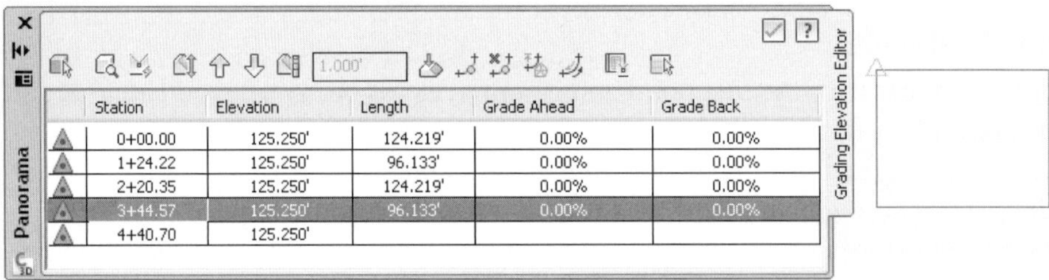


| Station | Elevation | Length | Grade Ahead | Grade Back |
|---|---|---|---|---|
| 0+00.00 | 125.250' | 124.219' | 0.00% | 0.00% |
| 1+24.22 | 125.250' | 96.133' | 0.00% | 0.00% |
| 2+20.35 | 125.250' | 124.219' | 0.00% | 0.00% |
| 3+44.57 | 125.250' | 96.133' | 0.00% | 0.00% |
| 4+40.70 | 125.250' | | | |

**FIGURE 10.10**

## ELEVATION POINT

Elevation points are not the beginning, intermediate, or end vertex of a feature line. Rather, they are a point along the feature line's path, where the elevation changes (see Figure 10.11). This point displays a green circle.

When converting an object to a feature line and sampling a surface assigning elevations to each vertex, the Insert intermediate grade break points option assigns elevation points to the feature line where the line crosses a TIN triangle (see Figure 10.8).

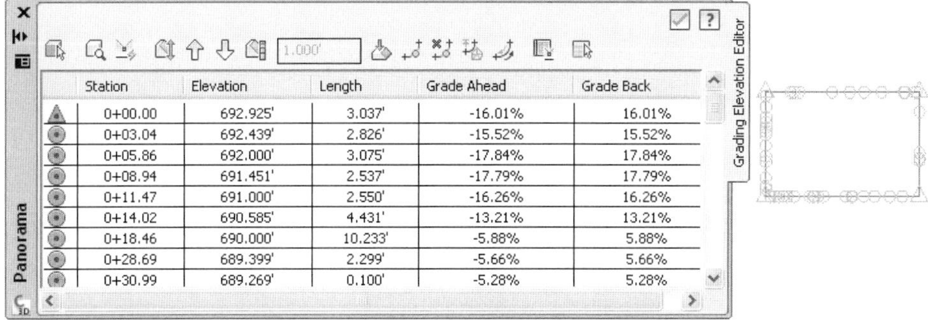


**FIGURE 10.11**

## ELEVATION EDITOR

Elevation Editor assigns and modifies feature line elevations in a Panorama vista (see Figure 10.9). Clicking in a cell puts the cell in edit mode, making the cell's value available for editing. Elevation and slopes/grades are editable. The Editor's rightmost icon deselects any selected row(s).

### Raise/Lower

The Raise/Lower icon assigns an elevation to all feature line points when no entry is selected. When you select an elevation cell and click the icon, the elevation value cell prompts you for an elevation. After you enter a value and press ENTER, the cell is updated.

### Raise and Lower Incrementally

Clicking the Raise or Lower incrementally icons changes the selected point's elevation by the increment set at the Grading Elevation Editor's top center.

### Set Increment

The default increment is 1 foot. Clicking this icon makes the current increment available for editing.

### Flatten to Elevation or Grade

Clicking the Flatten to Elevation or Grade icon opens the Flatten dialog box (see Figure 10.12).

If you select Flatten to constant elevation, all points are set to the first point's elevation. Flatten to grade sets the selected points at a constant grade based on the elevation difference between the first and last selected points.

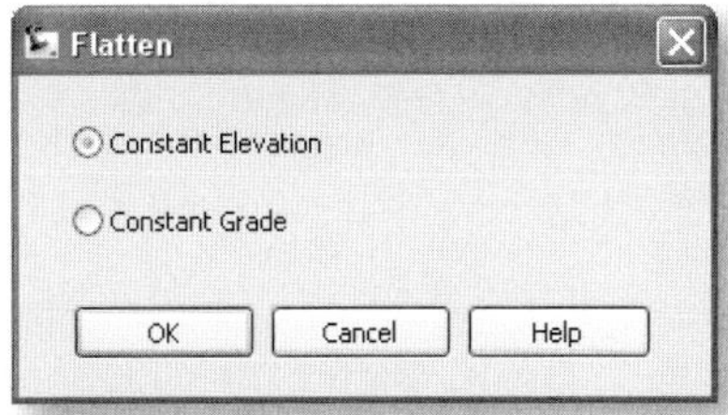


**FIGURE 10.12**

### Quick Editor

Quick Editor edits a feature line control or elevation point after you have selected the point in the drawing. The command line prompts you for a new elevation or grade. The new elevations and grades display tooltips and direction arrows.

### Edit Elevation

Edit Elevation uses the command line to edit each vertice's elevation or slope, to add new vertices, or to assign elevations from a surface. The command prompts are similar to AutoCAD's Pedit command.

The Feature Lines toolbar has alternative methods to assign elevations: grade, slope, and elevation difference. First, you select the feature, and then you select the starting and ending vertices, and set the method to change their elevation.

Set Grade/Slope between points changes elevations by changing elevations or grade/slope between two selected points on a feature line.

Insert High/Low Point places a point elevation between two control points and uses a grade or slope from each point.

Raise/Lower adjusts the entire feature line by a user-specified amount; positive raises and negative lowers.

Elevations from Surface does just that; it assigns elevations from a surface.

### Editing Feature Lines

The Modify tab's Edit Elevations and Geometry panels include tools to edit the feature line.

#### Insert Elevation Point/Delete Elevation Point

An elevation point is not a control point; it is an elevation along a feature line's geometry. This point can also represent a surface's triangle elevation where it intersects a feature line. See the earlier discussion on Converting objects to feature lines using surface elevations. Delete elevation points deletes a feature elevation point.

#### Insert PI/Delete PI

Adds or removes a control point from a feature line. After you select the feature line, Insert PI displays a jig that shows the control point pair between which the new control will be placed.

#### Break/Trim/Join

Break breaks a feature between two selected points. Trim uses a cutting edge and trims a feature line to it. Join joins two or more feature lines into a single feature line. Settings's Grading, Commands Join Feature sets a fuzzy factor for the command: default value 0.01.

#### Reverse

Reverse reverses the feature line's direction. Feature lines that do not travel in the same direction cannot be joined. This is similar to the rule that two alignment segments that do not travel in the same direction cannot be joined.

### Fillet

Adds a curve to a feature line. The command previews the curve, using the default radius, and does allow changing the radius.

### Edit Radius

Changes a feature line's curve. After selecting the curve, the Edit Feature Line Curve dialog box opens and prompts you for a new radius (see Figure 10.13).

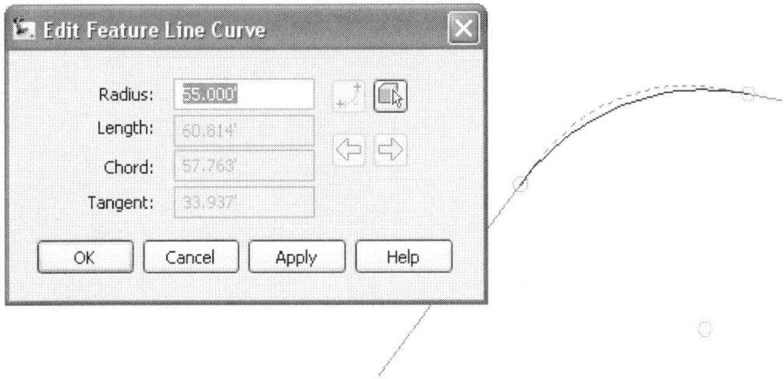


**FIGURE 10.13**

### Fit Curve and Smooth

Fit Curves creates a true curve from tessellated curves. Smooth performs a Bezier curve on a feature line.

### Weed

Weed removes redundant feature line vertices (see Figure 10.14). Weeding can be an individual factor or can be combinations of factors: angle, grade, or length. Close point is an additional option. Close point has a 3D distance tolerance and any two points within this distance have one removed. The Weed Vertices dialog box contains the vertices removal parameters. At its bottom, the dialog box reports how many vertices will be removed. The higher the values, the more vertices are weeded. While the values are adjusted, if visible, the removed vertices are highlighted in red and the remaining ones are highlighted in green.

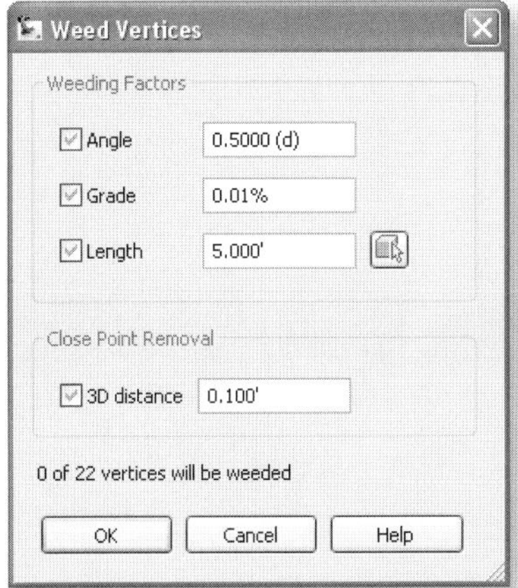


**FIGURE 10.14**

### Stepped Offset

Stepped Offset offsets a feature line and, if specified, modifies the elevations. You change all elevations by elevation difference, grade, slope, or specific elevation. The option list includes Variable, which individually adjusts each offset vertices' elevation by the methods mentioned earlier.

## LABELING FEATURE LINES

Feature Line labeling can be similar to labeling a parcel. However, interest in a feature line's properties includes grade, elevation change, and direction of slope. Each of these labels is dynamic and any change to an elevation on a feature line makes the label change, showing the new grade (see Figure 10.15).

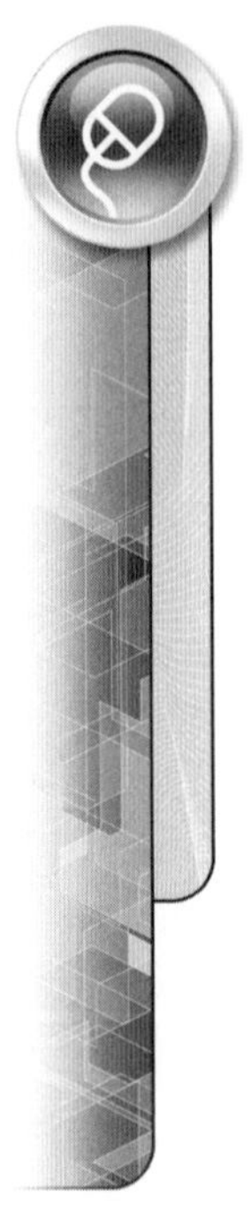

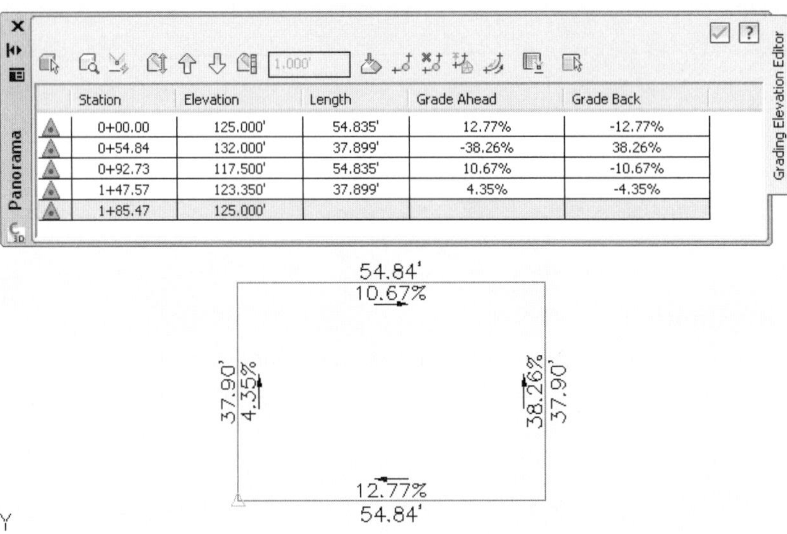

| Station | Elevation | Length | Grade Ahead | Grade Back |
|---|---|---|---|---|
| 0+00.00 | 125.000' | 54.835' | 12.77% | -12.77% |
| 0+54.84 | 132.000' | 37.899' | -38.26% | 38.26% |
| 0+92.73 | 117.500' | 54.835' | 10.67% | -10.67% |
| 1+47.57 | 123.350' | 37.899' | 4.35% | -4.35% |
| 1+85.47 | 125.000' | | | |



**FIGURE 10.15**

## POINTS FROM FEATURE LINES

All appropriate point routines interact with feature lines. For example, measure places points along a feature line at an interval.

## FEATURE LINE ANNOTATION

Add Labels adds grade, direction, elevation, and distance labels on feature lines. The labels are dynamic and change their values as the feature changes.

## EXERCISE 10-2

After completing this exercise, you will:

- Be familiar with Feature Line Edit Drawing Settings and Edit Feature Settings dialog boxes.
- Be able to adjust a profile PVI and Export Feature Lines from the profile.
- Be able to modify feature line elevations.
- Be able to use a feature line as surface data.
- Be able to convert polylines into feature lines.
- Be able to annotate a feature line.

## Exercise Setup

This exercise uses the *Chapter 10 – Unit 2.dwg* file. To find this file, browse to the CD that accompanies this textbook and open the *Chapter 10 – Unit 2.dwg* file.

1. If not in Civil 3D, double-click the ***Civil 3D*** desktop icon to start the application.
2. When you are at the command prompt, CLOSE the opening drawing and do not save the file.
3. At Civil 3D's top left, Quick Access Toolbar, click the ***Open*** icon, browse to the Chapter 10 folder of the CD that accompanies this textbook, and open the *Chapter 10 – Unit 2* drawing.
4. At Civil 3D's top left icon, click Civil 3D's drop-list arrow, from the Application Menu, highlight SAVE AS, from the flyout select AutoCAD Drawing, browse to the Civil 3D Projects folder, for the drawing name enter **Feature Lines**, and click ***Save*** to save the file.

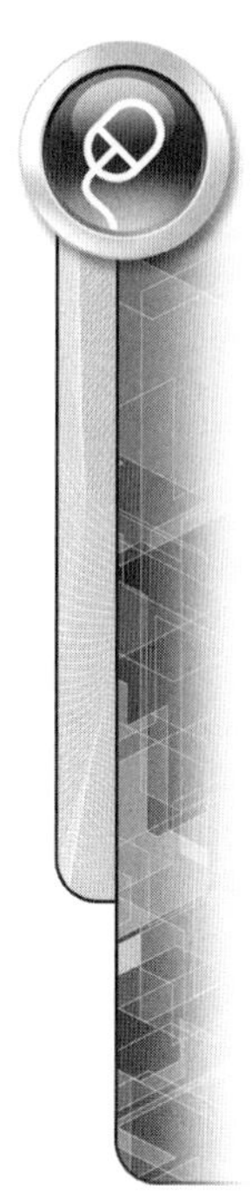

## Edit Drawing Settings

1. Click the ***Settings*** tab.
2. In Settings, select the drawing name, press the right mouse button, and from the shortcut menu, select EDIT DRAWING SETTINGS....
3. Click the ***Object Layers*** tab.
   Most objects have suffix modifiers that append each object with its name.
4. For **Feature Line, Grading, and Grading-Labeling**, set their modifiers to **Suffix** and their values to -*.
5. Click ***OK*** to exit the dialog box.

## Edit Feature Settings

1. In Settings, select the Grading heading, press the right mouse button, and from the shortcut menu, select Edit Feature Settings....
2. Expand the Default Styles and Name Format sections.

The Default section assigns a feature line style and naming templates.

3. Review the settings and then click ***OK*** to exit the dialog box.

## Drafting a Feature Line

Drafting a feature line also assigns elevations to it (see Figure 10.16).

1. Click the Ribbon's ***View*** tab. In the Views panel, select and restore the named view **PK-Island**.
2. Click the ***Home*** tab. On the Create Design panel, click the Feature Line icon, and from the shortcut menu, select ***Create Feature Line***.
3. In the Create Feature Lines dialog box, click ***OK*** to accept the defaults.
4. In the drawing use an *Endpoint* object snap to select the north island's gutter endpoint (see Figure 10.16). The command line prompts you for an elevation. Press ENTER to accept.
5. The command line prompts for the next point. For the next endpoint, in the command line, enter **@40 < 270** and press ENTER.
6. The command line prompts you for the new point's elevation. If not grade, in the command line, enter '**G**' and press ENTER. For the grade, enter **–2** and press ENTER.

7. The command line prompts for the next point. For the next endpoint, use the *Perpendicular* object snap, and in the drawing, select a point on the southern island flange line.
8. The command line prompts you for a grade. To enter an elevation in the command line, enter **'E'** and press ENTER. The command line prompts you with the selected point's elevation. Press ENTER to accept the elevation.
9. Press ENTER to exit the command.

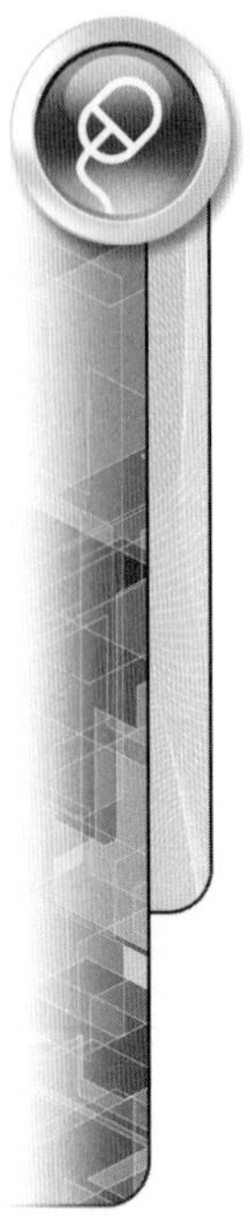

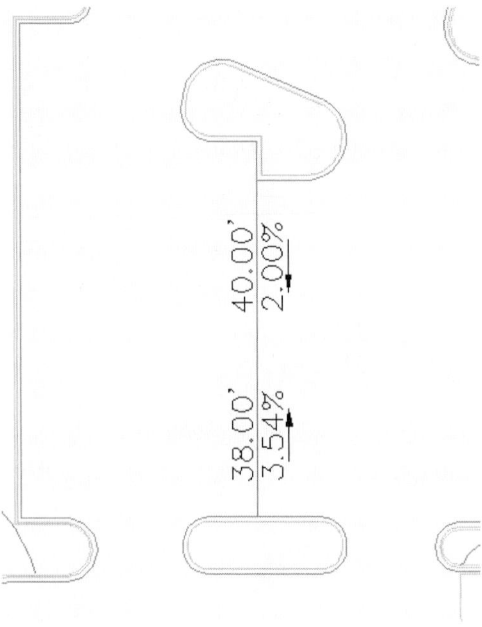


**FIGURE 10.16**

## Label Feature Line

1. Click the Ribbon's Annotate tab, on the Labels & Tables panel, select Add Labels.
2. In the Add Labels dialog box, change the Feature to **Line and Curve**, change the Label type to **Multiple Segment**, and set the Line label style to **Grade over Distance**. Click Add, and in the drawing, select the feature line that was just drawn.
3. Close the Add Labels dialog box.

## Elevation Editor...

Elevation Editor displays a feature line's elevations and grades in a vista.

1. PAN the feature line to the right side of the screen.
2. In the drawing, select the feature line. In the Feature Line tab, Modify panel, select the ***Edit Elevations*** icon to display the Edit Elevations panel. From the Edit Elevations panel, select ELEVATION EDITOR to display the Grading Elevation Editor vista.
3. Press ESC to remove the grips, place the cursor on the vista, and notice the triangles at the feature line control points.
4. In the editor, select a control point. Only its triangle is displayed on the feature line.
5. Select another entry in the vista and notice the redisplayed triangle.
6. Unselect the rows by selecting the ***Unselect All Rows*** icon at the vista's top center (rightmost icon).

7. Double-click in the Station 0+40 elevation cell, change it to **676**, and press ENTER.
8. Click the Panorama's ***green checkmark*** until it is closed.

## Quick Elevation Edit

Quick Elevation Edit adjusts a feature line's elevations and grades.

1. On the Ribbon's ***Modify*** tab. On the Edit Elevations panel's middle right, select the ***Quick Elevation Edit*** icon (first row, first icon).
2. In the drawing, slowly drag the cursor up and down the just drawn feature line.

Control points and grade directions and their values are displayed as you move the cursor over the feature line.

3. Place the cursor near the northern segment's midpoint to display its grade. Click the segment with the grade direction pointing south, and, for the new grade, enter **–2**, and press ENTER.
4. Move the cursor to the feature line's middle vertex. The elevation should not be 676.
5. Press ENTER to exit the command.

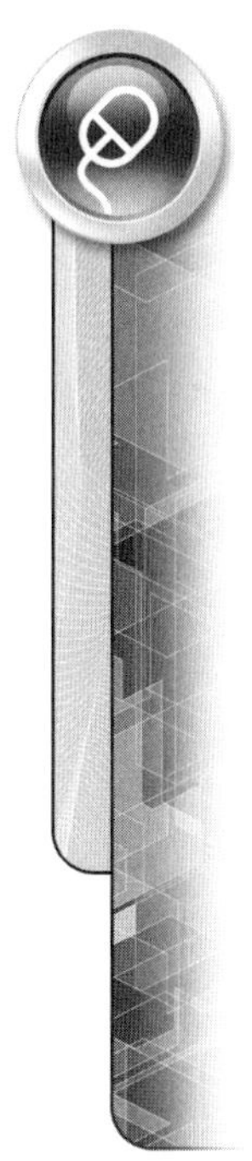

## Edit Elevations

Edit Elevations edits at the feature line's nearest vertex to the selection point and cycles through each feature line vertex.

1. On the Ribbon's ***Modify*** tab. On the Edit Elevations panel's middle right, select the ***Edit Elevations*** icon (first row, second icon).
2. In the drawing, select the feature line's northern end and press ENTER several times to cycle through the vertices.
3. Change the middle vertex's elevation to **676.00**, press ENTER, and enter '**X**' to exit the command.

## Converting a Feature Line

Converting an object to a feature line, if selected, assigns a style and an elevation.

1. Use the PAN and ZOOM commands to view the building footprint to the left of the parking lot.
2. On the Ribbon, click the ***Home*** tab. On the Create Design panel, click the ***Feature Line*** icon, and from the shortcut menu, select CREATE FEATURE LINES FROM OBJECTS.
3. The command line prompts you to select an object. In the drawing, select the building's footprint and press ENTER to continue.
4. The Create Feature Lines dialog box opens. For the feature line name, toggle **ON** Name, enter **470 Willow**, toggle **ON** Style, and from the list, select **Basic Feature Line**, toggle OFF Erase existing entities, toggle **ON** Assign elevations, and click ***OK*** to continue.
5. In the Assign Elevations dialog box, toggle **OFF** Insert intermediate grade break points, and click ***OK***.
6. In the drawing, select the feature line, press the right mouse button, and from the shortcut menu, select ELEVATION EDITOR....

Not inserting intermediate grade break points places elevations only at the feature vertices.

7. Click the vista's ***green checkmark*** to close it.
8. Use the AutoCAD Erase command to erase the feature line created.
9. On the Ribbon's Home tab, on the Create Design panel, click the ***Feature Line*** icon, and from the shortcut menu, select the CREATE FEATURE LINES FROM OBJECTS.
10. The command line prompts you to select an object. In the drawing, select the building's footprint, and press Enter to continue.
11. The Create Feature Lines dialog box opens. For the feature line name, toggle **ON** Name, enter **470 Willow**, toggle **ON** Style, and from the list, select **Basic Feature Line**, toggle **ON** Erase existing entities and Assign elevations, and click ***OK*** to continue.
12. In the Assign Elevation dialog box, toggle **ON** Insert intermediate grade break points and click ***OK***.
13. In the drawing, select the feature line to display its grips.

Each round grip represents a TIN intersection with the feature line.

14. In the drawing, select the feature line, press the right mouse button, from the shortcut menu, select ELEVATION EDITOR..., and review the elevation assignments.
15. Click the ***green checkmark*** to close the Panorama.

## Create a Surface from Feature Line

Feature Lines are surface breakline data.

1. Click the ***Prospector*** tab.
2. In Prospector, select the **Surfaces** heading, press the right mouse button, and from the shortcut menu, select CREATE SURFACE....
3. In Create Surface, change the style to **Contours and Triangles** and click ***OK*** to continue.
4. Expand the Surfaces branch until you are viewing the Surface1 Definition tree.
5. In the Definition tree, select **Breaklines**, press the right mouse button, and from the shortcut menu, select ADD....
6. For the Description enter **470 Willow**, click ***OK***. If necessary, in the drawing, select the building footprint.
7. In the drawing, select **Surface1**, right mouse click, and from the shortcut menu, select SURFACE PROPERTIES....
8. In Surface Properties, review the surface statistics.
9. Click ***OK*** to exit Surface Properties.
10. In Prospector, select **Surface1**, press the right mouse button, from the shortcut menu, select DELETE..., and click ***Yes*** to delete the surface.

## Create from Alignment

The Create from Alignment icon exports a feature line from an alignment's vertical profile.

1. In Layer Properties Manager, thaw the layer **C-ROAD-CORR-93rd - (1)** and click **X** to exit.
2. Use the ZOOM and PAN commands to better view the 93rd centerline just south of the entrance.
3. If necessary, click the ***Prospector*** tab.
4. Select the **Sites** heading, press the right mouse button, and from the shortcut menu, select NEW....

5. In the new Site Properties dialog box, for the name, enter **93rd Alignment** and click ***OK***.
6. In Prospector, select the Sites heading, press the right mouse button, and from the shortcut menu, select NEW....
7. In the Site Properties dialog box, for the site name enter Feature Lines from Corridor, and click OK to create the site.
8. On the Ribbon's Home tab, Create Design panel, click the ***Feature Line*** icon, and from the shortcut menu, select CREATE FEATURE LINES FROM ALIGNMENT.
9. In the drawing, select the 93rd centerline to display the Create Feature Line from Alignment dialog box.
10. Using Figure 10.17 as a guide, at the top of the dialog box, for the name, toggle ON Name, enter **93rd Centerline**, make sure the current profile is 93rd (3), toggle **ON** Style, and from the drop-list, select **Corridor Crown**, and click ***OK*** to open the Weed Vertices dialog box.
11. Click ***OK*** to accept the default weeding factors, thus creating the feature line.

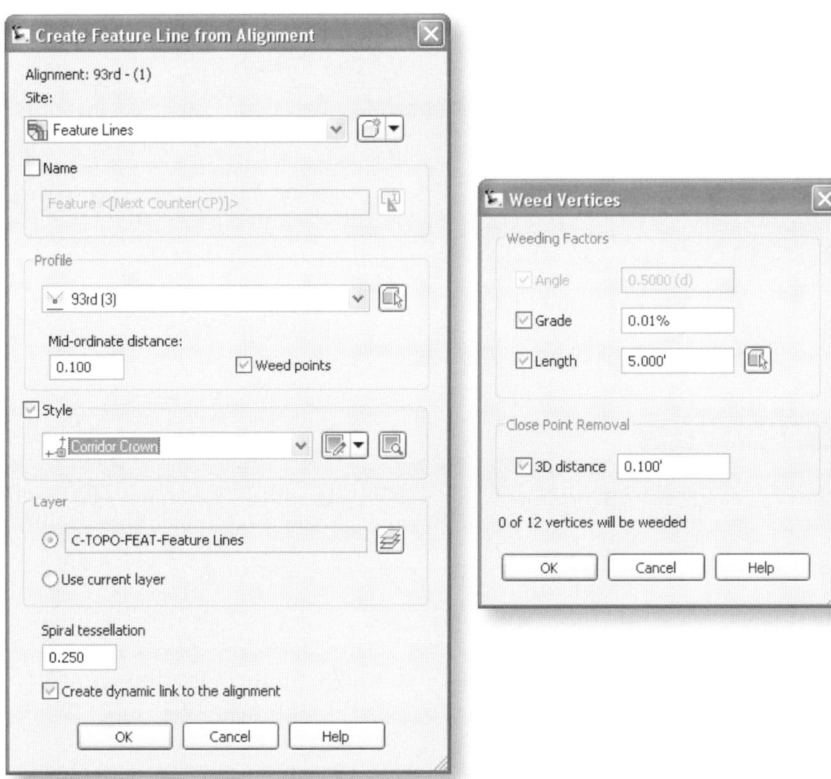


**FIGURE 10.17**

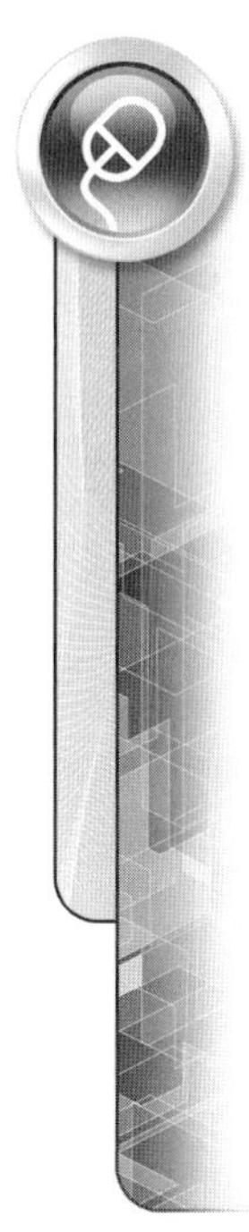

## Feature Lines from Corridor

1. Move your cursor around the corridor until you identify the ETW feature line.
2. On the Ribbon, click the ***Modify*** tab. On the Design panel, click the Corridor icon, and in the Corridor tab, Launch Pad panel's right side, select CREATE FEATURE LINES FROM CORRIDOR.

The command line prompts you to select a corridor feature line.

3. In the drawing, select the blue corridor ETW feature line, in the Select a Feature Line dialog box, select the feature line **ETW**, and click OK.

The Create Feature Line from Corridor dialog box opens.

4. In the dialog box, change the style to **Corridor Edge of Travel Way** and click ***OK*** to create the feature line.
5. Export a few more corridor feature lines. Some feature lines may be on top of one another. If this is the case, just export one of the feature lines. Press Enter to end the command.
6. Use the `LIST` command, select a feature line, and review the AutoCAD object report.
7. At Civil 3D's top left, Quick Access Toolbar, click the ***Save*** icon to save the drawing.

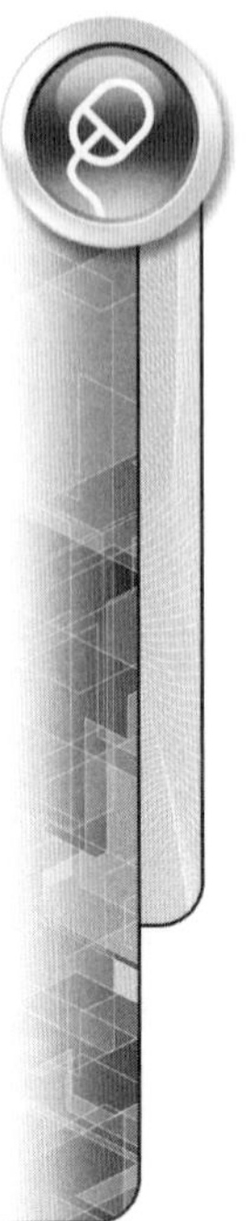

## SUMMARY

- Feature lines can be drawn or converted from other AutoCAD and Civil 3D objects and they can represent an alignment's vertical design or any other corridor point code.
- Each method used to create a feature line has a different method of assigning elevations.
- There are three feature line elevation editors: Elevation Editor..., Quick Elevation Edit, and Edit Elevations.
- Feature lines have control points: a vertex.
- Feature lines have elevation points: an elevation on a feature line segment that is not a vertex.
- The Modify - Feature Line tab edits, and annotates a feature line.

## UNIT 3: GRADING

Grading assigns feature lines grading criteria. A grading can be a task's solution, an elevation source, or a method of determining grades. Several settings affect grading and grading criteria.

### EDIT DRAWING SETTINGS

When you are working with grading objects, there will be multiple grading groups. It is a good practice to set a Modifier and its value. Edit Drawing Settings' Object Layers panel sets these values (see Figure 10.18).

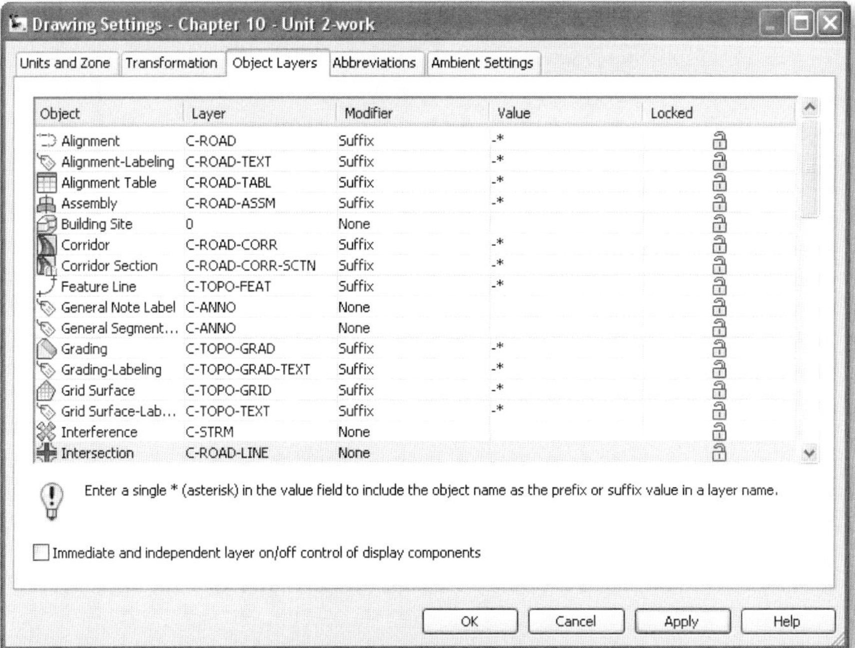


**FIGURE 10.18**

## EDIT FEATURE SETTINGS

Edit Feature Settings set default styles for grading and its naming format (see Figure 10.19).

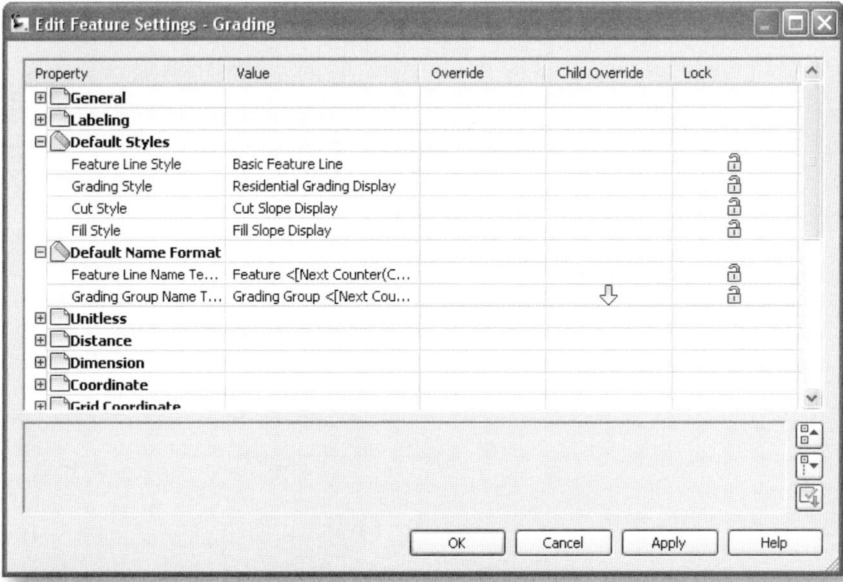


**FIGURE 10.19**

## GRADING STYLES

Grading styles affect how a grading object is displayed. A grading style names the style, sets the grading's center mark and its size, assigns a slope pattern, sets minimum and maximum slope values, and assigns components layers and their properties (see Figure 10.20). A grading marker is a grading solution selector.

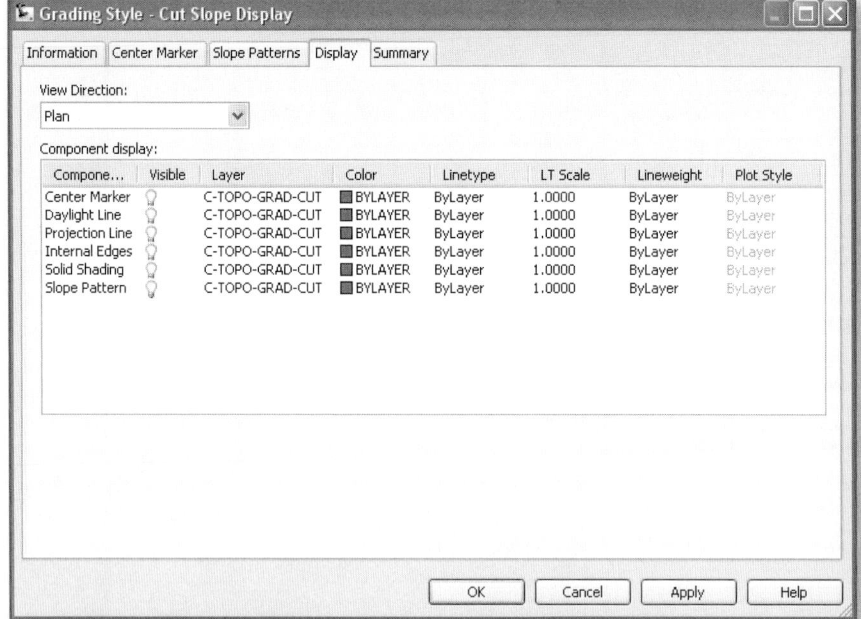


**FIGURE 10.20**

## GRADING CRITERIA SETS

Grading uses criteria sets. A criteria set is an alias for a grading method's collection (see Figure 10.21). Grading is a way to determine a solution by a slope or grade by an offset distance, by an elevation (absolute or relative), or by daylighting to a surface.

Grading criteria contains three or four sections: grading method; cut and/or fill slope projection; and conflict resolution. The grading method defines the target (surface, elevation, relative elevation, and distance), default elevation, and projection (cut, fill, both). Slope projection sections set default grades (slopes) and their default value. Conflict resolution defines how to clean up overlapping slope projections (use average slope, hold grade/slope maximum, and hold grade/slope as minimum).

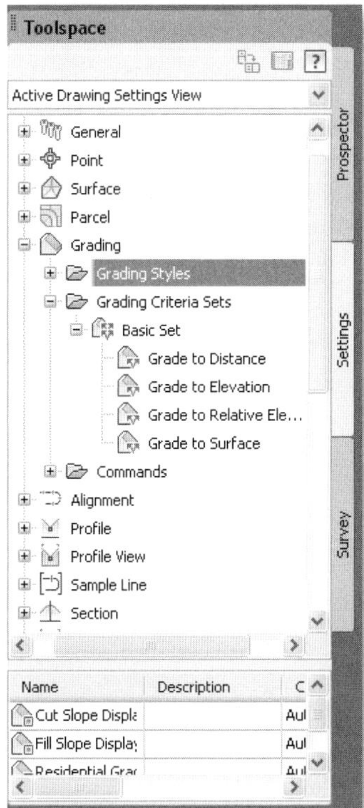


**FIGURE 10.21**

## GRADING TOOLS

Grading tools is a toolbar that defines a grading group by setting a target surface and the current criteria, creating grading and infills, editing grading group properties, and calculating volumes (see Figure 10.22). A grading group can have one or more grading objects. The volume utilities allow for an overall or individual volume calculation (see Figure 10.23).

**FIGURE 10.22**

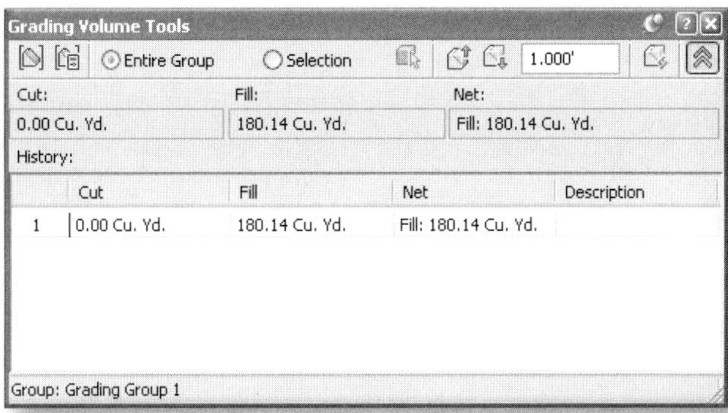


**FIGURE 10.23**

## Grading Group

Grading groups are a way to organize design areas and grading objects. Grading groups in different sites do not interact.

### EDIT GRADING

The Grading Editor changes a selected grading object's parameters. The grading editor lists the grading solution's parameters and allows for changes to their values (see Figure 10.24).

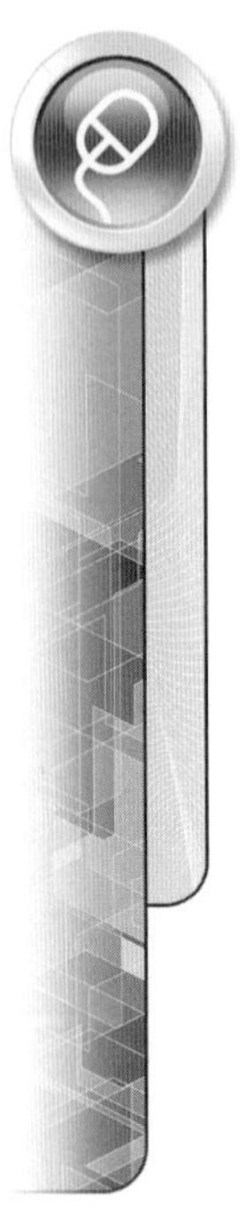

**FIGURE 10.24**

## EXERCISE 10-3

After completing this exercise, you will:

- Be able to convert a polyline to a feature line.
- Be able to modify a feature line.
- Be able to create a grading solution.
- Be able to edit a grading.

## Exercise Setup

This exercise uses the *Chapter 10 – Unit 3.dwg* file. To find this file, browse to the CD that accompanies this textbook and open the *Chapter 10 – Unit 3.dwg* file.

1. If not in Civil 3D, double-click the ***Civil 3D*** desktop icon to start the application.
2. Close the opening drawing and do not save the file.
3. At Civil 3D's top left, Quick Access Toolbar, click the ***Open*** icon, browse to the Chapter 10 folder of the CD that accompanies this textbook, and open the *Chapter 10 – Unit 3* drawing.
4. At Civil 3D's top left icon, click Civil 3D's drop-list arrow, from the Application Menu, highlight SAVE AS, from the flyout select AutoCAD Drawing, browse to the Civil 3D Projects folder, for the drawing name enter **Grading**, and click ***Save*** to save the file.

## Edit Drawing Settings

1. Click the ***Settings*** tab.
2. In Settings, select the drawing name, press the right mouse button, and from the shortcut menu, select EDIT DRAWING SETTINGS....

3. In Drawing Settings, click the ***Object Layers*** tab, toggle the **Suffix** modifier and set its value to a dash asterisk (**-***) for the objects **Feature Line, Grading**, and **Grading-Labeling**.
4. Click ***OK*** to exit and set the modifiers.
5. Click the ***Prospector*** tab.
6. In Prospector, select **Sites**, press the right mouse button, and from the shortcut menu, select NEW... to open the Site Properties dialog box.
7. In Site Properties, Information panel, for the site name enter **93rd Street** and click ***OK***.

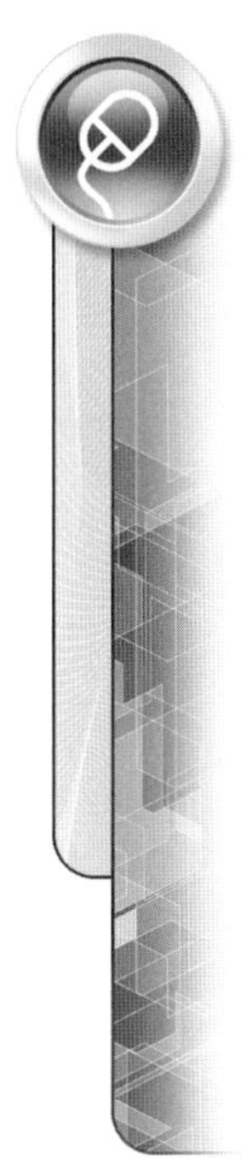

## Creating the Ditch Feature Line

Grading is ideal for ditch, pond, and surface design.

1. Click Ribbon's ***View*** tab. On the Views panel's left, select and restore the named view **South Ditch**. You may need to scroll down the list by selecting the down arrow.
2. On the Ribbon, click the ***Home*** tab. In the Create Design panel, click the ***Feature Line*** icon and from the shortcut menu, select CREATE FEATURE LINES FROM OBJECTS, select the ditch outline, and press ENTER to open the Create Feature Lines dialog box.
3. In Create Feature Lines, toggle **ON**, and for the name, enter **South Ditch**, keeping the counter, toggle **ON**, and set the style to **Grading Ditch**, and click ***OK*** to create the feature line.
4. Click the ***Modify*** tab. From the Edit Geometry panel, select ***Fillet*** (second row, the second icon), and in the drawing, select the new feature line.
5. The command line prompts you to specify a corner. Enter **'R'**, press ENTER to set the radius to **17.5**, and press ENTER to continue.
6. Move the cursor to the three corners to preview the resulting radius.
7. Select the three corners, except the lower left, to create the arc segments.
8. Still in the command, change the radius by entering **'R'**, and then press ENTER. For the radius, enter **5.0** and press ENTER.
9. In the lower left of the ditch, select the corner to create the new arc and press ENTER twice to exit the command.
10. At Civil 3D's top left, Quick Access Toolbar, click the ***Save*** icon to save the drawing.

## Creating the Ditch

1. In the drawing, select the **South Ditch** feature line, press the right mouse button, and from the shortcut menu, select ELEVATION EDITOR....
2. In the Grading Elevation Editor vista, click the ***Raise/Lower*** icon, in the elevation cell for the elevation, enter **671.00**, press ENTER to assign the elevation, and click the ***green checkmark*** until the vistas are closed.
3. Click the Ribbon's Home tab, Create Design panel, click the ***Grading*** icon and from the shortcut menu, select CREATE GRADING TOOLS to display the Grading Creation Tools toolbar.
4. On the toolbar's left, click the ***Set the Grading Group*** icon.
5. In the Site dialog box, select **93rd Street** and click ***OK*** to open the Create Grading Group dialog box.
6. In Create Grading Group, do not change the grading group Name, toggle **ON Automatic Surface Creation**, set the Surface style to **_No Display**, toggle **ON** and set the Volume base surface to **Existing Ground**, and click ***OK*** to close the dialog box (see Figure 10.25).

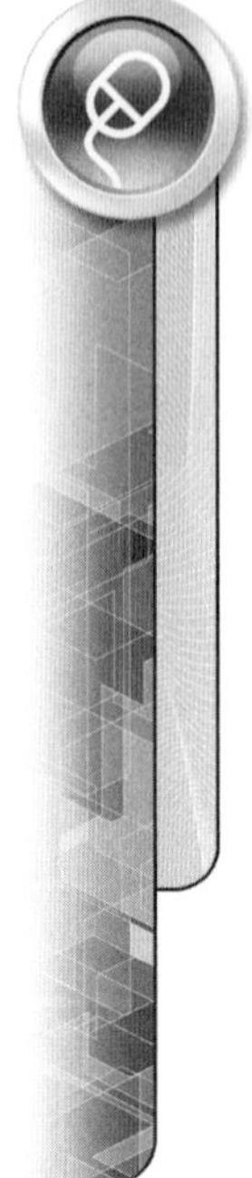

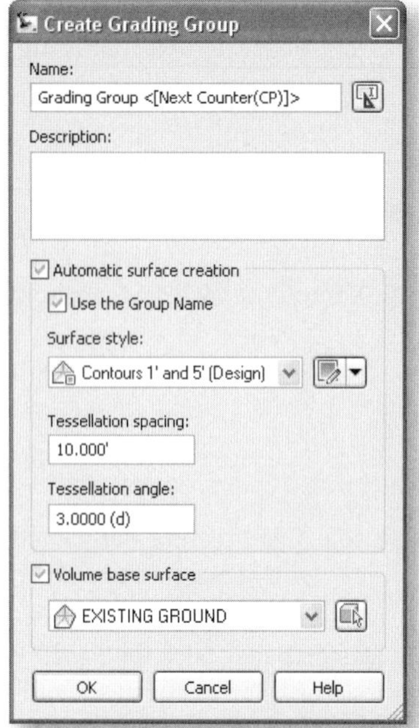


**FIGURE 10.25**

The Create Surface dialog box opens and has a surface name of Grading Group 1.

7. In Create Surface, for the surface description, enter **Grading Objects**, and click ***OK*** to exit the dialog box.
8. At the middle of the Grading Creation Tools toolbar, click the criteria drop-list arrow, and from the list, select **Grade to Relative Elevation**.
9. In the toolbar, to the right of criteria (the second icon to its right), click the drop-list arrow, and from the list, select CREATE GRADING.
10. In the drawing, select the southern ditch's feature line.
11. The command line prompts you for a grading side. In the drawing, select a point in the ditch's interior and press ENTER to apply the criteria to the entire feature line length.
12. The command line prompts you for a relative elevation. Enter **-5** and press ENTER.
13. The command line prompts you for the slope format. Press ENTER to accept the slope, for the slope, enter **2**, and press ENTER twice to create the grading and exit the command.

## Edit Grading

1. In the grading, select its diamond icon, press the right mouse button, and from the shortcut menu, select GRADING EDITOR....
2. In the Grading Editor vista, change the Relative Elevation to **-7** and press ENTER to change the grading solution.
3. Close the Panorama.

## North Ditch

1. Use the ZOOM and PAN commands to view the north ditch polyline.
2. On the Ribbon, click the ***Modify*** tab. In the Design panel, click the ***Feature Line*** icon. In the Feature Line tab's Modify panel, click the Edit Geometry panel icon.

On the Edit Geometry panel's right, select the ***Weed*** icon (third row, the first icon), select the northern ditch line, and review the weeding parameters.

3. In Weed Vertices, change the 3D distance to **15** and press ENTER.

The red vertices will be weeded from the feature line.

4. Click ***OK*** to weed the vertices.
5. Click the Ribbon's Home tab. On the Create Design panel, click the ***Feature Line*** icon, from the shortcut menu, select CREATE FEATURE LINES FROM OBJECTS, select the north ditch polyline, and press ENTER to open the Create Feature Lines dialog box.
6. In the Create Feature Lines dialog box, toggle **ON** name and enter **North Ditch**, toggle **ON** Style and set it to **Grading Ditch**, and click ***OK***.
7. From the Grading Creation Tools toolbar, make sure the criteria is **Grade to Relative Elevation**, click ***Create Grading***, and in the drawing, select the North Ditch feature line.
8. In the drawing, set the grading side by picking a point in the north ditch's interior, press ENTER to apply the grading to the entire length, for the relative elevation, enter **-6**, press ENTER to accept the Slope format, for the slope enter **2.5**, and press ENTER twice to create the grading and exit the routine.
9. At Civil 3D's top left, Quick Access Toolbar, click the ***Save*** icon to save the drawing.

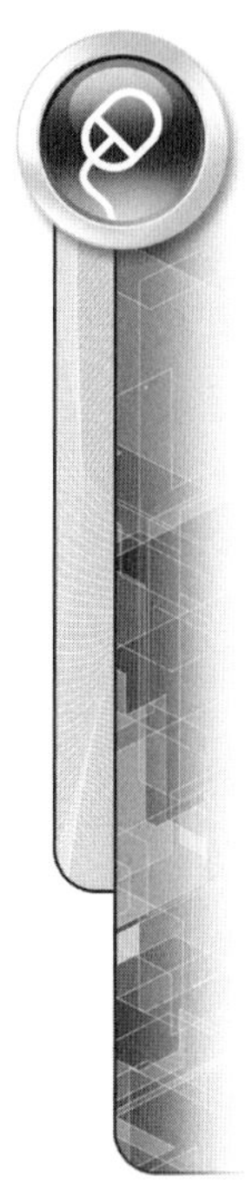

## Grading a Pond

A pond's outline is just to the northwest of the building. The first series of grading commands will use a grading solution to create a second solution. The interior of the pond will have the floor as an infill to put a floor in the pond.

1. Use the PAN and ZOOM commands to view the pond outline to the north of the building.
2. In the Grading Creation Tools toolbar, change the grading criteria to **Grade to Distance** and click ***Create Grading***.
3. In the drawing, select the pond's top contour (676), in the Create Feature Lines dialog box, toggle **ON** Style, set the style to **Basic Feature Line**, and click ***OK*** to continue.
4. The command line prompts you for a grading side. Select a point to the outside of the polyline.
5. The command line prompts you whether to apply the grading to the entire length. Press ENTER to accept.
6. The command line prompts you for a distance. For the distance, enter **6**, and press ENTER to continue.
7. The command line prompts you for a format. Press ENTER for Slope, for the slope value, enter **4**, and press ENTER twice to solve the grading and exit Create Grading.
8. In the Grading Creation Tools toolbar, change the grading criteria to **Grade To Surface** and click ***Create Grading***.
9. In the drawing, select the outer grading target line and press ENTER to apply the grading to the entire length.
10. The command line prompts you for a cut format. Press ENTER for **Slope**, for the slope value, enter **4**, and press ENTER to continue.
11. The command line prompts you for a fill slope. Press ENTER for Slope, for the slope value, enter **4**, and press ENTER twice to solve the grading and end the routine.

The grading solution intrudes on the building.

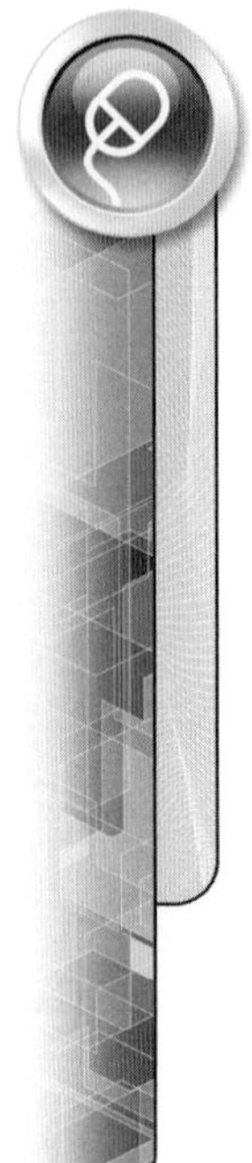

12. In the Grading Creation Tools toolbar, change the criteria to **Grade to Relative Elevation** and click ***Create Grading***.
13. In the drawing, select the pond edge feature line, select a point inside the pond for the offset side, and press ENTER to apply the grading to its entire length.
14. The command line prompts you for a relative elevation. Enter **-6** and press ENTER to continue.
15. The command line prompts you for a slope or grade. Press ENTER for slope, for the slope, enter **2.75**, and press ENTER twice to solve the grading and exit Create Grading.
16. In the Grading Creation Tools toolbar, to the right of Create Grading, click the drop-list arrow, from the list, select CREATE INFILL, and in the drawing, click in the pond's center.
17. Press ENTER to exit Create Grading.
18. At Civil 3D's top left, Quick Access Toolbar, click the ***Save*** icon to save the drawing.

## Edit Grading

The surface grading slope intrudes on the building and should be changed.

1. In the drawing, select the outer grading icon (diamond), press the right mouse button, and from the shortcut menu, select GRADING EDITOR....
2. In the Grading Editor vista, change the fill slope to **1.25**.

The grading is recomputed.

3. Close the Grading Editor vista.
4. At Civil 3D's top left, Quick Access Toolbar, click the ***Save*** icon to save the drawing.

## Viewing the Grading

1. In the drawing, select the pond grading objects, press the right mouse button, and from the shortcut menu, select OBJECT VIEWER....
2. After reviewing the grading, exit the Object Viewer.

## Grading Volume

1. In the Grading Creation Tools toolbar, select the ***Grading Volume Tools*** icon (the fifth icon in from the right) to display the Grading Volume Tools.

The dialog box reports the grading group volume.

2. Click the red **X** to close the toolbars.
3. On the Feature Line tab, click the Close icon.

This completes the grading unit. Grading is a powerful tool when used with other grading design methods.

## SUMMARY

- Grading objects are excellent for simple grading scenarios.
- Grading objects understand grading solutions intersections and if intersections do occur, the resulting solution shows the intersection.
- There are simple grading volume tools.
- A grading group can create a surface.

## UNIT 4: SITE AND BOUNDARY VOLUMES

Volume calculations are based on the differences in elevations between two surfaces. Surfaces used in a comparison can be any combination of two surface types (TIN or Grid). The Grid and TIN volume methods calculate volumes by different algorithms. Each method is directly dependent on the surface data quality for its results. The better the quality, the better the resulting volume estimate.

### SITE VOLUMES

Civil 3D uses two methods to calculate earthwork volumes: grid and composite. Prospector dynamically tracks the surface volume status and uses icons to represent surface references (the two comparison surfaces) and checks whether the volume is out of date. If you attempt to delete one of the comparison surfaces, Prospector will not allow its deletion because a volume surface depends on the comparison surfaces. To delete a dependent surface, you must first delete the volume surface.

#### Grid Volume Surface

A grid volume surface results from sampling elevation differences between two surfaces using a regularly spaced grid (X and Y). The sampling grid can have a rotation that reflects a rotation in one or both surfaces. While sampling the two surfaces, grid checks both surfaces for an elevation at each cell corner. If all four cell corners have elevations, each cell corner is assigned the difference in elevation between the surfaces (see Figure 10.26).

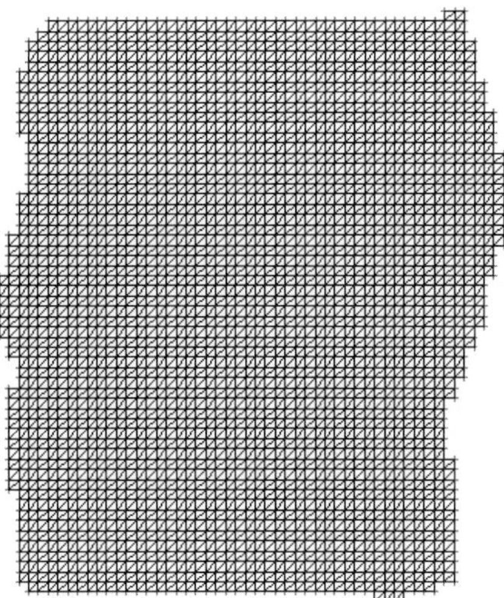

**FIGURE 10.26**

If there are varying cell sizes, then the sampling density varies and possibly changes the resulting volumes. Grid volumes' greatest problem, outside of bad data, is how densely to sample the two surfaces. The first issue of grid density is setting the grid spacing to an interval similar to the surface data spacing. In some cases, the data spacing can vary from 50 feet to as small as one-half foot. Sampling a site with a 20-foot grid is sampling the interpretations between 50-foot spaced data and missing the one-half foot spaced data.

The second issue is that if both surfaces have irregular borders and the grid spacing is too large, the volume surface may not include data around the surfaces' edges. This is because the spacing doesn't sample the irregular surface border shapes. Grid spacing should sample the greatest amount of intersecting area between surfaces with the fewest number of points. Optimum grid spacing varies based on surface size, shape, and relief. If the volume differences are severe, then you should question the sampling spacing or the surface quality.

### TIN Volumes

Volumes calculated from contour or breakline data limit the design's effect on the volume. The design usually starts by using the existing ground's contours as a seed for the design surface. By designing a second surface using Existing Ground's contours, the undisturbed contours limit volume calculations to the first undisturbed contour beyond the design. This is because the volume between all unchanged contours is 0 (zero).

The TIN volume surface results from the comparison of two surface triangulations (see Figure 10.27). Its elevations are the differences in elevations between two surfaces, and its triangles are a composite of both surfaces' triangles. This method samples the surface elevation difference at each triangle leg's end. TIN volume is the most comprehensive volume calculation method.

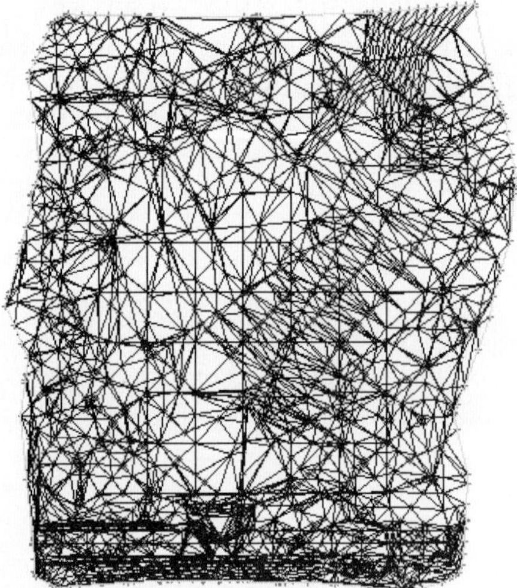

**FIGURE 10.27**

### Surface Utilities — Volumes

The Analyze panel's Volumes, Volumes command calculates a TIN volume without creating a volume surface (see Figure 10.28).

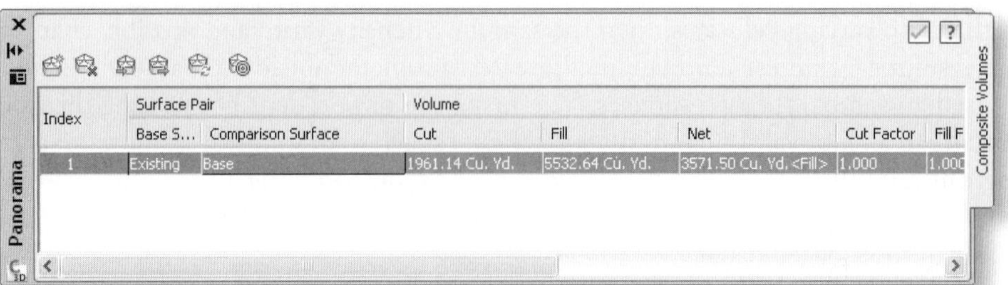


**FIGURE 10.28**

## BOUNDED VOLUMES

Bounded Volumes calculates a volume for an area defined by a polyline, polygon, or parcel. This method has to have a pre-existing volume surface.

## DESIGNING SURFACES

The best methods to moderate volume calculation errors are consistent surface design methodologies, consistent data densities, and an awareness of each volume calculation method's strengths and weaknesses. What is a good methodology for creating a design and what is the correct density of data are questions open to debate. What a user needs to do is develop a consistent approach to developing a surface and evaluate the calculated volume numbers by mentally figuring the "ball park" value.

Blending a second surface design into the first is best accomplished by starting with 3D polylines or contours from the first surface. This "seed" data is an effective way to create and blend a design surface into the existing conditions. The 3D polylines, feature lines, and contours also aid the process of new surface design. Blending a design into existing contours is straightforward and areas that are undisturbed by the design will produce no volume.

How does one create seed contours from a surface object? First, you apply a style to the surface to display the desired contours. Second, from the Modify - Surface panel, you use the Extract from Surface command to extract as lightweight polylines the major and minor contours. Third, you place the extracted contours on a new layer and start trimming, drafting, grading, and setting control points to define the proposed surface.

## SURFACE DESIGN TOOLS

Feature lines are the most powerful Civil 3D design tool. Its commands create feature lines as surface breakline data and grading objects.

### Stepped Offset

The Modify - Feature Line panel's Edit Geometry icon, displays the Stepped Offset icon. This routine offsets lines, survey figures, 3D polylines, or feature lines for a fixed distance. You adjust the offset object's elevation by specifying a change amount (1, 3, −1, or −3), a grade (25 or −33), or slope (8:1 or −8:1). When you enter a slope, you need only enter the sign (positive or negative) and the run. When entering a grade, you enter the sign and value, not the percent sign.

Optionally, the Multiple option repeats the offset if you need to make several offsets. In multiple mode, the command assumes the last offset object is the next object to offset.

When you change the offset's lines elevation, rather than uniformly changing it, you can use the Individual option to adjust each offset line's vertex elevations.

The following snippet demonstrates how to set the offset distance, change its elevation, and set the multiple option:

```
Command: Offset layer = Source
Specify offset distance or [Through/Layer]: 8 <Press ENTER>
Select a feature line, survey figure, 3d polyline, or
polyline to offset: <Select the object to offset>
```

```
Specify side to offset or [Multiple]: m <Press ENTER>
Specify side to offset: <Select the side to offset>
Specify elevation difference or [Grade/Slope] <0.000>: 1
<Press ENTER>
Specify side to offset: <Select the side to offset>
Specify elevation difference or [Grade/Slope] <1.000>:
<Press ENTER>
```

### Create Feature Line

Create Feature Line drafts a feature line. A feature line can contain tangent or curve segments. When you assign feature line vertex elevations, the choices are a specific elevation, a slope or grade, a difference in elevation between the previous and current vertex, or a surface elevation.

The following code snippet is from the Create Feature Line command:

```
Specify the next point or [Arc]: (5292.54 5351.15 732.4)
Distance 220.546', Grade -0.23, Slope -441.09: 1, Elevation
732.400'
Specify grade or [SLope/Elevation/Difference/SUrface]
<0.00>: e
Specify elevation or [Grade/SLope/Difference/SUrface]
<732.400>: <Press ENTER>
Specify the next point or [Arc/Length/Undo]:
```

### Insert Elevation Point

Insert Elevation Point adds elevation points to a feature line, but does not add a vertex to the feature line. Elevation points can be user-selected locations or an incremental value (every 10 or 20 units).

### Insert High/Low Elevation Point

Insert High/Low Elevation Point creates a new feature line vertex based on intersecting slopes or grades from two adjacent feature line vertices.

```
Command:
Select a feature line, survey figure, parcel line, or 3d
polyline:
Specify the start point: <move cursor to a vertex and select>
Specify the end point: <move cursor to opposite vertex and
select>
Start Elevation 732.900', End Elevation 732.750', Distance
206.176'
Specify slope ahead or [Grade]: 50 <press ENTER>
Specify slope back or [Grade]: 50 <press ENTER>
Select a feature line, survey figure, parcel line or 3d
polyline: <press ENTER to exit>
Command:
```

## VOLUME REPORT TOOLS

There are no surface Volume Report tools.

## CONTOUR DATA

See the discussion on Contour data in Chapter 4, Unit 2.

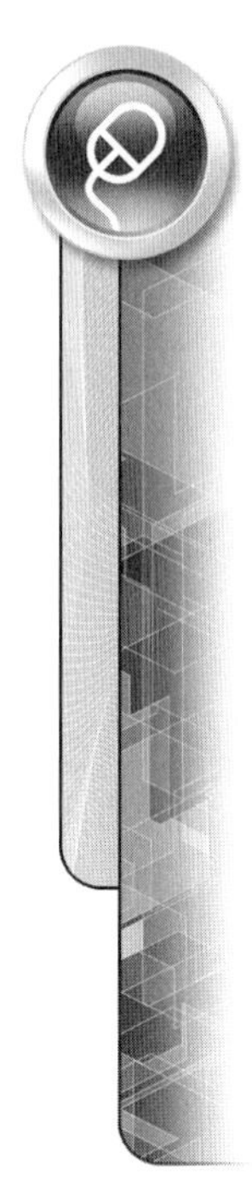

## EXERCISE 10-4

After completing this exercise, you will:

- Be able to build, evaluate, and calculate volumes.
- Be able to create a surface from contour and 3D polyline data.
- Be able to calculate a volume (in a panorama) between two surfaces.
- Be able to create a Triangular Irregular Network (TIN) and grid volume surface.
- Be able to review the volume surface statistics.

### Exercise Setup

This exercise has two parts. The first part involves creating a surface whose data is existing ground contours. This is done for two reasons: to compare the calculated volumes between the existing surface (points and breaklines); and to compare a contour data surface and the design surface.

The second part involves creating a design surface. The second surface sets the stage for the volume calculations.

This exercise uses the *Chapter 10 – Unit 4.dwg* file. To find this file, browse to the CD that accompanies this textbook and open the *Chapter 10 – Unit 4.dwg* file.

1. If you are not in ***Civil 3D***, double-click its desktop icon to start the application.
2. CLOSE the current drawing and do not save it.
3. At Civil 3D's top left, Quick Access Toolbar, click the ***Open*** icon, browse to the Chapter 10 folder of the CD that accompanies this textbook, and open the *Chapter 10 – Unit 4* drawing.
4. At Civil 3D's top left icon, click Civil 3D's drop-list arrow, from the Application Menu, highlight SAVE AS, from the flyout select AutoCAD Drawing, browse to the Civil 3D Projects folder, for the drawing name enter **Surface Volumes**, and click ***Save*** to save the file.

### Creating an EGCTR Surface

EGCTR uses the Existing surface contours.

1. If necessary, click the ***Prospector*** tab.
2. From the Ribbon's Home tab, open the Layer Properties Manager, thaw the layer **C-TOPO-Existing**, and click **X** to exit the palette.
3. On the Ribbon, click the ***Modify*** tab. On the Ground Data panel, click Surface and on the Surface Tools panel, select EXTRACT OBJECTS to open the Extract Objects From Surface dialog box.
4. In the dialog box, toggle **OFF** Border and click ***OK*** to extract the contours.
5. In Prospector, expand the Surfaces branch, select the surface **Existing**, press the right mouse button, and from the shortcut menu, select SURFACE PROPERTIES....

6. Click the ***Information*** tab, set the style to **_No Display**, and click ***OK*** to exit.
7. On the Surface tab, click Close.

The drawing contains extracted major and minor contours.

8. On the Ribbon, click the Home tab, open the Layer Properties Manager, create a **New** layer, for the layer's name enter **Existing Contours**, make it the current layer, assign it a color, and click **X** to exit the palette.
9. In the drawing, select the contours and place them on the Existing Contours layer.
10. In Prospector, select the Surfaces heading, press the right mouse button, and from the shortcut menu, select CREATE SURFACE....
11. In the Create Surface dialog box, for the name, enter **EGCTR**, for the description, enter **From EG Contours**, set the Surface Style to **Contours 1' and 5' (Design)**, and click ***OK*** to create the surface.
12. In Prospector, expand Surfaces until you are viewing the EGCTR Definition branch.
13. In the EGCTR Definition branch, select **Contours**, press the right mouse button, and from the shortcut menu, select ADD....
14. In Add Contour Data, for the description, enter **FROM EXISTING**, set the Supplemental distance to **35.000**, the Mid-ordinate to **0.100**, click ***OK***, and in the drawing, if necessary, select the contours (see Figure 10.29).
15. At Civil 3D's top left, Quick Access Toolbar, click the ***Save*** icon to save the drawing.

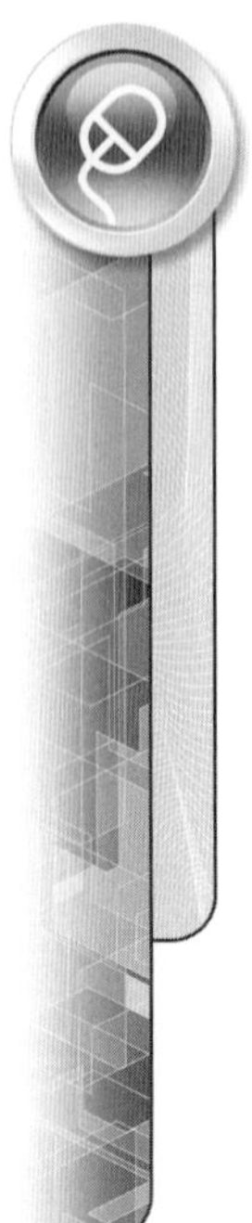

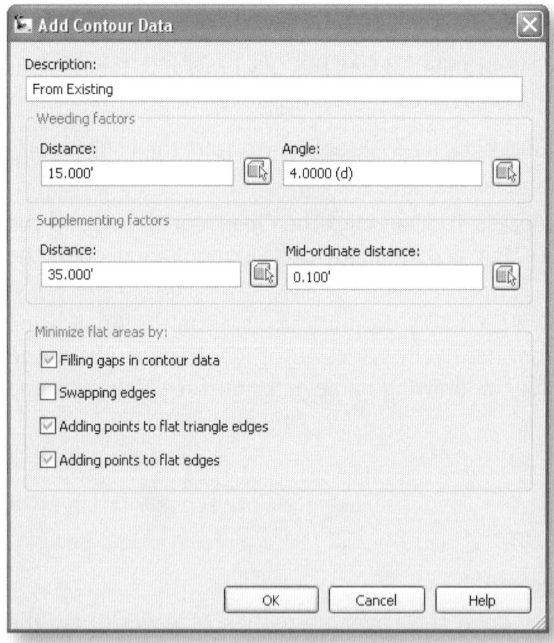


**FIGURE** 10.29

## Base Surface Seed Contours

The Base surface starts with extracted EGCTR contours.

1. On the Ribbon's Home tab, open the Layer Properties Manager, create a **New** layer, for the layer name, enter **Base Contours**, make it the current layer, freeze the Existing Contours layer, and click **X** to exit the palette.

2. On the Ribbon, click the ***Modify*** tab. On the Ground Data panel, click Surface. On the Surface tab's Surface Tools panel, select EXTRACT OBJECTS, and in the drawing select the surface to open the Extract Objects From Surface dialog box.
3. In the dialog box, toggle **OFF** Border and click ***OK*** to extract the contours.
4. In Prospector, the Surfaces branch, select the surface **EGCTR**, press the right mouse button, and from the shortcut menu, select SURFACE PROPERTIES....
5. Click the ***Information*** tab, set the style to **_No Display**, and click ***OK*** to exit.

The drawing contains extracted major and minor contours.

6. On the Surface tab, select Close.
7. In the drawing, select the contours and place them on the Base Contours layer.
8. At Civil 3D's top left, Quick Access Toolbar, click the ***Save*** icon to save the drawing.

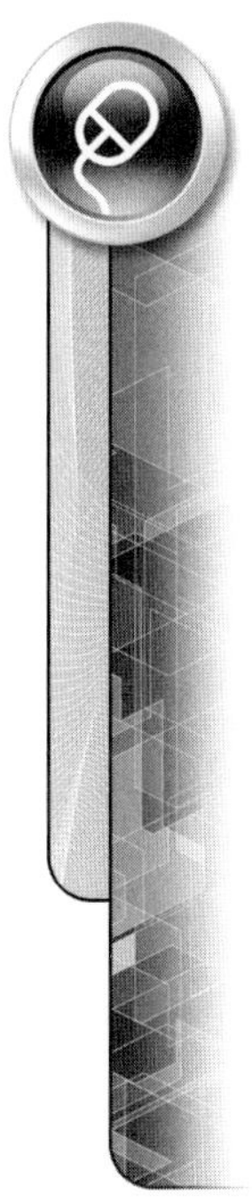

## Modifying Contours

The site's northeast corner is the first editing area. The priority regarding this area is filling in the swale and moving its contours northeasterly to the site's edge. If you are bringing fill into the site and placing it in the swale, the swale elevations rise to move the lower elevations to the northeast.

To create these changes, remove contours from the swale's interior and redraw new contours at the swale's northeastern edge (see Figure 10.30).

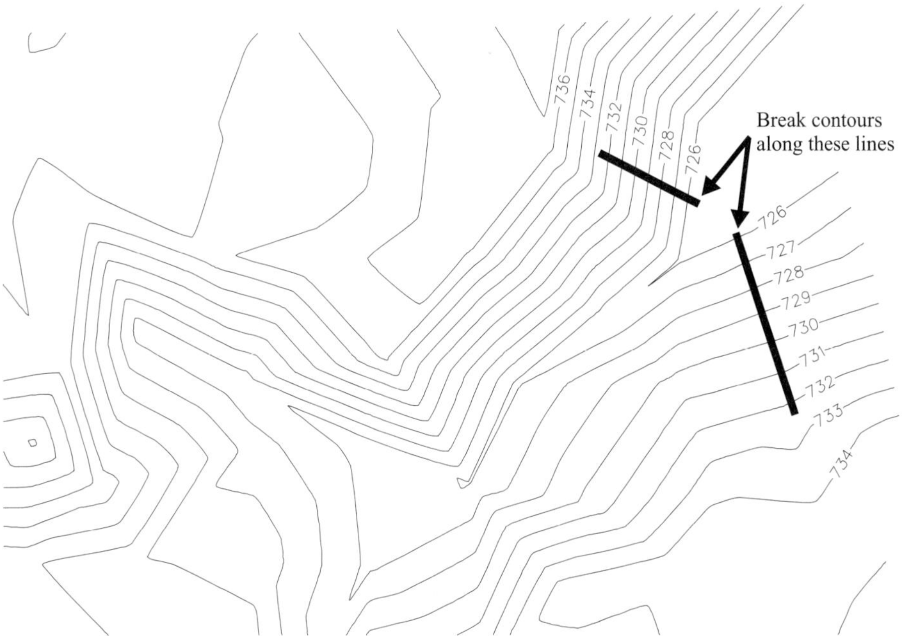


**FIGURE 10.30**

## Northeast Section:

1. Use the ZOOM and PAN commands and match your screen to Figure 10.30.
2. Using Figure 10.30 as a guide, use the LINE command and draw the lines as the figure indicates.

3. Using Figure 10.30 as a guide, start the TRIM command. In the drawing, at the cutting edge, select the lines just drawn, press ENTER, and trim the contours to the line's southwest side, removing the unwanted contours.
4. ERASE the two trimming lines.
5. At Civil 3D's top left, Quick Access Toolbar, click the ***Save*** icon to save the drawing.

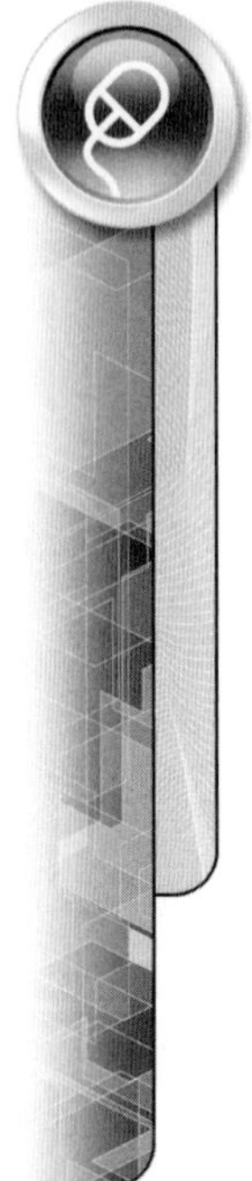

## Draw New Contours

1. In the status bar, toggle **ON** Object Snaps and set *Endpoint*.

Using Figure 10.31, draw a new 726' contour.

2. Start the POLYLINE command, and in the drawing, select the 726' contour's (the innermost contour) north end, toggle **OFF** object snap (F3), select a few points to represent the new 726' contour path, toggle **ON** object snaps (F3), select the 726' contour south end, and press ENTER to exit the polyline command.

**FIGURE 10.31**

3. Check the new contour's elevation by selecting it, pressing the right mouse button, from the shortcut menu, selecting PROPERTIES..., and reading the elevation value.

The polyline's Elevation value should match the connected exiting contour's elevation.

4. If the Elevation value does not match the 726' elevation, in Properties, change its elevation to **726**.

The engineer wants an 8:1 slope for the new contours. This means you must offset the contour 8' horizontally and raise its elevation 1'. You use Feature Line's Stepped Offset routine for this task. Stepped Offset's multiple mode creates several new contours from the one just drawn.

5. On the Ribbon, click the ***Modify*** tab. In the Edit Geometry panel, select STEPPED OFFSET (third row, the second icon). Set the Offset Distance to **8**, press ENTER, select the just-drawn polyline, in the command line, enter '**M**' for multiple, press ENTER, select a point to the southwest of the polyline, set the Elevation Difference to **1**, press ENTER, continue picking points to each new polyline's southwest, thus creating six additional contours with the same Elevation Difference, and press ENTER to exit the command.
6. To connect the polylines to the existing contours, activate the new polylines' grips, toggle **ON** object snaps (F3), and stretch the polylines until they are connected to their corresponding existing contours.
7. At Civil 3D's top left, Quick Access Toolbar, click the ***Save*** icon to save the drawing.

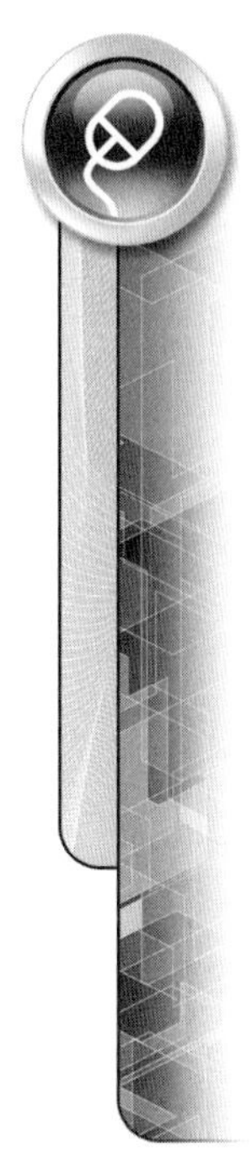

## Northwest Section:

Next, you modify the site's northwest side. This expands the central flat area by erasing the hill and redrafting contours (see Figures 10.32 and 10.33).

1. Use the ZOOM and PAN commands to make your view match Figure 10.32.
2. Use the ERASE command, and in the drawing, erase the hill contours around point 122.
3. Using Figure 10.32 as a guide, use the BREAK command to break the two contours at the indicated locations.
4. Press F3 to toggle **ON** Object Snaps.
5. Using Figure 10.33 as a guide, use the POLYLINE command and the **Endpoint** object snap, and redraw the contours connecting the southern endpoints to the northern endpoints.
6. In the drawing, select one of the new contours, press the right mouse button, and from the shortcut menu, select PROPERTIES....
7. If necessary, change the contour to the correct elevation.
8. Repeat the previous two steps and, if necessary, edit the remaining contour elevations.
9. At Civil 3D's top left, Quick Access Toolbar, click the ***Save*** icon to save the drawing.

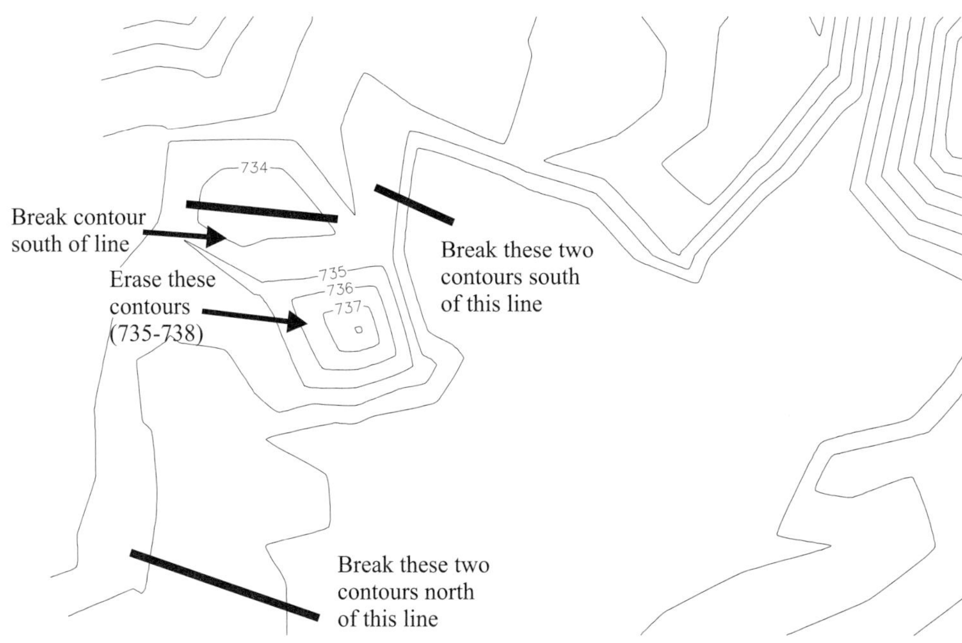


**FIGURE 10.32**

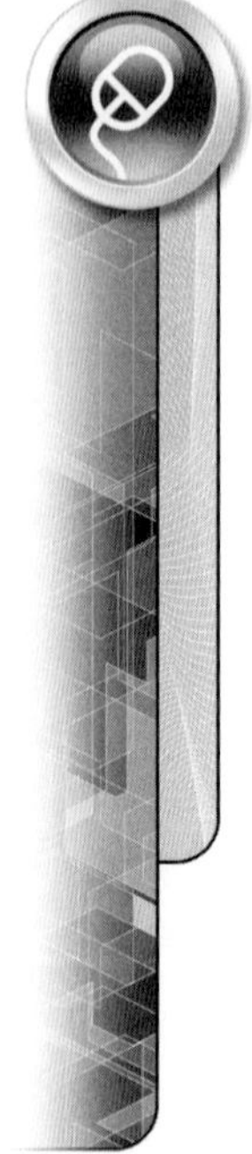

**FIGURE 10.33**

## Southeast Section:

Next, you edit contours in the site's southeast side. This expands the flat interior area to the southeast (see Figures 10.34 and 10.35). You also erase two closed contours.

1. Use the ZOOM and PAN commands to match your view to Figure 10.34.
2. Using Figure 10.34 as a guide, use the ERASE command to erase the two closed contours.
3. Using Figure 10.34 as a guide, use the BREAK command to break the 733 and 734 contours.
4. Toggle **ON** Object Snaps (F3).
5. Using Figure 10.35 as a guide, use the POLYLINE command to redraw the two contours.
6. In the drawing, select one of the new contours, press the right mouse button, and from the shortcut menu, select PROPERTIES....
7. If necessary, change its elevation to correct it.
8. If necessary, repeat the previous two steps to edit the remaining contour elevations.
9. At Civil 3D's top left, Quick Access Toolbar, click the ***Save*** icon to save the drawing.
10. Use the ZOOM EXTENTS (ZE) command to view the entire surface.

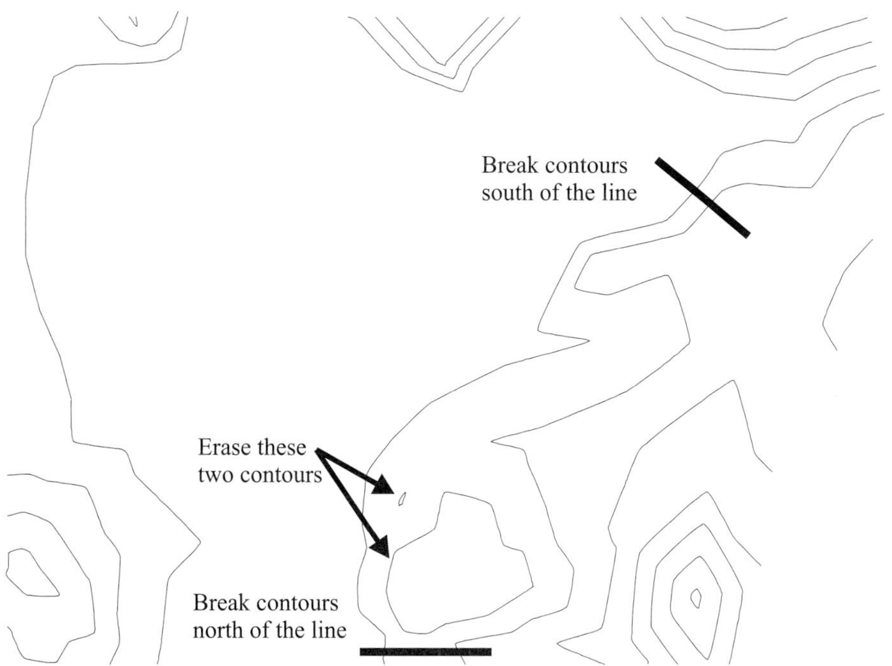


**FIGURE 10.34**

**FIGURE 10.35**

The drawing should be similar to Figure 10.36.

**FIGURE 10.36**

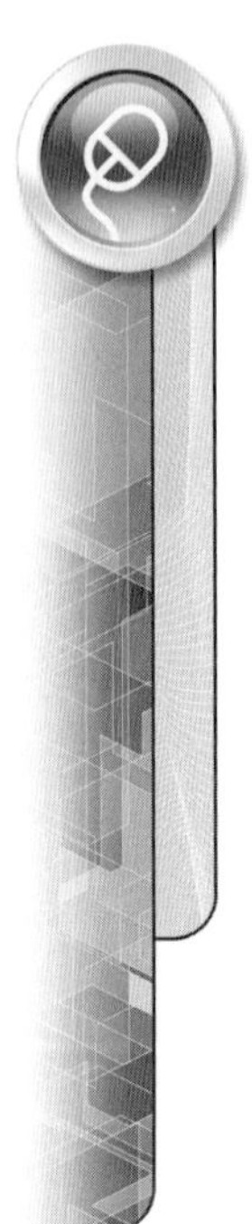

## Designing Site Drainage

The current design's problem is that the site's center is flat and water will not drain. One strategy to solve this problem is to use swales and a small berm to move the water to the northeast. The shallow swale uses points from the flat area's west side toward the northeast. Figure Lines define a second swale and a berm.

## Swale by Points

1. Click Ribbon's ***View*** tab. From the Views panel's left, select and restore the named view **Central**.
2. Toggle OFF Object Snaps (F3).
3. Click the Ribbon's Home tab, open the Layer Properties Manager, thaw the V-NODE layer, and click **X** to exit to close the palette.
4. In Prospector, select the **Points** heading, press the right mouse button, and from the shortcut menu, select CREATE....
5. At the toolbar's right side, click the Expand the Create Points dialog chevron.
6. Expand the Point Identity section and set the Next Point Number to **500**.
7. Expand the Points Creation section and set the following values:

   Prompt For Elevations: Manual
   Prompt For Descriptions: Automatic
   Default Description: SWCNTRL

8. At the toolbar's right side, click the Collapse the Create Points dialog chevron.
9. From the Create Points toolbar, click the Miscellaneous icon stack's drop-list arrow (first icon in from the left), and from the list, select MANUAL.
10. Using Figure 10.37 as a guide, place three points and use Table 10.1 to assign the listed elevations.
11. Press ENTER to exit the routine and return to the command line.

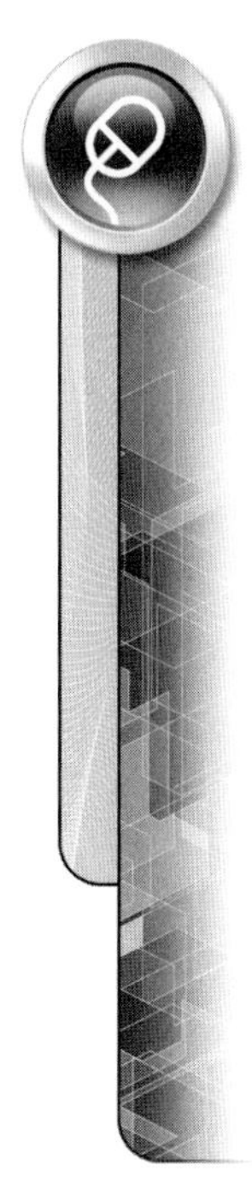

**TABLE 10.1**

| Point Number | Elevation |
|---|---|
| 500 | 732.90 |
| 501 | 732.40 |
| 502 | 732.10 |

**FIGURE 10.37**

These three points represent the upper, mid, and end swale points. To complete the swale, place points between the swale points, so the surface correctly triangulates the swale. The Interpolate routine is used to create these points.

12. In the Create Points toolbar, the Interpolate icon stack (the fifth icon in from the left), select its drop-list arrow, and from the list, select INTERPOLATE.
13. The command line prompts you to select a point object. In the drawing, select point **500**.
14. The command line prompts you to select a second point object. In the drawing, select point **501**.
15. The command line prompts you for the number of points between the two selected points. Enter **5** and press ENTER.
16. The command line prompts you for an offset value. Press ENTER for 0 (zero) to create the interpolated points.
17. Press ESC to end the routine and return to the command line.

18. Repeat Steps 11–16 and create interpolated points between points 501 and 502. For the number of interpolated points, enter **3**, for the offset press ENTER for 0 (zero), and press ESC to end the routine.
19. At Civil 3D's top left, Quick Access Toolbar, click the ***Save*** icon to save the drawing.

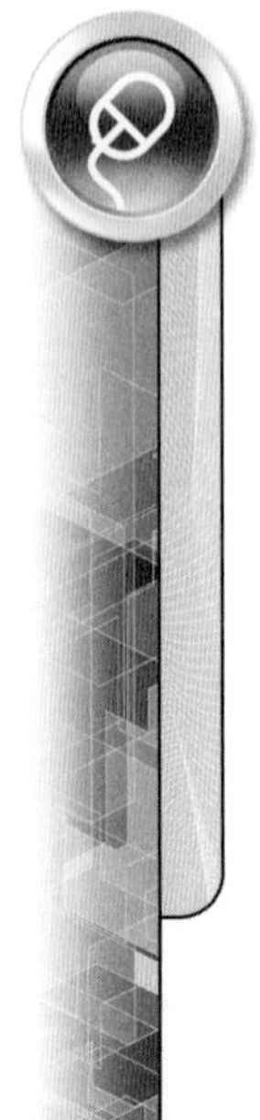

## Creating a Feature Line Swale

The second swale is a feature line with an elevation of 732.9 at its southern end and its northern end elevation (732.4) is from point 501.

1. Make sure the Transparent Commands toolbar is displayed and use Figure 10.37 as a guide for drawing the feature line.
2. On the Ribbon's ***Home*** tab. On the Create Design panel, click the ***Feature Line*** icon, from the shortcut menu, select the ***Create Feature Line*** to open the Create Feature Lines dialog box.
3. In the Create Feature Lines dialog box, toggle **ON** and name the feature line **Swale 1**, toggle **ON** Style, click the drop-list arrow, from the styles list, select **Grading Ditch**, and click ***OK*** to continue.
4. The command line prompts you for a starting point. In the drawing, select a point near the site's southern entrance.
5. The command line prompts you for an elevation. For the elevation, enter **732.90** and press ENTER to continue.
6. From the Transparent Commands toolbar, select the **Point Object** filter ('PO), and click anywhere on the point 501 label.
7. The command line may prompt you for a grade of <0.00>. If so, for elevation, enter **'E'** and press ENTER. The command line echoes point 501's elevation (732.40). Press ENTER to accept this elevation (see Figure 10.38).
8. The command line prompts you for a point object. To end the override and exit, press ESC twice.
9. At Civil 3D's top left, Quick Access Toolbar, click the ***Save*** icon to save the drawing.

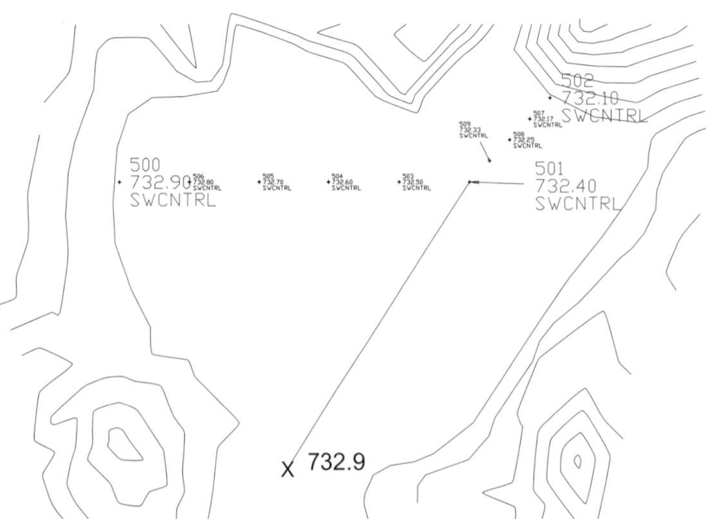


**FIGURE 10.38**

The following is the feature line command sequence:

```
Command:
Specify start point: <Select a point in the southern entrance
area>
Specify elevation or [Surface] <0.000>: 732.9 <Press ENTER>
Specify the next point or [Arc]: '_PO
>>
Select point object: <Select point 501>
Resuming DRAWFEATURELINE command.
Specify the next point or [Arc]: (5292.54 5351.15 732.4)
Distance 220.546', Grade -0.23, Slope -441.09:1,Elevation
732.400'
Specify grade or [SLope/Elevation/Difference/SUrface]
<0.00>: e
Specify elevation or [Grade/SLope/Difference/SUrface]
<732.400>: <Press ENTER>
Specify the next point or [Arc/Length/Undo]:
>>
Select point object: *Cancel* <Press ESC>
>>
Specify the next point or [Arc/Length/Undo]:
Resuming DRAWFEATURELINE command.
Specify the next point or [Arc/Length/Undo]: <Press ESC>
Command:
```

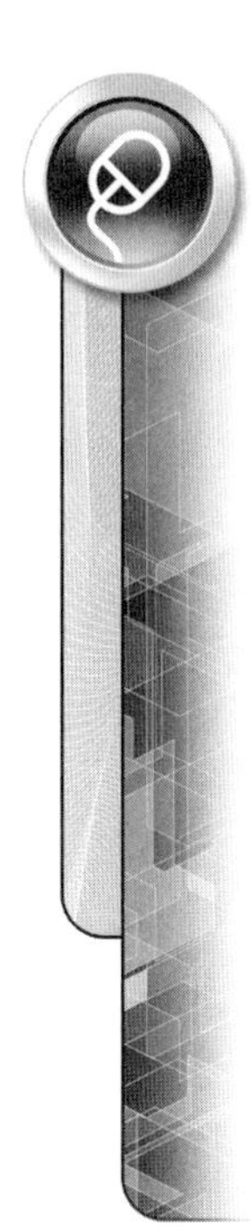

The feature line has only two vertices: its northern and southern ends. These two are the only swale data points. When you define the swale as a breakline, it can supplement the feature line's data.

## Creating a Berm

The berm sheds water to the point and feature line swale.

1. On the Home tab, Create Design panel, click the ***Feature Line*** icon, and from shortcut menu, select the ***Create Feature Line*** to open the Create Feature Lines dialog box.
2. In Create Feature Lines, toggle **ON** Name, for the name, enter **Berm 1**, toggle **ON** Style, click its drop-list arrow, from the styles list, select **Grading Ditch**, and click ***OK*** to continue.
3. The command line prompts you for a starting point. In the drawing, select a point in the southwest between the points and the feature line.
4. The command line prompts you for an elevation. For the elevation, enter **732.90**, and press ENTER to continue.
5. The command line prompts you for a second point. In the drawing, select a point between point number 507 and the feature line.
6. The command line prompts you for an elevation. For the elevation, enter **732.75** and press ENTER twice to set the elevation and exit the routine.

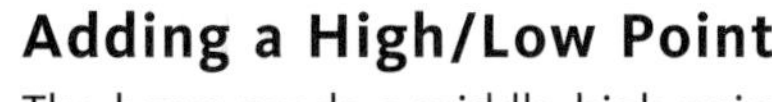

## Adding a High/Low Point

The berm needs a middle high point. Positive grades from each end vertex create a high point near its middle.

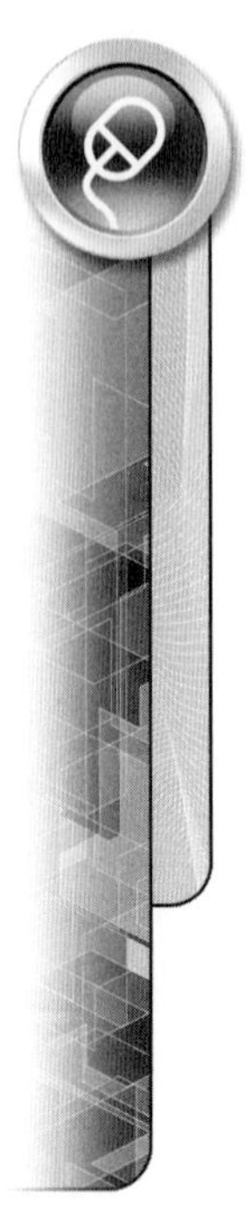

1. Click Ribbon's Modify tab. On the Edit Elevations panel's middle right, select ***Insert High/Low Elevation Point*** icon (first row, the fourth icon).
2. The command line prompts you to select an object (feature line, survey figure, parcel line, or 3D polyline). In the drawing, select the feature line that was just drawn.
3. The command line prompts you for the start point. In the drawing, select the feature line's southern end.
4. The command line prompts you for the end point. In the drawing, select the feature line's northern end.

The routine echoes Start Elevation 732.900', End Elevation 732.750', and Distance 206.176'.

5. The command line prompts you for the ahead slope or grade. Enter **'S'** and press ENTER, for the slope enter **25**, and press ENTER.
6. The command line prompts you for the back slope or grade. For the slope enter **25** and press ENTER twice to create the high point and exit the routine.
7. Select the feature line, press the right mouse button, and, from the shortcut menu, select ELEVATION EDITOR....

The feature line has an elevation point that is higher than either end.

8. Close the Grading Elevation Editor vista.
9. At Civil 3D's top left, Quick Access Toolbar, click the ***Save*** icon to save the drawing.

## Create the Base Surface

The base surface has points, breaklines, and contour data.

## Create Point Group

To assign points to a surface, they must belong to a point group.

1. If necessary, click the ***Prospector*** tab.
2. In Prospector, select the **Point Groups** heading, press the right mouse button, and from the shortcut menu, select NEW....
3. In the ***Information*** tab, for the point group Name, enter **Base**.
4. Click the Include tab, toggle ON With Numbers Matching, and for the range, enter **500–520**.
5. Click the ***Point List*** tab to review the selected points.
6. Click ***OK*** to exit the dialog box.

## Create Surface

1. In Prospector, select the Surfaces heading, press the right mouse button, and from the shortcut menu, select CREATE SURFACE....
2. In the Create Surface dialog box, set the surface name to **Base**, for the description, enter **Prelim Design**, set the Surface Style to **Border & Triangles & Points**, and click ***OK*** to exit.
3. In Prospector, expand the Surfaces branch until you are viewing the Base's Definition data type list.

4. From the data type list, select **Point Groups**, press the right mouse button, and from the shortcut menu, select ADD....
5. In Point Groups, select **Base**, and click ***OK***.
6. In the data type list, select **Breaklines**, press the right mouse button, and from the shortcut menu, select ADD....
7. In the Add Breaklines dialog box, for the description, enter **Breaklines**, set the type to **Standard**, toggle **ON** Supplementing factor's Distance, set it to **10**, set the Mid-Ordinate distance to **0.01**, click ***OK***, and in the drawing, select the swale and berm feature lines.
8. From the Home tab's, Layers panel, select ISOLATE. If necessary, set layers not isolated to Off and paperspace viewports to VPfreeze, and select a contour.
9. In the Definition tree type list, select Contours, press the right mouse button, and from the shortcut menu, select ADD....
10. In Add Contour Data, for the description, enter **EG AND NEW**, set the remaining values as shown in Figure 10.39, and when they match, click ***OK***.
11. In the drawing, select all contours and press ENTER to exit.
12. If the Event Viewer displays, close it by clicking the green check mark in the panorama's upper-right corner.
13. From the Home tab's, Layers panel, select UNISOLATE.
14. In Layer Properties Manager, for the current layer, set it to 0 (zero), freeze the layer **Base Contours**, **C-TOPO-Base** and **C-TOPO-FEAT**, and click ***OK*** to exit.

This completes building the new surface.

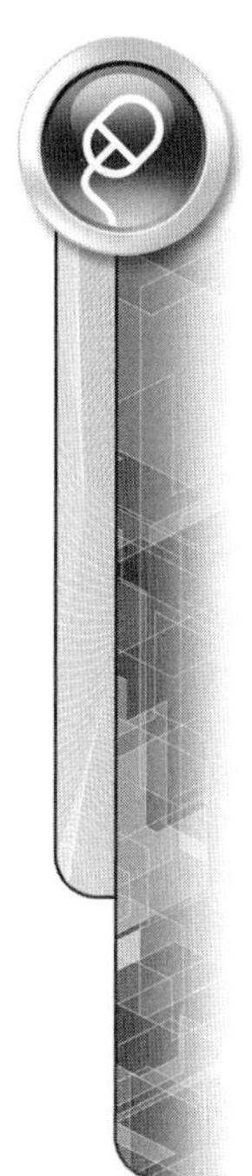

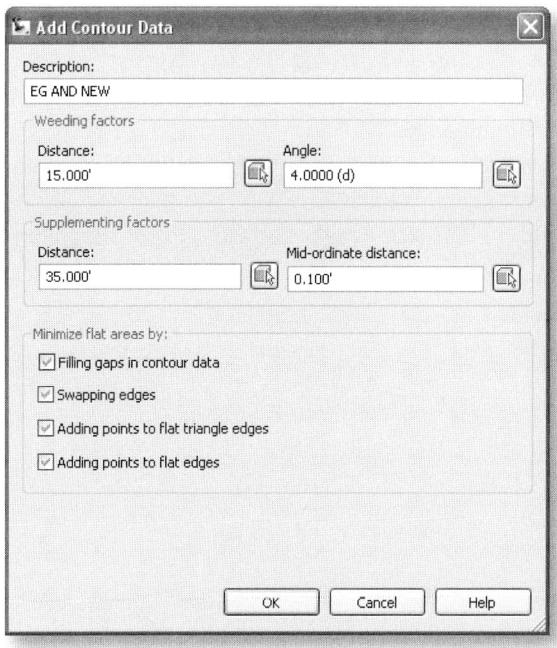


**FIGURE 10.39**

## Calculating a Volume Summary

1. On the Ribbon, click the Analyze tab. On the Volumes and Materials panel, click the Volumes icon, and from the shortcut menu, select VOLUMES to display the Composite Volumes vista.

2. In the vista, at its top left, click the ***Create new volume entry*** icon to create a calculation entry.
3. In the vista, click in the Index 1 Base Surface cell (<select surface>), click the drop-list arrow, and from the list, select **Existing.**
4. In the vista, click in the Index 1 Comparison Surface cell (<select surface>), click the drop-list arrow, and from the list, select **Base** (see Figure 10.40).
5. Repeat Steps 2 through 4 to calculate a second volume using **EGCTR** and **Base**.
6. Note the Cut, Fill, and Net volumes for each comparison.
7. In the Panorama, click its **X** to close it.

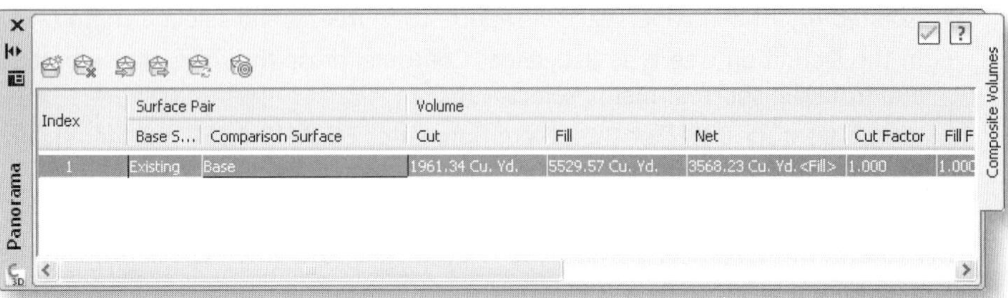


**FIGURE 10.40**

## Volume Surface Style

A volume surface uses style like any other surface object. For this exercise, you will define a style that shows a 2D view of contour ranges for cut and fill and in 3D as a model.

1. Click the ***Settings*** tab.
2. Expand the Surface branch until you are viewing the Surface Styles styles list.
3. From the styles list, select **Contours 1' and 5' (Design)**, press the right mouse button, and from the shortcut menu, select COPY....
4. In the Surface Style dialog box, click the ***Information*** tab, for the name, enter **Volume Contours**.
5. Click the ***Contours*** tab, expand the Contour Ranges section, and set the following values (see Figure 10.41):

   Major Color Scheme: Rainbow
   Minor Color Scheme: Land
   Group Values by: Quantile
   Number of Ranges: 12
   Use Color Scheme: True

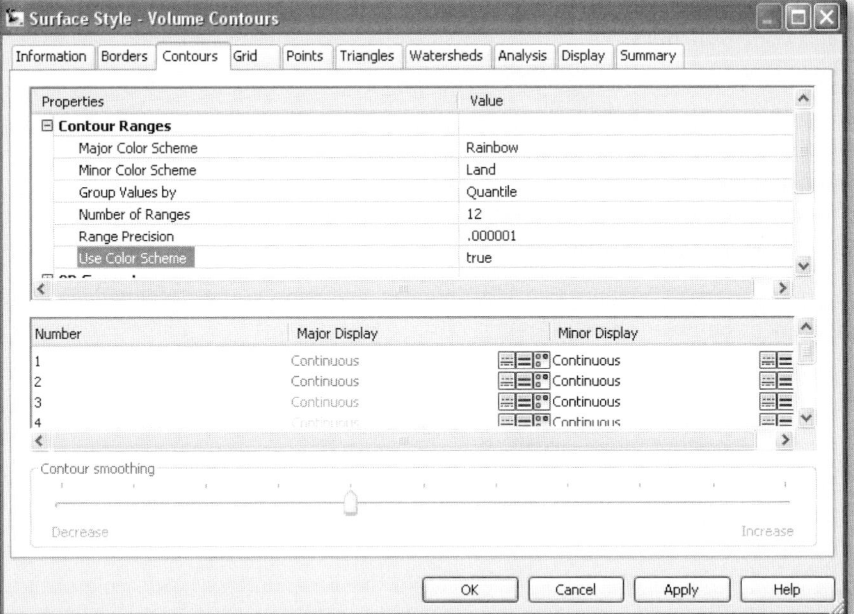


**FIGURE 10.41**

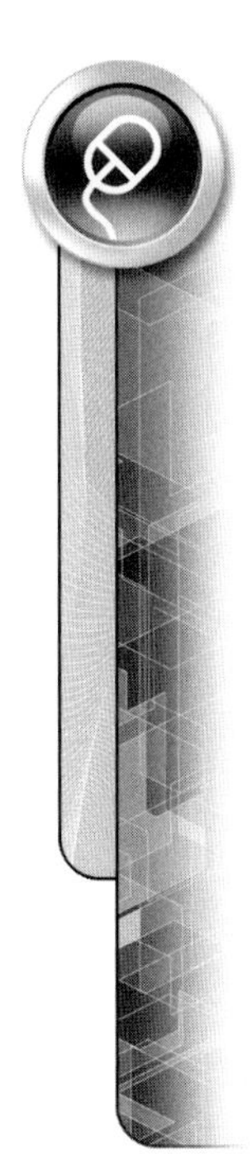

6. Click the ***Analysis*** tab, expand Elevations, and set the following values (see Figure 10.42):

   Scheme: Rainbow
   Group by: Quantile
   Number of Ranges: 12
   Display Type: 3D Faces
   Elevation Display Mode: Exaggerate Elevation
   Exaggerate Elevations by Scale Factor: 4

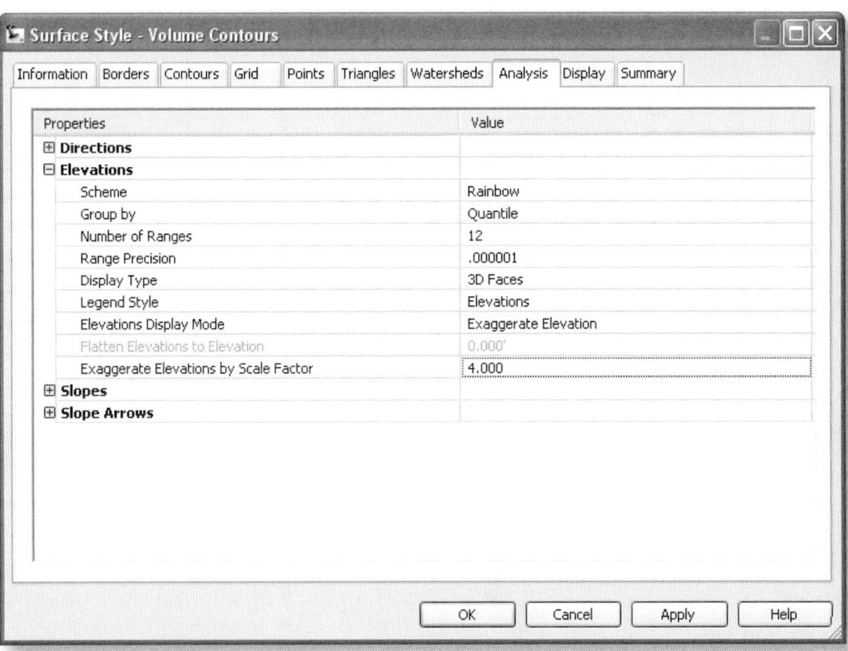

**FIGURE 10.42**

7. Click the ***Display*** tab, set the View Direction to **Plan**, toggle ON only **Border** and **User Contours**.
8. Set the View Direction to **Model** and toggle ON only **Elevations**.
9. Click ***OK*** to exit the dialog box.
10. At Civil 3D's top left, Quick Access Toolbar, click the ***Save*** icon to save the drawing.

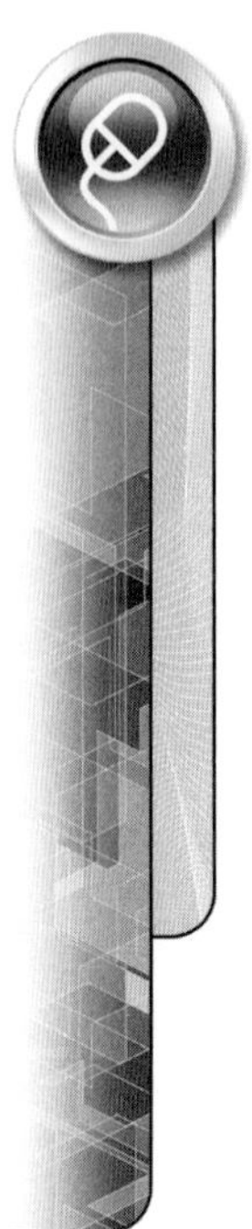

## Creating a TIN Volume Surface

Comparing two surfaces (Existing and Base) creates a volume surface whose elevations are the differences between the surfaces.

1. Click the ***Prospector*** tab.
2. In Prospector, click the **Surfaces** heading, press the right mouse button, and from the shortcut menu, select CREATE SURFACE....
3. In the Create Surface dialog box, set the surface Type to **TIN Volume Surface**, for the name, enter **EXISTING – BASE TIN VOL**, and enter the remainder of the information found in Figure 10.43.
4. At the bottom of the dialog box, click in the Base Surface value cell, click its ellipsis, in Select Base Surface, select **Existing**, and click ***OK***.
5. Repeat the previous step, but for the Comparison Surface select the surface **Base**.
6. Click ***OK*** to exit the dialog box and create the EXISTING – BASE TIN VOL surface.

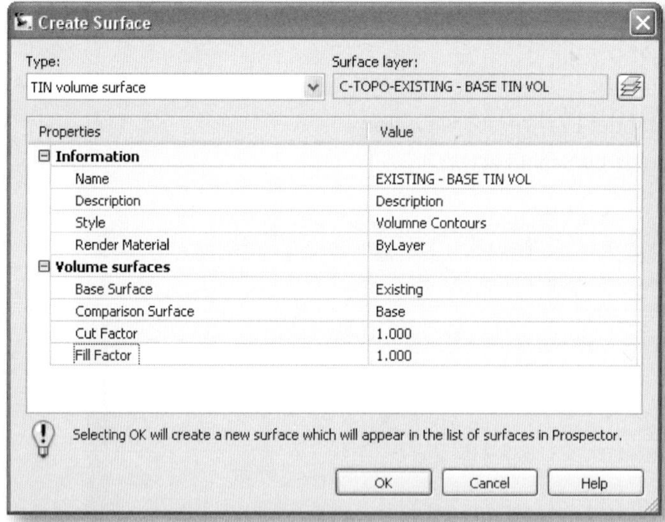


**FIGURE 10.43**

## Surface Properties

A volume surface's property dialog box reports the amount of cut and fill. The surface contours display the amounts of surface cut and fill.

1. From Prospector's Surfaces, select the EXISTING – BASE TIN VOL, press the right mouse button, and from the shortcut menu, select SURFACE PROPERTIES....

2. Click the ***Analysis*** tab, set the Analysis Type to **User-Defined Contours**, set the Number of Ranges to **12**, and click the ***Run Analysis*** icon.
3. In the Range Details section, change contour 6's elevation to **0.00** and click ***Apply***.
4. Change the Analysis Type to **Elevations**, set the Number of Ranges to **12**, and click the ***Run Analysis*** icon.
5. Select the ***Statistics*** tab and expand the Volume section to list the earthwork calculation.
6. Compare this volume to the volumes calculated previously.
7. Click ***OK*** to exit the Surface Properties dialog box.

The contours represent cut and fill amounts. The style displays a 3D view of the cut and fill amounts. The cut areas are the deepest holes, and the fill areas are the highest hills.

8. In the drawing, select any surface component, press the right mouse button, from the shortcut menu, select OBJECT VIEWER..., and view the surface from different angles.
9. Exit the Object Viewer.
10. In the drawing, select any surface component, press the right mouse button, and from the shortcut menu, select SURFACE PROPERTIES....
11. In Surface Properties, click the ***Information*** tab, set the Surface Style to **_No Display**, and click ***OK*** to exit the dialog box.
12. At Civil 3D's top left, Quick Access Toolbar, click the ***Save*** icon to save the drawing.

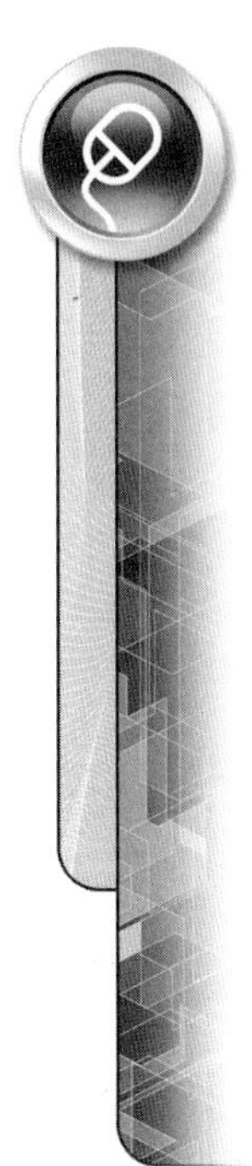

## Creating a Grid Volume Surface

Creating a Grid Volume surface is the same as creating a TIN Volume surface.

1. If necessary, click the ***Prospector*** tab.
2. In Prospector, select the **Surfaces** heading, press the right mouse button, and from the shortcut menu, select CREATE SURFACE....
3. In the Create Surface dialog box, set the surface type to **Grid Volume Surface** and enter the information found in Figure 10.43.
4. In the middle of the dialog box, set the X and Y Grid Spacing to **5**, and set the rotation to **0** (zero).
5. At the bottom of the dialog box, click in the Base Surface value cell. At its right side click the ellipsis, in the Base Surface dialog box, select **Existing**, and click ***OK***.
6. At the bottom of the dialog box, click in the Comparison Surface value cell, at its right side click the ellipsis, in the Comparison Surface dialog box, select **Base**, and click ***OK***.
7. Click ***OK*** to exit the dialog box and create the EXISTING – BASE GRID VOL surface.

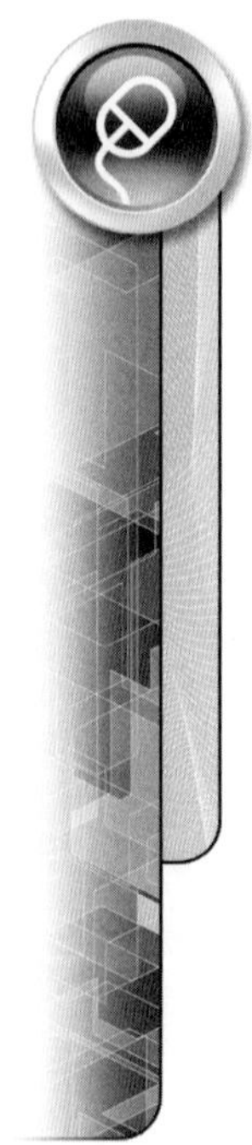

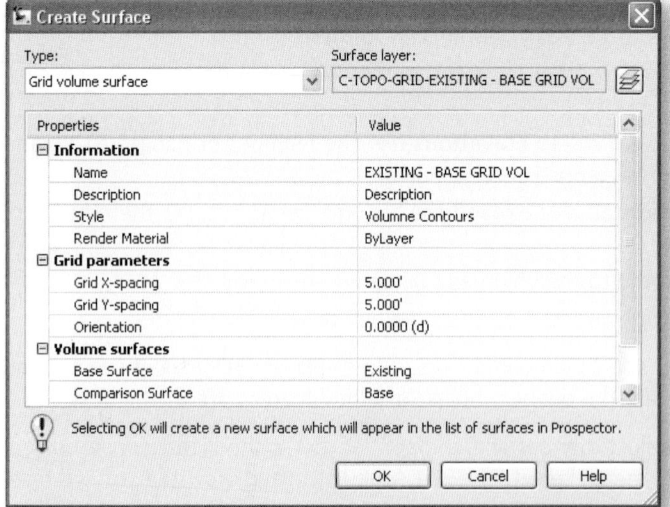


**FIGURE 10.44**

## Create User Contours

1. From Prospector's Surfaces list, select EXISTING – BASE GRID VOL, press the right mouse button, and from the shortcut menu, select SURFACE PROPERTIES....
2. Click the ***Analysis*** tab, set the Analysis type to **User-Defined Contours**, set the Number of Ranges to **12**, and click the ***Run Analysis*** icon.
3. Set the Analysis to **Elevations**, set the ranges to **12**, and click the ***Run Analysis*** icon.
4. Click the ***Statistics*** tab and expand the Volume section to list the cut and fill volumes.
5. Note the volume values and compare them to the previously calculated values.
6. Click ***OK*** to exit the Surface Properties dialog box.
7. In the drawing, select any surface component, press the right mouse button, from the shortcut menu, select OBJECT VIEWER..., and review the surface from different angles.
8. Exit Object viewer.
9. Select any surface component, press the right mouse button, and from the shortcut menu, select SURFACE PROPERTIES....
10. Click the Information tab, change the Surface Style to **Border & Triangles & Points**, and click ***OK*** to display the surface volume grid triangulation.
11. At Civil 3D's top left, Quick Access Toolbar, click the ***Save*** icon to save the drawing.

## SUMMARY

- Volume surface elevations are the difference in elevation between two compared surfaces.
- A volume surface property is an earthwork volume.
- Analysis surface styles evaluate a volume surface's data.

This ends the surface design tools and volumes chapter. As mentioned throughout this book, surfaces are fundamental to most projects. If a surface is not correct, this error affects many aspects of a design solution.

Next, you learn to develop pipe networks. The surface also plays a critical role in pipe network design.

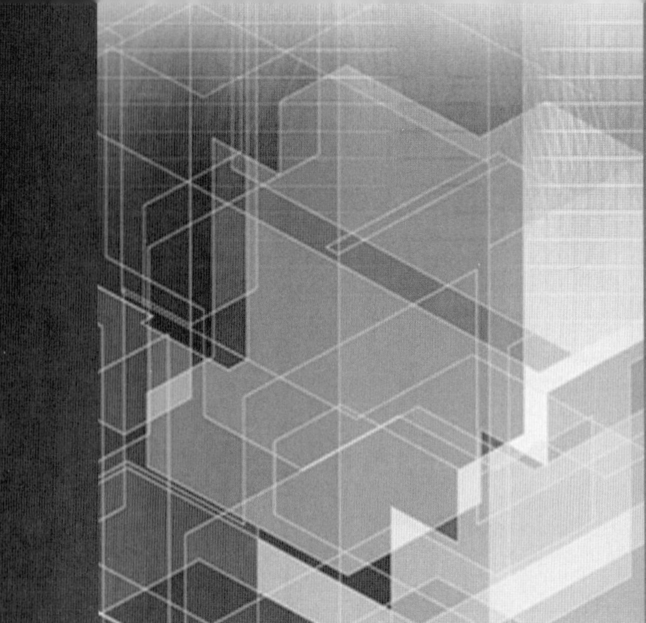

# CHAPTER 11

# Pipe Networks

## INTRODUCTION

Pipe networks are integral to a site-design solution. The piping system's complexity can vary from simple culverts to several storm and sanitary networks that service a residential area. Civil 3D creates plan, profile, section, and model network components. A pipe network can be a part of a roadway corridor's sample line group or can be a network of trunk lines and laterals. A pipe network can create profiles that display each segment's vertical design.

## OBJECTIVES

This chapter focuses on the following topics:

- Defining Pipe-Run Specific Settings
- Defining New Pipe-Run Structures
- Defining Pipe Runs
- Editing and Analyzing Pipe-Run Data
- Annotating Pipe Runs in Plan and Profile Views

### OVERVIEW

The Ribbon's Home and Modify, Pipe Networks panel commands create, edit, and annotate pipe and structure networks (see Figure 11.1). The pipe design toolset is in the Pipe Network Creation Tools toolbar (see Figure 11.2). Most piping designs are an interconnected network of trunk and branch lines. A pipe network can have several trunk lines and each trunk line can have any number of branches. The key is to include all of the appropriate trunks and branches with the correct network. You add to a network by editing it. Prospector lists all of the pipes and structures in a pipe network.

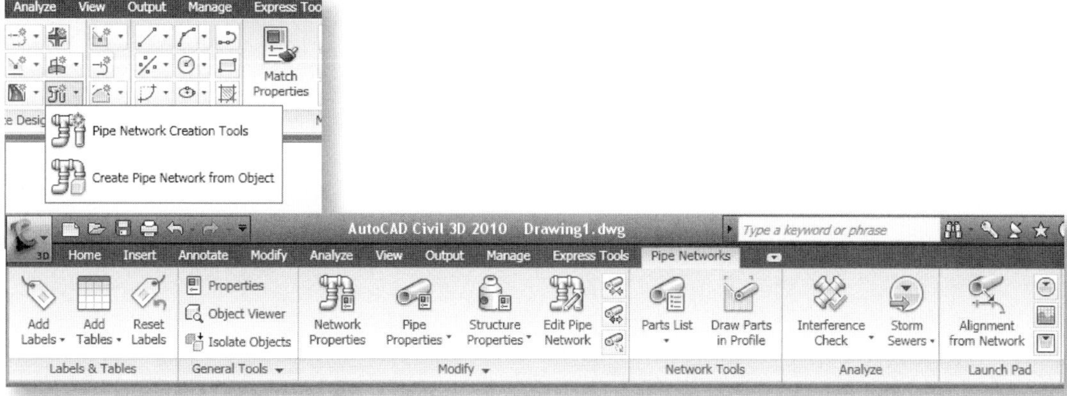


**FIGURE 11.1**

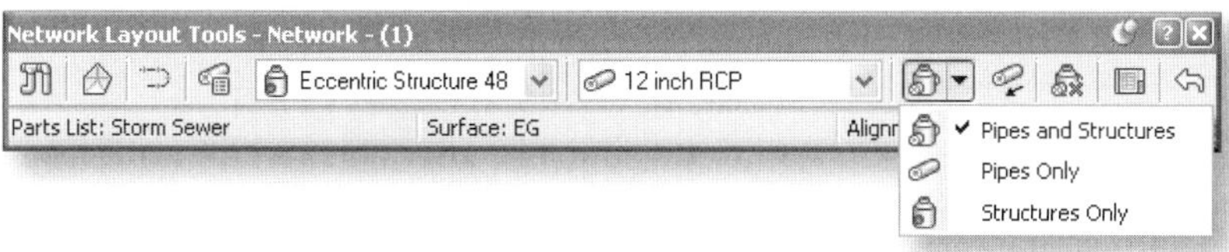


**FIGURE 11.2**

The pipe Network Layout Tools toolbar depends on a series of settings and styles. These settings and styles govern the structure type, pipe type and size, inverts, and a host of additional values.

A pipe network does not require an associated alignment. Each network stands on its own. If you want a pipe network alignment and profile, Modify's Pipe Networks panel's Launch Pad has tools you can use to create them. Other panel commands place all or a selection of pipes and structures in a profile view.

To edit a network, you select a pipe or structure, press the right mouse button, and from a shortcut menu, select Edit Pipe Network. Edit Pipe Network also is located in Modify's Pipe Networks panel or in a Prospector shortcut menu. Edit Pipe Network displays the pipe Network Layout Tools toolbar.

When you associate an alignment to a pipe network, you should do so when you start the network's design (see Figure 11.3). A pipe network can start or end beyond the associated alignment definition. If a pipe run is outside an alignment definition, you may have to edit structure and pipe values to produce correct elevations and slopes.

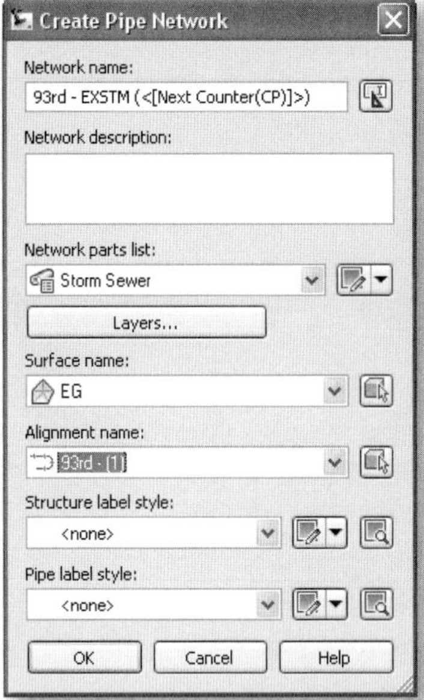


**FIGURE 11.3**

Pipe network review and editing is graphical and is done in Plan and/or Profile View, in a Pipe Network vista editor, or by editing values in the object's Properties dialog box. If a pipe network does not have an associated alignment, all editing is done in Plan View or in a panorama.

One issue is the pipe network layer assignment. When you set the Network parts list to Sanitary Sewer or Storm Sewer, the layer list does not switch; it stays set to the Storm layers. If you switch the parts list, the layers specified in the Create Pipe Network dialog box must be changed.

When you select a pipe or structure and press the right mouse button, a shortcut menu appears. The shortcut menu on the left side of Figure 11.4 shows the available options for selecting pipe, and the shortcut menu on the right side shows the options available for selecting a structure.

When you graphically edit a pipe or structure in Plan View, Civil 3D allows you to move them, disconnect them from the network, change their style, swap to a new part, or edit the network in a panorama (see Figure 11.4). When you are manipulating a pipe's grip, by moving it away from a structure, the pipe is disconnected from the structure. When you are moving the pipe back to the structure, the attach to structure icon must appear before you select and reattach the pipe. If you are selecting a structure with connected pipe, by moving it, you relocate the pipes.

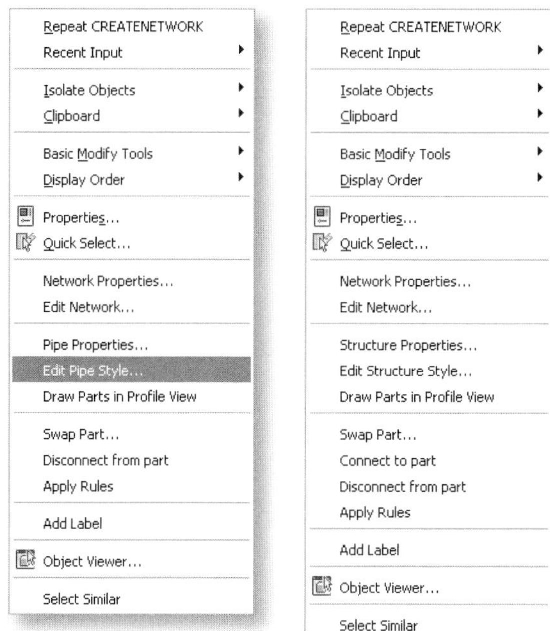


**FIGURE 11.4**

In Plan View, when you are breaking a pipe to add a structure or pipe, a break pipe icon is displayed and indicates the point at which the new structure or pipe attaches to the network.

**NOTE**

**Pipes and structures of different networks can now connect to each other.**

When you are editing a pipe network in Profile View, Civil 3D imports selected pipes and structures or the entire network into Profile View. Within Profile View's context, you can graphically change pipe locations, change pipe and structure sizes, and swap structure types.

You can use a panorama with Pipe and Structure vistas to edit an entire pipe network. Each vista presents the structure and pipe numbers. These vistas can be used to change almost any pipe or structure value. The biggest issue with these vistas is their size and complexity. Because the vistas contain numerous cells, they cannot be entirely displayed on the screen (you must scroll to see all their data). Also, the column placement can be difficult. It is necessary to move columns around to be able to view related values. In addition, changes made in the vista do not immediately affect the network; rather, they occur once you have exited it. Any changes made in a vista are permanent; there is no cancel or undo.

The pipe network and its objects have extensive settings and styles to automate many design tasks. These settings and styles affect drafting pipes and structures in plan, profile, and section; labeling; structures; and data values. Because there is a difference between sanitary and storm pipe networks pipes and structures, there are two parts lists. Each list specifies its respective typical structures and pipes.

Civil 3D includes an extensive pipe and structure parts catalogs. A catalog defines standard pipe and structure parts. A parts list is a subset of the catalog parts. A parts list contains typical structures and pipes, their descriptions, measurements, rules, and other settings.

### Unit 1

Unit 1 focuses on pipe network settings and styles. As mentioned earlier, the settings and styles list is extensive. This unit also reviews the assignment of pay items to items in a parts list.

### Unit 2

Unit 2 discusses drafting pipe networks.

### Unit 3

Unit 3 focuses on reviewing and editing pipe network data. Both review and edits can be done in Plan or Profile View, either graphically or from a panorama.

### Unit 4

Unit 4 covers pipe network annotation when the labeling is not a part of the originally drafted network. This unit also reviews pipes in cross-sections. The unit also reviews creating pipe network pay item reports.

## UNIT 1: PIPE SETTINGS AND STYLES

Settings' Pipe Network is an extensive list of settings and styles (see Figure 11.5). Pipe and Structure branches contain settings and values to define their shapes, rules, and labeling styles. Pipe Network defines parts lists for typical systems—storm and sanitary—from the Pipe and Structure branch definitions. A user can define other parts lists such as gas or water.

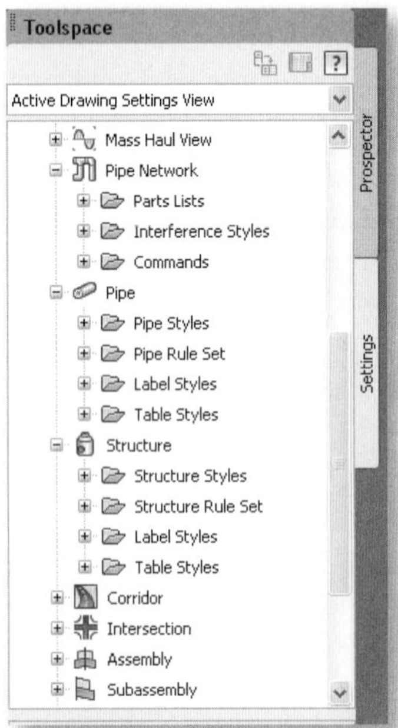


**FIGURE 11.5**

Parts lists for any one system should include typical structures and pipe sizes. A network branch may have structures and pipe sizes that are different from those of a trunk. In a storm sewer system, branches may be catch basins rather than manhole structures.

When developing parts lists, styles, and settings, it is important that you copy them into a template file. This way, every new job starts with the same settings and you are able to produce consistent drawings.

## EDIT DRAWING SETTINGS

Settings' Edit Drawing Settings contains base layer names for pipe, structure, and network objects (see Figure 11.6). The layer list includes entries for pipe networks in Plan, Profile, and Section Views. The problem with a modifier for pipes and structures is that a new layer is created for each structure and pipe. A pipe network does not have a single layer dedicated to its contents.

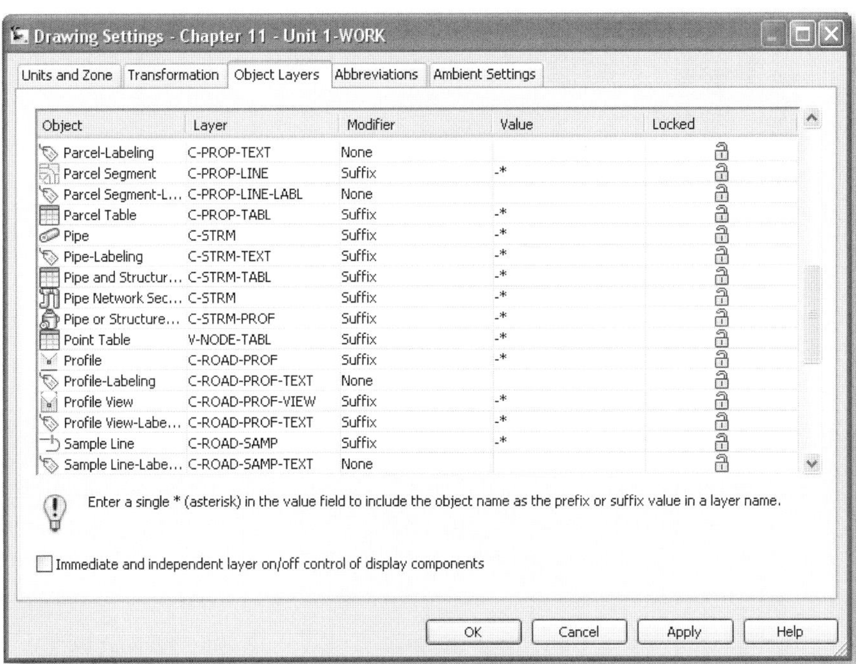


**FIGURE 11.6**

## PIPE AND STRUCTURE CATALOGS

The pipe and structure catalogs supply basic specifications for pipe network parts lists. A catalog populates many values in the myriad dialog boxes that open while you are developing storm and sanitary parts lists. Unfortunately, there isn't a menu call to display the catalogs.

Each catalog is in the operating system's All Users branch (XP) and is an HTML file that reads XML data files from folders below the catalog's location (see Figure 11.7). A catalog contains each type of pipe's and structure's sizes and other parameters. The only way to add types and parameters is to edit the XML files in an editor or use Part Builder.

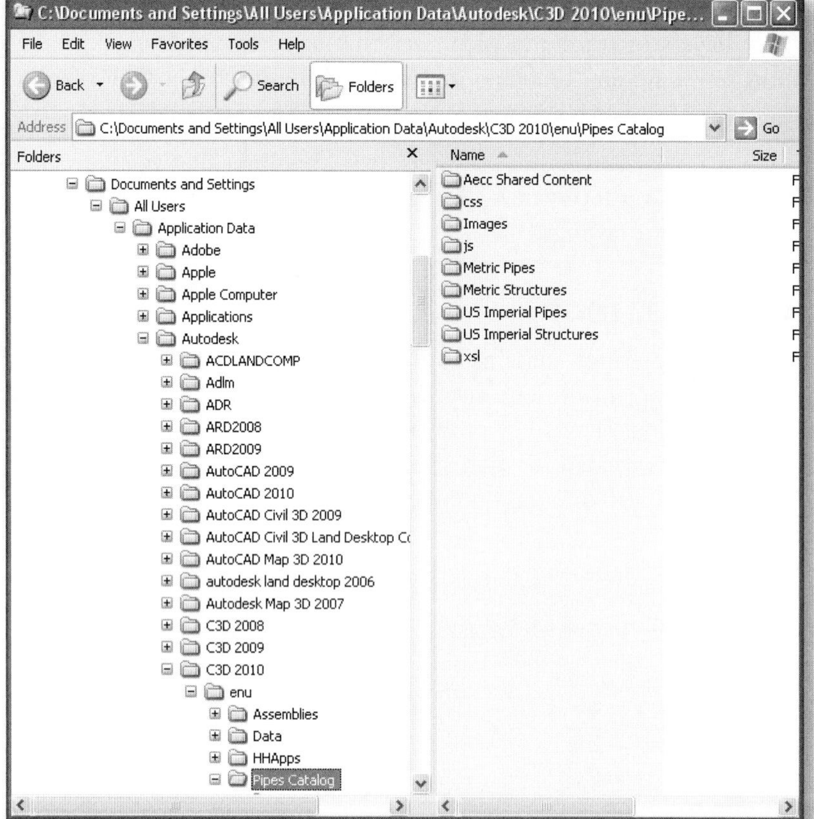


**FIGURE 11.7**

## PIPE NETWORK PARTS LIST

When you are developing or adding to a parts list, catalog values appear as content for the Network Parts List dialog box (see Figure 11.8). The Add Part Family command calls the Part Catalog dialog box, which then opens to display its different part categories.

A parts list can include every possible type and size. But like plotting scales, there will be a consistent set of typical pipe sizes and structures for every project. Users can swap a structure part to any size, even those not on the parts list. So, it is relatively easy to handle any exception occurring in any network.

Developing a parts list assigns the structure and pipe styles, rules, pay items, and rendering material. Initially, styles need to be defined prior to their parts list assignment. Rules affect error reporting, structures' positioning, and other crucial network values. When you change a parts list's pipes or structures, the change affects only those networks that are created after the change. If you need to change a style, make the change after you have completed the parts list modification.

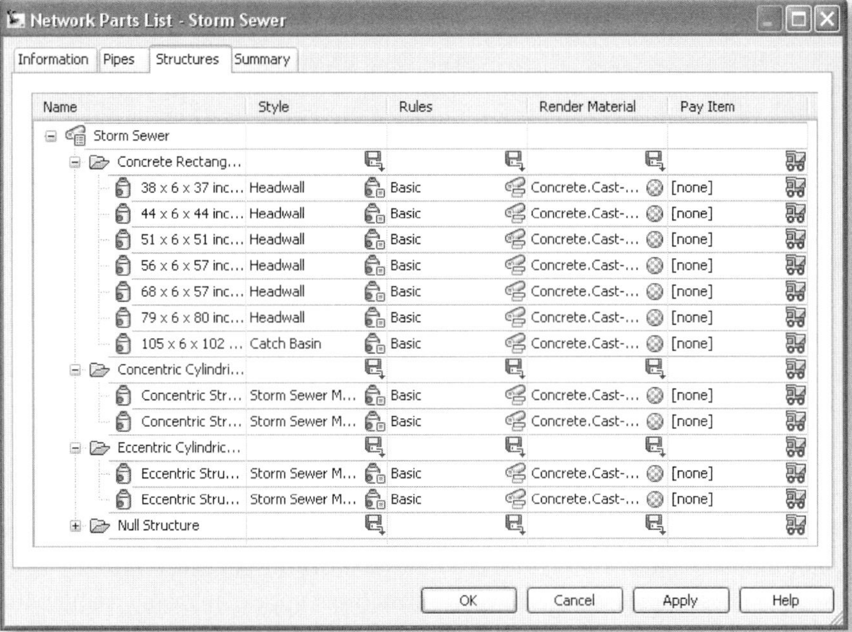


**FIGURE 11.8**

In the Part Catalog dialog box, to add the selected part family to the part list as a new family of potential part sizes (see Figure 11.9), you toggle ON a part family and click OK. When you select the new family name, after you press the right mouse button and select Add Part Size, a Part Size Creator dialog box opens with pipe or structure parameters (see Figure 11.10). Some pipe or structure parameters are from the catalog and are marked as lists in the source column. When you click in a List Source item value cell and then click its drop-list arrow, a list of selectable catalog values is displayed. Other values are calculated or are from a table that contains a range of values, and those values are identified in the Source column.

The Part Size Creator dialog box entries specify part material, frame, grate, and cover part numbers (see Figure 11.10).

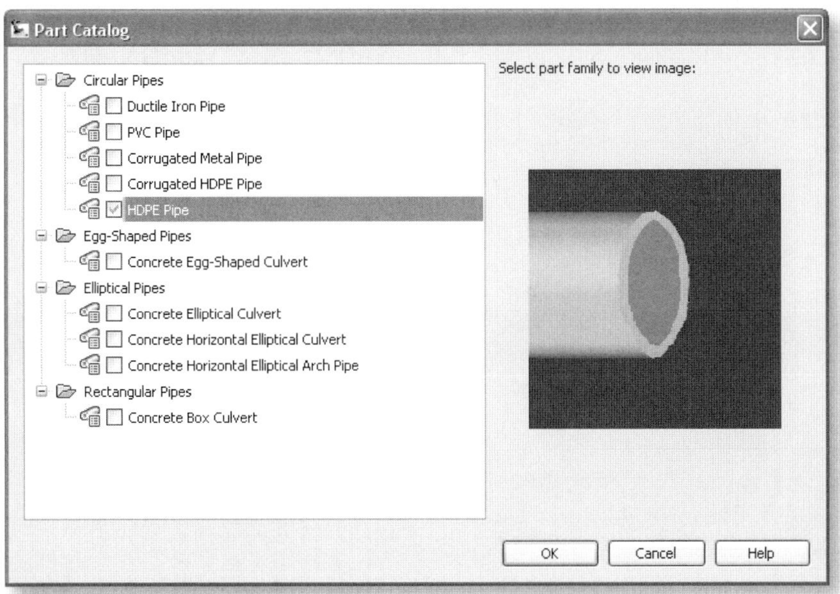


**FIGURE 11.9**

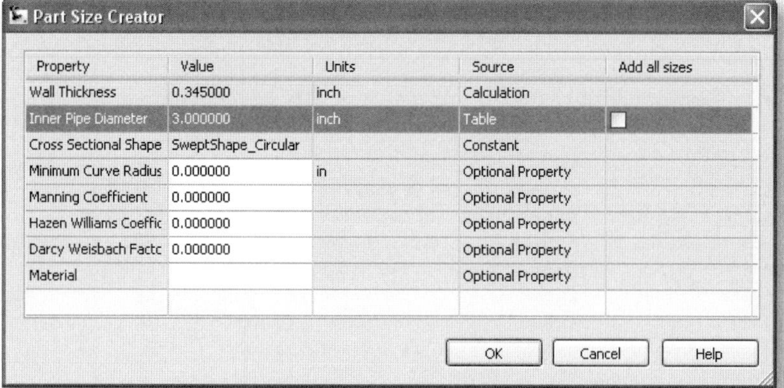


**FIGURE 11.10**

When you add parts to a parts list, Civil 3D assigns pipe and structure styles, rules, pay item, and render materials. Figure 11.8 shows a Storm parts list with assigned structure styles. These styles define how the structures appear in plan, profile, and section (see Figure 11.11). The default styles are set in the Pipe Network's Edit Feature Settings.

To assign a style, in the Style column, click the icon to the right of the Structure Category. A Structure Style dialog box opens and has a drop-list that contains all structure styles. After you select a style and exit, the selected style applies to *all* structures in that category. Or, to individually assign a style, you can click in the part sizes Style cell and select a style. Assigning Rules and Render Materials uses the same process. By selecting a Rule or a Render Material at the category level, you set the value for all of the instances in the category. When you set a Rule or Render Material for a single member, the value is set only for that instance.

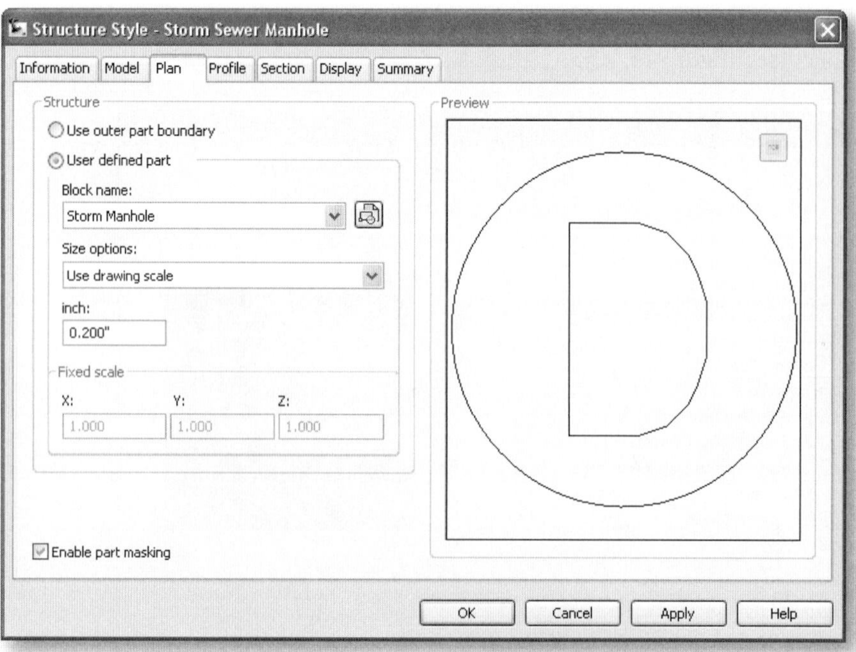


**FIGURE 11.11**

## Pay Items

Chapter 7, Unit 1, discusses defining and the structure of pay items. You assign pay item values to pipes and structures in a parts list. After creating the pipe networks, you can use reports (detailed or summary) from Quantity Take Manager to estimate costs.

# EDIT FEATURE SETTINGS

These settings affect the values found in the pipe network styles and commands, pipes, and structures.

## Pipe Network

Pipe network Edit Feature Settings sets default styles for the network parts list and the pipes and structures in the list. The Default Styles section sets default styles for structure and pipe types; their plan, profile, and section labels; render material; and parts list (see Figure 11.12). When you add a part to a parts list, these settings populate the parts values. If the settings are not correct, you will have to change them in the parts list.

The Default Name Format section defines the naming convention for a pipe network, structure, interferences, alignments from a pipe network, and pipe (see Figure 11.12).

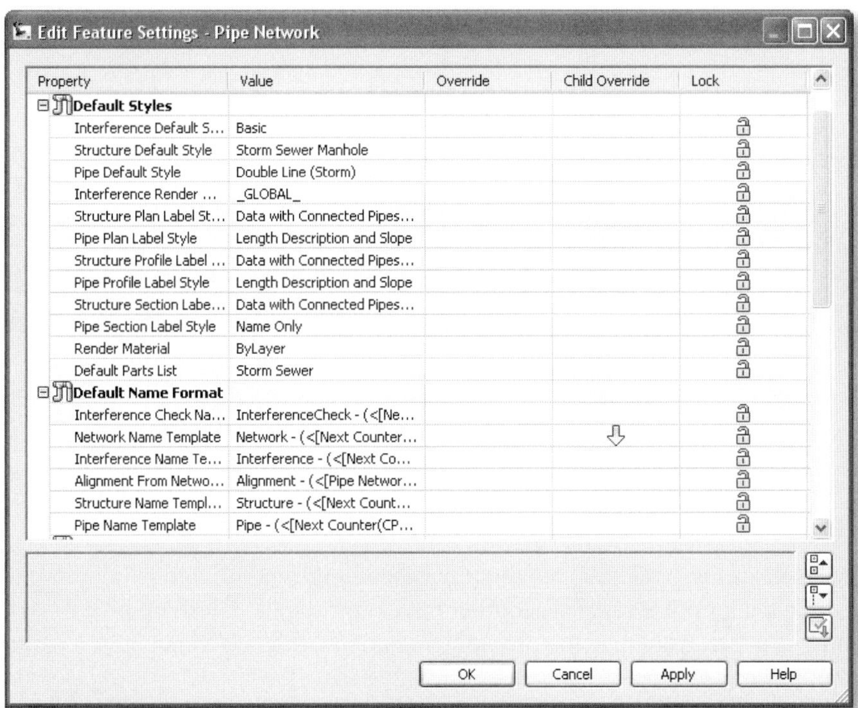


**FIGURE 11.12**

Default Rules sets the default pipe and structure rule. Default Profile and Default Section Label Placement set the label locations for pipes and structures. Each entry is an annotation anchor point. An anchor point can be above, below, in the middle, or on the pipe or structure. Each anchor point location affects where a label appears (see Figure 11.13). If you want labeling for storm systems to be different from sanitary systems, you need to adjust these values before you create the labeling.

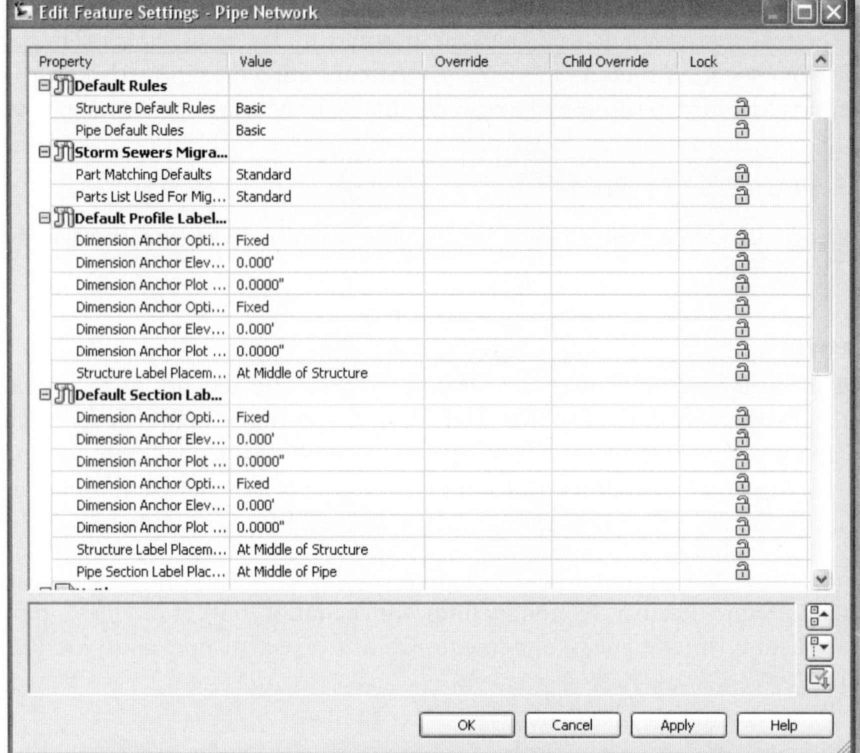


**FIGURE 11.13**

## PIPE STYLES

Pipe styles define a pipe segment's "look" in a pipe network.

### Plan

The Plan panel defines how a pipe is displayed in Plan View (see Figure 11.14). The panel's upper left defines the drafting of the pipe's walls. If you are not using the pipe's specs, the pipe size can have a user-defined size: sized by drawing scale; as a percentage of the screen size; or a fixed size.

The panel's upper right defines how to draft pipe ends. The panel's lower right defines options for drafting the pipe's centerline.

The bottom left of the panel defines the pipe's hatching method and its alignment. Under the hatch alignment toggle is another important toggle: Clean up pipe to pipe connections. When you are drafting pipe-only segments, this selection places a radius at each pipe vertex.

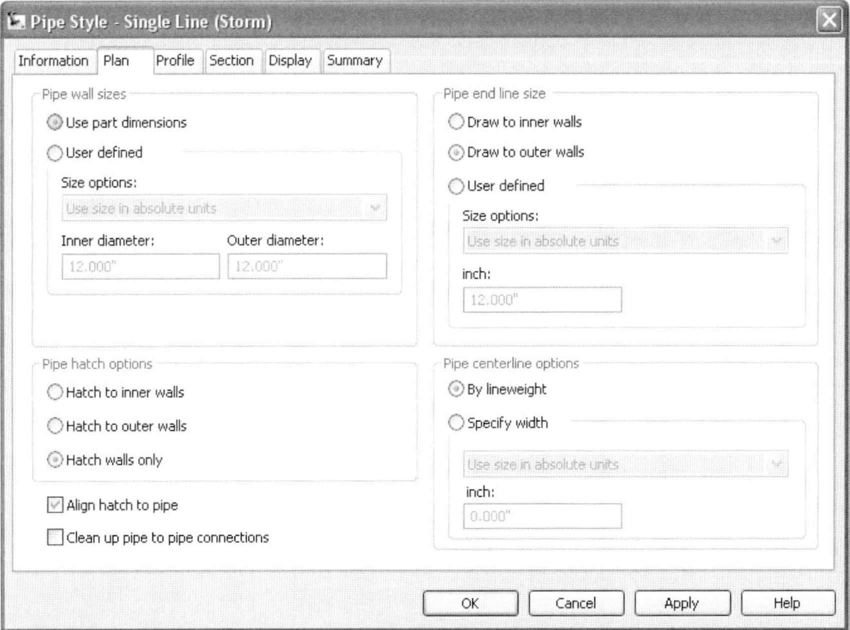


**FIGURE 11.14**

## Profile

This panel's settings are the same as those in the Plan panel, but apply to a pipe in Profile View (see Figure 11.15). The lower right contains crossing pipe hatching options.

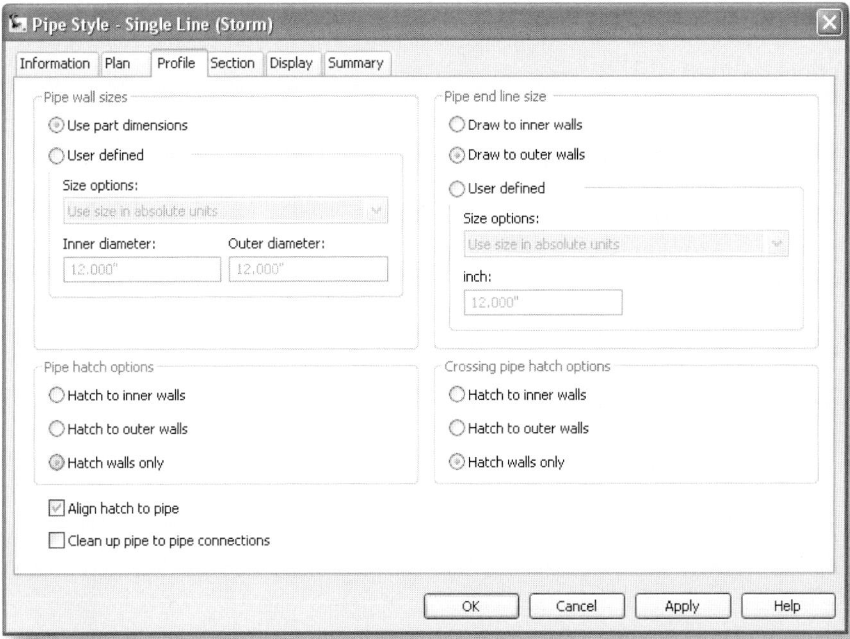


**FIGURE 11.15**

## Section

Section settings define how to hatch a pipe in a cross-section view (see Figure 11.16).

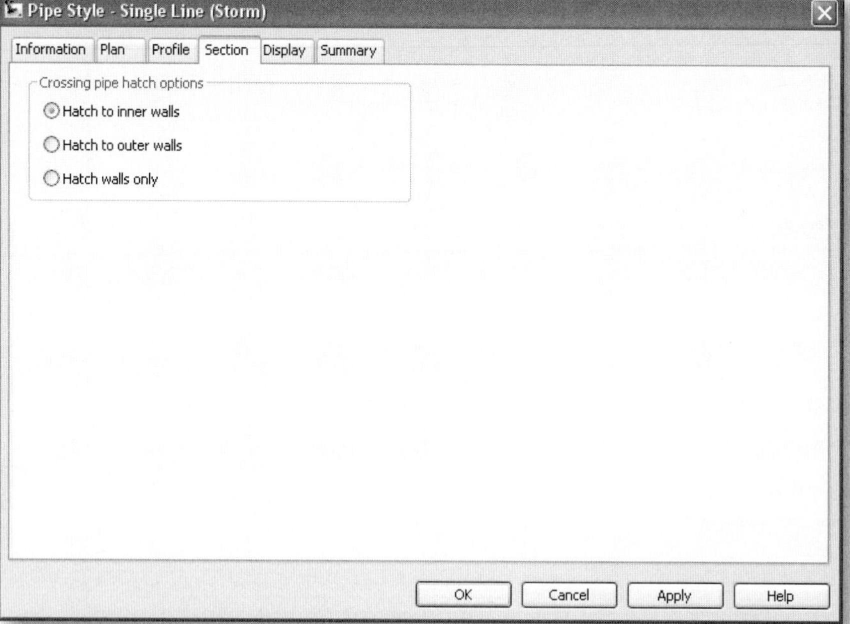


**FIGURE 11.16**

## Display

As with any style, you can set a value for each setting, but it is the display panel values that control what is visible in the drawing (see Figure 11.17). When you are using a Civil 3D content template, there are two pipe styles: single line and double line. It is the styles' visibility settings that differentiate them. The single line style draws pipes by displaying their centerline. The double line style draws two lines that represent the pipe's interior diameter.

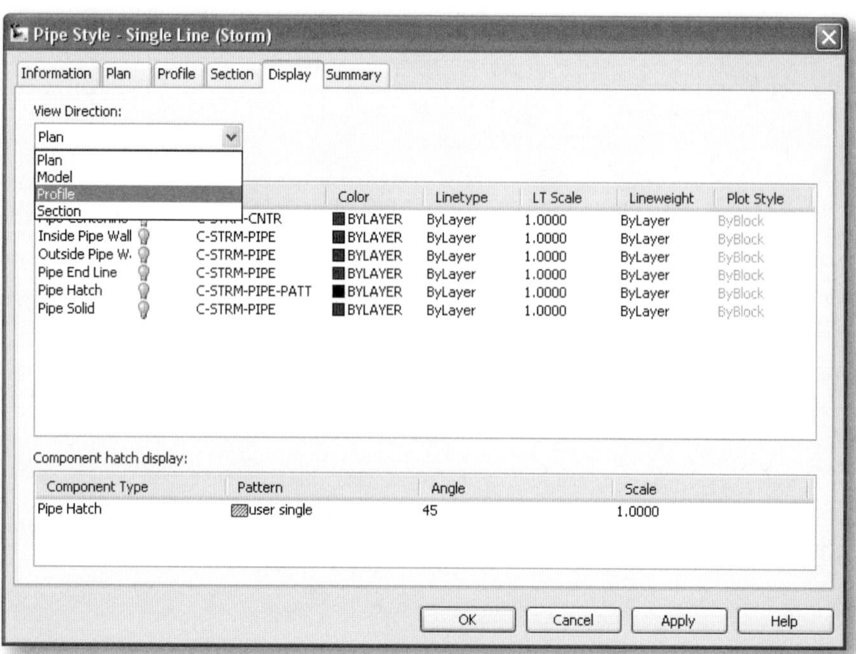


**FIGURE 11.17**

## PIPE RULES

Pipe rules define minimum/maximum slope, cover, and segment length rules for pipes (see Figure 11.18). Each pipe size or pipe family can have its own rule set. The Pipe to Pipe Match section allows the user to match pipe inverts by Crown, Centerline, or Invert.

The first pick of any run with pipes uses the minimum cover value to establish the pipe's invert. All subsequent selections use minimum slope for invert calculations. When you are connecting a pipe to a preexisting structure, use the Apply Rule command to resize the structure to accept the new pipe's invert.

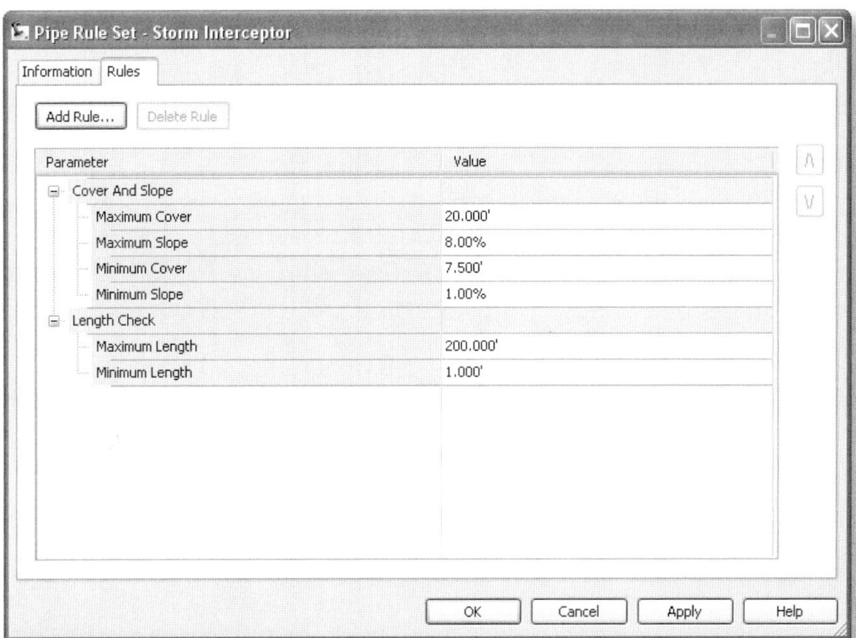


**FIGURE 11.18**

## PIPE LABELS

When you are using a Civil 3D content template, you can choose from several pipe label styles (see Figure 11.19). A pipe label style is the same as any other label style; it is anchored at a point on a pipe, can label multiple properties, and has a dragged state that can be different from or the same as the original label. The Text Component Editor defines the text contents and its formatting.

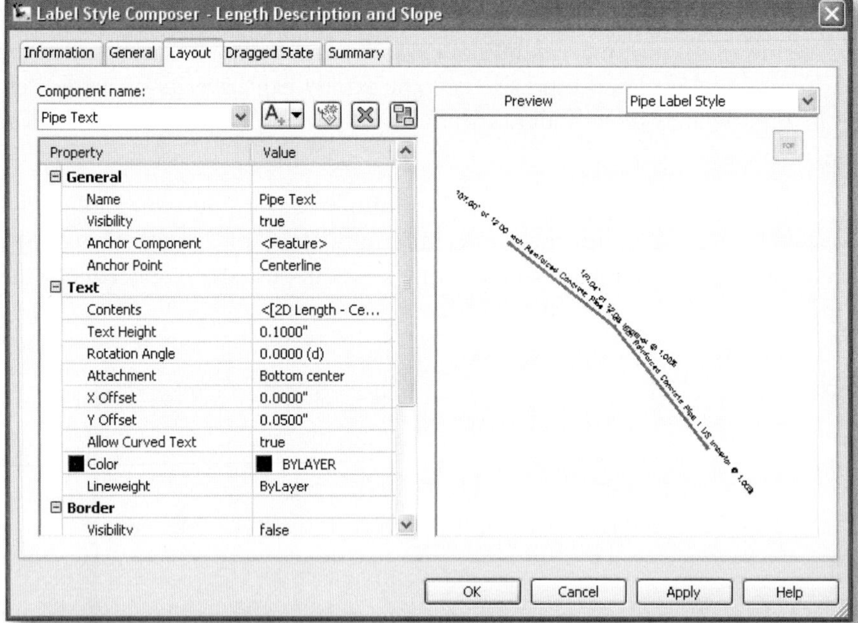


FIGURE 11.19

## STRUCTURE

Structure styles define junctions for pipe networks.

### Structure Style

These styles define a structure's form for plan, profile, section, and model.

### Model

Model defines a structure as a shape from the catalog (see Figure 11.20). If you are not using the catalog shape, other simpler shapes are available: box, cylinder, sphere, and a part-defined shape.

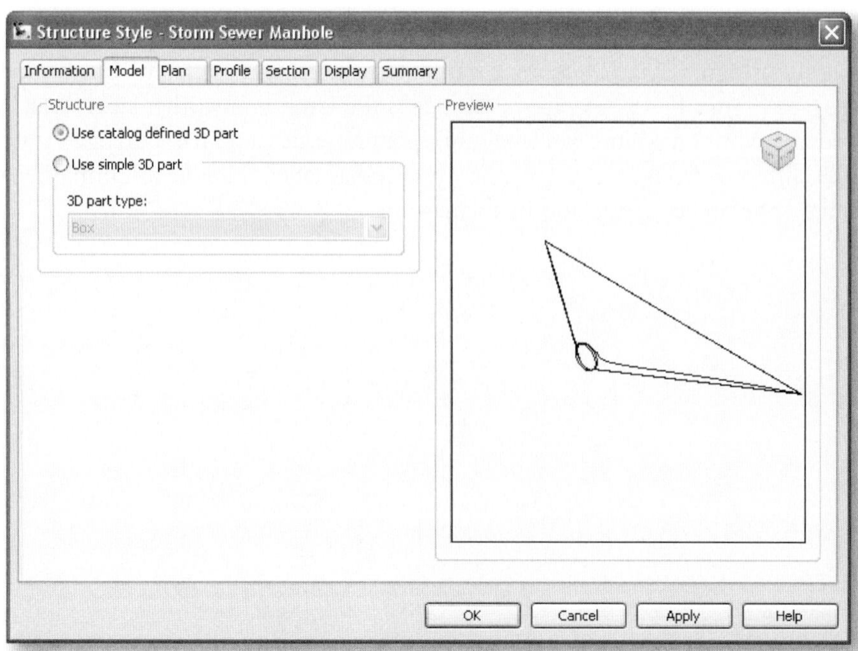


FIGURE 11.20

## Plan

Plan sets the symbol that identifies a structure in Plan View (see Figure 11.11). Size options control the symbol's size: by drawing scale; a fixed scale; in absolute units; size as a screen percentage; or as defined by the catalog parameters.

The Enable Part Masking toggle allows the structure to mask out connecting pipes.

## Profile and Section

Profile and Section defines how a structure is displayed in a Profile or Section view (see Figure 11.21). By default, the structure appears as a boundary in a Profile View. Or, the structure can be displayed as a block or solid. If you select block, sizing options become available.

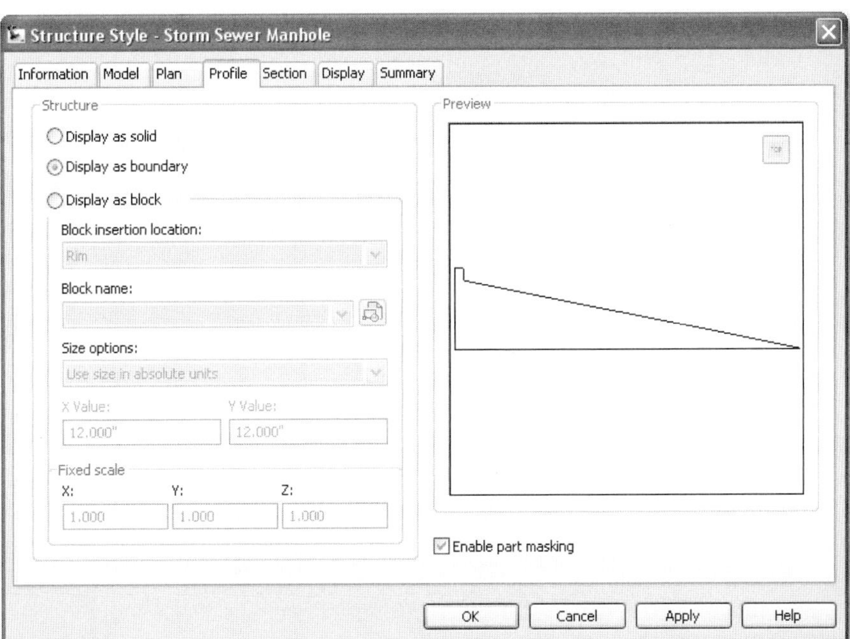


**FIGURE 11.21**

## Display

The Display panel controls what components are displayed and their layers when you are using a style (see Figure 11.22).

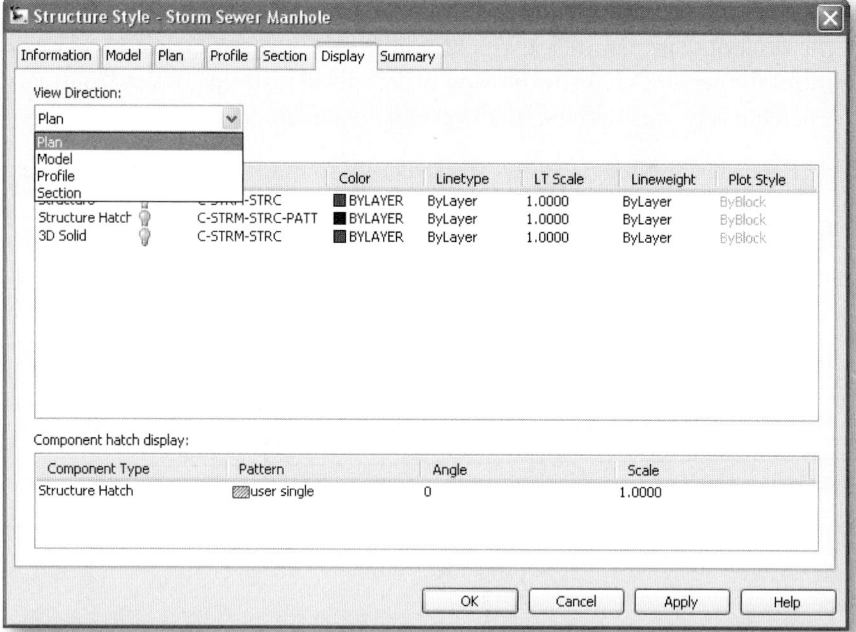


**FIGURE 11.22**

## Structure Rules

Structure rules affect the invert drop amount across a structure, sump depths, and maximum pipe diameter (see Figure 11.23).

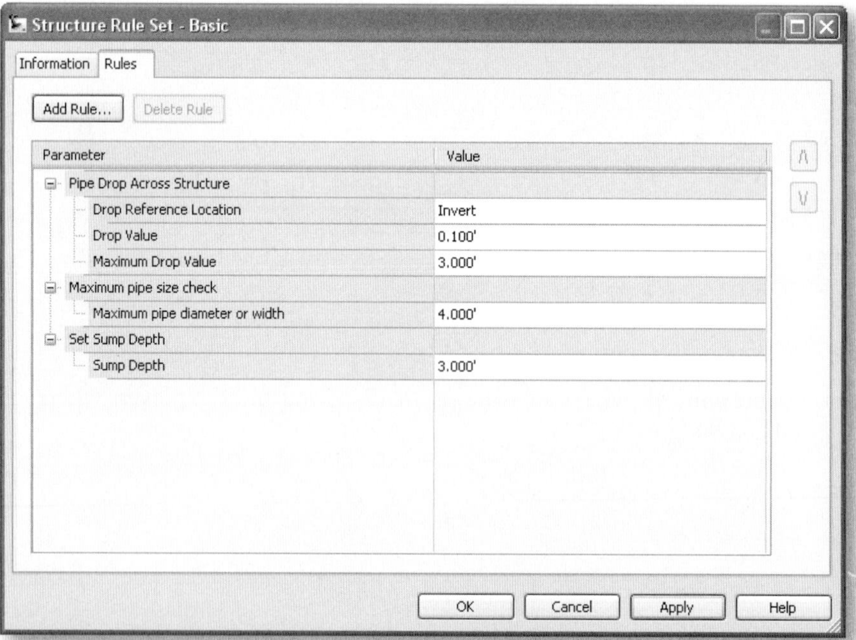


**FIGURE 11.23**

## Structure Labels

These label styles annotate structure information (see Figure 11.24). Labels include station/offset, name, rim, etc. Structure labels have a special label component for each pipe (in and out). This component type labels each connecting in and out pipe's invert.

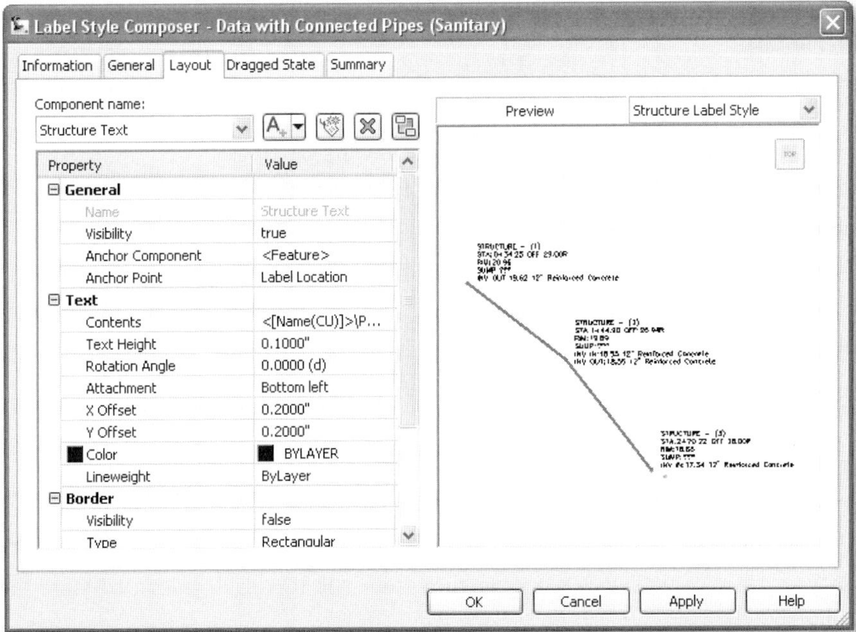


FIGURE 11.24

## INTERFERENCE STYLES

Pipe networks can check for interferences (colliding pipes). Interference styles define the markers at the interference (see Figure 11.25).

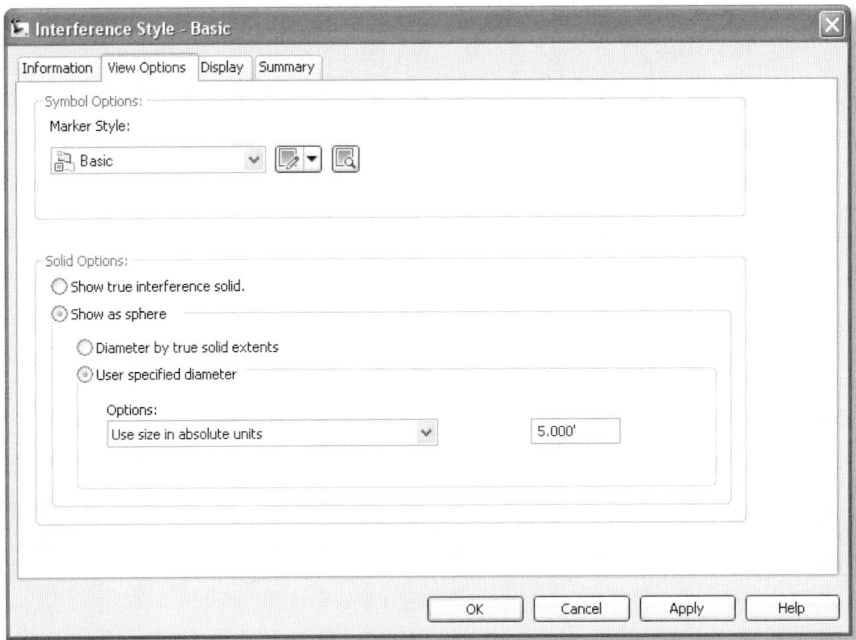


FIGURE 11.25

## EXERCISE 11-1

After completing this exercise, you will:

- Be familiar with the pipe network's Edit Drawing Settings.
- Be familiar with the pipe network's Edit Feature Settings.
- Be able to draft a pipe network.
- Be able to assign materials to pipes and structures.
- Be able to assign pipe rules.
- Be able to add structures to a parts list.
- Be able to assign pay item codes to part list items.

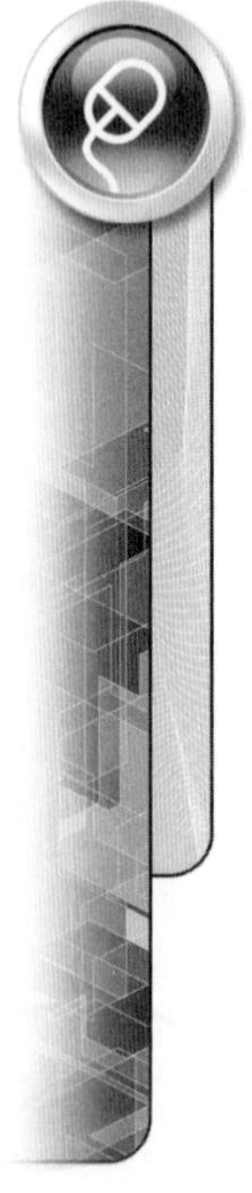

### Exercise Setup

This exercise uses an existing drawing. To find this file, browse to the Chapter 11 folder of the CD that accompanies this textbook, and open the *Chapter 11 - Unit 1.dwg* file.

1. If you are not in ***Civil 3D***, double-click the Civil 3D desktop icon to start the application.
2. When you are at the command prompt, close the opening drawing and do not save it.
3. At Civil 3D's top left, Quick Access Toolbar, click the ***Open*** icon, browse to the Chapter 11 folder of the CD that accompanies this textbook, and open the *Chapter 11 - Unit 1* drawing.
4. At Civil 3D's top left, click Civil 3D's drop-list arrow, and from the Application Menu, highlight SAVE AS, from the flyout select AUTOCAD DRAWING, browse to the Civil 3D Projects folder, for the drawing name enter **Pipe Network-1**, and click ***Save*** to save the file.

### Edit Drawing Settings

Edit Drawing Settings set several styles and values that affect pipe networks.

1. Click the ***Settings*** tab.
2. At Settings' top, select the drawing's name, press the right mouse button, and from the shortcut menu, select EDIT DRAWING SETTINGS....
3. Click the ***Object Layers*** tab and review its settings. If necessary, change your settings to match those in Figure 11.6. All Pipe and Structure layers except Structure and Pipe with a modifier (**suffix**) have a value of (**-***).
4. Click ***OK*** to exit the Edit Drawing Settings dialog box.

### Edit Feature Settings

Pipe Network's Edit Feature Settings values affect both pipes and structures. These values assign pipe segments and structures their initial styles and render materials (see Figures 11.12 and 11.13).

1. In Settings, select the **Pipe Network** heading, press the right mouse button, and from the shortcut menu, select EDIT FEATURE SETTINGS....
2. Expand and review the values for the Default Styles and Default Name Format sections.
3. In the Default Styles section, click in the Render Material's value cell to display an ellipsis. Click the ellipsis to open the Render Material dialog box.
4. In the Render Material dialog box, click the drop-list arrow, select **Concrete.Cast.In.Place.Flat.Grey.1**, and click ***OK*** to return to Edit Feature Settings.

5. Expand the section Storm Sewers Migration Default. Change the Parts List Used for Migration to **Storm Sewers**.
6. Expand the Default Rules, Default Profile Label Placement, and Default Section Label Placement sections and review their values.

Default Rules sets the pipes and structures default rule. The Default Profile Label Placement section sets initial label anchor points. The Default Section Label Placement section sets the initial labels of pipes and structures in a section view.

7. Click ***OK*** to exit the Edit Feature Settings dialog box.
8. At Civil 3D's top left, Quick Access Toolbar, click the ***Save*** icon to save the drawing.

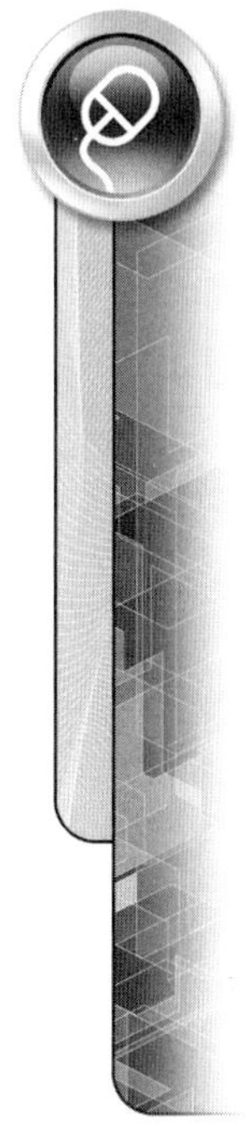

## Pipe and Structure Catalog

The Pipe and Structure Catalog contains basic pipe and structure shapes and parameters. There is a simple, but poorly documented, interface to add or modify the entries of this file: Partbuilder. The catalog is in the Documents and Settings, All Users path (XP), and for Vista, the path is ProgramData\Autodesk\C3D 2009\enu\Pipes Catalog\US Imperial Pipes. The Catalog is an HTML file located in the US Imperial Pipes and Structures folders: US Imperial Pipes.htm and US Imperial Structures.htm. Double-click the HTML file (see Figure 11.7) to review their values. You may get a message about scripts or ActiveX controls, click the message to allow blocked content, and click Yes in the security Warning dialog box.

1. Start Windows Explorer and make US Imperial Pipes the current folder.
2. Double-click the *US Imperial Pipes.htm* file, displaying Internet Explorer with the US Imperial Pipes Catalog.
3. Expand the various branches and review the pipe types and their values.
4. Exit the Catalog (Internet Explorer).
5. Change to the US Imperial Structures folder and double-click the *US Imperial Structures.htm* file displaying Internet Explorer with the US Imperial Structure Catalog.
6. Expand the various branches and review the structure types and their values.
7. Exit the Catalog (Internet Explorer).

## Parts List — Storm Sewer Parts Lists

Parts list contains typical pipe network pipe and structure specs. If you are using a Civil 3D content template, there will be two parts lists: storm and sanitary. You may have to add to the lists or change their part values to match your typicals.

1. In Settings, expand Pipe Network until you are viewing the Parts Lists list.
2. From the list, select **Storm Sewer**, press the right mouse button, and from the shortcut menu, select EDIT....
3. Click the Pipes tab to view and, if necessary, expand Storm Sewer and the Concrete Pipe list.

This panel lists more concrete pipes sizes than you would use in a typical project. But over several projects, the list may be complete. You could define a rule for each size and type of pipe. If you are doing so, this is where rules are assigned.

4. Click the ***Structures*** tab.
5. If necessary, expand Storm Sewer and each part family to view its contents.

Each structure has a descriptive name and has an appropriate object style (see Figure 11.8).

6. Click ***OK*** to exit the Storm Sewer Parts List.

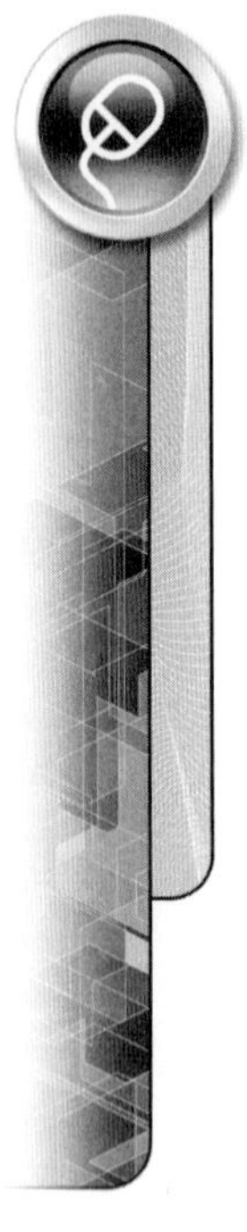

## Parts List — Sanitary Parts List

1. In the Pipe Network's Parts Lists, select **Sanitary Sewer**, press the right mouse button, and from the shortcut menu, select EDIT....
2. Click the ***Pipes*** tab and, if necessary, expand the Sanitary Sewer's PVC Pipe list.

The panel lists more PVC pipe sizes than would be typical for one project. But over several projects, the list may be complete. If you are defining a rule for each pipe size, this is the place to assign them.

3. Click the ***Structures*** tab and, if necessary, expand each Sanitary Sewer cylindrical parts family and review its contents.
4. Click ***OK*** to exit the Sanitary Sewer Parts List.

## Pipe Styles

1. In Settings, expand the Pipe branch until you are viewing the Pipe Styles list.
2. From the styles list, select **Single Line (Sanitary)**, press the right mouse button, and from the shortcut menu, select EDIT....
3. Click each tab to review its contents.
4. Click the ***Display*** tab to review component visibility settings.

This style draws only the pipe's centerline.

5. Click ***OK*** to exit the dialog box.
6. From the list, select **Double Line (Storm)**, press the right mouse button, and from the shortcut menu, select EDIT....
7. Click each tab to review its contents.
8. Click the ***Display*** tab to review component visibility settings.

The pipe's inside diameter draws the two lines.

9. Click ***OK*** to exit the dialog box.

## Pipe Rules

1. In Settings, expand the Pipe branch until you are viewing the Pipe Rule Set list.
2. From the list, select **Basic**, press the right mouse button, and from the shortcut menu, select EDIT....
3. Click the ***Rules*** tab.
4. If necessary, expand each rule section to review its values.

Cover and Slope set a pipe's minimum and maximum slope and cover. Pipe Length Check defines a pipe's shortest and longest length (see Figure 11.18).

5. Click ***OK*** to exit the dialog box.

## Pipe Label Styles

Pipe labels occur while you are drawing or after you have designed the network. Unit 4 of this chapter reviews these styles and how they are used.

## Structure Styles

Structure Styles define how storm and sanitary sewer systems display manholes, flared end sections, catch basins, etc., in Plan, Profile, and Section Views.

1. In Settings, expand the Structure branch until you are viewing the Structure Styles list.
2. From the styles list, select **Storm Sewer Manhole**, press the right mouse button, and from the shortcut menu, select EDIT....

3. Click the ***Model*** tab.

The catalog defines the structure's model shape, or you can choose a simpler shape.

4. Click the ***Plan*** tab.

This panel sets a structure's symbol. These settings also affect the symbol's size. A symbol can be an actual size, exaggerated, or resized each time the display area changes.

5. Click the ***Profile*** tab.

This panel sets how a structure is displayed in Profile View.

6. Click the ***Section*** tab to view its contents.

This panel sets how a structure is displayed in Section View.

7. Click the ***Display*** tab to review the component visibility settings.

This panel controls what components are visible for this structure in Plan, Model, Profile, or Section Views.

8. Click ***OK*** to exit the dialog box.

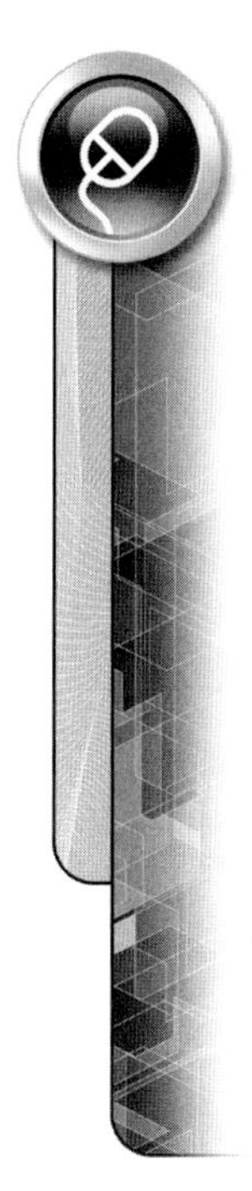

### Structure Rules

1. Expand Settings' Structure branch until you are viewing the Structure Rule Set list.
2. Select **Basic**, press the right mouse button, and from the shortcut menu, select EDIT....
3. Click the ***Rules*** tab.
4. If necessary, expand each section to view its contents.
5. Click ***OK*** to exit the dialog box.
6. At Civil 3D's top left, Quick Access Toolbar, click the ***Save*** icon to save the drawing file.

### Structure Label Styles

Label structures as you draw them or after you have created the network. Unit 4 of this chapter discusses these styles and their uses.

This completes the Pipe Networks settings review exercise. Pipe networks use an extensive styles library, rules, and catalog values. Next, you learn how to create pipe networks.

## SUMMARY

- Edit Drawing Settings sets the default layers, modifiers, and their values for pipes, structures, and pipe networks.
- Pipe Network's Edit Feature Settings values set initial styles, naming format, and labeling values for pipes and structures.
- If you want to label using different label anchor points, they must be changed in Pipe Network's Edit Feature Settings.
- A pipes and structures catalog provides the basic shapes and specs.
- Parts lists use values from a catalog, but those values are generally a subset of all possible sizes and types.
- A pipe rule affects slope, cover, and length.
- A structure rule affects invert drop, sump, and maximum connecting pipe sizes.
- Label pipe networks while you are drafting them or after you have drafted them.

## UNIT 2: CREATING A PIPE NETWORK

After defining styles, parts lists, and rules, you next create the pipe networks. You create pipe networks using two methods. Create Pipe Network From Objects creates a network from a selected drawing or Xref object. If you are selecting a line or arc, each selected line and arc becomes its own pipe network. If selecting a polyline, the entire polyline becomes the pipe network. Pipe Network Creation Tools displays a toolbar that has drafting and editing tools (see Figure 11.2).

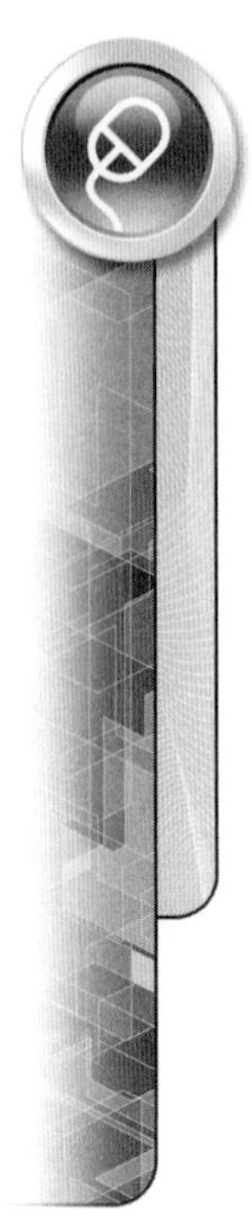

The Network Layout Tools toolbar lists the current parts list (storm, sewer, etc.), surface, and alignment (if referenced). Its middle portion lists the current structure type and pipe size. To the right of the pipe size is the drafting mode icon stack: Structure and Pipes, Pipes Only, and Structures Only. Structures Only locates structures with no connecting pipes and Pipes Only drafts pipes with no structures. Before you begin drafting, you must set the current slope. If it is set to up, then you are drafting from the lowest invert to the highest, and down from highest to lowest. All pipes and structures drafted with the toolbar active are members of the current network. If you have exited the toolbar, to continue working on the just-drawn network, you must edit it (Edit Network).

Pipes and structures receive names from the pipe network's Edit Feature Settings name template. It's easier to let Civil 3D name the pipes and structures and then edit their names in their Properties or vista. If the network is large, it may be confusing to edit in the Edit Network vista, because it shows all pipe or structure values. Without carefully tracking the network values, it is easy to make mistakes in the vistas.

### PIPE DRAFTING ICONS

When drafting a pipe network, you may want to connect a lateral pipe to an existing structure. When doing this, Pipes displays a starburst icon to the cursor's upper right. The starburst icon indicates, if you select the point, the pipe will attach to the structure.

When you want to break a pipe to attach to it, Pipes displays an "opposing fists" icon to the cursor's upper right. This icon indicates that selecting the point will break the pipe and either attach a pipe, the pipe (a pipe-pipe intersection), or place a structure at the selected point connecting the points.

### EXERCISE 11-2

After completing this exercise, you will:

- Be able to draft a pipe network.
- Be able to set structures and add pipes.

### Exercise Setup

This exercise continues with the previous exercise's drawing, Pipe Network-1. If you did not complete the previous exercise, browse to the Chapter 11 folder of the CD that accompanies this textbook and open the *Chapter 11 - Unit 2.dwg* file.

1. If not open, open the previous exercise's drawing or browse to the Chapter 11 folder of the CD that accompanies this textbook and open the *Chapter 11 - Unit 2* drawing.

## Existing Storm (93rd Street)

Along the centerline of 93rd Street is an existing storm sewer line (see Figure 11.26). Table 11.1 contains structure names and types for the network. The network starts at 93rd Street's eastern end and is drawn upslope.

**TABLE 11.1**

| Manhole # | Type | Material |
|---|---|---|
| EX-Stm-01 | Concentric 72" | Reinforced Concrete |
| EX-Stm-02 | Concentric 72" | Reinforced Concrete |
| EX-Stm-03 | Concentric 72" | Reinforced Concrete |
| EX-Stm-04 | Concentric 48" | Reinforced Concrete |

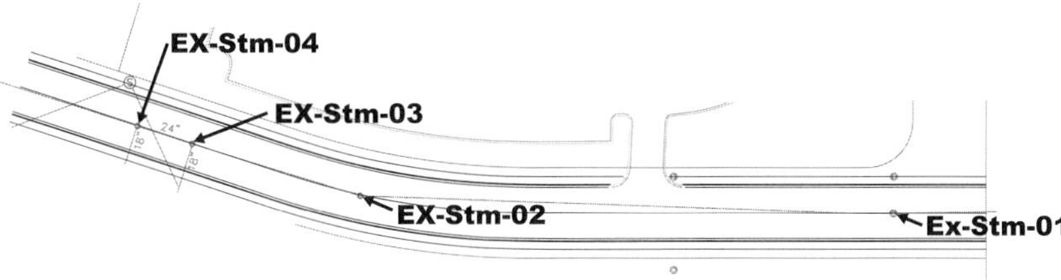


**FIGURE 11.26**

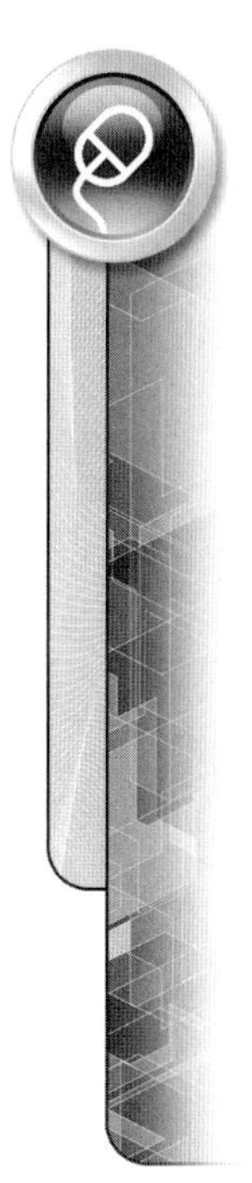

This pipe network has several different pipe sizes. The pipe size from EX-Stm-03 to EX-Stm-01 is 36". The pipe size between EX-Stm-04 and EX-Stm-03 is 24". Both EX-Stm-04 and EX-Stm-03 have 18" laterals feeding their barrels. Drawing this pipe network requires changing drafting modes and pipe sizes.

1. Use the ZOOM and PAN commands to view 93rd Street's easterly end and EX-Stm's first manhole.

The 93rd Street storm sewer uses a special rule, Storm Interceptor. This rule allows the network to be deeper than other storm systems (i.e., parking lot catch basins, etc.). You must set the rule before you define the next network. Otherwise, you have to apply a rule to each existing pipe.

2. If necessary, click the ***Settings*** tab.
3. In Settings, select the **Pipe Network** heading, press the right mouse button, and from the shortcut menu, select EDIT FEATURE SETTINGS....
4. If necessary, expand the Default Rules section. For Pipe Default Rules, click in the Value cell to display an ellipsis, and click the ellipsis to open the Pipe Default Rules dialog box.
5. In the Pipe Default Rules dialog box, click the drop-list arrow, from the list, select **Storm Interceptor**, and click ***OK*** until you have returned to the command line.
6. Click the ***Prospector*** tab.
7. On the Home tab, Create Design panel, click the ***Pipe Network*** icon, and from the shortcut menu, select PIPE NETWORK CREATION TOOLS to open the Create Pipe Network dialog box.
8. Using Figure 11.27 as a guide, for the pipe network name, enter **EX-STM**, keeping the counter. For the Network parts list, select the **Storm Sewer**, set the Surface name to **Proposed**, set the Alignment name to **93rd – (1)**, and click the ***OK*** button to exit the dialog box and display the Network Layout Tools toolbar.

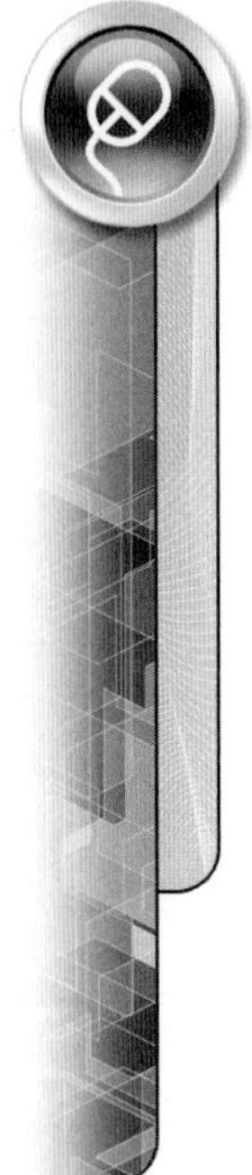

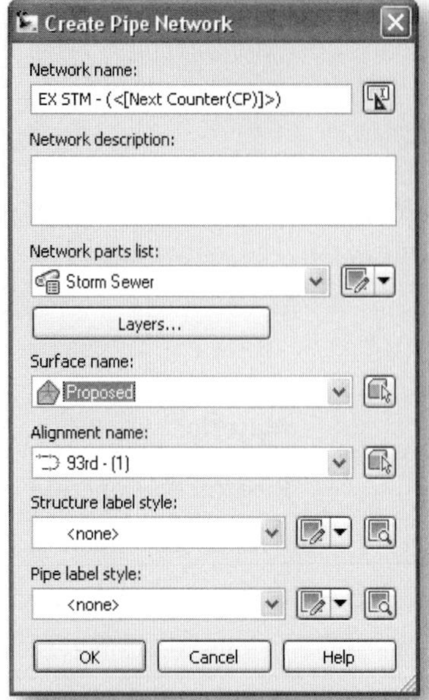


**FIGURE 11.27**

Along the toolbar's bottom, Parts List should be Storm Sewer, the Surface should be Proposed, and the Alignment should be 93rd - (1). Use Figure 11.26 as a guide for drafting this network.

9. At the toolbar's center, click the Structure drop-list arrow, and from the Concentric Cylindrical Structure parts list, select the **Concentric Structure 72" dia**.
10. At the toolbar's right of center, click the Pipe Size drop-list arrow, and from the list, select **36 inch RCP**.
11. In the toolbar, click the ***Upslope/Downslope*** icon (the second to the right of the pipe size) and set the design to **Upslope**.
12. In the toolbar, from the Drafting Mode icon stack, click the drop-list arrow, and from the list, select PIPES AND STRUCTURES.
13. Referencing Figure 11.26 and using a **Center** object snap, in the drawing, place the easternmost manhole (EX-Stm-01).
14. In the drawing, also using the Center object snap, locate the structures EX-Stm-02 and EX-Stm-03.
15. In the toolbar's center, click the Structure drop-list arrow, and from the Concentric Cylindrical Structure parts list, select **Concentric Structure 48" dia**.
16. In the toolbar, click the Pipe Sizes drop-list arrow, and from the pipe size list, select **24 inch RCP**.

Changing pipe sizes does not interrupt the drafting process.

17. Using the **Center** object snap, in the drawing, locate EX-Stm-04.
18. In the toolbar, click the Pipe Sizes drop-list arrow, and from the list, select **18 inch RCP**.
19. In the toolbar, click the Drafting Mode icon stack drop-list arrow, and from the list, select PIPES ONLY.

This breaks the cursor's connection to EX-Stm-04. To correctly draft the pipe, you need to reconnect the 18" pipe to EX-Stm-04 before you draft the pipe.

20. In the drawing, place the cursor near EX-Stm-04. A structure connection icon appears. Click near the EX-Stm-04 structure to start drafting the pipe.

21. In the drawing, click a point to the west of the structure as the pipe end.
22. To indicate that you are drafting a new pipe segment connecting to a different structure, click the ***Pipe Only*** drafting mode icon.
23. In the drawing, place the cursor near EX-Stm-04. The attach structure icon appears. Select a point near the structure and draw an 18" pipe EX-Stm-04 toward the south past the sidewalk polyline line.
24. To start a new pipe segment, click the ***Pipe Only*** drafting mode icon to place a new 18" pipe segment from EX-Stm-03 toward the south.
25. In the drawing, place the cursor at EX-Stm-03. A structure connection icon appears. Click near the structure to start drawing the pipe and select a second point south of the structure as the opposite end of the pipe.
26. Use the PAN command to view the EX-Stm-01 structure.
27. In the toolbar, click the Pipe Size drop-list arrow, and from the list, select **36 inch RCP**.
28. In the toolbar, click the ***Upslope/Downslope*** icon and set the mode to **Downslope**.
29. To draft a new pipe segment, in the toolbar, click the ***Pipe Only*** drafting mode icon to place a new 36" pipe segment from EX-Stm-01 toward the east.
30. In the drawing, place the cursor at EX-Stm-01. A structure connection icon appears. Click near the structure to start drawing the pipe, and select a second point east of the structure (near the end of the corridor's centerline) as the opposite end of the pipe.
31. Click the toolbar's red **X** to close it.
32. At Civil 3D's top left, Quick Access Toolbar, click the ***Save*** icon to save the drawing.

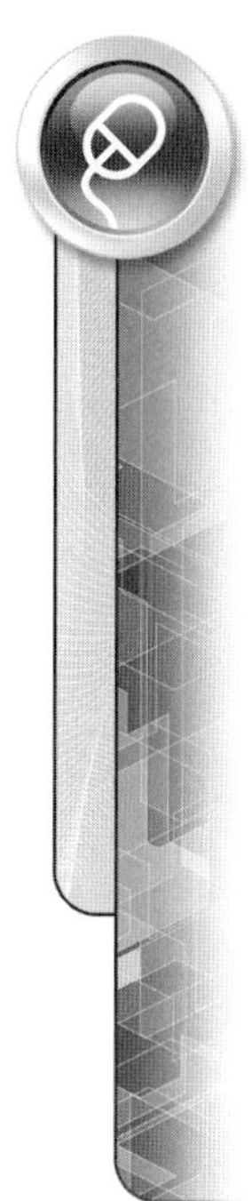

## Editing a Pipe Network

This simplest way to change structures and pipes names is by editing their properties, rather than trying the vista.

1. In the drawing, select each structure and edit its properties, changing each structure's name using the entries in Table 11.1.
2. In the drawing, select each pipe and edit its properties, changing each pipe's name using the entries in Table 11.2.
3. At Civil 3D's top left, Quick Access Toolbar, click the ***Save*** icon to save the drawing.

**TABLE 11.2**

| Pipe Name | From/to Manholes | Size |
|---|---|---|
| EX-Stm-P01 | East end of 93rd to EX-Stm-01 | 36 |
| EX-Stm-P02 | EX-Stm-01 to EX-Stm-02 | 36 |
| EX-Stm-P03 | EX-Stm-02 to EX-Stm-03 | 36 |
| EX-Stm-P04 | EX-Stm-03 to EX-Stm-04 | 24 |
| EX-Stm-P05 | EX-Stm-04 to West end of 93rd | 18 |
| EX-Stm-P06 | EX-Stm-03 to south of sidewalk | 18 |
| EX-Stm-P07 | EX-Stm-04 to south of sidewalk | 18 |

## Catch Basin Network

The last network is a series of catch basins and pipes with an outfall FES to a pond at the site's southeast corner. The drafting strategy for this network is to first place a 48"×48" rectangular structure at the parking lot's southeast corner next to the detention pond. After placing the structure, you next locate the three catch basins in the western part of the parking lot. After you locate the three western catch basins, you next locate a catch basin in the parking lot's northeast. All catch basins connect to the southeast manhole. Finally, you connect a pipe to the manhole and locate the FES in the detention pond. Use Figure 11.28 and Table 11.3 as guides for placing and naming the various network elements.

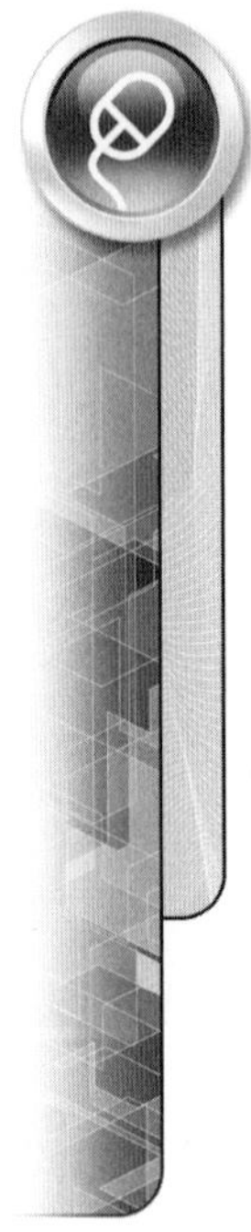

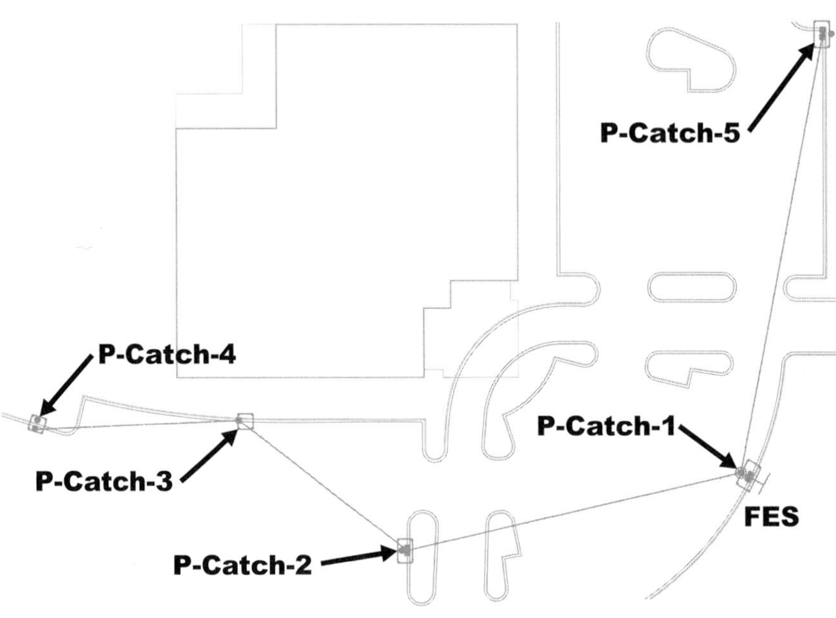


**FIGURE 11.28**

**TABLE 11.3**

| Structure | Type | Pipe Size | Pipe Name | Material |
|---|---|---|---|---|
| P-Catch-1 | 48"×48" Rectangular | 12" Concrete | P-Catch-P1 | Reinforced Concrete |
| P-Catch-2 | 24"×24" Rectangular | 12" Concrete | P-Catch-P2 | Reinforced Concrete |
| P-Catch-3 | 18"×18" Rectangular | 12" Concrete | P-Catch-P3 | Reinforced Concrete |
| P-Catch-4 | 15"×15" Rectangular | 12" Concrete | P-Catch-P4 | Reinforced Concrete |
| P-Catch-5 | 24"×24" Rectangular | 12" Concrete | P-Catch-P5 | Reinforced Concrete |
| FES | 44"×44" Rectangular Headwall | 24" Concrete | P-Catch-P6 | Reinforced Concrete |

## Adding a Part Family and Part to the Storm Sewer Parts List

The storm sewer parts list does not contain all the required structures. The parts need to be added to the storm sewer parts list.

1. Click the ***Settings*** tab.
2. In Settings, expand the Pipe Network branch until you are viewing the Parts Lists list.

3. Select the **Storm Sewer** parts list, press the right mouse button, and from the shortcut menu, select EDIT....
4. In the Network Parts List - Storm Sewer dialog box, click the ***Structures*** tab.
5. In the Structures tab, at the top left, select the **Storm Sewer** heading, press the right mouse button, and from the shortcut menu, select ADD PART FAMILY....
6. In the Part Catalog dialog box, toggle ON **Rectangular Structure Slab Top Rectangular Frame**, and click ***OK*** to return to the Network Parts List dialog box.
7. From the list of parts, select **Rectangular Structure Slab Top Rectangular Frame**, press the right mouse button, and from the shortcut menu, select ADD PART SIZE....
8. If necessary, change the Inner Structure Width and Length to **15"** and click ***OK*** to add the part. Use Figure 11.29 as a guide.
9. Repeat the previous two steps and add the following sizes: **18×18**, **24×24**, and **48×48**.

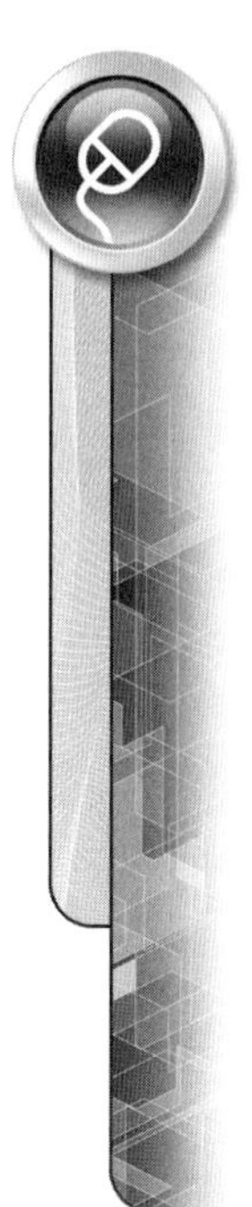

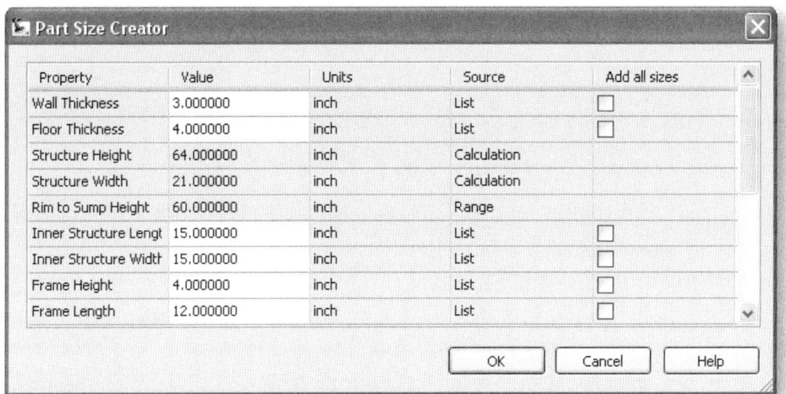

**FIGURE 11.29**

The style for the new structures is storm sewer manhole.

10. In the Part List dialog box, click the ***Style*** icon to the right of Rectangular Structure Slab Top Rectangular Frame to open the Structure Style dialog box. Click the drop-list arrow, from the style list, select **Catch Basin**, and click ***OK*** to return to the Network Parts List dialog box.
11. Click ***OK*** to exit the Network Parts List - Storm Sewer dialog box.

## Drafting the Catch Basin Network

1. Use the ZOOM and PAN commands to view the entire parking lot.
2. In the Layer Properties Manager, toggle **ON** the layer **PROP-STM-LIN** and click **X** to exit.
3. From the Ribbon's Home tab, Create Design panel, click the Pipe Network icon, and from the shortcut menu, select PIPE NETWORK CREATION TOOLS.
4. In the Create Pipe Network dialog box, for the Network name, enter **P-Catch**, leaving the counter, set the Network parts list to **Storm Sewer**, and set the Surface name to **Proposed**. Use Figure 11.30 as a guide.

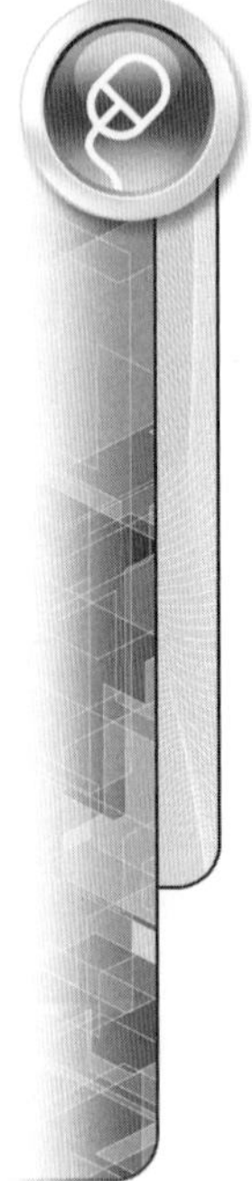

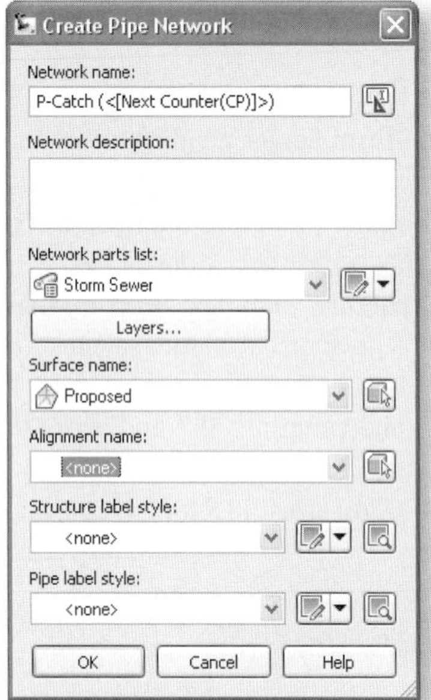


**FIGURE 11.30**

5. Click ***OK*** to exit the dialog box and display the pipe Network Layout Tools toolbar.
6. In the toolbar's center, click the Structure drop-list arrow, and from the Rectangular Structure Slab Top Rectangular Frame section, select the **48"×48" Rect Structure**.
7. In the toolbar, set ***Upslope/Downslope*** to **Upslope** (the second icon to the right of the pipe size).
8. In the toolbar, click the ***Drafting Mode*** icon stack drop-list arrow (to the right of the pipe size), and from the list, select STRUCTURES ONLY.
9. Using Figure 11.28 as a guide, in the drawing at the southeast parking lot's corner, place P-Catch - 1.
10. In the toolbar, click the Structure drop-list arrow, and from the Rectangular Structure Slab Top Rectangular Frame parts list, select the **24"×24" Rect Structure**.
11. In the toolbar, click the Pipe Size drop-list arrow, and from the list, select **18 inch RCP**.
12. In the toolbar, click the ***Drafting Mode*** icon stack drop-list arrow (to the right of the pipe size), and from the list, select PIPES AND STRUCTURES.
13. In the drawing, place the cursor near the structure you just placed. The connect to structure icon appears. Select the point to attach the pipe.
14. Use the PAN command toward the west and locate and place the P-Catch - 2 structure.
15. In the toolbar, click the **Structure** drop-list arrow, and from the Rectangular Structure Slab Top Rectangular Frame parts list, select the **18"×18" Rect Structure**.
16. In the toolbar, click the **Pipe Size** drop-list arrow, and from the list of sizes, select **15 inch RCP**.
17. In the drawing, PAN west and locate and place structure P-Catch - 3.
18. After placing the structure, in the toolbar, click the **Structure** drop-list arrow, from the Rectangular Structure Slab Top Rectangular Frame parts list, select

the **15"×15" Rect Structure**, and PAN west and locate and place structure P-Catch - 4.

19. In the toolbar, click the **Structure** drop-list arrow, and from the Rectangular Structure Slab Top Rectangular Frame parts list, select the **24"×24" Rect Structure**.
20. In the toolbar, click the **Pipe Size** drop-list arrow, and from the list of sizes, select **18 inch RCP**.
21. In the toolbar, click the ***Drafting Mode*** icon stack, and from the list, indicate the start of a new branch to the network by selecting PIPES AND STRUCTURES.
22. If necessary, use the PAN and ZOOM command to view the structure P-Catch - 1.
23. In the drawing, place the cursor near P-Catch – 1, and when the connect to structure icon appears, select the point.
24. Use the PAN command to pan north, locate and place the structure P-Catch - 5.
25. In the toolbar, click the **Structure** drop-list arrow, and from the list in the Concrete Rectangular Headwall part family, select **Headwall 44"×6"×44"**.
26. In the toolbar, click the **Pipe Size** drop-list arrow, and from the list of sizes, select **30 inch RCP**.
27. In the toolbar, click the ***Downslope/Upslope*** toggle to set it to **Downslope**.
28. In the toolbar, click the ***Drafting Mode*** icon.
29. In the drawing, place the cursor at P-Catch - 1. When the connect to structure icon appears, select the point and select a second point to the southeast in the detention pond to place the FES.
30. Click the Network Layout Tools toolbar's red **X** to close it.
31. At Civil 3D's top left, Quick Access Toolbar, click the ***Save*** icon to save the drawing.

This completes the pipe networks drafting exercise.

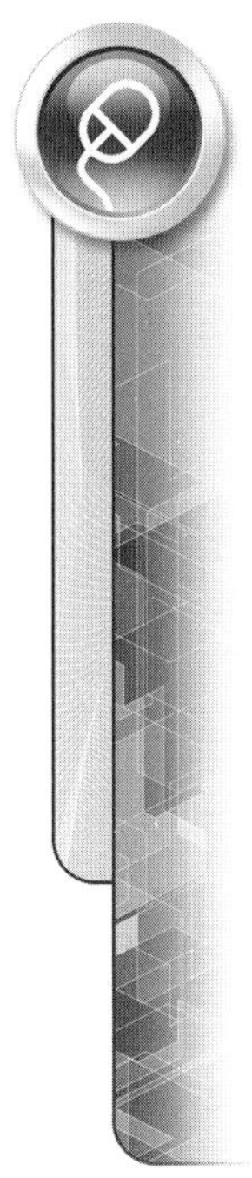

## SUMMARY

- A pipe network uses a parts list to set typical network structure and pipe sizes.
- If you have prematurely closed the pipe Network Layout Tools toolbar and you want to continue working with the pipe network, use Edit Network to continue working on the network.
- You can connect between different pipe networks.
- The current drafting mode remains active even when you change structure and/or pipe size or Upslope/Downslope.

## UNIT 3: PIPE NETWORKS REVIEW AND EDIT

Besides adding to an existing pipe network, other situations require network pipe or structure edits (for example, changing rules, resizing a pipe or structure, swapping structures or pipes, changing styles, etc.). Graphical editing includes moving, disconnecting and reconnecting elements, and deleting structures and pipes. Interference checks indicate areas that need to be edited.

When you select a pipe or a structure and press the right mouse button, a context-sensitive shortcut menu is displayed (see Figure 11.4). In Figure 11.4, the shortcut menu on the left is displayed when you select a pipe segment. Also in Figure 11.4, the right shortcut menu is displayed when you select a structure. Both menus include Edit Network…, which displays the pipe Network Layout Tools toolbar.

Reports Manager or Toolbox contain several pipe network reports: Pipes and Structures in HTML and CSV formats.

## NETWORK PROPERTIES

Pipe Network Properties include the original layout settings (see Figure 11.31), what labeling styles annotate the pipe network in a profile (see Figure 11.32), the layers the network uses in section views, and a review of the network statistics.

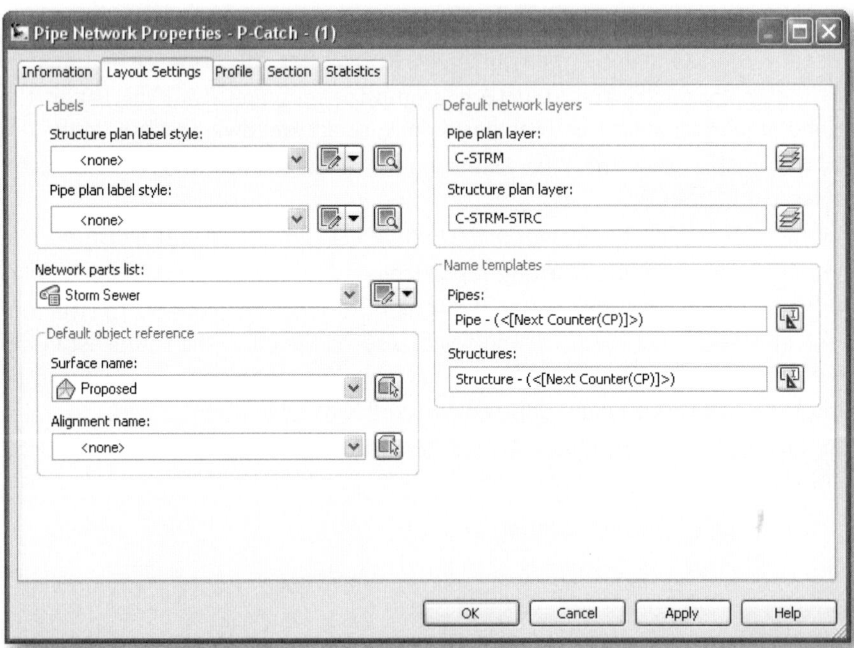


**FIGURE 11.31**

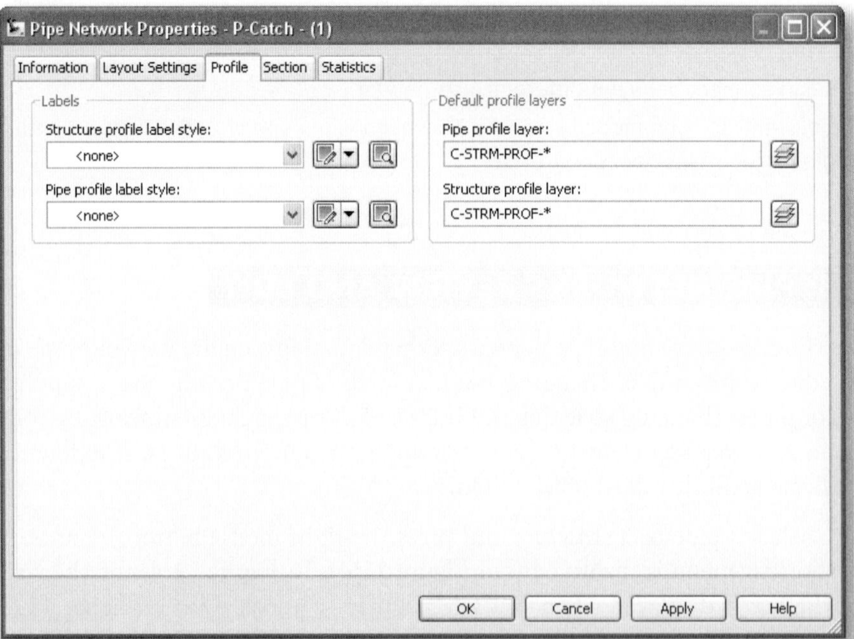


**FIGURE 11.32**

## PIPE PROPERTIES

When you select a pipe segment and, from a shortcut menu, you select Pipe Properties..., a Pipe Properties dialog box opens and has pertinent pipe segment values. The Information tab renames and reassigns the object style or the render material. The Part Properties panel displays inverts and part data, and sets the resize rule (see Figure 11.33). Any value in black is editable. The Rules panel displays the currently assigned rules and informs you if the pipe violates any rules (see Figure 11.34). If a rule is not met, an icon is displayed next to the broken rule.

Pipe Properties edits apply only to the selected pipe. Swap Part changes only the pipe's size. These edits do not allow changing pipe types and should be done as a network edit.

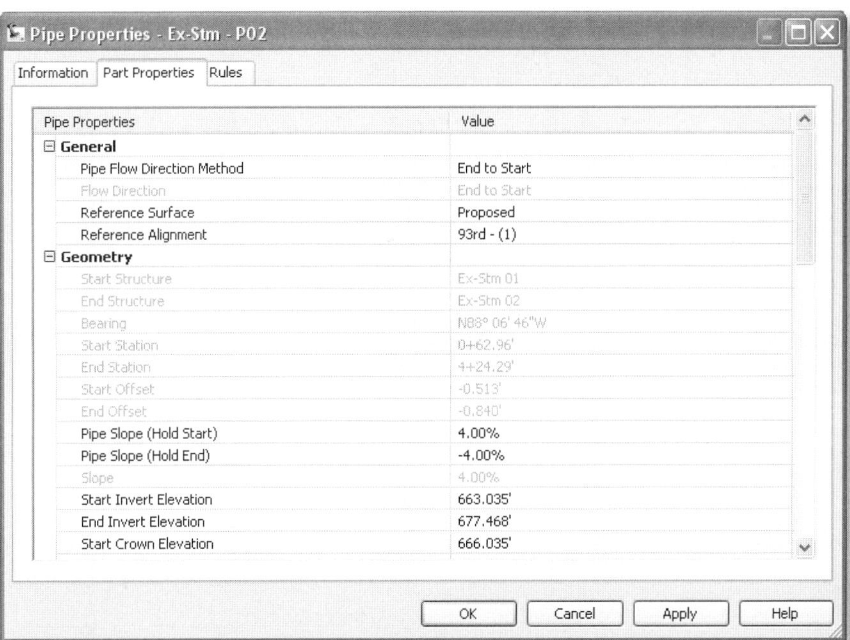


**FIGURE 11.33**

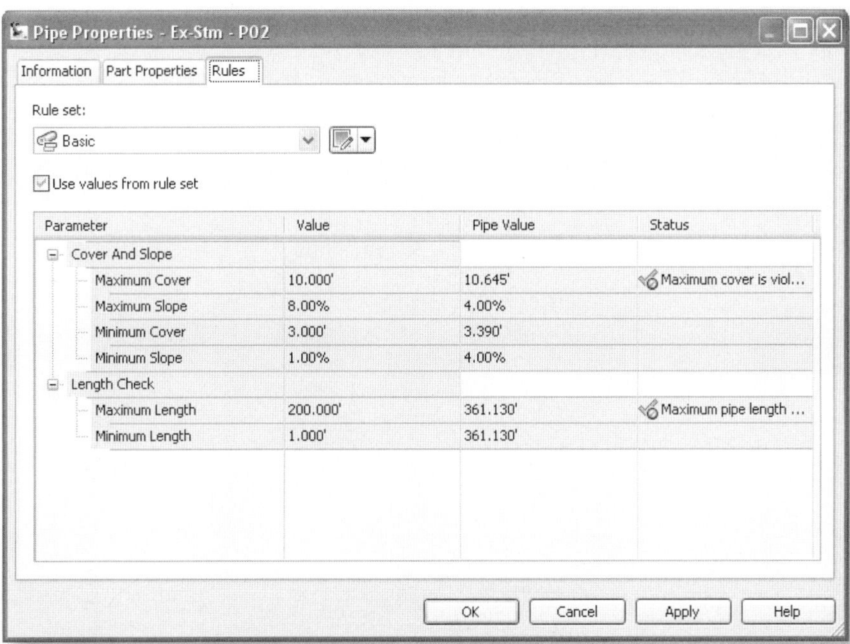


**FIGURE 11.34**

## STRUCTURE PROPERTIES

Structure Properties shows a structure's values. The Information tab renames or reassigns the object style or the render material. Part Properties displays information about the part and any calculated values (see Figure 11.35). Any value in black is editable.

Connected Pipes displays pipes connected to the structure, their rules violations, their inverts, and other critical information (see Figure 11.36).

Rules display the current rules and inform you if the structure violates any of them (see Figure 11.37). If a rule is not met, an icon is displayed next to the broken rule.

Structure Properties edits apply only to the selected structure. Swap Part changes the structure type and/or size.

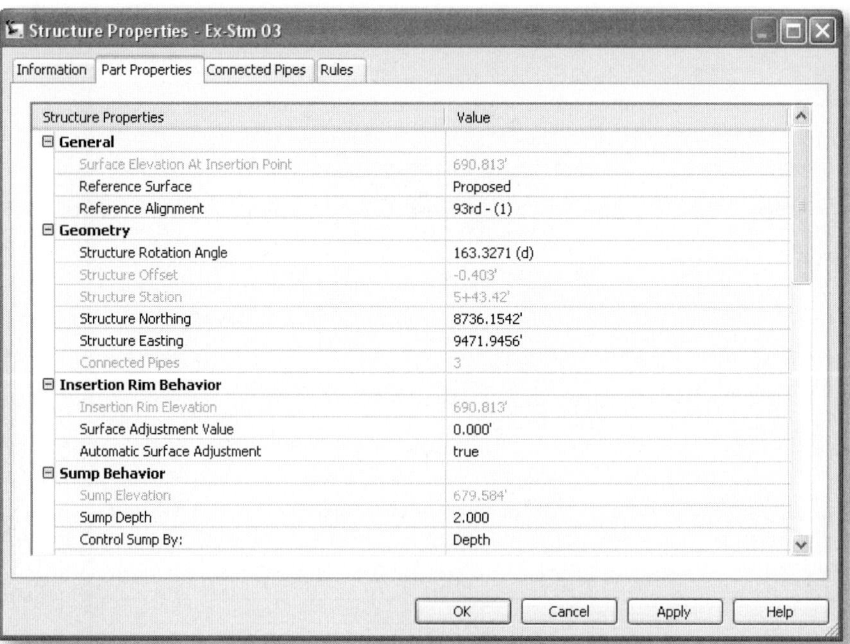


**FIGURE 11.35**

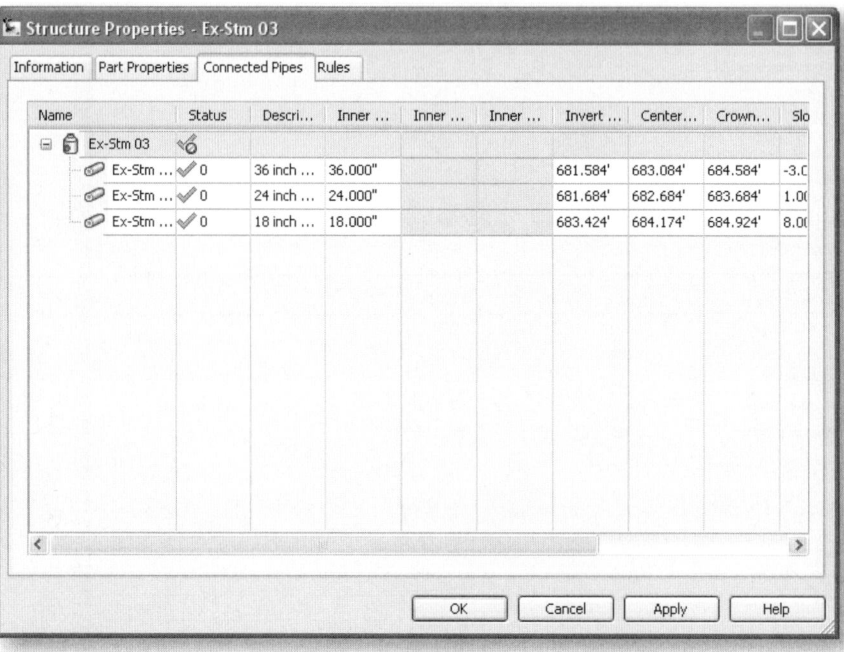


**FIGURE 11.36**

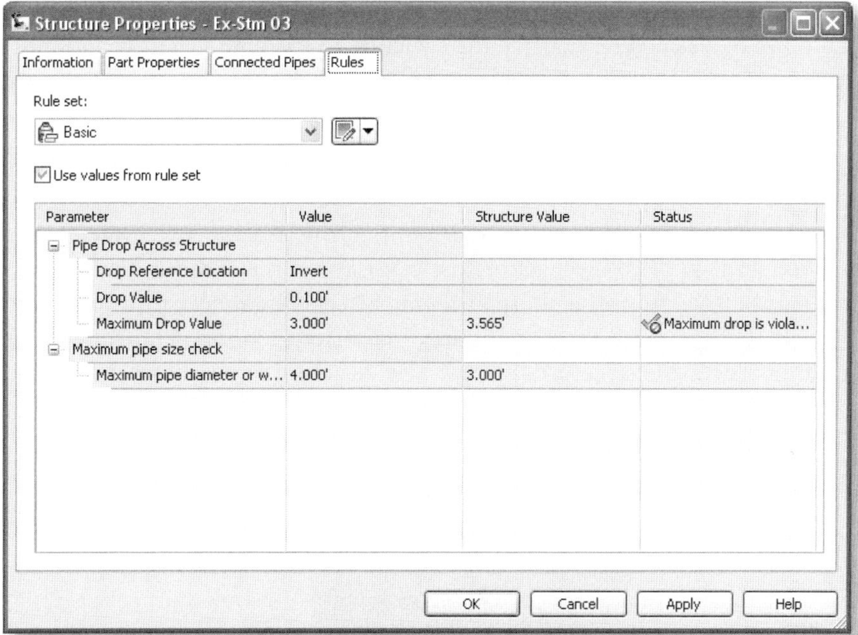


**FIGURE 11.37**

## SWAP PART: PIPE AND STRUCTURE

Swap Part exchanges the currently selected object with another object from a list of applicable choices (see Figure 11.38). Swap Part does not allow any changes to a pipe's type. Swap Part for a selected structure allows changes to the part type (headwall to eccentric cylindrical).

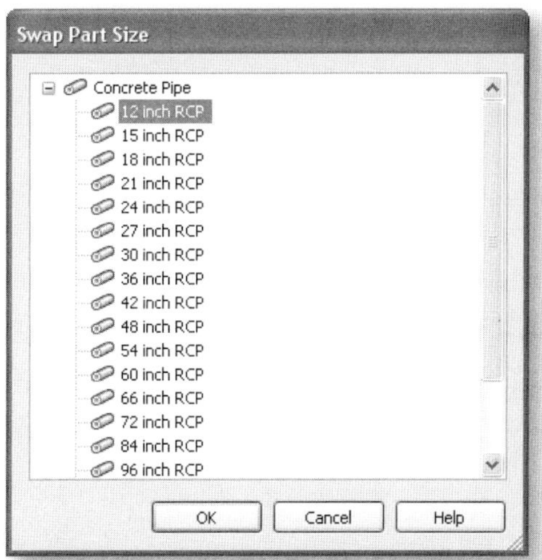


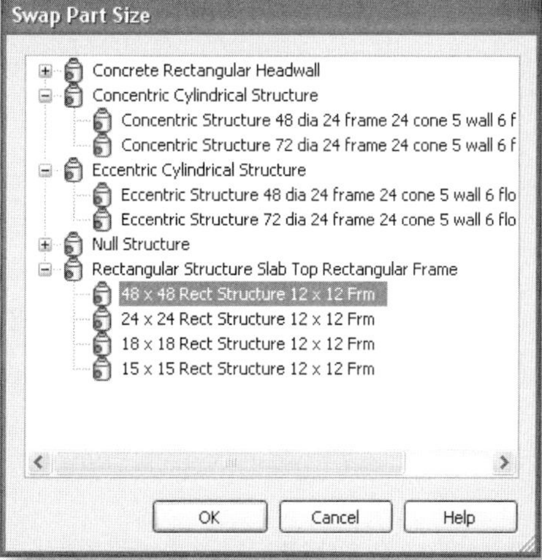


**FIGURE 11.38**

## GRAPHICAL EDITING

When you graphically edit a pipe network, pipes and structures are treated as connected objects. By selecting a structure, activating its grip, and relocating it, the attached pipes are relocated.

If you are grip editing pipes and you also want to move their attached structure (see Figure 11.39), the selection must include the structures. If you do not select the structures, and if the pipes are moved, then they are disconnected from the structure. To reattach pipes to a structure, relocate the structure to the pipe's vicinity and individually reconnect the pipes. To reattach a pipe, you click the pipe, press the right mouse button, and from the shortcut menu, select Connect to part. The routine prompts you for a connecting structure. In the drawing, select the appropriate structure.

To disconnect a pipe from a structure, activate its grips and move it away from a structure. If you want to disconnect a structure from a pipe, select the structure, press the right mouse button, from the shortcut menu, select Disconnect from part, and select a pipe. If there is more than one pipe, disconnect each pipe and then move the structure, or delete the structure, place a new structure in the desired location, and reconnect the pipes.

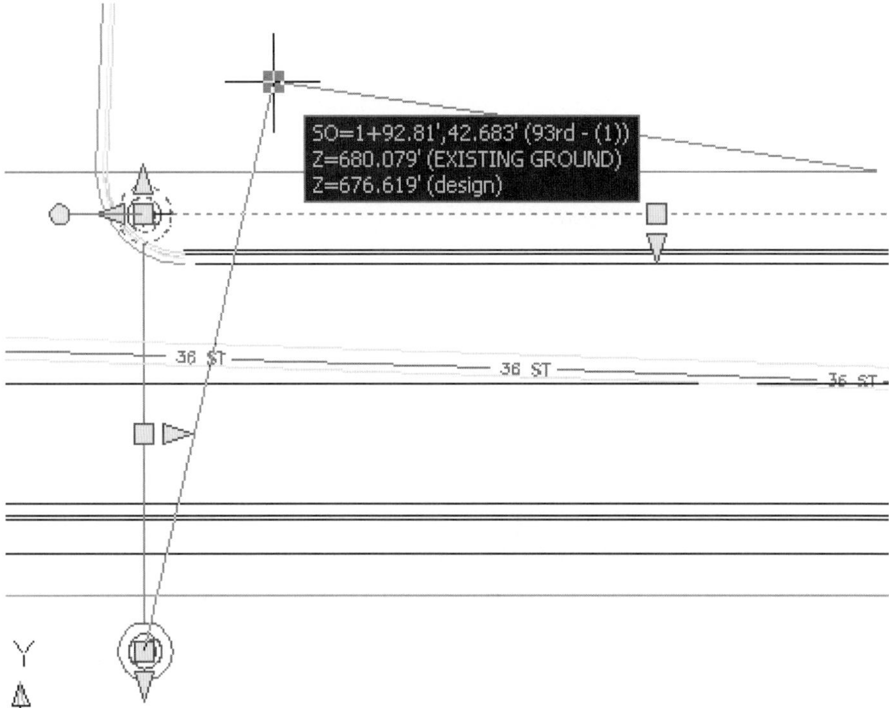


**FIGURE 11.39**

### CONNECT AND DISCONNECT PART

Connect and Disconnect Part separates a pipe or structure from adjacent pipe network objects. This is useful when you are inserting new structures and pipes in existing networks or when you are graphically editing network components.

### PROFILE EDITING

Pipes and structures that appear in a profile view can represent a selection set or an entire pipe network. All pipes and structures display their properties and can be graphically edited.

### INTERFERENCE CHECKING

When pipes cross, there is potential for them to intersect or not have enough vertical separation. Interference Check analyzes the separation between pipes of different networks or selected pipe segments (see Figure 11.40). Distance checking is a user-specified value or scale factor. Create Interference Check sets this value in the 3D proximity check Criteria dialog box (see Figure 11.41).

If you are editing pipe networks that are participating in an interference check, the interference goes out-of-date. To recalculate the interferences, click the interference, press the right mouse button, and from the shortcut menu, select Rerun Interference Check....

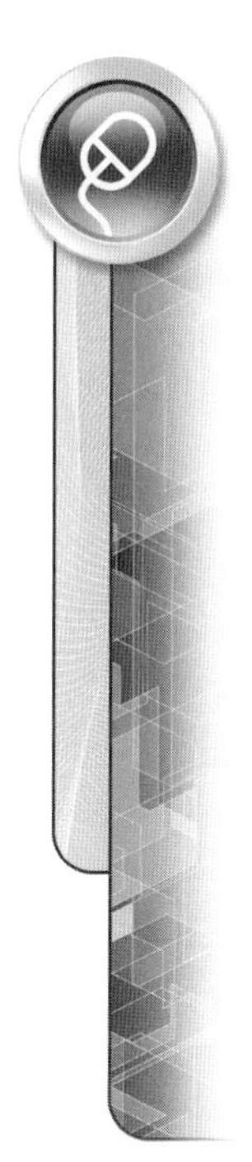

**FIGURE 11.40**

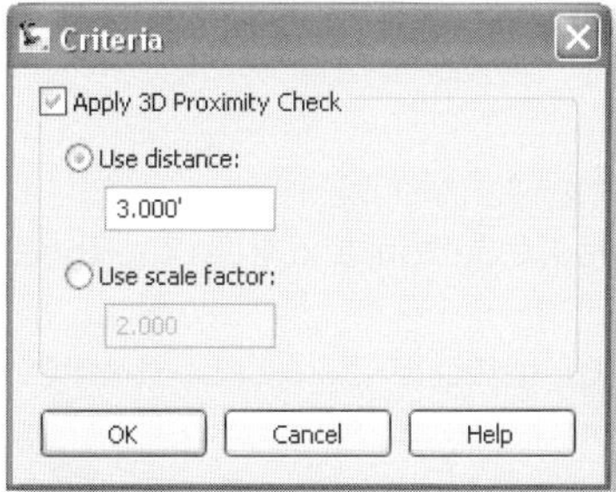


**FIGURE 11.41**

## EXERCISE 11-3

After completing this exercise, you will:

- Be able to review and edit pipe network properties.
- Be able to graphically edit pipe network segments.
- Be familiar with pipe or structure properties and be able to edit them.
- Be familiar with pipes or structures and be able to swap them out.

## Exercise Setup

This exercise continues with the previous exercise's drawing. If you did not complete the previous exercise, browse to the Chapter 11 folder of the CD that accompanies this textbook and open the *Chapter 11 - Unit 3.dwg* file.

1. If not open, open the previous exercise's drawing or browse to the Chapter 11 folder of the CD that accompanies this textbook, and open the *Chapter 11 - Unit 3* drawing.

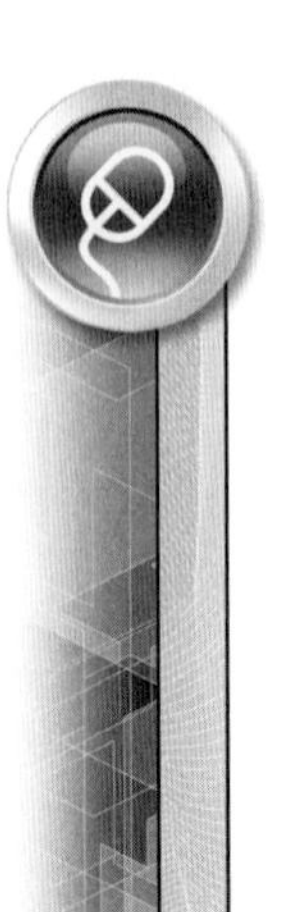

## Pipe Network Properties

Pipe Network Properties lists assigned values when the network is created.

1. Use the ZOOM and PAN commands until you are viewing the EX-Stm - (1) pipe network.
2. In the drawing, select a pipe or structure, press the right mouse button, and from the shortcut menu, select NETWORK PROPERTIES....
3. Click each tab and review its settings and values.
4. Click ***OK*** to exit the Pipe Network Properties dialog box.

## Structure Properties

Structure Properties displays structure information. You can adjust structure parameters, the structure's name, and review connected pipes.

1. If necessary, click the ***Prospector*** tab.
2. In Prospector, expand the Pipe Networks branch until you are viewing the EX-STM – (1) pipe network's Pipes and Structures.
3. Click **Structures** to list the structures in preview.
4. In Preview, select **EX-Stm-01**, press the right mouse button, and from the shortcut menu, select STRUCTURE PROPERTIES....
5. Click each tab to review the structure's settings and values, and click ***OK*** to exit the dialog box.

## Swap Parts

Swap pipes or structures to larger or smaller sizes.

1. In Preview, select **EX-Stm-01**, press the right mouse button, and from the shortcut menu, select ZOOM TO.
2. In the drawing, at the eastern side of EX-Stm-01, select the pipe (EX-Stm-P01), press the right mouse button, and from the shortcut menu, select SWAP PART... to display a pipe size list.
3. From the list, select a **48 inch RCP**, and click ***OK*** to swap the part.
4. At Civil 3D's top left, Quick Access Toolbar, click the ***Save*** icon to save the drawing.

## Profile Review of EX-STM – (1)

Civil 3D drafts some or all of a pipe network in Profile View. The 93rd Street storm system needs to be reviewed in Profile View.

1. Click Ribbon's ***Modify*** tab. In the Design panel, click the Pipe Network icon. From the Pipe Networks tab, on the Network Tools panel's right side, select DRAW PARTS IN PROFILE.

2. The command line prompts you for a network. In the drawing, select a pipe from the EX-STM – (1) network and press ENTER. The command line prompts you to select a Profile View. Use the Pan command until you are viewing the 93rd Street Profile View, and then select it.

The EX-STM – (1) pipe network is drawn in the Profile View.

## Edit Pipes in Profile View

Using Table 11.4 as a guide, edit each pipe's start and end inverts in each pipe's Pipe Properties dialog box.

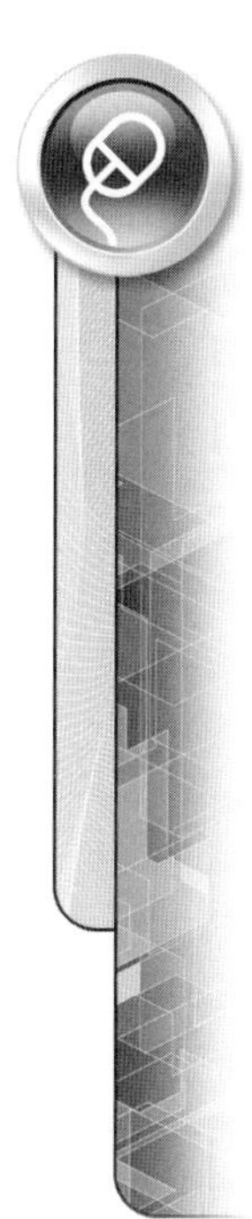

**TABLE 11.4**

| Pipe | Start Invert | End Invert |
|---|---|---|
| EX-Stm-P01 | 669.00 | 668.00 |
| EX-Stm-P02 | 669.50 | 675.00 |
| EX-Stm-P03 | 675.50 | 679.00 |
| EX-Stm-P04 | 680.00 | 681.00 |
| EX-Stm-P05 | 681.25 | 682.00 |
| EX-Stm-P06 | 681.50 | 684.00 |
| EX-Stm-P07 | 681.25 | 682.00 |

3. In the 93rd Street Profile View, select the pipe EX-Stm-P01, press the right mouse button, and from the shortcut menu, select PIPE PROPERTIES....
4. In the Pipe Properties dialog box, click the ***Part Properties*** tab, scroll down to the Start and End Invert Elevation entries, and using the values in Table 11.4, change the inverts for pipe EX-Stm-P01. Click ***OK*** to exit the Pipe Properties dialog box.
5. Repeat the previous two steps and edit the inverts for the EX-STM – (1) pipe network (EX-Stm-P02...P5).
6. In the drawing, select any entity that represents EX-STM – (1), press the right mouse button, and select from the shortcut menu, EDIT NETWORK....
7. In the Network Layout Tools toolbar, click the ***PIPE NETWORK VISTAS*** icon (right side).
8. Click the ***Pipes*** tab and edit the Start and End Inverts for EX-Stm-P06 and EX-Stm-P07 using the values in Table 11.4.
9. After editing the inverts, click the **X** to close the vistas and Network Layout Tools toolbar.
10. At Civil 3D's top left, Quick Access Toolbar, click the ***Save*** icon to save the drawing.

## Intersection Between EX-STM – (1) and EX-San – (1)

The EX-San – (1) pipe network crosses the EX-STM – (1) pipe network. The intersection needs to be checked for interferences.

1. Use the PAN and ZOOM commands to view the intersection of EX-San – (1) and EX-STM – (1).
2. If necessary, click the ***Prospector*** tab.
3. Expand the Pipe Networks branch until you view the Interference Checks heading.

4. Select **Interference Checks**, press the right mouse button, and from the shortcut menu, select CREATE INTERFERENCE CHECK....
5. The command line prompts you to select a part from the first network. In the drawing, select a pipe from the **EX-San – (1)** network.
6. The command line prompts you to select a part from the second network. In the drawing, select a pipe from the **EX-STM – (1)** network.

The Create Interference Check dialog box opens.

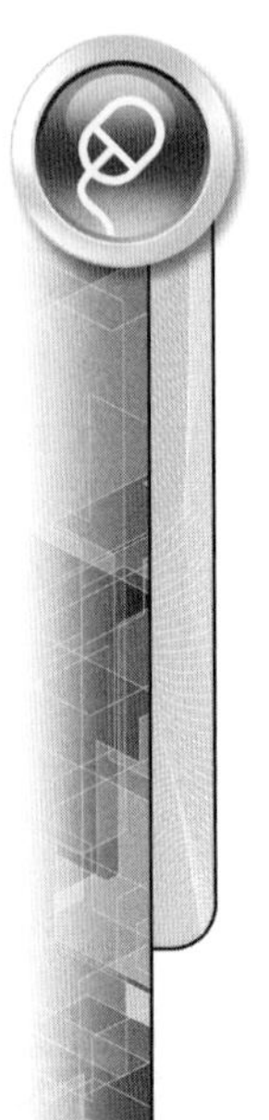

7. In the Create Interference Check dialog box, click ***3D Proximity Check Criteria...***.
8. In the Criteria dialog box, toggle **ON** Apply 3D Proximity Check, set the distance to **5.0**, and click ***OK*** to return to the Create Interference Check dialog box.
9. Click ***OK*** to exit Create Interference Check.

There are three interferences.

10. Click the ***OK*** button to close the interference report.

## Viewing a Pipe Network Model

Viewing a network with 3D Orbit or Object Viewer produces a 3D view of selected components.

1. In the drawing, select the interference symbols, pipes, and structures of the two pipe networks, press the right mouse button, and from the shortcut menu, select OBJECT VIEWER....
2. Rotate the model until you are viewing the network objects and interferences.
3. After reviewing the network and interferences, close the Object Viewer.
4. Use the ZOOM and PAN commands to view the P-Catch network in the parking lot.
5. In the drawing, select all of the P-Catch elements, press the right mouse button, and from the shortcut menu, select OBJECT VIEWER....
6. After reviewing the network, close the Object Viewer.
7. At Civil 3D's top left, Quick Access Toolbar, click the ***Save*** icon to save the current drawing.

## Graphical Editing

Graphically editing a pipe network relocates pipes or structures. Pipes and structures can be disconnected and reconnected by clicking the network part, and from the shortcut menu, selecting Reconnect or Disconnect.

1. Use the ZOOM and PAN commands to view the western end of the 93rd Street existing STM – (1) and San – (1) pipe networks.
2. In the drawing, select the **EX-SAN–1** structure and, using its grips, relocate some of its structure and its pipes.
3. Clear the grips by pressing ESC.
4. In the drawing, select one of the pipes connected to the EX-SAN–1 structure, click the grip nearest the structure, and move the pipe end to a new location.

The pipe disconnects from the structure.

5. In the drawing, select the pipe that was just disconnected, activate its northern grip, move it to the structure, and when the connect to structure icon appears, select it, connecting the pipe to the structure.
6. Close the drawing and **do not save** the changes.

This ends the exercise on editing a pipe network. The next unit reviews annotating a network's pipes and structures.

## SUMMARY

- To add branches to an existing network, you must edit it.
- You can attach a pipe from one network to a structure of a second network.
- When you are drafting upslope, the start invert is the lowest.
- When you are drafting downslope, the start invert is the highest.
- When you are attaching a pipe to a structure, a star icon (connect to structure) is displayed.
- When you are attaching a pipe to a pipe, a break at connection icon (two opposing fists) is displayed.
- If you are moving structure with pipes, you need only to select the structure. The pipes will relocate to the new structure location.
- To disconnect a pipe from a structure, select just the pipe and then relocate it.
- Editing pipe or structure parameters is best done in their Properties dialog box.

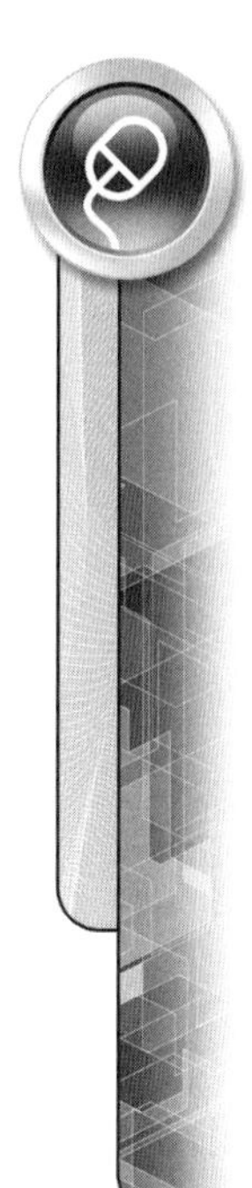

## UNIT 4: PIPE LABELS

Labeling a pipe network occurs during or after its creation. The Create Pipe Network dialog box is used to set structure and pipe label styles while drafting the network (see Figure 11.27). The Add Labels dialog box is used to add labels after you have created the network (see Figure 11.42).

There are two network label style types: pipes and structures (see Figure 11.5). Add Labels annotates individual parts or entire networks in plan, profile, or section. When you are labeling single parts, only the selected part is labeled. To label another part, select the next part.

By default, pipe labels do not use plan readability, but structure labels do.

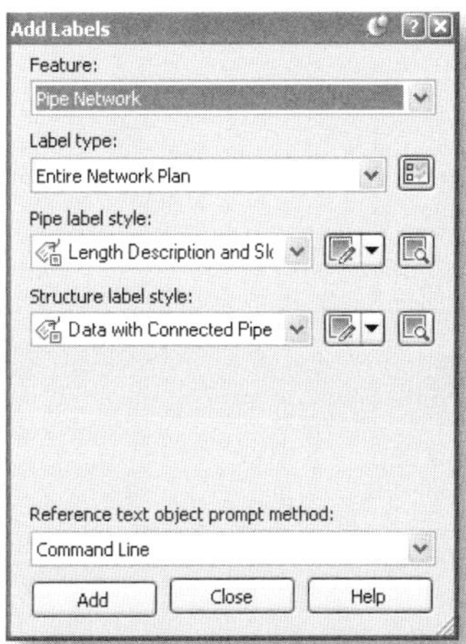


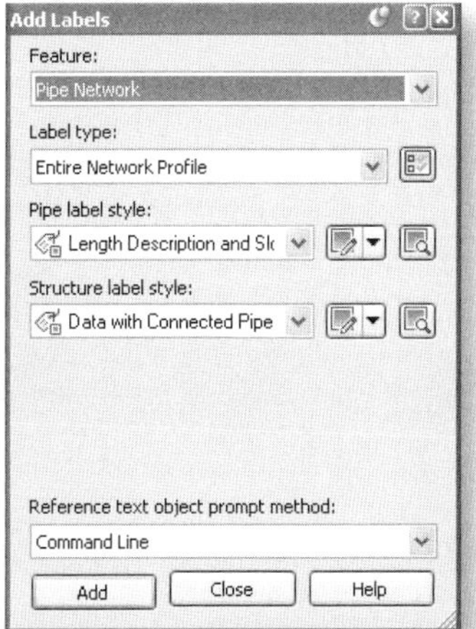


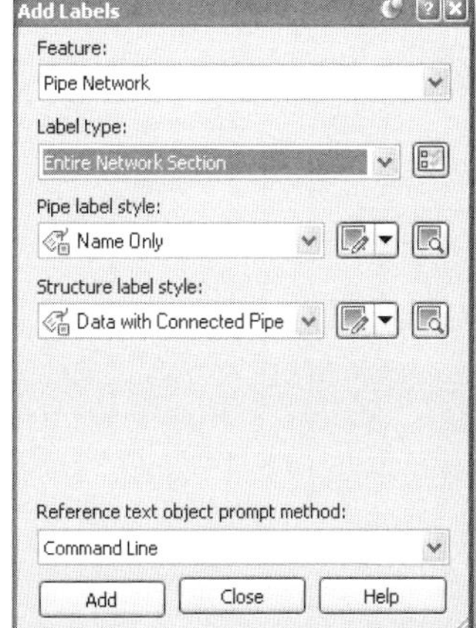


**FIGURE 11.42**

## STRUCTURE LABEL STYLES

A structure label annotates specified structure parameters. A structure label can include the structure name, its rim and sump elevation, in and out pipe inverts, clearances, and other parameters (see Figure 11.43).

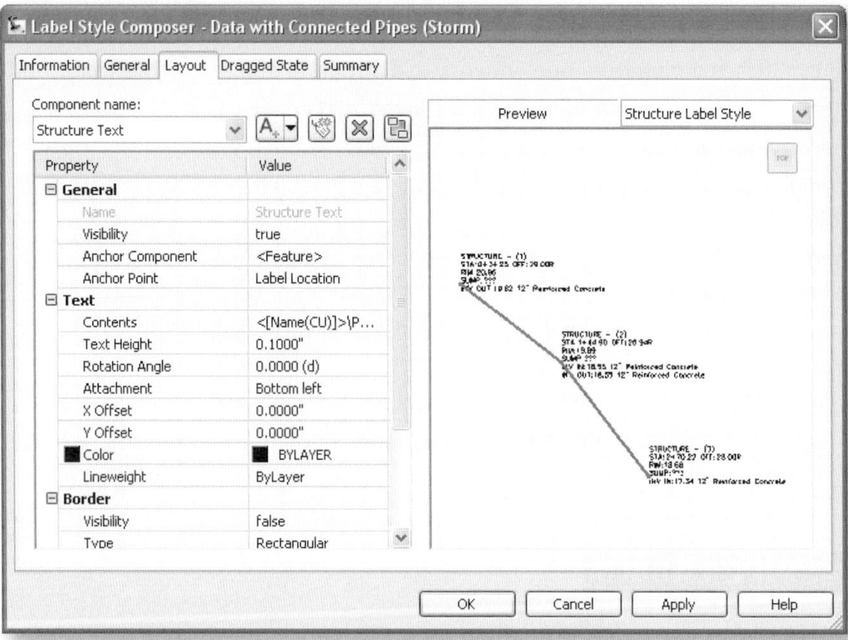


**FIGURE 11.43**

## PIPE LABEL STYLES

A pipe label annotates specified pipe parameters. A pipe label can include the pipe name, its starting and ending invert elevations, beginning and ending structure name, cover, and other parameters (see Figure 11.44).

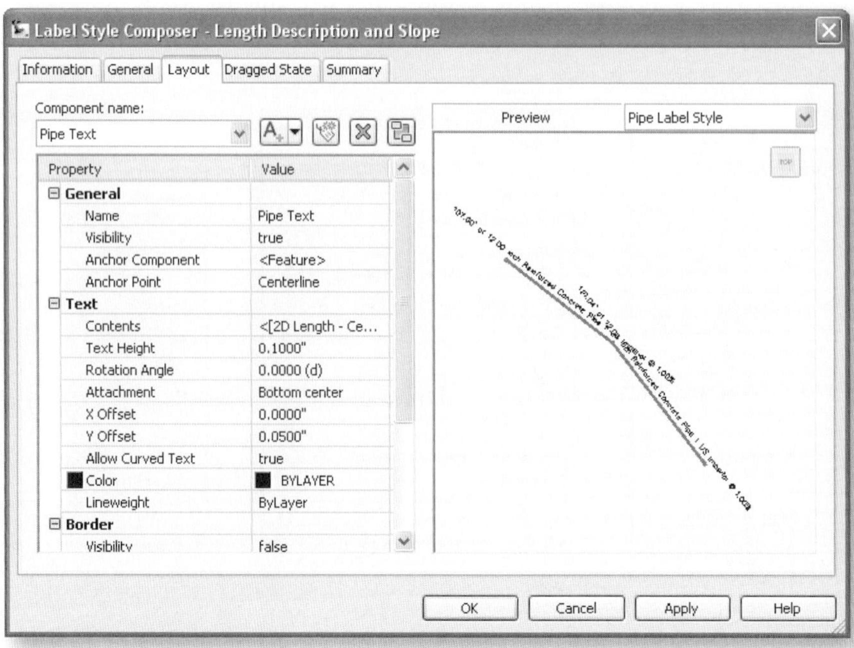


**FIGURE 11.44**

## PIPES IN SECTIONS

When you are creating a pipe network, you use the Create Pipe Network dialog box to set an association to an alignment (see Figure 11.27 or 11.30). If a network exists before you create a sample line group, its pipe and structure data is available for sampling (see Figure 11.45). If you do not want to sample a network, toggle it off before you create the sample line group. If you are developing a network after you have created a sample line group, the group needs to be resampled with the network(s) added to the sample list.

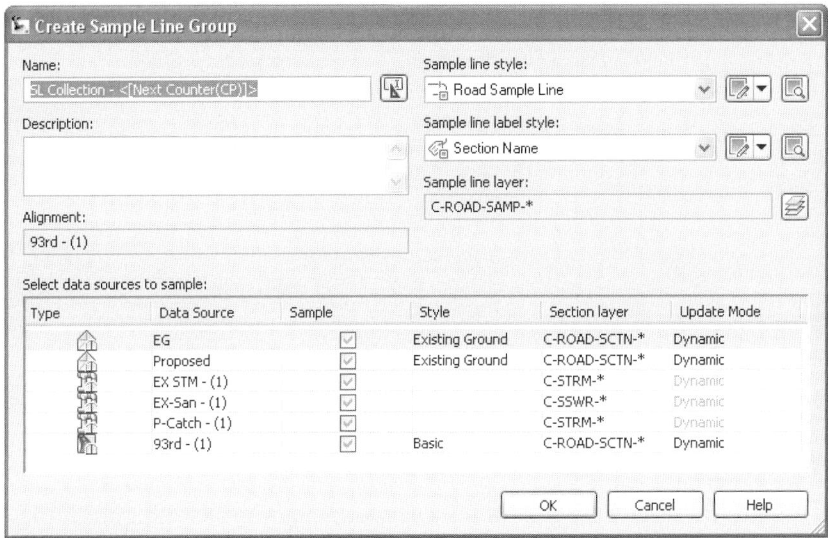


**FIGURE 11.45**

## CROSSING PIPES – PROFILE VIEW

Pipes and structures that intersect a profile view represent possible interferences and are a necessary part of any document. The Pipe Networks tab displays the drawn pipes and structures, but also displays all defined networks. Drawing the intersecting pipes and structures is a two-step process. First, you toggle on the pipe or structure to draw, and second, you choose what style will be used for your drawing (see Figure 11.46).

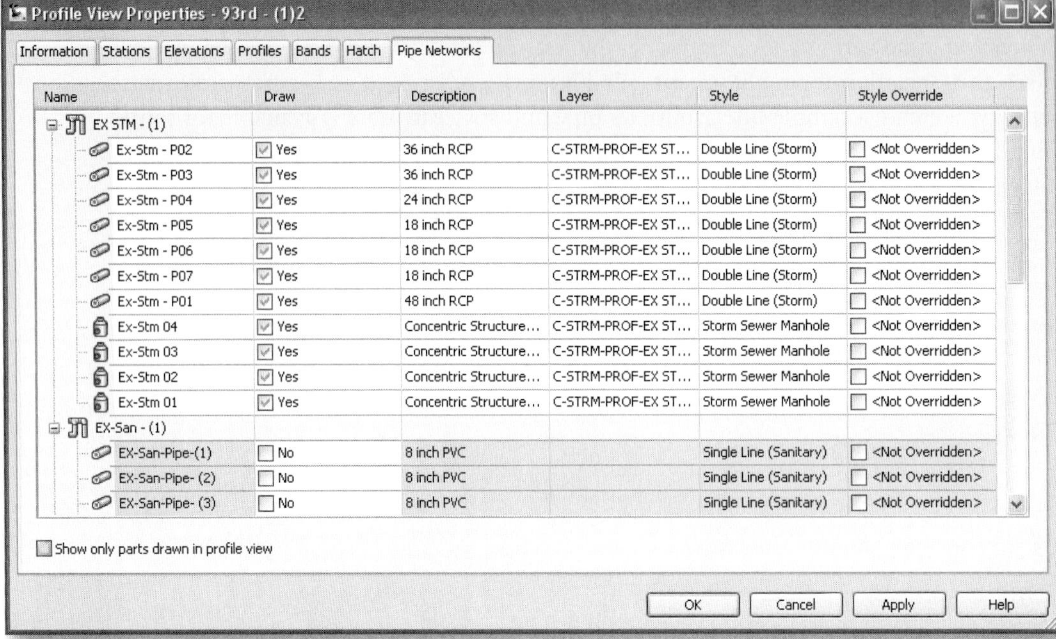


**FIGURE 11.46**

## PIPE NETWORK REPORTS

Civil 3D's Toolbox has several reports for pipe network quantities (see Figure 11.47). The Toolbox reports focus on quantities, not costs. The pay item reports focus on pipe and structure costs.

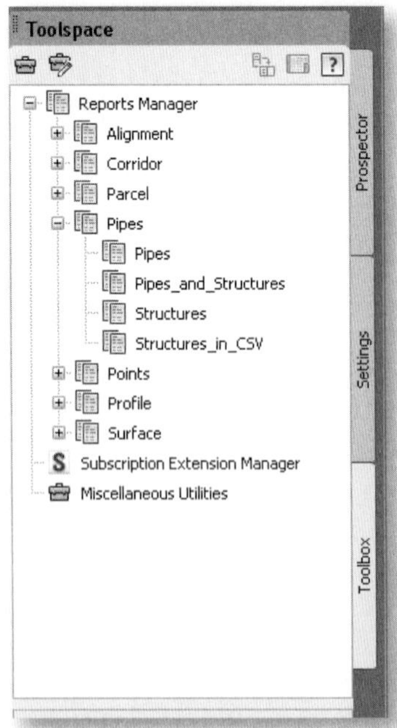


**FIGURE 11.47**

## Pay Item Reports

Objects with pay item assignments are available for reports using the item's values. There are two types of pay item reports: summary and detailed. A summary report lists the pay item ID, its description, total quantity, and unit of measure for each pay item. A detailed report contains a line of information for each selected object.

- A pay item report's scope is a drawing, sheet, or selected objects.
- A summary report does not calculate values for corridor codes.
- A detailed (itemized) report calculates Corridor codes assignments.
- If the pay item relates to an alignment, the report can be limited by alignment station values.

You assign pipe network pay items in the parts lists (see Figure 11.8).

For a discussion on Pay Items see Chapter 7, Unit 1.

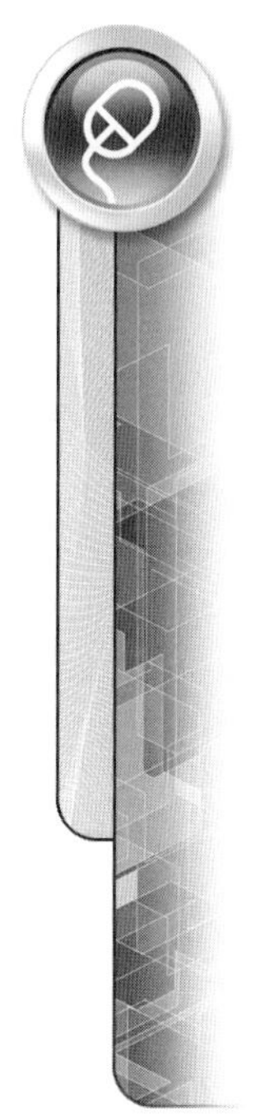

## EXERCISE 11-4

After completing this exercise, you will:

- Be able to annotate individual pipes and structures.
- Be able to annotate entire networks in plan and profile.

### Exercise Setup

This exercise continues with the previous exercise's drawing. If you did not complete the previous exercise, browse to the Chapter 11 folder of the CD that accompanies this textbook and open the *Chapter 11 - Unit 4.dwg* file.

1. Open the previous exercise's drawing or browse to the Chapter 11 folder of the CD that accompanies this textbook and open the *Chapter 11 - Unit 4* drawing.

### Annotating an Entire Pipe Network in Plan View

Add Labels annotates an entire pipe network in Plan View.

1. Use the ZOOM and PAN commands to view the entire 93rd Street storm network in Plan View.
2. Click the Ribbon's Annotate tab. In the Labels & Tables panel, click Add Labels to display the Add Labels dialog box.
3. In Add Labels, set the Feature to Pipe Network, set the Label Type to **Entire Network Plan**, set the Pipe Label Style to **Length Description and Slope**, and set the Structure Label Style to **Data with Connected Pipes (Storm)**.
4. Click ***Add*** and select a pipe or structure from the 93rd Street storm network.
5. Use the ZOOM and PAN commands to review the labels.

To make pipe labels respond to rotated views, a user must toggle ON plan readability.

6. Drag some labels to view their dragged state.

### Annotating an Entire Pipe Network in Profile View

Annotate an entire Pipe Network in Profile View by setting the correct Add Labels parameters.

1. Use the ZOOM and PAN commands to view the entire 93rd Street storm network in Profile View.

2. In Add Labels, set the Label Type to **Entire Network Profile**, set the Pipe Label Style to **Length Description and Slope**, set the Structure Label Style to **Data with Connected Pipes (Storm)**, click ***Add***, and in Profile View, select a pipe or structure from the 93rd Street storm network.
3. Use the ZOOM and PAN commands to review the labels.
4. Drag some labels to view their dragged state.
5. Click ***Close*** to exit Add Labels.
6. Close the drawing and **do not save** it.
7. Reopen the drawing file.

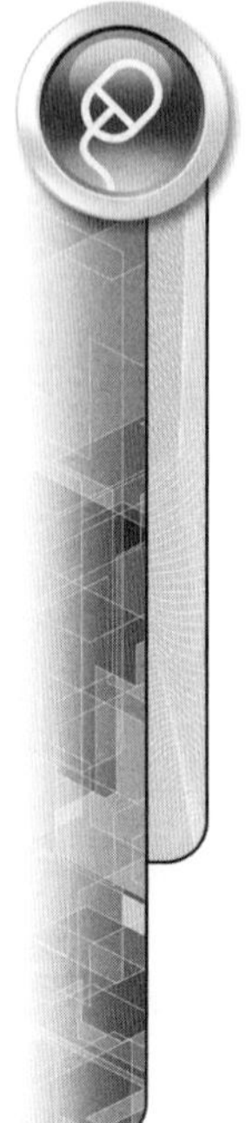

## Annotating Individual Parts in Plan View

Add Labels annotates individual pipes and structures in Plan View.

1. Use the ZOOM and PAN commands to view the EX-SAN-1 structure and its pipes at the 93rd Street storm sewer network's western end.
2. Click the Annotate tab. In the Labels & Tables panel, click Add Labels to display the Add Labels dialog box.
3. In the Add Labels dialog box, set the Feature to Pipe Network, set the Label Type to **Single Part Plan**, set the Pipe Label Style to **Length Description and Slope**, set the Structure Label Style to **Data with Connected Pipes (Storm)**, click ***Add***, and in the drawing, select a structure or a pipe.

Only the selected part receives a label.

4. Label additional pipes and structures by selecting a network pipe or structure.

## Annotating Single Parts in Profile View

Add Labels annotates individual pipes and structures in Profile View.

1. Use the ZOOM and PAN commands to view the 93rd Street Storm sewer network in Profile View.
2. In Add Labels, set the Label Type to **Single Part Profile**, set the Pipe Label Style to **Length Description and Slope**, set the Structure Label Style to **Data with Connected Pipes (Storm)**, click Add, and select a structure or pipe.

Only selected structures or pipes are labeled.

3. Label additional pipes and structures by selecting a part.
4. Click **Close** to exit Add Labels.
5. Close the drawing and **do not save** it.
6. Reopen the drawing file.

## Drawing Intersecting Pipes

1. In the drawing, select **Profile View**, press the right mouse button, and from the shortcut menu, select PROFILE VIEW PROPERTIES....
2. Click the ***Pipe Networks*** tab.
3. If necessary expand the EX-San – (1) section.
4. Toggle **ON** EX-San-Pipe-(6) and (7), for each pipe toggle **ON** Style Override, and in Pick Pipe Style, click the drop-list arrow, from the list, select **Pipe Crossing Pipe (Sanitary)**, and click ***OK***.
5. Click ***OK*** to exit Profile View Properties, and the crossing pipes are drawn in Profile View.

## Pipes in Road Section Views

1. In Prospector, expand the Corridors branch, if the **93rd – (1)** Corridor is out-of-date, rebuild it.
2. Use the AutoCAD ZOOM and PAN commands to view the entire 93rd Street storm network in Plan View.
3. Click the Home tab, on the Profile & Section Views panel, click the ***Sample Lines*** icon.
4. The command line prompts you for an Alignment. Press the right mouse button, from the list, select **93rd – (1)**, and click ***OK***.
5. In the Create Sample Line Group dialog box, uncheck the P-Catch – (1) pipe network, and click ***OK*** to accept the defaults.

The routine opens in select At a Station mode. To create a section centered on a structure, select its station using an AutoCAD Center object snap.

6. Using a **Center** object snap, select EX-Stm-01 to define the first section.
7. After selecting the structure, the command line prompts you for a swath width. Enter the right and left swath width as **60'**.
8. Repeat this three times for each EX-Stm manhole and press ENTER.

Note the sample line stations.

9. Use the PAN and ZOOM commands to view an empty area to the right of the site.
10. In the Home tab's Profile & Section Views panel, click the ***Section Views*** icon, and from the shortcut menu, select CREATE MULTIPLE VIEWS.
11. In Create Multiple Section Views' General panel, in the bottom right, click the Plot Group Style drop-list arrow, and select **Plot All**.
12. Click ***Next*** three times until you are viewing the Section Display Options panel.
13. In the Section Display Options panel, set Change Labels to **_No Labels**, set style for Proposed to **Finished Ground**, click ***Create Section Views***, and in the drawing, select a point to place the sections in the drawing (see Figure 11.48).

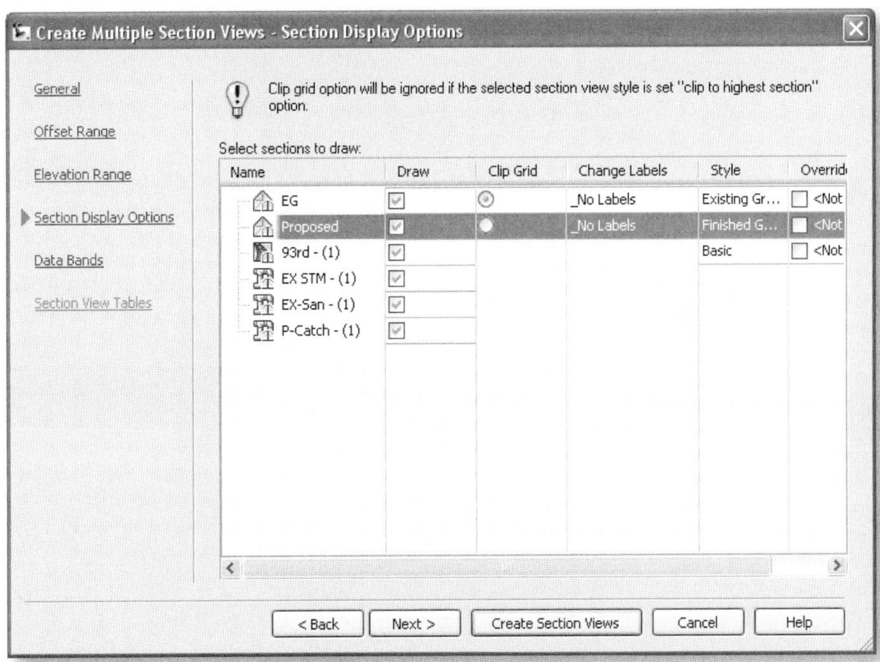

**FIGURE 11.48**

14. Use the ZOOM and PAN commands to better view the sections.
15. At Civil 3D's top left, Quick Access Toolbar, click the ***Save*** icon to save the current drawing.

Pipe networks have two labels types: pipe and structure. You can place labels while you are drafting a network or after you have completed its design.

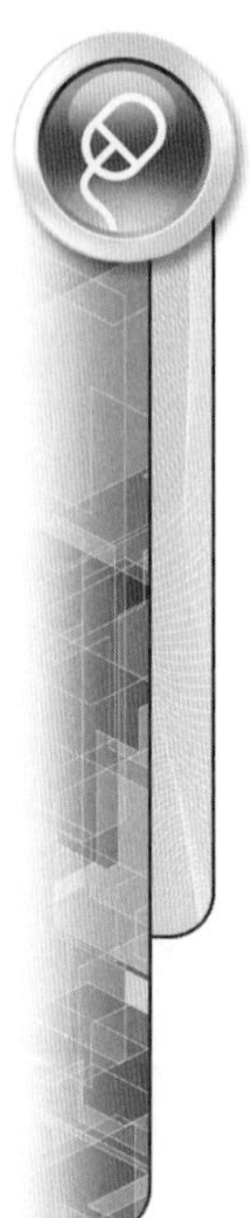

## SUMMARY

- Pipe network label types are pipe and structure.
- Civil 3D template label styles are simple and do not represent all pipe or structure properties.
- A pipe network should exist *before* you create the sample lines.
- A profile view can contain crossing pipe segments.

This ends the chapter on pipes. Chapter 12 reviews Civil 3D data-sharing methods. The chapter is a PDF on the CD that accompanies this textbook.

# CHAPTER 12

# Civil 3D Shortcuts

## INTRODUCTION

A project's data security and timeliness is always a concern, as is data sharing among design groups or contractors. Initially, Civil 3D provides minimal project support. However, in essence, a drawing is a project. Civil 3D has a simple data-sharing method: shortcuts. Shortcuts use an external LandXML file and, through this file, other drawings access object data. Shortcuts work very much like Xrefs and, like Xrefs, shortcuts can notify users of a data change.

## OBJECTIVES

This chapter focuses on the following topic:

- Data Shortcuts

### OVERVIEW

Initially, Civil 3D considered a drawing a project. All drawing objects represent a project's data. Considering a drawing a project is a radical departure from Autodesk Land Desktop's strategy. In LDT, an extensive external folder structure holds much of a project's data. As open and as vulnerable as it is, this strategy has saved many a project's data. The drawing as a project also made it difficult to share data between drawings.

A data shortcut is a LandXML file stored in a folder structure, and other drawings reference (use) the data shortcut to create their object instances. Civil 3D maintains links between the drawing and the shortcut. If the data shortcut definition changes, then Civil 3D notifies the user of the change and the drawing needs to be updated. A data shortcut expands data availability and ensures some degree of data quality.

A new way of thinking is also required when using shortcuts. Because data resides in a drawing, drawings should be thought of as sources of data. This means users will create drawings whose sole purpose is to serve data, and other drawings will use this data to create plan sets. When you use shortcuts, you should have a folder structure that organizes your company's implementation. Civil 3D provides project templates for such a strategy. A good starting place for understanding and planning an implementation is to read Civil 3D's Best Practices PDF (in Civil 3D's Help folder).

### Unit 1

This unit covers creating and using data shortcuts. When referencing shortcut data, Civil 3D uses icons to indicate data states, changes, and references.

## UNIT 1: DATA SHORTCUTS

A Civil 3D drawing serves as a repository for all project data. Shortcuts enable data sharing (see Figure 12.1). Data shortcuts are limited to surfaces, alignments, profiles, pipe networks, and view frame groups. If you are updating an object that is a shortcut, the reference receives a change notification.

At a minimum, shortcuts require a working folder and a shortcut folder. It can be more complex to implement shortcuts if you are viewing drawings as data sources. To implement this folder strategy, you create folders that contain drawings that, in turn, serve data to a project's design drawings.

After you determine where to locate shortcut files and drawings, you next create the shortcuts. There are two steps to creating and using a data shortcut. The first step is to create the shortcuts in a source drawing. The second step is to create references in a destination drawing. The source drawing defines a link between the drawing and the object(s) and exports those link(s) as an XML file (the shortcut file) to the appropriate folders. A destination drawing uses the working folder and its shortcut folder to reference the shortcut data and to create an object in the drawing.

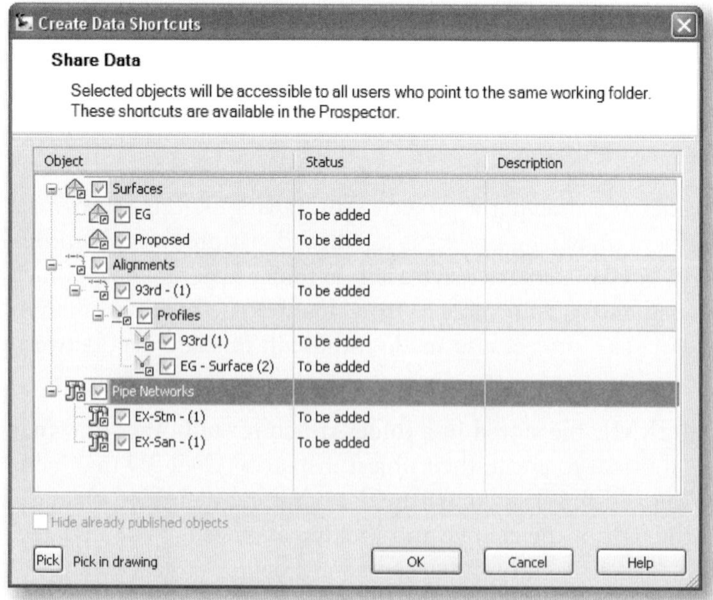


**FIGURE 12.1**

## SETTING UP WORKING AND SHORTCUT FOLDERS

Shortcuts require a working folder. This single folder can be a location for all projects and their shortcuts, or each project can have its own different folder. If you decide that each project will have its own working folder, you must reset the working folder and the shortcut folder locations whenever you switch to a different project.

After you have established a working folder, note that it must contain a shortcut folder. A shortcut folder can be a single folder that contains project shortcuts, or the shortcuts can be a part of a larger project folder structure that contains folders for source, working, and submission folders. These folders can contain more than drawings; they can contain supporting documents, e-mails, spreadsheets, survey data, etc.

Civil 3D installations include a sample template, which can be used for shortcuts and Vault. A shortcut folder structure can be migrated to a Vault project (see Figure 12.2). In this sample template, folders contain source drawings and their shortcuts. Whether you are using a project template or not, Civil 3D creates a _Shortcuts folder with Source Drawings subfolders.

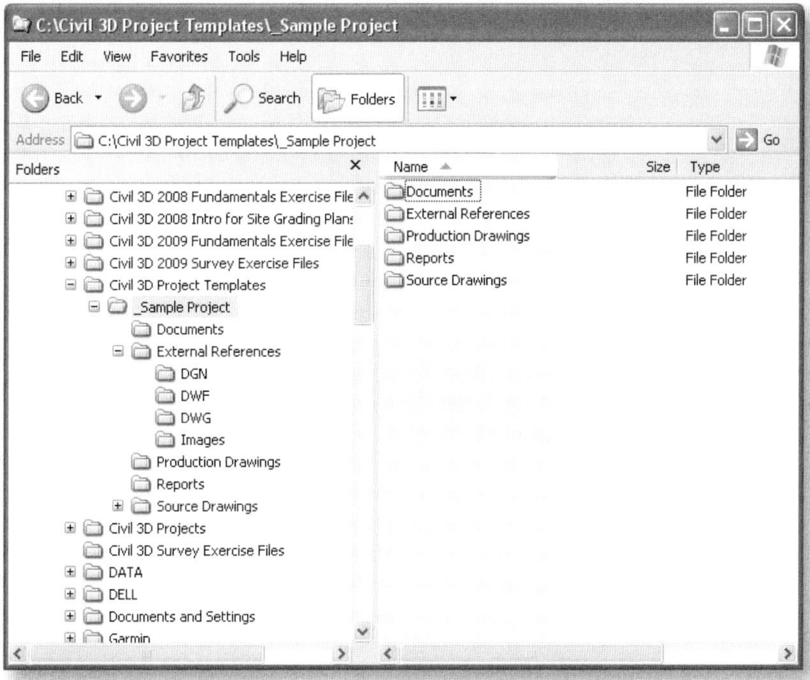


**FIGURE 12.2**

## CREATING MULTIPLE SHORTCUT FOLDERS IN ONE WORKING FOLDER

If the working folder has multiple shortcut folders, the correct shortcut folder must be set (see Figure 12.3). The left side of Figure 12.3 shows the shortcut menu listing Set Data Shortcuts Folder…. You select this command to open the Set Data Shortcuts Folder dialog box. The dialog box displays the current shortcut folder (with a green checkmark) and lists the other shortcut folders in the current working folder (see Figure 12.3).

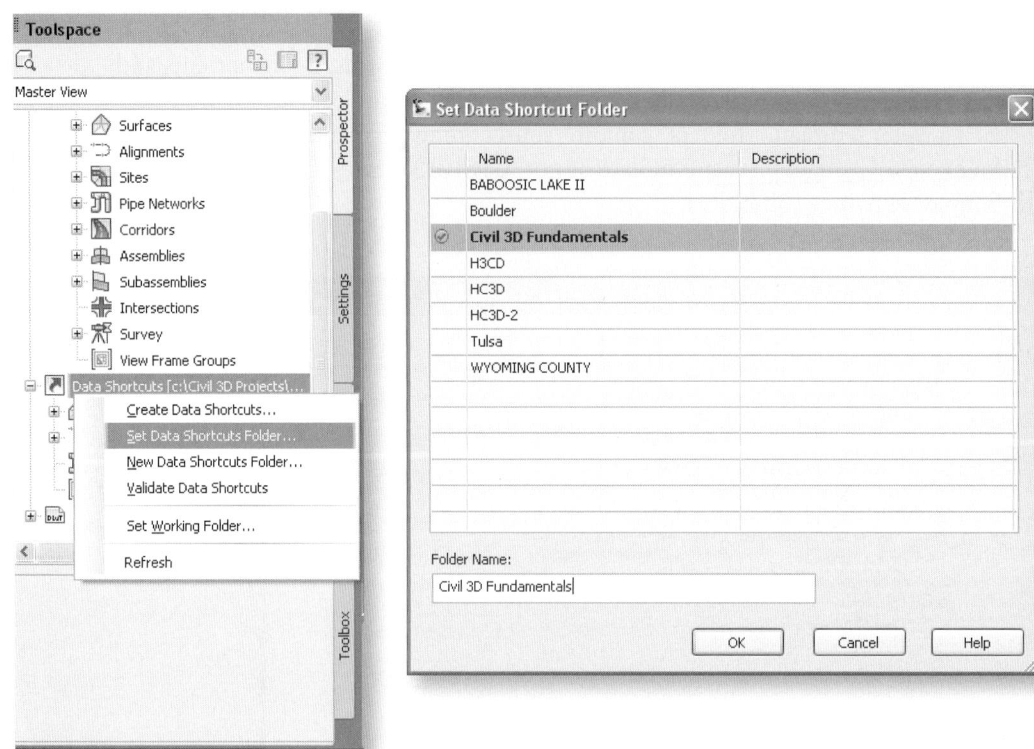


**FIGURE 12.3**

### Source Drawing — Create a Data Shortcut

Your first step is to create a data shortcut and export it to the shortcut subfolders. The Create Data Shortcuts dialog box creates links between the drawing's object and the drawing's name and location (see Figure 12.1). When you click OK, the selected shortcuts are exported to the working folder's appropriate shortcut subfolders.

You can toggle on object hiding. This prevents already-published objects from being referenced.

### Destination Drawing — Create a Reference

You first create a new drawing and save it. After you save the file, you set the working folder, listing the shortcuts. Next, you expand the object type, select the object, right mouse click, and from the shortcut menu, select Create Reference… (see Figure 12.4).

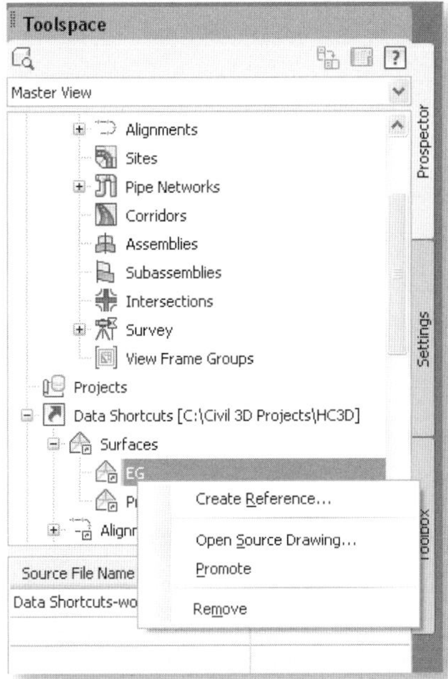


**FIGURE 12.4**

## OPEN SOURCE DRAWING

Open Source Drawing opens the drawing, thus creating the shortcut.

## SYNCHRONIZE

Synchronize updates the current drawing to the latest version of the referenced object. A balloon is used for out-of-date notification (like Xref's notification), or an out-of-date shield is displayed next to the object in Prospector.

## PROMOTE

Promote detaches a drawing's reference object from its shortcut object. When this is done, the promoted surface creates a snapshot. A snapshot prevents the object from going out of date and also can be modified.

## DATA SHORTCUT EDITOR

Data Shortcut Editor is an external program used to repair and edit shortcut paths and file locations (see Figure 12.5). The find and replace option quickly replaces paths, so the shortcut does not remain broken.

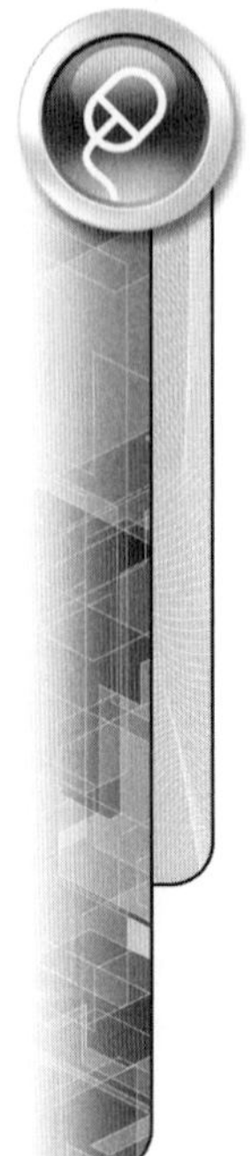

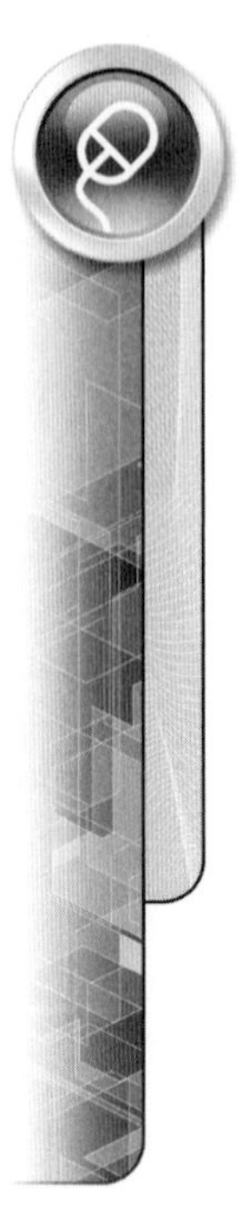

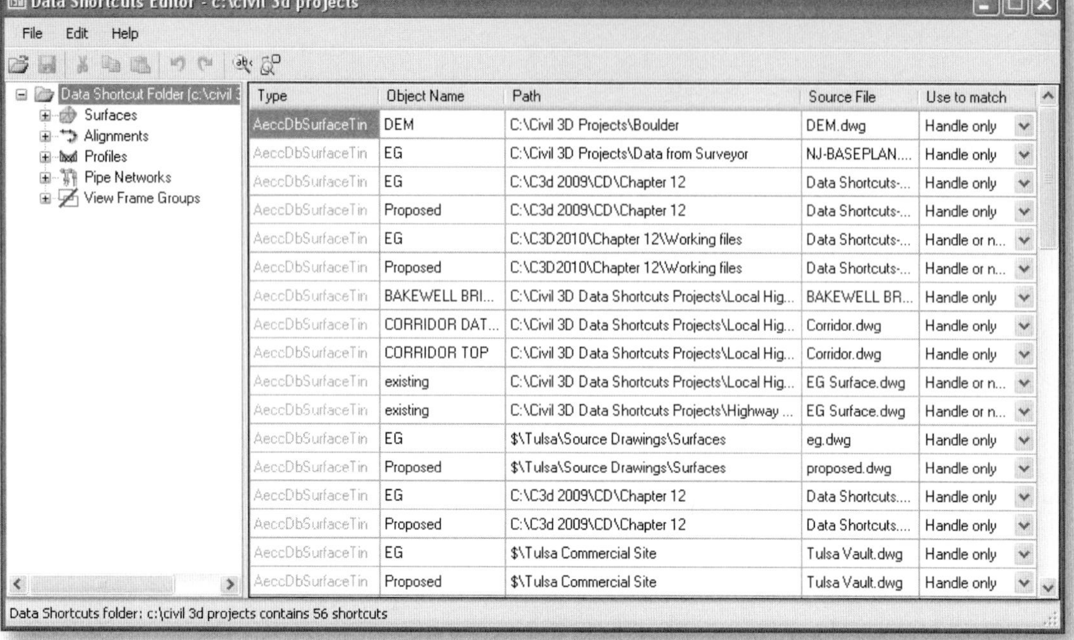


**FIGURE 12.5**

## EXERCISE 12-1

After completing this exercise, you will:

- Be able to set a working folder.
- Be able to create a new shortcut folder.
- Be able to switch between shortcut folders.
- Be familiar with creating Civil 3D data shortcuts.
- Be familiar with referencing data shortcuts.
- Be familiar with synchronizing drawing and data shortcut values.
- Be familiar with promoting references to drawing objects.

### Exercise Setup

This exercise uses the *Data Shortcuts.dwg* file. To find this file, browse to the Chapter 12 folder of the CD that accompanies this textbook, and open the *Data Shortcuts.dwg* file. This drawing has several Civil 3D objects. After you create the shortcuts, a new drawing references these shortcuts.

1. If you are not in ***Civil 3D***, start the application by double-clicking its desktop icon.
2. When you are at the command prompt, close the opening drawing and do not save it.
3. At Civil 3D's top left, Quick Access Toolbar, click the ***Open*** icon. In the Open dialog box, browse to the Chapter 12 folder of the CD that accompanies this textbook, select the *Data Shortcuts* drawing, and click ***Open***.
4. At Civil 3D's top left, click Civil 3D's drop-list arrow, and from the Application Menu, highlight Save As, from the flyout menu select AutoCAD Drawing, browse to the Civil 3D Projects folder, for the drawing name enter ***Data Shortcuts – work***, and click ***Save*** to save the file.

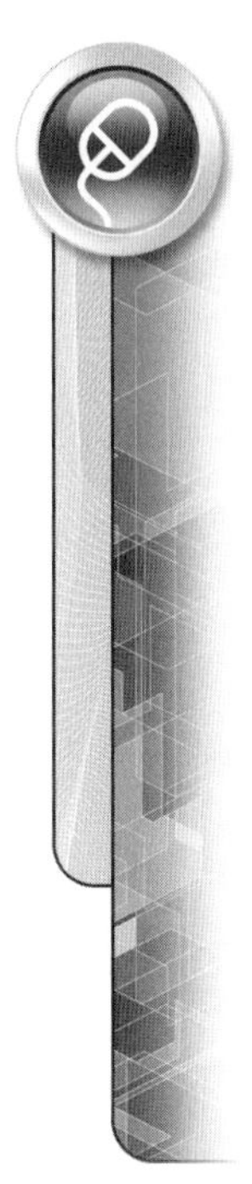

## Set a Working Folder

1. If necessary, click the ***Prospector*** tab.
2. If necessary, at the top of Prospector, click the View drop-list arrow, and from the list of views, select Master View.
3. If necessary, expand Data Shortcuts [].
4. Select **Data Shortcuts []**, press the right mouse button, and from the shortcut menu, select SET WORKING FOLDER... to open the Browse For Folder dialog box.
5. In the Browse For Folder dialog box, browse to and select the **C:\Civil 3D Projects** folder, and click ***OK***.

## Create a Shortcut folder

1. Select **Data Shortcuts []**, press the right mouse button, and from the shortcut menu, select NEW DATA SHORTCUTS FOLDER... to open the New Data Shortcut Folder dialog box.
2. In the New Data Shortcut Folder dialog box, for the shortcut folder name, enter **HC3D**; for the description, enter **Exercise folders for shortcuts**, and click ***OK***.
3. In Prospector, the Data Shortcuts heading now lists the working folder and the current shortcut folder: Data Shortcuts [C:\Civil 3D Projects\HC3D].

An HC3D folder now contains a _Shortcuts folder with five sub-folders: Alignments, PipeNetworks, Profiles, Surfaces, and ViewFrameGroups.

## Create Data Shortcuts

The current drawing contains several Civil 3D objects: surfaces, alignment and its profiles, and pipe networks.

1. In Prospector, select the Data Shortcuts heading, press the right mouse button, and from the shortcut menu select CREATE DATA SHORTCUTS....
2. In Create Data Shortcuts - Share Data panel, toggle **ON** all of the object types to share, **Surfaces, Alignments** and **Pipe Networks**, and click ***OK***.
3. At Civil 3D's top left, Quick Access Toolbar, click the ***Save*** icon to save the drawing.
4. Close the drawing.

## Create a Data Shortcut Reference

1. If necessary, click ***Prospector***.
2. Expand Drawing Templates and AutoCAD branches, from the template list, select *_AutoCAD Civil 3D (Imperial) NCS*, press the right mouse button, and from the shortcut menu, select CREATE NEW DRAWING.
3. Click the Civil 3D's icon drop-list arrow at the top left, from the Application Menu, highlight Save As, from the flyout select AutoCAD Drawing, browse to the Civil 3D Projects folder, for the drawing name enter **Data Shortcuts Destination**, and click ***Save*** to save the file.
4. In Prospector, expand Data Shortcuts' Surfaces branch. From the list of surfaces, select **EG**, press the right mouse button, and from the shortcut menu, select CREATE REFERENCE... to open the Create Surface Reference dialog box.
5. In the Create Surface Reference dialog box, review the settings, and click ***OK*** to create the surface reference.
6. In Prospector, expand the Data Shortcuts Destination's Surfaces branch and note the reference icon to the left of EG.

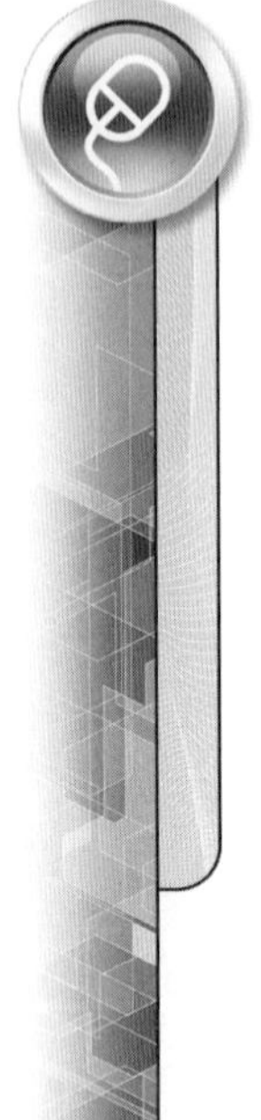

7. Repeat Steps 4 and 5 to create references for the surface **Proposed**, select **_No Display** for the surface style, alignment **93rd – (1)**, profile **93rd (1)**, and Pipe Network **EX-Stm – (1)**. When you create the references, the dialog boxes will be different for each object type.

## Create a Profile View from Reference Data

1. Use the PAN and ZOOM commands to move the site to the screen's left side.
2. On the Ribbon's Home tab, Create Design panel, click the ***Profile*** icon, and from the shortcut menu select CREATE SURFACE PROFILE.
3. In the Create Profile from Surface dialog box, select the EG surface and at the middle right, click ***Add>>***.

At the bottom of the dialog box, the profile list already includes the 93rd (1) vertical design profile.

4. Click ***Draw in Profile View***, set the Profile View Style to **Major Grids**, click ***Create Profile View***, and in the drawing, select a point to place the profile in the drawing.
5. At Civil 3D's top left, Quick Access Toolbar, click the ***Save*** icon to save the drawing.

## Edit the Data Shortcut Source Drawing

1. In Prospector, Data Shortcuts, select the *EG* surface, press the right mouse button, and from the shortcut menu, select OPEN SOURCE DRAWING....
2. In Prospector, under Data Shortcuts – works' Surfaces, expand the EG branch until you view the Definition heading's data list.
3. In the data list, select Edits, press the right mouse button, and from the shortcut menu, select RAISE/LOWER SURFACE.
4. The command line prompts you for the amount to add. Enter **1** and press ENTER to raise the surface 1 foot.
5. If necessary, REBUILD the EG surface.
6. At Civil 3D's top left, Quick Access Toolbar, click the ***Save*** icon to save the drawing.
7. Close the Data Shortcuts – work drawing.

At the Data Shortcuts Destination drawing's lower right, a change notification balloon should be displayed.

8. In Prospector, expand the Data Shortcuts Destination's Surfaces and the Alignments branches to view their out-of-date icons.
9. In the change balloon, click SYNCHRONIZE. A Panorama is displayed and notifies you that all shortcuts were updated.
10. Close the Panorama.
11. At Civil 3D's top left, Quick Access Toolbar, click the ***Save*** icon to save the drawing.

## Promote a Reference

1. In Prospector, select the **EG** surface reference, press the right mouse button, and in the shortcut menu, select PROMOTE.

This creates an object instance that breaks the shortcut link. The surface starts its definition with a snapshot. This allows the drawing to move to new locations and not go out-of-date.

2. Close the Data Shortcuts Destination drawing and do not save the changes.

## Multiple Shortcut Folders

A working folder can contain multiple shortcut folders. If this is the case, the shortcut folder needs to be set before making a reference.

1. In Prospector, Data Shortcuts, select the **EG** surface, right mouse click, and from the shortcut menu, select OPEN SOURCE DRAWING....
2. Select Data Shortcuts, press the right mouse button, and from the shortcut menu, select NEW DATA SHORTCUTS FOLDER... to open the New Data Shortcut Folder dialog box.
3. In the New Data Shortcut Folder dialog box, for the shortcut folder name, enter **HC3D-2**, for the description, enter **Exercise folders for shortcuts-2**, and click ***OK***.
4. In Prospector, the Data Shortcuts heading now lists the working folder and the current shortcut folder: Data Shortcuts [C:\Civil 3D Projects\HC3D-2].
5. Select **Data Shortcuts**, press the right mouse button, and from the shortcut menu, select **Set Data Shortcuts Folder...** to open the Set Data Shortcuts Folder dialog box.

The dialog box displays both shortcut folders; the current folder has a green checkmark.

6. In the dialog box, select the name **HC3D**, and click ***OK*** to change the current shortcut folder.
7. The Data Shortcuts heading updates accordingly.
8. Close all of the drawings.

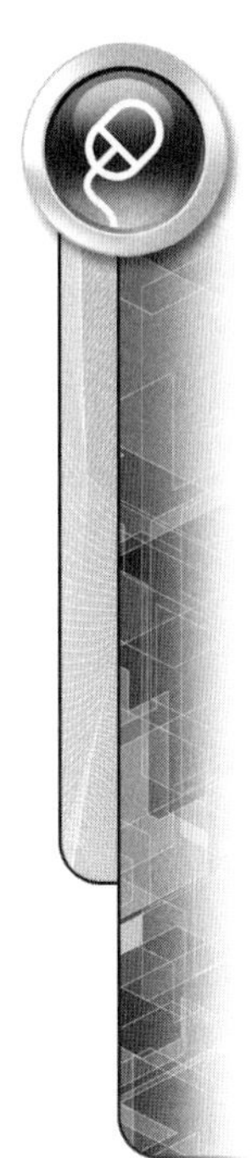

## Edit Shortcut Paths

1. Minimize Civil 3D.
2. Click Start, from All Programs, Autodesk, the AutoCAD Civil 3D 2010 flyout menu, select DATA SHORTCUTS EDITOR.
3. In the Data Shortcuts Editor, click the ***Open*** icon and the Browse to Folder dialog box opens.
4. In the dialog box, browse to the Civil 3D Projects folder, and click ***OK***.

The Editor displays the shortcuts and their values are available for editing.

5. Close the Data Shortcut Editor and maximize AutoCAD Civil 3D.

This ends the data shortcuts exercise.

## SUMMARY

- Data shortcuts link an object to an external LandXML file.
- When an object is a data shortcut and it changes, all drawings that reference the shortcut receive an out-of-date balloon notice and an out-of-date object icon.
- You can promote a referenced object to an object.
- When promoting a reference, the action breaks the data shortcut link.
- The Data Shortcut Editor resolves broken data links.
- Data shortcuts must have a working folder and a shortcut folder.
- If a working folder has more than one shortcut folder, before you use shortcuts, you must set the shortcut folder.

The next chapter reviews the Civil 3D hydrology tools.

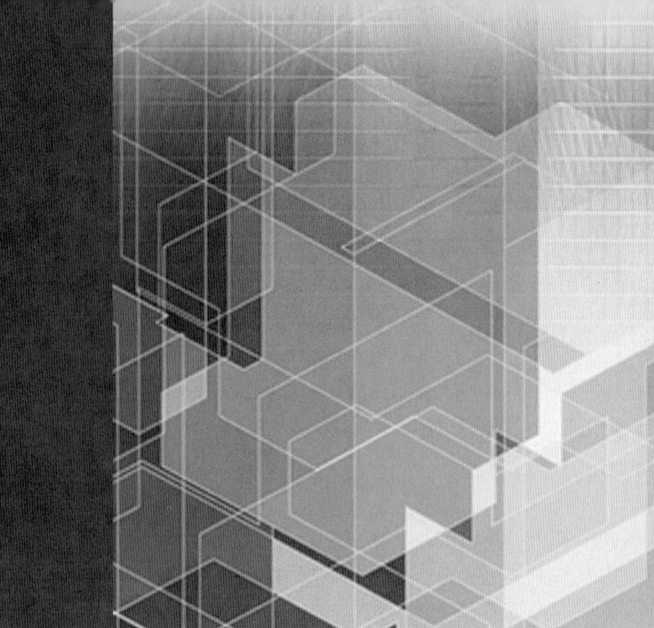

# CHAPTER 13

# Hydraulics and Pipe Design

## INTRODUCTION

A surface has slopes, and Civil 3D tools show the paths water takes across a uniform surface. However, most surfaces are not uniform. A site may have several surface types and each may affect the absorption, flow, and collection of surface water runoff. The hydrology extensions calculate runoff amounts, structures to handle runoff, and evaluate and design pipe networks.

## OBJECTIVES

This chapter focuses on the following topics:

- Establishing Rainfall Amounts
- Calculating Surface Runoff
- Using Structure Calculators
- Reviewing and Adjusting a Storm System

### OVERVIEW

Hydraulic extensions address water and how to handle it in a site. The interaction of hydraulic extensions with Civil 3D is limited, but it is helpful when you are creating and evaluating a design. When you work with the storm sewer tools, LandXML communicates a design between the extensions and the calculators.

The calculators use a project to transfer information between the calculators and the extensions. Most information from a drawing, area, location, etc., is not available directly within the calculators. A user must have notes and values available to work in the calculators.

### Unit 1

This unit focuses on the structure calculator. The calculator supports culverts, channels, inlets, and weirs.

### Unit 2

The calculation of runoff, its storage, and discharge is the focus of the second unit. The runoff calculations use IDF or SCS values and flow connection devices: ponds, reaches, culverts, orifices, and weirs.

### Unit 3

The focus of this unit is importing, reviewing, and revising a storm water pipe system. A design from Civil 3D must be imported and exported as a LandXML file. The calculators evaluate the design and then modify the design. The calculator can add and delete structures, if appropriate. A design can be developed in the extension and imported to Civil 3D.

### Unit 4

The focus of this unit is the plan and production tools that create plan and profile sheets from a pipe network design.

## UNIT 1: HYDRAFLOW EXPRESS

Hydraflow Express is a water-control structures calculator. Regardless of the type of structure under consideration, the interface is the same. It has basic settings that affect its calculations.

### USER INTERFACE

The Hydraflow Express interface has four sections: taskbar, input grid, graphic display, and results grid (see Figure 13.1).

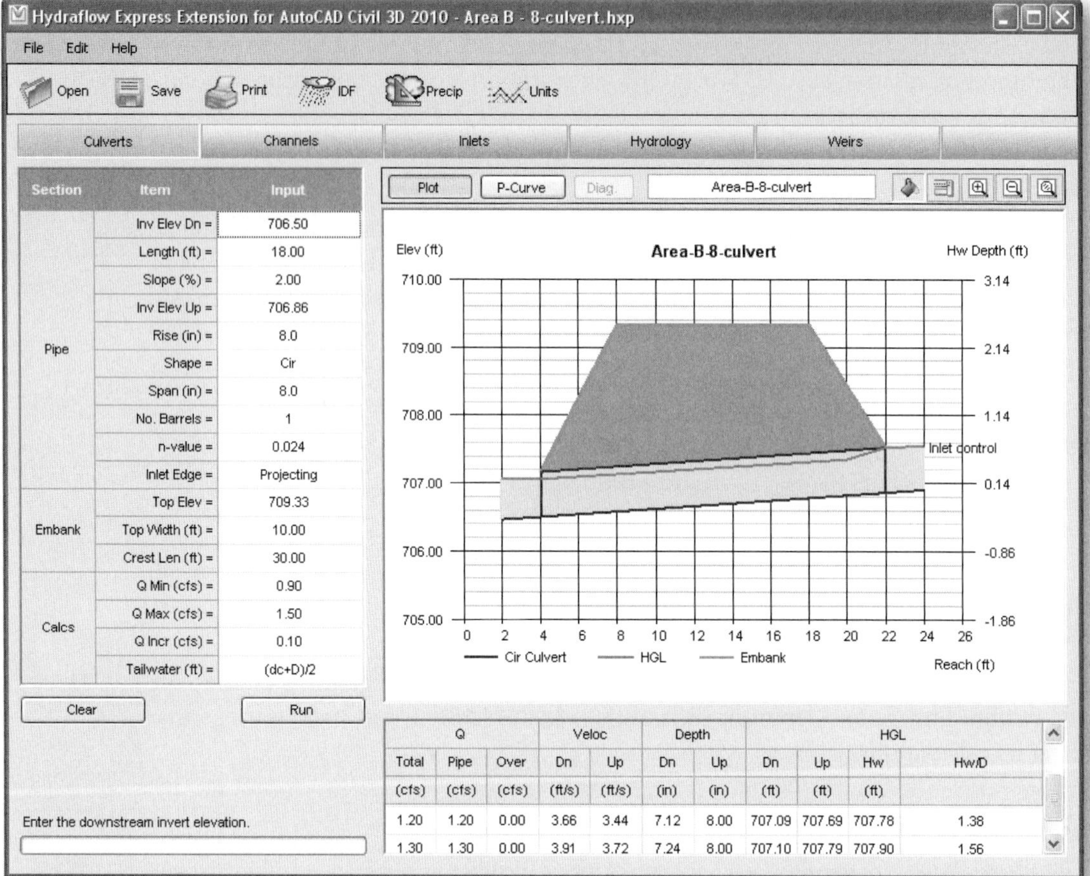


**FIGURE 13.1**

## Taskbar

Across the top, under the menus, you find the taskbar. You use the taskbar to select the structure type. There are four structure types: culverts, channels, inlets, and weirs. Express also includes a hydrology calculator that develops hydrographs for the structures to use.

An appropriate input grid (the left side of Figure 13.1), a graphic display (a structure schematic at the top right of Figure 13.1), and an output grid (bottom right of Figure 13.1) are displayed after you select a structure type. Channels, inlets, and weirs have a shapes subset. For example, a weir has the following shapes subset: rectangular, compound, circular, v-notch, trapezoidal, and proportional. Each shape type has varying input values (see Figure 13.2).

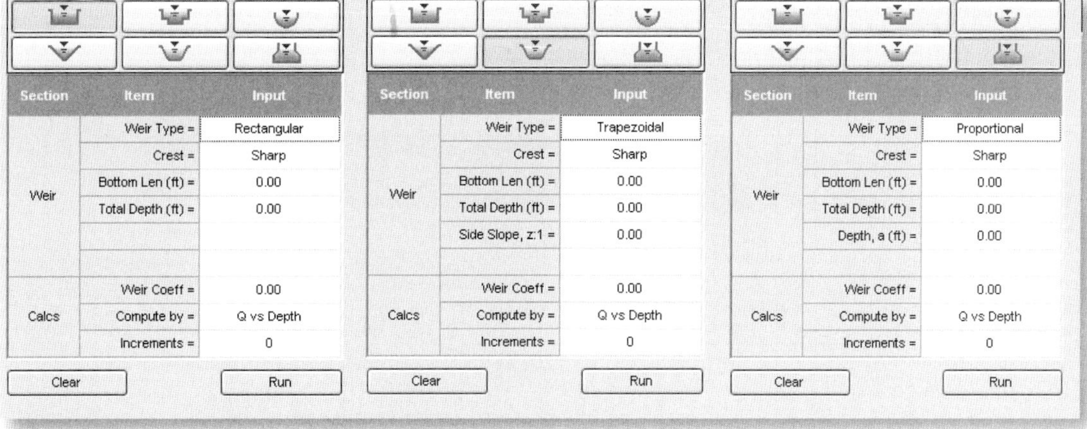


**FIGURE 13.2**

## Input Grid

The information in this section is required by the calculator. After you enter the values, click Run to perform the calculation using the currently entered values. You can change the values at any time and click Run again to compute new numbers. The values for which this section prompts reflect the selected structure type and shape (see Figure 13.2).

## Graphic Display and Results Grid

Calculation results appear in the Graphic Display as a schematic. The schematic shows the water level whose interval is set in the input grid, and that interval changes when you select an interval in the results grid (see Figures 13.3 and 13.4).

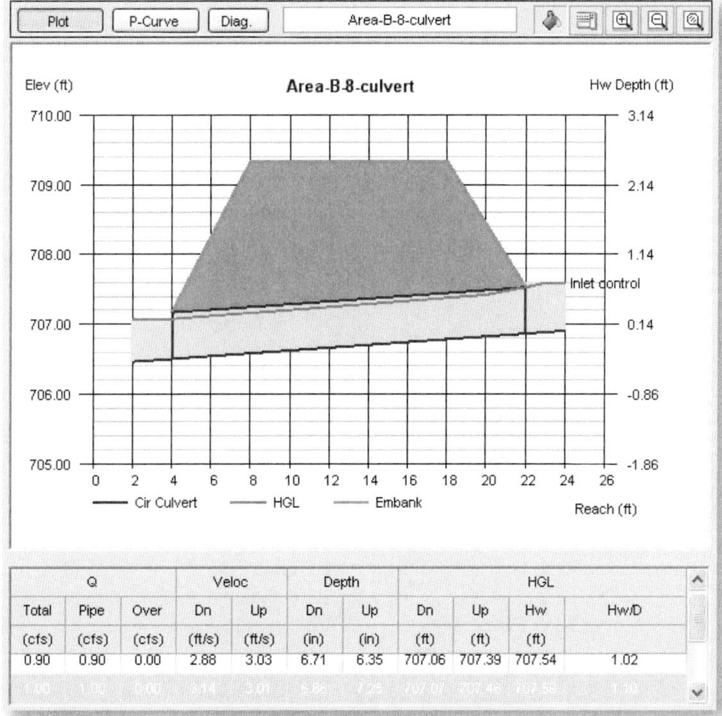


**FIGURE 13.3**

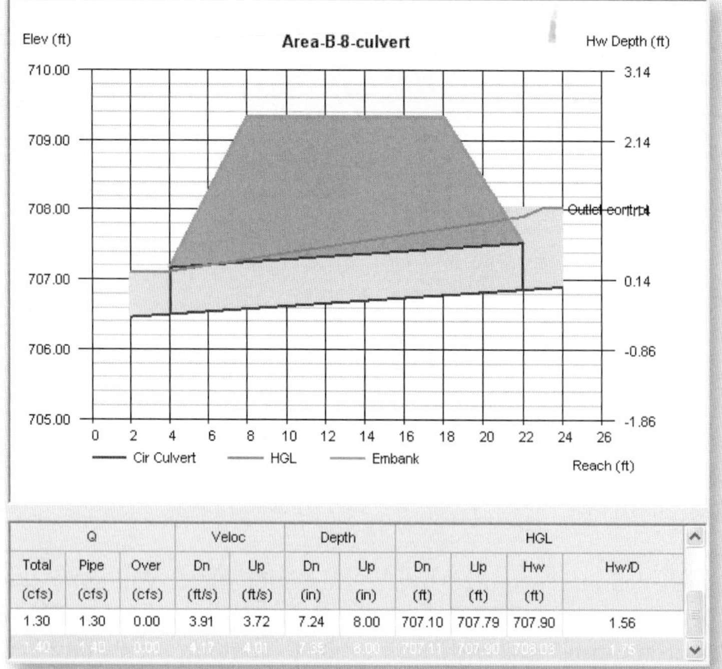


**FIGURE 13.4**

Graphics Display has three display modes that can be set by three buttons: Plot (the default), P-Curve, and diagram. Plot shows the water level for a selected result from the results grid. You select P-Curve to display the P-curve for the current structure definition (see Figure 13.5). With Plot toggled on, when you click diagram, a schematic is displayed that identifies the input variables (see Figure 13.6). With P-Curve toggled on, when you click diagram, a schematic P-Curve diagram is drawn (see Figure 13.7).

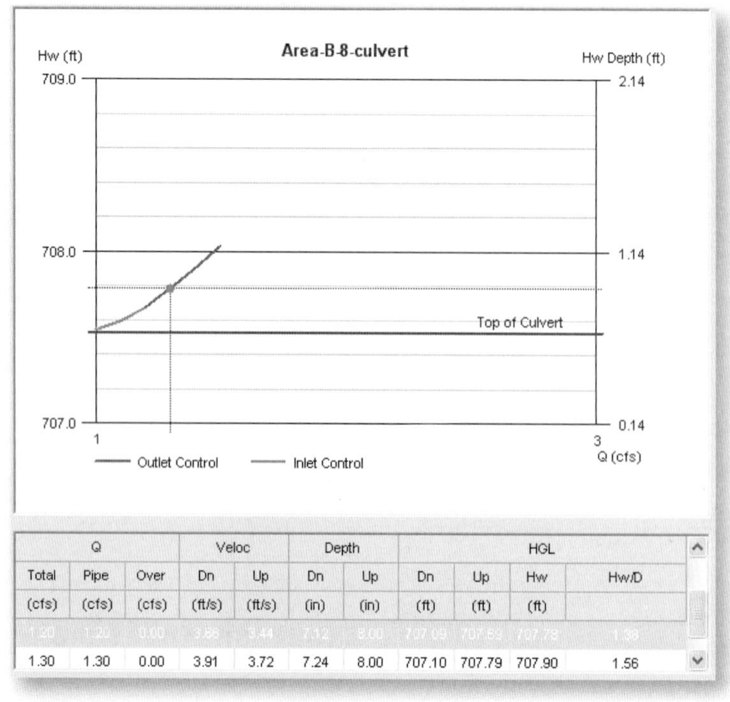


**FIGURE 13.5**

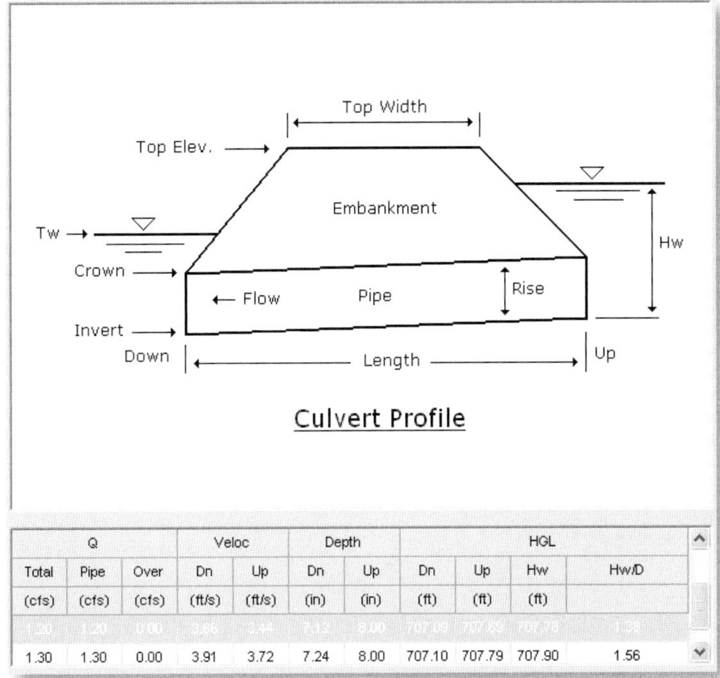


| Q | | | Veloc | | Depth | | HGL | | | |
|---|---|---|---|---|---|---|---|---|---|---|
| Total | Pipe | Over | Dn | Up | Dn | Up | Dn | Up | Hw | Hw/D |
| (cfs) | (cfs) | (cfs) | (ft/s) | (ft/s) | (in) | (in) | (ft) | (ft) | (ft) | |
| 1.20 | 1.20 | 0.00 | 3.66 | 3.44 | 7.12 | 8.00 | 707.09 | 707.69 | 707.78 | 1.38 |
| 1.30 | 1.30 | 0.00 | 3.91 | 3.72 | 7.24 | 8.00 | 707.10 | 707.79 | 707.90 | 1.56 |

**FIGURE 13.6**

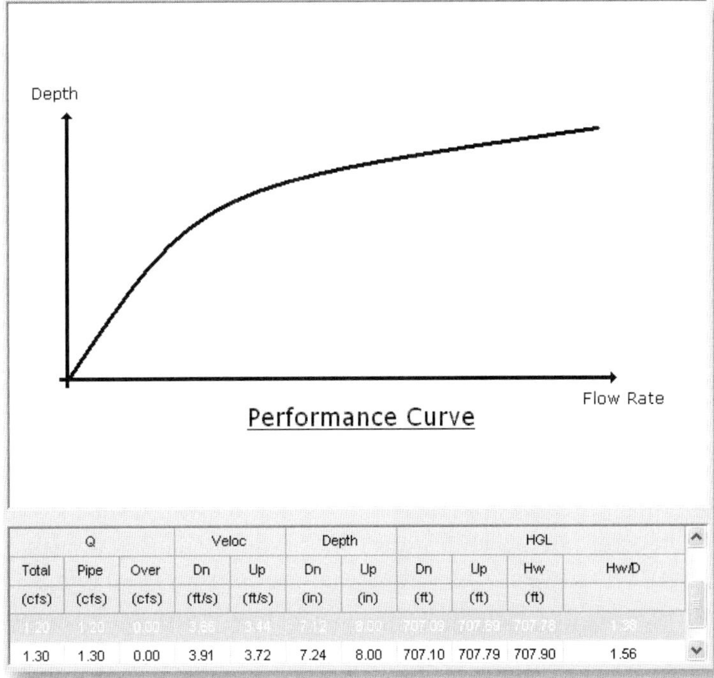


| Q | | | Veloc | | Depth | | HGL | | | |
|---|---|---|---|---|---|---|---|---|---|---|
| Total | Pipe | Over | Dn | Up | Dn | Up | Dn | Up | Hw | Hw/D |
| (cfs) | (cfs) | (cfs) | (ft/s) | (ft/s) | (in) | (in) | (ft) | (ft) | (ft) | |
| 1.20 | 1.20 | 0.00 | 3.66 | 3.44 | 7.12 | 8.00 | 707.09 | 707.69 | 707.78 | 1.38 |
| 1.30 | 1.30 | 0.00 | 3.91 | 3.72 | 7.24 | 8.00 | 707.10 | 707.79 | 707.90 | 1.56 |

**FIGURE 13.7**

### Results Grid

Results Grid displays by increment the calculated values of a structure. The increment is set in the Calcs' Grid section (Q incr). When Graphics Display is in plot mode, you click an interval in results grid to display its values graphically.

Graphics Display's center contains the structure's name.

If Graphics Display is in Plot mode, the icons to the right of the structure's name are active. These icons change colors to indicate if they are active or not (see Figure 13.8).

The bucket icon toggles between color-filled shapes or outlines and the icon to its right toggles on/off the EGL line. The remaining three icons are Zoom commands. The first icon on the left is a zoom window that shows more detail (less overall area), zooms back to show more area, and resets to return the graphics display back to its default centering.

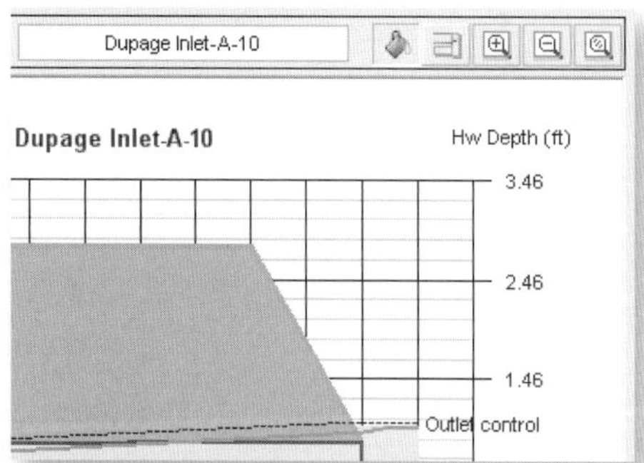


**FIGURE 13.8**

## PRECIPITATION

Runoff amounts are critical to many structure calculations. For example, can a weir discharge a 100-year storm safely, or can a culvert design handle a 50-year storm? How runoff amounts are determined falls into two categories: IDF curves (Intensity Duration Curves) and SCS rainfall amounts. The system used in your area depends on mandates by local governing boards. Some local governments have their own charts and amounts, and they may be different from those used in this textbook.

The Hydrology panel supports Rational, Modified Rational, and SCS runoff calculation methods. The next unit covers the use of the Hydrographs application.

### IDF Curves

IDF Curves describes rainfall intensity for a period of 60 minutes for several storm types, for example, 2-, 5-, 10-, or 25-year storms. If pertinent data is available, Express can calculate the IDF curves. This information came from the Hydro-35 and 40 circulars. However, NOAA has updated the data and publishes it on its rainfall Web site. The new data is also in NOAA's Volume 14 atlas.

The IDF precipitation calculator takes these intensities and produces the IDF curves. The Table tab of IDF is where a user inputs the intensities, and after their values have been input, the calculator displays the intensity curves (see Figures 13.9 and 13.10).

The Coefficients tab displays two types of standard intensities: FHA and Poly. Each is a different formula, yet obtains similar results. Again, it is your region that mandates the calculation to use.

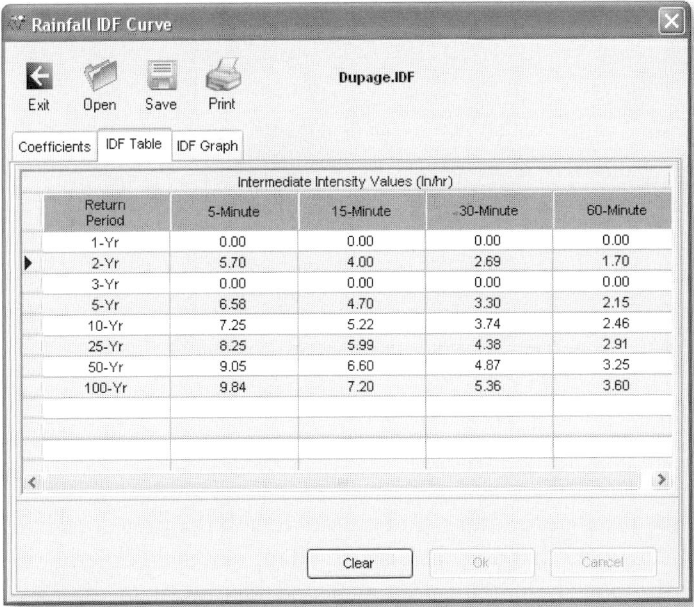


| Intermediate Intensity Values (In/hr) | | | | |
|---|---|---|---|---|
| Return Period | 5-Minute | 15-Minute | 30-Minute | 60-Minute |
| 1-Yr | 0.00 | 0.00 | 0.00 | 0.00 |
| 2-Yr | 5.70 | 4.00 | 2.69 | 1.70 |
| 3-Yr | 0.00 | 0.00 | 0.00 | 0.00 |
| 5-Yr | 6.58 | 4.70 | 3.30 | 2.15 |
| 10-Yr | 7.25 | 5.22 | 3.74 | 2.46 |
| 25-Yr | 8.25 | 5.99 | 4.38 | 2.91 |
| 50-Yr | 9.05 | 6.60 | 4.87 | 3.25 |
| 100-Yr | 9.84 | 7.20 | 5.36 | 3.60 |

**FIGURE 13.9**

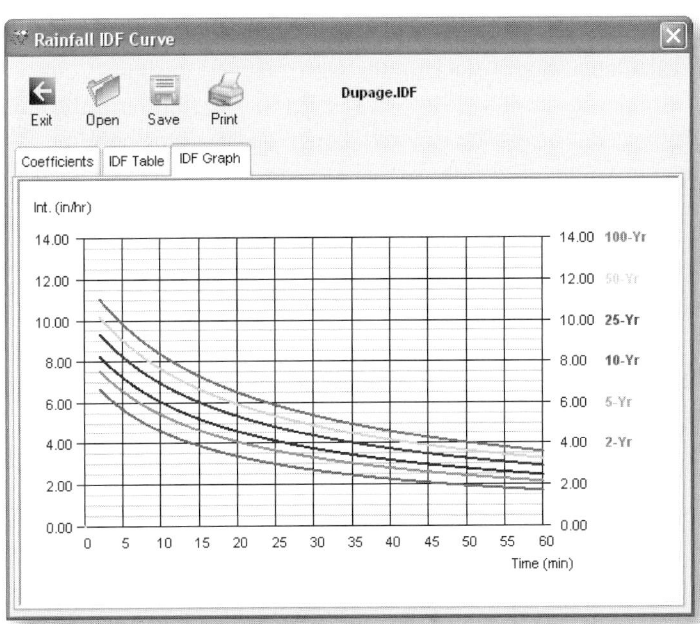


**FIGURE 13.10**

## SCS

The Soil Conservation Service also issues rainfall data. This data is the basis for TR-55 calculations: Rational and Modified Rational. This system uses a 24-hour amount for storms rated from 1- to 100-year storms (see Figure 13.11).

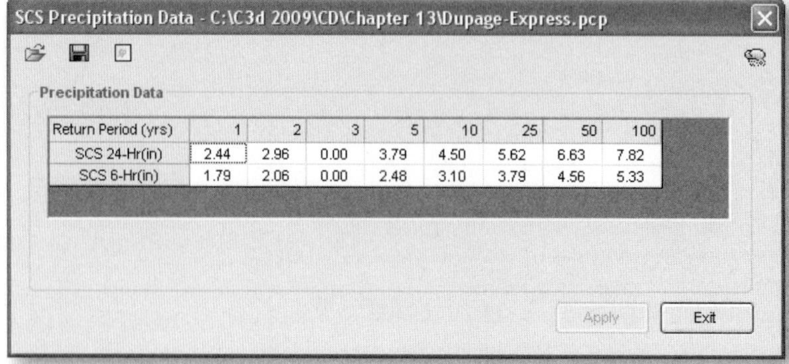

SCS Precipitation Data - C:\C3d 2009\CD\Chapter 13\Dupage-Express.pcp
Precipitation Data
Return Period (yrs) 1 2 3 5 10 25 50 100
SCS 24-Hr(in) 2.44 2.96 0.00 3.79 4.50 5.62 6.63 7.82
SCS 6-Hr(in) 1.79 2.06 0.00 2.48 3.10 3.79 4.56 5.33
Apply
Exit


**FIGURE 13.11**

## STRUCTURES

Many roadway projects require that runoff around or underneath the project be handled. Other sites accumulate water that must be discharged off-site in an orderly way. Structures are tools for the orderly handling of water and guiding it to discharge points. A structure may be how a site discharges its accumulated runoff. This calculator evaluates the structure over a range of flows. These same structures appear in Hydrographs as methods of routing water.

### Culverts

The Culverts calculator calculates water velocities through a culvert design. A culvert can be an arch, a box, a circular, or an Ell. There can be up to 4 barrels, and the culvert's inlet edge can be projecting or can be mitered, squared edge, and beveled. All coefficients are Manning's n values (see Figure 13.12). You must select the cell to expose a drop-list arrow when you are setting values for shape, n-values, edge, and number of barrels. When you click the drop-list arrow, the list displays the available selections.

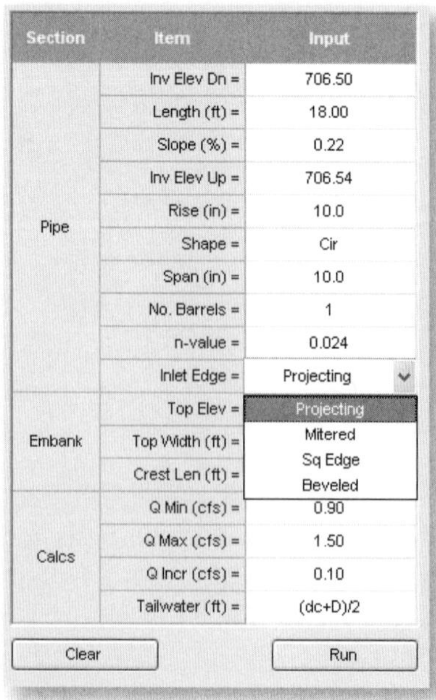


**FIGURE 13.12**

## Channel

Channels have six types: rectangular, triangular, trapezoidal, gutter, circular, and user-defined (see Figure 13.13). A user-defined channel can have up to 50 station, elevation, and manning's n numbers. Clicking the station-elevation cell opens a data-entry dialog box for the user values.

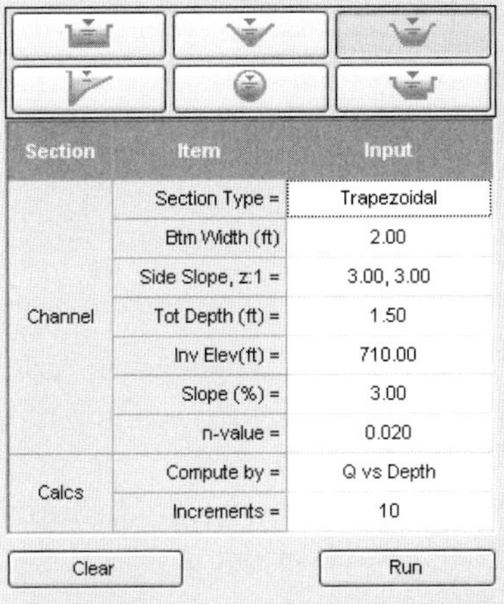


**FIGURE 13.13**

## Inlets

Inlets has six types: curb inlet, grate inlet, combination inlet, drop curb, grate, and slotted (see Figure 13.14). The values for which the Input grids prompt depend on the selected inlet type.

After you have selected the inlet type, the graphics area displays an inlet schematic. When you click in Input Grid's Gutter section, the graphics area displays a second schematic that explains the variables for the selected section (see Figure 13.15).

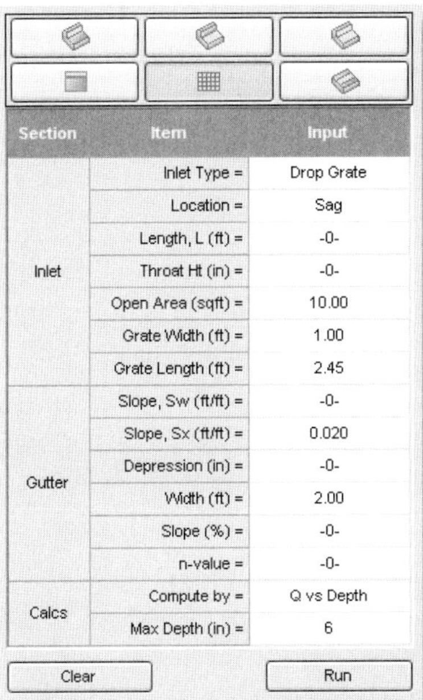


**FIGURE 13.14**

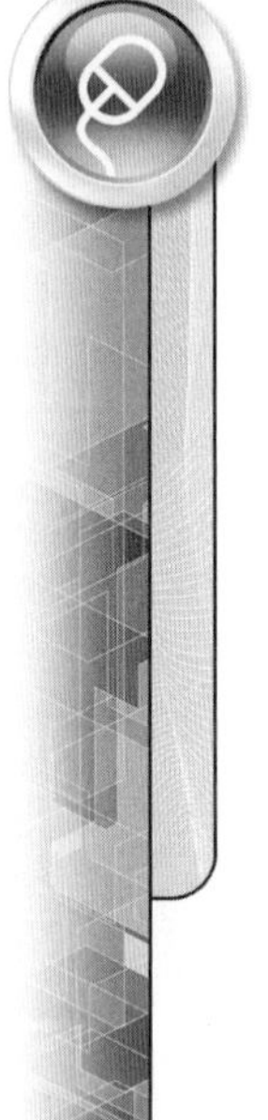

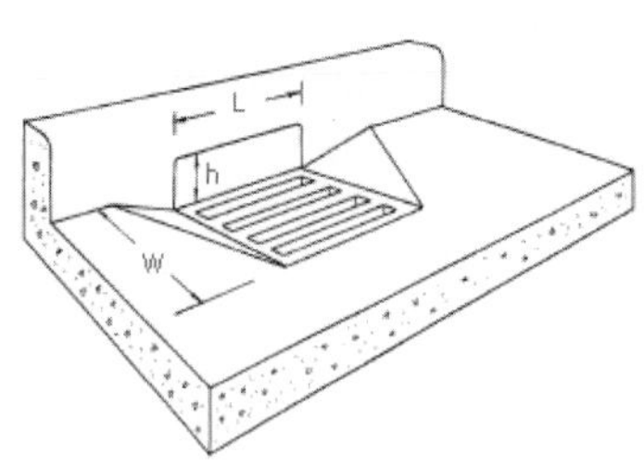


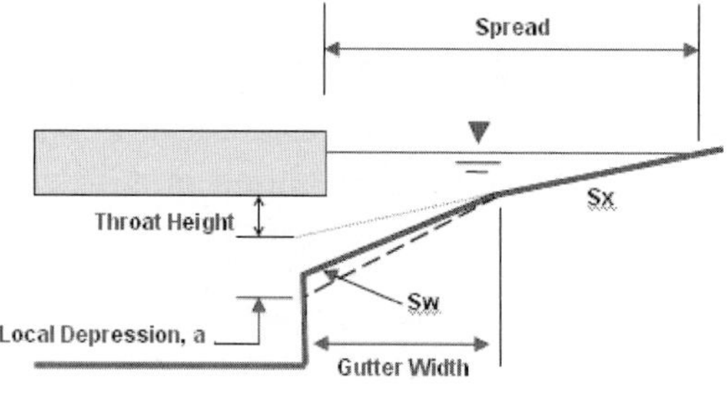


**FIGURE 13.15**

## Weirs

A weir allows variable flow as the water level varies behind it (less when lower, more when higher) (see Figure 13.16). There are six weir shapes: rectangular, compound, circular, v-notched, trapezoidal, and proportional. The type buttons at the panel's top left set the weir shape. A second value needs to be set. Click the crest cell in the Input grid to display a drop-list arrow listing the edge shapes: broad and sharp.

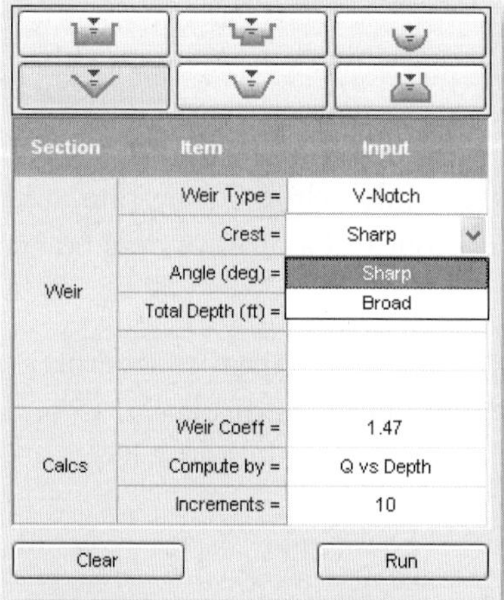


**FIGURE 13.16**

## REPORTS

Selecting the printer icon displays a shortcut menu listing Express's reports. Report prints only the selected values in the results grid. The Results Grid… report prints only the grid values and does not contain any structure name or data values.

### EXERCISE 13-1

After completing this exercise, you will:

- Be able to define an IDF chart.
- Be able to create an SCS chart.
- Be able to calculate culverts.
- Be able to calculate a weir.
- Be able to calculate a channel.

## Create an IDF Chart

1. If you are not in Civil 3D, start the application by double-clicking its desktop icon.
2. On the Ribbon, click the ***Analyze*** tab. On the Design panel, select LAUNCH EXPRESS.
3. In the Hydraflow Express application, at the top left, click the IDF icon to open the Rainfall IDF Curve dialog box.
4. In the dialog box's top left, click the ***Coefficients*** tab, and make sure that ***FHA*** is the active tab.
5. Click the ***IDF Table*** tab, and at its bottom, click ***Clear***.
6. Using Table 13.1 as a guide, enter the following values for the DuPage county IDF table:

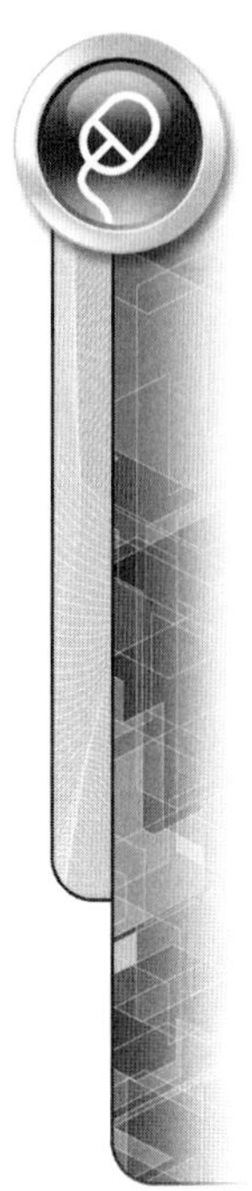

**TABLE 13.1**

| Year | 5-Minute | 15-Minute | 30-Minute | 60-Minute |
|---|---|---|---|---|
| 2 | 0.44 | 0.84 | 1.12 | 1.38 |
| 5 | 0.49 | 0.94 | 1.29 | 1.62 |
| 10 | 0.58 | 1.11 | 1.54 | 1.96 |
| 25 | 0.67 | 1.26 | 1.78 | 2.31 |
| 50 | 0.76 | 1.43 | 2.04 | 2.89 |
| 100 | 0.84 | 1.58 | 2.28 | 3.05 |

7. Click SAVE, browse to the Civil 3D Projects folder, for the IDF name, enter **DuPage**, click ***Save***, and click ***Exit*** to close the Rainfall IDF Curve.

## SCS Chart

1. In the Hydraflow Express application, at the top left, click the ***Precip*** icon to open the SCS Precipitation Data dialog box.
2. In SCS Precipitation Data, at its top left, click ***Clear***.
3. Using Table 13.2 as a guide, enter the following values for the SCS Precipitation Data for DuPage county:

**TABLE 13.2**

| Hr | 1 | 2 | 5 | 10 | 25 | 50 | 100 |
|---|---|---|---|---|---|---|---|
| 24 | 2.44 | 2.96 | 3.79 | 4.50 | 5.62 | 6.63 | 7.82 |
| 6 | 1.79 | 2.06 | 2.48 | 3.10 | 3.79 | 4.56 | 5.33 |

4. Click ***Apply***.
5. In the top left of SCS Precipitation Data, click SAVE, for the SCS name, enter **DuPage**, click ***Save***, and click ***Exit*** to close the SCS Precipitation Data dialog box.

## Culvert — Area-B 10-Inch Culvert

This exercise section calculates four culvert values: two for a 10-inch and two for an 8-inch culvert. The resulting values will be a part of the Unit 2 flow values. The target discharge from parking lot catch basins into a pond needs to be around 3 fps per catch basin. This target fps is for erosion control and for using the catch basins as delay structures in the event of heavy rains.

1. At the top left of Hydraflow Express, click ***Culverts*** to open the Culvert calculator.
2. Using Figure 13.17 as a guide, set the values in the Input grid. When you enter the culvert length, do not include the foot mark.

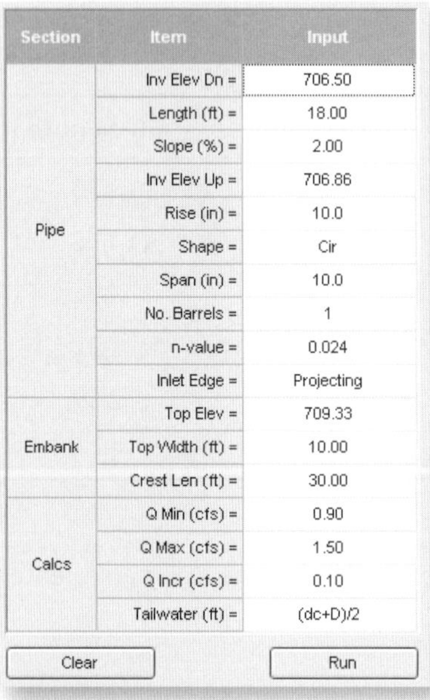

| Section | Item | Input |
|---|---|---|
| Pipe | Inv Elev Dn = | 706.50 |
| | Length (ft) = | 18.00 |
| | Slope (%) = | 2.00 |
| | Inv Elev Up = | 706.86 |
| | Rise (in) = | 10.0 |
| | Shape = | Cir |
| | Span (in) = | 10.0 |
| | No. Barrels = | 1 |
| | n-value = | 0.024 |
| | Inlet Edge = | Projecting |
| Embank | Top Elev = | 709.33 |
| | Top Width (ft) = | 10.00 |
| | Crest Len (ft) = | 30.00 |
| Calcs | Q Min (cfs) = | 0.90 |
| | Q Max (cfs) = | 1.50 |
| | Q Incr (cfs) = | 0.10 |
| | Tailwater (ft) = | (dc+D)/2 |



**FIGURE 13.17**

3. After entering the values, click ***Run*** to compute the culvert values.
4. At the top center, for the name, enter **Area-B-10-culvert**.
5. From the File menu, select SAVE AS..., save the computations as **Area B-10-culvert**, click ***Save***, and click ***OK*** to continue.
6. Review the culvert values in the results grid, select different results, and view the changes to the culvert diagram.
7. Click ***P-Curve*** to view the chart.
8. While in P-Curve mode, click ***Diag*** to view the P-Curve diagram.
9. Click ***Diag*** (now Plot) and then click ***Plot*** to return to the initial view.

The fps at the flow rate of 1cfs (the runoff rate Q) is different for the up and down velocity. The target is 3 fps for the up and down rate.

## Culvert — Area-B 8-Inch Culvert

1. Using Figure 13.18 as a guide, change the pipe rise to 8 (for an 8-inch pipe) and click ***Run*** to calculate the new flow rates for an 8-inch culvert.

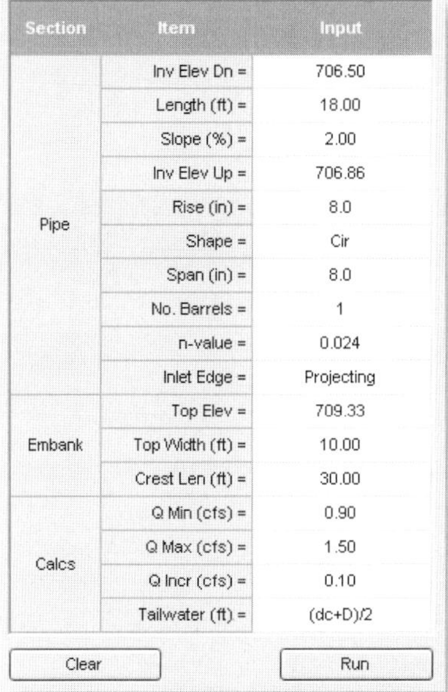

| Section | Item | Input |
|---|---|---|
| Pipe | Inv Elev Dn = | 706.50 |
| | Length (ft) = | 18.00 |
| | Slope (%) = | 2.00 |
| | Inv Elev Up = | 706.86 |
| | Rise (in) = | 8.0 |
| | Shape = | Cir |
| | Span (in) = | 8.0 |
| | No. Barrels = | 1 |
| | n-value = | 0.024 |
| | Inlet Edge = | Projecting |
| Embank | Top Elev = | 709.33 |
| | Top Width (ft) = | 10.00 |
| | Crest Len (ft) = | 30.00 |
| Calcs | Q Min (cfs) = | 0.90 |
| | Q Max (cfs) = | 1.50 |
| | Q Incr (cfs) = | 0.10 |
| | Tailwater (ft) = | (dc+D)/2 |

Clear Run

**FIGURE 13.18**

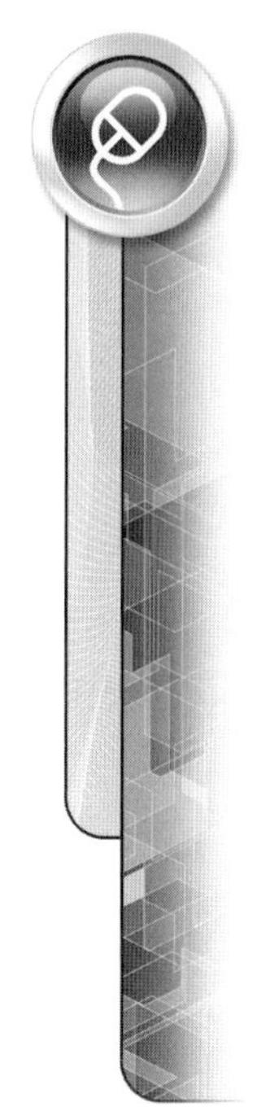

2. At the top center, for the name, enter **Area-B-8-culvert**.
3. From the File menu, select SAVE AS..., save the computations as **Area B-8-culvert**, click ***Save***, and click ***OK*** to continue.
4. Review the culvert values in the results grid and select different results to view the changes to the culvert diagram.
5. Click ***P-Curve*** to view the chart.
6. While in P-Curve mode, click ***Diag*** to view the P-Curve diagram.
7. Click ***Diag*** (now Plot) and then click ***Plot*** to return to the initial view.

The fps at the flow rate of 1cfs (the runoff rate Q) is approximately 3 fps, the target velocity. The Area-B catchbasin will also use an 8-inch outfall.

## Culvert — Area-C 10-Inch Culvert

1. Using Figure 13.19 as a guide, set the following values in the Culvert Design Input grid. When you enter the culvert length, do not include the foot mark.

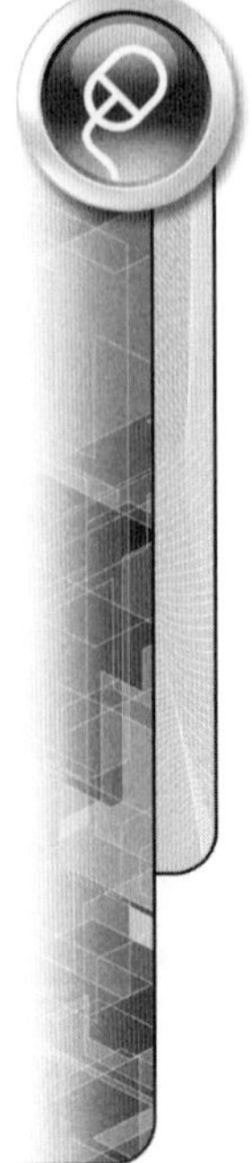

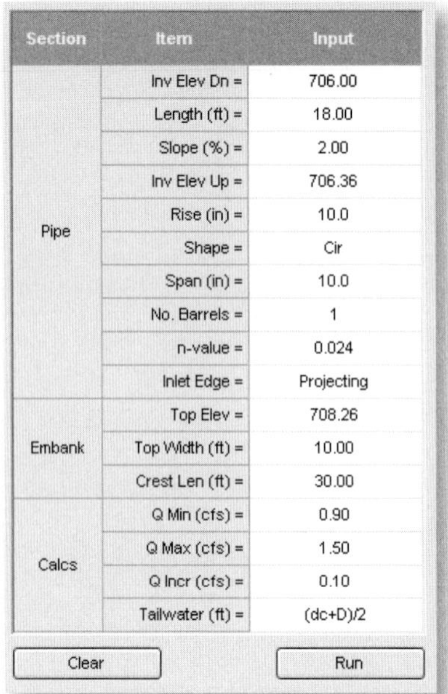

| Section | Item | Input |
|---|---|---|
| Pipe | Inv Elev Dn = | 706.00 |
| | Length (ft) = | 18.00 |
| | Slope (%) = | 2.00 |
| | Inv Elev Up = | 706.36 |
| | Rise (in) = | 10.0 |
| | Shape = | Cir |
| | Span (in) = | 10.0 |
| | No. Barrels = | 1 |
| | n-value = | 0.024 |
| | Inlet Edge = | Projecting |
| Embank | Top Elev = | 708.26 |
| | Top Width (ft) = | 10.00 |
| | Crest Len (ft) = | 30.00 |
| Calcs | Q Min (cfs) = | 0.90 |
| | Q Max (cfs) = | 1.50 |
| | Q Incr (cfs) = | 0.10 |
| | Tailwater (ft) = | (dc+D)/2 |



**FIGURE 13.19**

2. After entering the values, click ***Run*** to compute the culvert values.
3. At the top center, for the name, enter **Area-C-10-culvert**.
4. From the File menu, select SAVE AS..., save the computations as **Area C-10-culvert**, click ***Save***, and click ***OK*** to continue.
5. Review the culvert values in the results grid, and select different results to view the changes to the culvert diagram.
6. Click ***P-Curve*** to view the chart.
7. While in P-Curve mode, click ***Diag*** to view the P-Curve diagram.
8. Click ***Diag*** (now Plot) and then click ***Plot*** to return to the initial view.

The fps at the flow rate of 1cfs (the runoff rate Q) is different for the up and down velocity. The target is 3 fps for the up and down rate.

## Culvert — Area-C 8-Inch Culvert

1. Using Figure 13.20 as a guide, change the pipe rise to 8 (for an 8-inch pipe).

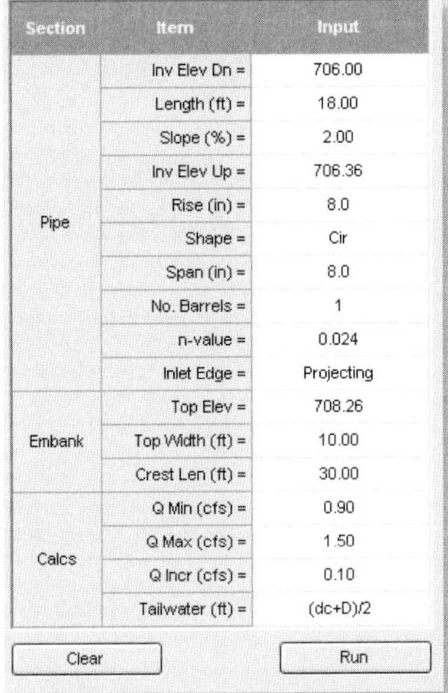

| Section | Item | Input |
|---|---|---|
| Pipe | Inv Elev Dn = | 706.00 |
| | Length (ft) = | 18.00 |
| | Slope (%) = | 2.00 |
| | Inv Elev Up = | 706.36 |
| | Rise (in) = | 8.0 |
| | Shape = | Cir |
| | Span (in) = | 8.0 |
| | No. Barrels = | 1 |
| | n-value = | 0.024 |
| | Inlet Edge = | Projecting |
| Embank | Top Elev = | 708.26 |
| | Top Width (ft) = | 10.00 |
| | Crest Len (ft) = | 30.00 |
| Calcs | Q Min (cfs) = | 0.90 |
| | Q Max (cfs) = | 1.50 |
| | Q Incr (cfs) = | 0.10 |
| | Tailwater (ft) = | (dc+D)/2 |

Clear    Run

**FIGURE 13.20**

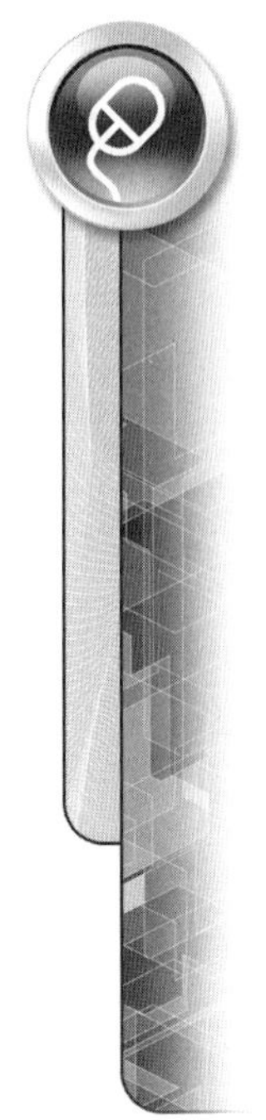

2. After entering the values, click ***Run*** to compute the culvert values.
3. At the top center, for the name, enter **Area-C-8-culvert**.
4. From the File menu, select SAVE AS..., save the computations as **Area C-8-culvert**, click ***Save***, and click ***OK*** to continue.
5. Review the culvert values in the results grid and select different results to view the changes to the culvert diagram.
6. Click ***P-Curve*** to view the chart.
7. While in P-Curve mode, click ***Diag*** to view the P-Curve diagram.
8. Click ***Diag*** (now Plot) and then click ***Plot*** to return to the initial view.

The fps at the flow rate of 1cfs (the runoff rate Q) is approximately 3 fps, the target velocity. The Area-C catchbasin will also use an 8-inch outfall.

## Culvert — Area-C-8-Culvert Report

1. From the File menu, Print flyout, select REPORT....
2. In the Print dialog box, select a printer and click PRINT.
3. If you are printing to a file, enter a report name.

The report is for the selected Results Grid item.

4. From the File menu, Print flyout, select RESULTS GRID....
5. In the Print dialog box, select a printer and click ***OK***.
6. If you are printing to a file, enter a report name.

The report is the current structure's results grid with no other annotation.

## Channel

The next exercise has a channel that empties Area-A into a pond. This section calculates the channel's flow into the pond.

1. Click ***Channels*** and, if necessary, click ***Clear*** to reset the values to zeros.
2. At the panel's top left, click the ***Trapezoidal*** icon.
3. Using Figure 13.21 as a guide, enter the channel's values.

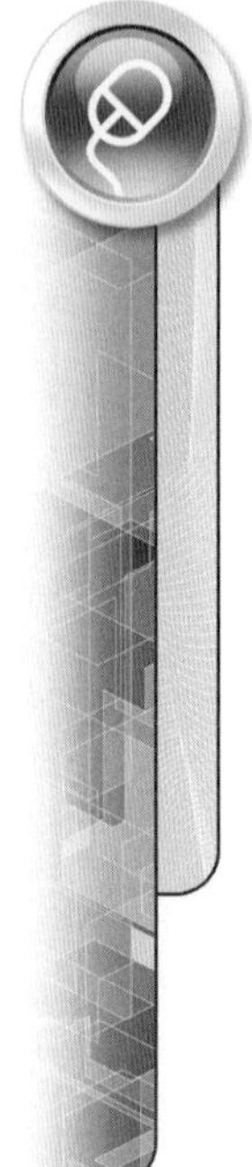

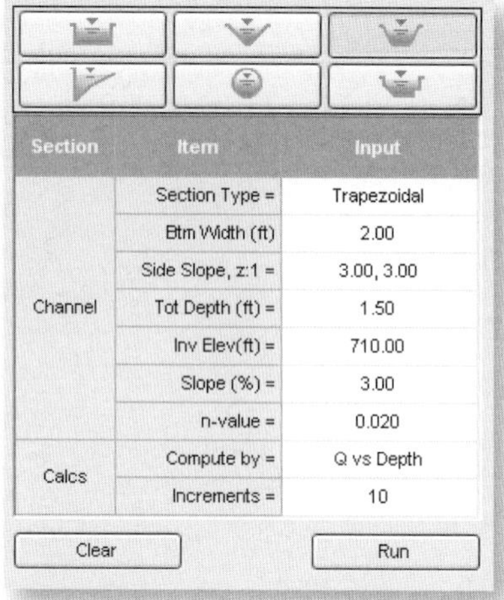

| Section | Item | Input |
|---|---|---|
| Channel | Section Type = | Trapezoidal |
| | Btm Width (ft) | 2.00 |
| | Side Slope, z:1 = | 3.00, 3.00 |
| | Tot Depth (ft) = | 1.50 |
| | Inv Elev(ft) = | 710.00 |
| | Slope (%) = | 3.00 |
| | n-value = | 0.020 |
| Calcs | Compute by = | Q vs Depth |
| | Increments = | 10 |



**FIGURE 13.21**

4. After entering the values, click ***Run*** to compute the channel values.
5. At the top center, for the name, enter **Area-A-channel**.
6. From the File menu, select SAVE AS..., save the computations as **Area A-channel**, click ***Save***, and click ***OK*** to continue.
7. Review the channel values in the results grid and select different results to view the changes to the culvert diagram.
8. Click ***P-Curve*** to view the chart.
9. While in P-Curve mode, click ***Diag*** to view the P-Curve diagram.
10. Click ***Diag*** (now Plot) and then click ***Plot*** to return to the initial view.

## Weir

1. At Express's top, click ***Weirs*** to open the Weir calculator.
2. At the panel's top left, click the ***V-Notch*** icon.
3. Using Figure 13.22 as a guide, enter the values into the calculator.

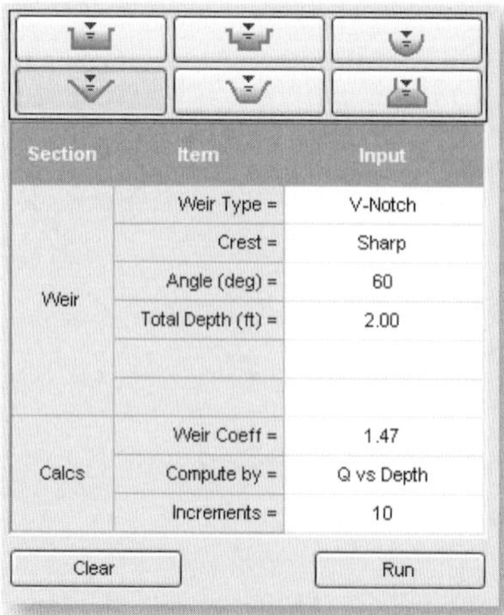

| Section | Item | Input |
|---|---|---|
| Weir | Weir Type = | V-Notch |
| | Crest = | Sharp |
| | Angle (deg) = | 60 |
| | Total Depth (ft) = | 2.00 |
| | | |
| | | |
| Calcs | Weir Coeff = | 1.47 |
| | Compute by = | Q vs Depth |
| | Increments = | 10 |



**FIGURE 13.22**

4. After entering the values, click ***Run*** to compute the channel values.
5. At the top center, for the name, enter **Seward's-Pond-Weir**.
6. From the File menu, select SAVE AS..., save the computations as **Sewards-Pond-Weir**, click ***Save***, and click ***OK***.
7. Review the weir values in the results grid and select different results to view the changes to the culvert diagram.

Select some results and notice that the plot changes to show the current values in schematic form.

8. Click ***P-Curve*** to view the chart.
9. While in P-Curve mode, click ***Diag*** to view the P-Curve diagram.
10. Click ***Diag*** (now Plot) and then click ***Plot*** to return to the initial view.
11. Close Hydraflow Express and return to Civil 3D.

This ends the Hydraflow Express extension exercise. Express's focus is hydrological structures and their performance. The next unit's focus is the Hydrograph extension.

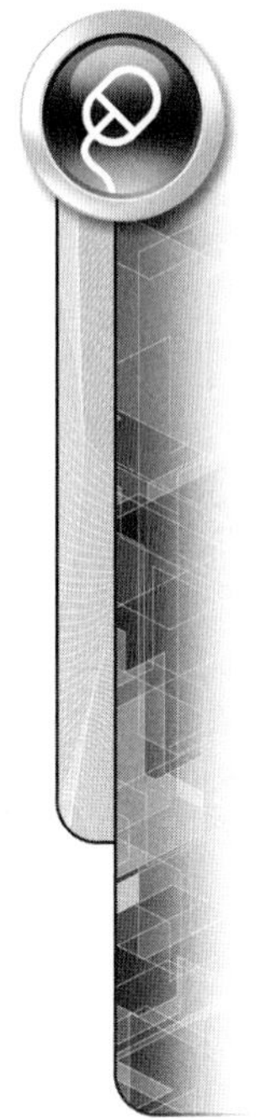

## SUMMARY

- Express supports channels, culverts, weirs, and inlets.
- Each structure type has type variations.
- Before you enter data, make sure the correct structure type is set.
- Express displays a results grid at the bottom right of the calculator.
- Selecting a result displays the selected result's values in the structure's plot, or P-Curve.
- There are two structure report types: Report (entered values) and Results grid (computation result values).

## UNIT 2: HYDROLOGY

Hydrographs model complex watersheds. To handle this complexity, modeling tools include drainage areas, combiners, diverters, reaches, and ponds.

### SCS RUNOFF CURVE NUMBER EDITOR

The Hydrograph application lists the common curve numbers in its help system. You should print a list of your most used values.

Many sites are not homogeneous; even a lot is a composite of pervious and impervious materials (house and lawn). Watersheds have differing conditions over their territory, and allowances need to be made to reflect their varying absorption rates. A Composite Runoff Curve Number calculator calculates a curve number that considers subarea characteristics (see Figure 13.23).

Composite CN

| Area 1 | | Area 4 | |
|---|---|---|---|
| Area (ac) ....... = | 0.07 | Area (ac) ....... = | 0.00 |
| Curve No. CN . = | 85 | Curve No. CN . = | 0 |
| **Area 2** | | **Area 5** | |
| Area (ac) ....... = | 0.07 | Area (ac) ....... = | 0.00 |
| Curve No. CN . = | 98 | Curve No. CN . = | 0 |
| **Area 3** | | **Area 6** | |
| Area (ac) ....... = | 0.00 | Area (ac) ....... = | 0.00 |
| Curve No. CN . = | 0 | Curve No. CN . = | 0 |
| **Composite CN** | | | |
| Curve No. CN . = | 92 | Ok / Clear / Exit | |

**FIGURE 13.23**

## TIME OF CONCENTRATION

Time of concentration has three components: sheet, shallow, and channel flow. Each watershed is different and may use only one time-of-concentration component, whereas other watersheds may use a combination of two or all three.

### Sheet Flow

Sheet flow is water flowing over a plane surface. The flow represents the farthest hydraulic distance in the watershed. The maximum sheet flow length is 300 feet.

Sheet flow calculation components are Manning's n, length of flow, two-year rainfall, and land slope (see Figure 13.24).

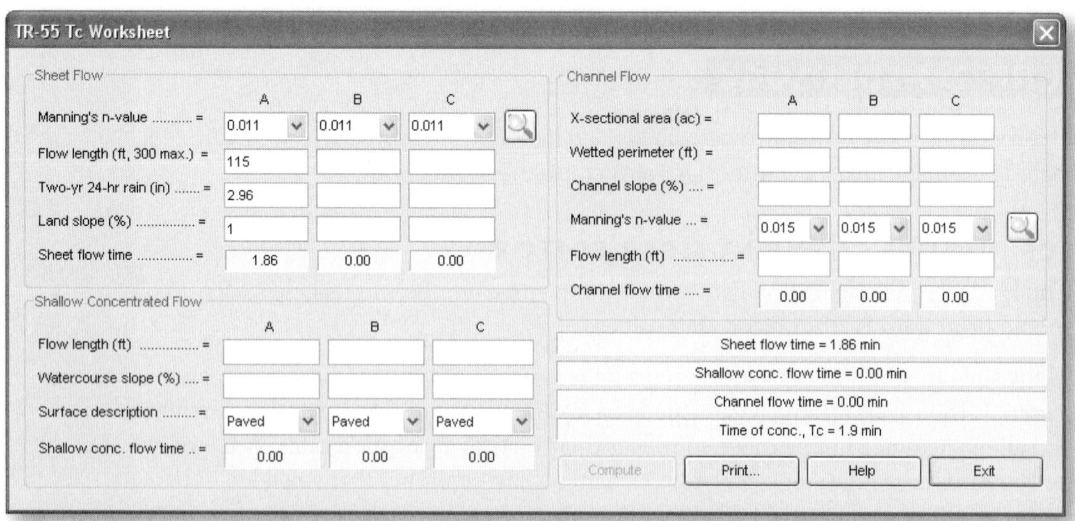


**FIGURE 13.24**

### Shallow Flow

Shallow flow is water that is starting to concentrate into organized structures. This flow occurs between the sheet and channel flow. Shallow flow components are surface, flow length, and slope (see Figure 13.24).

### Open Channel Flow

Channel flow is a structure, natural or manmade, that carries water in a defined space.

Channel flow components are cross-section flow area, wetter perimeter, channel slope, Manning's n, and flow length (see Figure 13.24).

## HYDROGRAPH COMPONENTS

The Hydrographs extension has three input interfaces: model, hydrographs, and pond. All components can be defined as a user creates a model or can be developed in their respective tabs (Hydrographs or Ponds). While building a model and defining each component, a user can review their properties by selecting the appropriate tab.

### Ponds

Pond definitions can be from contours, a trapezoid, manual, or chambers. Chambers represent underground storage and discharge (see Figure 13.25).

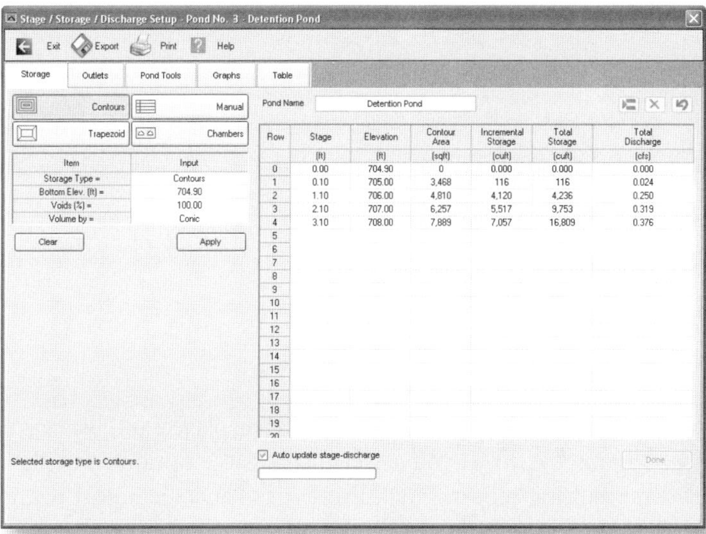


**FIGURE 13.25**

### Hydrographs

A hydrograph is a water flow that is from a runoff calculation or is outfall from a reach or pond. The Hydrographs tab defines or holds the model's defined runoffs and outfalls (see Figure 13.26). Select a hydrograph number and press the right mouse button to display a shortcut menu that lists runoff calculation methods or defines other hydrological components.

After a pond or hydrograph component has been defined in its own tab, it appears as an icon in the model tab. A user can relocate and attach components in the model tab.

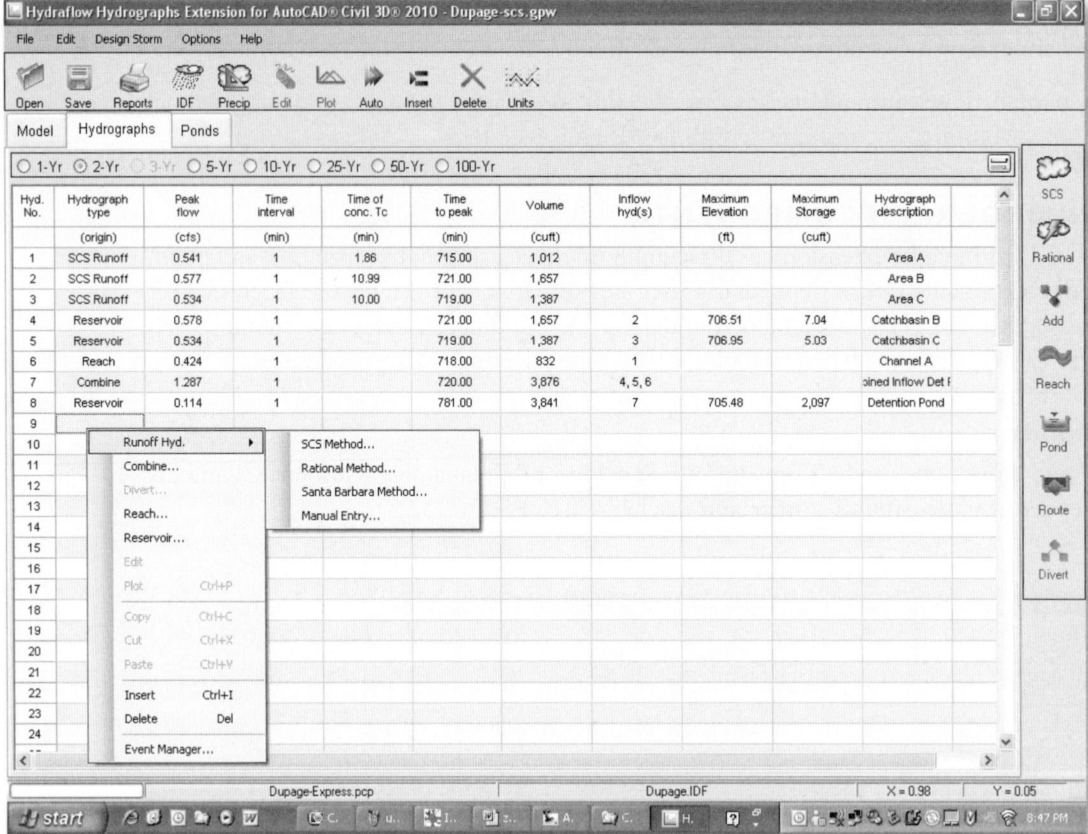

FIGURE 13.26

## REPORTS

When you select the Reports icon, the Reports Print Menu is displayed and lists what a report may contain (see Figure 13.27). If you select the possible options and click Preview, you can review the resulting report. Click Print… to send the report to the printer.

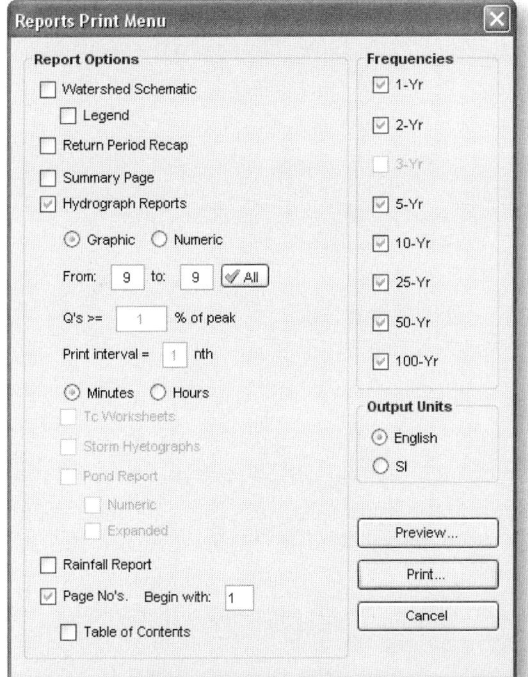


**FIGURE 13.27**

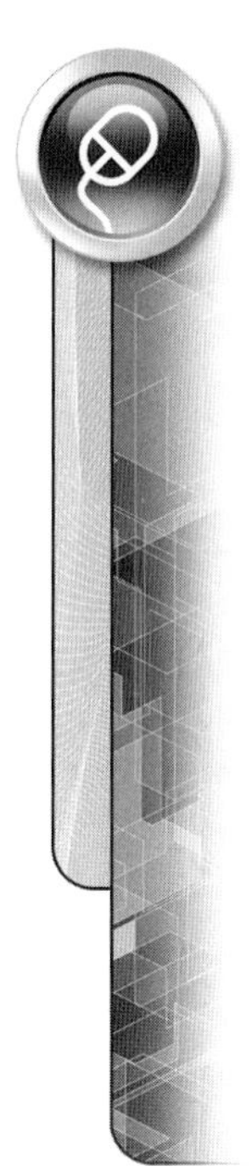

## EXERCISE 13-2

After completing this exercise, you will:

- Be able to calculate runoff.
- Be able to define a pond.
- Be able to build a hydrograph model.
- Be able to review and modify values.
- Print a report.

### Exercise Setup — SCS and IDF Charts

If you have not entered the values for the SCS and IDF charts for DuPage, you can complete the Create IDF and Create SCS chart of the previous exercise's sections. Or, the Chapter 13 folder of the CD that accompanies this textbook contains these files if you did not create them.

1. If you are not in Civil 3D, start the application by double-clicking its desktop icon.
2. On the Ribbon, click the ***Analyze*** tab. On the Design panel, select LAUNCH HYDROGRAPHS.
3. At Hydrograph's top, click ***IDF*** to open the Rainfall IDF Curve dialog box.
4. Click ***Open***, browse to and select *DuPage.idf*, click ***Open***, and at the top left, click ***Exit***.
5. At the top of Hydrograph, click ***Precip*** to open the Event Manager dialog box.
6. Click ***Open***, browse to and select *DuPage.pcp* (or DuPage-Express from the Chapter 13 folder of the CD that accompanies this textbook), click ***Open***, and at the bottom right, click ***Exit***.

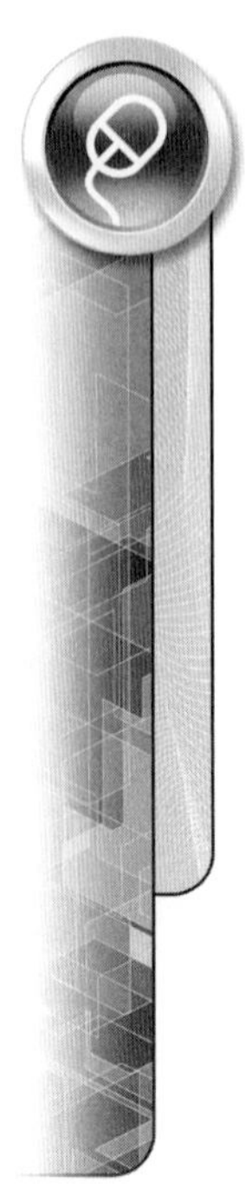

## Runoff Calculations — SCS — Area A

You first calculate runoff values for three zones. Area A is the entrance to the site and has the following statistics:

Name: Area A
Area: 0.14 Acres
Basin Data

| Type | Coefficient | Area 1 |
|---|---|---|
| Asphalt | 98 | 0.07 Acres |
| Heavy Soil | 85 | 0.07 Acres |

Time of Concentration: TR-55
Distance: 115 Ft.
Slope: 1

1. Select the ***Hydrographs*** tab, and change the rain period to **2-yr**.
2. For Hydrograph 1, select the Hydrograph Type cell, press the right mouse button, and from the Runoff Hyd. flyout, select `SCS METHOD...` to open the SCS Runoff Hydrograph dialog box.
3. For Description, enter **Area A** and check that the current SCS is DuPage by clicking ***Event Manager*** at the bottom right.
4. If necessary, change to DuPage.pcp (or DuPage-Express from the Chapter 13 folder of the CD that accompanies this textbook) by clicking the folder icon in the upper left, browse to and select *DuPage.pcp*, click ***Open***, and click ***Exit***.
5. For the Drainage Area (Ac) enter **0.14**.

The area's coefficient is a combination coefficient; it is the effect of the two types of zone materials.

6. In SCS Runoff Hydrograph, to the right of Curve Number (CN), click % (the percentage icon) to open the Composite CN dialog box.
7. Using Figure 13.28 as a guide and the Area A values from earlier in this unit, enter them, and when you are done, click ***OK*** to return to the SCS Runoff Hydrograph dialog box.

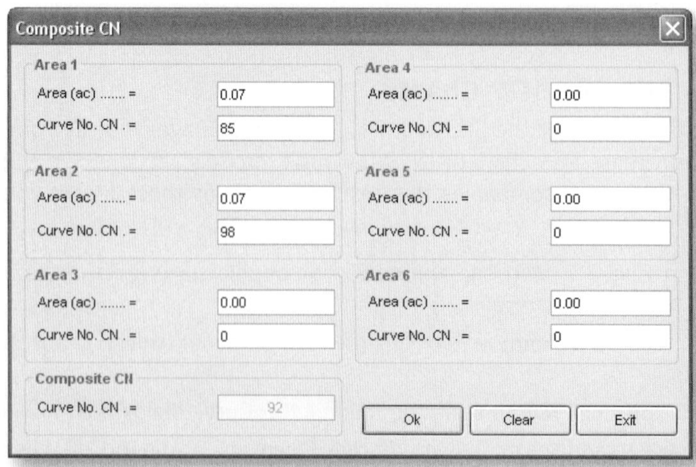


**FIGURE 13.28**

8. Change the Time of Concentration (Tc) Method to TR-55 by clicking the option button under the percent icon.
9. To the right of Time of Concentration, toggle TR55, and click ***TR-55*** to open the TR-55 Tc Worksheet dialog box.
10. Using Figure 13.29 as a guide, and the Area A values from earlier in this exercise section, enter **115** for the flow length and **1** for the slope. When you click in the two-year rainfall cell, the rainfall amount is automatically entered from the current SCS values.

TR-55 Tc Worksheet

Sheet Flow

| | A | B | C |
|---|---|---|---|
| Manning's n-value .......... = | 0.011 | 0.011 | 0.011 |
| Flow length (ft, 300 max.) = | 115 | | |
| Two-yr 24-hr rain (in) ....... = | 2.96 | | |
| Land slope (%) ............... = | 1 | | |
| Sheet flow time .............. = | 1.86 | 0.00 | 0.00 |

Shallow Concentrated Flow

| | A | B | C |
|---|---|---|---|
| Flow length (ft) ............... = | | | |
| Watercourse slope (%) .... = | | | |
| Surface description ......... = | Paved | Paved | Paved |
| Shallow conc. flow time .. = | 0.00 | 0.00 | 0.00 |

Channel Flow

| | A | B | C |
|---|---|---|---|
| X-sectional area (ac) = | | | |
| Wetted perimeter (ft) = | | | |
| Channel slope (%) .... = | | | |
| Manning's n-value ... = | 0.015 | 0.015 | 0.015 |
| Flow length (ft) ............... = | | | |
| Channel flow time .... = | 0.00 | 0.00 | 0.00 |

Sheet flow time = 1.86 min
Shallow conc. flow time = 0.00 min
Channel flow time = 0.00 min
Time of conc., Tc = 1.9 min

Compute | Print... | Help | Exit

**FIGURE 13.29**

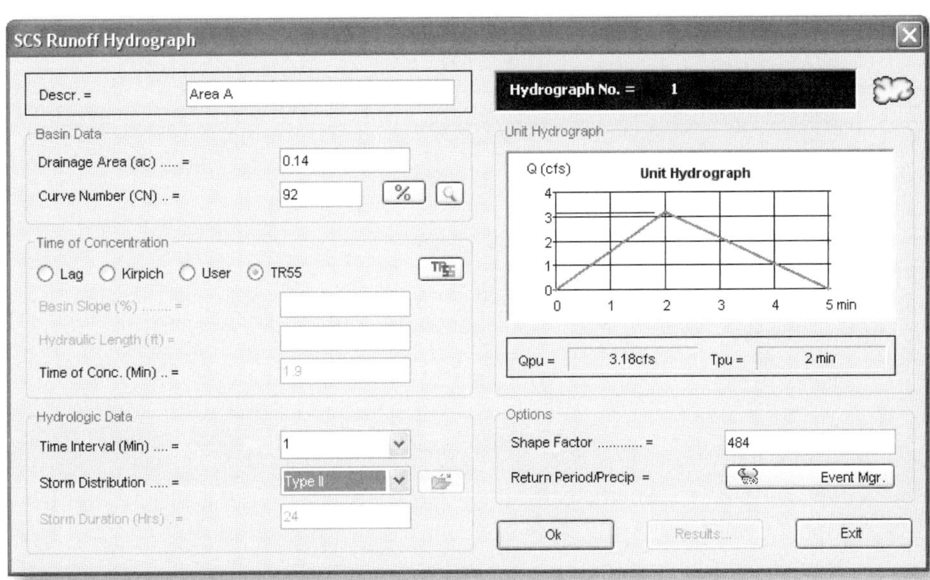


**FIGURE 13.30**

11. When you are done, click ***Compute*** to view the calculated values. Finally, click ***Exit*** to return to the SCS calculator.
12. In the dialog box's lower left, set the Time Interval (Min) to **1** and the Storm Distribution to **Type II**. Your calculator should look like Figure 13.30.
13. Review the calculated Hydrograph by clicking ***OK*** and then clicking ***Results...*** (see Figure 13.31).
14. Click ***Exit*** until you have returned to the Hydrographs page.
15. From the File menu, select SAVE PROJECT AS... and name the file **DuPage Lakewood**.

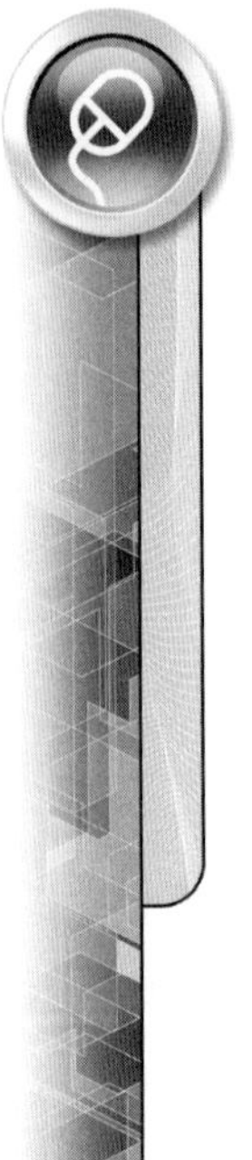

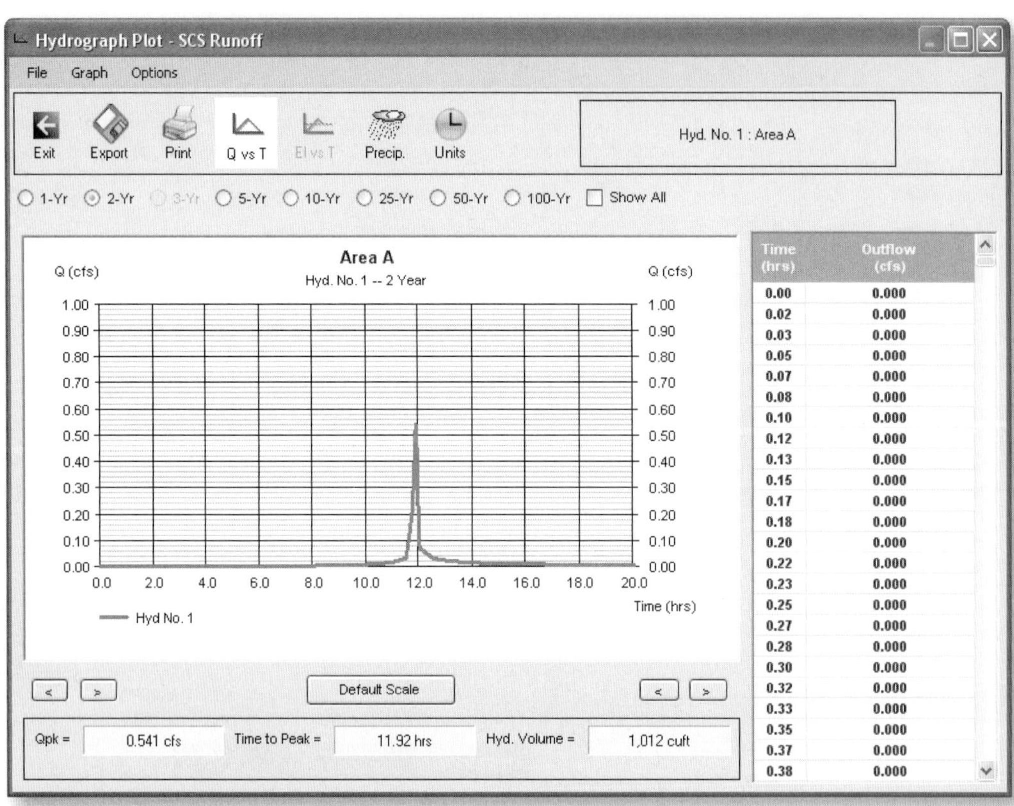


| Time (hrs) | Outflow (cfs) |
|---|---|
| 0.00 | 0.000 |
| 0.02 | 0.000 |
| 0.03 | 0.000 |
| 0.05 | 0.000 |
| 0.07 | 0.000 |
| 0.08 | 0.000 |
| 0.10 | 0.000 |
| 0.12 | 0.000 |
| 0.13 | 0.000 |
| 0.15 | 0.000 |
| 0.17 | 0.000 |
| 0.18 | 0.000 |
| 0.20 | 0.000 |
| 0.22 | 0.000 |
| 0.23 | 0.000 |
| 0.25 | 0.000 |
| 0.27 | 0.000 |
| 0.28 | 0.000 |
| 0.30 | 0.000 |
| 0.32 | 0.000 |
| 0.33 | 0.000 |
| 0.35 | 0.000 |
| 0.37 | 0.000 |
| 0.38 | 0.000 |

**FIGURE 13.31**

## Runoff Calculations — Area B

Area B is the eastern half of the entrance road and half of the parking lot and has the following statistics:

Name: Area B
Area: 0.17 Acres
Basin Data

| Type | Coefficient | Area 1 |
|---|---|---|
| Asphalt | 98 | 0.17 Acres |

Time of Concentration: TR-55
Distance: 150 Ft.
Slope: 2

1. Repeat the steps used to create Area A for Area B. Click in the Hydrograph Type cell for Hyd. No. 2.
2. Save the calculations after you have established Area B.

## Runoff Calculations — Area C

Area C is the parking lot's northern half and has the following statistics:

Name: Area C
Area: 0.16 Acres
Basin Data

| Type | Coefficient | Area 1 |
|---|---|---|
| Asphalt | 98 | 0.16 Acres |

Time of Concentration: TR-55
Distance: 115 Ft.
Slope: 1

1. Repeat the steps used to create Area A for Area C. Click in the Hydrograph Type cell for Hyd. No. 3.
2. Save the calculations after you have established Area C.
3. Your Hydrograph editor should look like Figure 13.32.

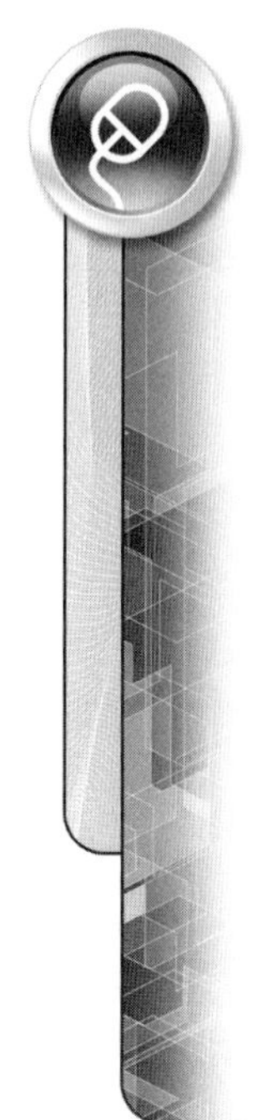

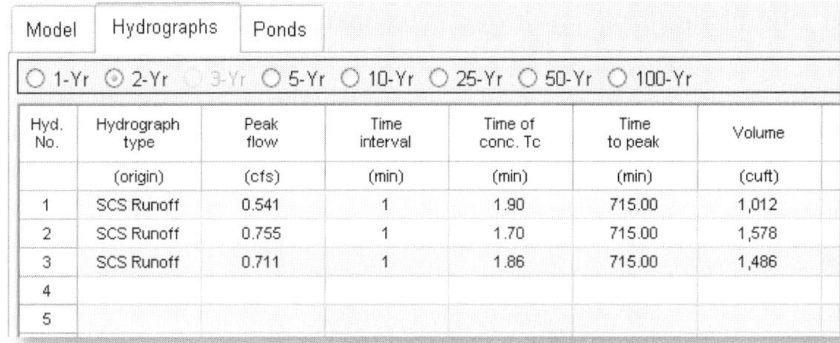

Model | Hydrographs | Ponds

○ 1-Yr ⊙ 2-Yr ○ 3-Yr ○ 5-Yr ○ 10-Yr ○ 25-Yr ○ 50-Yr ○ 100-Yr

| Hyd. No. | Hydrograph type | Peak flow | Time interval | Time of conc. Tc | Time to peak | Volume |
|---|---|---|---|---|---|---|
| | (origin) | (cfs) | (min) | (min) | (min) | (cuft) |
| 1 | SCS Runoff | 0.541 | 1 | 1.90 | 715.00 | 1,012 |
| 2 | SCS Runoff | 0.755 | 1 | 1.70 | 715.00 | 1,578 |
| 3 | SCS Runoff | 0.711 | 1 | 1.86 | 715.00 | 1,486 |
| 4 | | | | | | |
| 5 | | | | | | |

**FIGURE 13.32**

## Move Area A's Runoff to a Channel

Area A's Runoff drains to the detention pond through a channel. Channels are reaches; that is, they are concentrators and conveyors of runoff with no storage capacity. A reach can be defined in the Model or Hydrographs panel.

1. Select the ***Model*** tab, it should look like Figure 13.33. If the labels don't match, from the Options menu, Model flyout, select Show Hydrograph Numbers, and deselect Show Hydrograph Descriptions.

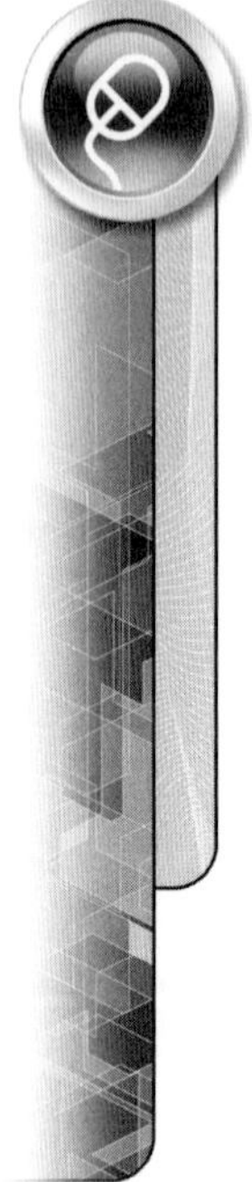

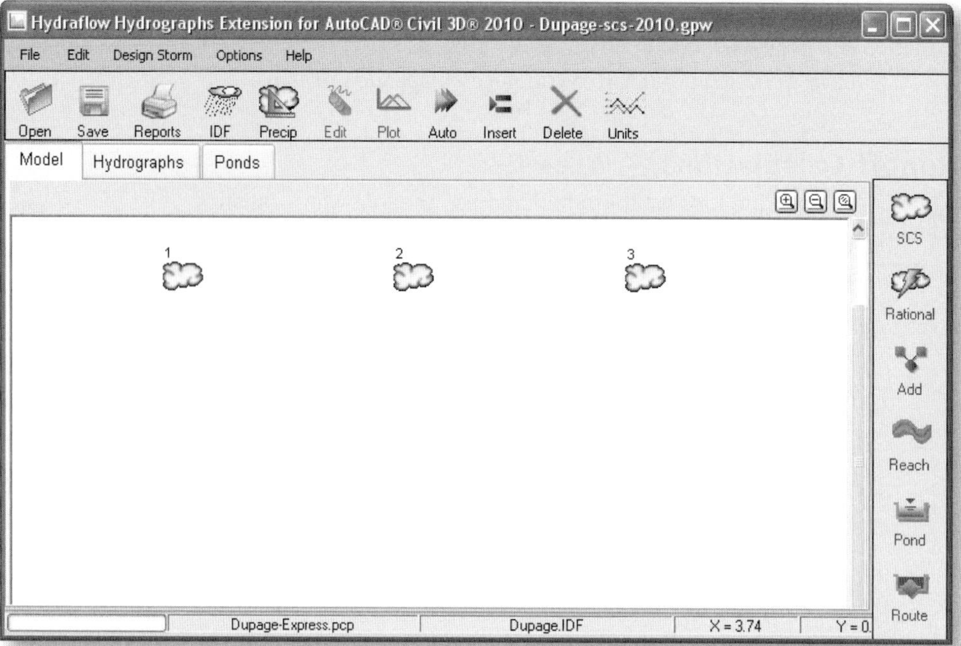


**FIGURE 13.33**

2. On the right side of the panel, click ***Reach*** to open the Reach dialog box. If Reach is grayed out, you will need to click in an open area of the panel to select Reach.
3. In the Reach dialog box, enter the values found in Figure 13.34. At the top left, name the reach **Channel A** and connect it to Area A's hydrograph.

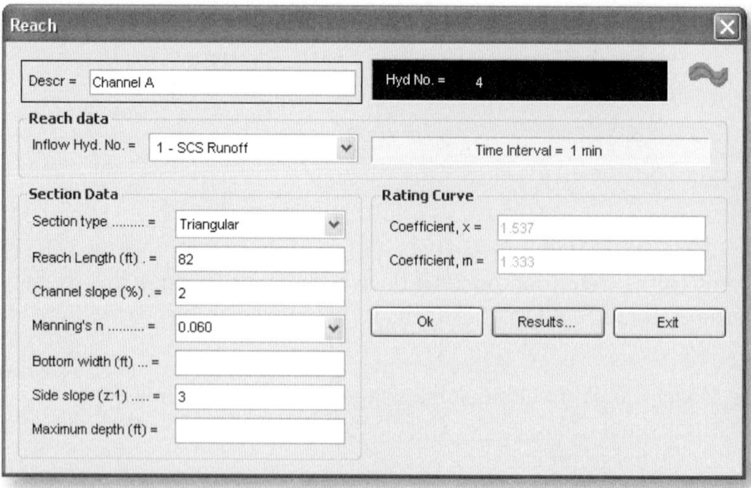


**FIGURE 13.34**

4. When you are done entering the data, click ***OK***, and then click ***Results...*** to view the hydrograph. After reviewing the hydrograph, click ***Exit*** to return to the Reach dialog box.
5. Click ***Exit*** to create the Reach.
6. Click ***Save*** to save the current model and hydrographs.

## Move Area B's Runoff to Catchbasin B

Runoff from Areas B and C flows through catch basins that discharge into a pond. The catch basins act like mini-culverts and take the water in from the parking lot and pass the flow to the pond. The catch basins are reservoirs with little or no detention capacity.

In the model, create a reservoir (catch basin) with little or no capacity. This, however, is not quite correct. If the rainfall intensity is sufficient, water will back up in the catch basin because of its restrictive outfall (8-inch pipe).

1. Click the ***Ponds*** tab.
2. Double-click the Pond1 Pond Name cell (<New Pond>) to open the Stage / Storage / Discharge Setup dialog box.
3. If necessary, at the top left, click ***Contours*** to set the Pond definition type.
4. For the pond's name, enter **Catchbasin B**.
5. Using Figure 13.35 as a guide, create the pond Catchbasin B.
6. At the top left, for Bottom Elevation, enter **705.9** and click ***Apply***.
7. Click in Row 1's elevation cell, and for the elevation enter **706.00**.
8. Click in Row 1's area cell, for the area enter **13**, and press ENTER.
9. Click in Row 2's elevation cell, and for the elevation enter **709.50**.
10. Click in Row 2's area cell, for the area enter **13**, and press ENTER.
11. In the bottom right, click ***Done***.

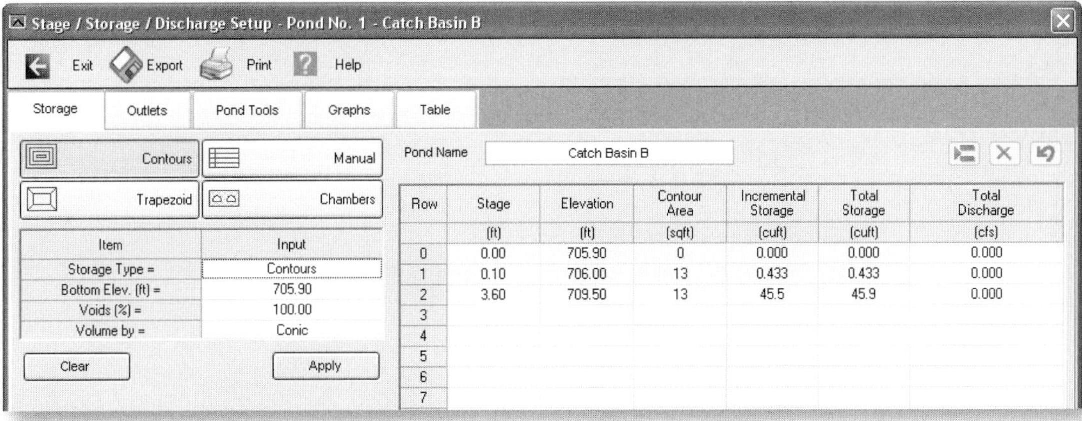


**FIGURE 13.35**

Catchbasin B has an 8-inch outfall.

12. Click the ***Outlets*** tab.
13. At the top left, Culv/Orifice, in column A, for the Rise enter **8**.
14. Click in column A's Span and it automatically sets to 8.
15. For the number of Barrels cell, set it to **1**.
16. Click in the Invert Elev. cell to fill in the value.
17. For Length enter **18**, and for slope enter **1**.
18. In the middle of the panel, click ***Compute*** to calculate the orifice's behavior.

Your Outlets panel should look similar to Figure 13.36.

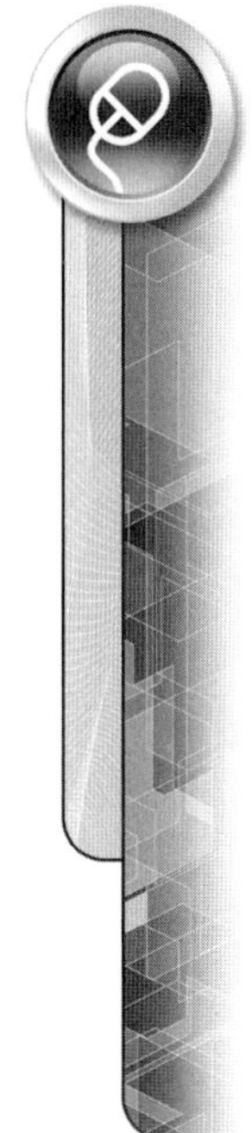

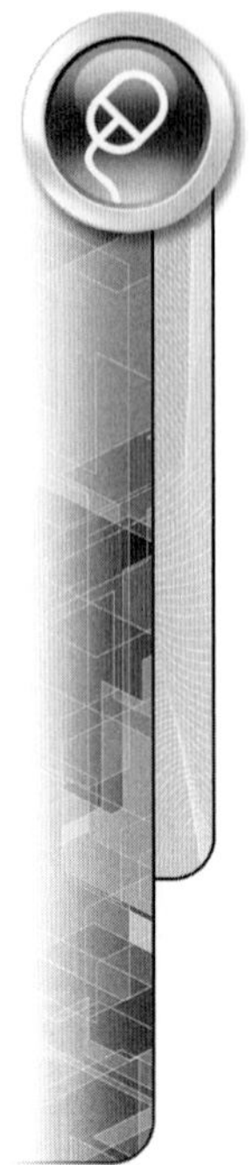

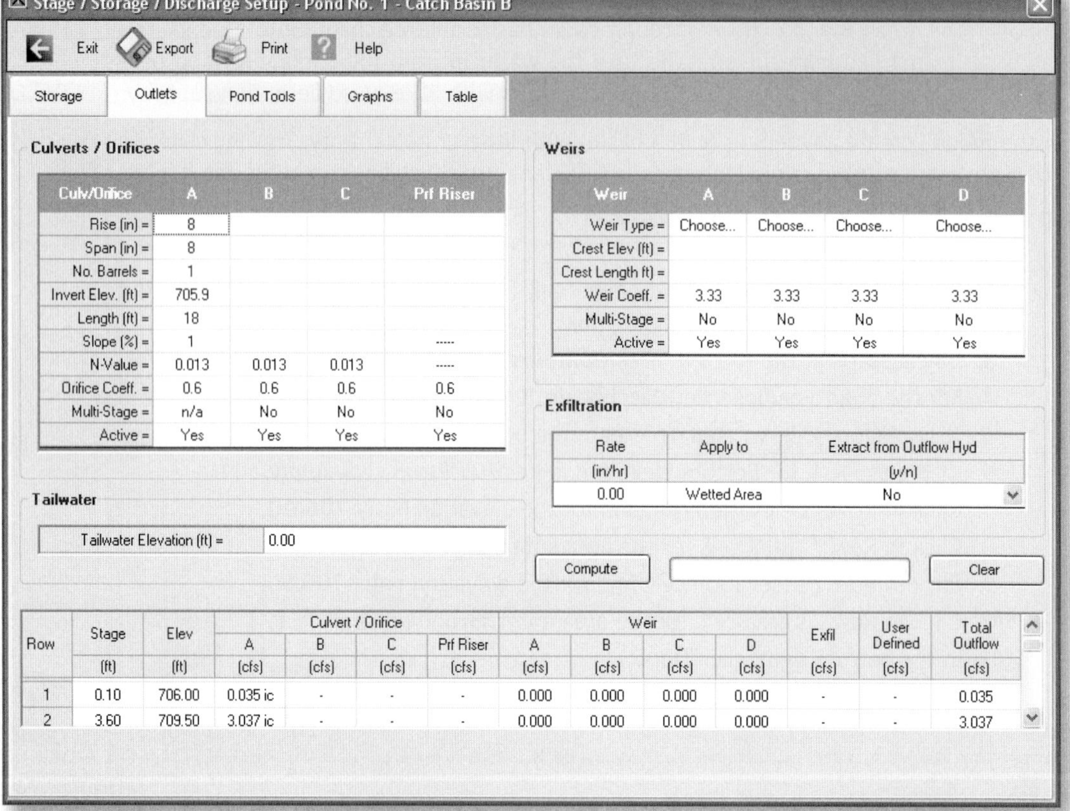


| Culv/Orifice | A | B | C | Prf Riser |
|---|---|---|---|---|
| Rise (in) = | 8 | | | |
| Span (in) = | 8 | | | |
| No. Barrels = | 1 | | | |
| Invert Elev. (ft) = | 705.9 | | | |
| Length (ft) = | 18 | | | |
| Slope (%) = | 1 | | | ----- |
| N-Value = | 0.013 | 0.013 | 0.013 | ----- |
| Orifice Coeff. = | 0.6 | 0.6 | 0.6 | 0.6 |
| Multi-Stage = | n/a | No | No | No |
| Active = | Yes | Yes | Yes | Yes |

| Weir | A | B | C | D |
|---|---|---|---|---|
| Weir Type = | Choose... | Choose... | Choose... | Choose... |
| Crest Elev (ft) = | | | | |
| Crest Length ft) = | | | | |
| Weir Coeff. = | 3.33 | 3.33 | 3.33 | 3.33 |
| Multi-Stage = | No | No | No | No |
| Active = | Yes | Yes | Yes | Yes |

| Rate (in/hr) | Apply to | Extract from Outflow Hyd (y/n) |
|---|---|---|
| 0.00 | Wetted Area | No |

| Row | Stage (ft) | Elev (ft) | Culvert / Orifice A (cfs) | B (cfs) | C (cfs) | Prf Riser (cfs) | Weir A (cfs) | B (cfs) | C (cfs) | D (cfs) | Exfil (cfs) | User Defined (cfs) | Total Outflow (cfs) |
|---|---|---|---|---|---|---|---|---|---|---|---|---|---|
| 1 | 0.10 | 706.00 | 0.035 ic | - | - | - | 0.000 | 0.000 | 0.000 | 0.000 | - | - | 0.035 |
| 2 | 3.60 | 709.50 | 3.037 ic | - | - | - | 0.000 | 0.000 | 0.000 | 0.000 | - | - | 3.037 |

**FIGURE 13.36**

19. Click ***Exit*** to return to Hydrographs.
20. At the top left, click ***Save*** to save the file.

## Catchbasin C

1. Double-click Pond2's Pond Name cell to open the Stage / Storage / Discharge Setup dialog box.
2. Repeat the steps from the previous section (Catchbasin B), but for Catchbasin C.
3. Using Figure 13.37 as a guide, enter the values for the Catchbasin C.
4. At the top left, for Bottom Elevation, enter **706.5** and click Apply.
5. Click in Row 1's elevation cell, enter **706.60** and set its area of **13**, and press ENTER.
6. Click in Row 2's elevation cell, enter **709.00** and set its area of **13**, and press ENTER.
7. After entering the values, in the dialog box's lower right, click ***Done***.

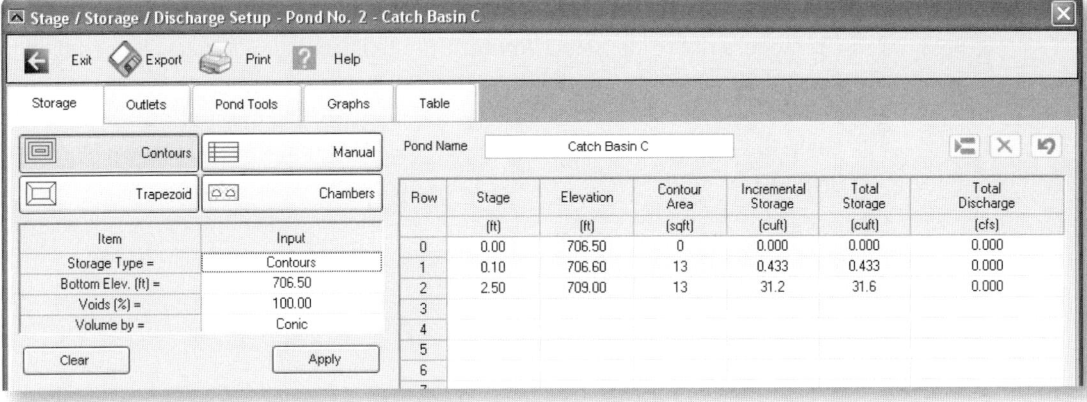

**FIGURE 13.37**

8. Click the ***Outlets*** tab.
9. At the top left, Culv/Orifice, in column A, for the Rise enter **8** and press ENTER.
10. Using the values shown in Figure 13.38, finish entering in the values for Catch-basin C – Outlets.

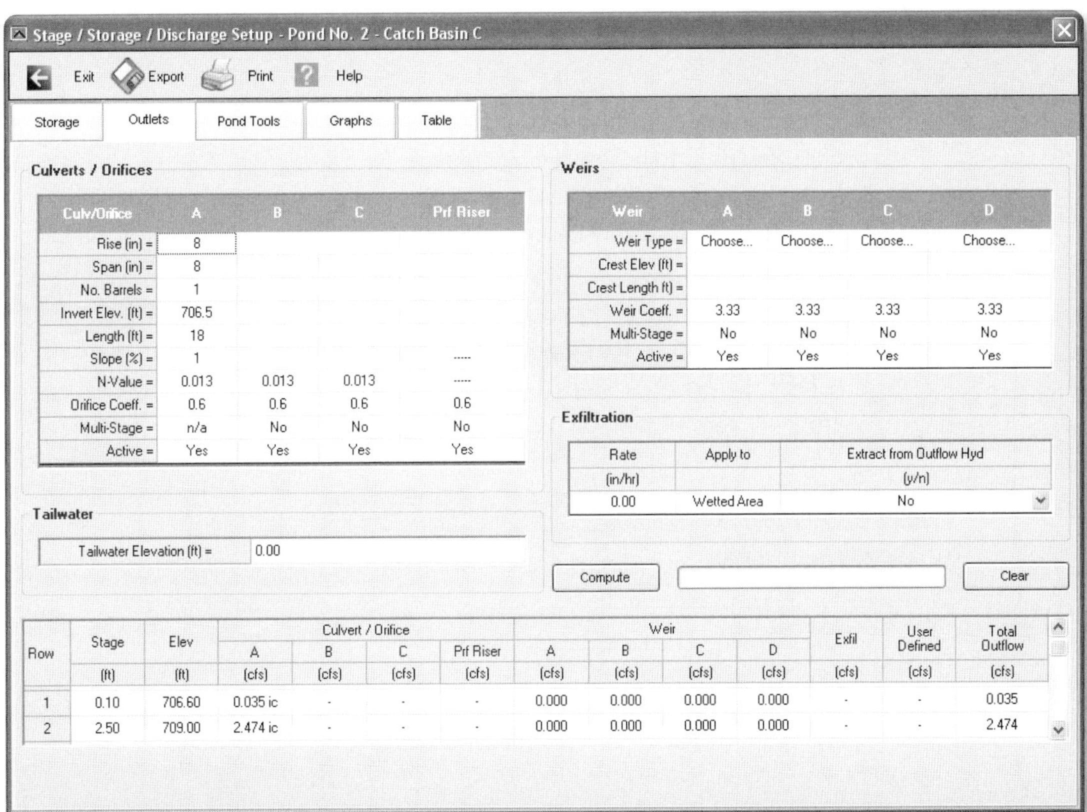

**FIGURE 13.38**

11. Click ***Exit*** to return to Hydrographs.
12. At the top left, click ***Save*** to save the file.
13. Click the ***Model*** Tab.

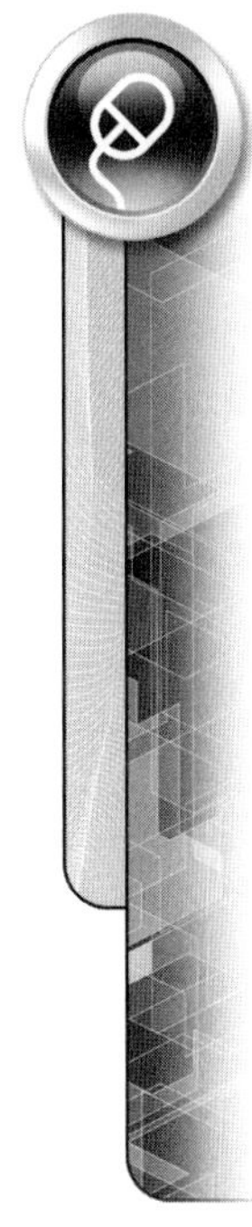

## Connect Area B to a Resevoir (Catchbasin B)

1. On Model's right side, select ***Route***. If Route is grayed out, you will need to click in an open area of the panel to select Route.
2. Using Figure 13.39, enter the necessary values. For the description, enter **Catchbasin B**, set the Inflow Hydrograph to 2 – SCS Runoff - Area B, and set the pond name to 1. Catchbasin B.

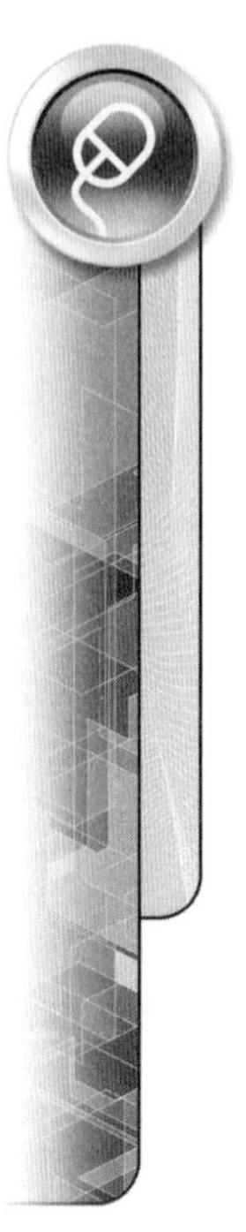

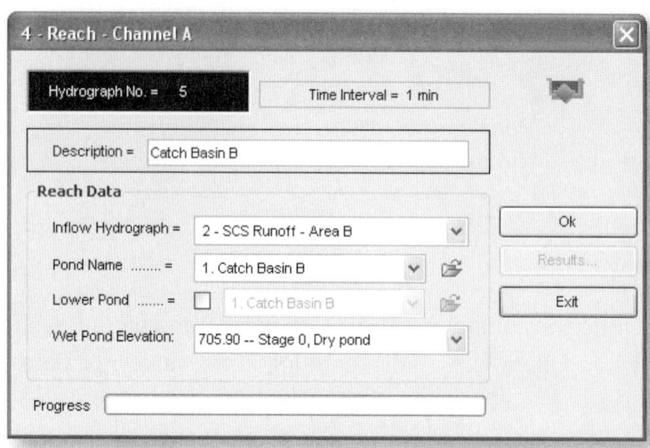


**FIGURE 13.39**

3. When the settings are correct, click ***OK***.
4. Click ***Results...*** to review the hydrograph.
5. In the Hydrograph Plot - Reservoir dialog box, at its top left, click ***Exit***.
6. Click ***Exit*** to close the Reservoir Route dialog box and return to the Model panel.

## Connect Area C to a Resevoir (Catchbasin C)

1. Repeat the previous Steps 1–6, and using Figure 13.40, connect Area C with Reservoir (Catchbasin C).

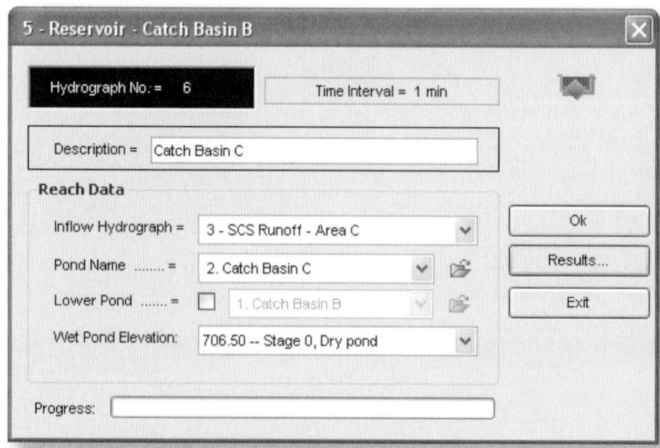


**FIGURE 13.40**

2. Click the ***Save*** icon to save the calculations.

## Combine the Inflow Hydrographs

All three areas discharge into the pond. The Model panel's Add icon merges hydrographs to a single model icon.

1. From the Model panel's right side, select ***Add***. If Add is grayed out, you will need to click in an open area of the panel to select Add.
2. Using Figure 13.41, enter the following values. For the description, enter **Combined Inflow Detention Pond**, select **Reach – Channel A, Reservoir - Catchbasin B**, and **Reservoir - Catchbasin C**.

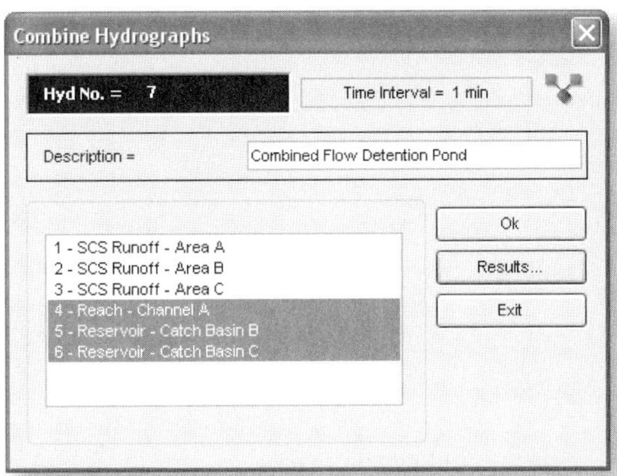


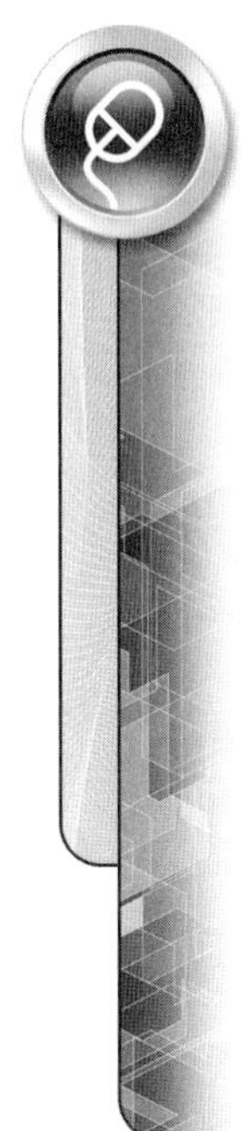

**FIGURE 13.41**

3. When the settings are correct, click ***OK***.
4. Click ***Results...*** to review the hydrograph.
5. In the Hydrograph Plot - Combine dialog box, at its top left, click ***Exit***.
6. Click ***Exit*** to close the Combine Hydrographs dialog box.
7. Click the ***Save*** icon to save the calculations.

Your model should look like Figure 13.42.

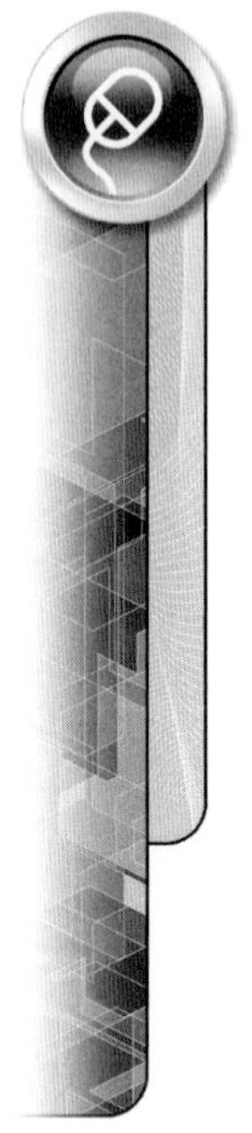

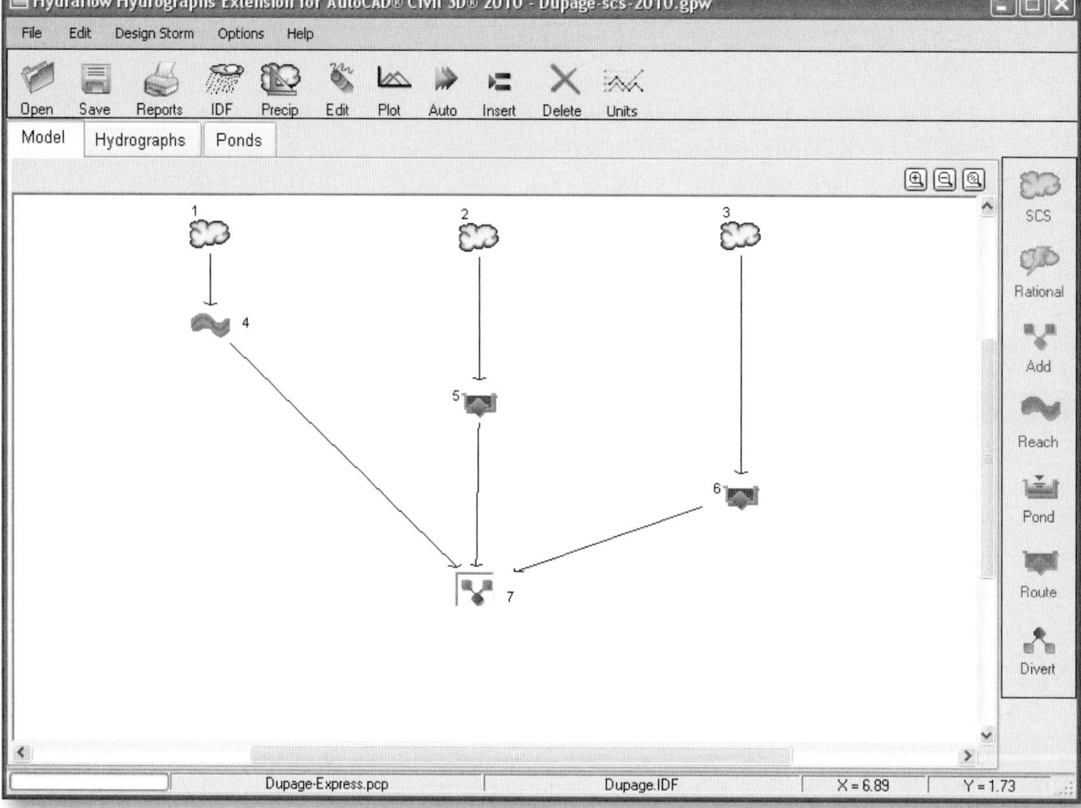


**FIGURE 13.42**

## Define the Detention Pond

1. Click the ***Ponds*** tab.
2. Double-click the Pond3 Pond name cell, to open the Stage / Storage / Discharge Setup dialog box.
3. If necessary, click ***Contours*** at the top left to set the Pond definition type.
4. For the pond name, enter **Detention Pond**.
5. At the top left, for Bottom Elevation, enter **704.9** and click ***Apply***.
6. Using Figure 13.43 as a guide to create the Detention Pond.
7. Use Table 13.3 to enter the elevation and area values. Fill in only the necessary cells.

**TABLE 13.3**

| Elevation | Area |
|---|---|
| 705 | 3468 |
| 706 | 4810 |
| 707 | 6257 |
| 708 | 7889 |

8. After carefully entering the elevation and area values, at the bottom right, click ***Done***.

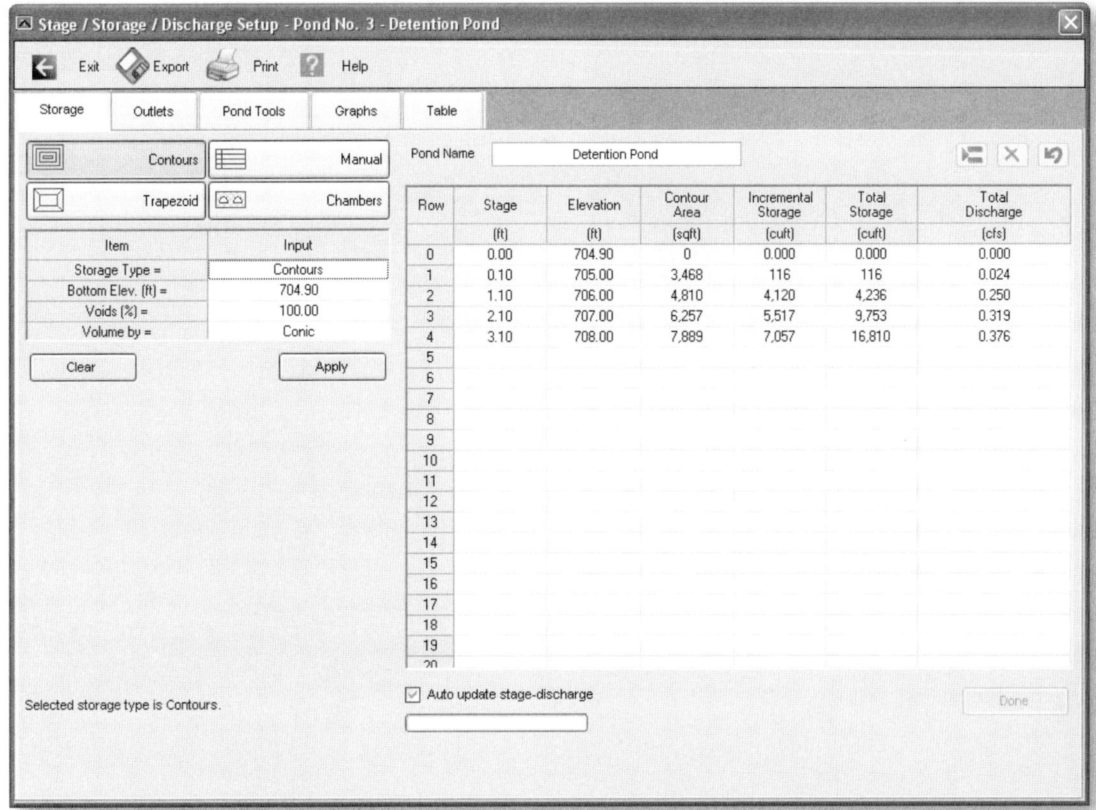


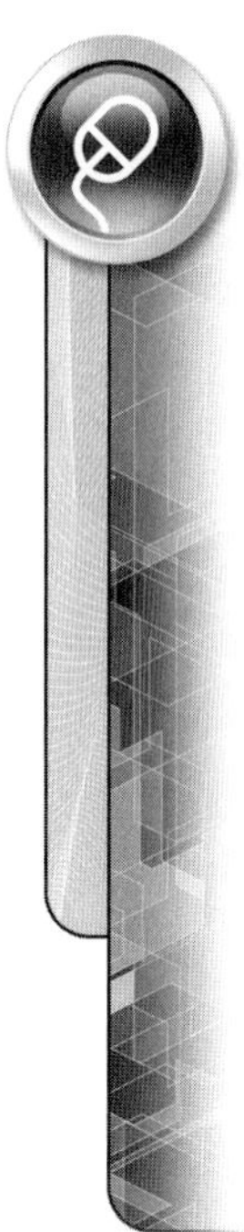

**FIGURE 13.43**

9. Click the ***Outlets*** tab.
10. At the top left, Culv/Orifices, in column A, for the Rise, enter **4**.
11. Click in column A's Span and No. Barrels cells and set their values to **4** and **1**.
12. For length enter **82**, and for slope, enter **1**.

Your Outlet should look similar to Figure 13.44.

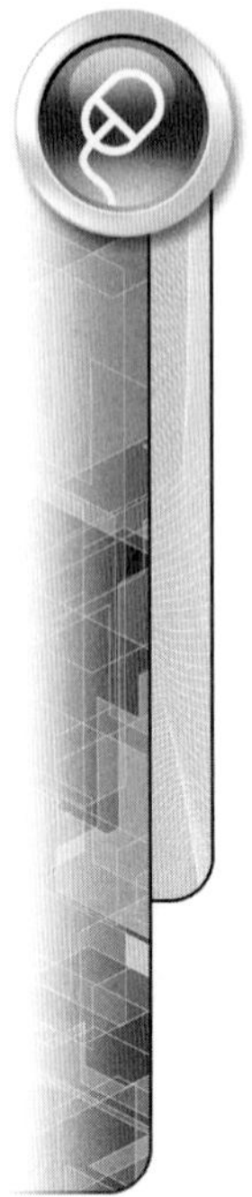

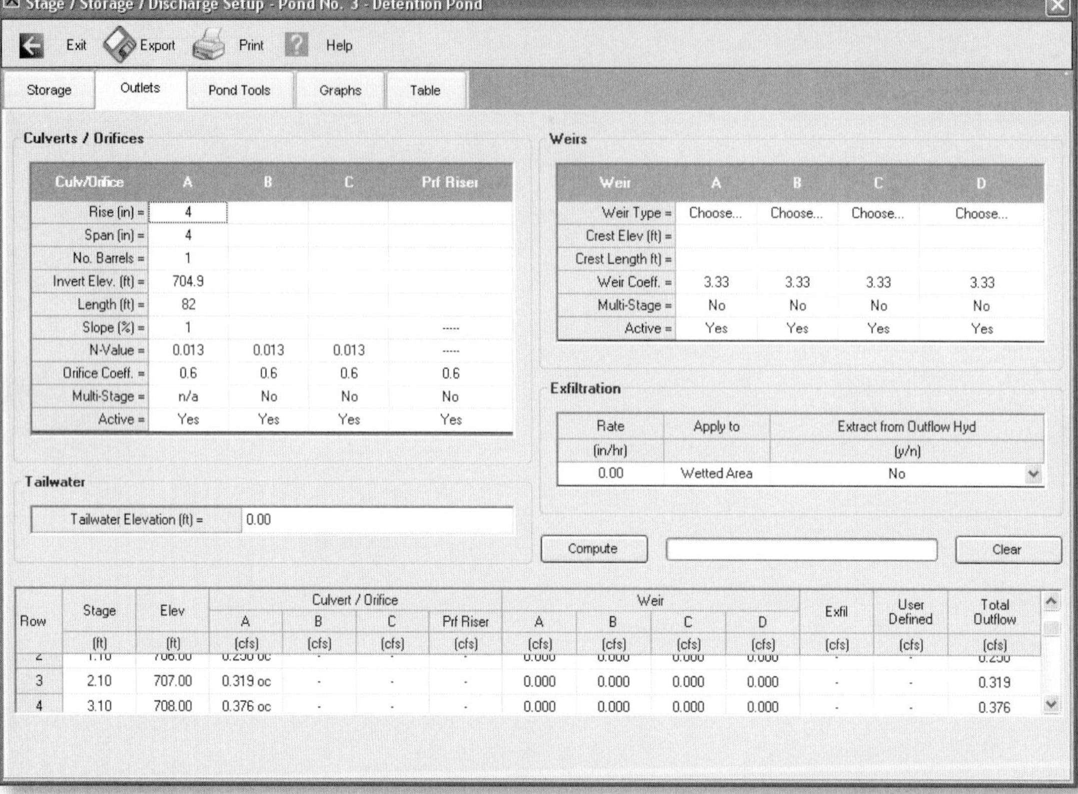


FIGURE 13.44

13. In the middle, click ***Compute*** to display the outfall values at the bottom.
14. Click ***Exit*** to return to Hydrographs.
15. At the top left, click the ***Save*** icon to save the file.

## Add the Inflow to a Detention Pond

1. At the top left, click the ***Model*** tab.
2. From the panel's right side, select ***Route***. If Route is grayed out, you will need to click in an open area of the panel to select Route.
3. Using Figure 13.45, enter the following values. For the description, enter **Detention Pond** and then select `COMBINE - COMBINED INFLOW DENTION POND`.

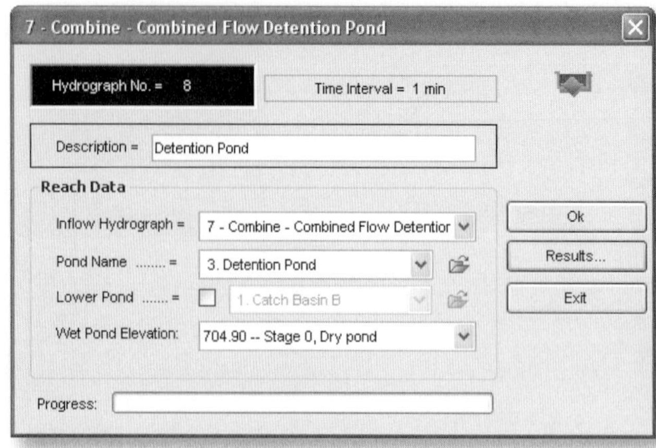


FIGURE 13.45

4. Set the Pond Name to **Detention Pond**.
5. When the settings are correct, click ***OK***.
6. Click ***Results...*** to review the hydrograph.
7. In the Hydrograph Plot – Reservoir – Detention Pond dialog box, at its top left, click ***Exit***.
8. Click ***Exit*** to close the Reservoir Route dialog box.
9. Click the ***Save*** icon to save the calculations.
10. Close Hydraflow Hydrographs and return to Civil 3D.

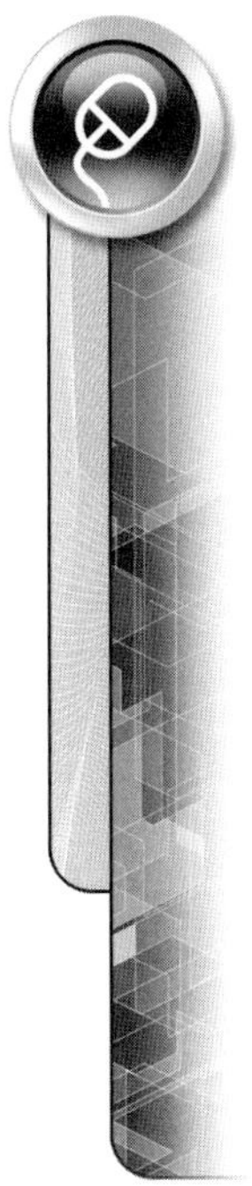

## SUMMARY

- Hydraflow Hydrographs has several components: hydrographs, reaches, ponds, and routes.
- The Model tab schematically connects the hydrographs and ponds.
- When completed, Hydrographs evaluates the routing's overall performance.

## UNIT 3: STORM PIPES

Civil 3D drafts storm water pipe networks. However, it does not have any design or evaluation capability. The Storm Sewers Extension provides design and evaluation tools. You access Storm Sewers through Modify's Pipe Network panel's Storm Sewers command. The command exports and imports storm sewer designs between Civil 3D and Storm Sewers.

When you are importing a pipe network, the import routine uses the default Pipe Network Edit Feature Settings. You must make sure they are appropriate for the system you are importing.

**NOTE**

**Draw all pipe network runs downslope.**

### DESIGN CODES

Design and design evaluation require minimum or typical standards. The Design Codes panel's values set minimums and maximums for pipes, inlets, and calculations.

#### Pipes

Design Code's Pipes panel sets minimum and maximum pipe values (see Figure 13.46). The left side has settings for minimum and maximum pipe sizes and slopes, design velocity, and a Manning's n. The upper-right setting excludes certain pipe sizes: 21, 27, and 33 inches. The lower right has settings for matching by crown or invert, the invert drop amount, and if smaller pipe sizes are allowed downstream.

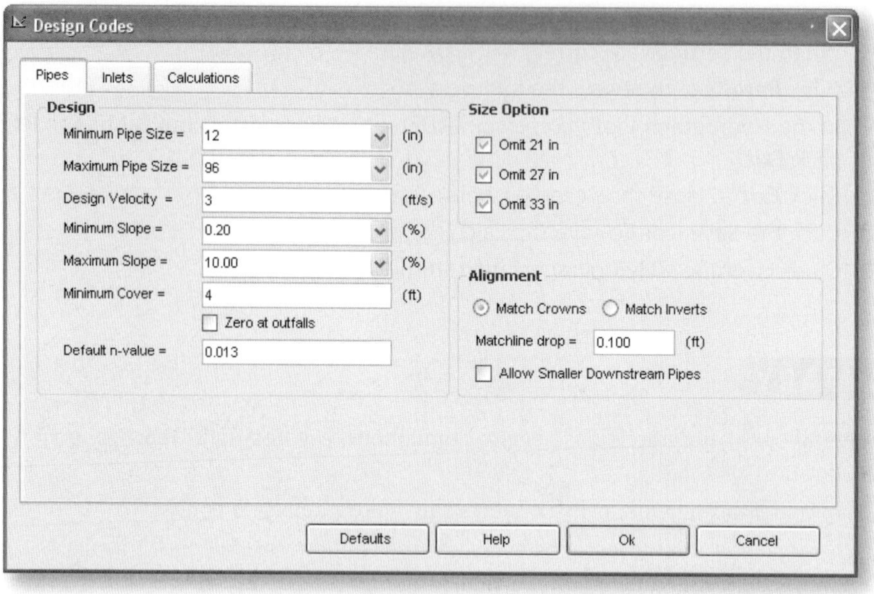

**FIGURE 13.46**

## Inlets

Inlets sets default values for curb, grate, and gutter inlets (see Figure 13.47). The lower-left side defines a schematic structure type. The upper-right grate design parameters and the lower-right parameters define composite runoff coefficient values.

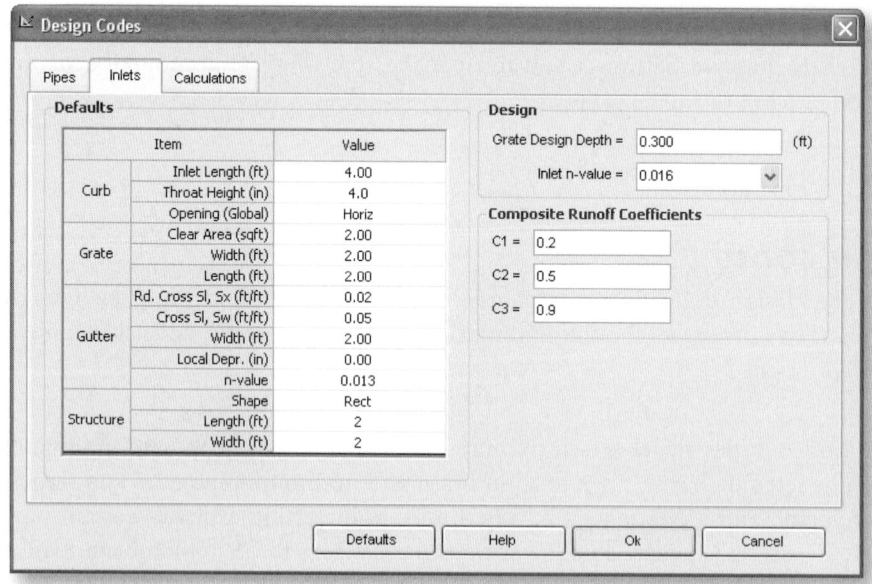


**FIGURE 13.47**

## Calculations

Calculations sets default values for HGL calculations, flow options, and junction loss calculations (see Figure 13.48).

## HGL Options

This group defines the minimum starting elevation of the HGL: crown, normal, (dc+D)/2, and critical. There is a toggle to resolve EGL discrepancies and inlet control checks. If the inlet control check is on, there are two value-evaluation methods: HDX-5 or Standard Orifice Calculations.

## Flow Options

Flow Options include values that affect accumulating known Qs, using captured inlet flows, whether to suppress travel times, setting the minimum Tc for intensity calculations, and how to calculate junction loss (manual entry or automatic).

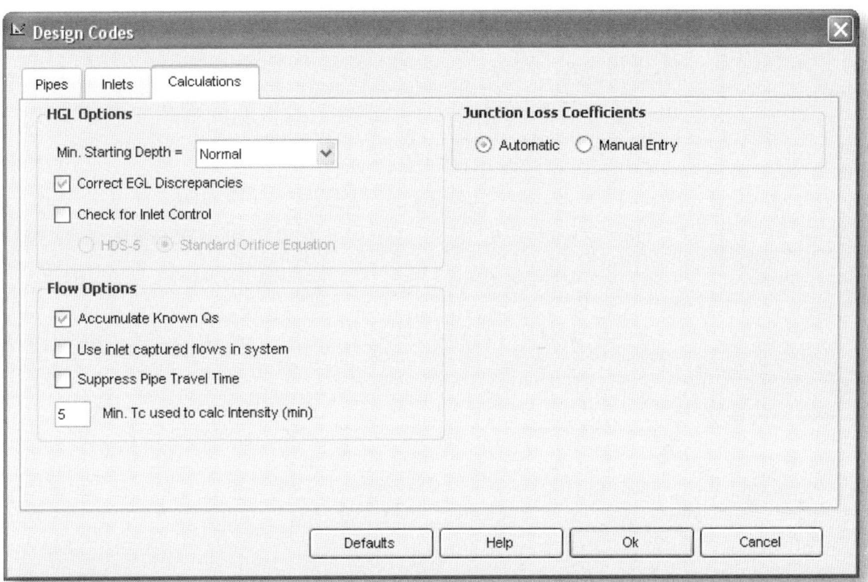


**FIGURE 13.48**

## XML Import/Export

This dialog box reads a LandXML file and builds a Storm Sewers model. After evaluating and adjusting its values, it then exports the design to a new or updated XML file (see Figure 13.49).

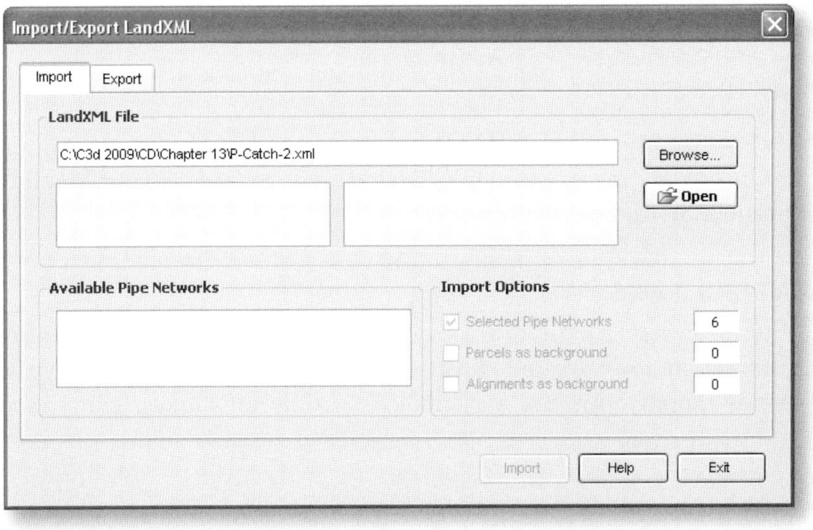

**FIGURE 13.49**

## Export and Import to Storm Sewers

When you use the Storm Sewers application to analyze and modify your design, you must modify the pipe network migration settings. These settings are a part of the Pipe Networks' Edit Feature Settings, Storm Sewers Migration Defaults. The migration settings have two parts. The first is to translate Civil 3D's pipes and structures to match those in Storm Sewers for import and export. The second is identifying the parts list to use.

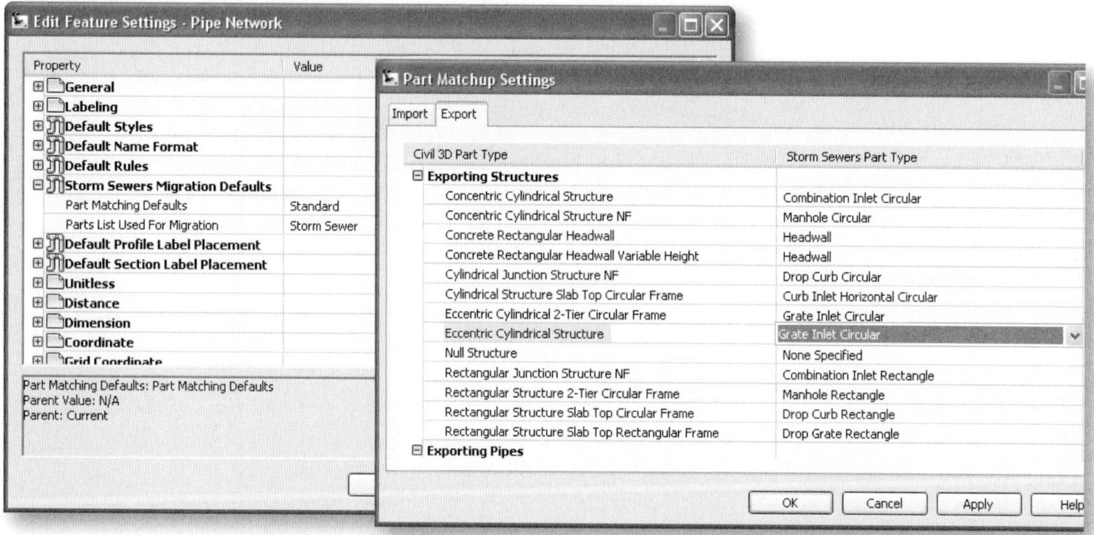


**FIGURE 13.50**

## Import/Export Parts Matching

The parts matching section sets equivalents between Civil 3D and Storm Sewers. For example, for export, an eccentric cylindrical junction with frame is a grate inlet – circular and on import a grate inlet – circular is an eccentric cylindrical junction with frame (see Figure 13.50).

The parts list setting is the active parts list creating the exported pipe network.

When you import a Storm Sewer project and you have an existing pipe network in the drawing, a dialog box displays asking if you want to modify the existing or want to create a new pipe network.

### DXF

The Storm Sewers extension imports DXFs as background images for its editor. A DXF can also be used to import a design. The preferred method is to use Modify's Pipe Network panel's Storm Sewer command.

### STORM WATERS TABS

When using Storm Sewer Extension to design or modify a pipe network, a user should start editing from downstream to upstream. This is because inlets have bypass runoff, and bypass cannot flow into a manhole, but only into inlets, grates, and outfalls.

## Plan

The Plan panel displays the design in Plan View. When you select a pipe and press the right mouse button, a shortcut menu is displayed with commands appropriate to the selected pipe (see Figure 13.51). You select Data Dialog to display all of the pipe's data. The Alignment section refers to the pipe's location; Flows is where inflow values are entered. Physical defines the invert elevations, pipe type, and other pertinent values.

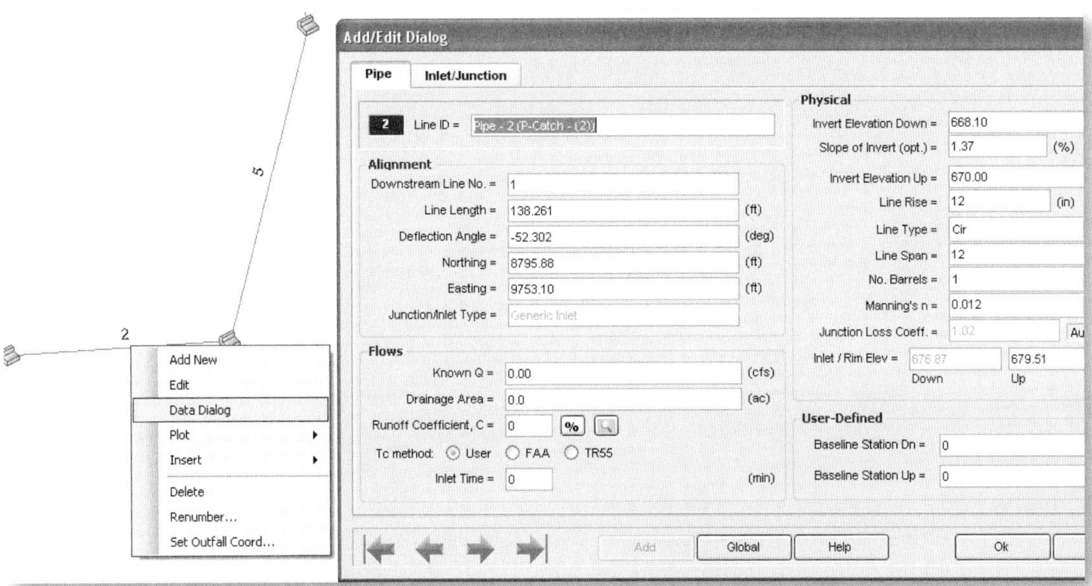


**FIGURE 13.51**

If you double-click a structure or click the Inlet/Junction tab, the Add/Edit dialog box opens and displays all of the structure's pertinent information (see Figure 13.52). General defines the type of inlet and, depending on the selected type, the lower left of the dialog box becomes active. In the upper right, Gutter Details define the road cross slope and other necessary values to evaluate or size the inlet. The lower-right section, Structure, defines the structure's shape.

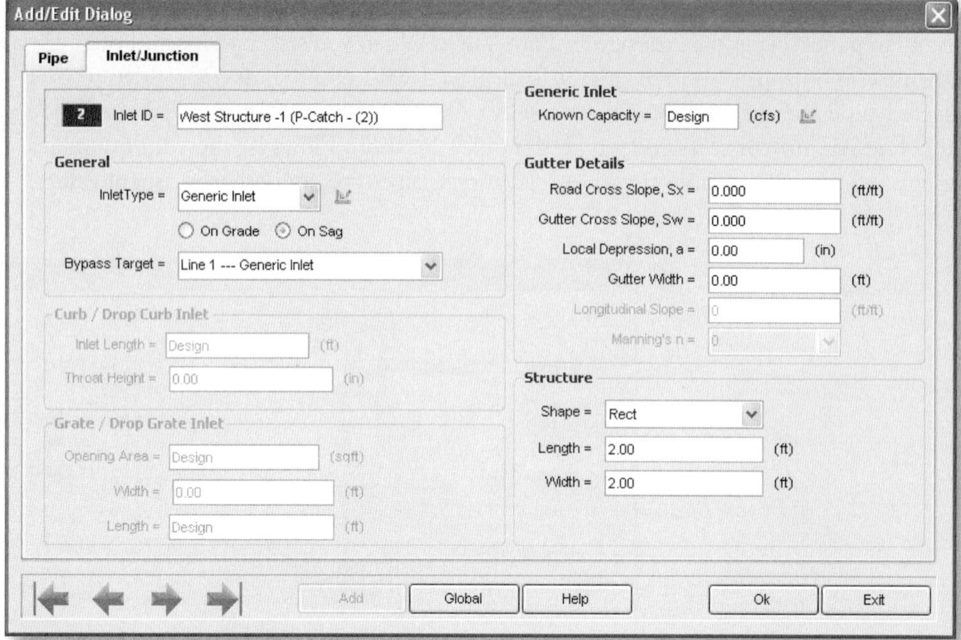


**FIGURE 13.52**

## Pipes

Pipes defines the inflow and surrounding slope, size, and basic parameters of an inlet (see Figure 13.53).

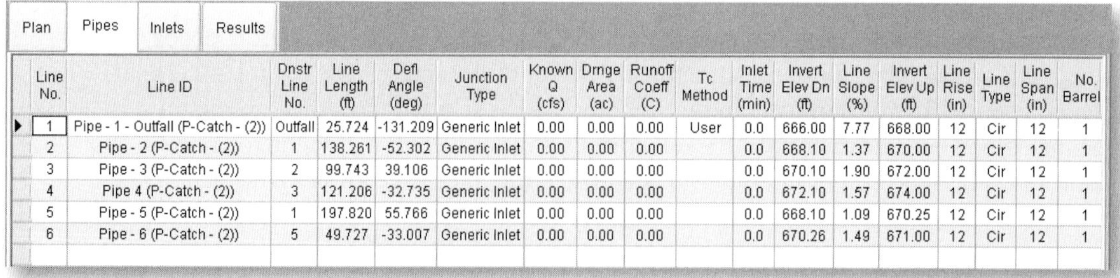


| Line No. | Line ID | Dnstr Line No. | Line Length (ft) | Defl Angle (deg) | Junction Type | Known Q (cfs) | Drnge Area (ac) | Runoff Coeff (C) | Tc Method | Inlet Time (min) | Invert Elev Dn (ft) | Line Slope (%) | Invert Elev Up (ft) | Line Rise (in) | Line Type | Line Span (in) | No. Barrel |
|---|---|---|---|---|---|---|---|---|---|---|---|---|---|---|---|---|---|
| 1 | Pipe - 1 - Outfall (P-Catch - (2)) | Outfall | 25.724 | -131.209 | Generic Inlet | 0.00 | 0.00 | 0.00 | User | 0.0 | 666.00 | 7.77 | 668.00 | 12 | Cir | 12 | 1 |
| 2 | Pipe - 2 (P-Catch - (2)) | 1 | 138.261 | -52.302 | Generic Inlet | 0.00 | 0.00 | 0.00 | | 0.0 | 668.10 | 1.37 | 670.00 | 12 | Cir | 12 | 1 |
| 3 | Pipe - 3 (P-Catch - (2)) | 2 | 99.743 | 39.106 | Generic Inlet | 0.00 | 0.00 | 0.00 | | 0.0 | 670.10 | 1.90 | 672.00 | 12 | Cir | 12 | 1 |
| 4 | Pipe 4 (P-Catch - (2)) | 3 | 121.206 | -32.735 | Generic Inlet | 0.00 | 0.00 | 0.00 | | 0.0 | 672.10 | 1.57 | 674.00 | 12 | Cir | 12 | 1 |
| 5 | Pipe - 5 (P-Catch - (2)) | 1 | 197.820 | 55.766 | Generic Inlet | 0.00 | 0.00 | 0.00 | | 0.0 | 668.10 | 1.09 | 670.25 | 12 | Cir | 12 | 1 |
| 6 | Pipe - 6 (P-Catch - (2)) | 5 | 49.727 | -33.007 | Generic Inlet | 0.00 | 0.00 | 0.00 | | 0.0 | 670.26 | 1.49 | 671.00 | 12 | Cir | 12 | 1 |

**FIGURE 13.53**

## Results

The Summary report is an overview of the network. Other reports can perform a cost estimate (see Figure 13.54).

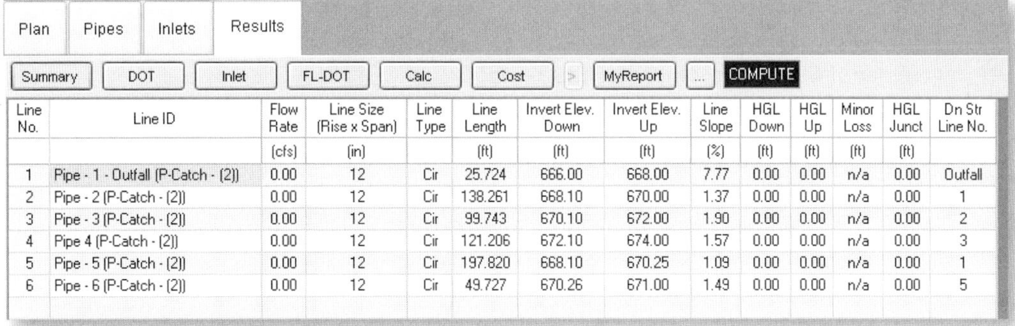

Plan | Pipes | Inlets | Results

Summary | DOT | Inlet | FL-DOT | Calc | Cost | MyReport | ... | COMPUTE

| Line No. | Line ID | Flow Rate | Line Size (Rise x Span) | Line Type | Line Length | Invert Elev. Down | Invert Elev. Up | Line Slope | HGL Down | HGL Up | Minor Loss | HGL Junct | Dn Str Line No. |
|---|---|---|---|---|---|---|---|---|---|---|---|---|---|
| | | (cfs) | (in) | | (ft) | (ft) | (ft) | (%) | (ft) | (ft) | (ft) | (ft) | |
| 1 | Pipe - 1 - Outfall (P-Catch - (2)) | 0.00 | 12 | Cir | 25.724 | 666.00 | 668.00 | 7.77 | 0.00 | 0.00 | n/a | 0.00 | Outfall |
| 2 | Pipe - 2 (P-Catch - (2)) | 0.00 | 12 | Cir | 138.261 | 668.10 | 670.00 | 1.37 | 0.00 | 0.00 | n/a | 0.00 | 1 |
| 3 | Pipe - 3 (P-Catch - (2)) | 0.00 | 12 | Cir | 99.743 | 670.10 | 672.00 | 1.90 | 0.00 | 0.00 | n/a | 0.00 | 2 |
| 4 | Pipe 4 (P-Catch - (2)) | 0.00 | 12 | Cir | 121.206 | 672.10 | 674.00 | 1.57 | 0.00 | 0.00 | n/a | 0.00 | 3 |
| 5 | Pipe - 5 (P-Catch - (2)) | 0.00 | 12 | Cir | 197.820 | 668.10 | 670.25 | 1.09 | 0.00 | 0.00 | n/a | 0.00 | 1 |
| 6 | Pipe - 6 (P-Catch - (2)) | 0.00 | 12 | Cir | 49.727 | 670.26 | 671.00 | 1.49 | 0.00 | 0.00 | n/a | 0.00 | 5 |

**FIGURE 13.54**

## Cost Codes

Cost Codes define the cost for each of the network items. Click Edit, and from the list, select Cost Codes… to open a dialog box where you can enter the per-unit values. After running an analysis, the Results tab displays a network cost estimate.

## Reports

Clicking the Reports icon opens the Print Reports dialog box. Print Reports allows a user to define a report and publish it to a printer (see Figure 13.55).

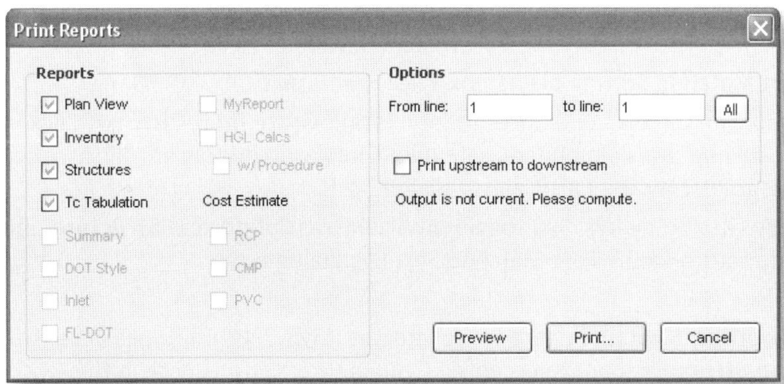


**FIGURE 13.55**

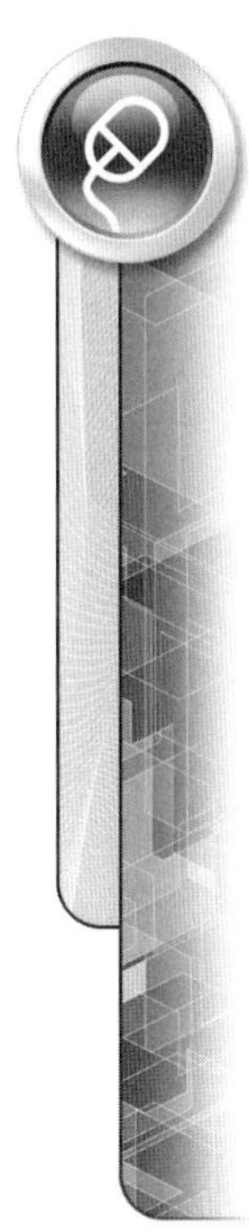

## EXERCISE 13-3

After completing this exercise, you will:

- Be able to create an IDF.
- Import and Export a pipe network to Storm Sewer.
- Be able to set Code Standards.
- Be able to add runoff to inlets.
- Be able to modify a design.

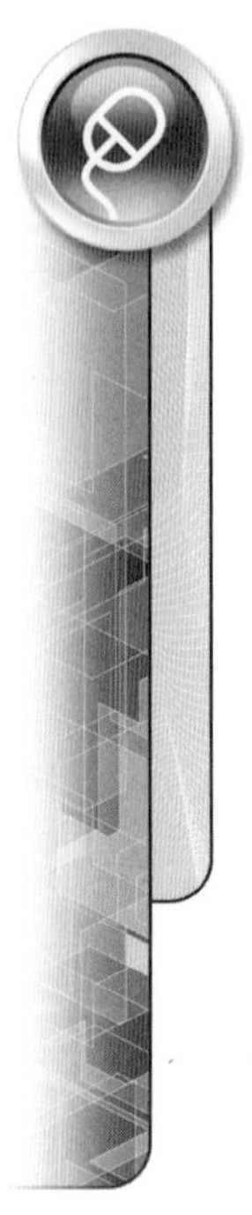

## Exercise Setup

This exercise uses the *Chapter 13 – Unit 3.dwg* file. To find this file browse to the Chapter 13 folder of CD that accompanies this textbook, and open the *Chapter 13 – Unit 3.dwg* file.

1. If it is not running, start **Civil 3D** by double-clicking its desktop icon.
2. When you are at the command prompt, close the open drawing and do not save it.
3. At Civil 3D's top left, Quick Access Toolbar, click the OPEN icon, browse to the Chapter 13 folder of the CD that accompanies this textbook, select the *Chapter 13 – Unit 3.dwg* file, and click ***Open***.
4. At Civil 3D's top left, click Civil 3D's drop-list arrow, in the Application Menu highlight SAVE AS, from the flyout select AutoCAD Drawing, browse to the Civil 3D Projects folder, for the drawing name enter **Tulsa Parking Lot – Work** and click ***SAVE*** to save the file.

## Edit Pipe Migration Settings

To export and import Civil 3D pipe networks to Storm Sewers, you must setup an equivalency chart. Civil 3D and Storm Sewers do not use the same terms for structures. The current pipe network uses eccentric cylindrical structures with a rectangular grate.

1. Click the Settings tab.
2. From the Settings list, select Pipe Network, right mouse click, and from the shortcut menu, select EDIT FEATURE SETTINGS....
3. Expand the Storm Sewers Migration Defaults section.
4. Click in the Part Matching Defaults value cell to display an ellipsis. Click the ellipsis and click it to display the Part Matchup Settings dialog box.
5. In the Part Matchup Setting dialog box, click the Import tab.
6. In the Importing Circular Structures section locate **Grate Inlet Circular**. Click in the Civil 3D Part Type column for Grate Inlet Circular to display an ellipsis. Click the ellipsis to display the Part Catalog dialog box.
7. In the Part Catalog dialog box, toggle on **Eccentric Cylindrical Structure** and click **OK** to return to the Part Matchup Settings dialog box.
8. Click the Export tab.
9. In the Exporting Structures section, locate the Civil 3D Part Type, **Eccentric Cylindrical Structure**. For the Eccentric Cylindrical Structure, click in the Storm Sewers Part Type column, click the drop-list arrow, and from the part's type list select, **Grate Inlet Circular**.
10. Click OK to return to Edit Feature Settings.
11. In the Storm Sewers Migration Defaults section, Parts List Used for Migration, click in the value cell to display an ellipsis.
12. Click the ellipsis and in the Parts List Used For Migration, select **Storm Sewer** and click **OK** to return to Edit Feature Settings.
13. Click **OK** to exit Edit Feature Settings.

## Export to Storm Sewers

1. In the Ribbon, click the ***Modify*** tab. On the Design panel, click the ***Pipe Network*** icon displaying the Pipe Networks tab.
2. On the Pipe Networks tab, Analyze panel, click the ***Storm Sewers*** icon and from the shortcut menu, select EDIT IN STORM SEWERS to display the Export to Storm Sewers dialog box.

3. From the pipe networks list, select PKLOT - PSTM – (1) and click ***OK*** to display Export Storm Sewers to File dialog box.
4. In Export Storm Sewers to File dialog box, browse to the folder C:\Civil 3D Projects, name the file *PKLOT - PSTM – (1)*, and click ***Save*** to display Hydraflow Storm Sewers application.

## Hydraflow Storm Sewers Setup — Set the Project to the Exported File

Even though the pipe network shows in Storm Sewers, if you were to save the project, it would not save to the file just exported. You need to open the exported file and all subsequent saves will be to the correct file.

1. At Storm Sewer's top left, from the File menu, select OPEN PROJECT....
2. In the Open Project dialog box, browse to the file just exported, select the file, and click ***Open*** to load the file as a project.

This sets the file as the project file and any subsequent saves will update this file. This file will in turn update your Civil 3D pipe network.

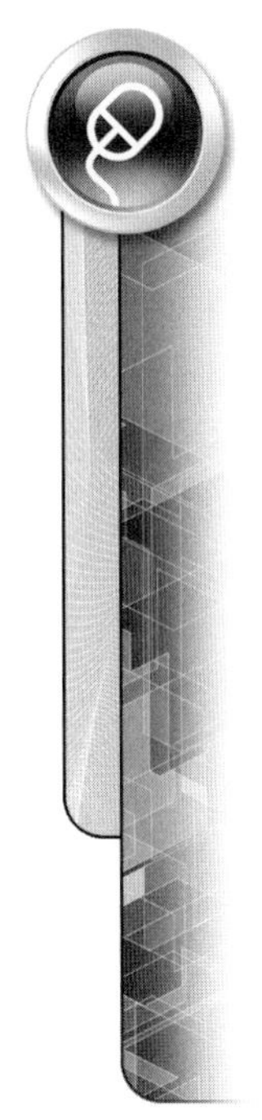

## Hydraflow Storm Sewers Setup — Tulsa IDF

The *Tulsa.IDF* and Codes must be set. The Chapter 13 folder of the CD that accompanies this textbook contains the Tulsa IDF.

1. At Storm Sewer's top, click the ***IDF*** icon to open the Rainfall IDF Curve dialog box.
2. Click the ***Open*** icon, browse to the Chapter 13 folder of the CD that accompanies this textbook, select the Tulsa.idf file and click Open. When the Tulsa.idf file has been displayed, click the ***Exit*** icon.

## Hydraflow Storm Sewers Setup — Codes

1. From the application's top, click the ***Codes*** icon to open the Design Codes dialog box.
2. If necessary, click the ***Pipes*** tab.
3. Adjust your settings to match those in Figure 13.56.
4. When the settings are correct, click the ***Inlets*** tab.
5. Adjust your settings to match those in Figure 13.57.
6. When the settings are correct, click the ***Calculations*** tab.
7. Adjust your settings to match those in Figure 13.58.
8. Click ***OK*** to close the dialog box.
9. From the Edit menu, select COST CODES....
10. Review the per-unit costs.
11. After reviewing the chart's values, click ***Exit***.
12. At Storm Sewers top left click ***Save***, and click OK to save the project.

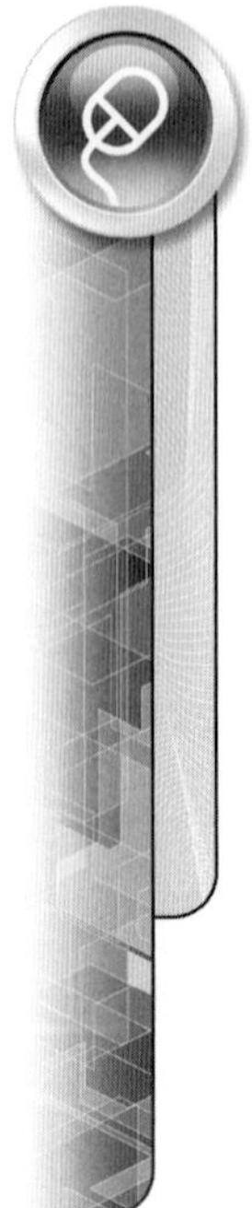

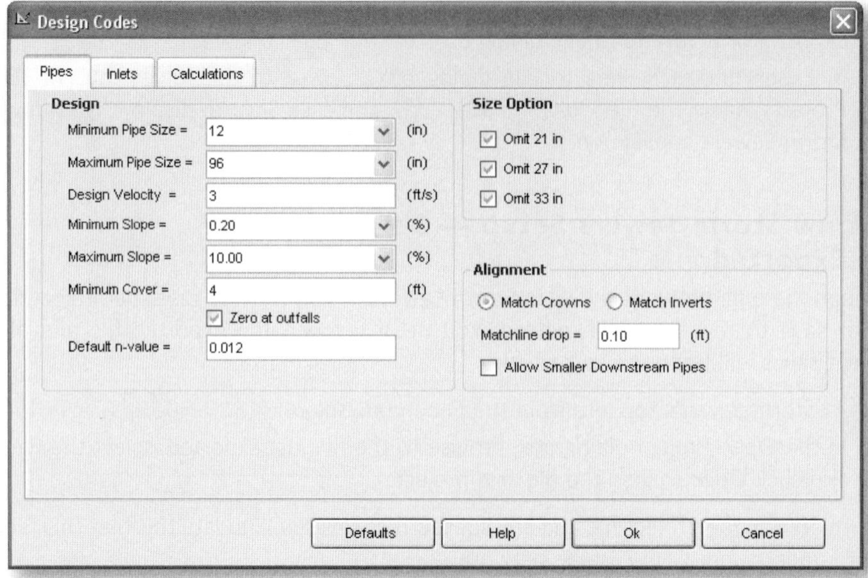


**FIGURE 13.56**

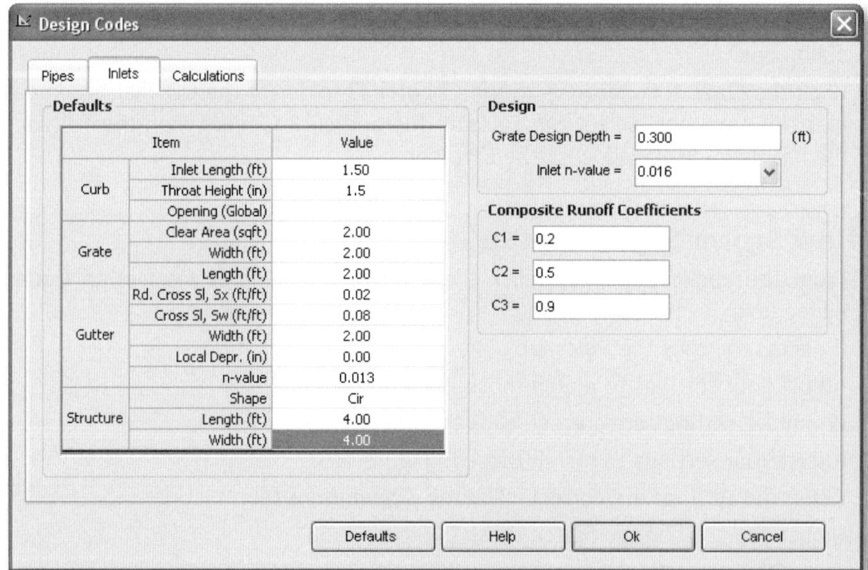


**FIGURE 13.57**

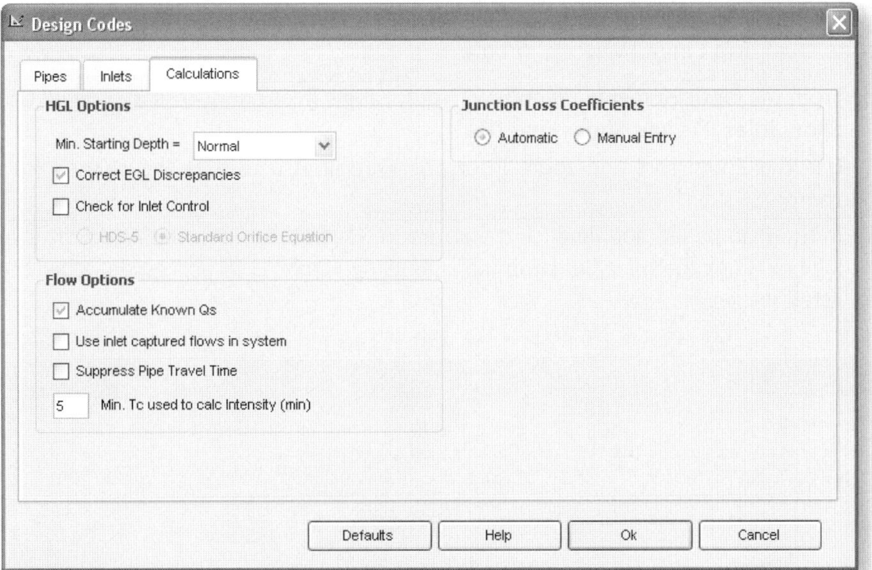


**FIGURE 13.58**

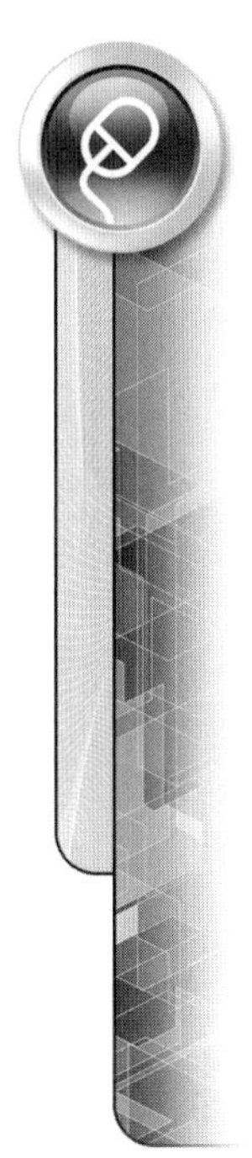

## Import a Background DXF

In the Chapter 13 folder of the CD that accompanies this textbook is a file ACAD_ CONTOURS.dxf. It contains contours that provide a background to the Hydraflow pipe network.

1. At Hydraflow Storm Sewers' top right, click the ***Import DXF*** icon, and from its shortcut menu, select BACKGROUND IMAGE FROM DXF....
2. In the Add DXF Background Image dialog box, browse to the Chapter 13 folder of the CD that accompanies this textbook, select the file *ACAD-CONTOURS.dxf*, and click ***Open*** to import the DXF.
3. Click the Save icon, and click OK to save the project.

## Pipe Network in Profile View

1. At the application's top, click the ***Profile*** icon.

This displays the Outfall structure and pipe. Storm Sewer can display only one branch at a time. A user must adjust the beginning and ending pipe segments at the top center of the panel.

2. At the Storm Sewer Profile's top, click To Line's drop-list arrow, from the list, select **4 – Pipe – W1**, and click the Update icon.

This displays the pipe network's western branch in Profile View. Write down the pipe sizes and note that they are all 12-inch.

3. At the Storm Sewer Profile's top, click To Line's drop-list arrow, from the list, select **8 – Pipe – E1**, and click the Update icon.

This displays the pipe network's eastern branch in Profile View. The pipes are also 12-inch pipes.

4. At the top left, click the ***Exit*** icon to return to plan view.

## Modify Structure Values

A structure must be an inlet to accept runoff.

1. From the Options menu, Plan View flyout, Labels flyout, if necessary, toggle **ON Show Inlet IDs** and **Show Line IDs**.
2. Use the Zoom tool and zoom in on the Outfall structure at the parking lot's southeast side.
3. In the Plan panel, double-click the structure **Outfall** to open the Add/Edit Dialog box. Click the ***Inlet/Junction*** tab, and using Figure 13.59 as a guide, enter the listed values.

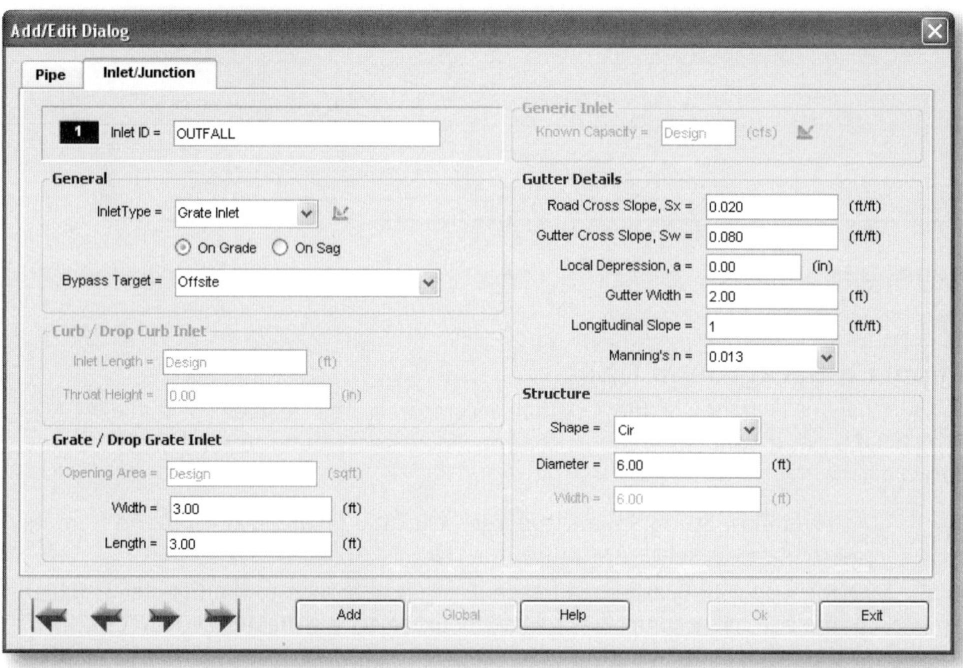


**FIGURE 13.59**

General
- Inlet Type = Grate Inlet
- On Grade

Grate / Drop Grate Inlet
- See Table 13.4

Gutter Details
- Road Cross Slope, Sx = 0.020
- Gutter Cross Slope, Sw = 0.080
- Local Depression, a = 0.000
- Gutter Width = 2.00
- Longitudinal Slope = 1
- Manning's n = 0.013

Structure
- Shape = Cir
- Diameter = 6.00

4. Click ***OK*** to set the values.
5. At the panel's lower left, click the **right-facing blue arrow** until you have reached structure **West 3**. You may have to click OK to reactivate the arrows.
6. Using Figure 13.60 as a guide, enter the listed values for West 3.

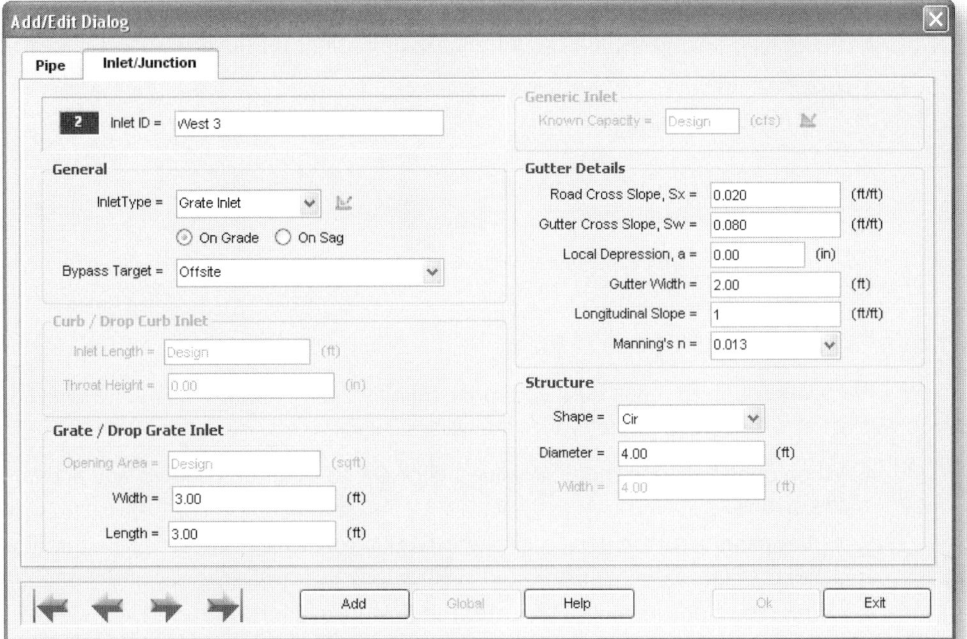

**FIGURE 13.60**

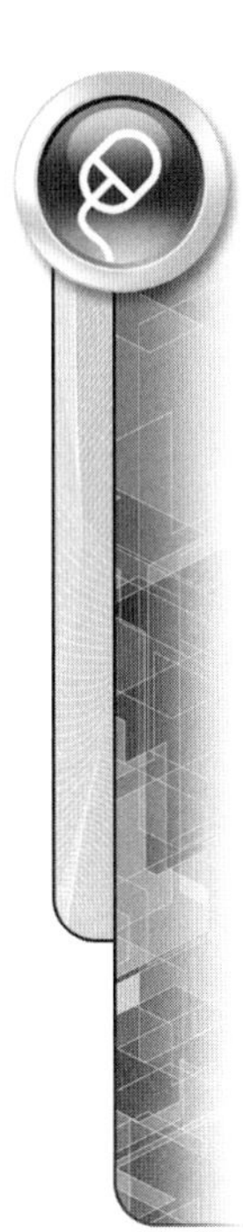

7. Click ***OK*** to set the values.
8. At the lower left of the Inlet/Junction panel, click the **RIGHT** arrow. You may have to click OK to reactivate the arrows. Use the same Gutter Details for all Inlet/Junction structures. However, the grate size changes for some of the structures (see Table 13.4). You will cycle from West 3 to West 1 and then East 4 to East 1.
9. After setting each structure's values, click ***Exit*** to return to the plan panel.
10. At Storm Sewer's top left, click the ***Save*** icon, and click OK to save the project.

**TABLE 13.4**

| Structure Name | Structure Size | Grate Type | Rectangular Grate |
|---|---|---|---|
| Outfall | 6 ft Circular | Grate Inlet | 3 ft × 3 ft |
| West 1 | 4 ft Circular | Grate Inlet | 2 ft × 2 ft |
| West 2 | 4 ft Circular | Grate Inlet | 2 ft × 2 ft |
| West 3 | 4 ft Circular | Grate Inlet | 3 ft × 3 ft |
| East 1 | 4 ft Circular | Grate Inlet | 1.5 ft × 1.5 ft |
| East 2 | 4 ft Circular | Grate Inlet | 1.5 ft × 1.5 ft |
| East 3 | 4 ft Circular | Grate Inlet | 2 ft × 2 ft |
| East 4 | 4 ft Circular | Grate Inlet | 2 ft × 2 ft |

## Setting Pipe Flows

After setting the structure values, next is setting all the flow values for each structure's pipe. This is done in each structure's Pipe panel.

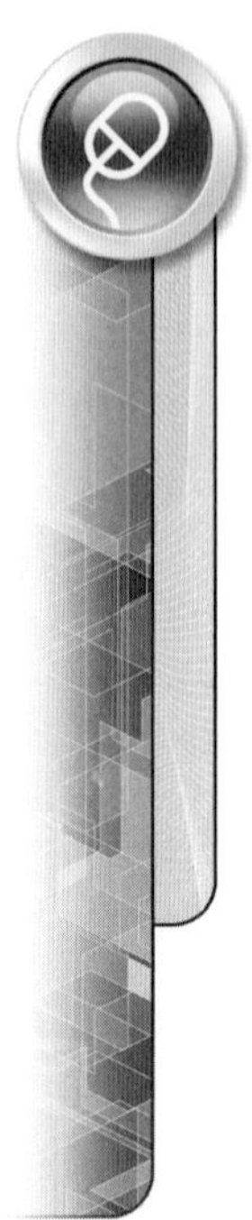

1. In the Storm Sewers plan panel, double click the **Outfall** structure to display the Inlet/Junction dialog box.
2. Click the Pipe tab.
3. In the Pipe tab's lower left enter the following values for flows (also in Table 13.5):

```
Flows
    Known Q: 1.594
    Drainage Area: 0.22
    Runoff Coefficient: 0.86
    Time of Concentration: 1.26
Physical
    Manning's n = 0.013
```

In the Physical section for each Pipe you **must** change the Manning's n value to 0.013 or you will get errors in the pipe network's analysis.

4. When done entering the flow values, click ***OK*** to save the changes and activate the arrows.
5. At the panel's bottom left, click the **RIGHT** arrow **three** times to view the pipe panel for Pipe – W1.
6. Using the values in Table 13.5, enter the flow values for **Pipe – W1**.
7. When done entering the values for Pipe – W1, click OK.
8. To move to Pipe – W2, at the panel's lower left, click the **LEFT** arrow.
9. Using the entries in Table 13.5, enter the flow values for **Pipe – W2**.
10. When done entering the values for Pipe – W2, click OK.
11. To move to Pipe – W3, at the panel's lower left, click the **LEFT** arrow.
12. Using the values in Table 13.5, enter the flow values for **Pipe – W3**.
13. When done entering the values for Pipe – W3, click OK.
14. After entering the values for Pipe – W3, click the **RIGHT** arrow **three** times to move to **Pipe – E4**.
15. If necessary, click **OK** to activate the panel's lower left arrows and click the **RIGHT** arrow **three** times to move to **Pipe – E1**.
16. Using the values in Table 13.5, enter the flow values for **Pipe – E1**.
17. To move to Pipe – E2, at the panel's lower left, click the **LEFT** arrow.
18. Using the entries in Table 13.5, enter the flow values for **Pipe – E2**.
19. Repeat steps 16 and 17 until arriving at Pipe – W1.
20. Click OK and then click ***Exit***.
21. At Storm Sewer's top left, click the ***Save***, and click OK to save the project.

**TABLE 13.5**

| Structure | Pipe Name | Known Q | Drainage Area | Runoff Coefficient | Inlet Time |
|---|---|---|---|---|---|
| Outfall | Pipe - Outfall | 1.594 | 0.22 | 0.86 | 1.26 |
| West 1 | Pipe - W1 | 3.564 | 0.45 | 0.96 | 1.60 |
| West 2 | Pipe – W2 | 4.515 | 0.57 | 0.96 | 2.19 |
| West 3 | Pipe – W3 | 1.549 | 0.21 | 0.92 | 1.67 |
| East 1 | Pipe – E1 | 1.561 | 0.22 | 0.86 | 1.97 |
| East 2 | Pipe – E2 | 1.535 | 0.20 | 0.90 | 1.55 |
| East 3 | Pipe – E3 | 1.594 | 0.20 | 0.92 | 1.46 |
| East 4 | Pipe – E4 | 1.077 | 0.15 | 0.87 | 1.65 |

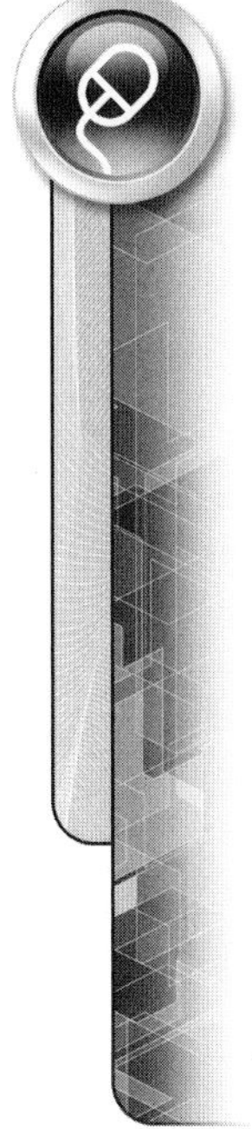

## Run the Analysis and Reset Pipe Sizes

Next, you run an analysis using the current pipe sizes. The pipe sizes are too small to handle the runoff amounts. When you run the analysis, you will allow the routine to resize them.

1. At the Plan panel's top, click the ***Run*** icon to open the Compute System dialog box.
2. In the Hydrology section, make sure it is set to **10 year**.
3. In the Design Options section, toggle **ON Reset Pipes Sizes**.
4. At Compute System's bottom, Starting HGLs, set the Starting HGL to **Normal**.
5. Click ***OK*** to evaluate and resize the pipes.

The Storm Sewer Design dialog box opens and displays the outfall structure.

6. At the lower right, click ***Up*** to review each structure and pipe in the system.
7. After reviewing the values, click ***Finish***.
8. At the top left, click the ***Save*** icon, and click ***OK*** to save the project.

## Review Pipe Results

1. At the top left, click the ***Pipes*** tab.

This opens the pipes panel that is displaying the adjusted results. A user can change any values in this worksheet.

2. Review the values for the pipe analysis.

## Review Inlet Results

1. At the top left, click the ***Inlets*** tab.

This opens the Inlets panel that is displaying the adjusted results. A user can change any values in this worksheet.

2. Review the values for the Inlet analysis.

## Review Results

1. At the top left, click the ***Results*** tab.

This opens the Results panel that is displaying the analysis in several different formats. One of the Results tabs is a cost estimate for the network.

2. Review the values for the various Results tabs.

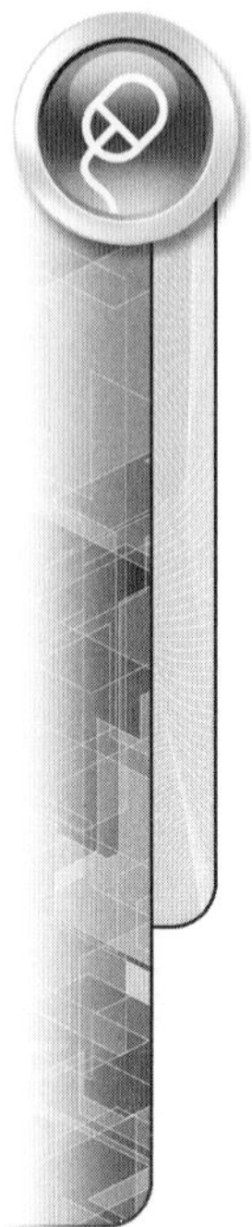

## Import the Revised Pipe Network

The last step is updating the PKLOT – PSTM – (1) pipe network in Civil 3D.

1. If necessary, restart **Civil 3D** and open the *Tulsa Parking Lot – Work* drawing.
2. If necessary, in the Ribbon, Click the ***Modify*** tab. On the Design panel, click the ***Pipe Network*** icon displaying the Pipe Networks tab.
3. In the Pipe Networks tab, the Analysis panel, click the ***Storm Sewers*** icon and from the shortcut menu, select `IMPORT FILE`.
4. In the Import Storm Sewers Files dialog box, browse to and select *PKLOT - PSTM – (1)* and click ***Open*** to read the file.
5. The Existing Networks Found dialog box displays. Select Update the existing pipe network to modify the network to the new values from Storm Sewers.
6. An Events panorama is displayed. Click the ***green checkmark*** to close the panorama.
7. In the Pipe Networks tab, click Close.
8. In the drawing select a few pipes to view their new sizes.
9. At Civil 3D's top left, Quick Access Toolbar, click the ***Save*** icon to save the drawing.
10. Close the drawing.

You will have to reset the structure types because they follow the Pipe Network defaults.

## SUMMARY

- Storm Water Extension evaluates and resizes pipes and, if desired, inlets.
- You can import a background DXF to give the pipe run a context.
- You export a design to Storm Sewers, but you must open the exported file to set the project correctly.
- After modifying a design in Storm Sewers, you import it to the Civil 3D file.
- An imported Storm Sewers project can modify an existing or create a new pipe network.

# UNIT 4: PLAN PRODUCTION TOOLS

Creating plan, profile, or plan and profile sheets is a Plan Production process. Creating these sheets takes two steps: creating view frames and producing the sheets. Several styles affect this process and a template is key to creating the sheets. The template defines the sheet size, viewport function (plan or profile), scale, and the appropriate border.

## PRODUCTION TEMPLATE

The Production Template contains several sheet size and layout examples. It is worth your time to review them. Each viewport has a type: plan or profile. This value is set in the Properties palette for the viewport (at the bottom).

## VIEW FRAME GROUP

The first set is used to create viewports. Viewports define the number of sheets and their organization (see Figure 13.61). A multi-tabbed wizard sets values for the view frame group.

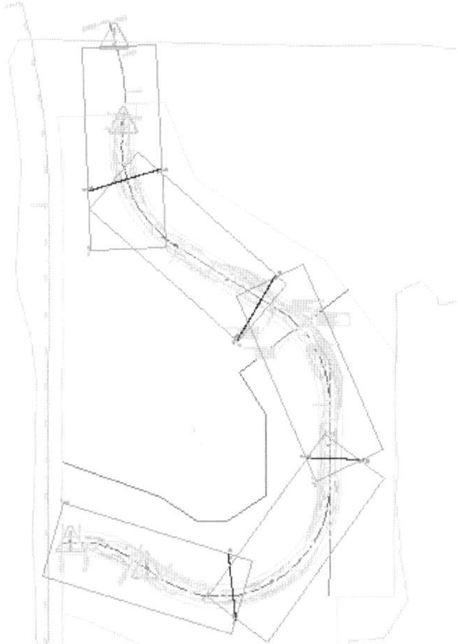

**FIGURE 13.61**

### Create View Frame Group Wizard

The Create View Frames wizard's first panel sets the alignment and its coverage (see Figure 13.62).

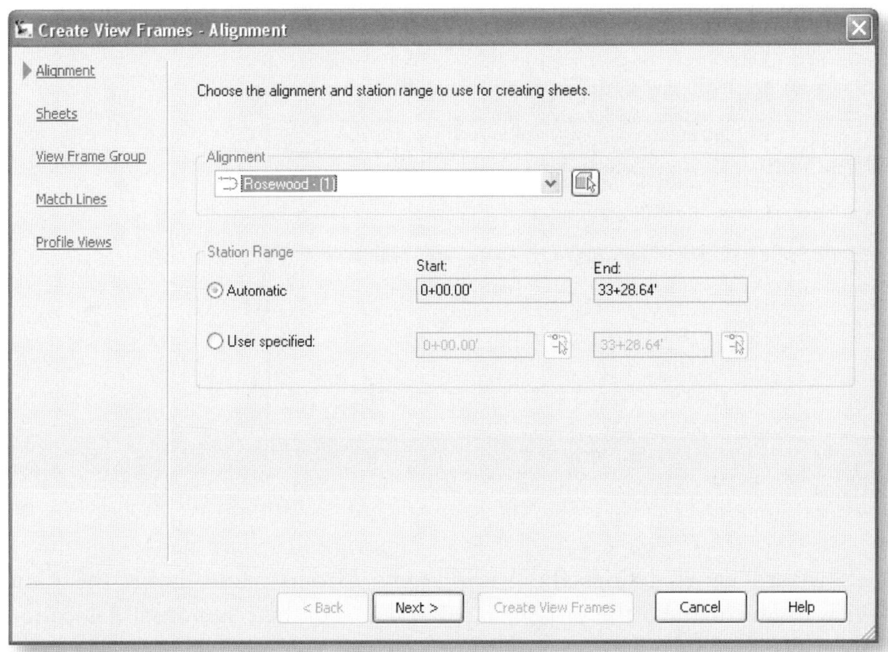


**FIGURE 13.62**

The second panel sets the sheets panel. This panel sets critical information for the set: the sheet type, its layout template, sheet size, and viewport scales (see Figure 13.63). The settings at the bottom affect the view frames' orientation along the alignment.

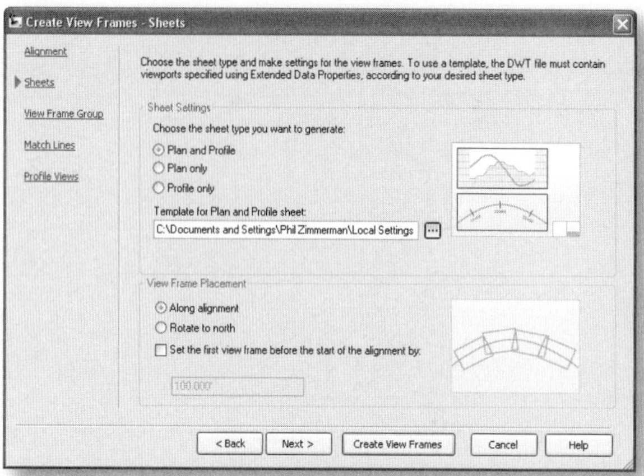


**FIGURE 13.63**

To identify the appropriate template file, you select the ellipsis at the right center of the panel. This opens the Select Layout as Sheet Template dialog box (see Figure 13.64). Each of the layouts listed is in a single drawing. The drawing's function is to define and provide the selected layout. Civil 3D installs with several drawings, each having a different grouping of layout: plan only, profile only, and plan and profile. You should adjust the layouts and add the correct border before using this process.

If no layouts appear, you must locate and select the drawing containing the layouts by clicking the ellipsis, browsing to, and then selecting the file.

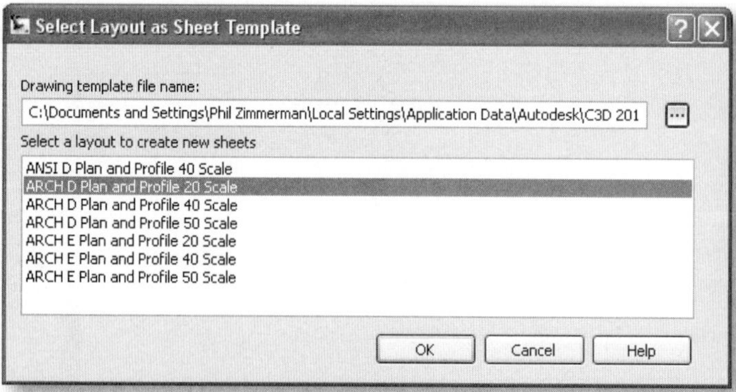


**FIGURE 13.64**

The next wizard panel defines the View Frame Group name, layer, the individual view frame names, their style, label style, and their annotation location (see Figure 13.65).

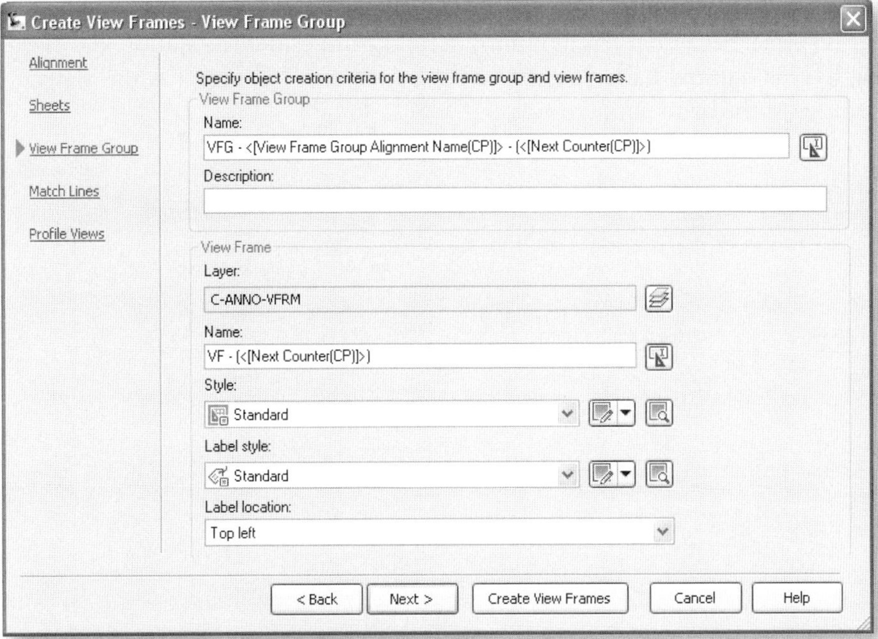


**FIGURE 13.65**

The Match Lines panel sets whether matchlines will appear along with the view frames, their style, and annotation (see Figure 13.66). Snap station value down to the nearest sets the matchline at a whole foot. Allow additional distance... defines the amount a view frame can move along an alignment if it needs to be repositioned.

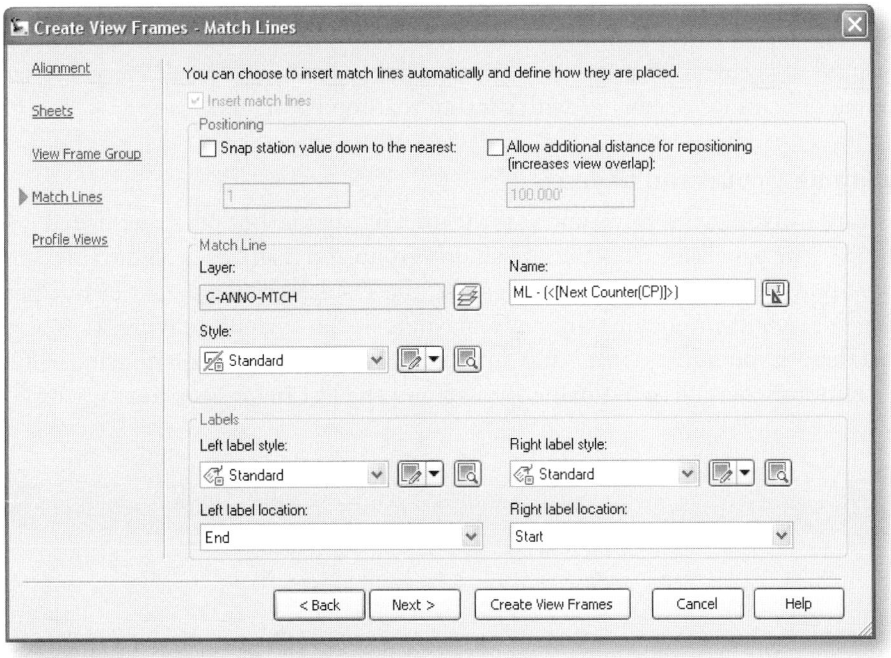

**FIGURE 13.66**

The Profile Views panel sets the profile view and its band set (see Figure 13.67). The band set is special because it has a dummy second band under the station and elevation band. This dummy band "centers" the profile view and prevents the viewport from cutting off the band annotation.

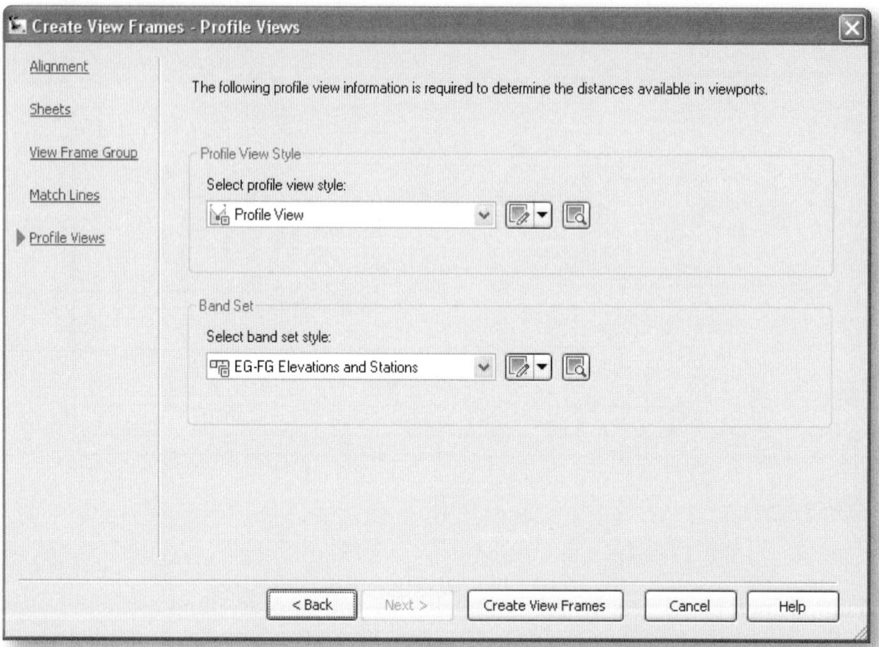


**FIGURE 13.67**

## CREATE PLAN PROFILE SHEETS

The second step in using Plan Production Tools is Create Sheets. This command also displays a wizard to step the user through the process. The result is a sheet set containing one or more drawings with a layout that represents the sheets.

### View Frame Group and Layouts

The wizard's first panel reviews and sets the drawing and layout tab values (see Figure 13.68). The panel's top sets the view frame group and if all or selected frames are created. Its middle defines how to organize the resulting layout: one layout per drawing, all layouts in one drawing, or all layouts in the current drawing. It is best to have these layouts in external drawing(s) and not in the current drawing. The panel's bottom sets the Layout name format, and the last item sets a north arrow.

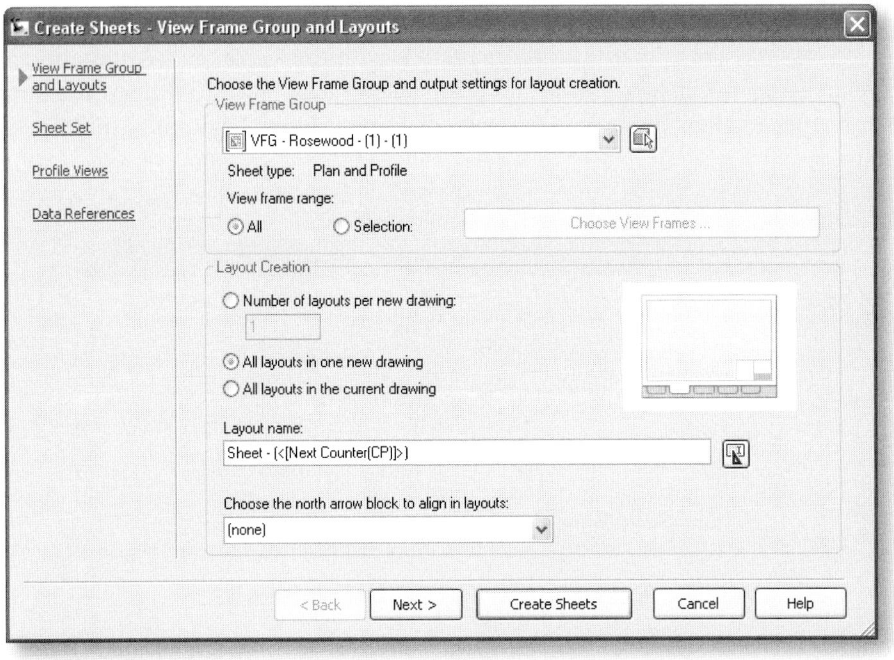

**FIGURE 13.68**

## Sheet Set

The result of creating a drawing set is a sheet set (see Figure 13.69). The panel's top portion sets the sheet set's name and location. The new sheets can be appended to an existing set, if desired. The panel's bottom sets the name for each drawing file containing the layouts. A sheet set can be assigned to a Vault project by logging in and assigning a project.

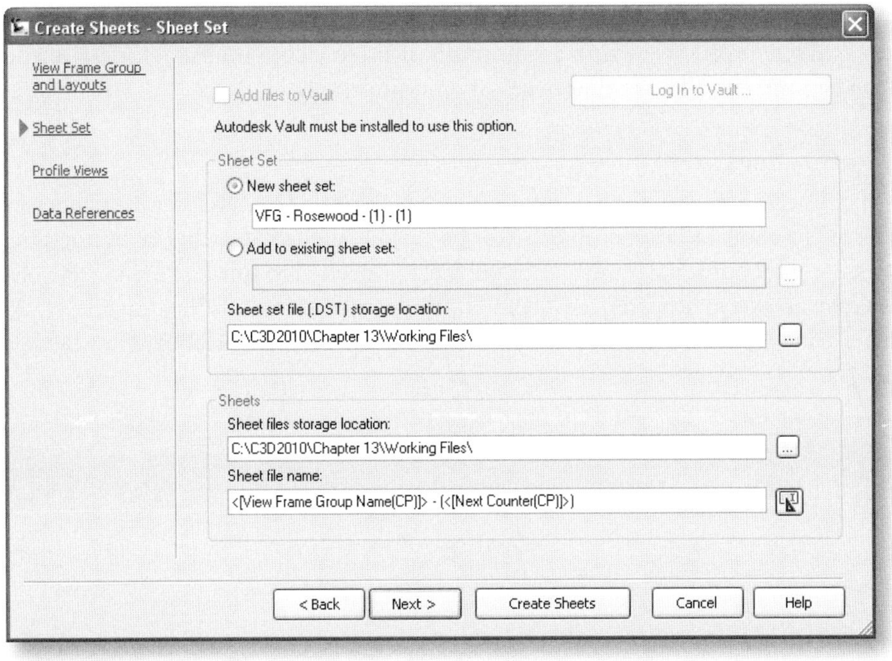

**FIGURE 13.69**

## Profile Views

Profile Views sets the annotation and orientation with the alignment. Annotation is set by referencing an existing profile and profile view (see Figure 13.70). At the bottom are toggles to align the profile with the alignment by its beginning, middle, or end point.

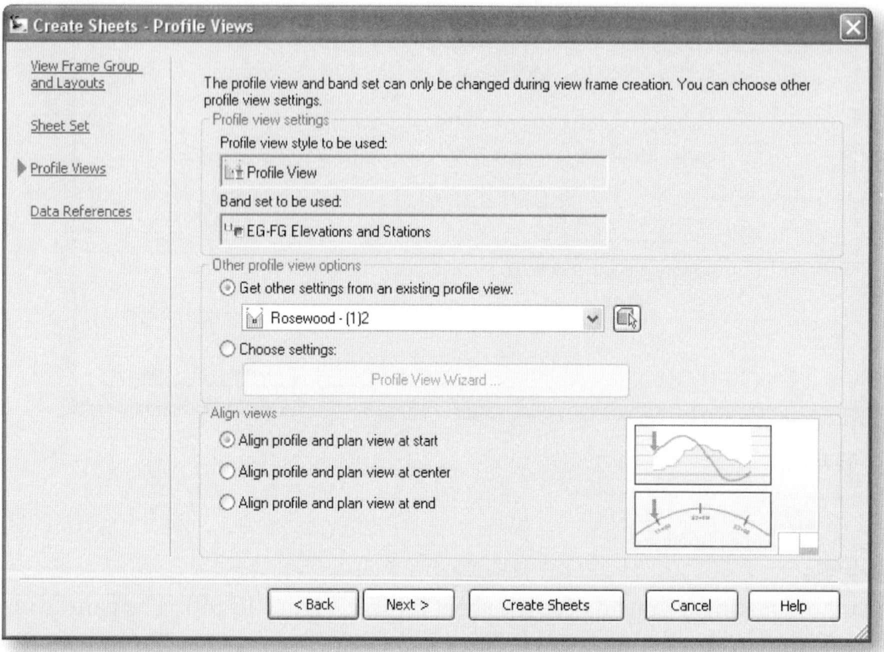


**FIGURE 13.70**

If you want different profile labeling, toggle Choose settings in the middle of the panel and click Profile View Wizard…; this opens the Profile View Wizard. The wizard can split the profile view into a minimum of three views; it allows for styles and labels reassignment and any other necessary modification.

## Data References

Data that is necessary for generating the sheets is set in the Data References panel (see Figure 13.71). Any additional objects can be put in the new drawing(s) by toggling them on here. If additional linework, not objects, is needed to finish the sheet, it must be added by an external reference (XREF) in the new plan and profile drawing(s). At the panel's bottom is a toggle; when on, it copies a pipe network's labeling to the new drawing(s).

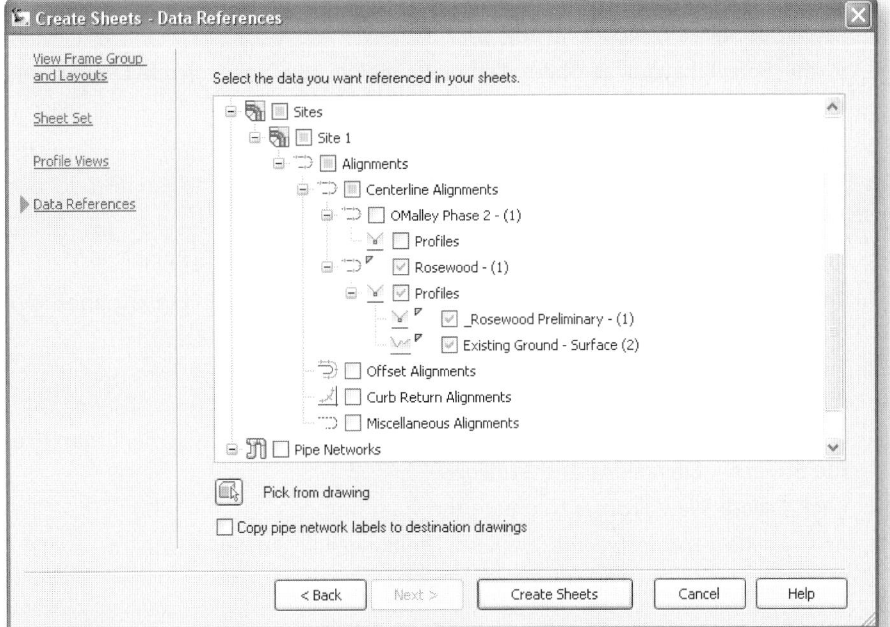


**FIGURE 13.71**

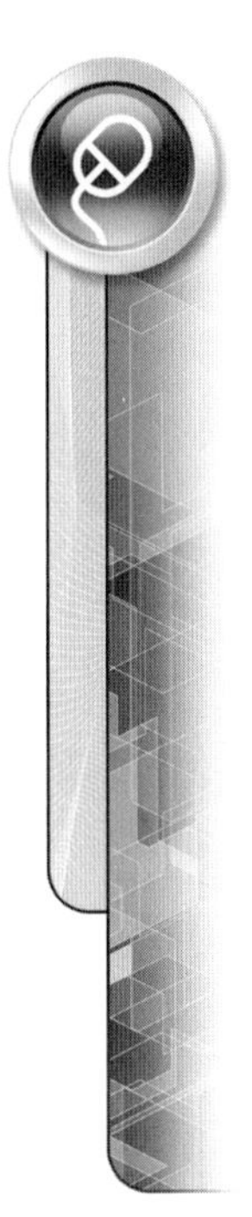

## EXERCISE 13-4

After completing this exercise, you will:

- Be able to create a View Frame Group.
- Be able to create a Plan and Profile sheet set.

### Exercise Setup

This exercise uses the *Chapter 13 – Unit 4.dwg* file. To find this file browse to the Chapter 13 folder of CD that accompanies this textbook, and open the *Chapter 13 – Unit 4.dwg* file.

1. If you are not in Civil 3D, start the program by double clicking on ***Civil 3D's*** desktop icon.
2. At Civil 3D's top left, Quick Access Toolbar, click the ***Open*** icon. In the Open dialog box, browse to the Chapter 13 folder of the CD that accompanies this textbook, select the file *Chapter 13 – Unit 4*, and click ***Open***.
3. At Civil 3D's top left, click the ***Civil 3D*** drop-list arrow, and from the Application Menu, highlight SAVE AS, from the flyout select AUTOCAD DRAWING, browse to the Civil 3D Projects folder, for the drawing name enter **Plan and Profile Sheets** and click ***Save*** to save the file.

### Create View Frames

1. On the Ribbon, click the ***Output*** tab. On the Plan Production panel, click the CREATE VIEW FRAMES icon displaying the Create View Frames wizard's Alignment panel.
2. Using Figure 13.62 as a guide, set the alignment to **Rosewood – (1)** and click ***Next*** to open the Sheets panel.
3. Using Figure 13.63 as a guide, if necessary, set the sheet type to **Plan and Profile** and, at the bottom, toggle the sheet orientation to **Along alignment**.

4. In the middle right of the Sheets panel, click the ellipsis to open the Select Layout as Sheet Template dialog box.
5. In the Select Layout as Sheet Template dialog box, select **Arch D Plan and Profile 20 scale** and click ***OK*** to return to the Sheets panel.

If no templates appear in the Select Layout as Sheet Template dialog box, browse to the Chapter 13 folder of the CD that accompanies this textbook and in the Plan Production folder, select from the template list.

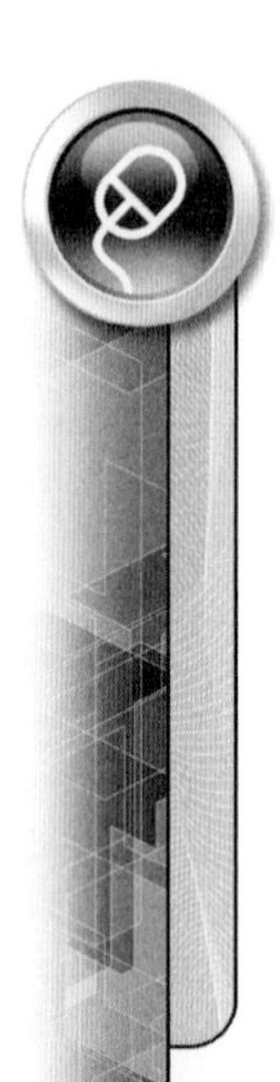

6. Click ***Next*** to open the View Frame Group panel (see Figure 13.65).
7. There are no changes, so click ***Next*** to open the Match Lines panel (see Figure 13.66).
8. There are no changes, so click ***Next*** to open the Profile Views panel (see Figure 13.67).
9. Click the Select band set style drop-list arrow and, from the list, select **Plan Profile Sheets – Elevations and Stations**.
10. Click ***Create View Frames*** to create the view frames.
11. At Civil 3D's top left, Quick Access Toolbar, click the ***Save*** icon to save the drawing.

## Create Plan Profile Sheets

The second step is to create a sheet set with the layouts defined by the view frame group.

1. On the Ribbon's ***Output*** tab. On the Plan Production panel, click CREATE SHEETS icon displaying the Create Sheets wizard.
2. In Create Sheets – View Frame Group and Layouts panel, in the middle, toggle **Layout Creation to All layouts in one new drawing**, set the north arrow block to **North**, and click ***Next*** to open the Sheet Set panel.

## Sheet Sets

The next panel defines the sheet set and its location.

1. Review the panel's current settings, in the middle of the panel, set the sheet set and sheet locations, and click ***Next*** to open the Profile Views panel.

## Profile Views

This panel defines the Profile and Profile View's annotation and alignment.

1. Review the panel's settings.
2. In the panel's middle, click ***Choose settings*** and then click PROFILE VIEW WIZARD....
3. In Profile View Wizard, click ***Next*** to open the Profile Display Options panel.
4. Scroll the profiles to the right until you are viewing their label assignments.
5. Double-click the _Rosewood Preliminary – (1) label assignment (_No Labels), in the Pick Profile Label Set dialog box, click the drop-list arrow, select **Complete Label Set**, and click ***OK*** to return to the Profile Display Options panel.
6. Click ***Next*** to open the Data Bands panel.

Notice the double band assignment: one for data, the dummy band for spacing.

7. Click ***Finish*** to return to the Profile Views panel.
8. Click ***Next*** to open the Data References panel.

## Data References

The new drawing(s) and their layouts can be linked to the data in the current drawing.

1. There is nothing to change in this panel.
2. Click ***Create Sheets*** to display the automatic save warning. Click ***OK*** to continue.
3. The command line prompts you for a profile view origin. Enter **0,0** and press ENTER.
4. An Events panorama is displayed, close the panorama.
5. Review the plan and profile sheets in the Sheet Set Manager palette.
6. In Sheet Set Manager, double-click each sheet to review its layout and annotation.
7. After reviewing the sheets, close the drawing and the Sheet Set Manager.
8. At Civil 3D's top left, Quick Access Toolbar, click the ***Save*** icon to save the drawing, and exit Civil 3D.

This ends the exercise on Hydrology and Plan and Profile Sheets.

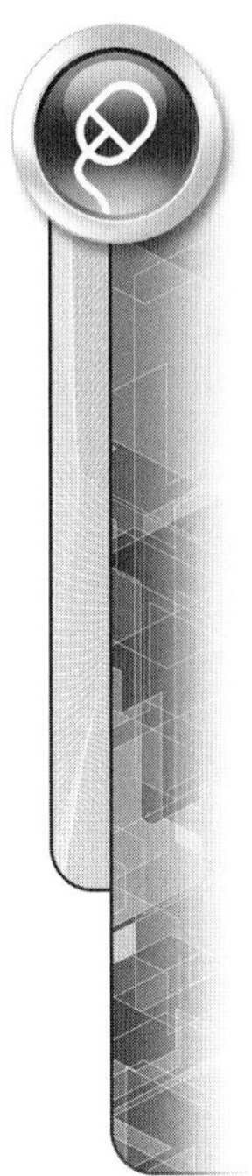

## SUMMARY

- Plan Production Tools create Plan, Profile, and Plan and Profile sheets.
- Each sheet is a layout in the current drawing, the layouts are in a single new drawing, or each single layout is in a new drawing.
- Layout viewports are either plan or profile, which is a viewport property.
- The resulting sheets are a sheet set.

The next chapter introduces Civil 3D's survey capabilities and concepts.

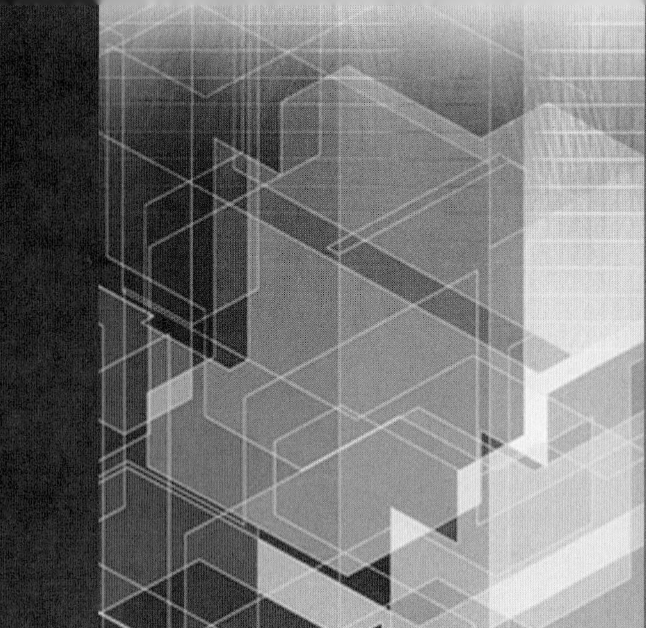

CHAPTER

# 14

# Survey Basics

## INTRODUCTION

Most projects start and end with point data. In the beginning, the field data describes the current site, its legal boundaries, and the utilities that serve it. The survey crew is the office's eyes, because only rarely does the office staff visit the site. After all planning and design is completed, the survey crew again enters the field to lay out the design for its construction.

## OBJECTIVES

This chapter focuses on the following topics:

- Becoming Familiar with the Field Book Language
- Converting Proprietary Data Files into the Field Book Language
- Becoming Familiar with the Survey Settings
- Importing a Field Book
- Reviewing Survey Data
- Editing Survey Data
- Creating Figure Styles
- Defining Figure Prefixes
- Editing Figure Commands in a Field Book
- Linework Code Sets

### OVERVIEW

Survey incorporates field-collected observations in a project. Field observations are obtained from a crew and its instruments observing critical site points (i.e., a survey). These observations come back to an office to be reduced to coordinates, line work, and symbols. The data collector holds the observations in the instrument manufacturer's proprietary data format. Survey needs to convert this proprietary format to one of two formats it can read: a field book or LandXML survey. One conversion program ships with Civil 3D: TDS link. Other manufacturers, such as Leica, Trimble, and Carlson, have programs you can download from their Web sites. You also can download programs from the manufacture's Web site. To communicate with the collectors, active sync must be installed.

A field crew is the office's eyes for the project. The crew members' site observations may be the only contact an engineer has with a project. These data points represent important site locations (for example, benchmarks, existing utilities, environmentally sensitive areas, etc.). These critical elements must be documented with a survey for legal and design reasons.

When the collector's proprietary file format is converted to the field book language, the data format changes, but not the data itself. A survey in the TDS format is included in the following code snippet:

```
SP,PN1,N 1095963.3900,E 747776.0520,EL652.638,--RBC
SP,PN8,N 1096024.7400,E 747359.1190,EL668.123,--PK
OC,OP1,N 1095963.3900,E 747776.0520,EL652.638,--RBC
LS,HI4.600000,HR4.940000
BK,OP1,BP8,BS278.2215,BC359.5958
--B
LS,HI4.600000,HR5.000000
SS,OP1,FP2000,AR187.5917,ZE91.2202,SD35.9700,--EP1
--B
SS,OP1,FP2001,AR185.5628,ZE91.1045,SD24.9500,--CL1
```

When the preceding data is converted to a field book, the result is the same data as the original file, but in field book coding:

```
NEZ  1 1095963.3900 747776.0520 652.6380 "RBC"
NEZ  8 1096024.7400 747359.1190 668.1230 "PK"
STN  1 4.600 "RBC"
PRISM  4.940
AZ  1 8 278.22150
BS  8 359.59580
BEG EP1
PRISM  5.000
F1 VA 2000 187.59170 35.970 91.22020 "EP1"
BEG CL1
F1 VA 2001 185.56280 24.950 91.10450 "CL1"
```

The field book language is a "neutral" language for all survey instrument manufacturers. In time, the field book language may be replaced by the LandXML survey format. Besides offering a neutral language, the field book language contains figure commands. Figure commands, when used properly, draw lines that represent linear survey features. Drawing lines from field observations help those who have not seen the site better understand how the line work connects to survey points. Having survey data draw line work is helpful when several points and lines converge at a common location.

Civil 3D Survey reads and sorts field book information and data into two categories: control points and non-control points. All points are survey points. Survey points are the initial observed points' coordinates. Control points have a known location and non-control points are points with known coordinates, but they are not used in the survey as stations (setups).

However, a survey and its coordinates may need adjustments to correct for observation errors. Survey errors come from misaligned optics, from damaged instruments, or from a field crew member having an unsteady hand. Having an instrument library with each instrument's error values is important for adjusting field book observations. An office's equipment library should have an entry for each instrument or at least one generic entry with reasonable error values.

Survey holds the initial survey point coordinates in a network's observation database. When reviewing or editing observation values, the user is editing the data in the observation database, not the field book. He or she may not have to edit the field book to correct a problem, but he or she must be aware that when correcting an observation database value, there is now a difference between the field book and the observation database. A likely reason for editing a field book is to correct minor figure-command errors. If you have major observations errors and are using figure commands, it is best to edit the original survey file. After you make the changes to the original survey file, convert it to a new field book that contains the corrections.

Survey's command window reviews values in the observation database and gives immediate answers to queries about observation values (for example, bearings, distances, slopes, inverses).

When you are satisfied with the data quality, it is time for adjustments. There are two types of adjustments: the entire survey (least squares) or traverse loop(s) within a network (compass, crandall, transit, or Least Squares). When you are adjusting at the survey level, the field data must contain redundant point observations within or outside the survey (water towers, cell towers, peaks, etc.). Having redundant data allows the adjustment to refine a point's location based on multiple point observations from different locations.

When a survey contains a traverse, traditional traverse methods create redundant data. Redundant traverse observations are only back to the last point occupied or forward to the next point to be occupied. A traditional traverse does not allow for cross-traverse or outside observations. This methodology calculates very precisely how to walk from the beginning to the end of a traverse path. However, using this methodology does not necessary work as well when you are walking a traverse loop from one side to the other. There is more data around the loop than across the loop.

A cursory knowledge of survey field techniques and definitions is desirable for this chapter and the next chapter. This knowledge helps when you encounter survey terms, values, and techniques.

Point and label styles and description key sets are important to know when you are creating documents from surveys. Styles and description key sets are both a part of the Chapter 2 point discussion.

With the advent of GPS, traditional surveys containing observations of angle, distance, and zenith angles change to point number, coordinates, and descriptions. Civil 3D allows a user to add line coding to a coordinate file to draw survey line work. These drafting commands are more powerful and easier to use than the commands for field book files.

### Unit 1

The first unit introduces the field book language. It is important to become comfortable with this language because Survey depends on its ability to communicate the field data.

### Unit 2

Survey settings are critical to correctly importing, adjusting, and resolving survey data. The Survey panel contains settings that affect the location of survey files and output. When you are creating a local survey database, this, too, has settings that affect the survey and its observation processing.

### Unit 3

The third unit reviews importing field books, describes the import results, and describes the tools for evaluating the resulting data.

### Unit 4

The fourth unit reviews defining figure styles, figure prefix library entries, and figure coding in field data files. Figures draw lines that represent linear features. Some figures represent naturally occurring lines swales, stream edges, or manmade lines, edges-of-pavement, centerlines, boundaries, etc. Many figures remain as line work that is important to site or project documentation. Other figures are critical to a surface. Survey has a tool for assigning figures to a surface as standard breaklines.

## UNIT 1: FIELD BOOK LANGUAGE

The field book language describes survey field techniques and observation data. This language provides Survey with a single language from which it can create points. The field book language is terse and assumes knowledge of and an understanding of surveying methods, data requirements, and methods for calculating point coordinates.

### FIELD BOOK LANGUAGE FORMAT

Field book language documentation is in Civil 3D's Help files and folder structures. The default path in Windows XP is C:\Documents and Settings\All Users\Application Data\Autodesk\C3D 2010\enu\Survey; for Vista it's located in C:\ProgramData\Autodesk\C3D 2010\enu\Survey. In this folder is a valuable file: *cmdhelp.ref*. This text file opens in Notepad and it lists all field book commands and their format.

Field book language formats survey and observation data. Each field book line is an observation or survey method (setup, backsight, prism adjustment). The survey crew's first task is establishing a station or setup. A station places the instrument over a point, which can have assumed or actual coordinates. If the survey has vertical measurements, the instrument height must be recorded.

When observing a point, there must be a way to record how much the instrument has turned and the distance from the instrument to the point. If vertical data is included, the point's location is measured by one of two methods: a vertical angle or a vertical distance. The following snippet is from the Cmdhelp.ref file and lists the commands for entering turn angles and distances (AD entries) and establishing a setup, including an azimuth and backsight:

```
AD (point) [angle] [distance] (descript)
AD [VA] (point) [angle] [distance] [vert angle] (descript)
```

```
AD [VD] (point) [angle] [distance] [vert distance]
(descript)
AZ [point 1] [point 2] [azimuth]
B [point1] [point2] [bearing] [quadrant]
BS [point] (orientation)
STN [point] (inst. height) (descript)
```

The capitalized letters (AD, AD [VD], AD [VA], etc.) are the command names, and the remaining values are the data entries. A mandatory value has a square bracket enclosing it (that is, AD must have an angle and a distance). An entry enclosed in parentheses is an optional value. A point number is not necessary when Autopoint number (AP (Number)) is on, and it is not mandatory to describe a point.

Each command and its data format reflect a field crew's method and resulting data. At the beginning of a survey, the crew establishes a station by setting up an instrument on a point and measuring the instrument's height above it. The activities define a station and the crew records this data as a station with an instrument height. The field book STN command represents the station process and its recorded data. Next, a (0) zero angle direction is determined. The BS command represents this process and its recorded data. From this zero direction, a field crew can determine and record the angular amount the instrument has turned to view a forward point. This activity becomes the data for the AD field book entries. The field book command language is an action-by-action translation of field activities.

## SURVEY BASICS

A surveyor needs several pieces of information to begin a survey, but the most important piece of information is a known location and possibly its elevation. If the surveyor does not have these control point values, then the surveyor makes assumptions about the point's coordinates and elevation. Traditionally, when the control point's actual coordinates are not known, a surveyor assumes a coordinate value: 5000, 5000 or 10000, 10000. If vertical measurements are included, a surveyor may assume an elevation (for example, 100). During the survey or after it has been completed, the surveyor researches the actual point coordinates and substitutes them in the survey data. If the surveyor changes the coordinates for a benchmark, all the points in the survey change, because their measurements are relative to this point. The same applies for the elevation. When the correct elevation has been determined and then substituted for the assumed value, all the points adjust their elevations.

The field book language commands for points with known coordinates and/or elevations are the following:

```
NE (point) [North] [East] (descript)
NEZ (point) [North] [East] (elev) (descript)
NE SS (point) [North] [East] (elev) (descript)
```

NE represents known coordinates only. NEZ represents coordinates and known or assumed elevations. NE SS is a special case. These are points with known locations and possibly elevations, but they are not used as control points (a location for the instrument) within the survey. When a field book containing these points types is imported, they appear as Network Non-Control Points.

When determining a station (STN) and turning an angle to observe a point, there has to be a way to measure the turn amount. A field crew uses a backsight to determine the 0 (zero) angle direction. This direction is expressed as an azimuth (an angle from 0 to 359.5959) or an observation to a second point. After establishing the zero angle direction, the instrument can measure the turn amount when viewing a point in the field. This is also known as the turned horizontal angle.

```
STN [point] (inst. height) (descript)
AZ [point 1] [point 2] [azimuth]
BS [point] (orientation)
```

The preceding commands occur in the listed sequence: establish a station, determine the 0 (zero) angle direction, and establish the 0 direction on the instrument. If there is no actual backsight point, the crew sights a direction and assigns a dummy point number. If an assumed azimuth is assigned, this can change in an editing session and as a result rotate the entire survey and its points.

When they are in the field observing, the crew uses a prism pole. A prism is an optical device that reflects the instrument beam back to the instrument. The delay between sending and receiving is how the instrument determines the angle and distance to the observed point. The prism attaches to a pole, and its height must be recorded to correctly calculate elevations. In field book language, this is the PRISM [height] command. During a survey, this height may never change, but in many surveys the prism height is frequently adjusted. When it is changed, a PRISM [height] appears in the field book.

Once all of the preliminaries are done and measured, it is time to survey (or make observations). A survey's purpose can be to determine a boundary's location or to inventory existing site conditions (a topographic survey). When a survey's focus is a boundary, the field crew may use a field data collection method known as a traverse. (Traverses are the focus of Chapter 15.) This chapter focuses on topographic surveys.

Existing site conditions include naturally occurring slopes, grades, and linear features. Naturally occurring linear features are streams, swales, berms, etc. Manmade linear features include edges-of-travelway, gutters, sidewalks, etc. A field crew collects points that represent each of these features (i.e., points along the edge-of-pavement, stream edge, sidewalk's edge, sign posts, utility poles, benchmarks, elevation, etc.). All points and their information can be used to create a property boundary legal description, a road right-of-way, a map of existing utilities, and a surface.

Vertical data is not required to observe points in the field. A surveyor's responsibility is 2D measurements, not 3D. However, when the surveyor is surveying for a topography and has a contour map or surface model as a deliverable, each observation has a vertical component. When surveying and collecting horizontal data, each observation needs only two pieces of information: a turned angle and a horizontal distance. When this data appears in a field book, it uses the AD command format:

```
AD (point) [angle] [distance] (descript)
```

Optional point numbers and descriptions are always filled in and only rarely are they automatically generated.

When collecting vertical data, the observations use one of two formats: zenith angle and vertical distance or difference. When using zenith angle, the angle of the scope relative to a normal passing through the instrument down to the station's point creates the recorded angle. When observing the prism through the instrument's scope, the crew records the scope's angle relative to the zenith normal. The angle ranges between 0 and

180 degrees, but for most surveys, the value is around 90 degrees (80–100). If the setup instrument height and the prism height are the same, a zenith angle of 90 implies the point is the same elevation as the station. If the angle is less than 90 degrees, the observation is upward and the point is higher than the station point. Conversely, with an angle greater than 90 degrees, the observed point is lower than the station.

Vertical distance, or difference, records the prism's height above or below instrument's eyepiece (the eyepiece height creates a horizontal plane). If the instrument's and the prism's heights are the same, they are at the same elevation. When the prism is higher (positive value), it is higher than the instrument's elevation, and if lower (negative value), it is lower than the instrument's elevation.

Using vertical measurements, all observed point elevations are dependent on the control point's elevation. If it is 100, all points are near this value. If you are researching and determining the true point elevation, simply replacing the assumed elevation with its researched value raises or lowers the entire survey.

The same can be said for an initial backsight azimuth. If you assume an azimuth and then research its actual value, you can change the field book's or Survey's azimuth to rotate the entire survey by the new azimuth value.

The following snippet is a hand-recorded survey containing control points (points with known coordinates), a stationing and its values, prism settings, and observations:

```
Benchmarks
POINT NUMBER N    E         Z         D
             1206 2265.3919 1552.9250        "+"
             1213 2601.6437 1497.8024 634.52 "IP"  IP SET
             1226 2623.0725 660.6777  636.43 "+"
INST @ 1226 INST HI 5.13    BACKSIGHT 1213
PRISM 4.65
STA  AR        SD     VD     DESC
200  92.1224   61.31  -0.89  BC/WK     BC AND F/WK
201  87.3738   61.13  -0.82  BC/WK
202  84.4515   17.12  -0.43  BC/WK
203  101.0837  17.49  -0.51  BC        BC AND F/WK
204  252.2341  13.08  -0.32  BC        BC AND F/WK
205  274.3455  12.68  -0.37  BC        E@ 10' GRVL
206  262.1906  33.27  -0.28  BC        BC AND F/WK
207  259.3437  33.45  -0.74  FLW LINE  INLET B
```

Creating a field book from the preceding data is a line-for-line translation. For example, benchmarks are NE or NEZ entries, and the instrument setup at point 1226 and backsight to point 1213 is an STN entry. Each survey line has a field book equivalent. The same known coordinates (benchmarks), stationing, and observations appear as the following in the field book:

```
NE 1206 2265.3919 1552.9250 "+"
NEZ 1213 2601.6437 1497.8024 634.52 "IP"
NEZ 1226 2623.0725 660.6777 636.43 "+"
```

```
STN 1226 5.13
BS 1213
PRISM 4.65
AD VD 200  92.1224 61.31 -0.89 "BC"
AD VD 201  87.3738 61.13 -0.82 "BW"
AD VD 202  84.4515 17.12 -0.43 "BW"
AD VD 203 101.0837 17.49 -0.51 "BC"
AD VD 204 252.2341 13.08 -0.32 "BC"
AD VD 205 274.3455 12.68 -0.37 "BW"
AD VD 206 262.1906 33.27 -0.28 "BC"
AD VD 207 259.3437 33.45 -0.74 "INB"
```

## ELECTRONIC CONVERSIONS

When you are using an application to convert a proprietary file format to the field book format, you are doing the same thing as manually transcribing a handwritten survey to a Notepad field book. TDS Link ships with Civil 3D, or you can visit the Trimble, Leica, or Carlson Web pages to download their translator for Survey.

### TDS Survey Link

TDS Survey Link has a format editor and conversion utilities. The TDS file format editor fixes any obvious collection errors. Once it is correct, convert the file.

First, from the Conversion menu, you select Convert File Format... to open the Convert dialog box (see Figure 14.1). The menu has additional routines for converting Survey Pro Job files, Nikon files, Trimble, and several other formats.

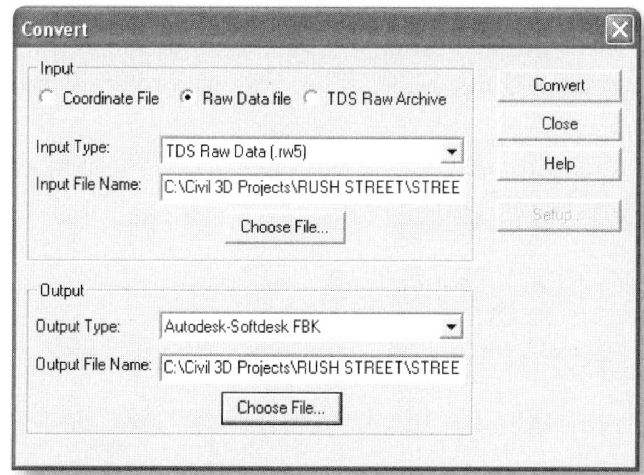


**FIGURE 14.1**

The Convert dialog box lists the raw and resulting file types (see Figure 14.2).

Convert
Input
Coordinate File Raw Data file TDS Raw Archive
Input Type: TDS Raw Data (.rw5)
Input File Name:
TDS Raw Data (.rw5)
TDS Raw Data (.raw)
Sokkia SDR Numeric 4
Sokkia SDR Alpha 14
Topcon FC-4
Leica GRE3
SMI
Topcon GTS210/220/310 Raw Data
Nikon AP800,700,DTM500,400,300,DR48
Topcon DL-100 Digital Level
Leica GSI Raw Data
Output
Output Type:
Output File Name:
Choose File...
Convert
Close
Help
Setup...

**FIGURE 14.2**

The following is a snippet from a TDS format file:

```
SP,PN1,N 1095963.3900,E 747776.0520,EL652.638,--RBC
SP,PN8,N 1096024.7400,E 747359.1190,EL668.123,--PK
OC,OP1,N 1095963.3900,E 747776.0520,EL652.638,--RBC
LS,HI4.600000,HR4.940000
BK,OP1,BP8,BS278.2215,BC359.5958
--B
LS,HI4.600000,HR5.000000
SS,OP1,FP2000,AR187.5917,ZE91.2202,SD35.9700,--EP1
--B
SS,OP1,FP2001,AR185.5628,ZE91.1045,SD24.9500,--CL1
```

After setting the Input file type, select the file. After setting the input file and its format, do the same for output (see Figure 14.3). The output format is the Autodesk-Softdesk FBK.

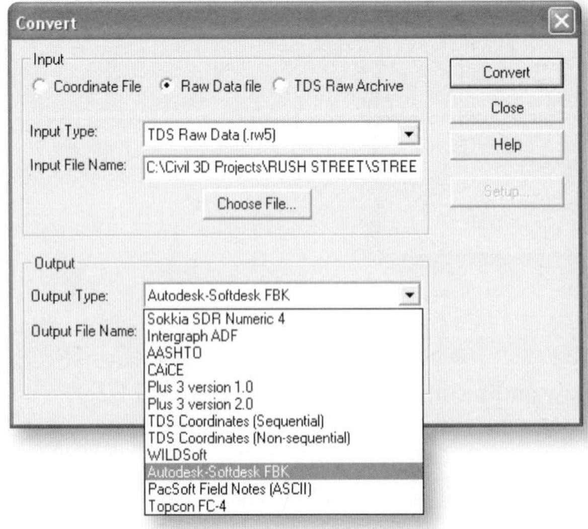


**FIGURE 14.3**

When converted to a field book, the TDS format looks like the following snippet:

```
NEZ 1 1095963.3900 747776.0520 652.6380 "RBC"
END
NEZ 8 1096024.7400 747359.1190 668.1230 "PK"
CONT RBC
STN 1 4.600 "RBC"
PRISM 4.940
AZ 1 8 278.22150
BS 8 359.59580
BEG EP1
PRISM 5.000
F1 VA 2000 187.59170 35.970 91.22020 "EP1"
BEG CL1
F1 VA 2001 185.56280 24.950 91.10450 "CL1"
```

### LINE WORK CODING SETS

See the discussion in Unit 4 of this Chapter.

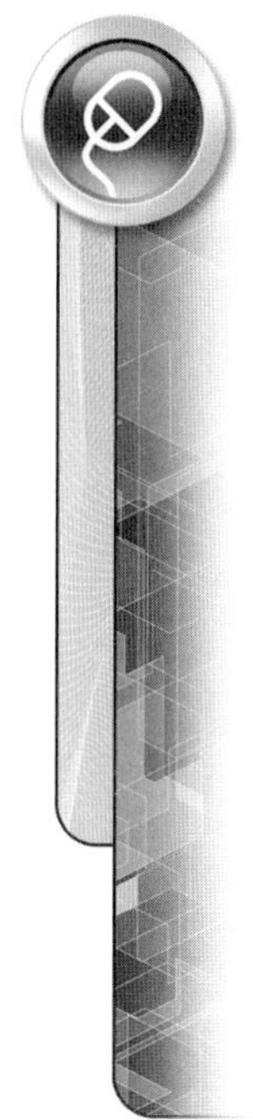

## EXERCISE 14-1

After completing this exercise, you will:

- Be familiar with the Cmdhelp.ref file.
- Be able to read field book files.
- Be able to write a field book file.
- Be able to review and print Cmdhelp.ref.

### Locate and Review Cmdhelp.ref

1. From the Windows All Programs, Accessories flyout, select Notepad.
2. In Notepad, from the File menu, select OPEN..., browse to the folder C:\Documents and Settings\All Users\Application Data\Autodesk\C3D 2010\enu\Survey (XP) (in Vista it is in C:\ProgramData\Autodesk\C3D 2010\enu\Survey), change the Files of Type to **All Files**, select *Cmdhelp.ref*, and click ***Open***.
3. From the File menu, select SAVE AS..., and save a copy to the desktop.
4. Review the command's language Point Creation and Point Location formats.
5. If possible, print the file and use it as a reference.

### Review and Print Nichol.txt

1. Open Windows Explorer, browse to the Chapter 14 folder of the CD that accompanies this textbook, locate and select the file *Nichols.doc* or *Nichols.txt*, and click ***Open***.
2. Review the survey information and, if possible, print the file for reference.

### Convert the Survey to a Field Book — Manually

1. Start a second Notepad file, and in the Notepad file containing the **Nichols.txt** file, review the Benchmark entries.

One benchmark entry does not have an elevation; the remaining entries do.

2. In the empty Notepad file, using the Field Book formats for points with known coordinates, for point 1206, enter its coordinates (the NE format), and enter the coordinates for points 1213 and 1226 (NEZ coordinates with elevation).

   After entering their values, your field book should look like the following:

   ```
   NE 1206 2265.3919 1552.9250 "+"
   NEZ 1213 2601.6437 1497.8024 634.52 "IP"
   NEZ 1226 2623.0725 660.6777 636.43 "+"
   ```

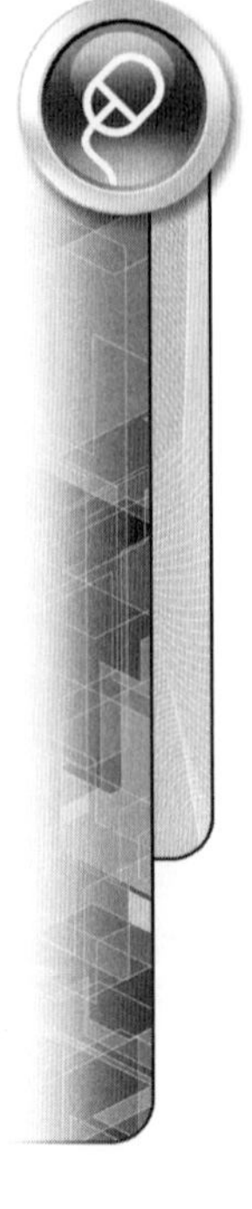

3. In field book Notepad (the entered the NE and NEZ values), from its File menu, select SAVE AS..., browse to the C:\Civil 3D Projects folder, change the file type to **All Files (*.*)**, enter *Nichols.fbk* for the filename, and click Save to save the file.

Next in the survey is setting the station at point 1213 and setting an instrument height.

4. Using STN, set the instrument at 1226 with the instrument height of 5.13.

The backsight (BS entry) is to a known port. All that needs to be done is to enter the backsight point number. Survey assumes it is the 0 (zero) direction.

5. Enter the backsight as BS 1213.

After setting the station and backsight values, you next set the prism height (PRISM).

6. For the prism height, enter 4.65 using the PRISM format.

After entering the values, your field book should look like the following:

```
NE 1206 2265.3919 1552.9250 "+"
NEZ 1213 2601.6437 1497.8024 634.52 "IP"
NEZ 1226 2623.0725 660.6777 636.43 "+"
STN 1226 5.13
BS 1213
PRISM 4.65
```

7. From Notepad's File menu, select SAVE.

## Enter Point Observations

Next, you enter point observations. These observations are angles and distances with a vertical difference. These observations use the AD VD command format.

1. Use the AD VD format and for points 200–207, code their observations.

After entering the values, your field book should look like the following:

```
NE 1206 2265.3919 1552.9250 "+"
NEZ 1213 2601.6437 1497.8024 634.52 "IP"
NEZ 1226 2623.0725 660.6777 636.43 "+"
STN 1226 5.13
BS 1213
PRISM 4.65
AD VD 200 92.1224 61.31 -0.89 BC
AD VD 201 87.3738 61.13 -0.82 BW
AD VD 202 84.4515 17.12 -0.43 BW
```

```
AD VD 203 101.0837  17.49 -0.51 BC
AD VD 204 252.2341  13.08 -0.32 BC
AD VD 205 274.3455  12.68 -0.37 BW
AD VD 206 262.1906  33.27 -0.28 BC
AD VD 207 259.3437  33.45 -0.74 INB
```

2. From Notepad's File menu, select SAVE.
3. Close all open Notepad files.

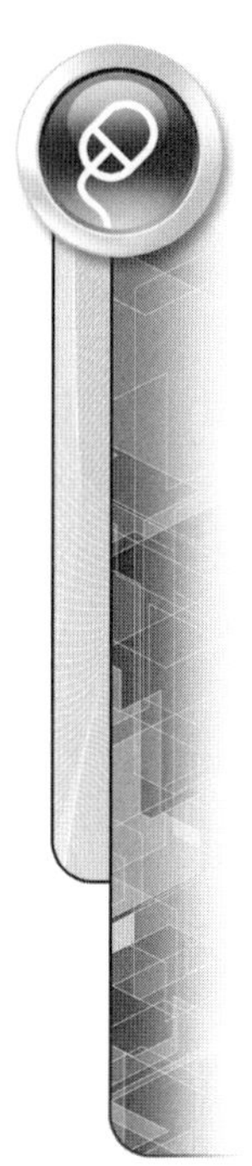

## Create a Local Survey Database

1. If you are not in Civil 3D, on the Windows Desktop double-click the ***Civil 3D*** icon.
2. At Civil 3D's top left, click the Civil 3D drop-list arrow, and from the Application Menu, highlight SAVE AS, from the flyout select AUTOCAD DRAWING, browse to the Civil 3D Projects folder, for the drawing name enter **Survey**, and click ***Save*** to save the file.
3. If necessary, in Ribbon's Home tab, Palettes panel, click the ***Survey Toolspace*** icon or click the Toolspace's ***Survey*** tab.
4. At the Toolspace's top, click the ***Survey User Settings*** icon.
5. In Survey User Setting, Import Defaults section, use Figure 14.7 as a guide, and set your values to match.
6. Click ***OK*** to exit the dialog box.
7. In Survey, place your cursor over Survey Databases, press the right mouse button, and from the shortcut menu, select NEW LOCAL SURVEY DATABASE....
8. In New Local Survey Database, for the name, enter **Road Survey**, and click ***OK***.
9. In Survey select the survey database Road Survey, right mouse click, and from the shortcut menu, select EDIT SURVEY DATABASE SETTINGS....
10. In Edit Survey Database Settings, if necessary, expand the Units section, set the distance value to US Foot by clicking the drop-list value and selecting US Foot.
11. Click OK to exit the Edit Survey Database dialog box.

## Create a Network

1. In Road Survey, select Networks, press the right mouse button, and from the shortcut menu, select NEW... to open the New Network dialog box.
2. In New Network, for the network name, enter **Road Survey**, enter today's date as its description, and click ***OK***.
3. Use Windows Explorer and from the Chapter 14 folder of the CD that accompanies this textbook, copy the file *10113f2.rw5* to the network folder Road Survey (C:\Civil 3D Projects\Road Survey\Road Survey).
4. Minimize Windows Explorer.

## Convert a TDS File to Autodesk–Softdesk FBK

1. On Ribbon's Home tab, click the Create Ground Data panel title to unfold the panel, and from the shortcut menu, select SURVEY DATA COLLECTION LINK to display TDS Survey Link.
2. From the File menu of Survey Link DC 7.5.5, select OPEN....
3. In Open, browse to, select, and open the *10113f2.rw5* file (C:\Civil 3D Projects\Road Survey\Road Survey).
4. In the editor, scroll through and review its data, and when you are done, exit the editor.

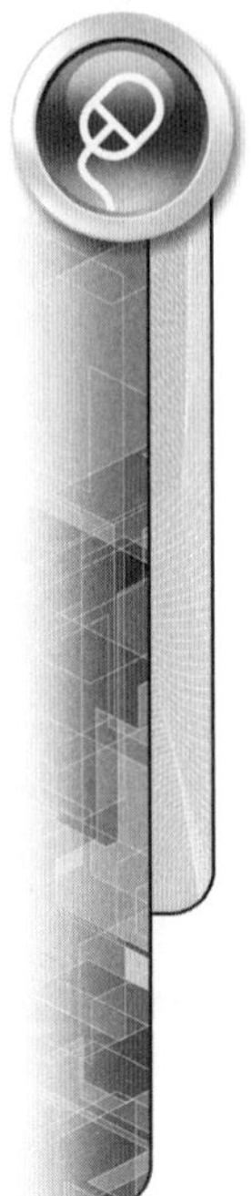

5. In Survey Link DC 7.5.5, from the Conversions menu, select CONVERT FILE FORMAT....
6. In the Convert dialog box, change the Input to **Raw Data File**, and set Input Type to **TDS Raw Data (.rw5)**.
7. Below Input File Name:, click ***Choose File...***, browse to C:\Civil 3D Projects\Road Survey\Road Survey, select the file *10113f2.rw5*, and click ***Open*** to return to the Convert dialog box.
8. In the Convert dialog box, change the Output Type to **Autodesk-Softdesk FBK**.
9. Below Output File Name..., click *CHOOSE FILE...*, browse to the C:\Civil 3D Projects\Road Survey\Road Survey, for the filename enter *10113f2*, and click ***Save*** to return to the Convert dialog box.
10. In the Convert dialog box, click ***Convert***, and when the Successful conversion dialog box appears, click ***OK*** to exit the message.
11. Click ***Close*** to exit the Convert dialog box.
12. At the upper right of Survey Link DC 7.5.5, click the **X** to close it.
13. In Survey, select the network Road Survey, right mouse click, and from the shortcut menu, select EDIT FIELD BOOK....
14. Browse to the Road Survey network folder, select *10113f2.fbk*, and click ***Open***.
15. In the editor, in the second line, add US to the FOOT entry.
    The second line should now have the following text:

    UNIT USFOOT DMS
16. Save the file and close Notepad.
17. At Civil 3D's top left, Quick Access Toolbar, select the ***Save*** icon to save the drawing.

## SUMMARY

- A survey is a systematic method for collecting data.
- A survey has required control and observation methods (i.e., a control point at the beginning with observations made from the control point and then occupying a foreward observed point and observing additional points from it).
- The field book command language can transcribe a written survey to an electronic field book.
- The field book command language is a neutral survey industry language.

This completes the field book language and converting manual or electronic surveys to a field book's unit.

## UNIT 2: SURVEY SETTINGS

Survey Toolspace contains most survey settings. Settings defines Network and Figure styles.

### NETWORK STYLE

A network style defines how a network object displays its components: known control points, unknown control points, sideshot points, network lines, direction lines, sideshot lines, and error ellipses. The Components panel sets the Marker Styles and Error Ellipses scale factor (see Figure 14.4).

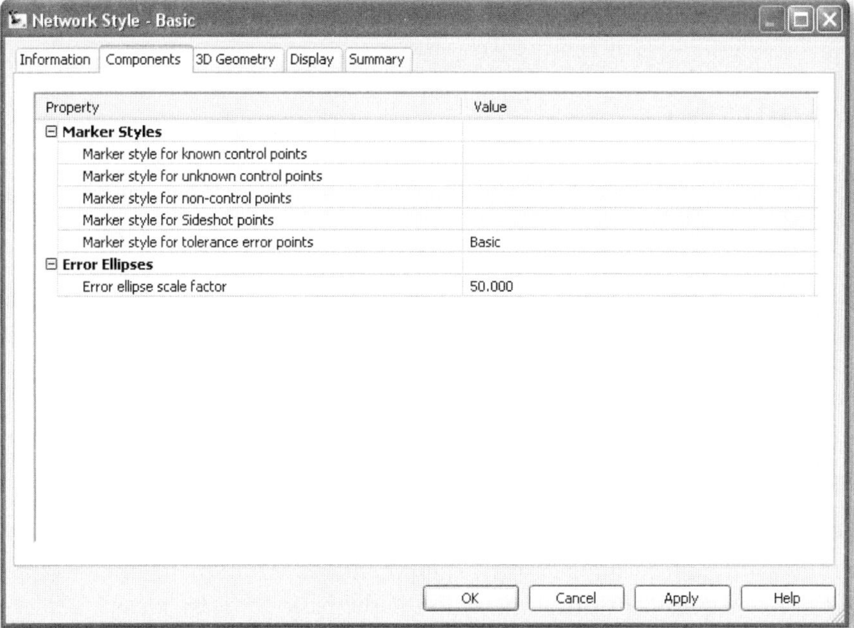


**FIGURE 14.4**

3D Geometry sets the network's vertical behavior. A network can have elevations, be flattened to a specified elevation, or have exaggerated elevations. Having a network at its actual elevations allows you to evaluate and locate any vertical blunders.

Display sets network object component layers and visibility.

## FIGURE STYLES

Figure styles define how a figure is displayed in a drawing. The Plan and Model tab defines figure marker symbol shapes and orientation (figure vertices) (see Figure 14.5).

Figure Style - Road Centerline

Information | Plan and Model | 3D Geometry | Profile | Section | Display | Summary

| Property | Value |
|---|---|
| **Vertex Markers** | |
| Marker style | |
| Align markers with figure | yes |
| **Midpoint Markers** | |
| Marker style | |
| Align markers with figure | yes |
| **Endpoint Markers** | |
| Start point marker style | |
| End point marker style | |
| Align markers with figure | yes |
| **Additional Markers** | |
| Marker style | |
| Additional marker placement method | None |
| Interval | 10.000' |
| Divide figure by | 2 |
| Place marker at figure start point | no |
| Place marker at figure end point | no |
| Align markers with figure | yes |

OK | Cancel | Apply | Help

**FIGURE 14.5**

3D Geometry controls the figure and its markers' elevations. Display assigns a component's layer and properties and visibility.

## SURVEY TOOLSPACE

Survey Toolspace settings affect the behavior, survey adjustments environment, and the display of survey information. There are three levels of settings: Survey User Settings, Local Survey Database, and Network.

### Survey User Settings

Survey User Settings affect initial values and behaviors of various survey commands.

#### Miscellaneous and Survey Database Defaults Sections

These sections set the survey data editor, survey setting, and extended survey properties paths.

#### Equipment and Figure Default Sections

These sections set the file locations for the equipment and figure prefix libraries (see Figure 14.6). The Equipment Defaults section is the basis for least squares adjustment error estimates. If you want to adjust surveys, it is important that there is a record for each instrument or a single general entry with reasonable values.

#### Linework Processing Defaults

This section sets the path to the linework code set. A linework code set defines how to parse descriptions to draw survey figures. The Process linework toggle allows linework processing during imports. The named code set is the current set used for creating survey figures. Process linework sequence has two values: By import order or By point number. By import order processes the import in the order the points are found in the import file. By point number sequentially imports the points by their point number, not the order of appearance in the file.

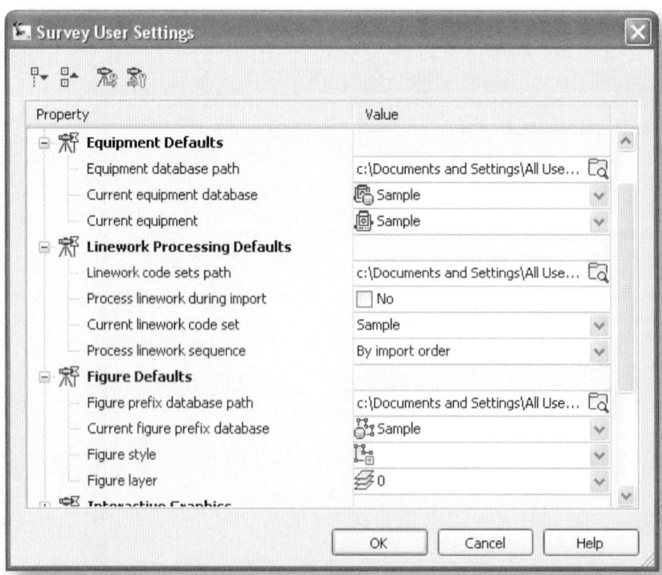


**FIGURE 14.6**

## Interactive Graphics

Interactive Graphics sets survey components' colors when you are displaying a survey network: foresight lines, backsight lines, prism and instrument colors, etc.

## Import Defaults

Import Defaults defines initial values when you are importing a field book. If you are importing a single field book, use the settings in Figure 14.7. The settings reset the observation database before reading in the field book data; remove any pre-existing points, network, figure objects; and does not insert any survey objects in the drawing when done reading the file. This is done so the user can choose what survey components he or she wants to view in a drawing.

When you import multiple field books, toggle off Erase survey points from drawing (if inserted), Reset network, and Delete network figures (if inserted).

The Export, Network, Setup, and Figure Preview sections control what is exported from a survey and if previews are available after importing a field book.

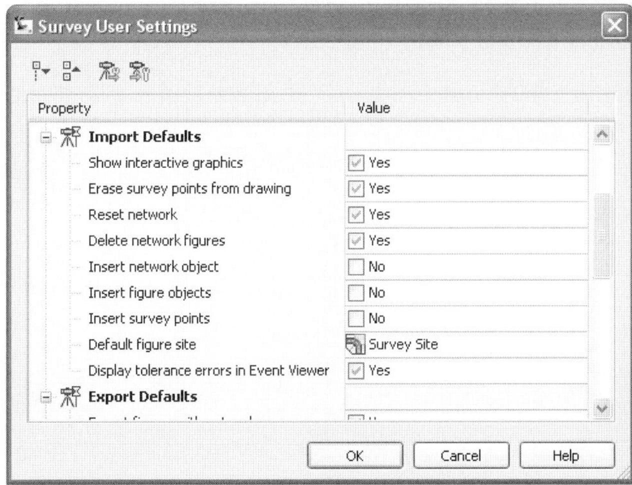


**FIGURE 14.7**

## Local Survey Settings

Two methods create a local survey database. The first method is to create a Vault project. When you create a Vault project, a local survey database also is created. The second method is to create a local survey database in the Survey panel. Select the Survey Databases heading, press the right mouse button, from the shortcut menu, select New local survey database…, and enter the survey's name. Click OK to create the folder structure in the project's working folder (default C:\Civil 3D Projects).

After you have created a local database, always check its initial settings. The settings include coordinate zone and unit defaults. By default, there is no assigned coordinate zone and the base unit is International foot (see Figure 14.8).

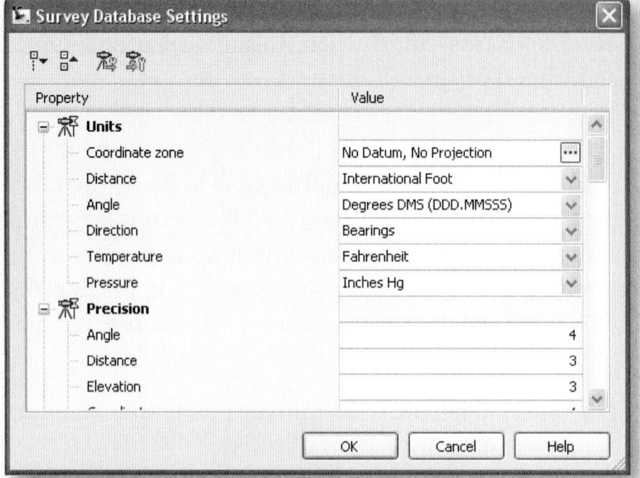


**FIGURE 14.8**

Next, you enter settings that affect default precision, types of data corrections, and measurements. These values affect manual survey data entry.

Traverse Analysis Defaults set initial horizontal and vertical adjustment methods and the quality level (see Figure 14.9).

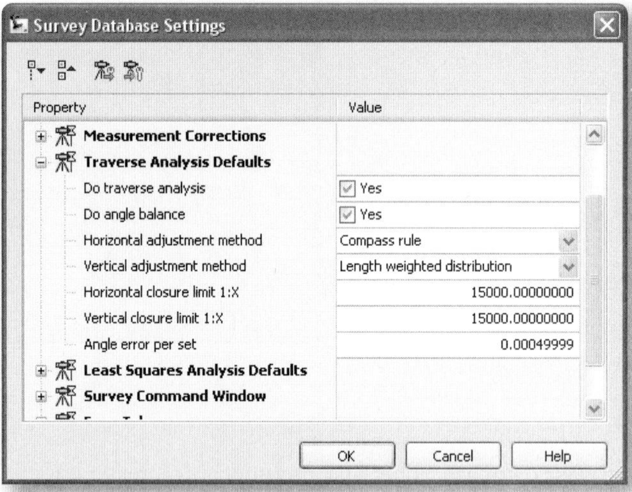


**FIGURE 14.9**

Least Squares Analysis Defaults set initial least squares adjustment values. A Least Squares can be 2D or 3D, with or without blunder detection, confidence level, and a convergence factor. The convergence factor is when two calculations for a point's location are less than the convergence value; the coordinate is then considered resolved.

Survey Command Window and its settings are discussed in the next unit.

The next to last section is Error Tolerance (see Figure 14.10). If an observation is above a tolerance value, the error triggers an event displayed in the Panorama.

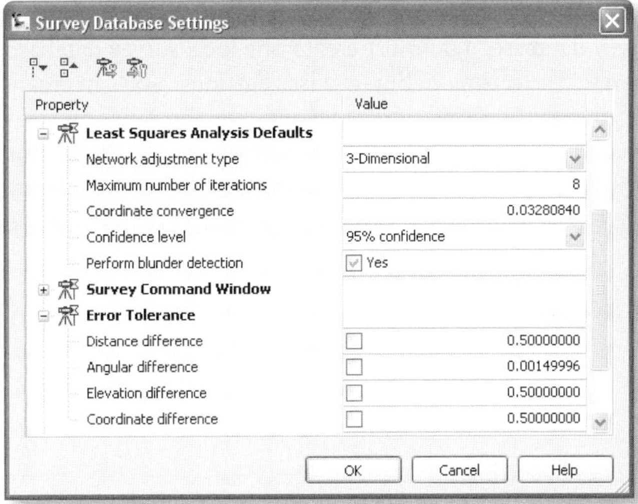


**FIGURE 14.10**

Point Protection controls how Survey handles redundant point numbers. By default the option is to average a point's coordinates, elevations, distance, or angular values only if those values fall within the specified maximum allowed value. If a measurement is over the value, it is not used, but is included in the observation data. Survey flags these values in the Edit Observations vista. The remaining options for the Overwrite method are the following:

### Notify

This method informs the user of the error and prompts for a solution.

### Check Shots

This option displays an error dialog box but does nothing to the data.

### Average All

This option averages all values associated with a point's observation. The user is not notified if any value is over the limits set in the Point Protection section.

### Overwrite All

This option uses the last value for a point as its actual values. All previous values are lost.

### Ignore All

This option uses the first value for a point and ignores any other values.

### Renumber All

This option renumbers any subsequent observation to a new point number.

## Exporting and Importing Survey Settings

Survey settings can be exported to a file. The exported settings can become defaults for a new local survey database or assigned to an existing survey database. The two rightmost icons in Survey User Settings and Survey Database Settings export and import these settings.

In Survey User Settings, saved Survey Database Settings can be assigned the default settings for any new local survey database. You must create one local survey database and set its values before you can set the setting in Survey User Settings. You can create several local survey settings, but only one can be the default. For example, most of your surveys are local coordinate based. If so, assign these settings as the default for new survey databases. If a survey is state plane based, you can change the local survey database settings and if desired export the settings for the next survey that will use state plane coordinates.

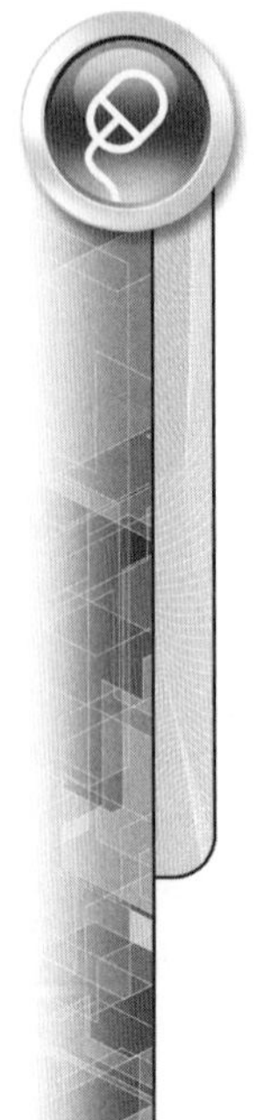

**EXERCISE 14-2**

After completing this exercise, you will:

- Be familiar with the styles in settings.
- Be familiar with the various settings for Survey.

## Exercise Setup

1. If you are not in **Civil 3D**, double-click its desktop icon.
2. When you are at the command prompt, close the open drawing and do not save it.
3. If necessary, click the ***Prospector*** tab and set Prospector to Master View.
4. In Prospector, expand the Drawing Templates and AutoCAD list, from the list, select **_Autodesk Civil 3D (Imperial) NCS**, right mouse click, and from the shortcut menu, select CREATE NEW DRAWING.
5. Click the ***Settings*** tab.
6. At the top of Settings, select the drawing name, press the right mouse button, and from the shortcut menu, select EDIT DRAWING SETTINGS....
7. Click the ***Object Layers*** tab and scroll to the object list's bottom.
8. In Object Layers for Survey Network and Survey Figure, set the Modifier to **Suffix** and its value to **-***.
9. Click ***OK*** to close the dialog box.
10. At Civil 3D's top left, click Civil 3D's drop-list arrow, from the Application Menu, highlight SAVE AS and from the flyout select AUTOCAD DRAWING, browse to the C:\Civil 3D Projects directory, for the drawing name enter **Nichols**, and click ***Save*** to save the file.

## Survey Database Settings

1. Click the ***Survey*** tab.
2. At Survey's top, select the ***Survey User Settings*** icon.
3. If necessary, Expand each section and review its settings.
4. In the Import Defaults section, make sure your settings match those in Figure 14.7.
5. Click ***OK*** to close the Survey User Settings dialog box.

## Create a Local Survey Database

1. In Survey, select the Survey Databases heading, press the right mouse button, and from the shortcut menu, select NEW LOCAL SURVEY DATABASE....
2. In New Local Survey Database, for the Name, enter **Nichols,** and click ***OK***.

A Nichols folder structure is built under Survey Databases.

3. From the Survey Databases list, select **Nichols**, press the right mouse button, and from the shortcut menu, select EDIT SURVEY DATABASE SETTINGS....
4. In the Survey Database Settings dialog box, expand the Units section and set the Distance to **US Foot**.
5. At the top of Survey Database Settings, click the ***export*** icon (the third from left), for the setting's name, enter Local coordinate survey, and click ***Save*** to save the settings.

When you create the next new survey database, simply import these saved settings.

6. Click ***OK*** to exit Survey Database Settings.

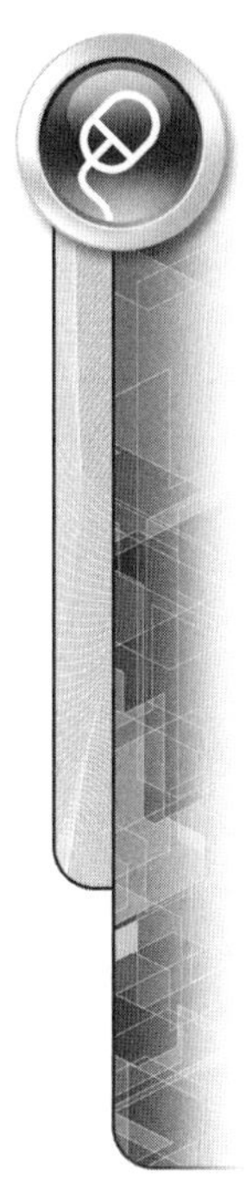

### Create a Named Network

1. In Survey, the Nichols survey, select the Networks heading, press the right mouse button, and from the shortcut menu, select NEW....
2. In the Network dialog box, for the Name, enter **Topo**, and click ***OK***.
3. Select the Nichols survey database, press the right mouse button, and from the shortcut menu, select CLOSE SURVEY DATABASE.

Survey creates a network folder, Topo, below the Nichols survey folder.

4. At Civil 3D's top left, Quick Access Toolbar, select the ***Save*** icon to save the drawing.

### Copy the Field Book File

1. Using Windows Explorer, from the C:\Civil 3D Projects folder or from the Chapter 14 folder of the CD that accompanies this textbook, copy the file *Nichols.fbk* and place it in the C:\Civil 3D Projects\Nichols\Topo folder.

## SUMMARY

- Settings and folders are important to the importing and adjusting of field book data.
- A local survey database is a folder in the current project's working folder.
- Survey User Settings affect the environment and Survey command behavior for all surveys.
- Local Survey settings affect only the selected Survey's settings.
- When you are setting up typical local survey settings, save them to use for the next survey.

The next unit imports a field book and reviews its resulting information.

## UNIT 3: IMPORTING AND EVALUATING FIELD BOOKS

There must be a Survey database and network before importing a field book. Also, check the settings before importing to make sure they are correct. To import a field book, select the network name and from its Import flyout menu, select Import field book... (see Figure 14.11). The network shortcut menu contains commands to edit a field book, export the survey as a field book, insert the network into the drawing, adjust the survey with Least Squares, reset the network observation database, etc. Also, this is where the user calls the Survey Command Window.

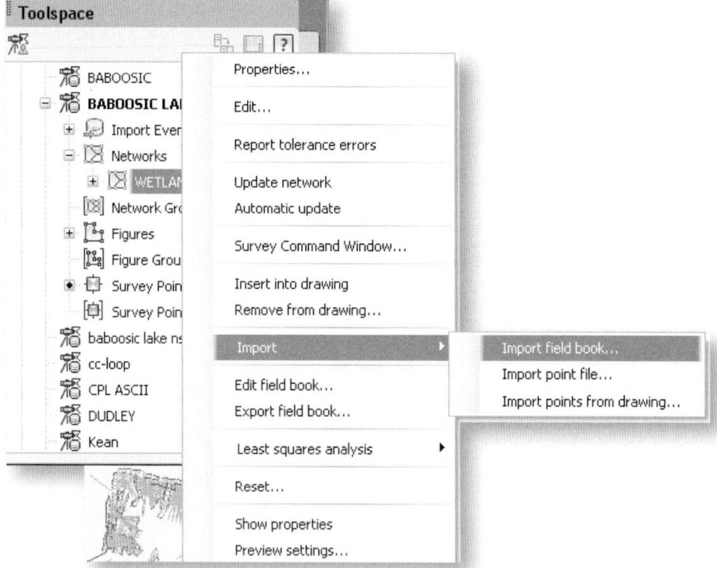


**FIGURE 14.11**

## IMPORT EVENTS

With the new linework coding, Survey now has other sources of survey figures: an ASCII file and a LandXML file. Each network has an Import Events branch. When selecting Import Events, pressing the right mouse button, and selecting Import survey data..., the Import Events wizard displays. Import Events is a wizard that allows a user to edit a network's settings; create networks; and import Field books, LandXML files, point (coordinate) files, and read points from a drawing.

After import, the Import Events heading lists the imported file. The Import Events branch has three sections: Networks, Figures, and Survey Points. An import event knows what network it was imported into and what figures and points it created.

### Specify Database

The Specify Database panel sets the current survey database (see Figure 14.12). A user can create a new survey database by clicking the Create New Survey Database... button at the panel's bottom. The button to its right allows a user to edit the survey database's settings. The default survey is the one that calls the wizard.

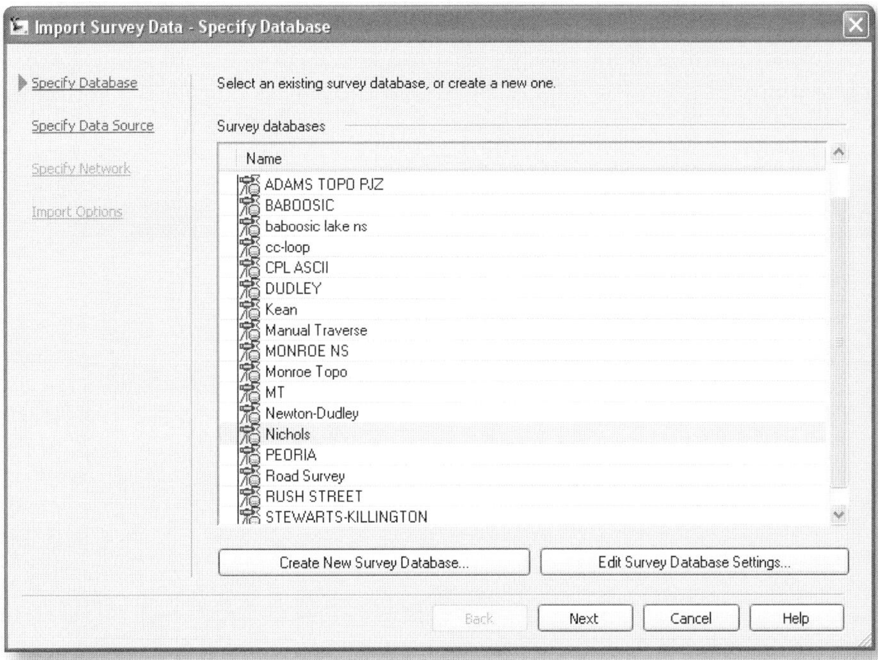


**FIGURE 14.12**

## Specify Data Source

The Specify Data Source panel identifies the file type and the file to import (see Figure 14.13).

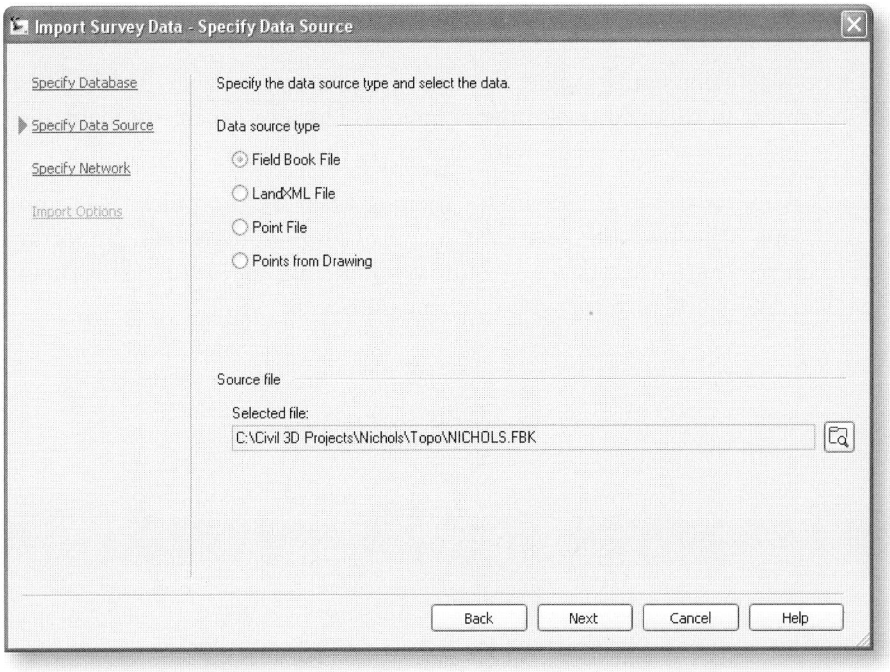

**FIGURE 14.13**

## Specify Network

Specify Network identifies of the survey's networks and allows a user to select the one for the current import event (see Figure 14.14). At the panel's bottom is a button to create a network.

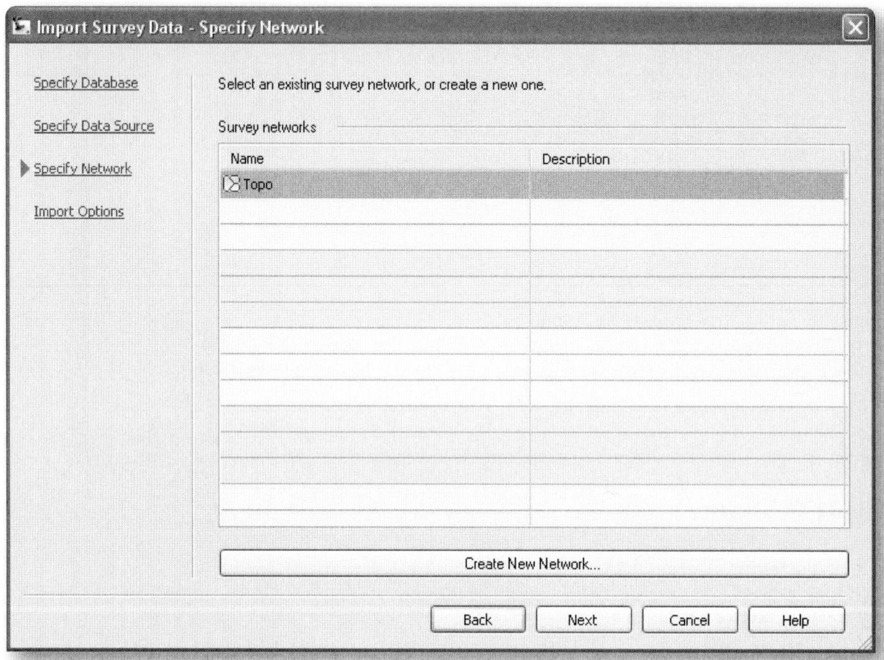


**FIGURE 14.14**

## IMPORT OPTIONS

Import Options contains settings values for the import event (see Figure 14.15). These settings are similar to the field book import options. However, there are many new options: linework and adding a point number offset to the imported points.

The Linework option and its values are for creating linework from files NOT containing traditional field book figure coding, i.e. ASCII coordinate files. Importing several files into a single network may require renumbering the points. This occurs when the field crew starts each file with the same numbers.

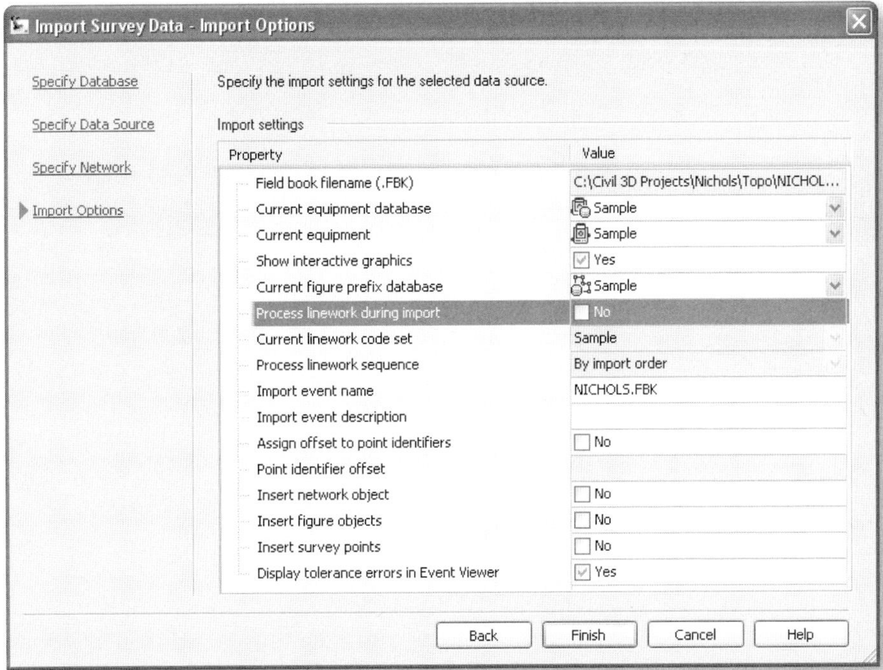


**FIGURE 14.15**

## NETWORK COMPONENTS

At a minimum, a survey network contains two components: control points and stations with observations (setups). However, any network may contain up to four components: Control Points, Non-Control Points, Directions, and Setups (see Figure 14.16). The Properties dialog box for each component displays the selected component's values. Edit its values in a Panorama by selecting Edit… from the shortcut menu (see Figure 14.17).

Any named network can produce figures and survey points. A figure is a group of points that define drawing line work, a surface breakline, or a parcel segment. Survey points are the coordinate list of observed points from each named network. A survey may contain one or more named networks, each defining new figures and points. The Figures and Survey Points lists represent all figures and points defined by the individually named networks.

Each network component can be inserted or removed from a drawing. This allows the user to edit and review numbers, but sometimes things must be seen to evaluate them. The component Insert… routine places the selected object in the drawing. This allows a visual inspection. An inserted network shows all of the setups, sideshots, and figures of a survey.

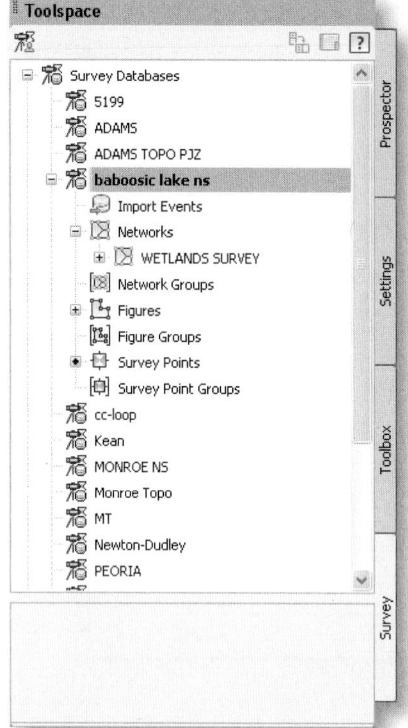


**FIGURE 14.16**

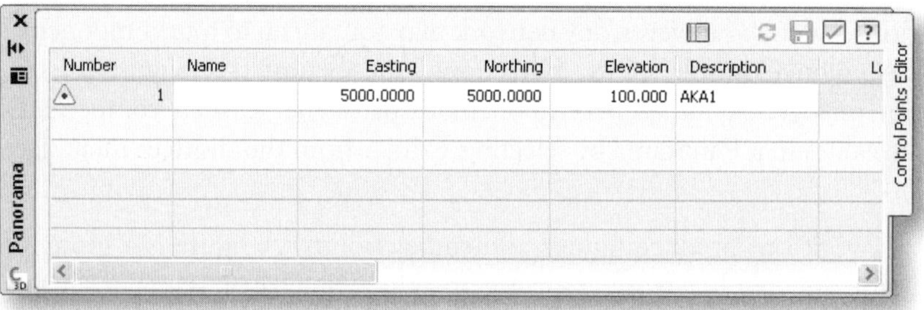


**FIGURE 14.17**

## Control Points

Control points are points with known elevations and are coded as NE and NEZ.

## Non-Control Points

Non-control points are points with known coordinates, but are coded as NESS. These points are not occupied during the survey.

## Directions

When establishing a station or setup without a known point as the backsight, surveyors use an azimuth to set the backsight direction (the 0 (zero) angle). Changing this value rotates the entire survey.

## Setups

Setups or stations in the field book are coded as STN. A setup represents the location of an instrument during the survey. It references a known point, a height of instrument over the known point, and can have an actual or assumed elevation.

### Figures

Figures are linear survey features (for example, centerlines, gutters, fences, edge-of-travelway, etc.).

### Survey Points

Survey points are the initial coordinates of observed points. These points have an icon that identifies them as unadjusted (see Figure 14.18). After adjusting the survey, the points' icons change to indicate their adjusted state.

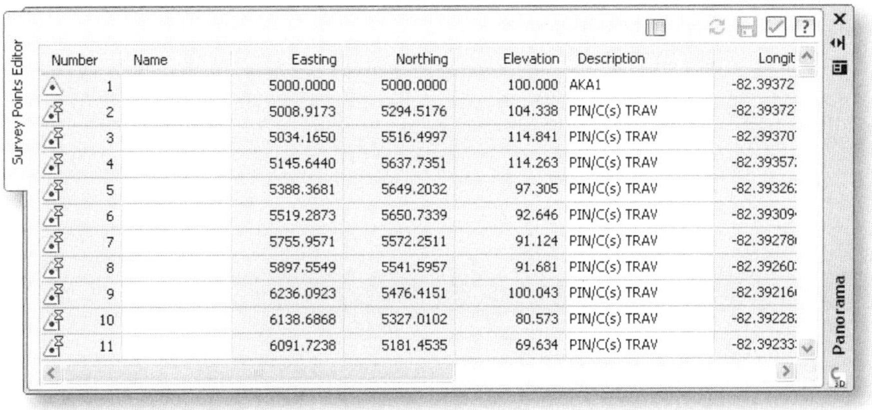

Survey Points Editor

| Number | Name | Easting | Northing | Elevation | Description | Longit |
|---|---|---|---|---|---|---|
| 1 | | 5000.0000 | 5000.0000 | 100.000 | AKA1 | -82.39372 |
| 2 | | 5008.9173 | 5294.5176 | 104.338 | PIN/C(s) TRAV | -82.39372 |
| 3 | | 5034.1650 | 5516.4997 | 114.841 | PIN/C(s) TRAV | -82.39370 |
| 4 | | 5145.6440 | 5637.7351 | 114.263 | PIN/C(s) TRAV | -82.39357 |
| 5 | | 5388.3681 | 5649.2032 | 97.305 | PIN/C(s) TRAV | -82.39326 |
| 6 | | 5519.2873 | 5650.7339 | 92.646 | PIN/C(s) TRAV | -82.39309 |
| 7 | | 5755.9571 | 5572.2511 | 91.124 | PIN/C(s) TRAV | -82.39278 |
| 8 | | 5897.5549 | 5541.5957 | 91.681 | PIN/C(s) TRAV | -82.39260 |
| 9 | | 6236.0923 | 5476.4151 | 100.043 | PIN/C(s) TRAV | -82.39216 |
| 10 | | 6138.6868 | 5327.0102 | 80.573 | PIN/C(s) TRAV | -82.39228 |
| 11 | | 6091.7238 | 5181.4535 | 69.634 | PIN/C(s) TRAV | -82.39233 |

Panorama

**FIGURE 14.18**

## SURVEY COMMAND WINDOW

Survey Command Window (SCW) is a floating dialog box that functions as a command-line coordinate geometry program (COGO) (see Figure 14.19). SCW accepts any correctly formed field book language command, records the entry in a file (batch.txt), displays the entry and its response in the middle of the dialog box, and at its top shows command results. The Batch.txt is an external file the user can edit and rerun to view the results (see Figure 14.20). A second file associated with the Survey Command Window, the Output.txt file, contains the results of the Survey Command Window entries (see Figure 14.21).

When you select a Survey Command Window menu routine, a dialog box opens and prompts you for the necessary points to execute the query. After entering the values and exiting the dialog box, the query answer appears at the command window's top.

The command window has a ditto option. This option reruns the same query by only entering in the appropriate point numbers (separated by a space(s)). For example, the distance query is D 1226 1213. After executing this command once, to execute it again, the user only needs to enter 1213 1206, to get the distance between points 1213 and 1206.

Survey Command Window has four types of data queries: Baseline, Centerline, Intersection, and Point Information.

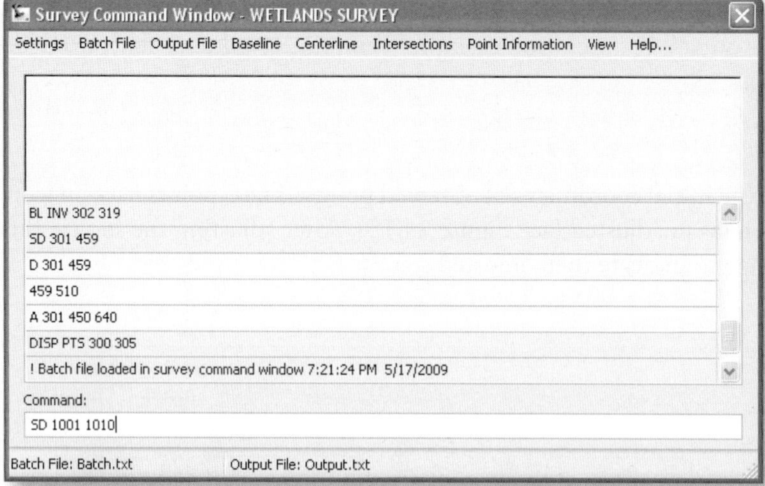


**FIGURE 14.19**

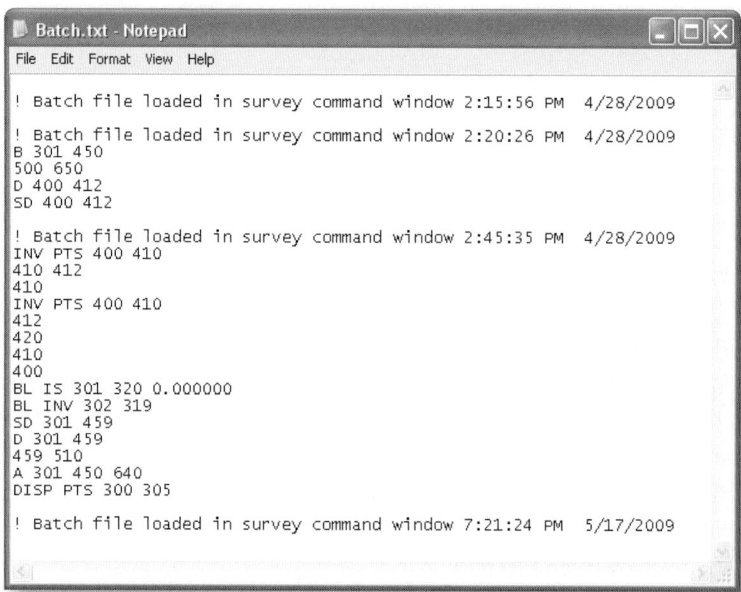


**FIGURE 14.20**

```
Output.txt - Notepad
File  Edit  Format  View  Help
|
!Page 1                              PROJECT: baboosic lake ns


B 301 450
!
!    Bearing: N 24-36-23 E
500 650
!
!    Bearing: N 57-09-38 E
D 400 412
!
!    Distance:  227.477
SD 400 412
!
!    Slope distance: 227.876

! Batch file loaded in survey command window 2:15:56 PM  4/28/2009
! Batch file loaded in survey command window 2:20:26 PM  4/28/2009
B 301 450
!
!    Bearing: N 24-36-23 E
500 650
!
!    Bearing: N 57-09-38 E
D 400 412
```

**FIGURE 14.21**

## Baseline

Baseline creates a temporary baseline (an imaginary line between two points). The baseline's beginning point has a user-defined station. Once the baseline is established, the baseline and user-defined station can be inversed to determine their station and offset relative to the baseline. This query type identifies point locations that are to one side or the other of the baseline (for example, fence posts over a property line, etc.)

## Centerline

The Centerline menu provides tools to define centerlines and inverse points relative to the centerline, create new stations, and create new points relative to a centerline (see Figure 14.22).

The station portion of the menu creates new points by Station/Offset with or without elevation and by setting a point with an elevation from a rod height or a vertical distance.

At the menu's bottom, the Set Cross Section routine defines the current centerline cross-section. The remaining routines place points by offset elevation, by offset and rod height, and by offset and a vertical distance.

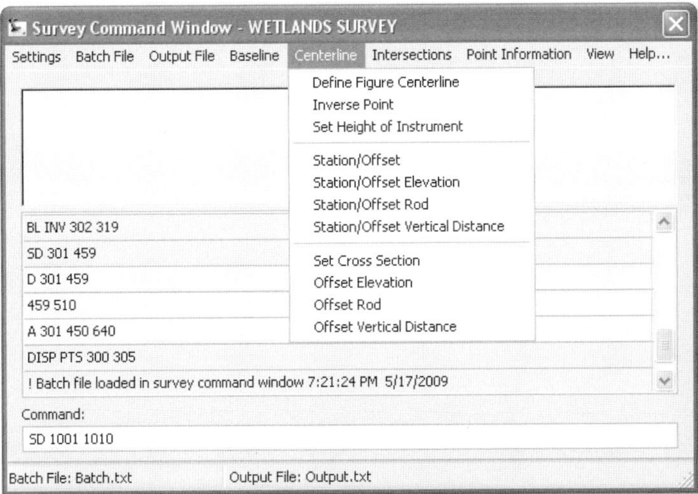


**FIGURE 14.22**

### Intersections

Intersection menu commands set points by defining intersections (for example, bearings, azimuths, lines, and several arc intersection methods using survey points) (see Figure 14.23).

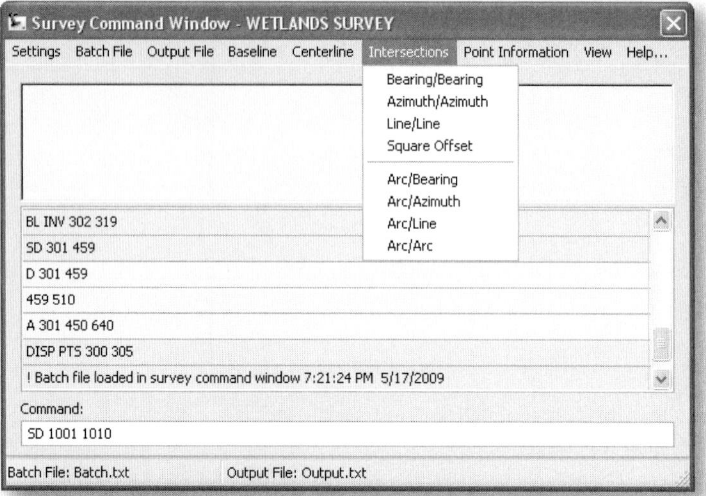


**FIGURE 14.23**

### Point Information

The Point Information menu routines return values from survey point queries. The routines include bearing; horizontal and slope distances; angles; inverse between points; and elevations, grades, or slope angles (see Figure 14.24).

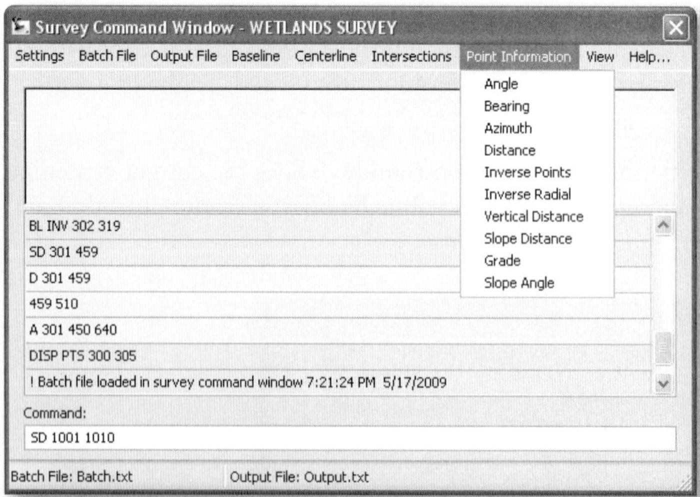


**FIGURE 14.24**

**EXERCISE 14-3**

After completing this exercise, you will:

- Be able to import a field book.
- Be able to insert network objects.
- Be able to review Survey component values.
- Be able to use the Survey command window.

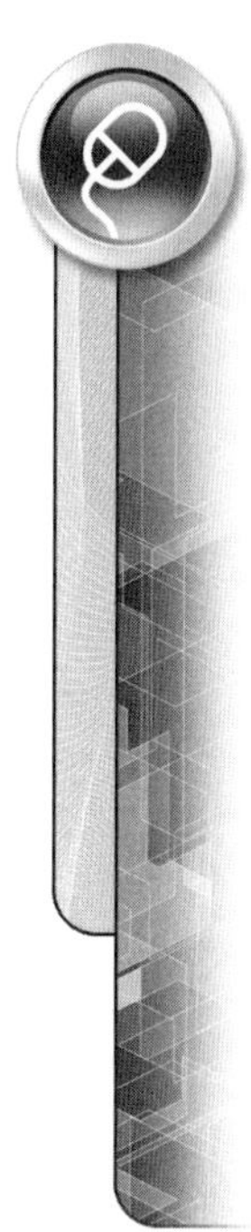

## Exercise Setup

This exercise continues with the field book from the previous exercise and its saved drawing.

1. If necessary, start **Civil 3D** by double-clicking its desktop icon.
2. When you are at the command prompt, close the open drawing and do not save it.
3. Click the ***Prospector*** tab.
4. If necessary, at Prospector's top, click the view drop-list arrow and from the list of views, select **Master View**.
5. At Civil 3D's top left, click ***Civil 3D's*** drop-list arrow, in the Application Menu highlight OPEN, in the flyout click AUTOCAD DRAWING, browse to the folder C:\Civil 3D Projects\Nichols, select the drawing named **Nichols**, and click ***Open***.
6. Click the ***Survey*** tab.

## Review Field Book Data

1. In Survey, select the Nichols survey, press the right mouse button, and from the shortcut menu, select ***Open***.
2. In the Nichols survey, Networks list, select **Topo**, press the right mouse button, from the shortcut menu, select EDIT FIELD BOOK..., browse to the C:\Civil 3D Projects\Nichols\Topo folder, select the *Nichols.fbk*, and click ***Open*** to edit the file.
3. Review the file's contents, and when you are done, close Notepad.

## Import the Field Book

1. In the Nichols survey, select Import Events, press the right mouse button, from the shortcut menu, and select IMPORT SURVEY DATA... to display the Import Survey Data wizard.
2. In Import Survey data, click ***Next***.
3. In the Specify Data Source panel, toggle on Field Book File, in the lower right click the ***identify source file*** icon, browse to the C:\Civil 3D Projects\Nichols\ Topo folder, select the *Nichols.fbk*, and click ***Open***.
4. Click the ***Next***.
5. The Network is Topo; click ***Next***.
6. In Import Options, match your settings to Figure 14.25 and when set, click ***Finish*** to import the file.

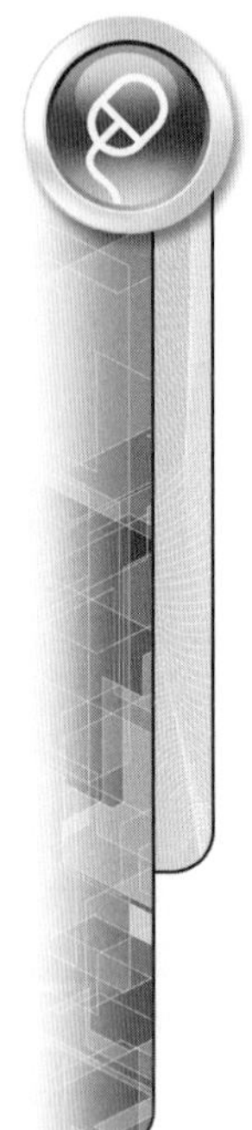

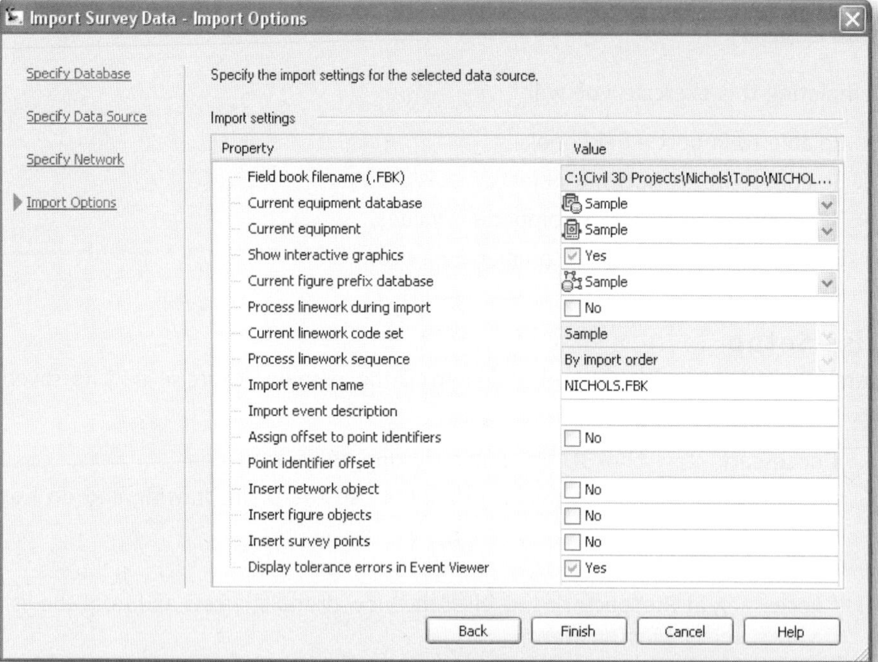


**FIGURE 14.25**

## Review Network Component Values

1. In the Nichols survey, Networks list, expand Topo until you are viewing the Control Points. Select the Control Points heading, press the right mouse button, and from the shortcut menu, select EDIT....
2. In the Control Points Editor vista, review control points values.
3. Close the Panorama.
4. In the Nichols survey, the Topo network, expand Setups, select the only setup, press the right mouse button, and from the shortcut menu, select PROPERTIES....
5. Review the setup's values and when you are done, click ***OK*** to close the Setup Properties dialog box.
6. If the Survey goes out-of-date, select the network name Topo, press the right mouse button, and from the shortcut menu, select UPDATE NETWORK.
7. In the Nichols survey, select the Survey Points heading, press the right mouse button, and from the shortcut menu, select EDIT....

The vista displays all of the survey points. A prism icon next to a point indicates that it is a sideshot. Control points have a triangular icon next to their number.

8. In the vista, review the points and click the ***green checkmark*** until closing the Panorama.

## Insert the Network Components

1. In the Nichols survey, select the network Topo, press the right mouse button, and from the shortcut menu, select INSERT INTO DRAWING.

The network is displayed in the drawing. The red line represents the backsight and the tentacles on the western end represent the observations to point 200–207.

2. Use the ZOOM and PAN commands to better view the observations on the western end.

3. In the Nichols survey, select the network Topo, press the right mouse button, from the shortcut menu, select REMOVE FROM DRAWING..., and click Yes to remove the network.
4. In the Nichols survey, select the Survey Points heading, press the right mouse button, and from the Points flyout menu, select INSERT INTO DRAWING.

Only markers appear in the drawing. The _All Points point group point labels style is set to None.

5. Click the ***Prospector*** tab and expand the Point Groups branch.
6. In Prospector, from the Points Groups list, select **_All Points**, press the right mouse button, and from the shortcut menu, select PROPERTIES....
7. In Point Group Properties, the Information tab, change the Point Label Style to **Point#-Elevation-Description**, and click ***OK***.

The points now have a marker and point label.

8. Click the ***Survey*** tab.
9. In the Nichols survey, select the Survey Points heading, press the right mouse button, and from the Points flyout menu, select REMOVE FROM DRAWING....
10. In the ***Are you sure*** dialog box, click ***Yes***.

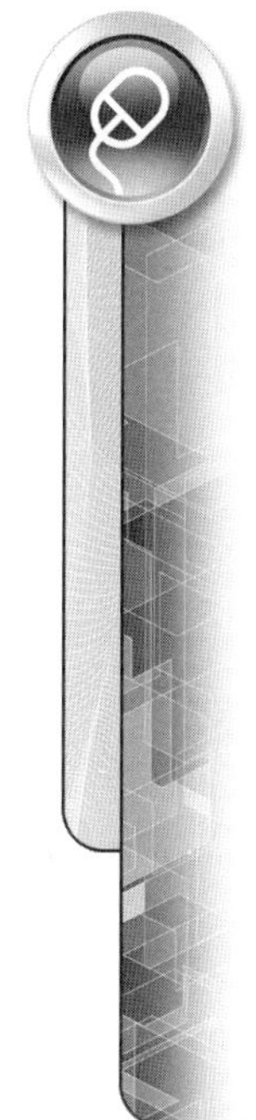

### Use the Survey Command Window

1. In the Nichols survey, select the network **Topo**, press the right mouse button, and from the shortcut menu, select SURVEY COMMAND WINDOW....
2. In the Survey Command Window, from the Point Information menu, select BEARING.
3. In Point Information – Bearing, for the Start Point, enter point number **1213**, for the Ahead Point, enter point number **1226**, and click ***OK*** to review the Survey Command Window results.
4. In the Survey Command Window Command line, enter **1206 1213** and press ENTER to query the bearing between points.
5. Use a few more commands from the Point Information menu to query the current point data.
6. Close the Survey Command Window.
7. Select the Nichols survey database, press the right mouse button, and from the shortcut menu, select CLOSE SURVEY DATABASE.
8. Close and save the Nichols drawing.

This ends the exercise on importing a field book and querying basic survey data.

## SUMMARY

- A survey has six components: Control Points, Non-Control Points, Directions, Setups, Figures, and Survey Points.
- At a minimum, a survey has to have a control point, a setup, a prism entry, and observations from the setup.
- Each component's Properties and Edits display the same data, but Properties is a dialog box and Edit is a vista.
- The Survey Command Window is an excellent data-review tool.
- The Survey Command Window defines centerlines and adds survey data.

The next unit reviews figure commands and styles.

## UNIT 4: SURVEY FIGURES

Figures represent surveyed linear features. Survey figures represent naturally occurring lines (streams, slope edges, toe of slopes, etc.) and manmade lines (boundaries, centerlines, edges-of-pavement, gutter lines, etc.). Civil 3D 2010 supports two figure creation methods. The first is the traditional method of field observations containing figure codes. This file type has to be converted from a instrument's format into a field book. The second method is for ASCII coordinate files and requires no field book conversion.

The field crew is the source of figure command coding. The field observations allow them to see and code linear features. Coding is relatively simple. If you are using the TDS format, a figure code is a Note followed by the letter B. This tells the conversion application that the following point's description is the figure name. The following snippet represents TDS figure coding:

```
BK,OP1,BP8,BS278.2215,BC359.5958
--B
LS,HI4.600000,HR5.000000
SS,OP1,FP2000,AR187.5917,ZE91.2202,SD35.9700,--EP1
--B
SS,OP1,FP2001,AR185.5628,ZE91.1045,SD24.9500,--CL1
--B
SS,OP1,FP2002,AR189.1856,ZE92.4909,SD12.9700,--EP2
--C3
SS,OP1,FP2003,AR150.1421,ZE90.0348,SD16.8050,--EP2
SS,OP1,FP2004,AR126.1635,ZE88.5429,SD20.0550,--EP2
SS,OP1,FP2005,AR100.4603,ZE87.5553,SD20.1950,--EP2
```

In the preceding snippet, the second line states that the next observation is a new figure beginning and whose name is the point's description (i.e., EP1). Later in the snippet, two more figures are started, CL1 and EP2, and a curve (C3) using points 2003–2005 is started.

The C3 note indicates that the next three shots are the PC (beginning of a curve), a point on the curve, and the PT (end of the curve). It does not matter what data format the user uses, a curve must have three sequential observations.

If you are using the SDR format, figure coding is a Note or an embedded figure code in a point's description. The following snippet represents an SDR file and coding figures by embedding the figure commands in the description:

```
09F100010002115.52000091.1144444303.352222GROUND
09F100011001110.70000090.1394444290.872777B FLL
09F100011002126.32000090.2061111291.630000FLL
09F100011003125.21000090.4394444295.942222FLL
09F100011004119.40000090.8550000300.730000FLL
09F100011005105.88000091.1561111303.967777FLL
13NMC3
09F10001100640.310000092.1244444316.698333FLL
09F10001100729.400000092.6344444323.123333FLL
```

```
09F10001100817.150000094.1900000336.901111FLL
09F100011027110.23000089.8938888291.182222B CL
09F100011028150.68000090.9855555312.382222B FLR
09F100011029150.61000090.8255555312.679444B BCR
09F100011030151.93000090.9727777308.789444B CL
09F100011031134.52000091.0300000308.256111CL
09F100011032134.25000091.0300000312.400555FLR
```

The preceding snippet's second line starts the figure FLL, and the fifth from the last line and the two lines after it start additional figures (FLR, BCR, and CL). The 13NMC3 line is a note indicating that the next three shots are the PC (beginning of a curve), a point on the curve, and the PT (end of the curve). It does not matter what format the user uses, a curve must have three sequential observations.

Figure coding works in surveys observing each line from its beginning to end, or by crisscrossing (zorroing) several lines in a single pass. Crisscrossing a site is similar to surveying a road at stations along its path. To draw the lines, the user needs to know only three figure commands: BEGIN, C3, and END.

There are some rules to making figure coding work. When you want to begin a new figure, use B, BEG, or BEGIN. When Survey encounters this code in the field book, it starts a new figure. When it encounters a description that matches a figure name, the conversion routine automatically adds the figure command, CONT (continue). The assumption is that a figure continues from its last described point to the next point because the points have the same description as the figure name. This method works whether crisscrossing and identifying several lines or surveying a single line. The following snippet is an example of the field data starting and continuing several lines in a crisscross survey:

```
09F100011027110.23000089.8938888291.182222BCL
09F100011028150.68000090.9855555312.382222B FLR
09F100011029150.61000090.8255555312.679444B BCR
09F100011030151.93000090.9727777308.789444B CL
09F100011031134.52000091.0300000308.256111 CL
09F100011032134.25000091.0300000312.400555 FLR
09F100011033134.22000090.8422222312.699444 BCR
09F100011034118.82000090.8633333313.568333 BCR
09F100011035118.65000091.0783333313.257222 FLR
09F100011036117.49000091.0783333308.633888 CL
09F100011037109.23000091.0783333314.283888 FLR
09F100011038109.33000090.8622222314.634444B CR
```

The goal is to draw a line with each similarly described point (for example, connecting the first FLR to the next, and so forth). When this file is converted, the conversion routine places CONTs in front of each line to tell Survey that this point continues the figure referenced in an earlier observation. The following snippet shows the converted file and the insertion of CONT in the field book:

```
F1 VA 1027 291.10560 110.230 89.53380 "BCL"
BEG FLR
F1 VA 1028 312.22560 150.680 90.59080 "FLR"
```

```
BEG BCR
F1 VA 1029 312.40460 150.610 90.49320 "BCR"
BEG CL
F1 VA 1030 308.47220 151.930 90.58220 "CL"
F1 VA 1031 308.15220 134.520 91.01480 "CL"
CONT FLR
F1 VA 1032 312.24020 134.250 91.01480 "FLR"
CONT BCR
F1 VA 1033 312.41580 134.220 90.50320 "BCR"
F1 VA 1034 313.34060 118.820 90.51480 "BCR"
CONT FLR
F1 VA 1035 313.15260 118.650 91.04420 "FLR"
CONT CL
F1 VA 1036 308.38020 117.490 91.04420 "CL"
CONT FLR
F1 VA 1037 314.17020 109.230 91.04420 "FLR"
CONT BCR
F1 VA 1038 314.38040 109.330 90.51440 "BCR"
END
```

The snippet starts by beginning three lines with points 1028, 1029, and 1030. Each line continues with later shots along their path. The conversion routine sees to it that each new line observation continues the line drafting.

Remaining disciplined and remembering figure names is the most difficult part of using figure codes. Each figure should have a unique name (for example, EP1, EP2, BOC1, BOC2, CL-Maple, etc.). If you are reusing EP1 and EP2, you must know how to stop and start the coding to draw the lines correctly. The best place to edit figure coding is in the field data file; the coding may become quite complex if it is done in the field book. Let the conversion routine create the field book with its potentially complex coding. Many times, it is easier to remove figure coding than it is to add it (see Figure 14.26).

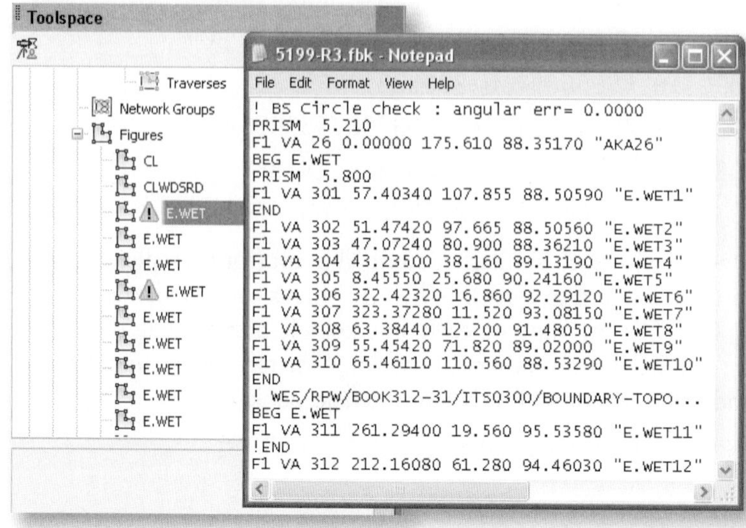


**FIGURE 14.26**

A continuous figure is not always desirable (for example, back-of-curb shots following a road and encountering a driveway). In this example, a sidewalk line intersects each back-of-curb point after the curb return. The sidewalk line cuts off the back-of-curb line, but across the driveway, the back-of-curb line restarts. In this situation, the back-of-curb line is several segments (figures) with the same name. Each jump from the back-of-curb line to the other side of the driveway is the beginning of a new back-of-curb, and no line should connect the back-of-curb from one side of the driveway to the other. To restart a new line with the same figure name, all you need to do is to enter a BEG before starting the next segment. This tells the conversion routine that the last back-of-curb observation does not connect to the current observation; rather, it starts a new figure with the same name. Importing this file with this type of coding creates a survey with several figures that have the same name. In Figure 14.26, the survey includes several individual edge-of-water figures, all with the same name.

Sometimes the conversion program puts too many ENDs in a field book. This creates problems with the resulting figures (i.e., a figure with one vertex). If a figure is malformed, Survey flags the offending figures with a yellow triangular shield that contains an exclamation point (!).

- Figure codes can be in data collector raw files or ASCII point files.

## FIGURE CURVES

There are three figure curve commands: PC, the next two points start and end an arc; C3, a three point curve; and MCS and MCE, a multiple observation curve. A C3 curve requires three sequential observations to make a curve. MCS and MCE bracket a series of observations defining a curve. There can be any number of observations between the codes. All observations between MCS and MCE are considered points on the curve.

## FIGURES, PARCELS, AND BREAKLINES

A figure can be parcel segments or a closed parcel boundary. If a figure is a parcel, figure properties assign a parcel style and site. If a figure is a breakline, it contains a breakline property. When you create figure breaklines, they become standard breaklines (3D). When you create breaklines from figures, the routine selects only those figures with the breakline property.

## LINEWORK CODE SETS

Line Code Sets are user-defined letters in a point's description denoting a figure action. Linework codes are for ASCII coordinate or LandXML survey files and extend the traditional field book language. For example, a B following a figure name starts drafting the figure, an H sets a horizontal offset to follow a figure, and a V sets a vertical offset to an offset figure. These codes are a part of a point's description and can contain several code letters; they should be separated with a space " " and placed AFTER the point's raw description (see the following code).

In an ASCII file:

```
175,466.0829,481.8260,100.0000,BC1 B
176,462.3638,538.8656,100.0200,BC1 BC
177,464.2371,553.2380,101.0500,BC1
178,469.7670,562.4632,101.0700,BC1
```

```
179,476.2186,569.2898,101.0400,BC1
180,490.4407,575.4662,101.0400,BC1 EC
181,517.6959,580.9047,101.0440,BC1
```

In the above snippet, point 175 starts the figure, 176 starts a multiple point arc, and point 181 ends the arc. A best fit curve connects the points. An arc coded by the above method does NOT need consecutively observed points.

In XML:

```
<RawObservation targetHeight="5.35" horizAngle="-3.5206"
     slopeDistance="10.6" zenithAngle="89.0604"
     directFace="true">
<TargetPoint name="1051" desc="TREE 2 CIR10">1000.715029044817
     1010.574548710302 100.266292027833</TargetPoint>
```

In the above snippet, point 1051 is a tree with a 2 inch trunk and a 10 foot diameter canopy.

### Description Keys and Linework Code Sets

When using Linework Code Sets (LCS) the Description Key Set's Full Description entries must be changed. Having the raw description values first allows the full description translation work like it always has. However, the use of $* puts all of the raw description and its linework coding to appear in the full description. For example, from the above ASCII snippet, the format of $* makes point 176's full description, BC1 BC and from the LandXML file, the tree's description, TREE 2 CIR10. To use only the necessary raw description for the full description, you must use parameter coding in the format column.

The coding for BC1 should be either $0 or BOC. The coding for TREE should be $1" $0 or $1" TREE.

### Line Codes

When a point is a line's beginning, continuance, and end, a code set uses the letters: B, C, or E. When a point's description contains multiple figure names, depending on the linework coding, the codes can start, continue, or end a line. In the following field book line, point 1001 starts two lines: FLL and BC1. The begin (B), continue (C), and end (E) codes can be before or after the figure name in the point's description. Again, the point's raw description should be the description's first entry.

```
F1 VA 1001 290.52220 110.700 90.08220 "GUTTER FLL B BC1 B"
```

In the following code, the point is a continuance of the two lines. The presence of the figure names in the description assumes they are a continuance of the figures.

```
F1 VA 1005 303.58040 105.880 91.09220 "GUTTER FLL BC1"
```

As with all line codes, starting a new figure with the name of an existing figure starts a new figure, without connecting to the last figure's point. For example, if a point's description contains "FLL B", the code starts a new line with no connection to the previous FLL point; in essence ending the previous FLL figure.

```
F1 VA 1007 323.07240 29.400 92.38040 "FLL B BC1"
```

### Lines with Offsets

When observing lines in the field, they may represent two or more lines having horizontal and vertical offsets. The coding is complex and can be added to the file after completing the field work. However, a single entry at the beginning of the figure, applies to the entire figure length. When beginning a new figure with the same name, you must add the offsets to the first figure point. The following code snippet creates a figure FLL and three offsets, two with a vertical offset of .5 (higher by a half foot) and the third with a vertical offset of .1 (higher by a tenth foot).

```
<RawObservation targetHeight="5.35" horizAngle="-12.2846"
    slopeDistance="110.7" zenithAngle="90.0822"
    directFace="true">
<TargetPoint name="1001" desc="FLL B H.5 V.5 H.1 H-1.5 V.1">
    1023.921019313217 1108.084236821268 99.830582515985
    </TargetPoint>
```

### Line with Connected Point

A figure may have other lines connecting to one or more of its points. In the snippet below the figure, EP1 connects points 11 and 12. By adding the linework code CPN13, connect point 11 to point 13, point 11 starts the EP1 figure and connects a line segment to point 13.

```
11,29.0546,4.2947,100.0000,EP1 B CPN13
12,29.0546,29.2947,100.0000,EP1
13,39.0546,4.2947,100.0000,HC1
```

### Pad Offset

Linework allows you to connect a figure between two points and then define a rectangular offset from the just drawn line. The following is a snippet drawing a line and offsetting the line to create a rectangle (rect).

```
100,500,500,100,BLDG1 B
101,500,550,100,BLDG1 RECT40
```

### Curve codes

Many linear field objects contain curves, curb returns, parking lots, islands, etc. Many of these curves are measured by more than one or two shots. The traditional figure codes of PC and C3 create curves from two or three points. Linework coding allows multiple points and they do not need to be consecutive. In a previous code snippet, BC starts the curve and EC ends the curve. Although the snippet shows consecutive points, sequential points are not necessary to create the curve. If coding contains only a BC the assumption is a three point curve.

A curve included in a line figure with offset coding will continue the offset around the curve. Also, a curve can start a figure.

### Point On Curve

The Point On Curve code forces the figure to create a three point arc with the point before and after the point on the arc. If the lines are not parallel, the resulting arc may not begin or end where the beginning and ending points are. The code for a point on a curve is OC. The following is a snippet using the OC linework code.

```
190,458.9067,608.9372,100.0100,BOC B
191,552.9917,660.8947,100.0200,BOC
192,567.7514,729.5742,100.0250,BOC OC
193,524.1334,769.6248,100.0200,BOC
194,423.9202,719.1105,100.0100,BOC
```

### Three Point Curve

The Three Point Curve code uses BC, OC, and EC to force a figure through three points. This would be like defining a cul-de-sac bulb. The following code snippet forces a three point curve.

```
190,458.9067,608.9372,100.0100,BOC B
191,552.9917,660.8947,100.0200,BOC BC
192,567.7514,729.5742,100.0250,BOC OC
193,524.1334,769.6248,100.0200,BOC EC
194,423.9202,719.1105,100.0100,BOC
```

### Multiple Point Curves

Figures with multiple point curves will create compound and reverse curves if they are in the data. The figure will pass through each point on the curve segment. The following is a snippet with multiple curves because of the point locations.

```
175,466.0829,481.8260,100.0000,BC1 B
176,462.3638,538.8656,100.0200,BC1 BC
177,464.2371,553.2380,101.0500,BC1
178,469.7670,562.4632,101.0700,BC1
179,476.2186,569.2898,101.0400,BC1
180,490.4407,575.4662,101.0400,BC1 EC
181,517.6959,580.9047,101.0440,BC1
```

### Linework Code Set File

The Linework Code Set File contains the figure creation codes (see Figure 14.27). The default values are the field book codes from Autodesk/Softdesk. However, you can edit them to match your coding and cause little retraining of your crews. The file is stored in the Survey folder pointed to by User Survey Settings and can be edited by selecting the file from the Survey panel's Linework Code Sets branch, pressing the right mouse button, and from the shortcut menu, selecting Edit....

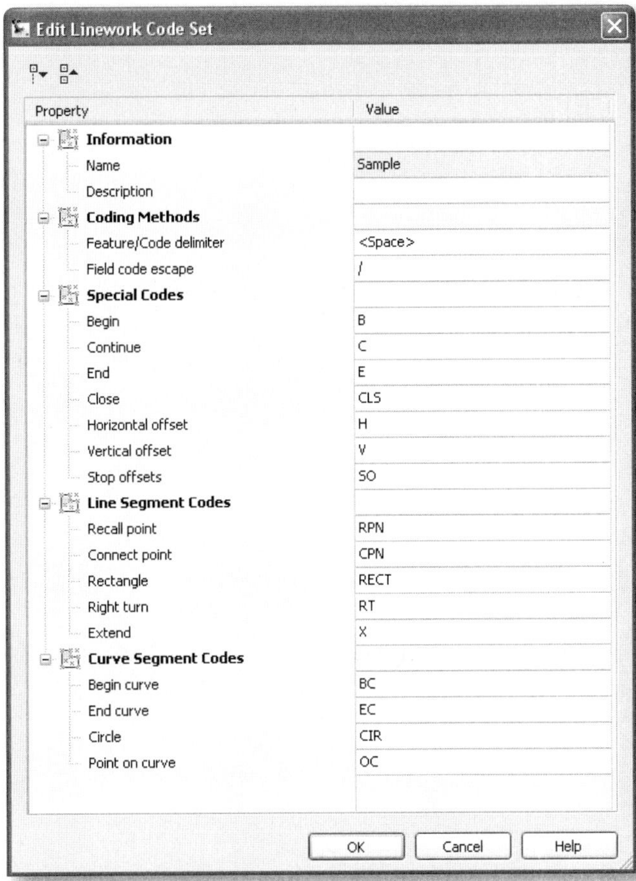


**FIGURE 14.27**

## FIGURE PREFIX LIBRARY

A figure prefix library contains definitions for all potential figures in a survey. A prefix definition assigns the figure use (breakline, parcel, or just a line), a figure style and possible parcel style, and a layer (see Figure 14.28). To create a new figure prefix, you select the Figure Prefix Databases heading, press the right mouse button, and from the shortcut menu, select New....

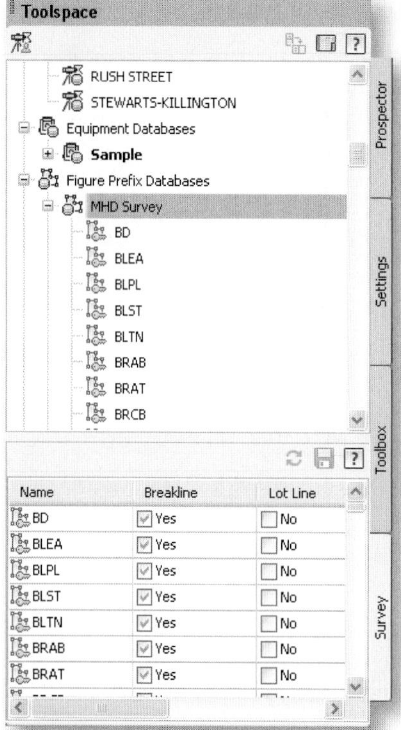


**FIGURE 14.28**

## FIGURE STYLE

A figure style defines a figure's markers, 3D Geometry, and component layers and their properties (see Figure 14.29).

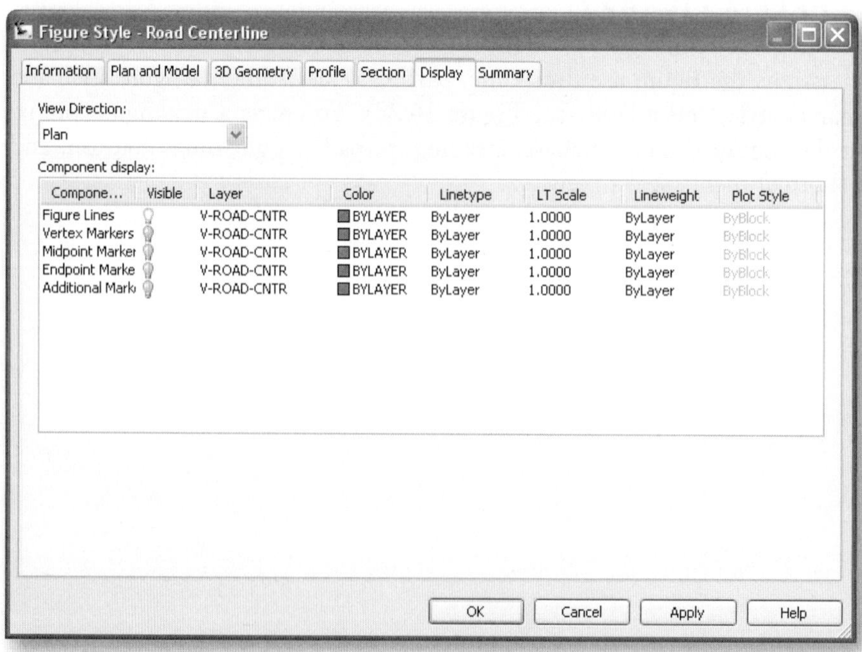


**FIGURE 14.29**

## EXERCISE 14-4

After completing this exercise, you will:

- Be able to create Survey figure prefixes.
- Be able to set up a local survey database.
- Be able to adjust the Survey settings.
- Be able to insert points.
- Be able to create a point group.
- Be able to create a surface.
- Be able to add figures as breakline data.

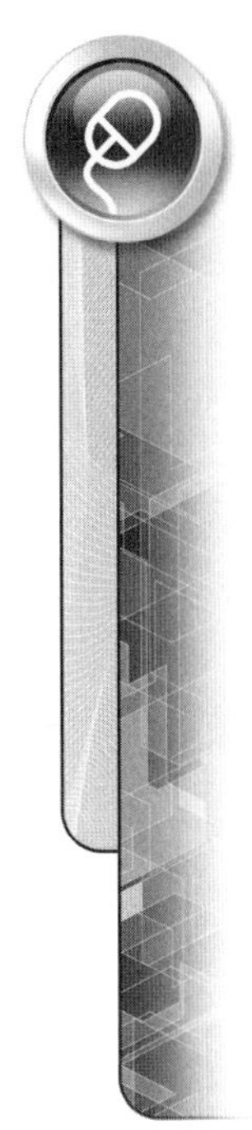

### Drawing Setup

1. If it is not open, start **Civil 3D** by double-clicking its desktop icon.
2. When you are at the command prompt, close the open drawing and do not save it.
3. At Civil 3D's top left, click Civil 3D's drop-list arrow, and from the Application Menu, click ***New***.
4. In the Select Template dialog box, browse to the Chapter 14 folder of the CD that accompanies this textbook, select the file *HC3D-Extended NCS (Imperial) template*, and click ***Open***.
5. If necessary, click the ***Survey*** tab.

### Create Figure Prefixes

1. In the Survey panel, expand the Figure Prefix Databases branch until you view the Sample figure prefix database.
2. In Figure Prefixes, select Sample, press the right mouse button, and from the shortcut menu, select NEW....
3. In New Figure Prefix, for the Name, enter **BC**, toggle on Breakline, assign the layer **C-ROAD-FIG**, for the style assign **Curb**, and click ***OK*** to create the prefix.
4. Repeat the previous two steps and create three additional figure prefixes. Use the values from Table 14.1 to create the figure prefix:

**TABLE 14.1**

| Name | Breakline | Layer | Style |
|---|---|---|---|
| FL | Yes | C-ROAD-FIG | Gutter |
| FN | Yes | C-ROAD-FIG | Fences |
| CL | Yes | C-ROAD-FIG | Road Centerline |

### Create the Local Survey Database and Network

1. From the Survey panel, select the Survey Databases heading, press the right mouse button, and from the shortcut menu, select NEW LOCAL SURVEY DATABASE....
2. In New Local Survey Database, for the Name enter **Peoria**, and click ***OK***.

A Peoria data structure is displayed under Survey Databases.

3. From the Survey Databases list, select Peoria, press the right mouse button, and from the shortcut menu, select EDIT SURVEY DATABASE SETTINGS....
4. In Survey Database Settings, if necessary, expand the Units section, set the Distance to **US Foot**, and click ***OK***.
5. At Civil 3D's top left, click Civil 3D's drop-list arrow, from the Application Menu, highlight SAVE AS, from the flyout select AUTOCAD DRAWING, browse to the C:\Civil 3D Projects\Peoria folder, for the drawing name enter **Peoria**, and click ***Save*** to save the file.
6. In Survey Databases, the Peoria survey, select its Networks heading, press the right mouse button, and from the shortcut menu, select NEW....
7. In New Network, if necessary, expand Network, for the name enter **Parking Lot Topo**, and click ***OK***.

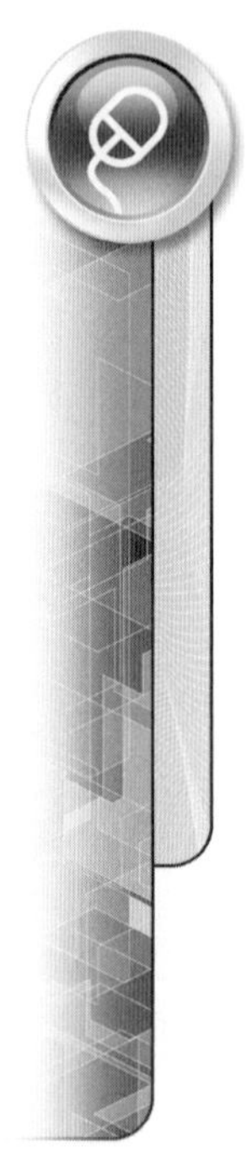

## Copy the Field Book to the Parking Lot Topo Folder

1. In Windows Explorer, browse to the Chapter 14 folder of the CD that accompanies this textbook, copy the file *Peoria.fbk*, and place it in the C:\Civil 3D Projects\Peoria\Parking Lot Topo folder.

## Review the Field Book File

1. In Survey, the Peoria survey, expand Networks, select Parking Lot Topo, press the right mouse button, and from the shortcut menu, select EDIT FIELD BOOK.... Browse to the C:\Civil 3D Projects\Peoria\Parking Lot Topo folder, select *Peoria.fbk*, and click ***Open*** to edit the file.
2. In Notepad, scroll to the line containing F1 VA for point 1001.

Note that the previous line starts a figure FLL and that all the points are the same description (FLL). This indicates that the survey crew observed the line from its beginning to end. The same is true for the BCL line. All of the points that describe the FLL and BCL line are sequential observations.

3. In Notepad, scroll to the beginning of the FLR, BCR, and CL figures.

In this section, the field crew crisscrossed three figures and observed (collected information on) additional items (trees). When converting the field data file, the conversion program added the CONT commands when it encountered a point with an active figure name.

4. Close Notepad and do not save any changes.

## Import the Field Book

1. In Survey, the Peoria survey, select the heading Import Events, press the right mouse button, and from the shortcut menu, select Import survey data....
2. In the Import Survey Data – Survey Database panel, check the current survey, it should have Peoria highlighted, and click ***Next***.
3. In the Specify Data Source panel, toggle on Field Book File, and at the lower right of the panel, click the ***identify source file*** icon, browse to the C:\Civil 3D Projects\Peoria\Parking Lot Topo folder, select *Peoria.fbk*, and click ***Open***.
4. Click ***Next*** to continue.
5. The Specify Network panel allows you to change the current network for the field book. Click ***Next*** to continue.

6. In the Import Options panel, make sure your toggles match those in Figure 14.30, and when set, click ***Finish*** to import the file.

The field book processes, but does not insert, objects in the drawing.

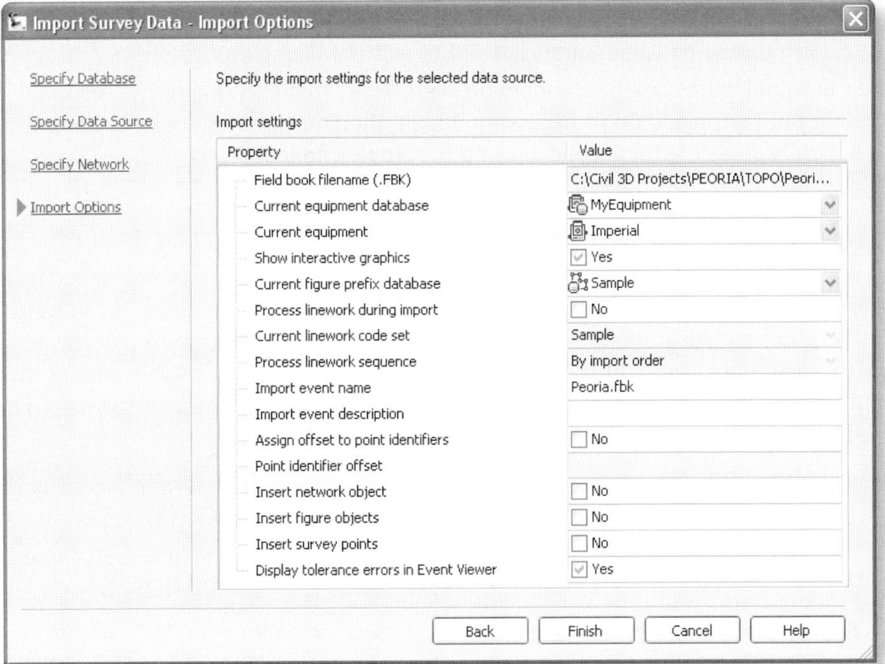


**FIGURE 14.30**

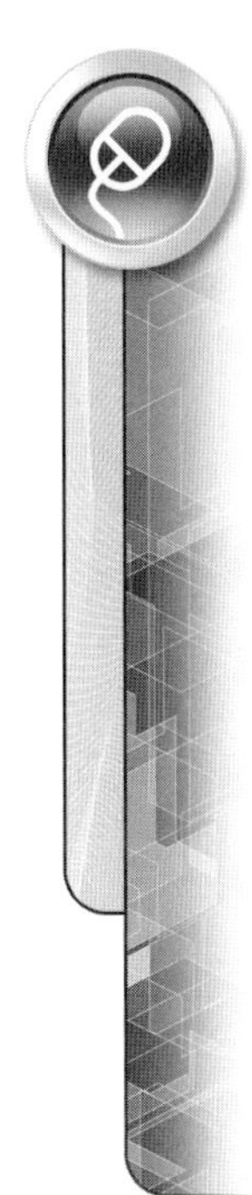

## Review Survey Data

1. In Survey, the Peoria survey, from the list of Networks, expand Parking Lot Topo until viewing Control Points. Select the heading Control Points, press the right mouse button, and from the shortcut menu, select EDIT....
2. Review values for each point number and click the ***green checkmark*** to close the Panorama.
3. In the Peoria survey, from Networks - Parking Lot Topo, expand Directions, from the list, select **1 to 2**, press the right mouse button, and from the shortcut menu, select EDIT....

A vista is displayed with the point 1 to 2 direction expressed as an azimuth.

4. Review the values and click the ***green checkmark*** to close the Directions Editor vista.
5. In the Peoria survey, the Network - Parking Lot Topo, expand Setups, from the list, select the first **Station:1, Backsight:2**, press the right mouse button, and from the shortcut menu, select EDIT OBSERVATIONS....

A vista appears with all observations made from the setup. In the drawing, Survey inserts the observations as a preview object.

6. Review the values and click the ***green checkmark*** to close the Observations Editor vista.

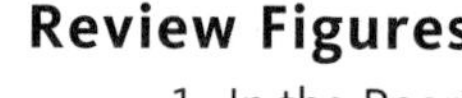

## Review Figures

1. In the Peoria survey, expand Figures, and from the list, select **FLL**. A figure preview is displayed in the drawing.
2. Click each figure to preview it in the drawing.
3. In the Peoria survey, from its figure list, select a figure, press the right mouse button, from the shortcut menu, select PROPERTIES..., expand the Figure section to review its values, and click OK to exit the dialog box.
4. Click the Figures heading, and, in preview, verify that each figure has the breakline option toggled on. If necessary, toggle the breakline option by clicking the figure's Breakline toggle and clicking the ***Apply changes*** icon.

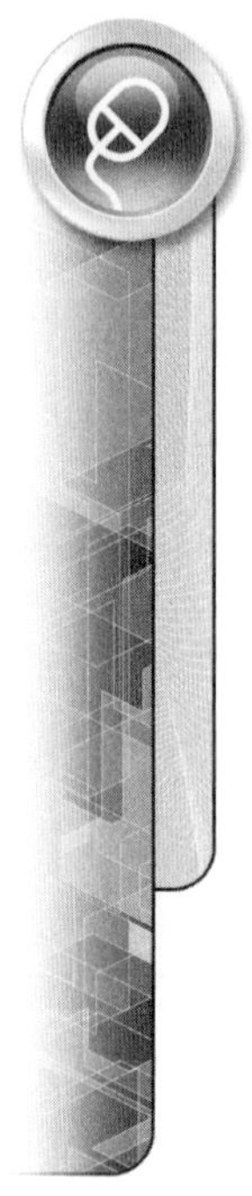

## Edit Field Book Figure Commands

There is a problem with two figures. FLL and BCL overlap on the survey's east side. They overlap because the field crew did not take the middle curve observations in the same place for both curves. Also, the crew forgot to add two C3 notes that make curves. You will edit the field book and add the C3 notes in the correct location. After adding the C3s, the field book is correct, but it needs to be reimported to create new survey data, changing the figures.

1. In the Peoria survey, from the Figures list, select **FLL**, press the right mouse button, and from the shortcut menu, select INSERT INTO DRAWING.
2. In the Peoria survey, from the Figures list, select **BCL**, press the right mouse button, and from the shortcut menu, select INSERT INTO DRAWING.

The two figures overlap at their eastern end.

3. In the Peoria survey, from the Figures list, select **FLL**, press the right mouse button, and from the Points flyout, select INSERT INTO DRAWING.
4. Use the ZOOM and PAN commands to better view the figure's eastern end.

Only markers appear in the drawing. The _All Points point group's point labels style is set in Point Group Properties.

5. Click the ***Prospector*** tab and expand its Point Groups branch.
6. In Prospector, from the Points Groups list, select _All Points, press the right mouse button, and from the shortcut menu, select PROPERTIES....
7. In Properties, if necessary, click the Information tab, change the Point Label Style to **Point#-Elevation-Description**, and click ***OK***.

The figure should have a curve starting at point 1003 and ending at 1005, but there are only line segments connecting the points. We need to review the figure coding at these point numbers to see if the C3 code is missing.

8. Click the ***Survey*** tab.
9. In the Peoria survey, from the Figures list, select **FLL**, press the right mouse button, and from the Points flyout, select REMOVE FROM DRAWING....
10. In the Are You Sure dialog box, click ***Yes***.
11. In the Peoria survey, from the Figures list, select **BCL**, press the right mouse button, and from the Points flyout, select INSERT INTO DRAWING.
12. Use the ZOOM and PAN commands to better view the figure's eastern end.

The figure should have a curve starting at point 1023 and ending at 1025, but there are only line segments connecting the points. We need to review the figure coding at these point numbers to see if the C3 code is missing.

13. In the Peoria survey, from the Figures list, select **BCL**, press the right mouse button, and from the Points flyout, select REMOVE FROM DRAWING....
14. In the Are You Sure dialog box, click ***Yes***.

## Edit a Field Book

The field book needs to be checked to verify if the C3s need to be added. You display the field book editor from the same shortcut menu you used to import it.

1. In the Peoria survey, from the list of Networks, select Parking Lot Topo, press the right mouse button, and select EDIT FIELD BOOK.... If necessary, browse to the C:\Civil 3D Projects\Peoria\Parking Lot Topo folder, select *Peoria.fbk*, and click ***Open***.
2. In Notepad, scroll down to the entries for points 1001–1005.
3. In Notepad, add a line before point 1003, and in the new line, enter C3.

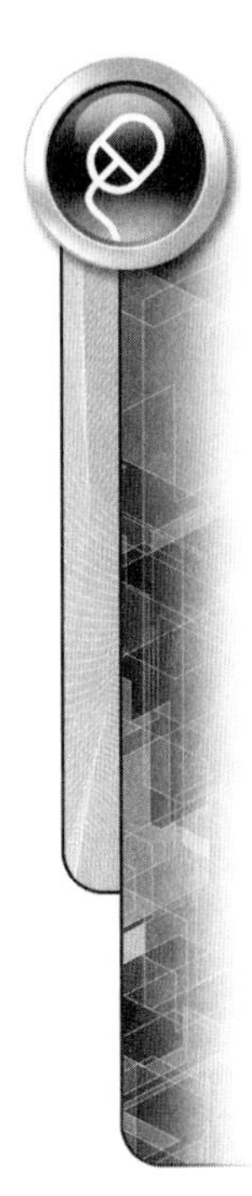

Your field book should look like the following around the entry for point 1002:

```
F1 VA 1002 291.37480 126.320 90.12220 "FLL"
C3
F1 VA 1003 295.56320 125.210 90.26220 "FLL"
F1 VA 1004 300.43480 119.400 90.51180 "FLL"
F1 VA 1005 303.58040 105.880 91.09220 "FLL"
```

The second edit takes place in the line before 1023.

4. In Notepad, scroll to the line containing the observation for point 1023.
5. In Notepad, add a line before point 1023, and in the new line, enter C3.

Your field book should look like the following around the entry for point 1023:

```
F1 VA 1022 317.07040 37.130 91.28120 "BCL"
C3
F1 VA 1023 303.42260 105.520 90.53320 "BCL"
F1 VA 1024 302.09220 113.880 90.46280 "BCL"
F1 VA 1025 296.12240 124.400 90.15540 "BCL"
F1 VA 1026 291.53000 125.680 89.59460 "BCL"
F1 VA 1027 291.10560 110.230 89.53380 "BCL"
```

6. Exit Notepad and save the changes.

## Reimport a Field Book

The current survey does not include the field book changes. You need to reimport the field book and remove all current data, replacing it with the edited field book values. To remove the current survey data, the survey must be reset. This is done from a shortcut menu, or from a toggle in the Import Field Book dialog box.

1. In Survey, the Peoria survey, from the list of Networks, select Parking Lot Topo, press the right mouse button, and from the shortcut menu, select RESET....
2. In the Are you sure dialog box, click ***Yes***.
3. In the Peoria survey, expand Import Events, select the Peoria.fbk import event, right mouse click, and from the shortcut menu, select Re-import.

4. Your toggles should match Figure 14.30, and when they do, click ***OK*** to re-import the file.

Survey processes the field book file, but does not insert anything in the drawing.

5. In the Peoria survey, from the Figures list, select **FLL**, press the right mouse button, and from the shortcut menu, select INSERT INTO DRAWING.
6. In the Peoria survey, from the Figures list, select **BCL**, press the right mouse button, and from the shortcut menu, select INSERT INTO DRAWING.

The two figures do not overlap at their eastern ends.

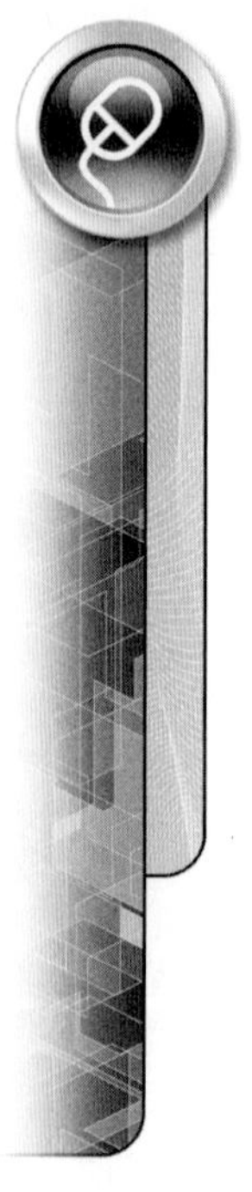

## Review Survey Data

Use the Survey Command Window to query the survey data.

1. From the Peoria survey database, select the network Parking Lot Topo, press the right mouse button, and from the shortcut menu, select SURVEY COMMAND WINDOW....
2. In the Survey Command Window, from the Point Information menu, select BEARING.
3. In Point Information – Bearing, for the Start Point, enter point number **1003**, for the Ahead Point, enter point number **1010**, and click ***OK*** to review the results.
4. In the Survey Command Window's Command line, enter **1006 1033** and press ENTER to query the bearing between the points.
5. Use a few more Point Information commands to query the current point data.
6. Close the Survey Command Window.

## Create a Surface from Survey Data

The survey contains points and figures that represent surface data and linear surface features. To use the data, you need to perform two steps. First, you place the points in a point group and assign the figures as surface breakline data.

1. In Survey, the Peoria survey, select the Survey Points heading, press the right mouse button, and from the Points flyout, select INSERT INTO DRAWING.
2. If a duplicate point warning appears, change the value to **overwrite**, apply it to all duplicate points, and click ***OK***.
3. In the Peoria survey, select the Figures heading, press the right mouse button, and from the Points flyout, select INSERT INTO DRAWING.
4. If a duplicate warning appears, overwrite the figures and click ***OK***.
5. Click the ***Prospector*** tab.
6. In Prospector, select the Point Groups heading, press the right mouse button, and from the shortcut menu, select NEW....
7. In Point Group Properties, the Information tab, for the name, enter **EG**.
8. Click the Include tab, toggle on With Numbers Matching, enter the point number range **1-1075**, and click the Point List tab to view the selected points list.
9. Click the ***Exclude*** tab, toggle on With Elevations Matching, and enter **<90**.
10. Click ***OK*** to create the point group.
11. In Prospector, select the Surfaces heading, press the right mouse button, and from the shortcut menu, select CREATE SURFACE....
12. In Create Surface, for the Name enter **EG**, for the description, enter **Parking Lot Topo – Peoria**, for the Style select **BTP**, and click ***OK*** until you have exited the dialog boxes.

13. In Prospector, expand Surfaces and EG branches until you are viewing EG's Definition data type list.
14. From EG's Definition data type list, select Point Groups, press the right mouse button, and from the shortcut menu, select ADD....
15. In Point Groups, from the list select **EG**, and click ***OK*** to assign the point group.
16. If the Event Viewer displays, close it by clicking the green check mark in the upper-right of the panorama.
17. Click the ***Survey*** tab.
18. In Survey, the Peoria survey, select the Figures heading, press the right mouse button, and from the shortcut menu, select CREATE BREAKLINES....
19. In Create Breaklines, make sure the Surface is **EG** and each breakline is toggled on to be a breakline, and click OK to continue.
20. In Add Breaklines, for the description, enter **Parking Lot Topo Figures**, make sure the Type is **Standard**, set the Mid-ordinate value to **0.01**, and click ***OK*** to continue.
21. In the drawing, select any triangle leg, press the right mouse button, and from the shortcut menu, select SURFACE PROPERTIES....
22. If necessary, click the Information tab, change the Surface Style to **Contours 1' and 5' (Design)**, and click ***OK*** to exit.
23. At Civil 3D's top left, Quick Access Toolbar, click the ***Save*** icon to save the drawing.
24. Close the drawing.

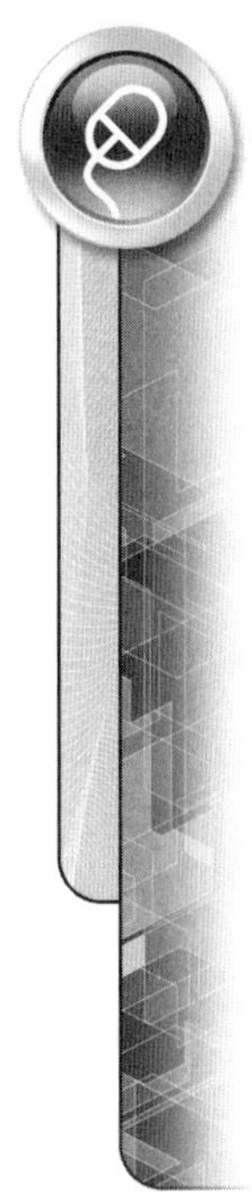

## Linework Coding

The Survey panel contains the linework coding values. All of the linework imports will be in a single survey and drawing. After reviewing the import you will delete the import event and update the network.

1. At Civil 3D's top left, Quick Access Toolbar, click the Open icon, browse to the CD that accompanies this textbook, and in the Linework folder for Chapter 14, select the drawing *Linework*, and click ***Open***.
2. If necessary, click the ***Survey*** tab.
3. In the Survey panel, select Survey Databases, press the right mouse button, and from the shortcut menu, select NEW LOCAL SURVEY DATABASE....
4. In the New Survey Database dialog box, for the survey name enter Linework.
5. Open Windows Explorer and from the Chapter 14 folder of the CD that accompanies this textbook, browse to and select the Linework folder, and copy the files to the Linework survey's folder (C:\Civil 3D Projects\Linework).

## Linework – Building Offset

Linework coding is for ASCII files or LandXML survey files. The following exercise sections show code snippets and how to import them in to a survey network.

1. In the Linework survey, select the Import Events heading, press the right mouse button, and from the shortcut menu, select IMPORT SURVEY DATA....
2. In the Survey Database panel, make sure Linework is highlighted and click ***Next***.
3. In the Specify Data Source panel, set the file type to Point File, at the bottom set the point file format to PNEZD (comma delimited).

4. At the panel's middle right, click the ***Browse*** icon. In the Select Source File dialog box browse to the C:\Civil 3D Projects\Linework folder, set the Files of type to *.txt, select the file *Bldg Offset.txt*, and click Open.

The panel displays the file path and its format.

5. Click ***Next***.
6. In the Specify Network panel, at the bottom, click Create New Network.... In the New Network dialog box, for the network name, enter Linework, and click ***OK*** to return to the Specify Network panel.
7. Click ***Next***.
8. In the Import Options panel, match your toggles to Figure 14.31, and when set click ***Finish*** to import the file and create the points and their linework.
9. In the Linework survey, expand the Figures list and select and review the figure BLDG1.
10. After reviewing the figure's values, in the Linework survey, select the *Bldg Offset.txt* import event, press the right mouse button, and from the shortcut menu, select ***Delete....***
11. In the Linework survey, select the Linework network, right mouse click, and from the shortcut menu, select ***Reset....*** In the Are you sure dialog box, click ***Yes***.

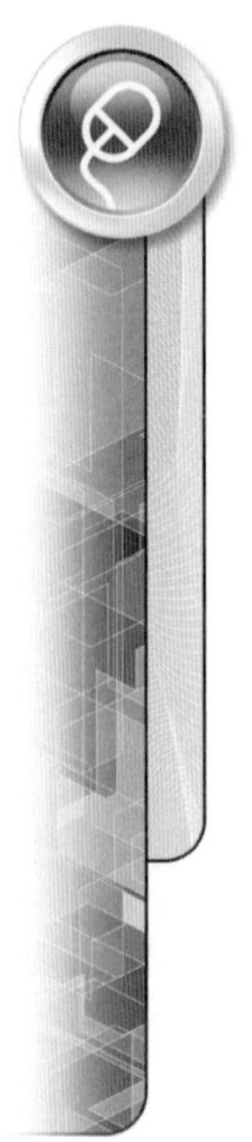

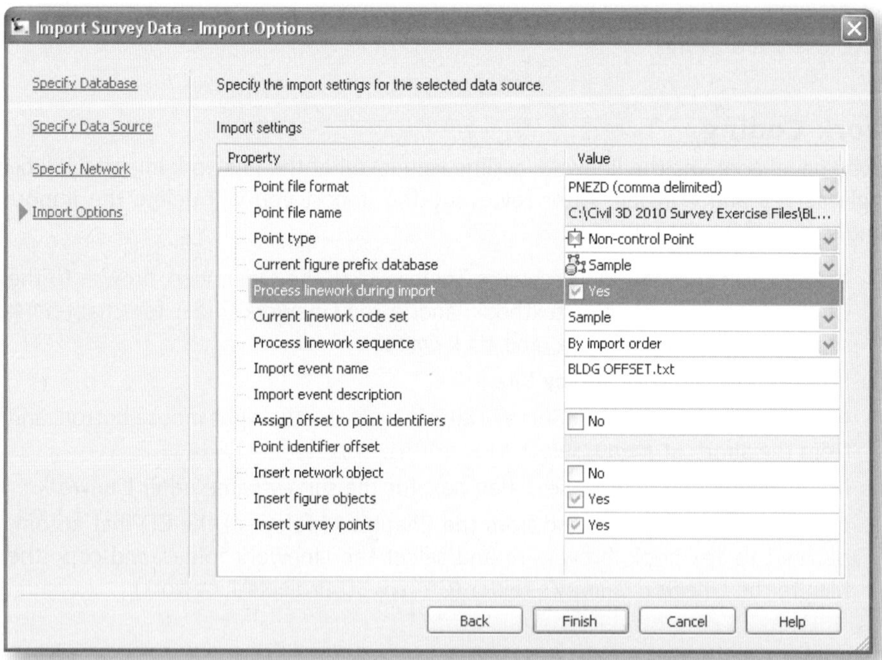


**FIGURE 14.31**

## Linework – Offset Lines

1. In the Linework survey, select the Import Events heading, right mouse click, and from the shortcut menu, select IMPORT SURVEY DATA....
2. In the Survey Database panel, make sure Linework is highlighted and then click ***Next***.

3. In the Specify Data Source panel, set the file type to Point File, at the bottom set the point file format to PNEZD (comma delimited).
4. At the panel's middle right, click the ***Browse*** icon. In the Select Source File dialog box browse to the C:\Civil 3D Projects\Linework folder, select the file *Curb.txt*, and click ***Open***.

The panel displays the file path and its format.

5. Click ***Next*** twice.
6. In the Import Options panel, match your settings to Figure 14.31 (except for the file name), and when set click ***Finish*** to import the file and create the points and their linework.
7. In the Survey, expand the Figures list and select and review the figure BC1.
8. After reviewing the figure's values, in the Linework survey, select *Curb.txt* import event, press the right mouse button, and from the shortcut menu, select ***Delete....***
9. In the Linework survey, select the Linework network, right mouse click, and from the shortcut menu, select ***Reset....*** In the Are you sure dialog box, click ***Yes***.

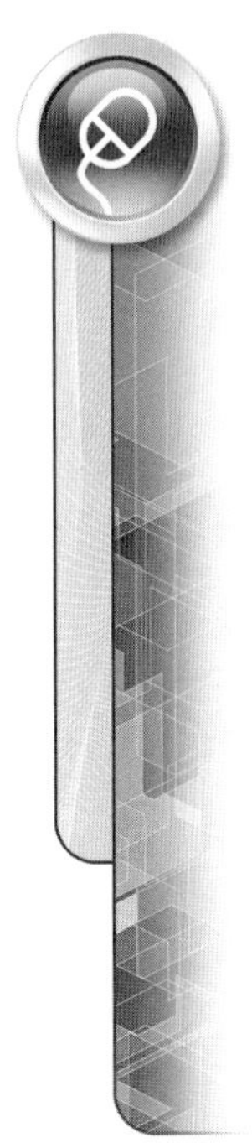

## Linework Curves

1. Using the same process from the previous two exercise sections, import the remaining text files: Line Continue.txt, Multi-point Curve.txt, Point on Curve.txt, and TOS.txt.

## Linework – Survey

Linework coding applies to unconverted LandXML survey files. These files do not need to be converted to a field book file.

1. Before importing the survey, delete all import events and reset the network (select the network name, right-mouse click, and from the shortcut menu select Reset...).
2. In the Linework survey, select the Import Events heading, right mouse click, and from the shortcut menu, select IMPORT SURVEY DATA....
3. In the Survey Databases panel, make sure Linework is highlighted and then click ***Next***.
4. In the Specify Data Source panel, set the file type to LandXML File.
5. At the panel's middle right, click the ***Browse*** icon. In the Select Source File dialog box browse to the C:\Civil 3D Projects\Linework folder, select the file *Peoria-Linework.xml*, and click ***Open***.
6. Click ***Next*** twice.
7. In the Import Options panel, match your settings to Figure 14.31 (except for the file name), and when set click ***Finish*** to import the file and create the points and their linework.
8. In the Survey, expand the Figures list and select and review the figures.
9. Close the survey database.
10. Close the drawing and do not save it.

This ends the exercise on field books, their command language, and their resulting data. Field books are powerful tools for taking handwritten or converted surveys and processing and evaluating their values in Civil 3D Survey.

## SUMMARY

- The best place to code figures is in the data collector.
- When you are converting a data file to a field book, the conversion program adds the BEGIN, CONT, and END figure commands.
- If you need to edit figure commands, simple edits should be done in the field book and extensive editing should be done in the collector data file.
- When you are importing an edited field book, in the Import Field Book dialog box, toggle on Erase points from drawing, reset network, and delete network figures.
- For figures to become surface breaklines, you must define the figure prefix as a breakline.

The next chapter reviews the survey and survey traverse adjustment routines.

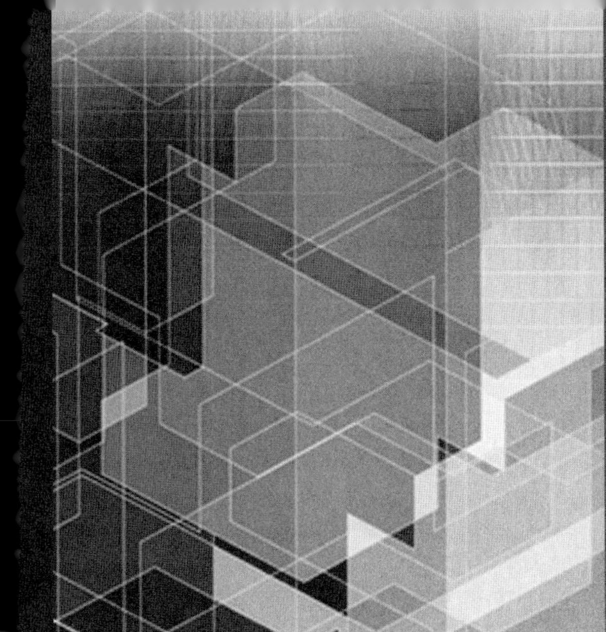

# CHAPTER 15

# Survey and Traverse Adjustments

## INTRODUCTION

The previous chapter focuses on topographic surveys. A topographic survey collects data on current site conditions (i.e., existing utilities, surface characteristics, and boundary determination). Surveying and establishing a site boundary can be a legal matter; a surveyor can be sued for errors and omissions found in his or her survey. If a site boundary is not correctly defined, a project can be in grave jeopardy. It is the survey crew's office's responsibility to define an accurate and precise boundary. The boundary survey's data-collection method is known as a traverse. A traverse has a prescribed field collection methodology. If accepted methodologies are not followed, the boundary data may be jeopardized.

A survey can contain a traverse, as well as observations among traverse points and possibly points external to the traverse path. If this is the case, you need to use the Least Squares analysis method to adjust the survey.

## OBJECTIVES

This chapter focuses on the following topics:

- Surveys Without Open Drawings
- Traditional Traverse Analysis Methods
- Traverses
- Closed Loop Traverse Adjustments
- Closed Connect Traverse Adjustments

### OVERVIEW

A traverse is a defined method to systematically measure straight lines and their directions. A point along the traverse path is a station (or "setup"), and each line is a traverse leg. The most common traverse type is the closed loop. A traverse starts at a point, visits other stations along a path (boundary), and returns to the beginning station. Other traverses types are closed, connected, and open. A closed connected traverse starts at a known point or at known points and ends at a known point or points. The last type of traverse is an open traverse. This traverse starts at a known point, but it ends at an observed point. An open traverse's only known point(s) is its starting station and possibly the backsight. An open traverse is the least desirable traverse type.

Traverse methodology starts with a beginning station and observes the next point to be occupied as a foresight. When the foresighted point (a station) is occupied, the field crew observes the last occupied point as a backsight. This continues around the loop until the crew returns to the beginning point. This method creates redundant observation data—looking forward to the next occupied point and looking back to the last point occupied.

A traverse has three traditional adjustment methods. The first and most accepted method is the Compass Rule adjustment. The Compass Rule adjustment assumes that both the observed angles and the distances are in error. Another method, the Transit Rule adjustment, assumes that the greatest amount of error is in the distances of the traverse. The Crandall Rule adjustment assumes that there are errors in distance measurements and distributes an angular error throughout the traverse.

As a final method, users can use the Least Squares adjustment on the traverse data because a traverse has redundant observations (forward and backward) along the traverse course. A Least Squares adjustment attempts to find the least amount of error for all observed traverse points.

A traverse can be transcribed from a written survey journal to a field book or can be entered in the survey Traverse Editor. Users can then export a field book containing the observation data, and thus preserve the traverse, should something happen to the application.

Users can also record a traverse in a survey. Many times, a traverse starts a survey. When the field book is imported, it contains the traverse, as well as all of the control points, stations, and survey points of the overall survey. A traverse can span several field books, as long as the field crew follows the prescribed traverse observation methods. When all of the field books are imported, the observation database represents an overall collection of survey data. To do this, the user must toggle off the Reset network in the Import Field Book dialog box.

### Unit 1

The first unit transcribes a manual survey using the Traverse Editor. The editor anticipates the data collected for the traverse and allows easy access to the data for editing errors. This unit will review the results of several adjustment methods.

### Unit 2

The second unit uses a traverse within a survey—a closed loop survey. The traverse noted by the surveyor is not found, and this unit's task is to identify what has happened in the survey. After the problem has been determined and corrected, this unit will review the results of several adjustment methods.

### Unit 3

The third unit looks into the closed connected traverse. This traverse type starts with known points and ends at known points. The question is, do the observation errors in the traverse affect the known ending points? This traverse has redundant and outside-of-traverse observations.

## UNIT 1: THE TRAVERSE EDITOR

To use the traverse editor, users must have a local survey database. Users define a local survey database as a part of a project or as a local folder. When a Vault project is created, it also creates a survey database with the same name as the project. When the survey database is a part of vault, the user must check in and out the survey to work with its data. When not using vault, all the user has to do is create the local survey database.

After a local survey database is created, a defined network is necessary. A network is a survey container. To start a traverse, there must be an initial station and backsight point. The initial station must be a control point. Users define control points in the named network that holds the traverse data. The backsight can be a second control point or a direction. When using a direction, the user uses the control point as the first point, types a dummy number for the second point, and enters a direction to the dummy backsight point number.

With these preliminaries finished, it is time to define a traverse and assign it the initial station and backsight points.

### LOCAL SURVEY DATABASE SETTINGS

Local Survey Database Settings contain traverse values that directly affect manual survey entries. It is easier to set the correct data types for the survey by entering the data in the Traverse Editor (see Figure 15.2). The settings in Figure 15.1 are for a horizontal traverse. These settings change what the Traverse Editor expects as data. If you are entering a survey, this survey settings section needs your attention. The Units section sets the format for angle entry. The default is the surveyor's shorthand entry, that is, decimal degrees (ddd.mmss). In this format, the "mm" and "ss" are the minutes and seconds values, and cannot exceed 59 minutes or seconds.

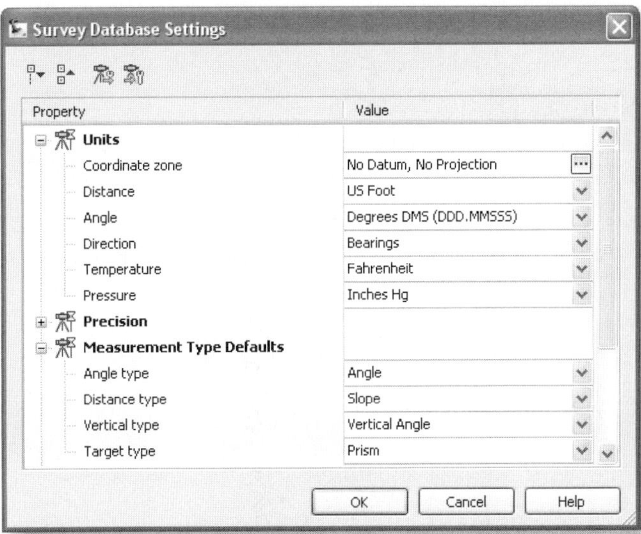


**FIGURE 15.1**

## TRAVERSE EDITOR

Users enter a manual traverse or review its data in the Traverse Editor (see Figure 15.2). When the data is floating, one side of the editor displays the station data and its backsight. The other side lists each traverse station and its observations. The editor displays only the traverse stations and their observations; no sideshots are shown. A traditional traverse does not use or acknowledge cross-traverse or outside-of-traverse observations.

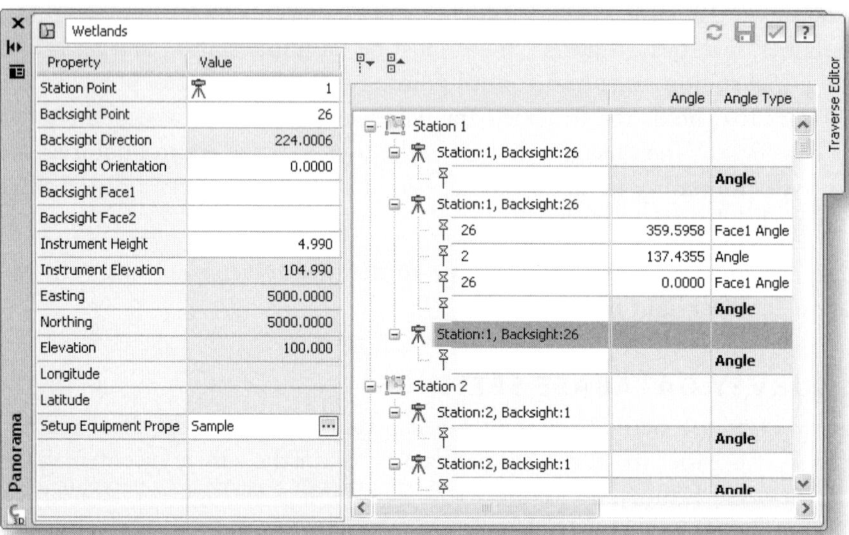


FIGURE 15.2

If you are manually entering traverse data, you first must define it (i.e., give it a name and a description). After you define the loop, you next enter the traverse data. The starting station and backsight points are entered in the New Traverse dialog box (see Figure 15.3).

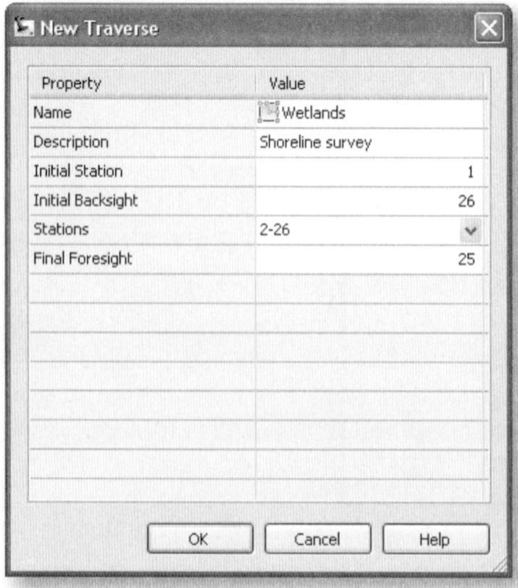


FIGURE 15.3

After you click OK in the Specify Initial Setup dialog box, the Traverse Editor opens and contains the first setup (see Figure 15.4).

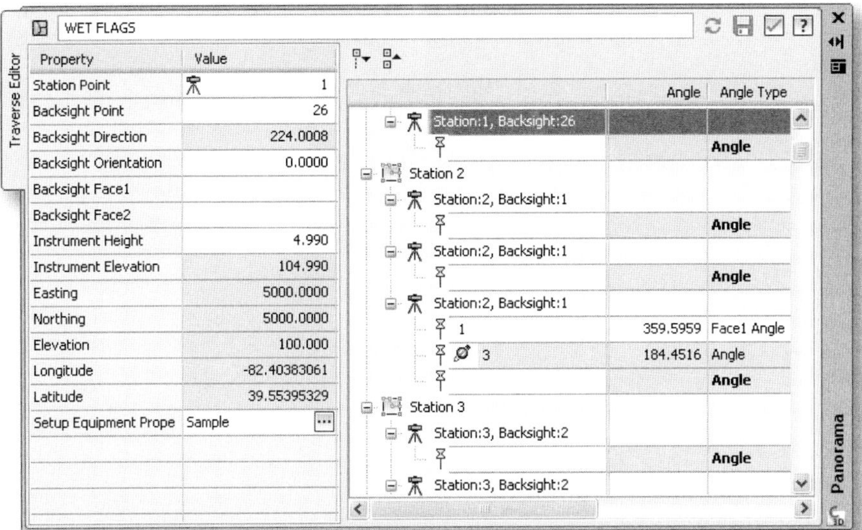

**FIGURE 15.4**

The editor assumes that the first foresight entry for a station is the point number of the next forward station. Then, using the traverse data, the user enters the remaining stations and observations.

When you are using the Traverse Editor to evaluate a defined traverse, all of the stations listed in its definition fill in the station and observation values. The editor is the easiest place to identify where a traverse may be broken (due to insufficient data or an absence of data).

## SURVEY AND VAULT

You need to define the project first when you are using Vault with a survey because defining a project also defines a survey database. When you view the Survey panel, you will see that the Local Survey has two icons: one for a project-controlled survey (a pick and shovel) and one for when it is locked (a pencil with a circle slash).

To unlock and use the survey folders, you must check out the survey from Vault. To check out a project's survey, you must go to Prospector's Master View project area, select the Survey heading (it must have a white circular icon), press the right mouse button, and select Check out from the shortcut menu. A white circular icon indicates that the survey is available for check out. If someone else has the survey checked out (shown by a white circle with a checkmark), you must contact that person to have him or her relinquish the check out. If you edit and change any files, you may jeopardize the survey's quality. You may overwrite someone else's efforts when you try to check in work.

The Vault survey folder is the current project working folder set in the Prospector Projects area.

## EQUIPMENT LIBRARY

Users should have at least one equipment definition that is general enough to describe the errors in most instruments used for surveys. There should be one entry for each instrument used for surveys and traverses. Civil 3D Survey uses the instrument's entry in the library to calculate observation errors (see Figure 15.5).

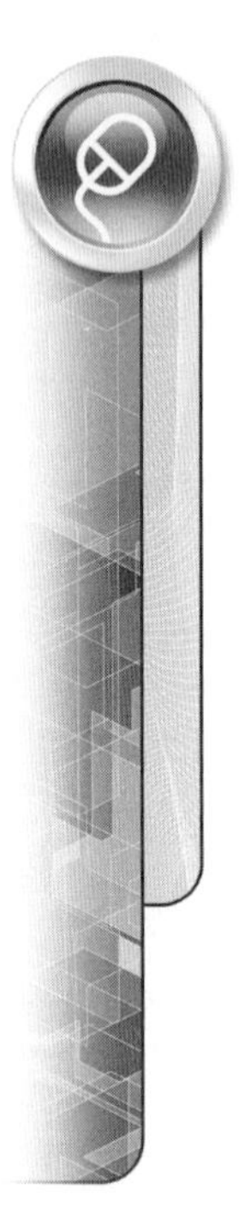

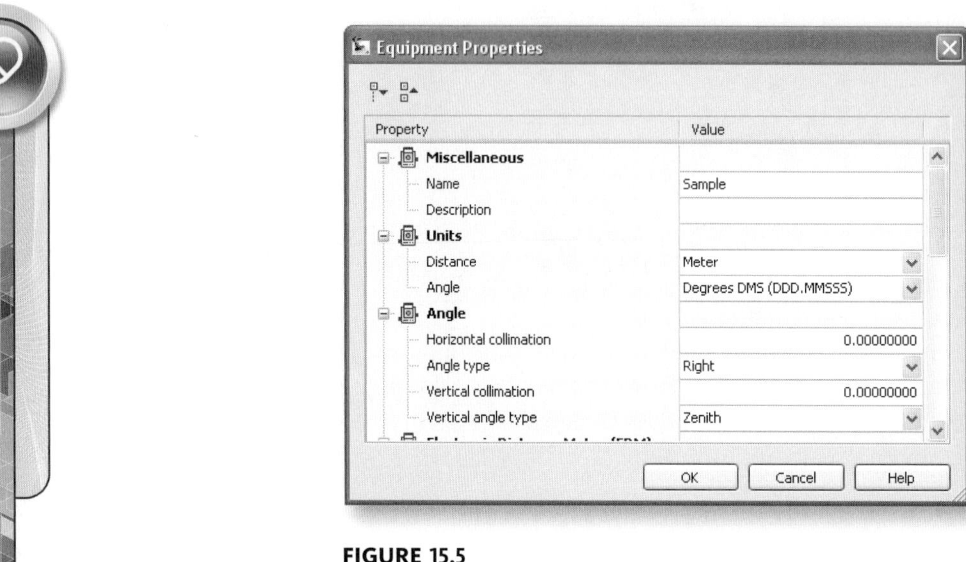


**FIGURE 15.5**

## EXERCISE 15-1

After completing this exercise, you will:

- Be able to create a new control point.
- Be able to create a direction.
- Be able to define a traverse.
- Be able to enter traverse data.
- Be able to review preliminary traverse adjustments.
- Be able to adjust a traverse.
- Be able to close a survey database.

## Drawing Setup

This exercise uses a local survey database.

1. If you are not in **Civil 3D**, on the Windows Desktop double-click the Civil 3D icon.
2. When you are at the command prompt, close the open drawing and do not save it.
3. If necessary, click the ***Prospector*** tab and set Prospector to **Master View**.
4. If necessary, in Ribbon's Home tab, Palettes panel, click the ***Survey Toolspace*** icon or click the ***Survey*** tab.

5. In the Survey panel, at its top select Survey Databases, press the right mouse button, and from the shortcut menu, select NEW LOCAL SURVEY DATABASE....
6. In the New Local Survey Database dialog box, for the survey name enter **Manual Traverse**, and click ***OK*** to create the survey.
7. From Start, Accessories, start Notepad, and from its File menu, select OPEN....
8. In Open, browse to the Chapter 15 folder of the CD that accompanies this textbook, select the *ManTrv.txt* file, and click ***Open***.
9. If possible, print the file.
10. Exit Notepad.

The file contains the same information as Table 15.1, which is found in this exercise.

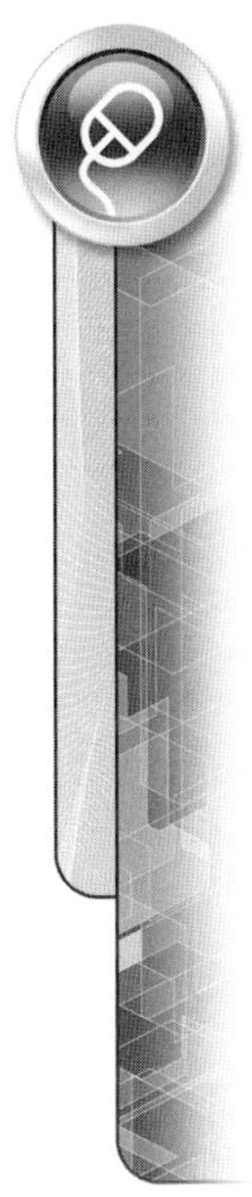

## Local Survey Database

1. In Survey, select Manual Traverse, press the right mouse button, and from the shortcut menu, select EDIT SURVEY DATABASE SETTINGS....
2. In Survey Database Settings, expand the Units section and change the Distance units to **US Foot**.
3. In the Measurement Type Defaults, set Angle Type to **Angle**, set Distance Type to **Horizontal**, set Vertical Type to **None**, and set Target Type to **Prism**.
4. Scroll down and if necessary, expand the Traverse Analysis Defaults. Set the Vertical Adjustment Method to **None**.
5. Scroll down and if necessary, expand the Least Squares Analysis Defaults. Set the Network Adjustment Type to **2-Dimensional**.
6. Click ***OK*** to set the values and exit the dialog box.

## Define the Manual PB Loop Network

1. In Survey, the Manual Traverse survey, select the Networks heading, press the right mouse button, and from the shortcut menu, select NEW....
2. In New Network, for the name enter *Manual PB Loop* and click OK.

## Define a Control Point and a Direction

To enter a survey, there must be at least one control and a backsight point.

1. In the Manual Traverse survey, expand Networks and then expand the network Manual PB Loop. Select the Control Points heading, press the right mouse button, and from the shortcut menu, select NEW....
2. In New Control Point, assign the following values:

   Point Number: 1

   Easting: 4464.1621

   Northing: 4715.0055

3. After entering the values, click ***OK*** to exit.
4. In the Manual Traverse survey, the Manual PB Loop network, select Directions, press the right mouse button, and from the shortcut menu, select NEW....
5. In New Direction, for the From Point enter **1**, for the To Point enter **25**, for the Azimuth (Direction) enter **275.0000**, set the measurement type to **Azimuth**, and click ***OK*** to exit.
6. Select the Manual PB Loop, press the right mouse button, and from the shortcut menu, select Automatic Update.

## Define the PB Loop Traverse

1. In the Manual Traverse survey, the network Manual PB Loop, select the Traverses heading, press the right mouse button, and from the shortcut menu, select NEW....
2. In the New Traverse dialog box, for the name enter *PB Loop*, for the description enter **Preliminary Loop – Manual**, set the Initial Station to 1, set the Initial Backsight to 25, and click ***OK***.

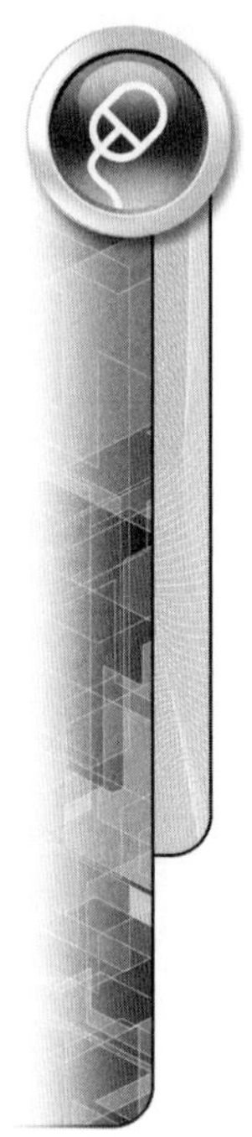

## Enter Traverse Data

Table 15.1 contains the traverse data. Each setup has two lines of information. The first line is the station point number, the backsight point number, and the angle. The second line is the foresight point number, horizontal angle, horizontal distance, and description.

The Local Survey Settings directly affect manual survey value prompting.

1. In the Manual Traverse survey, the network Manual PB Loop, expand Traverses. From the list, select PB Loop, press the right mouse button, and from the shortcut menu, select EDIT....
2. The Traverse Editor vista displays and shows the first setup and its backsight. The editor's cell-based side is where the turned angles and distance are entered.

   The foresight entry for the first station is the second line of Table 15.1.

**TABLE 15.1**

| STN | Pnt# | Angle | Hrz Dist | DescType |
|---|---|---|---|---|
| 1 | 25 | 0.0 | | BS |
| | 6 | 71.3033 | 429.7884 | TRV-1FS |
| 6 | 1 | 0.0 | | BS |
| | 102 | 214.0533 | 438.5154 | TRV-2FS |
| 102 | 6 | 0.0 | | BS |
| | 106 | 235.4233 | 451.0348 | TRV-3FS |
| 106 | 102 | 0.0 | | BS |
| | 107 | 205.2035 | 564.9924 | TRV-4FS |
| 107 | 106 | 0.0 | | BS |
| | 8 | 267.4337 | 369.7415 | TRV-5FS |
| 8 | 107 | 0.0 | | BS |
| | 108 | 170.3710 | 313.1525 | TRV-6FS |
| 108 | 8 | 0.0 | | BS |
| | 2 | 261.3654 | 590.6619 | TRV-7FS |
| 2 | 108 | 0.0 | | BS |
| | 1 | 180.1813 | 405.0081 | BM1FS |

The following step enters the data from the first line of Table 15.1.

3. In the Traverse Editor for Station 1, and below Station 1, Backsight 25, click in the cell to the right of the prism, and for the foresight point type **6**.

The editor presents a new station based on the foresight of point 6, backsighting point 1. The editor assumes that the first entry is the next forward station's point number.

4. Scroll the editor to the right. For the Angle enter **71.3033**, for the horizontal Distance enter **429.7884**, click the Apply changes icon, and for the Description enter **TRV-1FS**.
5. In the Traverse Editor for Station 6, and below Station 6, Backsight 1, click in the cell to the right of the prism, and for the foresight point enter **102**.
6. Scroll the editor to the right. For the angle enter **214.0533**, for the horizontal distance enter **438.5154**, click the Apply changes icon, and for the description enter **TRV-2FS**.
7. In Table 15.1, using the remaining data, finish entering traverse point numbers, angles, distances, and descriptions.
8. After entering the data, in the Traverse Editor, click the floppy icon to save the edits.
9. Click the green checkmark to close the Traverse Editor vista.
10. In the Manual Traverse survey, the Manual PB Loop network, expand Traverses. From the list of traverses, select PB Loop, press the right mouse button, and from the shortcut menu, select PROPERTIES....

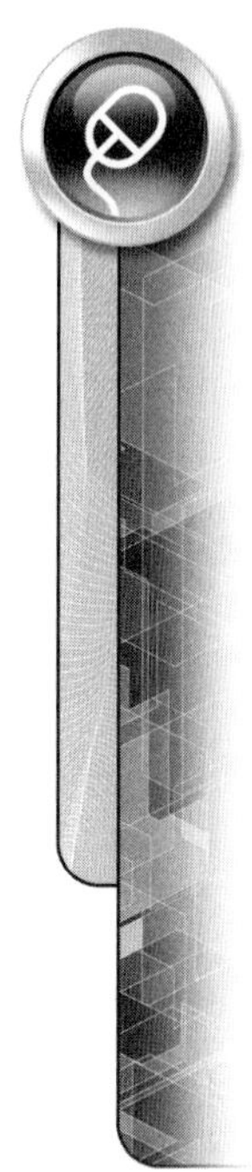

The traverse has the initial station, backsight, observed stations, and final foresight.

11. Click ***OK*** to exit the dialog box.
12. In the Manual Traverse survey, select the Survey Points heading, press the right mouse button, and from the shortcut menu, select Edit....
13. Review the points for the current traverse.

The only control point is point 1. The remaining points are survey points without an adjustment (prism icon).

14. Click the ***green checkmark*** to close the Survey Points Editor vista.

## Traverse Adjustments

1. In the Manual Traverse survey, the Manual PB Loop network, from the list of traverses, select PB Loop, press the right mouse button, and from the shortcut menu, select TRAVERSE ANALYSIS....
2. In Traverse Analysis, toggle **OFF** Update Survey Database and click ***OK*** to produce a preliminary adjustment report.

The Adjustment routine computes and displays the results as three reports in Notepad.

3. From the Notepad group, select the file *PB Loop Raw Closure.trv*.

This is an overall traverse quality report.

4. Close the file *PB Loop Raw Closure.trv*.
5. From the Notepad group, select the file *PB Loop Balanced Angles.trv*.

This report reviews the raw traverse calculations on the right and the potentially correct coordinates on the left with their deltas at the far right.

6. Close the *PB Loop Balanced Angles.trv* file.
7. From the Notepad group, select the *PB Loop.lso* file.

This report lists the traverse coordinates on the left, the adjustment-corrected coordinates on the left, and at the far left, the accumulated amount of change.

8. Close the *PB Loop.lso* file.

## Select an Adjustment and Update the Survey

1. In the Manual Traverse survey, the Manual PB Loop network, from the list of Traverses, select PB Loop, press the right mouse button, and from the shortcut menu, select TRAVERSE ANALYSIS....
2. In Traverse Analysis, for the Horizontal Adjustment Method, select Compass Rule, toggle **ON Update Survey Database**, and click ***OK*** to adjust the traverse.

The Adjustment routine computes and displays the results as three reports in Notepad.

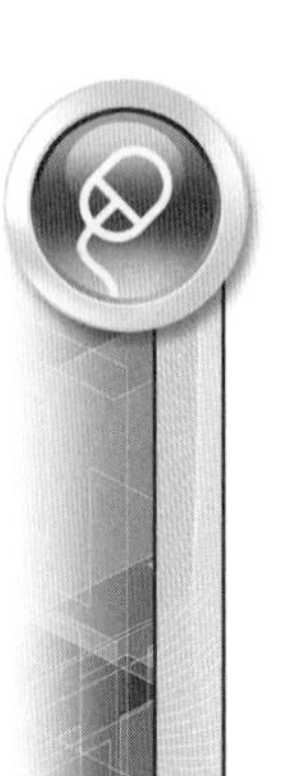

3. Review the reports and close them.
4. In the Manual Traverse survey, the Manual PB Loop network, select the Control Points heading and view the control points list.

The observed points are now adjusted and considered control points.

## Close the Survey Database

1. In Survey, select the survey database Manual Traverse, press the right mouse button, and from the shortcut menu, select CLOSE SURVEY DATABASE.
2. Close the drawing.

This ends the exercise on manually entering data in a traverse.

## SUMMARY

- All traverses must have at least one control point and a backsight.
- A backsight can be a second control point or a dummy point with an azimuth or bearing to it from the first station.
- The Measurement Type Defaults under the Local Survey Database settings directly affect what values the Traverse Editor expects while you are entering traverse data.
- The Traverse Editor uses the first entry for an ahead station as the next forward point in the traverse.
- Users can preview the adjustment values by toggling off Update Survey Database in the Traverse Analysis dialog box.

The next unit works with a topographic survey that contains a traverse.

## UNIT 2: SURVEY WITH A TRAVERSE

Many times, a traverse is a part of an overall survey. The field crew should record in a journal (also known as the field book) the traverse, its point numbers, and a drawing that represents the traverse loop. The crew should keep track of point numbers and their descriptions. However, sometimes things do not go correctly, or the crew does not know that the observations have not been done as noted. Civil 3D Survey tools help identify what may be wrong or where a survey breaks down. The Edit Observations and Traverse Editor are the two best places to identify issues with survey and/or traverse data.

The next exercise is such a survey. The field crew has a sketch in the journal that describes the traverse. However, when defining the traverse, an error message appears that states there is not enough data for the traverse analysis.

There are two ways to discover this error's cause, and both take time to rummage through the survey's setups and data. By knowing what a traverse is and what data it needs, users can review a setup and look for the necessary data and move on. If the necessary data is not in a setup, this is where the traverse breaks. Once the break is discovered, there is the issue of whether to continue the traverse or not. If no data is present or there is data that supports an alternate traverse definition, the user should consult with the surveyor about the errors and options.

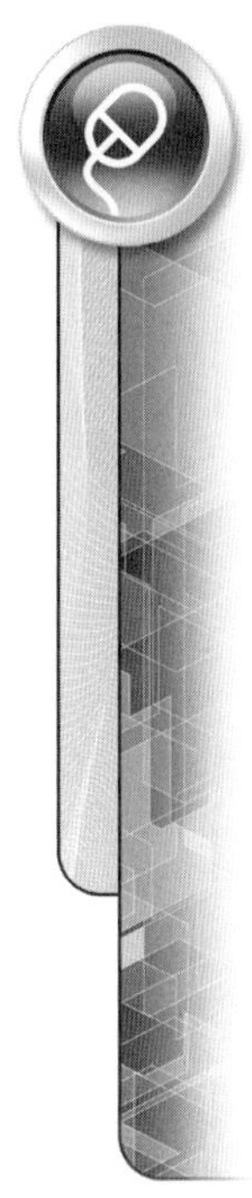

## EXERCISE 15-2

After completing this exercise, you will:

- Be able to set local survey settings.
- Be able to define a network.
- Be able to import a field book.
- Be able to define a traverse.
- Be able to identify a traverse error.
- Be able to adjust a traverse definition.

### Exercise Setup

This exercise uses a local survey database.

1. If you are not in **Civil 3D**, double-click its desktop icon.
2. When you are at the command prompt, close the open drawing and do not save it.
3. At Civil 3D's top left, click Civil3D's drop-list arrow, from the Application Menu, select New.
4. In the Select Template dialog box, browse to the Chapter 15 folder of the CD that accompanies this textbook, select the *HC3D-Extended NCS (Imperial)* template, and click ***Open***.
5. At Civil 3D's top left, click Civil 3D's drop-list arrow, from the Application Menu, highlight SAVE AS, from the flyout select AUTOCAD DRAWING, browse to the C:\Civil 3D Projects folder, for the drawing name enter **Oloop**, and click ***Save*** to save the file.

### Create a Local Survey Database

This survey is an example of a closed loop traverse.

1. If necessary, in Ribbon's Home tab, Palettes panel, click the ***Survey Toolspace*** icon or click the Toolspace's ***Survey*** tab.
2. In the Survey panel, at its top select Survey Databases, press the right mouse button, and from the shortcut menu, select NEW LOCAL SURVEY DATABASE....
3. In the New Local Survey Database dialog box, for the survey name enter **Oloop**, and click ***OK***.

### Set Values for the Oloop Survey Database

4. From the Survey Databases list, select Oloop, press the right mouse button, and from the shortcut menu, select EDIT SURVEY DATABASE SETTINGS....
5. In the Survey Database Settings, if necessary, expand the Units section, change the Distance units to **US Foot**, and set Angles to **Degrees DMS (DDD.MMSSS)**.

6. If necessary, expand the Traverse Analysis Defaults section, set the Horizontal Adjustment Method to **Compass Rule**, and set the Vertical Adjustment Method to **None**.
7. If necessary, expand the Least Squares Adjustment Defaults section, set the Network Adjustment Type to **2-Dimensional**, and toggle **ON Perform Blunder Detection**.
8. Click ***OK*** to exit the dialog box.

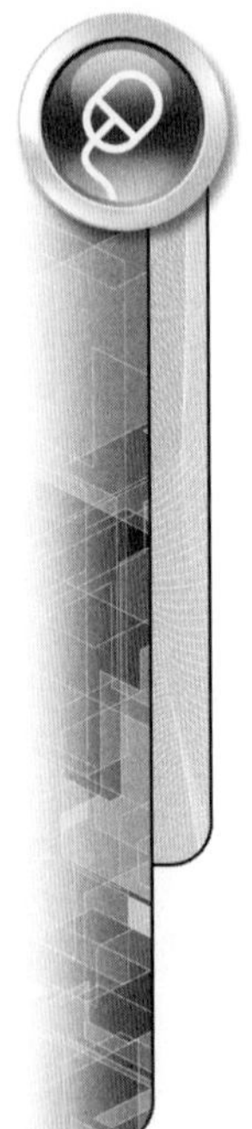

### Surveyor's Traverse Notes

The surveyor needs to have a survey brought into and adjusted by Civil 3D. The survey contains a topography and a traverse loop. The survey starts at point 10 and ends at point 11, foresighting to point 10. Figure 15.6 contains the traverse loop with stations and the traverse direction.

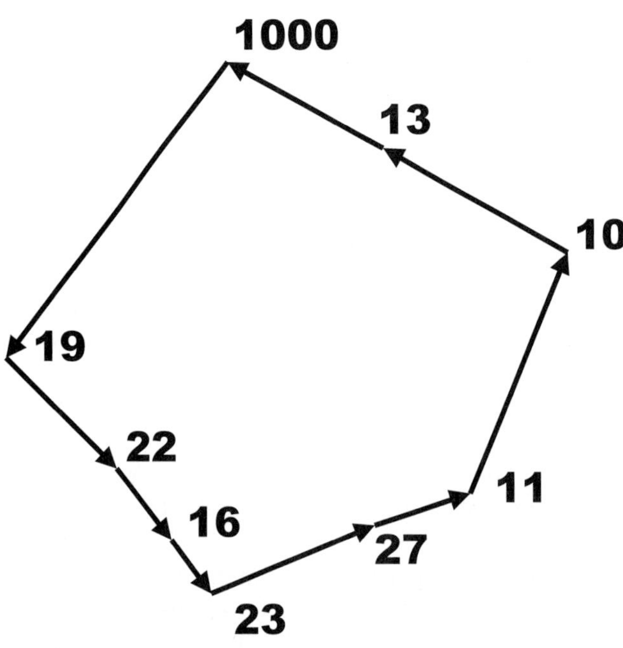


**FIGURE 15.6**

### Create the Network and Import the Field Book

1. In the Oloop survey, select Import Events, press the right mouse button, and from the shortcut menu, select Import Survey Data....
2. In the Import Survey Data dialog box, make sure the Oloop survey is selected and click ***Next***.
3. In the Specify Data Source panel, toggle the Data Source Type to **Field Book File**.
4. At the panel's middle right, click the ***Browse*** icon. In the Field Book Filename dialog box, browse to the Chapter 15 folder of the CD that accompanies this textbook, select the file *OLOOP.fbk*, and click ***Open*** to return to the Specify Data Source panel.
5. Click ***Next***.
6. In the Specify Network panel, at its bottom, click ***Create New Network....*** In the New Network dialog box, for the Network name enter **Boundary**, and click ***OK*** to return to the Specify Network panel.
7. Click ***Next***.

8. In the Import Options panel make your toggles match those in Figure 15.7, and click ***Finish*** to import the field book.

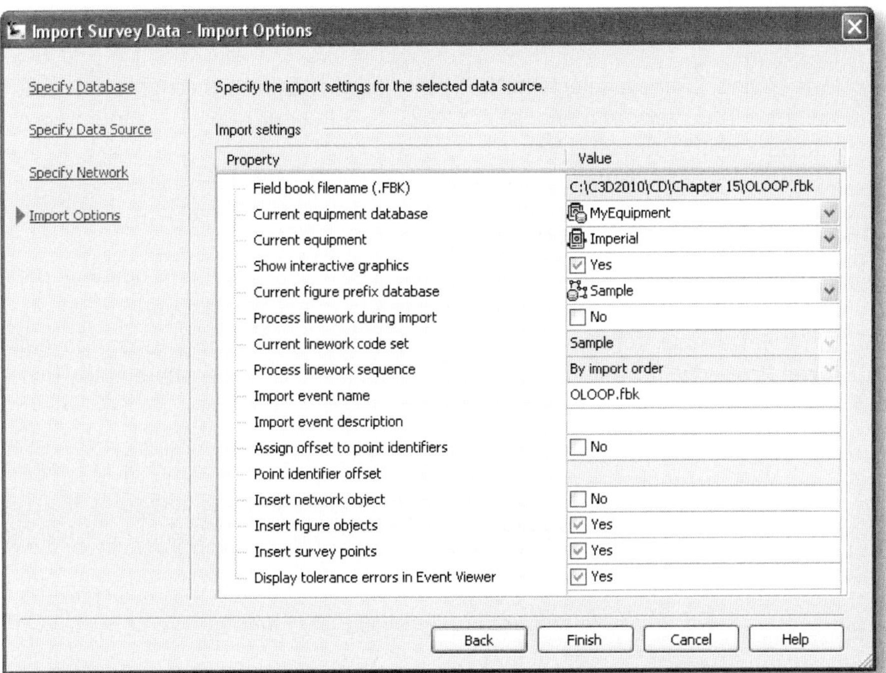


**FIGURE 15.7**

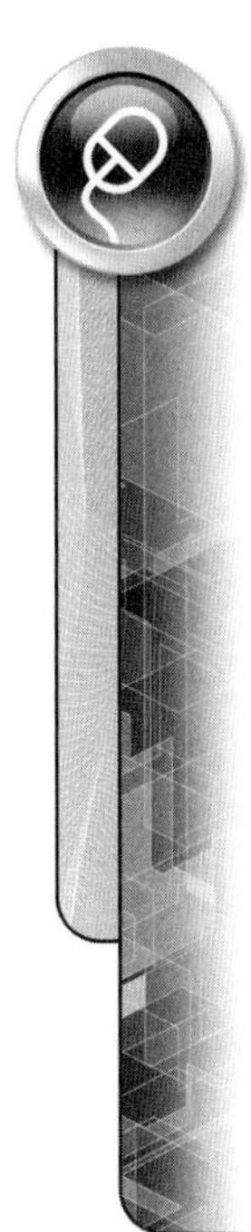

## Review the Survey Data

1. In the Oloop survey, expand the Boundary network, select the Control Points heading, press the right mouse button, and from the shortcut menu, select EDIT....
2. In the Control Points Editor, review the values for the only survey control point.
3. Click the ***green checkmark*** to exit the Control Points Editor vista.
4. In the Oloop survey, select the Survey Points heading, press the right mouse button, and from the shortcut menu, select EDIT....
5. In the Survey Points Editor vista, review the point list.

Each point has an icon that indicates its status: control, observed and occupied, and side shots.

6. In the Survey Points Editor vista, click the question mark and review the icon definitions.
7. Close Help and click the ***green checkmark*** to close the Survey Points Editor vista.

## Review the Setups

If the list of setups matches the survey sketch in Figure 15.6, things look good.

1. In the Oloop survey, the Boundary network, expand Setups. From the list, select Station: 10, Backsight: 11, press the right mouse button, and from the shortcut menu, select EDIT OBSERVATIONS....

2. Select and review a few more setups by editing their observations. When you are finished, click the ***green checkmark*** to close the Observations Editor.
3. In the Oloop survey, select the Boundary network, press the right mouse button, and from the shortcut menu, select INSERT INTO DRAWING.

The survey is displayed in the drawing.

4. In the Oloop survey, select the Boundary network, press the right mouse button, from the shortcut menu, select REMOVE FROM DRAWING..., and in the Are You Sure dialog box, click ***Yes***.

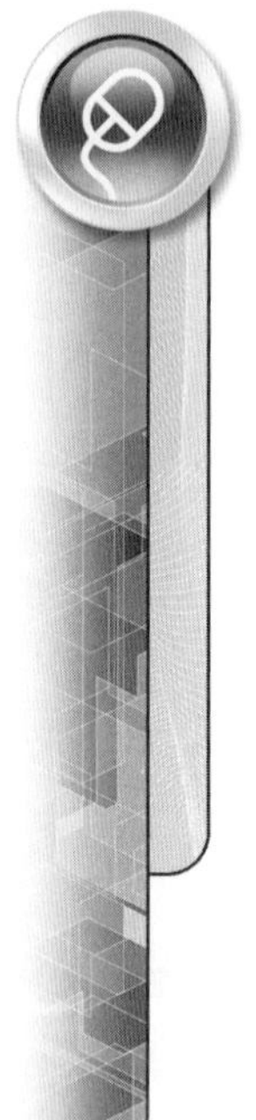

## Define Traverse per Sketch

1. In the Oloop survey, the Boundary network, select the Traverses heading, press the right mouse button, and from the shortcut menu, select NEW....
2. In New Traverse, for the name enter **Preliminary**, for the Description enter **Jewel Property**, for the Initial Station enter **10**, and click anywhere in the dialog box to have it fill in the remaining entries.

The current traverse definition does not contain point 16.

3. In the New Traverse dialog box, between points **22 and 23,** add point **16**. Your New Traverse dialog box should look like Figure 15.8. When complete, click ***OK***.

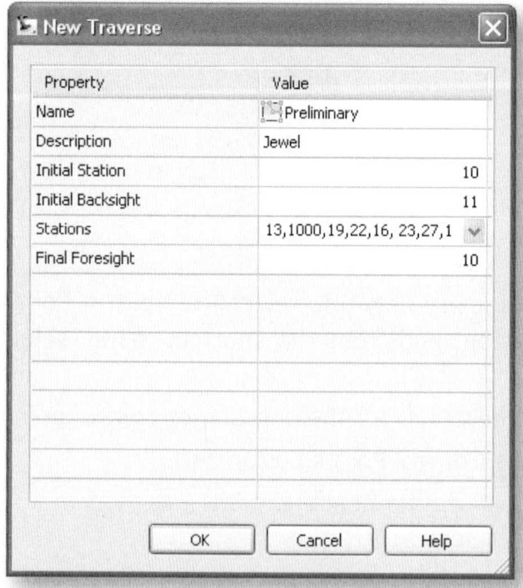


**FIGURE 15.8**

## Adjust the Traverse

1. In the Oloop survey, the Boundary network, expand Traverses, from the list of Traverses select Preliminary. Press the right mouse button, and from the shortcut menu, select TRAVERSE ANALYSIS....
2. In the Traverse Analysis dialog box, set the Horizontal Adjustment Method to **Compass Rule**, set the Vertical Adjustment Method to **None**, toggle **OFF Update Survey Database**, and click ***OK***.

Survey issues an error dialog box indicating that there is insufficient angle data to adjust the survey.

3. Click ***OK*** to exit.

Why does this error exist and where is this error coming from?

## Review the Traverse in the Traverse Editor

The traverse is broken. One setup's (station) observations break the ahead and back traverse loop observation method. You could find this by looking through each setup listed under the Boundary network. The easiest place to find broken data is by reviewing it in the Traverse Editor. If a traverse is broken, a setup will have no ahead and/or back loop observations in the Editor.

1. In the Oloop survey, the Boundary network, from the list of Traverses select Preliminary. Press the right mouse button, and from the shortcut menu, select Edit....
2. Scroll through the station and observations, making sure the correct backsight and foresight observations support the traverse definition.

When you arrive at the stations for 16 and 23 and you review their observations, you will see there are no observations at stationing on point 16 to point 23. This is where the traverse is broken.

The only thing you can do is find out if there are observations from the setup on point 22 to the next ahead point, 23. If there are, point 16 is a sideshot, not a traverse station. The loop would then go from 22 to 23 and bypass point 16 (see Figure 15.9).

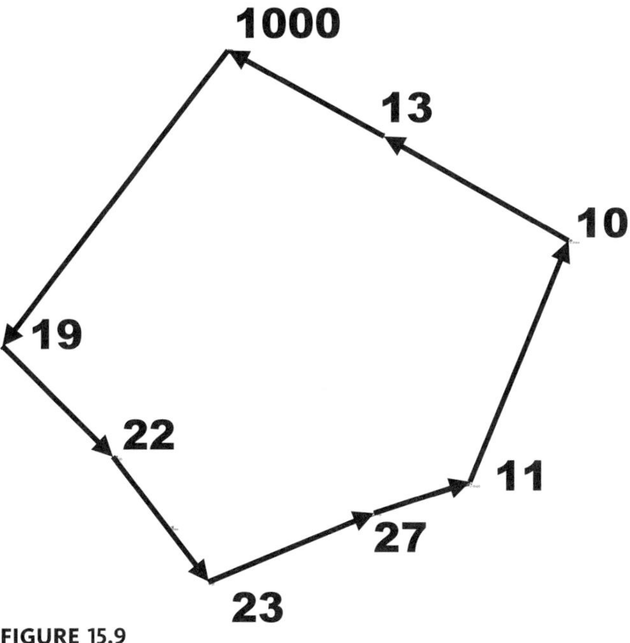


**FIGURE 15.9**

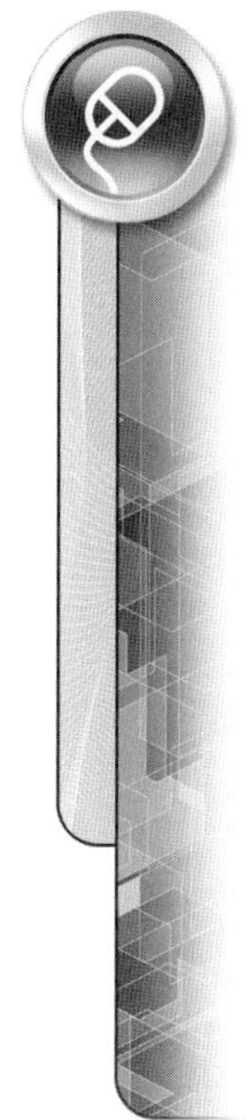

These alternative observations do not show in the Traverse Editor because the editor's focus is on the current loop definition.

You need to return to the Boundary network, Setups' stations list to look for possible alternative observations.

3. Click the ***green checkmark*** to close the Traverse Editor.
4. In the Oloop survey, expand the Boundary network's Setups list. From the list, select **Station: 22, Backsight: 19**, press the right mouse button, and from the shortcut menu, select EDIT OBSERVATIONS....

In the Observations Editor, note that, from this setup, the necessary foresight observation to point 23 is present.

There is also a station on point 23 that backsights to point 19. These two stations give you the look ahead and back observations you need to define the loop. With this information, point 16 is a sideshot, not a station along the traverse path.

5. Close the Observations Editor vista by clicking its green checkmark.

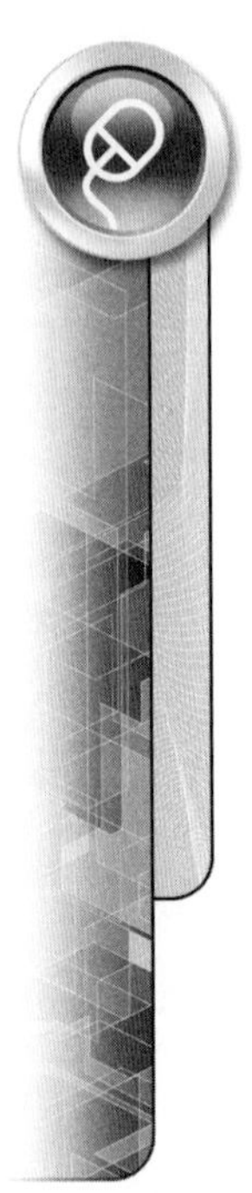

## Change a Traverse's Properties

The traverse definition needs to change to correctly define the traverse loop.

1. In the Oloop survey, from the Boundary network's Traverses list, select Preliminary. Press the right mouse button, and from the shortcut menu, select PROPERTIES....
2. In the Traverse Properties dialog box, change the station list to **13**, **1000**, **19**, **22**, **23**, **27**, and **11** (removing point 16 from the loop), and click ***OK*** to exit.

## Adjust the Traverse

1. In the Oloop survey, from the Boundary network's Traverses list, select Preliminary, press the right mouse button, and from the shortcut menu, select TRAVERSE ANALYSIS....
2. In Traverse Analysis, set the Horizontal Adjustment Method to **Compass Rule**, set the Vertical Adjustment Method to **None**, toggle **OFF Update Survey Database**, and click ***OK***.
3. Review the adjustment reports.
4. Repeat Steps 1 through 3, trying different adjustment methods.
5. In Traverse Analysis, set the Horizontal Adjustment Method to **Compass Rule**, set the Vertical Adjustment Method to **None**, toggle **ON Update Survey Database**, and click ***OK***.
6. Review the results and close the Notepad reports after you have reviewed their values.

## Insert Points into a Drawing

1. In the Oloop survey, select the Networks heading, press the right mouse button, and from the Points flyout menu, select INSERT INTO DRAWING.
2. Click the ***Prospector*** tab.
3. In Prospector, expand the Point Groups branch.
4. From the list of point groups, select _All Points, press the right mouse button, and from the shortcut menu, select PROPERTIES....
5. If necessary, click the Information tab and change the Point Label style to **Point#-Elevation-Description**, and click ***OK*** to exit.
6. At Civil 3D's top left, Quick Access Toolbar, click the ***Save*** icon to save the drawing.
7. Click the ***Survey*** tab.

## Close the Survey

1. In Survey, select the Oloop heading, press the right mouse button, and from the shortcut menu, select CLOSE SURVEY DATABASE.
2. Close and save the Oloop drawing.

This completes this close loop and adjustment exercise.

## SUMMARY

- The Traverse Editor is the best place to discover where a traverse is broken.
- Observations listed in the Traverse Editor relate only to the current traverse definition.
- When you are reviewing a station's complete observations, go to the Network's Setup list and review its values in Edit Observations.

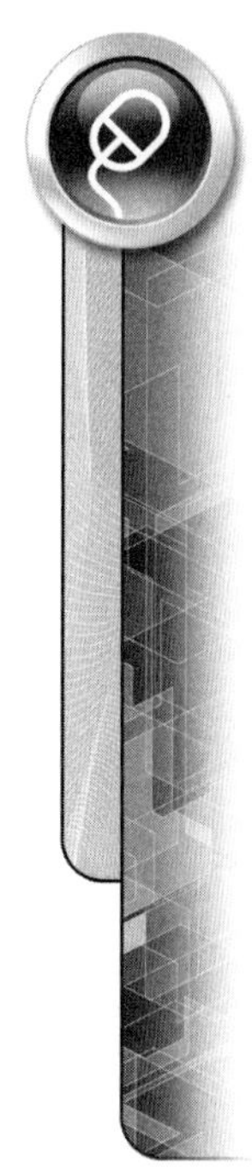

The next unit reviews a second traverse type, the closed connected traverse.

## UNIT 3: CLOSED CONNECTED SURVEY

What happens when the beginning and ending points of a survey are known coordinates? Does the Least Squares analysis method adjust these ending points even if they have known coordinates? How does Civil 3D Survey handle survey errors when it observes these known ending coordinates? Each question is legitimate. Points with known coordinates, even if they are observed with errors, remain at their original coordinates. These points are treated as control points even if they are a part of the survey observations. Observation errors place the points at a different set of coordinates, and the observations are adjusted to keep the points at their known coordinates at the survey's end.

Any external or redundant cross-traverse observations are ignored in traverse analysis. These methods analyze only traditional traverse data. To analyze a survey that contains additional observation data, users have to use the network Least Squares analysis.

## EXERCISE 15-3

After completing this exercise, you will:

- Be able to set local survey settings.
- Be able to define a network.
- Be able to import a field book.
- Be able to define a traverse.
- Be able to identify a traverse error.
- Be able to adjust a traverse definition.
- Be able to adjust a traverse with the Least Squares analysis method.
- Be able to review an adjustment ellipsis.

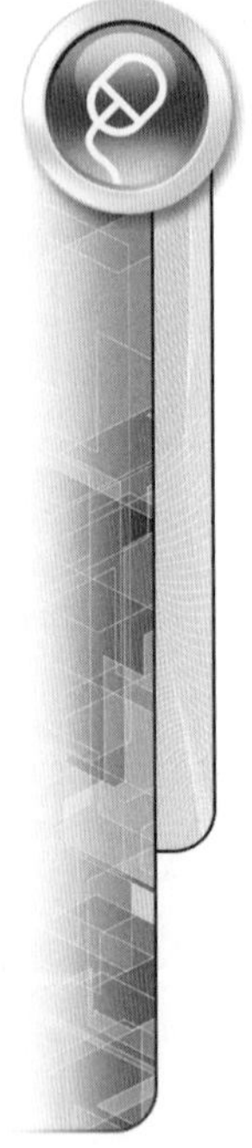

## Exercise Setup

This exercise uses a local survey database.

1. If you are not in **Civil 3D**, double-click its desktop icon.
2. If necessary, click the Survey tab.
3. In the Survey panel, select the Survey Databases heading, press the right mouse button, and from the shortcut menu, select NEW LOCAL SURVEY DATABASE....
4. In the New Local Survey Database dialog box, for the name enter **CC-Loop**, and click ***OK***.
5. At Civil 3D's top left, Quick Access Toolbar, click the ***Open*** icon, browse to the Chapter 15 folder of the CD that accompanies this textbook, select the *Property-CC.dwg* drawing, and click ***Open***.
6. At Civil 3D's top left, click ***Civil 3D's*** drop-list arrow, from the Application Menu, highlight SAVE AS, from the flyout select AUTOCAD DRAWING, browse to the C:\Civil 3D Projects\CC-Loop folder, for the drawing name enter *Property-CC-Work*, and click ***Save*** to save the file.

## Edit the CC-Loop Survey Database Settings

1. From the Survey Databases list, select the CC-Loop heading, press the right mouse button, and from the shortcut menu, select EDIT SURVEY DATABASE SETTINGS....
2. In Survey Database Settings, the Units section, change the Distance units to **US Foot**, and the Angles to **Degrees DMS (DDD.MMSSS)**.
3. If necessary, expand the Traverse Analysis Defaults, set the Horizontal Adjustment Method to **Compass Rule**, and set the Vertical Adjustment Method to **None**.
4. If necessary, expand the Least Squares Adjustment Defaults, set the Network Adjustment type to **2-Dimensional**, and toggle **ON Perform Blunder Detection**.
5. Click ***OK*** to exit the dialog box.

## Surveyor's Notes on the Traverse

The surveyor needs to have a survey brought into Civil 3D to be adjusted. The survey contains a closed connected traverse loop. The traverse is between two known points, points 1250 and 1375. There are redundant observations, as well as observations to an external point, point 1163. Figure 15.11 contains the traverse stations loop and the traverse direction.

The survey in this exercise can be treated as a traditional traverse or as a network Least Squares adjustment, because it contains cross-traverse and external point observations.

## Import the Field Book and Create the Network

1. In the CC-Loop survey, select Import Events, press the right mouse button, and from the shortcut menu, select Import Survey Data....
2. In the Import Survey Data dialog box, make sure the CC-Loop survey is selected and click ***Next***.
3. In the Specify Data Source panel, toggle the Data Source Type to **Field Book File**.
4. At the panel's middle right, click the ***Browse*** icon. In the Field Book Filename dialog box, browse to the Chapter 15 folder of the CD that accompanies this textbook, select the file *CC-Loop.fbk*, and click ***Open*** to return to the Specify Data Source panel.
5. Click ***Next***.

6. In the Specify Network panel, at its bottom, click ***Create New Network....*** In the New Network dialog box, for the Network name enter **Interior Boundary**, and click ***OK*** to return to the Specify Network panel.
7. Click ***Next***.
8. In the Import Options panel make your toggles match those in Figure 15.10, and then click ***Finish*** to import the field book.

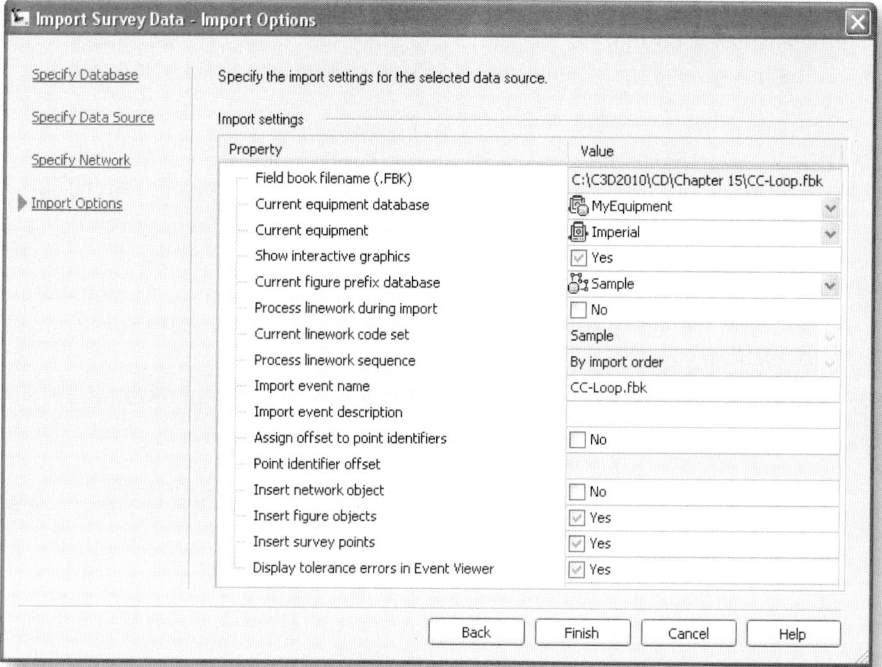

**FIGURE 15.10**

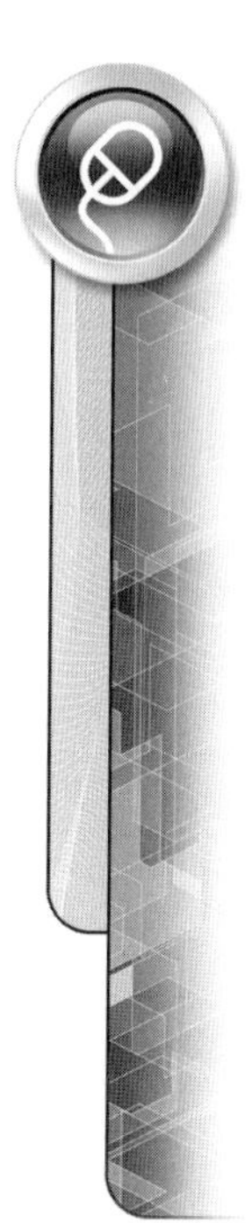

## Review the Survey Data

1. In the CC-Loop survey, expand the Interior Boundary network, and select the heading Control Points, press the right mouse button, and from the shortcut menu, select EDIT....
2. In the Control Points Editor vista, review the control point's values.
3. Click the ***green checkmark*** to exit the Control Points Editor vista.
4. In the CC-Loop survey, select the Survey Points heading. Press the right mouse button, and from the shortcut menu, select EDIT....
5. In the Survey Points Editor vista, review the point list.

Each point has an icon that indicates its status in the survey.

6. Click the ***green checkmark*** to close the Survey Points Editor vista.
7. In the CC-Loop survey, select the Interior Boundary network. Press the right mouse button, and from the shortcut menu, select INSERT INTO DRAWING.

The network shows the stations and foresights to the exterior reference point 1163.

## Review the Setups

The traverse follows the setups.

1. In the CC-Loop survey, expand Setups under the Interior Boundary network. From the list, select **Station: 100, Backsight: 1250**. Press the right mouse button, and from the shortcut menu, select EDIT OBSERVATIONS....

2. Select and review the remaining setups by editing their observations.
3. Click the ***green checkmark*** to close the Observations Editor vista.

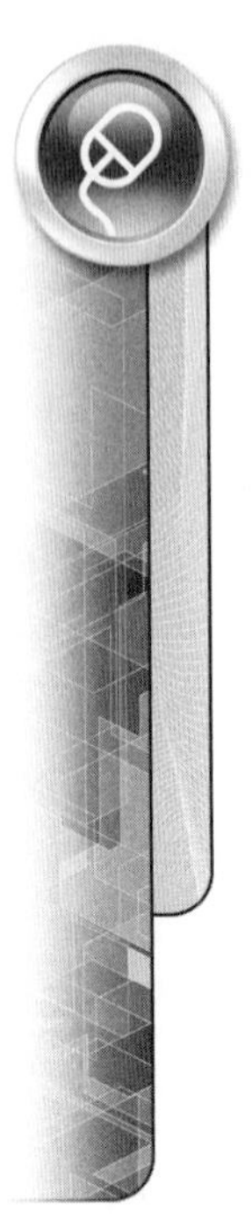

## Define Traverse per Sketch

1. In the CC-Loop survey, the Interior Boundary network, select the Traverses heading. Press the right mouse button, and from the shortcut menu, select NEW....
2. In New Traverse, for the name enter **Disputed**, for the description enter **Interior Disputed Bound**, for the Initial Station enter **1250**, click anywhere in the dialog box to let survey fill in the remaining stations, and click ***OK***.

## Preliminary Adjustment Review

1. In the CC-Loop survey, the Interior Boundary network, expand the Traverses branch. From the list of traverses, select Disputed, press the right mouse button, and from the shortcut menu, select TRAVERSE ANALYSIS....
2. In Traverse Analysis, set the Horizontal Adjustment Method to **Compass Rule**, set the Vertical Adjustment Method to **None**, toggle **OFF Update Survey Database**, and click ***OK***.
3. Review the preliminary results and close the Notepad reports after reading them.

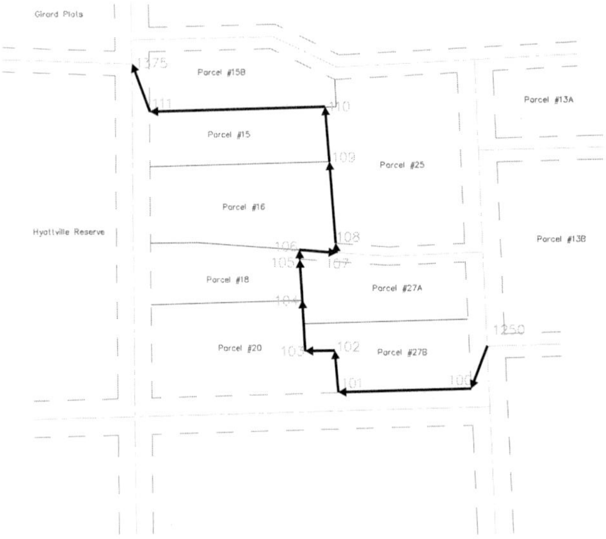


**FIGURE 15.11**

## Least Squares Analysis

Traverse adjustments do not take external observations into consideration. To use these observations, you must use the Least Squares adjustment from the named network shortcut menu.

1. In the CC-Loop survey, from the Networks list, select Interior Boundary. Press the right mouse button, and from the Least Squares Analysis flyout, select CREATE INPUT FILE.
2. In the CC-Loop survey, from the Networks list, select Interior Boundary. Press the right mouse button, and from the Least Squares Analysis flyout, select EDIT INPUT FILE....

The input file lists all cross-traverse observations and the observations to point 1163.

3. Close the Network.lsi file.
4. In the CC-Loop survey, from the Networks list, select Interior Boundary. Press the right mouse button, and from the Least Squares Analysis flyout, select PERFORM ANALYSIS....
5. In Least Squares Analysis, from the Procedure section, toggle **ON Update Survey Database**, and click ***OK*** to continue.
6. If Network file exists is displayed, click ***Yes*** to continue.
7. Close the Network.lsi file.
8. Use the ZOOM and PAN commands to view the error ellipses at stations 100–111. They are very small.

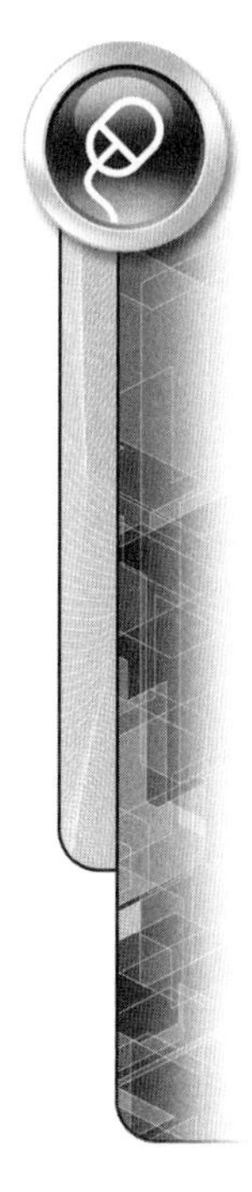

## Close the Survey and Save the Drawing

1. In the Survey Databases list, select the CC-Loop heading, press the right mouse button, and from the shortcut menu, select CLOSE SURVEY DATABASE.
2. Click the ***Prospector*** tab.
3. At Civil 3D's top left, Quick Access Toolbar, click the ***Save*** icon to save the drawing.
4. Close the drawing.

This completes the closed connected traverses exercise. Survey is the beginning and ending of every project. A survey's data is important and time review at each step of its analysis.

This chapter is the last chapter of this textbook. Each chapter highlights the tools available to the user to complete an engineering project from beginning to end.

## SUMMARY

- Traditional traverses use only foresight and backsight data.
- When adjusting a traverse by the Least Squares method, you cannot review the Least Squares input file.
- A survey containing a traverse with cross-traverse observations or outside-of-traverse observations must be analyzed by the Least Squares at the network level.

# INDEX



## A



## B



## C

## G



## H

## I



## K



## L



## M



## N



## O

# P

## S